Selected Works in Probability and Statistics

For further volumes:
http://www.springer.com/series/8556

Sandrine Dudoit
Editor

Selected Works of Terry Speed

Springer

Editor
Sandrine Dudoit, PhD
Professor of Biostatistics and Statistics
Chair, Graduate Group in Biostatistics
University of California, Berkeley
101 Haviland Hall, #7358
Berkeley, CA 94720-7358

ISBN 978-1-4614-1346-2 e-ISBN 978-1-4614-1347-9
DOI 10.1007/978-1-4614-1347-9
Springer New York Dordrecht Heidelberg London

Library of Congress Control Number: 2012930265

Printed on acid-free paper

Springer is part of Springer Science+Business Media (www.springer.com)

To Terry — teacher, colleague, and friend

NO
WAR

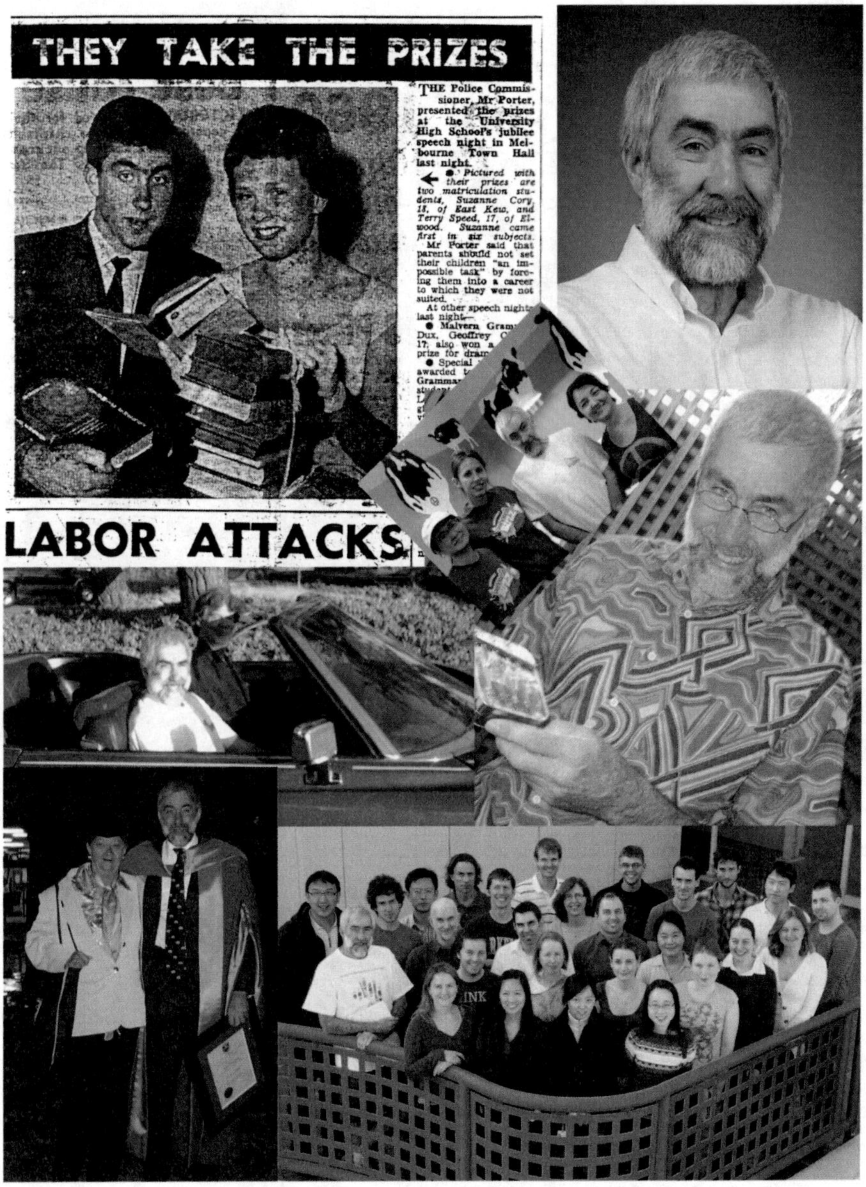
THEY TAKE THE PRIZES
THE Police Commissioner, Mr Porter, presented the prizes at the University High School's jubilee speech night in Melbourne Town Hall last night.
● Pictured with their prizes are two matriculation students, Suzanne Cory, 18, of East Kew, and Terry Speed, 17, of Elwood. Suzanne came first in six subjects.
Mr Porter said that parents should not set their children "an impossible task" by forcing them into a career to which they were not suited.
At other speech nights last night—
● Malvern Gram
Dux, Geoffrey
17, also won a
prize for dram
● Special
awarded
Gramma
LABOR ATTACKS

Preface to the Series

Springer's Selected Works in Probability and Statistics series offers scientists and scholars the opportunity of assembling and commenting upon major classical works in statistics, and honors the work of distinguished scholars in probability and statistics. Each volume contains the original papers, original commentary by experts on the subject's papers, and relevant biographies and bibliographies.

Springer is committed to maintaining the volumes in the series with free access of SpringerLink, as well as to the distribution of print volumes. The full text of the volumes is available on SpringerLink with the exception of a small number of articles for which links to their original publisher is included instead. These publishers have graciously agreed to make the articles freely available on their websites. The goal is maximum dissemination of this material.

The subjects of the volumes have been selected by an editorial board consisting of Anirban DasGupta, Peter Hall, Jim Pitman, Michael Sörensen, and Jon Wellner.

Preface

The purpose of this volume is to provide an overview of Terry Speed's contributions to statistics and beyond. Each of the fifteen chapters concerns a particular area of research and consists of a commentary by a subject-matter expert and selection of representative papers. Note that, due to space constraints, not all articles discussed in the commentaries are reprinted in this volume. The reader is referred to the book website for access to these papers (http://www.stat.berkeley.edu/~sandrine/Pubs/SelectedWorksTerrySpeed/). The chapters, organized more or less chronologically in terms of Terry's career, encompass a wide variety of mathematical and statistical domains, along with their application to biology and medicine. Accordingly, earlier chapters tend to be more theoretical, covering some algebra and probability theory, while later chapters concern more recent work in genetics and genomics. The chapters also span continents and generations, as they present research done over four decades, while crisscrossing the globe.

The commentaries provide insight into Terry's contributions to a particular area of research, by summarizing his work and describing its historical and scientific context, motivation, and impact. I've enjoyed reading the personal anecdotes, which remind us that one cannot always dissociate the scholar from the person and show how relationships beginning as professional collaborations can turn into long-lasting friendships. In addition to shedding light on Terry's scientific achievements, the commentaries reveal endearing aspects of his personality, such as his intellectual curiosity, energy, humor, and generosity. The title of Bin Yu's piece, "the $n \to \infty$ dimensions of Terry", says it all and captures Terry as an avid and tireless scholar and explorer.

Due to space constraints, this volume is only the tip of the iceberg, as it is clearly impossible to give a complete account of Terry's work. And it is certain that additional significant contributions are forthcoming — Terry's thirst for knowledge has not abated, and neither has his dynamic pace. For "coming attractions", one will have to wait for another such volume ...

Berkeley, CA
June 2011

Sandrine Dudoit

Acknowledgements for the Series

This series of selected works is possible only because of the efforts and cooperation of many people, societies, and publishers. The series editors originated the series and directed its development. The volume editors spent a great deal of time organizing the volumes and compiling the previously published material. The contributors provided comments on the significance of the papers. The societies and publishers who own the copyright to the original material made the volumes possible and affordable by their generous cooperation:

American Institute of Physics
American Mathematical Society
American Statistical Association
Applied Probability Trust
Bernoulli Society
Cambridge University Press
Canadian Mathematical Society
Danish Society for Theoretical Statistics
Elsevier
Finnish Statistical Society
Indian Statistical Institute
Institute of Mathematical Statistics
International Chinese Statistical Association
International Statistical Institute
John Wiley and Sons
New Zealand Statistical Association
Norwegian Statistical Society
Oxford University Press
Princeton University and the Institute for Advanced Studies
Royal Statistical Society

Statistical Society of Australia
Swedish Statistical Society
University of California Press
University of Illinois, Department of Mathematics
University of North Carolina Press

Acknowledgements

First and foremost, I would like to thank the fifteen contributors for their insightful and inspiring commentaries. This project would not have been possible without their hard work, support, generosity, and enthusiasm. In particular, they were remarkably gracious and efficient while working under pressure to meet tight publication deadlines. I have thoroughly enjoyed interacting with each of them and have learned much about Terry's career and life while reading their commentaries.

I am grateful to the reviewers for immediately and kindly agreeing to contribute to this volume and for their thoughtful reports.

I have greatly appreciated Jim Pitman's guidance and support throughout this project, regarding both the general editing process and technical aspects of bibliography management.

Matthew Watkins' work on Terry's bibliography was very valuable.

I am also thankful for Darlene Goldstein's encouragement and help during the final stages of this project.

Last but not least, editing this volume was an extraordinary opportunity to work with Terry again and "delve into his many lives". I have learned immensely from him, whether in a Berkeley classroom, sipping a milkshake on Bancroft Avenue, attending a performance at the San Francisco Opera, or trying to keep up with him on a morning jog in the mountains overlooking Lago Maggiore. His sharp intellect, vast culture, humanism, energy, enthusiasm, and humor never cease to inspire me. For this, I am most thankful.

Berkeley, CA
June 2011

Sandrine Dudoit

Contents

T. P. Speed. Iterative proportional fitting. In P. Armitage and T. Colton, editors, *Encyclopedia of Biostatistics*. John Wiley & Sons, New York, 1999. Reprinted with permission of John Wiley & Sons 373

9 History and Teaching Statistics 377
Deborah Nolan
T. P. Speed. Questions, answers and statistics. In R. Davidson and J. Swift, editors, *Proceedings: The Second International Conference on Teaching Statistics*, pages 18–28. University of Victoria, Victoria, BC, Canada, 1986. Reprinted with permission of the International Association for Statistical Education 380
D. Nolan and T. P. Speed. Teaching statistics: Theory through applications. *Am. Stat.*, 53(4):370–375, 1999. Reprinted with permission of the American Statistical Association 391

10 Genetic Recombination 397
Mary Sara McPeek
T. P. Speed, M. S. McPeek, and S. N. Evans. Robustness of the no-interference model for ordering genetic markers. *Proc. Natl. Acad. Sci. USA*, 89(7):3103–3106, 1992. Reprinted with permission of the National Academy of Sciences of the USA 402
M. S. McPeek and T. P. Speed. Modeling interference in genetic recombination. *Genetics*, 139:1031–1044, 1995. Reprinted with permission of the Genetics Society of America 406
H. Zhao, T. P. Speed, and M. S. McPeek. Statistical analysis of crossover interference using the chi-square model. *Genetics*, 139:1045–1056, 1995. Reprinted with permission of the Genetics Society of America 420
H. Zhao and T. P. Speed. On genetic map functions. *Genetics*, 142(4):1369–1377, 1996. Reprinted with permission of the Genetics Society of America 432

11 Molecular Evolution 441
Steven N. Evans
S. N. Evans and T. P. Speed. Invariants of some probability models used in phylogenetic inference. *Ann. Stat.*, 21(1):355–377, 1993. Reprinted with permission of the Institute of Mathematical Statistics 448

12 Statistical Genetics 471
Darlene R. Goldstein
S. Dudoit and T. P. Speed. A score test for linkage using identity by descent data from sibships. *Ann. Stat.*, 27(3):943–986, 1999. Reprinted with permission of the Institute of Mathematical Statistics 477

Biographical Sketch of Terry Speed

Terence Paul (Terry) Speed was born on March 14th, 1943, in Victor Harbor, South Australia. He grew up in Melbourne, attending Westgarth Central School and University High School. In his final year of high school, he decided that he wanted to pursue a career in medical research, influenced by the award that year (1960) of the Nobel Prize in Medicine to Sir Frank Macfarlane Burnet, the Director of Melbourne's Walter and Eliza Hall Institute (WEHI) of Medical Research. In 1961, Terry enrolled in a joint Medicine and Science degree at the University of Melbourne. By the end of the first term, his lack of enthusiasm for laboratory work prompted him to change his enrollment to Science alone, majoring in mathematics, while maintaining a strong interest in genetics. He graduated in 1964 with an honours degree in mathematics and statistics. In his final year, he edited the magazine *Matrix* of the mathematics students society and also attended lectures on algebra at Monash University, located in an outer suburb of Melbourne. At the end of that year, he married Freda Elizabeth (Sally) Pollard, whom he had met at a party at the home of Carl Moppert, then a Senior Lecturer in the Department of Mathematics at the University of Melbourne.

Although an attempt to join the PhD program in the Department of Statistics at the University of California, Berkeley (UCB) fell through, Terry was awarded an Australian Commonwealth Postgraduate Research Scholarship in the Department of Statistics at the University of Melbourne. He began his graduate studies in 1965, under the supervision of Professor Evan J. Williams. Rather than pursuing research in the area of his supervisor (Fisherian statistics), Terry developed an interest in probability theory, along the lines of Michel Loève's work at Berkeley. He did not however complete his doctoral degree at that point. In mid-1965, he took a job as a tutor in the Department of Mathematics at nearby Monash University and enrolled for a part-time PhD in mathematics under the supervision of Professor Peter D. Finch. With hindsight, it is interesting to note that several elements that were to feature prominently in Terry's later life had already manifested themselves: interests in medical and genetic research, the Walter and Eliza Hall Institute, and probability and statistics as practiced in the Department of Statistics at the University of California, Berkeley.

At Monash, Finch had eclectic interests in probability theory and mathematics and encouraged Terry to examine probability and measure theory on the class of lattices generalizing Boolean algebras that corresponded to the intuitionistic logic of L. E. J. Brouwer. This led to Terry's 1969 PhD thesis entitled *Some topics in the theory of distributive lattices*. In addition to working on his doctoral research, Terry taught introductory probability and statistics to large classes for four years running, and developed and presented undergraduate lecture courses on information theory (introductory and advanced), measure theory, projective geometry, and lattice theory. He also lectured on the theory of games to students in the Department of Mathematics at the new La Trobe University, located in another outer suburb of Melbourne. On top of this, he completed a Diploma of Education at Monash, reasoning that, if all else failed, he would be happy as a secondary school mathematics teacher and that it would be wise to be qualified.

While waiting for the examiners' reports on his thesis, Terry met Professor Joe Gani, then Director of the Manchester-Sheffield School of Probability and Statistics in the United Kingdom. Gani encouraged him to consider a lecturing position in the School. The presence in Manchester of Professor K. R. Parthasarathy — who carried out research on probability theory over algebraic structures such as locally compact abelian and Lie groups — proved to be the clincher. So off to Sheffield he and Sally went! Sheffield was an exciting place at that time, with excellent staff and lots of visitors. Equally important was its accessibility to other centers of probability and statistics such as Manchester and London. Initially, Terry travelled to Manchester weekly to attend Partha's seminar and went down to London to attend seminars at Imperial College, meetings of the Royal Statistical Society, and the like. There was lots of train travel. However, Terry's career in Partha-style probability theory did not take off and, in due course, he found himself collaborating with Elja Arjas on the topic of random walks, an experience that was both satisfying and productive. A later visitor to Manchester, Professor Debrabata Basu, re-kindled his interest in Berkeley-style statistics and led to a new obsession: sufficiency.

Terry returned to Australia to head the small group of statisticians in the Department of Mathematics at the University of Western Australia (UWA). He started at UWA as Associate Professor in 1974, became Professor in 1975, and spent a very happy and productive period there, culminating in being appointed Head of Department in 1982. From late 1977 until early 1979, he had his first sabbatical, spending time at the University of Copenhagen, Princeton University, Rothamsted Experimental Station, and UC Berkeley, all with Sally, and on his own at the Indian Statistical Institute in Calcutta.

In 1982, Terry was invited to apply for the position of Chief, Division of Mathematics and Statistics, at Australia's Commonwealth Scientific and Industrial Research Organization (CSIRO). He took up that appointment in 1983 and had a very hectic first year, being based in Canberra, but travelling to visit members of his division in every state capital and several other centers around Australia.

In 1984, he spent two enjoyable months visiting the Department of Statistics at the University of California, Berkeley, in a way, fulfilling his unrealized dream from 1964. While there, he was encouraged to apply for a permanent position, and three years later, in fall 1987, joined UCB as a tenured professor. On the basis of his administrative experience with CSIRO, he was appointed Department Chair 1989–94, after which he took a second sabbatical, encouraged by Sally to explore job opportunities back in Australia. Nothing happened on this front for two years, but in 1996, a former classmate from University High School, Professor Suzanne Cory, by then Director of the Walter and Eliza Hall Institute of Medical Research, invited him to start up bioinformatics at WEHI. Sally said "yes!" and so he half accepted. From August 1997 to July 2009, Terry split his time evenly between UCB and WEHI or, as he used to say, spent half his time in Berkeley, half in Melbourne, and the other half in the air in between.

Following yet more encouragement from Sally, Terry officially retired from teaching at UC Berkeley at the end of the US academic year 2008–9 and took on a full-time appointment at WEHI. At the time of writing, he still has four PhD students, three postdoctoral fellows, and a number of continuing collaborations at Berkeley. He visits there for short periods every 1–2 months and remains as active as ever.

To quote from one of Terry's recent e-mails: "Life has been hectic!"

Terry has (co-)authored over 300 refereed articles, in journals such as *Nature* and *The Annals of Statistics*, and on a wide variety of subjects, ranging from distributive lattices and ring theory in algebra, to pre-processing of high-throughput microarray and sequencing data in genomics. He contributes a regular column, *Terence's Stuff*, to the *Institute of Mathematical Statistics Bulletin*, with his unique and provocative opinions on the current state of statistical practice and education. His book *Stat Labs: Mathematical Statistics Through Applications* provides a glimpse into his teaching philosophy, which integrates the theory of statistics with its practice through case studies. As illustrated by his dizzying travel schedule, he is a much sought-after speaker worldwide. He has delivered prestigious lectures such as the 2001 Wald Memorial Lectures and 2006 Fisher Lecture, at the Joint Statistical Meetings, and the 2007 Hotelling Lectures, at the University of North Carolina, Chapel Hill.

Terry is an active and dedicated member of the main statistical and biological professional societies, journal editorial boards, and grant and peer review committees. He is also highly-solicited as a consultant and scientific advisory board member in industry. He is a Fellow of the Institute of Mathematical Statistics (1984), the American Statistical Association (1989), the American Association for the Advancement of Science (1990), and the Australian Academy of Science (2001). He has received various honors, including the 2002 Pitman Medal (Statistical Society of Australia), the 2003 Moyal Medal (Macquarie University), an Australian Government Centenary Medal (2003), the 2004 American Statistical Association Outstanding Statistical Application Award (for the paper Irizarry et al. (2003), *Biostatistics*, 4(2):249–264), as well as an Achievement Award for excellence in health and medical research (2007) and an Australia Fellowship (2009) from Australia's National Health and Medical Research Council (NHMRC).

In addition to his invaluable contributions to research, Terry is an extraordinary teacher, who has trained and influenced generations of students at Berkeley, in Australia, in the United Kingdom, and beyond. According to the Mathematics Genealogy Project (`http://genealogy.math.ndsu.nodak.edu/id.php?id=30979`), he has advised or co-advised 60 PhD students and has over 120 "descendants". He is a most inspiring and generous mentor. His contagious enthusiasm and intellectual curiosity have made him one of the most popular advisors in the UC Berkeley Department of Statistics and a great resource for students in other departments.

Berkeley, CA
June 2011

Sandrine Dudoit
with contributions from Terry Speed

PhD Students of Terry Speed

Name	Institution	Year
Michael Evans	Monash University	1973
Philip Pegg	University of Sheffield	1973
James (Jim) Pitman	University of Sheffield	1974
John Whitehead	University of Sheffield	1975
Anne Houtman	Princeton University	1980
Harri Kiiveri	University of Western Australia	1982
Matthew Knuiman	University of Western Australia	1983
Jens Breckling	University of Western Australia	1987
Bin Yu	University of California, Berkeley	1990
Sang Ho Lee	University of California, Berkeley	1991
Trang Nguyen	University of California, Berkeley	1991
Rudy Guerra, Jr.	University of California, Berkeley	1992
Darlene Goldstein	University of California, Berkeley	1993
Ferdinand Verweyen	University of California, Berkeley	1993
Mary Sara McPeek	University of California, Berkeley	1993
Steven Rein	University of California, Berkeley	1993
Ann Kalinowski	University of California, Berkeley	1995
David Nelson	University of California, Berkeley	1995
Hongyu Zhao	University of California, Berkeley	1995
Gregory Alexander	The American University	1996
Mark Grote	University of California, Berkeley	1996
Karl Broman	University of California, Berkeley	1997
Barathi Sethuraman	University of California, Berkeley	1997
William Forrest, III	University of California, Berkeley	1998
Lei Li	University of California, Berkeley	1998
Sandrine Dudoit	University of California, Berkeley	1999
Ru-Fang Yeh	University of California, Berkeley	1999
Shiying Ling	University of California, Berkeley	2000
Simon Cawley	University of California, Berkeley	2000

Name	Institution	Year
Alexandre Bureau	University of California, Berkeley	2001
Nicola Armstrong	University of California, Berkeley	2001
Yevgeniya (Jane) Fridlyand	University of California, Berkeley	2001
Fletcher (Hank) Ibser	University of California, Berkeley	2002
Von Bing Yap	University of California, Berkeley	2002
Yee Hwa (Jean) Yang	University of California, Berkeley	2002
Yongchao Ge	University of California, Berkeley	2003
Pratyaksha Wirapati	University of Melbourne	2003
Jacqueline Wicks	Australian National University	2003
Xiaoyue Zhao	University of California, Berkeley	2004
Natalie Thorne	University of Melbourne	2004
Benjamin Bolstad	University of California, Berkeley	2005
Yu Chuan Tai	University of California, Berkeley	2005
Ingileif (Inga) Hallgrímsdóttir	University of California, Berkeley	2005
Frédéric Schütz	University of Melbourne	2005
Ingrid Lönnstedt	Uppsala Universitet	2005
Tracey Wilkinson	University of Melbourne	2005
Richard Bourgon	University of California, Berkeley	2006
Jing Yi	University of California, Berkeley	2006
Yun Zhou	University of California, Berkeley	2006
Hui Tang	University of California, Berkeley	2007
Mark Robinson	University of Melbourne	2008
Tobias Sargeant	University of Melbourne	2008
Gregory Hather	University of California, Berkeley	2008
Margaret Taub	University of California, Berkeley	2009
Xin Wang	University of California, Berkeley	2009
Nancy Wang	University of California, Berkeley	2009
Frances Tong	University of California, Berkeley	2010
Oleg Mayba	University of California, Berkeley	2011
Alex Gout	University of Melbourne	2011

Contributors

Elja Arjas
University of Helsinki and National Institute for Health and Welfare, Finland
e-mail: elja.arjas@helsinki.fi

Karl W. Broman
Department of Biostatistics & Medical Informatics, University of Wisconsin–Madison
e-mail: kbroman@biostat.wisc.edu

Simon E. Cawley
Ion Torrent
e-mail: simon.cawley@lifetech.com

Anirban DasGupta
Department of Statistics, Purdue University and Indian Statistical Institute
e-mail: dasgupta@stat.purdue.edu

Brian A. Davey
Department of Mathematics and Statistics, La Trobe University, Australia
e-mail: b.davey@latrobe.edu.au

Sandrine Dudoit
Division of Biostatistics and Department of Statistics, University of California, Berkeley
e-mail: sandrine@stat.berkeley.edu

Steven N. Evans
Department of Statistics, University of California, Berkeley
e-mail: evans@stat.berkeley.edu

Jane Fridlyand
Department of Biostatistics, Genentech
e-mail: fridlyand.jane@gene.com

Darlene R. Goldstein
Institut de mathématiques d'analyse et applications, École Polytechnique Fédérale de Lausanne, and Swiss Institute of Bioinformatics, Switzerland
e-mail: darlene.goldstein@epfl.ch

Steffen L. Lauritzen
Department of Statistics, University of Oxford, United Kingdom
e-mail: steffen@stats.ox.ac.uk

Lei M. Li
Academy of Mathematics and Systems Science, Chinese Academy of Sciences, and Computational Biology and Bioinformatics, University of Southern California
e-mail: lilei@amss.ac.cn

Peter McCullagh
Department of Statistics, University of Chicago
e-mail: pmcc@galton.uchicago.edu

Mary Sara McPeek
Departments of Statistics and Human Genetics, University of Chicago
e-mail: mcpeek@uchicago.edu

Deborah Nolan
Department of Statistics, University of California, Berkeley
e-mail: nolan@stat.berkeley.edu

Terry Speed
Department of Statistics, University of California, Berkeley, and Division of Bioinformatics, Walter and Eliza Hall Institute, Australia
e-mail: terry@stat.berkeley.edu

Bin Yu
Departments of Statistics and Electrical Engineering & Computer Sciences, University of California, Berkeley
e-mail: binyu@stat.berkeley.edu

Chapter 1
Algebra

Brian A. Davey

It gives me great pleasure to present this brief commentary on some of T. P. Speed's papers on algebra. It may come as a surprise to many of Speed's colleagues to know that his 1968 PhD thesis was entitled *Some Topics in the Theory of Distributive Lattices*. Moreover, of his first 15 papers only one was in probability theory with the remainder in algebra. Nevertheless, this fruitful excursion into algebra has its roots in the foundations of probability theory. In the introduction to his PhD thesis, Speed writes:

> In July 1965, the author began to look at the lattices associated with intuitionistic logic which are called variously – relatively pseudo-complemented, brouwerian or implicative lattices. This was under the direction of Professor P. D. Finch and aimed towards defining probability measures over these lattices. It was hoped that a probability theory could be developed for the intuitionistic viewpoint similar to the Kolmogorov one for classical logic.

Speed never returned to the search for an intuitionistic probability theory for, as he says later in the introduction to his thesis, he became "*sold on distributive lattices*". In the summer of 1968–1969, between my third and honours years, I spent three months on a Monash University Graduate Assistantship during which I read Speed's PhD thesis. By the end of that summer I was also *sold on distributive lattices* and have been ever since [2].

Between 1969 and 1974, Speed published 17 papers on a range of algebraic topics: distributive lattices, including their topological representation (9), Baer rings (3), Stone lattices (2), semigroups (2), and ℓ-groups (1). In the commentary below, I will discuss five of these papers. Only one of these papers, the first discussed, comes from Speed's thesis.

B.A. Davey
Department of Mathematics and Statistics, La Trobe University, Australia
e-mail: b.davey@latrobe.edu.au

S. Dudoit (ed.), *Selected Works of Terry Speed*, Selected Works in Probability and Statistics,
DOI 10.1007/978-1-4614-1347-9_1, © Springer Science+Business Media, LLC 2012

Distributive lattices in general

Most of Speed's work on distributive lattices revolves around the role of particular sorts of prime ideals, with an emphasis on minimal prime ideals. In this section, we will look at two of the seven papers that fall into this category, namely, *On rings of sets* [10] and *On rings of sets. II. Zero-sets* [16].

In the first of these papers, Speed provides a unified approach to a number of representations of distributive lattices as rings of sets, that is, as lattices of subsets of some set in which the operations are set-theoretic union and intersection. Each of these characterisations was originally given in terms of the existence of enough elements of a special form, and their proofs looked quite different. Given cardinals $\mathfrak{m}$ and $\mathfrak{n}$, a lattice L is called $(\mathfrak{m},\mathfrak{n})$-complete if it is closed under the operations of least upper bound and greatest lower bound of sets of at most $\mathfrak{m}$ and $\mathfrak{n}$ elements, respectively. An $(\mathfrak{m},\mathfrak{n})$-complete lattice of sets is an $(\mathfrak{m},\mathfrak{n})$-*ring of sets* if $\mathfrak{m}$-ary least upper bounds and $\mathfrak{n}$-ary greatest lower bounds are given by set union and intersection, respectively. For example, the open sets of a topological space form an $(\mathfrak{m},2)$-ring of sets for every cardinal $\mathfrak{m}$. Speed introduces $\mathfrak{n}$-prime $\mathfrak{m}$-ideals and employs them to give natural necessary and sufficient conditions for an $(\mathfrak{m},\mathfrak{n})$-complete lattice to be isomorphic to an $(\mathfrak{m},\mathfrak{n})$-ring of sets. As Speed remarks in the introduction to the paper, *It is interesting to note that the elementary methods used in representing distributive lattices carry over completely and yield all these results, although this is hardly obvious when one considers special elements of the lattice.*

In *On rings of sets. II. Zero-sets* [16], Speed turns his attention to an important example of $(2,\omega)$-rings of sets, the *lattice* $\mathbf{Z}(X)$ *of zero-sets* of continuous real-valued functions on a topological space X. The paper, which is deeper and somewhat more technical than the first, includes lattice-theoretic characterisations of $\mathbf{Z}(X)$ in two important cases, when X is compact (Theorem 4.1) and when X is an arbitrary topological space (Theorem 5.9). In both cases, the characterisations involve minimal prime ideals. Along the way he proves a result (Theorem 3.1) that very nicely generalises Urysohn's Lemma for normal topological spaces and the fact that, in a completely regular space, disjoint zero-sets can be separated by a continuous function.

Distributive lattices—Priestley duality

About the same time that Speed was writing his PhD thesis at Monash University, H. A. Priestley was writing her DPhil at the University of Oxford. Speed was amongst the first to realise the importance of the new duality for bounded distributive lattices that Priestley established in her thesis (see Priestley [8, 9] and Davey and Priestley [2]).

In *On the order of prime ideals* [13], Speed addresses the question, raised by Chen and Grätzer [1], of characterising *representable* ordered sets, that is, ordered sets that arise as the ordered set of prime ideals of a bounded distributive lattice. By using Birkhoff's duality between finite distributive lattices and finite ordered sets, he shows that an ordered set is representable if and only if it is the inverse limit of an inverse system of finite ordered sets. Speed observes that, when combined with deep results of Hochster [5], this tells us that an ordered set is isomorphic to the ordered set of prime ideals of a commutative ring with unit if and only if it is isomorphic to an inverse limit of finite ordered sets. This cross fertilisation in Speed's work between commutative rings with unit and bounded distributive lattices will arise again in Section 1.

Soon after writing Speed [13], Speed became aware of Priestley's results. He quickly realised that, since an inverse limit of finite sets is endowed with a natural compact topology, his characterisation of representable ordered sets could be lifted to a characterisation of compact totally order-disconnected spaces, the ordered topological spaces that arise in Priestley duality (and are now referred to simply as *Priestley spaces*). In *Profinite posets* [12], he proved that an ordered topological space is a Priestley space if and only if it is isomorphic, both order theoretically and topologically, to an inverse limit of finite discretely topologised ordered sets.

Baer rings

Speed's PhD thesis was strongly influenced by the seminal paper *Minimal prime ideals in commutative semigroups* [6]. He took ideas from Kist's paper and reinterpreted them in the context of distributive lattices. Speed saw that there was some informal connection between the commutative Baer rings introduced and studied in Kist [6] and Stone lattices, a class of distributive lattices introduced by Grätzer and Schmidt [4]. A commutative ring R is a *Baer ring* if, for every element $a \in R$, the annihilator $\operatorname{ann}(a) := \{x \in R \mid xa = 0\}$ is a principal ideal generated by a (necessarily unique) idempotent a^*. A bounded distributive lattice L is a *Stone lattice* if, for every element $a \in L$, the annihilator $\operatorname{ann}(a) := \{x \in L \mid x \wedge a = 0\}$ is a principal ideal generated by an element a^*, and in addition the equation $a^* \vee a^{**} = 1$ is satisfied. While quite different looking, the requirements that a^* be an idempotent, in the ring case, and the identity $a^* \vee a^{**} = 1$, in the lattice case, guarantee that the elements a^* form a Boolean algebra and correspond precisely to the direct product factorisations of the ring or lattice.

While the proofs will typically be quite different, it is often true that a result about Baer rings will translate to a corresponding result about Stone lattices and vice versa. For example:

(i) Grätzer [3] proved that Stone lattices form an equational class; Speed and Evans [17] proved that Baer rings also form an equational class. (In both cases, * is added as an additional unary operation.)
(ii) Grätzer and Schmidt [4] proved that, in a Stone lattice, each prime ideal contains a unique minimal prime ideal; Kist [6] proved that precisely the same condition holds in a Baer ring.

In separate papers on Stone lattices [11] and Baer rings [14], Speed proves that there are broad classes of distributive lattices and rings, respectively, within which Stone lattices and Baer rings are characterised by the property that each prime ideal contains a unique minimal prime ideal.

In his third and final paper on Baer rings [15], Speed considers the question of embedding a commutative semiprime ring R into a Baer ring B. Two such embeddings had already been given: the first by Kist [6] and the second by Mewborn [7]. In both cases, the Baer ring B was constructed as a ring of global sections of a sheaf over a Boolean space. Speed shows that, in fact, there is a hierarchy of Baer extensions of R, the smallest being Kist's and the largest Mewborn's. Moreover, he is able to replace the sheaf-theoretic construction with a purely algebraic one similar in nature to one that had been used previously in the theory of lattice-ordered groups. The underlying lattice of a lattice-ordered group is distributive, so again we see Speed's fruitful use of the interplay between rings and distributive lattices.

References

[1] C. C. Chen and G. Grätzer. Stone lattices. II. Structure theorems. *Canad. J. Math.*, 21:895–903, 1969.

[2] B. A. Davey and H. A. Priestley. *Introduction to Lattices and Order*. Cambridge University Press, New York, second edition, 2002.

[3] G. Grätzer. Stone algebras form an equational class. Remarks on lattice theory. III. *J. Aust. Math. Soc.*, 9:308–309, 1969.

[4] G. Grätzer and E. T. Schmidt. On a problem of M. H. Stone. *Acta Math. Acad. Sci. Hungar.*, 8:455–460, 1957.

[5] M. Hochster. Prime ideal structure in commutative rings. *Trans. Amer. Math. Soc.*, 142:43–60, 1969.

[6] J. Kist. Minimal prime ideals in commutative semigroups. *Proc. Lond. Math. Soc. (3)*, 13:31–50, 1963.

[7] A. C. Mewborn. Regular rings and Baer rings. *Math. Z.*, 121:211–219, 1971.

[8] H. A. Priestley. Representation of distributive lattices by means of ordered stone spaces. *Bull. Lond. Math. Soc.*, 2:186–190, 1970.

[9] H. A. Priestley. Ordered topological spaces and the representation of distributive lattices. *Proc. Lond. Math. Soc. (3)*, 24:507–530, 1972.

[10] T. P. Speed. On rings of sets. *J. Aust. Math. Soc.*, 8:723–730, 1968.

[11] T. P. Speed. On Stone lattices. *J. Aust. Math. Soc.*, 9:297–307, 1969.

[12] T. P. Speed. Profinite posets. *Bull. Aust. Math. Soc.*, 6:177–183, 1972.

[13] T. P. Speed. On the order of prime ideals. *Algebra Univers.*, 2:85–87, 1972.

[14] T. P. Speed. A note on commutative Baer rings. *J. Aust. Math. Soc.*, 14:257–263, 1972.

[15] T. P. Speed. A note on commutative Baer rings. III. *J. Aust. Math. Soc.*, 15: 15–21, 1973.

[16] T. P. Speed. On rings of sets. II. Zero-sets. *J. Aust. Math. Soc.*, 16:185–199, 1973. Collection of articles dedicated to the memory of Hanna Neumann, II.
[17] T. P. Speed and M. W. Evans. A note on commutative Baer rings. *J. Aust. Math. Soc.*, 13:1–6, 1972.

ON RINGS OF SETS

T. P. SPEED

(Received 16 March 1967)

1. Introduction

In the past a number of papers have appeared which give representations of abstract lattices as rings of sets of various kinds. We refer particularly to authors who have given necessary and sufficient conditions for an abstract lattice to be lattice isomorphic to a complete ring of sets, to the lattice of all closed sets of a topological space, or to the lattice of all open sets of a topological space. Most papers on these subjects give the conditions in terms of special elements of the lattice. We thus have completely join-irreducible elements – G. N. Raney [7]; join prime, completely join prime, and supercompact elements – V. K. Balachandran [1], [2]; $\mathcal{N}$-sub-irreducible elements – J. R. Büchi [5]; and lattice bisectors – P. D. Finch [6]. Also meet-irreducible and completely meet-irreducible dual ideals play a part in some representations of G. Birkhoff & O. Frink [4].

What we do in this paper is define a new kind of prime ideal – called an $\mathfrak{n}$-prime $\mathfrak{m}$-ideal – and show that all the above concepts correspond to a particular kind of $\mathfrak{n}$-prime $\mathfrak{m}$-ideal. Here and throughout we mean $\mathfrak{m}$ and $\mathfrak{n}$ to be (possibly infinite) cardinals, always greater than 1. Also the symbol ∞ will be used to denote an arbitrarily large cardinal number. A class of lattices called $(\mathfrak{m}, \mathfrak{n})$-rings of sets is then defined and some theorems proved which cover all the representation theorems mentioned above. It is interesting to note that the elementary methods used in representing distributive lattices carry over completely and yield all these results, although this is hardly obvious when one considers special elements of the lattice.

I wish to express my gratitude to Professor P. D. Finch, whose paper [6] was the inspiration for this work.

2. Notations and Definitions

We assume a familiarity with the elementary notions of lattice theory as outlined in G. Birkhoff [3].

Definition 2.1. A lattice $\mathscr{L} = \langle L; \vee, \wedge \rangle$ is said to be $(\mathfrak{m}, \mathfrak{n})$-*complete* if the join of not more then $\mathfrak{m}$ elements of L belongs to L, and the meet of not more than $\mathfrak{n}$ elements of L belongs to L.

Thus an $(\mathfrak{m}, \mathfrak{n})$-complete lattice may be considered as an algebra with the $\mathfrak{m}$-ary operation of join and the $\mathfrak{n}$-ary operation of meet.

DEFINITION 2.2. An $(\mathfrak{m}, \mathfrak{n})$-complete lattice of sets $\mathscr{L} = \langle L; \vee, \wedge \rangle$ is called an $(\mathfrak{m}, \mathfrak{n})$-*ring of sets* if the $\mathfrak{m}$-ary operation of join corresponds to set union, and the $\mathfrak{n}$-ary operation of meet corresponds to set intersection.

EXAMPLE. The lattice of all open sets of a topological space is an $(\infty, 2)$-ring of sets.

DEFINITION 2.3. An ideal P of the $(\mathfrak{m}, \mathfrak{n})$-complete lattice $\mathscr{L} = \langle L; \vee, \wedge \rangle$ is called an $\mathfrak{n}$-prime $\mathfrak{m}$-ideal if

(i) For $\{x_\gamma : \gamma \in \Gamma\} \subseteqq L$ with $|\Gamma| \leqq \mathfrak{m}$ we have:

$$x_\gamma \in P \; \forall \gamma \in \Gamma \;\Leftrightarrow\; \bigvee_{\gamma \in \Gamma} x_\gamma \in P$$

(ii) For $\{y_\delta : \delta \in \Delta\} \subseteqq L$ with $|\Delta| \leqq \mathfrak{n}$ we have:

$$y_\delta \notin P \; \forall \delta \in \Delta \;\Leftrightarrow\; \bigwedge_{\delta \in \Delta} y \notin P.$$

REMARKS. 1. An ordinary prime ideal is a 2-prime 2-ideal in the above notation.

2. The definition is obviously not the most general possible but it will suffice for the purpose of this paper.

3. If P is an $\mathfrak{n}$-prime $\mathfrak{m}$-ideal then $L \setminus P$ is an $\mathfrak{m}$-prime $\mathfrak{n}$-dual ideal with the obvious (dual) definition of the latter.

DEFINITION 2.4. An homomorphism ψ between two $(\mathfrak{m}, \mathfrak{n})$-complete lattices is called an $(\mathfrak{m}, \mathfrak{n})$-*homomorphism* if ψ preserves joins of $\mathfrak{m}$ elements and meets of $\mathfrak{n}$ elements.

Note that a lattice of sets is not assumed to have set union and intersection as lattice operations unless stated, although the partial ordering is set inclusion.

3. $(\mathfrak{m}, \mathfrak{n})$-rings of sets

In this section we clarify the notion of $(\mathfrak{m}, \mathfrak{n})$-ring of sets.

PROPOSITION 3.1. *Let $\mathscr{L} = \langle L; \vee, \wedge \rangle$ be an $(\mathfrak{m}, \mathfrak{n})$-complete lattice of subsets of a set S. Then $\mathscr{L}$ is an $(\mathfrak{m}, \mathfrak{n})$-ring of sets if and only if for any $s \in S$*

(i) *$s \notin \bigvee\{l \in M : s \notin l\}$ for any $M \subseteqq L$ with $|M| \leqq \mathfrak{m}$*

(ii) *$s \in \bigwedge\{l \in N : s \in l\}$ for any $N \subseteqq L$ with $|N| \leqq \mathfrak{n}$.*

PROOF. If $\mathscr{L}$ is an $(\mathfrak{m}, \mathfrak{n})$-ring of sets, then the $\mathfrak{m}$-ary join and the $\mathfrak{n}$-ary meet operations correspond to set union and intersection respectively. It is thus clear that (i) and (ii) hold in this case.

For the converse we assume (i) and (ii). Observe that we must always have (for $M \subseteqq L$ with $|M| \leqq \mathfrak{m}$)

$$\bigvee \{l : l \in M\} \supseteqq \bigcup \{l : l \in M\}.$$

Now if $s \notin \bigcup\{l : l \in M\}$ then $s \notin l \; \forall l \in M$ and thus by (i) we see that $s \notin \bigvee \{l : l \in M\}$. The reverse inclusion is hence proved and we obtain $\bigvee \{l : l \in M\} = \bigcup\{l : l \in M\}$. Similarly $\bigwedge\{l : l \in N\} \subseteqq \bigcap \{l : l \in N\}$ always holds for $N \subseteqq L$ with $|N| \leqq \mathfrak{n}$, and (ii) implies the reverse inclusion giving

$$\bigwedge \{l : l \in N\} = \bigcap \{l : l \in N\}.$$

The proposition is thus proved.

Our next result is a direct generalisation of G. Birkhoff's theorem for distributive lattices (= (2, 2)-rings of sets), [3] p. 140.

PROPOSITION 3.2. *Let* $\mathscr{L} = \langle L; \vee, \wedge \rangle$ *be an* $(\mathfrak{m}, \mathfrak{n})$*-complete lattice. Then* $\mathscr{L}$ *is isomorphic to an* $(\mathfrak{m}, \mathfrak{n})$*-ring of sets if and only if* $\mathscr{L}$ *has a faithful representation as a subdirect union of a family* $\{\mathscr{L}_\alpha : \alpha \in A\}$ *of replicas of* $\mathbf{2}$ *in which each projection* $\pi_\alpha : \mathscr{L} \to \mathscr{L}_\alpha$ *is an* $(\mathfrak{m}, \mathfrak{n})$*-homomorphism.*

PROOF. Assume first that $\mathscr{L}$ has a sub-direct union representation with the stated properties. This is equivalent to the existence of an isomorphism ψ of $\mathscr{L}$ onto a lattice $\langle \mathscr{A}, \cup, \cap \rangle$ of subsets of the index set A; explicitly

$$\psi : l \to l\psi = \{\alpha \in A : l\pi_\alpha = 1\}, \quad \mathscr{A} = \{l\psi : l \in L\}.$$

It is clear that $\langle \mathscr{A}; \cup, \cap \rangle$ is a $(\mathfrak{m}, \mathfrak{n})$-complete lattice. We show it is a $(\mathfrak{m}, \mathfrak{n})$-ring of sets. Take $\mathscr{M} \subseteqq \mathscr{A}$ with $|\mathscr{M}| \leqq \mathfrak{m}$, and an arbitrary $\alpha \in A$.

$$\begin{aligned}
\text{Now} \quad & \bigvee \{K \in \mathscr{M} : \alpha \notin K\} \\
&= \bigvee \{l\psi \in \mathscr{M} : \alpha \notin l\psi] \text{ since every } K \in \mathscr{M} \text{ is of the form } l\psi, l \in L \\
&= [\bigvee \{l \in M : \alpha \notin l\psi\}]\psi \quad \text{where } M = \mathscr{M}\psi^{-1} \subseteqq L \\
&= [\bigvee \{l \in M : l\pi_\alpha = 0\}]\psi \text{ since } \alpha \notin l\psi \equiv l\pi_\alpha = 0.
\end{aligned}$$

Further, $[\bigvee \{l \in M : l\pi_\alpha = 0\}]\pi_\alpha = 0$ since $|M| \leqq \mathfrak{m}$ and the π_α are $(\mathfrak{m}, \mathfrak{n})$-homomorphisms, so that $\alpha \notin \bigvee\{K \in \mathscr{M} : \alpha \in K\}$ for $\mathscr{M} \subseteqq \mathscr{A}$ with $|\mathscr{M}| \leqq \mathfrak{m}$.

$$\begin{aligned}
\text{Similarly} \quad & \bigwedge \{K \in \mathscr{N} : \alpha \in K\} \\
&= \bigwedge \{l\psi \in \mathscr{N} : \alpha \in l\psi\} \\
&= [\bigwedge \{l \in N : \alpha \in l\psi\}]\psi \\
&= [\bigwedge \{l \in N : l\pi_\alpha = 1\}]\psi \text{ for } \mathscr{N} \subseteqq \mathscr{A} \text{ and } \alpha \in A.
\end{aligned}$$

This gives $$[\bigwedge \{l \in N : l\pi_\alpha = 1\}]\pi_\alpha = 1 \text{ if } |\mathscr{N}| = |N| \leqq \mathfrak{n}$$

since the π_α are $(\mathfrak{m}, \mathfrak{n})$-homomorphisms, so that $\alpha \in \bigwedge\{K \in \mathscr{N} : \alpha \in K\}$, and we have shown that (i) and (ii) of Proposition 3.1 are satisfied. Hence $\mathscr{L}$ is an $(\mathfrak{m}, \mathfrak{n})$-ring of sets.

For the converse assume $\mathscr{L}$ is isomorphic to an $(\mathfrak{m}, \mathfrak{n})$-ring of sets $\mathscr{L}'$. Then $\mathscr{L}'$ has a representation as a subdirect union of replicas of **2** and the working above readily reverses to establish the fact that the π_α are $(\mathfrak{m}, \mathfrak{n})$-homomorphisms.

4. $\mathfrak{n}$-prime $\mathfrak{m}$-ideals

We now discuss the notion of $\mathfrak{n}$-prime $\mathfrak{m}$-ideal. The first result is straightforward but the corollary is used to establish the equivalence between our ideals and the various concepts mentioned in the introduction. These concepts are not defined here – we refer to the papers concerned – for this reason the corollary is presented without proof.

PROPOSITION 4.1. *Let* $\mathscr{L} = \langle L; \vee, \wedge \rangle$ *be an* $(\mathfrak{m}, \mathfrak{n})$*-complete lattice. Then there is a one-one correspondence between*

(i) $\mathfrak{n}$*-prime* $\mathfrak{m}$*-ideals,*

(ii) $\mathfrak{m}$*-prime* $\mathfrak{n}$*-dual ideals,*

(iii) $(\mathfrak{m}, \mathfrak{n})$*-homomorphisms onto* **2**.

PROOF. It has already been remarked that (i) and (ii) are in one-one correspondence. Let $\psi : \mathscr{L} \to \mathbf{2}$ be an $(\mathfrak{m}, \mathfrak{n})$-homomorphism onto **2**. Then it is easy to see that $\{1\}\psi^{-1}$ is an $\mathfrak{m}$-prime $\mathfrak{n}$-dual ideal and $\{0\}\psi^{-1}$ is an $\mathfrak{n}$-prime $\mathfrak{m}$-ideal. Conversely if P is an $\mathfrak{n}$-prime $\mathfrak{m}$-ideal, we may define a map $\pi : \mathscr{L} \to \mathbf{2}$ by setting $l\pi = 0$ or 1 according as $l \in P$ or $l \notin P$. π may be checked to be an $(\mathfrak{m}, \mathfrak{n})$-homomorphism and our proposition is proved.

COROLLARY (Special Cases). Under the conditions of the proposition, with the appropriate values of $\mathfrak{m}$ and $\mathfrak{n}$, there is a one-one correspondence between the objects in the following groups.

A. $(\mathfrak{m} = 2, \mathfrak{n} = \infty)$

(i) prime principal dual ideals

(ii) join prime elements (V. K. Balachandran [2]); lattice bisectors (P. D. Finch [6]); $\mathscr{N}$-sub-irreducible elements for a certain $\mathscr{N}$ (J. R. Büchi [5]).

(iii) $(2, \infty)$-homomorphism onto **2**; lower complete homomorphisms onto **2** (P. D. Finch [6]).

B. $(\mathfrak{m} = \infty, \mathfrak{n} = 2)$

(i) ∞-prime dual ideals; completely prime dual ideals (G. Birkhoff & O. Frink [4]).

(ii) prime principal ideals

(iii) $(\infty, 2)$-homomorphisms onto **2**.

[5] On rings of sets

C. $(\mathfrak{m} = \infty, \mathfrak{n} = \infty)$

(i) completely prime principal dual-ideals

(ii) completely join prime elements (V. K. Balachandran [2]); supercompact elements (V. K. Balachandran [1]); completely join irreducible elements (G. N. Raney [7]).

(iii) (∞, ∞)-homomorphisms onto **2**; complete homomorphisms onto **2** (G. N. Raney [7]).

LEMMA 4.2. *Let $\mathscr{L} = \langle L; \vee, \wedge\rangle$ and $\mathscr{L}' = \langle L', \vee, \wedge\rangle$ be two $(\mathfrak{m}, \mathfrak{n})$-complete lattices. Suppose there is an $(\mathfrak{m}, \mathfrak{n})$-homomorphism*

$$\pi : \mathscr{L} \to \mathscr{L}'.$$

Then if P' is an $\mathfrak{n}$-prime $\mathfrak{m}$-ideal of $\mathscr{L}'$, $P = P'\pi^{-1}$ is an $\mathfrak{n}$-prime $\mathfrak{m}$-ideal of $\mathscr{L}$.

PROOF. P is well known to be an ideal of $\mathscr{L}$. We first show that P is an $\mathfrak{m}$-ideal. Let $\{l_\gamma : \gamma \in \Gamma\} \subseteqq P$ be such that $|\Gamma| \leqq \mathfrak{m}$. Then

$$\Big(\bigvee_{\gamma\in\Gamma} l_\gamma\Big)\pi = \bigvee_{\gamma\in\Gamma} l_\gamma\pi$$

and since $l_\gamma\pi \in P'$, $\forall\gamma \in \Gamma$, $\bigvee_{\gamma\in\Gamma} l_\gamma\pi \in P$, and we deduce that $\bigvee_{\gamma\in\Gamma} l_\gamma \in P = P'\pi^{-1}$.

Finally we show that P is $\mathfrak{n}$-prime. Suppose $\{l_\delta : \delta \in \Delta\} \subseteqq L$ is such that $|\Delta| \leqq \mathfrak{n}$ and $l_\delta \notin P$ $\forall\delta \in \Delta$.

Then $(\bigwedge_{\delta\in\Delta} l_\delta)\pi = \bigwedge_{\delta\in\Delta} l_\delta\pi$ and since $l_\delta \notin P$ $\forall\delta\in\Delta$ we have $l_\delta\pi \notin P'$ $\forall\delta\in\Delta$. Thus, since P' is $\mathfrak{n}$-prime, $\bigwedge_{\delta\in\Delta} l_\delta\pi \notin P'$ and so $\bigwedge_{\delta\in\Delta} l_\delta \notin P = P'\pi^{-1}$. The result is proved.

LEMMA 4.3. *Let $\mathscr{L} = \langle L; \vee, \wedge\rangle$ be an $(\mathfrak{m}, \mathfrak{n})$-ring of sets, subsets of a set $\mathscr{X}$. Then for any $x \in \mathscr{X}$, $P_x = \{l \in L : x \notin l\}$ is an $\mathfrak{n}$-prime $\mathfrak{m}$-ideal of $\mathscr{L}$.*

PROOF. P_x is clearly an ideal of $\mathscr{L}$. We show it is an $\mathfrak{m}$-ideal.

Let $\{l_\gamma : \gamma\in\Gamma\} \subseteqq P_x$ be such that $|\Gamma| \leqq \mathfrak{m}$. Since $x \notin l_\gamma$ for $\gamma \in \Gamma$, Proposition 3.1 (i) tells us that $x \notin \bigvee_{\gamma\in\Gamma} l_\gamma$ or $\bigvee_{\gamma\in\Gamma} l_\gamma \in P_x$.

Similarly let $\{l_\delta : \delta\in\Delta\} \subseteqq L$ be such that $|\Delta| \leqq \mathfrak{n}$ and $l_\delta \notin P_x \forall\delta\in\Delta$. Then Proposition 3.1 (ii) tells us that $x \in \bigwedge_{\delta\in\Delta} l_\delta$ or $\bigwedge_{\delta\in\Delta} l_\delta \notin P_x$.

We have thus proved P_x is $\mathfrak{n}$-prime and so it is an $\mathfrak{n}$-prime $\mathfrak{m}$-ideal.

5. Representation of lattices by $(\mathfrak{m}, \mathfrak{n})$-rings of sets

In this section we give a fundamental representation theorem and then show all such representations are of this form.

PROPOSITION 5.1. *Let $\mathscr{L} = \langle L; \vee, \wedge\rangle$ be an $(\mathfrak{m}, \mathfrak{n})$-complete lattice and $\mathscr{P} = \mathscr{P}(\mathscr{L}; \mathfrak{m}, \mathfrak{n})$ the set of all $\mathfrak{n}$-prime $\mathfrak{m}$-ideals of $\mathscr{L}$. We assume $\mathscr{P} \neq \Box$. Let $\mathscr{X}$ denote a non-empty subset of $\mathscr{P}$ and define a lattice $\mathscr{R}_{\mathscr{X}} = \langle R; \vee, \wedge\rangle$ by*

$\mathscr{R}_{\mathscr{X}} = \mathscr{L}\rho$ where $\rho = \rho_{\mathscr{X}}$ is defined by $\rho : \mathscr{L} \to \mathscr{R}_{\mathscr{X}}$, $l\rho = \{P \in \mathscr{P} : l \notin P\}$. Then $\mathscr{R}_{\mathscr{X}}$ is an $(\mathfrak{m}, \mathfrak{n})$-ring of sets and ρ is an $(\mathfrak{m}, \mathfrak{n})$-homomorphism.

PROOF. We show that ρ is an $(\mathfrak{m}, \mathfrak{n})$-homomorphism and it will then follow that $\mathscr{R}_{\mathscr{X}}$ is an $(\mathfrak{m}, \mathfrak{n})$-ring of sets. Take $\{l_\gamma : \gamma \in \Gamma\} \subseteqq L$ with $|\Gamma| \leqq \mathfrak{m}$.

Since

$$l_\gamma \leqq \bigvee_{\gamma \in \Gamma} l_\gamma$$

we deduce that

$$l_\gamma\rho \supseteqq (\bigvee_{\gamma \in \Gamma} l_\gamma)\rho$$

and hence

$$\bigcup_{\gamma \in \Gamma} l_\gamma\rho \supseteqq (\bigvee_{\gamma \in \Gamma} l_\gamma)\rho.$$

Now if $P \in \bigcup_{\gamma \in \Gamma} l_\gamma\rho$, then $l_\gamma \notin P$ for some $\gamma \in \Gamma$. Thus $\bigvee_{\gamma \in \Gamma} l_\gamma \notin P$ and hence $P \in (\bigvee_{\gamma \in \Gamma} l_\gamma)\rho$. We have proved ρ preserves joins of $\mathfrak{m}$ elements.

Next take $\{l_\delta : \delta \in \Delta\} \subseteqq L$ with $|\Delta| \leqq \mathfrak{n}$:

$$\bigwedge_{\delta \in \Delta} l_\delta \leqq l_\delta \quad \forall \delta \in \Delta$$

and so $(\bigwedge_{\delta \in \Delta} l_\delta)\rho \supseteqq l_\delta\rho$, giving

$$(\bigwedge_{\delta \in \Delta} l_\delta)\rho \supseteqq \bigcap_{\delta \in \Delta} l_\delta\rho.$$

For the reverse inclusion take $P \in (\bigwedge_{\delta \in \Delta} l_\delta)\rho$. Then $\bigwedge_{\delta \in \Delta} l_\delta \notin P$ and so, since P is $\mathfrak{n}$-prime, we must have $l_\delta \notin P \; \forall \delta \in \Delta$;

Thus $P \in \bigcap_{\delta \in \Delta} l_\delta\rho$ and we have

$$(\bigwedge_{\delta \in \Delta} l_\delta)\rho = \bigcap_{\delta \in \Delta} l_\delta\rho.$$

ρ is now proved to be an $(\mathfrak{m}, \mathfrak{n})$-homomorphism and the statements in the proposition all follow.

Our next result is basic.

PROPOSITION 5.2. *Let $\mathscr{L} = \langle L; \vee, \wedge\rangle$ be a $(\mathfrak{m}, \mathfrak{n})$-complete lattice, and ϕ a $(\mathfrak{m}, \mathfrak{n})$-homomorphism of $\mathscr{L}$ onto a $(\mathfrak{m}, \mathfrak{n})$-ring of sets $\mathscr{K} = \langle K; \cup, \cap\rangle$, subsets of some set $\mathscr{Y}$. Then there is a nonempty subset $\mathscr{X}$ of $\mathscr{P} = \mathscr{P}(\mathscr{L}; \mathfrak{m}, \mathfrak{n})$ and an isomorphism $\theta : \mathscr{K} \to \mathscr{R}_{\mathscr{X}}$ such that $\phi \circ \theta = \rho_{\mathscr{X}}$.*

PROOF. Let us first look at $\mathscr{K}$. Since $\mathscr{K}$ is a $(\mathfrak{m}, \mathfrak{n})$-ring of subsets of $\mathscr{Y}$, $P_y = \{k \in K : y \notin k\}$ is an $\mathfrak{n}$-prime $\mathfrak{m}$-ideal of $\mathscr{K}$ by Lemma 4.3. Also, since ϕ is a $(\mathfrak{m}, \mathfrak{n})$-homomorphism of $\mathscr{L}$ onto $\mathscr{K}$, $P_y\phi^{-1}$ is a $\mathfrak{n}$-prime $\mathfrak{m}$-ideal of $\mathscr{L}$ by Lemma 4.2.

Define $\mathscr{X} \subseteqq \mathscr{P}$ by $\mathscr{X} = \{P_y\phi^{-1} : y \in \mathscr{Y}\}$. In the statement of the proposition $\mathscr{R}_{\mathscr{X}}$ and $\rho = \rho_{\mathscr{X}}$ are defined as in Proposition 5.1. It remains to check that θ defined by $\phi \circ \theta = \rho_{\mathscr{X}}$ is an isomorphism of $\mathscr{K}$ onto $\mathscr{R}_{\mathscr{X}}$.

(i) θ is well defined. For suppose $l_1\phi = l_2\phi$ for $l_1, l_2 \in L$. Then

$$\{y \in \mathscr{Y} : l_1\phi \notin P_y\} = \{y \in \mathscr{Y} : l_2\phi \notin P_y\}$$

and so
$$\{y \in \mathscr{Y} : l_1 \notin P_y\phi^{-1}\} = \{y \in \mathscr{Y} : l_2 \notin P_y\phi^{-1}\}.$$

Thus
$$\{P \in \mathscr{X} : l_1 \notin P\} = \{P \in \mathscr{X} : l_2 \notin P\}$$

and so
$$l_1\rho = l_2\rho.$$

(ii) θ is an injection. For suppose $l_1\phi\theta = l_2\phi\theta$. Then $l_2\rho = l_2\rho$ by definition of θ, and the lines above reverse completely to prove $l_1\phi = l_2\phi$.

(iii) θ is clearly a surjection, for $\rho_{\mathscr{X}}$ is a surjection and so is ϕ.

(iv) We finally check that θ is an homomorphism. Take $k_1, k_2 \in K$ such that $k_i = l_i\phi$. Then

$$k_1 \vee k_2 = l_1\phi \vee l_2\phi = (l_1 \vee l_2)\phi$$

whence

$$\begin{aligned}(k_1 \vee k_2)\theta &= (l_1 \vee l_2)\phi \circ \theta = (l_1 \vee l_2)\rho = l_1\rho \vee l_2\rho \\ &= (l_1)\phi\theta \vee (l_2)\phi\theta = k_1\theta \vee k_2\theta.\end{aligned}$$

Similarly $(k_1 \wedge k_2)\theta = k_1\theta \wedge k_2\theta$ and θ is established to be an isomorphism. The proposition is thus proved.

We close with a theorem which determines when faithful representations exist. For the theorem, let $\mathscr{P}^d(\mathscr{L}; \mathfrak{m}, \mathfrak{n})$ denote the set of all $\mathfrak{m}$-prime $\mathfrak{m}$-dual ideals of $\mathscr{L}$.

THEOREM 5.3. *Let* $\mathscr{L} = \langle L; \vee, \wedge\rangle$ *be an* $(\mathfrak{m}, \mathfrak{n})$*-complete lattice. Then the following are equivalent*:

(i) $\mathscr{L}$ *is isomorphic with an* $(\mathfrak{m}, \mathfrak{n})$*-ring of sets.*

(ii) $(l] = \bigcap \{P \in \mathscr{P}(\mathscr{L}; \mathfrak{m}, \mathfrak{n}) : l \in P\}$ *for all* $l \in L$.

(iii) $[l) = \bigcap \{D \in \mathscr{P}^d(\mathscr{L}; \mathfrak{m}, \mathfrak{n}) : l \in D\}$ *for all* $l \in L$.

PROOF. Assume $\mathscr{L}$ is isomorphic with an $(\mathfrak{m}, \mathfrak{n})$-ring of sets. Then by Proposition 5.2 there must be a set $\mathscr{X} \subseteqq \mathscr{P}$ such that $\rho_{\mathscr{X}}$ is one-one. Thus we see that the map $l \to \{P \in \mathscr{P}(\mathscr{L}; \mathfrak{m}, \mathfrak{n}): l \in P\}$ is also one-one and hence

$$(l] = \bigcap \{P \in \mathscr{P}(\mathscr{L}; \mathfrak{m}, \mathfrak{n}): l \in P\}.$$

So (i) $\Rightarrow$ (ii).

It is clear that (ii) and (iii) are equivalent. Let us assume (ii). Then the map ρ is seen to be one-one and so $\mathscr{L}$ has a faithful representation as an $(\mathfrak{m}, \mathfrak{n})$-ring of subsets of $\mathscr{P}$. The proof of the theorem is now complete.

REMARK. We do not deduce all possible corollaries. It suffices to illustrate the method by taking $\mathfrak{m} = \mathfrak{n} = \infty$ and deducing the result.

COROLLARY. (G. N. Raney [7], V. K. Balachandran [1]). *A complete lattice $\mathscr{L}$ is isomorphic with a complete ring of sets if and only if $\mathscr{L}$ possesses a join basis of completely join irreducibles.*

PROOF. Take $\mathfrak{m} = \mathfrak{n} = \infty$ in Theorem 5.3 parts (i) and (ii). An ∞-prime ∞-dual ideal is equivalent to a completely prime principal dual ideal and its generator is thus a completely join irreducible element. Since the intersection of a family of principal dual ideals is the principal dual ideal generated by the join of the generators of the family, we see that (iii) tells us that for any $l \in L$

$[l) = \bigcap_{\nu \in V} [j_\nu) = [\bigvee_\nu j_\nu)$ where the j_ν are completely join irreducible. This is equivalent to $l = \bigvee j_\nu$ and our Corollary is proved.

References

[1] V. K. Balachandran, "A characterisation of $\Sigma\Delta$-rings of subsets', *Fundamenta Mathematica* 41 (1953), 38—41.

[2] V. K. Balachandran, 'On complete lattices and a problem of Birkhoff and Frink,' *Proc. Amer. Math. Soc.* 6 (1953), 548—553.

[3] G. Birkhoff, *Lattice Theory* (Amer. Math. Soc. Colloq. New York 1948).

[4] G. Birkhoff & O. Frink, 'Representations of lattices by sets', *Trans. Amer. Math. Soc.* 64 (1948), 299—316.

[5] J. R. Büchi, 'Representations of complete lattices by sets', *Portugaliae Mathematicae* 11 (1952), 151—167.

[6] P. D. Finch, 'On the lattice equivalence of topological spaces', *Journ. Aust. Math. Soc.* 6 (1966), 495—511.

[7] G. N. Raney, 'Completely distributive complete lattices', *Proc. Amer. Math. Soc.* 3 (1952), 677—680.

BULL. AUSTRAL. MATH. SOC. MOS 54F05, 06A35
VOL. 6 (1972), 177-183.

Profinite posets

T.P. Speed

The class of ordered topological spaces which are projective limits of finite partially ordered sets (equipped with the restriction of the product of the discrete topologies) is shown to coincide with the class of compact totally order-disconnected ordered topological spaces. Hence this is another category of spaces equivalent to the category of distributive lattices with zero and unit.

1. Introduction

In her papers [7], [8], Miss Priestley has discussed in detail the equivalence of the category of compact totally order-disconnected ordered topological spaces (with continuous monotone maps) and the category of distributive lattices with zero and unit (with zero and unit preserving lattice homomorphisms). More recently it has been shown [10] that the partially ordered set (= poset) of all prime ideals of such a lattice must be of the form $\varprojlim_{\alpha \in I} X_\alpha$ where each X_α $(\alpha \in I)$ is a finite poset. A synthesis of these two results immediately suggests itself, and we prove the following:

THEOREM. *Let X be an ordered topological space. Then X is compact and totally order-disconnected iff $X \cong \varprojlim_{\alpha \in I} X_\alpha$, where $\{X_\alpha, f_{\alpha\beta}\}$ is an inverse system of finite posets each equipped with the discrete topology.*

We prove this theorem in §§3, 4. An ordered topological space which

Received 8 October 1971. The author would like to thank Miss Hilary Priestley for the preprint of her paper [8].

is of the form $\varprojlim_{\alpha\in I} X_\alpha$ for an inverse system $\{X_\alpha, f_{\alpha\beta}\}$ of finite discretely topologised posets will be called a *profinite poset* by analogy with the group case. Thus the theorem above is an analogue of the well known characterization of profinite groups; see also [6] for other related results.

2. Preliminaries

The notation and terminology of [7], [8] will be adopted without further comment. Let us write $A \nleq B$ for subsets A, B of a poset $(X; \leq)$ iff for all $a \in A$, $b \in B$ we have $a \nleq b$.

LEMMA 1. *Let* $(X, \tau, \leq)$ *be a compact totally order-disconnected space. Then for disjoint closed sets* A, B *we have* $A \nleq B$ *iff there is an order-disconnection* $(U|L)$ *such that* $A \subseteq U$, $B \subseteq L$.

Proof. Assume $A \nleq B$. Then since X is totally order-disconnected, for any $x \in A$, $y \in B$ there is an order-disconnection $(U_{x,y}|L_{x,y})$ such that $x \in U_{x,y}$, $y \in L_{x,y}$. Fix x. Then the family $\{L_{x,y} : y \in B\}$ constitutes an open cover of B, and so there exists a finite sub-cover $\left\{L_{x,y_j} : j = 1, 2, \ldots, n\right\}$. Put $U_x = \bigcap_{j=1}^{n} U_{x,y_j}$ and $L_x = \bigcup_{j=1}^{n} L_{x,y_j}$ and we observe that $(U_x|L_x)$ is an order-disconnection with $x \in U_x$, $B \subseteq L_x$. Now the family $\{U_x : x \in A\}$ is an open cover of A and so has a finite subcover $\left\{U_{x_i} : i = 1, 2, \ldots, m\right\}$. Put $U = \bigcup_{i=1}^{m} U_{x_i}$ and $L = \bigcap_{i=1}^{n} L_{x_i}$ and we have an order-disconnection $(U|L)$ such that $U \supseteq A$, $L \supseteq B$ as required.

REMARK. This lemma shows that, as one would expect, compact subsets behave in much the same way as points in compact ordered spaces. For further evidence of this see Theorem 4, p. 46 of [4]. When the order is trivial, Lemma 1 reduces to a well known result for boolean algebras.

Let $(X; \leq)$ be a poset and $\rho \subseteq X \times X$ an equivalence relation on

X . Then one way of defining a quasi-order on X/ρ is to write $x/\rho \leq' y/\rho$ iff there exists $x_1 \equiv x\ (\rho)$, $y_1 \equiv y\ (\rho)$ such that $x_1 \leq y_1$. Unfortunately this relation $\leq'$ is not always a partial order on X/ρ ; when it is we say that ρ is *order compatible*. Thus the equivalence ρ on X is order compatible iff for any x_1, y_1 in X , if $x_1 \equiv x_2\ (\rho)$ and $y_1 \equiv y_2\ (\rho)$ and $x_1 \leq y_1$, $x_2 \geq y_2$ then $x_1 \equiv x_2 \equiv y_1 \equiv y_2\ (\rho)$. Equivalently, ρ is order compatible iff for any $x, y \in X$ such that $x \not\equiv y\ (\rho)$, we have either $\{x_1 : x_1 \equiv x\ (\rho)\} \nleq \{y_1 : y_1 \equiv y\ (\rho)\}$ or $\{x_1 : x_1 \equiv x\ (\rho)\} \ngeq \{y_1 : y_1 \equiv y\ (\rho)\}$.

3. First proof of the theorem

Suppose $X = \varprojlim_{\alpha \in I} X_\alpha$ where $\{X_\alpha, f_{\alpha\beta}\}$ is an inverse system of finite posets each equipped with the discrete topology, and I is a directed set. Then X is certainly a compact space ([1], Chapter I, §9.6, Proposition 8). For any $\alpha \in I$ and $x'_\alpha \in X_\alpha$ write $U_{x'_\alpha} = \{x \in X : x_\alpha \geq x'_\alpha\}$, $L_{x'_\alpha} = \{x \in X : x_\alpha \leq x'_\alpha\}$ and $T_{x'_\alpha} = \{x \in X : x_\alpha = x'_\alpha\}$, where $x = \langle x_\alpha \rangle_{\alpha \in I}$ denotes a typical element of X . Then $T_{x'_\alpha}$ is clopen, and (since each X_α is discrete) so are $U_{x'_\alpha}, L_{x'_\alpha}$. Further $U_{x'_\alpha}$ is increasing and $L_{x'_\alpha}$ is decreasing. We now prove that X is totally order-disconnected. Suppose $x \nleq y$ in X ; then for some $\alpha \in I$ we must have $x_\alpha \nleq y_\alpha$. Thus $\left(U_{x_\alpha} \middle| L_{y_\alpha}\right)$ is an order-disconnection and $x \in U_{x_\alpha}$, $y \in L_{y_\alpha}$, and so the result is proved.

For the converse we suppose that X is compact and totally order-disconnected. Let R denote the family of all clopen order compatible equivalences ρ on X , that is, all order compatible equivalences of the form $\rho = \bigcup_{i=1}^{m} V_i \times V_i$ for some finite partition $\{V_i\}$ of X into open sets. Then $X_\rho = X/\rho$ is a finite poset, and, when equipped with the discrete topology, is a continuous monotone image of X under the canonical projection $pr_\rho : X \to X/\rho$.

T.P. Speed

Now Lemma 1 implies that the equivalence ρ is order compatible iff $V_i \neq V_j$ implies that there exists an order disconnection $(U|L)$ such that $V_i \subseteq U$, $V_j \subseteq L$ or $V_j \subseteq U$, $V_i \subseteq L$. We now prove that the family of all clopen order compatible equivalences is directed, and that $\bigcap\{\rho : \rho \in R\} = \Delta$, the diagonal of $X \times X$. The last remark is easy, for if $x \neq y$ then either $x \nleq y$ or $y \nleq x$. Suppose $x \nleq y$; then there is an order-disconnection $(U|L)$ such that $x \in U$, $y \in L$. But it is easily checked that $\{U, L, U^c \cap L^c\}$ is a partition which induces an order compatible equivalence ρ, and hence $x \nleq y\ (\rho)$.

Suppose ρ and ρ' are two clopen order compatible equivalences induced by the partitions $\{V_i : i = 1, 2, \ldots, m\}$ and $\{V'_j : j = 1, 2, \ldots, n\}$ respectively. Then the partition

$$\{V_i \cap V'_j : i = 1, 2, \ldots, m,\ j = 1, 2, \ldots, n,\ V_i \cap V'_j \neq \emptyset\}$$

induces an order compatible equivalence $\rho \vee \rho'$. For if $V_i \cap V'_j \neq V_{i_1} \cap V'_{j_1}$, then either $V_i \neq V_{i_1}$ or $V'_j \neq V'_{j_1}$, say the former. Then either $V_i \nleq V_{i_1}$ or $V_{i_1} \nleq V_i$, again suppose the former. By Lemma 1 there is an order-disconnection $(U|L)$ such that $V_i \subseteq U$ and $V_{i_1} \subseteq L$. But now $V_i \cap V'_j \subseteq U$ and $V_{i_1} \cap V'_{j_1} \subseteq L$ which proves that $V_i \cap V'_j \nleq V_{i_1} \cap V'_{j_1}$ and so $\rho \vee \rho'$ is order compatible.

We now collect the foregoing results: the system $\{X_\rho : \rho \in R\}$ where for $\rho \subseteq \rho'$ the canonical map $f_{\rho\rho'} : X_{\rho'} \to X_\rho$ is continuous and monotone, and R is directed, becomes an inverse system $\{X_\rho, f_{\rho\rho'}\}$. The map $\phi : X \to \varprojlim_{\rho \in R} X_\rho$ given by $\phi(x) = \left\langle pr_\rho(x) \right\rangle_{\rho \in R}$ is continuous, bijective, and an order isomorphism, and so X and $\varprojlim_{\rho \in R} X_\rho$ are homeomorphic as required.

4. Second proof of the theorem

We quickly sketch an alternative, shorter, proof of the theorem. It does however, have the disadvantage of using results from [2], [5], [10] of a non-topological nature, but is the way the theorem was originally deduced.

Suppose $X = \varprojlim_{\alpha\in I} X_\alpha$ is a projective limit of finite, discretely topologised posets. Then $X_\alpha = \mathrm{Patch} A_\alpha$ for a unique distributive lattice A_α. Thus $X = \varprojlim X_\alpha \cong \varprojlim \mathrm{Patch} A_\alpha \cong \mathrm{Patch}(\varinjlim A_\alpha) = \mathrm{Patch}\, A$ where $A = \varinjlim A_\alpha$ is the direct limit of the direct system $\{A_\alpha, f^*_{\alpha\beta}\}$, and where $f^*_{\alpha\beta} : A_\alpha \to A_\beta$ is the dual map to $f_{\alpha\beta} : X_\beta \to X_\alpha$ for $\alpha \leq \beta$. By the main result of [7] and some remarks of [2], $X = \mathrm{Patch}\, A$ is compact and totally order-disconnected.

Conversely, suppose X is compact and totally order-disconnected. By the main result of [7], $X = \mathrm{Patch}\, A$ for a unique distributive lattice A. Write $A = \varinjlim A_\alpha$ as a direct limit of its finitely generated (finite) sublattices A_α. Then

$$\mathrm{Patch}\, A \cong \mathrm{Patch}(\varinjlim A_\alpha) \cong \varprojlim \mathrm{Patch}\, A_\alpha = \varprojlim X_\alpha$$

where $\{X_\alpha\}$ is a family of finite posets equipped with discrete topologies. The details of this proof can be reconstructed from [2], [5].

In a notice which appeared after this note was written, Joyal [3] states a theorem closely related to our main result. His proof is probably more like the one sketched above.

5. Final remarks

The theorem of this note and other results show that the following categories are equivalent:

(i) distributive lattices with zero and unit (with zero and unit preserving homomorphisms);

(ii) spectral spaces (with spectral maps);

(iii) compact totally order-disconnected spaces (with continuous monotone maps);

T.P. Speed

(iv) profinite posets (with continuous monotone maps).

The study of the relations between (i) and (ii) was begun by Stone in [11]; some further details are in [9] and the forthcoming part II, while much useful information is in [2]. The relation (i) ↔ (iii) is the object of [7], [8], and the connections between (i), (ii) and (iii) are being studied at the moment.

References

[1] Nicolas Bourbaki, *Elements of mathematics, General topology*, Part I (Hermann, Paris; Addison-Wesley, Reading, Massachussetts; Palo Alto; London; Don Mills, Ontario; 1966).

[2] M. Hochster, "Prime ideal structure in commutative rings", *Trans. Amer. Math. Soc.* 142 (1969), 43-60.

[3] André Joyal, "Spectral spaces and distributive lattices", *Notices Amer. Math. Soc.* 18 (1971), 393-394. "Spectral spaces. II", *Notices Amer. Math. Soc.* 18 (1971), 618.

[4] Leopoldo Nachbin, *Topology and order* (Van Nostrand, Princeton, New Jersey; Toronto; New York; London; 1965).

[5] A. Nerode, "Some Stone spaces and recursion theory", *Duke Math. J.* 26 (1959), 397-406.

[6] Katsumi Numakura, "Theorems on compact totally disconnected semigroups and lattices", *Proc. Amer. Math. Soc.* 8 (1957), 623-626.

[7] H.A. Priestley, "Representation of distributive lattices by means of ordered Stone spaces", *Bull. London Math. Soc.* 2 (1970), 186-190.

[8] H.A. Priestley, "Ordered topological spaces and the representation of distributive lattices", *Proc. London Math. Soc.* (to appear).

[9] T.P. Speed, "Spaces of ideals of distributive lattices. I: Prime ideals", *Bull. Soc. Roy. Sci. Liège* 38 (1969), 610-628.

[10] T.P. Speed, "On the order of prime ideals", manuscript.

[11] M.H. Stone, "Topological representations of distributive lattices and Brouwerian logics", *Čas. Mat. Fys.* 67 (1937), 1-25.

Department of Probability and Statistics,
The University of Sheffield,
Sheffield.

ON THE ORDER OF PRIME IDEALS

67u

T. P. SPEED

A poset X is isomorphic to the poset of all prime ideals of a (distributive) lattice with zero and unit if, and only if, X is the projective limit of an inverse system of finite posets.

1. Introduction

The problem of characterising posets of the form X_A where X_A denotes the set of all prime ideals of the (distributive) lattice A with zero and unit, ordered by inclusion, was raised by C. C. Chen and G. Grätzer [2]. In a similar context M. Hochster [3] discussed the same problem for commutative rings with identity, and gave a solution in terms of a certain family of order preserving maps. We note below that these problems have a common solution.

Let us call a poset X *profinite* if $X \cong \varprojlim X_\alpha$ for an inverse system $\{(X_\alpha), (\phi_{\alpha\beta})\}$ of *finite* posets defined over some directed set I. In terms of this notion we will prove the following:

THEOREM. *A poset X is isomorphic to the poset of all prime ideals of a (distributive) lattice with zero and unit if, and only if, X is profinite.*

COROLLARY 1. *A poset X is isomorphic to the poset of all prime ideals of a commutative ring with identity if, and only if, X is profinite.*

COROLLARY 2. *A poset X is isomorphic to the poset of all prime ideals of a (distributive) lattice [resp. lattice with zero, lattice with unit] if, and only if, X with largest and smallest [resp. with largest, with smallest] element adjoined, is profinite.*

2. Preliminary lemmas

For any (finite) distributive lattice A the set X_A of all prime ideals of A ordered by inclusion is a (finite) poset; further if $f: A \to A'$ is a zero and unit preserving lattice homomorphism between distributive lattices A and A' with zero and unit, there is an induced order preserving map $f^*: X_{A'} \to X_A$.

Also, if X is a finite poset, there is a finite distributive lattice $A_X = A$, unique up to isomorphism, such that $X \cong X_A$; again if $\phi: X \to X'$ is an order preserving map between finite posets X and X', there is an induced zero and unit preserving lattice homo-

Presented by G. Grätzer. Received November 24, 1971. Accepted for publication in final form April 4, 1972.

morphism $*\phi: A_{X'} \to A_X$. All these results are well known [1], and we summarise them in:

LEMMA 1. *The assignment $A \mapsto X_A$, $f \mapsto f^*$, defines a contravariant functor between the category $\mathscr{D}_F$ of all finite distributive lattices (with zero and unit preserving lattice homomorphisms) and the category $\mathscr{P}_F$ of all finite posets (with order preserving maps). This functor is a category equivalence.*

Let I be a directed set and suppose $\{(A_\alpha), (f_{\alpha\beta})\}$ is a direct system in $\mathscr{D}_F$ over I. Denote by X_α the poset of all prime ideals of A_α and by X the poset of all prime ideals of $A = \underrightarrow{\lim} A_\alpha$. Then we can see that each $x \in X$ defines a thread (x_α), $x_\alpha \in X_\alpha$, $\alpha \in I$, such that if $\alpha \leqslant \beta$, $x_\beta f_{\alpha\beta}^{-1} = x_\alpha$: we simply put $x_\alpha = x f_\alpha^{-1}$ where $f_\alpha: A_\alpha \to A$ is the canonical map into the direct limit. Conversely each such thread (x_α) can be readily seen to define an element $x \in X$: we put $x = \bigcup_\alpha x_\alpha f_\alpha$. This correspondence can be shown to be bijective and order preserving in both directions, and we then have

LEMMA 2. $X_A \cong \underleftarrow{\lim} X_\alpha$.

3. Proofs of the main results

We first prove the theorem. Let X be a profinite poset i.e. $X \cong \underleftarrow{\lim} X_\alpha$ where $\{(X_\alpha), (\phi_{\alpha\beta})\}$ is an inverse system in $\mathscr{P}_F$ relative to a directed set I. By Lemma 1 we then have a direct system $\{(A_\alpha), (f_{\alpha\beta})\}$ in $\mathscr{D}_F$ with $X_{A_\alpha} = X_\alpha$, $f_{\alpha\beta} = *\phi_{\alpha\beta}$. Put $A = \underrightarrow{\lim} A_\alpha$. By Lemma 2 $X_A \cong X$ and we have proved that X arises as X_A for a suitable distributive lattice A zero and unit.

Conversely, let A be a distributive lattice with zero and unit. Then we may write $A = \underrightarrow{\lim} A_\alpha$ where $\{(A_\alpha), (f_{\alpha\beta})\}$ is the direct system in $\mathscr{D}_F$ of all finite sublattices of A containing the zero and unit of A, with connecting maps $f_{\alpha\beta}: A_\alpha \to A_\beta$ when $A_\alpha \subseteq A_\beta$ being the canonical injections. By Lemma 1 we then have an inverse system $\{(X_\alpha), (\phi_{\alpha\beta})\}$ in $\mathscr{P}_F$ with $X_\alpha = X_{A_\alpha}$, $\phi_{\alpha\beta} = f_{\alpha\beta}^*$. Put $X = \underleftarrow{\lim} X_\alpha$. By Lemma 2 $X \cong X_A$ and we have proved that X_A is profinite.

This completes the proof of the theorem.

Corollary 1 can be proved using Proposition 12 of [3]; we omit the details.

If a distributive lattice fails to have a zero [resp. unit, zero and unit] we can simply add one, thereby adding a smallest [resp. largest, smallest and largest] prime ideal. By the theorem the poset of all prime ideals obtained must be profinite, and there is a natural converse. Thus we have Corollary 2.

4. Final remarks

Since a first draft of the above results was written, equivalent results were announced (without proof) by A. Joyal [4].

In conclusion I would like to thank Professor G. Grätzer for his remarks concerning the first draft, and also Brian Davey for his interest.

REFERENCES

[1] Garrett Birkhoff, Lattice Theory (A.M.S. Colloq. Publ. 25 Providence, R.I) 3rd Edition 1967.
[2] C.C. Chen and G. Grätzer, *Stone lattices II: Structure theorems*, Can. J. Math. *21* (1969), 895–903.
[3] M. Hochster, *Prime ideal structure in commutative rings*, Trans. Amer. Math. Soc. *142* (1969), 43–60.
[4] A. Joyal, *Spectral spaces and distributive lattices*, Abstract 71T A18 Notices A.M.S. *18* (1971), 292.

University
Sheffield
U.K.

ON RINGS OF SETS II. ZERO-SETS

Dedicated to the memory of Hanna Neumann

T. P. SPEED

(Received 2 May 1972)

Communicated by G. B. Preston

Introduction

In an earlier paper [11] we discussed the problem of when an (m, n)-complete lattice $\mathbf{L}$ is isomorphic to an (m, n)-ring of sets. The condition obtained was simply that there should exist sufficiently many prime ideals of a certain kind, and illustrations were given from topology and elsewhere. However, in these illustrations the prime ideals in question were all principal, and it is desirable to find and study examples where this simplification does not occur. Such an example is the lattice $\mathbf{Z}(X)$ of all zero-sets of a topological space X; we refer to Gillman and Jerison [5] for the simple proof that $\mathbf{Z}(X)$ is a $(2, \sigma)$-ring of subsets of X, where we denote aleph-zero by σ.

Lattices of the form $\mathbf{Z}(X)$ have occurred recently in lattice theory in a number of places, see, for example, Mandelker [10] and Cornish [4]. These writers have used such lattices to provide examples which illuminate a number of results concerning annihilators and Stone lattices. We also note that, following Alexandroff, a construction of the Stone-Čech compactification can be given using ultrafilters on $\mathbf{Z}(X)$; the more recent Hewitt realcompactification can be done similarly, and these topics are discussed in [5]. A relation between these two streams of development will be given below.

In yet another context, Gordon [6], extending some aspects of the work of Lorch [9], introduced the notion of a zero-set space $(X, \mathscr{Z})$. This is a structure abstracted from the system consisting of a set X and the family $\mathscr{Z}$ of zero-sets of the functions in a uniformly closed ring of real-valued functions defined on X. Gordon's axioms naturally embody some of the lattice-theoretic properties of $\mathbf{Z}(X)$ for a topological space X, but as we shall see below, they are more general.

We can now explain the contents of this paper. After listing our notation and terminology, we give some lattice-theoretic results which are necessary for subsequent analysis, but not without interest separately. We then give some constructions, similar to Urysohn's, of certain functions separating disjoint sets. They are more delicate than the usual since the family of sets used is closed under (finite unions and) countable intersections only, and hence the notion of closure is not available. Also these results enable us to give alternative proofs of some results of Gordon [6], thus avoiding the use of proximity spaces and the consequent application of Čech's difficult version of Urysohn's lemma, valid for uniformizable proximity spaces. In §§4,5 we turn to the main task which is find properties of $\mathbf{Z}(X)$ in addition to those which follow from its being a $(2,\sigma)$-ring of sets. Our results include algberaic characterisations of $\mathbf{Z}(X)$ for X a compact, respectively arbitrary, topological space.

To conclude this introduction we gratefully thank Drs. J. W. Baker and C. J. Knight for listening to, and helpfully commenting upon, early versions of the material presented below. Also the referee is to be thanked for pointing out an incorrect result stated in the first version, and for remarks leading to some shortening of proofs.

1. Notation and terminology

(1.1) *Lattice theory.* Most of the concepts from lattice theory we need are defined somewhere in Birkhoff [1], while the more special ones relating to rings of sets and special prime ideals are discussed in [11]. All our lattices will be assumed to possess a zero (least element) 0 and unit (greatest element) 1, and all sublattices will be assumed to contain the same zero and unit. The join and meet operations are denoted $\vee$ and $\wedge$ respectively, and thus a lattice can be considered as an abstract algebra $\mathbf{L} = (L; \vee, \wedge, 0, 1)$ with carrier L; we use the partial order on L without comment. Typical elements of L will be denoted a, b, c, $d, \cdots$; typical prime or minimal prime ideals will be denoted w, x, $y, \cdots$. We will abbreviate the term $(2,\sigma)$-prime (see [11]) to σ-prime, in accordance with usual practice. A lattice is said to have *enough* ideals of a specified type if distinct elements of the lattice can be separated by ideals of that type. The lattice $\mathbf{L}$ is said to be a $(2,\sigma)$-*regular sublattice* of the lattice $\mathbf{L}'$ if $\mathbf{L}$ is a sublattice of $\mathbf{L}'$ such that countable meets of elements in $\mathbf{L}'$ which exist in $\mathbf{L}'$ or $\mathbf{L}$ exist in both and coincide.

(1.2) *Topology.* Our general reference in this sphere is Bourbaki [2], while the reference for the less common concepts used below, such as zeroset, z-filter, realcompactification etc. is Gillman and Jerison [5]. We will reserve W, X, Y for topological spaces; generic elements will be denoted by the corresponding lower case letter; typical subsets will be written A, B, $C, \cdots$; typical open sets $G, \cdots$; typical closed sets $F, \cdots$.

(1.3) *General.* For subsets A, B of a set X we write $A \cup B$, $A \cap B$ for set union and intersection respectively, and $\complement A$ for the complement of A in X. The empty set is denoted ϕ. If $f: X \to Y$ is a map, we write fA for the direct image of

$A \subseteq X$ and $f^{-1}B$ for the inverse image of $B \subseteq Y$; parentheses will only be included where necessary. The unit interval $\{t \in \mathbb{R}: 0 \leqq t \leqq 1\}$ is denoted $[0, 1]$.

2. Some lattice-theoretic results

Our first definition is based on the work of Cornish [4]; see also Kerstan [7] §6, Definition 2 for a closely related definition.

DEFINITION 2.1. A lattice **L** is *normal* if for any pair a, $b \in L$ with $a \wedge b = 0$, there exists c, $d \in L$ such that $a \wedge c = b \wedge d = 0$ and $c \vee d = 1$.

It is not hard to see that a Hausdorff space X is normal if, and only if, the lattice $\mathbf{F}(X)$ of all closed subsets of X is a normal lattice. Further it has been known for some time that the lattice $\mathbf{Z}(X)$ of all zero-sets of a topological space X is a normal lattice.

A number of equivalent formulations of 2.1 in the case **L** a distributive lattice are given in [4], and although we need none of these, we note the following: a distributive lattice **L** is normal if, and only if, every prime ideal contains a unique minimal prime ideal. This last result is known for $\mathbf{Z}(X)$ in the form: a prime z-filter is contained in a unique z-ultrafilter, ([5] 2.13). We also refer to [4] for many consequences of normality. For later use we note that any Boolean lattice is normal.

Another topologically inspired concept we need is that of a G_δ-element of a lattice **L**, and again we note that a similar idea occurs in [7].

DEFINITION 2.2. An element $a \in L$ is a G_δ in the lattice **L** if there exists a sequence $\{a_n : n \geqq 1\}$ of (not necessarily distinct) elements of L with the following properties:

(α) $a \wedge a_n = 0$ for all n;
(β) if for $b \in L$ we have $b \wedge a_n = 0$ for all n, then $b \leq a$.

Our final definition in this section is the following abstraction of the analogous topological property.

DEFINITION 2.3. A lattice **L** is *perfectly normal* if (α) **L** is normal; and (β) every $a \in L$ is a G_δ.

Clearly a Hausdorff space X is perfectly normal if, and only if, the lattice $\mathbf{F}(X)$ is perfectly normal. Also it is easy to prove ([6] 2.3) that for any topological space X, the lattice $\mathbf{Z}(X)$ is perfectly normal.

We turn to some algebraic consequences of the definitions.

LEMMA 2.4. *A lattice* **L** *in which every* $a \in L$ *is a* G_δ *is disjunctive.*

PROOF: Take $a \nleqq b$ in L. By 2.2 (β) there must exist an n such that $a \wedge b_n \neq 0$ while by 2.2 (α) $b \wedge b_n = 0$. This proves the result.

T. P. Speed [4]

A deeper result which we use frequently below requires the characteristic property of a minimal prime ideal, Kist [8] viz: a prime ideal x of a distributive lattice **L** is minimal if, and only if, for any $a \in x$ there exists $b \notin x$ such that $a \wedge b = 0$.

LEMMA 2.5. *Let y be a σ-prime ideal in a $(2, \sigma)$-complete perfectly normal distributive lattice* **L**. *Then y is a minimal prime ideal.*

PROOF: Let $a \in y$; we must find $b \notin y$ such that $a \wedge b = 0$. Since a is a G_δ there exists a sequence $\{a_n: n \geqq 1\}$ with properties 2.2 (α), (β). Thus $a \wedge a_n = 0$, and so normality of L implies the existence of two sequences $\{c_n\}$, $\{d_n\}$ with: $a \wedge c_n = 0 = a_n \wedge d_n$ and $c_n \vee d_n = 1$ for all n. If, for some n, $c_n \notin y$, then we are through. Suppose now that $c_n \in y$ for all n; then $d_n \notin y$ for all n, and by the σ-prime property of y, $d = \bigwedge_n d_n \notin y$. But for all n, $a_n \wedge d \leqq a_n \wedge d_n = 0$ and so by 2.2 (β) $d \leqq a$ which contradicts $a \in y$, $d \notin y$.

Hence $a \wedge c_n = 0$ for some $c_n \notin y$ and y is minimal.

3. Constructions similar to Urysohn's

In this section we will be working with a $(2, \sigma)$-ring of subsets of a set X satisfying various conditions, and a careful analysis will enable us to extend the construction of a continuous function separating disjoint closed sets to this situation. We conclude by giving an alternative, direct, proof of a result of Gordon.

THEOREM 3.1. *Let* **H** *be a $(2, \sigma)$-ring of subsets of a set X. Then the following are equivalent:*

1) **H** *is a normal lattice.*

2) *For any A, $B \in H$ with $A \cap B = \phi$ there exists a function $f: X \to [0, 1]$ such that*

(α) $f^{-1}F \in H$ *for every closed subset F of* $[0, 1]$;

(β) $A \subseteq f^{-1}\{0\}$, $B \subseteq f^{-1}\{1\}$.

PROOF: 1) implies 2). We will explain the proof backwards thus motivating the construction. Let A, $B \in H$ with $A \cap B = \phi$ be given. Our aim is to define a system

(*) $$\mathscr{U} = \{U(t), F(t): 0 \leqq t \leqq 1\} \text{ where}$$

(i) $\complement U(t) \in H$, $F(t) \in H$, $0 \leqq t \leqq 1$;

(ii) $A \subseteq U(0)$, $B \subseteq \complement U(1)$;

(iii) If $0 \leqq t < t' \leqq 1$ then $U(t) \subseteq F(t) \subseteq U(t')$.

Then we will see that the well-known procedure of defining a map $f: X \to [0, 1]$ by writing, for $x \in X$:

(**) $$f(x) = \inf\{t: x \in U(t)\}$$

gives a function satisfying:

(iv) $f^{-1}[0,t] = F(t), f^{-1}[t,1] = \complement U(t)$.

Having done this we may take an arbitrary closed subset $[0,1]\setminus\bigcup_n(\alpha_n,\beta_n)$ of $[0,1]$ and find that

$$\begin{aligned} f^{-1}\,[0,1]\setminus\bigcup_n\ (\alpha_n,\beta_n) \\ &= f^{-1}\ \bigcap_n\ \{[0,\alpha_n]\cup[\beta_n,1]\} \\ &= \bigcap_n\ \{f^{-1}[0,\alpha_n]\cup f^{-1}[\beta_n,1]\} \\ &\in H \text{ as required.}\end{aligned}$$

Thus our function f so constructed satisfies (α) and (β) of (3.1)2) above.

An so we turn to defining the system $\mathscr{U}$. To do this we first define a subsystem $\mathscr{U}_\Delta$, where Δ is the set of binary rationals in $[0,1]$:

(*)′ $$\mathscr{U}_\Delta = \{U(\delta),\ F(\delta)\colon \delta\in\Delta\} \text{ where}$$

(i)′ $\complement U(\delta)\in H,\ F(\delta)\in H,\ \delta\in\Delta$;
(ii)′ $A\subseteq U(0),\ B\subseteq \complement U(1)$;
(iii)′ If $0\leqq\delta<\delta'\leqq 1$ then $U(\delta)\subseteq F(\delta)\subseteq U(\delta')$.

Let us suppose for the moment that $\mathscr{U}_\Delta$ is defined and satisfies (i)′, (ii)′ and (iii)′. Then if we write, for $0\leqq t\leqq 1$:

(†) $$U(t) = \bigcup_{\delta>t} U(\delta),\ F(t) = \bigcap_{\delta>t} F(\delta),$$

we clearly obtain a system $\mathscr{U}$ satisfying (i) and (ii). We check (iii). Take t, t' with $0\leqq t<t'\leqq 1$; there exists $\delta,\ \delta',\ \delta''\in\Delta$ with $t<\delta<\delta'<\delta''<t'$, and so by (iii)′ and (†)

$$U(t)\subseteq U(\delta)\subseteq F(\delta)\subseteq U(\delta')\subseteq F(\delta')\subseteq U(\delta'')\subseteq U(t').$$

Clearly $F(\delta)\subseteq F(t)\subseteq F(\delta')$ follows from (†) and so with the above we obtain

$$U(t)\subseteq F(\delta)\subseteq F(t)\subseteq F(\delta')\subseteq U(t') \text{ which implies (iii).}$$

Thus our problem reduces to constructing $\mathscr{U}_\Delta$ satisfying (i)′, (ii)′, (iii)′. This is done inductively, using the representation $\Delta=\{k2^{-m}\colon k=0,1,\cdots,2^m;\ m\geqq 0\}$; we define for $m\geqq 0$:

(*)″ $$\mathscr{U}_m = \{U(k2^{-m}),\ F(k2^{-m})\colon 0\leqq k\leqq 2^m\} \text{ where}$$

(i)″ $\complement U(k2^{-m})\in H,\ F(k2^{-m})\in H,\ 0\leqq k\leqq 2^m$;
(ii)″ $A\subseteq U(0),\ B\subseteq \complement U(1)$;
(iii)″ If $0\leqq k<l\leqq 2^m$ then $U(k2^{-m})\subseteq F(k2^{-m})\subseteq U(l2^{-m})$.

Then we put $\mathscr{U}_\Delta=\bigcup_{m\geqq 0}\mathscr{U}_m$.

Case $m = 0$. Put $\mathcal{U}_0 = \{U(0), F(0), U(1), F(1)\}$ where

$$U(1) = \complement B,\ F(1) = X, \text{ and}$$

$U(0)$, $F(0)$ are chosen so that $\complement U(0) \in H$, $F(0) \in H$, and

$$A \subseteq U(0) \subseteq F(0) \subseteq \complement B.$$

This can be done: for $A \cap B = \phi$ implies, by the assumed normality of H, the existence of V, $F \in H$ with:

$$A \cap V = \phi,\ B \cap F = \phi \text{ and } V \cup F = X.$$

But these relations imply $A \subseteq \complement V$, $F \subseteq \complement B$ and $\complement V \subseteq F$ so that putting $U(0) = \complement V$ with $F(0) = F$ satisfies our requirements.

Now suppose that for some $m \geqq 1$, $\mathcal{U}_{m-1}$ is defined and satisfies (i)″, (ii)″ and (iii)″, and let us consider $\mathcal{U}_m$. For even k we define $U(k2^{-m})$, $F(k2^{-m})$ in the obvious way. For odd $k \geqq 1$ we note that (iii)″ implies:

$$U((k-1)2^{-m}) \subseteq F((k-1)2^{-m}) \subseteq U((k+1)2^{-m}).$$

The last inclusion can be written $F((k-1)2^{-m}) \cap \complement U((k+1)2^{-m}) = \phi$ and so we may proceed as for the case $m = 0$ with F replacing A, $\complement U$ replacing B, and find elements V, F of H with

$$F((k-1)2^{-m}) \subseteq \complement V \subseteq F \subseteq U((k+1)2^{-m}).$$

Thus we may put $U(k2^{-m}) = \complement V$ and $F(k2^{-m}) = F$ and satisfy (iii)″ thus completing the inductive step.

And so we have constructed $\mathcal{U}_\Delta$ and thus $\mathcal{U}$, and it only remains to prove (iv) is valid in order to complete the proof of 1) implies 2). Recall the definition (**) of f.

If $f(x) \leqq t$ for some $t \in [0,1]$ then for any $\delta \in \Delta$ with $\delta > t$ we have $x \in U(\delta) \subseteq F(\delta)$ whence $x \in \bigcap_{\delta > t} F(\delta) = F(t)$.

On the other hand, if $x \in F(t)$, then for any $\delta \in \Delta$ with $\delta > t$ we may find $\delta' \in \Delta$ with $\delta > \delta' > t$ and so $x \in F(\delta') \subseteq U(\delta)$. Thus $f(x) < \delta$ for each $\delta > t$ whence $f(x) \leqq t$, and we have proved that $f^{-1}[0,t] = F(t)$. The other part of (iv) is proved similarly.

2) implies 1) Let us assume that $A \cap B = \phi$ for A, $B \in H$. By 2) there exists $f : X \to [0,1]$ such that $A \subseteq f^{-1}\{0\}$, $B \subseteq f^{-1}\{1\}$ and $f^{-1}F \in H$ for each closed $F \subseteq [0,1]$. If we take $F = [0,\frac{1}{2}]$ and $[\frac{1}{2},1]$ we obtain $D = f^{-1}[0,\frac{1}{2}] \in H$, $C = f^{-1}[\frac{1}{2},1] \in H$ such that $A \cap C = \phi$, $B \cap D = \phi$ and $C \cup D = X$, as required.

Remark 3.2. If $\mathbf{H} = \mathbf{F}(X)$ is the lattice closed subsets of a topology on X then 3.1 is just Urysohn's lemma. Another special case is when $\mathbf{H} = \mathbf{Z}(X)$ is the lattice of all zero-sets of a completely regular (Hausdorff) space X. In this case we have proved (cf. [5] 1.15) that disjoint zero-sets can be separated by a continuous function.

[7] On rings of sets

Our application of 3.1 is in the following result:

THEOREM 3.3. *Let* **H** *be a normal* $(2,\sigma)$*-ring of subsets of a set* X. *Then the following are equivalent for an element* $A \in H$:

1) *There exists a sequence* $\{A_n : n \geqq 1\}$ *of elements of* H *with the following properties*:

(α) $A \cap A_n = \phi$ *for all* n;
(β) *if for* $B \in H$ *we have* $B \cap A_n = \phi$ *for all* n, *then* $B \subseteq A$.

2) *There exists a function* $f: X \to [0,1]$ *such that*
(α) $f^{-1}F \in H$ *for every closed subset* F *of* $[0,1]$;
(β) $A = f^{-1}\{0\}$.

PROOF: 1) implies 2). By 3.1 there exists $f_n: X \to [0,1]$ satisfying 2) (α) such that $A \subseteq f_n^{-1}\{0\}$ and $A_n \subseteq f_n^{-1}\{1\}$. This uses only 1) ($\alpha$) and works for all n. Now consider the element $\bigcap_{n \geqq 1} f_n^{-1}\{0\}$ of H; clearly $A \subseteq \bigcap_{n \geqq 1} f_n^{-1}\{0\}$, and for any n,

$$A_n \cap \bigcap_{n \geqq 1} f_n^{-1}\{0\} \subseteq f_n^{-1}\{1\} \cap \bigcap_{n \geqq 1} f_n^{-1}\{0\} = \phi.$$

Thus by 1) (β) $A \supseteq \bigcap_{n \geqq 1} f_n^{-1}\{0\}$ and if we define $f: X \to [0,1]$ by

$$f = \sum_{n \geqq 1} 2^{-n} f_n$$

it is easy to see $f^{-1}\{0\} = \bigcap_{n \geqq 1} f_n^{-1}\{0\} = A$, and Lemma 2.4 of [6] implies that f satisfies 2) (α). This completes the proof of the first implication.

2) implies 1). If $A = f^{-1}\{0\}$ for a function f satisfying 2) (β), then we may define $A_n = f^{-1}[1/n, 1]$ for $n \geqq 1$, With this definition

$$\complement A = \bigcup_{n \geqq 1} A_n$$

and conditions 1) (α), 1) (β) are readily checked.

The following corollary can easily be proved using the two previous results.

COROLLARY 3.4. *Let* **H** *be a* $(2,\sigma)$*-ring of subsets of* X. *Then the following are equivalent*:

1) **H** *is a perfectly normal lattice.*
2) *For every* $A \in H$ *there exists a function* $f: X \to [0,1]$ *such that*:
(α) $f^{-1}F \in H$ *for every closed subset* F *of* $[0,1]$;
(β) $A = f^{-1}\{0\}$.

In the terminology we are using, a zero-set space is a pair $(X, \mathscr{Z})$ where X is a set and $\mathscr{Z}$ is a perfectly normal $(2,\sigma)$-ring of subsets of X which separates points of X. With any such space Gordon associates the set $S(X, \mathscr{Z})$ of all functions $f: X \to \mathbb{R}$ such that $f^{-1}F \in \mathscr{Z}$ for every closed subset F of $\mathbb{R}$; such functions are called zero-set functions. The lemma ([6] 2.4) used in the previous proof shows that $S(X, \mathscr{Z})$ is a uniformly closed ring of functions on X; also $S(X, \mathscr{Z})$ separates points of X and contains the constant functions.

We give a new proof of [6] 3.5 viz:

THEOREM 3.5. *Let $(X, \mathscr{Z})$ be a zero-set space and $S(X, \mathscr{Z})$ the family of all zero-set functions on X. Then*

$$\mathscr{Z} = \{Z(f): f \in S(X, \mathscr{Z})\}.$$

PROOF. By 3.4 every $A \in \mathscr{Z}$ is the zero-set of a suitable function of $S(X, \mathscr{Z})$; if $f \in S(X, \mathscr{Z})$ then $Z(f) = f^{-1}\{0\} \in \mathscr{Z}$ and the proof is complete.

4. The lattice Z (X) for X compact

It is well known that a completely regular space X is compact if, and only if, every z-ultrafilter is fixed. However, as in the case of the ring $C(X)$, the notion of fixed (resp. free) is not a lattice-theoretic invariant and so we must proceed slightly differently. At this point also, our treatment begins to differ from that in [6] since we only have the lattice $\mathbf{Z}(X)$ and not X itself.

The main result of this section is given a proof independently of the discussion in the next section, although it can also be derived from results there. We do this because the simplications which occur when X is compact allow quite different techniques to be used.

THEOREM 4.1. *Let* $\mathbf{L}$ *be a lattice. Then the following are equivalent:*

1) $\mathbf{L}$ *is isomorphic to the lattice* $\mathbf{Z}(X)$ *for a compact space* X.

2) (α) $\mathbf{L}$ *is a* $(2, \sigma)$*-complete lattice;*

(β) *Every minimal prime ideal of* $\mathbf{L}$ *is* σ*-prime;*

(γ) $\mathbf{L}$ *is perfectly normal.*

The space X of 1) *is NOT unique up to homeomorphism.*

PROOF. 1) implies 2). We will show that for any compact space X the lattice $\mathbf{Z}(X)$ has properties 2) (α), (β), (γ); these are obviously lattice invariants and so the implication will be proved. But we have already noted the validity of (α), (γ) for X general, and (β) follows since every minimal prime ideal of $\mathbf{Z}(X)$ is exactly those elements not belonging to a particular fixed ultrafilter $u_x = \{a \in \mathbf{Z}(X): x \in a\}$, where $x \in X$ is unique. Clearly such a minimal prime is σ-prime, completing the proof.

2) implies 1) Suppose we are given a lattice $\mathbf{L}$ satisfying (α), (β), (γ) of 2). Let X denote the set of all minimal prime ideals of $\mathbf{L}$, and equip X with the topology whose closed sets are intersections of the sets in $L' = \{X_a: a \in L\}$ where for $a \in L$, $X_a = \{x \in X: a \notin x\}$. We will prove that X so defined is a compact

(Hausdorff) space, and that the family of all zero-sets of X is exactly $\mathbf{L}'$, a lattice which will be shown to be isomorphic to $\mathbf{L}$.

We prove this last remark first. Condition 2) (γ) together with Lemma 2.5 above implies that $\mathbf{L}$ is disjunctive, and so by a result which is well known (see e.g. [8]) $a \to X_a$ is bijective. It can be readily checked that for a, $b \in L$, $X_a \cup X_b = X_{a \vee b}$, and since the ideals in X are all σ-prime, we find that for $\{a_n : n \geqq 1\}$ $L, \bigcap_{n \geqq 1} X_{a_n} = X_a$ where $a = \bigwedge_{n \geqq 1} a_n$. The latter exists, of course, by 2) (α).

The proof that X, so topologised, is a compact (Hausdorff) space given (in a dual form) in [4] Theorem 7.3 hence we omit it.

And so it remains to prove that L' is exactly the family of all zero-sets of X. Now X is normal and so it is enough to prove that L' is exactly the set of all closed G_δ-subsets of X. But every $a \in L$ is a G_δ and this is easily seen to imply

$$\complement X_a = \bigcup_{n \geqq 1} X_{a_n},$$

proving that X_a is a G_δ-subset, by definition, closed, of X. This proves half of what is required, and to complete the proof we take an arbitrary closed G_δ-subset F of X. By definition, there is a sequence M_n of subsets of L, and a subset $B \subseteq L$ such that

$$\bigcap_{b \in B} X_b = F = \bigcap_{n \geqq 1} \left\{ \bigcup_{a \in M_n} \complement X_a \right\}.$$

We concentrate on the right-hand equality first. Since F is compact, for any $n \geqq 1$ there is a finite subset $m_n \subseteq M_n$ such that

$$F \subseteq \bigcup_{a \in m_n} \complement X_a = \complement X_{a_n},$$

where $a_n = \bigwedge_{a \in m_n} a$. Thus we see that

$$\bigcap_{b \in B} X_b = F = \bigcap_{n \geqq 1} \complement X_{a_n}.$$

Now each X_{a_n} is compact, and so for each n there is a finite subset $B_n \subseteq B$ such that

$$X_{b_n} = \bigcap_{b \in B_n} X_b \subseteq \complement X_{a_n},$$

where $b_n = \bigwedge_{b \in B_n} b$. Putting these results together gives

$$F = \bigcap_{n \geqq 1} \complement X_{a_n} \supseteq \bigcap_{n \geqq 1} X_{b_n} \supseteq \bigcap_{b \in B} X_b = F,$$

whence $F = \bigcap_{n \geqq 1} X_{b_n} = X_b$, where $b = \bigwedge_{n \geqq 1} b_n$, and in this last step we have used the fact that $b \to X_b$ is a $(2, \sigma)$-homomorphism. The proof is now complete.

An interesting byproduct which will be explained in the next section is the following:

COROLLARY 4.2. *Let Y be a pseudocompact topological space. Then there is a compact space X such that* $\mathbf{Z}(Y)$ *and* $\mathbf{Z}(X)$ *are lattice isomorphic.*

PROOF. Putting together 5.8(b) and 5.14 of [5] we find that for Y pseudocompact $\mathbf{Z}(Y)$ satisfies (4.1) 2) (β) and so the result follows.

In particular, the corollary shows that non-homeomorphic pseudocompact spaces can have isomorphic lattices of zero-sets. We will see that this cannot happen when the spaces are both realcompact. Finally, we note that the space X in 4.2 can be taken to be βY, or any space $Y \subseteq X \subseteq \beta Y$.

5. The lattice $\mathbf{Z}(X)$ for a general X

In this section we characterise the lattice $\mathbf{Z}(X)$ algebraically, for a general topological space X. We begin with a reduction, relying heavily upon results from [5].

PROPOSITION 5.1. *For every topological space X there exists a completely regular space Y and a continuous map* τ *of X onto Y such that the map:*

$$Z_Y(g) \to Z_A(g \circ \tau)$$

is an isomorphism of $\mathbf{Z}(Y)$ *onto* $\mathbf{Z}(X)$.

PROOF. See [5] 3.9. The details are easy, and omitted.

The next stage of our reduction is again similar to the ring case.

PROPOSITION 5.2. *For every completely regular space X there exists a realcompact space* υX *and a continuous map* τ *of X into* υX *such that the map:*

$$a \to cl_{\upsilon X^a}$$

is an isomorphism of $\mathbf{Z}(X)$ *onto* $\mathbf{Z}(\upsilon X)$).

PROOF. See [5] 8.8.

From now on we will suppose, where appropriate, that X is realcompact. As a first attack on our characterisation problem we abstract the lattice-theoretic properties of a zero-set structure.

DEFINITION 5.3. A lattice $\mathbf{L}$ is a *z-lattice* if

(α) $\mathbf{L}$ is $(2, \sigma)$-complete;
(β) $\mathbf{L}$ has enough σ-prime minimal prime ideals;
(γ) $\mathbf{L}$ is perfectly normal.

For any topological space X the lattice $\mathbf{Z}(X)$ is a z-lattice. To see this we need only check (β) as (α) and (γ) of 5.3 have already been noted. Now for any $x \in X$ the family $j_x = \{a \in \mathbf{Z}(X) : x \notin a\}$ is easily seen to be a σ-prime ideal of $\mathbf{Z}(X)$ and there are certainly enough of these ideals to distinguish elements of $\mathbf{Z}(X)$. Thus (β) will be satisfied if we show that all the σ-prime ideals j_x are minimal. But this follows from 2.5 above; alternatively a direct proof can be given.

We will see below that although not every z-lattice is isomorphic to a lattice $\mathbf{Z}(X)$, such a lattice can be embedded as a sublattice of $\mathbf{Z}(X)$ for a suitable X in a particularly precise manner, which it is convenient to formulate separately. For any z-lattice $\mathbf{L}$ (possibly with superscripts) we denote by $X^{\mathbf{L}}$, or just X if no ambiguity is possible, (with the same superscripts) the set of all σ-prime minimal prime ideals of $\mathbf{L}$; for $a \in L$ we write $X^{\mathbf{L}}_a = X_c = \{x \in X^{\mathbf{L}} : a \notin x\}$.

DEFINITION 5.4. A z-lattice $\mathbf{L}$ is said to be a *z-sublattice* of the z-lattice $\mathbf{L}'$, equivalently, $\mathbf{L}'$ is a *z-extension* of $\mathbf{L}$, if

(α) $\mathbf{L}$ is a $(2, \sigma)$ regular sublattice of $\mathbf{L}'$;
(β) the map $x' \to x' \cap L$ is a bijection from X' onto X;
(γ) for any $b \in L'$, $X'_b = \bigcap\{X'_a : a \in L,\ a \geqq b\}$.

We will see that the property of being a z-sublattice is transitive, a fact needed below.

LEMMA 5.5. *If* $\mathbf{L}$ *is a z-sublattice of* $\mathbf{L}'$, *and* $\mathbf{L}'$ *is a z-sublattice of* $\mathbf{L}''$, *then* $\mathbf{L}$ *is a z-sublattice of* $\mathbf{L}''$.

PROOF. Clearly (α) and (β) are true so we need only prove (γ). Let $b \in L'$. We will show that

(*) $$X_b = \bigcap\{X_a : a \in L,\ a \geqq b\}$$

is true, and then the fact that for any $c \in L''$

$$X''_c = \bigcap\{X''_b : b \in L',\ b \geqq c\}$$

will complete the proof. Now suppose that $x'' \in X''$ is such that $a \notin x'$ for all $a \in L$ with $a \geqq b$. Then $a \notin x'' \cap L' = x'$ say, for all $a \in L$ such that $a \geqq b$, and so $b \notin x'$, since $\mathbf{L}'$ is a z-extension of $\mathbf{L}$. Thus we have proved $b \notin x''$ and the equality (*) is proved.

The following results is the main step in our characterisation theorem.

THEOREM 5.6. *Let* $\mathbf{L}$ *be a z-lattice. Then* $X = X^{\mathbf{L}}$ *is a realcompact space, and* $\mathbf{L}$ *is isomorphic to a z-sublattice of the z-lattice* $\mathbf{Z}(X)$.

PROOF. We give X the topology whose closed subsets are intersections of sets in $L' = \{X_a : a \in L\}$. Exactly as in 4.1 above we can prove that $\mathbf{L}$ is isomorphic to $\mathbf{L}'$ where $\mathbf{L}'$ is the set L' under the operations of finite set-union and countable set-intersection; the isomorphism is a $(2, \sigma)$-homomorphism.

T. P. Speed **[12]**

Thus $\mathbf{L}'$ is a z-lattice, and so the results of §3 above will apply. We prove that X is completely regular. Take a point $x \in X$ and a closed set $F = \bigcap\{X_a : a \in \mathbf{M}\}$ not containing x. Then there is $X_a \subseteq F$ with $x \notin X_a$; since a in L and hence X_a in L' is a G_δ, $x \in X_b$ with $X_a \cap X_b = \phi$ for a suitable $b \in L$. By Theorem 3.1 there is a continuous function $f : X \to [0,1]$ with $f(x) = 1$ and $f^{-1}\{0\} \supseteq F$. Now Corollary 3.4 shows that every element of L' is a zero-set of X and so $L \subseteq Z(X)$. Before we show that $\mathbf{L}'$ is a z-sublattice of $\mathbf{Z}(X)$ it will be necessary to prove that X is realcompact. Let η be a real z-ultrafilter on X; then $y = \{a \in Z(X) : a \notin \eta\}$ is a σ-prime minimal prime ideal on $\mathbf{Z}(X)$, and so $y' = y \cap L'$ is a σ-prime ideal of $\mathbf{L}'$. Lemma 2.5 implies that y' is in fact a minimal prime ideal, and so $y' = x$ for some unique $x \in X$. Now the intersection of all the zero-sets in η is, by the definition of the topology, an intersection of all the zero-sets of the form X_a in η and this intersection contains x; thus X is realcompact by [5] 5.15.

Having now established that $\mathbf{L}$ is isomorphic to the sublattice $\mathbf{L}'$ of the lattice $\mathbf{Z}(X)$ where X is a realcompact space, our proof is completed by proving that $\mathbf{Z}(X)$ is a z-extension of $\mathbf{L}'$. This is really quite easy once we observe that the z-lattice $\mathbf{Z}(X)$ has a space $X^{Z(X)}$ of σ-prime minimal prime ideals which is canonically homeomorphic to X under the map $x \to j_x = \{a \in Z(X) : x \notin a\}$. Referring to 5.4 we see that (α) is valid, (β) follows from Lemma 2.5 and the preceding remark, and (γ) simply expresses the fact that every zero-set in X (more precisely, its homeomorph $X^{Z(X)}$) is closed and hence an intersection of the basic closed sets in L' (more precisely, their copies inside $X^{Z(X)}$). Thus the theorem is proved.

It might have been hoped that in the previous construction, $\mathbf{L}'$ actually coincides with $\mathbf{Z}(X)$, but as already observed, this is not generally so. After examining an example which validates this assertion we formulate and prove the maximality property possessed by lattices $\mathbf{Z}(X)$, and our main characterisation theorem quickly follows.

EXAMPLE 5.7. Consider the z-lattice $\mathbf{B} = \mathbf{B}[0,1]$ of all Borel subsets of $[0,1]$. Then $\mathbf{B}$ is a z-sublattice of the power set $\mathbf{P} = \mathbf{P}[0,1]$.

PROOF. To see this we also need to refer to $\mathbf{F} = \mathbf{F}[0,1]$, the z-lattice of all closed sets (= zero-sets) of $[0,1]$ with the usual topology. Before the assertion can be proved we need to describe the σ-prime minimal prime ideals of each of $\mathbf{F}$, $\mathbf{B}$, and $\mathbf{P}$. Since $[0,1]$ is compact those of $\mathbf{F}$ are all fixed, i.e. of the form $j_x \cap F$ where $j_x = \{a \in P : x \notin a\}$, for $x \in [0,1]$. Also the non-measurability of the cardinal of $[0,1]$ implies that the σ-prime minimal prime ideals of $\mathbf{P}$ are all of the form j_x for $x \in [0,1]$. Now all three of $\mathbf{F}$, $\mathbf{B}$ and $\mathbf{P}$ are perfectly normal and so Lemma 2.6 implies that every σ-prime minimal prime ideal of $\mathbf{B}$ is of the form $j_x \cap B$ for some $x \in [0,1]$. This last result is also a consequence of 8.4 [6].

Turning now to proving that $\mathbf{B}$ is a z-sublattice of $\mathbf{P}$ we note that 5.4 (α) is obviously true, (β) has already been remarked upon, and so only (γ) remains. But each singleton $\{x\}$ belongs to B and so (γ) is easily seen to be equivalent to

$$[0,1]\backslash b = \bigcap_{x \in b} [0,1]\backslash\{x\}, \; b \subseteq [0,1];$$

where the sets in the right-hand intersection are all in B. Thus **B** is a proper z-sublattice of **P**.

Our next result shows that a zero-set lattice $\mathbf{Z}(X)$ can never be a proper z-sublattice of a z-lattice, and this is the point where we can see why Gordon's results [6] differ in some respect from the usual topological ones. Simply put, his zero-set structures are more general than those which can arise in the topological context, and so a result such as: a product of pseudo-compact zero-set spaces is pseudo-compact, can be valid in the former while failing in the latter. Put another way, z-lattices such as the **B** of 5.7 can never arise as $\mathbf{Z}(X)$ for a topological space X. We note that if this could happen, results of Mandelker [10] imply that X would be at least a P-space!

THEOREM 5.8. *Suppose X and Y are realcompact spaces and that $\mathbf{Z}(Y)$ is isomorphic to a z-sublattice $\mathbf{L}$ of $\mathbf{Z}(X)$. Then $\mathbf{Z}(Y)$ is isomorphic to $\mathbf{Z}(X)$.*

PROOF. We prove that X and Y are homeomorphic under the stated assumptions. It is easy to see that the space of σ-prime minimal prime ideals of $\mathbf{Z}(X)$ topologised as in 5.6 is canonically homeomorphic to X; we denote it X^* with points $j_x = \{a \in Z(X) \colon x \notin a\}$. Similarly for $\mathbf{Z}(Y)$. Thus we have the following diagram, where $L' = \{X_a^{Z(X)} \colon a \in L\}$ and $Z' = \{X_b^{Z(X)} \colon b \in Z(X)\}$:

$$\begin{array}{ccccc} \mathbf{Z}(Y) \xrightarrow{h} & \mathbf{L} & \subseteq & \mathbf{Z}(X) & \\ & \downarrow & & \downarrow & \searrow \\ & \mathbf{L}' & \subseteq & \mathbf{Z}' & \subseteq \; \mathbf{Z}(X^*) \end{array}$$

where all the maps which are not inclusions are isomorphisms; the vertical maps are as in the construction 5.6 and the diagonal map is defined using the homeomorphism $x \to j_x$.

Now (5.4) (β) states that the map

$$x^* = j_x \to j_x \cap L$$

is a bijection; by construction it is continuous from X^* onto $X^{\mathbf{L}}$. We show that it is a closed map. A typical closed subset F of X^* is of the form

$$F = \bigcap_{b \in B} X_b^*$$

where $B \subseteq \mathbf{Z}(X)$. Condition (5.4) ($\gamma$) states in this context that for each $b \in \mathbf{Z}(X)$, $X_b^* = \bigcap \{X_a^* \colon a \in L, \; a \geqq b\}$ whence

$$F = \bigcap_{a \in A} X_a^*$$

where $A \subseteq L$. But now we may apply the bijection above, and we find that $F \to \bigcap_{a \in A} X_a^{\mathbf{L}}$, a closed subset of $X^{\mathbf{L}}$. Thus X^* is homeomorphic to $X^{\mathbf{L}}$ and so we deduce that X is homeomorphic to Y. Finally we complete the proof by noting that the lattice of zero-sets of a topological space is a topological invariant.

We can now finish off with the characterisation theorem.

THEOREM 5.9. *The following are equivalent for a lattice* **L**.

1) **L** *is isomorphic to* $\mathbf{Z}(X)$ *for a topological space* X.

2) (α) **L** *is* $(2, \sigma)$*-complete*;

(β) **L** *has enough* σ*-prime minimal prime ideals*;

(γ) **L** *is perfectly normal*;

(δ) **L** *is isomorphic to every* z*-extension* $\mathbf{L}'$ *of* **L**.

PROOF. 1) implies 2). Properties (α), (β) and (γ) have already been observed Suppose that $\mathbf{Z}(X)$ is a z-sublattice of a z-lattice $\mathbf{L}'$. Noting that we may suppose X is realcompact by 5.2, we have the following diagram:

$$\begin{array}{ccc} \mathbf{Z}(X) & \subseteq & \mathbf{L}' \\ \downarrow & & \downarrow \\ \mathbf{Z}' & \subseteq & \mathbf{L}'' \subseteq \mathbf{Z}(X^{\mathbf{L}'}) \end{array}$$

where the vertical maps are isomorphisms as in the construction of 5.6, and the horizontal maps are inclusions. By the transitivity of the property of being a z-sublattice, $\mathbf{Z}'$ is a z-sublattice of $\mathbf{Z}(X^{\mathbf{L}'})$ isomorphic to $\mathbf{Z}(X)$, and so $\mathbf{Z}(X) \cong \mathbf{L}'$ follows from 5.8.

2) implies 1). We have already proved that if **L** satisfies 2) (α), (β), (γ), **L** is isomorphic to a z-sublattice $\mathbf{L}'$ of $\mathbf{Z}(X)$ for a completely regular space X, and so by (δ) we may conclude that $\mathbf{L} \cong \mathbf{Z}(X)$.

References

[1] G. Birkhoff, *Lattice Theory* Colloq. Publ. No. XXV, 3rd Ed., (Amer. Math. Soc. Providence, Rhode Island, 1967).

[2] N. Bourbaki, *Elements of mathematics, General topology, Part* 1 (Hermann, Paris: Addison-Wesley, Reading Massachussetts, Palo Alto, London, Don Mills Ontario; 1966).

[3] E. Čech, *Topological Spaces* (Interscience Publishers, John Wiley and Sons, London, New York; Sydney, 1966).

[4] W. H. Cornish, 'Normal lattices', *J. Austral Math. Soc.* 14 (1972), 200–215.

[5] L. Gillman and M. Jerison, *Rings of continuous functions* (D. Van Nostrand Princeton, New Jersey, Toronto, London, New York, 1960).

[6] H. Gordon, 'Rings of functions determined by zero sets', *Pacific. J. Math.* 36 (1971), 133–157.
[7] J. Kesttan, 'Eine Charakterisierung der vollständig regulären Raume', *Math. Nachr.* 17 (1958/1959), 27–46.
[8] J. Kist, 'Minimal prime ideals in commutative semigroups', *Proc. London Math. Soc.* (3) 13 (1963), 31–50.
[9] E. R. Lorch, 'Compactification, Baire functions, and Daniell integration', *Acta. Sci. Math. Szeged*, 24 (1963), 204–218.
[10] M. Mandelker, 'Relative annihilators in lattices', *Duke Math. J.* 37 (1970), 377–386.
[11] T. P. Speed, 'On rings of sets', *J. Austral. Math. Soc.* 8 (1968), 723–730.

Department of Probability of Statistics
University
Sheffield, S3 7RH, England

Chapter 2
Probability

Elja Arjas

Writing a brief commentary on three of Terry Speed's papers in probability brings to mind many memories from a time now almost forty years away. Two of these papers were written while Terry worked as a Lecturer in Sheffield, and during this period my encounters with Terry were very frequent. The third paper was written after Terry had already moved on to Perth.

These were times "when we were very young", and there was a great deal of excitement about new developments in probability. One of the main sources of inspiration was Volume 2 of *Introduction to Probability Theory and its Applications* by Feller [8], which had come out sixteen years after the publication of Volume 1 [7], and was then followed five years later by an expanded Second Edition. Feller was a master in making probability theory look like it were a collection of challenging puzzles, for which one, if only sufficiently clever, could find an elegant solution by some ingenious trick that actually made the original problem look like it had been trivial. Feller's books offered also a large number of examples leading to potentially important applications. This idea of making probability a tool for practical mathematical modeling was gaining ground in other ways, too. An important move in this direction, in 1964, was founding, at the initiative of Joe Gani, of the *Applied Probability* journals. The Department of Probability and Statistics in Sheffield, also Gani's creation, was a hub of these developments and it attracted a number of young talents to its circles from around the world, Terry being one of them.

Another source of inspiration at the time was 'the general theory of stochastic processes', which was represented, most importantly, by the French and the Russian schools of probability. The key figure behind this in France was Paul-André Meyer and his book *Probability and Potentials* [10] was one of the favorites in Terry's impressive home library in Sheffield. (A sign of Terry's interest in the works coming from the French school is that he translated into English J. Neveu's book *Martingales à temps discret* [11], which appeared in 1975 with the title *Discrete Pa-*

E. Arjas
University of Helsinki and National Institute for Health and Welfare, Finland
e-mail: elja.arjas@helsinki.fi

S. Dudoit (ed.), *Selected Works of Terry Speed*, Selected Works in Probability and Statistics,
DOI 10.1007/978-1-4614-1347-9_2, © Springer Science+Business Media, LLC 2012

rameter Martingales [12]. I remember Terry wondering why the French publishers did not seem to make any effort towards marketing their books outside France, or even making them available in the largest bookstores in UK.)

Chronologically, the earliest of the three papers on probability in this collection is the one entitled *Symmetric Wiener-Hopf factorisations in Markov additive processes*, which Terry and I submitted to the prestigious Springer journal 'ZW' in November 1972 [2]. For me, the background story leading to this is as follows: Not finding anyone in Finland to suggest a topic to work on for a PhD in probability, let alone to act as a supervisor, I had in desperation written to Professor Gani, asking him whether he would let me come and spend some time in his Department in Sheffield. I was immediately welcomed, and I stayed there for the winter and spring 1970–71. Sheffield turned out to be an excellent choice, with lots of academically interesting things going on all the time. There were many visitors, good weekly seminars, and if this wasn't sufficient, the Department paid train trips for us to go to London and Manchester to listen to more. But above all, there were people roughly of my age some of whom were working towards a PhD just like I was, and others who were already much beyond, like Terry. There I learned what doing research in probability might involve in practice. My contact with Terry, which grew into a friendship, was particularly important in this respect. During the first and longest stay in Sheffield in the spring of 1971 I lived next door from Terry and Sally, and on my later visits I enjoyed their hospitality as a guest in their home.

This paper on Wiener-Hopf factorizations was inspired, in particular, by the ideas on Random Walks in $\mathbb{R}^1$ that were contained in Chapter XII of Feller's Volume 2, with that same title. On the introductory page of this chapter Feller writes: "The theory presented in the following pages is so elementary and simple that the newcomer would never suspect how difficult the problems used to be before their natural setting was understood." The key to such elementary understanding offered by Feller is the concept of 'ladder point', a pair of random variables consisting of a 'ladder epoch' and 'ladder height'. Consecutive ascending (descending) ladder points make up the sequence of new maximal (minimal) record values of the random walk. The sample path of the random walk arising from its first n steps can now be divided into random excursions, each ending with a new maximal (minimal) record value, and finally including an incomplete excursion from such a record value to where the random walk is after n steps. Due to the assumed iid structure of the random walk, the differences between the successive ascending (descending) ladders are also iid, and therefore the distribution of the sum of any k of them can be handled by forming a k-fold convolution 'power' of the distribution of one. These convolution powers of the common distribution of the ascending ladder heights make up the 'positive part' of the Wiener-Hopf factorization. The 'negative part' stems from the incomplete excursion, by first noting that its distribution remains the same when the order of its steps is reversed and that, when considered in this manner 'backwards in time', the position at which the original random walk had its maximum now becomes a minimal record value. Therefore the distribution of this incomplete excursion gets a similar representation as the original sample path up to the maximal value, but now in terms convolution powers arising from the descending ladder points.

A second ingredient leading to our ZW paper was the emergence, in varying formulations and uses, of the concept of conditional independence. Conditional independence had been previously considered, for example, by Pyke [14] and Çinlar [4] in connection of semi-Markov and Markov renewal processes, and it was also an essential ingredient in Hidden Markov Models (HMMs) introduced by Baum and Petrie [3]. The general definitions and properties of conditional independence were expressed in measure theoretic terms in Meyer's book [10]. In statistics, it seems to have taken a few more years, to the well-known discussion paper of Dawid [6], until the fundamentally important ideas relating to conditional independence were fully appreciated and elaborated on. Presently, as is well known, conditional independence plays a major role particularly in Bayesian statistical modeling.

By replacing 'time' in Markov renewal processes by an additive real valued variable led us to consider, in a straightforward manner, a stochastic process called 'random walk defined on a Markov chain', or somewhat more generally, to Markov additive processes [5, 1]. It was relatively easy to see that the key ideas of Feller's treatment of random walks could be retained if the model was extended to include an underlying Markov chain, then assuming that the increments of the additive variable were conditionally independent given the states of this chain. In the case where the state space of the chain is finite, ordinary univariate convolutions used in the original random walk would be replaced by the corresponding matrix convolutions. Our paper in ZW adds a further level of generality to these results, by stating them in terms of transition kernels defined on a measurable state space. The technically most demanding aspect here was the construction of the dual or adjoint operators, corresponding to the time reversal in the original process. For the record, I should say that it was Terry who was primarily responsible for correctly adding all necessary mathematical bells and whistles to these general formulations.

The second paper, entitled *A note on random times* [13], provides the natural definition of, as it is called there, randomized stopping time in the case of processes of a discrete time parameter. In this brief note, Jim and Terry not only define this concept, but actually exhaust the topic completely by listing all its relevant properties and by linking it to different variants of essentially the same concept that existed in the literature at the time. Here, too, the key concept is conditional independence: Definition 1 says that a random time is a randomized stopping time relative to a family of histories if its occurrence, given the past, has no predictive value concerning the future. Of the properties derived, of most interest would seem to be the equivalence of (i) and (ii) of Proposition 2.5, and the intuitive explanation that is provided afterwards. To put it simply, a randomized stopping time is an 'ordinary' stopping time if it is considered relative to a family of bigger histories. What is required of these larger histories is that, at any given time point and given the past of the 'original' history, events in the past of this larger history do not help in predicting the future of the original. When expressed in this way, one can see how close it is to the concept of 'non-causality' of Granger [9], which is famous in the time series and econometrics literature, as well as, for example, to the property of local independence introduced by Schweder [15].

Looking at a result like this, one gets the feeling that the message it conveys should have been read, and understood, by generations of statisticians working in the area of survival analysis, in need of a natural definition of the concept of non-informative right censoring. They should have been thinking in terms of randomized stopping times! Instead, the common assumption stated in nearly all of the survival analysis literature is that of the 'random censoring model', which postulates for each considered individual the existence of two independent random variables, of which only the smaller is actually observed in the data. This model leads to strange events such as 'censoring of a person who is already dead'.

Terry is sole author of the third paper discussed here, entitled *Geometric and probabilistic aspects of some combinatorial identities* [16]. It is rather difficult to describe its contents in an understandable way in only a few sentences. In geometrical terms, it is concerned with certain hyperplanes in the positive orthant of the $(k+1)$-dimensional integer lattice. The main focus is on a particular combinatorial expression, which is shown to correspond to the number of minimal lattice paths from the origin to the considered hyperplane and such that the paths do not touch that plane until at the last point. This geometric interpretation then leads to concise derivations of some convolution type identities between the combinatorial expressions. Later on, the paper provides probabilistic interpretations, and corresponding proofs, for these results by considering the first passage time of a random walk from the origin to the hyperplane. There are also results on the associated moment generating functions, which have interesting analogues in the theory of branching processes. Although these combinatorial identities were not included in Feller's two books, one could say that Terry's approach to deal with them is very much Feller-like: when going through the mathematical derivations, at some point there is a phase transition from mysterious to intuitive and obvious. Another thing about this paper which I liked is its careful citing of the work of all authors who had earlier contributed, in various versions, to this same topic. But it looks like Terry just about exhausted this topic since, according to Google Scholar, to date this paper has been cited only once, and it isn't even listed in the ISI Web of Knowledge database.

Epilogue

When looking at the list of contents of this volume, which covers fifteen topics starting from algebra and ending with analysis of microarray data, one soon concludes that it would be hopeless to try to compete with Terry in terms of scientific output. In fact, competing with him in anything turned out to be a futile attempt. I once tried, in the late 1970s, when Terry visited me in Oulu and we went jogging. As we came back, I believe Terry was a bit more out of breath than I. Later on, however, Terry started practicing regularly by running up and down the steep hills surrounding Berkeley, and at some point I was told that he had run the marathon in less than three hours. My first marathon is still due. But luckily, there may be a sport where

I have a chance of beating him: cross-country skiing. This is an open invitation to Terry to try.

References

[1] E. Arjas and T. P. Speed. Topics in Markov additive processes. *Math. Scand.*, 33:171–192, 1973.
[2] E. Arjas and T. P. Speed. Symmetric Wiener-Hopf factorisations in Markov additive processes. *Z. Wahrscheinlichkeitstheorie und verw. Geb.*, 26(2):105–118, 1973.
[3] L. E. Baum and T. Petrie. Statistical inference for probabilistic functions of finite state Markov chains. *Ann. Math. Statist.*, 37:1554–1563, 1966.
[4] E. Çinlar. Markov renewal theory. *Adv. Appl. Prob.*, 1:123–187, 1969.
[5] E. Çinlar. Markov additive processes. *Z. Wahrscheinlichkeitstheorie und verw. Geb.*, 24:85–93, 1972.
[6] A. P. Dawid. Conditional independence in statistical theory. *J. Roy. Stat. Soc. B*, 41:1–31, 1979.
[7] W. Feller. *Introduction to Probability Theory and its Applications*, volume 1. Wiley, New York, 1950.
[8] W. Feller. *Introduction to Probability Theory and its Applications*, volume 2. Wiley, New York, 1966.
[9] C. W. J. Granger. Investigating causal relations by econometric models and cross-spectral methods. *Econometrica*, 37:424–438, 1969.
[10] P. A. Meyer. *Probability and Potentials*. Blaisdell, Waltham, MA, 1966.
[11] J. Neveu. *Martingales à temps discret*. Masson, Paris, 1972.
[12] J. Neveu. *Discrete-Parameter Martingales*, volume 10 of *North-Holland Mathematical Library*. North-Holland, Amsterdam, 1975. Translated from French by T. P. Speed.
[13] J. W. Pitman and T. P. Speed. A note on random times. *Stoch. Proc. Appl.*, 1: 369–374, 1973.
[14] R. Pyke. Markov renewal processes: Definitions and preliminary properties. *Ann. Math. Statist.*, 32:1231–1242, 1961.
[15] T. Schweder. Composable Markov processes. *J. Appl. Probab.*, 7:400–410, 1970.
[16] T. P. Speed. Geometric and probabilistic aspects of some combinatorial identities. *J. Aust. Math. Soc.*, 22:462–468, 1976.

Z. Wahrscheinlichkeitstheorie verw. Geb. 26, 105–118 (1973)

Symmetric Wiener-Hopf Factorisations in Markov Additive Processes

E. Arjas and T. P. Speed

The classical Wiener-Hopf factorisation of a probability measure is extended to an operator factorisation associated with a semi-Markov transition function. Some consequences of this factorisation are indicated including a set of duality relations.

1. Introduction

The classical Wiener-Hopf factorisation of a probability measure F on $(\mathbb{R}^1, \mathscr{R}^1)$ has been put in a symmetric form by Spitzer [14] and Feller [7] and can be written as follows:

$$\delta_0 - F = (\delta_0 - H^-) * (\delta_0 - \zeta\,\delta_0) * (\delta_0 - H^+) \tag{1.1}$$

where δ_0 is the unit mass at zero, $0 \leqq \zeta < 1$ and H^+, H^- are possibly defective probability measures concentrated on $(0, \infty)$ and $(-\infty, 0)$ respectively. In fact H^+ (resp. H^-) is identified as the distribution of the strict ascending (resp. descending) ladder variable.

In his very interesting extension of (1.1) Dinges [6] considered a substochastic transition function P on a measurable space $(E, \mathscr{E})$ with a total order, and constructed a factorisation:

$$I - \tau P = \left(I - \sum_1^\infty \tau^k P_k^-\right) \circ \left(I - \sum_1^\infty \tau^k P_k^{\cdot}\right) \circ \left(I - \sum_1^\infty \tau^k P_k^+\right) \tag{1.2}$$

where P_k^-, $P_k^{\cdot}$, and $P_k^+, k = 0, 1, \ldots$, are suitable operators or sub-stochastic transition functions, $0 \leqq \tau < 1$ and "$\circ$" denotes composition. Dinges' result gives (1.1) as a special case, but first a few rearrangements are required to do this. The reason is that although P_k^- and P_k^+ are notationally dual their constructions are not immediately seen to be so, and thus it is desirable to clarify this point. Further Presman [11, 12] has unsymmetric matrix factorisations which are similar to ones derived below, but these are obtained algebraically.

It is the purpose of this paper to obtain a symmetric factorisation which generalises (1.1) in two distinct ways: for we deal with Markov additive processes $\{(X_n, S_n) : n \geqq 0\}$, which reduce to the classical random walk by specialising the first component to a single value, or by suppressing the second component and specialising the first to be a random walk. Thus we can also obtain a result like (1.2) with the difference that our factorisation is manifestly symmetric. We formulate our results in an abstract way and the different results referred to are special cases. One aspect we emphasise throughout is the duality obtained from, and implicit in the proof of, our symmetric factorisations. In this respect our method is quite analogous to that of Feller's [7] Fourier analytic derivation of (1.1) in Chapter XVIII.

We now describe the contents of this paper. After some preliminaries concerning Markov additive processes we consider briefly Markov additive processes in duality. Next we formulate our abstract Wiener-Hopf factorisation and give its simple proof. The following two sections give concrete applications of this result and give a selection of corollaries. We close with some purely probabilistic duality results which are of some interest in themselves, and which can also be used to give alternative (probabilistic) proofs of our factorisations.

2. Markov Additive Processes

Our approach and notation will be based as far as possible upon Çinlar [4, 5] which in turn, is modelled upon Blumenthal and Getoor [3]. We recall some terminology. If $(G, \mathscr{G})$ and $(H, \mathscr{H})$ are measurable spaces and if $f: G \to H$ is measurable with respect to $\mathscr{G}$ and $\mathscr{H}$ then we write $f \in \mathscr{G}/\mathscr{H}$. If $H = \bar{\mathbb{R}}^1 = [-\infty, \infty]$ and $\mathscr{H} = \bar{\mathscr{R}}^1$, the Borel subsets of $\bar{\mathbb{R}}^1$, then we write $f \in \mathscr{G}$ instead of $f \in \mathscr{G}/\mathscr{H}$. Further $b\mathscr{G} = \{f \in \mathscr{G}: f \text{ is bounded}\}$, $\mathscr{G}_+ = \{f \in \mathscr{G}: f \geqq 0\}$ and $b\mathscr{G}_+ = b\mathscr{G} \cap \mathscr{G}_+$.

A mapping $N: F \times \mathscr{G} \to [0, 1]$ is called a *transition function* from $(F, \mathscr{F})$ into $(G, \mathscr{G})$ if a) $A \to N(x, A)$ is a measure on $\mathscr{G}$ for all fixed $x \in F$, and b) $x \to N(x, A)$ is in $b\mathscr{F}$ for all fixed $A \in \mathscr{G}$. Analogously, we define a mapping $Q: E \times (\mathscr{E} \times \bar{\mathscr{R}}^m) \to [0, 1]$ to be a *semi-Markov transition function* (abbrev. SMTF) on $(E, \mathscr{E}, \bar{\mathscr{R}}^m)$ if a) $x \to Q(x, A \times B)$ is in $b\mathscr{E}$ for every $A \in \mathscr{E}$, $B \in \bar{\mathscr{R}}^m$, b) $A \times B \to Q(x, A \times B)$ is a measure on $\mathscr{E} \times \bar{\mathscr{R}}^m$ for every $x \in E$.

If Q, R are two SMTF's on $(E, \mathscr{E}, \bar{\mathscr{R}}^m)$ we may define the *convolution product* $Q \circ R$ as the function,

$$(2.1) \qquad (x, A \times B) \to (Q \circ R)(x, A \times B) = \int_E \int_{\bar{\mathbb{R}}^m} Q(x, dx' \times ds)\, R(x', A \times (B - s)).$$

$Q \circ R$ is easily checked to be an SMTF. For any SMTF Q we define $Q^0 \equiv I$ where $I(x, A \times B) = \delta_x(A)\, \delta_0(B)$, and for $n \geqq 1$ $Q^n = Q^{n-1} \circ Q$.

There are many different ways of viewing a SMTF Q, and at various times we will be doing this. Thus Q may be viewed as a positive contraction valued measure defined on $(\bar{\mathbb{R}}^m, \bar{\mathscr{R}}^m)$ by the map $B \to Q(B)$, where $(Q(B) I_A)(x) = Q(x, A \times B)$; as a transition function on $(E \times \bar{\mathbb{R}}^m, \mathscr{E} \times \bar{\mathscr{R}}^m)$ which is homogeneous in the second component by the map $((x, s), A \times B) \to Q(x, A \times (B - s))$; as a transition function from $(E, \mathscr{E})$ to $(E \times \bar{\mathbb{R}}^m, \mathscr{E} \times \bar{\mathscr{R}}^m)$ by $(x, A \times B) \to Q(x, A \times B)$ (cf. Çinlar [4] (1.2)); and finally as giving a sequence $\{Q^n: n \geqq 0\}$ satisfying Definition (1.1) of Çinlar [5].

Any SMTF Q induces a family $\{Q(\theta): \theta \in \mathbb{R}^m\}$ of contractions on the Banach space $b\mathscr{E}$ by writing $(Q(\theta) f)(x) = \iint Q(x, dx' \times dy) \cdot f(x')\, e^{i(\theta, y)}$, where $(\cdot, \cdot)$ denotes the usual inner product in $\mathbb{R}^m$. We call $\{Q(\theta)\}$ the *Fourier transform* of Q.

We will consider a Markov process with state space $(E, \mathscr{E})$ to be a sextuple $X = (\Omega, \mathscr{M}, \mathscr{M}_n, X_n, \theta_n, P^x)$ $(x \in E)$, and all such processes will be assumed nonterminating (see Blumenthal and Getoor [3]). Following Çinlar [5] we have:

(2.2) **Definition.** Let X be a Markov process with state space $(E, \mathscr{E})$, write $(F, \mathscr{F}) = (\bar{\mathbb{R}}^m, \bar{\mathscr{R}}^m)$, and let $S = \{S_n: n \geqq 0\}$ be a family of functions from $(\Omega, \mathscr{M})$ into $(F, \mathscr{F})$. Then $(X, S) = (\Omega, \mathscr{M}, \mathscr{M}_n, X_n, S_n, \theta_n, P^x)$ is called a *Markov additive process* (abbrev. MAP) provided the following hold:

a) $S_0=0$ a.s.;

b) for each $n\geqq 0$, $S_n\in\mathcal{M}_n/\mathcal{F}$;

c) for each $n\geqq 0$, $A\in\mathcal{E}$, $B\in\mathcal{F}$, the mapping $x\to P^x\{X_n\in A, S_n\in B\}$ of E into $[0,1]$ is in $\mathcal{E}_+$;

d) for each $k, l\geqq 0$, $S_{k+l}=S_k+S_l\circ\theta_k$ a.s.;

e) for each $k, l\geqq 0$, $x\in E$, $A\in\mathcal{E}$, $B\in\mathcal{F}$

$$P^x\{X_l\circ\theta_k\in A, S_l\circ\theta_k\in B|\mathcal{M}_k\}=P^{X_k}\{X_l\in A, S_l\in B\}.$$

We follow Çinlar [5] in our notation for objects associated with the definition,

$$Q(x,C)=P^x\{(X_1,S_1)\in C\},\qquad C\in\mathcal{E}\times\mathcal{F}; \tag{2.3}$$

$$P(x,A)=Q(x,A\times F),\qquad A\in\mathcal{E}. \tag{2.4}$$

The action of $Q(B)$ mentioned above is as follows: for $f\in\mathcal{E}_+$

$$(Q(B)f)(x)=E^x[f(X_1);\, S_1\in B]. \tag{2.5}$$

Let N be a stopping time on Ω relative to $\{\mathcal{M}_n\}$; we define the (operator) transforms associated with (X_N, S_N) and with the behaviour of (X_n, S_n) for $n<N$: for $f\in b\mathcal{E}_+$ $\theta\in\mathbb{R}^m$, $0\leqq\tau<1$:

$$(Gf)(x)=E^x\left[\sum_0^{N-1}\tau^n e^{i(\theta,S_n)}f(X_n)\right], \tag{2.6}$$

$$(Hf)(x)=E^x[\tau^N e^{i(\theta,S_N)}f(X_N);\, N<\infty]. \tag{2.7}$$

A fundamental passage-time identity relating the transforms $G=G_N(\tau,\theta)$, $H=H_N(\tau,\theta)$ and $Q(\theta)$ is the following proved in Arjas and Speed [2] (I is the identity operator):

(2.8) **Proposition.** $G_N(\tau,\theta)[I-\tau Q(\theta)]=I-H_N(\tau,\theta)$.

3. Markov Additive Processes in Duality

Let us suppose that we are given a σ-finite measure π over our fixed state space $(E,\mathcal{E})$. We shall say that the MAP's

$$(X,S)=(\Omega,\mathcal{M},\mathcal{M}_n,X_n,S_n,\theta_n,P^x)\quad\text{and}\quad(\hat{X},\hat{S})=(\hat{\Omega},\hat{\mathcal{M}},\hat{\mathcal{M}}_n,\hat{X}_n,\hat{S}_n,\hat{\theta}_n,\hat{P}^x)$$

with SMTF's Q, $\hat{Q}$ respectively, are in duality relative to π if

a) for every $x\in E$, $P(x,\cdot)\ll\pi$, $\hat{P}(x,\cdot)\ll\pi$;

b) for every $B\in\bar{\mathcal{R}}^m$, $f,g\in\mathcal{E}_+$

$$\langle f, Q(B)g\rangle=\langle f\hat{Q}(-B), g\rangle \tag{3.1}$$

where, for $f_1, g_1\in\mathcal{E}_+$, we have $\langle f_1,g_1\rangle=\int f_1(x)\,g_1(x)\,\pi(dx)$. In this case we say also that Q and $\hat{Q}$ are in duality relative to π.

It can be proved (cf. Blumenthal and Getoor [3]) that π is P-excessive where $P=Q(\mathbb{R}^m)$ is the Markov transition function of X, and similar results hold for $\hat{P}$.

Thus (cf. Nelson [10]) the operators $Q(B)$ (resp. $\hat{Q}(B)$) defined by (2.5) act as linear contractions on $L^p(\pi)$ for $1 \leqq p \leqq \infty$. With this interpretation (3.1) expresses the fact that $\hat{Q}(-B)$, acting on $L^p(\pi)$, is the Banach space adjoint of $Q(B)$ acting on $L^q(\pi)$ where $p^{-1}+q^{-1}=1$. Slightly modifying this terminology we will speak of T and T^* being *adjoint* if $\langle f, T(B)g\rangle = \langle f T(-B)^*, g\rangle$ for every $B \in \bar{\mathscr{R}}^m$, $f, g \in \mathscr{E}_+$.

4. The Factorisation

In this section we present an axiomatic approach to symmetric Wiener-Hopf factorisations of SMTF's. A special case of our work is the unsymmetric matrix factorisation of Presman [12] whose derivation is abstract algebraic in nature. We would like to emphasise that while the discussion to follow is in a sense abstract, probabilistic considerations are used throughout and thus our arguments could hardly be termed algebraic.

Our formulation of the Wiener-Hopf factorisation will be in terms of the Fourier transforms of certain operator-valued measures. Explicitly, we will call a map $B \to T(B)$ from $\bar{\mathscr{R}}^m$ into the space of all bounded linear operators over $L^p(\pi)$ an operator-valued measure if for every $f \in L^p$, $g \in L^q$, the set function $B \to \langle f, T(B)g\rangle$ is countably additive. In this case the Fourier transform of the operator-valued measure is the operator-valued function $\theta \to T(\theta)$ from $\mathbb{R}^m$ into the space of all bounded linear operators over $L^p(\pi)$ where we write, for $f \in L^p$, $g \in L^q$, $\langle f, T(\theta)g\rangle = \int e^{i(\theta,y)}\langle f, T(dy)g\rangle$. It is easy to see that the functions $\theta \to G_N(\tau,\theta)$ and $\theta \to H_N(\tau,\theta)$ are Fourier transforms of suitable operator-valued measures. The space of all such Fourier transforms will be denoted $\mathscr{A}$, clearly an algebra over $\mathbb{C}$.

We make the following convention which shortens somewhat our statements: We say that a statement holds

(i) *symmetrically* (abbrev. s.) if it holds when all "+" symbols are replaced by "−" symbols and vice versa;

(ii) *dually* (abbrev. d.) if it holds when (X, S) and the possible other elements associated with it are replaced by $(\hat{X}, \hat{S})$ and the corresponding associated elements.

As we conceive them, symmetric Wiener-Hopf factorisations of transforms of SMTF's have three essential ingredients. We assume the following (I–III) throughout this section (almost surely):

I: A decomposition $\mathbf{A} = \mathbf{A}^- \oplus \mathbf{A}^{\cdot} \oplus \mathbf{A}^+$ of a subalgebra $\mathbf{A} \subset \mathscr{A}$ with

(i) $\mathbf{A}^-, \mathbf{A}^{\cdot}, \mathbf{A}^+$ all subalgebras of $\mathbf{A}$;

(ii) $\mathbf{A}^- \mathbf{A}^{\cdot} \subset \mathbf{A}^-$, $\mathbf{A}^{\cdot} \mathbf{A}^- \subset \mathbf{A}^-$, and s.;

(iii) $(\mathbf{A}^+)^* = \mathbf{A}^-$ and s., $(\mathbf{A}^{\cdot})^* = \mathbf{A}^{\cdot}$.

Here $\mathbf{A}^- \mathbf{A}^{\cdot} = \{ST\colon S \in \mathbf{A}^-, T \in \mathbf{A}^{\cdot}\}$ etc., and $(\mathbf{A}^+)^* = \{S^*\colon S \in \mathbf{A}^+\}$ and s.

We call a decomposition as in I a *symmetric W-decomposition*. The letter W is to stand for "Wendel" as there is a close relationship between the above and the so-called Wendel-projections of Kingman [9].

II: A system of stopping times N^+, $N^{\cdot +}$, N_+ relative to $\{\mathscr{M}_n\}$, and s. and d., such that almost surely

(i) $N_+ = N^{\cdot +} < N^+$ if $N^{\cdot +} < \infty$ and $N_+ = N^+$ if $N^{\cdot +} = \infty$, and s. and d.;

(ii) on $\{N^{\cdot +} < \infty\}$ $N^+ = N^{\cdot +} + N^+ \circ \theta_{N^{\cdot +}}$, and s. and d.

The stopping time N^+ will be sometimes described as a *strict ladder index* and N_+ as a *weak ladder index*, and s. and d.

We require that the above stopping times be adapted to the symmetric W-decomposition, by which we mean:

III: (i) $I \in \mathbf{A}^{\cdot}$;

(ii) $H_{N^+} \in \mathbf{A}^+$, $G_{N^+} \in \mathbf{A}^- \oplus \mathbf{A}^{\cdot}$, and s. and d.;

(iii) $H_{N_+} \in \mathbf{A}^{\cdot} \oplus \mathbf{A}^+$, $G_{N_+} - I \in \mathbf{A}^-$, and s. and d.;

where $\mathbf{A}^-$, $\mathbf{A}^{\cdot}$ and $\mathbf{A}^+$ stay fixed when statements are dualised.

We now prove two important preliminary lemmas, which give the desired factorisation as an almost immediate corollary. In the first lemma only II is used, whereas the second lemma is based on I and III.

(4.1) **Lemma** *(Relation between strict and weak ladder indices).*

$$I - H_{N_+} = (I - H_{N^{\cdot +}})(I - H_{N^+}), \text{ and s. and d.}$$

Proof. We note first that for $x \in E$, $0 \leqq \tau \leqq 1$, $\theta \in \mathbb{R}^m$, $f \in L^p$

$$E^x[\tau^{N^+} e^{i(\theta, S_{N^+})} f(X_{N^+}); \quad N^{\cdot +} < N^+ < \infty] = (H_{N^{\cdot +}} H_{N^+} f)(x). \tag{4.2}$$

To see this we write

$$E^x[\tau^{N^+} e^{i(\theta, S_{N^+})} f(X_{N^+}); N^{\cdot +} < N^+ < \infty]$$

$$= E^x[\tau^{N^{\cdot +}} e^{i(\theta, S_{N^{\cdot +}})} E^x[\tau^{N^+ \circ \boldsymbol{\theta}_{N^{\cdot +}}} e^{i(\theta, S_{N^+ \circ \boldsymbol{\theta}_{N^{\cdot +}}})} f(X_{N^+} \circ \boldsymbol{\theta}_{N^{\cdot +}});$$

$$N^+ \circ \boldsymbol{\theta}_{N^{\cdot +}} < \infty \,|\, \mathcal{M}_{N^{\cdot +}}]; N^{\cdot +} < \infty] \quad \text{by II and the general properties of conditional expectations}$$

$$= E^x[\tau^{N^{\cdot +}} e^{i(\theta, S_{N^{\cdot +}})} (H_{N^+} f)(X_{N^{\cdot +}}); N^{\cdot +} < \infty] \quad \text{by the (strong) Markov property}$$

$$= (H_{N^{\cdot +}} H_{N^+} f)(x).$$

Then, using II (i) and (4.2), we observe that

$$\begin{aligned}
(H_{N_+} f)(x) &= E^x[\tau^{N_+} e^{i(\theta, S_{N_+})} f(X_{N_+}); \ N_+ < \infty] \\
&= E^x[\tau^{N_+} e^{i(\theta, S_{N_+})} f(X_{N_+}); \ N_+ = N^{\cdot +} < \infty] \\
&\quad + E^x[\tau^{N_+} e^{i(\theta, S_{N_+})} f(X_{N_+}); \ N_+ = N^+ < \infty] \\
&= E^x[\tau^{N^{\cdot +}} e^{i(\theta, S_{N^{\cdot +}})} f(X_{N^{\cdot +}}); \ N^{\cdot +} < \infty] \\
&\quad + E^x[\tau^{N^+} e^{i(\theta, S_{N^+})} f(X_{N^+}); \ N^+ < \infty] \\
&\quad - E^x[\tau^{N^+} e^{i(\theta, S_{N^+})} f(X_{N^+}); \ N^{\cdot +} < N^+ < \infty] \\
&= (H_{N^{\cdot +}} f)(x) + (H_{N^+} f)(x) - (H_{N^{\cdot +}} H_{N^+} f)(x) \quad \text{by (4.2)}
\end{aligned}$$

which completes the proof. The symmetric and dual statements are proved similarly.

The second of the preliminary lemmas is

(4.3) **Lemma** *(Duality).*

(i) $G_{N^+} = (I - \hat{H}^*_{\hat{N}_+})^{-1}$, *and s. and d.*;

(ii) $G_{N_+} = (I - \hat{H}^*_{\hat{N}^+})^{-1}$, *and s. and d.*

Proof. By Proposition (2.8) applied to N_+, and its dual form applied to $\hat{N}^+$, for $0 \leqq \tau < 1$,

$$(I-\tau Q)^{-1} = (I - H_{N_+})^{-1} G_{N_+}$$

and

$$(I-\tau\hat{Q})^{-1} = (I - \hat{H}_{\hat{N}^+})^{-1} \hat{G}_{\hat{N}^+}.$$

These equations are mutually adjoint because $\hat{Q} = Q^*$, and so comparing the right hand sides we get

$$(I - H_{N_+})^{-1} G_{N_+} = \hat{G}^*_{\hat{N}^+} (I - \hat{H}^*_{\hat{N}^+})^{-1},$$

and further

$$G_{N_+} (I - \hat{H}^*_{\hat{N}^+}) = (I - H_{N_+}) \hat{G}^*_{\hat{N}^+}.$$

From I and III follows that the left hand side is of the form $I + K$ where $K \in \mathbf{A}^-$, and the right hand side is in $\mathbf{A}^{\cdot} \oplus \mathbf{A}^+$. Hence both sides must be I, giving (4.3)(ii) and the dual statement of (4.3)(i). Other symmetric and dual statements are proved similarly.

(4.4) **Corollary.** (i) $H_{N^{\cdot +}} = \hat{H}^*_{\hat{N}^{\cdot +}}$ *and s.;*

(ii) $H_{N^{\cdot +}} \in \mathbf{A}^{\cdot}$ *and s. and d.*

Proof. (i)

$$\begin{aligned} I - H_{N^{\cdot +}} &= (I - H_{N_+})(I - H_{N^+})^{-1} && \text{by (4.1)} \\ &= G_{N_+}(I - \tau Q)(I - \tau Q)^{-1} G_{N^+}^{-1} && \text{by (2.8)} \\ &= G_{N_+} G_{N^+}^{-1} && \text{cancelling} \\ &= (I - \hat{H}^*_{\hat{N}^+})^{-1} (I - \hat{H}^*_{\hat{N}_+}) && \text{by (4.3)} \\ &= [(I - \hat{H}_{\hat{N}_+})(I - \hat{H}_{\hat{N}^+})^{-1}]^* = I - \hat{H}^*_{\hat{N}^{\cdot +}} && \text{by (4.1).} \end{aligned}$$

(ii) $H_{N^{\cdot +}} \in \mathbf{A}^{\cdot} \oplus \mathbf{A}^+$ follows from the first line of the above proof when using III, and $\hat{H}^*_{\hat{N}^{\cdot +}} \in \mathbf{A}^{\cdot} \oplus \mathbf{A}^-$ can be proved similarly. The assertion then follows from (4.4)(i).

(4.5) **Theorem** *(Wiener-Hopf factorisation). Let (X, S) and $(\hat{X}, \hat{S})$ be in duality relative to π, and assume* I–III *to be valid. Then, for $0 \leqq \tau < 1$, $\theta \in \mathbb{R}^m$:*

(4.6) $$I - \tau Q(\theta) = [I - \hat{H}^*_{\hat{N}^+}(\tau, \theta)] [I - H_{N^{\cdot +}}(\tau, \theta)] [I - H_{N^+}(\tau, \theta)], \quad \textit{and s. and d.},$$

where the middle term is interchangeable with $I - \hat{H}^*_{\hat{N}^{\cdot +}}(\tau, \theta)$, and s. and d. Further, the factorisation (4.6) is unique in the sense that for a given W-decomposition there are no other factorisations with the non-unit term of the first (resp. second, third) factor in $\mathbf{A}^-$ (resp. $\mathbf{A}^{\cdot}$, $\mathbf{A}^+$), and s., and d.

Proof.

$$\begin{aligned} I - \tau Q(\theta) &= G_{N_+}^{-1}(\tau, \theta) [I - H_{N_+}(\tau, \theta)] && \text{by (2.8)} \\ &= [I - \hat{H}^*_{\hat{N}^+}(\tau, \theta)] [I - H_{N_+}(\tau, \theta)] && \text{by (4.3)(ii)} \\ &= [I - \hat{H}^*_{\hat{N}^+}(\tau, \theta)] [I - H_{N^{\cdot +}}(\tau, \theta)] [I - H_{N^+}(\tau, \theta)] && \text{by (4.1),} \end{aligned}$$

which is the required factorisation. The interchangeability of $I - H_{N^{\cdot +}}(\tau, \theta)$ with $I - \hat{H}^*_{\hat{N}^{\cdot +}}(\tau, \theta)$ follows from (4.4)(i).

We now prove uniqueness. To do this let us abbreviate the notation and assume that

$$I - \tau Q = K^- K^{\cdot} K^+ = L^- L^{\cdot} L^+$$

are two factorisations with factors invertible such that $I-K^-$, $I-L^- \in \mathbf{A}^-$; $I-K^\cdot, I-L^\cdot \in \mathbf{A}^\cdot$ and $I-K^+, I-L^+ \in \mathbf{A}^+$. Then

$$K^\cdot K^+ (L^+)^{-1} (L^\cdot)^{-1} = (K^-)^{-1} L^-,$$

and arguing as in the proof of (4.4)(ii) we see that both sides must be equal I, giving

$$K^- = L^- \quad \text{and} \quad K^\cdot K^+ = L^\cdot L^+.$$

A similar argument on the latter equation shows that $K^+ = L^+$ and $K^\cdot = L^\cdot$. (This proof followed a familiar pattern, cf. Dinges [6].)

We also state the factorisation in a measure form, allowing a direct comparison to the factorisation (1.2) of Dinges. Without going through the lengthy preliminaries (regarding the decomposition of the convolution algebra of operator-valued measures etc.) or making qualifications regarding uniqueness we simply describe the form of the factorisation and briefly explain some details of its components.

(4.7) **Theorem** *(Wiener-Hopf factorisation, measure form). For suitable operator-valued measures $H_n^+, H_n^{\cdot +} \hat{H}_n^+$, $n \geqq 1$, we have*

$$(4.8) \quad [I-\tau Q](B) = \left[I - \sum_1^\infty \tau^n (\hat{H}_n^+)^*\right] \circ \left[I - \sum_1^\infty \tau^n H_n^{\cdot +}\right] \circ \left[I - \sum_1^\infty \tau^n H_n^+\right](B),$$

and s. and d.

Interpretation. (i) "$\circ$" denotes the convolution product (see (2.1)) and "$*$" the adjoint as in § 3;

(ii) for $x \in E$, $B \in \bar{\mathscr{R}}^m$, $f \in L^p$ and $n \geqq 1$:

$$(H_n^+(B) f)(x) = E^x[f(X_n);\ N^+ = n,\ S_n \in B],$$
$$(H_n^{\cdot +}(B) f)(x) = E^x[f(X_n);\ N^{\cdot +} = n,\ S_n \in B],$$
$$(\hat{H}_n^+(B) f)(x) = \hat{E}^x[f(\hat{X}_n);\ \hat{N}^+ = n,\ \hat{S}_n \in B].$$

5. A Factorisation for Markov Chains with Totally Ordered State Space

We now specialise the results of the previous section to give a symmetrised factorisation for a transition function P, analogous to Dinges' [6] result. Recall however that we have assumed our process to be non-terminating, whereas in Dinges' case no extra assumptions of this kind are made save the necessary ones regarding order. These are that E has a reflexive, transitive binary relation, denoted $\leqq$, such that for any $x, x' \in E$ either $x \leqq x'$ or $x' \leqq x$. Further, if we write $x \sim x$ iff $x \leqq x'$ and $x' \leqq x$, and $x < x'$ if $x \leqq x'$ and $x \sim x'$ is false, then we require that $\{(x, x'): x' < x\}$ belong to the product σ-field $\mathscr{E} \times \mathscr{E}$.

For our algebra $\mathbf{A}$ (subalgebra of $\mathscr{A}$) we choose the real algebra generated by the set of all positive contractions on $L^p(\pi)$; this arises by putting $\theta = 0$ in each element of $\mathscr{A}$. Using the well-known equivalence between positive contractions and transition functions on $(E, \mathscr{E})$ we define the appropriate symmetric W-decomposition as follows: for $T \in \mathbf{A}$, $x \in E$, $A \in \mathscr{E}$ put

$$(5.1) \quad \begin{aligned} T^+(x, A) &= T(x, \{x': x < x'\} \cap A); \\ T^\cdot(x, A) &= T(x, \{x': x' \sim x\} \cap A); \\ T^-(x, A) &= T(x, \{x': x' < x\} \cap A); \end{aligned}$$

clearly $T=T^- + T^{\cdot} + T^+$ and this is easily seen to define a direct sum decomposition of **A** satisfying I(i), (ii) of § 4. To see how the decomposition can be defined directly in terms of its action on functions, we refer to Dinges [6].

The system of stopping times is the familiar one – the usual ladder indices:

$$N^+ = \inf\{n>0\colon X_0 < X_n\};$$

$$N_+ = \inf\{n>0\colon X_0 \leqq X_n\};\tag{5.2}$$

$$N^{\cdot +} = N_+ \quad \text{if } N_+ < N^+, \text{ and } N^{\cdot +} = \infty \text{ otherwise};$$

and s. and d.

We omit the verification of the fact that (5.2) satisfies II and III of § 4; II(ii) follows because on $\{N^{\cdot +} < \infty\}$ $X_{N^{\cdot +}} \sim X_0$ and $N^{\cdot +} < N^+$ so that $N^+ = \inf\{n > N^{\cdot +}\colon X_{N^{\cdot +}} < X_n\}$, and other requirements are satisfied quite obviously. Thus we can read off the following theorem, where we write $H_{N^+}(\tau) = H_{N^+}(\tau, 0)$ etc.:

(5.3) **Theorem.** *Let P and $\hat{P}$ be in duality relative to π, and consider the stopping times (5.2). Then as a relation between contractions on $L^p(\pi)$ for $0 \leqq \tau < 1$*

$$I - \tau P = [I - \hat{H}^*_{\hat{N}^+}(\tau)]\,[I - H_{N^{\cdot +}}(\tau)]\,[I - H_{N^+}(\tau)], \qquad \text{and s. and d.,}\tag{5.4}$$

where the middle term is interchangeable with $I - \hat{H}^*_{\hat{N}^{\cdot +}}(\pi)$, and s. and d. The uniqueness is as in Theorem (4.5).

(5.5) *Application 1.* The one-dimensional random walk. Suppose that $X_n = \sum_1^n Z_k$ where the $\{Z_k\}$ are i.i.d. random variables with law μ. Let λ denote Lebesgue measure on $(\mathbb{R}^1, \mathscr{R}^1)$; then it is easy to see that λ is P-excessive with $\hat{P}(x, A) = \hat{\mu}(A - x)$ where $\hat{\mu}$ is the measure μ reflected in the origin i.e. for $B \in \mathscr{R}^1$ $\hat{\mu}(B) = \mu(-B)$.

Now the operator P on $L^\infty(\lambda)$ is

$$(Pf)(x) = E^x[f(X_1)] = \int f(x + x')\,\mu(dx').\tag{5.6}$$

Following Dinges [6] we call this operator T_μ; note that if $e(x) = e^{i\theta x}$ for $\theta \in \mathbb{R}^1$ then $(T_\mu e)(x) = \phi(\theta)\,e(x)$ i.e. scalar multiplication by the characteristic function $\phi(\theta)$ of μ. The following expressions are readily checked: with notation as in Feller [7], Chapter XVIII (3.5)

$$(H_{N^+} e)(0) = \chi(\tau, \theta),$$

$$(H_{N^{\cdot +}} e)(0) = f(\tau),\tag{5.7}$$

$$(\hat{H}^*_{\hat{N}^+} e)(0) = \chi^-(\tau, \theta).$$

Note that in the last case the adjoint simply means complex-conjugation; the Eq. (3.5) of Feller is now seen to be an immediate consequence of (5.4) above acting on $e(x)$ and evaluating at $x=0$.

(5.8) *Application 2.* The m-dimensional random walk.

Here $X_n = \sum_1^n Z_k$ where $\{Z_k\}$ is a sequence of i.i.d. random variables with law μ. The dual process $\hat{X}$ is constructed as in the previous example, with respect to λ_m, m-dimensional Lebesgue measure. We order the state space $(\mathbb{R}^m, \mathscr{R}^m)$ by selecting a basis for $\mathbb{R}^m$ so that each Z_k can be written $Z_k=(Z_k^{(1)}, \ldots, Z_k^{(m)})$ and we then write:

$$(x'^{(1)}, \ldots, x'^{(m)}) \underset{[\sim]}{>} (x^{(1)}, \ldots, x^{(m)}) \quad \text{iff } x'^{(m)} \underset{[=]}{>} x^{(m)}.$$

In terms of this order the ladder indices N^+ etc. relate to the hyperplane $x^{(m)}=0$. Exactly as we found in the preceding example a factorisation arises by operating on $e(x)=e^{i(\theta,x)}$ for $\theta\in\mathbb{R}^m$.

(5.9) *Application 3*. A duality principle.

We now briefly describe a duality principle which is implicitly contained in Lemma (4.3). We express it as adjointness of two transition functions or rather, their associated contractions. For $x\in E$, $A\in\mathscr{E}$, $n\geqq 1$, define:

(5.10) (i) $D_n(x, A)=P^x\{X_1\leqq x, \ldots, X_n\leqq x,\ X_n\in A\}$,

(ii) $\hat{D}_n(x, A)=\hat{P}^x\{\hat{X}_1\leqq\hat{X}_n, \ldots, \hat{X}_{n-1}\leqq\hat{X}_n, \hat{X}_n\in A\}$.

Clearly these transition functions induce contractions D_n and $\hat{D}_n$ on $L^p(\pi)$ and $L^q(\pi)$ respectively, and the duality result is:

(5.11) **Propostion.** *$D_n^*=\hat{D}_n$ for all $n>0$.*

(5.12) *Remark.* The symmetric statements, where $\leqq$ in (5.10) is replaced systematically by $<$, $\geqq$ or $>$, and the dual statements hold also.

Proof. With the stopping times N^+ and $\hat{N}_+$ and the duality being used in this section we see that with definition (5.10)(i)

$$G_{N^+}(\tau)=\sum_0^\infty \tau^n D_n \qquad \textit{where } D_0=I.$$

Further, observing that

$$\hat{D}_n(x, A)=\hat{P}^x\{n \text{ is a weak ascending ladder index, } \hat{X}_n\in A\}$$

we readily find that

$$(I-\hat{H}_{\hat{N}_+}(\tau))^{-1}=\sum_0^\infty \tau^n \hat{D}_n \qquad \text{where } \hat{D}_0=I,$$

and the proof is an immediate consequence of Lemma (4.3)(ii).

(5.13) *Remark.* We can express Proposition (5.11) as follows: for $f\in L^p(\pi)$, $g\in L^q(\pi)$, $n>0$:

$$\begin{aligned}\langle f, D_n g\rangle &= \iint f(x)\, P^x\{X_1\leqq x, \ldots, X_n\leqq x, X_n\in(dx')\}\, g(x')\,\pi(dx)\\ &= \iint f(x)\, \hat{P}^{x'}\{\hat{X}_1\leqq\hat{X}_n, \ldots, \hat{X}_{n-1}\leqq\hat{X}_n, \hat{X}_n\in(dx)\}\, g(x')\,\pi(dx')\\ &= \langle f\hat{D}_n, g\rangle.\end{aligned}$$

In this form it is easy to give a direct probabilistic proof, and with this proof of Lemma (4.3), combined with a direct probabilistic proof of Lemma (4.1), we have an alternative method of obtaining Theorem (5.3).

6. A Factorisation Associated with the Second Component of a MAP

As a second specialisation we derive a factorisation using the ladder indices associated with the S-component of a MAP (X, S). This was our original aim and

E. Arjas and T. P. Speed:

amongst many possible applications, it gives an alternative way of deriving the result (1.1). Throughout we suppose the dimension $m=1$, see Remark (6.6).

The algebra which we decompose is the full algebra $\mathscr{A}$ of all Fourier transforms $T(\theta)$. For any such transform we have $T(\theta)=\int e^{i\theta y}\,T(dy)$, and we define

$$(6.1)\qquad \begin{aligned} T(\theta)^- &= \int_{-\infty}^{0-} e^{i\theta y}\,T(dy),\\ T(\theta)^{\cdot} &= T(\{0\}),\\ T(\theta)^+ &= \int_{0+}^{\infty} e^{i\theta y}\,T(dy), \end{aligned}$$

where the right sides can be interpreted formally or precisely, as operator integrals. For example, if $f\in L^p$, $g\in L^q$, $p^{-1}+q^{-1}=1$, then we define such integrals by

$$\langle f, T(\theta)^- g\rangle = \int_{-\infty}^{0-} e^{i\theta y}\langle f, T(dy)\,g\rangle$$

and similarly for $T(\theta)^+$. Clearly $T(\theta)=T(\theta)^-+T(\theta)^{\cdot}+T(\theta)^+$ and this decomposition induces a decomposition of $\mathscr{A}$ satisfying I(i), (ii) of § 4. The system of stopping times is the family of ladder indices for S:

$$(6.2)\qquad \begin{aligned} &N^+=\inf\{n>0\colon S_n>0\};\\ &N_+=\inf\{n>0\colon S_n\geqq 0\};\\ &N^{\cdot +}=N_+ \quad \text{if } N_+<N^+, \text{ and } N^{\cdot +}=\infty \text{ otherwise};\\ &\text{and s. and d.} \end{aligned}$$

We again omit the verification of the fact that (6.2) satisfies II and III of § 4; II(ii) now follows because $S_{N^{\cdot +}}=0$ on $\{N^{\cdot +}<\infty\}$. We have the following theorem, where $H_{N^{\cdot +}}(\tau)=H_{N^{\cdot +}}(\tau,0)$:

(6.3) **Theorem.** *Let Q and $\hat{Q}$ be in duality relative to π, and consider the stopping times* (6.2) *and s. and d. Then as a relation between contractions on $L^p(\pi)$, for $0\leqq\tau<1$, $\theta\in\mathbb{R}^1$:*

$$(6.4)\qquad \begin{aligned} &I-\tau Q(\theta)=[I-\hat{H}^*_{\hat{N}^+}(\tau,\theta)]\,[I-H_{N^{\cdot +}}(\tau)]\,[I-H_{N^+}(\tau,\theta)],\\ &\textit{and s. and d.}, \end{aligned}$$

where the middle term is interchangeable with $I-\hat{H}^*_{\hat{N}^{\cdot +}}(\tau)$, and s. and d. The uniqueness is as in Theorem (4.5).

We now suppose that the state space $E=\{1,2,\ldots,s\}$ and for a given SMTF Q the underlying chain P is ergodic. Thus there is a unique invariant measure π such that $\pi(i)>0$, $i\in E$. Put $\Delta=(\delta_{ij}\pi(i))$.

(6.5) **Corollary.** *In the finite-state case just described, if t denotes matrix transpose:*

$$I-\tau Q(\theta)=\Delta^{-1}[I-\hat{H}_{\hat{N}^+}(\tau,\theta)]^t\Delta\,[I-H_{N^{\cdot +}}(\tau)]\,[I-H_{N^+}(\tau,\theta)]$$

and s. and d.

This result is a symmetrised form of Theorem (2.1) of Presman [12], and if the last two factors are combined it becomes exactly his result.

(6.6) *Remark.* Before going on to give applications of Theorem (6.3) we will observe that the restriction to $m=1$ in this section is purely for simplicity. At least one interesting situation in $m>1$ dimensions is when N is the hitting time to a half-space through 0, as described in § 5. This topic can be treated exactly as the 1-dimensional case has been, giving rise to a generalised form of (6.3).

(6.7) *Application 1.* A duality principle.

The following discussion is a generalisation of the result Feller [7], p. 609, as indeed was the result (5.9). In a manner similar to our previous discussion we define SMTF's $D_n, \hat{D}_n$: for $x \in E$, $A \in \mathscr{E}$, $B \in \bar{\mathscr{R}}^1$ and $n \geq 1$

(6.8) (i) $D_n(x, A \times B) = P^x\{X_n \in A, S_1 \leqq 0, \ldots, S_n \leqq 0, S_n \in B\}$;
(ii) $\hat{D}_n(x, A \times B) = \hat{P}^x\{\hat{X}_n \in A, \hat{S}_1 \leqq \hat{S}_n, \ldots, \hat{S}_{n-1} \leqq \hat{S}_n, \hat{S}_n \in B\}$.

It is easy to see that these induce contractions on $L^p(\pi)$ and $L^q(\pi)$ respectively, and the duality result here is:

(6.9) **Proposition.** *$D_n^*(B) = \hat{D}_n(B)$ for all $B \in \bar{\mathscr{R}}^1$, $n>0$.*

Proof. The proof is almost identical to that given for Proposition (5.11).

Remark (5.12) applies here as well. Also as in § 5 we can give a direct proof of this result, but we refer to the final section for a fuller discussion.

We now discuss briefly the above duality in the context of the bivariate processes $(X, W) = \{(X_n, W_n): n \geqq 0\}$ and $(X, M) = \{(X_n, M_n): n \geqq 0\}$ where we define

(6.10)
$$(X_0, W_0) = (X_0, 0)$$
$$(X_n, W_n) = (X_n, (W_{n-1} + S_n - S_{n-1})^+), \quad n>0;$$

and

(6.11)
$$(X_n, M_n) = (X_n, \min(0, S_1, \ldots, S_n)), \quad n \geqq 0.$$

We now formulate this duality explicitly as:

(6.12) **Theorem.** *For (X, S) and $(\hat{X}, \hat{S})$ in duality the bivariate processes (X, W) and $(\hat{X}, \hat{M})$ are adjoint.*

Proof. As shown in Arjas and Speed [2] the resolvent of (X, W) is

$$\Lambda(\tau, \theta) = [I - H_{N_-}(\tau, 0)]^{-1} G_{N_-}(\tau, \theta)$$

and that of $(\hat{X}, \hat{M})$ is

$$\hat{\Phi}(\tau, \theta) = [I - \hat{H}_{\hat{N}^-}(\tau, \theta)]^{-1} \hat{G}_{\hat{N}^-}(\tau, 0),$$

where the stopping times are the ladder indices (6.2). Now if we take the adjoint of $\Lambda(\tau, \theta)$ we find

$$\begin{aligned} \Lambda^*(\tau, \theta) &= G_{N_-}^*(\tau, \theta)[I - H_{N_-}^*(\tau, 0)]^{-1} \\ &= [I - \hat{H}_{\hat{N}^-}(\tau, \theta)]^{-1} \hat{G}_{\hat{N}^-}(\tau, 0) \quad \text{by Lemma (4.3)} \\ &= \hat{\Phi}(\tau, \theta) \quad \text{as stated.} \end{aligned}$$

(6.13) *Application 2.* A moment identity.

In Feller [7] one of the more immediate consequences of the factorisation (1.1) is a relation between the expectations of the hitting times to half-lines (assuming both exist) which reads

$$-\tfrac{1}{2}\sigma^2 = E[S_{N^-}][1-\zeta]\,E[S_{N^+}]. \tag{6.14}$$

We now derive an analogue of (6.14) for the stopping times under discussion in this section. Let $E^\pi[f]$ be an abbreviation for $\langle 1, f\rangle = \int f(x)\,\pi(dx)$ and let us consider (when possible) the limited expansions:

$$\begin{aligned} Q(\theta) &= P + i\theta Q_1 - \tfrac{1}{2}\theta^2 Q_2 + o(\theta^2);\\ H_{N^+}(1,\theta) &= H^+ + i\theta M^+ + o(\theta);\\ H_{N^{\cdot +}}(1) &= H^{\cdot +};\\ &\text{and d.} \end{aligned} \tag{6.15}$$

(6.16) **Theorem.** *Let Q and $\hat{Q}$ be in duality relative to π, and consider the stopping times* (6.2). *Then, if S_{N^+}* (resp. $\hat{S}_{\hat{N}^+}$) *is proper and has a finite expectation irrespective of the starting point X_0 of X* (resp. $\hat{X}_0$ *of* $\hat{X}$),

$$Q_1 = 0, \qquad Q_2 < \infty$$

and

$$-\tfrac{1}{2}E^\pi[S_1^2] = \iint E^x[\hat{S}_{\hat{N}^+}]\,[I - H^{\cdot +}](x, dx')\,E^{x'}[S_{N^+}]\,\pi(dx).$$

Proof. We use the factorisation (6.4) at $\tau = 1$, giving

$$\begin{aligned} &\langle 1, [I - Q(\theta)]\,1\rangle\\ &\quad= \langle 1, [I - \hat{H}^*_{\hat{N}^+}(1,\theta)]\,[I - H_{N^{\cdot +}}(1)]\,[I - H_{N^+}(1,\theta)]\,1\rangle\\ &\quad= \langle [I - \hat{H}^+ - i\theta\hat{M}^+ + o(\theta)]\,1, [I - H^{\cdot +}]\,[I - H^+ - i\theta M^+ + o(\theta)]\,1\rangle\\ &\quad= -\theta^2\langle \hat{M}^+ 1, [I - H^{\cdot +}]\,M^+ 1\rangle + o(\theta^2), \end{aligned}$$

since, by the assumption of properness, $\hat{H}^+ 1 = 1$ and $H^+ 1 = 1$. On the other hand we can use the expansion

$$\begin{aligned} &\langle 1, [I - Q(\theta)]\,1\rangle\\ &\quad= \langle 1, [I - P - i\theta Q_1 + \tfrac{1}{2}\theta^2 Q_2 + o(\theta^2)]\,1\rangle\\ &\quad= -i\theta\langle 1, Q_1 1\rangle + \tfrac{1}{2}\theta^2\langle 1, Q_2 1\rangle + o(\theta^2), \end{aligned}$$

and the assertion follows by comparing the coefficients of θ and θ^2.

7. Two-Barrier Duality Relations in MAP's

In this final section we show that some general duality relations obtained recently by one of us in the case of one-dimensional random walks carry over to the present situation. In particular we can use them to give a direct probabilistic proof of (6.3).

Let (X, S) be as before, $m = 1$, and define the "reflected" process (X', S') with SMTF Q' by $Q'(B) = Q(-B)$, $B \in \bar{\mathscr{R}}^1$. Further, let (X, V) (resp. (X', V')) be the process obtained from (X, S) (resp. (X', S')) by placing two absorbing barries for the second component at specified positions, and (X, W) (resp. (X', W')) be the process obtained from (X, S) (resp. (X', S')) by placing two impenetrable barries for the second component at 0 and $a>0$. In the latter case we have inductively

$$W_0 = S_0; \quad W_n = \min(a, \max(W_{n-1} + S_n - S_{n-1}, 0)), \quad n>0.$$

The dual processes $(\hat{X}, \hat{S})$, $(\hat{X}', \hat{S}')$, $(\hat{X}, \hat{V})$, $(\hat{X}', \hat{V}')$, $(\hat{X}, \hat{W})$ and $(\hat{X}', \hat{W}')$ have their obvious meanings. We remark that the definition of an MAP can easily be extended to allow S to have a non-zero starting position.

Our duality relations are expressed in terms of the equality and adjointness of certain operators on $L^p(\pi)$. We define the following transition functions, where absorbing barriers are placed in braces following the expressions: for $x \in E$, $A \in \mathscr{E}$, an interval $I \in \mathscr{R}^1$, $y, z \in \mathbb{R}^1$, $n \geq 0$, $a>0$:

$$\begin{aligned}
&D_n(x, A, I, y, z) = P^x\{X_n \in A, W_n \leq z, S_n \in I + y \mid S_0 = y\};\\
(7.1)\quad &\hat{D}_n(x, A, I, y, z) = \hat{P}^x\{\hat{X}_n \in A, \hat{V}_n \leq a - y, \hat{S}_n \in I + a - z \mid \hat{S}_0 = a - z\}, \quad \{0, a+\};\\
&D'_n(x, A, I, y, z) = P^x\{X'_n \in A, W'_n \geq a - z, S'_n \in -I + a - y \mid S'_0 = a - y\};\\
&\hat{D}'_n(x, A, I, y, z) = \hat{P}^x\{\hat{X}'_n \in A, \hat{V}'_n \geq y, \hat{S}'_n \in -I + z \mid \hat{S}'_0 = z\}, \quad \{0-, a\}.
\end{aligned}$$

The associated operators are denoted by dropping the first two arguments e.g. $D_n(I, y, z)$ arises from $D_n(x, A, I, y, z)$.

(7.2) **Proposition.** *The following operators coincide*:

(1) $D_n(I, y, z)$,
(2) $\hat{D}_n^*(I, y, z)$,
(3) $D'_n(I, y, z)$,
(4) $\hat{D}'^*_n(I, y, z)$.

Further, if the inequalities on the right side of (7.1) are made strict and the barriers changed to $\{0-, a\}$ and $\{0, a+\}$ respectively, the above result is still true.

Proof. The result (1)=(2) follows from the corresponding result of Speed [13] by proving that for $f \in L^p$, $g \in L^q$:

$$\begin{aligned}
&\iint f(x) P^x\{X_n \in (dx'), W_n \leq z, S_n \in I + y \mid S_0 = y\} g(x') \pi(dx)\\
&\quad = \iint f(x) \hat{P}^{x'}\{\hat{X}_n \in (dx), \hat{V}_n \leq a - y, \hat{S}_n \in I + a - z \mid \hat{S}_0 = a - z\} g(x') \pi(dx').
\end{aligned}$$

All the other assertions are proved similarly.

Finally we remark that the case $a = \infty$ (one impenetrable or absorbing barrier only) can be formulated as (7.2) above using the analogous results in the i.i.d. case.

References

1. Arjas, E.: On a fundamental identity in the theory of semi-Markov processes. Advances Appl. Probab. **4**, 258–270 (1972).
2. Arjas, E., Speed, T. P.: Topics in Markov additive processes. To appear in Math. Scand.
3. Blumenthal, R. M., Getoor, R. K.: Markov Processes and Potential Theory. New York: Academic Press 1968.
4. Çinlar, E.: Markov additive processes. I. Z. Wahrscheinlichkeitstheorie verw. Gebiete **24**, 85–93 (1972).

Arjas and Speed: Symmetric Wiener-Hopf Factorisations in Markov Additive Processes

5. Çinlar, E.: Markov additive processes. II. Z. Wahrscheinlichkeitstheorie verw. Gebiete **24**, 95–121 (1972).
6. Dinges, H.: Wiener-Hopf-Faktorisierung für substochastische Übergangsfunktionen in angeordneten Räumen. Z. Wahrscheinlichkeitstheorie verw. Gebiete **11**, 152–164 (1969).
7. Feller, W.: An Introduction to Probability Theory and its Applications, Volume 2, Second Edition. New York: Wiley 1971.
8. Kemeny, J.G., Snell, J.L., Knapp, A.: Denumerable Markov Chains. Princeton: Van Nostrand 1966.
9. Kingman, J.F.C.: On the algebra of queues. J. Appl. Probab. **3**, 285–326 (1966).
10. Nelson, E.: The adjoint Markoff process. Duke Math. J. **25**, 671–690 (1958).
11. Presman, E.L.: A boundary value problem for the sum of lattice random variables given on a finite regular Markov chain. Theor. Probab. Appl. **12**, 323–328 (1967). (English translation.)
12. Presman, E.L.: Factorization methods and boundary problems for sums of random variables given on Markov chains. Math. USSR Izvestija **3**, 815–852 (1969). (English translation.)
13. Speed, T.P.: A note on random walks. II. To appear in J. Appl. Probab. **10**, 218–222 (1973).
14. Spitzer, F.: Principles of Random Walk. Princeton: Van Nostrand 1964.

E. Arjas
C.O.R.E.
de Croylaan 54
B-3030 Heverlee
Belgium

T.P. Speed
Department of Probability and Statistics
The University of Sheffield
Sheffield S3 7RH
England

(Received November 15, 1972)

Stochastic Processes and their Applications 1 (1973) 369–374. © *North-Holland Publ. Co.*

A NOTE ON RANDOM TIMES

J.W. PITMAN and T.P. SPEED

Department of Probability and Statistics, University of Sheffield, Sheffield, UK

Received 26 June 1973

Abstract. A generalisation of the notion of stopping time is stated, and related to similar generalisations introduced by Bahadur, Kemperman, Siegmund and others with a view to permitting auxiliary experimentation to enter into the definition of stopping rule. The main aim of this note is to draw attention to the conditional independence implicit in the definitions of these writers, and briefly indicate some consequences of this.

random time, conditional independence
stopping time

1. Introduction and description of results

Suppose that $(X_n, n = 1, 2, ...)$ is a sequence of random variables defined on a probability space $(\Omega, \mathcal{F}, \mathbf{P})$, and let $\mathcal{F}_n$ denote the σ-field generated by the random variables $X_1, X_2, ..., X_n$. In the theory of optional stopping of such a process (X_n), the random times considered are commonly assumed to be *stopping times of* $(\mathcal{F}_n)$, that is to say, extended positive integer-valued random variables t such that for each $n = 1, 2, ...$, the event $\{t > n\}$ lies in the σ-field $\mathcal{F}_n$ determined by the evolution of the process up to time n. Several authors have also considered stopping procedures involving the outcomes of random experiments auxiliary to the basic process (X_n); in this connection we mention Bahadur [2], Kemperman [6], Singh [9], Siegmund [8], Chow, Robbins and Siegmund [4], and Arjas and Speed [1].

In order to provide a unified approach to the work of these authors we make the definition which follows. Let $(\Omega, \mathcal{F}, \mathbf{P})$ be a probability space and let us refer to an extended positive integer-valued random variable defined on Ω as a *random time*. Suppose that $(\mathcal{F}_n, n = 1, 2, ...)$ is an increasing sequence of sub-σ-fields of $\mathcal{F}$, and let $\mathcal{F}_\infty$ denote the smallest σ-field containing every $\mathcal{F}_n$, $n = 1, 2, ...$.

Definition 1.1. A random time t is a *randomised stopping time* of $(\mathcal{F}_n)$ if for each $n = 1, 2, \ldots$, the event $\{t > n\}$ and the σ-field $\mathcal{F}_\infty$ are conditionally independent given $\mathcal{F}_n$.

The point of this note is to state Propositions 2.4 and 2.5 below which utilise some elementary properties of conditional independence to give various equivalent formulations of the above definition. These formulations show that the kinds of random times considered by Bahadur and Siegmund are essentially the same and just randomised stopping times according to the above definition, and it is clear that the random times considered by Kemperman and Singh are also included. We do not discuss here any applications of randomised stopping times, but refer the reader to the papers and books mentioned above.

When $\mathcal{F}_n$ is the σ-field generated by random variables $X_1, \ldots, X_n$, we refer to a (randomised) stopping time of $(\mathcal{F}_n)$ as a *(randomised) stopping time of* (X_n). It will be seen that a randomised stopping time of (X_n) can be thought of as being generated in the following way: an observer watches the evolution of the process (X_n) as time n increases, until a random time t when he stops observing the process; if at time k he has not yet stopped observing the process, the observer notes the value of X_k and then decides according to the outcome of some random experiment whether to stop at time k or to continue to observe the process. The random time t is a randomised stopping time of (X_n) if for each k, the outcome of the random experiment at time k and the as yet unobserved future $(X_n, k < n < \infty)$ are conditionally independent given the observed past $(X_n, 1 \leq n \leq k)$. The random time t is a stopping time of (X_n) if for each k, the decision at time k is made deterministically (and measurably) according to the past $(X_n, 1 \leq n \leq k)$.

A consequence of Proposition 2.5 is that properties of randomised stopping times associated with Markov processes or martingales can be immediately deduced from the well-known properties of stopping times of these processes. Indeed, let $(X_n, n = 1, 2, \ldots)$ be a sequence of random variables adapted to an increasing sequence of σ-fields $(\mathcal{F}_n, n = 1, 2, \ldots)$, and suppose that (X_n) is a Markov process (respectively, martingale) with respect to $(\mathcal{F}_n)$. If t is a randomised stopping time of $(\mathcal{F}_n)$ and $\mathcal{F}_n^t$ denotes the σ-field generated by $\mathcal{F}_n$ and the events $\{t = 1\}, \ldots$ $\ldots, \{t = n\}$, then it follows from the equivalence of (i) and (iv) in Proposition 2.5 that (X_n) is also a Markov process (respectively, martingale) with respect to $(\mathcal{F}_n^t)$, and since t is a stopping time of $(\mathcal{F}_n^t)$, all the standard results for stopping times of Markov processes and martingales can be applied at once to randomised stopping times of these processes.

For the sake of simplicity, we have only considered here random times associated with processes whose time set is the positive integers, but most of the discussion is easily adapted to the other usual time sets.

2. Details

Let N denote the set of natural numbers $\{1, 2, \ldots, n, \ldots\}$. Suppose throughout that $(\Omega, \mathcal{F}, \mathbf{P})$ is a probability space and that $(\mathcal{F}_n, n \in \mathbf{N})$ is an increasing sequence of sub-σ-fields of $\mathcal{F}$, with $\mathcal{F}_\infty$ the smallest sub-σ-field of $\mathcal{F}$ containing each $\mathcal{F}_n$. The reader is referred to [7] for a treatment of conditional independence.

Remark 2.1. Recalling from the introduction the definition of a randomised stopping time of $(\mathcal{F}_n)$, we observe that alternative but equivalent definitions are obtained by replacing the set $\{t > n\}$ apearing in the definition by any one of the sets $\{t \leq n\}$, $\{t \geq n\}$ and $\{t \neq n\}$.

Examples 2.2. Any random time independent of $\mathcal{F}_\infty$ is a randomised stopping time of $(\mathcal{F}_n)$, and so too is any stopping time of $(\mathcal{F}_n)$. For a less trivial example, consider a real-valued process (X_n) defined on $(\Omega, \mathcal{F}, \mathbf{P})$ and suppose that Y is a real-valued random variable independent of the process (X_n). Let $t = \inf\{n : X_n \geq Y\}$. Then it is easily seen that t is a randomised stopping time of (X_n). A similarly defined randomised stopping time of a continuous time process finds an application in [3, p. 276].

For another example, suppose that (X_n) is a Markov chain and let T_n be the time of the nth visit to state i. Let T be any stopping time of (X_n) and define a random time t by $t = \inf\{n : T_n \geq T\}$, so that $t - 1$ is the number of visits to state i before time T. Then t is a randomised stopping time of (T_n), as may be seen from the fact that $\{t > n\} = \{T_n < T\}$, [7, IV T41] and the strong Markov property (cf. [5, p. 27, proof of Theorem (76)]).

For sub-σ-fields $\mathcal{A}$ and $\mathcal{B}$ of $\mathcal{F}$, let us denote by $\mathcal{A} \vee \mathcal{B}$ the smallest sub-σ-field of $\mathcal{F}$ which contains both $\mathcal{A}$ and $\mathcal{B}$. Suppose we are given sub-σ-fields $\mathcal{E}_1$, $\mathcal{E}_2$ and $\mathcal{E}$ of $\mathcal{F}$.

Lemma 2.3. *The following statements are equivalent:*

(i) $\mathcal{E}_1$ *and* $\mathcal{E}_2$ *are conditionally independent given* $\mathcal{E}$;
(ii) $P[A \mid \mathcal{E} \vee \mathcal{E}_2] = P[A \mid \mathcal{E}]$ *a.s. for every set* A *in* $\mathcal{E}_1$;
(iii) $\mathcal{E} \vee \mathcal{E}_1$ *and* $\mathcal{E} \vee \mathcal{E}_2$ *are conditionally independent given* $\mathcal{E}$;
(iv) $E\{Y \mid \mathcal{E} \vee \mathcal{E}_2\} = E\{Y \mid \mathcal{E}\}$ *a.s. for every integrable* $\mathcal{E} \vee \mathcal{E}_1$*-measurable random variable* Y.

In (ii) and (iii) the subscripts 1 and 2 can be interchanged to give further statements equivalent to (i).

Proof. The equivalence of (i) and (ii) is proved in [7, II T 51]. The further equivalence of (iii) and (iv) follows by repeated application of this result.

With the aid of Lemma 2.3, the conditional independence condition in the definition of randomised stopping time can now be rephrased in a multitude of ways. Proposition 2.4 below displays some minimal conditions for a random time to be a randomised stopping time of $(\mathcal{F}_n)$, while in Proposition 2.5 the conditional independence is exploited to the full to give some strong properties of randomised stopping times.

Suppose that t is a random time on $(\Omega, \mathcal{F}, P)$.

Proposition 2.4. *The following statements are equivalent:*
(i) t *is a randomised stopping time of* $(\mathcal{F}_n)$;
(ii) *for all* $n \in N$, $P[t > n \mid \mathcal{F}_\infty] = P[t > n \mid \mathcal{F}_n]$ *a.s.*;
(iii) *for all* $n \in N$, $A \in \mathcal{F}_\infty$, $P[t > n, A] = \int_{\{t>n\}} P[A \mid \mathcal{F}_n] dP$.
Further statements equivalent to (i) *are obtained by replacing the set* $\{t > n\}$ *appearing in* (ii) *and* (iii) *by any of* $\{t \leq n\}$, $\{t = n\}$ *and* $\{t \neq n\}$.

Proof. Let $\mathcal{S}_n$ denote the sub-σ-field of $\mathcal{F}$ generated by the event $\{t > n\}$. By definition, t is a randomised stopping time of $(\mathcal{F}_n)$ if and only if for each n in N, the σ-fields $\mathcal{S}_n$ and $\mathcal{F}_\infty$ are conditionally independent given $\mathcal{F}_n$. The equivalence of (i), (ii) and (iii) now follows from the equivalence of (i) and (ii) in Lemma 2.3 since $\mathcal{F}_\infty \vee \mathcal{F}_n = \mathcal{F}_\infty$ and $\mathcal{S}_n \vee \mathcal{F}_n$ has an obvious simple structure. Using Remark 2.1 the remaining assertions can be proved in an identical manner.

Continuing to suppose that t is a random time on $(\Omega, \mathcal{F}, P)$, let $\mathcal{F}'_n$ denote the smallest sub-σ-field of $\mathcal{F}$ containing $\mathcal{F}_n$ and the events $\{t = 1\}, \ldots, \{t = n\}$.

§ 2. *Details*

Proposition 2.5. *The following statements are equivalent:*

(i) *t is a randomised stopping time of $(\mathcal{F}_n)$;*

(ii) *for each $n \in \mathbf{N}$ the σ-fields $\mathcal{F}^t_n$ and $\mathcal{F}_\infty$ are conditionally independent given $\mathcal{F}_n$;*

(iii) *for each $n \in \mathbf{N}$, $\mathbf{E}\{Y \mid \mathcal{F}_\infty\} = \mathbf{E}\{Y \mid \mathcal{F}_n\}$ a.s. for each integrable $\mathcal{F}^t_n$-measurable random variable Y;*

(iv) *for each $n \in \mathbf{N}$, $\mathbf{E}\{Z \mid \mathcal{F}^t_n\} = \mathbf{E}\{Z \mid \mathcal{F}_n\}$ a.s. for each integrable $\mathcal{F}_\infty$-measurable random variable Z.*

Proof. Let $\mathcal{F}_n$ denote the sub-σ-field of $\mathcal{F}$ generated by the events $\{t = 1\}, \ldots, \{t = n\}$, so that $\mathcal{F}^t_n = \mathcal{F}_n \vee \mathcal{F}_n$. The fact that $(\mathcal{F}_n)$ is an increasing sequence of σ-fields ensures that t is a randomised stopping time of $(\mathcal{F}_n)$ if and only if for each n in $\mathbf{N}$, the σ-fields $\mathcal{F}_n$ and $\mathcal{F}_\infty$ are conditionally independent given $\mathcal{F}_n$, and the proposition now follows by applying Lemma 2.3.

Put another way, the equivalence of (i) and (ii) in Proposition 2.5 means that t is a randomised stopping time of $(\mathcal{F}_n)$ if and only if there exists an increasing sequence of σ-fields $(\mathcal{G}_n)$ within $\mathcal{F}$ such that $\mathcal{F}_n \subseteq \mathcal{G}_n$, t is a stopping time of $(\mathcal{G}_n)$, and $\mathcal{G}_n$ and $\mathcal{F}_\infty$ are conditionally independent given $\mathcal{F}_n$. With this conditional independence criterion written in the form

$$\mathbf{P}[A \mid \mathcal{G}_n] = \mathbf{P}[A \mid \mathcal{F}_n] \quad \text{a.s.}$$

for all $n \in \mathbf{N}$ and $A \in \mathcal{F}_\infty$, this is just the property required by Siegmund of his 'randomised stopping variables for $(\mathcal{F}_n)$'.

Given an increasing sequence $(\mathcal{F}_n)$ of σ-fields in a probability space $(\Omega, \mathcal{F}, \mathbf{P})$, it may be that the probability space as it stands is not large enough to support many randomised stopping times of $(\mathcal{F}_n)$. As an extreme case, if $\mathcal{F} = \mathcal{F}_\infty$, then the only randomised stopping times of $(\mathcal{F}_n)$ are stopping times of $(\mathcal{F}_n)$. For this reason, it seems reasonable to consider the possibility of enlarging the original probability space in some way to allow room for experimentation auxiliary to $\mathcal{F}_\infty$. Consider, for example, the following procedure used by Bahadur [2]. Suppose that there is given for each n, an $\mathcal{F}_n$-measurable function a_n with $0 \leq a_n \leq 1$.

Observe the sequence of σ-fields $(\mathcal{F}_n)$ in succession, and given that the first m σ-fields have been observed, conduct an auxiliary random experiment with probability of success equal to the observed value of a_m, stopping at the time of the first success. This procedure will define a random time t on a probability space $(\Omega', \mathcal{F}', \mathbf{P}')$ constructed from $(\Omega, \mathcal{F}, \mathbf{P})$ and all necessary auxiliary experiments. This probability space will contain an isomorphic image of $\mathcal{F}$ in $\mathcal{F}'$ on which $\mathbf{P}'$ agrees with $\mathbf{P}$, and after identifying σ-fields and random variables defined on $(\Omega, \mathcal{F}, \mathbf{P})$ with their isomorphic copies in $(\Omega', \mathcal{F}', \mathbf{P}')$ it is being assumed that for each $n \in \mathbf{N}$, the event $\{t > n\}$ and the σ-field $\mathcal{F}_\infty$ are conditionally independent given $\mathcal{F}_n$ and $\{t \geq n\}$, and that

$$\mathbf{P}'[\{t = n\} \mid \mathcal{F}_n, \{t \geq n\}] = a_n \qquad \text{on } \{t \geq n\}.$$

The construction of the probability space $(\Omega', \mathcal{F}', \mathbf{P}')$ and random time t is easily formalised, and it may be shown that t is a randomised stopping time of $(\mathcal{F}_n)$ in $(\Omega', \mathcal{F}', \mathbf{P}')$, with

$$\mathbf{P}'[\{t > n\} \mid \mathcal{F}_\infty] = (1 - a_1) \dots (1 - a_n), \qquad n \in \mathbf{N}.$$

Moreover, if t^* is any randomised stopping time of $(\mathcal{F}_n)$ defined (in the obvious way) on an enlargement $(\Omega^*, \mathcal{F}^*, \mathbf{P}^*)$ of $(\Omega, \mathcal{F}, \mathbf{P})$, then a randomised stopping time of $(\mathcal{F}_n)$ having the same joint distribution with $\mathcal{F}_\infty$ as t^* can be constructed in the manner described above by taking

$$a_n = \mathbf{P}^*[t = n \mid \mathcal{F}_n] / \mathbf{P}^*[t \geq n \mid \mathcal{F}_n].$$

References

[1] E. Arjas and T.P. Speed, Markov additive processes (1972), manuscript.

[2] R.R. Bahadur, Sufficiency and statistical decision functions, Ann. Math. Statist. 25 (1954) 423–462.

[3] L. Breiman, Probability (Addison–Wesley, Reading Mass., 1968).

[4] Y.S. Chow, H. Robbins and D. Siegmund, Great Expectations: The Theory of Optimal Stopping (Houghton Mifflin, Boston, Mass., 1971).

[5] D. Freedman, Markov Chains (Holden–Day, San Francisco, Calif., 1971).

[6] J.H.B. Kemperman, The Passage Problem for a Stationary Markov Chain (Univ of Chicago Press, Chicago, Ill., 1961).

[7] P.A. Meyer, Probability and Potentials (Blaisdell, Waltham, Mass., 1966).

[8] D. Siegmund, Some problems in the theory of optimal stopping, Ann. Math. Statist. 38 (1967) 1627–1640.

[9] R. Singh, Existence of unbiased estimates, Sankhya Ser. A 26 (1964) 93–96.

J. Austral. Math. Soc. **22** (Series A) (1976), 462–468.

GEOMETRIC AND PROBABILISTIC ASPECTS OF SOME COMBINATORIAL IDENTITIES

T. P. SPEED

(Received June 9, 1975; revised October 20, 1975)

Communicated by W. D. Wallis

Abstract

For positive integers a, b and n define the combinational expression

$$A_n(a,b) = \frac{a}{a+bn}\binom{a+bn}{n}.$$

We give geometric and probabilistic interpretations of these expressions (and their multidimensional extensions) and find new, simple proofs of the convolution identities known to hold for such expressions.

1. Introduction

For non-negative integers a, b with $a + bn > 0$ let us define the combinatorial expression

$$A_n(a,b) = \frac{a}{a+bn}\binom{a+bn}{n}. \tag{1}$$

In two papers written some twenty years ago Gould (1956, 1957) discussed the above (and related) expressions. He obtained, amongst other results, the following convolution identity: for positive integral c

$$\sum_{m=0}^{n} A_m(a,b)A_{n-m}(c,b) = A_n(a+c,b). \tag{2}$$

Gould's two papers contain different approaches to this identity whilst in his recent article Gould (1974) gives yet another. We also note that Riordan (1958) presents an inductive proof of (2), as do Gould and Kaucky (1966) where further comments and extensions can be found. The proof of Blackwell and Dubins (1966) is perhaps closest in spirit to the one given below. Mohanty (1966a) extended the argument of Gould's first paper, stating and proving multinomial analogues of (2) and the related identities. He also gave a probabilistic interpretation of these facts. To formulate these results we use bold letters to denote k-tuples of non-negative integers, $\boldsymbol{b} = (b_1, b_2, \cdots, b_k)$ and $\boldsymbol{n} = (n_1, n_2, \cdots, n_k)$. Further we use the usual dot-product notation $\boldsymbol{b} \cdot \boldsymbol{n} = b_1 n_1 + b_2 n_2 + \cdots + b_k n_k$ and write $\mathbf{1} = (1, 1, \cdots, 1)$. With these preliminaries we can extend the notation above when $a + \boldsymbol{b} \cdot \boldsymbol{n} > 0$ writing

$$A_{\boldsymbol{n}}(a, \boldsymbol{b}) = \frac{a}{a + \boldsymbol{b} \cdot \boldsymbol{n}} \binom{a + \boldsymbol{b} \cdot \boldsymbol{n}}{\boldsymbol{n}} \tag{1'}$$

where $\binom{N}{\boldsymbol{n}} = N(N-1) \cdots (N - \mathbf{1} \cdot \boldsymbol{n} + 1) / \Pi_1^k n_i!$ denotes the usual multinomial coefficient. In this notation one of Mohanty's results (1966a, equation (9) p. 502), the generalisation of (2) above, can be written

$$\sum_{\boldsymbol{m}=\mathbf{0}}^{\boldsymbol{n}} A_{\boldsymbol{m}}(a, \boldsymbol{b}) A_{\boldsymbol{n}-\boldsymbol{m}}(c, \boldsymbol{b}) = A_{\boldsymbol{n}}(a + c, \boldsymbol{b}). \tag{2'}$$

Here is the summation from $m_1 = 0$ to $m_1 = n_1, \cdots, m_k = 0$ to $m_k = n_k$ as the notation suggests, and $\boldsymbol{n} - \boldsymbol{m} = (n_1 - m_1,\ n_2 - m_2, \cdots, n_k - m_k)$.

It is the purpose of this note to provide new proofs of these identities, the first, it is believed, that involve the geometrical interpretation of the expression (1'). After doing this we reconsider the probabilistic aspects of (2'), being somewhat more concrete than Mohanty in obtaining a random walk whose first passage probabilities to a certain hyperplane provide yet another interpretation and proof of (2').

2. Geometric interpretation of $A_{\boldsymbol{m}}(a, \boldsymbol{b})$

It is hoped that the notation will enable us to deal with the general case (arbitrary k) almost as as easily as one would the case $k = 1$, but this will involve some slightly unusual temporary usages. We will be working in the positive orthant of the integer lattice in $k + 1$ dimensions, the coordinate variables being denoted by $X_0,\ X_1, \cdots, X_k$ and an arbitrary element will be denoted by $(x_0, \boldsymbol{x})$ where $\boldsymbol{x} = (x_1, x_2, \cdots, x_k)$. The first coordinate will be treated differently, and all all bold letters will be k-*tuples* of non-negative integers.

Given any k-tuple $\boldsymbol{b}$ and non-negative integer a we can define a hyperplane by the equation

$$X_0 = a + (\boldsymbol{b} - \mathbf{1}) \cdot \boldsymbol{X}. \tag{P}$$

Clearly the point $(a + (\boldsymbol{b} - \mathbf{1}) \cdot \boldsymbol{n}, \boldsymbol{n})$ lies on (P) for any k-tuple $\boldsymbol{n}$, and we may now state the desired interpretation as follows:

PROPOSITION 1 (Mohanty). *The number of minimal lattice paths from* $(0, \mathbf{0})$ *to* $(a+(\mathbf{b}-\mathbf{1})\cdot\mathbf{n}, \mathbf{n})$ *which do not touch the plane* (P) *until the last point, is* $A_{\mathbf{n}}(a, \mathbf{b})$.

For the case $k = 1$ this result is in implicit in Mohanty and Narayana (1961) (following by duality from their Corollary on p. 256), and appears in the present generality in Mohanty (1972). The following proof is essentially Mohanty's but we include it for completeness.

PROOF. The minimal lattice paths from $(0, \mathbf{0})$ to $(a+(\mathbf{b}-\mathbf{1})\cdot\mathbf{n}, \mathbf{n})$ can be put into one-one correspondence with N-tuples $L = (\lambda_1, \lambda_2, \cdots, \lambda_N)$, where $N = a + \mathbf{b}\cdot\mathbf{n}$; for each $i, 1 \leqq i \leqq N$, λ_i is one of the symbols $S_0, S_1, \cdots, S_k$; for each j, $1 \leqq j \leqq k$ there are precisely n_j symbols S_j, and there are $a+(\mathbf{b}-\mathbf{1})\cdot\mathbf{n}$ symbols S_0.

Given such an N-tuple L we can build up a minimal lattice path, starting at either end, by interpreting a symbol S_j to mean 'move one unit along the X_j-axis towards the other end'. Conversely any minimal lattice path defines a unique such N-tuple in the obvious way.

It is also clear that there are precisely $\binom{N}{\mathbf{n}}$ such N-tuples and so this is the total number of minimal lattice paths connecting $(0, \mathbf{0})$ with $(a+(\mathbf{b}-\mathbf{1})\cdot\mathbf{n}, \mathbf{n})$. But we want the number of these which do not touch the plane (P) other than at the last point. To express this requirement as a property of the N-tuple L we need a little more notation. For each h, $1 \leqq h \leqq N$ and j, $0 \leqq j \leqq k$ define

$$\sigma_{hj} = \begin{cases} 1 & \text{if } \lambda_h = S_j \\ 0 & \text{otherwise.} \end{cases}$$

Clearly $\Sigma_{j=0}^{k} \sigma_{hj} = 1$. Also put $\xi_{ij} = \Sigma_{h=1}^{i} \sigma_{hj}$, this being the number of times the symbol S_j appears in the first i positions of L, and finally write $\boldsymbol{\xi}_i = (\xi_{i1}, \xi_{i2}, \cdots, \xi_{ik})$.

We will build up the lattice path by working backwards from the endpoint using L. After i steps have been incorporated, the X_j coordinate has reduced by $\xi_{ij} (0 \leqq j \leqq k)$ and so we are at the point $(a+(\mathbf{b}-\mathbf{1})\cdot\mathbf{n} - \xi_{i0}, \mathbf{n} - \boldsymbol{\xi}_i)$. This point lies in the half-space defined by (P) which contains the origin for all i, $1 \leqq i \leqq N$, if, and only if,

$$a+(\mathbf{b}-\mathbf{1})\cdot\mathbf{n} - \xi_{i0} < a+(\mathbf{b}-\mathbf{1})\cdot(\mathbf{n}-\boldsymbol{\xi}_i) \qquad (1 \leqq i \leqq N),$$

equivalently, upon expanding and using the fact that $\xi_{i0} + \mathbf{1}\cdot\boldsymbol{\xi}_i = i$, if and only if

$$\text{(C)} \qquad \mathbf{b}\cdot\boldsymbol{\xi}_i < i \qquad (1 \leqq i \leqq N).$$

Now $\mathbf{b}\cdot\boldsymbol{\xi}_i = \Sigma_{j=1}^{k} b_j \xi_{ij}$ is simply a partial sum along L of numerical terms if we replace S_j by the integer b_j, $1 \leqq j \leqq k$, and S_0 by 0. With this interpretation we can immediately recognise the condition (C) and use a well-known result to deduce that of the N cyclic permutations of the N-tuple L (with the numerical components just indicated), precisely $a = N - \boldsymbol{b} \cdot \boldsymbol{n}$ have the property (C); that is, satisfy the condition that for all i ($1 \leqq i \leqq N$) the partial sums of the first i terms are less than i.

We refer to Takacs (1967) p. 4 for this result; for a geometric proof more in the spirit of the present paper, see Mohanty (1966b).

This completes the proof that the number of minimal lattice paths from $(0, \mathbf{0})$ to $(a + (\boldsymbol{b} - \mathbf{1}) \cdot \boldsymbol{n}, \boldsymbol{n})$ not touching (P) before their endpoint is $\frac{a}{N}\binom{N}{\boldsymbol{n}}$ where $N = a + \boldsymbol{b} \cdot \boldsymbol{n}$.

3. Derivation of identities

Let us consider the hyperplane (P) defined above, and the parallel hyperplane (c being another non-negative integer)

(P′) $$X_0 = a + c + (\boldsymbol{b} - \mathbf{1}) \cdot \boldsymbol{X}.$$

Clearly any minimal lattice path from $(0, \mathbf{0})$ to $(a + c + (\boldsymbol{b} - \mathbf{1}) \cdot \boldsymbol{n}, \boldsymbol{n})$ on (P') must hit (P) *for the first time* at $\boldsymbol{X} = \boldsymbol{m}$ for some $\boldsymbol{m}$, $\mathbf{0} \leqq \boldsymbol{m} \leqq \boldsymbol{n}$. Indeed there are precisely $A_{\boldsymbol{m}}(a, \boldsymbol{b})$ such paths. Each can be completed in $A_{\boldsymbol{n}-\boldsymbol{m}}(c, \boldsymbol{b})$ ways, as can be seen by viewing (P') relative to the coordinate system $(X_0', \boldsymbol{X}')$ where $X_0' = X_0 - a$ and $\boldsymbol{X}' = \boldsymbol{X} - \boldsymbol{m}$. This, plus an obvious counting argument, completes the proof of (2′).

Another identity which can be derived in a similar way is:

(3) $$\sum_{\boldsymbol{m}=\mathbf{0}}^{\boldsymbol{n}} A_{\boldsymbol{m}}(a, \boldsymbol{b}) A_{\boldsymbol{n}-\boldsymbol{m}}(\boldsymbol{d} \cdot \boldsymbol{m}, \boldsymbol{b} + \boldsymbol{d}) = A_{\boldsymbol{n}}(a, \boldsymbol{b} + \boldsymbol{d}).$$

To get this one we consider the hyperplane (P) and the 'steeper' plane (P'') having the same X_0-intercept viz:

(P″) $$X_0 = a + (\boldsymbol{b} + \boldsymbol{d} - \mathbf{1}) \cdot \boldsymbol{X}.$$

Any minimal lattice path from $(0, \mathbf{0})$ to $(a + (\boldsymbol{b} + \boldsymbol{d} - \mathbf{1}) \cdot \boldsymbol{n}, \boldsymbol{n})$ on (P'') must hit the hyperplane (P) *for the first time* at $\boldsymbol{X} = \boldsymbol{m}$ for some $\boldsymbol{m}$, $\mathbf{0} \leqq \boldsymbol{m} \leqq \boldsymbol{n}$. Again there are $A_{\boldsymbol{m}}(a, \boldsymbol{b})$ such, and each can be completed in $A_{\boldsymbol{n}-\boldsymbol{m}}(\boldsymbol{d} \cdot \boldsymbol{m}, \boldsymbol{b} + \boldsymbol{d})$ ways, as we can see by viewing (P'') relative to the coordinate system $(X_0'', \boldsymbol{X}'')$ where $X_0'' = X_0 - a - (\boldsymbol{b} - \mathbf{1}) \cdot \boldsymbol{m}$, $\boldsymbol{X}'' = \boldsymbol{X} - \boldsymbol{m}$. Thus (3) follows in the same way as (2′).

The general identity in Mohanty (1966a) is seen to be a combination of (2′) and (3). Another identity derived in Gould's papers involves the expressions

(4) $$A_n(a, b) = \frac{a}{a + bn} \frac{(a + bn)^n}{n!}.$$

Gould (1957 equation 6) shows that (2) holds with this definition of $A_n(a, b)$ and one might wonder whether a geometric interpretation exists for the entities

(4) similar to that derived for (2). I have been unable to find such an interpretation although a probabilistic one exists, and Raney (1964) gives the combinatorial interpretations of closely related expressions which lead to the proof of (2) in this case. The definition (4) also suggests a generalisation not discussed by Mohanty, namely

$$(4') \qquad A_{\mathbf{n}}(a, \mathbf{b}) = \frac{a}{a + \mathbf{b} \cdot \mathbf{n}} \frac{(a + \mathbf{b} \cdot \mathbf{n})^{\mathbf{1} \cdot \mathbf{n}}}{\mathbf{n}!}$$

where $\mathbf{n}! = n_1!\, n_2! \cdots n_k!$ and $\mathbf{1} \cdot \mathbf{n} = n_1 + n_2 + \cdots + n_k$. The coefficients $A_{\mathbf{n}}(a, \mathbf{b})$ defined by (1′) approximate those defined in (4′) when a and $\mathbf{b}$ are large so it is reasonable to suppose that the convolution identity (2′) also holds in this case. This is indeed true, the result being deducible (with a little effort) from Raney (1964).

4. Associated probability distributions

Let $(p_0, \mathbf{p})$ be a $(k+1)$-tuple with $p_0 > 0$, $p_1 > 0, \cdots, p_k > 0$ and $p_0 + p_1 + \cdots + p_k = 1$. Then if $\mathbf{b} \cdot \mathbf{p} \leqq 1$, Mohanty (1966a) proved that

$$(5) \qquad \sum_{\mathbf{n}=0}^{\infty} A_{\mathbf{n}}(a, \mathbf{b})\, p_0^{a+(\mathbf{b}-\mathbf{1}) \cdot \mathbf{n}} \mathbf{p}^{\mathbf{n}} = 1$$

where $\mathbf{p}^{\mathbf{n}} = p_1^{n_1} p_2^{n_2} \cdots p_k^{n_k}$. We will offer an alternative derivation of (5) based upon a random walk interpretation. To do this we consider the random walk on the lattice points in the positive orthant in $(k+1)$-space which begins at $(0, \mathbf{0})$ and at each step moves along the X_j axis one unit in the positive direction with probability p_j $(0 \leqq j \leqq k)$, steps being mutually independent and identically constructed.

PROPOSITION 2. *The probability that the above random walk ever hits the hyperplane* (P) *is* π^a, *where* π *is the smallest positive root of*

$$(6) \qquad \sum_{1}^{k} p_j x^{b_j} - x + p_0 = 0.$$

PROOF. Let us define the function, in fact a probability generating function:

$$(7) \qquad f(x) = p_0 + \sum_{1}^{k} p_j x^{b_j}$$

We will see that the probability that the walk ever hits (P) is π^a where π is a probability that the walk ever hits (P) when $a = 1$, and that π is the smallest positive root of the equation $f(x) = x$. The first assertion is an immediate consequence of the assumed independence of the steps in the walk, as the passage from $(0, \mathbf{0})$ to (P) can be viewed as a succession of independent and probabilistically identical passages from $(0,\mathbf{0})$ to $X_0 = 1 + (\mathbf{b} - \mathbf{1}) \cdot \mathbf{X}$, from this plane to $X_0 = 2 + (\mathbf{b} - \mathbf{1}) \cdot \mathbf{X}$, and so on up to (P).

Let $\pi_a(m)$ denote the probability that the walk hits the hyperplane (P) in less than m steps. Clearly $\pi_1(m) \uparrow \pi$ as $m \to \infty$. If $m > 1$ we may condition upon the outcome of the first (random) step and find that

$$(8) \qquad \pi_1(m) = p_0 + \sum_1^k p_j \pi_{b_j}(m-1).$$

Now $\pi_{b_j}(m-1) \leqq \pi_{b_j}(m) \leqq [\pi_1(m)]^{b_j}$ and so we find that $\pi_1(m)$ satisfies the inequality

$$(9) \qquad 0 \leqq \pi_1(m) \leqq f(\pi_1(m)).$$

Letting $m \to \infty$ we see from (8) and the remarks opening this proof that $\pi = f(\pi)$ and it follows from (9) that π is the smallest such positive root.

COROLLARY. *$\pi = 1$ if and only if $\mathbf{b} \cdot \mathbf{p} \leqq 1$.*

PROOF. This is easily derived using methods well known in the theory of branching processes. See for example Harris (1963).

Let us define T to be random time, possibly infinite, which the walk takes to hit the hyperplane (P). Then we have the distribution of T involving our coefficients.

PROPOSITION 2. (i) *$P(T = a + \mathbf{b} \cdot \mathbf{n}) = A_{\mathbf{n}}(a, \mathbf{b}) p_0^{a + (\mathbf{b} - \mathbf{1}) \cdot \mathbf{n}} \mathbf{p}^{\mathbf{n}}$.*
(ii) *$P(T < \infty) = \pi^a$ where π is defined above.*

PROOF. Result (i) follows from Proposition 1 and the definition of the walk, whereas (ii) follows from the previous proposition.

COROLLARY 2. *Identity* (5) *holds if $\mathbf{b} \cdot \mathbf{p} \leqq 1$.*

If we denote by T_a the above random variable, then it is probabilistically obvious that the first passage time T_{a+c} should be distributed as the sum of a r.v. T_a and another, independent, r.v. T_c. This convolution property is equivalent to (2′) as is easily checked. Thus an alternative, probabilistic, proof of (2′) could be constructed. The details are left to the reader.

Finally we note that $E\{T_a\} = a/(1 - \mathbf{b} \cdot \mathbf{p})$ can be proved in a manner analogous to that used to obtain the equation for π. That is, by first deriving the equation $E\{T_a\} = aE\{T_1\}$, and then conditioning upon the outcome of the first step obtaining

$$E\{T_1\} = p_0 + \sum_1^k p_j(1 + E\{T_{b_j}\}).$$

The variance formula for T_a can be derived in a similar way.

5. Acknowledgements

I am indebted to H. W. Gould and G. N. Raney for providing references to some relevant work.

References

David Blackwell and Lester Dubins (1966), 'An elementary proof of an identity of Gould's', *Bol. Soc. Mat. Mexicana Ser.* 2 **11,** 108–110.

H. W. Gould (1956), 'Some generalisations of Vandermonde's convolution', *Amer. Math. Monthly* **63,** 84–91.

H. W. Gould (1957), 'Final analysis of Vandermonde's convolution', *Amer. Math. Monthly* **64,** 409–415.

H. W. Gould (1974), 'Coefficient identities for powers of Taylor and Dirichlet series', *Amer. Math. Monthly* **81,** 3–13.

H. W. Gould and J. Kaucky (1966), 'Evaluation of a class of binomial coefficient summations', *J. Combinatorial Theory* **1,** 233–247.

Theodore E. Harris (1963), *The theory of branching processes.* Springer-Verlag Berlin, Göttingen, Heidelberg.

S. G. Mohanty (1966a), 'Some convolutions with multinomial coefficients and related probability distributions', *SIAM Review* **8,** 501–509.

S. G. Mohanty (1966b), 'An urn problem related to the ballot problem', *Amer. Math. Monthly* **73,** 526–528.

S. G. Mohanty (1972), 'On queues involving batches', *J. Applied Probability* **9,** 430–435.

S. G. Mohanty and T. V. Narayana (1961), 'Some properties of compositions and their application to probability and statistics I', *Biom. Zeit.* **3,** 252–258.

George N. Raney (1964), 'A formal solution of $\Sigma_{i=1}^{\infty} A_i e^{B_i X} = X$', *Canad. J. Math.* **16,** 755–762.

John Riordan (1958), *An introduction to combinatorial analysis.* John Wiley and Sons, New York, London, Sydney.

Lajos Takács (1967) *Combinatorial methods in the theory of stochastic processes.* John Wiley and Sons, Inc., New York, London, Sydney.

Department of Mathematics,
University of Western Australia,
Nedlands, W.A. 6009.

Chapter 3
Sufficiency

Anirban DasGupta

It was the Fall of 1978. I had just finished my masters in statistics and started out as a PhD student in the stat-math division at the ISI in Calcutta. Teachers of the calibre of B.V. Rao and Ashok Maitra had taught me an enormous amount of mathematics and probability theory. But deep inside me I was curious to learn much more of statistical theory. Unfortunately, Basu had already left and moved to the US, and C.R. Rao was rarely seen in the Calcutta center. I considered following Basu to Tallahassee, but my friend Rao Chaganty warned me that the weather was so outlandishly good that I would probably never graduate. My other favorite teacher T. Krishnan was primarily interested in applied statistics, and J.K. Ghosh had only just returned from his visit to Pittsburgh. I remember being given a problem on admissibility; but, alas, that too turned out to be a modest extension of Karlin [30].

ISI allowed its students an unlimited amount of laziness and vagrancy, and I exploited this executive nonchalance gratuitously. I was not doing anything that I wanted to admit. Stat-Math was then located in an unpretentious, dark old building across from the central pond in the main campus. One day I was intrigued to see a new face; a visitor from Australia, someone whispered. In a week or so, the office sent out an announcement of a course on sufficiency by our visitor; the name was Terence P. Speed. That is how I first met Terry 34 years ago, and became one of his early students. Much later, I came to know that he was professionally and personally close to Basu, who had an enduring influence on my life. Together, Terry and Basu prepared a comprehensive bibliography of sufficiency [8]. They had intended to write a book, but communication at great distances was not such a breeze 40 years ago, and the book never came into being. Most recently, Terry and I worked together on summarizing Basu's work for the Selected Works series of Springer. I am deeply honored and touched to be asked to write this commentary on Terry's contributions to statistics, and particularly to sufficiency. Terry has worked on such an incredible variety of areas and problems that I will limit myself to just a few of his contributions that have directly influenced my own work and education. Sufficiency is certainly

A. DasGupta
Department of Statistics, Purdue University and Indian Statistical Institute
e-mail: dasgupta@stat.purdue.edu

S. Dudoit (ed.), *Selected Works of Terry Speed*, Selected Works in Probability and Statistics,
DOI 10.1007/978-1-4614-1347-9_3,

one of them. My perspective and emphasis will be rather different from other survey articles on it, such as Yamada and Morimoto [51].

For someone who does not believe in a probability model, sufficiency is of no use. It is also of only limited use in the robustness doctrine. I think, however, that the importance of sufficiency in inference must be evaluated in the context of the time. The idea of data summarization in the form of a low dimensional statistic without losing information must have been intrinsically attractive and also immensely useful when Fisher first formulated it [23]. In addition, we now know the various critical links of sufficiency to both the foundations of statistics, and to the elegant and structured theory of optimal procedures in inference.

For example, the links to the (weak and the strong) likelihood principle and conditionality principle are variously summarized in the engaging presentations in Barnard [3], Basu [6], Berger and Wolpert [10], Birnbaum [14], Fraser [26], and Savage [42]. And we are also all aware of such pillars of the mathematical theory of optimality, the Rao-Blackwell and the Lehmann-Scheffé theorem [12, 35], which are inseparably connected to sufficient statistics. At the least, sufficiency has acted as a nucleus around which an enormous amount of later development of ideas, techniques, and results have occurred. Some immediate examples are the theory of ancillarity, monotone likelihood ratio, exponential families, invariance, and asymptotic equivalence [5, 17, 18, 22, 33, 36, 38]. Interesting work relating sparse order statistics (e.g., a small fraction of the largest ones) to approximate sufficiency is done in Reiss [40], and approximate sufficiency and approximate ancillarity are given a direct definition, with consequences, in DasGupta [20]. We also have the coincidence that exact and nonasymptotic distributional and optimality calculations can be done precisely in those cases where a nontrivial sufficient statistic exists. The fundamental nature of the idea of sufficiency thus cannot be minimized; not yet.

Collectively, Kolmogorov, Neyman, Bahadur, Dynkin, Halmos, and Savage, among many other key architects, put sufficiency on the rigorous mathematical pedal. If $\{P, P \in \mathscr{P}\}$ is a family of probability measures on a measurable space $(\Omega, \mathscr{A})$, a sub σ-field $\mathscr{B}$ of $\mathscr{A}$ is *sufficient* if for each measurable set $A \in \mathscr{A}$, there is a (single) $\mathscr{B}$ measurable function g_A such that $g_A = E_P(I_A \,|\, \mathscr{B})$, a.e. $(P) \forall P \in \mathscr{P}$. This is rephrased in terms of a *sufficient statistic* by saying that if $T : (\Omega, \mathscr{A}) \longrightarrow (\Omega', \mathscr{A}')$ is a mapping from the original (measurable) space to another space, then T is a sufficient statistic if $\mathscr{B} = \mathscr{B}_T = T^{-1}(\mathscr{A}')$ is a sufficient sub σ-field of $\mathscr{A}$. In a classroom situation, the family $\mathscr{P}$ is often parametrized by a finite dimensional parameter θ, and we describe sufficiency as the conditional distribution of any other statistic given the sufficient statistic being independent of the underlying parameter θ. Existence of a fixed dimensional sufficient statistic for all sample sizes is a rare phenomenon for regular families of distributions, and is limited to the multiparameter exponential family (Barankin and Maitra [2], Brown [16]; it is also mentioned in Lehmann [34]). Existence of a fixed dimensional sufficient statistic in location-scale families has some charming (and perhaps unexpected) connections to the Cauchy-Deny functional equation [29, 32, 39].

Sufficiency corresponds to summarization without loss of information, and so the maximum such possible summarization is of obvious interest. A specific sub

σ-field $\mathscr{B}^*$ is a *minimal sufficient* sub σ-field if for any other sufficient sub σ-field $\mathscr{B}$, we have the inclusion that $\mathscr{B}^* \vee \mathscr{N}_{\mathscr{P}} \subseteq \mathscr{B} \vee \mathscr{N}_{\mathscr{P}}$, where $\mathscr{N}_{\mathscr{P}}$ is the family of all $\mathscr{P}$-null members of $\mathscr{A}$. In terms of statistics, a specific sufficient statistic T^* is minimal sufficient if given any other sufficient statistic T, we can write T^* as $T^* = h \circ T$ a.e. $\mathscr{P}$, i.e., a minimal sufficient statistic is a function of every sufficient statistic. A sufficient statistic that is also *boundedly complete* is minimal sufficient.

This fact does place completeness as a natural player on the scene rather than as a mere analytical necessity; of course, another well known case is Basu's theorem [4]. The converse is not necessarily true; that is, a minimal sufficient statistic need not be boundedly complete. The location parameter t densities provide a counterexample, where the vector of order statistics is minimal sufficient, but clearly not boundedly complete. It is true, however, that in somewhat larger families of densities, the vector of order statistics is complete, and hence boundedly complete [9]. If we think of a statistic as a partition of the sample space, then the partitions corresponding to a minimal sufficient statistic T^* can be constructed by the rule that $T^*(x) = T^*(y)$ if and only if the likelihood ratio $\frac{f_\theta(x)}{f_\theta(y)}$ is independent of θ. Note that this rule applies only to the dominated case, with $f_\theta(x)$ being the density (Radon-Nikodym derivative) of P_θ with respect to the relevant dominating measure.

Halmos and Savage [28] gave the *factorization theorem* for characterizing a sufficient sub σ-field, which says that if each $P \in \mathscr{P}$ is assumed to be absolutely continuous with respect to some P_0 (which we may pick to be in the convex hull of $\mathscr{P}$), then a given sub σ-field $\mathscr{B}$ is sufficient if and only if for each $P \in \mathscr{P}$, we can find a $\mathscr{B}$ measurable function g_P such that the identity $dP = g_P dP_0$ holds. Note that we insist on g_P being $\mathscr{B}$ measurable, rather than being simply $\mathscr{A}$ measurable (which would be no restriction, and would not serve the purpose of data summarization). Once again, in a classroom situation, we often describe this as T being sufficient if and only if we can write the joint density $f_\theta(x)$ as $f_\theta(x) = g_\theta(T(x))p_0(x)$ for some g and p_0. The factorization theorem took the guessing game out of the picture in the dominated case, and is justifiably regarded as a landmark advance. I will shortly come to Terry Speed's contribution on the factorization theorem.

Sufficiency comes in many colors, which turn out to be equivalent under special sets of conditions (e.g. Roy and Ramamoorthi [41]). I will loosely describe a few of these notions. We have *Blackwell sufficiency* [15] which corresponds to sufficiency of an experiment as defined via comparison of experiments [48, 50], *Bayes sufficiency* which corresponds to the posterior measure under any given prior depending on the data x only through $T(x)$, and *prediction sufficiency* (also sometimes called *adequacy*) which legislates that to predict an unobserved Y defined on some space $(\Omega'', \mathscr{A}'')$ on the basis of an observed X defined on $(\Omega, \mathscr{A})$, it should be enough to only consider predictors based on $T(X)$. See, for example, Takeuchi and Akahira [49], and also the earlier articles Bahadur [1] and Skibinsky [44]. I would warn the reader that the exact meaning of prediction sufficiency is linked to the exact assumptions on the prediction loss function. Likewise, Bayes sufficiency need not be equivalent to ordinary sufficiency unless $(\Omega, \mathscr{A})$ is a standard Borel space, i.e., unless $\mathscr{A}$ coincides with the Borel σ-field corresponding to some compact metrizable topology on Ω.

Consider now the enlarged class of probability distributions defined as $P_C(A) = P(X \in A \,|\, Y \in C), P \in \mathscr{P}, C \in \mathscr{A}''$. Bahadur leads us to the conclusion that prediction sufficiency is equivalent to sufficiency in this enlarged family of probability measures. A major result due to Terry Speed is the derivation of a factorization theorem for characterizing a prediction sufficient statistic in the dominated case [45]. A simply stated but illuminating example in Section 6 of Speed's article shows why the particular version of the factorization theorem he gives can be important in applications. As far as I know, a theory of partial adequacy, akin to partial sufficiency [7, 25, 27], has never been worked out. However, I am not sure how welcome it will now be, considering the diminishing importance of probability and models in prevalent applied statistics.

Two other deep and delightful papers of Terry that I am familiar with are his splendidly original paper on spike train deconvolution [37], and his paper on Gaussian distributions over finite simple graphs [47]. These two papers are precursors to what we nowadays call independent component analysis and graphical models. Particularly, the spike train deconvolution paper leads us to good problems in need of solution. However, I will refrain from making additional comments on it in order to spend some time on a most recent writing of Terry that directly influenced me.

In his editorial column in the *IMS Bulletin* [46], Terry describes the troublesome scenario of irreconcilable quantitative values obtained in bioassays conducted under different physical conditions at different laboratories (actually, he describes, specifically, the example of reporting the expression level of the HER2 protein in breast cancer patients). He cites an earlier classic paper of Youden [52], which I was not previously familiar with. Youden informally showed the tendency of a point estimate derived from one experiment to fall outside of the error bounds reported by another experiment. In Youden's cases, this was usually caused by an unmodelled latent bias, and once the bias was taken care of, the conundrum mostly disappeared.

Inspired by Terry's column, I did some work on reconcilability of confidence intervals found from different experiments, even if there are no unmodelled biases. What I found rather surprised me. Theoretical calculations led to the conclusion that in as few as 10 experiments, it could be quite likely that the confidence intervals would be nonoverlapping. In meta-analytic studies, particularly in clinical trial contexts, the number of experiments combined is frequently 20, 25, or more. This leads to the apparently important question: how does one combine independent confidence intervals when they are incompatible? We have had some of our best minds think about related problems; for example, Fisher [24], Birnbaum [13], Koziol and Perlman [31], Berk and Cohen [11], Cohen et al. [19], and Singh et al. [43]. Holger Dette and I recently collaborated on this problem and derived some exact results and some asymptotic theory involving extremes [21]. It was an exciting question for us, caused by a direct influence of Terry.

Human life is a grand collage of countless events and emotions, triumphs and defeats, love and hurt, joy and sadness, the extraordinary and the mundane. I have seen life from both sides now, tears and fears and feeling proud, dreams and schemes and circus crowds. But it is still my life's illusion of those wonderful years in the seventies that I recall fondly in my life's journey. Terry symbolizes that fantasy and uncomplicated part of my life. I am grateful to have had this opportunity to write a few lines about Terry; *prendre soin*, Terry, my teacher and my friend.

References

[1] R. R. Bahadur. Sufficiency and statistical decision functions. *Ann. Math. Statist.*, 25:423–462, 1954.
[2] E. W. Barankin and A. P. Maitra. Generalization of the Fisher-Darmois-Koopman-Pitman theorem on sufficient statistics. *Sankhyā Ser. A*, 25:217–244, 1963.
[3] G. A. Barnard. Comments on Stein's "A remark on the likelihood principle". *J. Roy. Stat. Soc. A*, 125:569–573, 1962.
[4] D. Basu. On statistics independent of a complete sufficient statistic. *Sankhyā*, 15:377–380, 1955.
[5] D. Basu. The family of ancillary statistics. *Sankhyā*, 21:247–256, 1959.
[6] D. Basu. Statistical information and likelihood. *Sankhyā Ser. A*, 37(1):1–71, 1975. Discussion and correspondence between Barnard and Basu.
[7] D. Basu. On partial sufficiency: A review. *J. Statist. Plann. Inference*, 2(1): 1–13, 1978.
[8] D. Basu and T. P. Speed. Bibliography of sufficiency. Mimeographed Technical report, Manchester, 1975.
[9] C. B. Bell, D. Blackwell, and L. Breiman. On the completeness of order statistics. *Ann. Math. Statist.*, 31:794–797, 1960.
[10] J. O. Berger and R. L. Wolpert. *The Likelihood Principle*, volume 6 of *Lecture Notes – Monograph Series*. Institute of Mathematical Statistics, Hayward, CA, 2nd edition, 1988.
[11] R. H. Berk and A. Cohen. Asymptotically optimal methods of combining tests. *J. Am. Stat. Assoc.*, 74(368):812–814, 1979.
[12] P. J. Bickel and K. A. Doksum. *Mathematical Statistics: Basic Ideas and Selected Topics*, volume I of *Holden-Day Series in Probability and Statistics*. Holden-Day, Inc., San Francisco, CA, 1977.
[13] A. Birnbaum. Combining independent tests of significance. *J. Am. Stat. Assoc.*, 49:559–574, 1954.
[14] A. Birnbaum. On the foundations of statistical inference. *J. Am. Stat. Assoc.*, 57:269–326, 1962.
[15] D. Blackwell. Comparison of experiments. In *Proceedings of the Second Berkeley Symposium on Mathematical Statistics and Probability, 1950*, pages 93–102, Berkeley and Los Angeles, 1951. University of California Press.
[16] L. Brown. Sufficient statistics in the case of independent random variables. *Ann. Math. Statist.*, 35:1456–1474, 1964.
[17] L. D. Brown. *Fundamentals of Statistical Exponential Families with Applications in Statistical Decision Theory*, volume 9 of *Lecture Notes – Monograph Series*. Institute of Mathematical Statistics, Hayward, CA, 1986.

[18] L. D. Brown and M. G. Low. Asymptotic equivalence of nonparametric regression and white noise. *Ann. Stat.*, 24(6):2384–2398, 1996.

[19] A. Cohen, J. I. Marden, and K. Singh. Second order asymptotic and nonasymptotic optimality properties of combined tests. *J. Statist. Plann. Inference*, 6(3): 253–276, 1982.

[20] A. DasGupta. Extensions to Basu's theorem, factorizations, and infinite divisibility. *J. Statist. Plann. Inference*, 137:945–952, 2007.

[21] A. DasGupta and H. Dette. On the reconcilability and combination of independent confidence intervals. Preprint, 2011.

[22] P. Dawid. Basu on ancillarity. In *Selected Works of Debabrata Basu*, Selected Works in Probability and Statistics, pages 5–8. Springer, 2011.

[23] R. A. Fisher. On the mathematical foundations of theoretical statistics. *Phil. Trans. R. Soc. Lond. A*, 222:309–368, 1922.

[24] R. A. Fisher. *Statistical Methods for Research Workers*. Oliver & Boyd, 14th, revised edition, 1970.

[25] D. A. S. Fraser. Sufficient statistics with nuisance parameters. *Ann. Math. Statist.*, 27:838–842, 1956.

[26] D. A. S. Fraser. *The Structure of Inference*. John Wiley & Sons Inc., New York, 1968.

[27] J. Hájek. On basic concepts of statistics. In *Proceedings of the Fifth Berkeley Symposium on Mathematical Statistics and Probability (Berkeley, Calif., 1965/66), Vol. I: Statistics*, pages 139–162, Berkeley, CA, 1967. University of California Press.

[28] P. R. Halmos and L. J. Savage. Application of the Radon-Nikodym theorem to the theory of sufficient statistics. *Ann. Math. Statist.*, 20:225–241, 1949.

[29] V. S. Huzurbazar. *Sufficient Statistics: Selected Contributions*. Marcel Dekker, New York, NY, 1976.

[30] S. Karlin. Admissibility for estimation with quadratic loss. *Ann. Math. Statist.*, 29:406–436, 1958.

[31] J. A. Koziol and M. D. Perlman. Combining independent chi-squared tests. *J. Am. Stat. Assoc.*, 73(364):753–763, 1978.

[32] K.-S. Lau and C. R. Rao. Integrated Cauchy functional equation and characterizations of the exponential law. *Sankhyā Ser. A*, 44(1):72–90, 1982.

[33] L. Le Cam. Sufficiency and approximate sufficiency. *Ann. Math. Statist.*, 35: 1419–1455, 1964.

[34] E. L. Lehmann. *Testing Statistical Hypotheses*. John Wiley & Sons Inc., New York, 1959.

[35] E. L. Lehmann and G. Casella. *Theory of Point Estimation*. Springer Texts in Statistics. Springer-Verlag, New York, second edition, 1998.

[36] E. L. Lehmann and J. P. Romano. *Testing Statistical Hypotheses*. Springer Texts in Statistics. Springer, New York, third edition, 2005.

[37] L. Li and T. P. Speed. Parametric deconvolution of positive spike trains. *Ann. Stat.*, 28(5):1279–1301, 2000.

[38] M. Nussbaum. Asymptotic equivalence of density estimation and Gaussian white noise. *Ann. Stat.*, 24(6):2399–2430, 1996.

[39] C. R. Rao and D. N. Shanbhag. Recent results on characterization of probability distributions: A unified approach through extensions of Deny's theorem. *Adv. Appl. Prob.*, 18(3):660–678, 1986.

[40] R. D. Reiss. A new proof of the approximate sufficiency of sparse order statistics. *Stat. Prob. Lett.*, 4:233–235, 1986.

[41] K. K. Roy and R. V. Ramamoorthi. Relationship between Bayes, classical and decision theoretic sufficiency. *Sankhyā Ser. A*, 41(1-2):48–58, 1979.

[42] L. J. Savage. The foundations of statistics reconsidered. In *Proceedings of the 4th Berkeley Symposium on Mathematical Statistics and Probability, Vol. I*, pages 575–586, Berkeley, CA, 1961. University of California Press.

[43] K. Singh, M. Xie, and W. E. Strawderman. Combining information from independent sources through confidence distributions. *Ann. Stat.*, 33(1):159–183, 2005.

[44] M. Skibinsky. Adequate subfields and sufficiency. *Ann. Math. Statist.*, 38: 155–161, 1967.

[45] T. P. Speed. A factorisation theorem for adequate statistics. *Aust. J. Stat.*, 20 (3):240–249, 1978.

[46] T. P. Speed. Enduring values. *IMS Bulletin*, 39(10):10, 2010.

[47] T. P. Speed and H. T. Kiiveri. Gaussian Markov distributions over finite graphs. *Ann. Stat.*, 14(1):138–150, 1986.

[48] C. Stein. Notes on the comparison of experiments. Technical report, University of Chicago, 1951.

[49] K. Takeuchi and M. Akahira. Characterizations of prediction sufficiency (adequacy) in terms of risk functions. *Ann. Stat.*, 3(4):1018–1024, 1975.

[50] E. Torgersen. *Comparison of Statistical Experiments*, volume 36 of *Encyclopedia of Mathematics and its Applications*. Cambridge University Press, Cambridge, 1991.

[51] S. Yamada and H. Morimoto. Sufficiency. In *Current Issues in Statistical Inference: Essays in Honor of D. Basu*, volume 17 of *Lecture Notes – Monograph Series*, pages 86–98. Institute of Mathematical Statistics, Hayward, CA, 1992.

[52] W. J. Youden. Enduring values. *Technometrics*, 14(1):1–11, 1972.

Sankhyā : The Indian Journal of Statistics
1976, Volume **38**, Series A, Pt. 2, pp. 194–196.

A NOTE ON PAIRWISE SUFFICIENCY AND COMPLETIONS

By T. P. SPEED
University of Western Australia, Nedlands

SUMMARY. The result of Halmos and Savage that pairwise sufficiency implies sufficiency for dominated families of measures is obtained here by a simple technique. We also use this technique to obtain an easy proof of the equality between the two notions of completion with respect to a dominated family of probability measures for sub-σ-fields in a measure space. The proofs have natural generalisations to the so called campact case.

1. Introduction

In their proof that pairwise sufficiency implies sufficiency for dominated families of measures Halmos and Savage (1949) use four technical Lemmas. One of the purposes of the present note is to describe a simple method of obtaining this result directly. We find that the technique used also provides a simpler method of deriving a result of Le Bihan, Littaye-Petit and Petit (1970) asserting the equality between two notions of completion with respect to a family of probability measures for sub-σ-fields in a measure space. We close the note with an indication of how simple generalisations of these two results to the so-called compact case can be obtained.

2. Pairwise sufficiency implies sufficiency in the dominated case

Let $\mathcal{B}$ be a sub-σ-field of $\mathcal{A}$ in a measure space $(\mathcal{X}, \mathcal{A})$ and suppose that $\mathcal{B}$ is pairwise sufficient for a countable family $\mathcal{P}_0$ of probability measures on $\mathcal{A}$; that is, for any bounded $\mathcal{A}$ measurable function f on $\mathcal{X}$ and a pair $\{P, Q\} \subseteq \mathcal{P}_0$, there exists a $\mathcal{B}$-measurable function $f_{P,Q}$ on $\mathcal{X}$ such that

$$f_{P,Q} = E_P^{\mathcal{B}} f \quad \text{a.s. } P$$

and

$$f_{P,Q} = E_Q^{\mathcal{B}} f \quad \text{a.s. } Q.$$

Put $f^* = \bigvee_P \bigwedge_Q f_{P,Q}$ the sup and inf extending over $\mathcal{P}_0$, and observe that for every $P \epsilon \mathcal{P}_0$ we have

$$\bigwedge_Q f_{P,Q} \leqslant f^* \leqslant \bigvee_Q f_{Q,P}.$$

We can now see that for every $P \epsilon \mathcal{P}_0$

$$f^* = E_P^{\mathcal{B}} f \quad \text{a.s. } P \qquad \dots \quad (1)$$

and the proof that $\mathcal{B}$ is sufficient for $\mathcal{P}_0$ on $\mathcal{A}$ is completed. For a dominated family $\mathcal{P}$ of probability measures on $\mathcal{A}$ take a countable equivalent subset $\mathcal{P}_0$ of $\mathcal{P}$, see

Halmos and Savage (1949) for this fact, and argue as above obtaining one f^* such that (1) holds. Now take $Q \in \mathscr{P} \setminus \mathscr{P}_0$ and consider $\mathscr{P}_0 \cup \{Q\}$. We can produce an f^{**} such that (1) holds with f^* replaced by f^{**}, and $f^{**} = E_Q^{\mathscr{B}} f$ a.s. Q. But then we see that $f^* = f^{**}$ a.s. P for all $P \in \mathscr{P}_0$, and hence $f^* = f^{**}$ a.s. Q; i.e. $f^* = E_Q^{\mathscr{B}} f$ a.s. Q.

3. Pairwise and strong completions of a sub-σ-field

In the notation used above let us define the *pairwise completion* of $\mathscr{B}$ with respect to $\mathscr{P}$ on $\mathscr{A}$, in symbols $\hat{\mathscr{B}}^{[\mathscr{P},\mathscr{A}]}$ or just $\hat{\mathscr{B}}$ where no confusion can result, to be all elements $A \in \mathscr{A}$ with the following property : for each pair $\{P, Q\} \subseteq \mathscr{P}$ there exists $B = B_{P,Q} \in \mathscr{B}$ such that $P(A \Delta B) = Q(A \Delta B) = 0$. We can also define the *strong completion* $\bar{\mathscr{B}} = \bar{\mathscr{B}}^{[\mathscr{P},\mathscr{A}]}$ of $\mathscr{B}$ with respect to $\mathscr{P}$ on $\mathscr{A}$ to be all elements $A \in \mathscr{A}$ for which there exists $B \in \mathscr{B}$ such that $P(A \Delta B) = 0$ for every $P \in \mathscr{P}$. Using analytic sets it was proved in Marie-Francoise Le Bihan *et al* (1970) that for a dominated family $\mathscr{P}$ we have $\bar{\mathscr{B}} = \hat{\mathscr{B}}$. Let us see that this result follows easily using the argument of Section 2. If $A \in \hat{\mathscr{B}}$, then for each pair $\{P, Q\} \subseteq \mathscr{P}_0$, a countable equivalent subfamily of $\mathscr{P}$, there exists $B_{P,Q} \in \mathscr{B}$ such that $P(A \Delta B_{P,Q}) = Q(A \Delta B_{P,Q}) = 0$. Put

$$B = \bigcup_P \bigcap_Q B_{P,Q},$$

the unions and intersections being taken over $\mathscr{P}_0$, and we see that $P(A \Delta B) = 0$ for all $P \in \mathscr{P}_0$, and hence for all $P \in \mathscr{P}$. This proves that $\hat{\mathscr{B}} \subseteq \bar{\mathscr{B}}$ and the reverse conclusion is immediate.

4. Compactness

Both of the above results generalise to the case of a compact family of probability measures $\mathscr{P}$ on $(\mathscr{X}, \mathscr{A})$. A family $\{f_P : P \in \mathscr{P}\}$ of bounded $\mathscr{B}$-measurable functions on $\mathscr{X}$ will be called *pairwise* [respectively finitely, countably, completely] *compatible* if for every subfamily $Q \subseteq \mathscr{P}$ consisting of two [respectively finitely, countably, arbitrarily many] measures, there exists a $\mathscr{B}$-measurable function f_Q such that for all $P \in Q$ we have $f_Q = f_P$ a.s. P. Pitcher (1965) defined the notion of compactness for families $\mathscr{P}$ of probabilities and an equivalent form of it is the following : $\mathscr{P}$ is compact on the sub-σ-field $\mathscr{B}$ of $\mathscr{A}$ if every countably compatible family $\{f_P, P \in \mathscr{P}\}$ of bounded $\mathscr{B}$-measurable functions is completely compatible.

Let us observe that a dominated family $\mathscr{P}$ of probabilities on $(\mathscr{X}, \mathscr{A})$ is compact on every sub-σ-field $\mathscr{B}$ of $\mathscr{A}$. As before we first take a countable equivalent subset $\mathscr{P}_0 \subseteq \mathscr{P}$. Any pairwise compatible family $\{f_P : P \in \mathscr{P}_0\}$ of $\mathscr{B}$-measurable functions is clearly completely compatible here : simply put $f^* = \bigvee_P \bigwedge_Q f_{P,Q}$ where $f_{P,Q}$ is the function defining the compatibility of f_P and f_Q, and the sup and inf are taken over $\mathscr{P}_0$. Now take $Q \in \mathscr{P} \setminus \mathscr{P}_0$. By considering $\mathscr{P}_0 \cup \{Q\}$ we find a

$\mathcal{B}$-measurable f^{**} such that $f^{**} = f_P$ a.s. P for all $P \in \mathcal{P}_0$, and $f^{**} = f_Q =$ a.s. Q. This first condition implies that $f^{**} = f^*$ a.s. P for all $P \in \mathcal{P}_0$ and so $f^{**} = f^* = f_Q$ a.s. Q and the proof is complete.

These remarks prove that the following result is a generalisation of those outlined in Sections 2, 3 and provides an alternative approach to them, although the details are not dissimilar.

Theorem : *Let $\mathcal{P}$ be a family of probabilities over $(\mathcal{X}, \mathcal{A})$ and $\mathcal{B}$ a sub-σ-field of $\mathcal{A}$ such that $\mathcal{P}$ is compact on $\mathcal{B}$. Then*

(i) $\bar{\mathcal{B}} = \hat{\mathcal{B}}$, *and further,*

(ii) *$\mathcal{B}$ is sufficient for $\mathcal{P}$ on $\mathcal{A}$ whenever it is pairwise sufficient for $\mathcal{P}$ on $\mathcal{A}$.*

Proof : (i) For $A \in \hat{\mathcal{B}}$ we take f_P to be the indicator of any $B \in \mathcal{B}$ for which $P(A \Delta B) = 0$; such exist and it follows from the assumption on A that $\{f_P : P \in \mathcal{P}\}$ is a pairwise compatible family of $\mathcal{B}$-measurable functions.

This easily extends to countable compatibility by using the sup-inf trick and so the compactness assumption on $\mathcal{B}$ ensures the existence of a $\mathcal{B}$-measurable f with $f = f_P$ a.s. P for each $P \in \mathcal{P}$. Clearly $B = \{f = 1\} \in \mathcal{B}$ satisfies $P(A \Delta B) = 0$, $P \in \mathcal{P}$ and the proof of (i) is complete.

(ii) For a bounded $\mathcal{A}$-measurable function f we put $f_P = E_P^{\mathcal{B}} f$. The system $\{f_P : P \in \mathcal{P}\}$ is again a pairwise compatible family of $\mathcal{B}$-measurable functions by the pairwise sufficiency assumption and again this lifts to countable compatibility using the sup-inf trick. The proof is completed by invoking the compactness of $\mathcal{P}$ on $\mathcal{B}$.

References

Paul R. Halmos and Savage, L. J. (1949) Application of the Radon-Nikodym theorem to the theory of sufficient statistics. *Ann. Math. Statist.*; **20**, 225-241.

Tokitake Kusama and Sakaturo Yamada (1972) : On compactness of the statistical structure and sufficiency. *Osaka J. Math.*, **9**, 11-18.

Marie-Françoise Le Bihan, Monique Littaye-Petit and Jean-Luc Petit (1970) : Exhaustivité par paire. *C. R. Acad. Sci. Paris Ser. A.*, **270**, 1753-1756.

Pitcher, T. S. (1965) : A more general property than domination for sets of probability measures, *Pacific J. Math.*, **15**, 597-611.

Paper received : July, 1975.

Austral. J. Statist., **20** (**3**), 1978, 240–249.

A FACTORISATION THEOREM FOR ADEQUATE STATISTICS[1]

T. P. Speed

Department of Mathematics, University of Western Australia

1. Introduction

In his interesting paper [3] Goro Ishii convincingly demonstrated the usefulness of the notion of adequacy in a number of situations by using this concept, together with completeness, to derive optimality properties of some known predictors. These proofs were all based upon analogues of the well known Rao–Blackwell and Lehmann–Scheffé theorems and to help recognise the adequacy of the statistics under discussion, Ishii cited an extension of the Fisher–Neyman factorisation criterion due to Sugiura and Morimoto [7]. I have not been able to consult this work.

The purpose of this paper is to state and prove a factorisation theorem of the type mentioned above. Our result is slightly more general than the one cited in [3]; we do not suppose a product structure for the underlying measure space and so in this respect our situation is more like that in the original paper by Skibinsky [5], and further, we do not suppose the dominating measure to be a probability measure. We begin by reviewing some simple facts regarding conditional expectations and then prove a factorisation theorem characterising conditional independence. A section is devoted to organising the known factorisation theorem characterising sufficiency in the dominated case, and may be of some independent interest. Finally a combination of these results gives the theorem of our title.

2. Conditional expectations

Our basic setting is a measurable space $(\mathscr{X}, \mathscr{A})$ equipped with a σ-finite measure μ; these will remain fixed throughout the paper. In this and the next section we will be considering a probability measure P on $\mathscr{A}$ which is absolutely continuous with respect to μ; let its Radon–Nikodym derivative be p and let us write this relationship as $P = p\,.\,\mu$.

For any sub-σ-field $\mathscr{C} \subseteq \mathscr{A}$ on which the restriction $\mu_{\mathscr{C}}$ of μ remains σ-finite we can define a conditional expectation operator $E_{\mu}^{\mathscr{C}}$

[1] Manuscript received January 6, 1977; revised November 22, 1977.

in the usual way: for any measurable non-negative or μ-integrable $f:\mathscr{X}\to\mathbb{R}$ there exists a $\mathscr{C}$-measurable function $E_\mu^\mathscr{C} f$ satisfying

(2.1) $$\int_C E_\mu^\mathscr{C} f\, d\mu_\mathscr{C} = \int_C f\, d\mu \quad (C\in\mathscr{C}).$$

If we suppose that $\mathscr{C}$ is $[\mu, \mathscr{A}]$-complete i.e. contains all the elements of $\mathscr{A}$ of zero μ-measure, (2.1) defines (see Neveu [6] p. 1) a unique μ-equivalence class of $\mathscr{C}$-measurable functions and we will adopt the usual procedure of denoting this class or a representative of this class by $E_\mu^\mathscr{C} f$. The relation between $E_\mu^\mathscr{C}$ and the usual operator $E_P^\mathscr{C}$ is as follows: for any measurable non-negative f

(2.2) $$E_P^\mathscr{C} f = E_\mu^\mathscr{C} fp / E_\mu^\mathscr{C} p \quad \text{a.s. } P$$

where $p = dP/d\mu$ satisfies $P(\{E_\mu^\mathscr{C} p = 0\}) = 0$. This is not hard to prove from the definitions and can be found in Loève [4] pp. 344–345; see also Neveu [6] pp. 16–17 for a brief discussion. Another relation which we need below is the following minor modification of the result just noted: if $\mathscr{B}\supseteq\mathscr{C}$ is another sub-σ-field of $\mathscr{A}$ and f *is also* $\mathscr{B}$-measureable, then

(2.3) $$E_P^\mathscr{C} f = E_\mu^\mathscr{C} fp_1 / E_\mu^\mathscr{C} p_1 \quad \text{a.s. } P$$

where

$$p_1 = dP_\mathscr{B}/d\mu_\mathscr{B} = E_\mu^\mathscr{B} p.$$

3. Conditional independence

Let $\mathscr{B}_1$, $\mathscr{B}_2$ and $\mathscr{C}$ be sub-σ-fields of $\mathscr{A}$. We say that $\mathscr{B}_1$ *and* $\mathscr{B}_2$ *are conditionally P-independent given* $\mathscr{C}$ if for all non-negative $\mathscr{B}_i$-measurable functions f_i $(i=1,2)$:

(3.1) $$E_P^\mathscr{C} f_1 f_2 = E_P^\mathscr{C} f_1 E_P^\mathscr{C} f_2 \quad \text{a.s. } P.$$

It is well-known that this is equivalent to: for all non-negative $\mathscr{B}_2\vee\mathscr{C}$-measurable functions f_2:

(3.2) $$E_P^{\mathscr{B}_1\vee\mathscr{C}} f_2 \quad \text{is } \mathscr{C}\text{-measurable, and so} = E_P^\mathscr{C} f_2$$

where we suppose that $\mathscr{C}$ is $[P, \mathscr{A}]$-complete. For a proof of essentially this assertion see Loève [4] pp. 563–564. It is also shown in Loève that if $\mathscr{B}_1$ and $\mathscr{B}_2$ are conditionally P-independent given $\mathscr{C}$ then so also are $\mathscr{B}_1\vee\mathscr{C}$ and $\mathscr{B}_2\vee\mathscr{C}$.

When $\mu_\mathscr{C}$ is σ-finite the operator $E_\mu^\mathscr{C}$ is well-defined and we can give an analogous definition of *conditional μ-independence.* It will then follow that (3.1) and (3.2) with μ instead of P are still equivalent.

Conditional P-independence of σ-fields has a formulation in terms of the factorisation of a density just as independence of σ-fields does, but I am unable to locate the following result in the literature. It is surely well known in its special case as the corollary. Let $(\mathscr{X}, \mathscr{A})$, μ, P and p be as in §2 and suppose that $\mathscr{B}_1$, $\mathscr{B}_2$ and $\mathscr{C}$ are sub-σ-fields of $\mathscr{A}$; for simplicity we will assume that $\mathscr{B}_1 \vee \mathscr{C} \vee \mathscr{B}_2 = \mathscr{A}$, and that $\mathscr{C}$ is $[\mu, \mathscr{A}]$-complete.

Proposition 1. *Let $\mu_{\mathscr{C}}$ be σ-finite and $\mathscr{B}_1$ and $\mathscr{B}_2$ be conditionally μ-independent given $\mathscr{C}$. If $\mathscr{B}_1$ and $\mathscr{B}_2$ are conditionally P-independent given $\mathscr{C}$ then we have the factorisation*

$$p = p_1 p_2 q^{-1} I_{\{q>0\}} \quad \text{a.s. } P \tag{3.3}$$

where $p_i = dP_{\mathscr{B}_i \vee \mathscr{C}}/d\mu_{\mathscr{B}_i \vee \mathscr{C}}$ $(i=1,2)$ and $q = dP_{\mathscr{C}}/d\mu_{\mathscr{C}}$. Conversely, if p can be factorised

$$p = f_1 f_2 \quad \text{a.s. } P \tag{3.4}$$

where f_i is $\mathscr{B}_i \vee \mathscr{C}$-measurable $(i=1,2)$, then $\mathscr{B}_1$ and $\mathscr{B}_2$ are conditionally P-independent given $\mathscr{C}$.

Loosely speaking the result asserts that when $\mathscr{B}_1$ and $\mathscr{B}_2$ are conditionally μ-independent given $\mathscr{C}$ and $P \ll \mu$, a necessary and sufficient condition for $\mathscr{B}_1$ and $\mathscr{B}_2$ to be conditionally P-independent given $\mathscr{C}$ is that the density p of P with respect to μ factorises into the product of a non-negative $\mathscr{B}_1 \vee \mathscr{C}$-measurable function and a non-negative $\mathscr{B}_2 \vee \mathscr{C}$-measurable function.

Proof. Suppose that $\mathscr{B}_1$ and $\mathscr{B}_2$ are conditionally P-independent given $\mathscr{C}$. We will prove (3.3) and it is here that we use the fact that $\mathscr{B}_1 \vee \mathscr{C} \vee \mathscr{B}_2 = \mathscr{A}$; trivial modifications would allow us to drop this assumption. Take $B_1 \in \mathscr{B}_1$, $C \in \mathscr{C}$ and $B_2 \in \mathscr{B}_2$. Then

$$\int_{B_1 C B_2} p_1 p_2 q^{-1} I_{\{q>0\}}\, d\mu$$

$$= \int_{C \cap \{q>0\}} p_1 I_{B_1} p_2 I_{B_2} q^{-1}\, d\mu$$

$$= \int_{C \cap \{q>0\}} E_\mu^{\mathscr{C}}(p_1 I_{B_1} p_2 I_{B_2}) q^{-1}\, d\mu_{\mathscr{C}} \qquad \text{as } q \text{ is } \mathscr{C}\text{-measurable,}$$

$$= \int_{C \cap \{q>0\}} E_\mu^{\mathscr{C}}(p_1 I_{B_1}) E_\mu^{\mathscr{C}}(p_2 I_{B_2}) q^{-1}\, d\mu_{\mathscr{C}} \qquad \text{using cond. } \mu\text{-independence}$$

$$= \int_{C \cap \{q>0\}} \frac{E_\mu^{\mathscr{C}}(p_1 I_{B_1})}{q} \frac{E_\mu^{\mathscr{C}}(p_2 I_{B_2})}{q}\, q\, d\mu_{\mathscr{C}}$$

$$= \int_{C \cap \{q>0\}} E_P^{\mathscr{C}}(I_{B_1}) E_P^{\mathscr{C}}(I_{B_2})\, dP_{\mathscr{C}}$$

using 2.3 and the fact that $q = dP_{\mathscr{C}}/d\mu_{\mathscr{C}} = E_\mu^{\mathscr{C}} p$,

$$= \int_{C \cap \{q>o\}} E_P^{\mathscr{C}}(I_{B_1 B_2})\, dP_{\mathscr{C}} \qquad \text{using cond. } P\text{-independence}$$

$$
\begin{aligned}
&= \int_{C\cap\{q>0\}} I_{B_1B_2}\, dP \\
&= P(C\cap\{q>0\}B_1B_2) \\
&= P(CB_1B_2) \quad \text{since} \quad P(\{q>0\}) = 1 \\
&= \int_{B_1CB_2} p\, d\mu \quad \text{as} \quad P = p\,.\,\mu.
\end{aligned}
$$

This implies the relation (3.3) as sets of the form B_1CB_2 form a π-system generating $\mathcal{B}_1 \vee \mathcal{C} \vee \mathcal{B}_2 = \mathcal{A}$.

For the converse suppose that p can be factorised as in (3.4). It is easy to check that $P(\{f_1=0\}) = P(\{f_2=0\}) = P(\{E_\mu^{\mathcal{B}_1\vee\mathcal{C}} f_1f_2 = 0\})$ follows from the fact that $P = f_1f_2\,.\,\mu$. Let f be a non-negative $\mathcal{B}_2\vee\mathcal{C}$-measurable function; then

$$
\begin{aligned}
E_P^{\mathcal{B}_1\vee\mathcal{C}} f &= E_\mu^{\mathcal{B}_1\vee\mathcal{C}} fp / E_\mu^{\mathcal{B}_1\vee\mathcal{C}} p \\
&= E_\mu^{\mathcal{B}_1\vee\mathcal{C}} ff_1f_2 / E_\mu^{\mathcal{B}_1\vee\mathcal{C}} f_1f_2 \\
&= f_1E_\mu^{\mathcal{B}_1\vee\mathcal{C}} ff_2 / f_1E_\mu^{\mathcal{B}_1\vee\mathcal{C}} f_2 \\
&= E_\mu^{\mathcal{B}_1\vee\mathcal{C}} ff_2 / E_\mu^{\mathcal{B}_1\vee\mathcal{C}} f_2 \\
&= E_\mu^{\mathcal{C}} ff_2 / E_\mu^{\mathcal{C}} f_2 \\
&= \text{a } \mathcal{C}\text{-measurable function,}
\end{aligned}
$$

provided we suppose that $\mathcal{C}$ is $[\mu, \mathcal{A}]$-complete. This completes the proof.

Corollary 1. *Suppose that* $\mathcal{X} = \mathcal{X}_1 \times \mathcal{X}_2$, $\mathcal{A} = \mathcal{A}_1 \otimes \mathcal{A}_2$, $\mu = \mu_1 \otimes \mu_2$ *where the* μ_i *are* σ*-finite on* $\mathcal{A}_i$ $(i = 1, 2)$, *and that* $\mathcal{B}_1$, $\mathcal{B}_2$ *are the* σ*-algebras generated by the coordinate projections. Let* $\mathcal{C} \subseteq \mathcal{A}_1$ *be a* $[\mu_1, \mathcal{A}_1]$*-complete sub-*σ*-field of* $\mathcal{A}_1$ *such that* μ_1 *remains* σ*-finite when restricted to* $\mathcal{C}$, *and let us also denote the sub-*σ*-field of* $\mathcal{A}$ *isomorphic to* $\mathcal{C}$ *by* $\mathcal{C}$. *Then for a probability measure* $P \ll \mu$ *we have* $\mathcal{B}_1$ *and* $\mathcal{B}_2$ *conditionally P-independent given* $\mathcal{C}$ *if and only if the density* $p = dP/d\mu$ *can be factorised*

$$
p = f_1f_2 \quad \text{a.s. } \mu
$$

where f_1 *is* $\mathcal{B}_1$*-measurable and* f_2 *is* $\mathcal{C}\vee\mathcal{B}_2$*-measurable.*

Proof. This is an immediate consequence of the proposition, for the hypotheses of the corollary imply that $\mathcal{B}_1$ and $\mathcal{B}_2$ are actually μ-independent and hence conditionally μ-independent given $\mathcal{C} \subseteq \mathcal{B}_1$.

4. Sufficiency

Let $\mathcal{P}$ be a family of probability measures on $\mathcal{A}$ and $\mathcal{B}$, $\mathcal{C}$ be sub-σ-field of $\mathcal{A}$. We say that $\mathcal{C}$ *is sufficient for* $\mathcal{P}$ *on* $\mathcal{B}$ if for any $B \in \mathcal{B}$

there exists a $\mathscr{C}$-measurable function φ_B such that for every $P \in \mathscr{P}$

(4.1) $$\varphi_B = E_{\mathscr{P}}^{\mathscr{C}} I_B \quad \text{a.s. } P$$

We will always suppose that $\mathscr{C}$ is $[\mathscr{P}, \mathscr{A}]$-complete i.e. contains all elements of $\mathscr{A}$ having zero P-measure for every $P \in \mathscr{P}$. In this section we may take $\mathscr{B} = \mathscr{A}$ without loss of generality, but in section 4 we will revert to the more general situation.

To understand the factorisation theorem of Fisher and Neyman let us consider the case $\mathscr{P} = \{P, Q\}$ with $P \ll Q$. We will denote dP/dQ by g and suppose $\mathscr{C} \subseteq \mathscr{A}$ to be $[Q, \mathscr{A}]$-complete.

Proposition 2. *$\mathscr{C}$ is sufficient for $\{P, Q\}$ on $\mathscr{A}$ if and only if g is $\mathscr{C}$-measurable.*

Proof. We choose an arbitrary $A \in \mathscr{A}$ and hope to prove that there exists a φ_A satisfying (4.1) if and only if g is $\mathscr{C}$-measurable. But (4.1) in this case implies that such a φ_A, if it exists, must be unique a.s. Q, and so must actually be $E_Q^{\mathscr{C}} I_A$. Thus we are really trying to prove: the necessary and sufficient condition that for all $A \in \mathscr{A} E_Q^{\mathscr{C}} I_A = E_P^{\mathscr{C}} I_A$ a.s. P, is $g = dP/dQ$ be $\mathscr{C}$-measurable.

Let us integrate $E_Q^{\mathscr{C}} I_A$ and $E_P^{\mathscr{C}} I_A$ over $C \in \mathscr{C}$ with respect to $P = g \, . \, Q$; we obtain

$$\int_C E_Q^{\mathscr{C}} I_A \, dP = \int_C E_Q^{\mathscr{C}} I_A g \, dQ = \int_C I_A E_Q^{\mathscr{C}} g \, dQ = \int_A I_C E_Q^{\mathscr{C}} g \, dQ;$$

$$\int_C E_P^{\mathscr{C}} I_A \, dP = \int_C I_A \, dP = \int_C I_A g \, dQ = \int_A I_C g \, dQ.$$

Now if $\mathscr{C}$ is sufficient for $\{P, Q\}$ on $\mathscr{A}$, equivalently, if for all $A \in \mathscr{A}$, $E_Q^{\mathscr{C}} I_A = E_P^{\mathscr{C}} I_A$ a.s. P, then the first terms in the two above equations coincide and hence so must the last. Taking $C = \mathscr{X}$ and varying $A \in \mathscr{A}$ we have proved that $E_Q^{\mathscr{C}} g = g$ a.s. Q; but $\mathscr{C}$ is $[Q, \mathscr{A}]$-complete and so g is $\mathscr{C}$-measurable.

On the other hand, if g is $\mathscr{C}$-measurable then the last terms in the above two equations, and hence the first, must coincide. By the remarks beginning the proof, this means that $\mathscr{C}$ is sufficient for $\{P, Q\}$ on $\mathscr{A}$. This completes the proof.

Now let us suppose that $\mathscr{P}$ is dominated by our σ-finite measure μ on $(\mathscr{X}, \mathscr{A})$. It is well known (see Halmos and Savage [2]) that this implies the existence of a countable subset $\mathscr{P}_0 \subseteq \mathscr{P}$ equivalent to $\mathscr{P}$—simply take (notation as in [6] p. 121) $\mathscr{P}_0$ for which

$$\sup_{P \in \mathscr{P}_0} I\{dP/d\mu > 0\} = \operatorname{ess\,sup}_{P \in \mathscr{P}} I\{dP/d\mu > 0\}$$

—*and so a probability measure* Q in the countable convex hull of $\mathscr{P}$ equivalent to $\mathscr{P}$. Furthermore, if $\mathscr{C}$ is sufficient for $\mathscr{P}$ on $\mathscr{B}$, $\mathscr{C}$ is also sufficient for $\mathscr{P} \cup \{Q\}$ on $\mathscr{B}$: the same φ_A works in (4.1) for Q since

$Q = \sum a_n P_n (a_n \geq 0, \sum a_n = 1, P_n \in \mathcal{P})$. Thus we have the following result assuming that $\mathcal{C}$ is $[\mathcal{P}, \mathcal{A}]$-complete, and thus also $[Q, \mathcal{A}]$-complete:

Corollary 1. *$\mathcal{C}$ is sufficient for $\mathcal{P}$ on $\mathcal{A}$ if and only if for every $P \in \mathcal{P}$, $g_P = dP/dQ$ is $\mathcal{C}$-measurable.*

The most frequently used form of the factorisation theorem is that of Bahadur [1] given in the next corollary.

Corollary 2. *Let $\mathcal{P}$ be dominated by μ on $(\mathcal{X}, \mathcal{A})$. Then for a $[\mathcal{P}, \mathcal{A}]$-complete sub-$\sigma$-field $\mathcal{C} \subseteq \mathcal{A}$ to be sufficient for $\mathcal{P}$ on $\mathcal{A}$ it is necessary and sufficient that there exists a non-negative function h, and for each $P \in \mathcal{P}$, non-negative functions g_P defined on $\mathcal{X}$ such that*
(i) h is $\mathcal{A}$-measurables;
(ii) for all $P \in \mathcal{P}$, g_P is $\mathcal{C}$-measurable;
(iii) for all $P \in \mathcal{P}$, $P = g_P h \,.\, \mu$ on $\mathcal{A}$.

Proof. This is an immediate consequence of the previous corollary; simply take Q as above, and $h = dQ/d\mu$.

5. Adequacy

Let $\mathcal{P}$ be a family of probability measures on $(\mathcal{X}, \mathcal{A})$ and $\mathcal{B}_1$, $\mathcal{B}_2$, $\mathcal{C}$ sub-σ-fields of $\mathcal{A}$. Slightly paraphrasing Skibinsky [5] we say that *$\mathcal{C}$ is adequate for $\mathcal{B}_1$ with respect to $\mathcal{B}_2$ and $\mathcal{P}$* if (i) $\mathcal{C}$ is sufficient for $\mathcal{P}$ on $\mathcal{B}_1$, and (ii) $\mathcal{B}_1$ and $\mathcal{B}_2$ are, for all $P \in \mathcal{P}$, conditionally P-independent given $\mathcal{C}$. In Skibinsky [5] $\mathcal{C} \subseteq \mathcal{B}_1$, but although we could replace $\mathcal{B}_1$ by $\mathcal{B}_1 \vee \mathcal{C}$ to achieve this, and no generality would be lost in doing so, we prefer the more symmetric situation natural to a formulation involving conditional independence. We will also suppose the $\mathcal{C}$ is $[\mathcal{P}, \mathcal{A}]$-complete, and that $\mathcal{B}_1 \vee \mathcal{C} \vee \mathcal{B}_2 = \mathcal{A}$.

As we did with sufficiency, it is instructive to formulate the factorisation theorem in the case $\mathcal{P} = \{P, Q\}$ where $P = g \,.\, Q$ on $\mathcal{A}$.

Proposition 3a. *If $\mathcal{C}$ is adequate for $\mathcal{B}_1$ with respect to $\mathcal{B}_2$ and $\{P, Q\}$ then g is $\mathcal{C} \vee \mathcal{B}_2$-measurable.*

Proof. The conditional independence of $\mathcal{B}_1$ and $\mathcal{B}_2$ given $\mathcal{C}$ with respect to P and Q gives, using Proposition 1 with Q replacing μ

(5.1) $$g = g_1 g_2 q^{-1} I_{\{q>0\}} \quad \text{a.s. } Q$$

where

$$g_i = dP_{\mathcal{B}_i \vee \mathcal{C}} / dQ_{\mathcal{B}_i \vee \mathcal{C}} = E_Q^{\mathcal{B}_i \vee \mathcal{C}} g \quad (i = 1, 2)$$

and

$$q = dP_{\mathcal{C}} / dQ_{\mathcal{C}} = E_Q^{\mathcal{C}} g.$$

But $\mathscr{C}$ is sufficient for $\{P, Q\}$ on $\mathscr{B}_1$, and hence $\mathscr{B}_1 \vee \mathscr{C}$, and so by Proposition 2

(5.2) $$g_1 = dP_{\mathscr{B}_1 \vee \mathscr{C}}/dQ_{\mathscr{B}_1 \vee \mathscr{C}} \text{ is } \mathscr{C}\text{-measurable.}$$

This fact combined with (5.1) implies that g is $\mathscr{C} \vee \mathscr{B}_2$-measurable.

To see that we cannot get a complete analogue of Proposition 2 by proving that the adequacy of $\mathscr{C}$ follows from the measurability of g with respect to $\mathscr{C} \vee \mathscr{B}_2$ we need only consider the case $P = Q$. For in this case $g = 1$ is certainly measurable $\mathscr{C} \vee \mathscr{B}_2$, the sufficiency part of the adequacy of $\mathscr{C}$ trivial but the conditional independence part false in general. In fact the converse requires a conditional independence assumption.

Proposition 3b. *If $\mathscr{B}_1$ and $\mathscr{B}_2$ are conditionally Q-independent given $\mathscr{C}$ and g is $\mathscr{C} \vee \mathscr{B}_2$-measurable, then $\mathscr{C}$ is adequate for $\mathscr{B}_1$ with respect to $\mathscr{B}_2$ and $\{P, Q\}$.*

Proof. The fact that $\mathscr{B}_1$ and $\mathscr{B}_2$ are conditionally P-independent given $\mathscr{C}$ is an immediate consequence of Proposition 1 now that we have assumed their conditional Q-independence given $\mathscr{C}$. To see that $\mathscr{C}$ is sufficient for $\{P, Q\}$ on $\mathscr{B}_1$ or, what is the same $\mathscr{B}_1 \vee \mathscr{C}$, we prove that $g_1 = dP_{\mathscr{B}_1 \vee \mathscr{C}}/dQ_{\mathscr{B}_1 \vee \mathscr{C}}$ is $\mathscr{C}$-measurable, and then use Proposition 2. But this is also a consequence of the assumed conditional Q-independence for, as g is $\mathscr{C} \vee \mathscr{B}_2$-measurable by hypothesis, $g_1 = E_Q^{\mathscr{B}_1 \vee \mathscr{C}} g$ is $\mathscr{C}$-measurable by (3.2).

Having considered the factorisation theorem for adequate sub-σ-fields in this very special case, the way is now clear to formulate an analogue of Bahadur's result (given as Corollary 2 to Proposition 2 above).

Theorem 1. *Let $(\mathscr{X}, \mathscr{A}, \mu)$ be a σ-finite measure space, $\mathscr{P}$ a family of probability measures on $\mathscr{A}$ dominated by μ, $\mathscr{B}_1$, $\mathscr{B}_2$, $\mathscr{C}$ sub-σ-fields of $\mathscr{A}$, such that $\mathscr{B}_1 \vee \mathscr{C} \vee \mathscr{B}_2 = \mathscr{A}$, $\mu_{\mathscr{C}}$ is σ-finite and $\mathscr{C}$ is $[\mathscr{P}, \mathscr{A}]$-complete. Suppose further that $\mathscr{B}_1$ and $\mathscr{B}_2$ are conditionally μ-independent given $\mathscr{C}$. Then a necessary and sufficient condition for $\mathscr{C}$ to be adequate for $\mathscr{B}_1$ with respect to $\mathscr{B}_2$ and $\mathscr{P}$ is that there exist a non-negative h, and for each $P \in \mathscr{P}$ non-negative functions g_P, on $\mathscr{X}$ such that*

(*i*) *h is $\mathscr{B}_1 \vee \mathscr{C}$-measurable,*
(*ii*) *for all $P \in \mathscr{P}$, g_P is $\mathscr{C} \vee \mathscr{B}_2$-measurable,*
(*iii*) *for all $P \in \mathscr{P}$, $P = g_P h \,.\, \mu$ on $\mathscr{A}$.*

Proof. Suppose that $\mathscr{C}$ is adequate for $\mathscr{B}_1$ with respect to $\mathscr{B}_2$ and $\mathscr{P}$. Then by Proposition 1 for each $P \in \mathscr{P}$ we can write

$$\frac{dP}{d\mu} = g_{P,1} g_{P,2} q_P^{-1} I_{\{q_P > 0\}} \quad \text{a.s. } \mu$$

where $g_{P,i}$ is $\mathscr{B}_i \vee \mathscr{C}$-measurable $(i = 1, 2)$ and q_P is $\mathscr{C}$-measurable.

A FACTORISATION THEOREM FOR ADEQUATE STATISTICS

Furthermore $g_{P,1} = dP_{\mathscr{B}_1 \vee \mathscr{C}}/d\mu_{\mathscr{B}_1 \vee \mathscr{C}}$. By Corollary 2 to Proposition 2 these can all be factorised

$$g_{P,1} = g'_P \,.\, h \quad \text{a.s.} \quad [P, \mathscr{B}_1 \vee \mathscr{C}]$$

where g'_P is $\mathscr{C}$-measurable and h is $\mathscr{B}_1 \vee \mathscr{C}$-measurable. Putting

$$g_P = g'_P g_{P,2} q_P^{-1} I_{\{q_P > 0\}},$$

a $\mathscr{C} \vee \mathscr{B}_2$-measurable function, and keeping the common h completes the proof of the necessity of this factorisation.

Conversely, suppose that functions h and g_P can be found satisfying (i), (ii) and (iii) of the theorem. Then by Proposition 1 $\mathscr{B}_1$ and $\mathscr{B}_2$ are, for all P, conditionally P-independent given $\mathscr{C}$. The sufficiency of $\mathscr{C}$ for $\mathscr{P}$ on $\mathscr{B}_1 \vee \mathscr{C}$ follows more or less as it did in the proof of Proposition 3b. We note that $g_{P,1} = d\mu_{\mathscr{B}_1 \vee \mathscr{C}}/d_{\mathscr{B}_1 \vee \mathscr{C}}$ can be obtained as $E_\mu^{\mathscr{B}_1 \vee \mathscr{C}}(dP/d\mu)$ and so

$$g_{P,1} = E_\mu^{\mathscr{B}_1 \vee \mathscr{C}}(g_P h) = (E_\mu^{\mathscr{B}_1 \vee \mathscr{C}} g)h = (E_\mu\, g)h$$

since h is $\mathscr{B}_1 \vee \mathscr{C}$ measurable, g is $\mathscr{C} \vee \mathscr{B}_2$-measurable, and (3.2) applies with μ instead of P. The proof is completed by invoking Corollary 2 to Proposition 2. This completes the proof of our main result.

Corollary 1. *Suppose that $\mathscr{X} = \mathscr{X}_1 \times \mathscr{X}_2$, $\mathscr{A} = \mathscr{A}_1 \otimes \mathscr{A}_2$, $\mu = \mu_1 \otimes \mu_2$ where the $\mathscr{B}_i$ are σ-finite on $\mathscr{A}_1$ $(i = 1, 2)$ and that $\mathscr{B}_1$, $\mathscr{B}_2$ are the σ-fields generated by the coordinate projections. Let $\mathscr{C} \subseteq \mathscr{A}_1$ be a sub-σ-field of $\mathscr{A}_1$ such that μ_1 remains σ-finite when restricted to $\mathscr{C}$ and let us also denote the sub-σ-field of $\mathscr{A}$ isomorphic to $\mathscr{C}$ by $\mathscr{C}$. Then a necessary and sufficient condition for $\mathscr{C}$ to be adequate for $\mathscr{B}_1$ with respect to $\mathscr{B}_2$ and a family $\mathscr{P}$ of probability measures dominated by μ is that there exist a non-negative function h, and for each $P \in \mathscr{P}$ non-negative functions g_P defined on $\mathscr{X}$ such that*

(i) *h is $\mathscr{B}_1$-measurable,*
(ii) *for all $P \in \mathscr{P}$, g_P is $\mathscr{C} \vee \mathscr{B}_2$-measurable,*
(iii) *for all $P \in \mathscr{P}$, $P = g_P h \,.\, \mu$ on $\mathscr{A}$.*

Proof. Once we observe that $\mathscr{B}_1$ and $\mathscr{B}_2$ are μ-independent, and hence conditionally μ-independent given $\mathscr{C} \subseteq \mathscr{B}_1$, there is nothing left to prove. This corollary includes the result cited as Theorem 2.2 in Ishii [3].

6. An Illustration

The examples in Ishii's paper [3] all concern cases in which the σ-fields $\mathscr{B}_1$ and $\mathscr{B}_2$ are actually P-independent for each $P \in \mathscr{P}$. Apart from a problem noted in the next section, the theorem of §5 above applies to Ishii's examples, and we now consider a simple situation in which a more general type of behaviour takes place.

Let $(X_1, X_2, \ldots, X_n, X_{n+1})$ be jointly normally distributed random variables such that for each i, $1 \le i \le n+1$, the distribution of X_1 given $X_{i-1}, X_{i-2}, \ldots, X_0$ is $N(\rho X_{i-1}, 1)$, where X_0 is given (i.e. held constant) and $|\rho| < 1$. Put $\mathscr{X}_1 = \mathbb{R}^n$, $\mathscr{X}_2 = \mathbb{R}$, and

$$\mu_1(dx_1 \ldots . dx_n) = (2\pi)^{-n/2} \exp\left(-\tfrac{1}{2}\sum_1^n x_i^2\right) dx_1 \ldots dx_n,$$

$$\mu_2(dx_{n+1}) = (2\pi)^{-1/2} \exp\left(-\tfrac{1}{2}x_{n+1}^2\right) dx_{n+1}.$$

The conditions on $(X_1, \ldots, X_{n+1})$ are then seen to be equivalent to the requirement that their joint distribution P has density with respect to $\mu = \mu_1 \otimes \mu_2$ given by

$$\frac{dP}{d\mu} = \exp\left[\rho \sum_1^{n+1} x_{i-1}x_i - \tfrac{1}{2}\rho^2 \sum_1^{n+1} x_i^2\right].$$

Defining the sub-σ-field $\mathscr{C} = \sigma(\sum_1^n X_{i-1}X_i, \sum_1^n X_i^2, X_n)$ of $\mathscr{B}_1 = \sigma(X_1, \ldots, X_n)$, and putting $\mathscr{B}_2 = \sigma(X_{n+1})$, we see that the conditions of Corollary 1 in §5 above are satisfied, viewing $\mathscr{P}$ as the class of probabilities indexed by ρ, $|\rho| < 1$. Thus the triple $(\sum_1^n X_{i-1}X_i, \sum_1^n X_i^2, X_n)$ is an *adequate* reduction of $(X_1, \ldots, X_n)$ as far as *estimation* of ρ and *prediction* of X_{n+1} is concerned. It is not hard to see that X_n could not be omitted from the triple, and this is, of course, intuitively obvious.

The reason why we chose to use the particular probability measure $\mu = \mu_1 \otimes \mu_2$ as the dominating measure, rather than the more natural dominating measure λ^{n+1}, $(n+1)$-dimensional Lebesgue measure, is explained below.

7. Limitations

In all of the results involving μ as a dominating measure we have had to suppose $\mu_{\mathscr{C}}$ σ-finite in order to be able to formulate conditional μ-independence given $\mathscr{C}$. This is a strong assumption and is violated in many simple examples.

Example. Let $\mathscr{X} = \mathbb{R}^3$, $\mathscr{A} = \mathscr{R}^3$ and $\mu = \lambda^3$ (Lebesgue measure) corresponding to three real random variables X_1, X_2, X_3 defined as the coordinate projections. One frequently considers a system such as

$$\mathscr{B}_1 = \sigma(X_1, X_2),\ \mathscr{B}_2 = \sigma(X_3) \text{ and } \mathscr{C} = \sigma(X_1 + X_2) \subseteq \mathscr{B}_1$$

and would certainly hope that since $\mathscr{B}_1$ and $\mathscr{B}_2$ are, in a sense, μ-independent, one would also have $\mathscr{B}_1$ and $\mathscr{B}_2$ conditionally μ-independent given $\mathscr{C}$. But μ is not σ-finite when restricted to $\mathscr{C}$, every set in $\mathscr{C}$ having zero or infinite μ-measure.

In hardly needs stating that in many examples of interest the μ will be exactly of the kind in the example, and a similar remark applies to the sub-σ-field $\mathscr{C}$, since many sufficient σ-fields are based upon sums.

Thus, although it seems very reasonable to assert that the theorem in §5 is in a sense a natural analogue of the Halmos–Savage formulation of the Fisher–Neyman theorem, it simply does not cover many examples of interest to statisticians. This also applies to the result of Sugiura and Morimoto cited in Ishii [3].

When the sample $(\mathscr{X}, \mathscr{A}, \mu)$ has a Euclidean structure it appears possible to prove a theorem which includes the examples mentioned but what seems difficult at the moment is the formulation of a result *generalising Theorem* 1 to cover all such cases. The problem is this: a satisfactory theory of conditional expectations, and hence of conditional independence, given sub-σ-fields on which the basic measure is not σ-finite, has yet to be developed. One might approach the problem via the notion of *set of σ-finiteness* Neveu [6] p. 16–17 but so far this has not been worked out. On the other hand there may be no alternative to formulating Euclidean problems in an Euclidean measure-theoretic framework cf. Tjur [8].

8. Acknowledgments

I would like to thank Professor D. Basu for his constant interest in this and related work, Professor J. F. C. Kingman for remarks which clarified some confusion in my mind, and Dr. M. L. Thornett, whose interest also contributed to this note. I am also indebted to Professor H. Morimoto for encouraging me to publish this extension of his result with M. Sugiura.

References

[1] Bahadur R. R. (1954) "Sufficiency and statistical decision functions." *Ann. Math. Statist.* 25, 423–462.

[2] Halmos, Paul R. and Savage, L. J. (1949) "Application of the Radon-Nikodym theorem to the theory of sufficient statistics." *Ann. Math. Statist.* 20, 225–241.

[3] Ishii, Goro (1969) "Optimality of unbiased predictors." *Ann. Inst. Statist. Math.* 21, 471–488.

[4] Loève, Michel (1963) *Probability Theory* Third Edition. Van Nostrand, Princeton N.J.

[5] Skibinsky, Morris (1967) "Adequate subfields and sufficiency." *Ann. Math. Statist.* 38, 155–161.

[6] Neveu, Jacques (1972) *Martingales a temps discret.* Masson et Cie, Paris.

[7] Sugiura, M. and Morimoto, H. (1969) "Factorisation theorem for adequate σ-field." (in Japanese). *Sūgaku* 21.

[8] Tjur, Tue (1974). *Conditional Probability Distributions.* Institute of Mathematical Statistics, University of Copenhagen.

Chapter 4
Interaction Models

Steffen L. Lauritzen

The articles in this bundle are all associated with the notion of *interaction* and represent the genesis of the subject of graphical models in its modern form, the origins of these being traceable back to Gibbs [11] and Wright [30] and earlier.

Around 1976, Terry was fascinated by the notion of conditional independence, along the lines later published in Dawid [6, 7]. In 1976, Terry invited me to Perth and we were running a daily research seminar with the theme of studying similarities and differences between Statistics and Statistical Mechanics. In particular, we wondered what the relations were between notions of interaction as represented in linear models, in multi-dimensional contingency tables, and in stochastic models for particle systems; in addition, the purpose was also to understand what was the relation between these concepts and conditional independence.

As we discovered that these were all essentially the same concepts, the similarity being obscured by very different traditions of notation, the term graphical model was coined. Our findings, also obtained in collaboration with John Darroch, were collected in Darroch et al. [4], and later expanded and published in Speed [24], Darroch et al. [5], and Darroch and Speed [3] as well as Lauritzen et al. [19] and to some extent Speed [25], the latter giving an overview of a number of different variants and proofs of what has become known as the Hammersley–Clifford theorem [14, 2].

Of these articles, Darroch et al. [5] rather quickly had a seminal impact and a small community of researchers in the area of graphical models gradually emerged. In a certain sense, the article does not contain much formally new material (if any at all), but for the first time a simple, visual description and interpretation of the class of log-linear models [12, 13], which otherwise could seem obscure, was available. The interpretation of a subclass of the models in terms of conditional independence had an immediate intuitive appeal. In addition, the article identified and emphasized models represented by *chordal* or *triangulated* graphs as those where estimation

S.L. Lauritzen
Department of Statistics, University of Oxford, United Kingdom
e-mail: steffen@stats.ox.ac.uk

S. Dudoit (ed.), *Selected Works of Terry Speed*, Selected Works in Probability and Statistics,
DOI 10.1007/978-1-4614-1347-9_4, © Springer Science+Business Media, LLC 2012

and other issues had a particularly simple solution, the combinatorial theory of these graphs being further studied in Lauritzen et al. [19].

Darroch and Speed [3] studied the notion of interaction from an algebraic point of view in terms of fundamental decompositions of the linear space of functions on a product of finite sets; indeed it essentially but implicitly uses the fundamental decomposition of this space into irreducible components which are stable under a product of symmetric groups [9] and thus gives an elegant algebraic perspective on the Hammersley–Clifford theorem.

Towards the end of 1976, Terry serendipitously came across Wermuth [29], which identified that a completely analogous theory could be developed for the Gaussian case, with chordal graphs playing essentially the same role as in the case of log-linear models; indeed, Dempster [8] had developed the basic computational and statistical theory for these under the name of models for *covariance selection*. This fact and the corresponding interpretation was emphasized and discussed in Darroch et al. [4] as well as in Speed [24, 25], but received otherwise relatively little attention at the time. Gaussian graphical models have had a remarkable renaissance in connection with the modern analysis of high-dimensional data, for example concerning gene expression [10, 23]. Out of this early work with Gaussian graphical models grew also the article by Speed and Kiiveri [26], which describes and unifies a class of iterative algorithms for fitting Gaussian graphical models of which special cases previously had been considered by e.g. Dempster [8]. Essentially, there are two fundamental types, of which one initially uses the estimate under no restrictions and iteratively ensures that restrictions of the model are satisfied; the other type initially uses a trivial estimator and iteratively ensures that the likelihood equations are satisfied. The article elegantly shows that an abundance of hybrids of these algorithms can be constructed and gives a unified proof of their convergence.

The last two articles [16, 17], represent the genesis of what today is probably the most prolific and well-known type of graphical models; these are based on *directed* acyclic graphs and admitting interpretation in *causal* terms similar to that of *structural equation models* [1]. At the time when these articles appeared they were (undeservedly) largely ignored both by the statistical and structural equation communities. Graphical models based on directed acyclic graphs—now mostly known as *Bayesian networks* [21]—have an unquestionable prominence in current scientific literature, but the surge of interest in these models was in particular generated by the prolific research activities in computer science, where work such as, for example, Lauritzen and Spiegelhalter [18], Pearl [22], Spirtes et al. [27], Heckerman et al. [15], and Pearl [20] established these models as objects worthy of intense study. In retrospect, it is clear that the global Markov property defined in Kiiveri et al. [17] was not the optimal one as there are independence relations true in any Bayesian network that cannot be derived from it, but fundamentally this article establishes the correct class of directed Markov models for the first time and thus yields a conditional independence perspective on structural equation models, as later elaborated, for example by Spirtes et al. [28].

References

[1] K. A. Bollen. *Structural Equations with Latent Variables*. John Wiley and Sons, New York, 1989.

[2] P. Clifford. Markov random fields in statistics. In G. R. Grimmett and D. J. A. Welsh, editors, *Disorder in Physical Systems: A Volume in Honour of John M. Hammersley*, pages 19–32. Oxford University Press, 1990.

[3] J. N. Darroch and T. P. Speed. Additive and multiplicative models and interactions. *Ann. Stat.*, 11:724–738, 1983.

[4] J. N. Darroch, S. L. Lauritzen, and T. P. Speed. Log-linear models for contingency tables and Markov fields over graphs. Unpublished manuscript, 1976.

[5] J. N. Darroch, S. L. Lauritzen, and T. P. Speed. Markov fields and log-linear interaction models for contingency tables. *Ann. Stat.*, 8:522–539, 1980.

[6] A. P. Dawid. Conditional independence in statistical theory (with discussion). *J. Roy. Stat. Soc. B*, 41:1–31, 1979.

[7] A. P. Dawid. Conditional independence for statistical operations. *Ann. Stat.*, 8: 598–617, 1980.

[8] A. P. Dempster. Covariance selection. *Biometrics*, 28:157–175, 1972.

[9] P. Diaconis. *Group Representations in Probability and Statistics*, volume 11 of *Lecture Notes–Monograph Series*. Institute of Mathematical Statistics, Hayward, CA, 1988.

[10] A. Dobra, C. Hans, B. Jones, J. R. Nevins, and M. West. Sparse graphical models for exploring gene expression data. *J. Multivariate Anal.*, 90:196–212, 2004.

[11] W. Gibbs. *Elementary Principles of Statistical Mechanics*. Yale University Press, New Haven, Connecticut, 1902.

[12] L. A. Goodman. The multivariate analysis of qualitative data: Interaction among multiple classifications. *J. Am. Stat. Assoc.*, 65:226–256, 1970.

[13] S. J. Haberman. *The Analysis of Frequency Data*. University of Chicago Press, Chicago, 1974.

[14] J. M. Hammersley and P. E. Clifford. Markov fields on finite graphs and lattices. Unpublished manuscript, 1971.

[15] D. Heckerman, D. Geiger, and D. M. Chickering. Learning Bayesian networks: The combination of knowledge and statistical data. *Mach. Learn.*, 20:197–243, 1995.

[16] H. Kiiveri and T. P. Speed. Structural analysis of multivariate data: A review. In S. Leinhardt, editor, *Sociological Methodology*. Jossey-Bass, San Francisco, 1982.

[17] H. Kiiveri, T. P. Speed, and J. B. Carlin. Recursive causal models. *J. Aust. Math. Soc. A*, 36:30–52, 1984.

[18] S. L. Lauritzen and D. J. Spiegelhalter. Local computations with probabilities on graphical structures and their application to expert systems (with discussion). *J. Roy. Stat. Soc. B*, 50:157–224, 1988.

[19] S. L. Lauritzen, T. P. Speed, and K. Vijayan. Decomposable graphs and hypergraphs. *J. Aust. Math. Soc. A*, 36:12–29, 1984.

[20] J. Pearl. Causality: Models, Reasoning, and Inference. Cambridge University Press, Cambridge, UK, 2000.

[21] J. Pearl. Fusion, propagation and structuring in belief networks. *Artif. Intell.*, 29:241–288, 1986.

[22] J. Pearl. Probabilistic Inference in Intelligent Systems. Morgan Kaufmann Publishers, San Mateo, CA, 1988.

[23] J. Schäfer and K. Strimmer. An empirical-Bayes approach to inferring large-scale gene association networks. *Bioinformatics*, 21:754–764, 2005.

[24] T. P. Speed. Relations between models for spatial data, contingency tables and Markov fields on graphs. *Adv. Appl. Prob.: Supplement*, 10:111–122, 1978.

[25] T. P. Speed. A note on nearest-neighbour Gibbs and Markov probabilities. *Sankhyā Ser. A*, 41:184–197, 1979.

[26] T. P. Speed and H. Kiiveri. Gaussian Markov distributions over finite graphs. *Ann. Stat.*, 14:138–150, 1986.

[27] P. Spirtes, C. Glymour, and R. Scheines. *Causation, Prediction and Search*. Springer-Verlag, New York, 1993. Reprinted by MIT Press.

[28] P. Spirtes, T. S. Richardson, C. Meek, R. Scheines, and C. Glymour. Using path diagrams as a structural equation modeling tool. *Sociol. Method. Res.*, 27:182–225, 1998.

[29] N. Wermuth. Analogies between multiplicative models in contingency tables and covariance selection. *Biometrics*, 32:95–108, 1976.

[30] S. Wright. The method of path coefficients. *Ann. Math. Statist.*, 5:161–215, 1934.

The Annals of Statistics
1980, Vol. 8, No. 3, 522–539

MARKOV FIELDS AND LOG-LINEAR INTERACTION MODELS FOR CONTINGENCY TABLES[1]

BY J. N. DARROCH, S. L. LAURITZEN AND T. P. SPEED

The Flinders University of South Australia, University of Copenhagen and University of Western Australia.

We use a close connection between the theory of Markov fields and that of log-linear interaction models for contingency tables to define and investigate a new class of models for such tables, graphical models. These models are hierarchical models that can be represented by a simple, undirected graph on as many vertices as the dimension of the corresponding table. Further all these models can be given an interpretation in terms of conditional independence and the interpretation can be read directly off the graph in the form of a Markov property. The class of graphical models contains that of decomposable models and we give a simple criterion for decomposability of a given graphical model. To some extent we discuss estimation problems and give suggestions for further work.

0. Introduction and summary. In the present paper we shall utilize some close connections between the theory of Markov fields and that of log-linear interaction models to define a new class of models for multidimensional contingency tables: *graphical models*. The graphical models have two important properties:

(i) they can be represented by an undirected, finite graph with as many vertices as the table has dimensions;

(ii) they can be interpreted in terms of conditional independence (in fact, a Markov property) and the interpretation can be read directly off the graph.

This class of models is a proper subclass of the so-called *hierarchical models*, but it strictly contains the *decomposable models* (Goodman (1970, 1971), Haberman (1970, 1974), Andersen (1974)). This implies that we can give a simple, visual representation of any decomposable model, thus making the interpretation easy.

We also characterise those graphs that correspond to decomposable models, thus giving an alternative to Goodman's algorithm for checking decomposability of a given hierarchical model: first, check whether it is graphical and then, if it is, check whether the graph is decomposable, i.e., whether there are any cyclic subgraphs of length $\geqslant 4$.

In Section 1 we introduce some notation and define the various classes of models for contingency tables. In Section 2 we review some basic elements of the theory of Markov fields and Gibbs states. In Section 3 we draw together the results in these

Received November 1978; revised March 1979.
[1]This research was supported in part by the Danish Natural Science Research Council.
AMS 1970 *subject classifications.* Primary 62F99; secondary 60K35.
Key words and phrases. Contingency tables, decomposability, Gibbs states, graphical models, triangulated graphs.

two sections, define the graphical models and discuss their interpretation. Section 4 contains the arguments needed to realise that all decomposable models are graphical and we also give the characterisation of decomposable graphs. Section 5 is devoted to maximum likelihood estimation in decomposable models. Although this is completely solved by Haberman (1974) we define an index directly interpretable from the graph and show how these indices are the powers of the marginal counts in the estimation formula. A combinatorial property of this index can also be used as a characterisation of decomposable graphs. Section 6 contains a list of all graphical models of dimension less than or equal to five together with their interpretation and these are divided into decomposables and nondecomposables. This is meant to both illustrate our theory and be an analogue of the tables in Goodman (1974) with all hierarchical models of dimension less than or equal to four together with an interpretation of the decomposables among them. Finally we give some suggestions regarding the use of the models and some directions for possible further work.

The present paper is almost without proofs. Most of our results are just "translations" of results from other areas. It is somewhat technical to establish the connection between graphical models and decomposable models. In fact, in our opinion these results are of a purely graph theoretic nature and the proofs and necessary formalism to derive the results can be found in Lauritzen, Speed and Vijayan (1978).

1. Preliminaries. We shall discuss log-linear interaction models for contingency tables. Since we want to use the analogies between the theory of Markov fields and that of such models, it will be convenient to introduce a notation that makes such analogies more apparent.

We shall consider a finite set C of *classification criteria* or *factors*. For each $\gamma \in C$ we let I_γ be the set of *levels* of the criterion or factor γ. The set of *cells* in our table is the set $I = \Pi_{\gamma \in C} I_\gamma$ and a particular cell will be denoted $\mathbf{i} = (i_\gamma, \gamma \in C)$. A set of n objects is classified according to the criteria and we let the *counts* $n(\mathbf{i})$ be the number of objects in cell $\mathbf{i}$.

For $a \subseteq C$, we consider the *marginal counts* $n(\mathbf{i}_a)$. $n(\mathbf{i}_a)$ is the number of objects in the marginal cell $\mathbf{i}_a = (i_\gamma, \gamma \in a)$ and is obtained as the sum of the $n(\mathbf{i})$ for all such $\mathbf{i}$ that agree with $\mathbf{i}_a$ on the coordinates corresponding to a. In other words, $n(\mathbf{i}_a)$ are the counts in the *marginal table*, where objects only are classified according to the criteria in a. Similarly we let $P(\mathbf{i})[P(\mathbf{i}_a)]$ denote the probability that any given object belongs to the [marginal] cell $\mathbf{i}[\mathbf{i}_a]$.

We consider the classifications of the n objects as n independent observations of the distribution P such that the distribution of the counts becomes a multinomial distribution:

$$P\{N(\mathbf{i}) = n(\mathbf{i}), \mathbf{i} \in I\} = \binom{n}{n(\mathbf{i}), \mathbf{i} \in I} \Pi_{\mathbf{i} \in I} P(\mathbf{i})^{n(\mathbf{i})}.$$

The general log-linear interaction model involves specification of the above unknown distribution P as follows: firstly we expand the logarithm of P as

$$\log P(\mathbf{i}) = \Sigma_{a \subseteq C} \xi_a(\mathbf{i}_a),$$

where ξ_a are functions of $\mathbf{i}$ that only depend on $\mathbf{i}$ via the coordinates in a, i.e., through $\mathbf{i}_a$. If $a = \emptyset$, $\xi_\emptyset$ is the constant vector.

Such an expansion can be made for any P with $P(\mathbf{i}) > 0$ for all $\mathbf{i} \in I$. If we are interested in having a one-to-one correspondence between the system of functions $\{\xi_a, a \subseteq C\}$ and P, we have to introduce standardising constraints as, e.g.,

$$\forall b \subset a : \Sigma_{\{\mathbf{i}'_a : \mathbf{i}'_b = \mathbf{i}_b\}} \xi_a(\mathbf{i}'_a) \equiv 0 \quad \text{for all } \mathbf{i}_b,$$

i.e., that summation over any factor gives a zero. This is all well known and standard although the notation is slightly unusual.

The functions ξ_a are called the *interactions* among the factors in a. If $|a| = 1$ we call ξ_a the *main effect*, if $|a| = 2$ a *first-order interaction* and, in general, if $|a| = m$, ξ_a is an interaction of order $m - 1$. A general log-linear interaction model involves specifying certain of these interactions to vanish and letting the remaining interactions be arbitrary and unknown. It is usually convenient to work with a smaller class of models, the *hierarchical models*.

A hierarchical model is an interaction model where the specifications of vanishing interactions satisfy the following property: *if ξ_a is specified to vanish and $b \supseteq a$ then ξ_b is specified to vanish*. In other words, if there is no interaction among factors in a then there is no interaction of higher order involving all the factors in a.

As is easily seen and well known, a hierarchical model can be specified via a so-called *generating class* being a set $\mathcal{C}$ of pair-wise incomparable (w.r.t. inclusion) subsets of C to be interpreted as the maximal sets of permissible interactions, i.e.,

$$\xi_a \equiv 0 \text{ iff } \textit{there is no } c \in \mathcal{C} \text{ with } a \subseteq c.$$

A probability P belonging to a hierarchical model with generating class $\mathcal{C}$ is uniquely determined by the marginal probabilities given by the elements of $\mathcal{C}$. The maximum likelihood estimate of P is obtained by equating these marginal probabilities to the marginal sample proportions.

A certain subclass of hierarchical models is of special interest: the *decomposable models*, introduced by Goodman (1970, 1971) and later defined formally by Haberman (1970, 1974). Following Haberman, a generating class is *decomposable* if either it has only one element or if it can be partitioned into generating classes $\mathcal{A}$ and $\mathcal{B}$ with $\mathcal{A} \cap \mathcal{B} = \emptyset$, $\mathcal{C} = \mathcal{A} \cup \mathcal{B}$ and such that

$$(\cup_{a \in \mathcal{A}} a) \cap (\cup_{b \in \mathcal{B}} b) = a^* \cap b^*$$

for some $a^* \in \mathcal{A}$, $b^* \in \mathcal{B}$. A slightly different definition was given by Lauritzen, Speed and Vijayan (1978) (henceforth referred to as LSV) but it is shown in the same paper that the definitions are equivalent.

MARKOV FIELDS AND LOG-LINEAR MODELS

As shown by Haberman (1970) these models have two fundamental properties

(i) the problem of maximum likelihood estimation has an explicit solution;
(ii) the models can be interpreted in terms of conditional independence, independence and equiprobability.

The basic idea in our work is that such an interpretation is most directly formulated as a Markov property. Goodman (1970), in fact, uses the terminology "models of Markov type" for decomposable models.

This leads us to consider Markov fields on finite graphs and from these considerations it turns out that it is natural to define a class of models, *graphical models* whose interpretation most elegantly is given as a Markov property of a certain random field associated with the model.

2. **Markov fields and Gibbs states.** In the theory of Markov fields, see, e.g., Kemeny, Snell and Knapp (1976), we operate with a set Γ of *sites* and here we assume Γ to be finite. Γ will correspond to the set of factors C. At each site $\gamma \in \Gamma$ there is a finite set I_γ of *elementary states*. The set $I = \Pi_{\gamma\in\Gamma} I_\gamma$ is the set of *configurations*. A given configuration is denoted by $\mathbf{i} = (\mathbf{i}_\gamma, \gamma \in \Gamma)$. Further there is an undirected *graph* $\mathbf{\Gamma}$ on Γ, i.e., a pair $\mathbf{\Gamma} = (V(\mathbf{\Gamma}), E(\mathbf{\Gamma}))$ consisting of the *vertex set* $V(\mathbf{\Gamma}) = \Gamma$ and *edge set* $E(\mathbf{\Gamma})$, where $E(\mathbf{\Gamma})$ is a set of unordered pairs of distinct elements of Γ. We say that α and β are *adjacent* or *neighbours* and write $\alpha \sim \beta$ iff $\{\alpha, \beta\} \in E(\mathbf{\Gamma})$.

If $a \subseteq \Gamma$, the *boundary* of a, ∂a, is the set of vertices in $\Gamma \setminus a$ that are adjacent to some vertex in a. The *closure* of a is $a \cup \partial a$ and is denoted by $\bar{a}$. When no confusion is possible we write $\partial\alpha$, $\bar{\alpha}$ instead of $\partial\{\alpha\}$, $\overline{\{\alpha\}}$. A *complete subset* is a subset $a \subseteq \Gamma$ where all elements are mutual neighbours. A *clique* is a maximal (w.r.t. inclusion) complete subset.

We now consider a probability P on I with $P(\mathbf{i}) > 0$ for all $\mathbf{i} \in I$ and the random variables defined by coordinate projections:

$$X_\gamma(\mathbf{i}) = i_\gamma, \qquad \gamma \in \Gamma$$

and

$$X_a(\mathbf{i}) = \mathbf{i}_a \quad \text{for } a \subseteq \Gamma, \qquad a \neq \emptyset.$$

The random field $(X_\gamma, \gamma \in \Gamma)$ is said to be *Markov* w.r.t. P *and* $\mathbf{\Gamma}$ (or P *is Markov* w.r.t. $\mathbf{\Gamma}$) if one of the following four equivalent properties hold:

(i) for all $\gamma \in \Gamma$, X_γ and $X_{\Gamma\setminus\bar{\gamma}}$ are conditionally independent given $X_{\partial\gamma}$;
(ii) for all $\alpha, \beta \in \Gamma$ with $\alpha \nsim \beta$, X_α and X_β are conditionally independent given $X_{\Gamma\setminus\{\alpha,\beta\}}$;
(iii) for all $a \subseteq \Gamma$, X_a and $X_{\Gamma\setminus\bar{a}}$ are conditionally independent given $X_{\partial a}$;
(iv) if two disjoint subsets $a \subseteq \Gamma$ and $b \subseteq \Gamma$ separated by a subset $d \subseteq \Gamma$ in the sense that all paths from a to b in $\mathbf{\Gamma}$ go via d, then X_a and X_b are conditionally independent given X_d.

That these four conditions in fact are equivalent for a probability with $P(\mathbf{i}) > 0$ is more or less well known, see, e.g., Pitman (1976) or Kemeny, Snell and Knapp (1976). It can be proved with quite elementary methods.

A *potential* is a real-valued function Φ on I of the form

$$\Phi(\mathbf{i}) = \Sigma_{a \subseteq \Gamma} \xi_a(\mathbf{i}_a)$$

where the functions ξ_a depend on $\mathbf{i}$ through $\mathbf{i}_a$ only and are called the *interaction potentials*. In fact, any real-valued function is a potential, see the remarks in the previous section, so this notion first gets interesting when we make restrictions on the ξ_a - functions.

A probability P on I is called a *Gibbs state with potential* Φ if

$$P(\mathbf{i}) = e^{\Phi(\mathbf{i})}.$$

Similarly, any probability on I with $P(\mathbf{i}) > 0$ for all $\mathbf{i}$ is a Gibbs state (with potential $\Phi(\mathbf{i}) = \log P(\mathbf{i})$). Φ is called a *nearest-neighbour potential* if it is built up from interactions only among mutual neighbours, i.e., if $\xi_a \equiv 0$ if not all vertices in a are mutual neighbours, i.e., if a is not a *complete subset* of Γ. P is called a *nearest-neighbour Gibbs state* iff P is a Gibbs state with potential Φ, where Φ is a nearest-neighbour potential.

One of the most basic results about Markov fields and nearest-neighbour Gibbs states asserts that, in fact, the two notions are identical: *P is a nearest-neighbour Gibbs state if and only if the corresponding random field is Markov*. A proof of this result can be found many places. In the case $I_\gamma = I_0$ there is, e.g., a proof in Kemeny, Snell and Knapp (1976), and the method of proof there easily extends to the case with I_γ depending on γ, see, e.g., Pitman (1976) or Speed (1976).

This theorem is in fact the key to our results: it establishes a connection between certain linear restrictions on the logarithm of a probability (being n.-n.-Gibbs) and a Markov property (an interpretation in terms of conditional independence). What remains to be done is to introduce the graphs in the contingency table framework.

3. Graphical models. Let us return to the contingency table set-up. Assume that we have given a graph $\mathbf{C}$ on our set of factors C, specified by the vertex set $V(\mathbf{C}) = C$ and edge set $E(\mathbf{C})$. Let $\mathcal{C}$ be the *cliques* of $\mathbf{C}$, i.e., the maximal complete subsets. The *graphical model* given by $\mathbf{C}$ is the hierarchical model with generating class $\mathcal{C}$. Note that $\mathcal{C}$ also uniquely defines the graph $\mathbf{C}$ by $\alpha \sim \beta$ iff $\exists c \in \mathcal{C}$ such that $\{\alpha, \beta\} \subseteq c$. In that sense our graph $\mathbf{C}$ is just another representation of the generating class $\mathcal{C}$.

Let us examine the restrictions on our interactions given by this generating class. By the definition of a hierarchical model we have $\xi_a \equiv 0$ *unless a is contained in a maximal complete subset*, i.e., *unless a is a complete subset*. In other words, the set of probabilities P in our model is *exactly the set of nearest-neighbour Gibbs states corresponding to* $\mathbf{C}$.

MARKOV FIELDS AND LOG-LINEAR MODELS

Consequently, by the fundamental theorem in the previous section, we have that the probabilities P, contained in our model are exactly *those making* $(X_\gamma, \gamma \in \mathbf{C})$ *a Markov field.* It is now clear that our model is given by conditional independence constraints involved in the four equivalent formulations of the Markov property. It is thus clear that if two sets of factors are in different connected components of the graph, they are independent. If two factors are not neighbours, they are conditionally independent given the other factors. If two sets of factors a and b are separated by a set of factors d, they are conditionally independent given those in d, etc.

We should like to point out, that not all hierarchical models are of the graphical type. It is, however, still possible to associate a graph with any generating class. The graph defines the interaction structure in part.

Let $\mathcal{C}$ be a generating class and assume that $C = \cup_{c \in \mathcal{C}} c$ (this assumption is merely of technical nature). Define a graph $\mathbf{C} = (V(\mathbf{C}), E(\mathbf{C}))$ by letting $V(\mathbf{C}) = C$ and $\{\alpha, \beta\} \in E(\mathbf{C})$ if and only if $\{\alpha, \beta\} \subseteq c$ for some $c \in \mathcal{C}$. We could call this graph the *first-order interaction graph* for $\mathcal{C}$ since it has all main effects as vertices and first-order interactions as edges. It is clear, that $\mathcal{C}$ corresponds to a graphical model if and only if $\mathcal{C}$ exactly is the set of cliques of this graph. If this is the case, we shall say that $\mathcal{C}$ is a graphical generating class. If there are cliques in the graph that are not in $\mathcal{C}$, which very well can be the case, then $\mathcal{C}$ is not graphical and the interaction structure in the model is not adequately described by the graph alone. Note that these remarks imply that the interaction structure in a graphical model is *determined by the first-order interactions,* since these interactions define the graph, which, in turn, gives us its cliques and thus its interactions of higher order.

The simplest example of a hierarchical model which is not graphical is that with $C = \{1, 2, 3\}$ and $\mathcal{C} = \{\{1, 2\}, \{2, 3\}, \{1, 3\}\}$. Its first-order interaction graph is

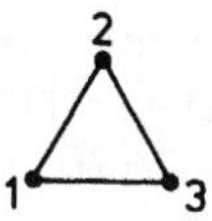

i.e., the complete 3-graph. If $\mathcal{C}$ had been graphical, $\mathcal{C}$ should have been $\{\{1, 2, 3\}\}$ which is not the case. The model in question, that of vanishing second-order interaction in a three-way table, is also known as the simplest nondecomposable hierarchical model, and it is well known, that it cannot be interpreted in terms of conditional independence.

In the next section we shall see that all decomposable models are graphical and characterise graphs corresponding to decomposable models.

4. Decomposable models and graphical models. Lauritzen, Speed and Vijayan (1978) (LSV) study properties of generating classes and their first-order interaction graphs, especially w.r.t. the notion of a decomposition. This is done in a purely graph-theoretic framework and they therefore use a slightly different terminology to be able to relate their results to other areas of mathematics.

A generating class is, in LSV, called a *generating class hyper graph* (g.c. hypergraph). The first-order interaction graph of a generating class is called the *2-section of the* g.c. *hypergraph*.

Here we shall quote some of the results from LSV of importance to us. For proofs and details, the reader is referred to that paper using the "translation key" just given. Corollary 4 in LSV asserts that *any decomposable model is graphical*. This fact was noted by Andersen (1974) in a somewhat disguised form (his Theorem 5).

We are now led to the following considerations: decomposability is a property of a generating class, a property which is not too easy to get hold of and verify directly. We have just seen that any decomposable model is graphical, i.e., is very well represented by its first-order interaction graph. Then decomposability must be a property of such a graph. Theorem 2 of LSV asserts (among other things) that: *the cliques of a graph form a decomposable generating class if and only if the graph is triangulated* (i.e., contains no cycles of length ≥ 4 without a chord). For the notion of a triangulated graph, see Berge (1973).

This result is definitely the main result of LSV and gives us a possibility of making an immediate visual check on the decomposability of a given graphical model, see our tables in Section 6.

Thus the smallest nondecomposable graphical generating class is given by the 4-cycle:

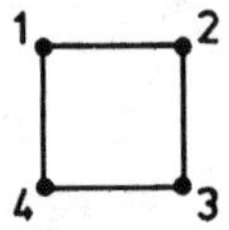

i.e., with $C = \{1, 2, 3, 4\}$, $\mathcal{C} = \{\{1, 2\}, \{2, 3\}, \{3, 4\}, \{1, 4\}\}$. In fact, Andersen (1974) gives this example of a nondecomposable model that can be interpreted in terms of conditional independence (1 and 3 are c.i. given 2 and 4, 2 and 4 are c.i. given 1 and 3).

The Markov interpretation originally made by Goodman, Haberman etc. is along the following lines: a generating class $\mathcal{C} = \{a_1, \cdots, a_k\}$ is decomposable iff its elements can be ordered so that

$$a_t \cap (a_1 \cup \cdots \cup a_{t-1}) = a_t \cap a_{r_t}, r_t \in \{1, \cdots, t-1\}, \quad t = 2, \cdots, k. \tag{4.1}$$

It follows that

$$b_t = a_t \setminus (a_1 \cup \cdots \cup a_{t-1}) = a_t \setminus a_{r_t} \neq \emptyset.$$

It is easy to see that, if P is hierarchical with generating class $\mathcal{C}$, that is

$$P(\mathbf{i}) = \exp \Sigma_{t=1}^{k} \Sigma_{a \subseteq a_t} \xi_a(\mathbf{i}_a),$$

then the conditional probability

$$P(\mathbf{i}_{b_k} | \mathbf{i}_{a_1 \cup \cdots \cup a_{k-1}})$$

simplifies to $P(\mathbf{i}_{b_k}|\mathbf{i}_{c_k})$ where

$$c_t = a_t \setminus b_t = a_t \cap a_{r_t},$$

and that the marginal probability $P_{a_1 \cup \cdots \cup a_{k-1}}$ satisfies the hierarchical model with generating class $\mathcal{C} = \{a_k\}$. It follows by induction that

$$P(\mathbf{i}) = P(\mathbf{i}_{a_1})\Pi_{t=2}^{k} P(\mathbf{i}_{b_t}|\mathbf{i}_{c_t})$$

and that the distribution of an $\mathbf{X}$ with probability P may be characterised by the sequence of Markov properties

$$\begin{aligned} &\text{conditional distribution of } \mathbf{X}_{b_t} \text{ given } \mathbf{X}_{a_1 \cup \cdots \cup a_{t-1}} \\ &= \text{conditional distribution of } \mathbf{X}_{b_t} \text{ given } \mathbf{X}_{c_t}, \qquad t = 2, \cdots, k. \end{aligned}$$

Further, (2) may be rearranged as

$$P(\mathbf{i}) = \frac{\Pi_{t=1}^{k} P(\mathbf{i}_{a_t})}{\Pi_{t=2}^{k} P(\mathbf{i}_{b_t})}$$

which is the explicit formula for P and includes as a special case the formula for the maximum likelihood estimate of P.

In order to arrive at this formula by the above method it is necessary to search for an ordering of the elements of $\mathcal{C}$ which satisfies (4.1). This search is helped by reference to the graph and also by the awareness that each element a_t must contain at least one element which is not in $a_1 \cup \cdots \cup a_{t-1}$. There are, generally, many orderings satisfying (4.1). Haberman proved that there are at least k by proving that any element of $\mathcal{C}$ may be chosen as initial member of some sequence. That there may be many more is illustrated by the example with $|\Gamma| = 6$ and

$$\mathcal{C} = \{\{1, 2\}, \{2, 3\}, \{4, 5\}, \{1, 5, 6\}\}$$

for which the graph is

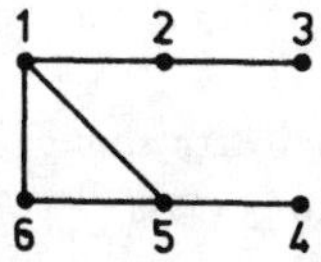

It turns out that 14 of the 4 ! = 24 possible orderings satisfy (4.1).

The description of the Markov property given by the graph seems more natural since it is immediate that the property does not involve an ordering of the elements of $\mathcal{C}$.

Theorem 2 in LSV also characterises decomposable graphs by a combinatorial property involving a certain counting index. Since this index is involved fundamentally in the estimation formula, we shall discuss this in the coming section.

5. The index and the estimation formula. Haberman (1974) introduces the *adjusted replication number* for subsets of sets in a generating class. In the decomposable case he shows that this number enters in the explicit formula for the maximum likelihood estimate $\hat{P}(\mathbf{i})$ of $P(\mathbf{i})$. In LSV a related quantity is defined. Whereas the adjusted replication number is defined recursively, this index is defined directly.

Let $\mathbf{C}$ be a connected graph $(C, E(\mathbf{C}))$ and $d \subseteq C$ be a complete subset. The *pieces* of C relative to d are defined as follows: remove d from C and form the subgraph $\mathbf{C} \setminus d$ with vertices $C \setminus d$ and edges which are those in $E(\mathbf{C})$ that do not involve vertices in d. $\mathbf{C} \setminus d$ now has one or more connected components $\mathbf{A}_t$, $t \in T$, say. Let $\mathbf{C}_t$ be the subgraphs of $\mathbf{C}$ obtained by readjoining d to the subgraphs $\mathbf{A}_t$, i.e., $\mathbf{C}_t$ has vertex set $A_t \cup d$ and edges which are those in $E(\mathbf{C})$ that only involve vertices in $A_t \cup d$. $\mathbf{C}_t$, $t \in T$ are *the pieces of C relative to d.*

Probably the procedure is best illustrated by an example:

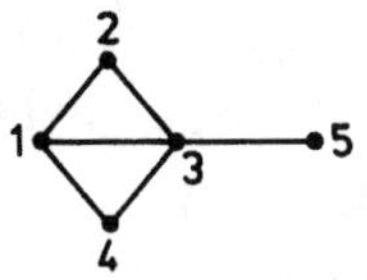

Consider this graph and let $d = \{3\}$. By removing d we get the following connected components:

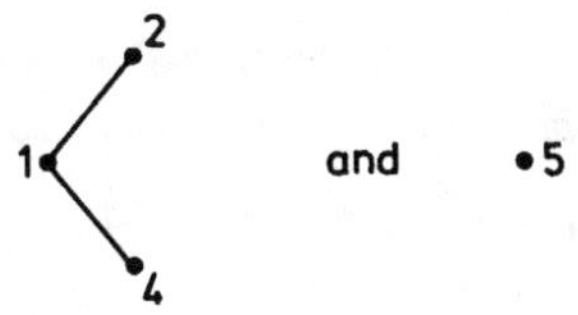

Readjoining d to these components we get the pieces:

For $d = \{1, 3\}$ we get components of $\mathbf{C} \setminus d$:

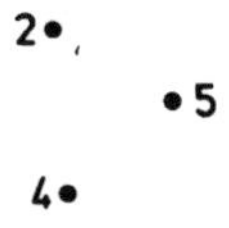

and thus pieces

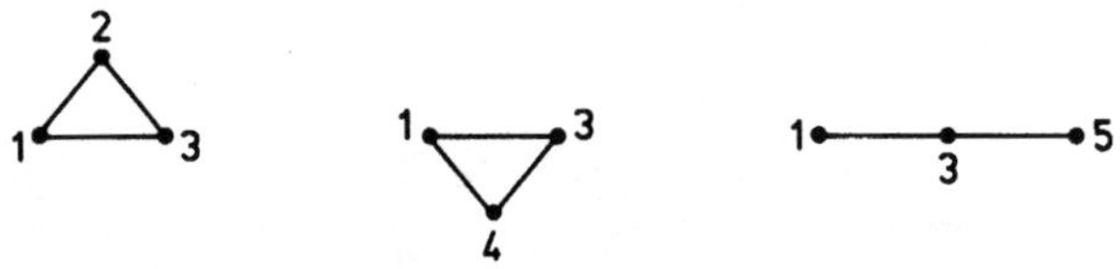

Clearly, since d was complete in $\mathbf{C}$, d is complete in all the pieces $\mathbf{C}_t$, but not necessarily a clique in $\mathbf{C}_t$ (i.e., maximal).

Let $\nu(d)$ be defined as

$$\nu(d) = 1 - \text{the number of pieces of } \mathbf{C} \text{ relative to } d \text{ in which } d \text{ is not a clique.}$$

In the example given above we have $\nu(\{3\}) = -1$, since $\{3\}$ is not a clique in any of the two pieces and $\nu(\{1, 3\}) = -1$ since $\{1, 3\}$ is a clique in 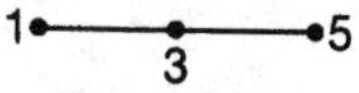 but not in the two remaining pieces.

Corollary 7 of LSV asserts that *for any connected graph* $\mathbf{C}$ *we have*

$$\Sigma_{d \text{ complete}} \nu(d) \geqslant 1$$

and Theorem 2 of LSV that $\mathbf{C}$ *is decomposable if and only if equality holds*. Thus we have a combinatorial identity characterising decomposable graphs.

If $\mathbf{C}$ is not connected itself but has connected components $\mathbf{C}_t$, $t \in T$ we define an index $\nu_t(d)$ for each of the components and have that $\mathbf{C}$ *is decomposable* iff

$$\Sigma_{t \in T} \Sigma_d \nu_t(d) = |T|,$$

which is an easy consequence of the inequality.

The index is primarily a tool for revealing combinatorial properties of decomposable graphs. However, it is worth noting that this index occurs in the estimation formula.

In a decomposable, and thus graphical model the maximum likelihood estimate $\hat{P}(\mathbf{i})$ *of* $P(\mathbf{i})$ *based upon n independent observations, is given by*

$$\hat{P}(\mathbf{i}) = \left[\Pi_{t \in T} \Pi_d n(\mathbf{i}_d)^{\nu_t(d)}\right] | n^{|T|},$$

provided that all $n(\mathbf{i}_d)$ *are positive.* (In this formula $\nu_t(d)$ is interpreted as zero if $d \nsubseteq \mathbf{C}_t$.)

To show this result we first realise that it is enough to consider connected graphs. For the various connected components correspond to independent sets of factors and their probabilities as well as their estimates multiply. Next we see that the formula is correct for a graph with just one clique. This is clear because such a graph corresponds to an unrestricted probability and in that case we have

$$\hat{P}(\mathbf{i}) = n(\mathbf{i})/n.$$

Noting that for such a graph we have $\nu(d) = 0$ unless $d = C$ in which case $\nu(d) = 1$, we see that our formula is correct in this case.

The final step in the proof is an induction argument using two basic facts:

(i) if a generating class $\mathcal{C}$ is decomposed into $\mathcal{A}$ and $\mathcal{B}$ such that $\mathcal{A} \cup \mathcal{B} = \mathcal{C}$, $\mathcal{A} \cap \mathcal{B} = \varnothing$ and $A \cap B = a^* \cap b^*$ for some $a^* \in \mathcal{A}$, $b^* \in \mathcal{B}$, where $A = \cup_{a \in \mathcal{A}} a$, $B = \cup_{b \in \mathcal{B}} b$, then

$$\hat{P}_{\mathcal{C}}(\mathbf{i}) = \frac{\hat{P}_{\mathcal{A}}(\mathbf{i}_A)\hat{P}_{\mathcal{B}}(\mathbf{i}_B)}{\hat{P}_{\{a^* \cap b^*\}}(\mathbf{i}_{a^* \cap b^*})},$$

which, e.g., follows directly from Theorem 2 of Andersen (1974);

(ii) if a generating class $\mathcal{C}$, where $\mathcal{C}$ is the maximal cliques of a connected graph **C** is decomposed as above, then both $\mathcal{A}$ and $\mathcal{B}$ are the cliques of the subgraphs **A** and **B**, these are both connected and the indices ν_A, ν_B and ν_C satisfy

$$\nu_C(d) = \nu_A(d) + \nu_B(d) \qquad \text{for } d \neq a^* \cap b^*$$

$$\nu_C(d) = \nu_A(d) + \nu_B(d) - 1 \quad \text{for } d = a^* \cap b^*.$$

This is Lemma 8 of LSV.

If we use these two facts and assume the result to be true for all graphical models with fewer than $|\mathcal{C}|$ cliques, we get

$$\hat{P}_{\mathcal{C}}(\mathbf{i}) = \frac{\hat{P}_{\mathcal{A}}(\mathbf{i}_A)\hat{P}_{\mathcal{B}}(\mathbf{i}_B)}{\hat{P}_{\{a^* \cap b^*\}}(\mathbf{i}_{a^* \cap b^*})} = \frac{\Pi_d n(\mathbf{i}_d)^{\nu_A(d)} \Pi_d n(\mathbf{i}_d)^{\nu_B(d)}}{n(\mathbf{i}_{a^* \cap b^*})} \Big/ n$$

$$= \Pi_d n(\mathbf{i}_d)^{\nu_C(d)} / n$$

where we again have let $\nu_A(d) = 0[\nu_B(d) = 0]$ if $d \nsubseteq A[d \nsubseteq B]$.

The estimation formula makes it possible for us to derive some further properties of our index. Let $n_\gamma = |I_\gamma|$ and suppose that we have $n = |I| = \Pi_{\gamma \in C} n_\gamma$ observations with exactly one observation in each cell, i.e., $n(\mathbf{i}) = 1$ for all $\mathbf{i}$. Then, clearly

$$\hat{P}(\mathbf{i}) = n^{-1}.$$

Using our formula for a connected graph C we also get

$$\hat{P}(\mathbf{i}) = n^{-1} \Pi_d n(\mathbf{i}_d)^{\nu(d)}$$

$$= n^{-1} \Pi_d (\Pi_{\gamma \notin d} n_\gamma)^{\nu(d)}$$

$$= n^{-1} \Pi_{\gamma \in C} \Pi_{d \not\ni \gamma} n_\gamma^{\nu(d)} = n^{-1} \Pi_{\gamma \in C} n_\gamma^{\Sigma_{d \subseteq C \setminus \{\gamma\}} \nu(d)}.$$

Since this expression is valid for all possible values of n_γ, we must have *for a connected, decomposable* graph **C**

$$\Sigma_{d \subseteq C \setminus \{\gamma\}} \nu(d) = 0 \quad \text{for all } \gamma \in C.$$

Since

$$\Sigma_d \nu(d) = 1 = \Sigma_{d \not\ni \gamma} \nu(d) + \Sigma_{d \ni \gamma} \nu(d),$$

we thus have, for all $\gamma \in \mathbf{C}$,

$$\Sigma_{d: \gamma \in d} \nu(d) = 1$$

for any connected, decomposable graph **C**.

A further identity is obtained by summation of the above identity for $\gamma \in C$:

$$|C| = \Sigma_{\gamma \in C} \Sigma_{d \ni \gamma} \nu(d) = \Sigma_d |d| \nu(d).$$

6. Graphical models of dimension less than or equal to five. Here, we shall give the graphical representation and the interpretation of all graphical models corresponding to an m-dimensional contingency table with $m \leqslant 5$. Apart from the interpretation column this is just a question of listing all graphs with less than five vertices. We do this both to illustrate the material in the previous sections and as a counterpart to the tables in Goodman (1970) of all hierarchical models of dimension $\leqslant 4$. We only list *connected* graphs since other models can be constructed by using these as connected components of other graphs. As remarked earlier, the various connected components in a graph of a graphical model correspond to independent sets of factors.

Giving the various interpretations in terms of conditional independence we shall use the notation of Goodman (1970), e.g.,

$$[1 \otimes 2|3]$$

meaning that, given 3, the factors 1 and 2 are conditionally independent. In Table 1 we list the decomposable graphical models and in Table 2 the nondecomposable models where we also indicate the critical $\geqslant 4$-cycle.

TABLE 1
Decomposable models of dimension less than or equal to five.

graph	interpretation
	unrestricted
	unrestricted
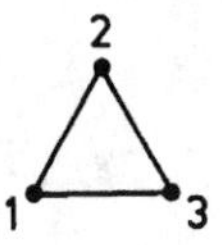	unrestricted
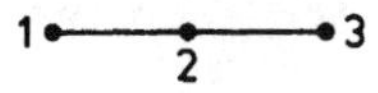	$[1 \otimes 3|2]$
	$[1 \otimes 3, 4|2] \cap [1, 2 \otimes 4|3]$
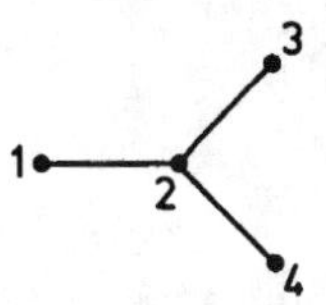	$[1 \otimes 3 \otimes 4|2]$
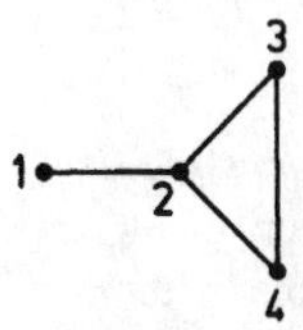	$[1 \otimes 3, 4|2]$

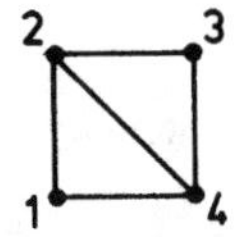

[1 ⊗ 3|2, 4]

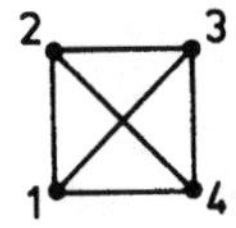

unrestricted

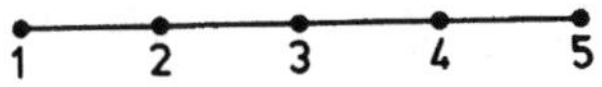

[1 ⊗ 3, 4, 5|2], etc.

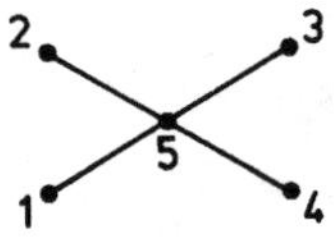

[1 ⊗ 2 ⊗ 3 ⊗ 4|5]

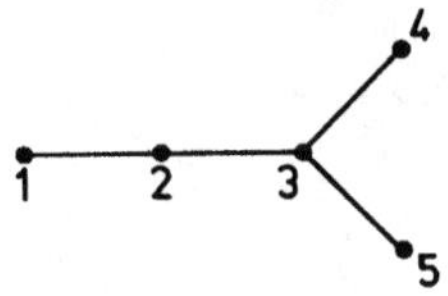

[1 ⊗ 3, 4, 5|2] ∩ [1, 2 ⊗ 4 ⊗ 5|3]

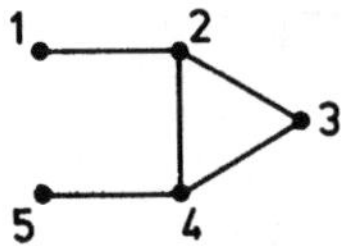

[1 ⊗ 5 ⊗ 3|2, 4] ∩ [1 ⊗ 3, 4, 5|2]
∩[5 ⊗ 1, 2, 3|4]

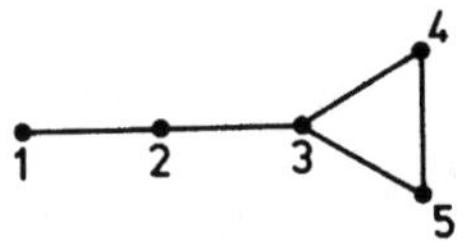

[1, 2 ⊗ 4, 5|3] ∩ [1 ⊗ 3, 4, 5|2]

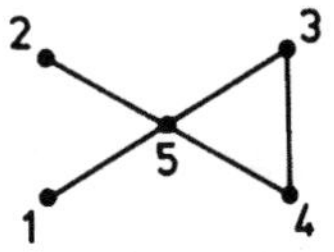

[1 ⊗ 2 ⊗ 3, 4|5]

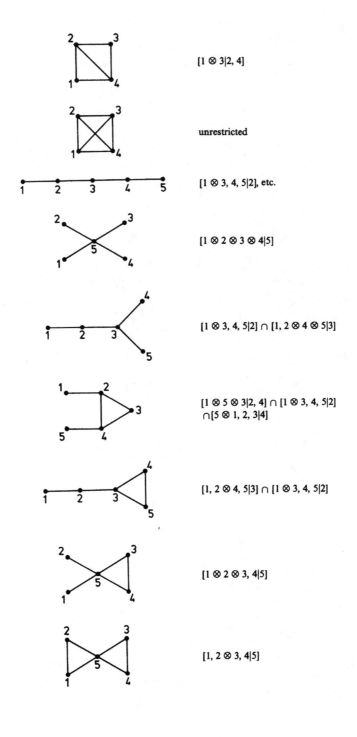

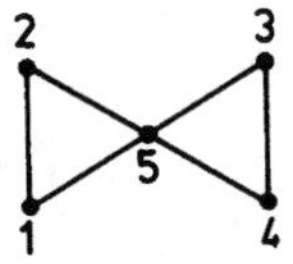

[1, 2 ⊗ 3, 4|5]

MARKOV FIELDS AND LOG-LINEAR MODELS

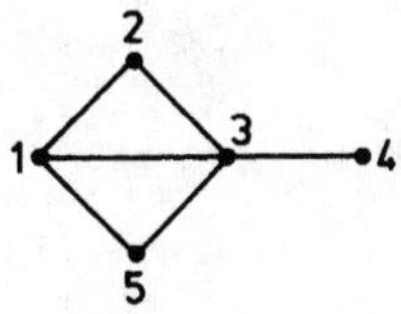

$[2 \otimes 5 \otimes 4|1, 3] \cap [1, 2, 5 \otimes 4|3]$

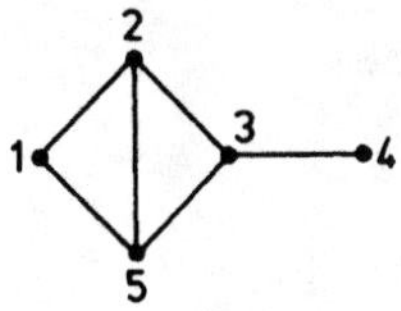

$[1, 2, 5 \otimes 4|3] \cap [1 \otimes 3, 4|2, 5]$

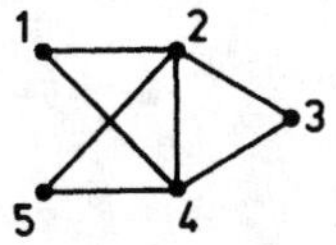

$[1 \otimes 3 \otimes 5|2, 4]$

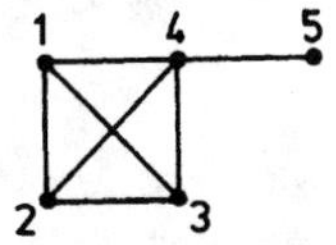

$[1, 2, 3 \otimes 5|4]$

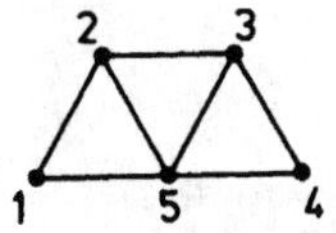

$[1 \otimes 3, 4|2, 5] \cap [1, 2 \otimes 4|3, 5]$

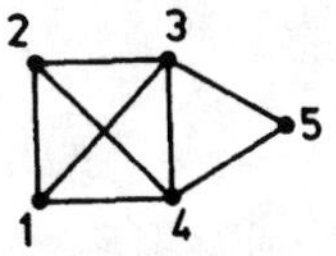

$[1, 2 \otimes 5|3, 4]$

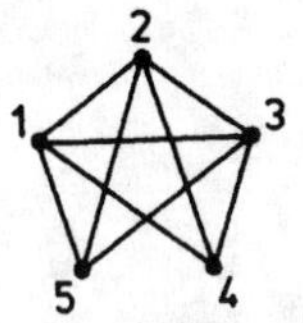

$[4 \otimes 5|1, 2, 3]$

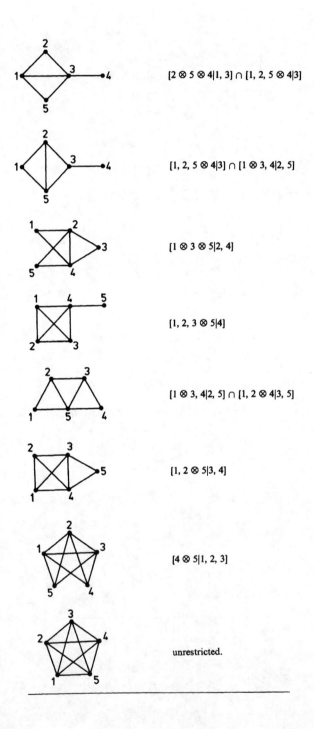

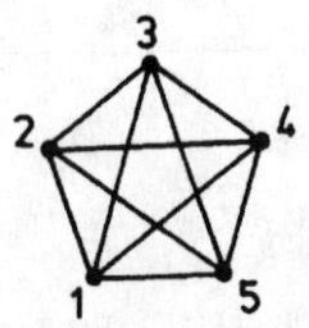

unrestricted.

TABLE 2

Nondecomposable models that are graphical of dimension less than or equal to five.

graph	> 4-cycle	interpretation
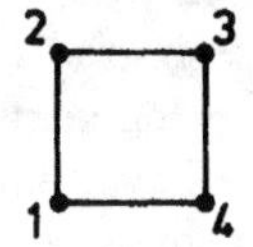	{1, 2, 3, 4}	[1 ⊗ 3|2, 4] ∩ [2 ⊗ 4|1, 3]
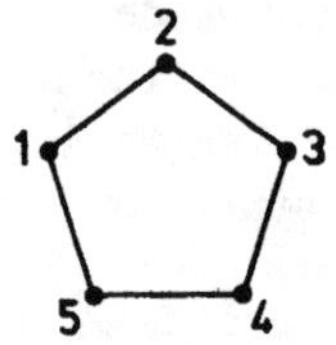	{1, 2, 3, 4, 5}	[1, 2 ⊗ 4|3, 5], etc.
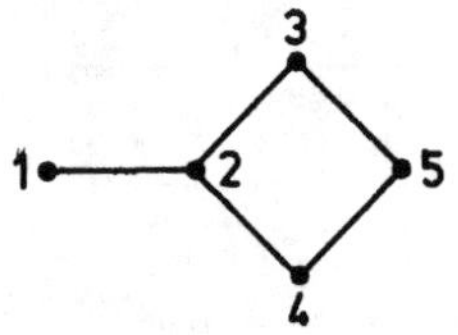	{2, 3, 4, 5}	[1, 2 ⊗ 5|3, 4] ∩ [1 ⊗ 3, 4, 5|2] ∩[3 ⊗ 1 ⊗ 4|2, 5]
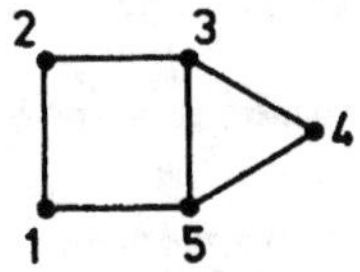	{1, 2, 3, 5}	[1, 2 ⊗ 4|3, 5] ∩ [1 ⊗ 3, 4|2, 5] ∩[2 ⊗ 4, 5|1, 3]
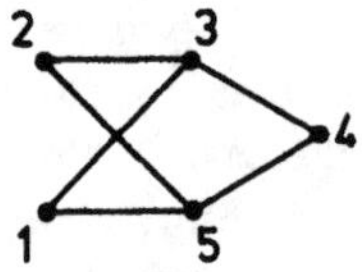	{1, 3, 4, 5} {2, 3, 4, 5} and {1, 2, 3, 5}	[1 ⊗ 2 ⊗ 4|3, 5] ∩[3 ⊗ 5|1, 2, 4]
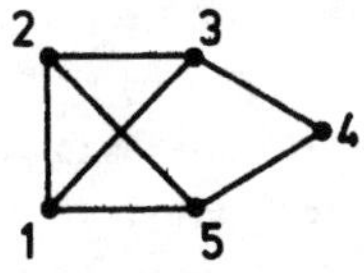	{1, 3, 4, 5} and {2, 3, 4, 5}	[1, 2 ⊗ 4|3, 5] ∩[3 ⊗ 5|1, 2, 4]
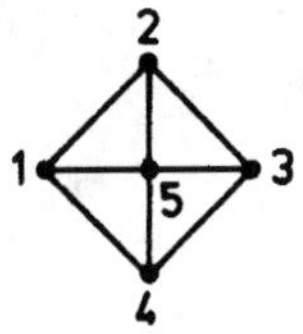	{1, 2, 3, 4}	[1 ⊗ 3|2, 4, 5] ∩[2 ⊗ 4|1, 3, 5].

Note that the last graph in Table 2 is *not* triangulated although it is made up by triangles. $\{1, 2, 3, 4\}$ is a cyclic subgraph without a chord. Thus the term "triangulated" is a bit misleading.

The interpretation column is made to give an interpretation in usual terms. Of course other conditional independence properties can be derived from those listed using rules of conditional independence. The most accurate interpretation will always be that the model consists of all Markov fields on the given graph.

To illustrate the complexity of the various types of models we have computed the number of possible models of any given type for a given contingency table of dimension $\leqslant 5$. The number of general log-linear interaction models is equal to 2^{2^n-1}. The number of graphical models is equal to $\sum_{i=0}^{n}\binom{n}{i}2^{\binom{i}{2}}$. The number of decomposable models does not seem to admit an explicit formula, but can be counted using the graphs in Tables 1 and 2. To count the number of hierarchical models is tedious for $n = 5$.

TABLE 3
Number of models of given type.

type \ dimension	1	2	3	4	5
Interaction	2	8	128	32,768	2,147,483,648
Hierarchical	2	5	19	167	7,580
Graphical	2	5	18	113	1,450
Decomposable	2	5	18	110	1,233

7. **Some final remarks.** Finally we shall give some suggestions as how to use the models and some possible directions for further work.

Searching for models. The graphical models are primarily relevant for the analysis of contingency tables of rather high dimension where it is difficult a priori to have very precise ideas about the relevant models and where one initially is looking for possible conditional independence among factors. We suggest that in such cases the graphs and their associated models be used directly in the search for possible models rather than the generating classes. It assures interpretability of any final model and it is in fact a very handy aid in visualising the features of the models. So, instead of trying gradually to remove interactions of high order, try to remove edges or throw in edges.

Estimation and test of hypotheses. At present, the graphs do not seem to be of great help in the numerical procedures of estimation and testing. There is something to be gained in discovering decomposability, thereby reducing the estimation problems. It might be the case that the graphs could be used in the estimation and testing problems. Consider for example the following model:

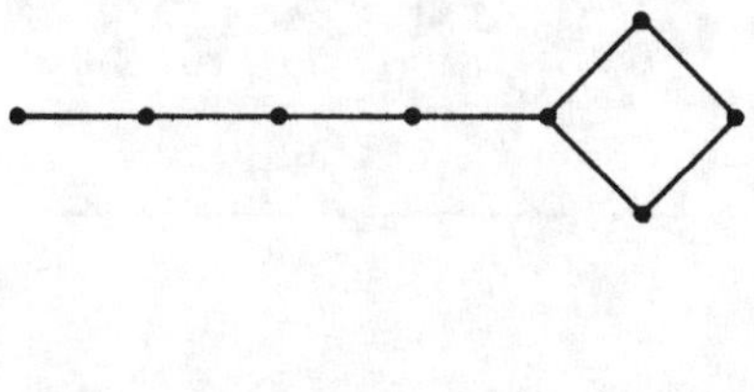

The model is not decomposable because of the 4-cycle to the right. On the other hand, the nondecomposability is isolated to that region. So, in fact, numerical iteration is only needed to find the marginal estimates in the table corresponding to these four factors. The estimate for the entire table can then be combined easily from this and an explicit formula for the marginal probability of the remaining factors using fact (i) in the proof of the basic estimation formula.

Similarly, we can get a simplification in a testing problem. Suppose that we want to find the likelihood ratio statistic for the hypothesis that the model

can be reduced to

Even though neither of the two models are decomposable, the difference between them is isolated to a decomposable region. Therefore, the likelihood ratio test statistic is nothing but that of testing independence in the two-way table involving the two factors at the left.

There is some work to be done in giving a good formulation of "local decomposability" and using such a notion in an efficient way in estimation and testing problems.

Exposition of the theory. Another possible use of the graphs is in an exposition of a theory of graphical models for contingency tables that uses the graphs *directly* instead of first relating these to generating classes and hierarchical models. This could have important pedagogical advantages.

We hope in the future to be able to give some more content to the vague remarks above.

Acknowledgments. We are grateful to M. L. Eaton, Minneapolis, for reading our manuscript and giving valuable suggestions.

REFERENCES

[1] ANDERSEN, A. H. (1974). Multidimensional contingency tables. *Scand. J. Statist.* **1** 115–127.

[2] BERGE, C. (1973). *Graphs and Hypergraphs*. Translated from French by E. Minieka. North-Holland, Amsterdam.

[3] GOODMAN, L. A. (1970). The multivariate analysis of qualitative data: Interaction among multiple classifications. *J. Amer. Statist. Assoc.* **65** 226–256.

MARKOV FIELDS AND LOG-LINEAR MODELS

[4] GOODMAN, L. A. (1971). Partitioning of chi-square, analysis of marginal contingency tables, and estimation of expected frequencies in multidimensional contingency tables. *J. Amer. Statist. Assoc.* **66** 339–344.

[5] HABERMAN, S. J. (1970). The general log-linear model. Ph. D. thesis, Depart. Statist. Univ. Chicago.

[6] HABERMAN, S. J. (1974). *The Analysis of Frequency Data*. IMS monographs, Univ. Chicago Press.

[7] KEMENY, J. G., SNELL, J. L. and KNAPP, A. W. (1976). *Denumerable Markov Chains 2nd edition*. Springer, Heidelberg, New York, Berlin.

[8] LAURITZEN, S. L., SPEED, T. P. and VIJAYAN, K. (1978). Decomposable graphs and hypergraphs. Preprint No. 9 Univ. Copenhagen, Inst. Math. Statist.

[9] PITMAN, J. W. (1976). Markov random fields. Lecture notes from a course given at the Univ. Copenhagen. Mimeographed.

[10] SPEED, T. P. (1976). Interaction. Unpublished manuscript.

SCHOOL OF MATH. SCIENCES
THE FLINDERS UNIVERSITY OF SOUTH AUSTRALIA
BEDFORD PARK, SOUTH AUSTRALIA 5042
AUSTRALIA 760511

DEPARTMENT OF MATHEMATICS
UNIVERSITY OF WESTERN AUSTRALIA
NEDLANDS, WESTERN AUSTRALIA 6009
AUSTRALIA

INSTITUTE OF MATHEMATICAL STATISTICS
UNIVERSITY OF COPENHAGEN
5 UNIVERSITETSPARKEN, DK-2100
COPENHAGEN Ø
DENMARK

The Annals of Statistics
1983, Vol. 11, No. 3, 724–738

ADDITIVE AND MULTIPLICATIVE MODELS AND INTERACTIONS

By J. N. Darroch and T. P. Speed

The Flinders University of South Australia and C.S.I.R.O.

A unified treatment is given of the classical additive models for complete factorial experiments and of multiplicative models and Lancaster-additive models for multi-dimensional contingency tables. The models are characterised by properties of being simplest subject to having a prescribed set of marginals. It is shown that, by using averaging operators and the notion of a generalised interaction, the interaction properties of these models can be derived very simply.

1. Introduction. Interaction models provide simplified structures for the arrays of unknown parameters which arise in *factorial experiments* and in multidimensional *contingency tables*. These two fields of application will be considered side by side, rather more attention being given to contingency tables.

In a factorial experiment there are, say, s factors $A_1, \cdots, A_s$ and a single response y. If the factors have $r_1, \cdots, r_s$ levels there are $r_1 \times \cdots \times r_s$ different combinations of levels called *cells*. The expected value $Ey = \eta$ of the response varies from cell to cell and inferential attention is focused on the array η of the $r_1 \times \cdots \times r_s$ values of η.

In a *pure response* s-dimensional contingency table there are s categorical variables $X_1, \cdots, X_s$ taking $r_1, \cdots, r_s$ values. This time the unknown parameter at each cell is the probability p of that particular combination of response values. The following discussion also applies to s-dimensional contingency tables in which some of the dimensions correspond to factors and the remainder to responses. The probability p is then the probability of the response values given the factor levels. There remains one further model for contingency tables. In it the $r_1 \times \cdots \times r_s$ frequencies are independent Poisson variables and the theory of this paper is applied to the array of their mean values μ.

The standard models for η, p, μ or some function of them are defined by *linear subspaces* of $\mathbb{R}^{\mathcal{I}}$, where

$$\mathcal{I} = \{i = (i_1, \cdots, i_s) : 1 \leq i_\sigma \leq r_\sigma, \sigma = 1, \cdots, s\}.$$

They are usually obtained by introducing a system of *interactions* and then requiring that a subset of these interactions vanish. This may be quite appropriate with additive models for factorial experiments, where the individual interactions can have a practical interpretation, but it is not necessarily so with multiplicative models for contingency tables. One of the aims of this paper is to give a simple account of an alternative approach in which we define models first (Section 2) and interactions later (Section 4). In doing so we take the opportunity to compare and contrast additive and multiplicative models, and to note the similarities and differences between two widely used parametrizations.

There is a certain amount of overlap in subject matter between this paper and the work of Haberman (1974, 1975) but the mathematical treatments of the common material are substantially different. Andersen (1975) gives a very clear summary of the general properties of interaction subspaces, applicable either to additive or multiplicative models, whilst other general treatments are by Mann (1949), Good (1958, 1963), Kurkjian and Zelen (1962), Grizzle, Starmer and Koch (1969), Goodman (1970) and Davidson (1973). Writings which concentrate upon multiplicative models for probabilities include several books: Haberman (1974, 1978, 1979), Bishop, Fienberg and Holland (1975), Fienberg (1977),

Received February 1980; revised December 1982.
AMS 1980 *subject classifications.* 62E10, 62E30.
Key words and phrases. Generalised interaction, Lancaster model, linear model log-linear model.

Gokhale and Kullback (1978), and Plackett (1981). Lancaster's theory of interaction and generalised correlation can be found in his book (1969), although the formulation given here (for finitely-valued random variables) is slightly different from his, being chosen to facilitate comparisons with other models. Further literature references are given in the body of the paper.

Inference matters are not discussed apart from a few comments on least-squares, sufficient reductions and maximum likelihood estimation. There is also no discussion of experimental design questions. A number of the results in this paper are new but in general the emphasis is on unifying existing results and on proving them by elementary methods.

2. Models and marginals. In this section we introduce the models which will be the main topic of the paper. The s factors or responses will be labeled by elements of $S = \{1, 2, \cdots, s\}$, subsets of which will be denoted by a, b, c, d. As in the introduction $\sigma \in S$ is supposed to have r_σ values (levels or response categories), and we write $\mathscr{I}$ for the set of cells i; precisely $\mathscr{I} = \{i = (i_\sigma) : i \leq i_\sigma \leq r_\sigma, \sigma \in S\}$. More generally we write i_a for the sub-tuple $i_a = (i_\sigma : \sigma \in a)$, $a \subseteq S$.

2.1 *The models.* Let $\mathscr{A}$ be a collection of subsets of S. The linear subspace $\Omega_{\mathscr{A}}$ of $\Omega = \mathbb{R}^{\mathscr{I}}$ is defined by the property that the function $f = (f(i) : i \in \mathscr{I})$ belongs to $\Omega_{\mathscr{A}}$ if and only if

$$f(i) = \sum_{a \in \mathscr{A}} \lambda_a(i_a) \tag{2.1}$$

for some functions $\{\lambda_a : a \in \mathscr{A}\}$. Having defined $\Omega_{\mathscr{A}}$, the *model* $M_{\mathscr{A}}$ for f is simply the property that f belongs to $\Omega_{\mathscr{A}}$. The collection $\mathscr{A}$ is called the *generating class* of the model. Given $\mathscr{A}$ let $\mathscr{A}^*$ denote the sub-collection of elements of $\mathscr{A}$ which are maximal with respect to inclusion. It is clear that $M_{\mathscr{A}^*}$ is the same model as $M_{\mathscr{A}}$ because if $b \subseteq a$, then $\lambda_a(i_a) + \lambda_b(i_b) = \mu_a(i_a)$. Whilst it is economical in practice to work with $\mathscr{A}^*$, the theory does not require us to do so.

EXAMPLE 2.1. All our examples will have $s \leq 4$ and for convenience we will write i, j, k and l instead of i_1, i_2, i_3 and i_4. Whenever no confusion is possible, we will use subscripts and omit the set describing the relevant indices. Thus we will write λ_{ijk} instead of $\lambda_{\{1,2,3\}}(i_1, i_2, i_3)$.

Suppose that $s = 3$ and $\mathscr{A} = \{\{1, 2\}, \{2, 3\}, \{3, 1\}\}$. Then $\Omega_{\mathscr{A}}$ consists of all arrays $f = (f_{ijk})$ representable in the form

$$f_{ijk} = \alpha_{ij} + \beta_{jk} + \gamma_{ki}$$

for some arrays (α_{ij}), (β_{jk}) and (λ_{ki}). □

Of the following interpretations of $M_{\mathscr{A}}$, the first is applicable mainly to a factorial experiment with observations $y = (y(i) : i \in \mathscr{I})$ and expected values $\eta = (\eta(i) : i \in \mathscr{I})$. The others are applicable to a contingency table with cell frequencies $n = (n(i) : i \in \mathscr{I})$ and probabilities $p = (p(i) : i \in \mathscr{I})$ or expected frequencies $\mu = (\mu(i) : i \in \mathscr{I})$.

Additive model: $\eta \in \Omega_{\mathscr{A}}$.
Multiplicative model: $\log p \in \Omega_{\mathscr{A}}$, p positive.
or $\log \mu \in \Omega_{\mathscr{A}}$, μ positive.
Lancaster-additive model I: $p/q \in \Omega_{\mathscr{A}}$.
Lancaster-additive model II: $P/Q \in \Omega_{\mathscr{A}}$.

Here the function $f = \log p$ is defined by $f(i) = \log p(i)$ whilst $f = p/q$ means $f(i) = p(i)/q(i)$ where $q(i) = p_1(i_1) \cdots p_s(i_s)$ is the product of the one-dimensional marginal probabilities from p. Finally $f = P/Q$ means $f(i) = P(i)/Q(i)$ where $P(i) = \sum_{j \leq i} p(i)$ and similarly for Q, where $j \leq i$ means $j_\sigma \leq i_\sigma$, $\sigma = 1, \cdots, s$. Additive and multiplicative models are commonly called *linear* and *log-linear* models, respectively. The general results below apply also to any generalised linear model; see Nelder and Wedderburn (1972), Baker and Nelder (1978).

Why should we study additive, multiplicative and Lancaster-additive models? In the first place, the way in which they combine linearity and economy has an obvious appeal. Less obvious is that they can be characterised by attractive properties relating them to their $\mathscr{A}$-marginal functions; these are given in the following section. Their best-known properties are the no-interaction ones by which they are usually characterised, and these are given in Section 4.

Suppose that f is known or assumed to satisfy $M_{\mathscr{A}}$ so that $f(i)$ is representable as the sum of parameters $\lambda_a(i_a)$. Leaving aside the trivial case when the generating class $\mathscr{A}$ contains only one element, it is always possible to choose more than one parametric representation of f. That is the parameters $\lambda_a(i_a)$ are not uniquely determined by f. The extent to which they are unique is discussed in Section 4.

Generally speaking, the parameters $\lambda_a(i_a)$ have little more than a mathematical existence but, on rare occasions, they also have a physical meaning.

EXAMPLE 2.2. Let i index the cities of a country, let j index age-categories of brides and let k index age-categories of bridegrooms. Let μ_{ijk} be the expected number of marriages, in a given year, in city i between brides of age j and bridegrooms of age k. Then

$$\mu_{ijk} = M_{ij} N_{ik} \rho_{ijk},$$

M_{ij}, N_{ik} being the numbers of eligible women of age j, men of age k in city i at the beginning of the year, and where ρ_{ijk} is the rate of marriages in city i between women of age j and men of age k. It may be very reasonable to assume that $\rho_{ijk} = \rho_{jk}$ so that

$$\log \mu_{ijk} = \log M_{ij} + \log \rho_{jk} + \log N_{ik}.$$

Thus we have an instance of the model of Example 2.1 in which the parameters (α_{ij}), (β_{jk}), (γ_{ik}) can be given a physical interpretation. □

2.2 *Marginals.* For an arbitrary element $\xi = (\xi(i) : i \in \mathscr{I}) \in \Omega$ and a subset $a \subseteq S$ we write $\xi_a(i_a) = \sum_{i_{a'}} \xi(i)$, the sum being over all $i_\sigma = 1, \cdots, r_\sigma$, $\sigma \in a' = S - a$, and call ξ_a the (unweighted) *a-marginal* of ξ. The $\mathscr{A}$-marginals of ξ are $\{\xi_a : a \in \mathscr{A}\}$. Now let $m(i)$ be a positive weight attached to cell i, where $\sum_i m(i) = 1$. It is necessary in much of what follows to work with weight functions different from the uniform weight function $m(i) = (\Pi_\sigma r_\sigma)^{-1}$. We define the ($m$-weighted) *a-marginal mean* $\bar{\eta}_a$ of $\eta \in \Omega$ by

$$\bar{\eta}_a(i_a) = \frac{1}{m_a(i_a)} \sum_{i_{a'}} m(i)\eta(i), \tag{2.2}$$

and the $\mathscr{A}$-*marginal* means of η are $\{\bar{\eta}_a : a \in \mathscr{A}\}$. The ($m$-weighted) *inner product* $\langle \xi, \eta \rangle_m$ of $\xi, \eta \in \Omega$ is defined by

$$\langle \xi, \eta \rangle_m = \sum_i m(i)\xi(i)\eta(i), \tag{2.3}$$

and its associated *norm* (length) is $\| \xi \|_m = \{\langle \xi, \xi \rangle_m\}^{1/2}$.

Additive models. In terms of these notions we can now characterise the additive model $M_{\mathscr{A}}$. We begin with a lemma.

LEMMA 2.1. *Fix $\xi, \eta_0 \in \Omega$ and consider the set of all η with the same $\mathscr{A}$-marginal means as η_0 and the squared distance $\| \eta - \xi \|_m^2$ of each such η from ξ. Suppose that, in this set, there exists η_1 satisfying $\eta_1 - \xi \in \Omega_{\mathscr{A}}$. Then η_1 uniquely minimizes $\| \eta - \xi \|_m^2$.*

PROOF. The condition that η_0 have the same $\mathscr{A}$-marginal means is $\eta - \eta_0 \perp_m \Omega_{\mathscr{A}}$, where orthogonality $\perp_m$ is with respect to the inner product (2.3). Therefore $\eta - \eta_1 \perp_m \Omega_{\mathscr{A}}$ since

MODELS AND INTERACTIONS

$\eta - \eta_0 \perp_m \Omega_{\mathscr{A}}$ and $\eta_1 - \eta_0 \perp_m \Omega_{\mathscr{A}}$. But $\eta_1 - \xi \in \Omega_{\mathscr{A}}$ and so $\langle \eta - \eta_1, \eta_1 - \xi \rangle_m = 0$. Rearrangement gives

$$\|\eta - \xi\|_m^2 - \|\eta_1 - \xi\|_m^2 = \|\eta - \eta_1\|_m^2$$

which establishes the truth of the Lemma. □

Note that if e is the unit function $e(i) = 1$ and $\xi = ke$, k constant, then $\xi \in \Omega_{\mathscr{A}}$ and it seems appropriate to describe ξ as *uniform*. The characterisation of the additive model can now be stated: any $\eta \in \Omega_{\mathscr{A}}$ is *simplest* in the sense that it is *closest* to being uniform amongst all arrays with the same $\mathscr{A}$-marginal means. Closeness is measured by $\|\cdot\|_m^2$ and simplest means that ξ is uniform. There is thus a separate characterisation for each positive weight function m.

The above discussion has not involved the question of existence, given η_0, ξ, of η_1 satisfying

$$\eta_1 - \eta_0 \perp_m \Omega, \quad \eta_1 - \xi \in \Omega_{\mathscr{A}} \tag{2.4}$$

but this question is well-known to have an affirmative answer. For $\eta_1 - \xi$ is the projection of $\eta_0 - \xi$ onto $\Omega_{\mathscr{A}}$ orthogonal with respect to $\langle \cdot, \cdot \rangle_m$; equivalently, $\eta_0 - \eta_1$ is the orthogonal projection of $\eta_0 - \xi$ onto $\Omega_{\mathscr{A}}^{\perp}$, the orthogonal complement of $\Omega_{\mathscr{A}}$.

Multiplicative models. The analogous characterisation of the multiplicative model $M_{\mathscr{A}}$, which is due to Good (1963) and Ku and Kullback (1968), closely resembles the previous one. Let the (unweighted) $\mathscr{A}$-marginals of the probability p be fixed at those of p_0 and measure the difference between p and a positive probability π by the Kullback discriminatory information

$$K(p, \pi) = \textstyle\sum_i p(i) \log p(i)/\pi(i) = \langle p, \log p/\pi \rangle \tag{2.5}$$

where $\langle \xi, \eta \rangle = \sum_i \xi(i)\eta(i)$ is the unweighted inner product.

LEMMA 2.2. *Suppose that, among all* p *with the same* $\mathscr{A}$*-marginals as* p_0*, there exists* p_1 *satisfying* $\log p_1/\pi \in \Omega_{\mathscr{A}}$. *Then* p_1 *uniquely minimises* $K(p, \pi)$.

PROOF. Since $p - p_0 \perp \Omega_{\mathscr{A}}$, $p_1 - p_0 \perp \Omega_{\mathscr{A}}$ and $\log p_1/\pi \in \Omega_{\mathscr{A}}$, we deduce that $\langle p - p_1, \log p_1/\pi \rangle = 0$. Rearranging this gives $\langle p, \log p/\pi \rangle - \langle p_1, \log p_1/\pi \rangle = \langle p, \log p/p_1 \rangle$, i.e. $K(p, \pi) - K(p_1, \pi) = K(p, p_1)$, from which the lemma follows. □

Taking π to be the uniform probability function gives the following characterisation: any p satisfying the multiplicative model $\log p \in \Omega_{\mathscr{A}}$ is simplest in the sense that it maximises $-\sum_i p(i) \log p(i)$ among all probabilities having the same $\mathscr{A}$-marginals. Assuming that $\cup \{a : a \in \mathscr{A}\} = S$, we may take $\pi = q_0$, the product of the one-dimensional marginals of p_0 and obtain the conclusion that any p satisfying the multiplicative model $M_{\mathscr{A}}$ is closest to being independent amongst all probabilities with the same $\mathscr{A}$-marginals, closeness being measured by K.

The existence of p_1 satisfying

$$p_1 - p_0 \perp \Omega_{\mathscr{A}}, \quad \log p_1/\pi \in \Omega_{\mathscr{A}} \tag{2.6}$$

is assured provided that the $\mathscr{A}$-marginals of p_0 admit a positive probability, see Haberman (1974), Barndorff-Nielsen (1978). Darroch and Ratcliff (1972) proved that, with this proviso and for any subspace ω of Ω, it is possible to construct p_1 given p_0, π and ω by generalised iterative scaling. When $\omega = \Omega_{\mathscr{A}}$ then iterative proportional scaling can be used.

Lancaster-additive Model I. The results concerning additive models can be adapted to provide a characterisation of the Lancaster-additive model I and because it is very similar to the two preceding ones, we only give a brief outline.

Suppose that the unweighted $\mathscr{A}$-marginals of p are held fixed at those of p_0, and that

$\cup \{a: a \in \mathscr{A}\} = S$. Then all of the univariate marginals p_σ are also held fixed and so too is $q = \Pi p_\sigma$, equal to q_0 say. If we put $m = q_0$ and $\eta = p/q = p/q_0$ in (2.2) we find that holding p_a fixed is equivalent to holding $\bar{\eta}_a$ fixed. With $\xi = e$ the difference $\|\eta - \xi\|_m^2$ simplifies to

$$\phi^2(p, q_0) = \textstyle\sum_i [p(i) - q_0(i)]^2/q_0(i), \tag{2.7}$$

the Pearson Chi squared measure of difference between p and q_0. Lemma 2.1 may be translated to apply in this context and using that we obtain the following characterisation: any p satisfying the Lancaster-additive model I with $\Omega_{\mathscr{A}}$ is simplest in the sense of being closest to independent among all probabilities having the same $\mathscr{A}$-marginals, closeness being measured by ϕ^2.

The equations that p_1 satisfies are

$$p_1 - p_0 \perp \Omega_{\mathscr{A}}, \quad p_1/q_0 \in \Omega_{\mathscr{A}} \tag{2.8}$$

where $\perp$ here denotes orthogonality with respect to the unweighted inner product. The existence of p_1 given p_0, that is, of a probability function having prescribed $\mathscr{A}$-marginals and satisfying the Lancaster additive model I is not now guaranteed; see Darroch (1974) for a counter-example when $\mathscr{A} = \{\{1, 2\}, \{2, 3\}, \{3, 1\}\}$, and for further comparisons between these models and the analogous multiplicative models.

2.3 *Fitting the models.* Let us suppose that data $y = (y(i): i \in \mathscr{I})$ from a factorial experiment has a normal distribution with mean $\eta \in \Omega_{\mathscr{A}}$ and covariance matrix $\sigma^2 \operatorname{diag}(m)^{-1}$, the diagonal matrix with value $m(i)^{-1}$ in the ith position. Then a *sufficient* reduction of y is to the pair $(\eta_1, \|\eta_1 - \eta_0\|_m^2)$ where η_1, the projection of $y = \eta_0$ onto $\Omega_{\mathscr{A}}$ orthogonal with respect to $\langle \cdot , \cdot \rangle_m$, satisfies (2.4) with $\xi = 0$. We have already seen that η_1 is completely determined by its $\mathscr{A}$-marginal means, and these coincide with those of y. If we further suppose that m is *completely multiplicative* in that it can be written

$$m(i) = k\, \Pi_\sigma m_\sigma(i_\sigma),$$

where for each $\sigma \in S$, $m_\sigma(i_\sigma) \geq 0$, $\sum_{i_\sigma} m_\sigma(i_\sigma) = 1$ and k is a constant, then we can express η_1 in terms of the $\mathscr{A}$-marginal means of y via formula (3.6) below. Thus (when m is completely multiplicative) the set of $\mathscr{A}$-marginal means is not only a sufficient reduction of y under the *additive model* $M_{\mathscr{A}}$, but also there is a closed-form solution of the least-squares (= maximum likelihood) estimation problem.

We turn now to the contingency table $n = (n(i): i \in \mathscr{I})$, supposing that n has a multinomial distribution with probability parameter p satisfying the *multiplicative model* $M_{\mathscr{A}}$ and total sample size $N = \sum_i n(i)$. The (unweighted) $\mathscr{A}$-marginal totals $\{n_a: a \in \mathscr{A}\}$ constitute a sufficient reduction of n and, provided these marginals admit a positive table, the log-likelihood $\langle n, \log p \rangle$ is maximised, or $K((1/N)n, p)$ is minimised, subject to $\log p/\pi \in \Omega_{\mathscr{A}}$ (normally π is uniform) when $p = p_1$ satisfies (2.6) with $p_0 = (1/N)n$. That these equations give the unique maximum likelihood solution is immediately verified on noting that $\langle \log p_1 - \log p, (1/N)n - p_1 \rangle = 0$ and on rearranging the term on the left-hand side of this equation to give $K((1/N)n, p) - K((1/N)n, p_1) = K(p_1, p)$. As was noted in 2.2 above, the equations (2.6) can be solved by the well-known iterative proportional scaling procedure.

To our knowledge there is no exact maximum-likelihood theory for the fitting of Lancaster-additive multinomial models to contingency tables, although a number of authors have discussed asymptotic theory for likelihood-ratio tests under the independence alternative, see Lancaster (1969) for details.

3. Generalised interactions. Denote by $M_{\mathscr{A}}$ the model for $f = (f(i): i \in \mathscr{I})$ defined by

$$M_{\mathscr{A}}: f \in \Omega_{\mathscr{A}}.$$

The function f will be variously interpreted as η, $\log p$, p/q or P/Q. In 3.2 below $M_{\mathcal{A}}$ will be formulated as imposing *zero generalised $\mathcal{A}$-interaction*, where generalised interactions are defined very simply by repeatedly averaging over the values $f(i)$ of f.

3.1 *Averaging operators.* Let w_σ be a weight function defined on $\{1, 2, \cdots, r_\sigma\}$, i.e. $\sum_{i_\sigma} w_\sigma(i_\sigma) = 1$. The numbers $w_\sigma(i_\sigma)$ will be thought of as non-negative although there is no strict need for them to be so. Write $S - \{\sigma\} = S - \sigma$. Then the *averaging operator* $T_{S-\sigma}$ operating on f is defined by

$$(T_{S-\sigma}f)(i) = \textstyle\sum_{i_\sigma} w_\sigma(i_\sigma) f(i).$$

Thus $T_{S-\sigma}$ takes weighted averages over the σth coordinate and leaves a function which depends on i through $i_{S-\sigma}$ only. For $a \subseteq S$ let T_a be the operator which takes averages over all coordinates with indices in $a' = S - a$. In other words,

$$T_a = \textstyle\prod_{\sigma \in a'} T_{S-\sigma}.$$

For example, if $S = \{1, 2, 3\}$, then $T_{\{1\}} = T_{S-2} T_{S-3} = T_{\{1,3\}} T_{\{1,2\}}$. When $a = S$ we define $T_S = I$, the *identity* operator. An alternative definition of T_a is possible via (2.2): $T_a \eta = \bar{\eta}_a$ where this average is weighted with respect to the completely multiplicative weight function $w(i) = \prod_{\sigma \in S} w_\sigma(i_\sigma)$. It is immediate that T_a is a linear operator on Ω, that $T_a^2 = T_a$ and, more generally that

$$T_a T_b = T_b T_a = T_{ab} \tag{3.1}$$

where for $a, b \subseteq S$ we write $a \cap b = ab$.

Two particular weight functions w are of special interest. One is the *uniform weight function* defined by

$$w_\sigma(i_\sigma) = 1/r_\sigma.$$

The other is the *substitution weight function* defined by

$$w_\sigma(i_\sigma) = \begin{cases} 0 & \text{if} \quad i_\sigma \neq r_\sigma, \\ 1 & \text{if} \quad i_\sigma = r_\sigma. \end{cases}$$

The resulting substitution operator T_a has the defining property $(T_a f)(i) = f(i_a r_{a'})$ where $j = i_a r_{a'}$ denotes the cell with $j_\sigma = i_\sigma$ if $\sigma \in a$ and $j_\sigma = r_\sigma$, if $\sigma \in a'$. Thus T_a substitutes r_σ for i_σ, $\sigma \in a'$. Of course any other fixed reference cell could be used instead of r. It will be convenient to denote $f(i_a r_{a'})$ by $f_a^r(i_a)$.

EXAMPLE 3.1. Let $s = 4$ and $a = \{1, 2\}$. When w is the uniform weight function the transformation $f \to T_a f$ replaces f_{ijkl} by $f_{ij\cdot\cdot}$ where, as usual, $\cdot$ denotes uniform average. When w is the substitution weight function f_{ijkl} is replaced under T_a by $f_{ijr_3r_4}$. □

Much of the theory in this paper is obtained using only the simple algebraic equipment of averaging operators. The same ground may be covered using sums and products of linear subspaces and their orthogonal projections. Little will be said about this approach here because it is part of this paper's aim to demonstrate the feasibility of the more elementary approach. It will suffice to show that T_a is an orthogonal projection operator.

We have already noted that $T_a^2 = T_a$ and so T_a is a projection operator. Since $T_a f = f$ iff $f(i) = \lambda(i_a)$ it follows that T_a projects onto the subspace Ω_a of Ω defined by this property. Further, T_a is self-adjoint with respect to $\langle \cdot, \cdot \rangle_w$ since

$$\begin{aligned} \langle f, T_a g \rangle_w &= \textstyle\sum_i w(i) f(i) [\sum_{i_{a'}} w_{a'}(i_{a'}) g(i)] \\ &= \textstyle\sum_{i_a} w_a(i_a) [\sum_{i_{a'}} (i_{a'}) f(i)][\sum_{i_{a'}} w_{a'}(i_{a'}) g(i)] = \langle T_a f, g \rangle_w. \end{aligned}$$

Finally, T_a is orthogonal with respect to $\langle \cdot , \cdot \rangle_w$ because $\langle (I - T_a)f, T_af\rangle_w = \langle T_a(I - T_a)f, f\rangle_w = \langle 0, f\rangle_w = 0$.

3.2 *Zero generalised interaction.* Given a generating class $\mathscr{A}$ of subsets of S, define the *generalised $\mathscr{A}$-interaction operator* $I - T_{\mathscr{A}}$ by

$$I - T_{\mathscr{A}} = \Pi_{a\in\mathscr{A}}(I - T_a). \tag{3.2}$$

By (3.1) the terms on the right-hand side of (3.2) can be multiplied together in any order and so, on expanding it, we find

$$T_{\mathscr{A}} = \textstyle\sum_a T_a - \sum\sum_{a\neq b} T_{ab} + \cdots \mp T_{\cap\mathscr{A}} \tag{3.3}$$

where the sums are over all $a \in \mathscr{A}$, distinct pairs $a, b \in \mathscr{A}$, etc. Another useful expression for $T_{\mathscr{A}}$ results from ordering the elements of $\mathscr{A}$ as $a_1, a_2, \cdots, a_m$, namely

$$T_{\mathscr{A}} = T_{a_1} + (I - T_{a_1})T_{a_2} + \cdots + \Pi_{l<m}(I - T_{a_l})T_{a_m}. \tag{3.4}$$

PROPOSITION 3.1. *The function f satisfies $M_{\mathscr{A}}$ if and only if*

$$T_{\mathscr{A}}f = f. \tag{3.5}$$

PROOF. If f satisfies $M_{\mathscr{A}}$ then for some functions $\{\lambda_a : a \in \mathscr{A}\}$ we can write $f = \sum_{a\in\mathscr{A}}\lambda_a$. Now $(I - T_a)\lambda_a = 0$ for each $a \in \mathscr{A}$, and so it follows that $\Pi_{a\in\mathscr{A}}(I - T_a)\sum_{a\in\mathscr{A}}\lambda_a = 0$; that is, $T_{\mathscr{A}}f = f$.

Conversely, if $T_{\mathscr{A}}f = f$ then, by (3.4),

$$f = T_{a_1}f + (I - T_{a_1})T_{a_2}f + \cdots + \Pi_{l<m}(I - T_{a_l})\cdot T_{a_m}f$$

which is of the form $\sum_{a\in\mathscr{A}}\lambda_a$. □

Since the $\{T_a\}$ are orthogonal projections onto the subspaces $\{\Omega_a\}$, it follows that $T_{\mathscr{A}}$ is the orthogonal projection onto $\Omega_{\mathscr{A}} = \sum_{a\in\mathscr{A}}\Omega_a$, although we do not use this fact in what follows.

The proposition formulates $M_{\mathscr{A}}$ as imposing zero generalised $\mathscr{A}$-interaction, in that $(I - T_{\mathscr{A}})f = 0$.

As foreshadowed in Section 2.3 above, when the weight function is completely multiplicative we have an explicit formula for an element satisfying the *additive model* $M_{\mathscr{A}}$ in terms of its $\mathscr{A}$-marginal means, namely

$$\eta = \textstyle\sum_a \bar{\eta}_a - \sum\sum_{a\neq b} \bar{\eta}_{ab} + \cdots \mp \bar{\eta}_{\cap\mathscr{A}}. \tag{3.6}$$

This result is an immediate consequence of (3.3) as soon as we recall that $T_a\eta = \bar{\eta}_a$. Using the substitution weight function we obtain the following special case of (3.6).

$$\eta(i) = \textstyle\sum_a \eta^r_a(i_a) - \sum\sum_{a\neq b} \eta^r_{ab}(i_{ab}) + \cdots \mp \eta^r_{\cap\mathscr{A}}(i_{\cap\mathscr{A}}).$$

From $(I - T_{\mathscr{A}})\log p = 0$ when $T_{\mathscr{A}}$ is based upon the substitution weight function, the multiplicative model is seen to be expressible as

$$\frac{p(i)\cdot \Pi\Pi_{a\neq b}\, p^r_{ab}(i_{ab}) \quad \cdots}{\Pi_a p^r_a(i_a)\cdot \Pi\Pi\Pi_{a\neq b\neq c}\, p^r_{abc}(i_{abc}) \quad \cdots}\,[p^r_{\cap\mathscr{A}}(i_{\cap\mathscr{A}})]^{\pm 1} = 1. \tag{3.7}$$

The left-hand side of (3.7) is a *generalised cross product ratio.*

EXAMPLE 3.2. As in Example 2.1 let $\mathscr{A} = \{\{1, 2\}, \{2, 3\}, \{3, 1\}\}$. Then (3.3) becomes

$$T_{\mathscr{A}} = T_{\{1,2\}} + T_{\{2,3\}} + T_{\{3,1\}} - T_{\{1\}} - T_{\{2\}} - T_{\{3\}} + T_{\phi}.$$

Using the uniform weight function, (3.6) expresses $M_{\mathscr{A}}$ as the familiar

$$\eta_{ijk} = \eta_{ij\cdot} + \eta_{\cdot jk} + \eta_{i\cdot k} - \eta_{i\cdot\cdot} - \eta_{\cdot j\cdot} - \eta_{\cdot\cdot k} + \eta_{\cdot\cdot\cdot}$$

while (3.7) becomes the equally familiar cross-product ratio formulation of no three-dimensional interaction, namely

$$\frac{p_{ijk}p_{ir_2r_3}p_{r_1jr_3}p_{r_1r_2k}}{p_{ijr_3}p_{r_1jk}p_{ir_2k}p_{r_1r_2r_3}} = 1.$$

□

Alternative formulations of the *Lancaster-linear models* $M_{\mathscr{A}}: p/q \in \Omega_{\mathscr{A}}$ and $P/Q \in \Omega_{\mathscr{A}}$, will now be given. First choose $w_\sigma = p_\sigma$. Then

$$T_a \frac{p(i)}{q(i)} = \textstyle\sum_{i_{a'}} q_{a'}(i_{a'}) \frac{p(i)}{q(i)} = \frac{1}{q_a(i_a)} \sum_{i_{a'}} p(i) = \frac{p_a(i_a)}{q_a(i_a)}.$$

Applying (3.3) the Lancaster-additive Model I is seen to be expressible was

(3.8) $$\frac{p}{q} = \textstyle\sum_a \frac{p_a}{q_a} - \sum\sum_{a \neq b} \frac{p_{ab}}{q_{ab}} + \cdots \mp \frac{p_{\cap\mathscr{A}}}{q_{\cap\mathscr{A}}}.$$

Turning now to the Lancaster-additive model II, let T_a be based on the substitution weight function. Then

$$T_a \frac{P(i)}{Q(i)} = \frac{P(i_a r_{a'})}{Q(i_a r_{a'})} = \frac{P_a(i_a)}{Q_a(i_a)}.$$

Consequently the model here is

(3.9) $$\frac{P}{Q} = \textstyle\sum_a \frac{P_a}{Q_a} - \sum\sum_{a \neq b} \frac{P_{ab}}{Q_{ab}} + \cdots \mp \frac{P_{\cap\mathscr{A}}}{Q_{\cap\mathscr{A}}}.$$

It is now easy to see that the two Lancaster-additive models are equivalent. After multiplication of (3.8) by $q(i)$ and (3.9) by $Q(i)$, each term in (3.9) is seen to be the distribution function of the corresponding term in (3.8).

3.3 *Marginals and generalised interactions.* A by-product of the model characterisations of 2.2 above is that, given $f \in \Omega_{\mathscr{A}}$, where f is η, $\log p$, p/q or P/Q, f is uniquely determined by its $\mathscr{A}$-marginals, suitably interpreted as weighted means or unweighted sums. This is a special case of the result which we now prove that given its $\mathscr{A}$-marginals and its generalised $\mathscr{A}$-interaction, f is uniquely determined.

There is almost nothing in the proof for η, p/q, P/Q. Thus, defining $T_{\mathscr{A}}$ with respect to any completely multiplicative weight function w, we can write $\eta = T_{\mathscr{A}}\eta + (I - T_{\mathscr{A}})\,\eta$ as the sum of the expansion (3.6), involving its $\mathscr{A}$-marginals, and its generalised $\mathscr{A}$-interaction. Similarly for p/q, except that we now define $T_{\mathscr{A}}$ with respect to $w = q$ and use (3.8), and for P/Q where the substitution operators are used.

There is no explicit demonstration of this uniqueness result for $\log p$ and it has to be proved using Lemma 2.2. Let us suppose that p is a positive probability and that $(I - T_{\mathscr{A}})\log p = u$. Define $\pi = k \exp u$ where k is the normalising constant making $\sum_i \pi(i) = 1$. Then $T_{\mathscr{A}}\log p/\pi = T_{\mathscr{A}}(\log p - \log k - u) = T_{\mathscr{A}}\log p - \log k = \log p - \log k - u$ by the definition of u and the fact that $T_{\mathscr{A}}u = 0$. But this means that $\log p/\pi \in \Omega_{\mathscr{A}}$ and by Lemma 2.2 there is only one p with this property having given $\mathscr{A}$-marginal sums, provided only that these marginals admit a positive probability.

A postscript on this result is the following: it does not matter which (completely multiplicative) weight function w is used to define the generalised $\mathscr{A}$-interaction function $(I - T_{\mathscr{A}})f$ because $(I - T_{\mathscr{A}})f$ defined with respect to one weight function is recoverable from $(I - T_{\mathscr{A}})f$ defined with respect to another. For, if $\{T_a\}$ and $\{\tilde{T}_a\}$ are defined with respect to w and $\tilde{w}$, we see from $\tilde{T}_{\mathscr{A}}T_a = T_a$, $a \in \mathscr{A}$, and (3.3) that $\tilde{T}_{\mathscr{A}}T_{\mathscr{A}} = T_{\mathscr{A}}$, i.e. that

$$(I - \tilde{T}_{\mathscr{A}})(I - T_{\mathscr{A}})f = (I - \tilde{T}_{\mathscr{A}})f.$$

Incidentally, this identity shows directly why $T_{\mathscr{A}}f = f$ iff $\tilde{T}_{\mathscr{A}}f = f$, a fact implicit in Proposition 3.1.

4. Interactions.

4.1 *Interaction operators.* In the previous section we saw that, given a weight function w and averaging operators T_a, the operators $T_{\mathscr{A}}$ and $I - T_{\mathscr{A}}$ arise naturally from consideration of the model $M_{\mathscr{A}}$. In the particular case $\mathscr{A} = \{S - \sigma: \sigma \in S\}$ the operator $I - T_{\mathscr{A}} = \Pi_{\sigma \in S}(I - T_{S-\sigma})$ will be denoted by U_S and called the S-interaction operator. Thus

$$U_S = \Pi_{\sigma \in S}(T_S - T_{S-\sigma}).$$

The definition is now extended to cover any subset b of S. Define $U_\phi = T_\phi$ and, otherwise

$$U_b = \Pi_{\sigma \in b}(T_b - T_{b-\sigma}).$$

The operator U_b will be called the b-interaction operator. Alternative ways of writing it are easily seen to be

$$U_b = \Pi_{\sigma \in b}(I - T_{b-\sigma}).\, T_b, \tag{4.1}$$

$$U_b = \Pi_{\sigma \in b}(I - T_{S-\sigma}).\, \Pi_{\sigma \in b'} T_{S-\sigma}, \tag{4.2}$$

$$U_b = \textstyle\sum_{c \subseteq b} (-1)^{|b-c|} T_c. \tag{4.3}$$

EXAMPLE 4.1. Again let $s = 3$. The interaction operator $U_{\{1,2,3\}}$ is identical to the operator $I - T_{\mathscr{A}}$, with $\mathscr{A} = \{\{1, 2\}, \{2, 3\}, \{3, 1\}\}$, discussed in Example 3.2. The interaction operator $U_{\{1,2\}}$ is expressible in various ways as

$$U_{\{1,2\}} = (I - T_{\{2\}})(I - T_{\{1\}})T_{\{1,2\}} = (I - T_{\{2,3\}})(I - T_{\{1,3\}})T_{\{1,2\}}$$

$$= T_{\{1,2\}} - T_{\{1\}} - T_{\{2\}} + T_\phi.$$

Thus, for the uniform weight function,

$$(U_{\{1,2\}}\eta)_{ijk} = \eta_{ij.} - \eta_{i..} - \eta_{.j.} + \eta_{...}\,. \quad \square$$

Interactions are usually introduced recursively and their recursive structure is clearly seen in the interaction operators. For example, when $s = 3$,

$$U_{\{1,2,3\}} = (I - T_{\{2,3\}})(I - T_{\{1,3\}}) - (I - T_{\{2,3\}})(I - T_{\{1,3\}})T_{\{1,2\}}.$$

The second term on the right side is $U_{\{1,2\}}$ and gives $\{1, 2\}$ interactions averaged over k. The first term gives $\{1, 2\}$ interactions within each level k. Thus $\{1, 2, 3\}$ interactions are clearly seen to be differences of $\{1, 2\}$ interactions.

Some basic results about interaction operators are collected together in the following lemma.

LEMMA 4.1 (i) $T_a U_b = 0$ *if* $b \not\subseteq a$.

(ii) $\sum_{i_\sigma} w_\sigma(i_\sigma) U_b f(i) = 0$ *if* $\sigma \in b$.

(iii) $T_a U_b = U_b$ *if* $b \subseteq a$.

(iv) $\sum_{b \subseteq a} U_b = T_a$.

(v) $U_b^2 = U_b$.

(vi) $U_a U_b = 0$ *if* $a \neq b$.

(vii) *Let* $b_1, \cdots, b_m$ *be distinct sets. Then*

$\sum_j k_j U_{b_j} f = 0$ *implies that* $k_j U_{b_j} f = 0$ *for all* j.

(viii) U_b *is self-adjoint with respect to the inner product* $\langle \cdot, \cdot \rangle_w$.

MODELS AND INTERACTIONS

Proof (i) Choose $\tau \in b - a$. Since $T_a = \Pi_{\sigma\in a'} T_{S-\sigma}$, it follows that $T_{S-\tau}(I - T_{S-\tau})$ is a factor of $T_a U_b$.
(ii) By (i) $T_{S-\sigma} U_b = 0$, $\sigma \in b$.
(iii) Apply (4.1) and (3.1).
(iv) First consider $a = S$. By (4.2) $\sum_{b\subseteq S} U_b = \sum_{b\subseteq S} [\Pi_{\sigma\in b}(I - T_{S-\sigma}) \Pi_{\sigma\in b'} T_{S-\sigma}]$ $= \Pi_{\sigma\in S}[(I - T_{S-\sigma}) + T_{S-\sigma}] = \Pi_{\sigma\in S} I = I$. Having established that $\sum_{b\subseteq S} U_b = I$, we now multiply by T_a to get $\sum_{b\subseteq S} T_a U_b = T_a$. Application of (i) and (iii) now gives (iv).
(v) U_b is a product of idempotent operators which commute and hence is itself idempotent.
(vi) Choose $\tau \in (b - a) \cup (a - b)$ and reason as in the proof of (i).
(vii) Multiply $\sum_j k_j U_{b_j} f = 0$ by U_b and apply (v) and (vi).
(viii) By (4.3) U_b is a linear combination of operators which are self-adjoint. □

We note that U_b is an orthogonal projection operator because it is idempotent and self-adjoint. Further $U_b f(i) = g(i_b)$ say and for each $\sigma \in b$, $\sum_{i_\sigma} w_\sigma(i_\sigma) g(i_b) = 0$. Moreover, if f is a function satisfying (a) $f(i) = h(i_b)$ and (b) $\sum_{i_\sigma} w_\sigma(i_\sigma) f(i) = 0$ for all $\sigma \in b$, then, by (4.1), $U_b f = f$. Thus U_b is the orthogonal projection operator onto the subspace Θ_b of all functions satisfying (a) and (b), although we will not use this interpretation in the sequel.

4.2 *Hierarchical no-interaction models.* Let the *closure* $\bar{\mathscr{A}}$ of a generating class $\mathscr{A}$ be defined by

$$\bar{\mathscr{A}} = \{b : b \subseteq a \quad \text{for some} \quad a \in \mathscr{A}\}.$$

The complement of $\bar{\mathscr{A}}$ is

$$\bar{\mathscr{A}}' = \{b : b \not\subseteq a \quad \text{for all} \quad a \in \mathscr{A}\}.$$

Note that the class $\bar{\mathscr{A}}'$ is hierarchical. That is, if $b_1 \in \bar{\mathscr{A}}'$ and $b_2 \supseteq b_1$, then $b_2 \in \bar{\mathscr{A}}'$.

Proposition 4.1. $T_{\mathscr{A}} = \sum_{b\in\bar{\mathscr{A}}} U_b$.

Proof. It is easier to prove that

$$I - T_{\mathscr{A}} = \sum_{b\in\bar{\mathscr{A}}'} U_b \tag{4.4}$$

from which the proposition follows. But this is a direct consequence of our definitions and Lemma 4.1. For

$$\begin{aligned} I - T_{\mathscr{A}} &= \Pi_{a\in\mathscr{A}}(I - T_a) \text{ by the definition (3.2)} \\ &= \Pi_{a\in\mathscr{A}}(\textstyle\sum_{b\not\subseteq a} U_b) \text{ by (iv) of Lemma 4.1} \\ &= \textstyle\sum_{b\in\bar{\mathscr{A}}'} U_b \text{ by (v) and (vi) of Lemma 4.1,} \end{aligned}$$

and the definition of $\bar{\mathscr{A}}'$. □

Thus the model $M_{\mathscr{A}}$ for f may now be expressed as

$$f(i) = \sum_{b\in\bar{\mathscr{A}}} U_b f(i) \tag{4.5}$$

or as

$$U_b f(i) = 0 \quad \text{for all} \quad b \in \bar{\mathscr{A}}'. \tag{4.6}$$

Formula (4.5) follows immediately from Proposition 4.1 and formula (4.6) by application of (iv) with $a = S$ and (vii) of Lemma 4.1. By virtue of (4.6), $M_{\mathscr{A}}$ may be called a *hierarchical no-interaction model.* Proposition 4.1 thus provides the link with the more common approach to models and interactions which starts with interactions and then defines models by requiring that a hierarchical set of interactions are zero.

Models with equal sized generating sets are frequently used in searches for parsimonious fits to data and, for such models, there is a simple formula relating $T_{\mathscr{A}}$ to $\{T_b: b \in \bar{\mathscr{A}}\}$.

EXAMPLE 4.2. Let $s = 5$ and consider $\mathscr{A} = \{12, 13, 14, 15, 23, 24, 25, 34, 35, 45\}$ where 12 denotes $\{1, 2\}$ etc. We shall prove that

$$T_{\mathscr{A}} = [T_{12} + \cdots + T_{45}] - 3[T_1 + \cdots + T_5] + 6T_\phi. \quad \square$$

The general result is given in Proposition 4.2 below. It really belongs in Section 3 but its proof uses results of this section.

PROPOSITION 4.2. *For* $0 \le t < s$ *let*

$$\mathscr{A} = \mathscr{A}_t^s = \{a \subseteq S: |a| = t\}.$$

Then

$$T_{\mathscr{A}} = \sum_{u=0}^t (-1)^{t-u} \binom{s-u-1}{t-u} \sum_{b:|b|=u} T_b.$$

PROOF.

$$\begin{aligned}
T_{\mathscr{A}} &= \sum_{b:|b|\le t} U_b = \sum_{b:|b|\le t} \Pi_{\sigma\in b}(I - T_{S-\sigma})\, \Pi_{\sigma\in b'} T_{S-\sigma} \\
&= \sum_{u=0}^t \text{coefficient of } z^u \text{ in } \Pi_{\sigma\in S}[z(I - T_{S-\sigma}) + T_{S-\sigma}] \\
&= \text{coefficient of } z^t \text{ in } (1-z)^{-1}\, \Pi_{\sigma\in S}[zI + (1-z)T_{S-\sigma}] \\
&= \text{coefficient of } z^t \text{ in } (1-z)^{-1} \sum_{u=0}^s z^u (1-z)^{s-u} \sum_{b:|b|=u} T_b \\
&= \text{coefficient of } z^t \text{ in } \sum_{u=0}^t z^u (1-z)^{s-u-1} \sum_{b:|b|=u} T_b \\
&= \sum_{u=0}^t (-1)^{t-u} \binom{s-u-1}{t-u} \sum_{b:|b|=u} T_b. \quad \square
\end{aligned}$$

4.3 *Dimensions of models.* Let us denote the rank of a linear operator T by $r(T)$, and the dimension of a subspace ω of Ω by $\dim \omega$. The following are immediate consequences of the relevant definitions.

$$\dim \Omega_a = r(T_a) = \Pi_{\sigma\in a} r_\sigma.$$

$$\dim \Theta_b = r(U_b) = \Pi_{\sigma\in b}(r_\sigma - 1).$$

Our next result is an immediate consequence of Propositions 3.1, 4.1, and the linearity of trace, as soon as we recall that $r(P) = \text{trace}\,(P)$ for a projection operator P.

PROPOSITION 4.3. (i) *For a generating class* $\mathscr{A}$

$$\dim \Omega_{\mathscr{A}} = \sum_a \Pi_{\sigma\in a} r_\sigma - \sum_{a\ne b}\sum \Pi_{\sigma\in ab} r_\sigma + \cdots \mp \Pi_{\sigma\in\cap\mathscr{A}} r_\sigma = \sum_{b\in\bar{\mathscr{A}}} \Pi_{\sigma\in b}(r_\sigma - 1).$$

(ii) *For any t satisfying* $0 \le t < |S| = s$

$$\dim \Omega_{\mathscr{A}_t^s} = \sum_{u=0}^t (-1)^{t-u} \binom{s-u-1}{t-u} \sum_{b:|b|=u} \Pi_{\sigma\in b} r_\sigma.$$

4.4 *Discussion.* As an illustration of the use of the interaction operators with the *additive model* $M_{\mathscr{A}}$, consider the following simple method of deriving the least-squares estimates of the interactions $U_b\eta(i)$ of η when we have data $y = (y(i,j): j = 1, \cdots, n(i), i \in \mathscr{I})$ with $n(i)$ observations made on cell i, and the cell frequencies $n(i)$ are proportional, that is, completely multiplicative

$$n(i) = \Pi_\sigma n_\sigma(i_\sigma)/N^{s-1} \tag{4.7}$$

MODELS AND INTERACTIONS

where $N = \sum_i n(i)$. Condition (4.7) is of course most likely to be realised in practice when $n(i)$ is constant. If we denote the mean of $y(i, j)$ over j by $y(i)$ then the sum of squared deviations of the observations from their expectations is $\sum_{i,j} (y(i, j) - \eta(i))^2 = \sum_{i,j} (y(i, j) - y(i))^2 + \sum_i n(i)(y(i) - \eta(i))^2$, so that the least value has to be found of

$$\text{(4.8)} \qquad \sum_i n(i)(y(i) - \eta(i))^2 = \sum_{b\subseteq S} \sum_{i_b} n_b(i_b)(U_b y(i) - U_b\eta(i))^2.$$

Identity (4.8) follows from the calculation

$$\langle z, z\rangle_w = \langle z, \sum_b U_b z\rangle_w = \sum_b \langle z, U_b^2 z\rangle_w = \sum_b \langle U_b z, U_b z\rangle_w,$$

using Lemma 4.1 (iv), (v) and (viii), with $z = y - \eta$ and $w(i) = n(i)/N$. Identity (4.8) shows that, for any no-interaction model (hierarchical or not), the least squares estimate of $U_b\eta$ is $U_b y$ (when the cell frequencies are proportional) for every model in which this interaction is not assumed zero.

Consider now the *multiplicative model* $M_{\mathscr{A}}$, i.e. $\log p \in \Omega_{\mathscr{A}}$. Two particular weight functions have been widely used in the literature. Since Birch's (1963) paper, most authors have used the uniform weight function. In this case

$$U_b \log p(i) = \sum_{c\subseteq b} (-1)^{|b-c|} \log p_c^*(i_c)$$

where $p_c^*(i_c)$ is the *geometric mean* of all $p(j)$ for which $j_c = i_c$, and we do not find these interactions easy to interpret. The system of interactions based upon the substitution weight function does seem easier to interpret with multiplicative models and has been used to effect by Plackett (1974). It was introduced by Mantel (1966), and is used more generally in GLIM, see Baker and Nelder (1978). Here

$$U_b \log p(i) = \sum_{c\subseteq b} (-1)^{|b-c|} \log p_c^r(i_c),$$

which is the logarithm of a cross product ratio. Thus if $d = 3$ and $b = \{1, 2\}$, the cross product ratio is

$$\frac{p(i, j, r_3)p(r_1, r_2, r_3)}{p(i, r_2, r_3)p(r_1, j, r_3)}.$$

Referring back to Section 2.3 above, we now turn to what may be called the estimated model interactions $U_b \log \hat{p}$, $b \in \mathscr{A}$, where $\hat{p}$ is the maximum likelihood estimate of p under $M_{\mathscr{A}}$. No matter which w is chosen, $U_b \log \hat{p}$ does not share the attractive properties of $U_b\hat{\eta}$ when the cell frequencies are proportional, properties which stem from the equation $U_b\hat{\eta} = U_b y$. Thus $U_b \log \hat{p}$ does not depend only on the b-marginal table n_b of $n = (n(i): i \in \mathscr{I})$ but, in general, on all $\mathscr{A}$-marginals. (An important exception occurs when the generating class is decomposable; see Haberman (1974), Darroch, Lauritzen and Speed (1980) and Lauritzen, Speed and Vijayan (1978).) Also it changes each time a different model (that is, a different $\mathscr{A}$) is fitted. This is one of the most important differences between the additive and multiplicative models. Of course when b is one of the maximal elements of $\mathscr{A}$, that is $b \in \mathscr{A}^*$, then $U_b \log \hat{p}$ can be put to use since its magnitude, relative to its standard deviation, indicates whether or not the model obtained from $M_{\mathscr{A}}$ by putting $U_b \log p = 0$ is likely to be acceptable; see Baker and Nelder (1978).

Finally we consider the implications of Proposition 4.1 for *Lancaster-additive models*. Using the weight function q (see Section 3.2) the b-interaction for model I is

$$\text{(4.9)} \qquad U_b \frac{p}{q} = \sum_{c\subseteq b} (-1)^{|b-c|} \frac{p_c}{q_c},$$

and using the substitution weight function the b-interaction for model II is

$$\text{(4.10)} \qquad U_b \frac{P}{Q} = \sum_{c\subseteq b} (-1)^{|b-c|} \frac{P_c}{Q_c}.$$

It is easy to see (cf. Section 3.2) that the two definitions of no b-interaction obtained from

(4.9) and (4.10) are equivalent to each other and to Lancaster's (1969, page 256) definition, namely

$$(4.11) \qquad \Pi_{\sigma\in b}(P_\sigma^*(i_\sigma) - P_\sigma(i_\sigma)) = 0,$$

where the P_σ^* are artificial functions multiplied according to the rule

$$\Pi_{\sigma\in c}P_\sigma^*(i_\sigma) = P_c(i_c).$$

Zentgraf (1975) proved that, if (4.11) holds for all b with $|b| > t$, then

$$(4.12) \qquad P(i) = \sum_{u=0}^{t} (-1)^{t-u}\binom{s-u-1}{t-u} \sum_{b:|b|=u} P_b(i_b)Q_{b'}(i_{b'}).$$

This result, when combined with its converse, amounts to a special case of Proposition 4.2 above.

4.5 *A uniqueness property of interactions.* The main purpose of this paper has been to show that many general properties linking models and interactions can be easily stated and proved using interaction operators. We have seen that given any model $M_{\mathscr{A}}$ and any multiplicative weight function w there corresponds a generalized interaction operator $T_{\mathscr{A}}$, that the interaction operators U_b provide a useful way of partitioning $T_{\mathscr{A}}$ and, finally, that $M_{\mathscr{A}}$ has the "hierarchical no-interaction" property by which it is usually characterised.

We conclude by returning to a question raised in Section 2.1, namely: given that f satisfies $M_{\mathscr{A}}$, to what extent are the parameters $\lambda_a(i_a)$ uniquely determined by f? The answer, as shown in the following proposition, is that interactions and only interactions of λ_a are uniquely determined.

PROPOSITION 4.4. *Assume*

$$(4.13) \qquad f(i) = \sum_{a\in\mathscr{A}} \lambda_a(i_a)$$

and let $c \in \mathscr{A}$. The extent to which λ_c is determined by f is defined by the equations

$$(4.14) \qquad U_b\lambda_c(i_c) = U_bf(i) \quad \text{for all} \quad b \in \bar{\mathscr{A}} - \bar{\mathscr{C}}$$

where $\mathscr{C} = \mathscr{A} - \{c\}$ and where the U_b are defined with respect to any multiplicative weight function.

PROOF. Since

$$f(i) - \lambda_c(i_c) = \sum_{a\in\mathscr{C}} \lambda_a(i_a)$$

therefore

$$U_b(f(i) - \lambda_c(i_c)) = 0 \quad \text{for all} \quad b \in \bar{\mathscr{C}}'.$$

However $U_bf(i) = U_b\lambda_c(i_c) = 0$ for all $b \in \bar{\mathscr{A}}'$. Thus, given (4.13), λ_c certainly satisfies (4.14).

We now prove that equations (4.14) define *all* that is uniquely determined about λ_c from a knowledge of f. This is done by showing that the information about λ_c contained in (4.14) is sufficient for us to construct a λ_c, λ_c^* say, such that

$$f(i) = \lambda_c^*(i_c) + \sum_{a\in\mathscr{C}} \lambda_a(i_a).$$

Simply define

$$\lambda_c^* = \sum_{b\in\bar{\mathscr{A}}-\bar{\mathscr{C}}} U_bf.$$

Then

$$f(i) - \lambda_c^*(i_c) = \sum_{b\in\bar{\mathscr{C}}} U_bf(i)$$

and, by Proposition 4.1, the right side can be written in the form

$$\textstyle\sum_{a\in\mathscr{C}}\lambda_a(i_a). \quad \square$$

EXAMPLE 2.2 (continued). We have $s = 3$ and $\mathscr{A} = \{\{1, 2\}, \{2, 3\}, \{3, 1\}\}$. Let $c = \{2, 3\}$ so that $\bar{\mathscr{A}} - \bar{\mathscr{C}} = \{\{2, 3\}\}$. Using the substitution weight function for convenience, we find that the total information about the marriage rates ρ_{jk} that can be determined from a knowledge of the expected numbers of marriages μ_{ijk} is contained in the equations

$$\frac{\rho_{jk}\rho_{r_2r_3}}{\rho_{jr_3}\rho_{r_2k}} = \frac{\mu_{r_1jk}\mu_{r_1r_2r_3}}{\mu_{r_1jr_3}\mu_{r_1r_2k}}.$$

Likewise, all that can be determined about the numbers M_{ij} of eligible women is contained in the equations

$$\frac{M_{ij}M_{r_1r_2}}{M_{ir_2}M_{r_1j}} = \frac{\mu_{ijr_3}\mu_{r_1r_2r_3}}{\mu_{ir_2r_3}\mu_{r_1jr_3}}. \quad \square$$

Acknowledgement. We are grateful to the referees and an Associate Editor for their helpful comments.

REFERENCES

ANDERSEN, A. H. (1974). Multidimensional contingency tables. *Scand. J. Statist.* **1** 115–127.

BAKER, R. J. and NELDER, J. A. (1978). *The GLIM System, Release* 3. *Generalised Linear Interactive Modelling.* The Numerical Algorithms Group, Oxford.

BARNDORFF-NIELSEN, O. (1978). *Information and Exponential Families.* Wiley, Chichester.

BIRCH, M. W. (1963). Maximum likelihood in three-way contingency tables. *J. Roy. Statist. Soc. Ser. B* **25** 220–233.

BISHOP, Y. M. M., FIENBERG, S. E. and HOLLAND, P. W. (1975). *Discrete Multivariate Analysis: Theory and Practice.* M.I.T. Press, Cambridge, Mass.

DARROCH, J. N. (1974). Multiplicative and additive interactions in contingency tables. *Biometrika* **61** 207–214.

DARROCH, J. N., LAURITZEN, S. L. and SPEED, T. P. (1980). Markov fields and log-linear interaction models for contingency tables. *Ann. Statist.* **8** 522–539.

DARROCH, J. N. and RATCLIFF, D. (1972). Generalized iterative scaling for log-linear models. *Ann. Math. Statist.* **43** 1470–1480.

DAVIDSON, R. (1973). Determination of confounding. *Stochastic Analysis.* Edited by D. G. Kendall and E. P. Harding. Wiley, New York.

FIENBERG, S. E. (1977). *The Analysis of Cross-Classified Data.* M.I.T. Press, Cambridge, Mass.

GOKHALE, D. V. and KULLBACK, S. (1978). *The Information in Contingency Tables.* Dekker, New York.

GOOD, I. J. (1958). The interaction algorithm and practical Fourier analysis. *J. Roy. Statist. Soc. Ser. B* **20** 361–372.

GOOD, I. J. (1963). Maximum entropy for hypothesis formulation, especially for multidimensional contengency tables. *Ann. Math. Statist.* **34** 911–934.

GOODMAN, L. A. (1970). The multivariate analysis of qualitative data: interaction among multiple classifications. *J. Amer. Statist. Assoc.* **65** 226–256.

GRIZZLE, J. E., STARMER, C. F. and KOCH, G. G. (1969). Analysis of categorical data by linear models. *Biometrics* **25** 489–504.

HABERMAN, S. J. (1974). *The Analysis of Frequency Data.* Univ. of Chicago Press, Chicago, Il.

HABERMAN, S. J. (1975). Direct products and linear models for complete factorial tables. *Ann. Statist.* **3** 314–333.

HABERMAN, S. J. (1978). *Analysis of Qualitative data,* Vol. 1. Academic, New York.

HABERMAN, S. J. (1979). *Analysis of Qualitative data,* Vol. 2. Academic, New York.

KU, H. H. and KULLBACK, S. (1968). Interaction in multidimensional contingency tables: an information theoretic approach. *J. Res. Nat. Bur. of Stand.* **72B** 159–199.

KURKJIAN, B., and ZELEN, M. (1962). A calculus for factorial arrangements. *Ann. Math. Statist.* **33** 600–619.

LANCASTER, H. O. (1969). *The Chi-Squared Distribution.* Wiley, New York.

LAURITZEN, S. L., SPEED, T. P. and VIJAYAN, K. (1978). *Decomposable Graphs and Hypergraphs.* Preprint No. 9. Institute of Mathematical Statistics, University of Copenhagen.

MANN, H. B. (1949). *Analysis and Design of Experiments.* Dover, New York.

MANTEL, N. (1966). Models for complex contingency tables and polychotomous dosage response curves. *Biometrics* **22** 83–95.

NELDER, J. A. and WEDDERBURN, R. W. M. (1972). Generalised linear models. *J. Roy. Statist. Soc. Ser.* A **135** 370–384.

PLACKETT, R. L. (1981). *Analysis of Categorical Data*, 2nd Ed. Griffin, London.

ZENTGRAF, R. (1975). A note on Lancaster's definition of higher-order interactions. *Biometrika* **62** 375–378.

SCHOOL OF MATHEMATICAL SCIENCES,
THE FLINDERS UNIVERSITY OF SOUTH AUSTRALIA,
ADELAIDE. S.A. 5042,
AUSTRALIA.

D.M.S., C.S.I.R.O.,
P.O. BOX 1965,
CANBERRA. A.C.T. 2601,
AUSTRALIA.

The Annals of Statistics
1986, Vol. 14, No. 1, 138–150

GAUSSIAN MARKOV DISTRIBUTIONS OVER FINITE GRAPHS

BY T. P. SPEED AND H. T. KIIVERI

CSIRO Division of Mathematics and Statistics, Canberra and Perth, Australia

Gaussian Markov distributions are characterised by zeros in the inverse of their covariance matrix and we describe the conditional independencies which follow from a given pattern of zeros. Describing Gaussian distributions with given marginals and solving the likelihood equations with covariance selection models both lead to a problem for which we present two cyclic algorithms. The first generalises a published algorithm for covariance selection whilst the second is analogous to the iterative proportional scaling of contingency tables. A convergence proof is given for these algorithms and this uses the notion of I-divergence.

1. Introduction. Most modelling of jointly Gaussian (normal) random variables involves the specification of a structure on the mean and the *covariance* matrix K. However, models which specify structure on K^{-1} have also been developed, although they are seemingly less popular. Our interest in this paper focuses on the covariance selection models, introduced by Dempster (1972) and studied by Wermuth (1976a, b), in which certain elements of K^{-1} are assumed to be zero.

In Section 2 we show how zeros in K^{-1} correspond to conditional independence statements and characterise all such statements consequent upon a given pattern of zeros. The characterisation is achieved by associating a simple graph [Behdzad et al. (1979)] with the elements of K^{-1} and providing rules for reading the graph. The results are a direct analogue of those given in Darroch et al. (1980) for contingency table models; see also Speed (1979).

The likelihood equations for covariance selection models lead naturally to a consideration of the problem of finding Gaussian distributions with prescribed margins. The results in Sections 3 and 4 provide a solution to this problem and a general algorithm for constructing the required distributions is given. Two special cases of this algorithm are considered. The first one is a generalisation of an algorithm in Wermuth and Scheidt (1977) whilst the second one has properties analogous to iterative proportional scaling for contingency tables [Haberman (1974)]. The notion of I-divergence [Csiszár (1975)] or discrimination information in the terminology of Kullback (1959), plays an important role in the convergence proof of this algorithm.

Finally, in Section 5 we show how the I-divergence geometry of Csiszár (1975) provides a framework in which both algorithms can be seen to be an iterated sequence of I-projections.

Received November 1983; revised September 1985.
AMS 1980 *subject classifications.* Primary 62F99; secondary 60K35.
Key words and phrases. Conditional independence, Markov property, simple graph, covariance selection, I-divergence geometry.

GAUSSIAN MARKOV DISTRIBUTIONS OVER FINITE GRAPHS

2. Conditional independence for Gaussian random variables. In the following we consider a random vector $\mathbf{X}$ having a Gaussian distribution with mean $\mathbf{0}$ and positive definite covariance matrix K. The components of $\mathbf{X}$ will be indexed by a finite set C and for $a \subset C$ we write $\mathbf{X}_a$ for the subset of the components of $\mathbf{X}$ indexed by a, namely $(X_\gamma\colon \gamma \in a)$. The covariance matrix $K = (K(\alpha, \beta)\colon \alpha, \beta \in C)$ on C is defined by $K(\alpha, \beta) = \mathbb{E}\{X_\alpha X_\beta\}$, $\alpha, \beta \in C$, where $\mathbb{E}$ denotes expected value. For subsets $a, b \subseteq C$, $K_{a,b} = \{K(\alpha, \beta)\colon \alpha \in a, \beta \in b\}$ denotes the cross covariance matrix of $\mathbf{X}_a$ and $\mathbf{X}_b$. When $a = b$ we write K_a instead of $K_{a,a}$. Note that care must be taken to distinguish between K_a^{-1} and $(K^{-1})_a$. The density $p(\mathbf{x})$ of $\mathbf{X}$ is, of course,

$$(1) \qquad p(\mathbf{x}) = (2\pi)^{-|C|/2}(\det K)^{-1/2}\exp\{-\tfrac{1}{2}\mathbf{x}^T K^{-1}\mathbf{x}\}, \qquad x \in \mathbb{R}^{|C|},$$

where $|\cdot|$ denotes the cardinality of the argument. Marginal densities are subscripted by their defining sets, e.g., $p_a(\mathbf{x}_a)$ or simply p_a, refers to the marginal density of $\mathbf{X}_a$, where a is an arbitrary subset of C.

Proposition 1 relates the conditional independence of two components of $\mathbf{X}$ to the structure of K. In the proposition and following we abbreviate the set intersection $a \cap b$ to ab and write $a \setminus b$ for the complement of b in a. The set $C \setminus b$ will be denoted b'.

PROPOSITION 1. *For subsets a, b of C with $a \cup b = C$ the following statements are equivalent.*

(i) $K_{a,b} = K_{a,ab}K_{ab}^{-1}K_{ab,b}$.
(i′) $K_{a\setminus b, b\setminus a} = K_{a\setminus b, ab}K_{ab}^{-1}K_{ab, b\setminus a}$.
(ii) $(K^{-1})_{a\setminus b, b\setminus a} = 0$.
(iii) $\mathbf{X}_a$ *and* $\mathbf{X}_b$ *are conditionally independent given* $\mathbf{X}_{ab}$.

PROOF. (i) and (i′) are easily seen to be equivalent by partitioning the rows of K over $a \setminus b$ and ab and the columns over $b \setminus a$ and ab. By partitioning over $a \setminus b$, $b \setminus a$, and ab, a straightforward use of the expression for the inverse of a partitioned matrix [Rao (1973, page 33)] proves that (i′) is equivalent to (ii). The standard formula (2) for the conditional covariance matrix gives the connection between (iii) and (i′),

$$(2) \qquad \operatorname{cov}(\mathbf{X}_{a\setminus b}, \mathbf{X}_{b\setminus a}|\mathbf{X}_{ab}) = K_{a\setminus b, b\setminus a} - K_{a\setminus b, ab}K_{ab}^{-1}K_{ab, b\setminus a}. \ \square$$

A useful special case of the above proposition is the following corollary, given by Wermuth (1976a).

COROLLARY 1. *For distinct elements α, β of C, X_α and X_β are conditionally independent given $X_{\{\alpha,\beta\}'}$ iff $K^{-1}(\alpha, \beta) = 0$.*

PROOF. Put $a = C \setminus \{\alpha\} = \{\alpha\}'$ and $b = \{\beta\}'$ in Proposition 1. □

Having shown that zeros in K^{-1} correspond to conditional independence statements we now describe all such statements which follow from a given pattern of zeros in K^{-1}. To do this we associate a simple undirected graph with the pattern of zeros and then give rules for reading the graph to obtain the independence relations.

To begin, some graph-theoretic notation and definitions are needed; for a general reference see Behdzad et al. (1979). Our simple undirected graph will be denoted by $\mathbf{C} = (C, E(C))$ where C is the vertex set, and $E(\mathbf{C})$ the edge set which consists of unordered pairs of distinct vertices. Pairs of vertices $\{\alpha, \beta\} \in E(\mathbf{C})$ are said to be *adjacent*. A maximal set of (≥ 2) vertices for which every pair is adjacent is called a *clique*. For any vertex γ we write $\partial\gamma = \{\alpha: \{\alpha, \gamma\} \in E(\mathbf{C})\}$ for the set of neighbours of γ. We also write $\bar{\gamma} = \gamma \cup \partial\gamma$.

An important notion is the separation of sets of vertices in $\mathbf{C}$. To define this we first need to define a *chain* which is a sequence $\gamma = \gamma_0, \gamma_1, \ldots, \gamma_m = \beta$ of vertices such that $\{\gamma_l, \gamma_{l+1}\} \in E(\mathbf{C})$ for $l = 0, 1, \ldots, m-1$. If $\gamma_0 = \gamma_m$ the chain is called a *cycle*. Two sets of vertices a, b are said to be separated by a third set d if every chain connecting an $\alpha \in a$ to a $\beta \in b$ intersects d.

The graph $\mathbf{C}$ is said to be *triangulated* [see Lauritzen et al. (1984)] iff all cycles $\gamma_0, \gamma_1, \ldots, \gamma_p = \gamma_0$ of length $p \geq 4$ possess a chord, where a *chord* is an edge connecting two nonconsecutive vertices of the cycle.

Finally, the graph $\tilde{\mathbf{C}}$ complementary to $\mathbf{C}$ has vertex set C and edge set $E(\tilde{\mathbf{C}})$ with the property that $\{\alpha, \beta\} \in E(\tilde{\mathbf{C}})$ iff $\alpha \neq \beta$ and $\{\alpha, \beta\} \notin E(\mathbf{C})$. Example 1 illustrates these ideas.

EXAMPLE 1. The graph $\mathbf{C}$ with vertex set $\{1,2,3,4\}$ and edge set $\{\{1,2\}, \{1,3\}, \{1,4\}, \{2,3\}, \{3,4\}\}$ could be depicted as in Figure 1. For this graph the set of neighbours of 1 is $\{2,3,4\}$; the cliques are $\{1,2,3\}$, $\{1,3,4\}$; a chain from $\{2\}$ to $\{4\}$ is 2, 3, 1, 4 and $\{2\}$ is separated from $\{4\}$ by $\{1,3\}$. Figure 2 shows the complementary graph.

As it stands the graph in Figure 1 is triangulated. However, if the edge $\{1,3\}$ were removed we would have the simplest example of a nontriangulated graph.

The characterisation of all conditional independence relations consequent upon a given pattern of zeros in K^{-1} is presented in Proposition 2.

PROPOSITION 2. *Let* $\mathbf{C}$ *be a simple graph with vertex set* C *indexing the Gaussian random variables* $\mathbf{X}$. *Then the following are equivalent.*

(i) $K^{-1}(\alpha, \beta) = 0$ *if* $\{\alpha, \beta\} \notin E(\mathbf{C})$ *and* $\alpha \neq \beta$;

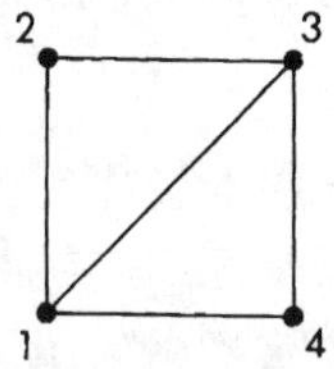

FIG. 1

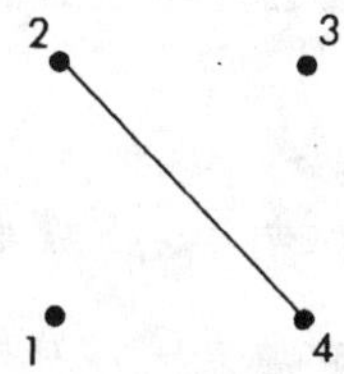

FIG. 2

The local Markov property:

(ii) *For every* $\gamma \in C$, X_γ *and* $\mathbf{X}_{\{\gamma\}'}$ *are conditionally independent given* $\mathbf{X}_{\partial\gamma}$;

The global Markov property:

(iii) *For every* a, b *and* d *with* d *separating* a *from* b *in* $\mathbf{C}$, $\mathbf{X}_a$ *and* $\mathbf{X}_b$ *are conditionally independent given* $\mathbf{X}_d$.

PROOF. To show the equivalence of (i) and (ii) we note that (i) is equivalent to $K^{-1}(\gamma, \overline{\{\gamma\}'}) = 0$. Putting $a = \overline{\{\gamma\}}$ and $b = \{\gamma\}'$ in Proposition 1 then proves the result.

The equivalence of (i) and (iii) for the case $a \cup b \cup d = C$ follows in a similar way if we put "a" $= a \cup d$ and "b" $= b \cup d$ in Lemma 1. When $a \cup b \cup d \neq C$ a simple maximality argument as in Vorobev (1963) shows that maximal sets a^*, b^* exist such that $a \subseteq a^*$, $b \subseteq b^*$, $a^* \cup b^* \cup d = C$, and a^* is separated from b^* by d. Proposition 1 then gives us $p = p_{a^*}p_{b^*}/p_d$ and integration to obtain the marginal density of $\mathbf{X}_{a \cup b \cup d}$ shows that (i) implies (iii).

The implication in the reverse direction follows on noting that if $(\alpha, \beta) \notin E(\mathbf{C})$ then α, β are separated by $\{\alpha, \beta\}'$. Hence by (iii) X_α and X_β are conditionally independent given $X_{\{\alpha,\beta\}'}$ and Corollary 1 shows that $K^{-1}(\alpha, \beta) = 0$. □

The results of Proposition 2 are illustrated in Example 2.

EXAMPLE 2. Suppose K^{-1} has the following pattern with $*$ denoting a nonzero element:

$$\begin{array}{c|ccccc} & 1 & 2 & 3 & 4 & 5 \\ \hline 1 & * & * & 0 & 0 & 0 \\ 2 & * & * & * & 0 & * \\ 3 & 0 & * & * & * & * \\ 4 & 0 & 0 & * & * & * \\ 5 & 0 & * & * & * & * \end{array}.$$

Then the corresponding graph $\mathbf{C}$ would be as shown in Figure 3. If we put $\gamma = \{2\}$, $\partial\gamma = \{1,3,5\}$, and use the local Markov property we deduce that X_2 and X_4 are conditionally independent given $\mathbf{X}_{\{1,3,5\}}$. Similarly with $a = \{1\}$, $b = \{4\}$, and $d = \{2\}$, the global Markov property can be used to assert that X_1 and X_4 are conditionally independent given X_2.

3. Gaussian Markov distributions with prescribed marginals. In this section we consider the problem of finding a Gaussian probability measure with prescribed marginals, i.e., we seek a joint probability density p whose marginals

$$p_{c_1}, \ldots, p_{c_n} \tag{3}$$

are known beforehand, $c_1, \ldots, c_n$ being proper subsets of C. (The notation is explained after (1) above.) Clearly if our marginal specifications are consistent it is necessary to give only the maximal c_i in (3).

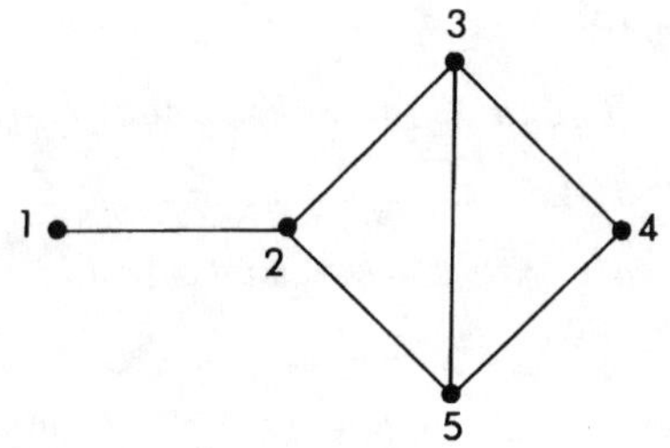

FIG. 3

As motivation for this problem consider the following. Suppose we have n independent and identically distributed observations $\mathbf{x}_1, \ldots, \mathbf{x}_n$ from (1) and we wish to find a maximum likelihood estimate of K subject to certain elements of K^{-1} being zero. When written in our notation, the likelihood equations for such a model (Dempster, 1972) are:

$$\begin{aligned} K(\alpha, \beta) &= S(\alpha, \beta) \quad \text{if } \{\alpha, \beta\} \in E(\mathbf{C}) \text{ or } \alpha = \beta, \\ K^{-1}(\alpha, \beta) &= 0 \quad \text{if } \{\alpha, \beta\} \notin E(\mathbf{C}) \text{ and } \alpha \neq \beta, \end{aligned} \tag{4}$$

where $nS = \sum_{i=1}^{n} \mathbf{x}_i \mathbf{x}_i^T$. The first equation in (4) is easily shown to be equivalent to

$$K_c = S_c \quad \text{if } c \in \mathscr{C}(\mathbf{C}), \tag{4'}$$

where $\mathscr{C}(\mathbf{C})$ is the class of cliques of $\mathbf{C}$. Since a Gaussian distribution with mean zero is completely specified by its covariance matrix, (4′) amounts to specifying the marginal distributions p_c for $c \in \mathscr{C}(\mathbf{C})$.

Theorem 1 can be used to describe the class of Gaussian measures with prescribed margins.

THEOREM 1. *Given positive definite matrices L and M defined on the vertices C of a graph* $\mathbf{C} = (C, E(\mathbf{C}))$ *there exists a unique positive definite matrix K such that*

(i) $K(\alpha, \beta) = L(\alpha, \beta)$ *if* $\{\alpha, \beta\} \in E(\mathbf{C})$ *or* $\alpha = \beta$,
(ii) $K^{-1}(\alpha, \beta) = M(\alpha, \beta)$ *if* $\{\alpha, \beta\} \notin E(\mathbf{C})$ *and* $\alpha \neq \beta$.

Equivalently

(i′) $K_c = L_c$ *if* $c \in \mathscr{C}(\mathbf{C})$;
(ii′) $K^{-1}(\tilde{c}, \tilde{c})$ *and* $M(\tilde{c}, \tilde{c})$ *agree except on the diagonals,* $\tilde{c} \in \mathscr{C}(\tilde{\mathbf{C}})$.

PROOF. The equivalence of (i) and (i′) follows from the relation

$$E(\mathbf{C}) = \bigcup_{c \in \mathscr{C}(\mathbf{C})} \bigcup_{\{\alpha, \beta\} \subseteq c} \{\alpha, \beta\}. \tag{5}$$

Replacing $\mathbf{C}$ by $\tilde{\mathbf{C}}$ in (5) enables the equivalence of (ii) and (ii′) to be demonstrated.

The main result of Theorem 1 can be established using the theory of exponential families [Barndorff-Nielsen (1978), Johansen (1979)] and such a proof is sketched by Dempster (1972, Appendixes A and B).

The results in Section 4 will show how to generate a sequence of matrices converging to the K of Theorem 1 and thus provide an alternative proof. We prefer this proof as it provides a basis for simple numerical algorithms which do not require Newton–Raphson type iterations or storage of large matrices to compute K. □

Replacing the L in Theorem 1 by the sample covariance matrix and setting $M = I$ shows that the estimation problem for covariance selection models has a well defined solution. When $M = I$, the K in Theorem 1 gives the Gaussian distribution with maximum entropy satisfying (i) or (i′) [see Dempster (1972)].

Note that varying the M in Theorem 1 gives the family of distributions with margins prescribed by L_c, $c \in \mathscr{C}(\mathbf{C})$.

In the next section we will make use of the notion of the *I-divergence* of two positive definite matrices. This is defined by

$$\mathscr{I}(P|R) = -\tfrac{1}{2}\{\log\det(PR^{-1}) + \mathrm{tr}(I - PR^{-1})\}. \tag{6}$$

The definition (6) results from evaluating the discrimination information measure of Kullback (1959), namely $\int p(\mathbf{x})\log\{p(\mathbf{x})/r(\mathbf{x})\}\,d\mathbf{x}$ for the two Gaussian distributions with densities $p(\mathbf{x})$, $r(\mathbf{x})$ defined by covariance matrices P, R. When it exists, the I-divergence behaves somewhat like a norm on a space of probability measures (Csiszár, 1975), although it is not.

Some properties of (6) which we will use later are given in Lemma 1. We write $\mathscr{P}$ for the set of $|C| \times |C|$ positive definite matrices and regard this as a (convex) subset of $\mathbb{R}^q$ where $q = |C|^2$. In the following a set of unordered pairs of (not necessarily distinct) elements of C will be denoted by E.

LEMMA 1. *The I-divergence* $\mathscr{I}(\cdot|\cdot)$ *has the following properties.*

(i) *If* $P, R \in \mathscr{P}$, $\mathscr{I}(P|R) \geq 0$ *with equality iff* $P = R$.

(ii) *Given* $P, R \in \mathscr{P}$, *if there exists a* $Q \in \mathscr{P}$ *such that*

(a) $Q(\alpha, \beta) = P(\alpha, \beta)$ *if* $(\alpha, \beta) \in E$, *and*

(b) $Q^{-1}(\alpha, \beta) = R^{-1}(\alpha, \beta)$ *if* $(\alpha, \beta) \notin E$, *then*

$$\mathscr{I}(P|R) = \mathscr{I}(P|Q) + \mathscr{I}(Q|R). \tag{7}$$

If such a Q *exists it is unique.*

(iii) *If* $\{K_n\}$ *and* $\{L_n\}$ *are sequences contained in compact subsets of* $\mathscr{P}$ *then* $\mathscr{I}(K_n|L_n) \to 0$ *implies* $K_n - L_n \to 0$.

PROOF. The first assertion is a well known property of the Kullback information measure so we focus on (ii) and (iii).

(ii) A simple calculation shows that for $Q \in \mathscr{P}$

$$\mathscr{I}(P|Q) + \mathscr{I}(Q|R) = \mathscr{I}(P|R) - \tfrac{1}{2}\operatorname{tr}\{(Q-P)\Delta\}, \tag{8}$$

where $\Delta = Q^{-1} - R^{-1}$. Conditions (a) and (b) then ensure that the trace term in (8) is zero.

To prove uniqueness suppose Q_1 and Q_2 satisfy (a) and (b) of (ii). Then setting $P = R = Q_1$ shows that

$$\mathscr{I}(Q_1|Q_1) = \mathscr{I}(Q_1|Q_2) + \mathscr{I}(Q_2|Q_1)$$

and since I-divergences are positive unless both arguments are equal we must have $Q_1 = Q_2$.

(iii) Suppose $\mathscr{I}(K_n|L_n) \to 0$ but $K_n - L_n \nrightarrow 0$. Then there exist convergent subsequences $K_{n'} \to K$ and $L_{n'} \to L$ with $K \neq L$. By continuity $\mathscr{I}(K_{n'}|L_{n'}) \to \mathscr{I}(K|L) \neq 0$, which is a contradiction. □

4. Algorithms. This section develops two algorithms for constructing the K of Theorem 1. The first algorithm preserves (i′) of Theorem 1 throughout the iterations and cycles through $\tilde{c} \in \mathscr{C}(\tilde{\mathbf{C}})$ forcing the off-diagonal elements of $K^{-1}(\tilde{c}, \tilde{c})$ to zero. The second algorithm preserves (ii′) whilst forcing $K_c = L_c$ as it cycles through $c \in \mathscr{C}(\mathbf{C})$. Both of these algorithms are special cases of a more general cyclic algorithm and we begin by presenting this algorithm. Throughout the discussion $E_1, E_2, \ldots, E_m$ denote sets of unordered pairs of (not necessarily distinct) elements of C whose union is denoted by E.

4.1. A general cyclic algorithm. The general cyclic algorithm is designed to solve the following problem. Given $G, H \in \mathscr{P}$ find an $F \in \mathscr{P}$ with the property that

$$F(\alpha,\beta) = G(\alpha,\beta) \quad \text{if } (\alpha,\beta) \in E, \tag{9}$$

$$F^{-1}(\alpha,\beta) = H(\alpha,\beta) \quad \text{if } (\alpha,\beta) \notin E. \tag{10}$$

The algorithm is defined as follows. Generate a sequence $\{F_n\}$ of positive definite matrices satisfying $F_0 = H^{-1}$ and, for $n \geq 1$,

$$F_n(\alpha,\beta) = G(\alpha,\beta) \qquad \text{if } (\alpha,\beta) \in E_{n'}, \tag{9′}$$

$$F_n^{-1}(\alpha,\beta) = F_{n-1}^{-1}(\alpha,\beta) \quad \text{if } (\alpha,\beta) \notin E_{n'}, \tag{10′}$$

where $n' = n (\operatorname{mod} m)$. Basically the idea is to maintain (10) throughout the sequence whilst cycling through the E_m and forcing (9). The crucial step in the algorithm involves going from F_{n-1} to F_n. Assuming for the moment that this step can be performed, a convergence proof for this algorithm, modelled upon that found in Csiszár (1975, Theorem 3.2), is given in Proposition 3. The two algorithms to be discussed are examples for which the sequence $\{F_n\}$ can be easily constructed. We write $\mathbb{N}$ for the set of nonnegative integers.

PROPOSITION 3. *The sequence $\{F_n\}$ generated by the general cyclic algorithm converges to the unique $F \in \mathscr{P}$ with the properties* (9) *and* (10).

PROOF. By (ii) of Lemma 1 we can write for $r \geqq 1$

$$\mathscr{I}(G|F_{r-1}) = \mathscr{I}(G|F_r) + \mathscr{I}(F_r|F_{r-1}). \tag{11}$$

Summing relations of the form (11) over r gives for $u \geqq 1$

$$\mathscr{I}(G|F_0) = \mathscr{I}(G|F_u) + \sum_{r=1}^{u} \mathscr{I}(F_r|F_{r-1}) \tag{12}$$

and from (12) we deduce that

$$\{F_n\} \in \{F\colon \mathscr{I}(G|F) \leqq \mathscr{I}(G|F_0)\} = A \quad \text{(say)}. \tag{13}$$

The set A is compact since $\mathscr{I}(G|F)$ is strictly convex (as a function of F^{-1}) with a unique minimum. From (12) it also follows that

$$\sum_{r=1}^{u} \mathscr{I}(F_r|F_{r-1}) \leqq \mathscr{I}(G|F_0). \tag{14}$$

Hence $\sum_{r=1}^{\infty} \mathscr{I}(F_r|F_{r-1})$ is convergent and $\mathscr{I}(F_r|F_{r-1}) \to 0$ as $r \to \infty$.

Now by (13) the vector sequence $\{F_{sm+1}, F_{sm+2}, \ldots, F_{sm+m}): s \geqq 0\}$ has a convergent subsequence, defined by $s \in \mathbb{N}_1 \subseteq \mathbb{N}$, with limit $(F_1^*, F_2^*, \ldots, F_m^*)$ say. For any $2 \leqq t \leqq m$ we can write

$$(F_t - F_{t-1}) = (F_t - F_{sm+t}) + (F_{sm+t} - F_{sm+t-1}) + (F_{sm+t-1} - F_{t-1}). \tag{15}$$

Letting $s \in \mathbb{N}_1 \to \infty$ and using (iii) of Lemma 1 with $L_n = K_{n-1}$ shows that $F_1^* = F_2^* = \cdots F_m^* = F$ (say). Note that (10) holds for each F_r and hence for the limit F. Similarly for each $s \in \mathbb{N}_1$ and t, $F_{sm+t}(\alpha, \beta) = G(\alpha, \beta)$ if $(\alpha, \beta) \in E_t$, so the same property holds for the limit F, i.e., (9) holds.

A similar argument for any other convergent subsequence shows that the limit point satisfies (9) and (10) of our proposition. Lemma 1, part (ii) then establishes that all convergent subsequences have the same limit and hence $\{F_n\}$ converges. □

The next lemma enables sequences $\{F_n\}$ satisfying (9′) or (10′) to be constructed when either

$$E_i = \{(\alpha, \beta)\colon \alpha, \beta \in a_i \subseteq C\} \tag{16}$$

or

$$E_i = \{(\alpha, \beta)\colon \alpha, \beta \in a_i \subseteq C, \alpha \neq \beta\}. \tag{17}$$

LEMMA 2. *Suppose Q, R, and $B \in \mathscr{P}$. Then*

(i) *for $a \subseteq C$ the matrix*

$$Q^{-1} = R^{-1} + \begin{bmatrix} B_a^{-1} - R_a^{-1} & 0 \\ 0 & 0 \end{bmatrix} \tag{18}$$

is positive definite and satisfies

(a) *$Q(\alpha, \beta) = B(\alpha, \beta)$ if $\alpha \in a$ and $\beta \in a$; and*
(b) *$Q^{-1}(\alpha, \beta) = R^{-1}(\alpha, \beta)$ if $\alpha \notin a$ or $\beta \notin a$.*

(ii) *The matrix Q is given by*

$$(19)\qquad Q = \begin{bmatrix} B_a & B_a R_a^{-1} R_{a,a'} \\ R_{a',a} R_a^{-1} B_a & R_{a'} - R_{a',a} R_a^{-1}(I - B_a R_a^{-1}) R_{a,a'} \end{bmatrix}$$

(iii) *We have the expression*:

$$(20)\qquad \mathscr{I}(Q|R) = -\tfrac{1}{2}\{\log\det B_a R_a^{-1} + \operatorname{tr}(I_a - B_a R_a^{-1})\}.$$

PROOF. (i) We use the density scaling of Kullback (1968). In the Gaussian case, given densities $b(\mathbf{x})$ and $r(\mathbf{x})$ corresponding to positive definite matrices B and R, scaling so that $r_a(\mathbf{x}_a)$ agrees with $b_a(\mathbf{x}_a)$ corresponds to computing

$$(21)\qquad q(\mathbf{x}) = \frac{r(\mathbf{x}) b_a(\mathbf{x}_a)}{r_a(\mathbf{x}_a)}.$$

Expanding the right-hand side of (21) gives

$$(22)\qquad \begin{aligned} q(\mathbf{x}) &= (2\pi)^{-|C|/2}\left(\frac{\det R \det B_a}{\det R_a}\right)^{-1/2} \\ &\quad \times \exp\left\{-\frac{1}{2}\mathbf{x}^T\left[R^{-1} + \begin{pmatrix} B_a^{-1} - R_a^{-1} & 0 \\ 0 & 0 \end{pmatrix}\right]\mathbf{x}\right\}, \end{aligned}$$

which by (18) is just

$$(23)\qquad (2\pi)^{-|C|/2}(\det Q)^{-1/2}\exp\{-\tfrac{1}{2}\mathbf{x}^T Q^{-1}\mathbf{x}\}.$$

The properties (a) and (b) are now immediate. A direct proof using matrix algebra can also be given.

The proofs of (ii) and (iii) are straightforward so we omit them. □

The two algorithms discussed below correspond to choosing the a_i in (16) and (17) to be the cliques of $\mathbf{C}$ or $\tilde{\mathbf{C}}$, respectively. In the following we will abbreviate the class of cliques of $\mathbf{C}$ by $\mathscr{C}$ and the class of cliques of $\tilde{\mathbf{C}}$ by $\tilde{\mathscr{C}}$. The notation $\operatorname{diag}(A)$ refers to a diagonal matrix whose diagonals are the same as those of A.

4.2. The first cyclic algorithm. List the cliques of the complementary graph $\tilde{\mathbf{C}}$ as $\tilde{c}_1, \dots, \tilde{c}_m$ and generate a sequence $\{K_u\}$ as follows: $K_0 = L$; for $s \in \mathbb{N}$, $1 \leq t \leq m$, $K_{sm+t} = Z_t(K_{sm+t-1})$, where $Z_t(K) = Q^{-1}$, Q being the matrix (18) of Lemma 2 with $R = K^{-1}$, $a = \tilde{c}_t$, and $B_a = \operatorname{diag}((K^{-1})_a^{-1})^{-1}$. The fact that this sequence converges to the required matrix K when $M = I$ follows from Proposition 3 on replacing a_i in (17) by $\tilde{c}_i$ and making the identifications $F_n = K_n^{-1}$, $G = M$, and $H = L$. It does not seem possible to give an explicit expression for B_a in the case when $M \neq I$.

For this algorithm the elements of the sequence $\{K_n\}$ are fixed over $\mathscr{C}$ whilst the elements of $\{K_n^{-1}\}$ vary over $\tilde{\mathscr{C}}$. From a computational point of view it is not necessary to compute the sequence $\{K_n\}$ by inverting K_n^{-1} at each step. The expression (18) provides a simple updating formula for K_n given K_{n-1}. Hence it is only necessary to invert $|\tilde{c}| \times |\tilde{c}|$ positive definite matrices when cycling through $\tilde{c} \in \tilde{\mathscr{C}}$.

The cyclic algorithm of Wermuth and Scheidt (1977) is also a special case of the general algorithm. Instead of using the cliques of $\tilde{\mathbf{C}}$ these authors cycle through the edges $\{\alpha, \beta\} \in E(\tilde{\mathbf{C}})$. The 2×2 matrix inversions required are explicitly performed and used to give a simple updating formula. Their algorithm is defined in the same way as above but they have $a \in E(\tilde{\mathbf{C}})$ and

$$B_a = \delta \begin{bmatrix} w^{-1} & 0 \\ 0 & u^{-1} \end{bmatrix},$$

where

$$(K^{-1})_a = \begin{bmatrix} u & v \\ v & w \end{bmatrix}$$

and $\delta = uw - v^2$. It is easily seen that at each step the current value of $K(\alpha, \beta)$ is changed by $-v/\delta$ so that $K^{-1}(\alpha, \beta) = 0$. A computer program for performing the adjustments is given in Wermuth and Scheidt's paper.

4.3. The second cyclic algorithm. Enumerate the cliques of $\mathbf{C}$ as $c_1, c_2, \ldots, c_m$ and define a sequence $\{K_r\}$ as follows: $K_0 = M^{-1}$; for $s \geqq 0$, $1 \leqq t \leqq m$, $K_{sm+t} = Y_t(K_{sm+t-1})$, where $Y_t(K) = Q$, Q being the matrix (6) of Lemma 1 with $R = K$, $a = c_t$, and $B = L$. Making the identifications $a_i = c_i$ in (16) and $F_n = K_n$, $G = L$, and $H = M$ in Proposition 3 shows that the second algorithm converges to the K of Theorem 1. This result also gives an alternative proof of Theorem 1. Note that $\{K_n^{-1}\}$ is held fixed over $\tilde{\mathscr{C}}$ whilst $\{K_n\}$ varies over $\mathscr{C}$.

That this second algorithm is analogous to iterative proportional scaling for contingency tables should be clear. At each step we "scale" the current covariance matrix to match the relevant "margin" L_c. We can also connect this algorithm with a general procedure in Kullback (1968) where, however, the proofs are incomplete. Using our notation, Kullback's procedure can be described as follows. Given the required marginal densities $g_{c_1}, \ldots, g_{c_m}$ and an initial density $\pi(\mathbf{x})$ construct the sequence $\{f_n\}$ (assumed to exist) defined by

$$f_0(\mathbf{x}) = \pi(\mathbf{x}),$$

and for $s \geqq 0$, $1 \leqq t \leqq m$

$$f_{sm+t}(\mathbf{x}) = \frac{f_{sm+t-1}(\mathbf{x}) g_{c_t}(\mathbf{x}_{c_t})}{(f_{sm+t-1})_{c_t}(\mathbf{x}_{c_t})}.$$

Note that this simply amounts to scaling the previous density to ensure the desired marginals and this is how we obtain the matrix Q of Lemma 2. Hence the second cyclic algorithm is a Gaussian version of Kullback's general procedure. It can also be shown to be a cyclic ascent algorithm.

4.4. Finite termination. When the graph $\mathbf{C}$ is triangulated and $M = I$ the second cyclic algorithm converges after one cycle if the cliques are suitably ordered. This result is completely analogous to the one cycle convergence of iterative proportional scaling for contingency tables when the generating class is decomposable [see Haberman (1974, Chapter 5)].

To demonstrate the result we need the following two lemmas. Without loss of generality we assume that the graph **C** is connected.

LEMMA 3. *If* **C** *is triangulated then there exists an enumeration* $c_1, \ldots, c_m$ *of the cliques such that for* $i = 2, \ldots, m$

$$c_i \setminus \bigcup_{l=1}^{i-1} c_l \neq \varnothing . \tag{24}$$

PROOF. The result is obtained by successively removing detachable cliques from **C** [see Lauritzen et al. (1984)]. □

Note that (24) states that for each i the clique c_i contains a vertex not in c_l for $l = 1, \ldots, i-1$.

The second lemma gives an expression for the determinant of the matrix K in Proposition 1 which is useful in proving the finite termination of the second algorithm.

LEMMA 4. *Suppose* $K \in \mathscr{P}$ *and* $K^{-1}_{a \setminus b,\, b \setminus a} = 0$ *for* a, b *with* $a \cup b = C$. *Then*

$$\det K = (\det K_a)(\det K_b)/\det K_{ab}. \tag{25}$$

PROOF. Note that (iii) of Proposition 1 implies $p = p_a p_b / p_{ab}$. Evaluation at $x = 0$ then gives the result. □

PROPOSITION 4. *If the cliques of* **C** *are ordered as in Lemma* 3 *and we start the second cyclic algorithm with* $K_0 = I$, *then*

(i) $(K_m)_c = L_c$ *for* $c \in \mathscr{C}$;
(ii) $(K_m^{-1})_{\tilde{c}}$ *is diagonal for* $\tilde{c} \in \tilde{\mathscr{C}}$.

PROOF. We will prove that $\mathscr{I}(K|K_m) = 0$ where K is the unique matrix of Theorem 1 with $M = I$. This will follow directly from (12) provided we can show that

$$\mathscr{I}(K|I) = \sum_{i=1}^{m} \mathscr{I}(K_i|K_{i-1}) \tag{26}$$

and we prove this by induction on m, the number of cliques. It is clearly true for $m = 1$ and so we assume that it is true for all $m \leqq q$ where $q \geqq 1$. If we can prove

$$\mathscr{I}(K|I) = \mathscr{I}(K_{q+1}|K_q) + \mathscr{I}(K_{\bar{c}}|I_{\bar{c}}), \tag{27}$$

where $\bar{c} = \bigcup_{i=1}^{q} c_i$, then (26) will follow for $m = q + 1$; q steps of the second algorithm starting from $K_0 = I$ generate matrices having the form

$$K_i = \begin{bmatrix} I & 0 \\ 0 & \tilde{K}_i \end{bmatrix}, \qquad i = 1, \ldots, q,$$

where $\tilde{K}_i$ is $|\bar{c}| \times |\bar{c}|$ and from the inductive hypothesis

$$\mathscr{I}(K_{\bar{c}}|I_{\bar{c}}) = \sum_1^q \mathscr{I}(\tilde{K}_i|\tilde{K}_{i-1}) = \sum_1^q \mathscr{I}(K_i|K_{i-1}).$$

Turning now to the proof of (27) we remark that it follows from Lemma 4 with $a = c_{q+1}$ and $b = \bar{c}$, the relationship (20) with $Q = K_{q+1}$, $R = K_q$, and $a = c_{q+1}$ as before, and the fact that

$$(K_q)_a = \begin{bmatrix} I & 0 \\ 0 & L_{ab} \end{bmatrix}.$$

The log det terms in the definition of $\mathscr{I}$ match up by Lemma 4 and the trace terms correspond by (20) and the fact just noted. □

We conclude this section with a few remarks comparing the two algorithms. When $M = I$, the main drawback of the first algorithm is the need to invert L at the beginning. It is possible that a numerical inversion of L could be difficult or impossible yet the second algorithm would work. This problem aside, it should be clear that the choice of which algorithm is to be favoured in any given situation is very much dependent on the number and sizes of the cliques in $\mathscr{C}$ and $\tilde{\mathscr{C}}$. However, if **C** is triangulated and $M = I$, the finite termination property of the second algorithm makes it attractive.

5. Some comments about the geometry. To give a geometric interpretation of the two algorithms it is convenient to define the "subspaces" $\mathscr{P}_{L,c} = \{P \in \mathscr{P}: P_c = L_c\}$, $\mathscr{Q}_{M,\tilde{c}} = \{Q \in \mathscr{P}: (Q^{-1})_{\tilde{c}}$ agrees with $M_{\tilde{c}}$ except on the diagonal$\}$, and $\mathscr{P}_{L,\mathscr{C}} = \cap\{\mathscr{P}_{L,c}: c \in \mathscr{C}\}$, $\mathscr{Q}_{M,\tilde{\mathscr{C}}} = \cap\{\mathscr{Q}_{M,\tilde{c}}: \tilde{c} \in \tilde{\mathscr{C}}\}$.

Equation (7) bears a resemblance to Pythagoras' theorem and clearly for all $P \in \mathscr{P}_{L,c}$ we have $\mathscr{I}(P|R) \geq \mathscr{I}(Q|R)$ with equality iff $Q = P$. Hence one can call the matrix Q the I-projection of R on to $\mathscr{P}_{L,c}$ [see Csiszár (1975)].

Viewing the adjustment defined by Q in Lemma 2 as an I-projection we can give an interpretation of the two cyclic algorithms as follows.

The first algorithm begins with a $K_0 \in \mathscr{P}_{L,\mathscr{C}}$ and cycles through $\tilde{c} \in \tilde{\mathscr{C}}$, I-projecting the current estimate of K onto $\mathscr{P}_{L,\mathscr{C}} \cap \mathscr{Q}_{I,\tilde{c}}$ in order to obtain the required element in $\mathscr{P}_{L,\mathscr{C}} \cap \mathscr{Q}_{I,\tilde{\mathscr{C}}}$. The fact that we are I-projecting follows from (ii) of Lemma 1. Using this, for all $K \in \mathscr{Q}_{I,c}$ we have

$$\mathscr{I}(K^{-1}|R^{-1}) = \mathscr{I}(K^{-1}|Q) + \mathscr{I}(Q|R^{-1})$$

or equivalently

$$\mathscr{I}(R|K) = \mathscr{I}(Q^{-1}|K) + \mathscr{I}(R|Q^{-1}),$$

and so $\mathscr{I}(R|K) \geq \mathscr{I}(R|Q^{-1})$ for all $K \in \mathscr{Q}_{I,c}$ with equality iff $K = Q^{-1}$.

For the second algorithm we begin with $K_0 \in \mathscr{Q}_{M,\tilde{\mathscr{C}}}$ and cycle through $c \in \mathscr{C}$, I-projecting the current estimate K onto $\mathscr{Q}_{M,\tilde{\mathscr{C}}} \cap \mathscr{P}_{L,c}$.

Both of the above algorithms are analogous to computing the projection onto the intersection of nonorthogonal (linear) subspaces by successively projecting onto each subspace [see for example von Neumann (1950, Chapter 13)].

Acknowledgment. The referees made many valuable suggestions and are warmly thanked for their contribution.

REFERENCES

BARNDORFF-NIELSEN, O. (1978). *Information and Exponential Families in Statistical Theory.* Wiley, Chichester.

BEHDZAD, M., CHARTRAND, B. and LESNIAK-FOSTER, L. (1979). *Graphs and Digraphs.* Prindle, Weber and Schmidt, Boston.

CSISZÁR, I. (1975). *I*-divergence geometry of probability distributions and minimisation problems. *Ann Probab.* **3** 146–158.

DARROCH, J. N., LAURITZEN, S. L. and SPEED, T. P. (1980). Log-linear models for contingency tables and Markov fields over graphs. *Ann. Statist.* **8** 522–539.

DEMPSTER, A. P. (1972). Covariance selection. *Biometrics* **28** 157–175.

HABERMAN, S. J. (1974). *The Analysis of Frequency Data.* Univ. Chicago Press.

JOHANSEN, S. (1979). *Introduction to the Theory of Regular Exponential Families.* Lecture Notes 3. Institute of Mathematical Statistics, Univ. Copenhagen.

KIIVERI, H. T. and SPEED, T. P. (1982). Structural analysis of multivariate data: a review. *In Sociological Methodology 1982* (S. Leinhardt, ed.). Jossey-Bass, San Francisco.

KULLBACK, S. (1959). *Information Theory and Statistics.* Wiley, New York.

KULLBACK, S. (1968). Probability densities with given marginals. *Ann. Math. Statist.* **39**: 1236–1243.

LAURITZEN, S. L., SPEED, T. P. and VIJAYAN, K. (1984). Decomposable graphs and hypergraphs. *J. Austral. Math. Soc. Ser. A* **36** 12–29.

RAO, C. R. (1973). *Linear Statistical Inference and its Applications.* 2nd ed. Wiley, New York.

SPEED, T. P. (1979). A note on nearest-neighbour Gibbs and Markov probabilities. *Sankhyā Ser. A* **41** 184–197.

VON NEUMANN, J. (1950). *Functional Operators: The Geometry of Orthogonal Spaces* **2**. Princeton Univ. Press.

VOROBEV, N. N. (1963). Markov measures and Markov extensions. *Theory Probab. Appl.* **8** 420–429.

WERMUTH, N. (1976a). Analogies between multiplicative models in contingency tables and covariance selection. *Biometrics* **32** 95–108.

WERMUTH, N. (1976b). Model search among multiplicative models. *Biometrics* **32** 253–263.

WERMUTH, N. AND SCHEIDT, E. (1977). Fitting a covariance selection model to a matrix. Algorithm AS105. *Appl. Statist.* **26** 88–92.

CSIRO
DIVISION OF MATHEMATICS AND STATISTICS
GPO BOX 1965
CANBERRA, ACT 2601
AUSTRALIA

CSIRO
DIVISION OF MATHEMATICS AND STATISTICS
PRIVATE BAG, P.O.
WEMBLEY, W.A. 6014
AUSTRALIA

Chapter 5
Last Words on Anova?

Terry Speed

Many people like to say the last words in an academic debate, and I am no exception. I have tried to do this on a few occasions, only to discover that when I came to say my piece, everyone had left the room. The analysis of variance is a case in point, and my comments on Tukey's contributions to anova explain the problem. If – as I believe to be true – people don't care much these days what Tukey thought about anova, they are going to care even less what I think. This is not said with any sense of bitterness. Indeed I regard myself as something of a student of fads, fashions and trends in statistics, so why should I expect otherwise? Nevertheless, I'm very happy to see these articles reprinted, as their easier availability may kindle the interest of someone, somewhere, sometime in what I still believe to be an important part of (the history of) our subject.

My main stimulus for work in this area came from the papers of six people: R.A. Fisher, Frank Yates, and John A. Nelder from the U.K, indeed all from Rothamsted, Alan T. James and Graham Wilkinson from Adelaide, Australia, and John W. Tukey from the U.S.A. Unpublished lecture notes by James were extremely helpful in getting me going. The anova program within GENSTAT, initially created by Wilkinson based on research by James, Wilkinson, James & Wilkinson and Nelder, was enormously influential. It was (and remains) truly brilliant in conception and execution, and I wanted to understand it. For a long time I was interested in – one might say obsessed with – the symmetries underlying much of anova, and that is reflected in some of the papers reprinted (thank you Rosemary Bailey)! But also I wanted to understand how users of anova saw things, including gory details such as the combination of information, the analysis of covariance and dealing with missing values, all topics with wonderful histories. I made one attempt to put it all together for general consumption, but that got rejected, and so I moved on to other things. As explained above, it is not clear how many people now care. I hope you enjoy the papers. There are several more if you do.

T. Speed
Department of Statistics, University of California, Berkeley, and Division of Bioinformatics, Walter and Eliza Hall Institute, Australia
e-mail: terry@stat.berkeley.edu

S. Dudoit (ed.), *Selected Works of Terry Speed*, Selected Works in Probability and Statistics, 141
DOI 10.1007/978-1-4614-1347-9_5, © Springer Science+Business Media, LLC 2012

References

[1] R. A. Bailey and T. P. Speed. On a class of association schemes derived from lattices of equivalence relations. In P. Schultz, C. E. Praeger, and R. Sullivan, editors, *Algebraic Structures and their Applications*. Marcel Dekker, New York, 1982.

[2] R. A. Bailey and T. P. Speed. A note on rectangular lattices. *Ann. Stat.*, 14: 874–895, 1986.

[3] T. P. Speed and R. A. Bailey. Factorial dispersion models. *Int. Stat. Rev.*, 55: 261–277, 1987.

[4] R. A. Bailey, C. E. Praeger, C. A. Rowley, and T. P. Speed. Generalized wreath products of permutation groups. *Proc. Lond. Math. Soc. (3)*, 47:69–82, 1983.

[5] A. Houtman and T. P. Speed. Balance in designed experiments with orthogonal block structure. *Ann. Stat.*, 11:1069–1085, 1983.

[6] A. Houtman and T. P. Speed. The analysis of multi-stratum designed experiments with incomplete data. *Aust. J. Stat.*, 26:227–246, 1984.

[7] T. P. Speed. John W. Tukey's contributions to analysis of variance. *Ann. Stat.*, 30(6):1649–1665, 2002.

[8] T. P. Speed. ANOVA models with random effects: An approach via symmetry. In J. Gani and M. B. Priestley, editors, *Essays in Time Series and Allied Processes : Papers in Honour of E. J. Hannan*, pages 355–368. Applied Probability Trust, Sheffield, UK, 1986.

[9] T. P. Speed. What is an analysis of variance? *Ann. Stat.*, 15:885–910, 1987.

GENERALIZED WREATH PRODUCTS OF PERMUTATION GROUPS

R. A. BAILEY, CHERYL E. PRAEGER, C. A. ROWLEY, *and* T. P. SPEED

[Received 19 July 1982]

1. *Introduction*

For $i = 1, 2$, let (G_i, Δ_i) be a permutation group. The *permutation direct product* $(G_1, \Delta_1) \times (G_2, \Delta_2)$ is $(G_1 \times G_2, \Delta_1 \times \Delta_2)$ with action defined by

$$(\delta_1, \delta_2)(g_1, g_2) = (\delta_1 g_1, \delta_2 g_2).$$

If $\Delta_1 \times \Delta_2$ is visualized as a rectangular array, an element of $G_1 \times G_2$ may be described as a permutation of rows by an element of G_1 followed by a permutation of columns by an element of G_2.

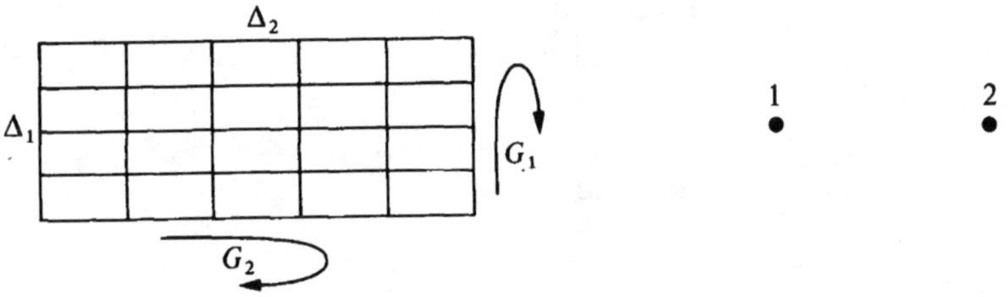

FIG. 1

The *permutation wreath product* $(G_1, \Delta_1) \operatorname{wr} (G_2, \Delta_2)$ is $(G_1{}^{\Delta_2} \times G_2, \Delta_1 \times \Delta_2)$, with action defined by

$$(\delta_1, \delta_2)(f, g_2) = (\delta_1(\delta_2 f), \delta_2 g_2),$$

where f is a function from Δ_2 to G_1. Thus an informal description of an element of $G_1 \operatorname{wr} G_2$ is 'independent permutations of the points within each column by elements of G_1, followed by a permutation of the columns by an element of G_2'.

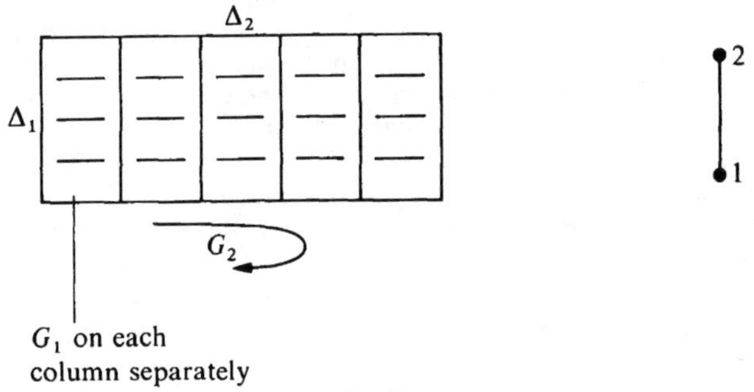

FIG. 2

In the first case, the indexes 1 and 2 play an equal role, and the *rows* of the array are significant. In the second case, the index 2 *dominates* the index 1, and the rows have no significance.

Sets with the two structures described above are frequently used in designed experiments. Nelder [**9**] described a class of structures obtained from these by successive *crossing* (corresponding to the direct product) and *nesting* (corresponding to the wreath product). He developed a large body of theory for these structures and asked whether only these structures satisfied his results.

However, there are many structures that are recognized as tractable by designers of statistical experiments, but which are not in Nelder's class: the simplest such was described by Throckmorton [**14**].

EXAMPLE 1. The set is divided into *rows* (index 1) crossed with *columns* (index 2). Each row is subdivided into *minirows* (index 3), which meet all columns. Within each *square* (row–column intersection), the fragments of minirows are crossed with *microcolumns* (index 4). Thus 1 dominates 3 and 4, while 2 dominates 4.

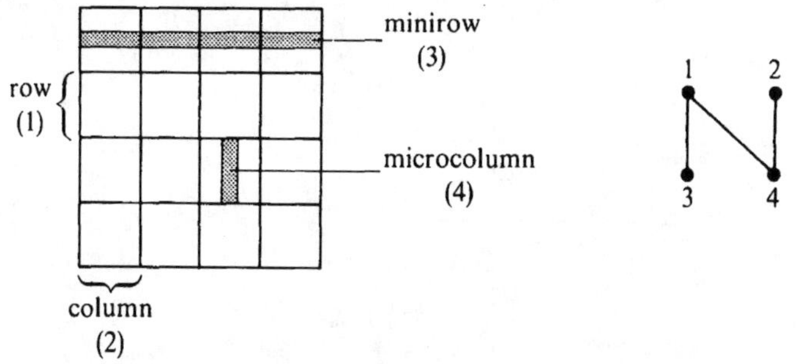

FIG. 3

Although this is not one of Nelder's structures, most statisticians working in the design of experiments could describe its automorphism group, in such terms as 'permute rows, permute columns, within each row separately permute minirows, within each square separately permute microcolumns'.

A more precise discussion of structures such as this is given by Speed and Bailey [**12**], and Bailey [**2**]: however, our concern here is with the associated *permutation groups*. In § 3 of this paper we introduce an explicit description of the elements of both the full automorphism group of such a structure and some of its subgroups. For Nelder's structures these groups can be obtained by successively forming direct products and wreath products of the appropriate permutation groups (group actions), but for our more general class we need to use a construction which Wells [**15**, § 7] described for actions of semigroups: he called it *the wreath product of an ordered set of actions.* The *ordered set* here is a partially ordered set, the partial order being given by the combinatorial structure. (The right-hand part of Figs 1–3 shows the appropriate partially ordered set.) We need to prove that if we start with *group* actions then Wells's wreath product action is also a group action.

Since statistical experiments are, necessarily, finite, our main interest is in structures defined by *finite* partially ordered sets: in this case, as is quite straightforward to check using the results of §4, the permutation group which we construct is identical with that constructed by Holland [6] and Silcock [11]. This is why we call it a generalized wreath product. However, finiteness is not essential to all of the theory: and so we have not confined our attention to this case.

Throughout this paper we rely heavily on Wells's explicit representation of the elements of the generalized wreath product. We use a slightly modified version based on the notation introduced in §2.

In §§5–6 we address two questions which are relevant to the use of these structures in designed experiments. What are the orbits on pairs of points (§5)? In particular, which subgroups have the 'expected' orbits? What are the irreducible constituents of the permutation linear representation (§6)? In particular, when is the centralizer algebra commutative?

2. *Notation and terminology*

The notation introduced in this section is used without comment in the rest of this paper. Throughout, (I, ρ) denotes a partially ordered set. We shall find it convenient to use both of the symbols ρ and $\leqslant$ for the partial order, the former in descriptive work and the latter in computations (when we shall also use the associated symbols $<$, $>$, and $\geqslant$ with their obvious meanings).

DEFINITIONS. Following Grätzer [4], we define a subset J of I to be
hereditary if, whenever $i < j$ and $j \in J$, then $i \in J$;
ancestral if, whenever $i > j$ and $j \in J$, then $i \in J$;
a *chain* if, whenever $i, j \in J$, then either $i \leqslant j$ or $j \leqslant i$;
an *antichain* if, whenever $i, j \in J$ and $i \neq j$, then neither $i \leqslant j$ nor $j \leqslant i$.
For $i \in I$ we define

$$A(i) = \{j \in I\colon j > i\}, \quad A[i] = \{j \in I\colon j \geqslant i\},$$

$$H(i) = \{j \in I\colon j < i\}, \quad H[i] = \{j \in I\colon j \leqslant i\},$$

and for $J \subseteq I$ we define

$$A(J) = \bigcup_{i \in J} A(i), \quad A[J] = \bigcup_{i \in J} A[i],$$

$$H(J) = \bigcup_{i \in J} H(i), \quad H[J] = \bigcup_{i \in J} H[i].$$

Note that all the A-subsets are ancestral and all the H-subsets are hereditary.

For $i \in I$, let Δ_i be a set with $|\Delta_i| \geqslant 2$ (this restriction is to avoid irritating special cases). For $J \subseteq I$, put $\Delta_J = \prod_{i \in J} \Delta_i$. If $K \subseteq J \subseteq I$, let π^J_K denote the natural projection from Δ_J onto Δ_K. If $K = \{k\}$, we shall often write π^J_k for π^J_K. We shall also abbreviate Δ_I to Δ and π^I_J to π_J. We shall need $\Delta_{A(i)}$ and $\pi_{A(i)}$ so often that we abbreviate them to Δ^i and π^i respectively.

We write elements of Δ as $\delta = (\delta_i)$ with $\delta_i \in \Delta_i$. For $J \subseteq I$ we define the equivalence relation $\underset{J}{\sim}$ on Δ by $\delta \underset{J}{\sim} \varepsilon$ if and only if $\delta\pi_J = \varepsilon\pi_J$.

For each $i \in I$ let G_i be a (faithful) permutation group on Δ_i with identity 1_i, and let F_i be the set of all functions from Δ^i into G_i. For $J \subseteq I$ put $F_J = \prod_{i \in J} F_i$, and let $F = F_I$. We write elements of F as $f = (f_i)$ with $f_i \in F_i$. If $K \subseteq J \subseteq I$, let φ^J_K denote the natural projection from F_J to F_K. The abbreviations φ^J_k and φ_J are used analogously to π^J_k and π_J.

Note that if J is empty then both Δ_J and F_J are singletons, and that if J is infinite then each is the full Cartesian product.

3. *Specification of the generalized wreath product*

Our aim is to identify F with a set of functions from Δ to Δ. We do this by defining an 'action' of F on Δ and showing that this action is, in a natural sense, faithful.

DEFINITION. For each $f = (f_i) \in F$, we define an *action* of f on Δ by the following rule: for each $\delta = (\delta_i) \in \Delta$,

$$\delta f = \varepsilon, \quad \text{where } \varepsilon = (\varepsilon_i) \in \Delta, \text{ and } \varepsilon_i = \delta_i(\delta\pi^i f_i);$$

that is, $(\delta f)_i = \delta_i(\delta\pi^i f_i)$.

LEMMA 1. *The action on Δ is faithful in the sense that if $f, h \in F$ and if $\delta f = \delta h$ for all $\delta \in \Delta$ then $f = h$.*

Proof. Let $i \in I$ and let $\gamma \in \Delta^i$. Put $x = \gamma f_i$ and $y = \gamma h_i$, so that $x, y \in G_i$. Let $\alpha \in \Delta_i$, and choose $\delta \in \Delta$ so that $\delta_i = \alpha$ and $\delta\pi^i = \gamma$. Then $(\delta f)_i = \alpha x$ and $(\delta h)_i = \alpha y$ so $\alpha x = \alpha y$. We can choose such a δ for every $\alpha \in \Delta_i$, and so $x = y$. Thus $\gamma f_i = \gamma h_i$ for all $\gamma \in \Delta^i$, and so $f_i = h_i$.

Lemma 1 shows that the above definition identifies F with a subset of Δ^Δ, and henceforth we shall regard F as this subset. Thus elements of F are 'multiplied' according to the composition of the corresponding functions from Δ to Δ. We wish to prove that this subset F of the semigroup Δ^Δ is a sub*group*. To do this we first need to investigate in some detail the relationship between the action of F on Δ and the partial order on I; these results (Lemmas 2, 3, and 6) are also used in subsequent sections. The proof that F is a submonoid (Lemmas 4 and 5) is immediate but the existence of inverses in F (Lemmas 7 and 8) requires a restriction on the partial order on I.

LEMMA 2. *Let J be an ancestral subset of I; let $\delta, \varepsilon \in \Delta$ and $f \in F$. If $\delta \sim_J \varepsilon$ then $\delta f \sim_J \varepsilon f$.*

Proof. Suppose that $\delta \sim_J \varepsilon$. Let $i \in J$. Then, since J is ancestral, $\delta_j = \varepsilon_j$ for all $j \geqslant i$. Thus $\delta_i = \varepsilon_i$ and $\delta\pi^i = \varepsilon\pi^i$. It follows that $(\delta f)_i = (\varepsilon f)_i$. Hence $\delta f \sim_J \varepsilon f$.

Lemma 2 shows that, if $f \in F$ and J is an ancestral subset of I, then f induces a map $\hat{f}_J : \Delta_J \to \Delta_J$ such that $f\pi_J = \pi_J \hat{f}_J$. It follows directly from the definitions that for J *any* subset of I and $f, g \in F$ we have $f\pi_J = g\pi_J$ if and only if $f\varphi_J = g\varphi_J$. Thus Lemma 2 gives us a way of defining an action of F_J on Δ_J, in the case when J is an ancestral subset, by identifying $f\varphi_J$ with $\hat{f}_J$. It is evident that this definition coincides with that obtained by defining the action of F_J on Δ_J analogously to that of F on Δ. If $J = \{j\}$ then $\Delta_J = \Delta_j$ and $\hat{f}_J = f_j$: there is therefore no ambiguity in writing f_J for $\hat{f}_J$ and f_j for $\hat{f}_{\{j\}}$.

LEMMA 3. *Let J, K be ancestral subsets of I with $K \subseteq J$, and let $f \in F$. Then $f_J\pi^J_K = \pi^J_K f_K$; in particular, $f\pi_J = \pi_J f_J$.*

LEMMA 4. *Let $f, h \in F$. Then $fh = t$, where*

$$t_i = f_i \times f_{A(i)}h_i$$

and the product of the functions f_i and $f_{A(i)}h_i$ from $\Delta_{A(i)}$ to G_i is defined pointwise.

Proof. Let $\delta \in \Delta$. Then

$$\begin{aligned}(\delta fh)_i &= (\delta f)_i(\delta f\pi^i h_i)\\ &= \delta_i(\delta\pi^i f_i)(\delta\pi^i f_{A(i)}h_i) \quad \text{(by Lemma 3)}\\ &= \delta_i(\delta\pi^i t_i).\end{aligned}$$

LEMMA 5. *If, for each $i \in I$, the function $z_i \in F_i$ is defined by $\gamma z_i = 1_i$ for all $\gamma \in \Delta^i$ then $z = (z_i)$ is the identity permutation on Δ.*

LEMMA 6. *Let J and K be ancestral subsets of I with $K \subseteq J$. Then $\varphi_K^J \colon F_J \to F_K$ is a semigroup homomorphism.*

DEFINITION. Let J be an ancestral subset of I and let $f \in F$. Then f is *invertible on J* if f_J has an inverse in F_J.

LEMMA 7. *Let $\mathscr{L}$ be a family of ancestral subsets of I; let $L = \bigcup \mathscr{L}$; let $f \in F$. If f is invertible on J for all J in $\mathscr{L}$ then f is invertible on the ancestral subset L.*

Proof. Since all the projections involved are semigroup homomorphisms, if $J, K \in \mathscr{L}$ and $i \in J \cap K$ then $(f_J)^{-1}\varphi_i^J = (f_K)^{-1}\varphi_i^K$. Hence we may define h in F_L by $h\varphi_i^L = (f_J)^{-1}\varphi_i^J$, using any $J \in \mathscr{L}$ such that $i \in J$. It is straightforward to check that $h = f_L^{-1}$.

LEMMA 8. *Suppose that (I, ρ) satisfies the maximal condition. If $f \in F$ and J is an ancestral subset of I then f is invertible on J.*

Proof. Let $X = \{i \in J : f$ is not invertible on $A[i]\}$. If X is not empty, then X contains a maximal element m. Then f is invertible on $A[i]$ for all $i > m$. By Lemma 7, f is invertible on $A(m)$, because $\bigcup_{i>m} A[i] = A(m)$. Define h in $F_{A[m]}$ by

$$h\varphi_i^{A[m]} = f_{A(m)}^{-1}\varphi_i^{A(m)} \quad \text{for all } i \in A(m)$$

and

$$\gamma(h\varphi_m^{A[m]}) = (\gamma f_{A(m)}^{-1} f_m)^{-1} \quad \text{for all } \gamma \in \Delta^m.$$

Then Lemma 4 shows that h is the inverse of $f_{A[m]}$, so f is invertible on $A[m]$. This contradiction shows that X is empty. Since $J = \bigcup_{i \in J} A[i]$, Lemma 7 shows that f is invertible on J.

The following example can be modified to show that, if (I, ρ) is any partially ordered set which contains an infinite ascending chain, then F contains an element which is not invertible on I. Thus the maximal condition in Lemma 8 is necessary.

EXAMPLE 2. Let $\mathbb{N}$ be the natural numbers with the usual ordering. For all $i \in \mathbb{N}$, let Δ_i be an arbitrary fixed 2-element set $\{a, b\}$, and let $G_i = \mathrm{Symm}(\Delta_i)$. Now, for all $i \in \mathbb{N}$, let f_i be the function in F_i which maps every element of Δ^i, with one exception, to the transposition in G_i: the exception being the element $\mathbf{a}$ in Δ^i, each of whose co-ordinates is equal to a; this element $\mathbf{a}$ is mapped to the identity of G_i. Then f is not a bijection on Δ, because its image does not include the element $\mathbf{b}$ with each co-ordinate equal to b.

Holland **[6]** and Silcock **[11]** avoid this problem by restricting the sets Δ and F; other approaches are worth investigation.

We summarize the results of this section in the following theorem.

THEOREM A. *Let (I, ρ) be a partially ordered set with the maximal condition. Then*

(i) *for all ancestral subsets J of I, (F_J, Δ_J) is a (faithful) permutation group;*

(ii) *if J and K are ancestral subsets of I with $J \supseteq K$ then (φ_K^J, π_K^J) is a permutation homomorphism from (F_J, Δ_J) onto (F_K, Δ_K) with kernel*

$$N_K^J = \{f \in F_J : f_j = z_j \text{ for } j \in K\};$$

(iii) *if J, K and L are ancestral subsets of I and $J \supseteq K \supseteq L$ then*

$$(\varphi_K^J, \pi_K^J)(\varphi_L^K, \pi_L^K) = (\varphi_L^J, \pi_L^J).$$

In particular, (F, Δ) is a permutation group, which we call the generalized wreath product of the permutation groups $(G_i, \Delta_i)_{i \in I}$. More formally, we write $\prod_{(I,\rho)}(G_i, \Delta_i)$ for this generalized wreath product.

We observe that if ρ is the identity relation on I then $\prod_{(I,\rho)}(G_i, \Delta_i)$ is the permutation Cartesian product of the (G_i, Δ_i); if I is finite, this is simply the permutation direct product. Similarly, if I is the disjoint union $I_1 \cup I_2$ and no element of I_1 is comparable with any element of I_2 then $\prod_{(I,\rho)}(G_i, \Delta_i)$ is the permutation direct product of the generalized wreath products $\prod_{(I_1,\rho)}(G_i, \Delta_i)$ and $\prod_{(I_2,\rho)}(G_i, \Delta_i)$.

At the other extreme, if (I, ρ) is the finite chain $1 \leqslant 2 \leqslant \ldots \leqslant n$ then $\prod_{(I,\rho)}(G_i, \Delta_i)$ is the permutation wreath product $(G_1, \Delta_1) \operatorname{wr} (G_2, \Delta_2) \operatorname{wr} \ldots \operatorname{wr} (G_n, \Delta_n)$. More generally, if I is the disjoint union $I_1 \cup I_2$ and, for all $i \in I_1$ and $j \in I_2$, $i \leqslant j$, then $\prod_{(I,\rho)}(G_i, \Delta_i)$ is the permutation wreath product of the generalized wreath products $\prod_{(I_1,\rho)}(G_i, \Delta_i)$ and $\prod_{(I_2,\rho)}(G_i, \Delta_i)$.

Although, as remarked above, we cannot apply our generalized wreath product construction to obtain a group if (I, ρ) is the natural numbers with the usual ordering, we can if we take the opposite ordering $\mathbb{N}^-$. This gives a class of potentially interesting examples of uncountable permutation groups on uncountable sets which are built up by an explicit construction from (possibly) finite permutation groups: for example, for all $i \in \mathbb{N}$ take $|\Delta_i| = 2$ and $G_i = \operatorname{Symm}(\Delta_i) \cong \mathbb{Z}_2$.

Of practical significance to statistics, Example 1 shows that if $|I| \geqslant 4$ then there are partial orders on I which cannot be decomposed into chains and identity relations: thus these generalized wreath products include more than the wreath and direct products and their iterated composites.

4. *Poset block structures and their automorphism groups*

DEFINITION. A *poset block structure* is a pair (Δ, S), where

(i) Δ is the Cartesian product over a partially ordered set (I, ρ) of sets Δ_i $(i \in I)$, with $|\Delta_i| \geqslant 2$,

(ii) S is the following set of equivalence relations on Δ:

$$S = \{ \underset{J}{\sim} : J \text{ an ancestral subset of } I\}.$$

These structures, and their relationship with association schemes and distributive lattices, are discussed by Speed and Bailey in [12], with the appellation 'distributive lattices of commuting uniform equivalence relations'. Their automorphism groups are described, without proof, by Bailey in [2]; the purpose of this section is to prove Theorem 1 of that paper (our Theorem B).

DEFINITION. An *automorphism* of a poset block structure (Δ, S) is a permutation t of Δ such that, for all $\sigma \in S$,

$$\delta\,\sigma\,\varepsilon \quad \text{if and only if} \quad (\delta t)\,\sigma\,(\varepsilon t) \quad (\delta, \varepsilon \in \Delta).$$

The following example shows that the weaker condition

$$\delta\,\sigma\,\varepsilon \quad \text{implies} \quad (\delta t)\,\sigma\,(\varepsilon t)$$

is not sufficient to ensure that the inverse of t also preserves the block structure.

EXAMPLE 3. Let (I, ρ) be the two-element chain $1 \leqslant 2$, let $\Delta_1 = \Delta_2 = \mathbb{N}$, and define t as follows:

$$(n,0)t = (2n,0), \quad (n,1)t = (2n+1,0),$$
$$(n,m)t = (n,m-1) \quad \text{for } m \geqslant 2.$$

THEOREM B. *Let (Δ, S) be a poset block structure with poset (I, ρ). Let F be the generalized wreath product $\prod_{(I,\rho)} \mathrm{Symm}(\Delta_i)$. If (I, ρ) satisfies the maximal condition then F is the group of automorphisms of (Δ, S).*

Proof. Let $f \in F$ and let J be an ancestral subset of I. Lemma 2 shows that

$$\delta \underset{J}{\sim} \varepsilon \quad \text{implies} \quad \delta f \underset{J}{\sim} \varepsilon f.$$

But f^{-1} is also in F, and application of Lemma 2 to f^{-1} shows that

$$\delta \underset{J}{\sim} \varepsilon \quad \text{is implied by} \quad \delta f \underset{J}{\sim} \varepsilon f.$$

Hence f is an automorphism of (Δ, S).

Now let t be an automorphism of (Δ, S). We need to prove that there are functions $t_i \in F_i$ such that $t = (t_i) \in F$.

Fix $i \in I$, and put $J = A[i]$, $K = A(i)$, so that $\Delta_K = \Delta^i$. Because t is an automorphism, there exist permutations t_J and t_K of Δ_J, Δ_K respectively such that $t\pi_J = \pi_J t_J$ and $t\pi_K = \pi_K t_K$. Also, as in Lemma 3, $\pi^J_K t_K = t_J \pi^J_K$.

Identifying Δ_J with $\Delta_i \times \Delta_K$, for $\beta \in \Delta^i$ define βt_i: $\Delta_i \to \Delta_i$ by

$$\alpha(\beta t_i) = (\alpha, \beta) t_J \pi^J_i \quad \text{for } \alpha \in \Delta_i.$$

We shall show that

(i) for all β in Δ^i, the function βt_i is a permutation of Δ_i, which shows that $t_i \in F_i$;

(ii) for all $\delta \in \Delta$, $\delta t \pi_i = \delta_i(\delta\pi^i t_i)$.

Since these results hold for all $i \in I$, we can complete the proof as follows: by (i), $(t_i) \in F$, and by (ii), $t = (t_i)$.

Proof of (i). We need a preliminary equality. For all $\alpha \in \Delta_i$ and all $\beta \in \Delta_K$,

$$(\alpha, \beta)t_J = ((\alpha, \beta)t_J\pi_i^J, (\alpha, \beta)t_J\pi_K^J) = (\alpha(\beta t_i), (\alpha, \beta)\pi_K^J t_K) = (\alpha(\beta t_i), \beta t_K).$$

To prove that βt_i is injective, suppose that $\alpha, \alpha' \in \Delta_i$ and $\alpha(\beta t_i) = \alpha'(\beta t_i)$. By the above equality, $(\alpha, \beta)t_J = (\alpha', \beta)t_J$. Since t_J is injective, $\alpha = \alpha'$.

To prove that βt_i is surjective, we observe that, because t_J is surjective, for each $\alpha \in \Delta_i$ there are $\alpha' \in \Delta_i$ and $\beta' \in \Delta_K$ such that $(\alpha', \beta')t_J = (\alpha, \beta t_K)$; then $(\alpha'(\beta' t_i), \beta' t_K) = (\alpha, \beta t_K)$. But t_K is injective, so $\beta = \beta'$ and $\alpha'(\beta t_i) = \alpha$.

Proof of (ii).

$$\delta t\pi_i = \delta t\pi_J\pi_i^J = \delta\pi_J t_J\pi_i^J = (\delta\pi_i, \delta\pi_K)t_J\pi_i^J.$$

By the definition of t_i this is $(\delta\pi_i)(\delta\pi_K t_i)$, which is $\delta_i(\delta\pi' t_i)$.

5. *Orbits on* $\Delta \times \Delta$

Throughout this section we assume that (I, ρ) satisfies the maximal condition, so that, as we established in Theorem A, (F, Δ) is a permutation group. We are interested only in the case when F is transitive on Δ, so we first state the following lemma.

LEMMA 9. *The generalized wreath product of the permutation groups* $(G_i, \Delta_i)_{i \in I}$ *is transitive if and only if* (G_i, Δ_i) *is transitive for all* $i \in I$.

When F *is* transitive on Δ, we are concerned with the orbits of F on $\Delta \times \Delta$. For subsets Γ_i of $\Delta_i \times \Delta_i$, we denote by $\bigotimes_{i \in I} \Gamma_i$ the subset of $\Delta \times \Delta$ which contains (δ, ε) if and only if $(\delta_i, \varepsilon_i) \in \Gamma_i$ for all $i \in I$. For each $i \in I$ we denote by D_i the diagonal subset $\{(\alpha, \alpha) \colon \alpha \in \Delta_i\}$ of $\Delta_i \times \Delta_i$, and put $E_i = \Delta_i \times \Delta_i$.

DEFINITIONS. Let J be an ancestral subset of I. The *border* of J, denoted $B(J)$, is the set of maximal elements of $I \backslash J$. The subset O_J of $\Delta \times \Delta$ is defined by

$$O_J = (\bigotimes_{i \in J} D_i) \otimes (\bigotimes_{i \in B(J)} (E_i \backslash D_i)) \otimes (\bigotimes_{i \in I \backslash J \backslash B(J)} E_i).$$

LEMMA 10. (i) *If J is an ancestral subset of I and* $(\delta, \varepsilon) \in O_J$ *then* $\delta \underset{J}{\sim} \varepsilon$ *and J is the maximal ancestral subset with this property.*

(ii) *The set* $\Delta \times \Delta$ *is the disjoint union of the subsets* O_J*, taken over all ancestral subsets J of I.*

We shall refer to the subsets O_J as *association sets.*

In work on the design of experiments, Δ is taken to be a set of random variables and a covariance model on Δ is specified in terms of the equivalence relations $(\underset{A[i]}{\sim})_{i \in I}$. Many authors (see, for example, John [7]), assume a model in which the value of the covariance of δ and ε depends only on the values of i for which $\delta \underset{A[i]}{\sim} \varepsilon$; that is, only on the association set O_J containing (δ, ε). Thus the decomposition of $\Delta \times \Delta$ into the association sets is a useful one to study. Other authors (see, for example, Nelder [9] and Bailey [1]) assume a model in which the covariance of δ and ε depends only on the orbit of F on $\Delta \times \Delta$ containing (δ, ε), where $F = \prod_{(I, \rho)}(G_i, \Delta_i)$ and the groups G_i are specified transitive subgroups of the Symm(Δ_i). Still other authors (see, for example, Yates [17] and Preece, Pearce, and Kerr [10]), say that *randomization* based on the group F is valid if and only if the orbits of F on $\Delta \times \Delta$ coincide with the association sets.

By Lemma 2, it is evident that each association set is a union of orbits of F on $\Delta \times \Delta$. Grundy and Healy [**5**] showed that, for the direct product $(G_1, \Delta_1) \times (G_2, \Delta_2)$, 2-transitivity of each (G_i, Δ_i) ensures that each association set is a single orbit. Bailey [**2**] proved a similar result for generalized wreath products. Here we prove the following stronger result.

THEOREM C. *The orbits of* $\prod_{(I,\rho)}(G_i, \Delta_i)$ *on* $\Delta \times \Delta$ *are precisely the association sets if and only if, for each* $i \in I$, *the permutation group* (G_i, Δ_i) *is 2-transitive.*

We shall prove Theorem C as a corollary to Theorem D. First we illustrate Theorem C by an example, and comment on its significance.

EXAMPLE 1. When F is transitive on Δ, as it is in this example, the orbits of F on $\Delta \times \Delta$ are more conveniently displayed as the orbits on Δ of the stabilizer F_δ in F of a fixed element $\delta \in \Delta$. In this case the ancestral subsets of I are $\varnothing$, $\{1\}$, $\{2\}$, $\{1,2\}$, $\{1,3\}$, $\{1,2,3\}$, $\{1,2,4\}$, and $\{1,2,3,4\}$. If each of the four groups (G_i, Δ_i) is 2-transitive, the orbits of F_δ on Δ are as shown in Table 1.

TABLE 1

Orbit	Corresponding ancestral set
$\{\delta\}$	$\{1,2,3,4\}$
$\{\varepsilon$: δ and ε are in the same micro-column, and $\delta \neq \varepsilon\}$	$\{1,2,4\}$
$\{\varepsilon$: δ and ε are in the same column and minirow, and $\delta \neq \varepsilon\}$	$\{1,2,3\}$
$\{\varepsilon$: δ and ε are in the same minirow but different columns$\}$	$\{1,3\}$
$\{\varepsilon$: δ and ε are in the same square but different minirows and different microcolumns$\}$	$\{1,2\}$
$\{\varepsilon$: δ and ε are in the same row but different minirows and different columns$\}$	$\{1\}$
$\{\varepsilon$: δ and ε are in the same column but different rows$\}$	$\{2\}$
$\{\varepsilon$: δ and ε are in different rows and different columns$\}$	$\varnothing$

The *significance* of Theorem C is that, by Theorem B, the orbits of the automorphism group of a poset block structure are precisely the combinatorially defined association sets. Nelder [**9**] showed that this is true for those poset block structures in which the partial order ρ is successively built up from chains and identity relations, and asked which other structures have this property. We have here a wider class of structures with this property, and hence a partial answer to the question. There are still other block structures, not based on posets, for which association schemes can be combinatorially defined (see Speed and Bailey [**12**]). The association sets of some of these structures are identical to the orbits of their automorphism groups, but only under fairly severe extra conditions (see Bailey [**2, 3**]). Thus a complete answer to the question does not yet seem to be known.

THEOREM D. *Suppose that, for* $i \in I$, (G_i, Δ_i) *is transitive. For* $i \in I$, *let* $M_{ij(i)}$ *for* $j(i) \in \Lambda_i$ *be the non-diagonal orbits of* G_i *on* $\Delta_i \times \Delta_i$. *Then, if* S *is any antichain in* I, *and, for* $i \in S, j(i)$ *is any element of* Λ_i, *the following subset* M *is an orbit of* F *on* $\Delta \times \Delta$:

$$M = \Big(\bigotimes_{i \in I \backslash H[S]} D_i\Big) \otimes \Big(\bigotimes_{i \in S} M_{ij(i)}\Big) \otimes \Big(\bigotimes_{i \in H(S)} E_i\Big).$$

Moreover, each orbit of F *on* $\Delta \times \Delta$ *has a unique representation of this form.*

Proof. Since $|\Delta_i| \geqslant 2$, none of the D_i or $M_{ij(i)}$ is E_i, nor is Λ_i ever empty; therefore distinct antichains S and distinct families $(j(i))_{i \in S}$ give distinct subsets of $\Delta \times \Delta$.

Now let $\delta, \varepsilon \in \Delta$. Then $(\delta, \varepsilon) \in M$ if and only if both $\delta_i = \varepsilon_i$ for all $i \in I \backslash H[S]$ and $(\delta_i, \varepsilon_i) \in M_{ij(i)}$ for all $i \in S$. Let $(\delta, \varepsilon) \in M$. If $i \in I \backslash H(S)$ then $A(i) \subseteq I \backslash H[S]$, so $\delta\pi^i = \varepsilon\pi^i$. Thus if $f \in F$ then $\delta\pi^i f_i$ and $\varepsilon\pi^i f_i$ are the *same* elements of G_i for $i \in I \backslash H(S)$, and hence $(\delta f)_i = (\varepsilon f)_i$ for $i \in I \backslash H[S]$ and $((\delta f)_i, (\varepsilon f)_i) \in M_{ij(i)}$ for $i \in S$. Therefore $(\delta f, \varepsilon f) \in M$.

Conversely, suppose that $\alpha, \beta, \gamma, \varepsilon \in \Delta$, that $(\alpha, \beta) \in M$ and $(\gamma, \varepsilon) \in M$. We must show that there is an element $f \in F$ such that $\alpha f = \gamma$ and $\beta f = \varepsilon$. We deal with coordinates in $I \backslash H(S)$ and $H(S)$ separately.

(i) If $i \in I \backslash H(S)$ then there is an element $g_i \in G_i$ such that $\alpha_i g_i = \gamma_i$ and $\beta_i g_i = \varepsilon_i$. Let $f_i \colon \Delta^i \to G_i$ be the constant function with image g_i.

(ii) If $i \in H(S)$ then there is an element $k \in S$ with $i < k$. Since $(\alpha_k, \beta_k) \in M_{kj(k)}$, we have $\alpha_k \neq \beta_k$. By transitivity, there are elements g_i and h_i in G_i such that $\alpha_i g_i = \gamma_i$ and $\beta_i h_i = \varepsilon_i$. Define $s_i \colon \Delta_k \to G_i$ by $\alpha_k s_i = g_i$ and $\omega s_i = h_i$ for all ω in $\Delta_k \backslash \alpha_k$. Let $f_i \colon \Delta^i \to G_i$ be the function $\pi_k^{A(i)} s_i$. Then $\alpha_i(\alpha\pi^i f_i) = \alpha_i g_i = \gamma_i$ and $\beta_i(\beta\pi^i f_i) = \beta_i h_i = \varepsilon_i$.

Now the element $f = (f_i)$ we have constructed maps α to γ and β to ε. Thus M is indeed an orbit of F on $\Delta \times \Delta$.

All that remains to show is that every orbit of F on $\Delta \times \Delta$ arises in this way. Let $\delta, \varepsilon \in \Delta$ and let J be the unique ancestral subset of I such that $(\delta, \varepsilon) \in O_J$. Now, $B(J)$ is an antichain. Moreover, $H(B(J)) = (I \backslash J) \backslash B(J)$, and so $I \backslash H[B(J)] = J$. By definition of O_J, for each i in $B(J)$ there is a unique $j(i)$ in Λ_i such that $(\delta_i, \varepsilon_i) \in M_{ij(i)}$. Thus, if M is defined as above for $S = B(J)$ and these $j(i)$, then $(\delta, \varepsilon) \in M$.

The proof of Theorem D gives the following, alternative, description of the orbits of F on $\Delta \times \Delta$. Each such orbit is specified by

(i) an ancestral subset J of I,

(ii) for each maximal element i of $I \backslash J$, a non-diagonal orbit M_i of G_i on $\Delta_i \times \Delta_i$.

The pair (α, β) is in the corresponding orbit if and only if

(i) $\alpha_i = \beta_i$ for all $i \in J$,

(ii) $(\alpha_i, \beta_i) \in M_i$ for each maximal element i of $I \backslash J$.

Proof of Theorem C. If (G_i, Δ_i) is 2-transitive then the only $M_{ij(i)}$ which occurs is $E_i \backslash D_i$: hence each antichain S gives just one orbit M_S. Let $J = I \backslash H[S]$. Then J is an ancestral subset of I, and $B(J) = S$; the last part of the proof of Theorem D shows that $M_S = O_J$.

To prove the converse, for $k \in I$ let $J = I \backslash H[k]$; then O_J is the union of $|\Lambda_k|$ orbits. If (G_k, Δ_k) is not 2-transitive then $|\Lambda_k| \geqslant 2$.

6. *Characters*

In this section we assume that I is finite and that, for each $i \in I$, (G_i, Δ_i) is finite and transitive. We investigate the permutation characters of the action of F on Δ and on Δ_J, for ancestral subsets J of I. Since we are using the letter π for projections, we denote the permutation character of F on Δ_J (that is, on equivalence classes of $\underset{J}{\sim}$) by ψ_J, and put $\psi = \psi_I$. Once again, the case in which every (G_i, Δ_i) is 2-transitive is particularly interesting and straightforward.

THEOREM E. *If I is finite and, for each $i \in I$, (G_i, Δ_i) is finite and 2-transitive then $\prod_{(I,\rho)}(G_i, \Delta_i)$ has distinct irreducible characters $\{\chi_J: J \subseteq I, J \text{ ancestral}\}$ such that, if K is any ancestral subset of I, the permutation character ψ_K of $\prod_{(I,\rho)}(G_i, \Delta_i)$ on the equivalence classes of $\underset{K}{\sim}$ is $\sum_{J \subseteq K} \chi_J$: in particular, ψ is the sum $\sum \chi_J$ over all ancestral subsets J. Moreover, if, for each $i \in I$, $|\Delta_i| = n_i$, the degree of χ_J is $\prod_{i \in m(J)}(n_i - 1)\prod_{i \in J\setminus m(J)} n_i$, where $m(J)$ is the set of minimal elements of J.*

We shall prove Theorem E as a corollary to Theorem F. Here we simply note that, since the ψ_J are very easy to compute in practice, so are the (irreducible) χ_J.

The permutation linear representations of (G_i, Δ_i) and (F, Δ) are afforded canonically by the vector spaces $\mathbb{R}^{\Delta_i}$ and $\mathbb{R}^{\Delta}$ respectively. Let $W_i = \mathbb{R}^{\Delta_i}$. We shall identify $\mathbb{R}^{\Delta}$ with $\bigotimes_{i \in I} W_i$ by regarding the tensor product of the functions w_i, with $i \in I$, to be the function which maps $\delta \in \Delta$ to $\prod_{i \in I} \delta_i w_i$. We shall also use the natural inner product on $\mathbb{R}^{\Delta}$ given by $v \cdot w = \sum_{\delta \in \Delta}(\delta v)(\delta w)$.

THEOREM F. *Let $F = \prod_{(I,\rho)}(G_i, \Delta_i)$. Suppose that I is finite and, for $i \in I$, (G_i, Δ_i) is finite and transitive. Let C_i be the subspace of constant functions in W_i, and let $V_{ij(i)}$ for $j(i) \in \Theta_i$ denote the other components of a direct decomposition of W_i into G_i-irreducible subspaces. Let S be an antichain in I, and, for each $i \in S$, let $j(i)$ be an element of Θ_i. Then*

(i) *the following subspace V of $\mathbb{R}^{\Delta}$ is F-irreducible:*

$$V = \Big(\bigotimes_{i \in A(S)} W_i\Big) \otimes \Big(\bigotimes_{i \in S} V_{ij(i)}\Big) \otimes \Big(\bigotimes_{i \in I\setminus A[S]} C_i\Big).$$

(ii) *Moreover, $\mathbb{R}^{\Delta}$ is the direct sum of such subspaces.*

Let

$$W = \Big(\bigotimes_{i \in A(T)} W_i\Big) \otimes \Big(\bigotimes_{i \in T} V_{ik(i)}\Big) \otimes \Big(\bigotimes_{i \in I\setminus A[T]} C_i\Big)$$

for some antichain T and some family $(k(i))_{i \in T}$ such that, for each $i \in T$, $k(i) \in \Theta_i$. Then

(iii) *$V = W$ if and only if $S = T$ and $j(i) = k(i)$ for each $i \in S$.*

(iv) *The subspaces V and W afford equivalent representations of F if and only if $S = T$ and, for each $i \in S$, the spaces $V_{ij(i)}$ and $V_{ik(i)}$ afford equivalent representations of G_i.*

Proof. (iii) Since, for each $i \in I$, $|\Delta_i| \geqslant 2$, none of the subspaces W_i or $V_{ij(i)}$ is equal to C_i and Θ_i is non-empty; therefore $V \cap W$ is the zero subspace unless $S = T$ and, for each $i \in S$, $j(i) = k(i)$.

(i) (a) Now we show that the subspace V given above is F-invariant. For each $\alpha \in \Delta_{A(S)}$ let $w_\alpha: \Delta_{A(S)} \to \mathbb{R}$ be the characteristic function of $\{\alpha\}$. Then V is spanned by the set of all functions $v: \Delta \to \mathbb{R}$ of the following form:

$$v = (\pi_{A(S)} w_\alpha) \prod_{i \in S} \pi_i v_i,$$

where $\alpha \in \Delta_{A(S)}$, and, for $i \in S$, $v_i \in V_{ij(i)}$; the product is pointwise. It is, therefore, sufficient to show that, for each $f \in F$ and each function v of this form, the function

$fv\colon \Delta \to \mathbb{R}$ is in V. Put $\beta = \alpha f_{A(S)}^{-1}$, and let $\delta \in \Delta$. Since $\pi_{A(S)} f_{A(S)} = f\pi_{A(S)}$, we have $\delta f \pi_{A(S)} = \alpha$ if and only if $\delta\pi_{A(S)} = \beta$.

Thus if $\delta\pi_{A(S)} \neq \beta$ then $\delta f v = 0$. On the other hand, if $\delta\pi_{A(S)} = \beta$ then, for each $i \in S$,

$$\delta f \pi_i v_i = \delta_i(\delta\pi^i f_i)v_i = \delta_i(\delta\pi_{A(S)}\pi^{A(S)}_{A(i)} f_i)v_i = \delta_i(\beta\pi^{A(S)}_{A(i)} f_i)v_i = \delta_i g_i v_i = \delta\pi_i g_i v_i,$$

where $g_i = \beta\pi^{A(S)}_{A(i)} f_i \in G_i$ and so $g_i v_i \in V_{ij(i)}$ because $V_{ij(i)}$ is G_i-invariant.

Thus $fv = (\pi_{A(S)} w_\beta)\prod_{i\in S}\pi_i g_i v_i$, which is in V, and so V is F-invariant.

(ii) To see that $\mathbb{R}^\Delta$ is the direct sum of such subspaces, for each $i \in I$ we let H_i be the orthogonal complement of C_i in W_i. Now for each antichain S let

$$Y_S = (\bigotimes_{i\in A(S)} W_i)\otimes(\bigotimes_{i\in S} H_i)\otimes(\bigotimes_{i\in I\setminus A[S]} C_i).$$

Then, since F acts orthogonally on $\mathbb{R}^\Delta$, Y_S is the direct sum

$$\bigoplus_{j\in\Sigma_S}((\bigotimes_{i\in A(S)} W_i)\otimes(\bigotimes_{i\in S} V_{ij(i)})\otimes(\bigotimes_{i\in I\setminus A[S]} C_i)), \tag{6.1}$$

where Σ_S is the set of families $(j(i))_{i\in S}$ such that, for all $i \in S$, $j(i) \in \Theta_i$. Moreover, if X_S is the subspace generated by $\{Y_T\colon T \text{ is an antichain}, T \subseteq A[S], \text{ and } T \neq S\}$, then $Y_S \cap X_S$ is the zero subspace. Thus the Y_S generate their direct sum. Now, $\mathbb{R}^\Delta$ is spanned by vectors of the form $w = \bigotimes_{i\in I} w_i$, where, for $i \in I$, either $w_i \in H_i$ or $w_i \in C_i$. Let $S(w) = m(\{i \in I\colon w_i \in H_i\})$. Then $S(w)$ is an antichain and $w \in Y_{S(w)}$. Thus

$$\mathbb{R}^\Delta = \bigoplus_{\substack{\text{antichains}\\ S}} Y_S.$$

(i)(b) and (iv). For each $i \in I$, denote by m_i the sum of the squares of the multiplicities of the distinct inequivalent G_i-irreducible components of H_i. Let m and m_S be the corresponding sums of squares for the F-irreducible components of $\mathbb{R}^\Delta$ and Y_S respectively. Then Proposition 29.2 of Wielandt [**16**] shows that $m_i = |\Lambda_i|$, where Λ_i is as defined in § 5; then, with Theorem D, it shows that

$$m = \sum_{\substack{\text{antichains}\\ S}} \prod_{i\in S} m_i. \tag{6.2}$$

Since $\mathbb{R}^\Delta$ is the direct sum of the F-subspaces Y_S,

$$m \geqslant \sum_{\substack{\text{antichains}\\ S}} m_S, \tag{6.3}$$

with equality if and only if, for $S \neq T$, no F-component of Y_S is equivalent to any F-component of Y_T.

Let $j, k \in \Sigma_S$. If, for all $i \in S$, the spaces $V_{ij(i)}$ and $V_{ik(i)}$ afford equivalent representations of G_i, then the direct summands in (6.1) corresponding to j and k afford equivalent representations of F. The sum of the squares of the numbers of these F-equivalent direct summands is calculated to be $\prod_{i\in S} m_i$. Since (6.1) gives Y_S, we have

$$m_S \geqslant \prod_{i\in S} m_i, \tag{6.4}$$

with equality if and only if *both* each V is F-irreducible *and* there are no more F-equivalences among the direct summands of (6.1) than those just described.

Now Equation 6.2 forces equality in Equations 6.3 and 6.4, and the result follows.

Proof of Theorem E. If (G_i, Δ_i) is 2-transitive then the orthogonal complement H_i of C_i in W_i is G_i-irreducible. Thus, for each antichain S, the set Σ_S contains just one element j, and so, by Theorem F, the subspace Y_S defined above is F-irreducible.

For each antichain S, the subset $A[S]$ is ancestral and $m(A[S]) = S$. For each ancestral subset J, the subset $m(J)$ is an antichain and $A[m(J)] = J$. Thus the maps $J \mapsto m(J)$ and $S \mapsto A[S]$ are mutually inverse bijections. For each ancestral subset J, let χ_J be the irreducible character of F afforded by $Y_{m(J)}$.

If K is any ancestral subset of I, the permutation character ψ_K of F is afforded by the subspace

$$V_K = \mathbb{R}^{\Delta_K} \otimes \bigotimes_{i \notin K} C_i.$$

Applying Theorem F to the partially ordered set K, and then tensoring the result with $\bigotimes_{i \notin K} C_i$, gives

$$V_K = \bigoplus_{\substack{\text{antichains} \\ S \subseteq K}} Y_S = \bigoplus_{\substack{\text{ancestral} \\ J \subseteq K}} Y_{m(J)},$$

and so $\psi_K = \sum_{\text{ancestral}\, J \subseteq K} \chi_J$.

From the point of view of the statistician, a permutation group (G, Γ) is useful only if the centralizer algebra $\mathscr{A}_G$ of G in $\mathbb{R}^\Gamma$ is commutative, or possibly if only the subset $\mathscr{S}_G$ of symmetric matrices in this centralizer algebra is commutative (and so forms a subalgebra) (see McLaren [**8**], and Speed, Bailey, Praeger, and Taylor [**13**]). Denote the permutation character of (G, Γ) by ψ_G.

LEMMA 11. (i) *$\mathscr{A}_G$ is commutative if and only if ψ_G is multiplicity-free.*

(ii) *$\mathscr{S}_G$ is commutative if and only if all irreducible quaternionic characters in ψ_G have multiplicity 2 and all other irreducible characters in ψ_G have multiplicity 1.*

Proof. Part (i) is well-known (see Wielandt [**16**, Theorem 29.3]), whilst (ii) is a slight modification of (i), and is proved by McLaren [**8**] and Speed et al. [**13**].

COROLLARY TO THEOREM F. *Under the hypotheses of Theorem F:*

(i) *$\mathscr{A}_F$ is commutative if and only if $\mathscr{A}_{G_i}$ is commutative for all $i \in I$;*

(ii) *$\mathscr{S}_F$ is commutative if and only if $\mathscr{S}_{G_i}$ is commutative for all $i \in I$ and there is no two-element antichain $\{i, j\}$ such that ψ_{G_i} includes quaternionic characters and ψ_{G_j} includes either non-real or quaternionic characters.*

For example, if I is the two-element antichain $\{1, 2\}$ and, for $i = 1, 2$, (G_i, Δ_i) is the regular representation of the quaternion group Q_8, then $\mathscr{S}_{G_1}$ and $\mathscr{S}_{G_2}$ are both commutative but $\mathscr{S}_F$ is not.

References

1. R. A. BAILEY, 'A unified approach to design of experiments', *J. Roy. Statist. Soc. Ser. A*, 144 (1981), 214–223.
2. R. A. BAILEY, 'Distributive block structures and their automorphisms', *Combinatorial mathematics VIII* (ed. K. L. McAvaney), Lecture Notes in Mathematics 884 (Springer, Berlin, 1981), pp. 115–124.
3. R. A. BAILEY, 'Latin squares with highly transitive automorphism groups', *J. Austral. Math. Soc. Ser. A*, 32 (1982), 18–22.
4. G. GRÄTZER, *Lattice theory. First concepts and distributive lattices* (W. H. Freeman, San Francisco, 1971).

WREATH PRODUCTS OF PERMUTATION GROUPS

5. P. M. Grundy and M. J. R. Healy, 'Restricted randomization and quasi-Latin squares', *J. Roy. Statist. Soc. Ser. B*, 12 (1950), 286–291.
6. W. C. Holland, 'The characterization of generalized wreath products', *J. Algebra*, 13 (1969), 152–172.
7. P. W. M. John, *Statistical design and analysis of experiments* (Macmillan, New York, 1971).
8. A. D. McLaren, 'On group representations and invariant stochastic processes', *Proc. Cambridge Philos. Soc.*, 59 (1963), 431–450.
9. J. A. Nelder, 'The analysis of randomized experiments with orthogonal block structure', *Proc. Roy. Soc. London Ser. A*, 283 (1965), 147–178.
10. D. A. Preece, S. C. Pearce, and J. R. Kerr, 'Orthogonal designs for three-dimensional experiments', *Biometrika*, 60 (1973), 349–358.
11. H. L. Silcock, 'Generalized wreath products and the lattice of normal subgroups of a group', *Algebra Universalis*, 7 (1977), 361–372.
12. T. P. Speed and R. A. Bailey, 'On a class of association schemes derived from lattices of equivalence relations', *Algebraic structures and applications* (eds. P. Schultz, C. E. Praeger, and R. P. Sullivan, Marcel Dekker, New York, 1981), pp. 45–54.
13. T. P. Speed, R. A. Bailey, C. E. Praeger, and D. E. Taylor, *The analysis of variance*, to appear.
14. T. N. Throckmorton, 'Structures of classification data', Ph.D. thesis, Iowa State University, 1961.
15. C. Wells, 'Some applications of the wreath product construction', *Amer. Math. Monthly*, 83 (1976), 317–338.
16. H. Wielandt, *Finite permutation groups* (Academic Press, New York, 1964).
17. F. Yates, 'The formation of Latin squares for use in field experiments', *Empire J. Exp. Agric.* (1933), 235–244.

R. A. Bailey
Statistics Department
Rothamsted Experimental Station
Harpenden
Herts. AL5 2JQ

C. A. Rowley
Mathematics Faculty
The Open University
Milton Keynes
MK7 6AA

C. E. Praeger *and* T. P. Speed
Mathematics Department
University of Western Australia
Nedlands 6009
Western Australia

The Annals of Statistics,
1983, Vol. 11, No. 4, 1069–1085

BALANCE IN DESIGNED EXPERIMENTS WITH ORTHOGONAL BLOCK STRUCTURE

BY A. M. HOUTMAN AND T. P. SPEED

A. C. Nielsen and CSIRO

The notion of general balance due to Nelder is discussed in relation to the eigenvectors of an information matrix, combinatorial balance and the simple combinability of information from uncorrelated sources in an experiment.

1. Introduction. This paper is about the notion of general balance (GB) introduced by Nelder (1965) in two papers on designed experiments with orthogonal block structure. Nelder defined (GB) as a relationship between the block structure or dispersion model for the data and the treatment structure or model for the expected value of the data. It embodies and unifies three important and apparently unrelated ideas concerning designed experiments: the usefulness of eigenvectors of the associated information matrices, the combinatorial and statistical notions of balance, and the simple combinability of information from different, uncorrelated, sources in the experiment. These ideas have been discussed independently by a number of authors including Yates (1936, 1939, 1940), Sprott (1956), Morley Jones (1959), Pearce (1963), Martin and Zyskind (1966), Corsten (1976) and many others. We will review the work of these authors in Section 3 and relate it to Nelder's (1965) work.

Nelder (1965, 1968) has shown how a simple and unified approach may be adopted to the analysis of multistratum designed experiments satisfying (GB), including the estimation of stratum variances and the combination of information across strata. We summarise these facts in Section 4 and also prove a useful supplementary result: that (GB) is not only a sufficient but also a necessary condition (assuming known stratum variances) for the simple recovery of all information on every contrast from every stratum in which it is estimable. Our definition of (GB) is slightly different from Nelder's in that we accommodate unequal treatment replications, but it has all the same consequences, and the broad scope of the notion so defined is underlined by the fact that all block designs with equal block size are then generally balanced (assuming the standard dispersion model). It will be seen from our examples and the associated discussion that essentially all designs with orthogonal block structure which have ever been recommended for use satisfy (GB). It also provides a convenient basis for the classification of designs, one which is connected with the simple and directly interpretable analysis.

Section 5 below is devoted to examples, beginning with the balanced incomplete block design (BIBD) which is the prototype of all designs satisfying (GB). Instead of going on to prove directly that partially balanced incomplete block designs (PBIBDs) all satisfy (GB), we obtain the same conclusion for their natural generalisations to more general block structures. Following a brief discussion of some further examples, we close the paper with a row-column design *not* satisfying (GB).

2. Basic framework.

2.1. *Treatment structure.* Our data will be viewed as a random array $y = (y_i)_{i \in \mathbf{I}}$ indexed by a set $\mathbf{I}$ of $n = |\mathbf{I}|$ *unit labels* and taking values in the vector space $\mathscr{D} = \mathbb{R}^{\mathbf{I}}$

Received May 1982, revised February 1983.

AMS 1980 *subject classifications.* Primary 62K05B.

Key words and phrases. Balance, designed experiment, recovery of information, simple combinability.

which has the inner product $\langle c \mid d \rangle = \sum_i c_i d_i$ and squared norm $\| c \|^2 = \langle c \mid c \rangle$. The models we consider for $\tau = \mathbb{E}y$, termed the *treatment structure*, will all be *linear*, i.e. of the form

$$\mathbb{E}y \in \mathscr{T} \tag{2.1}$$

where $\mathscr{T} \subseteq \mathscr{D}$ is a linear subspace of $\mathscr{D}$. In the theory of designed experiments this usually arises as follows: we have a set $\mathscr{X}$ of $v = |\mathscr{X}|$ *treatment labels*, a *design map* $x: \mathbf{I} \to \mathscr{X}$ which assigns a treatment to each unit, and a *design matrix* X satisfying $X(i, u) = 1$ if $x(i) = u$, $i \in I$, $u \in \mathscr{X}$, and $= 0$ otherwise. In this case $\mathscr{T} = \mathscr{R}(X)$, the range of X, and $\tau = X\alpha$ for some $\alpha \in \mathbb{R}^{\mathscr{X}}$. However none of the general discussion which follows assumes that $\mathscr{T}$ arises in this way. The (unweighted) orthogonal projection of $\mathscr{D}$ onto $\mathscr{T}$ will be denoted by T; if $\mathscr{T} = \mathscr{R}(X)$ then $T = X(X'X)^{-1}X'$.

A vector $c = (c_i) \in \mathscr{D}$ of constants satisfying $\sum c_i = 0$ is said to define (or be) a *contrast*; if $c \in \mathscr{T}$, then c defines (or is) a *treatment contrast*. This usage arises because least-squares estimation concentrates on the estimation of linear functions $\langle t \mid \tau \rangle$ of $\tau = \mathbb{E}y$ ($t \in \mathscr{T}$) based upon linear functions $\langle c \mid y \rangle$ of the data. Thus the term contrast refers in each case to the coefficients of these linear functions. In many analyses interest focuses on treatment contrasts $\langle t \mid \tau \rangle$ defined by elements t of specific *subspaces* of $\mathscr{T}$; for examples, we refer to Section 5 below. When $\mathscr{T} = \mathscr{R}(X)$ we say that *simple* treatment contrasts are those elements $t_{u,v} \in \mathscr{T}$ for which $\langle t_{u,v} \mid \tau \rangle$ is proportional to $\alpha_u - \alpha_v$, $u, v \in \mathscr{X}$, where $X\alpha = \tau$.

2.2. *Block structure.* Following Nelder (1965) we use the term *block structure* to mean the model for the dispersion matrix $V = \mathbb{D}y$, and all our models for V will have the form

$$\mathbb{D}y \in \mathscr{V} \tag{2.2}$$

where $\mathscr{V}$ is a suitably parameterized set of positive semi-definite (p.s.d.) matrices. We will say that we have *orthogonal block structure* (OBS) when $\mathscr{V}$ consists of all p.s.d. matrices $V(\xi) = \sum_\alpha \xi_\alpha S_\alpha$, where $\xi_\alpha \geq 0$ for all α, and the $\{S_\alpha\}$ are a family of known pairwise orthogonal projectors summing to the identity matrix, i.e. $S_\alpha = S'_\alpha = S^2_\alpha$, $S_\alpha S_\beta = S_\beta S_\alpha = 0$ if $\alpha \neq \beta$, and $\sum_\alpha S_\alpha = I$, the identity matrix. We call this representation of $V(\xi)$ its spectral form. In the theory of designed experiments such models usually arise in the following way: there is a system $\{A_a\}$ of *association matrices* defined over the set $\mathbf{I}$ of unit labels, and the dispersion matrix $V = \mathbb{D}y$ has the form $V = \sum_a \gamma_a A_a$ where $\{\gamma_a\}$ is a set of *covariances* varying freely subject only to the constraints ensuring that V is p.s.d. If the matrices $\{A_a\}$ satisfy the requirements of an *association scheme* then there always exist matrices $P = (p_{\alpha a})$ and $Q = (q_{a\alpha})$ of coefficients such that $S_\alpha = (1/n) \sum_a q_{a\alpha} A_a$ satisfies the properties listed above, and $\xi_\alpha = \sum_\alpha p_{\alpha a} \gamma_a$ constitutes an invertible linear reparametrization; see MacWilliams and Sloane (1978, Chapter 21, especially Section 2) for definitions and the results cited. Once more we remark that the general results which follow do not assume that our orthogonal block structure arose in this way although in practice the vast majority (block, row-column, split-plot designs etc.) do so. For example, any model $\mathscr{V}$ whose elements have the form $V = \sum_j \theta_j C_j$, where the $\{C_j\}$ are known symmetric idempotent matrices which *commute*, will be a submodel of a model of the form (OBS) above as the $\{C_j\}$ are simultaneously diagonalizable, but in general there will be more ξs than θs.

Summarising, we will be supposing that our data y is modeled by (2.1) and (2.2) where $\mathscr{T}$ is a linear subspace of $\mathscr{D}$ and $\mathscr{V}$ satisfies (OBS). The subspaces $\mathscr{S}_\alpha = \mathscr{R}(S_\alpha)$ are termed the *strata* of the dispersion model, the $\{S_\alpha\}$ are *strata projectors* and the $\{\xi_\alpha\}$ the *strata variances* (for it is easy to see that $\mathbb{D}S_\alpha y = \xi_\alpha S_\alpha$). *Multi-strata designs* are those with two or more strata variances in the dispersion model.

2.3. *Examples.*

EXAMPLE 1. The data y from an experiment consisting of v treatments applied across b blocks of k plots each are usually analysed under the *mixed model*

(2.3) $$y = X\alpha + Z\gamma + \varepsilon,$$

where X and Z are the $n \times v$ and $n \times b$ treatment and block incidence matrices, respectively, α is a $v \times 1$ vector of treatment parameters, and γ is a $b \times 1$ vector of zero-mean block effects having dispersion matrix $\sigma_b^2 I_b$ uncorrelated with the $n \times 1$ vector ε of errors which have dispersion matrix $\sigma^2 I_n$.

The dispersion matrix associated with such a model is $V = \sigma_b^2 ZZ' + \sigma^2 I_n$, and its spectral form is

(2.4) $$V = \xi_0 G + \xi_1(B - G) + \xi_2(I_n - B)$$

where $G = n^{-1}11'$ is the grand mean averaging operator (1 is the $n \times 1$ vector of ones), $B = k^{-1}ZZ'$ is the block averaging operator, $\xi_0 = \xi_1 = k\sigma_b^2 + \sigma^2$ and $\xi_2 = \sigma^2$. Note that here we have the constraint $\xi_0 = \xi_1 \geq \xi_2 > 0$.

A *randomisation model* for y, see Nelder (1954), would generate a dispersion matrix of the form (2.4).

In order to include both types of model, we will assume when analysing data from block designs with equal block size (which are the only sort we consider) that $S_0 = G$, $S_1 = B - G$ and $S_2 = I - B$ defines our block structure satisfying (OBS). It will be simpler, and necessary for most results, to assume $\xi_0 > 0$, $\xi_1 > 0$ and $\xi_2 > 0$ as well. □

EXAMPLE 2. The data y from an experiment in which v treatments are allocated to the $n = rc$ plots of a row-column design consisting of r rows and c columns are usually analysed under the *mixed model*

(2.5) $$y = X\alpha + Z_1\gamma_1 + Z_2\gamma_2 + \varepsilon$$

where X, Z_1 and Z_2 are the treatment, row and column incidence matrices, respectively, and γ_1, γ_2 and ε are uncorrelated zero-mean vectors having dispersion matrices $\sigma_r^2 I_r$, $\sigma_c^2 I_c$ and $\sigma^2 I_n$, respectively.

This time the dispersion matrix of y is $V = \sigma_r^2 Z_1 Z_1' + \sigma_c^2 Z_2 Z_2' + \sigma^2 I_n$ and its spectral form is

(2.6) $$V = \xi_0 G + \xi_1(R - G) + \xi_2(C - G) + \xi_3(1 - R - C + G)$$

where $G = (rc)^{-1}11'$, $R = c^{-1}Z_1Z_1'$ and $C = r^{-1}Z_2Z_2'$, $\xi_0 = c\sigma_r^2 + r\sigma_c^2 + \sigma^2$, $\xi_1 = c\sigma_r^2 + \sigma^2$, $\xi_2 = r\sigma_c^2 + \sigma^2$ and $\xi_3 = \sigma^2$. Again we have constraints: $\xi_1 \geq \xi_3 > 0$, $\xi_2 \geq \xi_3 > 0$ and $\xi_0 = \xi_1 + \xi_2 - \xi_3$.

A *randomisation model* for y would also generate a dispersion matrix of the form (2.6). Accordingly we will analyse row-column designs below with $S_0 = G$, $S_1 = R - G$, $S_2 = C - G$ and $S_3 = I - R - C + G$, a block structure satisfying (OBS). Again we will usually assume that $\xi_0 > 0$, $\xi_1 > 0$, $\xi_2 > 0$ and $\xi_3 > 0$. □

2.4. *Designed experiments.* The *design* of an experiment, i.e. the actual allocation of treatments to units, affects the least-squares analysis (under our model) of the data generated through the relationships it determines between the treatment subspace $\mathscr{T}$ and the strata subspaces $\{\mathscr{S}_\alpha\}$. For example, it is known that if T *commutes* with all the $\{S_\alpha\}$, then the analysis is easy; such designs are known as *orthogonal* designs, a class which includes completely randomised, randomised block, latin square and split-plot designs. For other designs, such as the balanced incomplete block designs (BIBDs), this commutativity fails, and a more elaborate analysis is required. Nelder's (1965) notion of general balance (GB) describes a relationship between T and the $\{S_\alpha\}$ which generalises, but in a sense is no more difficult than, that which arises with a BIBD, and as a consequence we find that essentially all designed experiments may be analysed in a manner almost identical to that of a BIBD. Note that the $\{C_i\}$ of Nelder (1965) correspond to our $\{S_\alpha\}$. Before giving any further details of these ideas, we devote the next section to reviewing the antecedents of general balance and clarifying its connections with similar notions which have appeared since 1965. See also Bailey (1981) for a related discussion.

3. Eigenvectors, balance and simple combinability.

3.1. *Eigenvectors of information matrices.* It has long been known in linear regression analysis that contrasts which are eigenvectors of the information matrix have special properties which make inference concerning them particularly straightforward; the analogy with principal components analysis explains why this is so. However it appears that Morley Jones (1959) was the first person to examine these ideas in some detail in the context of block experiments, and because of their relevance to general balance we will summarise his results within the framework introduced in Example 1 of the previous section.

Morley Jones analysed the data y under the "fixed block effects" model: $\mathbb{E}y \in \mathcal{T} + \mathcal{B}$, $\mathbb{D}y \in \mathcal{V}$ where $\mathcal{B} = \mathcal{R}(B)$ and $\mathcal{V} = \{\sigma^2 I : \sigma^2 > 0\}$, and he concentrated upon the *intra-block analysis*, i.e. that using the reduced data $\bar{B}y$ ($\bar{B} = I - B$) consisting of the observations adjusted by their block means. Clearly $\mathbb{E}\bar{B}y \in \bar{B}\mathcal{T}$ and $\mathbb{D}\bar{B}y \in \bar{B}\mathcal{V}\bar{B}$, and the task of minimising $\| \bar{B}y - \bar{B}\tau \|^2$ over $\tau \in \mathcal{T}$ is equivalent to solving the reduced normal equations ("eliminating blocks"):

$$T\bar{B}T\tau = T\bar{B}y$$

for $\tau \in \mathcal{T}$. In this context the eigenvectors and eigenvalues of the information matrix $T\bar{B}T$ are likely to be of interest. (In fact Morley Jones studied a closely-related matrix with the same eigenvectors but eigenvalues one minus those of $T\bar{B}T$.) He made the following observations: (a) an element $t \in \mathcal{T}$ is an eigenvector of $T\bar{B}T$ iff there exists a constant k such that for all $u \in \mathcal{T}$, $\langle u \mid (B - G)t \rangle = k\langle u \mid \bar{B}t \rangle$; (b) if one of two orthogonal treatment contrasts t and u is an eigenvector of $T\bar{B}T$, then their inter-block components Bt, Bu (resp. intra-block components $\bar{B}t$, $\bar{B}u$) are also orthogonal; (c) the best linear unbiased estimators (BLUEs) of contrasts $\langle t \mid \tau \rangle$ defined by eigenvectors of $T\bar{B}T$ are easy to compute, as are their precisions, and these are related to the corresponding eigenvalue; (d) the eigenvalues of $T\bar{B}T$ are directly related to the Fisher *efficiency factors* describing the relative loss of information occurring by restricting attention only to the intrablock analysis; and (e) normalised contrasts defined by eigenvectors of $T\bar{B}T$ corresponding to the same eigenvalue are estimated with the same precision; in particular, all contrasts are estimated with the same precision in BIBDs.

Although not explicitly referring to eigenvector contrasts, similar ideas can be found in Kurkjian and Zelen (1963). Their "property A" is equivalent to the spectral decomposition $T\bar{B}T = \sum_\beta \lambda_\beta T_\beta$ where the $\{T_\beta\}$ are the orthogonal projections decomposing $\mathcal{T}$ into subspaces $\{\mathcal{T}_\beta\}$ corresponding to main effects and interactions in a factorial experiment laid out in blocks. Their conclusions included (c) above, with the BLUE of $\langle t_\beta \mid \tau \rangle$ based upon $\bar{B}y$ being $\lambda_\beta^{-1}\langle t_\beta \mid \bar{B}y \rangle$ for an arbitrary $t_\beta \in \mathcal{T}_\beta$, having variance $\sigma^2\lambda_\beta^{-1} \| t_\beta \|^2$, and they observed that BLUEs of contrasts defined by elements of the different subspaces $\{\mathcal{T}_\beta\}$ are uncorrelated (cf. (b) above). They also applied their results to other types of incomplete block designs including group divisible and direct product designs. A further paper, Zelen and Federer (1964) extended the same ideas to row-column designs, but still only in the context of the lowest stratum analysis, i.e. that based upon $(I - R - C + G)y$; cf. Example 2 above.

In Pearce, Caliński and Marshall (1974) the eigenvectors of $T\bar{B}T$ are called "basic contrasts", and these authors note that those with eigenvalue 1 can be estimated with full efficiency in the intra-block analysis, those with eigenvalue 0 are "totally confounded" with blocks, whilst the remainder are "partially confounded". They recommend that the spectral decomposition of $T\bar{B}T$ be used by experimenters to ensure that the design permits contrasts of particular interest to be estimated with maximum efficiency in the intra-block analysis.

Corsten's (1976) *canonical analysis* is also equivalent to the spectral analysis of $T\bar{B}T$. He calls the eigenvectors (with non-zero eigenvalues) "identifiable contrasts" and views the corresponding eigenvalues as the squared cosines of the *canonical angles* between the subspaces $\mathscr{T}$ and $\mathscr{B}^{\perp}$ the orthogonal complement of $\mathscr{B}$; the same geometric approach is used by James and Wilkinson (1971).

3.2. *Balance.* BIBDs were introduced by Yates (1935) as incomplete block designs with equal block sizes, equal replications, and having the combinatorial property that every pair of distinct treatments appeared together in a block the same number of times. It followed that simple treatment contrasts were all estimated with the same precision, and as a consequence, that normalised treatment contrasts were also estimated with the same precision. Thus combinatorial balance was related to the property of sets of contrasts being estimated with the same precison.

Generalised forms of these ideas appeared soon afterwards: PBIBDs were introduced by Bose and Nair (1939); designs with unequally replicated treatments having a restricted form of balance were studied by Nair and Rao (1942); designs with supplemented balance by Hoblyn, Pearce and Freeman (1954), and Pearce (1960, 1963). Morley Jones (1959) continued this line of development.

Balance in block designs was first linked to the spectral properties of the intra-block information matrix (or a closely related matrix) by V. R. Rao (1958) and Morley Jones (1959). The latter proved that a block design is balanced with respect to a set of treatment contrasts iff those contrasts span a subspace of an eigenspace of $T\bar{B}T$. The combinatorial aspects of balance are reviewed in Raghavarao (1971), although we will see that the approach through general balance is more relevant to the problem of analysing data from an experiment with a design exhibiting the given type of balance.

3.3. *Simple combinability.* The term *recovery of interblock information* has come to mean the double task of estimating the relevant strata variances and the calculation of weighted combinations of the inter- and intra-block estimates (where this is appropriate) of a given treatment contrast. Following earlier work with cubic lattice designs, Yates (1939), Yates (1940) showed that the overall (weighted least squares) BLUE of any treatment contrast in a BIBD was the linear combination of its BLUE calculated using the intra-block data $(I - B)y$ and that calculated using the inter-block data $(B - G)y$, each weighted inversely according to its variance. We shall call this result, which assumes that the strata variances are known, the property of *simple combinability*, which is valid for *all* contrasts in a BIBD. Yates also gave a method of estimating the usually unknown strata variances from the anova table.

Conditions on a design which ensure the simple combinability in PBIBDs of certain sets of treatment contrasts were described by Sprott (1956) in a paper which gave great insight into the relation between combinability and combinatorial balance. In particular Sprott showed that the property of simple combinability holds for all contrasts in a PBIBD only if the design is actually a BIBD. This and other results along the same lines are special cases of a general theorem proved in the next section.

A link between the spectral properties of $T\bar{B}T$ and simple combinability in an incomplete block design was established by Zyskind and Martin (1966), who showed that a treatment contrast is simply combinable iff it is an eigenvector of $T\bar{B}T$. Thus these three topics: the eigenspaces of $T\bar{B}T$, balance, in either the combinatorial sense or in the statistical sense of contrasts being estimable with the same precision, and simple combinability are all seen to be intimately related. With this introduction to general balance we now turn to its definition and study.

4. General balance. As we have explained in Section 2 above, our model for the data $y = (y_i)_{i \in I}$ associated with our designed experiment is given by (2.1) $\mathbb{E}y \in \mathscr{T}$ and (2.2) $\mathbb{D}y \in \mathscr{V}$, where $\mathscr{T} \subseteq \mathscr{D}$ is a linear subspace and $\mathscr{V} = \{V(\xi): V(\xi) = \sum_\alpha \xi_\alpha S_\alpha,$

$\xi_\alpha > 0$ for all $\alpha\}$ is a dispersion model satisfying (OBS). General balance is a structural property relating $\mathcal{T}$ and the strata $\{\mathcal{S}_\alpha\}$.

4.1. *Definition of* (GB). We say that a design with (OBS) defined by $\{S_\alpha\}$ and treatment structure $\mathcal{T}$ is *generally balanced* with respect to the decomposition $\mathcal{T} = \oplus_\beta \mathcal{T}_\beta$ or just generally balanced if there exists a matrix $(\lambda_{\alpha\beta})$ of numbers such that for all α

$$\text{(GB)} \qquad TS_\alpha T = \textstyle\sum_\beta \lambda_{\alpha\beta} T_\beta,$$

where the $\{T_\beta\}$ are the orthogonal projectors onto the subspaces $\{\mathcal{T}_\beta\}$. It is clear that (GB) is equivalent to the requirement that the matrices $\{TS_\alpha T\}$ are simultaneously diagonalisible, with the $\{\mathcal{T}_\beta\}$ as their common eigenspaces. Another equivalent form is the following: there exists numbers $(\lambda_{\alpha\beta})$ such that for all α, β and β'

$$T_\beta S_\alpha T_{\beta'} = \begin{cases} \lambda_{\alpha\beta} T_\beta & \text{if } \beta = \beta', \\ 0 & \text{otherwise.} \end{cases}$$

Since the $\{S_\alpha\}$ and $\{T_\beta\}$ are all projectors, we must have $0 \le \lambda_{\alpha\beta} \le 1$ for all α and β, and it follows from $\sum_\alpha S_\alpha = I$ that for all β, $\sum_\alpha \lambda_{\alpha\beta} = 1$. A statistical interpretation of the $\lambda_{\alpha\beta}$ as *efficiency* factors will be explained in Section 4.3 below, and we refer to Fisher (1935) for the first use of such a two-way array. *Orthogonal designs* are just those for which each $\lambda_{\alpha\beta}$ is 0 or 1.

4.2. *Overall analysis assuming* (GB)*: known strata variances.* It is well known that the BLUE of $\tau = \mathbb{E}y$ based on y is given by the solution $\tau \in \mathcal{T}$ of the normal equation

$$\text{(NE)} \qquad TV^{-1}T\tau = TV^{-1}y;$$

equivalently, that it is given by $\hat{\tau} = Uy$ where $U = P^V_{\mathcal{T}}$ is projection of $\mathcal{D}$ onto $\mathcal{T}$ orthogonal with respect to the weighted inner product $\langle c \mid d\rangle_V := \langle c \mid V^{-1}d\rangle$. Yet one further statement of this (Gauss's) result is the following: $\langle t \mid \hat{\tau}\rangle$ is the unique BLUE of $\langle t \mid \tau\rangle$ for every $t \in \mathcal{T}$.

Now $TV^{-1}T = \sum_\beta \nu_\beta T_\beta$ under (GB), where we write $\nu_\beta = \sum_\alpha \lambda_{\alpha\beta}\xi_\alpha^{-1}$, and so the unique matrix inverse of $TV^{-1}T$ on the subspace $\mathcal{T}$ is $\sum_\beta \nu_\beta^{-1} T_\beta$. Consequently the solution $\hat{\tau} = Uy$ of (NE) is given by

$$(4.1) \qquad U = \textstyle\sum_{\alpha,\beta} w_{\alpha\beta}\lambda_{\alpha\beta}^{-1} T_\beta S_\alpha$$

where we have written $w_{\alpha\beta} = \nu_\beta^{-1}\xi_\alpha^{-1}\lambda_{\alpha\beta}$. This expression is called the *weight* for the treatment term β within stratum α, a name which we will shortly justify. Here and later all summations involving $\lambda_{\alpha\beta}^{-1}$ will be restricted only to those α or β for which $\lambda_{\alpha\beta} > 0$.

As we have already observed, the unique BLUE of $\langle t \mid \tau\rangle$ for $t \in \mathcal{T}$ is $\langle t \mid \hat{\tau}\rangle$ and by (4.1) this is just

$$(4.2) \qquad \langle t \mid \hat{\tau}\rangle = \textstyle\sum_{\alpha,\beta} w_{\alpha\beta}\lambda_{\alpha\beta}^{-1}\langle t \mid T_\beta S_\alpha y\rangle$$

with *variance* $\sum_\beta \nu_\beta^{-1} \| T_\beta t \|^2$. If $t = t_\beta \in \mathcal{T}_\beta$, the BLUE simplifies to

$$(4.3) \qquad \langle t_\beta \mid \hat{\tau}\rangle = \textstyle\sum_\alpha w_{\alpha\beta}\lambda_{\alpha\beta}^{-1}\langle t_\beta \mid S_\alpha y\rangle$$

with variance $\nu_\beta^{-1} \| t_\beta \|^2$.

Finally, the *covariance* between two BLUEs $\langle t_1 \mid \hat{\tau}\rangle$ and $\langle t_2 \mid \hat{\tau}\rangle$ is just

$$\textstyle\sum_\beta \nu_\beta^{-1}\langle T_\beta t_1 \mid T_\beta t_2\rangle,$$

and if $t_1 \in \mathcal{T}_\beta$, $t_2 \in \mathcal{T}_{\beta'}$, $\beta \neq \beta'$, this reduces to zero.

It is clear from the above that as long as the strata variances are known (up to a common scalar multiplier) and we can readily effect the projections $\{S_\alpha\}$ and $\{T_\beta\}$, the weighted least squares analysis of data from a designed experiment with generally balanced block structure is particularly simple. We will deal with the problem of unknown strata variances in the next subsection and in Section 4.5 below. On the issue of the ease of calculation and computation of the projections we can say this: the $\{S_\alpha\}$ are commonly built up from simple averaging operators such as G and B in Example 1 or R, C and G in Example 2 above, and rarely give any difficulties. The common decompositions $\{T_\beta\}$ relative to which designed experiments satisfy (GB) are also of this form, although there are some that are quite different, and in general the problem is not: "how do we compute the projections $\{T_\beta\}$?" but: "how do we discover them?" This is essentially a combinatorial problem, which needs to be done for each new design or class of designs. The usual mathematical skills (trial and error, ingenuity, etc.) help, as does the occasional computer-aided spectral analysis, and it is only the broader classes of block designs for which general solutions are unavailable; see Section 5.4.

4.3. *Within strata analysis assuming* (GB). A reduction of the full data y to its strata projections $S_\alpha y$ permits analyses *within strata* without knowledge of the strata variances, for $\mathbb{E}S_\alpha y \in S_\alpha \mathcal{T}$ and $\mathbb{D}S_\alpha y = \xi_\alpha S_\alpha$; in particular, the dispersion matrix of $S_\alpha y$ is known up to a scalar, and this is adequate for the usual least-squares analyses.

The least-squares *fitted value* $\hat{y}_\alpha$ of y in stratum α is $\hat{y}_\alpha = P_{S_\alpha \mathcal{T}} y$, the unweighted projection of y onto $S_\alpha \mathcal{T}$, unweighted because the subspace $S_\alpha \mathcal{T}$ is invariant under $\mathbb{D}S_\alpha y$ whence unweighted and weighted projectors coincide. The normal equation within $\mathcal{S}_\alpha$ is

$$(\mathrm{NE}_\alpha) \qquad TS_\alpha T\tau = TS_\alpha y$$

and its solution $\hat{\tau}_\alpha = U_\alpha y$ is given by (cf. Nelder (1965) equation 3.3)

$$(4.4) \qquad U_\alpha y = \textstyle\sum_\beta \lambda_{\alpha\beta}^{-1} T_\beta S_\alpha y$$

where the sum is only over those β for which $\lambda_{\alpha\beta} > 0$. We can readily prove that $P_{S_\alpha \mathcal{T}} = S_\alpha U_\alpha$. It follows from (4.4) that the unique BLUE of a contrast $\langle t \mid \tau \rangle$ which is estimable in $\mathcal{S}_\alpha$ (i.e. for which there exists a BLUE based on $S_\alpha y$) is

$$(4.5) \qquad \langle t \mid \hat{\tau}_\alpha \rangle = \textstyle\sum_\beta \lambda_{\alpha\beta}^{-1} \langle T_\beta t \mid S_\alpha y \rangle$$

with variance $\xi_\alpha \sum_\beta \lambda_{\alpha\beta}^{-1} \| T_\beta t \|^2$. If $t = t_\beta \in \mathcal{T}_\beta$ the BLUE simplifies to

$$(4.6) \qquad \langle t_\beta \mid \hat{\tau}_\alpha \rangle = \lambda_{\alpha\beta}^{-1} \langle t_\beta \mid S_\alpha y \rangle \quad \text{(provided } \lambda_{\alpha\beta} > 0)$$

with variance $\lambda_{\alpha\beta}^{-1} \xi_\alpha \| t_\beta \|^2$, and if $\lambda_{\alpha\beta} = 0$ then *no* contrast $\langle t_\beta \mid \tau \rangle$ is estimable in $\mathcal{S}_\alpha$. Finally, we remark that the covariance between two BLUEs $\langle t_1 \mid \hat{\tau}_\alpha \rangle$ and $\langle t_2 \mid \hat{\tau}_\alpha \rangle$ is $\xi_\alpha \sum_\beta \lambda_{\alpha\beta}^{-1} \langle T_\beta t_1 \mid T_\beta t_2 \rangle$ and if $t_1 \in \mathcal{T}_\beta$, $t_2 \in \mathcal{T}_{\beta'}$, $\beta \neq \beta'$, this again reduces to zero.

There are a number of points in the formulae above and in the corresponding ones in the previous sub-section which are worth noting. First, it is clear from both (4.6) and (4.3) that estimation *is* especially simple for contrasts which are eigenvectors of *all* the information matrices $TS_\alpha T$, cf. Section 3.1 point (c). Secondly, BLUEs of contrasts from distinct (common) eigenspaces of the $TS_\alpha T$ *are* orthogonal, cf. Section 3.1 point (b), and so the BLUEs of contrasts $\langle t \mid \tau \rangle$ for arbitrary $t \in \mathcal{T}$ are sums of the uncorrelated BLUEs of $\langle T_\beta t \mid \tau \rangle$ which have the simple form. And finally, the overall BLUE (4.3) of $\langle t_\beta \mid \tau \rangle$ for $t_\beta \in \mathcal{T}_\beta$ is quite clearly the simple combination of its BLUEs (4.6) in each stratum in which it is estimable, each weighted inversely according to its variance: $\langle t_\beta \mid \hat{\tau} \rangle = \sum_\alpha w_{\alpha\beta} \langle t_\beta \mid \hat{\tau}_\alpha \rangle$. This justifies our use of the term *weight* for $w_{\alpha\beta}$ introduced following equation (4.1). Similarly we can compare the variance of $\langle t_\beta \mid \hat{\tau}_\alpha \rangle$ to that of $\langle t_\beta \mid \hat{\tau} \rangle$ when the ξ_α are assumed equal, and see why $\lambda_{\alpha\beta}$ is termed the *efficiency factor* for treatment term β in stratum α, cf. point (d) in Section 3.1.

In a sense there is no single analysis of variance table which summarises all aspects of the least-squares analysis of a designed experiment satisfying (GB), but rather one for each stratum and one overall. See Table 1, the anova table within stratum α. Examples of

TABLE 1
Anova table within stratum α

Source	d.f.	*Sum of squares*	$\mathbb{E}$\{*Mean square*\}
⋮	⋮	⋮	⋮
Treatment term $\mathscr{T}_\beta$ (assuming $\lambda_{\alpha\beta} > 0$)	$\dim \mathscr{T}_\beta$	$\lambda_{\alpha\beta}^{-1} \| T_\beta S_\alpha y \|^2$	$\xi_\alpha + \frac{\lambda_{\alpha\beta}}{\dim \mathscr{T}_\beta} \| T_\beta \tau \|^2$
⋮	⋮	⋮	⋮
Residual	d_α:By difference ↑	By difference ↑ (may be zero)	ξ_α (if $d_\alpha > 0$)
Total	$\dim \mathscr{S}_\alpha$	$\| S_\alpha y \|^2$	

designs with residual degrees of freedom $d_\alpha = 0$ in some strata are quite common, e.g. symmetric BIBDs, double, triple, $\cdots$ lattice designs, rectangular lattice designs all have zero residual d.f. in the inter-block stratum, and the best general way to estimate ξ_α is certainly not via the anova table for stratum α. For further comments on the estimation of ξ_α, see Section 4.5 below.

4.4. *Simple combinability: a converse to* (GB). We now prove a result asserting that under certain general circumstances, if a set of contrasts spanning $\mathscr{T}$ is simply combinable, then the design satisfies (GB). The following lemma has its straightforward proof omitted. Our framework is that of Section 2.4 without assuming (GB).

LEMMA. *If the treatment contrast $\langle t \mid \tau \rangle$ is estimable in stratum α, then there exists a unique $c_\alpha = c_\alpha(t) \in \mathscr{R}(S_\alpha T)$ such that $Tc_\alpha = t$. Furthermore, the unique* BLUE *of $\langle t \mid \tau \rangle$ based on $S_\alpha y$ is then $\langle c_\alpha \mid y \rangle$.* □

PROPOSITION 4.1. *Let $\langle t \mid \tau \rangle$ be a treatment contrast such that for each stratum $\mathscr{S}_\alpha$ it is either estimable in or orthogonal to $\mathscr{S}_\alpha$, and suppose that there is a set $\{w_\alpha\}$ of non-negative weights summing to unity such that*

$$\langle t \mid \hat{\tau} \rangle = \textstyle\sum_\alpha w_\alpha \langle c_\alpha \mid y \rangle, \quad (y \in \mathscr{D}) \tag{4.7}$$

where $\langle c_\alpha \mid y \rangle$ is the BLUE *of $\langle t \mid \tau \rangle$ based on $S_\alpha y$, if $\langle t \mid \tau \rangle$ is estimable in $\mathscr{S}_\alpha$, and $w_\alpha = 0$ if t is orthogonal to $\mathscr{S}_\alpha$. Then for all α, t is an eigenvector of $TS_\alpha T$ with eigenvalue $\lambda_\alpha = \xi_\alpha w_\alpha (\sum_\alpha \xi_\alpha w_\alpha)^{-1}$.*

PROOF. It is not hard to prove that the transpose U' of $U = P_{\mathscr{T}}^V$ coincides with $V^{-1}UV$. It follows from equation (4.7) that $V^{-1}UVt = \sum_\alpha w_\alpha c_\alpha$ and so

$$UVt = (\textstyle\sum_\alpha \xi_\alpha S_\alpha)(\sum_\alpha w_\alpha c_\alpha) = \sum_\alpha \xi_\alpha w_\alpha c_\alpha. \tag{4.8}$$

Now $TU = U$ and since $Tc_\alpha = t$ for all α, (4.8) implies

$$UVt = (\textstyle\sum_\alpha \xi_\alpha w_\alpha)t. \tag{4.9}$$

On the other hand, (4.8) also implies that $S_\alpha UVt = \xi_\alpha w_\alpha c_\alpha$, and so

$$TS_\alpha UVt = \xi_\alpha w_\alpha t. \tag{4.10}$$

The conclusion now follows from (4.9) and (4.10). □

Now let us suppose that the subspace $\mathscr{T}$ has a basis consisting of vectors t satisfying the hypotheses of Proposition 4.1. Then for each such t there is a set $\{\lambda_{\alpha t}\}$ eigenvalues, and we can obtain a pairwise orthogonal system $\{\mathcal{T}_\beta\}$ of subspaces of $\mathcal{T}$ by grouping together all ts with a common set of eigenvalues, say $\{\lambda_{\alpha\beta}\}$ for each $t \in \mathcal{T}_\beta$. It is clear that the system $\{\mathcal{T}_\beta\}$ forms a complete set of eigenspaces common to all the matrices $\{TS_\alpha T\}$ and also that $\mathcal{T} = \oplus_\beta \mathcal{T}_\beta$. Thus we can obtain the following converse to (GB) implying equation (4.1).

Proposition 4.2. *If there exists an orthogonal decomposition $\mathcal{T} = \oplus_\beta \mathcal{T}_\beta$ of $\mathcal{T}$ and a set $\{w^*_{\alpha\beta}\}$ of weights such that for all $V \in \mathcal{V}$ the projection U onto $\mathcal{T}$ orthogonal with respect to $\langle \cdot \mid \cdot \rangle_V$ is $U = \sum_{\alpha,\beta} w^*_{\alpha\beta} T_\beta S_\alpha$, where $w^*_{\alpha\beta}\xi_\alpha$ is independent of α, then the design satisfies* (GB) *with respect to $\{\mathcal{T}_\beta\}$.* □

The proof will be omitted; it can be found in Houtman (1980). A stronger result can be obtained when there are only two effective strata, i.e. $\mathcal{V}$ is spanned by $S_0 = G$, S_1, S_2; for this case the hypothesis "for all $V \in \mathcal{V}$" in Proposition 4.2 is not required, as one suitable V leads to the same conclusion.

4.5. *The estimation of strata variances under* (GB). We remarked in Section 4.3 above that the residual operator $R_\alpha = S_\alpha - P_{S_\alpha \mathcal{T}}$ in stratum α may be zero, equivalently, that $d_\alpha = \operatorname{tr} R_\alpha = \dim \mathcal{S}_\alpha - \sum \{\dim \mathcal{T}_\beta : \lambda_{\alpha\beta} > 0\}$ may be zero. The reason for this is not hard to see: if $0 < \lambda_{\alpha\beta} < 1$, then treatment term $T_{\beta T}$ is being fitted and its full d.f. dim $\mathcal{T}_\beta$ removed not only in stratum α, but also in one or more other strata in which it is estimable. In a sense we should only remove that fraction $w_{\alpha\beta}(\dim \mathcal{T}_\beta)$ of the d.f. corresponding to the amount of information on $\mathcal{T}_\beta$ in $\mathcal{S}_\alpha$ and the approach of Nelder (1968) amounts to just this.

More precisely, Nelder's approach is based upon equating the observed with expected mean square of the *actual* residual $S_\alpha(I - U)y = S_\alpha\bar{U}y$ in stratum α rather than doing so with the *apparent* residual $R_\alpha y$ as is done if only the anova table is consulted. To illustrate the difference between the two we cite the following without proof:

Lemma (i) $\| S_\alpha \bar{U} y \|^2 = \| R_\alpha y \|^2 + \| (P_{S_\alpha \mathcal{T}} - S_\alpha U) y \|^2$.
(ii) $d'_\alpha = \operatorname{tr}(S_\alpha \bar{U}) = d_\alpha + \sum_\beta (1 - w_{\alpha\beta}) \dim \mathcal{T}_\beta$.
(iii) *When every treatment term is estimated in one of two strata, α and α' say, then*

$$\| (P_{S_\alpha \mathcal{T}} - S_\alpha U) y \|^2 = \textstyle\sum_\beta w_{\alpha'\beta} \lambda_{\alpha\beta} \| \Delta_\beta y \|^2$$

where $\Delta_\beta y = \lambda_{\alpha\beta}^{-1} T_\beta S_\alpha y - \lambda_{\alpha'\beta}^{-1} T_\beta S_{\alpha'} y$ is the difference between the estimates of treatment term β in the two strata, and a similar equation holds with the roles of α and α' reversed. □

Now both U and d'_α involve the weights $\{w_{\alpha\beta}\}$ so if we are to make use of the identity $\mathbb{E} \| S_\alpha \bar{U} y \|^2 = d'_\alpha \xi_\alpha$ in estimating ξ_α, an *iterative* approach must be used. We proceed as follows:

(0) Begin with initial estimates $\{\xi_\alpha^{(0)}\}$ or $\{w_{\alpha\beta}^{(0)}\}$ of the strata variances or weights, possibly making use of the strata anova tables;

(1) Given a set $\{\xi_\alpha\}$ and $\{w_{\alpha\beta}\}$ of *working* estimates of the strata variances and weights, calculate U and d'_α and obtain *revised* estimates $\{\xi^*_\alpha\}$ by solving for $\{\xi_\alpha\}$ in

$$\| S_\alpha \bar{U} y \|^2 = \xi_\alpha d'_\alpha, \quad \alpha = 0, 1, \cdots. \tag{4.11}$$

It is interesting to note that equation (4.11) is in fact the likelihood equation for $\{\xi_\alpha\}$ based upon $\| (I - T) y \|^2$ under the assumption that y has a multivariate normal distribution, see Patterson and Thompson (1971) for details. The information matrix corresponding to these *restricted* ML *estimates* $\{\hat{\xi}_\alpha\}$ under normality has elements

$$-2\mathbb{E}\left\{\frac{\partial^2 \log l}{\partial\xi_\alpha\xi_{\alpha'}}\right\} = \frac{1}{\xi_\alpha\xi_{\alpha'}} \times \begin{cases}[d_\alpha + \sum_\beta (1 - w_{\alpha\beta})^2(\dim \mathscr{T}_\beta)] & \text{if} \quad \alpha = \alpha' \\ [\sum_\beta w_{\alpha\beta}w_{\alpha'\beta}(\dim \mathscr{T}_\beta)] & \text{if} \quad \alpha \neq \alpha'\end{cases}$$

where the sums are over all β for which $\lambda_{\alpha\beta}$ (or $\lambda_{\alpha'\beta}$) > 0.

4.6. *Inferential difficulties under* (GB). Even when a designed experiment with orthogonal block structure defined by the strata $\{\mathscr{S}_\alpha\}$ and treatment structure $\{\mathscr{T}_\beta\}$ satisfies (GB), there remain difficulties with estimation and testing the model.

Although the formula (4.1) gives a precise expression for $\hat{\tau}$ when the strata variances $\{\xi_\alpha\}$ are *known*, these considerations no longer apply when we use the estimates $\{\hat{\xi}_\alpha\}$ obtained as in Section 4.5. The general problem of combining information on a common mean when the weights require estimation has a large literature; see Brown and Cohen (1974) for a general discussion and further references. In some of these papers the problem of combining information on treatment contrasts in BIBDs is considered and it would be of interest to extend these conclusions to multi-strata designs with a number of treatment terms.

A second difficulty arises when the analyst wishes to test the hypothesis $T_\beta\tau = 0$ for some β, say under a normality assumption. This can be done by an F-test in every stratum α for which $\lambda_{\alpha\beta} > 0$ and the stratum residual d.f. $d_\alpha > 0$, and although such tests would be *independent*, there appears to be no accepted procedure for combining the tests into a single one. On the other hand, an overall test might be sought, fitting to $\mathscr{T}$ first and then to the orthogonal complement $\mathscr{T} \ominus \mathscr{T}_\beta$, of $\mathscr{T}_\beta$ in $\mathscr{T}$ which still satisfies (GB). The problem here is the fact that the likelihood ratio test for such hypotheses does not appear to have been studied when information concerning $\mathscr{T}_\beta$ resides in more than one stratum.

Both of these problems would seem to warrant further research. Until straightforward exact or approximate solutions are found, most analysts will follow Yates (1940) and others in substituting the estimated weights into (4.1), and testing hypotheses $T_\beta\tau = 0$ in the stratum α for which $\lambda_{\alpha\beta}$ is largest.

5. Examples.

5.1. BIBDs. The basic notation for block designs was introduced in Section 2.3: b blocks of k plots each, and the term *balanced* means that the $v \geq k$ different treatments are applied to the plots in such a way that each pair of distinct treatments appears together in a block the same number of times, λ say. The strata projections are G, $B - G$ and $I - B$, all derived from simple averaging operators, whilst the treatment decomposition $T = G + (T - G)$ is similarly straightforward. We readily find that

$$TGT = G, \quad T(B - G)T = \bar{e}(T - G), \quad T(I - B)T = e(T - G) \tag{5.1}$$

where $e = (1 - k^{-1})/(1 - v^{-1}) = 1 - \bar{e}$ is the *efficiency factor* of the design; Yates (1936). The computation which establishes the (GB) conditions most easily is the checking that $(T - G)B(T - G) = \bar{e}(T - G)$ by applying $(T - G)B$ to a simple contrast $t_{u,v}$; in this form it is nothing more than checking the balance condition.

The overall BLUE of a treatment contrast $\langle t \mid \tau\rangle$ is given by $\langle t \mid \hat{\tau}\rangle = \xi_1^{-1}(\bar{e}\xi_1^{-1} + e\xi_2^{-1})^{-1}\langle t \mid (B - G)y\rangle + \xi_2^{-1}(\bar{e}\xi_1^{-1} + e\xi_2^{-1})^{-1}\langle t \mid (I - B)y\rangle$, the correctly weighted linear combination of the inter- and intra-block BLUEs $\bar{e}^{-1}\langle t \mid (B - G)y\rangle$, and $e^{-1}\langle t \mid (I - B)y\rangle$, respectively.

When we turn to the estimation of ξ_1 and ξ_2, we note that the residual d.f. $d_1 = (b - 1) - (v - 1)$ in the inter-block stratum is usually small and is zero if $v = b$. Nelder's iterative method or its Fisher scoring variant can be used with initial values $\xi_1^{(0)} = \xi_2^{(0)} = d_2^{-1}\|R_2 y\|^2$ on $d_2 = b(k - 1) - (v - 1)$ d.f. from the intra-block stratum. The only quantities needed for this calculation are the residual arrays

$$R_1 y = (B - G)y - \bar{e}^{-1}(B - G)T(B - G)y$$

$$R_2 y = \bar{B}y - e^{-1}\bar{B}T\bar{B}y$$

and the array of differences of effects estimated in the two strata:

$$\Delta_1 y = \bar{e}^{-1}T(B - G)y - e^{-1}T\bar{B}y.$$

The procedure generally converges quickly, and gives estimates which are close, although not identical, to those given by Yates' (1940) method based on anova tables, and the statistical properties of these estimates appear (by simulations) to be very similar to those of Yates' estimates.

5.2. *A natural generalisation of* PBIBDs. PBIBDs were introduced by Bose and Nair (1939) as generalisations of BIBDs and have been the subject of much study since then, mostly devoted to combinatorial aspects of the designs because the combinatorial objects now known as *association schemes* were first defined in this context, see MacWilliams and Sloane (1978) and Raghavarao (1971). The standard reference on the *analysis* of PBIBDs seems to be Clatworthy (1973). The idea behind PBIBDs is quite simple: where it is not possible for *every* pair of distinct treatment to be together in a block the same number λ of times, the pairs are partitioned into *association classes* forming an association scheme so that this can hold within classes, and the single number λ is replaced by a family λ_1, $\lambda_2, \cdots$ of numbers, one for each association class. Our generalisation carries this idea over to more general block structures than just blocks and plots such as nested BIBDs; Preece (1967).

Let us suppose that the orthogonal block structure of our design arises from a dispersion model based upon an association scheme $\{A_a\}$ over the set $\mathbf{I}$ of unit labels as described in Section 2.2. That is, the strata projections $\{S_\alpha\}$ are given by $S_\alpha = (1/n) \sum_a q_{a\alpha} A_a$ where $Q = (q_{a\alpha})$ is a matrix of structure constants. The association matrices $\{A_\alpha\}$ are defined in terms of the strata projections by $A_a = \sum_\alpha p_{\alpha a} S_\alpha$ where $P = (p_{\alpha a})$ is the "inverse" matrix of constants: $PQ = QP = nI$.

Similarly we suppose—as is customary with PBIBDs—that there is an association scheme $\{\tilde{B}_b\}$ defined over the set $\mathscr{X}$ of treatment labels, see Section 2.2, with corresponding orthogonal projectors $\{\tilde{T}_\beta\}$ given by $\tilde{T}_\beta = (1/v) \sum_b \tilde{q}_{b\beta} \tilde{B}_b$, where $\tilde{Q} = (\tilde{q}_{b\beta})$ and $\tilde{P} = (\tilde{p}_{\beta b})$ are the appropriate matrices of structure constants.

DEFINITION. A design map $x:\mathbf{I} \to \mathscr{X}$ is said to be $(\{A_a\}, \{\tilde{B}_b\})$*-balanced* if for all association classes a over $\mathbf{I}$ and b over $\mathscr{X}$ and $u_1, u_2 \in \mathscr{X}$ with $\tilde{B}_b(u_1, u_2) = 1$, the number $|\{(i, j) \in \mathbf{I} \times \mathbf{I}: A_a(i, j) = 1, x(i) = u_1, x(j) = u_2\}|$ depends only on b and not on the pair u_1, u_2 chosen. If we denote the number (of concurrences) in this definition by n_{ab} then, recalling the design matrix X introduced in Section 2.1 above, we see that an equivalent form of the definition is: there exists numbers n_{ab} such that for all a we have

$$X'A_aX = \textstyle\sum_b n_{ab}B_b. \tag{5.2}$$

In particular if we consider A_e and $\tilde{B}_e$ where e represents the identity association, we find that $n_{ee} = r$ defines the common replication number for the treatments of our design.

PROPOSITION 5.1. *An experiment with block structure arising from an association scheme* $\{A_a\}$ *over the set* $\mathbf{I}$ *of units, and having a design map which is* $(\{A_a\}, \{\tilde{B}_b\})$*-balanced with respect to an association scheme* $\{\tilde{B}_b\}$ *over the set* $\mathscr{X}$ *of treatments, satisfies* (GB). *In notation introduced above, the treatment decomposition is given by* $\{T_\beta\}$ *where* $T_\beta =$

$r^{-1}X\tilde{T}_\beta X'$, *and the matrix* $\Lambda = (\lambda_{\alpha\beta})$ *of efficiency factors is given by*

$$\lambda_{\alpha\beta} = (rn)^{-1} \textstyle\sum_a \sum_b q_{a\alpha} n_{ab} \tilde{p}_{\beta b}$$

where $\mathbf{n} = (n_{ab})$ *is the matrix of concurrences.*

Proof. We begin by noting that $T = r^{-1}XX'$. Then for all α

$$\begin{aligned} TS_\alpha T &= n^{-1} \textstyle\sum_a q_{a\alpha} TA_a T && \text{(definition of } S_\alpha) \\ &= (r^2n)^{-1} \textstyle\sum_a q_{a\alpha} X(X'A_aX)X' && \text{(definition of } T) \\ &= (r^2n)^{-1} \textstyle\sum_a \sum_b q_{a\alpha} n_{ab} X\tilde{B}_b X' && \text{(by (5.2))} \\ &= (r^2n)^{-1} \textstyle\sum_a \sum_b \sum_\beta q_{a\alpha} n_{ab} \tilde{p}_{\beta b} X\tilde{T}_\beta X' && \text{(definition of } \tilde{T}_\beta) \\ &= \textstyle\sum_\beta \{(rn)^{-1} \sum_a \sum_b q_{a\alpha} n_{ab} \tilde{p}_{\beta b}\} T_\beta && \text{(definition of } T_\beta) \end{aligned}$$

and the assertion is proved. □

Example 1. It is not hard to see that a BIBD is built over an association scheme on its units with associations which can be labeled e (equality), 1 (same block but different unit) and 2 (different block), whilst its treatments have the trivial association scheme with associations e (equality) and 1 (inequality). We readily find that $(rn)^{-1}Q'\mathbf{n}\tilde{P}'$ takes the form

$$(rn)^{-1}\begin{bmatrix} 1 & 1 & 1 \\ b-1 & b-1 & -1 \\ b(k-1) & -b & 0 \end{bmatrix}\begin{bmatrix} r & 0 \\ 0 & \lambda \\ r(r-1) & r^2-\lambda \end{bmatrix}\begin{bmatrix} 1 & 1 \\ v-1 & -1 \end{bmatrix} = \begin{bmatrix} 1 & 0 \\ 0 & 1-e \\ 0 & e \end{bmatrix}$$

making use of the relations $r(k-1) = \lambda(v-1)$ and $rv = bk = n$.

Example 2. Kshirsagar (1957) gave the very interesting 6×6 row-column design with 9 treatments A, B, C, D, E, F, G, H, I shown in Table 2. Let us consider the association scheme defined on the treatments by imposing a row-column *pseudo-structure* on them as shown in Table 3. If we let e, 1, 2 and 3 denote the associations of equality, same row (but unequal), same column (but unequal) and different row and column for both schemes, then we have what is shown in Table 4, with a similar result holding for $X'A_3X$ by differencing, since $A_1 + A_2 + A_3 = J - I$, where J is the matrix of all 1s. These clearly satisfy our balance condition with matrix $\mathbf{n} = (n_{ab})$ of concurrences, shown in Table 5. With these preliminaries we can readily get $\tilde{P}$ and Q and calculate the matrix $\Lambda = (\lambda_{\alpha\beta})$ of efficiency factors; this turns out to be as given in Table 6.

For many further such designs see Preece (1968, 1976) and Cheng (1981a, b).

Table 2
Treatment allocation to 36 units with a 6×6 row-column block structure

B	D	H	G	F	C
C	E	G	B	D	I
E	F	C	A	G	H
D	I	A	H	C	E
F	G	I	E	A	B
A	H	B	D	I	F

Table 3
3×3 row-column pseudostructure on 9 treatments

A	B	C
D	E	F
G	H	I

BALANCE IN DESIGNED EXPERIMENTS

TABLE 4

$X'A_1X =$

	A	B	C	D	E	F	G	H	I	
	0	2	2	2	3	3	2	3	3	A
		0	2	3	2	3	3	2	3	B
			0	3	3	2	3	3	2	C
				0	2	2	2	3	3	D
					0	2	3	2	3	E
by symmetry						0	3	3	2	F
							0	2	2	G
								0	2	H
									0	I

$X'A_2X =$

	A	B	C	D	E	F	G	H	I	
	0	3	3	3	2	2	3	2	2	A
		0	3	2	3	2	2	3	2	B
			0	2	2	3	2	2	3	C
				0	3	3	3	2	2	D
					0	3	2	3	2	E
by symmetry						0	2	2	3	F
							0	3	3	G
								0	3	H
									0	I

TABLE 5

$\mathbf{n} =$

	e	1	2	3	
	4	0	0	0	*e*
	0	2	2	3	1
	0	3	3	2	2
	12	11	11	11	3

TABLE 6

$\Lambda =$

Treatment pseudo-factor	gm	r	c	r · c		
	1	0	0	0	Grand mean	*Block stratum*
	0	1/8	1/8	0	Rows	
	0	0	0	1/8	Columns	
	0	7/8	7/8	7/8	Rows · Columns	

5.3. *Supplemented balance and related notions.* Pearce (1960) described a class of block designs possessing what he termed *supplemented balance*, and later Pearce (1963) extended the notion to row-column and more general designs. A typical example is a BIBD consisting of b blocks of k plots each and a standard balanced allocation of v treatments, which is supplemented by the addition of an extra plot to each block to which a *control* is applied. The resulting block design has b blocks each of $k + 1$ plots and $v + 1$ "treatments", but is readily found to satisfy (GB) for the "treatment" decomposition

$$\mathscr{T} = \mathscr{G} \oplus \mathscr{T}_* \oplus \mathscr{T}_c \tag{5.3}$$

where $\mathscr{G} = \mathscr{R}(G)$, $\mathscr{T}_*$ is the $(v-1)$-dimensional space of contrasts amongst the v original treatments, and $\mathscr{T}_c$ is the 1-dimensional subspace spanned by the contrast comparing the control to the average of the original treatments. This contrast is estimated with efficiency 1 in the intra-block stratum, whilst the contrasts in $\mathscr{T}_*$ are estimated intra-block with efficiency e^* where $1 - e^* = k(k+1)^{-1}(1-e)$, e being the efficiency factor of the original BIBD.

A similar analysis holds for block designs which only satisfy (GB) with more complicated treatment decompositions, and also for row-column and other designs with supplemented balance: in these cases $\mathscr{T}_*$ is replaced by the direct sum of the terms relative to which the original (unsupplemented) design satisfied (GB).

Pearce's block designs with supplemented balance are a special case of a class of block designs introduced by Nair and Rao (1942), which are themselves a variant on those described in the previous sub-section. They are analogous to PBIBDs with group-divisible association schemes defined on the treatments, but do not necessarily have equal group sizes, in which case they do not define an association scheme. Despite this fact, even when the group sizes are unequal the line of argument used in Proposition 5.1 carries over. We illustrate the results with the case of two groups, as discussed in Nair and Rao (1942), supposing that there are v_1 "rare" treatments each replicated r_1 times, and v_2 "frequent" treatments each replicated r_2 times. Each pair of "rare" (resp. "frequent") treatments occurs together in the same block n_{11} (resp. n_{22}) times, whilst pairs of treatments one of which is "rare" and the other "frequent" occur together in a block $n_{12} = n_{21}$ times. It is easy to establish that such designs are balanced with respect to the treatment decomposition

$$\mathscr{T} = \mathscr{G} \oplus \mathscr{T}_1 \oplus \mathscr{T}_2 \oplus \mathscr{T}_c$$

where $\mathscr{T}_1$ (resp. $\mathscr{T}_2$) is the space of dimension $n_1 - 1$ (resp. $n_2 - 1$) spanned by contrasts between the "rare" (resp. "frequent") treatments, and $\mathscr{T}_c$ is spanned by the single d.f. contrast comparing the average of the "rare" treatments with the average of the "frequent" treatments. The array of efficiency factors is shown in Table 7.

5.4. *Designs satisfying* (GB). Nelder (1965) observed that most of the common designs in use satisfied his definition of general balance. With our extension (GB) to designs in which treatments are not necessarily equally replicated, we can go further and assert that *all* block designs (with equal block sizes, and the usual dispersion model) satisfy (GB), since it is quite obvious that TGT, $T(B-G)T$ and $T(I-B)T$ all commute. All row and column designs which we have seen in the literature satisfy (GB), see Kshirsagar (1957), Pearce (1963, 1975), Zelen and Federer (1964a) for examples, and so also do all designs known to us with orthogonal block structure having three or more strata.

Knowing that a block design must satisfy (GB) is one thing; having explicit expressions for the orthogonal projections $\{T_\beta\}$ is quite another matter. There are a very large number of types of PBIBDs, and although it is generally not difficult to describe the structure of their Bose-Mesner algebra, see MacWilliams and Sloane (1978, Chapter 21), and hence obtain the $\{T_\beta\}$, most writers in statistics have not taken this viewpoint. Corsten (1976) is an exception.

For classes of block designs which are not PBIBDs, other methods must be used; the details concerning rectangular lattice designs, linked block and a number of other classes

TABLE 7

Treatment term:	gm	1	2	c	*Stratum*:
$\Lambda =$	1	0	0	0	grand mean
	0	$\dfrac{r_1 - n_{11}}{kr_1}$	$\dfrac{r_2 - n_{22}}{kr_2}$	$\dfrac{r_1 r_2 - bn_{12}}{r_1 r_2}$	blocks
	0	$\dfrac{r_1(k-1) + n_{11}}{kr_1}$	$\dfrac{r_2(k-1) + n_{22}}{kr_2}$	$\dfrac{bn_{12}}{r_1 r_2}$	plots

are available on request. Recently the class of α-designs was introduced, Patterson and Williams (1976), these being obtained in a particularly simple way from a basic generating array. This class seems to be so large, including BIBDs, PBIBDs, square and rectangular lattice designs as well as many others, that it does not seem to be possible to give a general description of the subspaces $\{\mathscr{T}_\beta\}$ relative to which the designs satisfy (GB). However this should be regarded as a challenging unsolved problem.

5.5. *Designs not satisfying* (GB).

A black sheep. Although all block designs satisfy (GB) this is not necessarily the case for row-column designs as the following 4 × 4 example with four equally-replicated treatments is shown in Table 8. To see that (GB) fails, one simply notes that the contrast which compares treatment 1 with the average of treatments 2, 3 and 4 is an eigenvector of $T(C - G)T$ (notation as in Section 2 above) and *not* of $T(R - G)T$.

Other designs. Some designs in common use which may not satisfy (GB) are those in which repeated measures are taken on a number of units, when both time (e.g. periods) and subjects (say) are assumed to contribute to the dispersion model, i.e. are regarded as "random effects", and "residual" as well as "direct" treatment effects are included in the model, see Cochran and Cox (1957) for a general discussion. The problem here is that there are no residual effects applying to the first period. In general both time and subjects are regarded as "fixed", in which case no problems arise because the dispersion model is then trivial.

Another class of designs whose structure and accepted analysis does not satisfy (GB) is the class of so-called *two-phase experiments*, McIntyre (1955, 1956), Curnow (1959). The explanation here appears to be simply the amount of structure in the experiment.

5.6. *Concluding discussion.* Throughout this paper we have discussed the notion of balance and its generalisations from a purely theoretical point of view, focusing upon contrasts with particular mathematical properties. It has not been our concern whether these contrasts are natural, or of possible scientific interest, although this is clearly the case in many common examples.

The designer of an experiment has a quite different perspective. Amongst other things, he tries to ensure that contrasts of primary interest are estimated with as high a precision as possible, subject to the constraints imposed by the experimental material. It by no means follows that he should always design his experiment so that such contrasts are eigenvectors of all the $\{TS_\alpha T\}$ of Section 4.1; indeed in many cases this is impossible.

If a designed experiment with orthogonal block structure satisfies (GB), then the coarsest decomposition $\mathscr{T} = \oplus\ \mathscr{T}_\beta$ with respect to which it does so is uniquely defined by the design. Other decompositions of $\mathscr{T}$ which satisfy (GB) can only arise by further decomposition of the individual $\{\mathscr{T}_\beta\}$ in the coarsest one. When the designer is able to arrange that all of the subspaces $\{\mathscr{T}_\beta\}$ consist of contrasts of interest, the analysis of data from the experiment and the display of the results will be particularly straightforward; examples here include BIBDs and the designs of Section 5.3. In general, however, not all

TABLE 8
Design not satisfying (GB).

2	1	1	1
1	3	3	2
2	2	4	3
4	4	4	3

contrasts of interest will belong to one of the $\mathcal{T}_\beta$, and it will be necessary in the analysis to use the more complicated formula (4.2) involving the projections $\{T_\beta\}$; examples here include unbalanced lattice designs.

A final point concerning the subspaces $\{\mathcal{T}_\beta\}$ in (GB) is worth making. Even when they do not consist of contrasts of scientific interest, they are frequently recognisable as arising from a pseudo-structure on the treatments, i.e. an artificial view of the treatments relative to which the $\{\mathcal{T}_\beta\}$ are natural or interpretable. Examples here include many PBIBDs, most lattice designs and Example 2 of Section 5.2. The most general design satisfying (GB)—and we need go no further than block α-designs to find examples—involves a decomposition of $\mathcal{T}$ into subspaces $\{\mathcal{T}_\beta\}$ which have neither scientific interest nor any natural or interpretable structure, however we care to view the treatments. Our general theory applies to such designs, although it may be an affront to some to describe them as balanced in any sense. We hope that our readers will appreciate the value of tracing the path from balance in BIBDs through to the notion of general balance, and conclude that the unity of outlook achieved outweighs any terminological problems met along the way.

6. Acknowledgements. This work was begun in the Department of Statistics at Princeton University whilst the first author was a graduate student and the second a visitor to the department. Many thanks are due to Professor Geoffrey S. Watson for his joint role as adviser and host, and to others in the department for the congenial work atmosphere they helped to create. The paper was written whilst the authors were at the Polytechnic Institute of New York and CORE, Belgium (A.M.H.) and the University of Western Australia (T.P.S.) and thanks are due to these institutions for their support. The referees are also to be thanked for their many helpful comments.

REFERENCES

BAILEY, R. A. (1981). A unified approach to design of experiments. *J. Roy. Statist. Soc. Ser. A* **144** 214–223.

BOSE, R. C. and NAIR, K. R. (1939). Partially balanced incomplete block designs. *Sankhyā* **4** 337–372.

BROWN, L. D. and COHEN, ARTHUR (1974). Point and confidence estimation of a common mean and recovery of interblock information. *Ann. Statist.* **2** 963–976.

CHENG, C.-S. (1981a). Optimality and construction of pseudo Youden designs. *Ann. Statist.* **9** 201–205.

CHENG, C.-S. (1981b). A family of pseudo Youden designs with row size less than the number of symbols. *J. Comb. Theory Ser. A* **31** 219–221.

CLATWORTHY, W. H. (1973). *Tables of Two-Associate-Class Partially Balanced Designs.* Appl. Math. Ser. 63, National Bureau of Standards, Washington, D.C.

COCHRAN, WILLIAM G. and COX, GERTRUDE M. (1957). *Experimental Designs.* Second Edition. Wiley, New York.

CORSTEN, L. C. A. (1962). Balanced block designs with two different numbers of replicates. *Biometrics* **18** 499–519.

CORSTEN, L. C. A. (1976). Canonical correlation in incomplete blocks. *Essays in Probability and Statistics.* Shinko Tsusho Co., Ltd. Japan, 125–154.

CURNOW, R. N. (1959). The analysis of a two phase experiment. *Biometrics* **15** 60–73.

FISHER, R. A. (1935). Discussion following Yates' "Complex Experiments". *Collected Papers of R. A. Fisher* **3** 332–333.

HOBLYN, T. N., PEARCE, S. C. and FREEMAN, G. H. (1954). Some considerations in the design of successive experiments in fruit plantations. *Biometrics* **10** 503–515.

HOUTMAN, ANNE M. (1980). The analysis of designed experiments. Ph.D. thesis, Princeton University.

JAMES, A. T. and WILKINSON, G. N. (1971). Factorization of the residual operator and canonical decomposition of non-orthogonal factors in analysis of variance. *Biometrika* **58** 279–294.

KSHIRSAGAR, A. M. (1957). On balancing in designs in which heterogeneity is eliminated in two directions. *Calc. Statist. Assoc. Bull.* **7** 161–166.

KSHIRSAGAR, A. M. (1966). Balanced factorial designs. *J. Roy. Statist. Soc. Ser. B* **28** 559–567.

KURKJIAN, B. and ZELEN, M. (1963). Application of the calculus of factorial arrangements, I. Block and direct product designs. *Biometrika* **50** 63–73.

MACWILLIAMS, F. J. and SLOANE, N. J. A. (1978). *The Theory of Error Correcting Codes.* North Holland, Amsterdam.

MARTIN, FRANK B. and ZYSKIND, GEORGE (1966). On combinability of information from uncorrelated linear models by simple weighting. *Ann. Math. Statist.* **37** 1338–1347.

MCINTYRE, G. A. (1955). Design and analysis of two-phase experiments. *Biometrics* **11** 324–334.

MCINTYRE, G. A. (1956). Query 123. *Biometrics* **12** 527–732.

MORLEY JONES, R. (1959). A property of incomplete blocks. *J. Roy. Statist. Soc. Ser. B* **21** 172–179.

NAIR, K. R. and RAO, C. R. (1942). Confounding in asymmetrtical factorial experiments. *J. Roy. Statist. Soc. Ser. B* **10** 109–131.

NELDER, J. A. (1954). The interpretation of negative components of variance. *Biometrika* **41** 544–548.

NELDER, J. A. (1965a). The analysis of randomized experiments with orthogonal block structure, I. Block structure and the null analysis of variance. *Proc. Roy. Soc. (London) Ser. A.* **273** 147–162.

NELDER, J. A. (1965b). The analysis of randomized experiments with orthogonal block structure, II. Treatment structure and the general analysis of variance. *Proc. Roy. Soc. (London) Ser. A* **273** 163–178.

NELDER, J. A. (1968). The combination of information in generally balanced designs. *J. Roy. Statist. Soc. Ser. B* **30** 303–311.

PATTERSON, H. D. and WILLIAMS, E. R. (1976). A new class of resolvable incomplete block designs. *Biometrika* **63** 83–92.

PATTERSON, H. D. and THOMPSON, R. (1971). Recovery of inter-block information when block sizes are unequal. *Biometrika* **58** 545–554.

PEARCE, S. C. (1960). Supplemented balance. *Biometrika* **47** 263–271.

PEARCE, S. C. (1963). The use and classification of non-orthogonal designs (with discussion). *J. Roy. Statist. Soc. Ser. A* **126** 353–377.

PEARCE, S. C. (1970). The efficiency of block designs in general. *Biometrika* **57** 339–346.

PEARCE, S. C. (1975). Row-and-column designs. *Applied Statistics* **24** 60–74.

PEARCE, S. C., CALINSKI, T. and MARSHALL, T. F. DE C. (1974). The basic contrasts of an experimental design with special referece to the analysis of data. *Biometrika* **61** 449–460.

PREECE, D. A. (1967). Nested balanced incomplete block designs. *Biometrika* **54** 479–486.

PREECE, D. A. (1968). Balanced 6 × 6 designs for 9 treatments. *Sankhyā Ser. B* **9** 201–205.

PREECE, D. A. (1976). A second domain of balanced 6 × 6 designs for 9 equally-replicated treatments. *Sankhyā Ser. B* **38** 192–194.

RAGHAVARAO, DAMARAJU (1971). *Construction and Combinatorial problems in Design of Experiments.* Wiley, New York.

RAO, V. R. (1958). A note on balanced designs. *Ann. Math. Statist.* **29** 290–294.

SPROTT, D. A. (1956). A note on combined interblock and intrablock estimation in incomplete block designs. *Ann. Math. Statist.* **27** 633–641.

YATES, F. (1936). Incomplete randomised blocks. *Ann. Eugenics London* **7** 121–140.

YATES, F. (1939). The recovery of inter-block information in variety trials arranged in three-dimensional lattices. *Ann. Eugenics* **9** 136–156.

YATES, F. (1940). The recovery of inter-block information in balanced incomplete block designs. *Ann. Eugenics* **10** 317–325.

ZELEN, M. and FEDERER, W. T. (1964a). Application of the calculus for factorial arrangements, II. Two-way elimination of heterogeneity. *Ann. Math. Statist.* **35** 658–672.

ZELEN, M. and FEDERER, W. T. (1964b). Application of the calculus for factorial arrangements, III. Analysis of factorials with unequal numbers of observations. *Sankhyā Ser. A* **25** 383–400.

56 AV. DES ARTS
BRUSSELS, BELGIUM

CSIRO
DIVISION OF MATHEMATICS AND STATISTICS
BOX 1965, GPO
CANBERRA, ACT 2601
AUSTRALIA

Austral. J. Statist., **26** (**3**), 1984, 227–246

THE ANALYSIS OF MULTISTRATA DESIGNED EXPERIMENTS WITH INCOMPLETE DATA[1]

A. Houtman

A. C. Nielsen, Belgium

and

T. P. Speed

CSIRO Division of Mathematics & Statistics, Australia

Summary

Building upon early work of E. A. Cornish we show that G. N. Wilkinson's version of Yates' approach to the analysis of designed experiments with a single error stratum carries over completely to designs with an arbitrary non-singular covariance matrix, initially assumed known. We show that the equations, corrections, adjustments and algorithms all have their more general analogues and that these can be solved, computed or executed quite readily if the design has orthogonal block structure and satisfies Nelder's condition of general balance. The results are illustrated with a split-plot and a simple (square) lattice design.

1. Introduction

The problem of analysing designed experiments with incomplete data—for example, missing or mixed-up values—has received a lot of attention when the designs are analysed with only a single error line. The corresponding questions for designs analysed with more than one error line (which we term multistrata designs), such as split-plot designs or block designs in which inter-block information is recovered, have rarely been raised, and in our opinion the accepted answers in these areas are not completely satisfactory. The most frequently adopted approach is to change the model back to one with only a single error line, that corresponding to the lowest stratum, and to carry out the analysis appropriate for incomplete data under the model in which all other terms (including the other errors) are fixed. Such an approach has the merit of simplicity, but it has no theoretical basis

[1] Manuscript received May 30, 1983.

and in our experience can give replacement values which are undesirably discordant with the remaining data. This paper reports an attempt to give an analysis of multistrata designed experiments with incomplete data which is closer to the exact one under the model usually assumed for such data.

We will build upon the early work of Cornish (1943, 1944, 1956), showing that Wilkinson's (1958a,b) version of Yates' (1933) approach carries over completely to experimental designs with an arbitrary non-singular covariance matrix **V**, initially assumed known. More precisely, we show that the equations, corrections, adjustments and algorithms associated with the analysis of an experiment with incomplete data but only one error line all have their more general analogues; the main problem is their solution, evaluation or execution. To simplify the discussion, as well as to make contact with the common multistrata designs, we then specialise to designs which are generally balanced in the sense of Nelder (1965a,b, 1968). This means that we suppose **V** to have a very specific relationship to the treatment model under discussion, and we remark that all common designs—e.g. all those in Cochran & Cox (1957)—possess this property, see Nelder (1965b) and Houtman & Speed (1983). Indeed most of the common multistrata designs have only two effective strata, i.e. all of the information concerning treatment contrasts lies in only two strata, and for such designs our results are simplified substantially.

Our results are all exact as long as the covariance matrix **V** is known, and in the case of generally balanced designs a natural extension of Nelder's (1968) method for estimating an unknown **V** suggests itself. The discussion is then illustrated by giving our analysis of a split-plot and a simple (square) lattice design, each having a single missing value.

We have not attempted in this paper to describe what we regard as the best way to carry out the associated calculations. One reason for this is our desire to outline a general approach and avoid concentrating on particular designs, but the main reason is the absence of widely-used general algorithms which perform multistrata analyses and are capable of the few modifications necessary to do the calculations we require. The ANOVA algorithm and the associated Macro facilities which can be found in GENSTAT, see Alvey *et al.* (1977), provide the most convenient framework known to us for doing the job.

2. Previous Work on the Subject

Formulae for replacing a single missing value in randomized complete block designs and in Latin squares were given by Allan & Wishart (1930), but it was Yates (1933) who laid down the general principles for replacing missing values in designed experiments and for correcting other aspects of the analysis of the completed data. Yates' method, suggested to him by Fisher, consists of using those replacement values which minimize the residual sum of squares when unknowns are substituted for the missing response values. When only one value is missing, this method leads to a simple direct formula for the replacement; when there are several missing values Yates suggests an iterative method for solving the equations. He notes that this method leads to the correct fitted values for the observed data, but with inflated treatment sums of squares, gives the correction for randomized blocks and for Latin squares and also computes the adjustment to the variance of a contrast for those two types of designs.

Yates' work was later generalized and expressed in a modern framework by Wilkinson (1958a,b, 1959) and a host of authors have made contributions to the formulation, interpretation, existence, uniqueness and solution of problems with incomplete data, see Hoyle (1971) for an extensive but incomplete bibliography. In particular we note the coordinate-free approach shown by Kruskal (1961) to include the estimation of mixed-up values and to provide an easy extension to the analysis of designs with extra observations.

A different approach to estimating missing values was introduced by Bartlett (1937). This method first assigns arbitrary response values to the missing plots and then adjusts the completed data by covariance upon pseudo-covariates, one being introduced for each missing plot and having value unity for that missing plot and zeroes elsewhere. It is easy to see that estimates so obtained are identical to those derived by Yates' method. Further contributions framed within the analysis of covariance approach with a single error can be found in Nair (1940), Truitt & Fairfield Smith (1956), Coons (1957), John & Prescott (1975), John & Lewis (1976) and P. L. Smith (1981).

Many authors have studied iterative methods to obtain estimates of missing values in single stratum experiments. The use of an iterative procedure was first recommended by Yates (1933). Later Healy & Westmacott (1956) gave a more general algorithm based on Yates' observation that the residuals after fitting the completed data must be zero in the cells corresponding to the missing values. Pearce (1965) improved the Healy–Westmacott algorithm by introducing an accelerating factor n/E where n is the total number of experimental units and E is the number of residual degrees of freedom for a complete experiment. This correction is also used in papers by Preece (1971) and Pearce & Jeffers (1971). More recently Rubin (1972), Haseman & Gaylor (1973), John & Prescott (1975) developed non-iterative methods involving $m+1$ uses of the same subroutine used for fitting the full model where m is the number of missing values. Jarrett (1978) describes the relationships between those various computing procedures.

The problem of mixed-up values was first considered by Nair (1940) using the analysis of covariance. Kruskal (1968) follows Yates' approach in a coordinate-free framework. Preece & Gower (1974) give an iterative procedure to deal with mixed-up values similar to the one advocated by Healy & Westmacott (1956) for missing values. John & Lewis (1976) give a direct procedure based on the appropriate analysis of variance.

Most of the literature on missing values concerns experiments with a single error stratum. The earliest efforts to adapt Yates' approach to designs with more than one error line, mainly lattices and BIB designs, are due to Cornish (1943, 1944, 1956) in three papers dealing with the recovery of interblock information. An influential early note of Anderson (1946) seems to end up recommending the lowest-stratum-only analysis for split-plot designs with missing data. Anderson's view has become accepted, see Cochran & Cox (1957), and is widely used to this day. We note in passing that little satisfaction can be gained by an appeal to the analysis of covariance, since, for multistrata designed experiments, this technique is not in much better state than the special cases which incomplete data pose. An exception is the unpublished report Cochran (1946) which discusses the analysis of covariance in split-plot designs and whose results may be modified to handle missing and mixed-up values. Recently Williams, Ratcliff & Speed (1981) showed how to get missing value estimates based on the information contained in the lowest two strata.

Finally we note that the EM algorithm described by Dempster, Laird & Rubin (1977) provides an iterative approach to the maximum likelihood estimation of parameters from incomplete data under quite general distributional assumptions. Under normality assumptions it can be shown that the recursion in the EM algorithm is the same as that satisfied by the estimates obtained at each step from Healy and Westmacott's algorithm, or from the extension we give of that algorithm for multistrata experiments. However the discussion below will only make the standard second-order assumptions usual in the analysis of designed experiments.

3. Derivation of the Basic Equations

We regard the observations as an array of numbers $\mathbf{y} = (y_i)_{i \in I}$ indexed by a set I of n unit labels and assume the following model for the expectation and dispersion of $\mathbf{y}$:

$$\mathbb{E}\mathbf{y} \in \mathcal{T} \tag{3.1}$$

$$\mathbb{D}\mathbf{y} = \mathbf{V}$$

where $\mathcal{T}$ is the subspace of arrays that are constant over treatments and $\mathbf{V}$ is a positive-definite matrix which is assumed to be known. The

n-dimensional vector space $\mathscr{D}$ of all possible arrays $\mathbf{y}$ is endowed with the natural inner product $\langle\mathbf{x}, \mathbf{y}\rangle = \sum_{i\in I} x_i y_i$ and with the inner product $\langle\mathbf{x}, \mathbf{y}\rangle_\mathbf{V} = \langle\mathbf{x}, \mathbf{V}^{-1}\mathbf{y}\rangle$ induced by $\mathbf{V}$; the associated norms are denoted respectively by $|.|$ and $|.|_\mathbf{V}$.

If the data are incomplete, the space $\mathscr{D}$ splits into the sum of two orthogonal sub-spaces

$$\mathscr{D} = \mathscr{D}_1 \oplus \mathscr{D}_2$$

reflecting the decomposition

$$\mathbf{y} = \mathbf{y}_1 + \mathbf{y}_2$$

of the data into the observed part $\mathbf{y}_1$ and the "missing" part $\mathbf{y}_2$. This notation was shown by Kruskal (1968) to include both the case of missing values and the case of mixed-up values: in the first case, $\mathbf{y}_1$ has zeroes in all units corresponding to missing observations and in the second case, $\mathbf{y}_1$ has a quantity $z = m^{-1}S$ in all m units corresponding to the observations whose sum S only was observed.

For the observed part $\mathbf{y}_1$ of the data, the model (3.1) now becomes

$$\begin{aligned} \mathbb{E}\mathbf{y}_1 &\in \mathscr{T}_1 \\ \mathbb{D}\mathbf{y}_1 &= \mathbf{V}_1 \end{aligned} \tag{3.2}$$

where $\mathscr{T}_1 = \mathbf{D}_1\mathscr{T}$, the orthogonal projection of $\mathscr{T}$ onto $\mathscr{D}_1$ and $\mathbf{V}_1 = \mathbf{D}_1\mathbf{V}\mathbf{D}_1$. (In this paper we will always use script letters to denote linear spaces and the corresponding capital letters to denote the orthogonal projections onto those spaces, with a superscript $\mathbf{V}$ if the projection is orthogonal with respect to $\langle.,.\rangle_\mathbf{V}$ rather than $\langle.,.\rangle$). Although the data could be fitted by estimating $\mathbb{E}\mathbf{y}_1$ by its BLUE $\hat{\boldsymbol{\tau}}_1 = \mathbf{T}_1^{\mathbf{V}_1}\mathbf{y}_1$, it is usually not straightforward to do so, since any special relationship that existed between $\mathbf{V}$ and $\mathscr{T}$ (e.g. orthogonality or general balance) would not usually continue to hold between $\mathbf{V}_1$ and $\mathscr{T}_1$. Accordingly, following Cornish (1956), we minimize

$$|\mathbf{y}_1 + \mathbf{y}_2 - \boldsymbol{\tau}|^2_\mathbf{V},$$

over $\boldsymbol{\tau} \in \mathscr{T}$ and $\mathbf{y}_2 \in \mathscr{D}_2$, and assume that $(\tilde{\boldsymbol{\tau}}, \tilde{\mathbf{y}}_2)$ is a pair at which the minimum is achieved. Arguing as Yates (1933) did in the single stratum case, we may minimize first over $\mathscr{T}$ and then over $\mathscr{D}_2$ to get

$$\tilde{\mathbf{y}}_2 = \mathbf{D}_2^\mathbf{V}(\tilde{\boldsymbol{\tau}} - \mathbf{y}_1),$$

and in the reverse order, obtaining

$$\tilde{\boldsymbol{\tau}} = \mathbf{T}^\mathbf{V}(\mathbf{y}_1 + \tilde{\mathbf{y}}_2).$$

Each of the two relations can be substituted into the other, leading to

$$\mathbf{D}_2^\mathbf{V}\bar{\mathbf{T}}^\mathbf{V}\tilde{\mathbf{y}}_2 = -\mathbf{D}_2^\mathbf{V}\bar{\mathbf{T}}^\mathbf{V}\mathbf{y}_1 \tag{3.3}$$

and

$$\mathbf{T}^{\mathbf{V}}\bar{\mathbf{D}}_2^{\mathbf{V}}\bar{\boldsymbol{\tau}} = \mathbf{T}^{\mathbf{V}}\bar{\mathbf{D}}_2^{\mathbf{V}}\mathbf{y}_1 \tag{3.4}$$

where $\bar{\mathbf{T}}^{\mathbf{V}} = \mathbf{I} - \mathbf{T}^{\mathbf{V}}$ and similarly $\bar{\mathbf{D}}_2^{\mathbf{V}} = \mathbf{I} - \mathbf{D}_2^{\mathbf{V}}$.

In the next section section we will show how to solve these equations. For the moment we simply state the result which justifies their solution as follows: the restriction $\mathbf{D}_1\mathbf{T}^{\mathbf{V}}(\mathbf{y}_1 + \bar{\mathbf{y}}_2)$ to the subspace corresponding to the observed data of the fitted values $\mathbf{T}^{\mathbf{V}}(\mathbf{y}_1 + \bar{\mathbf{y}}_2)$ obtained by analysing the observed data $\mathbf{y}_1$ completed with any solution $\bar{\mathbf{y}}_2$ of (3.3) coincides with the fitted values $\mathbf{T}_1^{\mathbf{V}'}\mathbf{y}_1$ of the observed data $\mathbf{y}_1$ to the appropriate submodel, i.e.

$$\hat{\boldsymbol{\tau}}_1 = \mathbf{D}_1\bar{\boldsymbol{\tau}}. \tag{3.5}$$

A proof of this result is given in the appendix.

From (3.3) or (3.4) it is easy to verify that the vector of residuals $\bar{\mathbf{r}} = \bar{\mathbf{T}}^{\mathbf{V}}(\mathbf{y}_1 + \bar{\mathbf{y}}_2)$ satisfies the equation

$$\mathbf{D}_2^{\mathbf{V}}\bar{\mathbf{r}} = \mathbf{0}, \tag{3.6}$$

which is similar but not equivalent to the property noticed by Yates in the single stratum case, that the residuals after fitting the completed data to $\mathcal{T}$ must be zeroes in the units corresponding to missing values.

4. Solutions of the Equations

In the simple case in which there is a single missing value (or only two mixed-up values) a direct formula may be obtained. Letting $\boldsymbol{\varepsilon}$ denote a dummy vector with unity in place of the missing value (or +1 and −1 in place of the two mixed-up values) and zeroes everywhere else, the unobserved vector has the form $\mathbf{y}_2 = a\boldsymbol{\varepsilon}$ where a is to be estimated, and $\mathbf{D}_2^{\mathbf{V}}\mathbf{z} = |\boldsymbol{\varepsilon}|_{\mathbf{V}}^{-2}\langle\boldsymbol{\varepsilon}, \mathbf{z}\rangle_{\mathbf{V}}\boldsymbol{\varepsilon}$. It follows then immediately from (3.3) that

$$\bar{\mathbf{y}}_2 = -\frac{\langle\boldsymbol{\varepsilon}, \bar{\mathbf{T}}^{\mathbf{V}}\mathbf{y}_1\rangle_{\mathbf{V}}}{|\bar{\mathbf{T}}^{\mathbf{V}}\boldsymbol{\varepsilon}|_{\mathbf{V}}^2}\boldsymbol{\varepsilon}. \tag{4.1}$$

A more manageable form of $\bar{\mathbf{y}}_2$ will be obtained in §6 for generally balanced designs and it will be illustrated with examples in §8. We will now suppose that $\dim \mathcal{D} > 1$, i.e. that there is more than one missing value or there are more than two mixed-up values, and study iterative methods for computing the solutions of (3.3) and (3.4) The following recursion formulae suggest themselves for $\bar{\mathbf{y}}_2$, $\bar{\mathbf{y}} = \mathbf{y}_1 + \bar{\mathbf{y}}_2$, $\bar{\boldsymbol{\tau}}$ and $\bar{\mathbf{r}} = \bar{\mathbf{T}}^{\mathbf{V}}\bar{\mathbf{y}}$:

(i) $\mathbf{y}_2^{(0)} = \mathbf{0}$; $\quad \bar{\mathbf{y}}_2^{(m+1)} = \mathbf{D}_2^{\mathbf{V}}[\mathbf{T}^{\mathbf{V}}(\mathbf{y}_1 + \bar{\mathbf{y}}_2^{(m)}) - \mathbf{y}_1]$, $\quad m \geq 0$.

(ii) $\bar{\mathbf{y}}^{(0)} = \mathbf{y}_1$; $\quad \bar{\mathbf{y}}^{(m+1)} = (\mathbf{I} - \mathbf{D}_2^{\mathbf{V}}\bar{\mathbf{T}}^{\mathbf{V}})\bar{\mathbf{y}}^{(m)}$, $\quad m \geq 0$.

(iii) $\bar{\boldsymbol{\tau}}^{(0)} = \mathbf{T}^{\mathbf{V}}\mathbf{y}_1$; $\quad \bar{\boldsymbol{\tau}}^{(m+1)} = \mathbf{T}^{\mathbf{V}}(\bar{\mathbf{D}}_2^{\mathbf{V}}\mathbf{y}_1 + \mathbf{D}_2^{\mathbf{V}}\bar{\boldsymbol{\tau}}^{(m)})$, $\quad m \geq 0$.

(iv) $\bar{\mathbf{r}}^{(0)} = \bar{\mathbf{T}}^{\mathbf{V}}\mathbf{y}_1$; $\quad \bar{\mathbf{r}}^{(m+1)} = \bar{\mathbf{T}}^{\mathbf{V}}\bar{\mathbf{D}}_2^{\mathbf{V}}\bar{\mathbf{r}}^{(m)}$, $\quad m \geq 0$.

All of these recursions are essentially the same, each being obtainable from the others by simple algebraic manipulations. Recursion (i) is a generalization of Healy and Westmacott's algorithm, and, under normality assumptions, (iii) can be shown to be equivalent to the EM algorithm of Dempster *et al.* (1977). It is the last recursion which most clearly indicates why convergence must take place, since $\bar{\mathbf{T}}^{\mathbf{V}}$ and $\bar{\mathbf{D}}_2^{\mathbf{V}}$ are projection operators, and so $|\bar{\mathbf{r}}^{(m+1)}|_{\mathbf{V}} \leq |\mathbf{r}^{(m)}|_{\mathbf{V}}$ for all $m = 0, 1, 2, \ldots$, with equality if and only if $\bar{\mathbf{r}}^{(m+1)} = \bar{\mathbf{r}}^{(m)}$, in which case the algorithm stops and $\bar{\mathbf{r}}^{(m)}$ is the solution $\bar{\mathbf{r}}$ by virtue of (3.6). An alternative proof of convergence uses a theorem of Von Neumann (1950, p. 55) showing that (iv) converges to the projection of $\mathbf{y}_1$ onto $\mathbf{V}\mathcal{T}^{\perp} \cap \mathbf{V}\mathcal{D}_1$, orthogonal with respect to $\langle .\, , . \rangle_{\mathbf{V}}$. We also notice that each algorithm is equivalent to a Taylor expansion.

The speed of convergence may be improved by the introduction of an appropriate acceleration factor ω. With the same initial values as before, the algorithms are modified as follows for $m \geq 0$:

(i)′ $\bar{\mathbf{y}}_2^{(m+1)} = \bar{\mathbf{y}}_2^{(m)} - \omega^{-1}\mathbf{D}_2^{\mathbf{V}}\bar{\mathbf{T}}^{\mathbf{V}}(\mathbf{y}_1 + \bar{\mathbf{y}}_2^{(m)})$.

(ii)′ $\bar{\mathbf{y}}^{(m+1)} = \bar{\mathbf{y}}^{(m)} - \omega^{-1}\mathbf{D}_2^{\mathbf{V}}\bar{\mathbf{T}}^{\mathbf{V}}\bar{\mathbf{y}}^{(m)}$.

(iii)′ $\bar{\boldsymbol{\tau}}^{(m+1)} = \bar{\boldsymbol{\tau}}^{(m)} - \omega^{-1}\mathbf{T}^{\mathbf{V}}\bar{\mathbf{D}}_2^{\mathbf{V}}(\bar{\boldsymbol{\tau}}^{(m)} - \mathbf{y}_1)$.

(iv)′ $\bar{\mathbf{r}}^{(m+1)} = \bar{\mathbf{r}}^{(m)} - \omega^{-1}\bar{\mathbf{T}}^{\mathbf{V}}\mathbf{D}_2^{\mathbf{V}}\bar{\mathbf{r}}^{(m)}$.

As all four algorithms are equivalent we study the convergence of (i)′. Using (3.3) one obtains

$$\bar{\mathbf{y}}_2 - \bar{\mathbf{y}}_2^{(m+1)} = (\mathbf{I} - \omega^{-1}\mathbf{D}_2^{\mathbf{V}}\bar{\mathbf{T}}^{\mathbf{V}}\mathbf{D}_2^{\mathbf{V}})(\bar{\mathbf{y}}_2 - \bar{\mathbf{y}}_2^{(m)}).$$

It follows that the algorithm converges to a solution of (3.3) for all $\mathbf{y}_2^{(0)}$ in $\mathcal{D}_2$ if and only if the spectral radius ρ of $I - \omega^{-1}\mathbf{D}_2^{\mathbf{V}}\bar{\mathbf{T}}^{\mathbf{V}}\mathbf{D}_2^{\mathbf{V}}$ is strictly smaller than unity. The solution is unique if we assume that $\mathcal{D}_2 \cap \mathcal{T} = \{0\}$, excluding, in particular, situations where all the observations on a treatment combination are missing. Under this assumption, the algorithm converges to the unique solution $\bar{\mathbf{y}}_2$ if and only if

$$\omega > \tfrac{1}{2}\lambda_{\max}(\mathbf{D}_2^{\mathbf{V}}\bar{\mathbf{T}}^{\mathbf{V}}\mathbf{D}_2^{\mathbf{V}})$$

where $\lambda_{\max}(\mathbf{A})$ is the largest eigenvalue of $\mathbf{A}$. The fastest convergence is obtained for the value ω_{opt} which minimizes ρ and hence

$$\omega_{\text{opt}} = \tfrac{1}{2}[\lambda_{\min}(\mathbf{D}_2^{\mathbf{V}}\bar{\mathbf{T}}^{\mathbf{V}}\mathbf{D}_2^{\mathbf{V}}) + \lambda_{\max}(\mathbf{D}_2^{\mathbf{V}}\bar{\mathbf{T}}^{\mathbf{V}}\mathbf{D}_2^{\mathbf{V}})],$$

where $\lambda_{\min}(\mathbf{A})$ is the smallest nonzero eigenvalue of $\mathbf{A}$.

5. Modifications to the Subsequent Analysis

The analysis performed on the data completed with a solution of (3.3) will produce the correct residual sum of squares (although the number of degrees of freedom must be reduced by the dimension of

$\mathscr{D}_2$) but treatment sums of squares (differences between residual sums of squares for a pair of nested treatment models) and variances of contrasts will need adjustment. Yates had already pointed this out in 1933 and he gave the corrections for the designs he studied. We now give the appropriate adjustments for multistrata experiments.

For a submodel $\mathbb{E}\mathbf{y} \in \mathscr{U} \subset \mathscr{T}$ of our original model, the treatment sum of squares $|\bar{\mathbf{U}}^{\mathbf{V}}\mathbf{y}|^2_{\mathbf{V}} - |\bar{\mathbf{T}}^{\mathbf{V}}\mathbf{y}|^2_{\mathbf{V}}$ will be inflated by the quantity

$$|\bar{\mathbf{U}}^{\mathbf{V}}\mathbf{y}(\mathscr{T})|^2_{\mathbf{V}} - |\bar{\mathbf{U}}^{\mathbf{V}}\mathbf{y}(\mathscr{U})|^2_{\mathbf{V}} = |\bar{\mathbf{U}}^{\mathbf{V}}[\mathbf{y}_2(\mathscr{T}) - \mathbf{y}_2(\mathscr{U})]|^2_{\mathbf{V}} \tag{5.1}$$

where $\mathbf{y}(\mathscr{T}) = \mathbf{y}_1 + \mathbf{y}_2(\mathscr{T})$ and $\mathbf{y}(\mathscr{U}) = \mathbf{y}_1 + \mathbf{y}_2(\mathscr{U})$ denote the completed data obtained by solving the missing values equations (3.3) respectively for the models $\mathbb{E}\mathbf{y} \in \mathscr{T}$ and $\mathbb{E}\mathbf{y} \in \mathscr{U}$. Equation (5.1) is simple the difference between the *apparent* sum of squares

$$|\bar{\mathbf{U}}^{\mathbf{V}}\mathbf{y}(\mathscr{T})|^2_{\mathbf{V}} - |\bar{\mathbf{T}}^{\mathbf{V}}\mathbf{y}(\mathscr{T})|^2_{\mathbf{V}}$$

and the *correct* sum of squares

$$|\bar{\mathbf{U}}^{\mathbf{V}}\mathbf{y}(\mathscr{U})|^2_{\mathbf{V}} - |\bar{\mathbf{T}}^{\mathbf{V}}\mathbf{y}(\mathscr{T})|^2_{\mathbf{V}},$$

the latter being smaller than the former since $\mathbf{y}(\mathscr{U})$ minimizes $|\bar{\mathbf{U}}^{\mathbf{V}}\mathbf{y}|^2_{\mathbf{V}}$ over $\mathscr{D}_2$. The algebra leading to the RHS of (5.1) is given in the appendix.

On the other hand, the variance of a contrast $\langle \mathbf{t}, \bar{\boldsymbol{\tau}} \rangle$ where $\bar{\boldsymbol{\tau}}$ satisfies (3.4) can be decomposed into the sum of the variance of that contrast when the data are complete and an adjustment due to the loss of precision encountered when estimating missing data. We have the identity

$$\operatorname{cov}(\langle \mathbf{t}, \bar{\boldsymbol{\tau}} \rangle, \langle \mathbf{u}, \bar{\boldsymbol{\tau}} \rangle) = \langle \mathbf{t}, \mathbf{T}^{\mathbf{V}}\mathbf{V}\mathbf{u} \rangle + \langle \mathbf{t}, \mathbf{T}^{\mathbf{V}}(\mathbf{D}_2^{\mathbf{V}}\bar{\mathbf{T}}^{\mathbf{V}}\mathbf{D}_2^{\mathbf{V}})^{-}\mathbf{D}_2^{\mathbf{V}}\mathbf{T}^{\mathbf{V}}\mathbf{V}\mathbf{u} \rangle. \tag{5.2}$$

If $\mathscr{D}_2$ is of dimension one only, this expression simplifies to

$$\operatorname{cov}(\langle \mathbf{t}, \bar{\boldsymbol{\tau}} \rangle, \langle \mathbf{u}, \bar{\boldsymbol{\tau}} \rangle) = \langle \mathbf{t}, \mathbf{T}^{\mathbf{V}}\mathbf{V}\mathbf{u} \rangle + \frac{\langle \mathbf{t}, \mathbf{T}^{\mathbf{V}}\boldsymbol{\varepsilon} \rangle \langle \mathbf{u}, \mathbf{T}^{\mathbf{V}}\boldsymbol{\varepsilon} \rangle}{|\bar{\mathbf{T}}^{\mathbf{V}}\boldsymbol{\varepsilon}|^2_{\mathbf{V}}} \tag{5.3}$$

where $\boldsymbol{\varepsilon}$ is the dummy vector introduced in §4. Again we leave the algebra to the appendix and illustrations to §8.

6. Generally Balanced Experiments

Following Nelder (1965a) we now assume that the design has an orthogonal block structure which determines the eigenstructure of the covariance operator

$$\mathbf{V} = \sum_{\alpha} \xi_\alpha \mathbf{S}_\alpha \tag{6.1}$$

where the $\{\xi_\alpha\}$ are (usually unknown) positive eigenvalues and the $\{\mathbf{S}_\alpha\}$ are known symmetric and idempotent projectors such that $\sum_\alpha \mathbf{S}_\alpha = \mathbf{I}$.

Designs with such dispersion models are said to be *generally balanced* for an orthogonal decomposition $\mathcal{T} = \bigoplus_\beta \mathcal{T}_\beta$ of the treatment space if for all α

$$\mathbf{T}\mathbf{S}_\alpha\mathbf{T} = \sum_\beta \lambda_{\alpha\beta}\mathbf{T}_\beta \tag{6.2}$$

for a set of eigenvalues $\{\lambda_{\alpha\beta}\}$ such that $0 \le \lambda_{\alpha\beta} \le 1$ and $\sum_\alpha \lambda_{\alpha\beta} = 1$. This condition will be assumed to hold for the rest of this section. The *effect* corresponding to treatment β in stratum α is then calculated by $\mathbf{Q}_{\alpha\beta} = \lambda_{\alpha\beta}^{-1}\mathbf{T}_\beta\mathbf{S}_\alpha$ (unless there is no information $\mathcal{T}_\beta$ in $\mathcal{S}_\alpha$, and $\lambda_{\alpha\beta} = 0$) and, assuming that the $\{\xi_\alpha\}$ are known, the overall effect of treatment term β is $\mathbf{Q}_\beta = \sum_\alpha w_{\alpha\beta}\mathbf{Q}_{\alpha\beta}$, a linear combination of the within strata effects with weights

$$w_{\alpha\beta} = \xi_\alpha^{-1}\lambda_{\alpha\beta}\left(\sum_{\alpha'} \xi_{\alpha'}^{-1}\lambda_{\alpha'\beta}\right)^{-1}$$

(the sum being over all α' such that $\lambda_{\alpha'\beta} \neq 0$). Further, the $\{\mathbf{Q}_\beta\}$ are mutually orthogonal and $\mathbf{T}^\mathbf{V} = \sum_\beta \mathbf{Q}_\beta$. We refer to Houtman & Speed (1983) for a fuller discussion.

For a single missing value the solution (4.1) may be written using (6.1) as

$$\tilde{\mathbf{y}}_2 = -\left[\frac{\sum_\alpha \xi_\alpha^{-1}\langle \mathbf{S}_\alpha\bar{\mathbf{T}}^\mathbf{V}\boldsymbol{\varepsilon}, \mathbf{S}_\alpha\bar{\mathbf{T}}^\mathbf{V}\mathbf{y}_1\rangle}{\sum_\alpha \xi_\alpha^{-1}|\mathbf{S}_\alpha\bar{\mathbf{T}}^\mathbf{V}\boldsymbol{\varepsilon}|^2}\right]\boldsymbol{\varepsilon}. \tag{6.3}$$

If the design is generally balanced and there are only two effective strata, the lowest, say α', and another, say α'', with eigenvalues $\lambda_{\alpha'\beta}$ and $\lambda_{\alpha''\beta}$ for treatment $\mathcal{T}_\beta$, and $\mathbf{Q}_{\alpha'\beta}$ and $\mathbf{Q}_{\alpha''\beta}$ as effects in those strata, we will write $\boldsymbol{\Delta}_\beta = \mathbf{Q}_{\alpha'\beta} - \mathbf{Q}_{\alpha''\beta}$ for their difference. Then

$$\tilde{\mathbf{y}}_2 = -\left[\frac{\sum_\alpha \xi_\alpha^{-1}\langle \mathbf{R}_\alpha\boldsymbol{\varepsilon}, \mathbf{R}_\alpha\mathbf{y}_1\rangle + \sum_\beta \mu_\beta\langle \boldsymbol{\Delta}_\beta\boldsymbol{\varepsilon}, \boldsymbol{\Delta}_\beta\mathbf{y}_1\rangle}{\sum_\alpha \xi_\alpha^{-1}|\mathbf{R}_\alpha\boldsymbol{\varepsilon}|^2 + \sum_\beta \mu_\beta|\boldsymbol{\Delta}_\beta\boldsymbol{\varepsilon}|^2}\right]\boldsymbol{\varepsilon}, \tag{6.4}$$

where $\mu_\beta = \xi_{\alpha'}^{-1}\xi_{\alpha''}^{-1}\lambda_{\alpha'\beta}\lambda_{\alpha''\beta}(\xi_{\alpha'}^{-1}\lambda_{\alpha'\beta} + \xi_{\alpha''}^{-1}\lambda_{\alpha''\beta})^{-1}$, and $\mathbf{R}_\alpha$ is the residual operator after fitting to $\mathcal{T}$ in stratum α.

If no treatment term is estimated in more than one stratum, then (6.4) simplifies to

$$\tilde{\mathbf{y}}_2 = -\left[\frac{\sum_\alpha \xi_\alpha^{-1}\langle \mathbf{R}_\alpha\boldsymbol{\varepsilon}, \mathbf{R}_\alpha\mathbf{y}_1\rangle}{\sum_\alpha \xi_\alpha^{-1}|\mathbf{R}_\alpha\boldsymbol{\varepsilon}|^2}\right]\boldsymbol{\varepsilon}. \tag{6.5}$$

This is the case for all orthogonal designs i.e. designs for which $\lambda_{\alpha\beta} = 0$ or 1. For example, the covariance operator of a complete randomized block design has spectral decomposition $\mathbf{V} = \xi_0\mathbf{G} + \xi_b(\mathbf{B} - \mathbf{G}) + \xi_w(\mathbf{I} - \mathbf{B})$ where $\mathbf{G}$ is the overall averaging operator (replacing all the components of $\mathbf{y}$ by the grand mean) and $\mathbf{B}$ is the block averaging operator (replacing all the components of $\mathbf{y}$ by the average of those components belonging to the same block). If there are b blocks and t treatments, a single missing value is estimated using (6.5) by

$$\frac{\xi_w^{-1}(t\sum_T+b\sum_B-\sum_G)+\xi_b^{-1}(\sum_G-b\sum_B)}{\xi_w^{-1}(t-1)(b-1)+\xi_b^{-1}(b-1)}$$

where $\sum_G$, $\sum_B$ and $\sum_T$ denote respectively the sum of all the observations, the sum of the observations in the block containing the missing observation and the sum of the observations that received the same treatment as the missing unit.

In a $t\times t$ Latin square with one missing value, the replacement using (6.5) is given by

$$\frac{\xi_p^{-1}[t(\sum_R+\sum_C+\sum_T)-2\sum_G]+\xi_r^{-1}(\sum_G-t\sum_R)+\xi_c^{-1}(\sum_G-t\sum_C)}{\xi_p^{-1}(t-1)(t-2)+\xi_r^{-1}(t-1)+\xi_c^{-1}(t-1)}.$$

Here $\mathbf{V}=\xi_0\mathbf{G}+\xi_r(\mathbf{R}-\mathbf{G})+\xi_c(\mathbf{C}-\mathbf{G})+\xi_p(\mathbf{I}-\mathbf{C}-\mathbf{R}+\mathbf{G})$ ($\mathbf{R}$ and $\mathbf{C}$ are the row and column averaging operators and $\mathbf{G}$ is as before), $\sum_R$ and $\sum_C$ are respectively the sum of the observations in the same row and same column as the missing observation and $\sum_G$ and $\sum_T$ are as before. The case of a split plot design will be discussed in §8.

In practice the operator $\mathbf{V}$ is only partially known: the projectors $\{\mathbf{S}_\alpha\}$ are determined by the structure of the design whilst the strata variances $\{\xi_\alpha\}$ need to be estimated, and we outline a method for doing so in the next paragraph. The $\{\xi_\alpha\}$ are not needed however if we assume that most of the information on treatments is concentrated in the lowest stratum in which they are estimable. If attention is restricted to that stratum only, we let $\xi_\alpha^{-1}\approx 0$ in all the other strata and write $\mathbf{Ry}$ for the vector of residuals in that stratum, then

$$\tilde{\mathbf{y}}_2=-\left[\frac{\langle\mathbf{Ry}_1,\mathbf{R}\boldsymbol{\varepsilon}\rangle}{|\mathbf{R}\boldsymbol{\varepsilon}|^2}\right]\boldsymbol{\varepsilon}.$$

This gives all the usual missing value estimates, see, for example, Cochran & Cox (1957).

7. Estimation of Strata Variances

When information on some or all treatment terms is available from more than one stratum, we saw in the previous paragraph that the missing value estimators involve the strata variances $\{\xi_\alpha\}$. We now outline a method for estimating the $\{\xi_\alpha\}$ in a generally balanced design. The method we propose is an extension of Nelder's way of handling the problem for a complete design. The main steps are as follows:

(i) complete the data with initial estimates $\mathbf{y}_2^{(0)}$ computed using lowest stratum information only;

(ii) calculate estimates $\{\xi_\alpha^{(0)}\}$ using the data completed with $\mathbf{y}_2^{(0)}$;

(iii) calculate $\mathbf{y}_2^{(1)}$ using one of the methods indicated in §§4 or 6;

(iv) calculate new estimates $\{\xi_\alpha^{(1)}\}$ and then go back to (iii), continuing as often as seems necessary.

Estimates of the $\{\xi_\alpha\}$ in (ii) and (iv) may be obtained in two ways. If, after estimating treatments, there are enough degrees of freedom left in stratum α, equating the error mean square in that stratum to its expectation easily provides an estimator of ξ_α. Indeed we have under (6.2)

$$\begin{aligned} \mathbb{E}\,|\mathbf{R}_\alpha \mathbf{y}|^2 &= \xi_\alpha \text{ trace } \mathbf{R}_\alpha \\ &= \xi_\alpha\Big(\text{trace } \mathbf{S}_\alpha - \sum_\beta \text{trace } \mathbf{T}_\beta\Big) \qquad (7.1) \\ &= \xi_\alpha d_\alpha \end{aligned}$$

where the sum is over all β such that $\lambda_{\alpha\beta} > 0$. But d_α is often small and may even be zero, and so we would rather use the *actual* residual in stratum α given by $\mathbf{S}_\alpha \bar{\mathbf{T}}^{\mathbf{V}}\mathbf{y}$. If (6.2) holds we have

$$\begin{aligned} \mathbb{E}\,|\mathbf{S}_\alpha \bar{\mathbf{T}}^{\mathbf{V}}\mathbf{y}|^2 &= \xi_\alpha \text{ trace } \mathbf{S}_\alpha \bar{\mathbf{T}}^{\mathbf{V}} \\ &= \xi_\alpha\Big\{d_\alpha + \sum_\beta (1 - w_{\alpha\beta}) \text{ trace } \mathbf{T}_\beta\Big\} \qquad (7.2) \\ &= \xi_\alpha d'_\alpha \end{aligned}$$

where the sum is again over all βs such that $\lambda_{\alpha\beta} > 0$. The "degrees of freedom" d'_α is larger than d_α in (7.1), but, like $|\mathbf{S}_\alpha \bar{\mathbf{T}}^{\mathbf{V}}\mathbf{y}|^2$, it involves the unknown $\{\xi_\alpha\}$ through the $w_{\alpha\beta}$. Nelder suggested an iterative method, choosing initial values $\{\xi_\alpha^{(0)}\}$, for example from (7.1) in the lowest stratum, obtaining $|\mathbf{S}_\alpha \bar{\mathbf{T}}^{\mathbf{V}(0)}\mathbf{y}|^2$ and $d_\alpha'^{(0)}$, and then revised estimators

$$\xi_\alpha^{(1)} = \frac{|\mathbf{S}_\alpha \bar{\mathbf{T}}^{\mathbf{V}(0)}\mathbf{y}|^2}{d_\alpha'^{(0)}}.$$

Again this needs to be continued as often as seems necessary.

In our experience the estimates of $\{\xi_\alpha\}$ do not change very much, and unless there is a lot of missing data, one would not expect them to. No result guaranteeing convergence is available even with complete data.

8. Examples

(i) *Split-plot:* Let us consider a general split-plot experiment with r replications of a plots (levels of A) each of b subplots (levels of B). The block structure determines the spectral form of $\mathbb{D}\mathbf{y} = \mathbf{V}$ as

$$\mathbf{V} = \xi_0 \mathbf{G} + \xi_r(\mathbf{R} - \mathbf{G}) + \xi_p(\mathbf{P} - \mathbf{R}) + \xi_s(\mathbf{I} - \mathbf{P}) \qquad (8.1)$$

where $\mathbf{G}, \mathbf{R}$ and $\mathbf{P}$ are respectively the overall averaging operator, the operator averaging over replications, and the operator averaging over plots. The factorial treatment structure determines the decomposition of the treatment space

$$\mathcal{T} = \mathcal{G} \oplus (\mathcal{A} \ominus \mathcal{G}) \oplus (\mathcal{B} \ominus \mathcal{G}) \oplus (\mathcal{T} \ominus (\mathcal{A} + \mathcal{B})) \tag{8.2}$$

corresponding to a decomposition of the vector of means

$$\mathbb{E}\mathbf{y} = \tau = \mathbf{G}\tau + (\mathbf{A} - \mathbf{G})\tau + (\mathbf{B} - \mathbf{G})\tau + \mathbf{T}_{\mathbf{A}.\mathbf{B}}\tau$$

for all $\tau \in \mathcal{T}$. The operator $\mathbf{G}$ is the same as above, $\mathbf{A}$ and $\mathbf{B}$ average respectively over the levels of A and over the levels of B, and $\mathbf{T}_{\mathbf{A}.\mathbf{B}} = \mathbf{T} - \mathbf{A} - \mathbf{B} + \mathbf{G}$ where $\mathbf{T}$ is the treatment averaging operator. We note that $\mathbf{G}$, $\mathbf{A}$, $\mathbf{B}$, $\mathbf{T}$ and $\mathbf{T}_{\mathbf{A}.\mathbf{B}}$ are orthogonal projectors with respective ranges $\mathcal{G}$, $\mathcal{A}$, $\mathcal{B}$, $\mathcal{T}$ and $\mathcal{T} \ominus (\mathcal{A} + \mathcal{B})$.

The experiment is generally balanced with respect to the treatment structure (8.2), with a set of eigenvalues all equal to zero or one, this always being the case for orthogonal designs. All the information on contrasts comparing levels of A (contrasts in $\mathcal{A} \ominus \mathcal{G}$) is contained in the main plot means adjusted by their replicate means $(\mathbf{P} - \mathbf{R})\mathbf{y}$, whilst all the information on contrasts comparing levels of B (contrasts in $\mathcal{B} \ominus \mathcal{G}$) and on those describing interaction between A and B (contrasts in $\mathcal{T} \ominus (\mathcal{A} + \mathcal{B})$) is contained in the stratum of subplot comparisons $(\mathbf{I} - \mathbf{P})\mathbf{y}$. And so a single missing value can be estimated using (6.5) by x where x is

$$\frac{\xi_s^{-1}(ra\sum_P + ab\sum_{AB} - a\sum_A) + \xi_p^{-1}(r\sum_R + a\sum_A - ra\sum_P - \sum_G) + \xi_r^{-1}(\sum_G - r\sum_R)}{\xi_s^{-1}[a(r-1)(b-1)] + \xi_p^{-1}[(r-1)(a-1)] + \xi_r^{-1}(r-1)}. \tag{8.3}$$

Here $\sum_P$ is the total of the observations in the plot containing the missing observation,

$\sum_R$ is the sum of all the observations in the replicate containing the missing observation,

$\sum_{AB}$ is the total of all the subplots that received the same treatment combination as the missing unit,

$\sum_A$ is the total of all the subplots that received the same level of treatment A as the missing one,

$\sum_G$ is the sum of all the observations.

This formula may be compared with the estimate obtained by Anderson (1946) by minimizing the subplot error only:

$$x' = \frac{r\sum_P + b\sum_{AB} - \sum_A}{(r-1)(b-1)}. \tag{8.4}$$

If ξ_p and ξ_r are both very large in comparison with ξ_s, then (8.3) reduces to (8.4). In the example treated in his paper, Anderson obtains a replacement of 763 whereas (8.3) gives 726 (using (7.1) to estimate the strata variances); the latter value is in better harmony with the rest of the data whose mean was 492.

The correction to the variance var $(\langle \mathbf{t}, \bar{\tau} \rangle)$ where $\bar{\tau}$ satisfies (3.4), will be zero whenever the missing value is at another treatment level than the levels compared in the contrast t. When this correction is not zero we will follow Cochran & Cox (1957, p. 303) and consider four simple kinds of contrasts. Let us write

$\mathbf{t_A}$ (resp. $\mathbf{t_B}$) = a difference between two A means (resp. B means),

$\mathbf{t_{Ba}}$ = a difference between two B means at the same level of A,

$\mathbf{t_{Ab}}$ = a difference between two A means at the same level of B, or at different levels of B.

With the denominator of the correction given in (5.3) equal to $|\bar{\mathbf{T}}^{\mathbf{V}}\boldsymbol{\varepsilon}|^2_{\mathbf{V}} = (r-1)(rab)^{-1}[\xi_r^{-1} + \xi_p^{-1}(a-1) + \xi_s^{-1}(b-1)a] = d$, we have

$$\text{var}\,(\langle \mathbf{t_A}, \bar{\tau} \rangle) = 2\xi_p(rb)^{-1} + (r^2b^2d)^{-1},$$
$$\text{var}\,(\langle \mathbf{t_B}, \bar{\tau} \rangle) = 2\xi_s(ra)^{-1} + (r^2a^2d)^{-1},$$
$$\text{var}\,(\langle \mathbf{t_{Ba}}, \bar{\tau} \rangle) = 2\xi_s r^{-1} + (r^2d)^{-1},$$
$$\text{var}\,(\langle \mathbf{t_{Ab}}, \bar{\tau} \rangle) = [2\xi_p(rb)^{-1} + 2\xi_s(b-1)(rb)^{-1}] + (r^2d)^{-1}.$$

Our corrections (second terms) reduce to the ones obtained by Cochran & Cox (1957, p. 303) and based on a lowest stratum estimate of the missing value by setting ξ_r^{-1} and ξ_p^{-1} equal to zero.

If we now consider the submodel having no AB-interaction term, namely $\mathbb{E}\mathbf{y} \in \mathcal{A} + \mathcal{B}$, the apparent sum of squares due to the interaction, $|\mathbf{T}^{\mathbf{V}}_{\mathbf{A.B}}\mathbf{y}(\mathcal{T})|^2_{\mathbf{V}} = \xi_s^{-1}\,|(\mathbf{T}-\mathbf{A}-\mathbf{B}+\mathbf{G})\mathbf{y}(\mathcal{T})|^2$ must be adjusted. This is done by subtracting from it the correction term

$$|\bar{\mathbf{P}}^{\mathbf{V}}_{\mathcal{A}+\mathcal{B}}(\mathbf{y}(\mathcal{T}) - \mathbf{y}(\mathcal{A}+\mathcal{B}))|^2_{\mathbf{V}} = (d-f)^2\,|\bar{\mathbf{P}}^{\mathbf{V}}_{\mathcal{A}+\mathcal{B}}\boldsymbol{\varepsilon}|^2_{\mathbf{V}}$$

where d is the replacement (8.3) under the full model, and f is the replacement under the submodel given by f where f is

$$\frac{\xi_s^{-1}(ra\Sigma_P + b\Sigma_B - \Sigma_G) + \xi_P^{-1}(r\Sigma_R + a\Sigma_A - ar\Sigma_P - \Sigma_G) + \xi_r^{-1}(\Sigma_G - r\Sigma_R)}{\xi_s^{-1}(ra-1)(b-1) + \xi_p^{-1}(a-1)(r-1) + \xi_r^{-1}(r-1)}. \tag{8.5}$$

The notation is as in (8.3) with Σ_B denoting the sum of the observations that received the same level of treatment B as the missing one, and finally

$$|\bar{\mathbf{P}}^{\mathbf{V}}_{\mathcal{A}+\mathcal{B}}\boldsymbol{\varepsilon}|^2_{\mathbf{V}} = (abr)^{-1}[\xi_s^{-1}(ra-1)(b-1) + \xi_p^{-1}(a-1)(r-1) + \xi_r^{-1}(r-1)].$$

(ii) *Simple (square) lattice:* We consider an experiment performed to compare k^2 treatments in two replicates of k blocks of k plots each. As in a split-plot experiment, the block structure here is doubly nested and so there are three strata (other than the grand mean): between replicates, within replicates between blocks, and within blocks. This defines the spectral decomposition of the dispersion matrix $\mathbb{D}\mathbf{y} = \mathbf{V}$ as

$$\mathbf{V} = \xi_0 \mathbf{G} + \xi_r(\mathbf{R}-\mathbf{G}) + \xi_b(\mathbf{B}-\mathbf{G}) + \xi_p(\mathbf{I}-\mathbf{B}) \tag{8.6}$$

where $\mathbf{G}$, $\mathbf{B}$ and $\mathbf{R}$ are respectively the overall, block and replicate averaging operators. This design does not satisfy the conditions (6.2) of general balance with respect to the natural treatment decomposition

$$\mathbf{T} = \mathbf{G} + (\mathbf{T}-\mathbf{G})$$

corresponding to "no structure" on treatments but general balance is obtained by introducing a factorial "pseudo-structure". This is determined by the following scheme: the treatments are arranged in a $k \times k$ square and treatments belonging to the same row (resp. column) of the square are allocated to the same block in the first (resp. second) replicate. Let us use M and N for the pseudo-factors corresponding to the rows and columns of the treatment array. The experiment is generally balanced with respect to the treatment decomposition

$$\boldsymbol{\tau} = \mathbf{G}\boldsymbol{\tau} + (\mathbf{M}-\mathbf{G})\boldsymbol{\tau} + (\mathbf{N}-\mathbf{G})\boldsymbol{\tau} + \mathbf{T}_{\mathbf{M}.\mathbf{N}}\boldsymbol{\tau} \tag{8.7}$$

where $\boldsymbol{\tau} = \mathbb{E}\mathbf{y} \in \mathcal{T}$. The notation here is as in (i) with $\mathbf{M}$ and $\mathbf{N}$ instead of $\mathbf{A}$ and $\mathbf{B}$. The relationships (6.2) are

$$\begin{aligned} \mathbf{T}(\mathbf{R}-\mathbf{G})\mathbf{T} &= 0, \\ \mathbf{T}(\mathbf{B}-\mathbf{R})\mathbf{T} &= \tfrac{1}{2}(\mathbf{M}-\mathbf{G}) + \tfrac{1}{2}(\mathbf{N}-\mathbf{G}), \\ \mathbf{T}(\mathbf{I}-\mathbf{B})\mathbf{T} &= \tfrac{1}{2}(\mathbf{M}-\mathbf{G}) + \tfrac{1}{2}(\mathbf{N}-\mathbf{G}) + \mathbf{T}_{\mathbf{M}.\mathbf{N}}; \end{aligned}$$

and so the effects are

$$\begin{aligned} \mathbf{Q}_{bM} &= (\tfrac{1}{2})^{-1}\mathbf{M}(\mathbf{B}-\mathbf{R}), & \mathbf{Q}_{pM} &= (\tfrac{1}{2})^{-1}\mathbf{M}(\mathbf{I}-\mathbf{B}), \\ \mathbf{Q}_{bN} &= (\tfrac{1}{2})^{-1}\mathbf{N}(\mathbf{B}-\mathbf{R}), & \mathbf{Q}_{pN} &= (\tfrac{1}{2})^{-1}\mathbf{N}(\mathbf{I}-\mathbf{B}), \\ \mathbf{Q}_{bM.N} &= \mathbf{0}, & \mathbf{Q}_{pM.N} &= \mathbf{T}_{M.N}(\mathbf{I}-\mathbf{B}) = \mathbf{T}_{M.N}, \end{aligned}$$

and the weights are

$$\begin{aligned} w_{bM} &= w_{bN} = w = \tfrac{1}{2}\xi_b^{-1}(\tfrac{1}{2}\xi_b^{-1} + \tfrac{1}{2}\xi_p^{-1})^{-1} = \xi_b^{-1}(\xi_b^{-1} + \xi_p^{-1})^{-1}, \\ w_{rM} &= w_{rN} = 1 - w = \xi_p^{-1}(\xi_b^{-1} + \xi_p^{-1})^{-1}. \\ w_{bM.N} &= 0, \quad w_{pM.M} = 1. \end{aligned}$$

Assuming known strata variances $\{\xi_\alpha\}$, a single missing value may be estimated using (6.4) by x where x is

$$\frac{\xi_p^{-1}(k^2\Sigma_T - 2\Sigma_R + \Sigma_G - 2kC - kC') - 2(\xi_b + \xi_p)^{-1}(2\Sigma_G - 4\Sigma_R - 2kC + kC')}{\xi_p^{-1}(k-1)^2 + 4(\xi_b + \xi_p)^{-1}(k-1)} \tag{8.8}$$

where Σ_G, Σ_R and Σ_T denote respectively the sum of all the observations, the total of the observations in the replicate containing the missing observation and the total of the observations that received the same treatment as the missing unit. We have borrowed from Cochran & Cox (1957) the notation

$C =$ [total (over replicates) of all treatments in the block to which the missing unit belongs] $- 2\sum_B$,

where $\sum_B$ is the sum of all the observations in the block containing the missing unit, and

$C' =$ sum of the C values for all blocks containing the treatment that was allocated to the missing unit.

Since there is no natural submodel of the treatment model assumed, corrections to sums of squares due to treatments will not usually be needed. And so we only compute the adjustments to be added to the variances of elementary contrasts (contrasts between pairs of treatments). For a complete experiment, the usual formulae for those variances are

$$\frac{2}{k}\left\{\frac{1}{\xi_b^{-1}+\xi_p^{-1}}+\frac{k-1}{2\xi_p^{-1}}\right\} \tag{8.9}$$

if the two treatments belong to the same block (in either replicate) and

$$\frac{2}{k}\left\{\frac{2}{\xi_b^{-1}+\xi_p^{-1}}+\frac{k-2}{2\xi_p^{-1}}\right\} \tag{8.10}$$

if the two treatments never appear in the same block. The correction to those variances, due to the estimation of a single missing value is given in (5.3) by $(\langle \mathbf{t}, \mathbf{T}^{\mathbf{V}}\boldsymbol{\varepsilon}\rangle)^2(|\bar{\mathbf{T}}^{\mathbf{V}}\boldsymbol{\varepsilon}|_{\mathbf{V}}^2)^{-1}$ with

$$\begin{aligned} D = |\bar{\mathbf{T}}^{\mathbf{V}}\boldsymbol{\varepsilon}|_{\mathbf{V}}^2 = \frac{1}{2k^4}[&4\xi_0^{-1}(k-1)^2+\xi_r^{-1}k^2 \\ &+\xi_b^{-1}\{4k^2(k-1)w^2+(k-1)(k-2)^2\} \\ &+\xi_p^{-1}\{4k^2(k-1)\bar{w}^2(k-1)k^3\}] \end{aligned} \tag{8.11}$$

where $\bar{w} = 1 - w$. Using D for the denominator, the correction to (8.9) is

$$\frac{1}{4k^2D}(k-2\bar{w})^2$$

for a comparison between the treatment allocated to the missing unit, say τ_m, and a treatment in the same block as the missing unit, and

$$\frac{1}{4k^2D}(k-2w)^2$$

for a comparison between τ_m and a treatment appearing in the same block as τ_m in the replicate that does not contain the missing unit. The correction to (8.10) is

$$\frac{1}{4D}$$

for a comparison between τ_m and a treatment never appearing in the same block as τ_m; it is

$$\frac{1}{k^2D}(1-2\bar{w})^2$$

for a comparison between a treatment appearing in the same block as τ_m in the same replicate as the missing unit and such a treatment in the other replicate; finally, it is

$$\frac{1}{D}\left(\frac{\bar{w}}{k}\right)^2\left(\text{resp.}\,\frac{1}{D}\left(\frac{w}{k}\right)^2\right)$$

for a comparison between a treatment appearing in the same block as τ_m in the same replicate as the missing unit (resp. in the replicate that does not contain the missing unit) and a treatment never appearing in the same block as τ_m. The corrections in all other cases are zero.

9. Appendix

(*i*) *Proof of* (3.5)

Step 1. We define the vector of residuals

$$\tilde{\mathbf{r}} = \bar{\mathbf{T}}^{\mathbf{V}}(\mathbf{y}_1 + \tilde{\mathbf{y}}_2) \qquad \text{(A1)}$$

where $\tilde{\mathbf{y}}_2$ satisfies the missing values equations (3.3), and first prove that

$$\tilde{\mathbf{r}} = (\mathbf{I} - \mathbf{P}^{\mathbf{V}}_{\mathscr{D}_2+\mathscr{T}})\bar{\mathbf{T}}^{\mathbf{V}}\mathbf{y}_1 \qquad \text{(A2)}$$

where $\mathbf{P}^{\mathbf{V}}_{\mathscr{D}_2+\mathscr{T}}$ is the projection onto the space $\mathscr{D}_2 + \mathscr{T} = \{\mathbf{y}_2 + \mathbf{t} \mid \mathbf{y}_2 \in \mathscr{D}_2, \mathbf{t} \in \mathscr{T}\}$, orthogonal w.r.t. $\langle .\,,.\rangle_{\mathbf{V}}$.

Using recursion (iv) of Section 4, we have

$$\begin{aligned} \tilde{\mathbf{r}} &= \lim_{n\to\infty}(\bar{\mathbf{T}}^{\mathbf{V}}\bar{\mathbf{D}}_2^{\mathbf{V}})^n\bar{\mathbf{T}}^{\mathbf{V}}\mathbf{y}_1 \\ &= \mathbf{P}^{\mathbf{V}}_{\mathbf{V}\mathscr{T}^{\perp}\cap\mathbf{V}\mathscr{D}_1}\bar{\mathbf{T}}^{\mathbf{V}}\mathbf{y}_1 \qquad \text{Von Neumann (1950, p. 55)} \\ &= (\mathbf{I} - \mathbf{P}^{\mathbf{V}}_{\mathscr{T}+\mathscr{D}_2})\bar{\mathbf{T}}^{\mathbf{V}}\mathbf{y}_1. \end{aligned}$$

This last relationship and the previous one both use the fact that the orthogonal complements w.r.t. $\langle .\,,.\rangle_{\mathbf{V}}$ of $\mathscr{T}$ and $\mathscr{D}_2$ are respectively $\mathbf{V}\mathscr{T}^{\perp}$ and $\mathbf{V}\mathscr{D}_1$.

Step 2.

$$\begin{aligned}
\mathbf{D}_1\tilde{\tau} &= \mathbf{D}_1\mathbf{T}^{\mathbf{V}}(\mathbf{y}_1+\tilde{\mathbf{y}}_2) && \text{where } \tilde{\mathbf{y}}_2 \text{ satisfies (3.3)}\\
&= \mathbf{D}_1(\mathbf{I}-\bar{\mathbf{T}}^{\mathbf{V}})(\mathbf{y}_1+\tilde{\mathbf{y}}_2)\\
&= \mathbf{y}_1-\mathbf{D}_1\tilde{\mathbf{r}} && \text{using (A1)}\\
&= \mathbf{y}_1-\mathbf{D}_1(\mathbf{I}-\mathbf{P}^{\mathbf{V}}_{\mathcal{T}+\mathcal{D}_2})\bar{\mathbf{T}}^{\mathbf{V}}\mathbf{y}_1 && \text{using (A2)}\\
&= \mathbf{y}_1-\mathbf{D}_1\mathbf{y}_1+\mathbf{D}_1\mathbf{T}^{\mathbf{V}}\mathbf{y}_1-\mathbf{D}_1\mathbf{P}^{\mathbf{V}}_{\mathcal{T}+\mathcal{D}_2}\mathbf{T}^{\mathbf{V}}\mathbf{y}_1+\mathbf{D}_1\mathbf{P}^{\mathbf{V}}_{\mathcal{D}_2+\mathcal{T}}\mathbf{y}_1\\
&= \mathbf{D}_1\mathbf{P}^{\mathbf{V}}_{\mathcal{D}_2+\mathcal{T}}\mathbf{y}_1. && \text{since } \mathbf{P}^{\mathbf{V}}_{\mathcal{T}+\mathcal{D}_2}\mathbf{T}^{\mathbf{V}}=\mathbf{T}^{\mathbf{V}}.
\end{aligned}$$

Step 3. We first observe that

$$\mathcal{D}_2+\mathcal{T}=\mathcal{D}_2+\mathcal{T}_1=\mathcal{D}_2\oplus\bar{\mathbf{D}}_2^{\mathbf{V}}\mathcal{T}_1$$

where the last sum is orthogonal with respect to $\langle .\,, .\rangle_{\mathbf{V}}$. Thus the weighted projection onto $\mathcal{D}_2+\mathcal{T}$ decomposes into the sum of the weighted projections onto $\mathcal{D}_2$ and $\mathcal{A}=\bar{\mathbf{D}}_2^{\mathbf{V}}\mathcal{T}_1$ respectively. We then have

$$\begin{aligned}
\mathbf{D}_1\tilde{\tau} &= \mathbf{D}_1\mathbf{D}_2^{\mathbf{V}}\mathbf{y}_1+\mathbf{D}_1\mathbf{A}^{\mathbf{V}}\mathbf{y}_1\\
&= \mathbf{D}_1\mathbf{A}^{\mathbf{V}}\mathbf{y}_1.
\end{aligned}$$

Now $\mathcal{R}(\mathbf{A}^{\mathbf{V}})=\mathcal{R}(\mathbf{V}(\mathbf{D}_1\mathbf{V}\mathbf{D}_1)^{-}\mathbf{T}_1)$, and so

$$\begin{aligned}
\mathbf{D}_1\tilde{\tau} &= \mathbf{D}_1\mathbf{V}(\mathbf{D}_1\mathbf{V}\mathbf{D}_1)^{-}\mathbf{T}_1[\mathbf{T}_1(\mathbf{D}_1\mathbf{V}\mathbf{D}_1)^{-}\mathbf{V}\mathbf{V}^{-1}\\
&\quad\times\mathbf{V}(\mathbf{D}_1\mathbf{V}\mathbf{D}_1)^{-}\mathbf{T}_1]^{-}\mathbf{T}_1(\mathbf{D}_1\mathbf{V}\mathbf{D}_1)^{-}\mathbf{V}\mathbf{V}^{-1}\mathbf{y}_1\\
&= \mathbf{T}_1[\mathbf{T}_1(\mathbf{D}_1\mathbf{V}\mathbf{D}_1)^{-}\mathbf{T}_1]^{-}\mathbf{T}_1(\mathbf{D}_1\mathbf{V}\mathbf{D}_1)^{-}\mathbf{y}_1\\
&= \mathbf{T}_1^{\mathbf{V}_1}\mathbf{y}_1 \qquad (\text{where } \mathbf{V}_1=\mathbf{D}_1\mathbf{V}\mathbf{D}_1)\\
&= \hat{\tau}_1
\end{aligned}$$

which proves (3.5).

(*ii*) *Verification of* (5.1)

$$\begin{aligned}
&|\bar{\mathbf{U}}^{\mathbf{V}}\mathbf{y}(\mathcal{T})|^2_{\mathbf{V}}-|\bar{\mathbf{U}}^{\mathbf{V}}\mathbf{y}(\mathcal{U})|^2_{\mathbf{V}}\\
&=\langle\bar{\mathbf{U}}^{\mathbf{V}}\mathbf{y}(\mathcal{T}),\bar{\mathbf{U}}^{\mathbf{V}}\mathbf{y}(\mathcal{T})\rangle_{\mathbf{V}}-\langle\bar{\mathbf{U}}^{\mathbf{V}}\mathbf{y}(\mathcal{U}),\bar{\mathbf{U}}^{\mathbf{V}}\mathbf{y}(\mathcal{U})\rangle_{\mathbf{V}}\\
&=\langle\mathbf{y}(\mathcal{T}),\bar{\mathbf{U}}^{\mathbf{V}}\mathbf{y}(\mathcal{T})\rangle_{\mathbf{V}}-\langle\mathbf{y}(\mathcal{U}),\bar{\mathbf{U}}^{\mathbf{V}}\mathbf{y}(\mathcal{U})\rangle_{\mathbf{V}}\\
&\quad+\langle\mathbf{y}(\mathcal{T}),\bar{\mathbf{U}}^{\mathbf{V}}\mathbf{y}(\mathcal{U})\rangle_{\mathbf{V}}-\langle\mathbf{y}(\mathcal{T}),\bar{\mathbf{U}}^{\mathbf{V}}\mathbf{y}(\mathcal{U})\rangle_{\mathbf{V}}\\
&=\langle\mathbf{y}(\mathcal{T}),\bar{\mathbf{U}}^{\mathbf{V}}[\mathbf{y}(\mathcal{T})-\mathbf{y}(\mathcal{U})]\rangle_{\mathbf{V}}+\langle\mathbf{y}(\mathcal{T})-\mathbf{y}(\mathcal{U}),\bar{\mathbf{U}}^{\mathbf{V}}\mathbf{y}(\mathcal{U})\rangle_{\mathbf{V}}.
\end{aligned}$$

Now, $\langle\mathbf{y}(\mathcal{T})-\mathbf{y}(\mathcal{U}),\bar{\mathbf{U}}^{\mathbf{V}}\mathbf{y}(\mathcal{U})\rangle_{\mathbf{V}}=\langle\mathbf{y}_2(\mathcal{T})-\mathbf{y}_2(\mathcal{U}),\bar{\mathbf{U}}^{\mathbf{V}}\mathbf{y}(\mathcal{U})\rangle_{\mathbf{V}}$

$$\begin{aligned}
&=\langle\mathbf{D}_2^{\mathbf{V}}[\mathbf{y}_2(\mathcal{T})-\mathbf{y}_2(\mathcal{U})],\bar{\mathbf{U}}^{\mathbf{V}}\mathbf{y}(\mathcal{U})\rangle_{\mathbf{V}}\\
&=\langle\mathbf{y}_2(\mathcal{T})-\mathbf{y}_2(\mathcal{U}),\mathbf{D}_2^{\mathbf{V}}\bar{\mathbf{U}}^{\mathbf{V}}\mathbf{y}(\mathcal{U})\rangle_{\mathbf{V}},
\end{aligned}$$

and, using (3.6) for the model $\mathcal{U}$, we have $\mathbf{D}_2^{\mathbf{V}}\bar{\mathbf{U}}^{\mathbf{V}}\mathbf{y}(\mathcal{U})=0$. Thus we may write

$$
\begin{aligned}
&|\bar{\mathbf{U}}^{\mathbf{V}}(\mathcal{T})|^2_{\mathbf{V}} - |\bar{\mathbf{U}}^{\mathbf{V}}\mathbf{y}(\mathcal{U})|^2_{\mathbf{V}} \\
&= \langle \mathbf{y}(\mathcal{T}), \bar{\mathbf{U}}^{\mathbf{V}}[\mathbf{y}(\mathcal{T}) - \mathbf{y}(\mathcal{U})]\rangle_{\mathbf{V}} - \langle \mathbf{y}(\mathcal{T}) - \mathbf{y}(\mathcal{U}), \bar{\mathbf{U}}^{\mathbf{V}}\mathbf{y}(\mathcal{U})\rangle_{\mathbf{V}} \\
&= \langle \mathbf{y}(\mathcal{T}), \bar{\mathbf{U}}^{\mathbf{V}}[\mathbf{y}_2(\mathcal{T}) - \mathbf{y}_2(\mathcal{U})]\rangle_{\mathbf{V}} - \langle \mathbf{y}(\mathcal{U}), \bar{\mathbf{U}}^{\mathbf{V}}[\mathbf{y}_2(\mathcal{T}) - \mathbf{y}_2(\mathcal{U})]\rangle_{\mathbf{V}} \\
&= \langle \mathbf{y}_2(\mathcal{T}) - \mathbf{y}_2(\mathcal{U}), \bar{\mathbf{U}}^{\mathbf{V}}[\mathbf{y}_2(\mathcal{T}) - \mathbf{y}_2(\mathcal{U})]\rangle_{\mathbf{V}} \\
&= |\bar{\mathbf{U}}^{\mathbf{V}}[\mathbf{y}_2(\mathcal{T}) - \mathbf{y}_2(\mathcal{U})]|^2_{\mathbf{V}}.
\end{aligned}
$$

(*iii*) *Verifications of* (5.2) *and* (5.3)

We start from equation (3.4) giving the vector of fitted values $\bar{\boldsymbol{\tau}}$ when the data are incomplete, writing the equation in the form

$$(\mathbf{I} - \mathbf{T}^{\mathbf{V}}\mathbf{D}_2^{\mathbf{V}}\mathbf{T}^{\mathbf{V}})\bar{\boldsymbol{\tau}} = \mathbf{T}^{\mathbf{V}}\bar{\mathbf{D}}_2^{\mathbf{V}}\mathbf{y}_1. \tag{A3}$$

We will only consider the case where $\mathcal{T} \cap \mathcal{D}_2 = \{0\}$ so that all the eigenvalues of $\mathbf{T}^{\mathbf{V}}\mathbf{D}_2^{\mathbf{V}}\mathbf{T}^{\mathbf{V}}$ are strictly smaller than one and hence all those of $\mathbf{I} - \mathbf{T}^{\mathbf{V}}\mathbf{D}_2^{\mathbf{V}}\mathbf{T}^{\mathbf{V}} = \mathbf{A}$ are strictly positive and the operator $\mathbf{A}$ is invertible. The unique solution of (A3) is thus

$$\bar{\boldsymbol{\tau}} = \mathbf{A}^{-1}\mathbf{T}^{\mathbf{V}}\bar{\mathbf{D}}_2^{\mathbf{V}}\mathbf{y}_1$$

and so we have

$$
\begin{aligned}
\operatorname{var} \bar{\boldsymbol{\tau}} &= \mathbf{A}^{-1}\mathbf{T}^{\mathbf{V}}\bar{\mathbf{D}}_2^{\mathbf{V}}\mathbf{D}_1\mathbf{V}\mathbf{D}_1(\bar{\mathbf{D}}_2^{\mathbf{V}})^*(\mathbf{T}^{\mathbf{V}})^*[\mathbf{A}^{-1}]^* \\
&= \mathbf{A}^{-1}\mathbf{T}^{\mathbf{V}}\bar{\mathbf{D}}_2^{\mathbf{V}}\mathbf{T}^{\mathbf{V}}\mathbf{A}^{-1}\mathbf{V} \\
&= \mathbf{A}^{-1}\mathbf{A}\mathbf{T}^{\mathbf{V}}\mathbf{A}^{-1}\mathbf{V} \\
&= \mathbf{T}^{\mathbf{V}}(\mathbf{I} - \mathbf{T}^{\mathbf{V}}\mathbf{D}_2^{\mathbf{V}}\mathbf{T}^{\mathbf{V}})^{-1}\mathbf{V}.
\end{aligned}
$$

We may re-express this as follows

$$
\begin{aligned}
\operatorname{var} \bar{\boldsymbol{\tau}} &= \mathbf{T}^{\mathbf{V}}\left[\sum_{i=0}^{\infty} (\mathbf{T}^{\mathbf{V}}\mathbf{D}_2^{\mathbf{V}}\mathbf{T}^{\mathbf{V}})^i\right]\mathbf{V} \\
&= \mathbf{T}^{\mathbf{V}}\mathbf{V} + \sum_{i=1}^{\infty} (\mathbf{T}^{\mathbf{V}}\mathbf{D}_2^{\mathbf{V}}\mathbf{T}^{\mathbf{V}})^i\mathbf{V} \\
&= \mathbf{T}^{\mathbf{V}}\mathbf{V} + \mathbf{T}^{\mathbf{V}}\mathbf{D}_2^{\mathbf{V}}\mathbf{T}^{\mathbf{V}}\left[\sum_{i=0}^{\infty} (\mathbf{T}^{\mathbf{V}}\mathbf{D}_2^{\mathbf{V}}\mathbf{T}^{\mathbf{V}})^i\right]\mathbf{V} \\
&= \mathbf{T}^{\mathbf{V}}\mathbf{V} + \mathbf{T}^{\mathbf{V}}\left[\sum_{i=0}^{\infty} (\mathbf{D}_2^{\mathbf{V}}\mathbf{T}^{\mathbf{V}}\mathbf{D}_2^{\mathbf{V}})^i\right]\mathbf{D}_2^{\mathbf{V}}\mathbf{T}^{\mathbf{V}}\mathbf{V} \\
&= \mathbf{T}^{\mathbf{V}}\mathbf{V} + \mathbf{T}^{\mathbf{V}}(\mathbf{D}_2^{\mathbf{V}}\bar{\mathbf{T}}^{\mathbf{V}}\mathbf{D}_2^{\mathbf{V}})^{-}\mathbf{D}_2^{\mathbf{V}}\mathbf{T}^{\mathbf{V}}\mathbf{V}\mathbf{u}.
\end{aligned}
$$

From this we have for all $\mathbf{t}, \mathbf{u} \in \mathcal{T}$

$$\operatorname{cov}(\langle \mathbf{t}, \bar{\boldsymbol{\tau}}\rangle, \langle \mathbf{u}, \bar{\boldsymbol{\tau}}\rangle) = \langle \mathbf{t}, \mathbf{T}^{\mathbf{V}}\mathbf{V}\mathbf{u}\rangle + \langle \mathbf{t}, \mathbf{T}^{\mathbf{V}}(\mathbf{D}_2^{\mathbf{V}}\bar{\mathbf{T}}^{\mathbf{V}}\mathbf{D}_2^{\mathbf{V}})^{-}\mathbf{D}_2^{\mathbf{V}}\mathbf{T}^{\mathbf{V}}\mathbf{V}\mathbf{u}\rangle$$

which is (5.2).

Now consider the case where $\dim \mathcal{D}_2 = 1$ so that $\mathcal{D}_2$ is spanned by the vector $\boldsymbol{\varepsilon}$ and $\mathbf{D}_2^{\mathbf{V}}\mathbf{z} = |\boldsymbol{\varepsilon}|_{\mathbf{V}}^{-2}\langle\boldsymbol{\varepsilon}, \mathbf{z}\rangle_{\mathbf{V}}\boldsymbol{\varepsilon}$. Let $c \in R$ be such that

$$(\mathbf{D}_2^{\mathbf{V}}\bar{\mathbf{T}}^{\mathbf{V}}\mathbf{D}_2^{\mathbf{V}})^{-}\mathbf{D}_2^{\mathbf{V}}\mathbf{T}^{\mathbf{V}}\mathbf{V}\mathbf{u} = c\boldsymbol{\varepsilon}. \tag{A4}$$

Since $\mathbf{D}_2^{\mathbf{V}}\mathbf{T}^{\mathbf{V}}\mathbf{V}\mathbf{u}$ is a vector in $\mathscr{D}_2$, premultiplying both sides of (A4) by $\mathbf{D}_2^{\mathbf{V}}\bar{\mathbf{T}}^{\mathbf{V}}\mathbf{D}_2^{\mathbf{V}}$ gives

$$\mathbf{D}_2^{\mathbf{V}}\mathbf{T}^{\mathbf{V}}\mathbf{V}\mathbf{u} = c\mathbf{D}_2^{\mathbf{V}}\bar{\mathbf{T}}^{\mathbf{V}}\mathbf{D}_2^{\mathbf{V}}\boldsymbol{\varepsilon} = c\mathbf{D}_2^{\mathbf{V}}\bar{\mathbf{T}}^{\mathbf{V}}\boldsymbol{\varepsilon},$$

and using the above expression for the projector $\mathbf{D}_2^{\mathbf{V}}$ we get

$$|\boldsymbol{\varepsilon}|_{\mathbf{V}}^{-2}\langle\boldsymbol{\varepsilon}, \mathbf{T}^{\mathbf{V}}\mathbf{V}\mathbf{u}\rangle_{\mathbf{V}}\boldsymbol{\varepsilon} = c\,|\boldsymbol{\varepsilon}|_{\mathbf{V}}^{-2}\langle\boldsymbol{\varepsilon}, \bar{\mathbf{T}}^{\mathbf{V}}\boldsymbol{\varepsilon}\rangle_{\mathbf{V}}.$$

Hence

$$\begin{aligned} c &= \langle\boldsymbol{\varepsilon}, \mathbf{T}^{\mathbf{V}}\mathbf{V}\mathbf{u}\rangle_{\mathbf{V}}[\langle\boldsymbol{\varepsilon}, \bar{\mathbf{T}}^{\mathbf{V}}\boldsymbol{\varepsilon}\rangle_{\mathbf{V}}]^{-1} \\ &= \langle\mathbf{T}^{\mathbf{V}}\boldsymbol{\varepsilon}, \mathbf{u}\rangle[|\bar{\mathbf{T}}^{\mathbf{V}}\boldsymbol{\varepsilon}|_{\mathbf{V}}^2]^{-1}. \end{aligned}$$

Using this value of c in (A4) and substituting (A4) into (5.2), we see that the latter expression simplifies in this particular case to

$$\operatorname{cov}(\langle\mathbf{t}, \bar{\tau}\rangle, \langle\mathbf{u}, \bar{\tau}\rangle) = \langle\mathbf{t}, \mathbf{T}^{\mathbf{V}}\mathbf{u}\rangle + \frac{\langle\mathbf{t}, \mathbf{T}^{\mathbf{V}}\boldsymbol{\varepsilon}\rangle\langle\mathbf{u}, \mathbf{T}^{\mathbf{V}}\boldsymbol{\varepsilon}\rangle}{|\bar{\mathbf{T}}^{\mathbf{V}}\boldsymbol{\varepsilon}|_{\mathbf{V}}^2}.$$

This completes the verification.

References

ALLAN, F. E. & WISHART, J. (1930). A method of estimating the yield of a missing plot in field experimental work. *Jour. Agric. Sci.* **20,** 399–406.

ALVEY, N. G. *et al.* (1977) GENSTAT. *A general statistical programme.* Rothamsted Experimental Station.

ANDERSON, R. L. (1946). Missing-plot techniques. *Biometrics* **2,** 41–47.

BARTLETT, M. S. (1937). Some examples of statistical methods of research in agriculture and applied biology. *J. Roy. Statist. Soc., Suppl.* **4,** 137–183.

COCHRAN, W. G. (1946). Analysis of covariance. *Institute of Statistics Mimeo Series No.* 6, University of North Carolina.

COCHRAN, W. G. & COX, G. M. (1957). *Experimental designs.* (2nd Edition) New York: Wiley.

COONS, I. (1957). The analysis of covariance as a missing-plot technique. *Biometrics* **13,** 387–405.

CORNISH, E. A. (1943). The recovery of inter-block information in quasi-factorial designs with incomplete data. I. Square, triple and cubic lattices. *C.S.I.R. Bull., No. 158.*

CORNISH, E. A. (1944). The recovery of inter-block information in quasi-factorial designs with incomplete data. II. Lattice squares. *C.S.I.R. Bull., No. 175.*

CORNISH, E. A. (1956). The recovery of inter-block information in quasi-factorial designs with incomplete data. III. Balanced incomplete blocks. *C.S.I.R.O. Divn. Math. Stats. Tech. Paper, No. 4.*

DEMPSTER, A. P., LAIRD, N. M. & RUBIN, D. B. (1977). Maximum likelihood from incomplete data via the EM algorithm. *J. Roy. Statist. Soc. B* **39,** 1–38.

HASEMAN, J. K. & GAYLOR, D. W. (1973). An algorithm for non-iterative estimation of multiple missing values for crossed classifications. *Technometrics* **15,** 631–636.

HEALY, M. J. R. & WESTMACOTT, M. H. (1956). Missing values in experiments analysed on automatic computers. *Appl. Statist.* **5,** 203–206.

HOUTMAN, A. M. (1980). *The analysis of designed experiments.* Ph.D. thesis, Princeton University.

HOUTMAN, A. M. & SPEED, T. P. (1983). Balance in designed experiments with orthogonal block structure. *Ann. Statist.,* **11,** 1069–1085.

HOYLE, M. H. (1971). Spoilt data—an introduction and bibliography. *J. Roy. Statist. Soc.* A **134,** 429–439.

JAMES, A. T. & WILKINSON, G. N. (1971). Factorization of the residual operator and canonical decomposition of nonorthogonal factors in the analysis of variance. *Biometrika* **58,** 279–294.

JARRETT, R. G. (1978). The analysis of designed experiments with missing observations. *Appl. Statist.* **27,** 38–46.

JOHN, J. A. & LEWIS, S. M. (1976). Mixed-up values in experiments. *Appl. Statist.* **25,** 61–63.

JOHN, J. A. & PRESCOTT, P. (1975). Estimating missing values in experiments. *Appl. Statist.* **24,** 190–192.

KRUSKAL, W. H. (1961). The coordinate-free approach to Gauss-Markov estimation, and its application to missing and extra observations. *Fourth Berkeley Symp. Math. Statist. Prob.* **1,** 435–451.

NAIR, K. R. (1940). The application of the technique of analysis of covariance to field experiments with several missing or mixed-up plots. *Sankhyā* **4,** 581–588.

NELDER, J. A. (1965a). The analysis of randomised experiments with orthogonal block structure. I. Block structure and the null analysis of variance. *Proc. Roy. Soc. (London) Ser. A.* **273,** 147–162.

NELDER, J. A. (1965b). The analysis of randomised experiments with orthogonal block structure II. Treatment structure and the general analysis of variance. *Ibid.* 163–178.

NELDER, J. A. (1968). The combination of information in generally balanced designs. *J. Roy. Statist. Soc. B* **30,** 303–311.

PEARCE, S. C. (1965). *Biological statistics: An Introduction.* New York: McGraw Hill.

PEARCE, S. C. & JEFFERS, J. R. N. (1971). Block designs and missing data. *J. Roy. Statist. Soc. B* **33,** 131–136.

PREECE, D. A. (1971). Iterative procedures for missing values in experiments. *Technometrics* **13,** 743–753.

PREECE, D. A. & GOWER, J. C. (1974). An iterative procedure for mixed-up values in experiments. *Appl. Statist.* **23,** 73–74.

RUBIN, D. B. (1972). A non-iterative method for L.S. estimation of missing values in any analysis of variance design. *Appl. Statist.* **21,** 136–141.

RUBIN, D. B. (1976). Non-iterative least squares estimates, standard errors and *F*-tests for analysis of variance with missing data. *J. Roy. Statist. Soc. B* **38,** 270–274.

RUBIN, D. B. (1976). Inference and missing data (with discussion). *Biometrika* **63,** 581–592.

SMITH, P. L. (1981). The use of Analysis of Covariance to analyse data from designed experiments with missing or mixed-up values. *Appl. Statist.* **30,** 1–8.

TRUITT, J. T. & FAIRFIELD SMITH, H. (1956). Adjustment by covariance and consequent tests of significance in split-plot experiments. *Biometrics* **12,** 23–39.

VON NEUMANN, J. (1950). *Functional operators, vol. II: The geometry of orthogonal spaces.* Princeton University Press.

WILKINSON, G. N. (1958a). Estimation of missing values for the analysis of incomplete data. *Biometrics* **14,** 257–286.

WILKINSON, G. N. (1958b). The analysis of variance and derivation of standard errors for incomplete data. *Biometrics* **14,** 360–384.

WILKINSON, G. N. (1960). Comparison of missing value procedures. *Austr. J. Statist.* **2,** 53–65.

WILLIAMS, E. R., RATCLIFF, D. & SPEED, T. P. (1981). Estimating missing values in multi-stratum experiments. *Appl. Statist.* **30,** 71–72.

YATES, F. (1933). The analysis of replicated experiments when the field results are incomplete. *Emp. Jour. Exp. Agric.* **1,** 129–142.

The Annals of Statistics
1986, Vol. 14, No. 3, 874–895

RECTANGULAR LATTICE DESIGNS: EFFICIENCY FACTORS AND ANALYSIS

BY R. A. BAILEY AND T. P. SPEED

Rothamsted Experimental Station and C.S.I.R.O.

Rectangular lattice designs are shown to be generally balanced with respect to a particular decomposition of the treatment space. Efficiency factors are calculated, and the analysis, including recovery of interblock information, is outlined. The ideas are extended to rectangular lattice designs with an extra blocking factor.

1. Introduction. The class of incomplete block designs known as *rectangular lattice* designs was introduced by Harshbarger (1946), with further details and extensions being given in a subsequent series of papers by Harshbarger (1947, 1949, 1951) and Harshbarger and Davis (1952). Apart from a contribution by Grundy (1950) concerning the efficient estimation of the stratum variances and the papers by Nair (1951, 1952, 1953) relating rectangular lattice designs to partially balanced designs, little further theoretical discussion of this class of designs seems to have occurred. Expositions of the basic results about rectangular lattice designs in two and three replicates, as well as tables of designs, can be found in Robinson and Watson (1949) and Cochran and Cox (1957). Discussions exist in other standard texts on the design and analysis of experiments, for example Kempthorne (1952), but, apart from recent contributions by Williams (1977) and Williams and Ratcliff (1980), the literature seems to end in the early 1950's. [In his recent note, Thompson (1983) uses the results in the present paper, as he acknowledges.] A possible explanation of this fact may be the observations of Nair (1951, 1953) that every 2-replicate rectangular lattice design is a partially balanced incomplete block design with four associate classes, whilst the obvious extension of the argument to r-replicate rectangular lattice designs for $r \geq 3$ fails in general, although the classes of rectangular lattice designs for $n(n-1)$ treatments in $n-1$ or n replicates again turn out to be partially balanced. Perhaps it was felt that, in not being partially balanced, rectangular lattice designs were rather too complicated.

In his fundamental papers on designed experiments with *simple orthogonal block structure* Nelder (1965a, b) introduced the notion of *general balance*, this being a relationship between the treatment structure and the block structure of the design. It is immediate from his definition that all block experiments (in the usual sense of the term) are generally balanced for some treatment structure [see Houtman and Speed (1983)], although here we might more properly use the term *treatment pseudo-structure*, and when this structure is elucidated for a given class of designs they can be regarded as understood and readily analysed. In a

Received June 1985; revised September 1985.

AMS 1980 *subject classifications.* Primary 62K10; secondary 62J10, 05B15.

Key words and phrases. Analysis of variance, block structure, combination of information, efficiency factor, general balance, Latin square, rectangular lattice, resolvable design, stratum, treatment decomposition.

later paper, Nelder (1968) showed the importance of general balance in permitting the straightforward estimation of stratum variances, introducing a method equivalent to that which has come to be known as *restricted maximum likelihood estimation* of variances [see Patterson and Thompson (1971) and Harville (1977)]. The definition of general balance in block designs is intimately connected with the eigenspaces of a certain linear transformation, denoted by $\mathbf{L}_B$ in this paper, and in this form a number of other authors have recently emphasised the same concept [see, for example, Pearce, Caliński, and Marshall (1974), who called the eigenvectors of $\mathbf{L}_B$ *basic contrasts*, and Corsten (1976)].

In Sections 3 and 4 of this paper we obtain an orthogonal decomposition of the space of all treatment contrasts associated with a general r-replicate rectangular lattice design. In Section 5 we use this decomposition to identify all the eigenspaces of the linear transformation $\mathbf{L}_B$. An equivalent description of our results is that we determine the treatment pseudo-structure relative to which the designs are generally balanced; equivalently again, we describe the basic contrasts of the design. Using these results, a full analysis, modelled on Nelder's (1965b, 1968) general approach, of rectangular lattice designs is given in Section 6, involving the derivation of a fully orthogonal analysis of variance and estimates of the stratum variances, and the calculations of estimates of treatment contrasts, together with their standard errors. A recursive analysis along the lines of Wilkinson (1970) is most satisfactory, as the eigenspaces are orthogonal complements of subspaces each of which has a simple formula for its orthogonal projection in terms of averaging operators, and so these subspaces can be swept out successively in a quite straightforward manner. Our general approach to the analysis of designed experiments is framed in vector space terms, similar to that used by James and Wilkinson (1971) and Bailey (1981), but in the multistratum framework of Nelder's papers.

Finally, we use the foregoing ideas to sketch the design and analysis of an experiment in which an extra blocking factor was imposed on a rectangular lattice design. Two examples are used throughout the paper to illustrate the theory.

EXAMPLE 1. This is a rectangular lattice for 20 treatments in three replicates of five blocks of four plots. Although this is an entirely abstract example, there being no associated experiment, it illustrates the general theory well because it has no special features: the design is *not* partially balanced, and its construction does *not* use a complete set of mutually orthogonal Latin squares. Tables 1, 3–5, 7, and 12–15 refer to Example 1.

EXAMPLE 2. In an experiment into the digestibility of stubble, 12 feed treatments were applied to sheep. There were 12 sheep, in three rooms of four animals each. There were three test periods of four weeks each, separated by two-week recovery periods. Each sheep was fed three treatments, one in each test period. During the recovery periods all animals received their usual feed, so that they would return to normal conditions before being subjected to a new treatment.

TABLE 1
Transversal of a 5×5 *Latin square*

①	2	3	4	5
2	1	④	5	3
3	⑤	1	2	4
4	3	5	1	②
5	4	2	③	1

It was desired that each treatment should be fed once in each room and once in each period. If periods are ignored, a suitable design is a rectangular lattice design in which sheep are blocks and rooms are replicates. We shall ignore the periods until Section 7, where we show how to deal with this extra blocking factor. Tables 9–11 and 18–19 refer to Example 2.

2. Construction. In this section we review the construction of rectangular lattice designs, partly in order to establish our terminology and notation.

A rectangular lattice design is a resolvable incomplete block design for t treatments in r replicates of n blocks of size $n - 1$, where $t = n(n - 1)$ and $2 \leq r \leq n$, for some integer n. We write b for rn, the total number of blocks, and N for $b(n - 1)$, the total number of plots. The design has the property that any pair of treatments occur together in at most one block. The design is constructed from a set of $r - 2$ mutually orthogonal $n \times n$ Latin squares $\Lambda_1, \ldots, \Lambda_{r-2}$.

A *transversal* of such a set of Latin squares is defined [see Dénes and Keedwell (1974), pages 28 and 331] to be a set of n cells with one cell in each row and one in each column, which between them have all the letters of all the squares $\Lambda_1, \ldots, \Lambda_{r-2}$. In Table 1 a transversal of a single 5×5 Latin square is indicated with circles. Transversals do not always exist: Table 2 shows a 4×4 Latin square with no transversal. A sufficient condition for the existence of a transversal is the existence of a Latin square Λ_{r-1} orthogonal to each of $\Lambda_1, \ldots, \Lambda_{r-2}$, for then each letter of Λ_{r-1} corresponds to a transversal. Such a set of mutually orthogonal $n \times n$ Latin squares $\Lambda_1, \ldots, \Lambda_{r-1}$ exists whenever n is a prime or prime power and r is less than or equal to n [see Dénes and Keedwell (1974), page 165]. However, this condition is not necessary, because the square in Table 1 has no orthogonal mate.

It is convenient (although not essential) to permute the rows and columns of $\Lambda_1, \ldots, \Lambda_{r-2}$ simultaneously so that the transversal lies down the main diagonal.

TABLE 2
A 4×4 *Latin square with no transversal*

1	2	3	4
4	1	2	3
3	4	1	2
2	3	4	1

RECTANGULAR LATTICE DESIGNS

TABLE 3a
Table 1 with rows permuted

1	2	3	4	5
3	5	1	2	4
2	1	4	5	3
5	4	2	3	1
4	3	5	1	2

TABLE 3b
Table 3a with letters permuted

1	5	4	3	2
4	2	1	5	3
5	1	3	2	4
2	3	5	4	1
3	4	2	1	5

This is achieved by moving the ith row to the jth row if the unique transversal cell in row i is in column j. It is also convenient to rename the "letters" of each square independently so that the letters on the main diagonal are in natural order. Tables 3a and 3b show the results of applying these processes to the square in Table 1.

An $n \times n$ square array is drawn. The diagonal cells are left blank, and the t treatments are allocated to the remaining cells, as in Table 4. In this example we have labelled the treatments $A, B, \ldots, T$, but we shall usually use ω to denote a general treatment, to avoid confusion with other symbols. We denote the n diagonal cells by $i, j, \ldots$ and the r *classifications* (that is, rows, columns, letters of $\Lambda_1, \ldots$, letters of Λ_{r-2}) by $a, b, \ldots$.

We define subsets of the treatments called *spokes* and *fans*. A *1-spoke* is the set of $n-1$ treatments in any row; a *2-spoke* is the set of $n-1$ treatments in any column. For $a = 3, \ldots, r$, an *a-spoke* is the set of $n-1$ treatments in the positions of any one letter of square Λ_{a-2}. For $a = 1, \ldots, r$ and $i = 1, \ldots, n$ we denote by $\mathscr{S}_{ai}$ the unique a-spoke which would naturally go through the ith diagonal cell if the diagonal cells were not excluded. For each fixed i, the *fan* $\mathscr{F}_i$ through the ith diagonal cell is defined to be the union of all spokes through that

TABLE 4
Treatment array for Example 1

*	*A*	*B*	*C*	*D*
E	*	*F*	*G*	*H*
I	*J*	*	*K*	*L*
M	*N*	*O*	*	*P*
Q	*R*	*S*	*T*	*

R. A. BAILEY AND T. P. SPEED

TABLE 5
Rectangular lattice block design (Example 1)
(blocks are columns)

replicate 1					replicate 2					replicate 3				
A	E	I	M	Q	E	A	B	C	D	F	D	C	B	A
B	F	J	N	R	I	J	F	G	H	J	K	H	E	G
C	G	K	O	S	M	N	O	K	L	P	M	N	L	I
D	H	L	P	T	Q	R	S	T	P	T	S	Q	R	O

diagonal cell; that is,

$$\mathscr{F}_i = \mathscr{S}_{1i} \cup \mathscr{S}_{2i} \cup \cdots \cup \mathscr{S}_{ri}.$$

The terminology is suggested by the fact that all spokes in a fan have the corresponding diagonal cell in common, while no two spokes in the same fan have any further cells in common. In the example given by Tables 3b and 4, we have

$$\mathscr{S}_{11} = \{A, B, C, D\},$$
$$\mathscr{S}_{24} = \{C, G, K, T\},$$
$$\mathscr{S}_{32} = \{D, K, M, S\},$$
$$\mathscr{F}_5 = \{Q, R, S, T, D, H, L, P, A, G, I, O\}.$$

The design is now constructed very easily. For $a = 1, \ldots, r$, the blocks of the ath replicate are just the a-spokes. Table 5 shows the (unrandomized) design which emerges in this way from Tables 3b and 4. Thus spokes have a genuine statistical meaning, as each spoke gives a block of the design. Fans have no direct statistical meaning, but they are a combinatorial consequence of the spokes which prove useful for the analysis of the design.

Orthogonal cyclic Latin squares may be constructed by the automorphism method of Mann (1942), which is described in Section 7.2 of Dénes and Keedwell (1974). If p is the smallest prime divisor of n then $p - 1$ orthogonal squares are obtained, and hence rectangular lattice designs may be constructed for $r \leq p$ (reserving one of the squares for the transversal). The same designs may also be constructed as α-designs [Patterson and Williams (1976)]. Let $q_1, q_2, \ldots, q_{r-1}$ be any integers such that no two are congruent modulo p and none is divisible by p. Without loss of generality we may take $q_1 = 1$. The generating α-array is in Table 6, in the format used by Patterson and Williams (1976), whose series I, II, and IV are all examples of the array shown here.

3. Decomposition of the treatment space. Let $\mathbb{R}^t$ be the real vector space of vectors indexed by the t treatments. We need to find an orthogonal decomposition of $\mathbb{R}^t$ that will enable us to analyse data from experiments with the rectangular lattice design. To this end, we define certain special vectors in and subspaces of $\mathbb{R}^t$.

Let $\mathbf{u}$ be the vector $(1, 1, \ldots, 1)$. For $a = 1, \ldots, r$ and $i = 1, \ldots, n$ let $\mathbf{v}_{ai}$ be the characteristic vector of the spoke $\mathscr{S}_{ai}$; that is, the ω-entry $(\mathbf{v}_{ai})_\omega$ of $\mathbf{v}_{ai}$ is 1 if

RECTANGULAR LATTICE DESIGNS

TABLE 6
Generators for α-designs which are also rectangular lattice designs (entries in the array should be reduced modulo n)

0	0	0	⋯	0
0	1	q_2	⋯	q_{r-1}
0	2	$2q_2$	⋯	$2q_{r-1}$
⋮	⋮	⋮		⋮
0	$n-2$	$(n-2)q_2$	⋯	$(n-2)q_{r-1}$
0	$n-1$	$(n-1)q_2$	⋯	$(n-1)q_{r-1}$

$\omega \in \mathscr{S}_{ai}$ and 0 otherwise. Similarly, for $i = 1, \ldots, n$, let $\mathbf{w}_i$ be the characteristic vector of the fan $\mathscr{F}_i$, so that

$$\mathbf{w}_i = \mathbf{v}_{1i} + \mathbf{v}_{2i} + \cdots + \mathbf{v}_{ri}.$$

Let U_μ be the subspace spanned by $\mathbf{u}$; let U_f be the subspace spanned by the fan vectors $\mathbf{w}_i$; let U_s be the subspace spanned by the spoke vectors $\mathbf{v}_{ai}$; and let U_ε be the whole space $\mathbb{R}^t$. [Our conventions for labelling the first and last of these spaces agree with those used by Throckmorton (1961) and Kempthorne (1982).] Then

$$U_\mu \subseteq U_f \subseteq U_s \subseteq U_\varepsilon.$$

For Example 1 we display each vector in $\mathbb{R}^{20}$ in a two-dimensional array corresponding to Table 4. Tables 7a and 7b give examples of vectors in $U_s \setminus U_f$ and in U_f respectively.

The dimension of U_μ is 1. The space $\mathbb{R}^t$ has an inner product $\langle \, , \rangle$ on it defined by

$$\langle \mathbf{z}, \mathbf{z}' \rangle = \sum_{\omega=1}^{t} z_\omega z'_\omega.$$

TABLE 7a
The vector $\mathbf{v}_{11} - 2\mathbf{v}_{24} + 5\mathbf{v}_{32}$

*	1	1	−1	6
0	*	0	−2	0
0	0	*	3	0
5	0	0	*	0
0	0	5	−2	*

TABLE 7b
The vector $\mathbf{w}_1 + 3\mathbf{w}_5$

*	4	1	1	4
1	*	1	3	3
4	1	*	0	3
1	0	3	*	4
4	3	3	4	*

We use this to find the dimensions of the spaces U_f and U_s. Note that

$$\langle \mathbf{v}_{ai}, \mathbf{v}_{bj} \rangle = |\mathscr{S}_{ai} \cap \mathscr{S}_{bj}|$$

$$(3.1) \qquad = \begin{cases} n-1 & \text{if } a = b \text{ and } i = j, \\ 0 & \text{if } a = b \text{ and } i \neq j, \\ 0 & \text{if } a \neq b \text{ and } i = j, \\ 1 & \text{if } a \neq b \text{ and } i \neq j, \end{cases}$$

so that

$$\langle \mathbf{w}_i, \mathbf{w}_j \rangle = |\mathscr{F}_i \cap \mathscr{F}_j|$$

$$(3.2) \qquad = \begin{cases} r(n-1) & \text{if } i = j, \\ r(r-1) & \text{if } i \neq j. \end{cases}$$

Moreover, $\Sigma_i \mathbf{w}_i = r\mathbf{u}$. Suppose that $\Sigma_i \lambda_i \mathbf{w}_i = \mathbf{0}$ for some real numbers λ_i. If $r \neq n$, taking inner products with individual $\mathbf{w}_i$ shows that $\lambda_1 = \cdots = \lambda_n$, and hence that $\lambda_1 = \cdots = \lambda_n = 0$: thus the fan vectors are linearly independent and so U_f has dimension n. On the other hand, if $r = n$ then $\mathbf{w}_i = \mathbf{u}$ for $i, \ldots, n$: thus $U_f = U_\mu$. Now suppose that $\Sigma_a \Sigma_i \lambda_{ai} \mathbf{v}_{ai} = \mathbf{0}$ for some real numbers λ_{ai}. Taking inner products with individual $\mathbf{v}_{ai}$ shows that there are real numbers θ_a and ϕ_i such that $\lambda_{ai} = \theta_a + \phi_i$ for all a and i. Since

$$\mathbf{v}_{a1} + \mathbf{v}_{a2} + \cdots + \mathbf{v}_{an} = \mathbf{u}$$

for $a = 1, \ldots, r$, this implies that $(\Sigma_a \theta_a)\mathbf{u} + \Sigma_i \phi_i \mathbf{w}_i = \mathbf{0}$. Hence U_s has dimension $nr - (r-1)$ if $r \neq n$, and $nr - (r-1) - (n-1)$ if $r = n$.

For Example 1, Equations (3.1) and (3.2) are demonstrated in Tables 7a and 7b, respectively. For example, the six entries equal to 4 in Table 7b correspond to the elements of $\mathscr{F}_1 \cap \mathscr{F}_5$. In this case the five fan vectors form a basis for U_f; while a basis of U_s consists of $\mathbf{u}$ and all but three spoke vectors, one being omitted for each classification.

We can form the orthogonal complements of the U-subspaces, and thus obtain the subspaces that really interest us. Specifically, we put

$$V_\mu = U_\mu,$$

$$V_f = \text{the orthogonal complement of } U_\mu \text{ in } U_f,$$

$$V_s = \text{the orthogonal complement of } U_f \text{ in } U_s,$$

$$V_\varepsilon = \text{the orthogonal complement of } U_s \text{ in } U_\varepsilon.$$

Then V_f is spanned by vectors of the form $\mathbf{w}_i - \mathbf{w}_j$; while V_s is spanned by vectors of the form $\mathbf{v}_{ai} - \mathbf{v}_{bi}$. Now $\mathbb{R}^t$ is the orthogonal direct sum

$$\mathbb{R}^t = V_\mu \oplus V_f \oplus V_s \oplus V_\varepsilon.$$

We record the important facts about this decomposition in Table 8.

In two special cases this decomposition can be described in simpler terms. If $r = n$ then the set $\{\Lambda_1, \ldots, \Lambda_{r-2}\}$ is only one square short of a complete set of mutually orthogonal Latin squares. Thus there exists a (unique) Latin square

RECTANGULAR LATTICE DESIGNS

TABLE 8
Decomposition of the treatment subspace

subspace	V_μ	V_f	V_s	V_ε
description	mean	contrasts between fans	contrasts between spokes within fans	orthogonal to spokes
dimension ($r < n$)	1	$n - 1$	$(n - 1)(r - 1)$	$(n - r)(n - 1) - 1$
dimension ($r = n$)	1	0	$(n - 1)^2$	$n - 2$

Λ_{n-1} orthogonal to all the others, by Theorem 1.6.1 of Rhagavarao (1971). One letter of Λ_{n-1} must correspond to the transversal. Each other letter of Λ_{n-1} occurs just once in each a-spoke, for each classification a. Hence the contrasts between these $n - 1$ other letters are orthogonal to spokes, and so they form the whole space V_ε. Since V_f is null in this case, V_s must consist of all treatment contrasts which are orthogonal to the letters of Λ_{n-1}. Thus the treatments have the simple nested structure $(n - 1) \to n$ [in the notation of Nelder (1965a)], and the treatment space decomposition is the familiar one into mean, between letters of Λ_{n-1} and within letters.

If $r = n - 1$ and $n \neq 4$, the results of Shrikhande (1961) and Bruck (1963) show that there is a unique complete orthogonal set $\{\Lambda_1, \ldots, \Lambda_{n-1}\}$ containing the original set $\{\Lambda_1, \ldots, \Lambda_{n-3}\}$ and that the original transversal corresponds to a letter of one of the two extra squares, say Λ_{n-2}. The same result is true even when $n = 4$, because the existence of the original transversal prevents Λ_1 from being isotopic to the square in Table 2, which is the only 4×4 Latin square (up to isotopy) which is not uniquely embeddable in a complete set of mutually orthogonal Latin squares [isotopy classes are also called transformation sets (see Fisher and Yates (1934))]. The treatments now have the simple crossed factorial structure $Q_1 \times Q_2$, where the levels of Q_1 are the $n - 1$ other letters of Λ_{n-2} and the levels of Q_2 are the n letters of Λ_{n-1}. Now V_ε is the main effect of Q_1; while V_f is the main effect of Q_2 and V_s is the Q_1Q_2 interaction.

Example 2 has $r = n - 1 = 3$. The rectangular lattice design is constructed from the set of mutually orthogonal 4×4 Latin squares in Table 9 : the rows, columns, and letters of Λ_1 are the three classifications; letter 1 of Λ_2 gives the transversal; the remaining letters of Λ_2 and Λ_3 give the 3×4 factorial treatment structure described above and shown in Table 10. The design is that shown in Table 11, ignoring periods.

In both these special cases the factorial treatment decomposition has no direct statistical meaning, but is merely an aid to the analysis. The factors Q_1 and Q_2 are entirely analogous to the pseudo-factors used in the construction and analysis of square lattice designs [Yates (1936)].

4. Treatment projection. Let $\mathbf{z}$ be a vector in $\mathbb{R}^t$. In order to use the spaces V_μ, V_f, V_s, and V_ε in the analysis of an experiment we need to know how to calculate the projections of $\mathbf{z}$ onto these spaces. This is done in terms of the

R. A. BAILEY AND T. P. SPEED

TABLE 9a
Three mutually orthogonal 4 × 4 *Latin squares*

Λ_1 (gives 3rd replicate)				Λ_2 ("1" gives transversal; other letters are levels of Q_1)				Λ_3 (letters are levels of Q_2)			
1	4	2	3	1	2	3	4	1	2	3	4
3	2	4	1	2	1	4	3	3	4	1	2
4	1	3	2	3	4	1	2	4	3	2	1
2	3	1	4	4	3	2	1	2	1	4	3

TABLE 9b
Array of twelve treatments for Example 2

*	*A*	*B*	*C*
D	*	*E*	*F*
G	*H*	*	*I*
J	*K*	*L*	*

TABLE 10
3 × 4 *factorial structure for Example* 2

treatment	*A*	*B*	*C*	*D*	*E*	*F*	*G*	*H*	*I*	*J*	*K*	*L*
level of Q_1	2	3	4	2	4	3	3	4	2	4	3	2
level of Q_2	2	3	4	3	1	2	4	3	1	2	1	4

TABLE 11
Design which is not generally balanced

room		1				2				3			
sheep		1	2	3	4	5	6	7	8	9	10	11	12
time period	**1**	*B*	*D*	*I*	*L*	*K*	*E*	*F*	*G*	*A*	*J*	*C*	*H*
	2	*C*	*E*	*H*	*K*	*A*	*L*	*I*	*J*	*G*	*B*	*D*	*F*
	3	*A*	*F*	*G*	*J*	*H*	*B*	*C*	*D*	*E*	*I*	*K*	*L*

following totals:

$$\text{grand total } G(\mathbf{z}) = \sum_{\omega} \mathbf{z}_{\omega},$$

$$\text{spoke total } S_{ai}(\mathbf{z}) = \sum \{\mathbf{z}_{\omega}\colon \omega \in \mathscr{S}_{ai}\} = \langle \mathbf{z}, \mathbf{v}_{ai} \rangle,$$

$$\text{fan total } F_i(\mathbf{z}) = \sum \{\mathbf{z}_{\omega}\colon \omega \in \mathscr{F}_i\} = \langle \mathbf{z}, \mathbf{w}_i \rangle.$$

RECTANGULAR LATTICE DESIGNS

TABLE 12
A particular vector $\mathbf{z}$ *in* $\mathbb{R}^{20}$

*	7	3	2	3
6	*	5	9	4
5	2	*	6	7
4	5	8	*	1
2	4	2	5	*

It is immediate that

(4.1) $$\sum_i S_{a\iota}(\mathbf{z}) = G(\mathbf{z}),$$

(4.2) $$\sum_a S_{ai}(\mathbf{z}) = F_i(\mathbf{z}),$$

(4.3) $$\sum_i F_i(\mathbf{z}) = rG(\mathbf{z}).$$

Define the *fan totals vector* $\mathbf{f(z)}$ and the *spoke totals vector* $\mathbf{s(z)}$ by

$$\mathbf{f}(\mathbf{z}) = \sum_i F_\iota(\mathbf{z})\mathbf{w}_\iota,$$

$$\mathbf{s}(\mathbf{z}) = \sum_a \sum_i S_{a\iota}(\mathbf{z})\mathbf{v}_{a\iota}.$$

We also need the *grand totals vector* $\mathbf{g(z)}$, all of whose entries are equal to $G(\mathbf{z})$.

Continuing our Example 1, a vector $\mathbf{z}$ is shown in Table 12. Its spoke totals are in Table 13: the column margins are the fan totals, and the row totals are all the grand total. The vectors $\mathbf{f(z)}$ and $\mathbf{s(z)}$ are shown in Table 14.

We aim to give the projections of $\mathbf{z}$ onto the spaces V_μ, V_f, V_s, and V_ε in terms of $\mathbf{f(z)}$, $\mathbf{s(z)}$, and $\mathbf{g(z)}$. The necessary calculations are contained in the following two lemmas.

LEMMA 1.

(i) $$\langle \mathbf{s}(\mathbf{z}), \mathbf{v}_{a\iota}\rangle = nS_{ai}(\mathbf{z}) + (r-1)G(\mathbf{z}) - F_\iota(\mathbf{z}),$$

(ii) $$\langle \mathbf{f}(\mathbf{z}), \mathbf{v}_{a\iota}\rangle = (n-r)F_i(\mathbf{z}) + r(r-1)G(\mathbf{z}),$$

(iii) $$\langle \mathbf{f}(\mathbf{z}), \mathbf{w}_i\rangle = r(n-r)F_i(\mathbf{z}) + r^2(r-1)G(\mathbf{z}).$$

TABLE 13
Spoke totals of $\mathbf{z}$

i	**1**	**2**	**3**	**4**	**5**	**total**
row ($a = 1$)	15	24	20	18	13	90
column ($a = 2$)	17	18	18	22	15	90
letter ($a = 3$)	13	15	13	20	29	90
fan totals	45	57	51	60	57	270

TABLE 14

fan totals vector f(z)					spoke totals vector s(z)				
*	159	156	156	159	*	62	53	50	45
162	*	153	174	165	61	*	55	75	52
153	153	*	168	168	66	51	*	57	55
162	168	168	*	162	50	49	65	*	46
153	174	165	162	*	43	51	46	48	*

PROOF. To simplify the expressions, we omit "(z)", the vector $\mathbf{z}$ being understood.

(i)
$$\begin{aligned}\langle \mathbf{s}, \mathbf{v}_{ai}\rangle &= \sum_b \sum_j S_{bj}\langle \mathbf{v}_{bj}, \mathbf{v}_{ai}\rangle \\ &= (n-1)S_{ai} + \sum_{b\neq a}\sum_{i\neq j} S_{bj} \quad (\text{by } (3.1)) \\ &= (n-1)S_{ai} + \sum_{b\neq a}(G - S_{bi}) \quad (\text{by } (4.1)) \\ &= nS_{ai} + (r-1)G - \sum_b S_{bi} \\ &= nS_{ai} + (r-1)G - F_i \quad (\text{by } (4.2)).\end{aligned}$$

(ii)
$$\begin{aligned}\langle \mathbf{f}, \mathbf{v}_{ai}\rangle &= \sum_j F_j\langle \mathbf{w}_j, \mathbf{v}_{ai}\rangle \\ &= (n-1)F_i + (r-1)\sum_{j\neq i} F_j \quad (\text{by } (3.1)) \\ &= (n-r)F_i + (r-1)\sum_j F_j \\ &= (n-r)F_i + r(r-1)G \quad (\text{by } (4.3)).\end{aligned}$$

(iii) Summing the equation in (ii) over all the spokes in $\mathscr{F}_i$ gives

$$\langle \mathbf{f}, \mathbf{w}_i\rangle = r(n-r)F_i + r^2(r-1)G. \qquad \square$$

LEMMA 2. *The orthogonal projections of* $\mathbf{z}$ *onto* $U_\mu, U_f, U_s, U_\varepsilon$, *respectively, are*

$$\frac{\mathbf{g}(\mathbf{z})}{n(n-1)}, \quad \frac{\mathbf{f}(\mathbf{z})}{r(n-r)} - \frac{(r-1)\mathbf{g}(\mathbf{z})}{(n-1)(n-r)},$$
$$\frac{\mathbf{s}(\mathbf{z})}{n} + \frac{\mathbf{f}(\mathbf{z})}{n(n-r)} - \frac{(r-1)\mathbf{g}(\mathbf{z})}{(n-1)(n-r)}, \quad \mathbf{z}$$

when $r \neq n$. *When* $r = n$ *then* $U_f = U_\mu$ *and the orthogonal projection of* $\mathbf{z}$ *onto* U_s *is*

$$\frac{\mathbf{s}(\mathbf{z})}{n} - \frac{(n-2)\mathbf{g}(\mathbf{z})}{n(n-1)}.$$

Proof. Put $\mathbf{x} = [r(n-r)]^{-1}\mathbf{f} - (r-1)[(n-1)(n-r)]^{-1}\mathbf{g}$ when $r \neq n$. Since $\mathbf{f}$ and $\mathbf{g}$ are both sums of fan vectors, $\mathbf{x} \in U_f$. Thus it suffices to show that $\mathbf{z} - \mathbf{x}$ is orthogonal to U_f. This is so if $\langle \mathbf{z} - \mathbf{x}, \mathbf{w}_i \rangle = 0$ for each fan $\mathscr{F}_i$. By Lemma 1(iii) and (3.2),

$$\langle \mathbf{x}, \mathbf{w}_i \rangle = \frac{r(n-r)F_i + r^2(r-1)G}{r(n-r)} - \frac{r(n-1)(r-1)G}{(n-1)(n-r)} = F_i = \langle \mathbf{z}, \mathbf{w}_i \rangle.$$

Similarly, put $\mathbf{y} = n^{-1}\mathbf{s} + [n(n-r)]^{-1}\mathbf{f} - (r-1)[(n-1)(n-r)]^{-1}\mathbf{g}$. Then $\mathbf{y} \in U_s$, because $\mathbf{s}$, $\mathbf{f}$, and $\mathbf{g}$ are all sums of spoke vectors, so it suffices to show that $\langle \mathbf{z} - \mathbf{y}, \mathbf{v}_{ai} \rangle = 0$ for all spokes $\mathscr{S}_{ai}$. Lemmas 1(i) and (ii) show that $\langle \mathbf{y}, \mathbf{v}_{ai} \rangle$ is equal to

$$\frac{nS_{ai} + (r-1)G - F_i}{n} + \frac{(n-r)F_i + r(r-1)G}{n(n-r)} - \frac{(n-1)(r-1)G}{(n-1)(n-r)},$$

which is S_{ai}, which is $\langle \mathbf{z}, \mathbf{v}_{ai} \rangle$.

Now let $r = n$ and put $\mathbf{y} = n^{-1}\mathbf{s} - (n-2)[n(n-1)]^{-1}\mathbf{g}$. Then

$$\langle \mathbf{y}, \mathbf{v}_{ai} \rangle = \frac{nS_{ai} + (n-2)G}{n} - \frac{(n-2)(n-1)G}{n(n-1)} = S_{ai} = \langle \mathbf{z}, \mathbf{v}_{ai} \rangle$$

so that $\mathbf{y} \in U_s$ and $\mathbf{z} - \mathbf{y}$ is orthogonal to U_s. □

Now subtraction gives the orthogonal projection of $\mathbf{z}$ onto $V_\mu, V_f, V_s, V_\varepsilon$.

Theorem 1. *Let $\mathbf{T}_\mu, \mathbf{T}_f, \mathbf{T}_s, \mathbf{T}_\varepsilon$ be the operators of orthogonal projection from $\mathbb{R}^t$ onto $V_\mu, V_f, V_s, V_\varepsilon$, respectively. Then, for all $\mathbf{z}$ in $\mathbb{R}^t$,*

$$\mathbf{T}_\mu \mathbf{z} = \frac{\mathbf{g}(\mathbf{z})}{n(n-1)},$$

$$\mathbf{T}_f \mathbf{z} = \frac{\mathbf{f}(\mathbf{z})}{r(n-r)} - \frac{r\mathbf{g}(\mathbf{z})}{n(n-r)} \quad \textit{when } r \neq n \textit{ and zero otherwise},$$

$$\mathbf{T}_s \mathbf{z} = \frac{\mathbf{s}(\mathbf{z})}{n} - \frac{\mathbf{f}(\mathbf{z})}{rn},$$

$$\mathbf{T}_\varepsilon \mathbf{z} = \mathbf{z} - (\mathbf{T}_\mu \mathbf{z} + \mathbf{T}_f \mathbf{z} + \mathbf{T}_s \mathbf{z}).$$

In Example 1 we have $n = 5$ and $r = 3$, so $\mathbf{T}_\mu \mathbf{z} = \mathbf{g}(\mathbf{z})/20$; $\mathbf{T}_f(\mathbf{z}) = \mathbf{f}(\mathbf{z})/6 - 3\mathbf{g}(\mathbf{z})/10$; $\mathbf{T}_s \mathbf{z} = \mathbf{s}(\mathbf{z})/5 - \mathbf{f}(\mathbf{z})/15$, and $\mathbf{T}_\varepsilon \mathbf{z}$ is best obtained by subtraction. For the particular vector $\mathbf{z}$ shown in Table 12, these four components of $\mathbf{z}$ are shown in Table 15. The orthogonality of the decomposition may be verified by noting that

$$\|\mathbf{T}_\mu \mathbf{z}\|^2 + \|\mathbf{T}_f \mathbf{z}\|^2 + \|\mathbf{T}_s \mathbf{z}\|^2 + \|\mathbf{T}_\varepsilon \mathbf{z}\|^2$$
$$= 405 + 24 + 47.2 + 21.8 = 498 = \|\mathbf{z}\|^2.$$

TABLE 15

$\mathbf{T}_\mu \mathbf{z}$					$\mathbf{T}_f \mathbf{z}$				
*	4.5	4.5	4.5	4.5	*	−0.5	−1.0	−1.0	−0.5
4.5	*	4.5	4.5	4.5	0.0	*	−1.5	2.0	0.5
4.5	4.5	*	4.5	4.5	−1.5	−1.5	*	1.0	1.0
4.5	4.5	4.5	*	4.5	0.0	1.0	1.0	*	0.0
4.5	4.5	4.5	4.5	*	−1.5	2.0	0.5	0.0	*
$\mathbf{T}_s \mathbf{z}$					$\mathbf{T}_\varepsilon \mathbf{z}$				
*	1.8	0.2	−0.4	−1.6	*	1.2	−0.7	−1.1	0.6
1.4	*	0.8	3.4	−0.6	0.1	*	1.2	−0.9	−0.4
3.0	0.0	*	0.2	−0.2	−1.0	−1.0	*	0.3	1.7
−0.8	−1.4	1.8	*	−1.6	0.3	0.9	0.7	*	−1.9
−1.6	−1.4	−1.8	−1.2	*	0.6	−1.1	−1.2	1.7	*

5. General balance. The block structure of a rectangular lattice design is the double nested classification of plots within blocks within replicates. This is one of the *simple orthogonal block structures* defined by Nelder (1965a). In what follows we retain the notation of Nelder (1965a, b, 1968) and Bailey (1981) as far as possible.

Let $\mathbb{R}^N$ be the real vector space associated with the N plots. Each grouping of the plots according to the block structure defines an averaging operation $\mathbf{P}$ on $\mathbb{R}^N$. In our case there are four averaging operators: the grand mean averaging operator $\mathbf{P}_\mu = \mathbf{J}/N$, where $\mathbf{J}$ is the all-1's matrix; the replicates averaging operator $\mathbf{P}_R$; the blocks averaging operator $\mathbf{P}_B$; and the identity $\mathbf{P}_\varepsilon = \mathbf{I}$. Nelder (1965a) showed that there is an orthogonal direct sum decomposition $\oplus_\alpha W_\alpha$ of $\mathbb{R}^N$ such that each W_α is an eigenspace of every $\mathbf{P}$. Let $\mathbf{C}_\alpha$ be the operator of orthogonal projection from $\mathbb{R}^N$ onto W_α. Nelder (1965a) showed that each $\mathbf{C}_\alpha$ is a linear combination of the $\mathbf{P}$'s with integer coefficients: Speed and Bailey (1982) gave explicit formulae for these coefficients. In our case we have

$$\mathbf{C}_\mu = \mathbf{P}_\mu, \qquad \mathbf{C}_R = \mathbf{P}_R - \mathbf{P}_\mu,$$

$$\mathbf{C}_B = \mathbf{P}_B - \mathbf{P}_R, \qquad \mathbf{C}_\varepsilon = \mathbf{P}_\varepsilon - \mathbf{P}_B.$$

The spaces W_α are called *strata*: they play an important role in analysis of variance [see Nelder (1965b) and Bailey (1981)]. Our covariance model for the data vector $\mathbf{y}$ is

$$\operatorname{Cov}(\mathbf{y}) = \xi_\mu \mathbf{C}_\mu + \xi_R \mathbf{C}_R + \xi_B \mathbf{C}_B + \xi_\varepsilon \mathbf{C}_\varepsilon \tag{5.1}$$

for unknown scalars ξ_μ, ξ_R, ξ_B, and ξ_ε.

Denote by $\mathbf{X}$ the $N \times t$ design matrix; that is, $\mathbf{X}_{p\omega}$ is 1 if plot p receives treatment ω and 0 otherwise. For each stratum W_α, the matrix $\mathbf{L}_\alpha$ defined by $\mathbf{L}_\alpha = \mathbf{X}'\mathbf{C}_\alpha\mathbf{X}$ is called the *information matrix* for that stratum. For designs with equal replication r, we have $\mathbf{L}_\mu = r\mathbf{T}_\mu$. If $\mathbf{L}_\alpha = \mathbf{0}$ there is no information about treatments in stratum W_α. Strata, other than W_μ, for which $\mathbf{L}_\alpha \neq \mathbf{0}$, are called *effective strata.*

Suppose that $\oplus_\theta V_\theta$ is an orthogonal direct sum decomposition of $\mathbb{R}^t$. Nelder (1965b) defined an equally replicated design to be *generally balanced* with respect to this treatment decomposition if each V_θ is an eigenspace of every information matrix; that is, there are numbers $\lambda_{\alpha\theta}$ such that $\mathbf{L}_\alpha = \Sigma_\theta \lambda_{\alpha\theta}\mathbf{T}_\theta$, where $\mathbf{T}_\theta$ denotes orthogonal projection onto V_θ. We have $0 \leq \lambda_{\alpha\theta} \leq r$ for all α and θ; and $\Sigma_\alpha \lambda_{\alpha\theta} = r$ for all θ. The quantity $\lambda_{\alpha\theta}/r$ is the *efficiency factor* for treatment term V_θ in the stratum W_α. In a simple block design with blocks stratum W_B, examination of the trace of $\mathbf{L}_B$ shows that $\Sigma_\theta \lambda_{B\theta}\dim(V_\theta)/r = b/r - 1$, the so-called *loss of information due to blocks.*

Houtman and Speed (1983) have shown that in any design with only two effective strata there must be *some* decomposition $\oplus V_\theta$ of $\mathbb{R}^t$ with respect to which the design is generally balanced. However, the decomposition may not be easy to find, use or interpret. Our claim is that a rectangular lattice design is generally balanced with respect to the treatment decomposition given in Section 3.

LEMMA 3. *For $a = 1,\ldots, r$ and $i = 1,\ldots, n$,*

$$\mathbf{X}'\mathbf{P}_B\mathbf{X}\mathbf{v}_{ai} = (n\mathbf{v}_{ai} - \mathbf{w}_i + (r-1)\mathbf{u})/(n-1).$$

PROOF. If $\mathscr{B}$ is any block and $\mathbf{v}$ is any vector in $\mathbb{R}^t$ then the entries of $\mathbf{P}_B\mathbf{X}\mathbf{v}$ for the plots in $\mathscr{B}$ are all equal to the average of the entries of $\mathbf{v}$ for those treatments which occur in $\mathscr{B}$. If $\mathbf{v} = \mathbf{v}_{ai}$ and $\mathscr{B}$ consists of $\mathscr{S}_{bj}$ then this average is equal to $\langle \mathbf{v}_{ai}, \mathbf{v}_{bj}\rangle/(n-1)$. Denote the characteristic vector of this block by $\mathbf{x}_{bj}$. Then

$$(n-1)\mathbf{P}_B\mathbf{X}\mathbf{v}_{ai} = \sum_b \sum_j \langle \mathbf{v}_{ai}, \mathbf{v}_{bj}\rangle \mathbf{x}_{bj}.$$

Since $\mathbf{X}'\mathbf{x}_{bj} = \mathbf{v}_{bj}$ we have

$$\begin{aligned}(n-1)\mathbf{X}'\mathbf{P}_B\mathbf{X}\mathbf{v}_{ai} &= \sum_b \sum_j \langle \mathbf{v}_{ai}, \mathbf{v}_{bj}\rangle \mathbf{v}_{bj} \\ &= (n-1)\mathbf{v}_{ai} + \sum_{b \neq a} (\mathbf{u} - \mathbf{v}_{bi}) \quad \text{(by (3.1))} \\ &= n\mathbf{v}_{ai} + (r-1)\mathbf{u} - \mathbf{w}_i. \qquad \square\end{aligned}$$

THEOREM 2. *Rectangular lattice designs are generally balanced with respect to the treatment decomposition given in Section 3.*

PROOF. We always have $\mathbf{L}_\mu\mathbf{u} = r\mathbf{u}$, and $\mathbf{L}_\mu\mathbf{z} = \mathbf{0}$ whenever $\mathbf{z}$ is orthogonal to $\mathbf{u}$. By definition of replicate, $\mathbf{X}'\mathbf{P}_R\mathbf{X}\mathbf{z} = r\mathbf{g}(\mathbf{z})/n(n-1) = \mathbf{X}'\mathbf{P}_\mu\mathbf{X}\mathbf{z}$, so $\mathbf{L}_R = \mathbf{0}$. Moreover, $\mathbf{L}_B = \mathbf{X}'\mathbf{P}_B\mathbf{X} - \mathbf{X}'\mathbf{P}_R\mathbf{X}$, and so

$$\mathbf{L}_B(\mathbf{v}_{ai} - \mathbf{v}_{bi}) = n(n-1)^{-1}(\mathbf{v}_{ai} - \mathbf{v}_{bi})$$

by Lemma 3. Since V_s is spanned by vectors of the form $\mathbf{v}_{ai} - \mathbf{v}_{bi}$, this shows that

TABLE 16
Efficiency factors of a rectangular lattice design

	treatment subspace			
	V_μ	V_f	V_s	V_ε
stratum				
mean W_μ	1	0	0	0
replicates W_R	0	0	0	0
blocks W_B	0	$\frac{n-r}{r(n-1)}$	$\frac{n}{r(n-1)}$	0
plots W_ε	0	$\frac{n(r-1)}{r(n-1)}$	$\frac{rn-r-n}{r(n-1)}$	1

V_s is an eigenspace of $\mathbf{L}_B$ with eigenvalue $\lambda_{Bs} = n/(n-1)$. Similarly, Lemma 3 shows that

$$\mathbf{L}_B(\mathbf{w}_i - \mathbf{w}_j) = (n-r)(n-1)^{-1}(\mathbf{w}_i - \mathbf{w}_j),$$

so V_f is an eigenspace of $\mathbf{L}_B$ with eigenvalue $\lambda_{Bf} = (n-r)/(n-1)$. Whether or not $r = n$, Table 8 now shows that $\lambda_{Bs}\dim(V_s) + \lambda_{Bf}\dim(V_f) = b - r$, so there can be no further nonzero eigenvalues in the blocks stratum. Thus V_ε must be an eigenspace of $\mathbf{L}_B$ with $\lambda_{B\varepsilon} = 0$.

By the result of Houtman and Speed (1983), the spaces V_f, V_s, V_ε are also eigenspaces of $\mathbf{L}_\varepsilon$. □

The eigenvalues in stratum W_ε are calculated by subtraction. Division by r gives the efficiency factors, which are shown in Table 16, which is laid out like the table in Section 4.2 of Nelder (1968).

Block designs are often classified by a single measure of efficiency: the harmonic mean of the efficiency factors (taking account of multiplicity) in stratum W_ε. It follows from Tables 8 and 16, that, whether $r = n$ or $r < n$, the harmonic mean efficiency factor for a rectangular lattice design is

$$\frac{n(r-1)(rn-r-n)(n^2-n-1)}{(r-1)^2n^2(n^2-n-1) - r^2(n-1)^2 + rn(r-1)}.$$

This efficiency factor is proportional to the reciprocal of the average variance of the intrablock estimates of simple treatment differences, and so may also be obtained from this average variance, which is given by Williams (1977, page 413).

6. Analysis. Since rectangular lattice designs are generally balanced, their analysis follows the pattern described by Nelder (1965b, 1968), Wilkinson (1970), and James and Wilkinson (1971). In this section we specialize their results to rectangular lattice designs, retaining most of Nelder's notation. We outline the procedure for fitting the model, deriving a complete analysis of variance, estimating the stratum variances ξ_R, ξ_B, and ξ_ε, and obtaining minimum variance unbiased linear estimates (with estimated weights) of arbitrary treatment contrasts, together with their estimated variances.

Let $\mathbf{t}$ be the $t \times 1$ vector of individual treatment effects and let $\mathbf{y}$ be the $N \times 1$ vector of observations. If $\lambda_{\alpha\theta} \neq 0$, the treatment *effect* $\mathbf{T}_\theta \mathbf{t}$ is estimated in stratum W_α by $\mathbf{h}_{\alpha\theta}$, where $\mathbf{h}_{\alpha\theta} = \mathbf{T}_\theta \mathbf{X}' \mathbf{C}_\alpha \mathbf{y}/\lambda_{\alpha\theta}$. The contribution of treatment term V_θ to the *fitted value* in stratum W_α is $\mathbf{C}_\alpha \mathbf{X} \mathbf{h}_{\alpha\theta}$, with the sum of squares $\lambda_{\alpha\theta}\|\mathbf{h}_{\alpha\theta}\|^2$. Thus the overall fitted value in stratum W_α is $\Sigma'_\theta \mathbf{C}_\alpha \mathbf{X}\mathbf{h}_{\alpha\theta}$, where Σ'_θ denotes summation over those θ for which $\lambda_{\alpha\theta} \neq 0$. The residual sum of squares, RSS_α, in stratum W_α, and its number of degrees of freedom, d_α, are obtained by subtraction:

$$\mathrm{RSS}_\alpha = \mathbf{y}'\mathbf{C}_\alpha \mathbf{y} - \sum_\theta{}' \lambda_{\alpha\theta}\|\mathbf{h}_{\alpha\theta}\|^2, \tag{6.1}$$

$$d_\alpha = \dim(W_\alpha) - \sum_\theta{}' \dim(V_\theta). \tag{6.2}$$

Thus we obtain the analysis of variance shown in Tables 17a ($r < n$) and 17b ($r = n$).

If the stratum variances ξ_α are known, we put $w_\theta = \Sigma_\alpha \lambda_{\alpha\theta}/\xi_\alpha$ and define *weights* $w_{\alpha\theta}$ by $w_{\alpha\theta} = \lambda_{\alpha\theta}/\xi_\alpha w_\theta$. The *weighted effect* corresponding to treatment term V_θ is $\Sigma_\alpha w_{\alpha\theta}\mathbf{h}_{\alpha\theta}$, and the overall *weighted fitted value* $\hat{\mathbf{t}}$ is $\Sigma_\theta \Sigma_\alpha w_{\alpha\theta}\mathbf{h}_{\alpha\theta}$. If $\mathbf{x}$ is any treatment contrast (that is, $\mathbf{x} \in \mathbb{R}^t$ and $\langle \mathbf{x}, \mathbf{u}\rangle = 0$) then the minimum variance unbiased linear estimate of $\langle \mathbf{x}, \mathbf{t}\rangle$ is $\langle \mathbf{x}, \hat{\mathbf{t}}\rangle$, with variance $\Sigma_\theta \|\mathbf{T}_\theta \mathbf{x}\|^2/w_\theta$.

TABLE 17a

Analysis of variance when $r < n$

stratum	source of variation	df	SS	EMS
mean		1	$\mathbf{y}'\mathbf{C}_\mu \mathbf{y}$	$r\|\mathbf{T}_\mu \mathbf{t}\|^2 + \xi_\mu$
replicates		$r-1$	$\mathbf{y}'\mathbf{C}_R \mathbf{y}$	ξ_R
blocks	V_f	$n-1$	$\lambda_{Bf}\|\mathbf{h}_{Bf}\|^2$	$\dfrac{\lambda_{Bf}\|\mathbf{T}_f \mathbf{t}\|^2}{n-1} + \xi_B$
	V_s	$(n-1)(r-1)$	$\lambda_{Bs}\|\mathbf{h}_{Bs}\|^2$	$\dfrac{\lambda_{Bs}\|\mathbf{T}_s \mathbf{t}\|^2}{(n-1)(r-1)} + \xi_B$
	total	$r(n-1)$	$\mathbf{y}'\mathbf{C}_B \mathbf{y}$	
plots	V_f	$n-1$	$\lambda_{\epsilon f}\|\mathbf{h}_{\epsilon f}\|^2$	$\dfrac{\lambda_{\epsilon f}\|\mathbf{T}_f \mathbf{t}\|^2}{n-1} + \xi_\epsilon$
	V_s	$(n-1)(r-1)$	$\lambda_{\epsilon s}\|\mathbf{h}_{\epsilon s}\|^2$	$\dfrac{\lambda_{\epsilon s}\|\mathbf{T}_s \mathbf{t}\|^2}{(n-1)(r-1)} + \xi_\epsilon$
	V_ϵ	$(n-r)(n-1)-1$	$\lambda_{\epsilon\epsilon}\|\mathbf{h}_{\epsilon\epsilon}\|^2$	$\dfrac{\lambda_{\epsilon\epsilon}\|\mathbf{T}_\epsilon \mathbf{t}\|^2}{(n-r)(n-1)-1} + \xi_\epsilon$
	error	$n(rn-2r-n+1)+1$	RSS_ϵ	ξ_ϵ
	total	$rn(n-2)$	$\mathbf{y}'\mathbf{C}_\epsilon \mathbf{y}$	

TABLE 17b
Analysis of variance when $r = n$

stratum	source of variation	df	SS	EMS
mean		1	$\mathbf{y}'\mathbf{C}_\mu\mathbf{y}$	$r\Vert\mathbf{T}_\mu\mathbf{t}\Vert^2 + \xi_\mu$
replicates		$r-1$	$\mathbf{y}'\mathbf{C}_R\mathbf{y}$	ξ_R
blocks	V_s	$(n-1)^2$	$\lambda_{Bs}\Vert\mathbf{h}_{Bs}\Vert^2$	$\dfrac{\lambda_{Bs}\Vert\mathbf{T}_s\mathbf{t}\Vert^2}{(n-1)^2} + \xi_B$
	error	$n-1$	RSS_B	ξ_B
	total	$n(n-1)$	$\mathbf{y}'\mathbf{C}_B\mathbf{y}$	
plots	V_s	$(n-1)^2$	$\lambda_{\varepsilon s}\Vert\mathbf{h}_{\varepsilon s}\Vert^2$	$\dfrac{\lambda_{\varepsilon s}\Vert\mathbf{T}_s\mathbf{t}\Vert^2}{(n-1)^2} + \xi_\varepsilon$
	V_ε	$n-2$	$\lambda_{\varepsilon\varepsilon}\Vert\mathbf{h}_{\varepsilon\varepsilon}\Vert^2$	$\dfrac{\lambda_{\varepsilon\varepsilon}\Vert\mathbf{T}_\varepsilon\mathbf{t}\Vert^2}{n-2} + \xi_\varepsilon$
	error	$(n-1)(n^2-2n-1)$	RSS_ε	ξ_ε
	total	$n^2(n-2)$	$\mathbf{y}'\mathbf{C}_\varepsilon\mathbf{y}$	

Usually the stratum variances ξ_α are not known. If $d_\alpha \neq 0$ then $\text{RSS}_\alpha/d_\alpha$ provides an unbiased estimate of ξ_α, but in general such estimates are based on too few degrees of freedom, because one or more treatment terms have been fitted and removed in more than one stratum. For a rectangular lattice design with $r < n$ there is no such estimate of ξ_B, because $d_B = 0$.

The solution to this difficulty is to estimate the stratum variances and the weights simultaneously. With the weighted fitted value $\hat{\mathbf{t}}$ given above, the sum of squares, R_α, for the residual in stratum W_α is given by

$$R_\alpha = \text{RSS}_\alpha + \sum_\theta \lambda_{\alpha\theta} \sum_\beta \sum_\gamma w_{\beta\theta} w_{\gamma\theta} \langle \mathbf{h}_{\alpha\theta} - \mathbf{h}_{\beta\theta}, \mathbf{h}_{\alpha\theta} - \mathbf{h}_{\gamma\theta} \rangle, \tag{6.3}$$

with expected value $d'_\alpha \xi_\alpha$, where

$$d'_\alpha = \dim(W_\alpha) - \sum_\theta{}' w_{\alpha\theta} \dim(V_\theta). \tag{6.4}$$

Equating observed and expected values of the R_α gives a set of equations in the ξ_α. As Nelder (1968) observed, (6.3) simplifies considerably when there are only two effective strata. Thus for rectangular lattice designs we obtain the following equations for ξ_B and ξ_ε:

$$\text{RSS}_B + \sum_\theta \lambda_{B\theta} w_{\varepsilon\theta}^2 \Vert\mathbf{h}_{B\theta} - \mathbf{h}_{\varepsilon\theta}\Vert^2 = \xi_B\Big[r(n-1) - \sum_\theta{}' w_{B\theta}\dim(V_\theta)\Big],$$

$$\text{RSS}_\varepsilon + \sum_\theta \lambda_{\varepsilon\theta} w_{B\theta}^2 \Vert\mathbf{h}_{\varepsilon\theta} - \mathbf{h}_{B\theta}\Vert^2 = \xi_\varepsilon\Big[rn(n-2) - \sum_\theta{}' w_{\varepsilon\theta}\dim(V_\theta)\Big].$$

Note that RSS_B is zero when $r < n$, and that the weights $w_{\alpha\theta}$ also involve the unknown ξ_α. However, these equations may be solved, iteratively if necessary, to give us estimates ξ_B and ξ_ε, which, under normality, correspond to the so-called *restricted maximum likelihood estimates*, and these may be used to give the best available estimates of linear combinations $\langle \mathbf{x}, \mathbf{t} \rangle$ and the estimated variances of those estimates.

It is clear that the analysis depends on the availability of the projection operators $\mathbf{C}_\alpha$ and $\mathbf{T}_\theta$. The former are quite standard, and correspond to fitting and removing the grand mean, replicate means, and block means. The latter are given by the fan and spoke totals, and so are straightforward to calculate, even by hand. If the statistical programming language GENSTAT is used, spoke totals are automatically calculated if r treatment pseudo-factors are declared, one for each classification: the levels of the ath pseudo-factor are the a-spokes. An alternative strategy is to input r copies of the data and use just two treatment pseudo-factors, FAN and SPOKE. In the ath copy, treatments in spoke $\mathcal{S}_{ai}$ are declared to have level i of FAN and level a of SPOKE. The treatment declaration FAN/SPOKE ensures that all the correct major calculations are done, using the *sweeps* of Wilkinson (1970), although minor adjustments have to be made to the output to allow for the multiple copies. Thompson (1983) explains this method, and its difficulties, in more detail, using the general methods of Thompson (1984), and shows that this type of pseudo-factorial structure is also useful for diallel experiments.

Thus, apart from the use of estimated weights because the stratum variances are in general not known, a completely satisfactory analysis of any rectangular lattice design can be made once the operators $\mathbf{T}_\theta$ are available. Given these, the analysis is analogous to that of a balanced incomplete block design with recovery of interblock information.

Williams and Ratcliff (1980) gave a procedure for the analysis of rectangular lattice designs which differs from ours in two respects. In the first place, their covariance model is of the form

$$\operatorname{Cov}[(\mathbf{I} - \mathbf{P}_R)\mathbf{y}] = \gamma_B \mathbf{P}_B + \gamma_\varepsilon \mathbf{I},$$

which differs from our equation (5.1). Secondly, our iterative analysis ensures that the final estimates of ξ_B, ξ_ε and the treatment effects are consistent with each other, while the Williams–Ratcliff procedure, which is based on that given by Yates (1940) and Cochran and Cox (1957, Section 1.3), is, roughly speaking, only the first cycle of the restricted maximum likelihood analysis of Patterson and Thompson (1971). The differences between these methods, which apply not only to rectangular lattice designs, will be discussed in more detail elsewhere.

7. Rectangular lattices with cross-blocking. The foregoing ideas may be extended to a more complicated block structure.

In Example 2 we have so far ignored the periods. However, it was desirable that each treatment should be fed once in each period. The experimenter concerned found that, for the rectangular lattice design constructed at the end of Section 3, the treatments could be permuted within sheep so that each treatment occurred once in each period: his proposed design is shown in Table 11.

Unfortunately, this design takes no account of the grouping of the 36 experimental units into nine room-periods: each room-period consists of the four observations made in the same test period in the same room. In the notation of Nelder (1965a), the block structure is

$$3 \text{ periods} \times (3 \text{ rooms} \to 4 \text{ sheep}).$$

The stratum projection matrices are given by

$$\begin{aligned} \mathbf{C}_\mu &= \mathbf{P}_\mu, \\ \mathbf{C}_R &= \mathbf{P}_R - \mathbf{P}_\mu, \\ \mathbf{C}_P &= \mathbf{P}_P - \mathbf{P}_\mu, \\ \mathbf{C}_{RP} &= \mathbf{P}_{RP} - \mathbf{P}_P - \mathbf{P}_R + \mathbf{P}_\mu, \\ \mathbf{C}_S &= \mathbf{P}_S - \mathbf{P}_R, \\ \mathbf{C}_\varepsilon &= \mathbf{P}_\varepsilon - \mathbf{P}_S - \mathbf{P}_{RP} + \mathbf{P}_R, \end{aligned}$$

where, for example, $\mathbf{P}_{RP}$ is the averaging matrix for room-periods. Although V_μ, V_f, V_s, and V_ε are eigenspaces of $\mathbf{C}_\mu$, $\mathbf{C}_R$, $\mathbf{C}_P$, and $\mathbf{C}_S$, they are *not* eigenspaces of $\mathbf{C}_{RP}$ and $\mathbf{C}_\varepsilon$, because the block design given by the room-periods alone is not in any sense balanced with respect to the treatment decomposition $V_\mu \oplus V_f \oplus V_s \oplus V_\varepsilon$. Thus the design is not generally balanced.

However, it is possible to permute the treatments given to each sheep so that each treatment occurs once in each period and the design is generally balanced. This may be done for $n(n-1)$ treatments in the simple orthogonal block structure

$$(n-1) \text{ periods} \times [(n-1) \text{ rooms} \to n \text{ sheep}]$$

as follows. Ignoring periods, the design is constructed from a set of mutually orthogonal Latin squares $\Lambda_1, \ldots, \Lambda_{n-1}$, as in Section 2. A supplementary $(n-1) \times (n-1)$ Latin square Δ is needed, whose letters are the remaining letters of Λ_{n-2}. Let δ_{ap} be the letter in row a and column p of Δ. Then the treatment in the pth period and the ith animal of the ath room is the unique treatment which is in spoke $\mathscr{S}_{ai}$ and in letter δ_{ap} of Λ_{n-2}. In our particular example we may take the supplementary square Δ shown in Table 18: the resulting design is in Table 19.

In the notation of Section 3, V_ε is the main effect of Q_1, where the levels of Q_1 are the remaining letters of Λ_{n-2}. By our construction, Q_1 is completely confounded with room-periods, while all treatment vectors which are orthogonal to Q_1 are also orthogonal to room-periods. Hence the efficiency factors for this extension of the rectangular lattice design are those shown in Table 20.

TABLE 18
Supplementary Latin square

2	3	4
3	4	2
4	2	3

RECTANGULAR LATTICE DESIGNS

TABLE 19
Generally balanced design for [periods × (rooms → sheep)]

room		1				2				3			
sheep		1	2	3	4	5	6	7	8	9	10	11	12
time period	1	*A*	*D*	*I*	*L*	*G*	*K*	*B*	*F*	*H*	*J*	*C*	*E*
	2	*B*	*F*	*G*	*K*	*J*	*H*	*E*	*C*	*L*	*I*	*D*	*A*
	3	*C*	*E*	*H*	*J*	*D*	*A*	*L*	*I*	*F*	*B*	*K*	*G*

TABLE 20
Efficiency factors of an extended rectangular lattice design

	treatment subspace			
	V_μ	$V_f = Q_2$	$V_s = Q_1Q_2$	$V_\varepsilon = Q_1$
stratum				
mean W_μ	1	0	0	0
rooms W_R	0	0	0	0
periods W_P	0	0	0	0
room-periods W_{PR}	0	0	0	1
sheep W_s	0	$\frac{1}{(n-1)^2}$	$\frac{n}{(n-1)^2}$	0
units W_ε	0	$\frac{n(n-2)}{(n-1)^2}$	$\frac{n^2-3n+1}{(n-1)^2}$	0

Acknowledgments. The contribution of T.P.S. to the work reported here was carried out whilst he was a visitor at the Indian Statistical Institute in Calcutta, and he would like to thank the then Acting Director, Dr. A. Maitra, and other staff and scholars, for providing such an enjoyable environment for study. Part of R.A.B.'s contribution was made while she was visiting the Mathematics Department of the University of Western Australia, to whose staff she would like to extend similar thanks. We are grateful to A. Grassia of C.S.I.R.O., Perth, for drawing our attention to the problem described in Example 2.

REFERENCES

BAILEY, R. A. (1981). A unified approach to design of experiments. *J. Roy. Statist. Soc. Ser. A* **144** 214–223.

BRUCK, R. H. (1963). Finite nets. II. Uniqueness and embedding. *Pacific J. Math.* **13** 421–457.

COCHRAN, W. G. and COX, G. M. (1957). *Experimental Designs*, 2nd ed. Wiley, New York.

CORSTEN, L. C. A. (1976). Canonical correlation in incomplete blocks. In *Essays in Probability and Statistics. A Volume in Honour of Prof. Junjiro Ogawa* (S. Ikeda et al., eds.) 125–154. Shinko Tsusho Co. Ltd., Tokyo.

DÉNES, J. and KEEDWELL, A. D. (1974). *Latin Squares and Their Applications*. English Universities Press Limited, London.

FISHER, R. A. and YATES, F. (1934). The 6 × 6 Latin squares. *Proc. Cambridge Philos. Soc.* **30** 492–507.

GRUNDY, P. M. (1950). The estimation of error in rectangular lattices. *Biometrics* **6** 25–33.

HARSHBARGER, B. (1946). Preliminary report on the rectangular lattices. *Biometrics* **2** 115–119.

HARSHBARGER, B. (1947). *Rectangular Lattices.* Virginia Agricultural Experiment Station Memoir 1.

HARSHBARGER, B. (1949). Triple rectangular lattices. *Biometrics* **5** 1–13.

HARSHBARGER, B. (1951). Near balance rectangular lattices. *Virginia J. Sci.* **2** 13–27.

HARSHBARGER, B. and DAVIS, L. L. (1952). Latinized rectangular lattices. *Biometrics* **8** 73–84.

HARVILLE, D. A. (1977). Maximum likelihood approaches to variance component estimation and to related problems (with discussion). *J. Amer. Statist. Assoc.* **72** 320–340.

HOUTMAN, A. M. and SPEED, T. P. (1983). Balance in designed experiments with orthogonal block structure. *Ann. Statist.* **11** 1069–1085.

JAMES, A. T. and WILKINSON, G. H. (1971). Factorization of the residual operator and canonical decomposition of nonorthogonal factors in the analysis of variance. *Biometrika* **58** 279–294.

KEMPTHORNE, O. (1952). *The Design and Analysis of Experiments.* Wiley, New York.

KEMPTHORNE, O. (1982). Classificatory data structures and associated linear models. In *Statistics and Probability: Essays in Honor of C. R. Rao* (G. Kallianpur, P. R. Krishnaiah and J. K. Ghosh, eds.) 397–410. North-Holland, Amsterdam.

MANN, H. B. (1942). The construction of orthogonal Latin squares. *Ann. Math. Statist.* **13** 418–423.

NAIR, K. R. (1951). Rectangular lattices and partially balanced incomplete block designs. *Biometrics* **7** 145–154.

NAIR, K. R. (1952). Analysis of partially balanced incomplete block designs illustrated on the simple square and rectangular lattices. *Biometrics* **8** 122–155.

NAIR, K. R. (1953). A note on rectangular lattices. *Biometrics* **9** 101–106.

NELDER, J. A. (1965a). The analysis of randomized experiments with orthogonal block structure. I. Block structure and the null analysis of variance. *Proc. Roy. Soc. London Ser. A* **283** 147–162.

NELDER, J. A. (1965b). The analysis of randomized experiments with orthogonal block structure. II. Treatment structure and the general analysis of variance. *Proc. Roy. Soc. London Ser. A* **283** 163–178.

NELDER, J. A. (1968). The combination of information in generally balanced designs. *J. Roy. Statist. Soc. Ser. B* **30** 303–311.

PATTERSON, H. D. and THOMPSON, R. (1971). Recovery of inter-block information when block sizes are unequal. *Biometrika* **58** 545–554.

PATTERSON, H. D. and WILLIAMS, E. R. (1976). A new class of resolvable incomplete block designs. *Biometrika* **63** 83–92.

PEARCE, S. C., CALIŃSKI, T. and MARSHALL, T. F. DE C. (1974). The basic contrasts of an experimental design with special reference to the analysis of data. *Biometrika* **61** 449–460.

RAGHAVARAO, D. (1971). *Constructions and Combinatorial Problems in Design of Experiments.* Wiley, New York.

ROBINSON, H. F. and WATSON, G. S. (1949). Analysis of simple and triple rectangular lattice designs. North Carolina Agricultural Experimental Station Tech. Bull. 88.

SHRIKHANDE, S. S. (1961). A note on mutually orthogonal latin squares. *Sankhyā Ser. A* **23** 115–116.

SPEED, T. P. and BAILEY, R. A. (1982). On a class of association schemes derived from lattices of equivalence relations. In *Algebraic Structures and Applications* (P. Schultz, C. E. Praeger and R. P. Sullivan, eds.) 55–74. Marcel Dekker, New York.

THOMPSON, R. (1983). Diallel crosses, partially balanced incomplete block designs with triangular association schemes and rectangular lattices. *Genstat Newsletter* **10** 16–32.

THOMPSON, R. (1984). The use of multiple copies of data in forming and interpreting analysis of variance. In *Experimental Design, Statistical Models, and Genetic Statistics. Essays in Honor of Oscar Kempthorne* (K. Hinkelmann, ed.) 155–173. Marcel Dekker, New York.

THROCKMORTON, T. N. (1961). Structures of classificatory data. Ph.D. thesis, Iowa State Univ.

WILKINSON, G. N. (1970). A general recursive algorithm for analysis of variance. *Biometrika* **57** 19–46.

RECTANGULAR LATTICE DESIGNS

WILLIAMS, E. R. (1977). A note on rectangular lattice designs. *Biometrics* **33** 410–414.

WILLIAMS, E. R. and RATCLIFF, D. (1980). A note on the analysis of lattice designs with repeats. *Biometrika* **67** 706–708.

YATES, F. (1936). A new method of arranging variety trials involving a large number of varieties. *J. Agric. Sci.* **26** 424–455.

YATES, F. (1940). The recovery of inter-block information in balanced incomplete block designs. *Ann. Eugenics* **10** 317–325.

STATISTICS DEPARTMENT
ROTHAMSTED EXPERIMENTAL STATION
HARPENDEN
HERTFORDSHIRE AL5 2JQ
ENGLAND

C.S.I.R.O.
DIVISION OF MATHEMATICS AND STATISTICS
G.P.O. Box 1965
CANBERRA ACT 2601
AUSTRALIA

The Annals of Statistics
1987, Vol. 15, No. 3, 885–910

SPECIAL INVITED PAPER

WHAT IS AN ANALYSIS OF VARIANCE?

BY T. P. SPEED

CSIRO, Canberra, Australia

The analysis of variance is usually regarded as being concerned with sums of squares of numbers and independent quadratic forms of random variables. In this paper, an alternative interpretation is discussed. For certain classes of dispersion models for finite or infinite arrays of random variables, a form of generalized spectral analysis is described and its intuitive meaning explained. The analysis gives a spectral decomposition of each dispersion in the class, incorporating an analysis of the common variance, and an associated orthogonal decomposition of each of the random variables. One by-product of this approach is a clear understanding of the similarity between the spectral decomposition for second-order stationary processes and the familiar linear models with random effects.

"... the analysis of variance, which may perhaps be called a statistical method, because the term is a very ambiguous one—is not a mathematical theorem, but rather a convenient method of arranging the arithmetic."

R. A. Fisher (1934)

1. Introduction. To most of us the expression analysis of variance or anova conjures up a subset of the following: multiindexed arrays of numbers, sums of squares, anova tables with lines; perhaps, somewhat more mathematically, independent quadratic forms of random variables, chi-squared distributions, and F-tests. We would also think of linear models and the associated notions of main effects and interactions of various orders; indeed the standard text on the subject, Scheffé (1959, page 5) essentially defines the analysis of variance to be regression analysis where the regressor variables (x_{ij}) take only the values 0 or 1, although he mentions in a footnote that -1 and 2 have also arisen. What is anova? Is there a variance being analysed? Is there a mathematical theorem, contrary to Fisher's assertion? Or is it just a body of techniques, a statistical method, ..., a convenient method of arranging the arithmetic?

Signs that there might be an underlying mathematical structure began to appear in the late 1950s and early 1960s. James (1957) emphasised the role of the algebra of projectors in the analysis of experimental designs, Tukey (1961) outlined the connection between anova and spectrum analysis [something which was made more explicit by Hannan (1961, 1965), who focussed on the decomposition of permutation representations of groups], whilst Graybill and Hultquist

Received February 1986; revised August 1986.

AMS 1980 *subject classification*. 62J10.

Key words and phrases. Analysis of variance, dispersion model, sums of squares, association scheme, group, symmetry, manova.

(1961) gave a definition of anova (assuming joint normality of all random variables concerned) which incorporated many of the same ideas as the others mentioned: the commuting of projectors and the spectral decomposition of a covariance matrix.

Of course, anova is just a word (or three) and people can give it any meaning they wish, so there is no sense in which the definition I offer in the following text has any greater claim to be the correct one than any other. What I do believe is that it is a mathematically fruitful definition, that it covers most if not all situations which statisticians would regard as being instances of anova and that its generality and simplicity are both pedagogically and scientifically helpful. And yes, I believe there are relevant mathematical theorems, although as we will see it is perhaps unreasonable to expect a single theorem to cover all existing cases.

2. Two simple examples. Let us begin with an array $y = (y_{ij})$ of mn random variables where $i = 1, \dots, m$ and $j = 1, \dots, n$ is nested within i, i.e., j only has meaning within the values of i. The following decomposition of the sum of squares is familiar to all who have met anova:

$$\sum_k \sum_l y_{kl}^2 = mny_{\cdot\cdot}^2 + n\sum_h (y_{h\cdot} - y_{\cdot\cdot})^2 + \sum_i \sum_j (y_{ij} - y_{i\cdot})^2, \tag{2.1}$$

and we denote the three terms on the right by SS_0, SS_1 and SS_2. Here $y_{i\cdot} = n^{-1}\Sigma_j y_{ij}$, $y_{\cdot\cdot} = m^{-1}\Sigma_i y_{i\cdot}$, etc. It is not hard to derive (2.1) by the standard juggling which many believe characterises anova. Of what interest or use is this decomposition? To answer this question, we must make some assumptions about the y_{ij}, and one set—the ones Fisher (1934) probably had in mind when he made the remark quoted—is the following: $Ey_{ij} = \mu_i$, where (μ_i) is a set of m unknown parameters, the (y_{ij}) are pairwise uncorrelated and they have a common variance σ^2; i.e., the dispersion matrix Dy of y is just $\sigma^2 I$. Under these assumptions we can prove (see the following text) that $E\{SS_0\} = mn\mu_{\cdot}^2 + \sigma^2$, $E\{SS_1\} = (m-1)\sigma^2 + n\Sigma_i(\mu_i - \mu_\cdot)^2$ and $E\{SS_2\} = m(n-1)\sigma^2$. It is here that we can see the point of Fisher's remark about "the arithmetic," for when the (y_{ij}) are jointly normal, SS_0/σ^2, SS_1/σ^2 and SS_2/σ^2 are mutually independent with chi-squared distributions on 1, $m-1$ and $m(n-1)$ degrees of freedom, respectively, and the ratio $F = m(n-1)SS_1/(m-1)SS_2$ permits a test of the hypothesis H: $\mu_1 = \mu_2 = \cdots = \mu_m$, having a central F-distribution with $(m-1, m(n-1))$ degrees of freedom when H is true. The F-test of this hypothesis has many desirable properties [Hsu (1941, 1945), Wald (1942), Wolfowitz (1949), Herbach (1959) and Gautschi (1959)] and the decomposition (2.1) is indeed a convenient method of arranging the arithmetic.

But all of this is just sums of squares—quadratic forms in normal variates if you wish; the only variance in sight is the common σ^2 and that does not appear to be undergoing any analysis. However, let us look closely at the proof of some of the foregoing assertions. How do we see that the quadratic forms SS_0, SS_1 and SS_2 are independent under the assumption $Dy = \sigma^2 I$ and joint normality? One approach, owing to Tang (1938), uses the fact that their (unsquared and un-summed) components $y_{\cdot\cdot}$, $y_{h\cdot} - y_{\cdot\cdot}$ and $y_{ij} - y_{i\cdot}$ are uncorrelated, and hence, by the joint normality, independent, and this property is retained when the components are squared and summed.

How do we see that these components are uncorrelated? Each is a linear combination of elements in the array y with easily calculated coefficients and, with the assumption that $Dy = \sigma^2 I$, their covariances are simply σ^2 times the sums of the products of these coefficients. For example, the coefficient of y_{kl} in $y_{h\cdot} - y_{\cdot\cdot}$ is $-1/mn$ if $k \neq h$ and $1/n - 1/mn$ if $k = h$, whilst that of y_{kl} in $y_{ij} - y_{i\cdot}$ is 0 if $k \neq i$, $-1/n$ if $k = i$ and $l \neq j$ and $1 - 1/n$ if $k = i$ and $l = j$. Thus if $h = i$,

$$\operatorname{cov}(y_{h\cdot} - y_{\cdot\cdot}, y_{ij} - y_{i\cdot})$$
$$= \sigma^2\left[-\frac{1}{mn}0(m-1)n + \left(\frac{1}{n} - \frac{1}{mn}\right)\left(-\frac{1}{n}\right)(n-1) + \left(\frac{1}{n} - \frac{1}{mn}\right)\left(1 - \frac{1}{n}\right)1\right],$$

which is zero as stated; the case $h \neq i$ is dealt with similarly. Similar calculations prove that $\operatorname{cov}(y_{\cdot\cdot}, y_{h\cdot} - y_{\cdot\cdot}) = \operatorname{cov}(y_{\cdot\cdot}, y_{ij} - y_{\cdot\cdot}) = 0$ and, further, that $E\{y_{\cdot\cdot}^2\} = \mu_{\cdot\cdot}^2 + (1/mn)\sigma^2$, $E\{(y_{h\cdot} - y_{\cdot\cdot})^2\} = ((m-1)/mn)\sigma^2 + (\mu_h - \mu_{\cdot})^2$ and $E\{(y_{ij} - y_{i\cdot})^2\} = (m(n-1)/mn)\sigma^2$.

It has just been proved that the three components in the sum

$$y_{ij} = y_{\cdot\cdot} \quad + \quad y_{i\cdot} - y_{\cdot\cdot} \quad + \quad y_{ij} - y_{i\cdot} \tag{2.2}$$

are uncorrelated; their variances thus add and we may write this as

$$\sigma^2 = \frac{1}{mn}\sigma^2 + \frac{m-1}{mn}\sigma^2 + \frac{m(n-1)}{mn}\sigma^2. \tag{2.3}$$

Here at last is a variance being analysed! But before we examine this any further let us see with a minimum of further algebra how the sums of squares of the components in (2.2) must add up and give (2.1). Denoting the coefficients of y_{kl} in $y_{\cdot\cdot}$, $y_{i\cdot} - y_{\cdot\cdot}$ and $y_{ij} - y_{i\cdot}$ by $S_0(ij, kl)$, $S_1(ij, kl)$ and $S_2(ij, kl)$, respectively, we can easily check that the $mn \times mn$ matrices S_0, S_1 and S_2 so defined are symmetric, idempotent, pairwise orthogonal and sum to the $mn \times mn$ identity matrix I. Symmetry is quickly apparent from their definition; orthogonality is implicit in the calculation which proved the components in (2.2) uncorrelated, whilst idempotence is proved by a similar calculation; and clearly they sum to the identity. Thus we can write $y = S_0 y + S_1 y + S_2 y$ as

$$(y_{ij}) = (y_{\cdot\cdot}) + (y_{i\cdot} - y_{\cdot\cdot}) + (y_{ij} - y_{i\cdot}), \tag{2.4}$$

where the S_α act on arrays $u = (u_{ij})$ of real numbers as follows $(S_\alpha u)_{ij} = \Sigma_k \Sigma_l S_\alpha(ij, kl) u_{kl}$, $\alpha = 0, 1, 2$. But then (2.4) is a decomposition of the array into component arrays which are orthogonal with respect to the inner product $\langle u, v\rangle = \Sigma_i \Sigma_j u_{ij} v_{ij}$, whilst (2.1) is simply the Pythagorean relationship

$$|y|^2 = |S_0 y|^2 + |S_1 y|^2 + |S_2 y|^2,$$

where $|y|^2 = \langle y, y\rangle$ is the associated squared norm.

An unexpected bonus. Without any further calculations we may assert that (2.2) remains an orthogonal decomposition of y_{ij} when the dispersion matrix $Dy = \Gamma$ has the form

$$\Gamma = \xi_0 S_0 + \xi_1 S_1 + \xi_2 S_2, \tag{2.5}$$

where the eigenvalues ξ_0, ξ_1 and ξ_2 are positive real numbers. A modified version of (2.3) also holds, namely

$$\operatorname{var}(y_{ij}) = \frac{1}{mn}\xi_0 + \frac{m-1}{mn}\xi_1 + \frac{m(n-1)}{mn}\xi_2. \tag{2.6}$$

These assertions are readily checked. For example,

$$\operatorname{cov}(y_{i\cdot} - y_{\cdot\cdot}, y_{ij} - y_{i\cdot}) = (S_1 \Gamma S_2)(ij, ij) = 0,$$

and

$$\operatorname{var}(y_{ij} - y_{i\cdot}) = (S_2 \Gamma S_2)(ij, ij) = \xi_2 S_2(ij, ij) = \frac{m(n-1)}{mn}\xi_2.$$

The question this observation now raises is: How wide is the class of matrices of the form (2.5)? Perhaps unexpectedly, it coincides with a class which arises frequently, namely the set of all matrices Γ having the form

$$\Gamma = \gamma_2 A_2 + \gamma_1 A_1 + \gamma_0 A_0, \tag{2.7}$$

where $A_2 = I$ is the identity matrix, $A_1(ij, kl) = 1$ if $i = k$, $j \neq l$ and 0 otherwise, $A_0(ij, kl) = 1$ if $i \neq k$ and 0 otherwise, and γ_2, γ_1 and γ_0 are a variance and two covariances constrained only to ensure that Γ is positive definite. The easiest way to see that Γ's of the form (2.5) and (2.7) coincide is to list the index ij lexicographically and write the matrices in tensor product form. We find that $A_2 = I_m \otimes I_n$, $A_1 = I_m \otimes (J_n - I_n)$ and $A_0 = (J_m - I_m) \otimes J_n$, whilst $S_0 = (1/m)J_m \otimes (1/n)J_n$, $S_1 = (I_m - (1/m)J_m) \otimes (1/n)J_n$ and $S_2 = I_m \otimes (I_n - (1/n)J_n)$, where I_m and J_m are the $m \times m$ identity and matrix of 1's, respectively. The eigenvalues ξ and the entries γ correspond in the following way:

$$\begin{bmatrix} \xi_0 \\ \xi_1 \\ \xi_2 \end{bmatrix} = \begin{bmatrix} 1 & n-1 & n(m-1) \\ 1 & n-1 & -n \\ 1 & -1 & 0 \end{bmatrix} \begin{bmatrix} \gamma_2 \\ \gamma_1 \\ \gamma_0 \end{bmatrix}, \tag{2.8a}$$

$$\begin{bmatrix} \gamma_2 \\ \gamma_1 \\ \gamma_0 \end{bmatrix} = \frac{1}{mn} \begin{bmatrix} 1 & m-1 & m(n-1) \\ 1 & m-1 & -m \\ 1 & -1 & 0 \end{bmatrix} \begin{bmatrix} \xi_0 \\ \xi_1 \\ \xi_2 \end{bmatrix}. \tag{2.8b}$$

Where have we gotten to? We have exhibited a set of covariance matrices (2.7) for a random array $y = (y_{ij})$ which are simultaneously diagonalisable, cf. (2.5); their eigenvalues are invertible linear combinations (2.8) of their entries; their common eigenspace projectors decompose the elements of the array into statistically orthogonal (i.e., uncorrelated) components (2.2) whilst also decomposing the arrays themselves into geometrically orthogonal arrays (2.4). Pythagoras' theorem applied to the decomposition of array elements gives an analysis of variance qua variance (2.6), whilst it gives the sum of squares decomposition (2.1) of an anova table when applied to the decomposition of arrays. We might also add that these decompositions all make "statistical sense."

How special is this example? Before answering this question let us look at a second example, which is not normally regarded as being an instance of anova. This time our array has a circular nature: A sequence $y = (y_t\colon t = 0, 1, \ldots, n-1)$ of $n = 2m+1$ random variables with $\operatorname{cov}(y_s, y_t) = \gamma_{|t-s|}$, $0 \le s, t < n$, i.e., $\Gamma = Dy$ is a symmetric circulant with first row $(\gamma_0\gamma_1 \cdots \gamma_m\gamma_m \cdots \gamma_1)$. To emphasize the similarity with (2.7) we write it as

$$\Gamma = \sum_0^m \gamma_a A_a, \tag{2.9}$$

where A_a is the symmetric circulant having first row $(0 \cdots 010 \cdots 010 \cdots 0)$ with 1's in the ath and $(n-a)$th position, $1 \le a \le m$, and $A_0 = I$, the $n \times n$ identity matrix. It is well known that the class of all such matrices is simultaneously diagonalisable with common projectors $S_0 = (1/n)J_n$ and $S_\alpha(s,t) = (2/n)\cos(2\pi(s-t)\alpha/n)$, $0 < \alpha \le m$, $0 \le s$, $t < n$, whilst their eigenvalues are linear combinations of their entries

$$\xi_\alpha = \gamma_0 + 2\sum_1^m \gamma_a \cos\left(\frac{2\pi}{n} a\alpha\right), \qquad \alpha = 0, \ldots, m, \tag{2.10a}$$

with inverses

$$\gamma_a = \frac{1}{n}\xi_0 + \frac{2}{n}\sum_1^m \xi_\alpha \cos\left(\frac{2\pi}{n} a\alpha\right), \qquad a = 0, \ldots, m. \tag{2.10b}$$

Further, we have an orthogonal decomposition of the random variables similar to (2.2):

$$y_t = y_{\cdot} + \sum_1^m S_\alpha y_t, \tag{2.11}$$

where $S_\alpha y_t = (2/n)\sum_0^{n-1} y_s \cos(2\pi(s-t)\alpha/n)$, $1 \le \alpha \le m$, cf. Hannan (1960, I.2), and the variances of each component add, corresponding to $a = 0$ in (2.10b).

Finally, we remark that a decomposition of the n-dimensional vector space analogous to (2.4) and its associated sum of squares decomposition may also be derived; it is just the (real form of the) discrete Fourier transform. The analogy with the view of the classical anova we have just presented is complete.

3. Sums of squares. Let $y = (y_t\colon t \in T)$ be a finite array of random variables with mean $Ey = 0$ and dispersion matrix $Dy = \Gamma \in \mathbf{V}$, where $\mathbf{V}$ is a family of positive definite matrices over T. The formal definition of anova given by Graybill and Hultquist (1961) refers to a decomposition of $|y|^2$ into a sum of quadratic forms under an assumption of joint normality of y. It had two aspects which we will recall shortly: one which in essence refers to properties of the individual matrices $\Gamma \in \mathbf{V}$, and one which was clearly a property of the model as a whole. Later writers on the same topic include Albert (1976), Brown (1984) and Harville (1984), and in all of these papers the role of anova as a property of a model **V** has tended to get emphasised less than the consequences of the definition for arrays y with $Dy \in \mathbf{V}$. In what follows we modify the Graybill and Hultquist (1961) definition slightly, removing some details without, we hope, losing its essence. We also express the definition solely in terms of the class **V** of dispersion matrices, removing the joint normality assumption. Finally we argue that the definition is most fruitful when applied to a particular parametrization of **V**, one which is not usual in this context, although as we will see it coincides with that used in developing the spectral theory of second-order stationary processes over index sets of various kinds.

Initially we will suppose that **V** is a class of positive definite matrices having the form

$$\Gamma(\theta) = \sum_{a=1}^{s} \theta_a A_a, \tag{3.1}$$

where the $\{A_a\}$ are known symmetric matrices and $\theta = (\theta_a)$ is an s-dimensional real parameter belonging to $\Theta \subset R^s$. It will be convenient to suppose that the $\{A_a\}$ are linearly independent matrices over T and that **V** contains s linearly independent elements. Dispersion models of this form have been studied by a number of authors over the years including Anderson (1969, 1970, 1973) and Jensen (1975), but our emphasis is quite different from theirs. Essentially following Graybill and Hultquist (1961) we say that an anova exists for **V** if there exists a family $\{S_\alpha\}$ of s known pairwise orthogonal symmetric idempotent matrices summing to the identity matrix I over T such that

(a) for every $\theta \in \Theta$ and α there exists $\xi_\alpha(\theta)$ such that

$$\Gamma(\theta)S_\alpha = \xi_\alpha(\theta)S_\alpha; \tag{3.2}$$

(b) the map $\theta = (\theta_a) \to \xi(\theta) = (\xi_\alpha(\theta))$ is linear and invertible.

Condition (a) replaces the condition that for each $\theta \in \Theta$ the s quadratic forms $\{|S_\alpha y|^2\}$ are mutually independent scale multiples of chi-squares under the assumption $y \sim N(0, \Gamma(\theta))$ [see Albert (1976, Theorem 1(a))], whilst condition (b) asserts that the multipliers $\xi_\alpha(\theta) = E\{d_\alpha^{-1}|S_\alpha y|^2\}$, where $d_\alpha = \operatorname{rank} S_\alpha$, are independent linear functions of the $\{\theta_a\}$.

It is clear from (a) that the matrices $\{S_\alpha\}$ simultaneously reduce all $\Gamma \in \mathbf{V}$, i.e., that $\Gamma = \Sigma_\alpha \xi_\alpha S_\alpha$, where we omit the dependence on θ if no confusion can result, and thus every element of **V** commutes with every other. As long as **V** contains s linearly independent elements, these conclusions extend to *all* matrices of the form $\Sigma_a \theta_a A_a$ with $\theta \in R^s$ and in particular we deduce that the $\{A_a\}$ commute. It also follows from (b) that, in general, $\Gamma(\theta)$ has s distinct eigenvalues.

Conversely, if the $\{A_a\}$ all commute, a well known theorem in linear algebra tells us that there is a family $\{S_\alpha\}$ of t (say) pairwise orthogonal symmetric idempotent matrices summing to I such that $A_a S_\alpha = p_{\alpha a} S_\alpha$ for constants $p_{\alpha a}$, $\alpha = 1, \ldots, t$, $a = 1, \ldots, s$. It follows that an element $\Gamma \in \mathbf{V}$ will have spectral decomposition $\Gamma = \Sigma_\alpha \xi_\alpha S_\alpha$, where $\xi_\alpha = \Sigma_a p_{\alpha a}\theta_a$, and if, in general, such a Γ has s distinct eigenvalues, then we deduce that $t = s$ and that $P = (p_{\alpha a})$ is an invertible $s \times s$ matrix.

Where have we gotten to? Without giving full details we have seen the reason why the preceding (a) and (b) are jointly equivalent to the two conditions

(c) the matrices $\{A_a\}$ commute,
(d) in general, $\Gamma(\theta)$ has s distinct eigenvalues.

This is in essence the content of Graybill and Hultquist (1961, Theorem 6). Note that under (c) and (d) we can write $A_a = \Sigma_\alpha p_{\alpha a} S_\alpha$ and $S_\alpha = (1/n)\Sigma_a q_{a\alpha} A_a$, where we have inserted a scale factor $n = |T|$ for later convenience, and where $\Sigma_a p_{\alpha a} q_{a\beta} = n\delta_\beta^\alpha$ and $\Sigma_\alpha q_{a\alpha} p_{\alpha b} = n\delta_b^a$, δ here being Kronecker's delta. These equations combine to give

$$A_a A_b = A_b A_a = \sum_c \left\{(1/n)\sum_\alpha p_{\alpha a} p_{\alpha b} q_{c\alpha}\right\} A_c, \tag{3.3}$$

implying that $\mathbf{V}$ may be extended to the linear algebra generated by the $\{A_a\}$ without invalidating anything we have said to date.

If the $\{A_a\}$ all have the property that all their row (column) sums are the same, i.e., if for each a there exists k_a such that $\Sigma_s A_a(s, t) = \Sigma_t A_a(s, t) = k_a$, then the matrix $S_0 = (1/n)J$, where J is the matrix of 1's over T, is always one of the $\{S_\alpha\}$.

Let us leave the matrices $\Gamma \in \mathbf{V}$ for a moment and turn to the elements y_t of random array $y = (y_t\colon t \in T)$ with $Dy = \Gamma \in \mathbf{V}$, still assuming that $\mathbf{V}$ satisfies (c) and (d). The prescription $S_\alpha y_t = \Sigma_s S_\alpha(s, t) y_s$ defines a family of random variables such that

$$y_t = \sum_\alpha S_\alpha y_t. \tag{3.4}$$

Now $\operatorname{cov}(S_\alpha y_t, S_\beta y_u) = (S_\alpha \Gamma S_\beta)(t, u) = \xi_\alpha S_\alpha(t, u)\delta_\beta^\alpha = 0$ if $\alpha \neq \beta$ and so the different terms on the R.H.S. of (3.4) are uncorrelated. Further $\operatorname{var}(S_\alpha y_t) = \xi_\alpha S_\alpha(t, t)$. Next suppose that $\operatorname{var}(y_t) = \sigma^2$ is the same for all $t \in T$, i.e., that the matrices $\{A_a\}$ are all constant down their diagonals. Then $S_\alpha(t, t) = n^{-1} d_\alpha$, where $d_\alpha = \operatorname{rank}(S_\alpha) = \operatorname{trace}(S_\alpha)$, and we can sum the variances in (3.4) obtaining

$$\sigma^2 = \sum_\alpha \phi_\alpha, \tag{3.5}$$

where $\phi_\alpha = n^{-1} d_\alpha \xi_\alpha = \operatorname{var}(S_\alpha y_t)$, independent of $t \in T$. Clearly this is an analysis of variance. The connection between it and the sum of squares decomposition

$$|y|^2 = \sum_\alpha |S_\alpha y|^2 \tag{3.6}$$

resulting from the geometric orthogonality of the terms in

$$y = \sum_\alpha S_\alpha y \tag{3.7}$$

is clear: The eigenvalues ξ_α are the expected mean squares:

$$\xi_\alpha = E\{d_\alpha^{-1}|S_\alpha y|^2\}. \tag{3.8}$$

Is this the correct anova? Does it have all the properties one might hope for? I would like to suggest that the answer to these questions is no, and that although the definition is basically correct, it is really only appropriate for a particular class $\{A_a\}$ of basis matrices and parameters $\{\theta_a\}$, namely, when the entries of the basis matrices are either 0 or 1 and the parameters are covariances. With this class we will find that we have a notion that extends fruitfully far beyond sums of squares.

4. Anova: Finite arrays. In this section we will sketch the most natural framework within which the special properties of our examples hold generally. The restriction to finite arrays is vital because there are many sorts of infinities and, perhaps surprisingly, no single mathematical framework is yet available which covers all the cases.

As before we begin with an array $y = (y_t \colon t \in T)$ of random variables indexed by a finite set T with $Ey = 0$ and we will consider a very special sort of parametrization of its dispersion matrix $\Gamma = Dy$, namely that defined by equality constraints among the elements of Γ. More fully, we will suppose that

$$\Gamma = \sum_a \gamma_a A_a, \tag{4.1}$$

where $\{A_a \colon a \in X\}$ is a class of matrices over T whose elements are 0 and 1 only satisfying (i) each matrix A_a is symmetric; (ii) $\Sigma_a A_a = J$, the matrix of 1's over T; (iii) one of these matrices, A_e say, is the identity matrix I over T; and (iv) there exist integers (n_{abc}), $a, b, c \in X$ such that $A_a A_b = \Sigma_c n_{abc} A_c$. Finally, $\{\gamma_a \colon a \in X\}$ is a set of covariances which are such that Γ given by (4.1) is positive definite.

Such matrices $\{A_a\}$ are the adjacency matrices of the association scheme over T defined by saying that s and t are a-associates, $a(s,t) = a$, say, if $A_a(s,t) = 1$, $s, t \in T$, $a \in X$; see MacWilliams and Sloane (1977, Chapter 21) for fuller background and the theory which follows.

We proceed to analyse the class of all Γ of the form (4.1). From (i) all such Γ are symmetric; from (ii) the $\{A_a\}$ are linearly independent and hence the dimension of the vector space $\mathbf{A}$ of all such Γ (forgetting positive definiteness for the moment) is $s = |X|$; from (iii) $\mathbf{A}$ contains the identity and from (iv) we deduce that $\mathbf{A}$ is a commutative algebra. The theorem in linear algebra already cited tells us that there exists a unique basis of $\mathbf{A}$ of primitive idempotents $\{S_\alpha \colon \alpha \in Z\}$, where $S_\alpha = S_\alpha^2 = S_\alpha'$, $S_\alpha S_\beta = S_\beta S_\alpha = 0$, $\alpha \neq \beta$, $\Sigma_\alpha S_\alpha = I$, containing $(1/n)J = S_0$, say. Further the transformation from this basis to the original one consisting of the $\{A_a\}$ is linear and invertible:

$$S_\alpha = \frac{1}{n} \sum_a q_{a\alpha} A_a, \tag{4.2a}$$

$$A_a = \sum_\alpha p_{\alpha a} S_\alpha, \tag{4.2b}$$

where $P = (p_{\alpha a})$ and $Q = (q_{a\alpha})$ are matrices of coefficients satisfying $PQ = QP = nI$, $n = |T|$ and I here is the identity matrix of order $s = |X| = |Z|$. Since the eigenvalues of A_a are $(p_{\alpha a})$ from (4.2b), those of $\Gamma = \Sigma_a \gamma_a A_a = \Sigma_\alpha \xi_\alpha S_\alpha$ are

(4.3a) $$\xi_\alpha = \sum_a p_{\alpha a} \gamma_a$$

whilst the entries γ_a of Γ in (4.1) are recoverable from the eigenvalues via

(4.3b) $$\gamma_a = (1/n) \sum_\alpha q_{a\alpha} \xi_\alpha.$$

Writing $k_a = |\{t \in T\colon A_a(s, t) = 1\}|$, independent of $s \in T$, and $d_\alpha = \operatorname{rank}(S_\alpha)$, we summarise some basic facts concerning these numbers and the matrices P and Q. Here δ denotes the Kronecker delta.

THEOREM (cf. MacWilliams and Sloane 1977, Chapter 21, Section 2).

(i) $p_{\alpha e} = q_{a0} = 1$; $p_{0a} = k_a$; $q_{e\alpha} = d_\alpha$; $d_\alpha p_{\alpha a} = k_a q_{a\alpha}$.
(ii) $\Sigma_\alpha d_\alpha p_{\alpha a} p_{\alpha b} = n k_a \delta_b^a$, $\Sigma_a k_a q_{a\alpha} q_{a\beta} = n d_\alpha \delta_\beta^\alpha$.
(iii) $p_{\alpha a} p_{\alpha b} = \Sigma_c n_{abc} p_{\alpha c}$.

All of these facts give us great insight into the structure of matrices of the form (4.1) and many examples can be found in the literature; see MacWilliams and Sloane (1977) and references therein. Speed and Bailey (1982) show that all standard ("balanced complete," "orthogonal") anova models arise from such schemes where X is a modular lattice of equivalence relations on T, and the Möbius function on X (together with the number of levels of each index) determines the matrices P and Q. These results are summarized in Section 6. For most but not all classical anova models, results equivalent to the preceding were given by Nelder (1954, 1965) when Γ is induced by randomisation; see Speed (1985) for more details concerning the connexions. Early forms of (3.4) and (3.6) can be found in Kempthorne (1952, Chapter 8), again with a randomisation distribution defining Γ.

Let us turn now to the elements y_t of the array y. As in Section 3 we write $S_\alpha y_t = \Sigma_u S_\alpha(t, u) y_u$, and find that (3.4) is a decomposition of y_t into uncorrelated components which in this context satisfies

(4.4) $$E\{(S_\alpha y_t)(S_\beta y_u)\} = n^{-1} \xi_\alpha q_{a(t,u)\alpha} \delta_\beta^\alpha,$$

and in particular this equals $n^{-1} d_\alpha \xi_\alpha = \phi_\alpha$ say, if $t = u$ and $\alpha = \beta$. Here $a(t, u)$ is the unique $a \in X$ such that $A_a(t, u) = 1$. With this notation we may write (4.3b) in the form

(4.5) $$\gamma_a = \sum_\alpha (d_\alpha^{-1} q_{a\alpha}) \phi_\alpha,$$

noting that the special case $a = e$ (the identity association) gives us the analysis of variance (3.5) corresponding to the decomposition (3.4). The index α labels the "lines" of the anova table—we call them strata—and the projectors S_α will be termed stratum projectors.

Summarising, we have seen that if $\Gamma = Dy$ has the form (4.1) where the $\{A_a\}$ satisfy conditions (i), (ii), (iii) and (iv) following (4.1), then, from Section 3, our variants (a) and (b) [equivalently, (c) and (d)] of Graybill and Hultquist's (1961) definition are certainly satisfied. Do we get anything extra which might justify our belief that it is only with these sorts of basis matrices and corresponding parameters that the term anova is appropriate? I believe we do, and make the following supporting observations:

(i) the present framework has a common variance (that to be analysed) as part of its formulation;

(ii) the $\{A_a\}$ matrices already have the property that their row (column) sums are the same, which implies that $S_0 = (1/n)J$ is one of the $\{S_\alpha\}$;

(iii) the $\{A_a\}$ matrices are all constant down their diagonal, a property which combines with (i) to give the analysis of the common variance;

(iv) we have the compact and extremely useful formula (4.4).

In the more general discussion of Section 3 each of the preceding (i), (ii) and (iii) had to be assumed in order to obtain the desired consequences, whilst (iv) shows the great simplification which results from covariance parametrization: With it, we need only know $\{A_a\}$, $\{d_\alpha\}$, $\{k_a\}$ and the function $s_\alpha(a) = k_a^{-1}p_{\alpha a} = d_\alpha^{-1}q_{a\alpha}$; without it (cf. Section 3) we need the entries of the $\{A_a\}$, the $\{S_\alpha\}$ and the change-of-basis matrices $(p_{\alpha a})$ and $(q_{a\alpha})$.

In a sense the reasons just given for selecting this formulation as the one deserving the title anova are mere details; the real reason is the fact that almost all examples and the natural generalisations and variants all derive from the present and no other approach. This will become more apparent in the next section, but first we give an example.

Example. Suppose that $T = \prod_1^r\{1,\ldots,n_j\}$ and that the indices are nested in a hierarchical structure t_1 nesting t_2 which nests t_3, etc. If we write $t = (t_1,\ldots,t_r)$ then there is an obvious way to define a set of matrices $\{A_a: a = 0,\ldots,r\}$ satisfying (i), (ii), (iii) and (iv), namely, $A_a(s,t) = 1$ if $s_h = t_h$, $h = 1,\ldots, a$, $s_{a+1} \neq t_{a+1}$, $A_a(s,t) = 0$ otherwise, $0 \le a < r$; $A_r = I$ $(= A_e)$. When working with this example it is helpful to introduce the equivalence matrices $\{R_a: a = 0,\ldots,r\}$ defined by $R_a(s,t) = 1$ if $s_h = t_h$, $h = 1,\ldots, a$, $R_a(s,t) = 0$ otherwise; clearly $R_a = A_a + \cdots + A_r$, $0 \le a \le r$, while $A_a = R_a - R_{a+1}$, $0 \le a < r$, and $A_r = R_r = I$. This is because the primitive idempotents $\{S_\alpha\}$ are now readily defined by

$$S_0 = (n_1 \cdots n_r)^{-1}R_0,$$

$$S_\alpha = (n_{\alpha+1} \cdots n_r)^{-1}R_\alpha - (n_\alpha \cdots n_r)^{-1}R_{\alpha-1}, \qquad 1 \le \alpha < r,$$

$$S_r = I - n_r^{-1}R_{r-1}.$$

It is easy to calculate that $k_r = 1 = d_0$, $k_a = (n_{a+1} - 1)n_{a+2} \cdots n_r$, $0 \le a < r$,

WHAT IS AN ANALYSIS OF VARIANCE?

$d_\alpha = n_1 \cdots (n_\alpha - 1)$, $0 < \alpha \le r$, and

$$d_\alpha^{-1} q_{a\alpha} = \begin{cases} 0, & a = 0, \ldots, \alpha - 2, \\ -(n_\alpha - 1)^{-1}, & a = \alpha - 1, \\ 1, & a = \alpha, \ldots, r. \end{cases}$$

The decomposition of $y_t = y_{t_1 \cdots t_r}$ is totally straightforward:

$$y_{t_1 t_2 \cdots t_{r-1} t_r} = y_{\cdot\cdot \cdots \cdot\cdot} + (y_{t_1 \cdot \cdots \cdot\cdot} - y_{\cdot\cdot \cdots \cdot\cdot}) + \cdots + (y_{t_1 t_2 \cdots t_{r-1} t_r} - y_{t_1 t_2 \cdots t_{r-1} \cdot})$$

and the other results follow immediately. This is one of the examples where X (and hence Z) have a lattice structure, namely the $(r+1)$-chain $\{\phi, \{1\}, \{1,2\}, \ldots, \{1,2,\ldots,r\}\}$; see Section 6.

5. Anova for infinite arrays. From the viewpoint presented in this paper one of the earliest instances of anova in statistics was the spectral representation of weakly stationary time series $y = (y_t \colon t \in \mathbb{Z})$, essentially put in its modern form by Cramér (1940) following earlier work by Khinchin (1934). Here the covariance matrix $\Gamma(s,t) = \operatorname{cov}(y_s, y_t)$ satisfies $\Gamma(s,t) = \Gamma(u,v)$ whenever $t - s = v - u$ and so may, formally at least, be written

$$\Gamma = \sum_0^\infty \gamma_a A_a, \tag{5.1}$$

where $A_0 = I$ is the doubly infinite identity matrix and A_a is the doubly infinite symmetric circulant having zeroth row $(\cdots 010 \cdots 0 \cdots 010 \cdots)$ with a 1 in the ath and $-a$th position, $a = 1, 2, \ldots$. Because Γ is positive definite, a theorem of Herglotz tells us that for such a matrix there exists a uniquely defined positive measure on $[-\pi, \pi)$ whose Fourier coefficients are the $\{\gamma_a\}$. Since $\gamma_{-a} = \gamma_a$, this measure must be symmetric about 0 and so we can obtain the real spectral representation

$$\gamma_a = \int_{[0,\pi)} \cos(a\alpha)\phi(d\alpha), \qquad a \in \mathbb{Z}, \tag{5.2}$$

a formula which can readily be compared with (2.10b). The corresponding (real) representation of y_t with $E\{y_t\} \equiv 0$ takes the form

$$y_t = y_\cdot + 2\int_{(0,\pi)} [\cos(t\alpha)u(d\alpha) + \sin(t\alpha)v(d\alpha)], \tag{5.3}$$

where u and v are additive and mean-square continuous random set functions defined on the Borel subsets of $(0, \pi)$, spanning the Hilbert space generated by $y = (y_t \colon t \in \mathbb{Z})$ having zero means and satisfying

$$\begin{aligned} E\{u(A)u(B)\} &= E\{v(A)v(B)\} = \phi(A \cap B), \\ E\{u(A)v(B)\} &= 0, \end{aligned} \tag{5.4}$$

for A, B Borel subsets of $(0, \pi)$. Finally $y_\cdot$ is the mean-square limit of $T^{-1}\Sigma_1^T y_t$ as

$T \to \infty$, which is easily shown to exist. To compare (5.3) and (2.11) one simply expands the $\cos(2\pi(s-t)\alpha/n)$ and separates out random variables from non-random coefficients.

This is one kind of "infinite anova"; there are many similar ones in the literature of stochastic processes; see Hannan (1970, Chapter 1) and references therein.

At this point we do not stop to consider the method of proof of (5.3); in essence it reduces to the spectral decomposition of a unitary operator in Hilbert space and this will be covered by the discussion in Section 6. Rather we turn to another kind of infinite array.

Our original example $y = (y_{ij}\colon i = 1, \dots, m;\ j = 1, \dots, n)$ with j nested within i and having $\Gamma = Dy$ of the form (2.7) makes perfect sense if m or n (or both) is (are) countably infinite. Indeed one such example is the "random effects model"

$$y_{ij} = \varepsilon_0 + \varepsilon_i + \varepsilon_{ij}, \tag{5.5}$$

where (ε_i) and (ε_{ij}) are uncorrelated infinite sequences of uncorrelated random variables with zero means and variances σ_1^2 and σ_2^2, respectively, and ε_0 is a zero mean random variable uncorrelated with the ε_i and the ε_{ij} with variance σ_0^2. In this case the parameters γ_2, γ_1 and γ_0 of (2.7) are

$$\gamma_2 = \sigma_0^2 + \sigma_1^2 + \sigma_2^2, \qquad \gamma_1 = \sigma_0^2 + \sigma_1^2, \qquad \gamma_0 = \sigma_0^2. \tag{5.6}$$

What is the analogue of (2.4), (2.5) and (2.6) for an array $y = (y_{ij})$ with $\Gamma = Dy$ satisfying (2.7) for $m = n = \infty$? Clearly we can truncate i and j (within i) to the ranges $1, \dots, m$ and $1, \dots, n$, respectively, and see what results as $m, n \to \infty$, and doing this leads to some simple and interesting conclusions. Denoting the parameters and other objects associated with the truncated array by a superscript (m, n), we can prove directly that $\phi_\alpha^{(m,n)} = (mn)^{-1} d_\alpha^{(m,n)} \xi_\alpha^{(m,n)}$ and $[d_\alpha^{(m,n)}]^{-1} q_{a\alpha}^{(m,n)}$ both converge as m and $n \to \infty$ to ϕ_α and $s_\alpha(a)$ say, $\alpha = 0, 1, 2$ and $a = 0, 1, 2$. It follows that the terms $\xi_\alpha S_\alpha$ in the spectral representation (2.5) also converge as m and $n \to \infty$, since $\xi_\alpha^{(m,n)} S_\alpha^{(m,n)}(ij, kl) = \xi_\alpha^{(m,n)} (mn)^{-1} q_{a(ij,kl)\alpha}^{(m,n)}$, and we find that the limiting form of (2.5) is

$$\Gamma = \phi_0 J \otimes J + \phi_1 I \otimes J + \phi_2 I \otimes I, \tag{5.7}$$

where I and J are the infinite identity matrix and matrix of all 1's, respectively. Although (5.7) is not a spectral representation in any obvious sense, it can be proved that the most general positive definite matrix of the form

$$\Gamma = \gamma_2 I \otimes I + \gamma_1 I \otimes (J - I) + \gamma_0 (J - I) \otimes J \tag{5.8}$$

has a unique representation in the form (5.7) with ϕ_0, ϕ_1 and ϕ_2 all positive. The relations between ϕ's and γ's are simple enough:

$$\gamma_2 = \phi_0 + \phi_1 + \phi_2, \qquad \gamma_1 = \phi_0 + \phi_1, \qquad \gamma_0 = \phi_0 \tag{5.9a}$$

with inverse

$$\phi_2 = \gamma_2 - \gamma_1, \qquad \phi_1 = \gamma_1 - \gamma_0, \qquad \phi_0 = \gamma_0. \tag{5.9b}$$

In an obvious notation we can also prove that for $m' \geq m$ and $n' \geq n$,

$$\text{(5.10a)} \quad \| y_{\cdot\cdot}^{(m,n)} - y_{\cdot\cdot}^{(m',n')} \|^2 = \phi_2 \left[\frac{1}{mn} - \frac{1}{m'n'} \right] + \phi_1 \left[\frac{1}{m} - \frac{1}{m'} \right],$$

$$\text{(5.10b)} \quad \begin{aligned} &\| (y_{i\cdot}^{(m,n)} - y_{\cdot\cdot}^{(m,n)}) - (y_{i\cdot}^{(m',n')} - y_{\cdot\cdot}^{(m',n')}) \|^2 \\ &= \phi_2 \left[\left(\frac{1}{n} - \frac{1}{n'} \right) \left(1 - \frac{1}{m} \right) + \frac{1}{n'} \left(\frac{1}{m} - \frac{1}{m'} \right) \right] + \phi_1 \left[\frac{1}{m} - \frac{1}{m'} \right], \end{aligned}$$

$$\text{(5.10c)} \quad \| (y_{ij}^{(m,n)} - y_{i\cdot}^{(m,n)}) - (y_{ij}^{(m',n')} - y_{i\cdot}^{(m',n')}) \|^2 = \phi_2 \left[\frac{1}{n} - \frac{1}{n'} \right],$$

from which it follows that $y_{\cdot\cdot}^{(m,n)}$, $y_{i\cdot}^{(m,n)} - y_{\cdot\cdot}^{(m,n)}$ and $y_{ij}^{(m,n)} - y_{i\cdot}^{(m,n)}$—the components in (2.2)—all converge in mean square as $m, n \to \infty$. Denoting their limits by ε_0, ε_i and ε_{ij}, respectively, it can also be proved that not only are ε_0, ε_h and ε_{ij} pairwise orthogonal—they come from different strata in the limiting form of (2.2)—but also ε_h and ε_i are orthogonal if $h \neq i$, and similarly ε_{ij} and ε_{kl} are orthogonal if $i \neq k$ or $i = k$ and $j \neq l$. But all this has proved that (5.5) is (up to second order) the most general form for an array $y = (y_{ij})$ with $Dy = \Gamma$ satisfying (5.8), and that (5.7) is the most general form for such Γ. In this sense the standard random effects models arise naturally as the spectral decompositions of infinite arrays of multiindexed random variables with the appropriate dispersion models. For further details including a proof of this general result we refer to Speed (1986).

For our final illustration of an anova for an infinite array we return to the Example at the end of Section 4 and suppose that the repeated nesting goes on ad infinitum, i.e., that $T = \prod_1^\infty \{1, \ldots, n_j\}$ with each index of $t = (t_1, t_2, \ldots) \in T$ nesting all subsequent ones. As with the finite version, we can define association matrices $\{A_a\colon a = 0, 1, \ldots\}$ to which we must add $A_\infty = I \,(= A_e$ in our general notation). The relationship matrices $\{R_a\colon a = 0, 1, \ldots, \infty\}$ are defined in the same way as we did earlier and the passage from A-matrices to R-matrices is as before. We now look for a spectral representation for the positive definite matrices of the form

$$\text{(5.11)} \quad \Gamma = \sum_{a=0}^{a=\infty} \gamma_a A_a.$$

As with our previous discussion, it is instructive to look at a truncated version of T, and the obvious candidate here is $T^{(r)} = \{t \in T\colon t_{r+1} = t_{r+2} = \cdots = 1\}$.

Denoting parameters and other expressions associated with the subarray $y^{(r)} = (y_t\colon t \in T^{(r)})$ with a superscript (r), we note that $s_\alpha(a) = [d_\alpha^{(r)}]^{-1} q_{\alpha a}^{(r)}$ does not depend upon r as long as $0 \leq a$, $a \leq r$. Furthermore, a straightforward calculation proves that $\phi_\alpha^{(r)} = (n_1 \cdots n_r)^{-1} d_\alpha^{(r)} \xi_\alpha^{(r)}$ satisfies

$$\text{(5.12)} \quad \phi_\alpha^{(r)} - \phi_\alpha^{(r+1)} = (1 - n_\alpha^{-1}) n_{\alpha+1}^{-1} \cdots n_r^{-1} (1 - n_{r+1}^{-1})(\gamma_e - \gamma_r),$$

which is nonnegative since $\gamma_a \leq \gamma_e$ for all a. Since $0 \leq \phi_\alpha^{(r)} \leq \gamma_e$ for all $r \geq 1$ and $\alpha \leq r$, we deduce that $\phi_\alpha^{(r)}$ converges, to ϕ_α say, as $r \to \infty$. Thus the elements of

$\xi_\alpha^{(r)} S_\alpha^{(r)}$ also converge as $r \to \infty$ and so we conjecture a unique representation for Γ in (5.11) taking the form of an ordinary infinite series

$$\Gamma = \sum_\alpha \phi_\alpha S_\alpha, \tag{5.13}$$

where the ϕ_α are positive (summing to γ_e—the anova) and the S_α satisfy $S_\alpha(s, t) = s_\alpha(a(s, t))$, i.e.,

$$S_\alpha = \sum_{a=0}^{a=\infty} s_\alpha(a) A_a. \tag{5.14}$$

These facts are readily proved and are perhaps most easily seen by using formal infinite tensor products. In an obvious notation $S_0 = J = J_{n_1} \otimes J_{n_2} \otimes \cdots$, whilst for $\alpha > 0$ we can use the expression for $s_\alpha(a)$ to get

$$\begin{aligned} S_\alpha &= \sum_{a \geq \alpha} A_a - (n_\alpha - 1)^{-1} A_{\alpha - 1} \\ &= \frac{n_\alpha}{n_\alpha - 1} I_{n_1} \otimes \cdots \otimes I_{n_{\alpha-1}} \otimes \left(I_{n_\alpha} - \frac{1}{n_\alpha} J_{n_\alpha} \right) \otimes J_{n_{\alpha+1}} \otimes J_{n_{\alpha+2}} \otimes \cdots . \end{aligned}$$

This completes our discussion of the spectral decomposition of Dy and we turn to that of y_t, $t \in T$. As with our previous example, its components are defined as mean-square limits, and in this case it is perhaps no surprise to see that these exist for

$$S_\alpha^{(r)} y_t = y_{t_1 \cdots t_\alpha \cdot\cdot \cdots \cdot t_{r+1} t_{r+2} \cdots} - y_{t_1 \cdots t_{\alpha-1} \cdot\cdot \cdots \cdot t_{r+1} t_{r+2} \cdots}$$

as $r \to \infty$. Indeed $\|S_\alpha^{(r)} y_t - S_\alpha^{(r')} y_t\|^2 = \phi_\alpha^{(r)} - \phi_\alpha^{(r')}$ for $1 \leq r \leq r'$, and by (5.12) this converges to zero as $r, r' \to \infty$ (assuming $n_r \geq 2$ for all r). Of course the mean-square limit $S_\alpha y_t$, say, of $S_\alpha^{(r)} y_t$, satisfies $\|S_\alpha y_t\|^2 = \phi_\alpha$, and so the spectral representation of y_t is the infinite sum, defined as a mean-square limit

$$y_t = \sum_\alpha S_\alpha y_t, \tag{5.15}$$

with associated anova $\gamma_e = \Sigma_\alpha \phi_\alpha$. Note that (5.15) is not the same as the expression

$$y_{t_1 t_2 t_3 \cdots} = \varepsilon_0 + \varepsilon_{t_1} + \varepsilon_{t_1 t_2} + \varepsilon_{t_1 t_2 t_3} + \cdots,$$

where $\{\varepsilon_0\}, \{\varepsilon_{t_1}\}, \{\varepsilon_{t_1 t_2}\}, \{\varepsilon_{t_1 t_2 t_3}\}, \ldots$ are uncorrelated sets of uncorrelated effects having variances $\phi_0, \phi_1, \phi_2, \phi_3, \ldots$; to get such a representation we would also need to let $n_1 \to \infty$, $n_2 \to \infty$, $n_3 \to \infty, \ldots$ in the preceding discussion.

These three examples of anovas for infinite arrays give a good idea of the range of possibilities. With the finite cyclic structure going over to the infinite one, we obtain a "continuous infinity" of strata; with the classical anova models illustrated by our second example, we simply recover standard random effects models, the number of strata remaining constant; whilst our final example shows how limits can be taken along infinite chains in the partially-ordered subset defining the nesting relationships on the set of indices, with the number of strata going to a countable infinity.

In none of these infinite examples does there appear to be a full analogue of the geometrically orthogonal decomposition of arrays y of real numbers, nor any associated sum of squares decompositions. Given that we never observe an infinite array of real numbers, this is no real limitation of the theory, and for many examples—most importantly the standard anova models in statistics—these decompositions for finite subarrays give useful information concerning aspects of the full array. Some details are sketched in Speed (1985) in a discussion relating the anova of a subarray, where it exists, to the anova of a full array.

The conclusion we come to after this discussion is that there is more to anova than sums of squares. Our view, already stated in the previous section, is that anova is a feature of certain models $\mathbf{V}$ which impose *equality constraints* on the covariances between pairs of elements of arrays of random variables.

6. Classical anova: Factorial dispersion models. The historically important anovas with multiply indexed arrays are the random effects models, dating back beyond Fisher (1925) to the last century, the randomization or permutation models following those discussed by Neyman, Iwaskiewicz and Kolodziejczyk (1935) and the more recent generalisations of de Finetti's exchangeability, studied by Aldous (1981) and others. Because of the importance of these ideas in statistics, I will sketch their common second-order theory.

We begin with a set F of factors $f_1, f_2, \ldots$, and a partial order $\leq$ on F where $f_1 \leq f_2$ means that the factor f_1 is nested within the factor f_2; cf. Nelder (1965). A subset $a \subseteq F$ is said to be a filter if $f_1 \in a$ and $f_1 \leq f_2$ implies that $f_2 \in a$, the need for such subsets arising because it is frequently necessary, when referring to the levels of a given factor f, to refer at the same time to all factors within which f is nested. The set of all filters of the partially ordered set $(F; \leq)$ forms a distributive lattice $L(F)$ under the operations of set union and intersection [see Aigner (1979, page 33)] and we refer to this book for all other order-theoretic terminology and results used in what follows. We remark in passing that our use of partially ordered sets in this context is closely related to, but does not coincide with, that of Throckmorton (1961), adopted by Kempthorne and Folks (1971, Section 16.11).

Next we suppose that the set of levels of factor f is T_f, $f \in F$, and we write $T = \prod_f T_f$ for the set of all combinations of levels of factors in F, denoting a typical element by $t = (t_f\colon f \in F)$. For any pair $s, t \in T$ we write $a(s, t)$ for the largest filter $a \in L(F)$ such that $s_f = t_f$ for all $f \in a$; e.g., if $s = ijk$ and $t = i'j'k'$, where we have three factors whose levels are denoted by the usual ijk rather than (s_1, s_2, s_3), and the second factor j is nested within the first i, then $a(s, t) = \{1, 2\}$ if $i = i'$, $j = j'$ and $k \neq k'$, whereas $a(s, t) = \{3\}$ if $i \neq i'$, $j = j'$ and $k = k'$, for $\{2, 3\}$ is not a filter of the partially ordered set of factors.

With these preliminaries we turn to the definition of factorial dispersion models. These are for arrays $y = (y_t\colon t \in T)$ of real random variables indexed by the set T of all combinations of levels of a set F of factors whose nesting relationships are defined by the partially ordered set $(F; \leq)$. The factorial dispersion model $\mathbf{V} = \mathbf{V}(F, T)$ is the class of all covariance matrices $\Gamma = Dy$ over

T which satisfy

$$\operatorname{cov}(y_s, y_t) = \operatorname{cov}(y_u, y_v) \tag{6.1}$$

whenever $a(s, t) = a(u, v)$, $s, t, u, v \in T$. Such classes are slightly more general than ones introduced by Nelder (1965), and we note that it has not yet been necessary to state whether or not the sets T_f are finite. For our summary of the structure of these models, we consider the two cases $|T_f| < \infty$ for all $f \in F$, and $|T_f| = \infty$ for all $f \in F$.

Finite factorial dispersion models. If $|T_f| = n_f < \infty$ for all $f \in F$, and we write $n = \prod_f n_f$, then $\mathbf{V}(F, T)$ is a class of $n \times n$ matrices whose structure is readily exhibited; see Speed and Bailey (1982) for full details. First we define the family $\{A_a\colon a \in L(F)\}$ of matrices over T by writing $A_a(s, t) = 1$ if $a(s, t) = a$ and $A_a(s, t) = 0$ otherwise, $s, t \in T$, $a \in L(F)$. Each element $\Gamma \in \mathbf{V}(F, T)$ satisfying (6.1) may then be represented uniquely in the form $\Gamma = \Sigma_a \gamma_a A_a$, the sum being over $L(F)$, with the parameters $\{\gamma_a\colon a \in L(F)\}$ being covariances.

It can be shown that the $\{A_a\}$ so defined form an association scheme, i.e., that (i), (ii), (iii) and (iv) of Section 4 and hence the consequences of these conditions hold, but here we can construct the structure constants $\{k_a\}, \{d_a\}$ and the functions $\{s_\alpha(a)\}$ directly. To do this we introduce a second representation of $\mathbf{V}(F, T)$ involving relationship matrices $\{R_b\colon b \in L(F)\}$, where $R_b(s, t) = 1$ if $s_f = t_f$ for all $f \in b$ and $R_b(s, t) = 0$ otherwise, $s, t \in T$ and $b \in L(F)$. Clearly $R_b = \Sigma_{a \supseteq b} A_a$ and the representation we refer to is

$$\Gamma = \sum_b f_b R_b, \tag{6.2}$$

where the parameters $\{f_b\colon b \in L(F)\}$ have been called canonical components of variance by Fairfield-Smith (1955), Σ-quantities by Wilk and Kempthorne (1956), and f-quantities by Nelder (1965), although he later called them components of excess variance [Nelder (1977)]. Unfortunately it would take us too far afield to explain fully the frameworks of these other writers and the correspondence of the different parameters.

Relating the $\{f_b\}$ to the $\{\gamma_a\}$ requires the zeta function of the lattice $L(F)$, defined by $\zeta(a, b) = 1$ if $a \subseteq b$, $\zeta(a, b) = 0$ otherwise, and the associated Möbius function μ defined by $\Sigma\zeta(a, b)\mu(b, c) = \Sigma\mu(a, b)\zeta(b, c) = \delta(a, c) = 1$ if $a = c$ and 0 otherwise; here a, b and $c \in L(F)$ and the sums are over all $b \in L(F)$; see Aigner (1979, page 141) for further details. In this notation

$$f_b = \sum_a \mu(a, b)\gamma_a \tag{6.3a}$$

and

$$\gamma_a = \sum_a \zeta(b, a) f_b = \sum_{b \subseteq a} f_b. \tag{6.3b}$$

It can be shown that for all lattices of the form $L(F)$ the Möbius function μ takes only the values 1, -1 or 0; indeed the following concise formula for μ can be proved:

(6.4) $$\mu(a, b) = \begin{cases} (-1)^{|b \setminus a|}, & \text{if } b \supset a \text{ and } b \setminus a \subseteq b_m, \\ 0, & \text{otherwise,} \end{cases}$$

where b_m denotes the set of minimal elements of $b \subseteq F$.

The final representation of elements of $\mathbf{V}(F, T)$ we present is an explicit form of their common spectral decomposition. If we write $\bar{n}_a = \prod\{n_f\colon f \notin a\}$ for an element $a \in L(F)$, then the formula

(6.5) $$S_\alpha = \sum_a \mu(a, \alpha)\bar{n}_a^{-1}R_a, \qquad \alpha \in L(F)$$

defines a set of pairwise orthogonal symmetric idempotent matrices summing to the identity matrix I over T. Further the formula

(6.6) $$\xi_\alpha = \sum_b \zeta(a, b)\bar{n}_b f_b$$

gives the eigenvalues of $\Gamma = \Sigma_b f_b R_b$ and its spectral decomposition is then $\Gamma = \Sigma_\alpha \xi_\alpha S_\alpha$. Thus the eigenvalues $\{\xi_\alpha\colon \alpha \in L(F)\}$ constitute a third set of parameters whose positivity succinctly defines the parameter space, and there are two related sets of parameters which also have been used: the specific components of variance $\{\sigma_\alpha^2\colon \alpha \in L(F)\}$ of Cornfield and Tukey (1956), given by $\sigma_\alpha^2 = \bar{n}_\alpha^{-1}\xi_\alpha$, and the spectral components of variance $\{\phi_\alpha\colon \alpha \in L(F)\}$, cf. Daniels (1939), given by $\phi_\alpha = n^{-1}d_\alpha\xi_\alpha$, where $d_\alpha = \text{rank}(S_\alpha)$.

If we combine the relationships between the $\{\gamma_a\}$ and the $\{f_b\}$ with those connecting the $\{f_b\}$ and the $\{\xi_\alpha\}$ we can obtain (4.3a) and (4.3b) where a and $\alpha \in L(F)$ and the sums are over $L(F)$, and of course (4.2a) and (4.2b) also hold with the same coefficients $(p_{a\alpha})$ and $(q_{a\alpha})$. The following formulas give expressions for the key quantities:

(6.7) $$d_\alpha = \prod_{f \in \alpha \setminus \alpha_m} n_f \times \prod_{f \in \alpha_m} (n_f - 1),$$

where α_m denotes the set of minimal elements of α,

(6.8) $$k_a = \prod_{f \in \bar{a} \setminus \bar{a}^m} n_f \prod_{f \in \bar{a}^m} (n_f - 1),$$

where $\bar{a}^m$ denotes the set of maximal elements of $\bar{a} = F \setminus a$, and the common value $s_\alpha(a)$ of $d_\alpha^{-1}q_{a\alpha} = k_a^{-1}p_{a\alpha}$ is

(6.9) $$s_\alpha(a) = \begin{cases} \displaystyle\prod_{f \in \alpha_m \setminus a} \{-1/(n_f - 1)\}, & \text{if } \alpha \setminus \alpha_m \subseteq a, \\ 0, & \text{otherwise,} \end{cases}$$

where an empty product is defined to be unity.

The foregoing discussion enables a fairly complete analysis of finite factorial dispersion models to be given and we now indicate the changes necessary when $|T_f| = n_f = \infty$ for all $f \in F$. The main conclusion is the fact that the first two representations, $\Gamma = \Sigma_a \gamma_a A_a$ and $\Gamma = \Sigma_b f_b R_b$, continue to apply because we never need to multiply these matrices. After a suitable normalization and limiting argument, the third representation turns out to coincide with the second. In particular the limiting forms of the two parametrizations, which are essentially normalized eigenvalues $\{\sigma_\alpha^2\}$ and $\{\phi_\alpha\}$, coincide with the corresponding $\{f_\alpha\}$. Finally, the limiting form of the function $s_\alpha(a)$ is just the zeta function $\zeta(\alpha, a) = 1$ if $\alpha \subseteq a$ and 0 otherwise.

We turn now to the spectral decompositions (3.4) and (3.6) in our classical anova context. It is easy to see that for finite arrays the matrices $\{\bar{n}_a^{-1}R_a$: $a \in L(F)\}$ act on $y_t(t \in T)$ by simply averaging out all indices t_f with $f \notin a$, and so by (6.4) the expression (6.5) for S_α reduces to an alternating sum of averaging operators starting with $\bar{n}_\alpha^{-1}R_\alpha$. For infinite arrays it all carries through using mean-square limits; cf. Section 5. In the finite case this is just the familiar anova decomposition of multi-indexed arrays into admissible main effects and interactions termed the population identity by Kempthorne (1952, Chapter 8) (his arrays having permutation or sampling distributions) and called the yield identity by Nelder (1965). For infinite arrays we recover the standard random effects linear models appropriate to the nesting structure on the indices: the components $S_\alpha y_t$ are not only uncorrelated across strata but (when $n_f \equiv \infty$) also, when distinct, within strata. Again we refer to Speed (1986) for more details.

7. Anova and groups. In all the particular examples we have given so far, and in the vast majority of those which occur in practice, there is an underlying group G acting transitively on the index set T, denoted $(g, t) \to t^g$, in such a way that the class of covariance matrices $\Gamma = Dy$ of $y = (y_t: t \in T)$ which we consider for our anovas coincides with the class of positive definite functions Γ on $T \times T$ which are G-invariant in the sense that

$$(7.1) \qquad \Gamma(s, t) = \Gamma(s^g, t^g), \qquad (s, t) \in T \times T, g \in G.$$

It will follow from a few simple manipulations that the mathematical parts of our anovas, getting the spectral representation of the matrices Γ and the corresponding orthogonal decompositions of the array elements y_t $(t \in T)$, are only a slightly disguised form of a standard problem in harmonic analysis. This should hardly come as a surprise given the earlier discussion of finite and infinite circular arrays $(y_t: t = 0, 1, \ldots, n-1)$ and $(y_t: t \in \mathbb{Z})$.

We will only sketch the connexion here; the interested reader is referred to Hannan (1965, Section 5) and Dieudonné (1978) for further details. Choosing and fixing an arbitrary $t_0 \in T$, we define the subgroup $K = \{g \in G: t_0^g = t_0\}$ of G and observe that the homogeneous space G/K of cosets of G modulo K corresponds naturally with T, gK corresponding to t iff $t^g = t_0$. Now a function Φ on T is said to be spherically symmetric (relative to K) if $\Phi(t) = \Phi(t^k)$, $t \in T$, $k \in K$; similarly a function Ψ on G is said to be bi-invariant (relative to K) if $\Psi(kgk') = \Psi(g)$, $g \in G$, $k, k' \in K$, whilst we have called a function Γ on $T \times T$ G-invariant if it satisfied (7.1). The simple manipulations previously referred to show that these three classes of functions are essentially the same one, e.g., if Γ is G-invariant on $T \times T$, then $\Psi(g) = \Gamma(t_0^g, t_0)$ is bi-invariant on G whilst $\Phi(t) = \Gamma(t, t_0)$ is spherically symmetric on T. Conversely, if Ψ is bi-invariant on G and g_s, g_t are elements g and $h \in G$ for which $s^g = t_0$, $t^h = t_0$, respectively, then $\Gamma(s, t) = \Psi(g_s^{-1}g_t)$ is G-invariant on $T \times T$. Finally, we let Y denote the space of all orbits of G over $T \times T$; clearly functions γ over Y correspond in an obvious way to G-invariant functions Γ on $T \times T$ and hence to the other classes previously mentioned. With this background our initial anova problems take the form: Describe the class of all functions γ on Y, in particular those for which $\Gamma(s, t) = \gamma_{b(s,t)}$ is positive definite over T, where $b(s, t)$ is the unique element of Y containing $(s, t) \in T \times T$.

Solutions to the problem just posed exist for many group actions, the most elegant case apparently being when (G, K) is a Gel'fand pair [Dieudonné (1978, page 55)] usually discussed when G is a unimodular separable metrizable locally compact group and K a compact subgroup. When (G, K) is a Gel'fand pair there is a class Z of functions called zonal spherical functions which plays a prominent role and in our terms these are the functions on Y defined by $s_\alpha(a) = d_\alpha^{-1}q_{a\alpha}$, $a \in Y$, $\alpha \in Z$. We note in passing that this class includes all characters of locally compact abelian groups, so our anova decomposition of the matrix Γ is a form of generalised Bochner–Godement theorem.

In his expositions Letac (1981, 1982) presents a wide range of applications of the theory of Gel'fand pairs in probability theory and we can clearly add anova to his list. The example in Letac (1982) which he calls the infinite symmetric tree is just the third example we discussed in the previous section—the infinitely nested hierarchical anova model—and so we have given an alternative approach to its harmonic analysis. It is also of interest to note that the theory of discrete Gel'fand pairs which Letac summarises in his paper is included within the theory of association schemes: All of his formulas can be found in the theorem we cited in Section 4, e.g., $\pi(a) = k_a$ is the measure on X induced by the uniform measure on T, the spherical functions are $s_\alpha(a) = d_\alpha^{-1}q_{a\alpha}$ as has already been noted and the Plancherel measure on Z is $\nu(\alpha) = n^{-1}d_\alpha$.

What of the spectral decompositions for the elements y_t $(t \in T)$ of the arrays? These arise from the decomposition of the permutation representation $g \to U_g$ of G into its irreducible constituents, where U_g is defined on the Hilbert space H spanned by the $(y_t\colon t \in T)$ [using the inner product $\langle y_s, y_t\rangle = \Gamma(s, t)$] by extending the assignment $U_g y_t = y_{tg}$, $t \in T$, $g \in G$ to the whole of H. In seeking to derive the decomposition in any particular case there are issues concerning the compactness of K, separability and local compactness of G, the nature of the representation $\{U_g\}$ and so on, which must be verified before general theory can be applied; we refer to Dieudonné (1978, 1980) for details. Perhaps surprisingly, none of the simple (infinite) classical anova models gives rise to pairs (G, K) for which these conditions hold, and so the ad hoc approach adopted in Speed (1986) still seems to be necessary. Even defining the groups for these classical anova models is a formidable task; see Bailey, Praeger, Rowley and Speed (1983) for details of the finite cases and Speed (1986) for some remarks on their infinite analogues.

8. Manova. The multivariate analysis of variance or manova does for arrays of random vectors what anova does for arrays of (real-valued) random variables, that is, gives suitable spectral decompositions of their dispersion matrices, orthogonal decompositions of both the elements of the arrays and the arrays themselves; associated with these are analysis of the variances and covariances and decompositions of the sums of squares and products. There are some twists, however, which require us to generalise slightly our earlier formulation involving association matrices. For example, suppose that $w = (w_t\colon t = 0, \dots, n-1)$ is a circular array of zero mean random vectors $w_t = (x_t, y_t)'$ with dispersion matrix

$$\Gamma = D\begin{bmatrix} x \\ y \end{bmatrix} = \begin{bmatrix} \Gamma^{xx} & \Gamma^{xy} \\ \Gamma^{yx} & \Gamma^{yy} \end{bmatrix}.$$

We assume that $\Gamma^{xx} = Dx$ and $\Gamma^{yy} = Dy$ both have the form (2.9) whilst $\Gamma^{xy} = \text{cov}(x, y)$ satisfies $\Gamma^{xy}(s, t) = \Gamma^{xy}(u, v)$ if $t - s = v - u$, i.e., Γ^{xy} is a circulant, although not necessarily a symmetric one. Indeed $\text{cov}(x_s, y_t)$ and $\text{cov}(y_s, x_t)$ are in general different. What is the decomposition of Γ^{xy} analogous to the diagonalisation of Γ^{xx} and Γ^{yy}?

The solution in this case is easy enough because the structure of arbitrary circulants is as transparent as that of symmetric circulants: Write $\Gamma^{xy} = \sum_0^{n-1}\gamma_b^{xy}B_b$, where B_b is the $n \times n$ circulant having a single 1 in the bth position and 0's elsewhere in its first row. Assuming that $n = 2m + 1$ as before—the case $n = 2m$ is just as readily dealt with—we recover our earlier association matrices by noting that $A_0 = B_0$, whilst $A_a = B_a + B_a'$, $a = 1, \dots, m$. The $(m+1) \times (m+1)$ structural matrices $P = (p_{a\alpha})$ and $Q = (q_{a\alpha})$ are best described by the equations

$$k_a^{-1}p_{a\alpha} = d_\alpha^{-1}q_{a\alpha} = \cos\left(\frac{2\pi}{n}a\alpha\right), \tag{8.1}$$

where $k_0 = d_0 = 1$, $k_a = d_\alpha = 2$, $1 \le a$, $\alpha \le m$. We now need to introduce another inverse pair of $m \times m$ matrices of structural constants, namely $T = (t_{b\alpha})$ and $L = (l_{\alpha b})$:

$$t_{b\alpha} = l_{\alpha b} = 2\sin\left(\frac{2\pi}{n}b\alpha\right), \qquad 1 \le \alpha, b \le m. \tag{8.2}$$

It is not hard to prove that $TL = LT = nI_m$. With these constants defined, we supplement the $\{S_\alpha\}$ defined following (2.9) with T_α defined by $T_\alpha(s, t) = (1/n)t_{b(s,t)\alpha}$ where $b(s, t) = (t - s) \pmod n$. This is equivalent to

$$T_\alpha = (1/n)\sum_1^m t_{b\alpha}(B_b - B_b'), \qquad \alpha = 1, \dots, m. \tag{8.3}$$

In these terms we have

$$B_b = S_0 + \tfrac{1}{2}\sum_1^m (p_{\alpha b}S_\alpha + l_{\alpha b}T_\alpha), \qquad b = 1, \dots, m, \tag{8.4}$$

which, incidentally, agrees with our earlier notation since

$$A_a = B_a + B_a' = 2S_0 + \sum_1^m p_{a\alpha}S_\alpha = \sum_0^m p_{a\alpha}S_\alpha, \qquad a = 1, \dots, m.$$

Also we see that $B_b - B_b' = \sum_1^m l_{b\alpha}T_\alpha$, a consequence of the relation $LT = TL =$

nI_m. It is not hard to check that $T'_\alpha = -T_\alpha$, $T^2_\alpha = -S_\alpha$, $\alpha = 1,\ldots m$, and with all these preliminaries we can write the real form of the spectral decomposition of Γ^{xy} as

$$\Gamma^{xy} = c_0^{xy}S_0 + \sum_1^m (c_\alpha^{xy}S_\alpha + q_\alpha^{xy}T_\alpha), \tag{8.5}$$

where c_α^{xy} and q_α^{xy} are given by

$$c_\alpha^{xy} = \gamma_0^{xy} + \sum_{a=1}^m \cos\left(\frac{2\pi}{n}a\alpha\right)\left[\gamma_a^{xy} + \gamma_{n-a}^{xy}\right], \tag{8.6a}$$

$$q_\alpha^{xy} = \sum_{a=1}^m \sin\left(\frac{2\pi}{n}a\alpha\right)\left[\gamma_{n-a}^{xy} - \gamma_a^{xy}\right], \tag{8.6b}$$

with inverse

$$\gamma_b^{xy} = \frac{1}{n}c_0^{xy} + \frac{2}{n}\sum_{\alpha=1}^m \left[c_\alpha^{xy}\cos\left(\frac{2\pi}{n}b\alpha\right) + q_\alpha^{xy}\sin\left(\frac{2\pi}{n}b\alpha\right)\right]. \tag{8.6c}$$

In fact $c_\alpha^{xy} = \mathrm{Re}(\xi_\alpha^{xy})$ and $q_\alpha^{xy} = -\mathrm{Im}(\xi_\alpha^{xy})$, $\alpha = 0, 1, \ldots, m$, where ξ_α^{xy}, $\alpha = 0, \ldots, n$, are the eigenvalues of Γ^{xy}, in general complex, although they do satisfy the reality constraint $\bar{\xi}_\alpha^{xy} = \xi_{n-\alpha}^{xy}$.

The element γ_b^{xy} can be viewed as the bth entry in Γ^{xy} or as the xy entry in Γ_b, the lag b cross covariance matrix of the two sequences (x_t) and (y_t):

$$\Gamma_b = \begin{bmatrix} \gamma_b^{xx} & \gamma_b^{xy} \\ \gamma_b^{yx} & \gamma_b^{yy} \end{bmatrix}.$$

Grouping the c_α and q_α into matrices we may combine (8.6c) with the corresponding results for γ_b^{xx} and γ_b^{yy} to get

$$\Gamma_b = \frac{1}{n}C_0 + \frac{2}{n}\sum_\alpha \left[C_\alpha\cos\left(\frac{2\pi}{n}b\alpha\right) + Q_\alpha\sin\left(\frac{2\pi}{n}b\alpha\right)\right]. \tag{8.7}$$

This is the real spectral representation of Γ_b with $\{C_\alpha\}$ and $\{Q_\alpha\}$ being termed the cospectral and quadrature spectral matrices, respectively. The former are positive definite and the latter antisymmetric, as we will see in due course. Either (8.5) (together with the corresponding result for Γ^{xx} or Γ^{yy}) or (8.7) leads to the real spectral representation of a Γ having the form

$$\Gamma = A_e \otimes \Gamma_e + \sum_1^m \left[B_b \otimes \Gamma_b + B_{n-b} \otimes \Gamma_{n-b}\right], \tag{8.8}$$

which is

$$\Gamma = S_0 \otimes C_0 + \sum_1^m \left[S_\alpha \otimes C_\alpha + T_\alpha \otimes Q_\alpha\right]. \tag{8.9}$$

Now that we have the equivalent of the relations (4.2a) and (4.3a) for this class of covariance matrices, we can consider the corresponding decomposition of the elements w_t and the arrays w. The orthogonal decomposition of elements is just what one would expect, namely

$$\begin{bmatrix} x_t \\ y_t \end{bmatrix} = \sum_\alpha \begin{bmatrix} S_\alpha x_t \\ S_\alpha y_t \end{bmatrix}, \tag{8.10}$$

where $S_\alpha x_t = \Sigma_u S_\alpha(t, u) x_u$ are similar for $S_\alpha y_t$; cf. (2.11). The terms are of course orthogonal across strata and obey the following rules within strata:

$$\operatorname{cov}(S_\alpha x_t, S_\alpha y_t) = n^{-1} d_\alpha c_\alpha^{xy}, \qquad \operatorname{cov}(T_\alpha x_t, S_\alpha y_t) = n^{-1} d_\alpha q_\alpha^{xy}. \tag{8.11}$$

We can combine (8.11) with the corresponding results for x_t and y_t alone and obtain the formulas

$$D\begin{bmatrix} S_\alpha x_t \\ S_\alpha y_t \end{bmatrix} = n^{-1} d_\alpha C_\alpha, \qquad D\left[\begin{bmatrix} T_\alpha x_t \\ T_\alpha y_t \end{bmatrix}, \begin{bmatrix} S_\alpha x_t \\ S_\alpha y_t \end{bmatrix}\right] = n^{-1} d_\alpha Q_\alpha, \tag{8.12}$$

from which it is clear that C_α is positive definite; since $T_\alpha' = -T_\alpha$, $T_\alpha x_t$ is orthogonal to $S_\alpha x_t$ and so Q_α is antisymmetric.

The preceding discussion gives a good illustration of the extra difficulties encountered when nonsymmetric elements B_b appear in the class of basis matrices describing the cross covariances between different components of a vector element of a random array. How general can the class of $\{B_b\}$ of matrices be and still permit a satisfactory manova? Condition (i) of symmetry on our family of adjacency matrices can be modified—the matrices would then be described as the adjacency matrices of a homogeneous coherent configuration [Higman (1975, 1976)], but more is needed to give a reasonable theory. The appropriate conditions on a class $\{B_b: b \in Y\}$ of matrices over a set T with entries 0 and 1 only are the following:

(i) the transpose B_b' belongs to the class $\{B_b\}$, i.e., there exists $b^\vee$ such that $B_b' = B_{b^\vee}$;
(ii) $\Sigma_b B_b = J$, the matrix of 1's over T;
(iii) one of the matrices, B_e say, is the identity matrix over T;
(iv) $B_b B_c = \Sigma_d n_{bcd} B_d$ for suitable integers (n_{bcd});
(v) the symmetric elements of the algebra **B** of all linear combinations of the $\{B_b\}$ commute, i.e., $(B_b + B_b')(B_c + B_c') = (B_c + B_c')(B_b + B_b')$.

The last condition was introduced in a similar context by McLaren (1963).

Some of the B_b may already be symmetric: Let us list them first and write them as A_a; the remaining A-matrices are the symmetrized B-matrices $A_a = B_a + B_a'$, and we can list the remaining B-matrices in transpose pairs.

A dispersion model for an array $w = (w_t: t \in T)$ of random vectors which has the form

$$\Gamma = \sum_a A_a \otimes \Gamma_a + \sum_b [B_b \otimes \Gamma_b + B_{b^\vee} \otimes \Gamma_{b^\vee}], \tag{8.13}$$

where the first sum is over the symmetric relations and the second over the appropriate half of the nonsymmetric relations will have a manova decomposition provided that (v) is satisfied as well as (i), (ii), (iii) and (iv). The general spectral decomposition of such a Γ then takes the form

$$\Gamma = \sum_\alpha^{(1)} S_\alpha \otimes C_\alpha + \sum_\alpha^{(2)} [S_\alpha \otimes C_\alpha + T_\alpha \otimes Q_\alpha] \tag{8.14}$$
$$+ \sum_\alpha^{(3)} [S_\alpha \otimes C_\alpha + T_\alpha \otimes Q_\alpha + U_\alpha \otimes D_\alpha + V_\alpha \otimes E_\alpha],$$

where the sums $\Sigma^{(1)}$, $\Sigma^{(2)}$ and $\Sigma^{(3)}$ are over what we term the real, complex and quaternionic types of strata, respectively; $T_\alpha' = -T_\alpha$, $U_\alpha' = -U_\alpha$, $V_\alpha' = -V_\alpha$, $T_\alpha^2 = U_\alpha^2 = V_\alpha^2 = -S_\alpha$, $T_\alpha U_\alpha = V_\alpha$, $U_\alpha V_\alpha = T_\alpha$ and $V_\alpha T_\alpha = U_\alpha$. In the representation (8.14) the parameter matrices $\{C_\alpha\}$ are positive definite whilst $\{Q_\alpha\}$, $\{D_\alpha\}$ and $\{E_\alpha\}$ are all antisymmetric; cf. (8.12). There are further sets of structure matrices beyond $P = (p_{\alpha a})$ and $Q = (q_{a\alpha})$ which continue to relate the $\{S_\alpha\}$ and the $\{A_a\}$; where complex strata occur we need matrices $T = (t_{b\alpha})$ and $L = (l_{\alpha b})$ to pass from the $\{B_b\}$ to the $\{T_\alpha\}$ as we did in the cyclic example; and where quaternionic strata arise we also need two further pairs of mutually inverse structure matrices to permit the passage between the $\{B_b\}$ and the $\{U_\alpha\}$ and $\{V_\alpha\}$. The details are straightforward but lengthy and will not be given here; they will appear in Chapter 11 of Bailey, Praeger, Speed and Taylor (1987).

When the structure of the vector space $\mathbf{B}$ spanned by the $\{B_b\}$ is fully exhibited, the decompositions of w_t and (w_t) follow as before. We have the familiar expression

$$w_t = \sum_\alpha S_\alpha w_t, \tag{8.15}$$

where, as usual, $S_\alpha w_t = \Sigma_u S_\alpha(t, u) w_u$ (i.e., S_α effectively acts componentwise) and the terms in (8.15) are orthogonal across strata and satisfy relations similar to (8.12) within complex or quaternionic strata. For example, if α is quaternionic we have

$$D(S_\alpha w_t) = \frac{d_\alpha}{n} C_\alpha, \qquad D(T_\alpha w_t, S_\alpha w_t) = \frac{d_\alpha}{n} Q_\alpha,$$

$$D(U_\alpha w_t, S_\alpha w_t) = \frac{d_\alpha}{n} D_\alpha, \qquad D(V_\alpha w_t, S_\alpha w_t) = \frac{d_\alpha}{n} E_\alpha,$$

whereas $D(U_\alpha w_t, V_\alpha w_t)$ must be worked out from (8.14) using the formulas given after it. The anova in this context is simply

$$\Gamma_e = \sum_\alpha \Phi_\alpha, \tag{8.16}$$

where $\Phi_\alpha = n^{-1} d_\alpha C_\alpha$ is the (matrix) spectral component of variance of stratum $\alpha \in X$.

9. What is an anova? It must be abundantly clear by now that we regard anova as a property of certain special classes of dispersion models for arrays of random variables, or vectors, namely, for certain models defined by equality constraints amongst (co)variances. There should be an appropriate (real) spectral decomposition for all the dispersion matrices in the model, and a corresponding orthogonal decomposition for elements of the array. The components in these decompositions have interpretations which range from the notions of (random) main effects and interactions, in the classical anovas, through to harmonics at different wavelengths, wave numbers, etc., in the more classical harmonic analyses. For finite arrays there are also decompositions of sums of squares.

All of this is in marked contrast to the current use of the term in regression analysis and variance component analysis, where analysis of variance decompositions is more-or-less arbitrary orthogonal decomposition of sums of squares relating to "fixed" or "random" effects in assumed linear models. At this point it is worth explaining why our theory concerns only those structures described as "balanced" or "orthogonal." The reason is simple: Arrays with an anova as we use the term—one might add unique and complete—all have a high degree of symmetry, and in a sense the underlying index set is "complete." By comparison, the so-called "unbalanced" or "nonorthogonal" (random effects) anova models are in general rather messy subarrays of arrays with anova, and do not have an anova in their own right. For some further discussion of these points, see Speed (1985).

Although the vast majority of anova decompositions—of the matrices (or functions) and the random variables—are associated with a group action, and hence could be viewed as a part of a theory of generalised harmonic analysis, this line of thinking is by no means the best or the most general approach. For many arrays of random variables, including the standard multi-indexed ones of classical anova, the permutation groups are extremely complicated, whilst a direct combinatorial approach by-passing all representation theory is quite efficient; see Speed and Bailey (1982). Also in the reference just cited, an example of an association scheme which is *not* induced by a group action is given which shows that there are cases without an underlying group action.

Is there a single general theorem? It is hard to believe that one theorem will ever be formulated which covers all the examples mentioned so far. It would have to include all homogeneous coherent configurations satisfying condition (v) of Section 8, all limits of finite association schemes such as those illustrated in Section 5, the theory of Gel'fand pairs mentioned in Section 7, and much more. For example James (1982) has discussed the classical diallel cross in genetics from essentially our viewpoint; the triallel, double cross and other genetic structures give further interesting examples.

In closing we state what must be quite obvious to the reader: This paper has concentrated on the question, "What is an anova?" We have not discussed any of the many questions, which are both mathematically and statistically interesting, which arise when the array of random variables has an anova.

Acknowledgments. I am very pleased to be able to acknowledge the assistance given to me over the years by my collaborators Rosemary Bailey, Cheryl Praeger and Don Taylor. Much of the material outlined in this paper is or has arisen out of joint work with one or more of them, and without their skills and enthusiasm this paper could not have been written. I would also like to thank the Associate Editor for his very careful reading of the manuscript.

REFERENCES

AIGNER, M. (1979). *Combinatorial Theory.* Springer, New York.

ALBERT, A. (1976). When is a sum of squares an analysis of variance? *Ann. Statist.* 4 775–778.

ALDOUS, D. J. (1981). Representations for partially exchangeable arrays of random variables. *J. Multivariate Anal.* 11 581–598.

ANDERSON, T. W. (1969). Statistical inference for covariance matrices with linear structure. In *Proc. Second Internat. Symp. on Multivariate Anal.* (P. R. Krishnaiah, ed.) 55–66. Academic, New York.

ANDERSON, T. W. (1970). Estimation of covariance matrices which are linear combinations or whose inverses are linear combinations of given matrices. In *Essays in Probability and Statistics* (R. C. Bose et al., eds.) 1–24. Univ. of North Carolina Press, Chapel Hill, N.C.

ANDERSON, T. W. (1973). Asymptotically efficient estimation of covariance matrices with linear structure. *Ann. Statist.* 1 135–141.

BAILEY, R. A., PRAEGER, C. E., ROWLEY, C. A. and SPEED, T. P. (1983). Generalised wreath products of permutation groups. *Proc. London Math. Soc.* (3) **47** 69–82.

BAILEY, R. A., PRAEGER, C. E., SPEED, T. P. and TAYLOR D. E. (1987). *Analysis of Variance.* To appear.

BROWN, K. G. (1984). On analysis of variance in the mixed model. *Ann. Statist.* **12** 1488–1499.

CORNFIELD, J. and TUKEY, J. W. (1956). Average values of mean squares in factorials. *Ann. Math. Statist.* **27** 907–949.

CRAMÉR, H. (1940). On the theory of stationary random processes. *Ann. of Math.* **41** 215–230.

DANIELS, H. (1939). The estimation of components of variance. *J. Roy. Statist. Soc. (Suppl.)* **6** 186–197.

DIEUDONNÉ, J. (1978). *Treatise on Analysis* **6**. Academic, New York.

DIEUDONNÉ, J. (1980). *Special Functions and Linear Representations of Lie Groups* Amer. Math. Soc., Providence, R.I.

FAIRFIELD-SMITH, H. (1955). Variance components, finite populations and experimental inference. North Carolina Institute of Statistics Mimeo Series No. 135.

FISHER, R. A. (1925). *Statistical Methods for Research Workers.* Oliver and Boyd, Edinburgh.

FISHER, R. A. (1934). Discussion in Wishart, J. Statistics in agricultural research. *J. Roy. Statist. Soc. Suppl.* 1 26–61 (see especially page 52).

GAUTSCHI, W. (1959). Some remarks on Herbach's paper "Optimum nature of the *F*-test for model II in the balanced case." *Ann. Math. Statist.* **30** 960–963.

GRAYBILL, F. A. and HULTQUIST, R. A. (1961). Theorems concerning Eisenhart's model II. *Ann. Math. Statist.* **32** 261–269.

HANNAN, E. J. (1960). *Time Series Analysis.* Methuen, London.

HANNAN, E. J. (1961). Application of the representation theory of groups and algebras to some statistical problems, Research Reports, Part 1, Summer Research Institute, Australian Math. Soc.

HANNAN, E. J. (1965). Group representations and applied probability. *J. Appl. Probab.* **2** 1–68.

HANNAN, E. J. (1970). *Multiple Time Series.* Wiley, New York.

HARVILLE, D. A. (1984). A generalised version of Albert's theorem, with applications to the mixed linear model. In *Experimental Design Statistical Models and Genetic Statistics, Essays in Honor of Oscar Kempthorne* (K. Hinkelmann, ed.) 231–238. Dekker, New York.

HERBACH, L. H. (1959). Properties of model II—type analysis of variance tests, *A*: Optimum nature of the *F*-test for model II in the balanced case. *Ann. Math. Statist.* **30** 939–959.

HIGMAN, D. G. (1975). Coherent configuration I. *Geom. Dedicata* **4** 1–32.

HIGMAN, D. G. (1976). Coherent configurations II. *Geom. Dedicata* **5** 413–424.

HSU, P. L. (1941). Analysis of variance from the power function standpoint. *Biometrika* **32** 62–69.

HSU, P. L. (1945). On the power function of the E^2 test and the T^2 test. *Ann. Math. Statist.* **16** 278–286.

T. P. SPEED

JAMES, A. T. (1957). The relationship algebra of an experimental design. *Ann. Math. Statist.* **28** 993–1002.

JAMES, A. T. (1982). Analysis of variance determined by symmetry and combinatorial properties of zonal polynomials. In *Statistics and Probability: Essays in Honor of C. R. Rao* (G. Kallianpur, P. R. Krishnaiah and J. K. Ghosh, eds.) 329–341. North-Holland, Amsterdam.

JENSEN, S. T. (1975). Covariance hypotheses which are linear in both the covariance and the inverse covariance. Preprint No. 1, Institute of Mathematical Statistics, Univ. Copenhagen.

KEMPTHORNE, O. (1952). *The Design and Analysis of Experiments.* Wiley, New York.

KEMPTHORNE, O. and FOLKS, L. (1971). *Probability, Statistics and Data Analysis.* Iowa State Univ. Press, Ames, Iowa.

KHINCHIN, A. YA. (1934). Korrelationstheorie der Stationären stochastischen Prozesse. *Math. Ann.* **109** 604–615.

LETAC, G. (1981). Problèmes classiques de probabilité sur un couple de Gel'fand. *Analytical Methods in Probability Theory. Lecture Notes in Math.* **861** 93–120. Springer, Berlin.

LETAC, G. (1982). Les fonctions sphériques d'un couple de Gel'fand symmétrique et les chains de Markov. *Adv. in Appl. Probab.* **14** 272–294.

MACWILLIAMS, F. J. and SLOANE, N. J. A. (1077). *The Theory of Error-Correcting Codes.* North-Holland, Amsterdam.

MCLAREN, A. D. (1963). On group representations and invariant stochastic processes. *Proc. Camb. Philos. Soc.* **59** 431–450.

NELDER, J. A. (1954). The interpretation of negative components of variance. *Biometrika* **41** 544–548.

NELDER, J. A. (1965). The analysis of randomised experiments with orthogonal block structure. I. Block structure and the null analysis of variance. *Proc. Roy. Soc. London Ser. A* **283** 147–162.

NELDER, J. A. (1977). A reformulation of linear models (with discussion). *J. Roy. Statist. Soc. Ser. A* **140** 48–76.

NEYMAN, J., IWASKIEWICZ, K. and KOLODZIEJCZYK, ST. (1935). Statistical problems in agricultural experimentation (with discussion). *J. Roy. Statist. Soc. Suppl.* **2** 107–189.

SCHEFFÉ, H. (1959). *The Analysis of Variance.* Wiley, New York.

SPEED, T. P. (1985). Dispersion models for factorial experiments. *Bull. Internat. Statist. Inst.* **51** 24.1-1–24.1-16.

SPEED, T. P. (1986). Anova models with random effects: An approach via symmetry. In *Essays in Time Series and Allied Processes: Papers in Honor of E. J. Hannan* (J. Gani and M. B. Priestley, eds.) 355–368. Applied Probability Trust, Sheffield.

SPEED, T. P. and BAILEY, R. A. (1982). On a class of association schemes derived from lattices of equivalence relations. In *Algebraic Structures and Applications* (P. Schultz, C. E. Praeger and R. P. Sullivan, eds.) 55–74. Dekker, New York.

TANG, P. C. (1938). The power function of the analysis of variance tests with tables and illustrations of their use. *Statist. Res. Mem.* **2** 126–149.

THROCKMORTON, T. N. (1961). Structures of classification data. Ph.D. thesis, Iowa State Univ.

TUKEY, J. W. (1961). Discussion, emphasising the connection between analysis of variance and spectrum analysis. *Technometrics* **3** 191–219. (Follows papers by G. M. Jenkins and E. Parzen on spectrum analysis.)

WALD, A. (1942). On the power function of the analysis of variance test. *Ann. Math. Statist.* **13** 434–439.

WILK, M. B. and KEMPTHORNE, O. (1956). Some aspects of the analysis of factorial experiments in a completely randomized design. *Ann. Math. Statist.* **27** 950–985.

WOLFOWITZ, J. (1949). The power of the classical tests associated with the normal distribution. *Ann. Math. Statist.* **20** 540–551.

DEPARTMENT OF STATISTICS
UNIVERSITY OF CALIFORNIA
BERKELEY, CALIFORNIA 94720

International Statistical Review (1987), **55**, 3, pp. 261–277. Printed in Great Britain

Factorial Dispersion Models

T.P. Speed[1] and R.A. Bailey[2]

[1]*CSIRO Division of Mathematics and Statistics, Canberra, Australia ACT 2601.* [2]*Statistics Department, Rothamsted Experimental Station, Harpenden, Herts., AL5 2JQ, UK.*

Summary

A class of dispersion models for multi-indexed arrays of random variables is introduced and discussed. These models generalize the second-order properties of variance component, randomization and exchangeability models, and lead naturally to general techniques for calculating the orthogonal decompositions, expected mean squares and other aspects of the analysis of variance of such arrays.

Key words: Analysis of variance; Association scheme; Canonical component of variance; Exchangeability; Linear model; Permutation model; Randomization; Sample; Symmetry; Variance components.

1 Introduction

The analysis of variance of multi-indexed arrays, i.e. data from factorial experiments, interpreting this expression in the widest possible sense, had its origins in the quantitative genetic research of R.A. Fisher. By the time of the publication of Fisher (1925) these ideas had also been applied to comparative experiments in agriculture and in the following 15 years the range of applications was broadened to include sampling (Yates & Zacopanay, 1935; Youden & Mehlich 1937; Cochran, 1939) and industrial statistics (Daniels, 1938, 1939). Over the same period the models and assumptions underlying the analysis of variance were closely scrutinized: see especially Eden & Yates (1933), who examined the z-test using nonnormal data, and the later work of Pitman (1938) and Welch (1937) on the same topic, and the critical study by Neyman et al. (1935) of the use of Fisher's methods in agricultural experiments. Somewhat different problems were being tackled within a similar framework in psychometrics (Spearman, 1910; Brown, 1913) and animal breeding (Lush, 1931; Lush et al., 1933). In both of these fields there were measurements with two components of error; in modern terms they were concerned with the estimation in the presence of random effects, a topic whose origins can be found in nineteenth century astronomy: see Scheffé (1956) for further details.

Many modern writers on what has come to be called *variance component* analysis take as their starting point a *linear model* for their data array built up from independent sets of independent random effects, with one set of effects for each appropriate index or set of indices: some of these effects are termed *main effects,* the remainder *interactions*. Such effects induce a variety of covariances between elements of the array, although it is not common to regard the estimation of these covariances as an issue of particular statistical interest. This linear model approach is not appropriate if the underlying distribution of the array is a permutation distribution, a viewpoint adopted by a number of writers from Fisher onwards, including Kempthorne (1952) and Nelder (1965), who have chosen to emphasize the *randomization* aspects of analysis of variance. Nor is it appropriate if the effects or indeed the whole array are to be regarded as *randomly sampled* without replacement from one or more finite populations, an approach also adopted by Fairfield

Smith (1955), Cornfield & Tukey (1956) and Hooke (1956a,b). Finally, we mention that a statistician may wish to assume nothing more than a certain amount of symmetry, invariance or stationarity, such as the generalization to multi-indexed arrays of de Finetti's exchangeability: see, for example, Dawid (1977) and Aldous (1981). Do these symmetry assumptions still permit us to carry out the usual analysis of variance calculations in a meaningful way?

All of the different views or models mentioned above lead to the same structure for the covariance matrix of the data, although the form and interpretation of the parameters and the problems attacked naturally differ between models. Our approach is therefore based upon this common *dispersion model.* We shall see that all of the second-order calculations associated with analysis of variance can be derived straightforwardly from the relevant aspects of the dispersion model. Here we are taking the point of view of Cox (1960) and Speed (1987) that the term 'analysis of variance' means the decomposition of the common variance of several random variables into variance components which are of intrinsic interest, rather than the calculations required to analyse data from a so-called 'fixed effects model' with a single variance which is a nuisance parameter. Of course, many of the calculations are the same for both cases, because they are merely the mathematical decompositions of quadratic forms or geometric projections, as Bryant (1984) and Saville & Wood (1986) have recently pointed out. However, the underlying philosophy is quite different.

The work reported in this paper had as its starting point the paper by Nelder (1965), which concerns the second-order properties of the class of multi-indexed arrays which can be built up by successively nesting and crossing simpler ones, starting from a single unstructured factor. Although ostensibly set within a randomization framework, Nelder's results have a broader applicability, and § 2 below refines them somewhat and extends them to random arrays with more general (not necessarily permutation) distributions, to a wider class of index sets, and to the case where the number of values, or levels, of the indices, or factors, may be countably infinite. Nelder's (1965) work was primarily motivated by the need to systematize analysis of variance techniques so that a general computer program could be written to replace large collections of subroutines, each appropriate for a particular 'design'. An independent stream of work, initiated by Kempthorne (1952) and continued throughout the 1950's and early 1960's, see for example Wilk (1955), Wilk & Kempthorne (1956a,b; 1957), was concerned with the objective development and interpretation of linear models for randomized experiments. This body of research, from what we shall call the Iowa (State University) school, includes a number of valuable techniques for calculating the averages of certain quadratic forms over random sampling and randomly selecting designs. At around the same time Cornfield & Tukey (1956) reported on work done by them some years earlier addressing essentially the same problem: the calculation of expected mean squares in analysis of variance tables, or, as they term it, 'average values of mean squares in factorials'. One of our aims was to derive the main results of these authors within the modified Nelder framework outlined above.

Much of the paper is devoted to the broad problem of relating the characteristics of a subset of a multi-indexed array, which we can think of as a *sample,* to those of the full array, thought of as the *population.* The results of Cornfield & Tukey (1956) and the Iowa school do come out naturally, as do some less familiar ones concerning the prediction of unobserved from observed random variables, a topic usually referred to in this context as the estimation of random effects; see, for example, Harville (1976).

We hope that our methods, which attempt to treat finite and infinite populations, the different models or approaches noted above, as well as various kinds of samples, in a uniform manner, will lead to:

(i) an understanding of the different parameterizations of factorial dispersion models, together with their interpretations;
(ii) the various orthogonal decompositions of random variables, of arrays of numbers, and of sums of squares, including the associated numbers of degrees of freedom;
(iii) techniques for calculating expected values of mean squares under a range of assumptions including linear models, over randomization, random sampling, and symmetry, obtaining answers in terms of the desired parameters;
(iv) procedures for getting 'best' estimates of all parameters;
(v) formulae for obtaining 'best' linear predictors of key unobserved random variables.

Throughout the paper we illustrate our results with a triply indexed array $y = (y_{ijk})$ and its associated dispersion matrix $\Gamma = Dy$, assuming that the second index, j, is nested within the first, i, and that these two are crossed with the third, k. As well as being complicated enough to exhibit most of the possibilities, this example allows the reader to specialize the results to a simple nesting, by suppressing k, and to a simple crossing, by suppressing j.

A good deal of the new work reported here is closely related to joint work with C.E. Praeger and D.E. Taylor which we hope will appear soon in a monograph entitled *Analysis of Variance.* We should like to thank them both for their collaboration over the years.

2 Factorial dispersion models

2.1 Preliminaries

We will suppose given a set Π of factors $p, q, \ldots$ and a partial order $\leqslant$ on Π, where $q \leqslant p$ means that the factor q is *nested* in the factor p, in the sense of Nelder (1965). It is helpful to draw the partially ordered set $(\Pi; \leqslant)$, which we term the *nesting poset,* with p above q if $q \leqslant p$ and connected to q if there is no r distinct from p and q for which $q \leqslant r$ and $r \leqslant p$; this is the so-called Hasse diagram; see Fig. 1. We refer to Aigner (1979) for terminology and further details concerning ordered sets. A subset $a \subseteq \Pi$ is said to be a *filter* if $p \in a$ whenever both $q \leqslant p$ and $q \in a$; such subsets have also been termed *admissible* by the Iowa school, but we shall adhere to (one) standard order-theoretic terminology. The need for such subsets arises because, in referring to the levels of a factor, it is frequently necessary to refer at the same time to all factors within which that factor is nested. The class $L(\Pi)$ of all filters of a poset $(\Pi; \leqslant)$ is readily found to be a *distributive lattice* under the operations of set union and set intersection, containing the empty set $\varnothing$ and the whole set Π (Aigner, 1979, p. 33). The lattice $L(\Pi)$ is also a poset under set-inclusion and so we can draw its Hasse diagram as well. It is convenient for our purposes to draw the subset lattice diagram 'upside down', using the reverse ordering from set inclusion. Figures 1 and 2 depict an *example* of a simple poset of three factors and its associated distributive lattice of filters; note that $2 < 1$ (meaning $2 \leqslant 1$ and $2 \neq 1$) is the sole nontrivial nesting relationship. This example will be used to illustrate much of what follows. Note that our use of Hasse diagrams in this context is quite different from that of Throckmorton (1961).

In a completely general factorial model, if factor q is nested in factor p then there is no need for q to have the same number of levels within each level of p. However, models

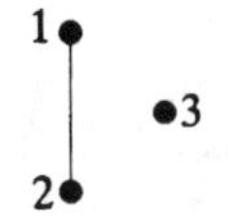

Figure 1. *The poset Π in the example.*

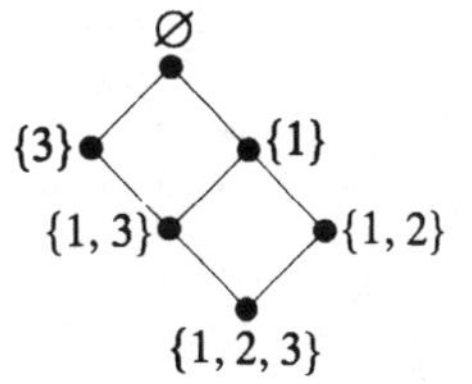

Figure 2. *The lattice L(Π) in the example.*

based on assumptions of exchangeability or randomization do imply that q has the same number of levels with each level of p, and § 4 makes it clear that models based on random sampling also imply this balance condition if all the random variables are to have the same variance. Of all the viewpoints discussed in § 1, only the linear model approach permits q to have different numbers of levels within different levels of p. Since this paper is concerned with the theory that is common to all the approaches in § 1, we may assume that q has the same number of levels within each level of p; this paper has nothing to say about so-called 'unbalanced' data.

Next we suppose that the set of *levels* of factor p is T_p for p in Π. Of course, if q is nested in p then the levels of q at different levels of p bear no relation to one another at all even if the number of levels is constant. Nevertheless, it is extremely common to use the same set T_q to denote the levels of q within each level of p; see, for example, John (1971) or Kempthorne (1952). Although this might appear somewhat confusing, there are two good reasons for this convention: it facilitates both the algorithm for analysis of variance calculations (Nelder, 1965) and some of the formal mathematics (Bailey et al.,1983). We write $T = \prod_p T_p$ for the set of all *combinations* of levels of factors in Π, denoting a typical element of T by $t = (t_p : p \in \Pi)$. For any pair t, u in T we write $a(t, u)$ for the largest filter a in $L(\Pi)$ such that $t_p = u_p$ for all p in a.

Example. With Π as in Fig. 1, let $t = ijk$ and $u = i'j'k'$. Then $a(t, u) = \{1, 2\}$ if $i = i'$, $j = j'$ and $k \neq k'$; whilst $a(t, u) = \{3\}$ if $i \neq i'$, $j = j'$ and $k = k'$, since $\{2, 3\}$ is not a filter of Π in this case. Here and below, when discussing our example, it is convenient to write (t_1, t_2, t_3) as (i, j, k) and abbreviate this to ijk.

With these preliminaries we can now define the *dispersion models* of our title. They are for arrays $y = (y_t : t \in T)$ of real random variables indexed by the set T of combinations of levels of a set Π of factors whose nesting relationships are described by the partially ordered set $(\Pi; \leqslant)$. The covariance matrix Dy is defined over T and is said to be *factorial* if $\operatorname{cov}(y_t, y_u) = \operatorname{cov}(y_v, y_w)$ whenever $a(t, u) = a(v, w)$ for t, u, v, w in T, and the class of all such covariance matrices is denoted by $V(\Pi, T)$; briefly, a covariance matrix is factorial if the covariance between any two elements y_t and y_u depends only on the (largest) subset (filter) of the factors corresponding to the components on which the indices t and u agree. This class is more general than that introduced by Nelder (1965), and we note that it has not yet been necessary to state whether or not the sets T_p are *finite* for p in Π. See Tjur (1984) for a discussion of an even wider class of models, and Bailey (1984) for a discussion of the relationship between Tjur's work and the present paper: factorial dispersion models correspond to Bailey's *poset block structures*.

2.2 *Finite index sets: Algebraic theory*

If $|T_p| = n_p < \infty$ for p in Π, and we write $n = \prod_p n_p$, then $V(\Pi, T)$ is a class of $n \times n$ positive-definite matrices whose structure is readily exhibited: see Speed & Bailey (1982)

Factorial Dispersion Models

for fuller details and proofs. Firstly, we can define the *association matrices* $\{A_a : a \in L(\Pi)\}$ over T by writing $A_a(t, u) = 1$ if $a(t, u) = a$, and $A_a(t, u) = 0$ otherwise. A general element Γ of $V(\Pi, T)$ thus has the form $\Gamma = \sum_a \gamma_a A_a$, the sum being over $L(\Pi)$, with the parameters $\{\gamma_a : a \in L(\Pi)\}$ being *covariances.*

For reference in § 4, we note that the set of matrices $\{A_a : a \in L(\Pi)\}$ forms an *association scheme*. This means that the following conditions are satisfied:

(i) for all a in $L(\Pi)$, every entry in A_a is equal to 0 or 1, but A_a is not the zero matrix;
(ii) for all a in $L(\Pi)$, the matrix A_a is symmetric;
(iii) the sum $\sum_{a \in L(\Pi)} A_a$ is the matrix J with every entry equal to 1;
(iv) one of the matrices (in this case A_Π) is equal to the identity matrix I;
(v) there are integers n_{abc} for a, b, c in $L(\Pi)$ such that, for all a, b in $L(\Pi)$,

$$A_a A_b = \sum_{c \in L(\Pi)} n_{abc} A_c.$$

See, for example, MacWilliams & Sloane (1977, Ch. 21) for a good discussion of the general theory of association schemes.

A second, useful representation of elements of $V(\Pi, T)$ involves the *relationship matrices* $\{R_b : b \in L(\Pi)\}$, where $R_b(t, u) = 1$ if $t_p = u_p$ for all p in b, and $R_b(t, u) = 0$ otherwise. Clearly

$$R_b = \sum_{a \supseteq b} A_a$$

and the representation we refer to is $\Gamma = \sum_b f_b R_b$, where the parameters $\{f_b : b \in L(\Pi)\}$ are called *canonical components of variance* by Fairfield Smith (1955), *Σ-quantities* by Wilk & Kempthorne (1956a); and simply *f-quantities* by Nelder (1965), although later he called them *components of excess variance* (Nelder, 1977). Relating the f's to the γ's requires the *zeta* function of the lattice $L(\Pi)$ given by $\zeta(a, b) = 1$ if $a \subseteq b$ and $\zeta(a, b) = 0$ otherwise, and the associated *Möbius function* μ defined by

$$\sum_b \zeta(a, b)\mu(b, c) = \sum_b \mu(a, b)\zeta(b, c) = \delta(a, c),$$

where $\delta(a, c) = 1$ if $a = c$, and $\delta(a, c) = 0$ otherwise; the sums are over all b in $L(\Pi)$: see Aigner (1979, p. 141) for further details. In this notation $f_b = \sum_a \mu(a, b)\gamma_a$.

Because it plays such a prominent role in the discussion which follows, we explain briefly how the Möbius function μ of a lattice L is calculated from a Hasse diagram. An easy reformulation of the definition is the following: $\mu(a, c) = 0$ unless $a \subseteq c$; $\mu(a, a) = 1$; and for $a \subset c$:

$$\mu(a, c) = -\sum_{a \subseteq b \subset c} \mu(a, b);$$

equivalently,

$$\mu(a, c) = -\sum_{a \subset b \subseteq c} \mu(b, c).$$

Example (cont.). Let us calculate some values of μ. From $\mu(\emptyset, \emptyset) = \mu(\{1\}, \{1\}) = 1$ and either of the above, we deduce that $\mu(\emptyset, \{1\}) = -1$. A similar argument applies to any pair connected by an edge in the Hasse diagram. Turning to $\mu(\emptyset, \{1, 3\})$ we can use the first of the above formulae to find that

$$\mu(\emptyset, \{1, 3\}) = -[\mu(\emptyset, \emptyset) + \mu(\emptyset, \{1\}) + \mu(\emptyset, \{3\})] = +1,$$

Table 1

The matrices of the zeta and Möbius functions for the example.

(a) Z

	$\varnothing$	{1}	{1, 2}	{3}	{1, 3}	{1, 2, 3}
$\varnothing$	1	1	1	1	1	1
{1}	0	1	1	0	1	1
{1, 2}	0	0	1	0	0	1
{3}	0	0	0	1	1	1
{1, 3}	0	0	0	0	1	1
{1, 2, 3}	0	0	0	0	0	1

(b) Z^{-1}

	$\varnothing$	{1}	{1, 2}	{3}	{1, 3}	{1, 2, 3}
$\varnothing$	1	−1	0	−1	1	0
{1}	0	1	−1	0	−1	1
{1, 2}	0	0	1	0	0	−1
{3}	0	0	0	1	−1	0
{1, 3}	0	0	0	0	1	−1
{1, 2, 3}	0	0	0	0	0	1

whilst similar reasoning shows that

$$\mu(\varnothing, \{1, 2\}) = -[\mu(\varnothing, \varnothing) + \mu(\varnothing, \{1\})] = 0.$$

Alternatively, the Möbius function may be calculated by matrix inversion. Let Z be the $L(\Pi) \times L(\Pi)$ matrix with entries $\zeta(a, b)$: Table 1(a) shows Z for our example. Since Z is upper triangular (if the elements of $L(\Pi)$ are written in a suitable order), it is easily inverted. The values of μ are simply the entries of Z^{-1}. For example, Table 1(b) shows that $\mu(\{1\}, \{1, 2, 3\}) = 1$.

It can be shown that, for all lattices of the form $L(\Pi)$ that we are considering, μ takes only the values 0, +1 or −1. A concise formula for the values of μ can be given but we do not need it here.

The final representation of elements of $V(\Pi, T)$ is the explicit form of their common spectral decomposition. If we write $\bar{n}_a = \Pi(n_p : p \notin a)$ for a in $L(\Pi)$, then the formula

$$S_\alpha = \sum_a \mu(a, \alpha)\bar{n}_a^{-1}R_a, \quad \alpha \in L(\Pi),$$

gives a set of pairwise orthogonal, that is $S_\alpha S_\beta = 0 = S_\beta S_\alpha$ if $\alpha \neq \beta$, idempotent ($S_\alpha^2 = S_\alpha$) symmetric matrices which sum to the identity ($I = \Sigma_\alpha S_\alpha$). Further, the formula

$$\xi_\alpha = \sum_b \zeta(\alpha, b)\bar{n}_b f_b, \quad \alpha \in L(\Pi),$$

gives the eigenvalues of Γ, whose spectral decomposition is then $\Gamma = \Sigma_\alpha \xi_\alpha S_\alpha$. Thus the *eigenvalues* $\{\xi_\alpha : \alpha \in L(\Pi)\}$ constitute a third set of parameters for $V(\Pi, T)$, and there are two related sets of parameters which have also been used: the *specific components of variance* $\{\sigma_\alpha^2 : \alpha \in L(\Pi)\}$ of Cornfield & Tukey (1956), where $\sigma_\alpha^2 = \bar{n}_\alpha^{-1}\xi_\alpha$, and the *spectral components of variance* $\{\phi_\alpha : \alpha \in L(\Pi)\}$, of Daniels (1939), where $\phi_\alpha = n^{-1} d_\alpha \xi_\alpha$ and $d_\alpha = \text{rank}(S_\alpha)$. Table 2 summarizes the main representations of a factorial covariance matrix.

The nonnegativity of the eigenvalues $\{\xi_\alpha : \alpha \in L(\Pi)\}$ or, equivalently, $\{\sigma_\alpha^2 : \alpha \in L(\Pi)\}$ or $\{\phi_\alpha : \alpha \in L(\Pi)\}$, succintly defines the parameter space for $V(\Pi, T)$. There is no such simple characterization in terms of the covariances $\{\gamma_a : a \in L(\Pi)\}$, nor, in general, of the $\{f_b : b \in L(\Pi)\}$. In the linear model approach the latter parameters are the variances of

Table 2

Representations of a factorial covariance matrix

	$\sum_{a \in L(\Pi)} \gamma_a A_a$	$\sum_{b \in L(\Pi)} f_b R_b$	$\sum_{\alpha \in L(\Pi)} \xi_\alpha S_\alpha$
Using	Association matrices	Relationship matrices	Orthogonal projectors

independent sets of random variables, and so are constrained to be nonnegative. This constraint is stronger than the nonnegative-definiteness of Γ, and so the linear model approach is, in general, a proper subset of $V(\Pi, T)$. We shall show in § 2.4 that the classes coincide when T_p is infinite for all p in Π.

It is convenient to combine the relationships between γ's and f's and f's and ξ's to give

$$\xi_\alpha = \sum_a p_{\alpha a}\gamma_a, \quad \gamma_a = n^{-1}\sum_\alpha q_{a\alpha}\xi_\alpha, \tag{1}$$

where a, $\alpha \in L(\Pi)$ and the sums are over $L(\Pi)$. It is also true that

$$S_\alpha = n^{-1}\sum_a q_{a\alpha}A_a, \quad A_a = \sum_\alpha p_{\alpha a}S_\alpha \tag{2}$$

and the matrices $P = (p_{\alpha a})$ and $Q = (q_{a\alpha})$ thus hold the key to the solution of many later problems.

The $\varnothing$-row of the matrix P consists of the elements $p_{\varnothing a} = k_a = |\{u : a(t, u) = a\}|$, independent of t, whilst the Π-row of Q consists of $q_{\Pi\alpha} = d_\alpha = \text{rank}\,(S_\alpha)$, the number of so-called *degrees of freedom* in the *stratum* α: see below for an explanation of this terminology.

It can be shown that $k_a q_{a\alpha} = d_\alpha p_{\alpha a}$ and the simplest way to describe the entries of P and Q is via formulae for d_α, k_a and the common value $s_\alpha(a)$ of $d_\alpha^{-1}q_{a\alpha} = k_a^{-1}p_{\alpha a}$. These are as follows:

$$d_\alpha = \prod_{p\in\alpha\backslash\alpha_m} n_p \times \prod_{p\in\alpha_m} (n_p - 1),$$

where α_m denotes the set of *minimal* elements of α;

$$k_a = \prod_{p\in\bar{a}\backslash\bar{a}^m} n_p \times \prod_{p\in\bar{a}^m} (n_p - 1),$$

where $\bar{a} = \Pi\backslash a$ and $\bar{a}^m$ denotes the set of *maximal* elements of $\bar{a}$; and

$$s_\alpha(a) = \begin{cases} \prod_{p\in\alpha_m\backslash a} \{-1/(n_p - 1)\} & \text{if } \alpha\backslash\alpha_m \subseteq a, \\ 0 & \text{otherwise,} \end{cases}$$

where an empty product is defined to be unity.

For our *example* given in Fig. 1 and 2 the values of d_a, k_a and $s_\alpha(a)$ are shown in Table 3.

2.3 *Finite index sets: Decomposition of arrays*

The preceding approach permits a full discussion of the structure of matrices in the class $V(\Pi, T)$. We now turn to the random array $y = (y_t : t \in T)$ with dispersion matrix Γ in $V(\Pi, T)$. The matrices $\{S_\alpha : \alpha \in L(\Pi)\}$ are pairwise orthogonal projectors summing to the identity and so define an orthogonal decomposition of the n-dimensional space $\mathbb{R}^T$ of T-indexed arrays of real numbers, and hence also of the space of random arrays taking values in $\mathbb{R}^T$. Thus the decomposition

$$y = \sum_{\alpha\in L(\Pi)} S_\alpha y \tag{3}$$

of the array y into component arrays $S_\alpha y$ is orthogonal with respect to the standard inner product $\langle x, y\rangle = \sum_t x_t y_t$. Therefore we have the sum of squares decomposition $|y|^2 = \sum_\alpha |S_\alpha y|^2$, where $|y|^2 = \langle y, y\rangle = \sum_t y_t^2$.

T.P. Speed and R.A. Bailey

Table 3

(a) Values of d and k in the Example

a	d_a	k_a
$\varnothing$	1	$(n_1-1)n_2(n_3-1)$
$\{1\}$	n_1-1	$(n_2-1)(n_3-1)$
$\{1,2\}$	$n_1(n_2-1)$	(n_3-1)
$\{3\}$	n_3-1	$(n_1-1)n_2$
$\{1,3\}$	$(n_1-1)(n_3-1)$	n_2-1
$\{1,2,3\}$	$n_1(n_2-1)(n_3-1)$	1

(b) Values of $s_\alpha(a)$ in the Example

α	$a=\varnothing$	$a=\{1\}$	$a=\{1,2\}$	$a=\{3\}$	$a=\{1,3\}$	$a=\{1,2,3\}$
$\varnothing$	1	1	1	1	1	1
$\{1\}$	$\frac{-1}{n_1-1}$	1	1	$\frac{-1}{n_1-1}$	1	1
$\{1,2\}$	0	$\frac{-1}{n_2-1}$	1	0	$\frac{-1}{n_2-1}$	1
$\{3\}$	$\frac{-1}{n_3-1}$	$\frac{-1}{n_3-1}$	$\frac{-1}{n_3-1}$	1	1	1
$\{1,3\}$	$\frac{1}{(n_1-1)(n_3-1)}$	$\frac{-1}{n_3-1}$	$\frac{-1}{n_3-1}$	$\frac{-1}{n_1-1}$	1	1
$\{1,2,3\}$	0	$\frac{1}{(n_2-1)(n_3-1)}$	$\frac{-1}{n_3-1}$	0	$\frac{-1}{n_2-1}$	1

By taking components of equation (3) and writing $S_\alpha y_t$ for $(S_\alpha y)_t$ we obtain the decomposition

$$y_t = \sum_{\alpha \in L(\Pi)} S_\alpha y_t \tag{4}$$

of the element y_t, for t in T, into components $S_\alpha y_t$ which depend only on indices in α. This is the *population identity* of Kempthorne (1952, Ch. 8), his arrays having permutation or sampling distributions, which we discuss below; it is also called the *yield identity* by Nelder (1965).

Example (cont.). This decomposition is the familiar one:

$$y_{ijk} = y_{...} + (y_{i..} - y_{...}) + (y_{ij.} - y_{i..}) + (y_{..k} - y_{...}) + (y_{i.k} - y_{i..} - y_{..k} + y_{...}) \\ + (y_{ijk} - y_{ij.} - y_{i.k} + y_{i..}), \tag{4a}$$

where the terms correspond to $\alpha = \varnothing$, $\{1\}$, $\{1,2\}$, $\{3\}$, $\{1,3\}$ and $\{1,2,3\}$ respectively, and we denote the *averaging* over an index by a dot in that position.

Factorial Dispersion Models

For simplicity, suppose that $Ey=0$. Since $\Gamma=\sum_\alpha \xi_\alpha S_\alpha$, the components of (3) are uncorrelated, as are the components of (4); thus the $S_\alpha y_t$ are the *principal components* of y. We find that $E(S_\alpha y_t)^2=\phi_\alpha$ for α in $L(\Pi)$ and so $\operatorname{var}(y_t)=\sum_\alpha \phi_\alpha$. Moreover, since $\phi_\alpha=n^{-1}d_\alpha\xi_\alpha$, we have

$$E\{d_\alpha^{-1}|S_\alpha y|^2\}=\xi_\alpha. \tag{5}$$

Example (cont.) As an illustration we consider $\alpha=\{1,2,3\}$. From Table 3 we have $d_{\{1,2,3\}}=n_1(n_2-1)(n_3-1)$ whilst (4a) gives $S_{\{1,2,3\}}y=(y_{ijk}-y_{ij.}-y_{i.k}+y_{i..})$; hence for any array $y=(y_{ijk})$ with zero mean and dispersion matrix in $V(\Pi, T)$ we have

$$E\left\{\frac{1}{n_1(n_2-1)(n_3-1)}\sum_i\sum_j\sum_k(y_{ijk}-y_{ij.}-y_{i.k}+y_{i..})^2\right\}=\xi_{\{1,2,3\}}.$$

From (1) and the values of k_a and $s_\alpha(a)$ in Table 3 we find that

$$\xi_{\{1,2,3\}}=\gamma_{\{1,2,3\}}-\gamma_{\{1,3\}}-\gamma_{\{1,2\}}+\gamma_{\{1\}}.$$

It should now be apparent from our Example, if not the general discussion, that we are providing an alternative interpretation of the *analysis of variance* of a multi-indexed array $y=(y_t)$. We have indicated how the components $S_\alpha y_t$ of (4) are the principal components of the random array y, provided that $Dy\in V(\Pi, T)$, and seen that their variances ϕ_α are, when suitably normalized by their multiplicities d_α and the array size n, the eigenvalues ξ_α of Dy. Because of the double role of the projectors S_α these components are also the terms which, when squared and summed, give the *sums of squares* decompositions that are such a familiar feature of analysis of variance *tables*. The *lines* of the analysis of variance table, termed *strata* by analogy with stratified sampling, are labelled by the filters of Π, that is by the elements α of the lattice $L(\Pi)$, and the number of *degrees of freedom* for the line labelled α is d_α, coinciding with the multiplicity of the corresponding eigenvalue ξ_α. And, finally, the *expected mean square* in line α given by (5) is the link between the principal components and the sum of squares decompositions. All this has been done by assuming only that $Dy\in V(\Pi, T)$; we have not *assumed* any linear model for the array y, although (4) is in a sense an *implicit* linear model. Note that we have assumed throughout that $Ey=0$ and so our discussion is truly an analysis of *variance* qua variance; the introduction of structured *mean values* is an additional complication which we do not discuss here. All of this seems very similar to the discussion by Hannan (1965, § 5.2) and indeed the connection with spectral analysis can be made complete.

2.4 *Infinite index sets*

Most of the foregoing extends to the situation where some or all of the factors have countably infinitely many levels, and for simplicity we discuss the case that T_p is countably infinite for *all* p in Π. The representations $\Gamma=\sum_a\gamma_a A_a=\sum_b f_b R_b$ continue to hold (as these matrices are never multiplied), and we find that the spectral representation $\Gamma^{(\mathbf{n})}=\sum_\alpha \xi_\alpha^{(\mathbf{n})}S_\alpha^{(\mathbf{n})}$ of the truncation $\Gamma^{(\mathbf{n})}$ of Γ, converges, after a suitable normalization, to the representation $\sum_\alpha f_\alpha R_\alpha$. Here we use the superscript $(\mathbf{n})$ to denote the truncation to $t_p\leqslant n_p$ for all p in Π and our limits are all as $n_p\to\infty$ for all p in Π. Indeed $\phi_\alpha^{(\mathbf{n})}=n^{-1}d_\alpha^{(\mathbf{n})}\xi_\alpha^{(\mathbf{n})}$ converges to f_α, as does $\bar{n}_\alpha^{-1}\xi_\alpha^{(\mathbf{n})}=(\sigma_\alpha^{(\mathbf{n})})^2$, and we find that the components $S_\alpha^{(\mathbf{n})}y_t$ of y_t, where $t_p\leqslant n_p$ for all p in Π, also converge in mean square. The decomposition (4) of y_t continues to hold in the limit with the additional property that $E\{(S_\alpha y_t)(S_\alpha y_u)\}=0$ if $(t_p:p\in\alpha)\neq(u_p:p\in\alpha)$, and so in this case (4) is (to second-order) just the often *assumed linear model* with random effects used in variance component analysis with the set Π of factors exhibiting the nesting structure characterized by $(\Pi;\leqslant)$.

Example (cont.). The above implies that the linear model

$$y_{ijk} = \mu + \alpha_i + \beta_{ij} + \gamma_k + \delta_{ik} + \varepsilon_{ijk},$$

where μ, $\{\alpha_i\}$, $\{\beta_{ij}\}$, $\{\gamma_k\}$, $\{\delta_{ik}\}$ and $\{\varepsilon_{ijk}\}$ are uncorrelated sets of uncorrelated effects with zero means and variances $\sigma^2_{\varnothing}$, $\sigma^2_{\{1\}}$, $\sigma^2_{\{1,2\}}$, $\sigma^2_{\{3\}}$, $\sigma^2_{\{1,3\}}$ and $\sigma^2_{\{1,2,3\}}$ respectively is, to second order, the most general array y with $Ey = 0$, $Dy \in V(\Pi, T)$ with Π as in Fig. 1 and $T_1 = T_2 = T_3 = \{1, 2, \dots\}$.

3 Permutation distributions

Suppose that $\eta = (\eta_t : t \in T)$ is a finite array of real numbers indexed by T as in § 2 above and that we define an array of random variables $y = (y_t : t \in T)$ by the rule $y_t = \eta_{\pi(t)}$ for t in T, where π is a *random permutation* of the index set T which respects the nesting relationships; see Bailey et al. (1983) for full details of the group G of all such permutations. Following Nelder (1954, 1965) we ask: What are the covariances induced by this randomization? It is not hard to see that to answer this question we do not need to know anything about the group G other than the following facts:

$$P(y_t = \eta_v) = n^{-1}, \quad P(y_u = \eta_w \mid y_t = \eta_v) = k^{-1}_{a(t,u)}\delta(a(t, u), a(v, w)),$$

where t, u, v and w are in T. With this information it is clear that $Ey_t = n^{-1}\sum_v \eta_v$, and, if $a(t, u) = a$,

$$Ey_t y_u = (nk_a)^{-1} \sum_v \sum_w A_a(v, w)\eta_v \eta_w.$$

By using the relations between the matrices $\{A_a\}$ and $\{S_\alpha\}$ given in (2) above, we find that $\Gamma = Dy = \sum_{\alpha \neq \varnothing} \xi_\alpha S_\alpha$, where $\xi_\alpha = d_\alpha^{-1} |S_\alpha \eta|^2$. Since $\sum_t y_t$ has the constant value $\sum_t \eta_t$ and $S_\varnothing y = y_{.}$, the eigenvalue $\xi_\varnothing$ is equal to zero. Using (1) we can obtain the covariances $\{\gamma_a\}$ of y in terms of the $\{\xi_\alpha\}$.

Example (cont.). We might ask for the covariance $\gamma_{\{3\}}$, which is cov $(y_{ijk}, y_{i'j'k})$ with $i \neq i'$. Since $\gamma_{\{3\}} = n^{-1}\sum_\alpha q_{\{3\},\alpha}\xi_\alpha$, Table 3 shows that

$$n_1 n_2 n_3 \gamma_{\{3\}} = \xi_\varnothing - \xi_{\{1\}} + (n_3 - 1)(\xi_{\{3\}} - \xi_{\{1,3\}}).$$

We therefore need expressions for $\xi_{\{1\}}$, $\xi_{\{3\}}$ and $\xi_{\{1,3\}}$, as $\xi_\varnothing = 0$. From (4a) and (5) we see that $(n_1 - 1)\xi_{\{1\}} = n_2 n_3 \sum_i (\eta_{i..} - \eta_{...})^2$, and the corresponding expressions for $\xi_{\{3\}}$ and $\xi_{\{1,3\}}$ are as readily obtained.

4 Restricting to subsets

In many situations, including all of those for which T is infinite, we can observe only a finite part $y^{(U)} = (y_u : u \in U)$ of our random array $y = (y_t : t \in T)$, where U is some subset of T. What can we learn from $y^{(U)}$ about the various sets of parameters $\{\gamma_a\}$, $\{f_b\}$ and $\{\xi_\alpha\}$ of Dy? It is evident from the simplest examples that the restriction of a factorial dispersion model $V(\Pi, T)$ to a subset U of T does not necessarily result in a factorial dispersion model over U, so we are led into some broader aspects of analysis of variance which it is beyond our scope to cover fully here. For a in $L(\Pi)$, let $A_a^{(U)}$ be the restriction $A_a|_{U \times U}$ of the association matrix A_a over T to the subset U. The restriction $A_a^{(U)}$ may be zero for some values of a: however, if $A_a^{(U)}$ and $A_b^{(U)}$ are both nonzero for distinct filters a,

b in $L(\Pi)$, then $A_a^{(U)} \neq A_b^{(U)}$. Let $M = \{a \in L(\Pi): A_a^{(U)} \neq 0\}$. It is clear that $\{A_a^{(U)}: a \in M\}$ satisfies conditions (i)–(iv) of § 2.2. It may happen that condition (v) is also satisfied, so that $\{A_a^{(U)}: a \in M\}$ forms an association scheme over U: in this case we shall call U a *tractable* subset of T. We note that, even if U is a tractable subset of T, the association scheme over U is not necessarily of the same kind as that over T: indeed, it may not even be a factorial association scheme.

Example (cont.). Let U be the subset $\{(i, j, j): 1 \leq i \leq r,\ 1 \leq j \leq v\}$, where r and v are integers with $r \leq n_1$, $v \leq n_2$ and $v \leq n_3$. Then $A_{\{1,2\}}^{(U)} = A_{\{1,3\}}^{(U)} = 0$ and $M = \{\varnothing, \{1\}, \{3\}, \{1, 2, 3\}\}$. It may be checked that $\{A_a^{(U)}: a \in M\}$ satisfies condition (v), and so U is tractable. In fact, the association scheme over U is the factorial one corresponding to the simple crossed structure obtained from the poset Π in Fig. 1 omitting the factor 2. In § 5 we shall give an example where the association schemes on U and T are given by the *same* poset Π.

For simplicity let us suppose T to be finite; the extension to infinite T is quite straightforward. The standard theory of association schemes (MacWilliams & Sloane, 1977, p. 655) shows that, if U is a tractable subset of T, the set $\{A_a^{(U)}: a \in M\}$ of association matrices can be simultaneously diagonalized, with $|M|$ distinct common eigenspaces. Thus there are orthogonal projectors $\{S_\lambda^{(U)}: \lambda \in \Lambda\}$, where $|\Lambda| = |M|$, and, as in § 2.2, a $\Lambda \times M$ matrix $P^{(U)}$ and $M \times \Lambda$ matrix $Q^{(U)}$ such that

$$S_\lambda^{(U)} = |U|^{-1} \sum_{a \in M} q_{a\lambda}^{(U)} A_a^{(U)}, \quad A_a^{(U)} = \sum_{\lambda \in \Lambda} p_{\lambda a}^{(U)} S_\lambda^{(U)}.$$

We shall comment later on how the coefficients $q_{a\lambda}^{(U)}$ and $p_{\lambda a}^{(U)}$ may be found explicitly.

Just as in § 2.3, the projectors $\{S_\lambda^{(U)}: \lambda \in \Lambda\}$ define orthogonal decompositions of random arrays $y^{(U)}$ and of elements y_u of $y^{(U)}$ for u in U, and sum of squares decompositions with known expressions for expected mean squares. In short, when U is tractable then the dispersion model $\{\Gamma|_{U \times U}: \Gamma \in V(\Pi, T)\}$ exhibits the main features of an analysis of variance.

Write $\xi_\lambda^{(U)}$ for the eigenvalue of $Dy^{(U)}$ corresponding to the projector $S_\lambda^{(U)}$, where $\lambda \in \Lambda$. Then $\xi_\lambda^{(U)}$ is not, in general, equal to an eigenvalue of Dy. However, since $Dy^{(U)} = Dy|_{U \times U}$, we know that every covariance $\gamma_a^{(U)}$ in $Dy^{(U)}$ must appear, as γ_a in Dy. Thus we can combine the expression $\xi_\lambda^{(U)} = \sum_{a \in M} p_{\lambda a}^{(U)} \gamma_a^{(U)}$ for $\xi_\lambda^{(U)}$ with the formula (1) for γ_a to relate the expected mean square parameters $\{\xi_\lambda^{(U)}: \lambda \in \Lambda\}$ in the analysis of variance of $y^{(U)}$ to the analogous parameters $\{\xi_\alpha: \alpha \in L(\Pi)\}$ of y. This relationship may be expressed in matrix form as

$$\xi^{(U)} = P^{(U)} I^{(U,T)} |T|^{-1} Q\xi, \tag{6}$$

where $I^{(U,T)}$ is an $M \times L(\Pi)$ matrix with $I^{(U,T)}(a, b) = \delta_{ab}$.

A sufficient condition for tractability is that the restricted relationship matrices $\{R_a^{(U)}: a \in L(\Pi)\}$, where $R_a^{(U)} = R_a|_{U \times U}$, arise from a lattice of commuting uniform equivalence relations on U; see Speed & Bailey (1982) for definitions and fuller details. We note that in such cases, if at least one of $A_a^{(U)}$, $A_b^{(U)}$ is zero, then $R_a^{(U)}$ may be equal to $R_b^{(U)}$ even if $a \neq b$. For convenience let L^* be any subset of $L(\Pi)$ containing M such that $\{R_a^{(U)}: a \in L^*\} = \{R_a^{(U)}: a \in L(\Pi)\}$ but that $R_a^{(U)} \neq R_b^{(U)}$ whenever a and b are distinct elements of L^*. Then, for a in $L(\Pi)$, let a^* be the unique element of L^* such that $R_a^{(U)} = R_{a^*}^{(U)}$. In many cases L^* is a (not necessarily distributive) lattice of the type considered by Speed & Bailey (1982, § 2), who proved that analogues of all the earlier formulae and results hold with μ and ζ replaced by the Möbius and zeta functions of the lattice L^*. In particular, we have the following simple relationship between the

f-parameters of the two systems:

$$f_a^{(U)} = \sum_{b^*=a} f_b \quad (a \in L^*). \tag{7}$$

In most examples $M = L^*$ and the theory of § 2 applies to $y^{(U)}$ with no difficulty: we shall give an example of such a subset U in § 5. However, there are minor technical difficulties when $M \subset L^*$. One example where this happens is a 3×3 Graeco–Latin square viewed as a subset of a full $3 \times 3 \times 3 \times 3$ array. The difficulties are caused by the facts that $A_a^{(U)} = 0$ for $a \in L^* \backslash M$, and, correspondingly, some of the projectors defined in terms of the $R_a^{(U)}$ and the Möbius function of L^* are zero.

It should be clear that the foregoing discussion covers classical variance component estimates, at least in principle, although we do not discuss any general ways of disentangling estimates of population (that is, T) parameters from, say, quadratic forms in observed subsets. Equations (6) and (7) are most useful in the so-called 'balanced' or 'orthogonal' cases, where the observed subarray has a high degree of symmetry closely related to that of the full array. In most recent literature the full array is taken to be infinite, arising from an assumed linear model rather than an assumed covariance structure, but the results quoted in § 2.4 show that these two sets of assumptions are equivalent for our present purposes; see also Speed (1986).

5 Random sampling from structured populations

As in § 3, suppose that $\eta = (\eta_t : t \in T)$ is a finite population of real numbers indexed by T, and let U be an arbitrary subset of T. Let us consider *sampling* a random copy of U within T; that is, obtaining a random subset $\psi(U)$ of T which is labelled by U via the random injection ψ. Note that different choices of ψ may give rise to the same *set* $\psi(U)$ and yet must be considered different *samples*, because the labelling by the elements of U is a crucial feature of the sample. Thus random sampling amounts to random choice of ψ from some set Ψ of injections from U to T. We can now define a random array $y^{(U)} = (y_u : u \in U)$ indexed by U by putting $y_u = \eta_{\psi(u)}$ for u in U. The randomness underlying the distribution of $y^{(U)}$ is provided by the random sampling of ψ.

If we take Ψ to be (the restriction to U of) the group G mentioned in § 3, then $y^{(U)}$ is identical to the array obtained by restricting to U the random array $y = (y_t : t \in T)$ having the permutation distribution defined in § 3. Since all our results depend only on second-order properties, it follows from § 3 that all we require of Ψ is that it satisfy the following condition: for all u, v in U and all t, w in T,

$$P(\psi(u) = t) = n^{-1},$$
$$P(\psi(u) = t \mid \psi(v) = w) = k^{-1}_{a(t,w)} \delta(a(t, w), a(u, v)). \tag{*}$$

Thus *-random sampling an array of numbers is (to second order) the same as restricting to a subset of an array of random variables having an appropriate permutation distribution. Indeed the array η could well consist itself of random variables; provided that $D\eta \in V(\Pi, T)$, a sampling procedure satisfying (*) has no effect on the form of the dispersion matrix of the random variables selected, although the values of the individual covariances will change if $E\eta$ is not constant. If η has zero mean and dispersion matrix in $V(\Pi, T)$ then we may restrict at the outset to the subarray $\eta^{(U)} = (\eta_u : u \in U)$ of the desired form.

With this background we can carry out calculations concerning the sampled array using the structure on U derived from that on T. If U is a tractable subset of T then we may use

Factorial Dispersion Models

(6) and (7) to derive many expressions for average values of mean squares over sampling distributions: compare this with the work of Cornfield & Tukey (1956), Wilk (1955) and Wilk & Kempthorne (1956a,b; 1957).

Example. A triply-indexed population of numbers,

$$\eta = (\eta_{IJK} : I = 1, \ldots, N_1, J = 1, \ldots, N_2, K = 1, \ldots, N_3),$$

where the three factors 'rows', 'subrows' and 'columns' have the nesting relationship shown in Fig. 1, may be sampled as follows. Using simple random sampling without replacement, obtain n_1 rows $\psi_1(1), \ldots, \psi_1(n_1)$, and, independently within each of these, n_2 subrows $\psi_2(1, 1), \ldots, \psi_2(1, n_2), \ldots, \psi_2(n_1, 1), \ldots, \psi_2(n_1, n_2)$, and, independently of all the foregoing, n_3 columns $\psi_3(1), \ldots, \psi_3(n_3)$. We then form the array

$$(y_{ijk}) = (\eta_{\psi_1(i)\psi_2(i,j)\psi_3(k)} : i = 1, \ldots, n_1, j = 1, \ldots, n_2, k = 1, \ldots, n_3).$$

By the equivalence above, this is no different from restricting to the *first* n_1 rows, the *first* n_2 subrows within each row, and the *first* n_3 columns of the array (y_{IJK}) having the appropriate permutation distribution, for condition (*) is easily checked in this case. For example, if $I \neq I'$ and $i \neq i'$ then

$$P(\psi_1(i') = I', \psi_2(i', j) = J', \psi_3(k) = K \mid \psi_1(i) = I, \psi_2(i, j) = J, \psi_3(k) = K)$$

is equal to $1/(N_1 - 1)N_2$, which is as it should be since Table 3 shows that $k_{\{3\}} = (N_1 - 1)N_2$.

In this case it is clear that the sample $y = (y_{ijk})$ has the same lattice structure as the population: we may therefore use (6) to relate the two sets of parameters. For example, if we write the ξ-parameters for the population as $(\Xi_\alpha : \alpha \in L(\Pi))$ and those for the sample as $(\xi_\alpha : \alpha \in L(\Pi))$, Table 3 shows that

$$\xi_{\{1,3\}} = N_2^{-1}[n_2 \Xi_{\{1,3\}} + (N_2 - n_2)\Xi_{\{1,2,3\}}]. \tag{8}$$

In other words

$$E\left\{\frac{1}{(n_1 - 1)(n_3 - 1)} \sum_i \sum_j \sum_k (y_{i.k} - y_{i..} - y_{..k} + y_{...})^2\right\}$$

$$= \frac{n_2}{N_2(N_1 - 1)(N_3 - 1)} \sum_I \sum_J \sum_K (\eta_{I-K} - \eta_{I--} - \eta_{--K} + \eta_{---})^2$$

$$+ \frac{(N_2 - n_2)}{N_2 N_1(N_2 - 1)(N_3 - 1)} \sum_I \sum_J \sum_K (\eta_{IJK} - \eta_{IJ-} - \eta_{I-K} + \eta_{I--})^2,$$

where we are using . and – to denote the sample and population averages respectively. If the array η were random with zero mean, we would simply enclose an expectation operator around the right-hand side. Similarly (7) shows that $\xi_{\{1,2\}} = n_3 f_{\{1,2\}} + f_{\{1,2,3\}}$, sample and population f's, and γ's, coinciding because $M = L^* = L(\Pi)$ and neither set of parameters directly involves the sizes of the arrays.

The foregoing example is a *regular* sample in the sense that $U = \prod_p U_p$, where $U_p \subseteq T_p$ for p in Π. Using the results and notation of § 2.2 with $n_p = |U_p|$ and $N_p = |T_p|$ for p in Π, it can be shown that, for a regular sample, the coefficient of ξ_β in the expansion (6) of $\xi_\alpha^{(U)}$ is $b_{\beta\alpha}$, which is equal to

$$(\bar{n}_\beta / \bar{N}_\beta) \prod_{p \in \beta_m \setminus \alpha} (1 - n_p/N_p) \tag{9}$$

if $\alpha \subseteq \beta$, and zero otherwise. This formula is given (in words) by Cornfield & Tukey

(1956) and may be used to derive the coefficients in (8). The first proof was given by Haberman (1975, Th. 2).

Example (cont.). Suppose that our rows and sub-rows are *blocks* and *plots*, respectively, and that our columns correspond to *treatments*. Then we may regard our initial sampling ψ as the selection of $n_1 n_2$ experimental units, in n_1 blocks of n_2 plots per block, together with n_3 treatments. Now put $n_1 = r$ and suppose that $n_2 = n_3 = v$, and allocate treatments to plots in such a way that plots in the same block receive different treatments, all such allocations being equally probable. There are many ways of doing this; we are in effect choosing a complete block design at random. As before, the labelling of the sample by i, j, k is important, and we do not want this labelling to destroy the relationship of having the same treatment. Thus we denote a complete block design by the function θ, where $\theta(i, k) = j$ if treatment k is assigned to plot j of block i. This procedure now defines a doubly-indexed array $\eta_{\psi_1(i)\psi_2(i,\theta(i,k))\psi_3(k)}$, which is a *random* $(1/v)$th *fraction* (subject to certain constraints) of our originally selected sample. However, it is easy to check that this (combined) sampling procedure still satisfies (*): for example,

$$P(\theta(i, k) = j \mid \theta(i', k') = j') = v^{-1}$$

whenever $i \neq i'$, and so

$$P(\psi_1(i') = I', \ \psi_2(i', \theta(i', k)) = J', \ \psi_3(k) = K \mid \psi_1(i) = I, \ \psi_2(i, \theta(i, k)) = J, \ \psi_3(k) = K)$$

is equal to $1/(r-1)v$ whenever $I \neq I'$ and $i \neq i'$. Note that, if we had replaced θ by the more natural function θ^* allocating treatments to plots, so that $\theta^*(i, j) = k$ whenever $\theta(i, k) = j$, and obtained the array

$$\eta_{\psi_1(i)\psi_2(i,j)\psi_3(\theta^*(i,j))},$$

then our combined sampling procedure would not have satisfied (*). This illustrates the care that is necessary in considering a random sample as a labelled subset.

By the equivalence given at the beginning of the section, we may regard the sample as being the *first* r blocks, the *first* v plots within each of these blocks, and the *first* v treatments, with the fraction selected being given by any one complete block design. It is convenient to use the complete block design in which the treatments have the *same* labels as the plots, in every block. Thus, our sampled array is $y^{(U)}$, where $U = \{(i, j, j): i = 1, \ldots r, j = 1, \ldots, v\}$ and $(y_t : t \in T)$ has the permutation distribution based on η. We showed in § 4 that U is a tractable subset of T and that $M = \{\varnothing, \{1\}, \{3\}, \{1, 2, 3\}\}$. Moreover, $R^{(U)}_{\{1,2\}} = R^{(U)}_{\{1,3\}} = R^{(U)}_{\{1,2,3\}}$, whilst $R^{(U)}_{\varnothing}$, $R^{(U)}_{\{1\}}$, $R^{(U)}_{\{3\}}$ and $R^{(U)}_{\{1,2,3\}}$ are distinct: thus $L^* = M$. Equation (7) gives

$$f^{(U)}_{\varnothing} = f_{\varnothing}, \quad f^{(U)}_{\{1\}} = f_{\{1\}}, \quad f^{(U)}_{\{3\}} = f_{\{3\}}, \quad f^{(U)}_{\{1,2,3\}} = f_{\{1,2\}} + f_{\{1,3\}} + f_{\{1,2,3\}}.$$

These identities were first derived by Wilk (1955).

The techniques of this and §§ 3 and 4 allow us to re-derive the results of Throckmorton (1961) and White (1963, 1975) in a unified and direct way which fully exploits the underlying combinatorial structure; but see also the Appendix of Neymann et al. (1935). We make no comments here on the relative merits of these as compared with other approaches to the analysis of experimental data; a discussion which did justice to the topic would take us too far from the main subject of this paper.

6 Prediction

Our final topic is the best (that is, minimum mean-squared error linear) *prediction* of linear combinations of elements of an array $y = (y_t : t \in T)$ with $Ey = 0$, and known Dy in

Factorial Dispersion Models

$V(\Pi, T)$, based upon the observation of a finite subarrary $y^{(U)}$ where $U \subseteq T$. As an illustration from many possible results we shall suppose that $U = \prod_p \{1, \ldots, n_p\}$ is a regular sample from $T = \prod_p \{1, \ldots, N_p\}$ where $n_p \leq N_p \leq \infty$ for p in Π and even here we shall consider the prediction of only the components $(S_\alpha y_t : \alpha \in L(\Pi), t \in U)$ in (4) above. In this case it can be shown that the best predictor of $S_\alpha y_t$ based upon $y^{(U)}$ when $t \in U$ is

$$\xi_\alpha^{(T)} \sum_\beta [\xi_\beta^{(U)}]^{-1} b_{\alpha\beta} S_\beta^{(U)} y_t, \tag{10}$$

where, for $\beta \in L(\Pi)$, $S_\beta^{(U)}$ denotes the matrix introduced in § 2 above for the array $y^{(U)}$. If $t \notin U$ a more complicated expression can be derived. Predictors of more general linear combinations of elements of y are best derived using (10) and linearity. If any of the N_p is infinite we must pass to a limit in (10) by combining $\xi_\alpha^{(T)}$ with the $b_{\alpha\beta}$ which, of course, depend upon the N_p for p in Π.

Example (cont.). Let us compute the best predictor of $y_{i--} - y_{---}$; that is, let us evaluate (10) with Π as in Fig. 1 and $\alpha = \{1\}$. Note that the averages we are predicting are in the population (T-indexed) array, and this will be done using averages from the sample (U-indexed) array. From (9) we find that $b_{\{1\},\{1\}} = n_2 n_3 / N_2 N_3$ whilst $b_{\{1\},\varnothing} = n_2 n_3 (1 - n_1/N_1)/N_2 N_3$ and so

$$\hat{y}_{i--} - \hat{y}_{---} = \xi_{\{1\}}^{(T)} \frac{n_2 n_3}{N_2 N_3} \left[(y_{i..} - y_{...})/\xi_{\{1\}}^{(U)} + \left(1 - \frac{n_1}{N_1}\right) y_{...}/\xi_{\varnothing}^{(U)} \right].$$

If we let N_1, N_2 and N_3 all tend to infinity and expand the ξ's in terms of f's, this expression simplifies to

$$\frac{n_2 n_3 f_{\{1\}}}{n_2 n_3 f_{\{1\}} + n_3 f_{\{1,2\}} + n_2 f_{\{1,3\}} + f_{\{1,2,3\}}} (y_{i..} - y_{...})$$
$$+ \frac{n_2 n_3 f_{\{1\}}}{n_1 n_2 n_3 f_\varnothing + n_2 n_3 f_{\{1\}} + n_3 f_{\{1,2\}} + n_1 n_2 f_{\{3\}} + n_2 f_{\{1,3\}} + f_{\{1,2,3\}}} y_{...} .$$

It is clear that the above discussion is essentially what some writers term the *estimation of random effects* (Harville, 1976). Our approach places it firmly within a prediction framework, of unobserved random variables by observed ones, but the two are, of course, equivalent. As we remarked earlier, matters become more complicated in the presence of a structured mean value (fixed effects), and we shall say nothing further about this topic here.

7 Closing remarks

There are many more aspects of this topic which could be addressed if space permitted. For example, concise *rules* for forming analysis of variance tables, that is sums of squares, degrees of freedom and expected mean squares and so on, can be formulated and proved in the above framework. Other randomization analyses can be obtained with a minimum of effort, for example, for split plot and other more complex designs. These topics and a number of others will be expounded together with a fuller exposition of the material we have surveyed in a forthcoming monograph mentioned in § 1.

References

Aigner, M. (1979). *Combinatorial Theory*. New York: Springer Verlag.
Aldous, D.J. (1981). Representations for partially exchangeable arrays of random variables. *J. Mult. Anal.* **11**, 581–598.

T.P. Speed and R.A. Bailey

Bailey, R.A. (1984). Discussion of paper by T. Tjur. *Int. Statist. Rev.* **52,** 65–77.

Bailey, R.A., Praeger, C.E., Rowley, C.A. & Speed, T.P. (1983). Generalized wreath products of permutation groups. *Proc. Lond. Math. Soc.* (3) **47,** 69–82.

Brown, W. (1913). The effects of "observational errors" and other factors upon correlation coefficients in psychology. *Br. J. Psychol.* **6,** 223–238.

Bryant, P. (1984) Geometry, statistics, probability: variations on a common theme. *Am. Statistician* **38,** 38–48.

Cochran, W.G. (1939). The use of the analysis of variance in enumeration by sampling. *J. Am. Statist. Assoc.* **34,** 492–510.

Cornfield, J. & Tukey, J.W. (1956). Average values of mean squares in factorials. *Ann. Math. Statist.* **27,** 907–949.

Cox, D. (1960). Discussion of paper by R. L. Plackett. *J. R. Statist. Soc.* B **22,** 209–210.

Daniels, H.E. (1938). Some problems of statistical interest in wool research (with discussion). *J. R. Statist. Soc. Suppl.* **5,** 89–128.

Daniels, H.E. (1939). The estimation of components of variance. *J.R. Statist. Soc. Suppl.* **6,** 186–197.

Dawid, A.P. (1977). Invariant distributions and analysis of variance models. *Biometrika* **64,** 291–297.

Eden, T. & Yates, F. (1933). On the validity of Fisher's z test when applied to an actual sample of non-normal data. *J. Agric. Sci.* **23,** 6–17.

Fairfield Smith, H. (1955). Variance components, finite populations and experimental inference. North Carolina Institute of Statistics Mimeo Series No 135.

Fisher, R.A. (1925). *Statistical Methods for Research Workers.* Edinburgh: Oliver & Boyd.

Haberman, S.J. (1975). Direct products and linear models for complete factorial tables. *Ann. Statist.* **3,** 314–333.

Hannan, E.J. (1965). *Group Representations and Applied Probability.* London: Methuen.

Harville, D. (1976). Extension of the Gauss–Markov theorem to include the estimation of random effects. *Ann. Statist.* **4,** 384–395.

Hooke, R. (1956a). Symmetric functions of a two-way array. *Ann. Math. Statist.* **27,** 55–79.

Hooke, R. (1956b). Some applications of bipolykays to the estimation of variance components and their moments. *Ann. Math. Statist.* **27,** 80–98.

John, P.W.M. (1971). *Statistical Design and Analysis of Experiments.* New York: Macmillan.

Kempthorne, O. (1952). *The Design and Analysis of Experiments.* New York: Wiley.

Lush, J.L. (1931). The number of daughters necessary to prove a sire. *J. Dairy Sci.* **14,** 209–220.

Lush, J.L., Henderson, A.L., Culbertson, C.C. & Hammond, W.E. (1933). The reliability of some measures of the productiveness of individual brood sows. *Record of the Proceedings of the Annual Meeting of the American Society of Animal Production* **26,** 282–298.

MacWilliams, F.J. & Sloane, N.J.A. (1977). *The Theory of Error-Correcting Codes.* Amsterdam: North Holland.

Nelder, J.A. (1954). The interpretation of negative components of variance. *Biometrika* **41,** 544–548.

Nelder, J.A. (1965). The analysis of randomized experiments with orthogonal block structure. I. Block structure and the null analysis of variance. *Proc. R. Soc. Lond.* A **283,** 147–162.

Nelder, J.A. (1977). A reformulation of linear models (with discussion). *J. R. Statist. Soc.* A **140,** 48–76.

Neyman, J. with co-operation of K. Iwaszkiewicz and St. Kołodziejczyk (1935). Statistical problems in agricultural experimentation (with discussion). *J. R. Statist. Soc. Suppl.* **2,** 107–180. See also *A Selection of Early Statistical Papers of J. Neyman* (1967). Cambridge University Press.

Pitman, E.J.G. (1938). Significance tests which may be applied to samples from any populations. III. The analysis of variance test. *Biometrika* **29,** 322–335.

Saville, D.J. & Wood, G.R. (1986). A method for teaching statistics using N-dimensional geometry. *Am. Statistician* **40,** 205–214.

Scheffé, H. (1956). Alternative models for the analysis of variance. *Ann. Math. Statist.* **27,** 251–271.

Spearman, C. (1910). Correlation calculated from faulty data. *Br.J. Psychol.* **3,** 271–295.

Speed, T.P. (1986). ANOVA models with random effects: An approach via symmetry. In *Essays in Time Series and Allied Processes: Papers in Honour of E.J. Hannan,* Ed. J. Gani and M.B. Priestly, pp. 355–368. Sheffield: Applied Probability Trust.

Speed, T.P. (1987). What is an analysis of variance? *Ann. Statist.* **15.** To appear.

Speed, T.P. & Bailey, R.A. (1982). On a class of association schemes derived from lattices of equivalence relations. In *Algebraic Structures and Applications,* Ed. P. Schultz, C.E. Praeger and R.P. Sullivan, pp. 55–74. New York: Marcel Dekker.

Throckmorton, T.N. (1961). Structures of classification data. Iowa State University PhD thesis.

Tjur, T. (1984). Analysis of variance models in orthogonal designs (with discussion). *Int. Statist. Rev.* **52,** 33–81.

Welch, B.L. (1937). On the z-test in randomized blocks and Latin squares. *Biometrika* **29,** 21–52.

White, R.F. (1963). Randomization analysis of the general experiment. Iowa State University PhD thesis.

White, R.F. (1975). Randomization and the analysis of variance. *Biometrics* **31,** 552–571.

Wilk, M.B. (1955). The randomization analysis of a generalized randomized block design. *Biometrika* **42,** 70–79.

Wilk, M.B. & Kempthorne, O. (1956a). Some aspects of the analysis of factorial experiments in a completely randomized design. *Ann. Math. Statist.* **27,** 950–985.

Wilk, M.B. & O. Kempthorne (1956b). Analysis of variance: Preliminary tests, pooling, and linear models, **2**. Wright Air Development Center Tech. Rep. 55-244.
Wilk, M.B. & Kempthorne, O. (1957). Non-additivities in a Latin square design. *J. Am. Statist. Assoc.* **52,** 218–236.
Yates, F. & Zacopanay, I. (1935). The estimation of the efficiency of sampling, with special reference to sampling for yield in cereal experiments. *J. Agric. Sci.* **25,** 545–577.
Youden, W.J. & Mehlich, A. (1937). Selection of efficient methods for soil sampling. *Contrib. Boyce Thompson Inst.* **9,** 59–70.

Résumé

On étudie un ensemble de modèles de structure des covariances pour des tableaux multi-indexés (i.e. indexés par les éléments d'un produit cartésien) de variables aléatoires. Par leurs propriétés au second ordre, ces modèles généralisent les modèles de composantes de la variance, les modèles de randomisation, ainsi que les modèles d'échangeabilité. Ils conduisent de façon naturelle à des techniques générales pour effectuer des décompositions orthogonales, calculer les espérances des carrés moyens et évaluer les autres quantités intervenant dans l'analyse de variance de ce type de tableaux.

[*Received April* 1986, *revised March* 1987]

The Annals of Statistics
2002, Vol. 30, No. 6, 1649–1665

JOHN W. TUKEY'S CONTRIBUTIONS TO ANALYSIS OF VARIANCE

BY T. P. SPEED

University of California, Berkeley

John Tukey connected the theory underlying simple random sampling without replacement, cumulants, expected mean squares and spectrum analysis. He gave us one degree of freedom for nonadditivity, and he pioneered finite population models for understanding ANOVA. He wrote widely on the nature and purpose of ANOVA, and he illustrated his approach. In this appreciation of Tukey's work on ANOVA we summarize and comment on his contributions, and refer to some relevant recent literature.

1. Introduction. Most (9/15) of John Tukey's contributions to analysis of variance (hereafter ANOVA) can be found in Volume 7 of *The Collected Works of John W. Tukey* [17]. Also in that volume are two items which will be of interest to readers of this paper. One is a six-page foreword to the nine collected papers by John Tukey himself. The other is an historical introduction to and remarks on the roles of analysis of variance, and some brief comments on the individual papers by the volume editor, David R. Cox. However, Tukey being Tukey, there is no substitute for reading the papers themselves. Every one of them advances our knowledge, at times dramatically, while seeming to be no more than a lucid exposition from first principles of some well-established part of our subject. There are exceptions to this last statement.

John Tukey's main published contributions to ANOVA were made in a little over a decade, from 1949 to 1961. They constitute approximately 20% of his output over this period, and so about 5% of his total output. In subject matter these papers range from the foundational to the computational, from the algebraic to the interpretational, and contain some strikingly original views of the topics he discusses. How many of us see a clear connection between finite-population simple random sampling as in books on sampling, Fisher's k-statistics and cumulants for calculating moments of sample moments, the moments of mean squares in ANOVA tables and the arithmetic of spectrum analysis? At the same time as he was clarifying the analysis of variance qua variance, he highlighted the importance of scale to the notion of interaction in the analysis of means, and gave us a tool for identifying and removing removable nonadditivity. He also showed us how to analyze a complex multifactorial data set; indeed in no fewer than four of the

Received January 2002; revised March 2002.
AMS 2000 subject classifications. Primary 62J10; secondary 94A20.
Key words and phrases. Odoffna, ANOVA, moments, cumulants, k-statistics, polykays, variances, components of variance, mean squares, factorials, interactions, pigeonhole model.

papers below we get his views on the nature and purpose of ANOVA. It was much broader than the usual one which focusses on testing.

In my opinion much of Tukey's work on ANOVA is underappreciated, and much of that which was appreciated at the time has been forgotten. He laments [17, page lii], wrongly as it turns out, "Perhaps regrettably, I am not aware of very much that extends papers 5, 6, 7, and 9" (of [17]). Some of his work on ANOVA, for example, his "dyadic ANOVA" and his "components in regression," was never followed up. Neither of these titles scores a hit (with Tukey's meaning) in *Current Index to Statistics*. Fashions change, and the foundational worries or solutions of one generation of statisticians can cease to be of interest to a later generation. It is for this reason as well as to celebrate Tukey's genius that it is a real pleasure to be able to remind readers of his wonderful contributions to ANOVA, including creating the abbreviation itself.

2. ODOFFNA. How we will miss Tukey's neologisms. His one degree of freedom for nonadditivity (ODOFFNA) paper [2] is perhaps his best-known and most striking contribution to the analysis of variance and needs little introduction here. Whereas others had paid attention to nonconstancy of the variance or nonnormality of the "errors" in ANOVA, Tukey was concerned with nonadditivity. Explaining his ideas in the context of a singly replicated row-by-column table, he showed how to isolate a single degree of freedom from the "residual," "error" or "interaction" sum of squares ("call it what you will" he said), and so test the null hypothesis of additivity using a statistic which gave power against a restricted class of multiplicative alternatives. The statistic was motivated by the idea of a power transformation; it was illustrated graphically through three examples, and some elegant distribution theory was presented. This is a gem of a paper and amply deserves its place in the texts [see, e.g., Scheffé (1959) or Seber (1977)]. Tukey's later papers [5, 13] on the same topic present no new ideas; rather they illustrate the earlier ideas in more general contexts, something he pointed out was possible in [2].

What has happened to ODOFFNA since the 1960s? These days most people concerned about the possibility that their linear model might better satisfy the standard assumptions of additivity, homoscedasticity and normality of errors after a transformation will make use of the Box and Cox (1964) theory. However, their approach to transformations is not a complete substitute for ODOFFNA, as can be seen in Tukey's [14] discussion of additive and multiplicative fits to two-way tables (see especially [14], Section 10F). It is likely that we will continue to extract ODOFFNA in new contexts in the future, and for more on this, see Tukey's own comments on the follow-up to ODOFFNA in his foreword to [17].

3. Complex analyses of variance: general problems [11]. In [11] Green and Tukey made a number of general points concerning complex analyses of variance in the course of analyzing a specific experimental data set. Some of the points are familiar, some were new at the time but most are still of interest today. The authors explain that the purpose of their analysis is "to provide a simple summary of the variation in the experimental data, and to indicate the stability of means and other meaningful quantities extracted from the data." They intended their approach to be in opposition to the view that the sole purpose of ANOVA is to provide tests of significance. It was aimed at researchers in psychology and followed a review of the use of ANOVA in that field a few years earlier.

The experiment is from psychophysics and involves six factors: sex (S, two levels), sight (I, two levels), persons (P, eight levels), rate (R, four levels), weight (W, seven levels) and date (D, two levels). All of S, I, R, W and D are crossed, while P is nested in a balanced way within $S \times I$ so we may describe the factor relationships by the formula $((S \times I)/P) \times R \times W \times D$. The response was a difference limen, a kind of threshold of perception, which could be expressed as a difference in weights, a squared difference in weights, a ratio of weights, a logarithm of a ratio of weights or even a response time.

One novel aspect of this paper is that the authors discuss not only what scale to use for the dependent variable; that is, possible transformations, but also just what the dependent variable should be in that context: a difference, a squared difference, a ratio, a log ratio, etc. After an initial analysis with one response variable, they choose another and obtain a new, and to their minds better, analysis. Another novelty at that time was the careful discussion of the nesting and crossing between factors and their implications for the analysis. This was no doubt inspired by the discussion of these matters Tukey and Cornfield gave in [8], which was published some four years before [11].

Perhaps the most interesting part of this paper is the extended section "Variance components and the proper error term" and the section "Variance components in the illustrative example" which follows it. The first of these discusses an example simpler than the actual experiment and draws heavily on material concerning the pigeonhole model in [8]; see Section 4.3. Then they turn to the experiment and things get interesting when they seek to impose a sampling model on the factors. The four levels of rate (50, 100, 150 and 200 g/s) and the seven levels of weight $(100, 150, \ldots, 400$ g) are admitted to present a problem for their pigeonhole model. Are they exhaustive samples from finite populations, that is, fixed; are they small samples from large populations of levels, that is, random; or are they something else? Whereas it was easy for them to view sex and sight (blind or not) as fixed, and person as random, the choice for R and W was far less obvious. After some discussion of various options, including a mention of using polynomials to fit responses to rate and weight, they decide to regard R and W as random "although we recommend against this procedure [for scaled variables] in general." The ideal that one ANOVA theory fits all cases seems hard to live up to, even when you are the creator of the theory.

As soon as all factors are assigned the category fixed or random, it is possible to write out all 39 expected mean square lines of the ANOVA table, and this they do. Next follows an illuminating discussion of "aggregation and pooling" of lines in the table, which, when implemented with the illustrative data, reduces the 39 lines to 15. They make use of a modified version of a procedure of Paull (1950) which Tukey highlights in his Introduction to [17] and seems to be of interest today. There are two useful graphical representations of the relative contributions of the different sources of variability, one in two dimensions which is especially appealing, but on the whole there is relatively little plotting of the data, a large contrast with Tukey's later work, for example, in [14].

A later analysis of this same data set can be found in Johnson and Tukey [15]. Looking back on this paper after four decades, and bearing in mind all that Tukey wrote on ANOVA before and after that time, one cannot help but be struck by how little use he made in this paper of the processes and procedures he recommended when considering such an analysis. Referring to matters to be discussed in Section 4, he made no attempt to assign standard errors to his estimated variance components, under either normality or any other assumptions, the scientific purpose of the experiment was nowhere mentioned, the situations or populations to which inference was to be made were nowhere mentioned, even the means he calculated and plotted were not assigned any measures of their stability, something that was stated at the beginning of the paper to be one of the major purposes of ANOVA. Granted this was an expository paper with a limited objective, and probably already long by the standards of the journal, but I think the point remains that it is hard to put Tukey's ANOVA theory into practice, even for Tukey himself.

4. Some moment calculations. Tukey wanted to derive average values and variances and later a third moment of consider later. He tells us [17, page liv] that his first attempt at deriving the variance of the between variance component in an unbalanced one-way design took five or six full days "using old-fashioned clumsy methods." He was "convinced that it ought not to be so hard" and so "went looking for better tools, and eventually came out with the polykays." Polykays are generalizations of Fisher's k-statistics and we now outline the main points from the papers in which they were introduced.

4.1. *Some sampling simplified; keeping moment-like computations simple* [3, 6]. In 1929 Fisher introduced k-statistics as unbiased estimators of cumulants and a computational technique which radically simplified much previous research on moments of moments. It would take us too far astray to describe his technique in detail [see Speed (1986a)], but we can describe the simplest of his results in this area as soon as we recall the following well-known facts. If $X_1, \ldots, X_n$ are i.i.d. random variables with common first two cumulants κ_1 and κ_2 (the mean and variance, respectively), then

$$k_1 = \frac{1}{n}\sum X_i = \bar{X} \quad \text{and} \quad k_2 = \frac{1}{n-1}\sum (X_i - \bar{X})^2$$

satisfy

$$\mathbf{E}k_1 = \kappa_1 \quad \text{and} \quad \mathbf{E}k_2 = \kappa_2.$$

Now the k's are Fisher's k-statistics, that is, unbiased estimates of the corresponding cumulants. The key results of Fisher (1929) were the general definition of k-statistics and a procedure for calculating their joint cumulants whose core was a rule for calculating the coefficients of lower order k-statistics in an expansion for the product of two k-statistics. The relationships above are the simplest relevant results: the expected values or first cumulants of the first two k-statistics. Next would come the results which come from replacing **E** by var or covar; that is, replacing first by second cumulant in the sample-population calculation.

We all know that $\text{var}(k_1) = \kappa_2/n$, but what about $\text{var}(k_2)$? This result, first derived by Gauss, is not quite so well known, but turns out to be

$$\text{var}(k_2) = \frac{2}{n-1}\kappa_2^2 + \frac{1}{n}\kappa_4.$$

Deriving this last fact is already messy enough to warrant thinking very carefully about the algebraic formulation one adopts, and any desire to obtain more general expressions of the same kind focusses the mind greatly on the same issue. Fisher had his approach, Tukey simplified it as we shall see and it can be simplified yet again; see Speed (1983) and McCullagh (1987).

Tukey's main aim in [3] and [6] was to extend these results (and others like them) to the finite population case. Apparently unknown to Tukey, this task had been begun by Neyman in 1923 [see Neyman (1925)], though far less elegantly or generally. To achieve his aim Tukey extended Fisher's entire machinery. He named the tool he developed polykays—multiply-indexed generalizations of k-statistics—later noting that these same functions had been introduced earlier by Dressel (1940) in a paper that was not noticed at the time. For Tukey polykays of order or weight r are indexed by partitions of the natural number r. For example, there are two of order 2, indexed by $(1, 1)$ and (2); three of order 3, indexed by $(1, 1, 1)$, $(1, 2)$ and (3); four of order 4, indexed by $(1, 1, 1, 1)$, $(1, 1, 2)$, $(1, 3)$ and (4); and so on. Fisher's k-statistics are the single subscript versions of the polykays, (1), (2), (3), (4) etc., hence Tukey's name. In what follows we drop the commas and parentheses in the partition notation, writing $1, 11, 2$, etc.

How are polykays defined in general? To do this Tukey made use of an auxiliary class of symmetric functions also labelled by partitions, which he called symmetric means or, more simply, brackets, denoted by $\langle 1\rangle$, $\langle 11\rangle$, $\langle 2\rangle$, etc. These functions had the appealing property of rather transparently being "inherited on the average," which means that the average of the sample function over simple random sampling without replacement from a finite population was just the corresponding population function. Tukey avoided using the term "unbiased" as (so he said) "there are now so many kinds of unbiasedness!" The sample mean

$$\langle 1\rangle = \frac{1}{n}\sum x_i$$

is clearly inherited on the average, as is

$$\langle 11 \rangle = \frac{1}{n(n-1)} \sum_{i \neq j} x_i x_j .$$

The value of brackets lies in the fact that [3, page 111] "every expression which is (i) a polynomial, (ii) symmetric, (iii) inherited in the average, can be written as a linear combination of brackets with coefficients which do not depend on the size of the set of numbers involved." As one illustration we give the following well-known and useful representation:

$$\frac{1}{n-1} \sum (x_i - \bar{x})^2 = \frac{1}{n} \sum x_i^2 - \frac{1}{n(n-1)} \sum_{i \neq j} x_i x_j ,$$

where the last two terms are transparently inherited in the average, neatly proving that the first term is also, a standard fact from sampling theory. Tukey would write this last relationship $(2) = \langle 2 \rangle - \langle 11 \rangle$, and in general he needed a rule giving the coefficients of brackets in the expansion of his parentheses (polykays). As he said ([3], page 124) "the single-index brackets have the coefficients for moments in terms of cumulants (given numerically by Kendall [(1943), Section 3.13] up to the 10th moment). The coefficients of brackets with several indices can be found by formal multiplication."

How do we use all this machinery? Elegant though it is, there is still some hard work: the multiplication tables need to be derived. Tukey derived his own, but by the time of publication of [3, 6] comprehensive tables had independently appeared [Wishart (1952a, b)]. A simple instance of a multiplication rule is

$$(*) \qquad (2)^2 = (22) + \frac{1}{n}(4) + \frac{2}{n-1}(22)$$

Let us see how this leads very painlessly to the main result of Neyman (1925). First, note that the preceding identity has a version connecting population k-statistics which is of the same form, but with n replaced by N. Next recall that the polykays (22), (4), etc. are all "inherited on the average." We now take the expectation (i.e., average) of $(*)$ over all samples and subtract from the result the population version of $(*)$. This leaves us with

$$\operatorname{var}((2)) = \left[\frac{1}{n} - \frac{1}{N}\right](4) + 2\left[\frac{1}{n-1} - \frac{1}{N-1}\right](22),$$

which is the formula Neyman worked hard to obtain. This was indeed "sampling simplified." Note also that if we let $N \to \infty$ (so-called infinite population) and use the easily proved fact that, in this case, (22) is just $(2)^2$, we obtain Gauss' result.

Tukey certainly simplified sampling. He demonstrated clearly that indeed finite populations are simpler to deal with, and more powerful, and he now had the machinery to carry out certain calculations in ANOVA.

Later developments cast Tukey's work in the framework of tensors [cf. Kaplan (1952) and, most recently within the general theory of symmetric functions, Speed (1986a)]. The gains from so doing are modest, but I think definitely worthwhile. One consequence of the tensor formulation is that some of Tukey's formal calculations (e.g., his symbolic o-multiplication) cease to be "tricks." Another is the greater simplicity which comes from allowing all random variables to be potentially different. For example, instead of calculating variances of variances, we calculate covariances of distinct covariances, and obtain variances by appropriately equating arguments. With this slightly greater generality, $(*)$ above becomes [Speed (1986a), page 43]

$$(12) \otimes (34) = (12|34) + \frac{1}{n}(1234) + \frac{1}{n-1}[(13|24) + (14|23)],$$

where $1, 2, 3$ and 4 all label distinct variables. This simplification removes certain multiplicity factors and then reveals the coefficients defining polykays to be values of the Möbius function over a partition lattice, which I think is a real step forward; see Speed (1983) and McCullagh (1987).

Where are polykays now? There was a little theoretical development of them after Tukey's work, but he left no major problems unaddressed. I extended them to multiply-indexed arrays in Speed (1986a, b) and Speed and Silcock (1988a), and used the extensions to generalize the calculations of Tukey discussed in the next section. Apart from my own work the most recent references to polykays are Tracy (1973) and, an application of them, McCullagh and Pregibon (1987). To my knowledge there have been no other publications concerning polykays since then. In short, it seems that after about 25 years of life, polykays have been dead or sleeping for 25 years. Apparently they have served their purpose, though I have no doubt that they will be resurrected, awakened or reborn at some time in the future, when another problem comes along for whose solution they are the natural tool.

4.2. *Variances of variance components* [7, 9, 10]. Why did Tukey go to all the trouble of inventing polykays and their calculus, and what did he learn from so doing? Giving as one purpose of the analysis of variance "to estimate the sizes of the various components contributed to the overall variance from the corresponding sources," he wanted "to obtain formulas for the variances of the natural estimates of these variance components." Along with Gauss, Fisher and many others, Tukey wanted to go beyond normality, but almost uniquely he did so in dispensing with infinite populations. He regretted ([7], page 157) that he still had to leave "the customary (and dangerous) independence assumptions" concerning the terms in his linear population models. This answers the question "Why?" Let us now see some of what he learned in a simple case: the balanced single (or one-way) classification. Tukey's model for this takes the form

$$x_{ij} = \mu + \eta_i + \omega_{ij}, \qquad i = 1, \ldots, c,\ j = 1, \ldots, r,$$

where the $\{\eta_i\}$ are sampled from a population of size n with k-statistics $k_1, k_2, \ldots,$ the $\{\omega_{ij}\}$ are from a population of size N with k-statistics $K_1, K_2, \ldots$ and the samplings are independent and order randomized. If we denote by B and W the usual between-class and within-class mean squares, respectively, with expectations k_2 and K_2, then Tukey showed, among other results, that

$$\begin{aligned}
\operatorname{var}(B) &= \left(\frac{1}{c} - \frac{1}{n}\right)k_4 + 2\left[\frac{1}{c-1} - \frac{1}{n-1}\right]k_{22} \\
&\quad + \frac{4}{r(c-1)}k_2K_2 + \frac{2(rc-1)}{r^2c(r-1)(c-1)}K_{22}, \\
\operatorname{cov}(B, W) &= -\frac{2}{rc(r-1)}K_{22}, \\
\operatorname{var}(W) &= \left[\frac{1}{rc} - \frac{1}{N}\right]K_4 + 2\left[\frac{1}{c(r-1)} - \frac{1}{N-1}\right]K_{22}.
\end{aligned}$$

The remainder of [7] consists of more formulae of this kind, derived for other variance component models: row-by-column classifications, Latin squares, balanced incomplete blocks and more general balanced models.

Paper [9] considers the special, more complicated case of an unbalanced one-way classification. One novelty here is that there is no single compelling estimate of the between-class component of variance, and so Tukey considers a class of estimates involving weights which need to be specified. He then derives the variances and covariances as before, generalizing those just given, and presents numerical examples. Lastly, paper [10] does what its title says: it presents the third moment about the mean, that is, the third cumulant of the quantity W given above.

What can we learn from or do with such formulae? In the first place, we can obtain qualitative insights by comparing the general finite population results with the special case of infinite normal populations. There k_4 and K_4 vanish, while k_{22} and K_{22} are k_2^2 and K_2^2, respectively, and of course $N = \infty$. In this case the results are familiar, and the extent to which the normal variances for the estimated variance components are too small or too large could, in principle, be examined. Interestingly, Tukey does not present formulae giving unbiased estimates of either the individual terms in his expressions for the variances of the estimated variance components, or for the variance expression as a whole. I would be very surprised if he did not have such formulae, for example, for k_4 and K_4 and k_{22} and K_{22} above, but he makes no mention of them. Without them, his aim of calculating estimates of the precision of estimated variance components under these more general assumptions must remain unfulfilled.

What has been done since the 1950s in this area? There has been more work on the topic of variances of estimated components and variance; see, for example, Harville (1969), but there, as in all other such cases that I know, the calculations are carried out under an assumption of normality. In some of my own work [Speed

(1986a, b), Speed and Silcock (1988a, b)] I have tried to extend Tukey's work to ANOVA models which are not built up additively from independent components.

4.3. *Average values of mean squares in factorials* [8]. This is an interesting and important paper: broad in coverage, profound in its analysis, beautifully written and elegant in its dealing with messy algebraic details. It is arguably Tukey's most important contribution to ANOVA. By the early to mid-1950s it was becoming clear that the concise description in Eisenhart (1947) of models for ANOVA did not provide a foundation for all uses of ANOVA. The now well-known mixed-model ambiguity concerning the interaction component of variance when (say) rows are "fixed" and columns "random" had emerged: in some linear model formulations this component appeared in the expected mean square line for both rows and columns, while in others it did not. It was apparent to many that the combining of linear models and ANOVA was not as simple as might have seemed at first. Neyman and his Polish colleagues found this out the hard way in 1935, but made no later attempt at a broad synthesis. Kempthorne (1952) in Ames, building on the work of Neyman and co-workers, Fairfield Smith in Raleigh, Tukey in Princeton, Cornfield at the National Institutes of Health in Bethesda and no doubt others elsewhere all sought to devise models of differing breadth and flexibility which would specialize appropriately under different assumptions, and lead to the desired analyses and inferences. Throughout all this, Fisher was silent on the topic, apparently holding to his view that "the analysis of variance is ... a convenient method of arranging the arithmetic."

Anyone who reads the five sections comprising the Initial discussion of [8] quickly realizes that providing a general framework for ANOVA is no mean task. The subsequent six sections spelling out Cornfield and Tukey's approach prior to their presenting any average values shows that theirs is not an easy resolution. So it should come as no surprise when I say that the situation today is hardly any better than it was then in the mid-1950s. Cornfield and Tukey's paper should be essential reading for all those who care about these matters. But it is not read, and neither their approach nor any other has taken root among the legions of users of ANOVA and linear models. No treatment of the issues that prompted them to write that paper has yet gained acceptance; see below.

What are the issues? Although in most of his writings on ANOVA Tukey emphasized estimation of variance components above significance testing, this paper is very much motivated by testing. Expressions for average values of mean squares in factorials are the primary basis for testing: they tell us which mean squares can usefully be compared with which; that is, they dictate the choice of error term. So attention focuses sharply on the model assumptions leading to these averages. As Tukey and Cornfield point out in Section 2 of their paper, the choice among assumptions is important and is not simple. It includes but goes beyond empirical questions about the behavior of the experimental material. Assumptions must also depend on the nature of the sampling and randomization involved in obtaining the data, and the purpose of the analysis, as expressed by the situations or populations to which one wishes to make statistical inference.

Cornfield and Tukey's way ahead is by the use of what they call a pigeonhole model, in which combinations of experimental factors (rows, columns, etc.) define pigeonholes containing a finite or infinite population. If, like them, we illustrate ideas with the replicated row-by-column classification, then their assumption is that a sample of r rows is drawn from a possible R, and a sample of c columns is drawn from a possible C. These rc intersections define the pigeonholes which are the cells of the actual experiment, and from each of the rc cells a sample of n elements is drawn. "All the samplings—of rows, of columns, and within pigeonholes—are at random and independent of one another." This is their approach. They discuss at considerable length the way in which an equivalent linear model can be defined, making it clear just how different their linear model was from those previously used (and used today). Of particular significance was their notion of "tied" interaction, their avoidance of what they term the "special and dangerous" assumption of independence of the variation of interaction terms of main effects terms.

After their lengthy preliminaries it is almost a relief to get to the algebraic part of the paper: definitions of components of variance and rules for calculating what we now term expected mean squares. They discuss two-way and three-way designs in detail and give rules for designs with factors nested or crossed in arbitrary ways. There is an interesting discussion of the nature of the various proofs then extant of the formulae. At that time there were two types: "Proofs using special machinery or indirect methods (e.g., symmetry arguments and equating of coefficients for special assumptions)," the approach preferred by Tukey, and "proofs using relatively straightforward algebra," which was the preferred way of Cornfield. Neither of these was particularly effective in full generality.

The mathematical content of [8] has been revisited at least twice since 1956. The first time was by Haberman (1975), in a dense paper which does not seem to have been widely read. He makes effective use of the calculus of tensor products of vector spaces to give very concise proofs of the main results. A quite different approach was used in Speed (1985) [see also Speed and Bailey (1987)] (also not widely read), where the discussion was expressed in terms of the eigenvalues of the associated dispersion matrices. Other, less general formulations can be found in books on linear models and ANOVA, for example, Searle, Casella and McCulloch (1992).

As suggested earlier, all attempts at providing a general framework for ANOVA since 1956 should have come to terms with the material in [8]: they should either incorporate it or suggest an alternative approach. There have been many such attempts over the last 45 years, with Nelder (1977) providing the most far-reaching alternative, building on Nelder (1965a, b). This paper and especially the discussion of it are well worth reading, especially today. The most recent discussions of the "mixed models controversy" [see, e.g., Schwarz (1993) and Voss (1999) and references therein] refer to neither Cornfield and Tukey, Kempthorne, Nelder nor any other of the earlier generation of researchers in this area. Plackett (1960) gives an excellent review of this early work.

Tukey's contribution to the discussion of Nelder (1977) is particularly interesting, in part because it reveals so clearly his distrust in models. It should be read in full, but here are some tantalizing excerpts, all the more relevant when one bears in mind that all recent discussion of this issue is a discussion of models:

> I join with the speaker in hoping for an eventual and agreed-upon description. I hope the present paper will help us approach this ideal state, but I must say that it has not brought us there.
> Three types of variability arise in almost any question about a set of comparative measurements, experimental or not: measurement variability, sampling variability and contextual variability.
> A major point, on which I cannot yet hope for universal agreement, is that our focus must be on questions, not models.
> One conclusion I draw from such examples is this: Models can—and will—get us into deep trouble if we expect them to tell us what the unique proper questions are.

I close this section with some personal comments, but before I do so, I should confess that I too have attempted to publish a description of ANOVA which I had hoped might have become "agreed-upon." It did not even get accepted for publication. However, I think I represent more than myself when I say that, for all my admiration of [8] and what it attempted to do, that solution was simply too far away from the world of linear models most of us inhabit. In my view, and I suspect that of many others, linear models are most readily specified through a model for the expected values and a model for the variances and covariances of the observables. After all, we are simply specifying (apart from the values of certain unknown parameters) the first two moments of our observables. Had their approach been in these terms, I believe it might still be discussed. Nelder (1977) had a related objection when he pointed out that randomization models (involving finite populations but random effects) could not be seen as a special case of the approach in [8]. The matter of providing linear unbiased estimates of quantities of interest figured nowhere in [8], and I believe this reduces many people's willingness to see its solution as general and relevant to their use of linear models and ANOVA. But perhaps the real reason that the description in [8] is not yet agreed upon is this: the majority of statisticians these days (perhaps even 50 years ago) are not interested in the issues that concerned Tukey, Cornfield, Kempthorne, Fairfield Smith and Neyman and co-workers, before them, and Nelder and others, including me, after them. Perhaps it is just too hard, connecting assumptions and models to the subject matter, to the data collection process, to the questions one is asking and the kinds of answers one seeks. "Does it really matter? Does it make any practical difference?" I get asked. It is so much easier discussing models and parameterizations.

5. Other ANOVA papers by Tukey.

5.1. *Dyadic ANOVA* [1]. This paper was based on a talk Tukey gave in November 1946, and is more interesting for what it tells us about the development of his thinking concerning ANOVA than for the material related to its title. Ostensibly about ANOVA for vectors, that is, what we would now call multivariate analysis of variance (MANOVA), the paper also contains a wealth of interesting material only marginally related to that topic. The reason he wrote it, he says, was that other accounts of MANOVA concentrate too much on tests and too little on that which is most useful and revealing in ordinary ANOVA. It is impossible to resist passing on one of his introductory remarks, presumably aimed at the average reader of *Human Biology*. He writes:

> It is a maxim of arithmetic that it is not proper to add 2 oranges to 1 apple; this is good arithmetic but may be poor vector algebra. For
>
> $$(2\,\text{oranges}, 0) + (0, 1\,\text{apple}) = (2\,\text{oranges}, 1\,\text{apple})$$
>
> is a meaningful and useful statement.

Later, he goes on:

> If we are to have an analysis of variance, we must have squares, and the solution is
>
> $$(2\,\text{oranges}, 1\,\text{apple})^2 = \begin{bmatrix} 4\,\text{orange}^2 & 2\,(\text{orange})(\text{apple}) \\ 2\,(\text{orange})(\text{apple}) & 1\,\text{apple}^2 \end{bmatrix}$$

The paper includes a concise discussion of components of variance, initially in the context of Eisenhart's (1947) models, but also including the finite population pigeonhole models which were to play such a big role in his later work. Rather surprisingly in view of his later disdain for F-tests, and his stated motivation for writing the paper, he makes a start on tests of significance for dyadic ANOVA, that is, the distribution of eigenvalues in 2×2 MANOVA. He even attempts to give fiducial intervals for quantities of interest, but concludes that more distribution theory is required.

A topic not obviously related to dyadic ANOVA is what he calls choice of terms, that is, choice of the response variable to be analyzed in a given experiment. He castigates Fisher for not paying more attention to this point, illustrating it dramatically by carrying out the same analysis on some hydrogen spectrum data using both wavelength and its reciprocal, wave number, as responses. In a fascinating analysis foreshadowing the power transformation underpinnings of ODOFFNA, he uses his newly developed dyadic ANOVA to find that linear combination of a response variate and the variate squared which minimizes the ratio of row plus column sums of squares to interaction sum of squares in an unreplicated row-by-column array. Illustrating the method on one of the data sets which he uses in his later paper on ODOFFNA, Tukey shows the considerable gain in efficiency he achieves with his transformation. The eigenvalue problem he solves is reminiscent of canonical variate analysis, and he ends that discussion with some interesting speculations on alternative criteria to optimize in the definition of discriminant functions.

A further point of interest in this paper can be found in the Appendix, headed "Two identities and a lemma." The lemma gives the variance of the average and the expectation of the sample variance of a set of variates which have different means and different variances, but a common covariance λ, a simple enough variant on the result which is well known for i.i.d. variates. He goes on to apply this result to his pigeonhole models, illustrating once more what was to be a recurrent theme in his statistical research: a desire to weaken standard assumptions wherever possible. He finds that, under these more general assumptions, the formulae are essentially unchanged, with a common variance σ^2 being replaced by the average variance $\sigma_{\cdot}^2 - \lambda$.

5.2. *Components in regression* [4]. This paper is about simple linear regression when both variates are subject to "error," and the use of instrumental variates in this context. The fields of application discussed include precision of measurement, psychology and econometrics, and, as is so often the case with Tukey, the paper demonstrates the prodigious breadth of his knowledge. The connection with ANOVA is slight, really only arising because he discusses an example in which measurements are taken in replicate. As he says, "We could have avoided mention of variance components . . . since we only deal with the simplest sorts . . . between-vs-within or regression-vs-balance. However, we have chosen to bring them in for two reasons. Mainly to set the analysis in terms which can easily be carried over to more complicated analyses where the correct procedure might otherwise be a mystery. Secondarily, to stress the analogy with variance components for a single variate." The paper is not easy reading and, since its connection to other material here is not great, we do not discuss it any further.

5.3. *ANOVA and spectral analysis* [12]. As might be expected from its context—the discussion of two papers on the spectral analysis of time series—[12] is much more about spectral analysis than ANOVA. It was placed in one of the time series volumes [21], not in [17], yet I want to mention it here, in part for its influence on me personally. What Tukey makes very clear in this discussion is that spectrum analysis, with a *line* for each frequency, *is* ANOVA. More fully, he says "the spectrum analysis of a single time series is just a branch of variance component analysis." This was one of his inspired connections which proved illuminating in both directions. It is clear from his remarks that Tukey supposed that his statistical audience knew something about ANOVA and could read [8] if they wished, and that this would enlarge their understanding of spectrum analysis, the topic of the papers. What was probably not apparent at the time was that there were people, myself included, for whom spectrum analysis was straightforward, but variance component analysis a mystery, and that his connection would be helpful to such people in the other direction. For evidence of the impact of this paper on me, see Speed (1987); for a valuable introduction to this paper, see the comments by Brillinger in [21].

5.4. *Toward robust ANOVA* [16]. This paper offers "a recipe for robust/resistant analysis of variance of data from factorial experiments in which all factors have three or more versions." Its motivation is eloquently explained as follows:

> Analysis of variance continues to be one of the most widely used statistical methods. Not only the form of the analysis of variance table with its lines of mean squares and degrees-of-freedom associated with each of several sorts of variation, but the entire analysis, including confidence statements, is classically supposed to be determined by the design—the hierarchical structure, conduct, and the intent of the experiment—alone. The behaviour of the data itself is, classically, not supposed to influence how its description is formatted. Hardly an exploratory attitude. ... In this account, rather than using a data-free structure to define our procedure, we provide a further stage of responding to the data's behaviour, one where summarization is based on a robust alternative to the mean.

The recipe is explained by its application to a particular $5 \times 3 \times 8$ array of data from an experiment concerning the hardness of gold alloy fillings. It begins with a *pre-decomposition*, this being a multiway analogue of median polish, and proceeds through the *identification* of so-called exotic entries, to a *re-decomposition* dealing with these, and a *robust analysis of variance* with the familiar sums of squares and degrees-of-freedom calculated from the re-decomposition. Next, a process of *downsweeping* is carried out, this being a variant of the pooling of mean squares which we met in Section 3 above, and the recipe concludes with the calculation of error mean squares, standard errors and confidence statements.

5.5. *Methods, comments, challenges* [18–20]. Tukey expounded and discussed ANOVA in a number of his many overview papers, and I will single out three of these for brief mention.

In [18] he goes over "some methods that form sort of a general core of the statistical techniques" that were used at that time. He aimed "to supply background: statistical, algebraic and perhaps intuitive," and he succeeded admirably. The exposition could hardly be improved upon, indeed is better than most we see today, in that it contains possibly the first instance of the "analysis of variance diagram" mentioned in the discussion of paper [11] in Section 3 above. This diagram surely deserves to be more widely used. Also noteworthy is a remark which may well be the first appearance in print of the abbreviation ANOVA.

In [19] Tukey offers 37 methodological comments about statistics on topics ranging from *exploration versus confirmation*, *re-expression* and *causation*, to *spectrum analysis*, and naturally he has something so say about ANOVA. Relevant comments concern regression and analysis of variance, nonorthogonal analysis and

MANOVA, and can only be described as stimulating and provocative. For example, in seeking a replacement of conventional MANOVA: "We could calculate principal components, but they are not likely to be simply interpretable. So let us not"; and: "Much the same could be said of 'dust bowl empiricists factor analysis'."

In Section 21 of the last of these three overview papers [20], Tukey foreshadows the issues dealt with more fully in [16] discussed above. We see clearly how keen Tukey was to unify his understanding of and approach to ANOVA with his robust/resistant and exploratory data analysis paradigms. While [16] is a fine start, it seems clear that there is much more to be said on this unification.

6. Concluding remarks. John Tukey was an extraordinarily able and creative statistician. He made a number of lasting contributions to ANOVA: to our understanding of what it is and what it can do for us; to the algebraic and computational aspects of the subject; and, perhaps most important and characteristic to showing us how to go beyond the usual assumptions. The impact of all this work on the subject today is less than it should be, perhaps in part because Tukey set his standards rather high. However, his papers are all there for anyone to read, and if this appreciation of them encourages one person who would not otherwise, to do so, its purpose will have been achieved.

Acknowledgments. I thank Gordon Smyth, for reading this manuscript in draft and offering many helpful comments, and David Brillinger, for many valuable references and for his enthusiasm for this project. Thanks also to the referees and editors for their comments and assistance.

JOHN W. TUKEY'S PUBLICATIONS ON ANOVA

[1] (1949a). Dyadic ANOVA. An analysis of variance for vectors. *Human Biology* **21** 65–110. (Paper [21].)

[2] (1949b). One degree of freedom for non-additivity. *Biometrics* **5** 232–242. (Paper [29].)

[3] (1950). Some sampling simplified. *J. Amer. Statist. Assoc.* **45** 501–519. (Paper [38].)

[4] (1951). Components in regression. *Biometrics* **7** 33–69. (Paper [41].)

[5] (1955). Answer to query 113. *Biometrics* **11** 111–113. (Paper [51].)

[6] (1956a). Keeping moment-like sampling computations simple. *Ann. Math. Statist.* **27** 37–54. (Paper [52].)

[7] (1956b). Variances of variance components. I. Balanced designs. *Ann. Math. Statist.* **27** 722–736. (Paper [53].)

[8] (1956c). Average values of mean squares in factorials. *Ann. Math. Statist.* **27** 907–949. (With J. Cornfield. Paper [55].)

[9] (1957a). Variances of variance components. II. The unbalanced single classification. *Ann. Math. Statist.* **28** 43–56. (Paper [57].)

[10] (1957b). Variances of variance components. III. Third moments in a balanced single classification. *Ann. Math. Statist.* **28** 378–384. (Paper [58].)

[11] (1960). Complex analyses of variance: General problems. *Psychometrika* **25** 127–152. (With B. F. Green, Jr. Paper [76].)

[12] (1961). Discussion, emphasizing the connection between analysis of variance and spectrum analysis. *Technometrics* **3** 191–219. (Paper [79].)

T. P. SPEED

[13] (1962). Handout for Meeting of Experimental Station Statisticians "Tests for Non-additivity." ([17], Paper 3).

[14] (1977). *Exploratory Data Analysis*. Addison-Wesley, Reading, MA. (Book [307].)

[15] (1987). Graphical exploratory analysis of variance illustrated on a splitting of the Johnson and Tsao data. In *Design, Data and Analysis by Some Friends of Cuthbert Daniel* (C. Mallows, ed.) 171–244. Wiley, New York. (With E. G. Johnson. Paper [282].)

[16] (2001). Towards robust analysis of variance. In *Data Analysis from Statistical Foundations* (A. K. Md. E. Saleh, ed.) 217–244. Nova Science Publishers, Huntington, NY. (With A. H. Seheult. Paper [213].)

[17] (1992). *The Collected Works of John W. Tukey VII. Factorial and ANOVA: 1949–1962*. Wadsworth, Belmont, CA.

The following are not entirely about ANOVA, but contain relevant material.

[18] (1951). Standard methods of analyzing data. In *Proc. Computation Seminar, December 1949* 95–112. IBM, Armonk, NY. (Paper [216].)

[19] (1980). Methodological comments focused on opportunities. In *Multivariate Techniques in Human Communication Research* (P. R. Monge and J. Cappella, eds.) 489–528. Academic Press, New York. (Paper [194].)

[20] (1982). An overview of techniques of data analysis, emphasizing its exploratory aspects. In *Some Recent Advances in Statistics* (J. Tiago de Oliveira and B. Epstein, eds.) 111–172. Academic Press, New York. (With C. L. Mallows. Paper [273].)

[21] (1984). *The Collected Works of John W. Tukey I. Time Series: 1949–1964*. Wadsworth, Belmont, CA.

REFERENCES

BOX, G. E. P. and COX, D. R. (1964). An analysis of transformations (with discussion). *J. Roy. Statist. Soc. Ser. B* **26** 211–252.

DRESSEL, P. L. (1940). Statistical semi-invariants and their estimates with particular emphasis on their relation to algebraic invariants. *Ann. Math. Statist.* **11** 33–57.

EISENHART, C. (1947). The assumptions underlying the analysis of variance. *Biometrics* **3** 1–21.

FISHER, R. A. (1929). Moments and product moments of sampling distributions. *Proc. London Math. Soc.* **30** 199–238.

HABERMAN, S. J. (1975). Direct products and linear models for complete factorial tables. *Ann. Statist.* **3** 314–333.

HARVILLE, D. A. (1969). Variances of variance-component estimators for the unbalanced 2-way cross classification with application to balanced incomplete block designs. *Ann. Math. Statist.* **40** 408–416.

KAPLAN, E. L. (1952). Tensor notation and the sampling cumulants of k-statistics. *Biometrika* **39** 319–323.

KEMPTHORNE, O. (1952). *Design and Analysis of Experiments*. Wiley, New York.

KENDALL, M. G. (1943). *The Advanced Theory of Statistics* **1**. Griffin, London.

MCCULLAGH, P. (1987). *Tensor Methods in Statistics*. Chapman and Hall, London.

MCCULLAGH, P. and PREGIBON, D. (1987). k-statistics and dispersion effects in regression. *Ann. Statist.* **15** 202–219.

NELDER, J. A. (1965a). The analysis of randomized experiments with orthogonal block structure. I. Block structure and the null analysis of variance. *Proc. Roy. Soc. London Ser. A* **283** 147–162.

NELDER, J. A. (1965b). The analysis of randomized experiments with orthogonal block structure. II. Treatment structure and the general analysis of variance. *Proc. Roy. Soc. London Ser. A* **283** 163–178.

NELDER, J. A. (1977). A reformulation of linear models (with discussion). *J. Roy. Statist. Soc. Ser. A* **140** 48–76.

NEYMAN, J. S. (1925). Contributions to the theory of small samples drawn from a finite population. *Biometrika* **17** 472–479. [Reprinted from *La Revue Mensuelle de Statistique* **6** (1923) 1–29.]

PAULL, A. E. (1950). On a preliminary test for pooling mean squares in the analysis of variance. *Ann. Math. Statist.* **21** 539–556.

PLACKETT, R. L. (1960). Models in the analysis of variance (with discussion). *J. Roy. Statist. Soc. Ser. B* **22** 195–217.

SCHEFFÉ, H. (1959). *The Analysis of Variance*. Wiley, New York.

SCHWARZ, C. J. (1993). The mixed-model ANOVA: The truth, the computer packages, the books. Part I. Balanced data. *Amer. Statist.* **47** 48–59.

SEARLE, S. R., CASELLA, G. and MCCULLOCH, C. E. (1992). *Variance Components*. Wiley, New York.

SEBER, G. A. F. (1977). *Linear Regression Analysis*. Wiley, New York.

SPEED, T. P. (1983). Cumulants and partition lattices. *Austral. J. Statist.* **25** 378–388.

SPEED, T. P. (1985). Dispersion models for factorial experiments. *Bull. Inst. Internat. Statist.* **51** 1–16.

SPEED, T. P. (1986a). Cumulants and partition lattices. II. Generalized k-statistics. *J. Austral. Math. Soc. Ser. A* **40** 34–53.

SPEED, T. P. (1986b). Cumulants and partition lattices. III. Multiply-indexed arrays. *J. Austral. Math. Soc. Ser. A* **40** 161–182.

SPEED, T. P. (1986c). Cumulants and partition lattices. IV. A.s. convergence of generalized k-statistics. *J. Austral. Math. Soc. Ser. A* **41** 79–94.

SPEED, T. P. (1987). What is an analysis of variance? (with discussion). *Ann. Statist.* **15** 885–941.

SPEED, T. P. and BAILEY, R. A. (1987). Factorial dispersion models. *Internat. Statist. Rev.* **55** 261–277.

SPEED, T. P. and SILCOCK, H. L. (1988a). Cumulants and partition lattices. V. Calculating generalized k-statistics. *J. Austral. Math. Soc. Ser. A* **44** 171–196.

SPEED, T. P. and SILCOCK, H. L. (1988b). Cumulants and partition lattices. VI. Variances and covariances of mean squares. *J. Austral. Math. Soc. Ser. A* **44** 362–388.

TRACY, D. S. and GUPTA, B. C. (1973). Multiple products of polykays using ordered partitions. *Ann. Statist.* **1** 913–923.

VOSS, D. T. (1999). Resolving the mixed models controversy. *Amer. Statist.* **53** 352–356.

WILK, M. B. and KEMPTHORNE, O. (1955). Fixed, mixed, and random models. *J. Amer. Statist. Assoc.* **50** 1144–1167.

WISHART, J. (1952a). Moment coefficients of the k-statistics in samples from a finite population. *Biometrika* **39** 1–13.

WISHART, J. (1952b). The combinatorial development of the cumulants of k-statistics. *Trabajos Estadistica* **3** 13–26.

DEPARTMENT OF STATISTICS
UNIVERSITY OF CALIFORNIA, BERKELEY
367 EVANS HALL
BERKELEY, CALIFORNIA 94720-3860
E-MAIL: terry@bilbo.berkeley.edu

Chapter 6
Cumulants and Partition Lattices

Peter McCullagh

This is the first paper to appear in the statistical literature pointing out the importance of the partition lattice in the theory of statistical moments and their close cousins, the cumulants. The paper was first brought to my attention by Susan Wilson, shortly after I had given a talk at Imperial College on the Leonov-Shiryaev result expressed in graph-theoretic terms. Speed's paper was hot off the press, arriving a day or two after I had first become acquainted with the partition lattice from conversations with Oliver Pretzel. Naturally, I read the paper with more than usual attention to detail because I was still unfamiliar with Rota [18], and because it was immediately clear that Möbius inversion on the partition lattice $\mathscr{E}_n$, partially ordered by sub-partition, led to clear proofs and great simplification. It was a short paper packing a big punch, and for me it could not have arrived at a more opportune moment.

The basic notion is a partition σ of the finite set $[n] = \{1,\ldots,n\}$, a collection of disjoint non-empty subsets whose union is $[n]$. Occasionally, the more emphatic term set-partition is used to distinguish a partition of $[n]$ from a partition of the integer n. For example 135|2|4 and 245|1|3 are distinct partitions of [5] corresponding to the same partition $3+1+1$ of the integer 5. Altogether, there are two partitions of [2], five partitions of [3], 15 partitions of [4], 52 partitions of [5], and so on. These are the Bell numbers $\#\mathscr{E}_n$, whose exponential generating function is $\exp(e^t - 1)$. The symmetric group acting on $\mathscr{E}_n$ preserves block sizes, and each integer partition is a group orbit. There are two partitions of the integer 2, three partitions of 3, five partitions of 4, seven partitions of 5, and so on.

It turns out that, although set partitions are much larger, the additional structure they provide is essential for at least two purposes that are fundamental in modern probability and statistics. It is the partial order and the lattice property of $\mathscr{E}_n$ that simplifies the description of moments and generalized cumulants in terms of cumulants. This is the subject matter of Speed's paper. At around the same time, from the late 1970s until the mid 1980s, Kingman was developing the theory of partition structures, or partition processes. These were initially described in terms of inte-

P. McCullagh
Department of Statistics, University of Chicago
e-mail: pmcc@galton.uchicago.edu

S. Dudoit (ed.), *Selected Works of Terry Speed*, Selected Works in Probability and Statistics, 277
DOI 10.1007/978-1-4614-1347-9_6, © Springer Science+Business Media, LLC 2012

ger partitions [3, 10], but subsequent workers including Kingman and Aldous have found it simpler and more natural to work with set partitions. In this setting, the simplification comes not from the lattice property, but from the fact that the family $\mathscr{E} = \{\mathscr{E}_1, \mathscr{E}_2, \ldots\}$ of set partitions is a projective system, closed under permutation and deletion of elements. The projective property makes it possible to define a process on $\mathscr{E}$, and the mutual consistency of the Ewens formulae for different n implies an infinitely exchangeable partition process.

In his 1964 paper, Rota pointed out that the inclusion-exclusion principle and much of combinatorics could be unified in the following manner. To any function f defined on a finite partially-ordered set, there corresponds a cumulative function

$$F(\sigma) = \sum_{\tau \le \sigma} f(\tau).$$

The mapping $f \mapsto F$ is linear and invertible with inverse

$$f(\sigma) = \sum_{\tau} m(\tau, \sigma) F(\tau),$$

where the Möbius function is such that $m(\tau, \sigma) = 0$ unless $\tau \le \sigma$. In matrix notation, $F = Lf$, where L is lower-triangular with inverse M. The Möbius function for the Boolean lattice (of sets, subsets and complements) is $(-1)^{\#\sigma - \#\tau}$, giving rise to the familiar inclusion-exclusion rule. For the partition lattice, the Möbius function relative to the single-block partition is $m(\tau, \{[n]\}) = (-1)^{\#\tau - 1}(\#\tau - 1)!$, where $\#\tau$ is the number of blocks. More generally, $m(\tau, \sigma) = \prod_{b \in \sigma} m(\tau[b], b)$ for $\tau \le \sigma$, where $\tau[b]$ is the restriction of τ to the subset b.

Although they have the same etymology, the word 'cumulative' in this context is unrelated semantically to 'cumulant', and in a certain sense, the two meanings are exact opposites: cumulants are to moments as f is to F, not vice-versa.

Speed's paper is concerned with *multiplicative* functions on the partition lattice. To understand what this means, it is helpful to frame the discussion in terms of random variables $X^1, X^2, \ldots, X^n$, indexed by $[n]$. The joint moment function μ associates with each subset $b \subset [n]$ the number $\mu(b)$, which is the product moment of the random variables $X[b] = \{X^i \colon i \in b\}$. Any such function defined on subsets of $[n]$ can be extended multiplicatively to a function on set partitions by $\mu(\sigma) = \prod_{b \in \sigma} \mu(b)$. Likewise, the joint cumulant function κ associates with each non-empty subset $b \subset [n]$ a number $\kappa(b)$, which is the joint cumulant of the random variables $X[b]$. The extension of κ to set partitions is also multiplicative over the blocks. It is a property of the partition lattice that if $f \equiv \kappa$ is multiplicative, so also is the cumulative function $F \equiv \mu$. In particular, the full product moment is the sum of cumulant products

$$\mu([n]) = \sum_{\sigma} \prod_{b \in \sigma} \kappa(b).$$

For zero-mean Gaussian variables, all cumulants are zero except those of order two, and the above expression reduces to Isserlis's theorem [5] for $n = 2k$, which is a the sum over $n!/(2^k k!)$ pairings of covariance products. Wick's theorem, as it is known in the quantum field literature, is closely associated with Feynman diagrams. These are not merely a symbolic device for the computation of Gaussian moments, but also an aid for interpretation in terms of particle collisions [4, Chapter 8]. For an account that is accessible to statisticians, see Janson [8] or the AMS feature article by Phillips [17].

The moments and cumulants arising in this way involve distinct random variables, for example $X^2X^3X^4$, never $X^3X^3X^4$. However, variables that are given distinct labels may be equal, say $X^2 = X^3$ with probability one, so this is not a limitation. As virtually everyone who has worked with cumulants, from Kaplan [9] to Speed and thereafter, has noted, *the general results are most transparent when all random variables are taken as distinct.*

The arguments put forward in the paper for the combinatorial lattice-theoretic approach are based on the simplicity of the proof of various known results. For example, it is shown that the ordinary cumulant $\kappa([n])$ is zero if the variables can be partitioned into two independent blocks. Subsequently, Streitberg [25] used cumulant measures to give an if and only if version of the same result. To my mind, however, the most compelling argument for Speed's combinatoric approach comes in Proposition 4.3, which offers a simple proof of the Leonov-Shiryaev result using lattice-theoretic operations. To each subset $b \subset [n]$ there corresponds a product random variable $X^b = \prod_{i\in b} X^i$. To each partition σ there corresponds a set of product variables, one for each of the blocks $b \in \sigma$, and a joint cumulant $\kappa^\sigma = \text{cum}\{X^b\colon b \in \sigma\}$. One of the obstacles that I had encountered in work on asymptotic approximation of mildly non-linear transformations of joint distributions was the difficulty of expressing such a generalized cumulant in terms of ordinary cumulants. The lattice-theoretic expression is remarkable for its simplicity:

$$\kappa^\sigma = \sum_{\tau:\tau\vee\sigma=\mathbf{1}_n} \prod_{b\in\tau} \kappa(b),$$

where the sum extends over partitions τ such that the least upper bound $\sigma \vee \tau$ is the single-block partition $\mathbf{1}_n = \{[n]\}$. Tables for these connected partitions are provided in McCullagh [14]. For example, if $\sigma = 12|34|5$ the third-order cumulant κ^σ is a sum over 25 connected partitions. If all means are zero, partitions having a singleton block can be dropped, leaving nine terms

$$\kappa^{12,34,5} = \kappa^{1,2,3,4,5} + \kappa^{1,2,3}\kappa^{4,5}[4] + \kappa^{1,3,5}\kappa^{2,4}[4]$$

in the abbreviated notation of McCullagh [13]. Versions of this result can be traced back to James [6], Leonov and Shiryaev [11], James and Mayne [7], and Malyshev [12].

A subject such as statistical moments and cumulants that has been thoroughly raked over by Thiele, Fisher, Tukey, Dressel and others for more than a century, might seem dry and unpromising as a topic for current research. Surprisingly, this is not the case. Although the area has largely been abandoned by research statisticians, it is a topic of vigorous mathematical research connected with Voiculescu's theory of non-commutative random variables, in which there exists a notion of freeness related to, but distinct from, independence. The following is a brief idiosyncratic sketch emphasizing the parallels between Speicher's work and Speed's paper.

First, Speed's combinatorial theory is purely algebraic: it does not impose positive definiteness conditions on the moments or cumulants, nor does it require them to be real-valued, but it does implicitly require commutativity of the variables. In a theory of non-commutative random variables, we may think of $X^1, \ldots, X^n$ as orthogonally invariant matrices of unspecified order. For a subset $b \subset [n]$, the *scalar product* $X^b = \operatorname{tr} \prod_{i \in b} X^i$ is the trace of the matrix product, which depends on the cyclic order. The first novelty is that $\mu(b) = E(X^b)$ is not a function on subsets of $[n]$, but a function on cyclically ordered subsets. Since every permutation $\sigma\colon [n] \to [n]$ is a product of disjoint cycles, every function on cyclically ordered subsets can be extended multiplicatively to a function on permutations $\mu(\sigma) = \prod_{b \in \sigma} \mu(b)$. Given two permutations, we say that τ is a *sub-permutation* of σ if each cycle of τ is a sub-cycle of some cycle of σ — in the obvious sense of preserving cyclic order [1]. For $\tau \leq \sigma$, the crossing number $\chi(\tau, \sigma)$ is the number of 4-cycles (i, j, k, l) below σ such that i, k and j, l are consecutive in τ: $\chi(\tau, \sigma) = \#\{(i, j, k, l) \leq \sigma\colon \tau(i) = k,\ \tau(j) = l\}$, and τ is called *non-crossing in* σ if $\chi(\tau, \sigma) = 0$. For a good readable account of the non-crossing property, see Novak and Sniady [16].

Although it is not a lattice, the set Π_n of permutations has a lattice-like structure; each maximal interval $[0_n, \sigma]$, in which 0_n is the identity and σ is cyclic, is a lattice. With sub-permutation as the partial order, $[0_n, \sigma] \cong \mathscr{E}_n$ is isomorphic with the standard partition lattice; with non-crossing sub-permutation as the partial order, each maximal interval is a partition lattice of a different structure. Speicher's combinatorial theory of moments and cumulants of non-commutative variables uses Möbius inversion on this lattice of non-crossing partitions [24]. If $f \equiv \kappa$ is multiplicative, so also is the cumulative function $F \equiv \mu$, and vice-versa. The function $\kappa(b)$ on cyclically ordered subsets is called the free cumulant because it is additive for sums of freely independent variables. Roughly speaking, freeness implies that the matrices are orthogonally or unitarily invariant of infinite order. For further discussion on this topic, see Nica and Speicher [15] or Di Nardo et al. [2].

The partition lattice simplifies the sampling theory of symmetric functions, leading to a complete account of the joint moments of Fisher's k-statistics and Tukey's polykays [19]. It led to the development of an extended theory of symmetric functions for structured and nested arrays associated with a certain subgroup [20, 21, 22, 23]. Elegant though they are, these papers are not for the faint of heart. With some limitations, it is possible to develop a parallel theory of spectral k-statistics and polykays — polynomial functions of eigenvalues having analogous finite-population inheritance and reverse-martingale properties. Simple expressions are easily obtained for low-order statistics, but the general theory is technically rather complicated.

Acknowledgement

I am grateful to Elvira Di Nardo, John Kolassa, Lek Heng Lim and Domenico Senato for helpful comments on an earlier draft. Support for this work was provided by NSF grant DMS-0906592.

References

[1] P. Biane. Some properties of crossings and partitions. *Discrete Math.*, 175: 41–53, 1997.
[2] E. Di Nardo, P. Petrullo, and D. Senato. Cumulants and convolutions via Abel polynomials. *Eur. J. Combin.*, 31(7):1792–1804, 2010.
[3] W. J. Ewens. The sampling theory of selectively neutral alleles. *Theor. Popul. Biol.*, 3:87–112, 1972.
[4] J. Glimm and A. Jaffe. *Quantum Physics*. Springer, 1987.
[5] L. Isserlis. On a formula for the product moment coefficient of any order of a normal frequency distribution in any number of variables. *Biometrika*, 12: 134–139, 1918.
[6] G. S. James. On moments and cumulants of systems of statistics. *Sankhyā*, 20:1–30, 1958.
[7] G. S. James and A. J. Mayne. Cumulants of functions of random variables. *Sankhyā*, 24:47–54, 1962.
[8] S. Janson. *Gaussian Hilbert Spaces*. Cambridge University Press, 1997.
[9] E. L. Kaplan. Tensor notation and the sampling cumulants of k-statistics. *Biometrika*, 39:319–323, 1952.
[10] J. F. C. Kingman. The representation of partition structures. *J. Lond. Math. Soc.*, 18:374–380, 1978.
[11] V. P. Leonov and A. N. Shiryaev. On a method of calculation of semi-invariants. *Theor. Probab. Appl.*, 4:319–329, 1959.
[12] V. A. Malyshev. Cluster expansions in lattice models of statistical physics and the quantum theory of fields. *Russ. Math. Surv.*, 35:1–62, 1980.
[13] P. McCullagh. Tensor notation and cumulants of polynomials. *Biometrika*, 71: 461–476, 1984.
[14] P. McCullagh. *Tensor Methods in Statistics*. Chapman and Hall, London, 1987.
[15] A. Nica and R. Speicher. *Lectures on the Combinatorics of Free Probability*. Cambridge University Press, London, 2006.
[16] J. Novak and P. Sniady. What is a free cumulant? *Notices of the American Mathematical Society*, 58:300–301, 2011.
[17] T. Phillips. Finite-dimensional Feynman diagrams. *American Math. Society feature column*, 2001.
[18] G.-C. Rota. On the foundations of combinatorial theory I. *Z. Wahrscheinlichkeitstheorie und verw. Geb.*, 2:340–386, 1964.
[19] T. P. Speed. Cumulants and partition lattices II: Generalised k-statistics. *J. Aust. Math. Soc. A*, 40:34–53, 1986.
[20] T. P. Speed. Cumulants and partition lattices III: Multiply-indexed arrays. *J. Aust. Math. Soc. A*, 40:161–182, 1986.
[21] T. P. Speed. Cumulants and partition lattices IV: A.S. convergence of generalised k-statistics. *J. Aust. Math. Soc. A*, 40:79–94, 1986.
[22] T. P. Speed and H. L. Silcock. Cumulants and partition lattices V. Calculating generalized k-statistics. *J. Aust. Math. Soc. A*, 44(2):171–196, 1988.

[23] T. P. Speed and H. L. Silcock. Cumulants and partition lattices VI. Variances and covariances of mean squares. *J. Aust. Math. Soc. A*, 44(3):362–388, 1988.
[24] R. Speicher. Multiplicative functions on the lattice of non-crossing partitions and free convolution. *Math. Ann.*, 296:611–628, 1994.
[25] B. Streitberg. Lancaster interactions revisited. *Ann. Stat.*, 18:1878–1885, 1990.

Austral. J. Statist., **25** (2), 1983, 378–388

CUMULANTS AND PARTITION LATTICES[1]

T. P. SPEED

CSIRO Division of Mathematics and Statistics, Canberra

Summary

The (joint) cumulant of a set of (possibly coincident) random variables is defined as an alternating sum of moments with appropriate integral coefficients. By exploiting properties of the Möbius function of a partition lattice some basic results concerning cumulants are derived and illustrations of their use given.

1. Introduction

Cumulants were first defined and studied by the Danish scientist T. N. Thiele (1889, 1897, 1899) who called them *half-invariants* (halvinvarianter); see Hald (1981) for a review of this early work. The ready interpretability and descriptive power of the first few cumulants was evident to Thiele, as was their role in studying non-linear functions of random variables, and these aspects of their use have continued to be important to the present day, see Brillinger (1975, Section 2.3). In a sense which it is hard to make precise, all of the important aspects of (joint) distributions seem to be simpler functions of cumulants than of anything else, and they are also the natural tools with which transformations (linear or not) of systems of random variables (independent or not) can be studied when exact distribution theory is out of the question.

The definition of multivariate cumulant most commonly used today involves moment-generating functions. If $X_1, \ldots, X_m$ is a system of m random variables and $\mathbf{r}=(r_1, \ldots, r_m)$ is an m-tuple of non-negative integers, then the cumulants $\{\kappa_{\mathbf{r}}\}$ of $X_1, \ldots, X_m$ are defined by $\kappa_{0\ldots0}=0$ and the identity

$$\sum_{\mathbf{r}} \kappa_{\mathbf{r}} \frac{\boldsymbol{\theta}^{\mathbf{r}}}{\mathbf{r}!} = \log \sum_{\mathbf{r}} \mathbb{E}\{\mathbf{X}^{\mathbf{r}}\} \frac{\boldsymbol{\theta}^{\mathbf{r}}}{\mathbf{r}!}. \tag{1.1}$$

where we have written $\boldsymbol{\theta}^{\mathbf{r}}=\theta_1^{r_1} \ldots \theta_m^{r_m}$, $\mathbf{X}^{\mathbf{r}}=X_1^{r_1} \ldots X_m^{r_m}$ and $\mathbf{r}!= r_1! \ldots r_m!$, and summed over $r_1 \geqslant 0, \ldots, r_m \geqslant 0$. Here and below all relevant moments are assumed to exist. An alternative approach which

[1] Manuscript received September 22, 1982; revised February 1, 1983.

is in some respects more convenient defines the joint cumulant $\mathscr{C}(X_1, \ldots, X_m)$ of $X_1, \ldots, X_m$ ($\kappa_{1\ldots1}$ in the notation above) directly:

$$\mathscr{C}(X_1, \ldots, X_m) = \sum_{\sigma} (-1)^{b(\sigma)-1}(b(\sigma)-1)! \prod_{a=1}^{b(\sigma)} \mathbb{E}\Big\{\prod_{i\in\sigma_a} X_i\Big\} \tag{1.2}$$

the sum being over all *partitions* σ of $\{1, \ldots, m\}$ into $b = b(\sigma) \geq 1$ *blocks* $\sigma_1, \sigma_2, \ldots, \sigma_b$. For example, if $m = 3$ we have

$$\mathscr{C}(X_1, X_2, X_3) = \mathbb{E}\{X_1X_2X_3\} - \mathbb{E}\{X_1X_2\}\mathbb{E}\{X_3\}$$
$$- \mathbb{E}\{X_1X_3\}\mathbb{E}\{X_2\} - \mathbb{E}\{X_1\}\mathbb{E}\{X_2X_3\} + 2\mathbb{E}\{X_1\}\mathbb{E}\{X_2\}\mathbb{E}\{X_3\}.$$

Note that we have not required that the random variables $X_1, \ldots, X_m$ are all *distinct*. If $X_1 = X_2 = X_3 = X$ in the last formula, we obtain an expression which in the notation of Kendall & Stuart (1969) we recognise to be the formula $\kappa_3 = \mu_3' - 3\mu_1'\mu_2' + 2(\mu_1')^3$. The general multivariate cumulant $\kappa_{\mathbf{r}}$ can be defined via (1.2) in a similar manner.

The purpose of this expository note is to derive some basic results concerning (joint) cumulants from definition (1.2) and give illustrations of their use. Our approach is based upon the fact that (1.2) is an instance of Möbius inversion over the lattice $\mathscr{P}(\underline{m})$ of all partitions of the set $\underline{m} = \{1, \ldots, m\}$, and further use of this technique leads to some new proofs. None of the results we prove are new; our aim is simply to show how a small investment in modern algebra—in this instance the theory of Möbius functions—helps us to step our way elegantly through some potentially messy classical algebra.

It is a great pleasure to be able to contribute to this number honouring Evan Williams. Amongst many other things he introduced me to cumulants and showed me their usefulness, and I hope that this note can convey some of the enjoyment I have found working with them.

2. Lattice Preliminaries

A *partition* σ of a non-empty set S is simply a family of non-empty subsets $\sigma_1, \ldots, \sigma_b$—called the *blocks* of σ—whose union is S. For example, the family $\sigma = \{\{1, 2\}, \{3\}, \{4\}\}$ is a partition of $S = \{1, 2, 3, 4\}$ and we denote it by $\sigma = 12|3|4$. If σ and τ are two partitions of the same set S and every block of σ is contained in a block of τ, then we say that σ is *finer* than τ (τ is *coarser* then σ) and write $\sigma \leq \tau$ ($\tau \geq \sigma$). In this way we find that the collection $\mathscr{P}(S)$ of all partitions of S becomes a *partially-ordered set* and it is in fact a *lattice*, for every pair $\sigma, \tau \in \mathscr{P}(S)$ has a least upper bound and a greatest lower bound in the partial order. The greatest lower bound $\sigma \wedge \tau$ of σ and τ is easy to describe directly: its blocks are just the non-empty intersections of blocks of σ with blocks of τ. For example, $123|4 \wedge 12|34 = 12|3|4$ and $12|34 \wedge 13|24 = 1|2|3|4$ hold in $\mathscr{P}(\underline{4})$. An excellent general

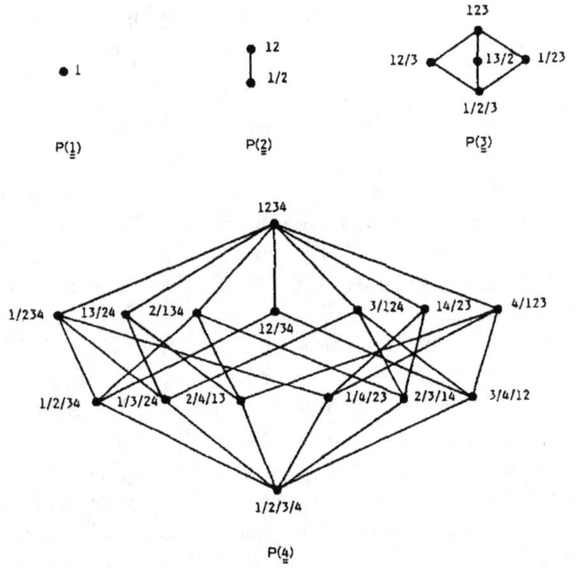

Fig. 1

reference for lattice theory and those results we quote below is Aigner (1979). We illustrate the foregoing with Hasse diagrams of the small partition lattices, see Figure 1.

In these diagrams each element of the partially-ordered set is denoted by a vertex, and an edge is drawn between the vertices corresponding to ρ and τ if $\rho<\tau$ (or $\rho>\tau$) and there is *no* element σ with $\rho<\sigma<\tau$ (or $\rho>\sigma>\tau$).

Associated with *any* finite partially ordered set $(\mathcal{P}, \leq)$ are two important numerical functions defined on $\mathcal{P}$: its *zeta function* $\zeta_{\mathcal{P}}$ given by $\zeta(\sigma,\tau)=1$ if $\sigma\leq\tau$, and 0 otherwise; and its Möbius function $\mu=\mu_{\mathcal{P}}$ which can be defined in many ways, one simple one being the following:

$$\mu(\rho,\tau)=\begin{cases}1 & \text{if} \quad \rho=\tau;\\ -\sum\limits_{\rho\leq\sigma<\tau}\mu(\rho,\sigma) & \text{if} \quad \rho<\tau;\\ 0 & \text{otherwise.}\end{cases}$$

It is not hard to prove that $\sum_\sigma \mu(\rho,\sigma)\zeta(\sigma,\tau)=\sum_\sigma \zeta(\rho,\sigma)\mu(\sigma,\tau)=\delta(\rho,\tau)$ where $\delta(\rho,\tau)=1$ if $\rho=\tau$ and 0 otherwise, i.e. the *matrices* $Z=(\zeta(\sigma,\tau))$ and $M=(\mu(\sigma,\tau))$ over $\mathcal{P}$ are mutually inverse.

Let us suppose that f is a real-valued function on $\mathcal{P}$ and that we define another function F on $\mathcal{P}$ by

$$F(\tau)=\sum_{\sigma\leq\tau} f(\sigma).$$

Thinking of f as a column vector this is saying that $F = Zf$. *Möbius inversion* is just the recovery of f from F: formally, $f = Z^{-1}F = MF$, and more fully

$$f(\tau) = \sum_{\sigma} \mu(\sigma, \tau) F(\sigma).$$

The power of Möbius inversion rests in the fact that for many familiar partially ordered sets, there is a simple formula for μ. Indeed it can be quite a useful technique without even having a formula! These basic ideas apply to any finite partially ordered set and we refer to Aigner (1979) for many illustrations.

It is clear from the definition of μ that if $\rho < \tau$ and there is no σ with $\rho < \sigma < \tau$, then $\mu(\rho, \tau) = -\mu(\rho, \rho) = -1$. Referring to the diagram of $\mathcal{P}(\underset{\sim}{3})$ we can readily calculate that $\mu(1|2|3, 123) = 2$, whilst all other μ-values there are $+1$ or -1. Similarly we find that in $\mathcal{P}(\underset{\sim}{4})$ the following are true: $\mu(12|3|4, 1234) = 2$ whilst $\mu(1|2|3|4, 1234) = -6$. It can be shown that for any $\sigma \in \mathcal{P}(\underset{\sim}{m})$ we have $\mu(\sigma, \underset{\sim}{m}) = (-1)^{b-1}(b-1)!$ where $b = b(\sigma)$ is the number of blocks of σ; a product of such expressions gives a formula for $\mu(\sigma, \tau)$ in $\mathcal{P}(\underset{\sim}{m})$ but we will have no occasion to use it. We refer to Rota (1964), and Aigner (1979) for a proof.

3. Equivalence of the Two Definition

We will begin the proof of the equivalence of the two definitions by seeking an expression for $\mathbb{E}\{X_1 \ldots X_m\}$ in terms of the $\{\kappa_{\mathbf{r}}\}$, and to this end we introduce some notation which plays a fundamental role in what follows. For a partition $\sigma = \sigma_1 | \ldots | \sigma_b$ of $\{1, \ldots, m\}$ let us write

$$\kappa_\sigma = \prod_{a=1}^{b(\sigma)} \kappa_{\mathbf{r}(\sigma_a)}$$

where $\mathbf{r}(\sigma_a) = (r_1, \ldots, r_m)$ is defined by $r_i = 1$ if $i \in \sigma_a$, $r_i = 0$ otherwise, $a = 1, \ldots, b(\sigma)$. For example, if $\sigma = 1234$, then $\kappa_\sigma = \kappa_{1111}$, whilst if $\sigma = 12|34$, then $\kappa_\sigma = \kappa_{1100}\kappa_{0011}$.

Now let us exponentiate both sides of (1.1) and calculate the coefficient of $\theta_1 \ldots \theta_m$ on the left-hand side. It is really quite straightforward to see that the answer is $\sum_\sigma \kappa_\sigma$, where the κ_σ have just been defined and the sum is over all partitions σ of $\{1, \ldots, m\}$. For example, $\mathbb{E}\{X_1X_2X_3X_4\}$ is the sum of 15 terms beginning with $\kappa_{1234} = \kappa_{1111}$ and ending with $\kappa_{1|2|3|4} = \kappa_{1000}\kappa_{0100}\kappa_{0010}\kappa_{0001}$. More generally, if τ is an arbitrary partition of $\{1, \ldots, m\}$ with blocks $\tau_1, \ldots, \tau_b$, then we can multiply expressions of the form just derived to obtain the identity

$$\prod_{a=1}^{b(\tau)} \mathbb{E}\left\{\prod_{i \in \tau_a} X_i\right\} = \prod_{a=1}^{b(\tau)} \sum_{\sigma_a \in \mathcal{P}(\tau_a)} \kappa_{\sigma_a} = \sum_{\sigma \leq \tau} \kappa_\sigma. \tag{3.1}$$

For example, $\mathbb{E}\{X_1X_2\}\mathbb{E}\{X_3\}\mathbb{E}\{X_4\} = (\kappa_{12}+\kappa_{1|2})\kappa_3\kappa_4 = \kappa_{12|3|4}+\kappa_{1|2|3|4}$. Now equation (3.1) can be inverted by Möbius inversion and doing so gives us the fundamental relationship:

$$\kappa_\tau = \sum_{\sigma} \mu(\sigma, \tau) \prod_{a=1}^{b(\sigma)} \mathbb{E}\left\{\prod_{i\in\sigma_a} X_i\right\}. \tag{3.2}$$

When $\tau = \underline{m}$ this reduces to (1.2), apart from the identification of $\mu(\sigma, \underline{m})$ as $(-1)^{b(\sigma)-1}(b(\sigma)-1)!$, and we have proved the equivalence of the definitions.

A more abstract and general theory including this equivalence can be found in Doubilet *et al.* (1972).

Example. Putting $m=4$ we see from (3.2) and Figure 1 that $\kappa_{1234} = \kappa_{1111}$ is an alternating sum of 15 terms with coefficients $+1$, -1, $+2$ and -6. If we identify two or more of the random variables $X_1, \dots, X_4$, additional numerical factors enter because the same expression appears more than once in the 15 terms. At the extreme, when $X_1 = X_2 = X_3 = X_4 = X$, we find cf. Kendall & Stuart (1969, p. 70l) the traditional expression

$$\kappa_4 = \mathbb{E}\{X^4\} - 4\mathbb{E}\{X^3\}\mathbb{E}\{X\} - 3(\mathbb{E}\{X^2\})^2 + 12\mathbb{E}\{X^2\}(\mathbb{E}\{X\})^2 - 6(\mathbb{E}\{X\})^4.$$

Here the factors of -4, -3 and 12 are a combination of multiplicities and Möbius function values.

It is a long standing observation of workers with cumulants that the general results are most transparent when all random variables under discussion are taken as distinct. The identification of some or all at a later stage merely introduces extra factors, and at times these multiplicities are not particularly easy to calculate.

4. Properties of Cumulants

Cumulants of order 2 are just variances and covariances and a number of properties which are familiar in this case seem much less well known in general. Our first result provides a good illustration of the way in which Möbius inversion may be used in this context although its proof using (1.1) is also easy. We take as given a set $X_1, \dots, X_m$ of random variables, and write $\underline{m} = \{1, \dots, m\}$.

Proposition 4.1. If there is a subset $s \subseteq \underline{m}$ such that the random variables $\{X_i : i \in s\}$ and $\{X_i : i \in t\}$ are independent, $t = \underline{m}\setminus s$, then $\mathscr{C}(X_1, \dots, X_m) = 0$.

Proof. For each $\pi \in \mathscr{P}(\underline{m})$ we denote the partition induced on s, i.e. that partition having as blocks the non-empty numbers of $\pi_1 \cap s, \dots, \pi_b \cap s$, by $\pi \cap s$, and similarly for $\pi \cap t$. The proof makes crucial use of the following simple fact: for any $\pi \in \mathscr{P}(\underline{m})$, $\sigma \in \mathscr{P}(s)$

and $\tau \in \mathcal{P}(t)$ we have

$$\pi \geq \sigma \mid \tau \quad \text{iff} \quad \pi \cap s \geq \sigma \quad \text{and} \quad \pi \cap t \geq \tau. \tag{4.1}$$

We can now go to (1.2) and calculate:

$$\begin{aligned}
\mathscr{C}(X_1, \ldots, X_m) &= \sum_\pi \mu(\pi, \underline{m}) \prod_{a=1}^{b(\pi)} \mathbb{E}\Big\{\prod_{i\in\pi_a} X_i\Big\} \\
&= \sum_\pi \mu(\pi, \underline{m}) \prod_{a=1}^{b(\pi\cap s)} \mathbb{E}\Big\{\prod_{i\in\pi_a\cap s} X_i\Big\} \prod_{a=1}^{b(\pi\cap t)} \mathbb{E}\Big\{\prod_{i\in\pi_a\cap t} X_i\Big\}
\end{aligned}$$

by independence,

$$= \sum_\pi \mu(\pi, \underline{m}) \Big\{\sum_\sigma \zeta(\sigma, \pi\cap s)\kappa_\sigma\Big\}\Big\{\sum_\tau \zeta(\tau, \pi\cap t)\kappa_\tau\Big\}$$

by (3.1),

$$= \sum_\pi \sum_\sigma \sum_\tau \mu(\pi, \underline{m})\zeta(\sigma, \pi\cap s)\zeta(\tau, \pi\cap t)\kappa_\sigma\kappa_\tau)$$

$$= \sum_\pi \sum_\sigma \sum_\tau \mu(\pi, \underline{m})\zeta(\sigma \mid \tau, \pi)\kappa_\sigma\kappa_\tau \quad \text{by (4.1)},$$

$$= \sum_\sigma \sum_\tau \delta(\sigma \mid \tau, \underline{m})\kappa_\sigma \tau_\sigma \quad \text{by Möbius inversion}$$

and this expression is zero since $\underline{m} \neq \sigma \mid \tau$ for any $\sigma \in \mathcal{P}(s)$, $\tau \in \mathcal{P}(t)$.

For the next two propositions we consider an array $(X_{ij} : j \in \underline{n}_i, i \in \underline{m})$ of real random variables and a similarly indexed array (a_{ij}) of real numbers. The following result also generalises a well known one for variances and covariances: it states that $\mathscr{C}$ is a *multi-linear* operator.

Proposition 4.2.

$$\mathscr{C}\Big(\sum_{j_1} a_{1j_1}X_{1j_1}, \ldots, \sum_{j_m} a_{mj_m}X_{mj_m}\Big) = \sum_{j_1} \cdots \sum_{j_m} a_{1j_1} \cdots a_{mj_m} \mathscr{C}(X_{1j_1}, \ldots, X_{mj_m}).$$

Proof. From (1.2) and the distributive law

$$\begin{aligned}
&\sum_\sigma \mu(\sigma, \underline{m}) \prod_{a=1}^{b} \mathbb{E}\Big\{\prod_{i\in\sigma_a} \sum_{j_i\in\underline{n}_i} a_{ij_i}X_{ij_i}\Big\} \\
&= \sum_\sigma \mu(\sigma, \underline{m}) \prod_{a=1}^{b} \mathbb{E}\Big\{\sum_{j_i\in\underline{n}_i, i\in\sigma_a} \prod_{i\in\sigma_a} a_{ij_i}X_{ij_i}\Big\} \\
&= \sum_\sigma \mu(\sigma, \underline{m}) \sum_{j_i\in\underline{n}_i, i\in\underline{m}} \prod_i a_{ij_i} \prod_{a=1}^{b} \mathbb{E}\Big\{\prod_{i\in\sigma_a} X_{ij_i}\Big\} \\
&= \sum_{j_1} \cdots \sum_{j_m} a_{1j_1} \cdots a_{mj_m} \sum_\sigma \mu(\sigma, \underline{m}) \prod_{a=1}^{b} \mathbb{E}\Big\{\prod_{i\in\sigma_a} X_{ij_i}\Big\}
\end{aligned}$$

which is the stated result.

Corollary. If $X_i = Y_i + Z_i$, $i = 1, \ldots, m$, where $\{Y_i\}$ and $\{Z_i\}$ are independent sets of m real random variables, then

$$\mathscr{C}(X_1, \ldots, X_m) = \mathscr{C}(Y_1, \ldots, Y_m) + \mathscr{C}(Z_1, \ldots, Z_m).$$

Proof. This is an immediate consequence of Propositions 4.1 and 4.2.

The following proposition is the core of the main result of Leonov & Shiryaev (1959). Our proof is much more direct than theirs and highlights the power of Möbius inversion. Any partition $\pi = \pi_1 | \ldots | \pi_b$ of the row labels $\underline{m}$ of the (X_{ij}) induces a partition $\bar{\pi}$ of the full set $S = \{(i, j) : j \in \underline{n}_i, i \in \underline{m}\}$ of labels in a natural way: $\bar{\pi}$ has blocks $\{(i, j) : j \in \underline{n}_i, i \in \pi_a\}$, $a = 1, \ldots, b(\pi)$. We say that a partition σ of S is *decomposable relative to* a partition π of $\underline{m}$ when $\sigma \leq \bar{\pi}$, where $\bar{\pi}$ has just been defined, and we call σ *indecomposable* if no such relation holds other than $\sigma \leq \underline{m}$. Brillinger (1975, p. 20) gives some equivalent formulations, and states without proof the following.

Proposition 4.3.

$$\mathscr{C}\Big(\prod_{j_1 \in \underline{n}_1} X_{1j_1}, \ldots, \prod_{j_m \in \underline{n}_m} X_{mj_m}\Big) = \sum_{\sigma}{}^{*} \prod_{a=1}^{b(\sigma)} \mathscr{C}(X_{ij} : (i, j) \in \sigma_a)$$

where $\sum^*$ denotes the sum over all *indecomposable* partitions σ of S.

Proof. For any $\pi \in \mathscr{P}(\underline{m})$ we have by (3.1)

$$\prod_{a=1}^{b(\pi)} \mathbb{E}\Big\{\prod_{i \in \pi_a} \prod_{j_i \in \underline{n}_i} X_{ij_i}\Big\} = \sum_{\sigma \leq \bar{\pi}} \prod_{a=1}^{b(\sigma)} \mathscr{C}(X_{ij} : (i, j) \in \sigma_a).$$

The sum on the right, which we denote by $F(\pi)$, is over all $\sigma \in \mathscr{P}(S)$ which are decomposable relative to π. Such σ may also be decomposable relative to some $\rho < \pi$, and so we can use Möbius inversion over $\mathscr{P}(\underline{m})$ to write $f(\pi) = \sum_\rho \mu(\rho, \pi) F(\rho)$ for the corresponding sum over all σ which are decomposable relative to π and no finer partition. With this notation we use (1.2) and Möbius inversion over $\mathscr{P}(\underline{m})$ once more to obtain

$$\mathscr{C}\Big(\prod_{j_1 \in \underline{n}_1} X_{1j_1}, \ldots, \prod_{j_m \in \underline{n}_m} X_{mj_m}\Big) = \sum_{\pi} \mu(\pi, \underline{m}) F(\pi) = f(\underline{m})$$

and the proof is complete.

This proposition provides easy access to a number of results due to Isserlis (1918–19a, b) Bergström (1918–19), Wishart (1928–29, 1929) and others.

Example. Let us take $S = \{(1, 1), (1, 2), (2, 1), (2,2)\}$ which we simplify to $\{1, 2, 3, 4\}$. Then we may refer to the lattice $\mathscr{P}(\underline{4})$ and, by omitting the decomposable partitions $12|34$, $12|3|4$, $1|2|34$ and

$1|2|3|4$ we readily see that

$$\begin{aligned}\mathscr{C}(X_1X_2, X_3X_4) &= \mathscr{C}(X_1, X_2, X_3, X_4)\\ &+ \mathscr{C}(X_1)\mathscr{C}(X_2, X_3, X_4) + \mathscr{C}(X_2)\mathscr{C}(X_1, X_3, X_4)\\ &+ \mathscr{C}(X_3)\mathscr{C}(X_1, X_2, X_4) + \mathscr{C}(X_4)\mathscr{C}(X_1, X_2, X_3) + \mathscr{C}(X_1, X_3)\mathscr{C}(X_2, X_4)\\ &+ \mathscr{C}(X_1, X_4)\mathscr{C}(X_2, X_3) + \mathscr{C}(X_1)\mathscr{C}(X_3)\mathscr{C}(X_2, X_4)\\ &\qquad + \mathscr{C}(X_1)\mathscr{C}(X_4)\mathscr{C}(X_2, X_3)\\ &+ \mathscr{C}(X_2)\mathscr{C}(X_3)\mathscr{C}(X_1; X_4) + \mathscr{C}(X_2)\mathscr{C}(X_4)\mathscr{C}(X_1, X_3).\end{aligned}$$

If X_1, X_2, X_3 and X_4 have a joint normal distribution, then cumulants of order exceeding two all vanish, and in this case if their means are all zero we have

$$\operatorname{cov}(X_1X_2, X_3X_4) = \operatorname{cov}(X_1, X_3)\operatorname{cov}(X_2, X_4) + \operatorname{cov}(X_1, X_4)\operatorname{cov}(X_2, X_4).$$

As a further illustration of this result, let us suppose that $X_1, \ldots, X_n$ are mutually independent and identically distributed random variables with cumulants $\kappa_1 = 0$, κ_2, κ_3, $\kappa_4, \ldots$ (traditional notation). Then for any matrix (a_{ij}) of coefficients, we have

$$\operatorname{var}\left(\sum_i \sum_j a_{ij}X_iX_j\right) = \kappa_4 \sum_i a_{ii}^2 + \kappa_2^2 \sum_i \sum_j (a_{ij}^2 + a_{ij}a_{ji}).$$

The proof is almost immediate once we observe that we require

$$\mathscr{C}\left(\sum_i \sum_j a_{ij}X_iX_j, \sum_i \sum_j a_{ij}X_iX_j\right) = \sum_{i_1} \sum_{i_2} \sum_{i_3} \sum_{i_4} a_{i_1i_2}a_{i_3i_4}\mathscr{C}(X_{i_1}X_{i_2}, X_{i_3}X_{i_4}).$$

Of the 15 possible combinations of equality and inequality on i_1, i_2, i_3 and i_4, each corresponding to an element of $\mathscr{P}(\underline{4})$ in an obvious way, only three give a non-zero cumulant, namely those corresponding to 1234, $13|24$ and $14|23$. Now

$$s^2 = \frac{1}{n-1}\sum_i (X_i - \bar{X})^2 = \frac{1}{n}\sum_i X_i^2 - \frac{1}{n(n-1)}\sum\sum_{i \neq j} X_iX_j$$

and so if we put

$$a_{ii} = \frac{1}{n} \quad \text{and} \quad a_{ij} = a_{ji} = \frac{-1}{n(n-1)}, \quad i \neq j,$$

in the preceding result we obtain the formula which goes back to Gauss (1823);

$$\operatorname{var}(s^2) = \frac{1}{n}\kappa_4 + \frac{2}{n-1}\kappa_2^2.$$

Our final result, due to Brillinger (1969), and generalizes to higher-order cumulants the familiar identity

$$\mathrm{Var}(X) = \mathbb{E}\{\mathrm{Var}(X \mid Y)\} + \mathrm{Var}(\mathbb{E}\{X \mid Y\})$$

for real random variables X and Y. We will use an obvious notation for conditional cumulants.

Proposition 4.4.

$$\mathscr{C}(X_1, \ldots, X_m) = \textstyle\sum_\pi \mathscr{C}(\mathscr{C}(X_i : i \in \pi_a \mid Y) : a \in \underline{b}(\pi)),$$

the sum being over all partitions π of $\{1, \ldots, m\}$.

Proof. The proof which follows is not as simple as Brillinger's, which uses moment-generating functions. A typical term in the expansion (1.2) for $\mathscr{C}(X_1, \ldots, X_m)$ is a product of terms of the form $\mathbb{E}\{\prod_{i \in \pi_a} X_i\} = \mathbb{E}\{\mathbb{E}\{\prod_{i \in \pi_a} X_i \mid Y\}\}$, and we expand the inner term on the right-hand side of this using (3.1), switch the sum and the outer expectation, and use (3.1) once more. Most terms cancel and the simple result is derived. The notational details are somewhat messy, but we proceed.

$$\begin{aligned}\mathbb{E}\left\{\mathbb{E}\left\{\prod_{i \in \pi_a} X_i \mid Y\right\}\right\} &= \sum_{\sigma_a \in \mathscr{P}(\pi_a)} \mathbb{E}\left\{\prod_{k \in \underline{b}(\sigma^a)} \mathscr{C}(X_i : i \in \sigma_a^k \mid Y)\right\} \\ &= \sum_{\sigma_a \in \mathscr{P}(\pi_a)} \sum_{\tau_a \in \mathscr{P}(\underline{b}(\sigma^a))} \prod_{l \in \underline{b}(\tau_a)} \mathscr{C}((\sigma_a^k \mid Y) : k \in \tau_a^l)\end{aligned}$$

where we have abbreviated $\mathscr{C}(X_i : i \in \sigma_a^k \mid Y)$ by $(\sigma_a^k \mid Y)$. Putting this expression into (1.2) we obtain

$$\mathscr{C}(X_1, \ldots, X_m) = \sum_\pi \mu(\pi, \underline{m}) \sum_{\sigma \le \pi} \sum_{\substack{\tau_a \in \mathscr{P}(\underline{b}(\sigma_a)) \\ a = 1, \ldots, b(\pi)}} \prod_{a \in \underline{b}(\pi)} \prod_{l \in \underline{b}(\tau_a)} \mathscr{C}((\sigma_a^k \mid Y) : k \in \tau_a^l)$$

where in the third sum we write $\sigma_a = \sigma \cap \pi^a$, $a = 1, \ldots, b(\pi)$. Our result is proved if we can show that only terms involving $\pi = \underline{m}$, i.e. $b(\pi) = 1$, survive.

To this end suppose that $\sigma \in \mathscr{P}(\underline{m})$ and $\tau \in \mathscr{P}(\underline{b}(\sigma))$ and write

$$P(\sigma, \tau) = \prod_{l \in \underline{b}(\tau)} \mathscr{C}((\sigma^k \mid Y) : k \in \tau^l)$$

$$\rho(\sigma, \tau) = \bigcup_{k \in \tau_1} \sigma^k \,\Big|\, \bigcup_{k \in \tau_2} \sigma^k \,\Big|\, \cdots .$$

Noting that $\rho(\sigma, \tau) \ge \sigma$, we find that the last sum can be written as

$$\sum_\pi \sum_\sigma \sum_\tau \mu(\pi, \underline{m}) \zeta(\pi, \rho(\sigma, \tau)) P(\sigma, \tau) = \sum_\sigma \sum_\tau \delta(\underline{m}, \rho(\sigma, \tau)) P(\sigma, \tau)$$

by Möbius inversion and the result is proved.

Example. For $m = 3$ this result asserts that

$$\begin{aligned}\mathscr{C}(X_1, X_2, X_3) = {} & \mathscr{C}(\mathscr{C}(X_1, X_2, X_3 \mid Y)) + \mathscr{C}(\mathscr{C}(X_1 \mid Y), \mathscr{C}(X_2, X_3 \mid Y)) \\ & + \mathscr{C}(\mathscr{C}(X_2 \mid Y), \mathscr{C}(X_1, X_3 \mid Y)) \\ & + \mathscr{C}(\mathscr{C}(X_3 \mid Y), \mathscr{C}(X_1, X_2 \mid Y)) \\ & + \mathscr{C}(\mathscr{C}(X_1 \mid Y), \mathscr{C}(X_2 \mid Y), \mathscr{C}(X_3 \mid Y)).\end{aligned}$$

If $X_1 = X_2 = X_3 = X$ and we adopt a suggestive notation, the previous expression simplifies to a formula similar to the well-known one for $\mathrm{Var}(X)$:

$$\kappa_3(X) = \mathbb{E}\{\kappa_3(X \mid Y)\} + 3 \operatorname{cov}(\mathbb{E}\{X \mid Y\}, \operatorname{Var}(X \mid Y)) + \kappa_3(\mathbb{E}\{X \mid Y\}).$$

We note in closing that Proposition 4.4 has been used to obtain the cumulants of random sums of (iid) random variables, see e.g. Lange *et al.* (1981).

5. Closing Remarks

The theory of *k-statistics* developed by Fisher (1928–29) and its generalized form involving the so-called *polykays* due to Tukey (1950) is also simplified greatly by a recognition of the role played by the underlying partition lattices and their Möbius functions. For example, it is possible to give a fairly compact proof of a generalization of Fisher's famous result concerning the joint cumulants of sample k-statistics along the lines of that of Proposition 4.3 above.

In a quite different direction, (joint) cumulants of another kind can be defined for arrays of random variables labelled by multiple indices as in a complex experimental design. Here the second order cumulants turn out to be components of variance, and many interesting generalizations of anova notions appear. We leave this and other work to another time.

References

AIGNER, M. (1979). *Combinatorial Theory*. New York: Springer-Verlag.

BERGSTRÖM, S. (1918–19). Sur les moments de la function de correlation normale de n variables. *Biometrika* **12,** 177–183.

BRILLINGER, D. R. (1969). The calculation of cumulants via conditioning. *Ann. Inst. Statist. Math.* **21,** 375–390.

BRILLINGER, D. R. (1975). *Time Series. Data analysis and theory.* New York: Holt, Rinehart and Winston.

DOUBILET, PETER, ROTA, GIAN-CARLO & STANLEY, RICHARD, P. (1972). On the foundations of combinatorial theory VI: The idea of a generating function. *Proceedings of the 6th Berkeley Symposium on Probability and Mathematical Statistics* 267–318. Edited by L. Le Cam, University of California Press.

FISHER, R. A. (1928–29). Moments and product moments of sampling distributions. *Proc. Lond. Math. Soc.* (2) **30,** 199–238.

T. P. SPEED

GAUSS, C. F. (1823). Theoria Combinationis Observationum Erroribus Minimis Obnoxiae Pars Posterior. *Comm. Soc. Reg. Scient. Gotting.* 5. See also: *Werke* Band IV.

HALD, A. (1981). T. N. Thiele's contributions to statistics. *Int. statist. Rev.* **49,** 1–20.

ISSERLIS, L. (1918–19a). On a formula for the product moment coefficient of any order of a normal frequency distribution in any number of variables. *Biometrika* **12,** 134–139.

ISSERLIS, L. (1918–19b). Formulae for determining the mean value of products of deviations of mixed moment coefficients in two to eight variables in samples from a limited population. *Biometrika* **12,** 183–184.

KENDALL, MAURICE G. & STUART, ALAN (1969). *The Advanced Theory of Statistics.* Volume 1 (Third Edition). London: Griffin.

LANGE, KENNETH, BOEHNKE, MICHAEL & CARSON, RICHARD (1981). Moment computations for subcritical branching processes. *J. Applied Probability* **18,** 52–64.

LEONOV, V. P. & SHIRYAEV, A. N. (1959). On a method of calculation of semi-invariants. *Theor. Prob. Appl.* **4,** 319–329.

ROTA, GIAN-CARLO (1964). On the foundations of combinatorial theory I. Theory of Möbius functions. *Zeit. f. Warsch.* **2,** 340–368.

THIELE, T. N. (1889). Forelaesninger over Almindelig Iagttagelseslaere: Sandsynlighedregning og mindste Kvadraters Methode. København: Reitzel.

THIELE, T. N. (1897). *Elementaer Iagttagelseslaere.* København: Gyldendalske. English translation: *Theory of Observations.* London 1903: Layton. Reprinted in *Ann. Math. Statist.* (1931) **2,** 165–308.

THIELE, T. N. (1899). Om Iagttagelseslaerens Halvinvarienter. *Overs. Vid. Sels. Forh. Nr.* **3,** 135–141.

TUKEY, JOHN W. (1950). Some sampling simplified. *J. Amer. Statist. Assoc.* **45,** 501–519.

WISHART, JOHN (1928–29). A problem in combinatorial analysis giving the distribution of certain moment statistics. *Proc. Lond. Math. Soc.* (2) **30,** 309–321.

WISHART, JOHN (1929). The correlation between product moments of any order in samples from a normal population. *Proc. Roy. Soc. Edin.* **49,** 1.

Chapter 7
Asymptotics and Coding Theory: One of the $n \to \infty$ Dimensions of Terry

Bin Yu

Terry joined the Berkeley Statistics faculty in the summer of 1987 after being the statistics head of CSIRO in Australia. His office was just down the hallway from mine on the third floor of Evans. I was beginning my third year at Berkeley then and I remember talking to him in the hallway after a talk that he gave on information theory and the Minimum Description Length (MDL) Principle of Rissanen. I was fascinated by the talk even though I did not understand everything. Terry pointed me to many papers, and before long Terry started to co-advise me (with Lucien Le Cam) as his first PhD student at Berkeley. It was truly a great privilege to work with Terry, especially as his first student at Berkeley since I had the luxury of having his attention almost every day – he would knock on my door to chat about research and to take me to the library to find references. Every Saturday I was invited to have lunch with him and his wife Sally at his rented house in the Normandy Village on Spruce Street, a cluster of rural European styled houses near campus (the most exotic part to me about the lunch was the avocado spread on a sandwich). Through my interactions with Terry, I was molded in $n \to \infty$ dimensions. In particular, I was mesmerised by the interplay shown to me by Terry of data, statistical models, and interpretations – it was art with rigor! I am able to pursue and enjoy this interplay in my current research, even though I ended up writing a theoretical PhD thesis.

The four papers under "asymptotics and coding theory" in this volume represent the MDL research done during my study with Terry (and Rissanen) and a paper after my PhD on Information Theory proper: lossy compression.

The Minimum Description Length (MDL) Principle was invented by Rissanen [7] to formalize Occam's Razor. Based on a foundation of the coding theory of Shannon, its most successful application to date is model selection, now a hot topic again under the new name of sparse modeling or compressed sensing in the high-dimensional situation. An idea closely related to MDL was Minimum Message Length (MML) first articulated in the context of clustering in Wallace and Boulton

B. Yu
Departments of Statistics and Electrical Engineering & Computer Sciences,
University of California, Berkeley
e-mail: binyu@stat.berkeley.edu

S. Dudoit (ed.), *Selected Works of Terry Speed*, Selected Works in Probability and Statistics, 295
DOI 10.1007/978-1-4614-1347-9_7, © Springer Science+Business Media, LLC 2012

[13]. In a nutshell, MDL goes back to Kolmogorov's algorithmic complexity, a revolutionary concept, but not one that is computable. By rooting MDL in Shannon's information theory, Rissanen made the complexity (or code length) of a statistical or probabilistic model computable by corresponding a probability distribution to a prefix code via Kraft's inequality. At the same time, this coding interpretation of probability distribution removed the necessity of postulating a true distribution for data, since it can be viewed operationally as a code-generating device. This seemingly trivial point is fundamental for statistical inference. Moreover, Rissanen put MDL on solid footing by generalizing Shannon's order source coding theorem to the second order to support the coding forms valid for use in MDL model selection. That is, he showed in Rissanen [8] that, for a nice parametric family of dimension k with n iid observations, they have to achieve a $\frac{k}{2}\log n$ lower bound asymptotically beyond the entropy lower bound when the data generating distribution is in the family. More information on MDL can be found in the review articles Barron et al. [3] and Hansen and Yu [5], and books Rissanen [6, 9] and Grünwald [4].

Not long before he and I started working on MDL in the late 1987, Terry had met Jorma Rissanen when Jorma visited Ted Hannan at the Australia National University (ANU). Hannan was a good friend of Terry. Jorma's homebase was close by, the IBM Almaden Research Center in San Jose, so Terry invited him to visit us almost every month. Jorma would come with his wife and discuss MDL with us while his wife purchased bread at a store in Berkeley before they headed home together after lunch. We found Rissanen's papers original, but not always easy to follow. The discussions with him in person were a huge advantage for our understanding of MDL.

After catching up with the literature on MDL and model selection methods such as AIC [1] and BIC [11], we were ready to investigate MDL from a statistical angle in the canonical model of Gaussian regression and became among the first to explore MDL procedures in the nonparametric case, using the convenient and canonical histogram estimate (which is both parametric and nonparametric). This line of research resulted in the first three papers on asymptotics and coding in this volume.

The research in Speed and Yu [12] started in 1987. The paper was possibly written in 1989, with many drafts including extensive comments by David Freedman on the first draft and it was a long story regarding why it took four years to publish. By then, it was well-known that AIC is prediction optimal and inconsistent (unless the true model is the largest model), while BIC is consistent when the true model is finite and one of the sub-regression models considered. Speed and Yu [12] addresses the prediction optimality question with refitting (causal or on-line prediction) and without refitting (batch prediction). A new lower bound on the latter was derived with sufficient achievability conditions, while a lower bound on the former had been given by Rissanen [8]. Comparisons of AIC, stochastic complexity , BIC, and Final Prediction Error (FPE) criteria [1] were made relative to the lower bounds and in terms of underfitting and overfitting probabilities. A finite-dimensional (fixed p to use modern terms) Gaussian linear regression model was assumed, as was common in other works around that time or before. The simple but canonical Gaussian regression model assumption made the technical burdens minimal, but it was sufficient to reveal useful insights such as the orders of bias-variance trade-off when there was underfit or overfit, respectively. Related trade-offs are seen in analysis of modern model selection (sparse modeling) methods such as Lasso under high-dimensional regression models (large p large n). In fact, Speed and Yu [12] entertained the idea of a high-dimensional model through a discussion of finite dimensional models vs infinite dimensional models. In fact, much insight from this paper is still relevant today: BIC does well both in terms of consistency and prediction when the bias term drastically decreases to a lower level at a certain point (e.g. a "cliff" bias decrease when there is a group of major predictors and rest marginal). Working with Terry on this first paper of mine taught me lessons that I try to practice to this day: mathematical derivations in statistics should have meanings and give insights, and a good formulation of a problem is often more important than solving it.

The next two papers, Rissanen et al. [10] and Yu and Speed [14], are on histograms and MDL. They extend the MDL paradigm to the nonparametric domain. Around the same time Barron and Cover were working on other nonparametric MDL procedures through the resolvability index [2]. Rissanen spearheaded the first of the two papers, Rissanen et al. [10], to obtain a (properly defined) code length almost sure lower bound in the nonparametric case in the same spirit as the lower bound in the parametric case of his seminal paper [7]. This paper also showed that a histogram estimator achieve this lower bound. The second paper [14] introduced the minimax framework to address both the lower and upper code length bound questions for Lipschitz nonparametric families. Technically the paper was quite involved with long and refined asymptotic derivations, a Poissonization argument, and multinomial/Poisson cumulant calculations for which Terry showed dazzling algebraic power. A surprising insight from the second paper was that predictive MDL seemed a very flexible way to achieve the minimax optimal rate for expected code length. Working on the two histogram/MDL papers made me realize that there is no clear cut difference between parametric and nonparametric estimation: the so-called infinite dimensional models such as the Lipschitz family actually correspond to parametric estimation problems of dimensions increasing with the sample size. This insight holds for all nonparametric estimation problems and the histogram is a concrete example of sieve estimation.

The last of the four paper was on lossy compression of information theory proper. MDL model selection criteria are based on lossless code (prefix code) lengths. The aforementioned lower bound in Rissanen [7] was also fundamental for universal source (lossless) coding when the underlying data generating distribution has to be estimated, in addition to being the cornerstone of the MDL theory in the parametric case. It was natural to ask whether there is a parallel result for lossy compression where entropy is replaced by Shannon's rate-distortion function. Yu and Speed [15] showed it was indeed the case and there are quite a few follow-up papers in the information theory literature including Zhang et al. [16].

During my study with Terry, starting in the late 1987, Terry was moving full steam into biology as a visionary pioneer of statistical bioinformatics. To accommodate my interest in analysis and asymptotic theory and possibly pursue his other love for information theory rather than biology, Terry was happy to work with me on theoretical MDL research and information theory, an instance of Terry's amazing intellectual versatility as amply clear from this volume.

References

[1] H. Akaike. A new look at the statistical model identification. *IEEE Trans. AC*, 19:716–723, 1974.

[2] A. R. Barron and T. M. Cover. Minimum complexity density estimation. *IEEE Trans. Inform. Theory*, 37:1034–1054, 1991.

[3] A. R. Barron, J. Rissanen, and B. Yu. The minimum description length principle in coding and modeling. *IEEE Trans. Inform. Theory*, 44:2743–2760, 1998.

[4] P. D. Grünwald. *The Minimum Description Length Principle*. MIT Press, Boston, 2007.

[5] M. H. Hansen and B. Yu. Model selection and the principle of minimum description length. *J. Am. Stat. Assoc.*, 96:746–774, 2001.

[6] J. Rissanen. *Information and Complexity in Statistical Modeling*. Springer, New York, 2007.

[7] J. Rissanen. Modeling by shortest data description. *Automatica*, 14:465–471, 1978.

[8] J. Rissanen. Stochastic complexity and modeling. *Ann. Stat.*, 14:1080–1100, 1986.

[9] J. Rissanen. *Stochastic Complexity and Statistical Inquiry*. World Scientific, Singapore, 1989.

[10] J. Rissanen, T. P. Speed, and B. Yu. Density estimation by stochastic complexity. *IEEE Trans. Inform. Theory*, 38:315–323, 1992.

[11] G. Schwarz. Estimating the dimension of a model. *Ann. Stat.*, 6:461–464, 1978.

[12] T. P. Speed and B. Yu. Model selection and prediction: Normal regression. *Ann. Inst. Stat. Math.*, 45(1):35–54, 1993.

[13] C. S. Wallace and D. M. Boulton. An information measure for classification. *Computer J.*, 11:185–194, 1968.

[14] B. Yu and T. P. Speed. Data compression and histograms. *Probab. Theory Relat. Fields*, 92:195–229, 1992.

[15] B. Yu and T. P. Speed. A rate of convergence result for a universal D-semifaithful code. *IEEE Trans. Inform. Theory*, 39:813–820, 1993.

[16] Z. Zhang, E. Yang, and V. K. Wei. The redundancy of source coding with a fidelity criterion. *IEEE Trans. Inform. Theory*, 43:71–91, 1997.

IEEE TRANSACTIONS ON INFORMATION THEORY, VOL. 38, NO. 2, MARCH 1992

Density Estimation by Stochastic Complexity

Jorma Rissanen, *Senior Member, IEEE*, Terry P. Speed, Bin Yu

***Abstract*—The results by Hall and Hannan on optimization of histogram density estimators with equal bin widths by minimization of the stochastic complexity, are extended and sharpened in two separate ways. As the first contribution, two generalized histogram estimators are constructed. The first has unequal bin widths which, together with the number of the bins, are determined by minimization of the stochastic complexity with help of dynamic programming. The other estimator consists of a mixture of equal bin width estimators, each of which is defined by the associated stochastic complexity. As the main contribution in this paper, two theorems are proved, which together extend the universal coding theorems to a large class of data generating densities. The first gives an asymptotic upper bound for the code redundancy in the order of magnitude, achieved with a special predictive type of histogram estimator, which sharpens a related bound. The second theorem states that this bound cannot be improved upon by any code whatsoever.**

***Index Terms*—MDL Principle, universal coding, histograms, asymptotic bounds, variable bin widths.**

I. Introduction

The MDL (minimum description length) principle to nonparametric density estimation is applied in this paper. This principle permits us to compare any two density estimators based upon a finite set of observed data by the codelength with which the data together with the estimator itself can be encoded. We prefer an estimator that achieves a short total codelength, which means that the best estimators are such that they assign high probabilities to clusters of the data points while at the same time the estimators themselves are not too complex to describe. Hence, for example, a histogram estimator with a large number of bins will not necessarily be good, because we have to describe, one way or another, the large number of counts of the observations falling in these bins. Similarly, the usual kernel estimators, which are formed as a sum of functions, one centered at each observed data point, are bad, because to describe them we need at least as many bits as for the description of the data points themselves. However, such estimators can be greatly simplified by retaining just enough functions to permit a good fit to the data, the number of them being subject to optimization. Such pruned-down kernel estimators turn out to have quite short codelengths [21].

In [11], an idealized codelength, the stochastic complexity, based upon the class of histogram estimators with equal-width bins was computed, which when minimized gave the optimal number of the bins and the associated density estimator. This estimator turns out to be good for data that are roughly uniformly distributed. However, when the distribution is strongly nonuniform, for instance having a long sparse tail, then many of the optimized number of bins may have very few data points or none at all, and one may then say that for the sparse portion of the data the density function is described with unnecessary detail. For such reasons, we extend the Hall–Hannan stochastic complexity calculation to the class of histogram estimators with variable-width bins, which can be calculated with dynamic programming. Despite an increased number of additional parameters to be encoded, the resulting codelength can be shorter than the Hall–Hannan stochastic complexity, while never exceeding it by more than about three bits for the entire data string. For small data sets histogram estimators lack smoothness. However, by constructing an estimator as a mixture of many equal bin width histograms we achieve a degree of "smoothness," not locally in terms of continuity or differentiability, but in a broader sense, without sacrificing efficiency. The analysis of the new estimators appears to be difficult, and we compare their performance with the equal bin width histogram estimator in an example.

It was shown in [11] that the optimal number of bins is also asymptotically of the correct magnitude to minimize the largest absolute deviation of the histogram estimator from the data generating density, in a fairly large family of nonparametric densities. Although such a result lends support to the idea of MDL principle providing good estimators, the support is somewhat indirect: the optimality is in terms of a sensible but still arbitrary distance measure. As the main analytic result in this paper, we prove another, stronger optimality property of a complexity based estimator, denoted $f^*(y \mid x^t)$, which is extended to a family of densities $f_n^*(x^n)$ for sequences $x^n = x_1, \cdots, x_n$ predictively by multiplication. In broad terms, we show that this estimator gives asymptotically the shortest codelength for the data in the order of magnitude that can be achieved by any density estimator, be it of histogram type or not, relative to the class $\mathcal{M}$ of densities $f(y)$ defined on the unit interval, which are uniformly bounded and also bounded away from zero and from infinity, and each having a bounded first derivative. These are extended to sequences by independence with the result $f^n(x^n)$. Moreover, we spell out the shortest mean

Manuscript received December 6, 1989. T. P. Speed and B. Yu are supported in part by NSF Grant DMS 88-02378.
J. Rissanen is with IBM Almaden Research Center, 650 Harry Road, Room K52/802, San Jose, CA 95120-6099.
T. P. Speed is with the Department of Statistics, Evans Hall, 3rd Floor, University of California, Berkeley, CA 94720.
B. Yu is with the Department of Statistics, University of Wisconsin, Madison, WI 53706.
IEEE Log Number 9104812.

0018-9448/92$03.00

IEEE TRANSACTIONS ON INFORMATION THEORY, VOL. 38, NO. 2, MARCH 1992

codelength, which we define to be the asymptotic stochastic complexity of the data, relative to the nonparametric model class $\mathcal{M}$ in question. More specifically, our first theorem states that

$$\frac{1}{n} E_f \log \frac{f^n(x^n)}{f_n^*(x^n)} \le A_f(n^{-2/3}), \tag{1.1}$$

while the second theorem states that for any family of density estimators $\{g_t(x^t)\}$

$$\frac{1}{n} E_f \log \frac{f^n(x^n)}{g_n(x^n)} \ge Kn^{-2/3} \tag{1.2}$$

holds for some positive constant K and all densities $f \in \mathcal{M}$, except for a set that is asymptotically ignorable in a suitable sense. We recall [17], [18] that for a model class with k free parameters, the right-hand side in (1.2), which represents the shortest codelength required to encode the optimal model with which the code for the data is designed, is $\frac{k}{2n} \log n$. Hence, we see that it takes a longer code to describe the estimator in the non-parametric family, as it should.

In [1], density estimators $\hat{p}$, minimizing a codelength criterion of the form $L(x^n, q) = -\log q(x^n) + L_n(q)$ were studied, where the second term denotes a prefix codelength for the estimator. Instead of providing an explicit construction for this length the authors specify it abstractly by certain properties. As their main contribution, the authors define the *index of resolvability* and show it to provide an upper bound both for the code redundancy, as well as for the Hellinger distance between the "true" density and its estimator of the form $\hat{p}(y \mid x^n)$, in probability. Further, an asymptotic formula is given for the index of resolvability. There are three main differences between their work and ours. First, we give an explicit construction for a density estimator, obtained with the MDL principle. Second, the class of estimators, provided by the two-stage codelength in [1], excludes those which do not satisfy the imposed condition that $L_n(\cdot)$ depends only on n as well as the important estimators obtained by a predictive coding process or by stochastic complexity. This is because the codelength for the data, resulting from these estimators, cannot be separated into codelengths for the model and the data, and hence the index of resolvability is inapplicable to them. By contrast, our second theorem does apply, not only to predictively constructed estimators but to estimators of any kind.

II. Histogram Estimators and Universal Codes

In [11], the *stochastic complexity* (for a general definition, see [21, Section 3.2]) of a set of observed data, relative to the class of histogram densities with equal size bins, their number to be optimized, was derived and the associated density estimator constructed. Another paper with similar ideas is [6], based on an earlier paper [5]. In the former, Dawid considers what is in effect a density estimate obtained from the equal bin width stochastic complexity in predictive form, and he demonstrates through simulations some of its desirable properties.

Although the histogram-like density estimators with the number of bins optimized will be shown to have strong asymptotic optimality properties among all density estimators whatsoever, other estimators may well perform better for small and medium size data or have other desirable properties such as a great degree of smoothness. For example, the various kernel estimators can be designed to provide any desired degree of smoothness, and the number of functions can be optimized with the MDL principle even though we cannot calculate the stochastic complexity for them in a closed form. In this section we study two generalizations of the usual histogram estimators, in both of which we can take advantage of the closed form solution for the stochastic complexity. The first class of estimators have variable-length bins, the lengths as well as the number of the bins determined by optimizing the stochastic complexity. This class shares the asymptotic optimality properties of the equal bin width histogram estimators. The other class consists of a mixture of a collection of ordinary equal bin width histograms, aimed at providing increased over-all "smoothness."

We begin by generalizing the Hall–Hannan complexity and the associated density estimators to histograms with variable-width bins. For convenience of notation we index the observed data so that $x_1 \le x_2 \le \cdots \le x_n$, and note that the indexes need have no bearing on the time order of their arrival. Without loss of generality, we take all the observed data points as integers with the smallest $x_1 = 0$, and we write $x^n = x_1, \cdots, x_n$. Let $a = (a_1, \cdots, a_{m-1})$ denote an increasing sequence of the end points of m bins $[0, a_1], (a_1, a_2], \cdots, (a_{m-1}, R]$, partitioning the range $[0, R]$; let $R_i = a_i - a_{i-1}$, $a_0 = 0$ and $a_m = R$, denote the length of the ith bin. Next, consider parametric histogram densities defined by $f(y \mid p, R, m, a) = \frac{p_i}{R_i}$, if y falls in the ith bin, where $p = (p_1, \cdots, p_m)$ denotes nonnegative parameters with sum unity.

With the uniform prior $\pi(p) = (m-1)!$ on the simplex defined by the parameters, we can evaluate the integral

$$f(x^n \mid R, m, a) = \int \prod_{t=1}^{n} f(x_t \mid p, R, m, a) \pi(p) \, dp$$

$$= \left(\prod_{i=1}^{m} R_i^{-n_i} \right) \frac{(m-1)! \Pi_i n_i!}{(m+n-1)!}. \tag{2.1}$$

Then the stochastic complexity, $I(x^n \mid R, m, a) = -\log f(x^n \mid R, m, a)$, fixing n, R, m and a, is given by

$$I(x^n \mid R, m, a) = \sum_{i=1}^{m} n_i \log R_i + \log \binom{n}{n_1, \cdots, n_m} + \log \binom{n+m-1}{n}, \tag{2.2}$$

in terms of the multinomial

$$\binom{n}{n_1, \cdots, n_m} = \frac{n!}{\Pi_i n_i!}$$

and the binomial

$$\binom{n+m-1}{n} = \frac{(n+m-1)!}{n!(m-1)!}$$

coefficients. Here, n_i denotes the number of observations that fall in the ith bin. In the special case $R = 1$ and $R_i = 1/m$ we get the Hall–Hannan complexity, which we write as $I(x^n \mid m)$.

We may interpret the three terms in (2.2) as codelengths corresponding to a particular encoding process. (Although a codelength is an integer-valued quantity it is convenient to regard the negative logarithm of a probability as a kind of idealized codelength, and with a further idealization we call even the negative logarithm of a density a codelength [17].) The last term is the length required to encode the m nonnegative integers n_i, when m is given; this is a special case of (2.4), derived next. The first two terms give the codelength for the observations when we imagine each to be specified by a pair (i, y), where i gives the bin (the ith) in which the observation falls, and y gives the position of the observation within this bin. Encoding of y, when we know that it belongs to the ith bin, clearly takes about $\log R_i$ bits, and the first term in (2.2) is the sum of these over all the n observations. Finally, the second term in (2.2) is the codelength required to encode the bin numbers (the first component in (i, y)) of all the n observations, for it is the logarithm of the number of all strings of length n in m symbols with the given counts. Another, predictive encoding process is defined by the conditional densities in (2.7), and taking the sum of their negative logarithms gives exactly the same codelength (2.2) for the same parameter values.

We can find the optimal sequence of the bins by dynamic programming. However, since the codelength required to encode the sequence of the bins, which must be added to (2.2), may be large, we generally get a shorter overall codelength if the end points of the bins are suitably restricted. We do this by introducing two parameters. The first is the precision, an integer d, with which the break points a_i are expressed; in other words, $a_i = k_i d$ and $\boldsymbol{a} = d\boldsymbol{k}$, where $\boldsymbol{k} = (k_1, \cdots, k_{m-1})$. The second new parameter is the minimum bin width we permit, say κd, also expressed as a multiple of d. To apply the dynamic programming argument, subdivide the interval $[0, R]$, where R is taken as a variable multiple of d, into $m+1$ bins with the break points $\boldsymbol{a} = a_1, \cdots, a_{m-1}, \tau$ and write $\boldsymbol{a}' = a_1, \cdots, a_{m-1}$. Then from (2.2) we get by a straightforward calculation a decomposition of the form $I(x^{n(R)} \mid R, m+1, \boldsymbol{a}) = I(x^{n(\tau)} \mid \tau, m, \boldsymbol{a}') + \cdots$, where $x_1, \cdots, x_{n(R)}$ denotes the portion of the observed data falling within $[0, R]$, and similarly for $x^{n(\tau)}$. The remaining term, represented by the dots, is given by the last three terms in (2.3). Next, let

$$L_m(R) = \min_{\boldsymbol{a}} I(x^{n(R)} \mid R, m, \boldsymbol{a}),$$

where $\boldsymbol{a} = d\boldsymbol{k}$ with $k_{i+1} - k_i \geq \kappa$ and $k_{m-1} \leq k - \kappa$, and $k = R/d$. By the dynamic programming argument, we then get the recursion

$$L_{m+1}(R) = \min_{\tau} \left\{ L_m(\tau) + (n(R) - n(\tau)) \log(R - \tau) + \log\binom{n(R)+m}{n(\tau)+m} + \log\frac{n(\tau)+m}{m} \right\}, \tag{2.3}$$

where τ, besides being less than R, is also restricted to be a multiple of d as well as by the requirement of the minimum bin width. The recursive equations are solved for $m \geq 1$ and for $R = d\kappa, d(\kappa+1), \cdots$, until the desired range including all the observations is reached. The initial value is $L_1(R) = n(R) \log R$ for all R. A recursive evaluation of (2.3) for the desired value of m and the range gives both the minimized stochastic complexity and the optimal sequence of the bin boundaries $\hat{\boldsymbol{a}}$ with about $(R/d)^2$ operations.

We need the codelengths required to encode the various integer-valued parameters, of which we first consider the increasing sequence $\boldsymbol{k}$ with $k_i - k_{i-1} \geq \kappa$ for $i = 1, \cdots, m-1$, $k_0 = 0$, and $k_m = k$. To get this length, associate with the sequence in a one-to-one fashion a binary string as follows. Begin with $k_1 - \kappa$ 0's and a 1, followed by $k_2 - k_1 - \kappa$ 0's and a 1, and so on until $k - k_{m-1} - \kappa$ 0's are added, followed by the last 1. The string has $k - m\kappa$ 0's and m 1's, and it always ends with a 1. Hence, the codelength required for such a sequence is to within one bit

$$L(\boldsymbol{k}) = \log\binom{k - m(\kappa-1) - 1}{m-1}. \tag{2.4}$$

This (nonprefix) length estimate is valid provided that m, $k = R/d$, and κ are given. In fact, we need to encode the four parameters m, d, κ, and k, since in general the range R cannot be regarded as given. The code for these four integers must be a prefix code, for we must be able to decode them from a preamble in the entire code string without a separating comma. We recall that a positive integer i can be encoded in a prefix manner with about $L^*(i) = 1.5 + \log i + \log\log i + \cdots$ bits, where the series includes all the positive terms [8], [16]. Hence, we can encode the four parameters with the length $L(d, m, \kappa, k) = L^*(d) + L^*(m) + L^*(\kappa) + L^*(\max\{1, k - m\kappa\})$ bits. The best codelength we can get for the data sequence using variable-length bins by this procedure is then

$$L_V(x) = \min_{m,k,d,\kappa} \{ I(x \mid \boldsymbol{k}, m, d\boldsymbol{k}) + L(\boldsymbol{k}) + L(d, m, \kappa, k) \}. \tag{2.5}$$

For each m, d, κ, and k only the first term in (2.5) depends on the sequence $\boldsymbol{k}$, and the minimization is done by the recursion (2.3). The minimization with respect to the remaining three bounded integer-valued parameters (k being determined by d and the range, which is not subject to optimization) is to be done by exhaustive search.

Consider the choice $d = 1$ and $\kappa = \lceil (k/m)$, where $\lceil x$ denotes the least integer upper bound for x. This forces the bins to have equal lengths, which means that $L(\boldsymbol{k}) = 0$, and

IEEE TRANSACTIONS ON INFORMATION THEORY, VOL. 38, NO. 2, MARCH 1992

with $k = R$ we have

$$L(1, m, \kappa, R) = L(m, R) + 2L^*(1),$$

where the first term denotes the prefix codelength for m and the range R, both to be added to the Hall–Hannan complexity to make it complete. We then see that the codelength $L_V(x^n)$ of the optimal variable bin width code never exceeds the length, say $L_E(x^n)$, of the optimal equal bin width code by more than about $2L^*(1) = 3$ bits. A further (trivial) subclass of uniform densities results from the choice $d = 1$ and $m = 1$.

Once the optimal parameters $\hat{m}$, $\hat{d}$, $\hat{\kappa}$, and $\hat{k}$, minimizing the stochastic complexity are found, we generally wish to construct a density estimate. One way is to calculate the natural histogram estimator

$$\hat{f}_V(y \mid x^n) = \frac{n_i}{n} \hat{R}_i^{-1}, \tag{2.6}$$

for y in the ith bin with length $\hat{R}_i = (\hat{k}_i - \hat{k}_{i-1})\hat{d}$. Another is defined by (2.1) as

$$\hat{f}(y \mid x^n) = \frac{f(x^n y \mid R, \hat{m}, \hat{a})}{f(x^n \mid R, \hat{m}, \hat{a})} = \frac{n_i + 1}{n + \hat{m}} \hat{R}_i^{-1}, \tag{2.7}$$

for y also in the ith bin; the pair $x^n y$ denotes the string $x_1, \cdots, x_n, y$ of length $n + 1$.

We next describe the estimator obtained as a mixture of the equal bin width histograms. Writing first

$$f(y \mid x^n, m) = \frac{n_i + 1}{n + m} \frac{m}{R}, \tag{2.8}$$

for the special case of (2.7) with equal bin widths, we define the mixture density estimator as

$$\hat{f}_M(y \mid x^n) = \frac{1}{M} \sum_{m=1}^{M} f(y \mid x^n, m)$$
$$= \frac{1}{RM} \sum_{m=1}^{M} m \frac{n_i + 1}{n + m}, \tag{2.9}$$

where, again, i is the index of the bin in which y falls, and n_i is the number of the data points that fall within this bin. The number M is taken as a parameter to be optimized. With this estimator the data sequence can be encoded with the codelength

$$-\log \hat{f}_M(x^n) = -\sum_{t=0}^{n-1} \log f_M(x_{t+1} \mid x^t). \tag{2.10}$$

Example: We calculated the optimal codelength $L_V(x^n) = 572$, obtained with the parameters $\hat{m} = 4$, $\hat{d} = 6$, $\hat{\kappa} = 18$, $k = 18$, and $\hat{k} = 12, 14, 16$ for the set of 76 integers 0, 7, 18, 39, 49, 50, 61, 80, 82, 82, 82, 82, 84, 84, 85, 86, 88, 89, 89, 91, 91, 92, 92, 92, 92, 92, 93, 95, 96, 96, 101, 101, 101, 101, 105, 107, 107, 111, 112, 115, 115, 116, 117, 117, 118, 119, 119, 121, 122, 123, 124, 124, 125, 125, 125, 129, 129, 129, 130, 131, 196, 201, 201, 203, 212, 232, 236, 241, 241, 243, 243, 243, 245, 246, 248, 248. We also calculated the "fit," $-\log f(x^n \mid R, \hat{m}, \hat{a}) = 549$. Fig. 1 shows the corresponding density estimator marked "variable bin width." The optimal number of equal-length bins is $\hat{m} = 13$, which gives the total codelength $L_E(x^n) = 577$ and the "fit" 554. In Fig. 1, the associated density estimator is marked "equal bin width." Finally, we calculated the total codelength for the mixture estimator with the optimized number $\hat{M} = 16$ as $L_M(x^n) = 578$, and the "fit" $-\log \hat{f}_M(x^n) = 557$. The dependence of the codelength on the number of terms M in the mixture is very slight, and we can pick it in the form of an integer power of two. The associated density estimator is marked with "mixture" in Fig. 1.

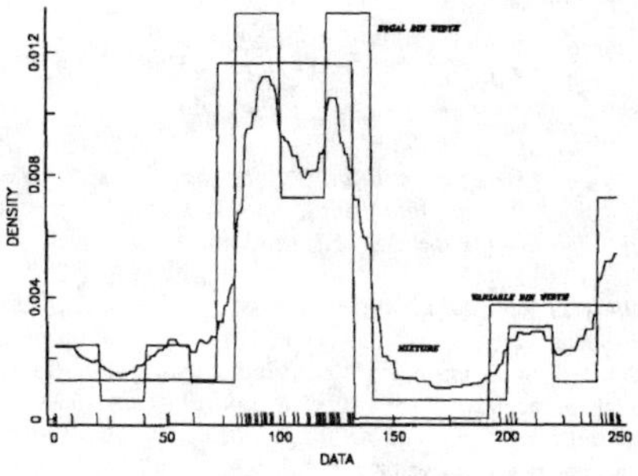

Fig. 1. Three density estimators.

Due to the relatively small data set the codelengths obtained with all the three estimators are virtually the same, despite the fact that the estimators differ considerably. We see in Fig. 1 that the large optimal number of bins, 13, in the equal bin width density estimator makes it somewhat "jumpy" in creating perhaps needlessly many local maxima and minima. The four bins in the variable bin width estimator, by contrast, give a less ragged density function. By far the best looking estimator, however, is the mixture density, in which, unlike in the usual kernel and spline estimators, "smoothness" is achieved without imposing analytic continuity. As a practical matter, both the equal bin width and the mixture estimators are easy to calculate, requiring only $O(n)$ number of operations for n observations, which makes them feasible to compute even for multidimensional data. By contrast, the variable bin width estimator requires $O(R^2)$ operations, which just about confines their calculations for scalar observations only.

III. Asymptotic Optimality

Ordinarily the goodness of a density estimator is expressed in terms of a suitably chosen distance measure between the estimated density and an assumed data generating one in some class. In this paper, in accordance with the MDL principle, we have taken the codelength with which the data and the estimator itself can be encoded as the yardstick for the quality of an estimator. The purpose of this section is to derive optimum asymptotic rates, in the order of magnitude, for the expected codelengths, relative to a class of smooth data generating densities, and to demonstrate an estimator and the associated universal code that achieves the optimal rate in the order of magnitude. The same measure also translates into the Kullback distance between the estimated and the data generating densities, which provides further support for the codelength criterion. Specifically, the class $\mathcal{M}$ consists of densities $f(y)$, defined on the unit interval such that $0 < c_0 \le f(y) \le c_1$, $c_0 < 1$, $c_1 > 1$, and each density having a bounded first derivative, say $|f'(y)| \le c_f$. This class is larger than the class $\mathcal{M}'$ considered in [11], in which the absolute derivative was required to be bounded uniformly over the densities. It was shown in [11] that for each density function f in the class $\mathcal{M}'$ the number of bins minimizing (2.2) for $R = 1$ satisfies $\hat{m}(x^n)/(n/\log n)^{1/3} \to C_f$ in probability, where C_f is a constant for each f. Moreover, the corresponding optimal bin width $1/\hat{m}(x^n)$ is also of the correct order of magnitude for minimizing the largest absolute deviation of the histogram estimator from any data generating density in the same class. With this number of bins, the stochastic complexity (2.2), denoted now by $I(x^n \mid m_n)$, behaves asymptotically like

$$\frac{1}{n} I(x^n \mid m_n) \approx -\frac{1}{n} \log f(x^n) + K\left(\frac{\log n}{n}\right)^{2/3} \quad (3.1)$$

in probability, and the second term gives also the amount by which the mean-per-symbol stochastic complexity exceeds the entropy. Since the codelength provided by the variable bin width estimator, constructed in the previous section, exceeds the optimal equal bin width estimator $I(x^n \mid m_n)$ only by at most three bits, its mean-per-symbol length, too, is asymptotically no greater than the right-hand side.

However, we get a smaller excess term for a different density estimator, constructed from the stochastic complexity $I(x^n \mid m)$ in a predictive way. This estimator is defined in terms of the conditional densities (2.8), rewritten here for $R = 1$ and $0 \le t$,

$$f(x_{t+1} \mid x^t, m) = \frac{t_i + 1}{(t + m)} m, \quad (3.2)$$

where we let the number of bins grow with t as $m_t^* = \lceil t^{1/3}$ to take advantage of an increasing information. Writing

$$f^*(y \mid x^t) = f(y \mid x^t, m_{t+1}^*), \quad (3.3)$$

where x^0 is the empty string, we obtain for any string $f_n^*(x^n) = \Pi_{t=1}^n f^*(x_t \mid x^{t-1})$, regardless of the ordering of the observations. Notice that the negative logarithm of $f_n^*(x^n)$ is not the stochastic complexity $I(x^n \mid m)$ for any single value of m, but rather it is the sum of the increments $I(x^{t+1} \mid m_t^*) - I(x^t \mid m_t^*)$. It is interesting that here the mean predictive codelength is asymptotically strictly shorter than the nonpredictive one. This is in contrast with all the parametric model classes studied, where the two mean lengths are asymptotically equal.

Theorem 1: For all $f \in \mathcal{M}$,

$$\frac{1}{n} E_f \log \frac{f^n(x^n)}{f_n^*(x^n)} \le A_f n^{-2/3}, \quad (3.4)$$

where $f^n(x^n) = \Pi_{t=1}^n f(x_t)$ and A_f a number dependent on f. Also,

$$E_f \int_0^1 f(y) \log \frac{f(y)}{f^*(y \mid x^n)} dy \le B_f n^{-2/3}, \quad (3.5)$$

B_f is a number dependent of f. The expectation E_f is taken with respect to f^n over the data sequences x^n.

The proof is given in Appendix A.

The question arises whether any code exists with shorter mean length in the order of magnitude than given by the right-hand side of (3.4). Just as for parametric model classes [17], [18], we cannot expect this to be the greatest lower bound for *all* data generating densities, since one designed with f clearly reaches the entropy, but what we can expect is the right-hand side of (3.4) to represent, in the order of magnitude, the shortest possible mean codelength for all but a negligible subset of the densities. This turns out to be true, although the lack of nonsingular measures in function spaces forces us to invent a plausible way to capture the intuitive idea of "negligible subset" of densities. For this we need some notation. Consider a partition of the unit interval into m_n equal size bins, where

$$m_n = \lceil (n^{1/3}/\log n). \quad (3.6)$$

For a density function f in $\mathcal{M}$ let p_i denote the probability of the ith bin. Write $f_i = m_n p_i$ and denote by $\theta_f = (f_1, \cdots, f_{m_n})$ the collection of such linear functionals that act as parameters although they do not determine the density function completely. Further, write $\Omega_n = \{\theta_f \in R^{m_n} \mid f \in \mathcal{M}\}$.

Theorem 2: Let $g = \{g_n(x^n)\}$ be any family of densities on I^n, where I is the unit interval, such that the Kolmogorov consistency conditions are satisfied, and each member is positive except in a set of measure zero. Then, there exists a positive constant K such that for all sufficiently large values of n and all f in $\mathcal{M}$,

$$\frac{1}{n} E_f \log \frac{f^n(x^n)}{g_n(x^n)} \ge K n^{-2/3}, \quad (3.7)$$

except for f in a set $\{f \mid (f_1, \cdots, f_{m_n}) \in A_{g,n} \subset \Omega_n \subset R^{m_n}\}$ such that the ratio of the volume of $A_{g,n}$ to that of the entire set $\Omega_n = \{\theta_f \in R^{m_n} \mid f \in \mathcal{M}\}$ satisfies

$$\frac{V(A_{g,n})}{V(\Omega_n)} \to 0,$$

as $n \to \infty$.

The proof is given in Appendix B.

Remarks: The requirement that the family $\{g_n\}$ satisfies the consistency conditions for a random process is not really needed in this version of the theorem. However, in universal coding the main interest is in encoding sequences modeled as samples from random processes, for which the consistency requirement provides a collection of Kraft-inequalities for the

IEEE TRANSACTIONS ON INFORMATION THEORY, VOL. 38, NO. 2, MARCH 1992

symbols and hence, prefix codes. Further, just as in the case with parametric model classes we may interpret the right-hand side in (3.4) as the optimal model cost per observation in order of magnitude; i.e, the codelength per observation required to encode the density estimator itself. Since the estimator is defined predictively, this cost does not appear explicitly in the total codelength, but it may be visualized as resulting from the cumulative effect of the errors in the estimated counts n_i. This cost is greater than in the case with parametric models, namely, $(k \log n)/2n$, where k denotes the number of free parameters, reflecting the fact that the nonparametric model class here is richer and its members more difficult to estimate. The choice of $m_t^* = \lceil t^{1/3}$ is seen to be appropriate for the model class $\mathcal{M}$ with its specific smoothness conditions. For a class with different smoothness conditions and hence different ϵ-entropy, [7], another choice would be better leading to a different optimal rate. Extensions and variations of Theorems 1–2 have already been proved, including an a.s. approximation for the codelength and a minimax form of Theorem 2. These will be published separately. Finally, the second bound (3.5) serves to indicate that not only does the codelength obtained with the estimator (3.3) converge to the entropy, but also the estimator itself converges to the data generating density at the same rate, when the distance is measured in terms of the Kullback distance.

We may regard the theorems as the latest step in the series of statements about universal codelength, relative to model classes of steadily increasing generality. The very first such result is Shannon's coding theorem for the singelton class $\{P(x)\}$. It was followed by the theorems in [3], [4], [14], establishing worst case bounds for independent and Markov sources as well as for some gaussian classes. In [17], a sharper inequality of the type in Theorem 2, valid for all but a vanishing subset of parameters, were proved for general parametric classes, which was further strengthened for the Markov sources in [19]. A further generalization of the latter to the ARMA class became possible through the works [9] and [18].

The reachability of the lower bound with predictive coding has important implications in prediction theory. Indeed, the bound for the codelength translates naturally to a bound for the mean prediction error. Here, the early results in [17] and [20] have been vastly generalized in [12], [10], and [13].

Acknowledgment

The authors thank A. Barron for letting them see the two papers cited prior to their publication and for his valuable comments on our draft manuscript.

Appendix A

Partition the unit interval into m equal-length bins and let $I_k = \left(\frac{k}{m}, \frac{k+1}{m}\right]$; here $0 < m < n$ and $k = 0, \cdots, m$. To simplify the notation in (3.2) slightly we write

$$f(y \mid x^n, m) \equiv f_{n,m}(y) = \frac{1}{1 + m/n}\left(\frac{n_k}{n} m + \frac{m}{n}\right),$$

for $y \in I_k$, while bearing in mind that this density depends on the data x^n. Notice that $0 < \frac{1}{n+1} \le f_{n,m}(y) \le m$.

Lemma 1: For every $f \in \mathcal{M}$, $0 < c_0 \le f(y) \le c_1$, $c_0 < 1$, $c_1 > 1$,

$$E_f \int_0^1 f(x) \log \frac{f(x)}{f_{n,m}(x)}\, dx$$
$$\le \frac{2}{c_0} E_f \int_0^1 (f(x) - f_{n,m}(x))^2\, dx + 4m(n+1)(c_1^2 + m^2) e^{-Bc_0^2 n/(4m)},$$

where B is a positive constant.

Proof of Lemma 1: Put $p_k = \int_{I_k} f(x)\, dx$, and we have $mp_k \ge c_0$. Further, [2, p. 10],

$$E \int_0^1 f(x) \log \frac{f(x)}{f_{n,m}(x)}\, dx \le E \int_0^1 \frac{(f(x) - f_{n,m}(x))^2}{f_{n,m}(x)}\, dx$$
$$= E\left\{1_{\{f_{n,m} > c_0/2\}} \int_0^1 \frac{(f(x) - f_{n,m}(x))^2}{f_{n,m}(x)}\, dx\right\} + E\left\{1_{\{f_{n,m} \not> c_0/2\}} \int_0^1 \frac{(f(x) - f_{n,m}(x))^2}{f_{n,m}(x)}\, dx\right\}.$$

The first term in the sum is bounded from above by the first term in the right-hand side of the inequality in the lemma. As to the second term, using the inequality $(f(x) - f_{n,m})^2 \le 2(f^2(x) + f_{n,m}^2(x))$ together with the bounds for f and $f_{n,m}$ we get the upper bound

$$2(n+1)(c_1^2 + m^2) P\{f_{n,m}(x) \not> c_0/2\} \tag{A.1}$$

for it. Now, any sequence x^n for which $\{f_{n,m}(x^n) \le c_0/2\}$ for some k (and, hence, $\{f_{n,m}(x^n) \not> c_0/2\}$ is true) also satisfies $\frac{n_k}{n} m \le \frac{c_0}{2} - \frac{m}{n}\left(1 - \frac{c_0}{2}\right) < \frac{c_0}{2}$, and since $mp_k \ge c_0$ it further satisfies $mp_k - \frac{n_k}{n} m > \frac{c_0}{2}$. Therefore, $P\{f_{n,m}(x) \not> c_0/2\} \le P\{n_k m/n - mp_k \ge c_0/2\}$ for some k. Further, by Bennett's inequality, [15],

$$P\left\{\frac{n_k}{n} m - mp_k \ge cm/\sqrt{n}\right\} \le P\left\{\left|\frac{n_k}{n} m - mp_k\right| \ge cm/\sqrt{n}\right\} \le 2e^{-Bmc^2}, \tag{A.2}$$

for any c, where B is a positive constant, independent of m, n, and k. Putting $\frac{cm}{\sqrt{n}} = \frac{c_0}{2}$ the upper bound (A.1) with (A.2) gives the second term in the right-hand side of the inequality in Lemma 1, which completes the proof. □

Lemma 2: For every $f \in \mathcal{M}$

$$E_f \int_0^1 (f(x) - f_{n,m}(x))^2\, dx \le \frac{2m}{n} + 4\frac{c_f^2}{m^2} + 4(1 + c_1^2)\frac{m^2}{n^2}, \tag{A.3}$$

where $c_f = \max_y |f'(y)|$.

Proof of Lemma 2: We have first

$$\begin{aligned}
&E\int_0^1 (f(x)-f_{n,m}(x))^2\,dx \\
&= E\sum_{k=1}^{m}\int_{I_k}\left(\frac{(n_k+1)m/n}{1+m/n}-f(x)\right)^2 dx \\
&= E\sum_{k=1}^{m}\int_{I_k}\left(\frac{n_k m/n - mp_k}{1+m/n} + \frac{mp_k+m/n}{1+m/n}-f(x)\right)^2 dx \qquad \text{(A.4)}\\
&\le E\sum_{k=1}^{m}\left[\int_{I_k} 2(n_k m/n - mp_k)^2 + 2\left(\frac{mp_k+m/n}{1+m/n}-f(x)\right)^2\right] dx.
\end{aligned}$$

For the first term in the right-hand side of the inequality we get

$$\begin{aligned}
&E\sum_{k=1}^{m}\int_{I_k} 2(n_k m/n - mp_k)^2\,dx \\
&= 2\sum_{k=1}^{m} E(n_k m/n - mp_k)^2 \times \frac{1}{m} \\
&= 2\sum_{k=1}^{m} m^2\,\frac{p_k(1-p_k)}{nm} \le \frac{2m}{n}. \qquad \text{(A.5)}
\end{aligned}$$

For the second term in the right-hand side of the inequality in (A.4), which does not depend on x^n, we get, again using $(a+b)^2 \le 2(a^2 + b^2)$ and the upper bounds for the densities and their derivatives

$$4\sum_{k=1}^{m}\left[\int_{I_k}(mp_k - f(x))^2\,dx + \frac{1}{m}\left(\frac{mp_k+m/n}{1+m/n}-mp_k\right)^2\right] \le 4\left(\frac{c_f^2}{m^2}+c_1^2\,\frac{m^2}{n^2}\right). \qquad \text{(A.6)}$$

This completes the proof. □

Returning to the proof of Theorem 1, we verify that the right-hand side of the inequality in Lemma 2 is minimized for m approximately $n^{1/3}$. At any rate, with the choice $m_{n-1}^* = \lceil n^{1/3}$ for m the right-hand side is bounded from above by

$$2(1+2c_f^2)n^{-2/3} + O(n^{-4/3}).$$

By Lemma 1, then, we get with this bound

$$E_f\int_0^1 f(x)\log\frac{f(x)}{f_{t,m_{t-1}^*}(x)}\,dx \le A_f t^{-2/3} + O(t^{5/3}e^{-Bt^{4/3}}) + O(t^{-4/3}),$$

where $A_f = 2(1+2c_f^2)$, proving (3.5). Further, with the notation (3.3) and the subsequent convention

$$\begin{aligned}
\frac{1}{n}E\log\frac{f_n(x^n)}{f_n^*(x^n)} &= \frac{1}{n}\sum_{t=1}^{n}E\int_0^1 f(x)\log\frac{f(x)}{f_{t,m_{t-1}^*}(x)}\,dx \\
&\le \frac{1}{n}(B_f n^{1/3} + O(1)),
\end{aligned}$$

for a constant B_f, which concludes the proof. □

Appendix B

We begin with an estimate of the rate with which $\hat{f}_i = m_n \frac{n_i}{n}$ converges to f_i in probability, where m_n is given in (3.6). By Bennett's inequality, (A.2), we get

$$P\left\{\bigcup_i |\hat{f}_i - f_i| \ge \frac{cm_n}{\sqrt{n}}\right\} \le 2m_n e^{-Bm_n c^2}. \qquad \text{(B.1)}$$

We wish to select c so that the right-hand side gets smaller than some number α, $0<\alpha<1$, say $\alpha = 2/3$, to be specific. This is true with the choice

$$c = \frac{1}{\sqrt{B}}\sqrt{\frac{\log(3m_n)}{m_n}}.$$

This value, in turn, determines the threshold $cm_n/\sqrt{n}$ in (B.1), for which we pick

$$r_n = (3B)^{-1/2} n^{-1/3}, \qquad \text{(B.2)}$$

which for large n is slightly larger than what required. With these choices (B.1) gives the inequality

$$P\left\{\bigcup_i |\hat{f}_i - f_i| \ge r_n\right\} \le 2/3. \qquad \text{(B.3)}$$

Next, we generalize an inequality in [17], [21], valid for parametric classes of models, which links the Kullback distance and the estimation rate for parameters. Consider a partition of the compact set Ω_n into m_n-dimensional hypercubes of edge length r_n, given in (B.2). Write $\hat{\Omega}_n$ for the finite set of the centers of these cubes, and let $C(\theta)$ denote the cube with its center at θ. Further, let $X_n(\theta) = \{(x_1,\cdots,x_n)\,|\,\hat{\theta}\in C(\theta)\}$, where $\hat{\theta} = \left(\frac{m_n n_1}{n},\cdots,\frac{m_n n_{m_n}}{n}\right)$. From (B.3)

$$P_n(\theta) \equiv P\{X_n(\theta)\} \ge 1/3. \qquad \text{(B.4)}$$

Next, consider the density function g_n, as specified in the theorem, and let $Q_n(\theta) = P_g(X_n(\theta))$ denote the probability mass g_n assigns to the set $X_n(\theta)$. Notice that for any two distinct points in $\hat{\Omega}_n$ these sets of strings are disjoint. The ratio $f^n(x^n)/P_n(\theta)$ defines a distribution on $X_n(\theta)$, as does of course $g_n(x^n)/Q_n(\theta)$. By the nonnegativity of the mutual information, applied to these two distributions, we get

$$T_n(\theta) \equiv \int_{X_n(\theta)} f^n(x^n)\log\frac{f^n(x^n)}{g_n(x^n)}\,dx^n \ge P_n(\theta)\log\frac{P_n(\theta)}{Q_n(\theta)}. \qquad \text{(B.5)}$$

Also,

$$E_f\log\frac{f^n(x^n)}{g_n(x^n)} \ge T_n(\theta) - 1, \qquad \text{(B.6)}$$

where we used the inequality $\log z \ge 1 - 1/z$ for $z = f^n(x^n)/g_n(x^n)$, whenever $g_n(x^n) > 0$, to get

$$\int_{\overline{X}_n(\theta)} f^n(x^n)\log\left[f^n(x^n)/g_n(x^n)\right]dx^n \ge Q_n(\theta) - P_n(\theta) > -1;$$

IEEE TRANSACTIONS ON INFORMATION THEORY, VOL. 38, NO. 2, MARCH 1992

here $\bar{X}$ denotes the complement of X. Notice that for each hypercube with its center θ in $\hat{\Omega}_n$ we have a set of density functions associated with that θ, any one of which by (B.4) assigns a $O(1)$ probability to the cube. Let $f_{n,\theta}$ denote one of them. Now, if a single density function g_n succeeds in approximating all these density functions $f_{n,\theta}$ well, as θ runs through all the centers, then the probability mass it assigns to each cube cannot go to zero as n grows. But since there is just so much probability mass available for this density function, there can be only so many cubes where the approximation can be very good. A quantitative evaluation of the number of the cubes, where a very good approximation is possible, is what gives the desired inequality.

Putting the just sketched plan to work let K be a positive number and let $A_{g,n}$ be the set of θ's such that the left-hand side of (B.5) satisfies the inequality

$$\frac{1}{n} T_n(\theta) < Kn^{-2/3}, \tag{B.7}$$

which means that for these θ's we are trying to force the codelength $-\log Q_n(\theta)$ to be close to the ideal $-\log P_n(\theta)$. This with (B.4) and (B.5) implies

$$-\log Q_n(\theta) < T_n(\theta)\left[P_n^{-1}(\theta) - \frac{\log P_n(\theta)}{T_n(\theta)}\right] < 2Kn^{1/3}, \tag{B.8}$$

which holds for $\theta \in A_{g,n}$ and for all sufficiently large n. This gives a lower bound for $Q_n(\theta)$, which we write as $q_n(\theta)$ for short; in other words, forcing (B.7) causes us to "spend" a certain minimum amount of the available probability mass. Next, let $B_{g,n}$ be the smallest set of the centers of the hypercubes which cover $A_{g,n}$; and let ν_n be the number of the elements in $B_{g,n}$. Since the sets $X_n(\theta)$, $\theta \in \Omega_n$, are disjoint, we have

$$1 \geq \sum_{\theta \in B_{g,n}} Q_n(\theta) \geq \nu_n q_n, \tag{B.9}$$

which with (B.8) gives the inequality $\log \nu_n < 2Kn^{1/3}$. The volume of $A_{g,n}$ is then bounded from above by

$$V(A_{g,n}) \leq \nu_n r_n^{m_n}, \tag{B.10}$$

which holds for all sufficiently large n.

We next calculate a lower bound for the volume of the m_n-dimensional set $\Omega_n = \{\theta_f = f_1, \cdots, f_{m_n} | f \in \mathcal{M}\}$. To do it, let $C = \frac{1}{3}\min\{1 - c_0, c_1 - 1\}$, and consider the set

$$D = \left\{\theta \in R^{m_n} \middle| \sum_{i=1}^{m_n} \theta_i = m_n, |\theta_j - 1| < C, \text{all } j, \right\}$$

which has the volume $(2C)^{m_n}$. This will be the sought-for lower bound after we show that D is a subset of Ω_n. Hence, we must demonstrate that for each $\theta = (\theta_1, \cdots, \theta_{m_n})$, $\Sigma_i \theta_i = m_n$, in D there is a density function in $\mathcal{M}$ such that $f_i = \theta_i$. In fact, define a density function f_θ successively on $I_0, \cdots, I_{m_n - 1}$, where $I_{i-1} = \left(\frac{i-1}{m_n}, \frac{i}{m_n}\right]$, as follows:

$$f_\theta(x) = \theta_i + (\theta_i - \theta_1)\sin\left[2\pi m_n\left(x - \frac{i-1}{m_n}\right) - \frac{\pi}{2}\right], \tag{B.11}$$

for $x \in I_{i-1}$. By a direct verification

$$m_n \int_{I_{i-1}} f_\theta(x)\, dx^n \equiv f_i = \theta_i, \tag{B.12}$$

so that the integral over the unit interval is unity. Further, the values of this function at the bin boundaries all equal θ_1, so the function is continuous. Its derivative at the bin boundaries vanishes, and the function has a first derivative in the entire unit interval. Also,

$$|f_\theta'(x)| \leq \max_i |2\pi m_n(\theta_i - \theta_1)| < \infty. \tag{B.13}$$

Finally,

$$f_\theta(x) \leq \max_i |\theta_i| + \max_j |\theta_j - \theta_1| \leq C + 1 + 2C \leq c_1$$

$$f_\theta(x) \geq -\max_i |\theta_i| - \max_j |\theta_j - \theta_1| \geq 1 - C - 2C \geq c_0. \tag{B.14}$$

By (B.12)–(B.14) f_θ belongs to $\mathcal{M}$.

The volume of Ω_n is then at least as large as the volume of D, or $(2C)^{m_n}$. Hence, with (B.10) we get

$$\log \frac{V(A_{g,n})}{V(\Omega_n)} \leq \log \nu_n - m_n \log(2C/r_n)$$

$$\leq n^{1/3}\left[2K - \frac{1}{3} - O\left(\frac{1}{\log n}\right)\right],$$

which goes to $-\infty$ for all K smaller than $1/6$. Hence, for each such K we get by (B.7) $\frac{1}{n} T_n(\theta) \geq Kn^{-2/3}$, except for $\theta \in A_{g,n}$, and by (B.6) the claim in the theorem follows.

References

[1] A. R. Barron and T. M. Cover, "Minimum complexity density estimation," submitted for publication, 1989.

[2] A. R. Barron and C. Sheu, "Approximation of density functions by sequences of exponential families," submitted to *Ann. Statist.*, 1988.

[3] L. D. Davisson, "Minimax noiseless universal coding for Markov sources," *IEEE Trans. Inform. Theory*, vol. IT-29, pp. 211–215, Mar. 1983.

[4] L. D. Davisson, R. J. McEliece, M. A. Pursley, and M. S. Wallace, "Efficient universal noiseless source code," *IEEE Trans. Inform. Theory*, vol. IT-27, pp. 269–279, May 1981.

[5] A. P. Dawid, "Present position and potential developments: Some personal views, statistical theory, The prequential approach," *J. Royal Statist. Soc. A*, vol. 147, pt. 2, pp. 278–292, 1984.

[6] —, "Prequential data analysis," to appear in *Issues and Controversies in Statistical Inference; Essays in Honor of D. Basu's 65th Birthday*, M. Ghosh and P. K. Pathak, Eds. 1989.

[7] L. Devroye, *A Course in Density Estimation*. Boston, MA: Birkhauser.

[8] P. Elias, "Universal codeword sets and representations of the integers," *IEEE Trans. Inform. Theory*, vol. IT-21, pp. 194–203, Mar. 1975.

[9] L. Gerencse'r, "On a class of mixing processes," *Stochastics*, vol. 26, pp. 165–191, 1989.

[10] L. Gerencse'r and J. Rissanen, "A prediction bound for Gaussian ARMA processes," *Proc. 25th CDC*, vol. 3, Athens, Greece, 1986, pp. 1487–1490.

[11] P. Hall and E. J. Hannan, "On stochastic complexity and nonparametric density estimation," *Biometrika*, vol. 75, pp. 705–714, 1988.

[12] E. J. Hannan, A. J. McDougall, and D. S. Poskitt, "Recursive estimation of autoregressions," *J. Roy. Statist. Soc., Ser. B*, vol. 51, no. 2, pp. 217–233, 1989.

[13] E. M. Hemerly and M. H. A. Davis, "Strong consistency of the PLS criterion for order determination of autoregressive processes," *Ann. Statist.*, vol. 17, no. 2, pp. 941-946, 1989.
[14] R. E. Krishevskii and V. K. Trofimov, "The performance of universal encoding," *IEEE Trans. Inform. Theory*, vol. IT-27, pp. 199-207, Mar. 1981.
[15] D. Pollard, *Convergence of Stochastic Processes*. New York: Springer-Verlag, 1984, 215 pages.
[16] J. Rissanen, "A universal prior for integers and estimation by minimum description length," *Ann. Statist.*, vol. 11, no. 2, pp. 416-431, 1983.
[17] ——, "Universal coding, information, prediction, and estimation," *IEEE Trans. Inform. Theory*, vol. IT-30, pp. 629-636, July 1984.
[18] ——, "Stochastic complexity and modeling," *Ann. Statist.*, vol. 14, pp. 1080-1100, 1982.
[19] ——, "Complexity of strings in the class of Markov sources," *IEEE Trans. Inform. Theory*, vol. IT-32, pp. 526-532, July 1986.
[20] ——, "A predictive least squares principle," *IMA J. Math. Contr. and Inform.*, vol. 3, no. 2-3, pp. 211-222, 1986.
[21] ——, *Stochastic Complexity in Statistical Inquiry*. Singapore: World Scientific Publ. Co., 1989, 175 pages.

Ann. Inst. Statist. Math.
Vol. 45, No. 1, 35–54 (1993)

MODEL SELECTION AND PREDICTION: NORMAL REGRESSION*

T. P. SPEED[1] AND BIN YU[2]

[1] *Department of Statistics, University of California at Berkeley, CA 94720, U.S.A.*
[2] *Department of Statistics, University of Wisconsin-Madison, WI 53706, U.S.A.*

(Received September 27, 1991; revised April 27, 1992)

Abstract. This paper discusses the topic of model selection for finite-dimensional normal regression models. We compare model selection criteria according to prediction errors based upon prediction with refitting, and prediction without refitting. We provide a new lower bound for prediction without refitting, while a lower bound for prediction with refitting was given by Rissanen. Moreover, we specify a set of sufficient conditions for a model selection criterion to achieve these bounds. Then the achievability of the two bounds by the following selection rules are addressed: Rissanen's accumulated prediction error criterion (APE), his stochastic complexity criterion, AIC, BIC and the FPE criteria. In particular, we provide upper bounds on overfitting and underfitting probabilities needed for the achievability. Finally, we offer a brief discussion on the issue of finite-dimensional vs. infinite-dimensional model assumptions.

Key words and phrases: Model selection, prediction lower bound, accumulated prediction error (APE), AIC, BIC, FPE, stochastic complexity, overfit and underfit probability.

1. Introduction

This paper discusses the topic of model selection for prediction in regression analysis. We compare model selection criteria according to the quality of the predictions they give. Two types of prediction errors, prediction with and without refitting, will be considered. A lower bound on the former type of error was given by Rissanen (1986*a*), and in this paper (Section 2) we provide a lower bound for the latter. Moreover, also in Section 2 we specify a set of sufficient conditions for a model selection criterion to achieve these bounds. Roughly speaking, to achieve these bounds, a model selection criterion has to be consistent and satisfy some underfitting and overfitting probability constraints. Section 3 concerns the following model selection criteria: Rissanan's predictive "minimum description length"

* Support from the National Science Foundation, grant DMS 8802378 and support from ARO, grant DAAL03-91-G-007 to B. Yu during the revision are gratefully acknowledged.

(accumulated prediction error, or predictive least squares), stochastic complexity, AIC, BIC and FPE. We consider bounds on their overfitting and underfitting probabilities, and therefore their achievability of the prediction lower bounds. In particular, the selection rule based on the accumulated prediction error and BIC achieve the two prediction lower bounds, but AIC does not unless the largest model considered is the true model.

Detailed proofs are relegated to the last section 5. All of our results are obtained under the assumption that a finite dimensional normal model generates the data under discussion. This contrasts greatly with most previous discussions, notably Shibata (1983*a*, 1983*b*) and Breiman and Freedman (1983), where the "true" model is infinite-dimensional. More discussion on finite-dimensional models vs. infinite-dimensional models can be found in Section 4.

2. Model selection and prediction in regression

In order to compare model selection procedures a number of choices need to be made; these can be critical. Two objectives of regression analysis are data description and prediction. The focus will be on the second, prediction.

Write $y = (y_1, \ldots, y_n)'$ for the n-dimensional column vector of observations, and $X = (x_{ij})$ for the $n \times K$ matrix of covariates or regressors. Inner products and squared norms are denoted by $\langle y, z\rangle = \sum y_t z_t$ and $|y|^2 = \langle y, y\rangle$, respectively. For $1 \le t \le n$, $1 \le k \le K$, denote by $y(t)$ and $X_k(t)$ that $t \times 1$ and $t \times k$ subvector and submatrix of y and X respectively, consisting of the first t rows and, in the case of X, of the first k columns. The subscript k or the parenthetical t will be omitted when they are clear from the context, or when $k = K$ or $t = n$. The t-th row of X is denoted by x_t' and the j-th column by ξ_j, whilst $x_t'(k)$ denotes the t-th row of X_k, with an analogous convention regarding the dropping of t or k. Parameter vectors are denoted by $\beta = (\beta_1, \ldots, \beta_k)'$, written $\beta(k)$ when necessary.

The class of models to be discussed will be denoted by $\{M_k : 1 \le k \le K\}$, where M_k is the model prescribing that y is $N(X_k\beta, \sigma^2 I)$ for some $\beta \in \boldsymbol{R}^k$ and $\sigma^2 > 0$. The number K of models is supposed known, and for the present discussion is held fixed as the sample size $n \to \infty$.

One framework for prediction involving regression is the following: (y_1, x_1), $(y_2, x_2), \ldots, (y_t, x_t)$ are given. The object is to predict y_{t+1} from x_{t+1}. An obvious approach is to select a model on the basis of the data available at time t, and predict y_{t+1} from this model with $t+1$ replacing t. The response y_t at time t is known before predicting y_{t+1}, so this framework is called *prediction with repeated refitting* because it allows model selection at each time.

A quite different framework assumes the existence of an initial data set $\{(y_1, x_1), \ldots, (y_n, x_n)\}$, often called a training sample, and the regressors $\tilde{x}_1, \ldots, \tilde{x}_m$ associated with a number of other units, the requirement being to predict the corresponding responses $\tilde{y}_1, \ldots, \tilde{y}_m$. A familiar variant on this would be when the "prediction" is in fact the *allocation* of units into predetermined groups. The standard solution to this problem is to select a model on the basis of the initial data set, and then predict or allocate using the model selected. This framework will be called *prediction without refitting*.

MODEL SELECTION AND PREDICTION: NORMAL REGRESSION

In this section, the above two frameworks for prediction will be discussed in detail: lower bounds are given in each case, and sufficient conditions for a model selection procedure to achieve them are obtained. However, we leave to Section 3 the achievability of these lower bounds by common selection procedures.

2.1 *Prediction with repeated refitting*

A natural measure of the quality of a sequence of predictions in the repeated refitting framework is the sum

$$(2.1) \qquad \mathrm{APE}_n = \sum_{t=1}^{n} (y_t - \hat{y}_{t|t-1})^2$$

where $\hat{y}_{t|t-1}$ denotes a predictor of y_t made on the basis of data up to and including time $t-1$, and any covariates available at time t. Model selection is thus permitted at every stage. The predictors which we consider below are $\hat{y}_{t|t-1} = x_t'\hat{\beta}_{t-1}(\hat{k}_{t-1})$, where $\hat{\beta}_{t-1}(\hat{k}_{t-1})$ is the least squares estimator based on model $M_{\hat{k}_{t-1}}$ at time t, and we will compare selection procedures leading to different $\hat{k}_s$ according to the average size of APE which is achieved for large n. For the purposes of our asymptotic analysis, it is not necessary to specify how we define $\hat{k}_t$ for $t \leq K$. In practice a number of reasonable approaches exist.

Our comparison is based upon a general inequality derived by Rissanen ((1986*a*), p. 1087). As in Sections 3 and 4 we denote by k^* the dimension associated with the true model, and $\hat{y}_{t|t-1}$ is *any* predictor of y_t which is a measurable function of $y_1, \ldots, y_{t-1}$, and $x_1, \ldots, x_t$. Although all our discussions so far have supposed that the error variance σ^2 is known and equal to unity, we will state the inequality for an arbitrary unknown σ^2. It asserts that for all k^* there is a Lebesgue null subset $A(k^*)$ of $\boldsymbol{R}^{k^*}$ such that for $\beta^* \notin A(k^*)$:

$$(2.2) \qquad \liminf_{n\to\infty} \frac{\boldsymbol{E}_{\beta^*}\{\sum_1^n (y_t - \hat{y}_{t|t-1})^2 - n\sigma^2\}}{k^* \log n} \geq \sigma^2.$$

We say that the lower bound (2.2) is achieved by a model selection criterion if it is achieved by the corresponding predictor $y_{t|t-1}$.

We need some assumptions before we can state our results on the achievability of the prediction lower bound (2.2).

Assume (cf. Lai *et al.* (1979)) that there exists a positive definite $K \times K$ matrix $C = C_K$ such that

$$(2.3) \qquad \lim_{N\to\infty} N^{-1} \sum_{t=M+1}^{M+N} x_t x_t' = C$$

uniformly in $M \geq 0$. If $M = 0$, the left-hand side is just $\lim_N N^{-1}X(N)'X(N)$. A further specialization gives $\lim_N N^{-1}X_k(N)'X_k(N) = C_k$, where C_k denotes the principal $k \times k$ submatrix of C. Assume also that

(2.4) $\quad M_{k^*} \subseteq M_K$ is the smallest true model, and $\beta(k^*)$ the true parameter.

With this background we can now state the following result, proved in Section 5 below.

THEOREM 2.1. *Suppose that* (2.3) *and* (2.4) *hold and that* $\hat{k}_n$, *the dimension defined by a model selection procedure, satisfies*:

(i) $\mathrm{pr}(\hat{k}_n < k^*) = O(n^{-2}(\log n)^{-c})$ *as* $n \to \infty$, *for some* $c > 1$, *and*

(ii) $\mathrm{pr}(\hat{k}_n > k^*) \le O((\log n)^{-\alpha})$ *as* $n \to \infty$, *for some* $\alpha > 2$.

Then the predictor $\hat{y}_{t|t-1} = x_t'\hat{\beta}_{t-1}(\hat{k}_{t-1})$ *achieves the lower bound* (2.2).

2.2 *Prediction without refitting*

Now let us suppose that we have observed $(y_1, x_1), \ldots, (y_n, x_n)$ and are required to predict the responses $\tilde{y}_1, \ldots, \tilde{y}_m$ corresponding to units with covariate vectors $\tilde{x}_1, \ldots, \tilde{x}_m$. In most discussions of this aspect of model selection, see e.g. Nishi (1984) and Shibata (1986*a*), $m = n$ and $x_i = \tilde{x}_i$, $1 \le i \le n$. Our framework is more realistic and although the general conclusions do not seem to be different from Shibata's, this was not obvious a priori.

Our predictors will all be of the form $\tilde{x}_u'\hat{\beta}(\hat{k})$, $u = 1, \ldots, m$ where $\hat{k}$ corresponds to a model selected on the basis of $\{(y_t, x_t) : t = 1, \ldots, n\}$. Given that $\hat{k} = k$, a natural measure of the quality of our set of m predictions is given by the *prediction error*

$$\mathrm{PE}(k) = \boldsymbol{E}\{|\tilde{y} - \tilde{X}_k\hat{\beta}(k)|^2 \mid y\} = m\sigma^2 + |\tilde{X}_{k^*}\beta(k^*) - \tilde{X}_k\hat{\beta}(k)|^2,$$

which averages over the new observations and conditions on the initial data. Following this line of thought, an equally natural measure of the effectiveness of the model selection procedure leading to $\hat{k}$ is $\boldsymbol{E}\{\mathrm{PE}(\hat{k}) - m\sigma^2\}$, where this time the expectation is over the possible initial data sets. What we now do is give some results on the behaviour of this quantity under a range of assumptions about $\tilde{X}$.

Our results are asymptotic in both n, the size of the initial sample, and m, the number of predictions being made. For this reason we need to supplement assumption (2.3) with an analogous, but weaker hypothesis concerning $\tilde{X}$ namely: that there exists a $K \times K$ positive definite $\tilde{C} = \tilde{C}_K$ such that

$$\lim_{M \to \infty} M^{-1} \sum_{u=1}^{M} \tilde{x}_u \tilde{x}_u' = \tilde{C}. \tag{2.5}$$

In the theorems which follow, $\hat{k} = \{\hat{k}_n\}$ is the index resulting from a procedure selecting from the models $\{M_k : 1 \le k \le K\}$.

The components of condition (B) below are defined by the partitioning

$$C_{k+1} = \begin{bmatrix} C_k & D_{k,k+1} \\ D_{k,k+1} & E_{k,k+1} \end{bmatrix},$$

where C_k, $k \le K$ is defined following (2.3).

THEOREM 2.2. *Assume conditions* (2.3), (2.4) *and* (2.5). *Then under any of the following conditions*:

MODEL SELECTION AND PREDICTION: NORMAL REGRESSION

(A) $\lim_{n\to\infty} \mathrm{pr}(\hat{k}_n < k^*) > 0$;

(B) $C_k^{-1} D_{k,k+1} = \tilde{C}_k^{-1} \tilde{D}_{k,k+1}$, $k^* \le k < K$;

(C) $\hat{k} = \hat{k}_{\mathrm{FPE}_\alpha}$ *for a sequence* $\alpha = (\alpha_n)$ *with* $n^{-1}\alpha_n \to 0$ *where* FPE_α *is the Final Prediction Error criterion defined in Section* 3, *we may conclude*

$$\lim_{m,n\to\infty} nm^{-1}\boldsymbol{E}\{\mathrm{PE}(\hat{k}_n) - m\sigma^2\} \ge \mathrm{tr}\{C_{k^*}^{-1}\tilde{C}_{k^*}\}\sigma^2. \tag{2.6}$$

The proof will be given in Section 5. It can be seen from the proof of this theorem that there will be other "symmetric" selection rules other than FPE_α for which the conclusion holds.

The next question of interest is the following: what kinds of selection rules attain the lower bound (2.6)?

THEOREM 2.3. *The lower bound* (2.6) *is attained for any consistent selection rule whose underfitting probability* $\mathrm{pr}(\hat{k}_n < k^*)$ *is* $o(n^{-2})$ *as* $n \to \infty$.

3. APE, stochastic complexity, and FPE

In this section, we consider the achievability of the two lower bounds in Section 2 of some commonly-used model selection criteria. We derive upper bounds on the underfitting and overfitting probabilities of these criteria and then use Theorem 2.1 or Theorem 2.3.

First, we consider the criterion based upon accumulated (one-step) prediction errors (APE) (or predictive least squares). This criterion is the predictive MDL criterion introduced in Rissanen (1984, 1986*b*). Many authors have discussed this criterion as detailed in the remark after Theorem 3.1.

We now introduce the definition of APE. Only ordinary least squares estimates will be used. For $1 \le k \le K$, $k+1 \le s \le n$, write

$$\hat{\beta}_s(k) = (X_k(s)'X_k(s))^{-1}X_k(s)'y(s)$$

and $\hat{\beta}(k) = \hat{\beta}_n(k)$. All of the matrices $X_k(t)$ will be assumed to have rank k when $t > k$. The *recursive residuals*, also called one-step prediction errors, based on M_k are $e_t(k) = y_t - x_t(k)'\hat{\beta}_{t-1}(k)$. The ordinary residuals are $r_{t,n}(k) = y_t - x_t(k)'\hat{\beta}_n(k)$. The parenthetical k will be dropped if its value is clear from the context.

For any fixed $k \le K$, consider the accumulated squared prediction error $\mathrm{APE}_n(k) = \sum_{t=k+1}^n e_t(k)^2$. Obviously, $\mathrm{APE}_n(k)$ is the same as the prediction error with refitting (2.2) when the model M_k is fixed through time t.

Expression $\mathrm{APE}_n(k)$ will lead us to a model selection criterion: choose that k which minimizes $\mathrm{APE}_n(k)$ over all $k \le K$.

For the remainder of this section σ^2 is supposed known and so, for simplicity, is taken to be 1. This is possible because, unlike many model selection criteria, the one based on APE does not require knowledge or an estimate of σ^2. The numbers $\{b_k\}$ which appear in the following theorem are normalized limiting (squared) bias terms defined by

$$b_k = \mathrm{tr}\{(E_{k,k^*} - D'_{k,k^*}C_k^{-1}D_{k,k^*})\zeta(k)\zeta(k)'\}$$

where for $k < k^*$ the principal submatrices C_k and C_{k^*} of C are written

$$C_{k^*} = \begin{bmatrix} C_k & D_{k,k^*} \\ D_{k,k^*} & E_{k,k^*} \end{bmatrix},$$

and $\beta(k^*) = (\beta(k)' \mid \varsigma(k)')'$ is the corresponding partitioning of $\beta(k^*)$. It is shown in Section 5 (Lemma 5.3) that $b_1 \geq b_2 \geq \cdots \geq b_{k^*-1} > 0$.

THEOREM 3.1. *Under assumptions* (2.3) *and* (2.4), *as* $n \to \infty$, *let* $\hat{k}_n$ *denote the dimension selected by minimizing* $\mathrm{APE}_n(k)$. *Then we have the following bounds*:

(i) $\mathrm{pr}(\hat{k}_n < k^*) \leq O(\exp(-bn))$ *as* $n \to \infty$, *for* $b = \min(b_{k^*-1}/3, b_{k^*-1}^2/18)$.

(ii) $\mathrm{pr}(\hat{k}_n > k^*) \leq O(n^{-1/6})$ *as* $n \to \infty$.

Remark. The upper bound in (i) shows the interplay between the bias term b_k and the sample size n; the product of them determines the underfitting probability, not the sample size n alone.

COROLLARY 3.1. *The lower bounds* (2.2) *and* (2.6) *are attained for the* APE *selection rule.*

PROOF. Straightforward from Theorems 2.1, 2.2 and 3.1.

Remark. (a) Convergence in probability of the APE selection rule was established by Rissanen (1986*b*) under essentially the same conditions as we have used here. Other writers who have suggested the use of APE or a related criterion to select regression models include Hjorth (1982) and Dawid (1984, 1992). The latter describes a generalization of the use of APE as the prequential approach to statistical analysis. (b) There is no doubt that our assumptions could be weakened, but the derivations of the same results are expected to be much more involved. In the context of time series, Wax (1988) derived the weak consistency of an analogous estimator of the order of an autoregressive process without the Gaussian assumption, and Hemerly and Davis (1989) strengthened it to the a.s. consistency. Moreover, Wei (1992) obtained the a.s. consistency and asymptotic expansions of APE under stochastic regression models.

Now we turn to selection rules based on the residual sum of squares, which is $\mathrm{RSS}_n(k) = \sum_1^n r_{t,n}(k)^2$ where the ordinary residuals $r_{t,n}(k)$ are defined above. When $\sigma^2 = 1$ in the regression models M_k the *final prediction error* (FPE) criterion is $\mathrm{FPE}_{\alpha_n}(k) = \mathrm{RSS}_n(k) + \alpha_n k$ where (α_n) is a sequence of positive numbers. For AIC, $\alpha_n \equiv 2$. For BIC (Schwartz (1978)), $\alpha_n = \log n$. When σ^2 is not known, we may replace it by its usual estimate from the largest model M_K. Our results should still hold in that case.

Rissanen (1986*a*) introduced stochastic complexity (SC) of a set of data relative to a model as variant of his MDL and PMDL expressions, and in many cases it is asymptotically equivalent to the latter, whilst being easier to calculate. We refer to his paper for definitions of these quantities. For our regression models with error variance equal to unity, SC takes a particularly simple form if the prior distribution for the parameter $\beta(k)$ is taken to be $N(0, \tau I_k)$ where $\tau > 0$ is a scale parameter, $k = 1, \ldots, K$. A simple calculation yields the expression

$$(3.1) \quad \mathrm{SC}_n(k) = \frac{1}{2} n \log 2\pi + \frac{1}{2} \log \det(I_n + \tau X_k X_k') + \frac{1}{2} y'(I_n + \tau X_k X_k')^{-1} y.$$

From Lemma 5.5 in Section 5 we see that as $n \to \infty$,

$$\mathrm{SC}_n(k) - \frac{1}{2} n \log 2\pi = k \log n + \mathrm{RSS}_n(k) + O(1) \quad \text{a.s.}$$

and so any discussion of model selection based upon stochastic complexity is subsumed under that of BIC.

The FPE criterion has been discussed by Akaike (1970, 1974), Bhansali and Downham (1977), Atkinson (1980), and Shibata (1976, 1986*a*) amongst others. Geweke and Meese (1981) discuss the problem quite generally, but with random regressors, whilst Kohn (1983) considers selection in general parametric models. Shibata (1984) may be consulted for further details on some cases of FPE. The consistency of FPE's, with α_n's satisfying $\lim n^{-1}\alpha_n = 0$ and $\underline{\lim}(2 \log\log n)^{-1}\alpha_n > 1$, was established in a time-series context by Hannan and Quinn (1979). Moreover, the equivalence of BIC and APE has been shown by Hannan *et al.* (1989) for the finite-dimensional autoregressive models and by Wei (1992) for finite-dimensional stochastic regression models.

THEOREM 3.2. *Let $\hat{k}_n$ denote the dimension selected by* FPE_{α_n} *for some sequence α_n such that $n^{-1}\alpha_n \to 0$ as $n \to \infty$. Then*

(i) *$\hat{k}_n$ overfits with probability approaching unity as $n \to \infty$. More precisely, for any constant $0 < b < b_{k^*-1}/4$, $\mathrm{pr}(\hat{k}_n < k^*) \le O(\exp(-bn))$ as $n \to \infty$.*

(ii) *If $k^* < K$, and $\liminf(2 \log\log n)^{-1}\alpha_n > 2$, we have, for some $\gamma > 2$, $\mathrm{pr}(\hat{k}_n > k^*) \le O((\log n)^{-\gamma})$ as $n \to \infty$.*

We omit the proof of this theorem in this paper because Woodroofe (1982) and Haughton (1989) contain smilar bounds for BIC under more general models. Moreover, a lower bound, instead of an upper one, on the overfit probability (ii) is given in the Appendix II of Merhav *et al.* (1989) for BIC. Their result suggests that the overfit probability of BIC tends to zero slower than exponentially as n tends to infinity.

COROLLARY 3.2. (i) *The selection rules defined by* BIC *and* SC *all lead to predictors which achieve the lower bounds* (2.2) *and* (2.6);

(ii) *If* $\lim(2 \log\log n)^{-1}\alpha_n < 1$, *the selection rules defined by* FPE_{α_n} *do* not *achieve the lower bounds* (2.2) *and* (2.6) *unless $k^* = K$; in particular,* AIC *does* not *achieve the lower bounds unless $k^* = K$.*

4. Discussion

The results presented seem to suggest that if prediction is part of the objective of a regression analysis, then model selection carried out using APE, BIC, SC or an equivalent procedure has some desirable properties. Of course there is a qualification: in deriving these theorems we have assumed that the model generating our data is (i) fixed throughout the asymptotics; (ii) finite-dimensional; and (iii) belongs to the class of models being examined. Before commenting on these assumptions, let us see that our theorems are at least in general agreement with a number of analyses and simulations in the literature. The first paper to point out clearly that consistent model selection gives better predictions seems to be Shibata (1984), although he does not emphasize this conclusion. Atkinson's (1980) results also suggest the conclusion we have reached, but again this is not emphasized. The simulation results of Clayton *et al.* (1986) led them to conclude "that if the 'true' or 'approximately true' model is included among the alternatives considered, all reasonable model selection procedures will possess rather similar predictive capabilities". We feel that this conclusion is more a reflection of the limited scope of the simulations conducted rather than the true state of affairs. Indeed a close examination of the sample sizes and models these authors studied suggests that there was little opportunity for the procedures (not the models) to be distinguished, as far as the squared prediction error of the resulting choices is concerned. More recently, Rissanen (1989) reported clear differences between cross validation and SC, and to the extent that cross-validation and AIC perform similarly, Stone (1977), this is explained by Corollary 3.2.

Shibata (1981, 1983*a*, 1983*b*, 1984, 1986*a*, 1986*b*) presents a number of theorems demonstrating the optimality of AIC or other forms of FPE_{α_n} with bounded sequences (α_n), as well as arguments rebutting the criticisms that such procedures are unsatisfactory by virtue of their inconsistency under assumptions (i), (ii) and (iii). Shibata (1981), and Breiman and Freedman (1983) using random regressors, suppose the true model to be *infinite*-dimensional rather than *finite*-dimensional. Shibata (1981) also offers an optimality result for AIC valid under a "moving truth" assumption.

Clearly, the prediction optimality of BIC and its analogues like APE depend on the assumption that the true model is finite-dimensional, i.e., the bias term $b_k = 0$ for $k \geq k^*$. When the true model is assumed to be infinite-dimensional, i.e., $b_k > 0$ for all k, Breiman and Freedman (1983) showed that AIC's equivalent is optimal in terms of one-step further prediction. We now show by the following three simple examples that the decay rate of the bias term plays a determining role in the battle of AIC vs. BIC.

For simplicity, let us take the framework of Breiman and Freedman (1983) where an infinite-dimensional model with Gaussan $N(0,1)$ independent regressors is assumed with the error variance $\sigma^2 = 1$. Then the one-step ahead prediction error for the $(n+1)$-st observation based on model M_k is roughly $\text{PE}(k) = b_k + kn^{-1}$. Moreover, AIC approximately minimizes $b_k + kn^{-1}$, while BIC minimizes $b_k + kn^{-1}\log n$. By the result of Breiman and Freedman (1983), asymptotically, $\text{PE}(\hat{k}_{\text{BIC}})/\text{PE}(\hat{k}_{\text{AIC}}) \geq 1$, where $\hat{k}_{\text{AIC}}$ is the model selected by AIC, and similarly for $\hat{k}_{\text{BIC}}$.

Example 1. Assume $b_k = k^{-\alpha}$. Straightforward calculation shows that, as $n \to \infty$, $\text{PE}(\hat{k}_{\text{BIC}})/\text{PE}(\hat{k}_{\text{AIC}}) \to \infty$.

Example 2. Assume $b_k = e^{-k}$. Then as $n \to \infty$, $\text{PE}(\hat{k}_{\text{BIC}})/\text{PE}(\hat{k}_{\text{AIC}}) \to 2$.

Example 3. Assume $b_k = e^{-e^k}$. Then as $n \to \infty$, $\text{PE}(\hat{k}_{\text{BIC}})/\text{PE}(\hat{k}_{\text{AIC}}) \to 1$.

To summarize, as the decay rate of the bias term increases, the prediction performance of BIC catches up with that of AIC. And, as we have seen, BIC out-performs AIC when $b_k = 0$ for $k > k^*$, i.e. when the model is finite.

Finally, all three of APE, BIC and SC derive from general approaches to the model selection problem and have extensions to situations where one or more of (i), (ii) and (iii) are dropped, see Sawa (1978) for some remarks about this situation. When something is known about these extensions, it will be of interest to compare them with AIC or, more generally FPE_{α_n}.

5. Proofs

Most of the arguments given below are straightforward. We have tried to be explicit wherever possible, and have included some proofs which may be found elsewhere in order to keep this paper self-contained.

The proofs are presented in the following order: Theorem 3.1, Corollaries 3.1 and 3.2, Theorem 2.2, Theorem 2.3 and Theorem 2.1. We continue to use the notation introduced in Section 2 above. It is straightforward to show

LEMMA 5.1. *For $k < s < t \le n$ and $c \in R(X_k(t))$, we have* $\text{cov}(e_{s+1}(k), c'y(t)) = 0$.

It follows from the lemma that

COROLLARY 5.1. (a) *For all $k < s < t \le n$, we have* $\text{cov}(e_s(k), e_t(k)) = 0$. (b) *For all $k < t \le n$, and $c \in R(X_k)$,* $\text{cov}(e_t(k), c'y) = 0$.

Let us write $\lambda_t(k) = \boldsymbol{E}\{e_t(k)\}$ and $\mu_t(k) = \text{Var}\{e_t(k)\} - 1$, $\epsilon_t = y_t - \boldsymbol{E}\{y_t\}$ and $H_n(k) = X_n(k)(X_n(k)'X_n(k))^{-1}X_n(k)'$, and define the following quantities:

$$V_n(k) = \sum_{t=k+1}^{n} \mu_t(k), \qquad B_n(k) = \sum_{t=k+1}^{n} \lambda_t(k)^2, \qquad N_n(k) = |H_n(k)\epsilon|^2,$$

$$N_n^\dagger(k) = \sum_{t=k+1}^{n} \mu_t(k) \left[\frac{(e_t(k) - \lambda_t(k))^2}{\mu_t(k) + 1} - 1 \right],$$

$$B_n^\dagger(k) = 2 \sum_{t=k+1}^{n} (e_t(k) - \lambda_t(k))\lambda_t(k).$$

It is clear from the proof of the result we state shortly that V is a *variance* term, B is a *bias* term, and N is a *noise* term, whilst $N^{\dagger}$ is a second noise term and $B^{\dagger}$ a part-noise part-bias term.

LEMMA 5.2. *With the above notation*

$$\sum_{t=k+1}^{n} e_t(k)^2 - \sum_{t=1}^{n} \epsilon_t^2 = V_n(k) + B_n(k) - N_n(k) + B_n^{\dagger}(k) + N_n^{\dagger}(k). \tag{5.1}$$

PROOF. It follows from Corollary 5.1 that $\{e_{k+1}(k), \ldots, e_n(k)\}$ are pairwise uncorrelated, and uncorrelated with $c'y$ for all $c \in R(X_k)$. Thus we can make an orthogonal transformation and obtain

$$|\epsilon|^2 = |H(k)\epsilon|^2 + \sum_{t=k+1}^{n} \frac{[e_t(k) - \boldsymbol{E}\{e_t(k)\}]^2}{\mathrm{Var}\{e_t(k)\}}. \tag{5.2}$$

The lemma then follows from this equation and the comparing two sides of (5.1). □

In the lemmas which follow, (2.1) and (2.2) will be assumed without comment. Moreover, to state our next result we need a little further notation. For $k < k^*$, write the principal $k \times k$ submatrix C_k of C given by (2.4) in the form

$$C_{k^*} = \begin{bmatrix} C_k & D_{k,k^*} \\ D'_{k,k^*} & E_{k,k^*} \end{bmatrix}$$

and we write $\beta(k^*) = (\beta(k)' \mid \zeta(k)')'$ and $X_{k^*}(n) = [X_k(n) \mid Z_k(n)]$.

LEMMA 5.3. $n^{-1}B_n(k) \to b_k$ *as* $n \to \infty$, *where*

$$b_k = \mathrm{tr}\{(E_{k,k^*} - D'_{k,k^*}C_k^{-1}D_{k,k^*})\zeta(k)\zeta(k)'\}$$

satisfies $b_1 \geq b_2 \geq \cdots \geq b_{k^*-1} > 0$.

PROOF. We begin by observing that for $k < k^*$, $\lambda_t(k) = A_k(t)'\zeta(k)$, where

$$A_k(t)' = z_t(k)' - x_t(k)'(X_k(t-1)'X_k(t-1))^{-1}X_k(t-1)'Z_k(t-1).$$

It follows that $\lambda_t(k)^2 = \mathrm{tr}\{A_k(t)A_k(t)'\zeta(k)\zeta(k)'\}$ and so

$$n^{-1}\sum_{t=k+1}^{n} \lambda_t(k)^2 = \mathrm{tr}\left\{n^{-1}\sum_{t=k+1}^{n} A_k(t)A_k(t)'\zeta(k)\zeta(k)'\right\}.$$

Using (2.4) and the notation introduced above, $t^{-1}X_k(t)'X_k(t) \to C_k$, $t^{-1}X_k(t)' \cdot Z_k(t) \to D_{k,k^*}$, and $t^{-1}Z_k(t)'Z_k(t) \to E_{k,k^*}$ as $t \to \infty$, and so it follows that

$$n^{-1}\sum_{t=k+1}^{n} A_k(t)A_k(t)' \to E_{k,k^*} - D_{k,k^*}C_k^{-1}D_{k,k^*}$$

as $n \to \infty$, giving the expression for b_k stated. The monotonicity of b_k can then be checked using the partial order of positive definite matrices. □

For the next lemma we need some notation paralleling that used in Lemma 5.2 above. Write $\bar{\lambda}_t(k) = \boldsymbol{E}\{r_t(k)\}$ and $\bar{B}_n(k) = \sum_1^n \bar{\lambda}_t(k)^2$. Furthermore, put $\bar{B}_n^\dagger(k) = 2\sum_1^n \bar{\lambda}_t(k)\epsilon_t$. By variants of the proofs of Lemmas 5.2 and 5.3 and by the law of iterative algorithm, we obtain

LEMMA 5.4.

$$\sum_1^n r_t(k)^2 - \sum_1^n \epsilon_t^2 = \bar{B}_n(k) - N_n(k) + \bar{B}_n^\dagger(k) \tag{5.3}$$

where for $k < k^*$, $n^{-1}\bar{B}_n(k) \to b_k$, *and* $\bar{B}_n^\dagger(k) = O((n \log\log n)^{1/2})$ *a.s. as* $n \to \infty$.

LEMMA 5.5. *In the notation introduced prior to equation* (3.1)

$$\begin{aligned}\log\det(I_n + \tau X_k(n)X_k(n)') &+ y(n)'(I_n + \tau X_k(n)X_k(n)')^{-1}y(n)\\ &= k\log n + \sum_1^n r_t(k)^2 + O(1) \qquad a.s.\ n \to \infty.\end{aligned}$$

PROOF. Straightforward from assumption (2.3) and Rao ((1973), p. 33). □

In the following lemmas we use the notation $\rho_k = \xi_{k+1} - X_k\gamma_k$, $\tilde{\rho}_k = \tilde{\xi}_{k+1} - \tilde{X}_k\gamma_k$ and $\eta_k = X_k(X_k'X_k)^{-1}\tilde{X}_k'\tilde{\rho}_k$, where $\gamma_k = (X_k'X_k)^{-1}X_k'\xi_{k+1}$. It is evident that γ_k is the regression coefficient of the $(k+1)$-st variable on the previous k, and so ρ_k and $\tilde{\rho}_k$ are essentially residuals when the current model is M_k, whereas η_k is part residual and part fitted value.

LEMMA 5.6.

$$\tilde{X}_{k+1}(X_{k+1}'X_{k+1})^{-1}X_{k+1}\epsilon = \tilde{X}_k(X_k'X_k)^{-1}X_k\epsilon + |\rho_k|^{-2}\langle\rho_k, \epsilon\rangle\tilde{\rho}_k.$$

PROOF. This is a straightforward consequence of the formula for the inverse of a partitioned matrix, see e.g. Rao ((1973), p. 33). □

If we write $N_{m,n}(k) = |\tilde{X}_k(X_k'X_k)^{-1}X_k'\epsilon|^2$ by analogy with the noise term introduced just before Lemma 5.2, then we have

COROLLARY 5.2.

$$N_{m,n}(k+1) = N_{m,n}(k) + 2|\rho_k|^{-2}\langle\eta_k, \epsilon\rangle\langle\rho_k, \epsilon\rangle + |\rho_k|^{-4}|\tilde{\rho}_k|^2\langle\rho_k, \epsilon\rangle^2.$$

Now let us write $\tilde{X}_{k^*} = [\tilde{X}_k \mid \tilde{Z}_k]$ and $\tilde{R}_k = \tilde{Z}_k - \tilde{X}_k(X_k'X_k)^{-1}X_k'Z_k$. Furthermore, for $k > k^*$, write

$$C_{k+1} = \begin{bmatrix} C_k & D_{k,k+1} \\ D_{k,k+1} & E_{k,k+1} \end{bmatrix}$$

and similarly for $\tilde{C}_{k+1}$. Finally, denote by $\Delta_{k,k+1}$ and Δ_k, the differences $\tilde{C}_k^{-1}\cdot\tilde{D}_{k,k+1} - C_k^{-1}D_{k,k+1}$ and $\tilde{C}_k^{-1}\tilde{D}_{k,k^*} - C_k^{-1}D_{k,k^*}$, respectively.

The following formulae bear a close resemblance to ones obtained in a similar context by Box and Draper (1959, 1963). There, however, the emphasis is on design: the choice of x vectors. It should be clear from the context whether or not $k \le k^*$ is required to give a non-trivial result.

LEMMA 5.7. *As $m, n \to \infty$ we have*
(i) $m^{-1}\tilde{X}_k'\tilde{R}_k \to \tilde{C}_k\Delta_k$.
(ii) $m^{-1}\tilde{R}_k'\tilde{R}_k \to \tilde{E}_k - \tilde{D}_{k,k^*}'\tilde{C}_k^{-1}\tilde{D}_{k,k^*} + \Delta_k'\tilde{C}_k^{-1}\Delta_k$.
(iii) $m^{-1}|\tilde{\rho}_k|^2 \to \tilde{E}_{k,k+1} - \tilde{D}_{k,k+1}'\tilde{C}_k^{-1}\tilde{D}_{k,k+1} + \Delta_{k,k+1}'C_k^{-1}\Delta_{k,k+1}$.
(iv) $n^{-1}|\rho_k|^2 \to E_{k,k+1} - D_{k,k+1}'C_k^{-1}D_{k,k+1}$.
(v) $nm^{-2}|\eta_k|^2 \to \Delta_{k,k+1}'\tilde{C}_kC_k^{-1}\tilde{C}_k\Delta_{k,k+1}$.

PROOFS. These are all straightforward consequences of the relevant definitions. □

Next we extend some earlier notation, writing $B_{m,n}(k) = \mathrm{tr}\{\tilde{R}_k'\tilde{R}_k\zeta(k)\zeta(k)'\}$, and $S_{m,n}(k) = 2\langle\tilde{R}_k\zeta(k), \tilde{X}_k(X_k'X_k)^{-1}X_k'\epsilon\rangle$. Clearly the first term is the analogue of the bias term introduced prior to Lemma 5.2, and reduces to it if $m = n$ and $\tilde{X} = X$. For the definition of PE(k), see Section 2 above.

LEMMA 5.8. *In the notation just introduced, we have*

$$\mathrm{PE}(k) - m\sigma^2 = B_{m,n}(k) + N_{m,n}(k) - S_{m,n}(k).$$

PROOF. $\mathrm{PE}(k) - m\sigma^2 = |\tilde{X}_{k^*}\beta(k^*) - \tilde{X}_k\hat{\beta}(k)|^2$, where we may write

$$\begin{aligned}\tilde{X}_{k^*}\beta(k^*) - \tilde{X}_k\hat{\beta}(k) &= \tilde{X}_{k^*}\beta(k^*) - \tilde{X}_k(X_k'X_k)^{-1}X_k'(X_{k^*}\beta(k^*) + \epsilon)\\ &= (\tilde{Z}_k - \tilde{X}_k(X_k'X_k)^{-1}X_k'Z_k)\zeta(k) - \tilde{X}_k(X_k'X_k)^{-1}X_k'\epsilon.\end{aligned}$$

The result now follows upon taking the squared norm of this vector. □

LEMMA 5.9. *As $m, n \to \infty$ we have*
(i) $m^{-1}B_{m,n}(k) \to \mathrm{tr}\{(\tilde{E}_k - \tilde{D}_{k,k^*}'\tilde{C}_k^{-1}\tilde{D}_{k,k^*} + \Delta_k'\tilde{C}_k^{-1}\Delta_k)\zeta(k)\zeta(k)'\}$.
(ii) $m^{-1}n\boldsymbol{E}\{N_{m,n}(k)\} \to \mathrm{tr}(\tilde{C}_kC_k^{-1})$.
(iii) $m^{-1}nN_{m,n}(k) = O(\log\log n)$ *a.s.*
(iv) $m^{-1}nS_{m,n}(k) \to 0$ *a.s. if* $\Delta_k = 0$.
(v) $m^{-1}S_{m,n}(k) = O((n^{-1}\log\log n)^{1/2})$ *a.s. if* $\Delta_k \ne 0$.

PROOF. (i) is an immediate consequence of Lemma 5.7(iv); (ii) and (iii) are straightforward calculations; (iv) follows from the definitions, whilst (v) is a now-familiar form of the law of the iterated logarithm. □

MODEL SELECTION AND PREDICTION: NORMAL REGRESSION

PROOF OF THEOREM 3.1. (i) We begin by obtaining some probability inequalities concerning the terms in $\text{APE}_n(k)$, cf. Lemma 5.2. Since $N_n(k) = |H_n(k)\epsilon|^2$ is a chi-squared r.v.,

$$\text{pr}(N_n(k) > \beta_n) \leq O(\exp(-\beta_n)) \quad \text{as} \quad n \to \infty.$$

Similarly, $B_n^\dagger(k)$ is a sum of independent zero mean normal r.v.'s whose variance is $O(n)$, and so $\text{pr}(|B_n^\dagger(k)| > \gamma_n) \leq O(\gamma_n^{-1} n^{1/2} \exp(-\gamma_n^2/2n))$.

Finally, $W_n(k) = V_n(k) + N_n^\dagger(k)$ is a sum of $n-k$ independent squared normals, the t-th of which is scaled by $\mu_t(k)$, and so

$$\begin{aligned}
\text{pr}(W_n(k) > \delta_n) &\leq \exp(-\delta_n) \prod_{k+1}^{n} (1 - 2\mu_t(k))^{-1/2} \leq \exp\left\{-\delta_n + \sum_{k+1}^{n} \mu_t(k)\right\} \\
&= \exp\{-\delta_n + k \log n + o(\log n)\} \\
&\leq n^{k+1} \exp(-\delta_n), \quad \text{as} \quad n \to \infty.
\end{aligned}$$

We now put these inequalities together, select (β_n), (γ_n) and (δ_n), and obtain (i). For simplicity, we drop subscripts n where no confusion will result. If $k < k^*$,

$$\begin{aligned}
\text{pr}(\hat{k} = k) &\leq \text{pr}\{\text{APE}(k) < \text{APE}(k^*)\} \\
&= \text{pr}\{B(k) - N(k) + W(k) + B^\dagger(k) \\
&\qquad < B(k^*) - N(k^*) + W(k^*) + B^\dagger(k^*)\} \\
&\leq \text{pr}\{W(k^*) \geq B(k) + B^\dagger(k) - N(k)\} \\
&\qquad\qquad \text{since} \quad W(k) > 0 \quad \text{and} \quad N(k^*) > 0, \\
&\leq \text{pr}\{W(k^*) \geq nb_k + o(n) - \gamma_n - B_n\} \\
&\qquad + P\{N(k) > B_n\} + P\{|B^\dagger(k)| > \gamma_n\} \\
&\leq n^{k+1} \exp(-nb_k + o(n) + \gamma_n + \beta_n) \\
&\qquad + O(\exp(-\beta_n)) + O(\gamma_n^{-1} n^{1/2} \exp(-\gamma_n^2/2n)).
\end{aligned}$$

We now see that if $\beta_n = b_k n/3$ and $\gamma_n = b_k n/3$, the desired conclusion follows since b_k decreases as k increases to $k^* - 1$.

(ii) For the overfitting probability, we estimate $\text{pr}(\hat{k} = k)$ for $k > k^*$, noting that in this case $\text{APE}(k) = V(k) - N(k) + N^\dagger(k)$, i.e. the bias terms disappear. In this proof we bound $-N^\dagger(k)$ and $N^\dagger(k^*)$ from below by the same quantity, β_n say, and calculate the tail probability as in the first part of the proof. We find that

$$\begin{aligned}
\text{pr}(N^\dagger(k) < -\beta_n) &= \text{pr}(-N^\dagger(k) > \beta_n) \\
&\leq \exp(-\beta_n) \prod_{k+1}^{n} \{(1 + 2\mu_t(k))^{-1/2} \exp \mu_t(k)\} \\
&\leq O(\exp(-\beta_n)).
\end{aligned}$$

Similarly we have $\text{pr}(N^\dagger(k^*) > \beta_n) \leq O(\exp(-\beta_n))$, and since $N(k) - N(k^*)$ is a chi-squared r.v. on $k - k^*$ degrees of freedom,

$$\text{pr}(N(k) - N(k^*) > \gamma_n) \leq O(\gamma_n^{-1+(k-k^*)/2} \exp(-\gamma_n/2)).$$

Thus if $k > k^*$,

$$\begin{aligned}\operatorname{pr}(\hat{k} = k) &= \operatorname{pr}\{\mathrm{APE}(k) < \mathrm{APE}(k^*)\} \\ &= \operatorname{pr}\{V(k) - N(k) + N^\dagger(k) < V(k^*) - N(k^*) + N^\dagger(k^*)\} \\ &\leq \operatorname{pr}\{V(k) - \beta_n - (N(k) - N(k^*)) < V(k^*) + \beta_n\} \\ &\quad + \operatorname{pr}\{N^\dagger(k) < -\beta_n\} + \operatorname{pr}\{N^\dagger(k) > \beta_n\} \\ &\leq O(\gamma_n^{-1+(k-k^*)/2} \exp(-\gamma_n/2)) + 2O(\exp(-\beta_n)),\end{aligned}$$

where $\gamma_n = (k - k^*)\log n + o(\log n) - 2\beta_n$, since $V(k) = k\log n + o(\log n)$, and similarly for $V(k^*)$. If we take $\beta_n = \beta \log n$ for $\beta = 6^{-1}$, say, then we deduce that $\operatorname{pr}(\hat{k} > k) \leq O(n^{-1/6})$. □

Corollary 3.2 can be shown by an argument similar to Theorems 2.1 and 2.3. Note that when the selection rule is not consistent, the inequality is sharp since the prediction error based on M_k for some $k > k^*$ is strictly larger than the one based on M_{k^*}, and underfitting does not cause any problem since all FPE's underfit with a probability vanishing exponentially fast (Theorem 3.1(i)).

Let $\{H_j : j = 1, \ldots, n\}$ be a set of pairwise orthogonal rank 1 projectors summing to the identity, such that for all $k = 1, \ldots, K$ we have $\sum_{p=1}^{k} H_p = H(k)$, where $R(H(k)) = R(X_k(n))$. Let $\epsilon = (\epsilon_i)$ be an n-tuple of iid $N(0,1)$ random variables, F any function of $|H_i\epsilon|^2$ for a fixed $i \in \{1, \ldots, n\}$, and ξ, η fixed vectors.

LEMMA 5.10. $\boldsymbol{E}\{\langle x_i, H_i\epsilon\rangle F(|H_i\epsilon|^2)\} = 0.$

PROOF. The lemma is an immediate consequence of the symmetry of the normal distribution. □

COROLLARY 5.3. *Let f be a function of $|H_1\epsilon|^2, \ldots, |H_k\epsilon|^2$. Then if $1 \leq i, j \leq k$, we have*

$$\boldsymbol{E}\{\langle \xi, H_i\epsilon\rangle f(|H_1\epsilon|^2, \ldots, |H_k\epsilon|^2)\} = 0,$$
$$\boldsymbol{E}\{\langle \xi, H_i\epsilon\rangle \langle \eta, H_j\epsilon\rangle f(|H_1\epsilon|^2, \ldots, |H_k\epsilon|^2)\} = 0.$$

PROOF. The identities follow from the lemma by a suitable conditioning. □

In the lemma which follows we use the expressions ρ_k and η_k defined prior to Lemma 5.6 above.

LEMMA 5.11. *Let $\hat{k}_n$ denote the dimension selected by FPE_{α_n} and suppose that $l > k \geq k^*$. Then we have*

$$\lim_{m,n} m^{-1} n |\rho_k|^{-2} \boldsymbol{E}\{\langle \rho_k, \epsilon\rangle \langle \eta_k, \epsilon\rangle 1_{\{\hat{k}_n = l\}}\} = 0. \tag{5.4}$$

PROOF. We begin by replacing $\hat{k}_n$ by $\tilde{k}_n$, that k which minimizes $\mathrm{FPE}(k)$ over the range $\{k^*, k^*+1, \ldots, K\}$. From Theorem 3.2 we know that $\operatorname{pr}(\hat{k}_n \neq \tilde{k}_n) \to 0$ as $n \to \infty$.

MODEL SELECTION AND PREDICTION: NORMAL REGRESSION

Now recall the definition of FPE(k) and note that if $k < l$, FPE(k) $\leq$ FPE(l) if and only if $\sum_{k+1}^{l} |H_p\epsilon|^2 \leq (l-k)\alpha$. Thus the event $\{\tilde{k} = l\}$ is the intersection of the two events: $\{\sum_{p=h+1}^{l} |H_p\epsilon|^2 \geq (l-h)\alpha; k^* \leq h < l\}$ and $\{\sum_{p=l+1}^{h} |H_p\epsilon|^2 \leq (h-l)\alpha, l < h \leq K\}$ whose indicators we denote by f_l and g_l respectively. Our aim is to show that

$$\boldsymbol{E}\{\langle\rho_k, \epsilon\rangle\langle\eta_k, \epsilon\rangle f_l g_l\} = 0 \tag{5.5}$$

and then deduce the conclusion of the lemma.

Since $\eta_k \in R(X_k)$, we may write $\langle\eta_k, \epsilon\rangle = \sum_{i=1}^{k}\langle\eta_k, H_i\epsilon\rangle$. Similarly, $\rho_k \in R(X_k)^\perp$ and so $\langle\rho_k, \epsilon\rangle = \sum_{j=k+1}^{n}\langle\rho_k, H_j\epsilon\rangle$. Thus our interim objective will be achieved if we can prove that for all i, j, $1 \leq i \leq k$, $k+1 \leq j \leq n$, we have

$$\boldsymbol{E}\{\langle\eta_k, H_i\epsilon\rangle\langle\rho_k, H_j\epsilon\rangle f_l g_l\} = 0. \tag{5.6}$$

Note that f_l is a function of $\{|H_p\epsilon|^2 : k^* < p \leq l\}$ whilst g_l is a function of $\{|H_p\epsilon|^2 : l < p \leq K\}$, and so if $i \leq k^*$ or $j > k$, (5.6) is trivially zero. If we take the case $k^* \leq i, j \leq l$, we can split off g_l by independence and use Corollary 5.3 to get the conclusion. Similarly if $k^* \leq i \leq l$ and $l < j \leq K$, we can again use independence this time splitting off $\langle\eta_k, H_i\epsilon\rangle f_l$, and again getting zero by the same corollary. Thus (5.6) and hence (5.5) are established.

The proof is completed by noting that $\lim_{m,n} m^{-1}n|\rho_k|^{-2}\boldsymbol{E}|\langle\eta_k, \epsilon\rangle\langle\rho_k, \epsilon\rangle|$ is finite, and so we can combine the result $\mathrm{pr}(\tilde{k}_n \neq \hat{k}_n) \to 0$ as $n \to \infty$ with (5.5) to obtain (5.4). □

PROOF OF THEOREM 2.2. We obtain (2.6) under each of the three conditions in turn; in all cases making use of Lemmas 5.8 and 5.9. Then by Lemma 5.8, the left-hand side of (2.6) will be $O(n)$ as $m, n \to \infty$, since the bias terms $nB_{m,n}(k)$ for $k < k^*$ are not all eliminated, and these are $O(n)$ as $m, n \to \infty$, and cannot be canceled by either of the noise terms. Thus (2.6) is trivially true. Now let us assume (B). By virtue of the result just established, we may also suppose that $\mathrm{pr}(\hat{k}_n < k^*) \to 0$ as $n \to \infty$. Otherwise we make no assumptions concerning the selection procedure $\hat{k}$. On the set $\{\hat{k} > k^*\}$, $B_{m,n}(\hat{k}) = S_{m,n}(\hat{k}) = 0$, and so $\mathrm{PE}(\hat{k}) - m\sigma^2 = N_{m,n}(\hat{k})$.

Our proof begins by observing that

$$\begin{aligned}
&\lim_{m,n} nm^{-1}\boldsymbol{E}||\rho_k|^{-2}\langle\eta_k, \epsilon\rangle\langle\rho_k, \epsilon\rangle| \\
&\quad \leq \lim_{m,n} nm^{-1}|\rho_k|^{-2}\{\boldsymbol{E}\langle\eta_k, \epsilon\rangle^2\boldsymbol{E}\langle\rho_k, \epsilon\rangle^2\}^{1/2} \\
&\quad = \lim_{m,n} nm^{-1}|\rho_k|^{-2}\{|\eta_k|^2|\rho_k|^2\}^{1/2},
\end{aligned}$$

and this limit is zero by Lemma 5.7 and (B).

Repeated application of this result and Corollary 5.2 give a series of inequalities, which imply that for $k > k^*$:

$$\lim_{m,n} nm^{-1}\boldsymbol{E}\{N_{m,n}(k)1_{\{\hat{k}=k\}}\} \geq \lim_{m,n} nm^{-1}\boldsymbol{E}\{N_{m,n}(k^*)1_{\{\hat{k}=k\}}\},$$

whence $\lim_{m,n} nm^{-1}\boldsymbol{E}\{N_{m,n}(\hat{k})1_{\{\hat{k}\geq k^*\}}\} \geq \lim_{m,n} nm^{-1}\boldsymbol{E}\{N_{m,n}(k^*)1_{\{\hat{k}\geq k^*\}}\}$. Since $\mathrm{pr}(\hat{k}_n \geq k^*) \to 1$ as $n \to \infty$, and $N_{n,m}(k^*) \geq 0$, $\lim_{m,n} nm^{-1}\boldsymbol{E}\{N_{m,n}(k^*)\} = \mathrm{tr}\{\tilde{C}_{k^*}C_{k^*}^{-1}\}$ implies (2.6) in case (B).

Finally we consider case (C). The proof goes as for case (B), and in particular the selection rules $\hat{k}$ based on FPE_{α_n} for α_n such that $n^{-1}\alpha_n \to 0$ as $n \to \infty$, overfit with probability approaching unity by Theorem 3.2. The chain of inequalities leading to the final conclusion is also true, but this time the individual steps are justified by Theorem 3.1, and the proof is completed exactly as it was in case (B). Any other selection rule for which the same symmetry argument is valid also has the lower bound. □

PROOF OF THEOREM 2.3. (i) We begin by proving that the underfitting contribution to the left-hand side of (2.6) is asymptotically negligible. This follows from the readily checked fact that when $k < k^*$, $nm^{-1}\boldsymbol{E}\{(\mathrm{PE}(k) - m\sigma^2)\} \leq O(n)$ as $m, n \to \infty$. Thus for all $k < k^*$,

$$nm^{-1}\boldsymbol{E}\{(\mathrm{PE}(\hat{k}) - m\sigma^2)1_{\{\hat{k}=k\}}\} \leq O(n)\sqrt{\mathrm{pr}(\hat{k}_n = k)} \to 0$$

as $m, n \to \infty$, and so $nm^{-1}\boldsymbol{E}\{\mathrm{PE}(\hat{k}) - m\sigma^2)1_{\{\hat{k}<k^*\}}\} \to 0$ as $n, m \to \infty$.

Turning now to the overfitting contribution, we begin by proving that in the chain of inequalities used to prove the lower bound in cases (B) and (C), the terms dropped—the second and third terms of the right-hand side of Corollary 5.2—all have absolute expectations which are $O(mn^{-1})$. The argument at the beginning of the proof of case (B) of Theorem 2.2 shows this for the second term, for even without the hypothesis (B) we get a constant at that stage by Lemma 5.7(v). Similarly for the third terms,

$$\lim_{m,n} nm^{-1}\boldsymbol{E}\{|\rho_k|^{-4}|\tilde{\rho}_k|^2\langle\rho_k, \epsilon\rangle^2\} = O(1)$$

by Lemma 5.7. Thus we may use the consistency hypothesis and get

$$\begin{aligned}
&\lim_{m,n} nm^{-1}\boldsymbol{E}\{(\mathrm{PE}(\hat{k}) - m\sigma^2)1_{\{\hat{k}>k^*\}}\} \\
&= \sum_{k=k^*+1}^{K} \lim_{m,n} nm^{-1}\boldsymbol{E}\{(\mathrm{PE}(\hat{k}) - m\sigma^2)1_{\{\hat{k}=k\}}\} \\
&= \sum_{k=k^*+1}^{K} \lim nm^{-1}\boldsymbol{E}\{(\mathrm{PE}(k^*) - m\sigma^2)1_{\{\hat{k}=k\}}\} \\
&= \lim_{m,n} nm^{-1}\boldsymbol{E}(\mathrm{PE}(k^*) - m\sigma^2) = \mathrm{tr}\{\tilde{C}_{k^*}C_{k^*}^{-1}\},
\end{aligned}$$

the second last step following from our assumption that $\mathrm{pr}(\hat{k}_n = k) \to 0$ as $n \to \infty$ for all $k > k^*$. This completes the proof of (i).

(ii) Now we suppose that $\hat{k}$ is obtained by minimizing FPE_{α_n} for a sequence $\alpha_n < 2\log\log n$. We know from Theorem 3.2 that $\mathrm{pr}(\hat{k} < k^*) = o(n^{-1})$ and so

MODEL SELECTION AND PREDICTION: NORMAL REGRESSION

need only consider overfitting. By Shibata (1984), $\liminf \mathrm{pr}(\hat{k}_n = k^* + 1) > 0$. We next simplify $\lim_{m,n} nm^{-1}\boldsymbol{E}\{(\mathrm{PE}(\hat{k}) - m\sigma^2)\}$ in the now familiar way, noting that (as in the proof of Theorem 2.2) it coincides with

$$\begin{aligned}\lim_{m,n} nm^{-1}\boldsymbol{E}\{(\mathrm{PE}(\hat{k}) - m\sigma^2)1_{\{\hat{k}\geq k^*\}}\} \\ \geq \mathrm{tr}\{\tilde{C}_{k^*}C_{k^*}^{-1}\} + \lim_{m,n} nm^{-1}\boldsymbol{E}\{|\rho_{k^*}|^{-4}|\tilde{\rho}_{k^*}|^2\langle\rho_{k^*},\epsilon\rangle^2 1_{\{\hat{k}=k^*+1\}}\}.\end{aligned}$$

Now the second term above is zero only if $\rho_{k^*} = 0$, which implies $k^* = K$, since we have assumed all design matrices to be of full rank. Thus the inequality (2.6) is strict for selection rules based on FPE_{α_n} with $\liminf(2\log\log n)^{-1}\alpha_n < 1$. □

PROOF OF THEOREM 2.1. Since ϵ_t is independent of $\hat{k}_{t-1}$ and $\hat{\beta}_{t-1}$ for all $t > 1$,

$$\boldsymbol{E}\left\{\sum_1^n (y_t - x_t'\hat{\beta}_{t-1}(\hat{k}_{t-1}))^2\right\} = n\sigma^2 + \sum_1^n \boldsymbol{E}(x_t'\beta^* - x_t'\hat{\beta}_{t-1}(\hat{k}_{t-1}))^2.$$

Write

$$\begin{aligned}U_n &= \sum_1^n \boldsymbol{E}\{(x_t'\beta^* - x_t'\hat{\beta}_{t-1}(\hat{k}_{t-1}))^2 1_{\{\hat{k}_{t-1}<k^*\}}\},\\ V_n &= \sum_1^n \boldsymbol{E}\{(x_t'\beta^* - x_t'\hat{\beta}_{t-1}(\hat{k}_{t-1}))^2 1_{\{\hat{k}_{t-1}=k^*\}}\},\\ W_n &= \sum_1^n \boldsymbol{E}\{(x_t'\beta^* - x_t'\hat{\beta}_{t-1}(\hat{k}_{t-1}))^2 1_{\{\hat{k}_{t-1}>k^*\}}\}.\end{aligned}$$

We deal with each of these three components in turn. Let us temporarily denote $x_t'(X_k(t-1)'X_k(t-1))^{-1}X_k(t-1)'\epsilon(t-1)$ by $d'\epsilon$. Then

$$\begin{aligned}U_n &= \sum_{k=1}^{k^*-1}\sum_{t=1}^n \boldsymbol{E}\{(x_t'\beta^* - x_t'\hat{\beta}_{t-1}(\hat{k}_{t-1}))^2 1_{\{\hat{k}_{t-1}=k\}}\}\\ &= \sum_{k=1}^{k^*-1}\sum_{t=1}^n \boldsymbol{E}\{(\lambda_t(k) - d'\epsilon)^2 1_{\{\hat{k}_{t-1}=k\}}\}\\ &\leq 2\sum_{k=1}^{k^*-1}\sum_{t=1}^n [\lambda_t(k)^2\mathrm{pr}(\hat{k}_{t-1}=k) + 2\boldsymbol{E}\{(d'\epsilon)^2 1_{\{\hat{k}_{t-1}=k\}}\}].\end{aligned}$$

Now for $k < k^*$, $\sum_1^n \lambda_t(k)^2 = b_k n + o(1)$ as $n \to \infty$, whilst $\mathrm{pr}(\hat{k}_{t-1} = k) \leq O(t^{-2}(\log t)^{-c})$ as $n \to \infty$, $c > 1$. Summing by parts we thus conclude that

$$\sum_{k=1}^{k^*-1}\sum_{t=1}^n \lambda_t(k)^2\mathrm{pr}(\hat{k}_{t-1}=k) = O(1) \quad \text{as} \quad n \to \infty.$$

Furthermore, $\boldsymbol{E}\{(d'\epsilon)^4\} = 3\boldsymbol{E}\{(d'\epsilon)^2\}$, and since $\boldsymbol{E}(d'\epsilon)^2 = |d|^2\sigma^2 = \mu_t(k)\sigma^2$,

$$\begin{aligned}\sum_{k=1}^{k^*-1}\sum_{t=1}^{n} \boldsymbol{E}\{(d'\epsilon)^2 1_{\{\hat{k}_{t-1}=k\}}\} &\le \sum_{k=1}^{k^*-1}\sum_{t=1}^{n} \sqrt{3}\sigma^2\mu_t(k)\{\mathrm{pr}(\hat{k}_{t-1}=k)\}^{1/2} \\ &= O(1) \quad \text{as} \quad n\to\infty,\end{aligned}$$

as argued above, but this time using $\sum_1^n \mu_t(k) = k\log n(1+o(1))$ as $n\to\infty$. Thus $U_n = O(1)$ as $n\to\infty$.

Turning now to the overfitting term V_n, we find only the quadratic form $(d'\epsilon)^2$, as the bias term vanishes. Thus we can argue as above, giving

$$\begin{aligned}W_n &= \sum_{k=k^*+1}^{K}\sum_{t=1}^{n} \boldsymbol{E}\{(d'\epsilon)^2 1_{\{\hat{k}_{t-1}=k\}}\} \\ &\le \sqrt{3}\sigma^2 \sum_{k=k^*+1}^{K}\sum_{t=1}^{n} \mu_t(k)\{\mathrm{pr}(\hat{k}_{t-1}=k)\}^{1/2} = O(1),\end{aligned}$$

since $\mathrm{pr}(\hat{k}_{t-1}=k) \le O(\log t^{-\alpha})$ as $t\to\infty$, where $\alpha > 2$.

Finally, we examine the term corresponding to getting the model correct. Since $\mathrm{pr}(\hat{k}_{t-1}\ne k^*) \le A(t^{-2}(\log t)^{-c}) + B(\log t)^{-\alpha}$ for large t,

$$\begin{aligned}V_n &= \sum_{t=1}^{n} \boldsymbol{E}\{(x_t'\beta^* - x_t'\hat{\beta}_{t-1}(k^*))^2 1_{\{\hat{k}_{t-1}=k^*\}}\} \\ &= \sum_{t=1}^{n} \boldsymbol{E}\{(d'\epsilon)^2\} - \sum_{t=1}^{n} \boldsymbol{E}\{(d'\epsilon)^2 1_{\{\hat{k}_{t-1}\ne k^*\}}\} \\ &= k^*\log n(1+o(1)) + O(1) \quad \text{as} \quad n\to\infty. \qquad \square\end{aligned}$$

Acknowledgements

We would like to thank Jorma Rissanen for his inspiration and for many useful discussions. Special thanks are due to David Freedman for his criticisms of the manuscript.

References

Akaike, H. (1970). Statistical predictor identification, *Ann. Inst. Statist. Math.*, **22**, 202–217.

Akaike, H. (1974). A new look at the statistical model identification, *IEEE Trans. Automat. Control*, **19**, 716–723.

Atkinson, A. C. (1980). A note on the generalized information criterion for choice of a model, *Biometrika*, **67**, 413–418.

Bhansali, R. H. and Downham, D. Y. (1977). Some properties of the order of an autoregressive model selected by a generalization of Akaike's FPE criterion, *Biometrika*, **64**, 547–551.

Box, G. E. P. and Draper, N. R. (1959). A basis for the selection of a regression surface design, *J. Amer. Statist. Assoc.*, **54**, 622–654.

MODEL SELECTION AND PREDICTION: NORMAL REGRESSION

Box, G. E. P. and Draper, N. R. (1963). The choices of a second order rotatable design, *Biometrika*, **50**, 335–352.

Breiman, L. A. and Freedman, D. F. (1983). How many variables should be entered in a regression equation?, *J. Amer. Statist. Assoc.*, **78**, 131–136.

Clayton, M. K., Geisser, S. and Jennings, D. (1986). A comparison of several model selection procedures, *Bayesian Inference and Decision* (eds. P. Goel and A. Zellner), 425–439, Elsevier, New York.

Dawid, A. P. (1984). Present position and potential developments: some personal views, Statistical theory—The prequential approach (with discussion), *J. Roy. Statist. Soc. Ser. A*, **147**, 278–292.

Dawid, A. P. (1992). Prequential data analysis, *Current Issues in Statistical Inference: Essays in Honor of D. Basu, Institute of Mathematical Statistics, Monograph*, **17** (eds. M. Ghosh and P. K. Pathak).

Geweke, J. and Meese, R. (1981). Estimating regression models of finite but unknown order, *Internat. Econom. Rev.*, **22**, 55–70.

Hannan, E. J. and Quinn, B. G. (1979). The determination of the order of an autoregression, *J. Roy. Statist. Soc. Ser. B*, **41**, 190–195.

Hannan, E. J., McDougall, A. J. and Poskitt, D. S. (1989). Recursive estimation of autoregressions, *J. Roy. Statist. Soc. Ser. B*, **51**, 217–233.

Haughton, D. (1989). Size of the error in the choice of a model to fit data from an exponential family, *Sankhyā Ser. A*, **51**, 45–58.

Hemerly, E. M. and Davis, M. H. A. (1989). Strong consistency of the predictive least squares criterion for order determination of autoregressive processes, *Ann. Statist.*, **17**, 941–946.

Hjorth, U. (1982). Model selection and forward validation, *Scand. J. Statist.*, **9**, 95–105.

Kohn, R. (1983). Consistent estimation of minimal dimension, *Econometrica*, **51**, 367–376.

Lai, T., Robbins, H. and Wei, C. Z. (1979). Strong consistency of least squares estimates in multiple regression II, *J. Multivariate Anal.*, **9**, 343–361.

Merhav, N., Gutman, M. and Ziv, J. (1989). On the estimation of the order of a Markov chain and universal data compression, *IEEE Trans. Inform. Theory*, **39**, 1014–1019.

Nishi, R. (1984). Asymptotic properties of criteria for selection of variables in multiple regression, *Ann. Statist.*, **12**, 758–765.

Rao, C. R. (1973). *Linear Statistical Inference*, 2nd ed., Wiley, New York.

Rissanen, J. (1984). Universal coding, information prediction, and estimation, *IEEE Trans. Inform. Theory*, **30**, 629–636.

Rissanen, J. (1986*a*). Stochastic complexity and modeling, *Ann. Statist.*, **14**, 1080–1100.

Rissanen, J. (1986*b*). A predictive least squares principle, *IMA J. Math. Control Inform.*, **3**, 211–222.

Rissanen, J. (1989). *Stochastic Complexity in Statistical Inquiry*, World Books, Singapore.

Sawa, T. (1978). Information criteria for discriminating among alternative regression models, *Econometrica*, **46**, 1273–1291.

Schwartz, G. (1978). Estimating the dimension of a model, *Ann. Statist.*, **6**, 461–464.

Shibata, R. (1976). Selection of the order of an autoregressive model by Akaike's information criterion, *Biometrika*, **63**, 117–126.

Shibata, R. (1981). An optimal selection of regression variables, *Biometrika*, **68**, 45–54.

Shibata, R. (1983*a*). Asymptotic mean efficiency of a selection of regression variables, *Ann. Inst. Statist. Math.*, **35**, 415–423.

Shibata, R. (1983*b*). A theoretical view of the use of AIC, *Times Series Analysis: Theory and Practice 4* (ed. O. D. Anderson), 237–244, Elsevier, Amsterdam.

Shibata, R. (1984). Approximate efficiency of a selection procedure for the number of regression variables, *Biometrika*, **71**, 43–49.

Shibata, R. (1986*a*). Selection of the number of regression variables; a minimax choice of generalized FPE, *Ann. Inst. Statist. Math.*, **38**, 459–474.

Shibata, R. (1986*b*). Consistency of model selection and parameter estimation, *Essays in Time Series and Allied Processes: Papers in Honour of E. J. Hannan, J. Appl. Probab.*, **23A**, 127–141.

Stone, M. (1977). An asymptotic equivalence of choice of model by cross-validation and Akaike's criterion, *J. Roy. Statist. Soc. Ser. B*, **39**, 44–47.

Wax, M. (1988). Order selection for AR models by predictive least squares, *IEEE Trans. Acoust. Speech Signal Process.*, **36**, 581–588.

Wei, C. Z. (1992). On the predictive least squares principle, *Ann. Statist.*, **20**, 1–42.

Woodroofe, M. (1982). On model selection and the arc sine laws, *Ann. Statist.*, **10**, 1182–1194.

IEEE TRANSACTIONS ON INFORMATION THEORY, VOL. 39, NO. 3, MAY 1993

A Rate of Convergence Result for a Universal D-Semifaithful Code

Bin Yu and T. P. Speed

***Abstract*—The problem of optimal rate universal coding in the context of rate-distortion theory is considered. A D-semifaithful universal coding scheme for discrete memoryless sources is given. The main result is a refined covering lemma based on the random coding argument and the method of types. The average codelength of the code is shown to appraoch its lower bound, the rate-distortion function, at a rate $O(n^{-1} \log n)$, and this is conjectured to be optimal based on a result of Pilc. Issues of constructiveness and universality are also addressed.**

***Index Terms*— Discrete memoryless source, rate-distortion, D-semifaithful, universal coding, optimal rate, random coding, method of types.**

I. Introduction

ENTROPY has a central position in information theory, in part because in the limit it gives the shortest possible per-symbol average length of a noiseless code. If we consider a discrete memoryless source with distribution P_0, the entropy $H(P_0)$ serves as a nonasymptotic lower bound to the average expected codelength for data strings from this source. Moreover, the entropy lower bound can be achieved asymptotically at the rate $O(n^{-1})$ when the source distribution P_0 is known, and at the rate $O(n^{-1} \log n)$ when P_0 is not known.

Rissanen [19] improved the entropy lower bound by showing that entropy $\frac{1}{2} kn^{-1} \log n$ is an asymptotic lower bound to the average expected codelength. His bound holds for data strings from parametric statistical models satisfying mild regularity conditions, and the k in the lower bound is the dimension of the model. Discrete memoryless sources are covered by his result, with k there being the cardinality of the source alphabet minus one, and the rate $O(n^{-1}\log n)$ is optimal in this case, when P_0 is not known. The rate $O(n^{-1}\log n)$ has been shown to be achievable for various other statistical models, see for example Davisson [8], Rissanen [19], [20], Hannan and Kavalieris [10], Hemerly and Davis [11], Gerenscer and Rissanen [9], Clarke and Barron [5], and Weinberger, Lempel, and Ziv [27]. Extensions to nonparametric models can be found in Barron and Cover [1], Rissanen, Speed, and Yu [21], and Yu and Speed [28].

Rate-distortion theory was started by Shannon [23], and in that context we consider block-codes with a fidelity criterion,

Manuscript received June 10, 1991; revised August 17, 1992. This work was supported in part by the Wisconsin Alumni Research Foundation and the Army Research Office Grant DAAL03-91-G-007, and the National Science Foundation Grant DMS 8802378.

B. Yu is with the Department of Statistics, University of Wisconsin, Madison, WI 53706.

T. P. Speed is with the Department of Statistics, University of California, Berkeley, CA 94720.

IEEE Log Number 9206227.

or semifaithful codes to use the term from a recent paper of Ornstein and Shields [15]. Instead of the expected codelength used in noiseless coding it is natural in rate-distortion theory to consider the log of the number of D-balls required to cover the n-tuple space of the source alphabet under some single-letter distance measure. The role of entropy in noiseless coding is then taken by the rate-distortion function, in the following sense: the rate-distortion function gives a lower bound to the log of the covering number, which we may also refer to as the expected code length of a D-semifaithful code, and this lower bound can be achieved in the limit by certain D-semifaithful codes. In particular, Ornstein and Shields [15] obtain D-semifaithful codes which achieve the rate-distortion function lower bound almost surely, for ergodic sequences, and Shields [22] uses Markov types for similar results. Earlier work for other classes of sources include Neuhoff, Gray, and Davisson [14], Mackenthun and Pursley [13] and Kieffer [12]. In the case of memoryless sources, the achievability proof can be found in standard texts, see for example, Cover and Thomas [7] for a recent exposition using the random coding argument. However, no results have yet been provided on the rate at which this lower bound is approached.

In this paper, we describe a D-semifaithful universal coding scheme of memoryless sources and obtain an associated rate result. We show, for a discrete memoryless source with a source alphabet of J elements and an unknown distribution P_0, that under some mild smoothness conditions on the rate-distortion function, a universal D-semifaithful code can be constructed such that the average expected length of this code tends to the rate-distortion function at the rate $n^{-1}\log n$. The techniques used are the method of types and random coding. The main result will be based on a refined coding lemma (Theorem 1) for type classes. It is "refined" because it improves the $o(1)$ term in the covering lemma in Csiszár and Körner [6] to an $O(n^{-1}\log n)$ term. In other words, we are able to give a better upper bound on (the log of) the number of D-balls needed to cover a type class, equivalently, on the number of D-semifaithful code words required to encode a type class. Then a two-stage code is constructed as the D-semifaithful code for all strings: first we encode the type class, and next we encode the elements of each type class using the refined covering lemma. The above results are contained in Section II.

In Section III, we conjecture that the rate $n^{-1} \log n$ is asymptotically optimal. Our conjecture is based on a result of Pilc [16], [17], which is expressed in terms of the inverse of the rate-distortion function: the distortion-rate function. Pilc has upper bounds and lower bounds for noiseless channels and

IEEE TRANSACTIONS ON INFORMATION THEORY, VOL. 39, NO. 3, MAY 1993

for noisy channels, but we use his results only for noiseless channels. Unfortunately, although the rate $n^{-1}\log n$ in the upper bound of our two-stage code matches that in Pilc's lower bound, his lower bound is on the log cardinality of an *expected* D-semifaithful code (cf. the forthcoming definition) while our code is *pointwise* D-semifaithful with an upper bound on the expected codelength. Hence, we do not know at this stage if the rate $n^{-1}\log n$ is indeed optimal in terms of expected codelength. Moreover, his bound does not include Rissanen's since it holds only for nonzero distortion levels.

In Section IV, we compare our code with the code corresponding to Pilc's upper bound. The main point made there is that our code is universal, while the other one is not. In addition, the issue of construction versus pure existence is addressed in relation to our code and the one corresponding to Pilc's upper bound.

We start with some preliminaries on rate-distortion theory and the method of types. Our main reference on rate-distortion theory is Berger [2], and that on the method of types is Csiszár and Körner [6].

II. Preliminaries

Let $\mathcal{A}_0 = \{1, 2, \cdots, J-1, J\}$ be the source alphabet, and let $\mathcal{B}_0 = \{1, 2, \cdots, K\}$ be the reproducing alphabet. $\mathcal{B}_0$ could be the same as or a subset of $\mathcal{A}_0$. We assume our source is memoryless, i.e., that the letters $x_1, \cdots, x_n$, which make up our strings are mutually independent and identically distributed (i.i.d.) with distribution P_0 on $\mathcal{A}_0$. Without loss of generality we assume $P_0(j) > 0$ for all $j \in \mathcal{A}_0$. We use a single-letter fidelity criterion to measure the distortion between any nth order source string $x^n = (x_1, \cdots, x_n) \in \mathcal{A}_0^n$, and its code word $y^n \in \mathcal{B}_0^n$. More precisely, let

$$d_n(x^n, y^n) = n^{-1}\sum_{t=1}^{n} d(x_t, y_t),$$

where d is a bounded real nonnegative function on $\mathcal{A}_0 \times \mathcal{B}_0$, with maximum d_M and minimum d_m. Then the rate distortion function $R_n(P_0, D)$ for the distribution of $x_1, \cdots, x_n$ equals $nR(P_0, D)$ where the rate-distortion function $R(P_0, D)$ of P_0 can be formally defined as follows:

$$R(P_0, D) = \min_W I(W, P_0)$$
$$= \min_W \sum_{j=1}^{J}\sum_{k=1}^{K} P_0(j)W(k|j)\log\frac{W(k|j)}{Q(k)},$$

where the minimum is taken over the set of matrices W from $\mathcal{A}_0$ to $\mathcal{B}_0$ such that for any $j, k, W(k|j) \geq 0$, for all j, $\sum_{k=1}^{K} W(k|j) = 1$,

$$\sum_{j=1}^{J}\sum_{k=1}^{K} P_0(j)W(k|j)\, d(j,k) \leq D,$$

and Q is the marginal distribution on $\mathcal{B}_0$ induced by P_0 and W, i.e., for $k \in \mathcal{B}_0$,

$$Q(k) = \sum_{j=1}^{J} P_0(j)W(k|j).$$

The following properties of $R(P, D)$ can be found in Berger [2].

1) $R(P, \cdot)$ is convex, monotonically decreasing on $[0, D_{\max}]$ where $D_{\max} = \min_k \sum_{j=1}^{J} P(j)\, d(j,k)$. Moreover, for $D \geq D_{\max}$, $R(P, D) = 0$, and $R(P, 0) = H(P) = -\sum_{j=1}^{J} P(j)\log P(j)$. Hence $R'_D(P, D) \leq 0$ for any D, where $'$ denotes differentiation with respect to D.

2) If $I(W, P) = R(P, D)$, then for any j:

$$\frac{\partial I(W,P)}{\partial W(k|j)}|W = P(j)\log\frac{W(k|j)}{Q(k)},$$

where for any j, k:

$$W(k|j) = \frac{Q(k)e^{sd(j,k)}}{\sum_{\ell} Q(\ell)e^{sd(j,\ell)}},$$

with $s = R'_D(P, D) \leq 0$.

Definition (D-semifaithful code): A map $M_n : \mathcal{A}_0^n \to \mathcal{B}_0^n$ is called a *pointwise* D-semifaithful code if for any $x^n \in \mathcal{A}_0^n$

$$d_n(x^n, M_n(x^n)) \leq D.$$

Similarly a map M_n is called *expected* D-semifaithful with respect to a source distribution P_0 if whenever $(x_1, \cdots, x_n)$ are i.i.d. with common distribution P_0,

$$E_{P_0} d_n(x^n, M_n(x^n)) \leq D.$$

Since our main argument will be based on the method of types, we next introduce the definition of type and some of its properties. We will follow the notation of Csiszár and Körner [6], with $1\{A\}$ denoting the indicator of the event A and $\equiv$ meaning equal by definition.

Definition (Type): The type of a sequence $x^n \in \mathcal{A}_0^n$ is the distribution P_{x^n} on $\mathcal{A}_0$ defined for $j \in \mathcal{A}_0$ by

$$P_{x^n}(j) \equiv \frac{1}{n}N(j|x^n) \equiv \frac{1}{n}\sum_{t=1}^{n} 1\{x_t = j\}$$

that is, the empirical distribution of x^n on $\mathcal{A}_0$. We write $T_P^n = \{x^n : x^n \text{ has type } P\}$ for any P on $\mathcal{A}_0$ such that $\{nP(j)\}$ are integers.

For any given $x^n \in \mathcal{A}_0^n$, and a stochastic matrix $W : \mathcal{A}_0 \to \mathcal{B}_0$, we next define conditional types.

Definition (Conditional type): The conditional type W of a sequence $y^n \in \mathcal{B}_0^n$ given $x^n \in \mathcal{A}_0^n$ is defined for $j \in \mathcal{A}_0$, $k \in \mathcal{B}_0$ by

$$N(j|x^n)W(k|j) \equiv N(j,k|x^n,y^n)$$
$$\equiv \sum_{t=1}^{n} 1\{x_t = j \text{ and } y_t = k\}.$$

We denote the set of sequences $y^n \in \mathcal{B}_0^n$ having the conditional type W given x^n by $T_W^n(x^n)$.

The cardinality of a type class, or a conditional type class can be bounded above and below as in the following results from Csiszár and Körner [6].

Lemma 1: For any type P of sequences in $\mathcal{A}_0^n$,

$$(n+1)^{-J}\exp(nH(P)) \leq |T_P^n| \leq \exp(nH(P)). \quad (1.1)$$

Lemma 2: For every $x^n \in \mathcal{A}_0^n$, and stochastic matrix $V : \mathcal{A}_0 \to \mathcal{B}_0$ such that $T_V^n(x^n)$ is nonempty,

$$(n+1)^{-JK} \exp\{nH(V|P_{x^n})\} \le |T_V^n(x^n)| \le \exp\{nH(V|P_{x^n})\}, \quad (1.2)$$

where $H(V|P) = \sum_{j=1}^J P(j)H(V(\cdot|j)) = -\sum_{j=1}^J P(j) \sum_{k=1}^K V(k|j) \log V(k|j)$.

Lemma 3: The total number of type classes is at most $(n+1)^J$.

III. A Universal Pointwise D-Semifaithful Code

In this section, we first use the random coding argument in Csiszár and Körner [6] to prove a refined covering lemma (Theorem 1). Then, we go on to give a two-stage universal D-semifaithful coding scheme (Theorem 2) with the rate $n^{-1} \log n$. We begin with a proposition extracted from Csiszár and Körner [6].

For a given type P on $\mathcal{A}_0^n$, positive constant D in $(0, D_{\max})$, and a subset B of $\mathcal{B}_0^n$, write

$$U_D(B) = \{x^n \in T_P^n : d_n(x^n, B) > D\},$$

where $d_n(x^n, B) := \min_{y^n \in B} d_n(x^n, y^n)$.

Proposition 1: Suppose $Z^{(m)} = \{Z_1, \cdots, Z_m\}$ are i.i.d. and uniform over a subset $G \subset \mathcal{B}_0^n$. If for some m_n we have $E|U_D(Z^{(m_n)})| < 1$, then there is a set $B_{P,D}$ such that $|B_{P,D}| \le m_n$ and $|U_D(B_{P,D})| < 1$. This implies that $U_D(B_{P,D}) = \phi$. In other words, that $B_{P,D}$ "covers" T_P^n within distance D.

See Csiszár and Körner [6] for the proof.

Moreover, note that,

$$\left|U_D\left(Z^{(m)}\right)\right| = \sum_{x^n \in T_P^n} 1\left\{U_D\left(Z^{(m)}\right)\right\}(x^n)$$

implying

$$E\left|\left(U_D\left(Z^{(m)}\right)\right| = \sum_{x^n \in T_P^n} \mathbf{P}_0\left(x^n \in U_D\left(Z^{(m)}\right)\right). \quad (2.1)$$

For any fixed $x^n \in T_P^n$, because the Z are i.i.d.,

$$\mathbf{P}_0\left(x^n \in U_D\left(Z^{(m)}\right)\right) = [\mathbf{P}_0(d_n(x^n, Z) > D)]^m, \quad (2.2)$$

where Z is uniformly distributed over G.

Furthermore, if we can find a subset $G_1(x^n) \subset G$ such that, for any $y^n \in G_1(x^n)$ we have $d_n(x^n, y^n) \le D$, then

$$\begin{aligned}\mathbf{P}_0(d_n(x^n, Z) > D) &= 1 - \mathbf{P}_0(d_n(x^n, Z) \le D)\\ &\le 1 - \mathbf{P}_0(Z \in G_1(x^n))\\ &= 1 - |G_1(x^n)|/|G|. \quad (2.3)\end{aligned}$$

Combining (2.1), (2.2), and (2.3), we get

$$\begin{aligned}E\left|U_D\left(Z^{(m)}\right)\right| &\le \sum_{x^n \in T_P^n} \left(1 - \frac{|G_1(x^n)|}{|G|}\right)^m\\ &\le \sum_{x^n \in T_P^n} \exp\left(-\frac{|G_1(x^n)|}{|G|} m\right). \quad (2.4)\end{aligned}$$

The last inequality holds because $(1-t)^m \le \exp(-tm)$ for any $t > 0$. Next we choose a conditional type class as $G_1(x^n)$, and a type class as G. For the chosen $G_1(x^n)$ and G, we select m_n using (2.4) such that $E|U_D(Z^{(m)})| < 1$. For any type P and constant D in $(0, D_{\max})$, take W such that

$$\sum_{j,k} W(k|j)P(j)d(j,k) \le D^*,$$

and $I(W,P) = R(P, D^*)$, where $D^* = D - n^{-1}JKd_M$.

Note that this W depends on both P and D^*, but for simplicity, we do not indicate this dependence in any way. Because the $nP(j)$ are integers, we can find a stochastic matrix $[W]$, a truncation of W, such that for all j and k, $n[W](k|j)P(j)$ are integers, and

$$|W(k|j) - [W](k|j)| \le \frac{1}{nP(j)}, \qquad \text{for } j = 1, \cdots, J.$$

Let $[Q] = [W] \cdot P$, i.e., $[Q](k) = \sum_{j-1}^J [W](k|j)P(j)$. Then, the $n[Q](k)$ are also integers. Therefore, the type class $T_{[Q]}^n$ and the conditional type class $T_{[W]}^n(x^n)$ are well defined for all $x^n \in T_P^n$.

Let us take $G = T_{[Q]}^n$ and $G_1(x^n) = T_{[W]}^n(x^n)$. Then, for any $y^n \in G_1(x^n)$,

$$N(j,k|x^n, y^n) = [W](k|j)P(j)n,$$

and

$$\begin{aligned}N(k|y^n) &= \sum_j N(j,k|x^n,y^n)\\ &= \sum_j [W](k|j)P(j)n = [Q](k)n,\end{aligned}$$

that is, $y^n \in G = T_{[Q]}^n$. Hence, $G_1(x^n) \subseteq G$.

In addition, for any $y^n \in G_1(x^n)$, since $D^* = D - n^{-1}JKd_M$,

$$\begin{aligned}d_n(x^n, y^n) &= \sum_{j,k} [W](k|j)P(j)d(j,k)\\ &\le \sum_{j,k} (W(k|j)P(j) + n^{-1})d(j,k)\\ &\le \sum_{j,k} W(k|j)P(j)d(j,k) + n^{-1}JKd_M\\ &\le D^* + n^{-1}JKd_M = D.\end{aligned}$$

Recalling the bounds in (1.1) and (1.2), we have

$$\begin{aligned}|G| &\le \exp(nH(Q)),\\ |G_1(x^n)| &\ge (n+1)^{-JK} \exp\{nH([W]|P)\}.\end{aligned}$$

Thus, since $I([W], P) = H([Q]) - H([W]|P)$,

$$-\frac{|G_1(x^n)|}{|G|} \le -(n+1)^{-JK} \exp(-nI([W], P)). \quad (2.5)$$

IEEE TRANSACTIONS ON INFORMATION THEORY, VOL. 39, NO. 3, MAY 1993

Putting (2.5) into (2.4), we find

$$E\left|U_D\left(Z^{(m)}\right)\right| \le \sum_{x^n\in T_P^n} \cdot \exp\{-(n+1)^{-JK}\exp(-nI([W],P))m\} = |T_P^n|\exp\{-(n+1)^{-JK} \cdot \exp(-nI([W],P))m\}$$

Finally, we get by Lemma 1:

$$E\left|U_D\left(Z^{(m)}\right)\right| \le \exp(nH(P))\exp\{-(n+1)^{-JK} \cdot \exp(-nI([W],P))m\}. \quad (2.6)$$

Now we can choose $m = m_n$ as an integer such that

$$\exp\{nI([W],P)+(JK+2)\log(n+1)\} \le m_n \le \exp\{nI([W],P)+(JK+4)\log(n+1)\}.$$

Then for such an m_n, (2.6) gives

$$E|U_D(Z^{(m)})| \le \exp(n\log J)\exp\left(-(n+1)^2\right) < 1, \quad \text{for } n \text{ large}.$$

Applying Proposition 1, we obtain the following theorem.

Theorem 1 (Refined Covering Lemma for Type Classes): Given a type P on $\mathcal{A}_0$ and D in $(0, D_{\max})$, there is a subset $B_{P,D} \subset \mathcal{B}_0^n$ such that for any $x^n \in T_P^n$, $d_n(x^n, B_{P,D}) \le D$ and

$$|B_{P,D}| \le \exp\{nI([W],P)+(JK+4)\log(n+1)\},$$

where for any j,k

$$|[W](k|j) - W(k|j)| \le \frac{1}{nP(j)},$$

and $I(W,P) = R(P,D^*)$ for $D^* = D - n^{-1}JKd_M$.

Next, we show that we can replace $[W]$ by W in Theorem 1. Since $[W]$ is close to W, and D^* is close to D, we expect $I([W],P)$ to be close to $I(W,P) = R(P,D^*)$, hence close to $R(P,D)$. Formally, we expand $I([W],P)$ around $I(W,P)$ as follows:

$$I([W],P) = I(W,P) + \sum_{j=1}^{J}\sum_{k=1}^{K-1}\frac{\partial I}{\partial W(k|j)} \cdot (\cdot,P)|W_{(k|j)}([W](k|j) - W(k|j)) + \cdots, \quad (2.7)$$

where $(\cdots)$ denotes smaller order terms. Since $I(W,P) = R(P,D^*)$, by property 2) in Section I, for any k,j

$$\frac{\partial I(\cdot,P)}{\partial W(k|j)}|W = P(j)\log\frac{W(k|j)}{Q(k)}.$$

Note that for any j,k, $|[W](k|j) - W(k|j) \le (nP(j))^{-1}$, so we have from (2.7):

$$I([W],P) = I(W,P) + \left\{\sum_{j=1}^{J}\sum_{k=1}^{K-1}\log\frac{W(k|j)}{Q(k)}\right\}n^{-1} + o(n^{-1}).$$

However, again by property 2) in Section I, for all j,k we have

$$W(k|j) = \frac{Q(k)e^{sd(j,k)}}{\sum_{\ell} Q(\ell)e^{sd(j,\ell)}},$$

where $s = R'_D(P,D^*) < 0$. Hence,

$$\frac{W(k|j)}{Q(k)} = \frac{e^{sd(j,k)}}{\sum_{\ell} Q(\ell)e^{sd(j,\ell)}} \le \frac{e^{-|s|d_m}}{\sum_{\ell} Q(\ell)e^{-|s|d_M}} = e^{|s|(d_M-d_m)}.$$

Similarly $W(k|j)/Q(k) \ge e^{-|s|(d_M-d_m)}$, and hence,

$$|\log\frac{W(k|j)}{Q(k)}| \le |s|(d_M - d_m).$$

Without loss of generality, assume $d_m = 0$. We then get

$$I([W],P) \le I(W,P) + n^{-1}JK|s|d_M + 0(n^{-1}) = R(P,D^*) + n^{-1}JK|s|d_M + O(n^{-1}).$$

On the other hand,

$$I(W,P) = R(P,D^*) = R(P,D) + R'_D(P,D)(D^*-D) + \cdots = R(P,D) + |s|JKd_Mn^{-1} + \cdots.$$

Thus,

$$I([W],P) \le R(P,D) + 2JK|s|d_Mn^{-1} + o(n^{-1}), \quad (2.8)$$

where $s = R'_D(P,D)$. We have proved the following

Corollary 1: Under the assumptions of Theorem 1

$$\log|B_{P,D}| \le nR(P,D) + 2JK|s|d_M + o(1) + (KJ+4)\log(n+1).$$

Theorem 2 (Universal Pointwise D-Semifaithful Coding): Let $\mathcal{F}$ be a class of distributions on $\mathcal{A}_0$ such that for some $D \in (0, D_{\max})$, the derivatives $\{\partial^2 R(P,D)/\partial P_j \partial P_{j'} : j,j' = 1,\cdots,J\}$ are uniformly bounded over $\mathcal{F}$ by a constant C, and $|E_{P_0}R'_D(P_{x^n},D)| < \infty$ for all $P_0 \in \mathcal{F}$. Then there exists a two-stage code $M_n : \mathcal{A}_0^n \to \mathcal{B}_0^n$ such that

$$d_n(x^n, M_n(x^n)) \le D,$$

and for all $P_0 \in \mathcal{F}$, as $n \to \infty$,

$$n^{-1}E_{P_0}L(M_n(x^n)) \le R(P_0,D) + (KJ+J+4)n^{-1} \cdot \log(n+1) + O(n^{-1}).$$

Proof of Theorem 2: For any $x^n \in \mathcal{A}_0^n$, our coding scheme has two stages: First, we encode P_{x^n}, which by Lemma 3 takes at most $J\log(n+1)$ bits. Next, we use Corollary 1, which asserts that for the type class $T_{P_{x^n}}^n$, there is a $B_{P_{x^n},D}$ which covers $T_{P_x^n}^n$ with radius D. We then take $M_n : T_{P_{x^n}}^n \to B_{P_{x^n},D}$ where $M_n(x^n) = y^n \in B_{P_{x^n},D}$ is such that $d_n(x^n,y^n) \le D$. This takes at most $\log|B_{P_{x^n},D}|$ bits, which, since $R'_D < 0$, is bounded by

$$nR(P_{x^n},D) + (KJ+4)\log(n+1) - 2KJR'_D(P_{x^n},D) + o(1).$$

For simplicity, denote P_{x^n} by $\overline{P}$. Taking the expectation of this last expression gives a bound on the expected codelength of

$$E_{P_0}R(\overline{P},D)+(KJ+4)n^{-1}\ \log(n+1) \\ -\ 2KJE_{P_0}R'_D(\overline{P},D)n^{-1}+o(n^{-1}).$$

To prove the theorem, it suffices to show

$$E_{P_0}R(\overline{P},D)=R(P_0,D)+O(n^{-1}), \qquad (2.9)$$

because $|E_{P_0}R'_D(\overline{P},D)|<\infty$ by assumption.

The idea to show (2.9) is simply a Taylor expansion of $R(P,D)$ around $P=P_0$, but because the Taylor expansion holds only in an $o(1)$ neighborhood of P_0, some effort has to be made to give a rigorous proof.

We split the set of types into two disjoint subsets:

$$\Omega_n=\{P:|P(j)-P_0(j)| \\ \le n^{-1/2}\log n,\ \text{for all } j\in\mathcal{A}_0\},$$

and

$$\Omega_n^c=\{P:|P(j)-P_0(j)| \\ > n^{-1/2}\log n,\ \text{for some } j\in\mathcal{A}_0\},$$

and we break the expectation $E_{P_0}R(\overline{P},D)$ up similarly, defining

$$E_1=\sum_{P\in\Omega_n}\mathbf{P}_0(x^n\in T_P^n)R(P,D)$$

$$E_2=\sum_{P\in\Omega_n^c}\mathbf{P}_0(x^n\in T_P^n)R(P,D),$$

where $\mathbf{P}_0$ denotes the probability measure on sequences defined by P_0. Before we go further, we need a good bound on $\sum_{P\in\Omega_n^c}\mathbf{P}_0(x^n\in T_P^n)$. Hoeffding's inequality, cf. Pollard [18, p. 191], implies that for all j,

$$\mathbf{P}_0(x^n:|\overline{P}(j)-P_0(j)|>n^{-1/2}\ \log\ n) \\ \le \exp(-2[\log n]^2\cdot n/4n) \\ =\exp\left(-\frac{1}{2}\,(\log n)^2\right).$$

As a result,

$$\sum_{P\epsilon\Omega_n^c}\mathbf{P}_0(x^n\epsilon T_P^n)$$

$$\le\sum_{j=1}^{J}\mathbf{P}_0\left(x^n:|\overline{P}(j)-P_0(j)|>n^{-1/2}\log n\right)$$

$$\le J\ \exp\left(-\frac{1}{2}\,(\log n)^2\right)=Jn^{-\frac{1}{2}\log n}. \qquad (2.10)$$

Then (2.10) and the inequality $R(P,D)\le\log J$ together yield

$$E_2=\sum_{P\in\Omega_n^c}\mathbf{P}_0(x^n\in T_P^n)R(P,D) \\ \le(\log J)Jn^{-\frac{1}{2}\log n} \\ =O(n^{-1}),\qquad \text{for } n \text{ large.}$$

On Ω_n, we can expand $R(P,D)$ as

$$R(P,D)=R(P_0,D)+\sum_{j=1}^{J-1}\frac{\partial R}{\partial P_j}(P_0,D)\cdot((P(j)-P_0(j))$$

$$+\frac{1}{2}\sum_{j,j'}(P(j')-P_0(j'))\frac{\partial^2 R}{\partial P_j\partial P_{j'}}(P_0',D) \\ \cdot((P(j)-P_0(j)),$$

where P_0' is in between P_0 and P. Because the partial derivatives around P_0 are bounded by a constant C, the third term on the right is bounded by

$$2JC\sum_{j=1}^{J}(P(j)-P_0(j))^2,$$

and so its expectation is $O(n^{-1})$ by a known result concerning the multinomial variance.

Moreover, the fact that $E_{P_0}(P(j)-P_0(j))=0$ for all j, implies that

$$\left|\sum_{P\in\Omega_n}\mathbf{P}_0(x^n\in T_P^n)(P(j)-P_0(j))\right|$$

$$=\left|\sum_{P\in\Omega_n^c}\mathbf{P}_0(x^n\in T_P^n)(P(j)-P_0(j))\right|$$

$$\le 2\sum_{P\in\Omega_n^c}\mathbf{P}_0(x^n\in T_P^n)=O(n^{-1}). \qquad (2.11)$$

Similarly, we can show

$$\sum_{P\in\Omega_n}P_0(x^n\epsilon T_P^n)R(P_0,D)=R(P_0,D)+O(n^{-1}). \qquad (2.12)$$

Hence,

$$E_{P_0}R(\overline{P},D)\le R(P_0,D)+O(n^{-1}).$$

This completes the proof that

$$n^{-1}E_{P_0}L(M_n(x^n))\le R(P_0,D) \\ +(KJ+J+4)n^{-1} \\ \cdot\log(n+1)+O(n^{-1}). \qquad \square$$

Remark: Note that it is very easy to check that the boundedness conditions on the derivatives of the rate-distortion funciton are satisfied by a Bernoulli(p) source with distortion measured by Hamming distance. In that case, the rate-distortion function is known, cf. Cover and Thomas [7], to be

$$R(p,D)=\begin{cases}H(p)-H(D), & \text{if } 0\le D\le\min(p,1-p),\\ 0, & \text{if } D>\min(p,1-p).\end{cases}$$

Then,

$$R'_D(p,D)=\log\frac{1-D}{D}$$

and

$$\frac{\partial^2}{\partial p^2}R(p,D)=-\frac{1}{p(1-p)}$$

IEEE TRANSACTIONS ON INFORMATION THEORY, VOL. 39, NO. 3, MAY 1993

It is clear that the boundedness conditions are satisfied in this case if we choose $\mathcal{F}$ as the set of binary distributions with uniform bounds on p away from both 0 and 1. In general, if those derivatives exist, they are likely to be bounded. Thus, our conditions do not appear to be too stringent.

IV. Lower Bound

For a string of i.i.d. discrete source letters from $\mathcal{A}_0$, we have shown in the last section that under some smoothness conditions there is a pointwise D-semifaithful code with its average expected code length tending to $R(P_0, D)$ at the rate $n^{-1}\log n$. Recalling that $R(P_0, D)$ is a lower bound on the average codelength of such D-semifaithful codes, we may ask: is the rate $n^{-1}\log n$ the best possible?

Unfortunately, we have not been able to show that $n^{-1}\log n$ is the optimal rate, though we conjecture it is the case. The main reason for our conjecture is a lower bound due to Pilc [16], [17] in terms on the distortion-rate function $D(P_0, R)$, which is the inverse function of $R(P_0, D)$ in the variable D. Note that $D(P_0, \cdot)$ is defined on $[0, H(P_0)]$.

Theorem 3 (Pilc [17]): Assume that $x_1, \cdots, x_n$ are i.i.d. with distribution P_0 on $\mathcal{A}_0$, and that M_n is a map from $\mathcal{A}_0^n \to \mathcal{B}_0^n$. Given $R \in (0, H(P_0))$, if $|M_n(\mathcal{A}_0^n)| \le 2^{nR}$, then

$$E_{P_0} d_n(x^n, M_n(x^n)) \ge D(P_0, R) + \frac{1}{2}\frac{\log n}{|s_0(R)|n}(1 + o(1)), \quad (3.1)$$

where s_0 satisfies

$$\mu(s_0, P_0) - s_0\mu'(s_0, P_0) = -R$$

with

$$\mu(s, P_0) = \sum_{j=1}^{J} P_0(j)\log\left(\sum_{k=1}^{K} Q(k)\exp\,(sd(j,k))\right),$$
$$Q = WP_0,$$

and

$$I(W, P_0) = R(P_0, D_R) = R.$$

Moreover, for any $\epsilon > 0$, there exists $N(\epsilon) > 0$, such that if $n > N(\epsilon)$,

$$\min E_{P_0} d_n(x^n, M_n(x^n)) \le D(P_0, R) + \frac{1}{2}(1+\epsilon)\frac{\log n}{|s_0|n}(1 + o(1)), \quad (3.2)$$

where the minimum is taken over all codes (maps) M_n such that $|M_n(\mathcal{A}_0^n)| \le 2^{nR}$.

It is worth noting that the constant $\frac{1}{2}$ in front of the rate $n^{-1}\log n$ does not depend on the dimension J of source distribution P_0, whereas in the noiseless case the corresponding constant is $\frac{1}{2}(J-1)$.

Applying the function $R(P_0, \cdot)$ to both sides of (3.1), we get the following corollary.

Corollary 2 (Lower Bound): Under the assumptions of Theorem 3, for any *expected* D-semifaithful code M_n, if $|M_n| = 2^{nR}$, then as $n \to \infty$,

$$R \ge R(P_0, D) + \frac{1}{2}n^{-1}\log n(1 + o(1)). \quad (3.3)$$

Proof: By (3.1) and the fact that $R(P_0, \cdot)$ is decreasing,

$$\begin{aligned} R(P_0, E_{P_0} d_n(x^n, M_n(x^n)) &\le R(P_0, D(P_0, R) \\ &\quad + \frac{1}{2}\frac{\log n}{2|s_0|n}(1 + 0(1))) \\ &= R(P_0, D(P_0, R)) \\ &\quad + R'_D(P_0, D(P_0, R)) \\ &\quad \cdot \frac{\log n}{2|s_0|n}(1 + o'(1)), \end{aligned}$$

where the $o'(1)$ term represents the sum of $o(1)$ term in the previous expression and the smaller order terms from the Taylor expansion. From the parametric representation of $R(P_0, \cdot)$ (Berger [2]), it is easy to see that $s_0(R) = R'_D(P_0, D(P_0, R)) < 0$.

Also note that $R'_D < 0$, $E_{P_0} d_n(x^n, M_n(x^n)) \le D$, and $R(P_0, D(P_0, R)) = R$, so we have

$$R \ge R(P_0, E_{P_0} d_n(x^n, M_n(x^n)) + \frac{1}{2}n^{-1}\log n(1 + o'(1)).$$

Since $R(P_0, \cdot)$ is decreasing,

$$R(P_0, E_{P_0}(d_n(x^n, M_n(x^n)) \ge R(P_0, D).$$

This completes the proof of (3.3). □

Remark 1: Pilc's lower bound in Theorem 3 relies on some large deviation bounds from Shannon and Gallager [25], [26]. Those bounds are for tails of sums of i.i.d. variables and are accurate to the order $n^{-1/2}\exp(-cn)$ with the best constant c. Moreover, Pilc's original lower bound does not hold for noiseless coding because at $R = H(P_0)$, $s_0(R) = R'_D(P_0, D) = -\infty$. Hence, his lower bound does not include Rissanen's lower bound in the noiseless coding case as a special case.

Remark 2: From the previous section, we have a universal code which is *pointwise* D-semifaithful and $R(P_0, D) + O(n^{-1}\log n)$ in expected codelength. It would have been perfect if Pilc's result was in terms of expected code length and for pointwise D-semifaithful codes. However, Pilc's lower bound in the form of Corollary 2 is something like a dual to the result we seek; it says that for any *expected* D-semifaithful code, the log of the cardinality of the set of its code words is bounded below by $R(P_0, D) + O(n^{-1}\log n)$. Note that this log cardinality is not a random quantity, unlike the codelength of our universal code. For a pointwise D-semifaithful code, the log cardinality is likely to be bigger than the expected codelength.

V. Discussion

In this section, we compare the proofs of our Theorem 2 and Pilc's Theorem 3 from the points of view of constructiveness and universality. Both Pilc's upper bound (3.2) and our Theorem 2 involve a random coding argument. We might think that neither of them can give a code constructively, and that we can not really say which code is universal, since neither result looks constructive. On the other hand, we observe that the random coding argument in Theorem 2 does not need the true distribution P_0, since we proved the existence of a D-semifaithful code for each type class, while Pilc's random coding argument used a knowledge of P_0. This difference in random coding seems to suggest that our code might be universal, whereas Pilc's coding might not be so. We now show this to be the case.

A. Construction of a Universal D-Semifaithful Code when P_0 is Unknown

For each type class T_P^n, let $[Q] = [W] \cdot P$ as in Section II. $[W]$ can be obtained to any precision numerically (if not analytically) by Blahut's algorithm, Blahut [4]. We require a precision of order $(nP(j))^{-1}$. Then, the $n[Q](k)$ are integers, i.e., $[Q]$ is a type. Take m_n to be the integer part of $\exp\{nI([W], P) + 3JK\log(n+1)\}$. For this m_n, we could in principle search through all subsets of size m_n in $T_{[Q]}^n$ in order to find a subset $B_{P,D}$ that covers T_P^n within D distance. That is, for each subset B, we check whether $d(x^n, B) \leq D$ for all $x^n \in T_P^n$. For the m_n previously chosen, Theorem I guarantees the existence of such a pointwise D-semifaithful code. In other words, through exhaustive search, we can find at least one set $B_{P,D} \subset T_{[Q]}^n$ satisfying our D-covering requirement. We take the first such $B_{P,D}$ found as our codebook for T_P^n, and we have "constructed" a universal pointwise D-semifaithful code. Note that the code we just described has its code length approach the rate-distortion function lower bound at the rate $n^{-1}\log n$, and this rate is optimal in the noiseless coding case.

B. Construction of a D-Semifaithful Code when P_0 is Known

Pilc's upper bound (3.2) says that for any $R \in (0, H(P_0))$ and $\epsilon > 0$, we can use a random coding argument to find a map $M_n : \mathcal{A}_0^n \to \mathcal{B}_0^n$ such that as $n \to \infty$,

$$E_{P_0} d_n(x^n, M_n(x^n)) \leq D(P_0, R) + \left(\tfrac{1}{2} + \epsilon\right) \cdot \frac{\log n}{|s_0| n}(1 + o(1)).$$

When P_0 is known, for any fixed $D \in (0, D_{\max})$, we can take R_n to be $R(P_0, D) + \frac{1}{2}(1+\epsilon)n^{-1}\log n$. For this R_n, we can search through all subsets in $\mathcal{B}_0^n$ of size less than or equal to 2^{nR_n}. We choose the codebook $\overline{B}_{P_0}$ as the set such that

$$E_{P_0} d(x^n, \overline{B}_{P_0}) = \min_B E_{P_0} d_n(x^n, B), \tag{4.1}$$

where the min is taken over all B with $|B| \leq 2^{nR_n}$.

Pilc's result guarantees this codebook $\overline{B}_{P_0}$ satisfies

$$\begin{aligned} E_{P_0} d_n(x^n, \overline{B}_{P_0}) &\leq D(P_0, R_n) + \frac{1}{2}(1+\epsilon)\frac{\log n}{|s_0| n}(1 + o(1)) \\ &= D(P_0, R(P_0, D)) - \frac{1}{2}(1+\epsilon)\frac{\log n}{|s_0| n} + \frac{1}{2}(1+\epsilon)\frac{\log n}{|s_0| n} + o(n^{-1}\log n). \end{aligned}$$

The last equality holds because of the Taylor expansion of $D(P_0, \cdot)$, the fact that $D'(P_0, R(P_0, D)) = s_0^{-1}$, and $D' < 0$. It follows that

$$E_{P_0} d_n(x^n, \overline{B}_{P_0}) = D + o(n^{-1}\log n).$$

Without knowledge of the $o(1)$ term in Pilc's upper bound, $D + o(n^{-1}\log n)$ is the best level of distortion we can establish; we cannot deduce that the code is expected semifaithful at the exact level D.

The code $\overline{B}_{P_0}$ clearly depends on P_0 as we need to know P_0 to check (4.1). The $D + o(n^{-1}\log n)$-semifaithful code obtained from Pilc's upper bound is, therefore, not universal.

When P_0 is not known, a natural remedy would be to use the empirical distribution $\overline{P}$ instead of P_0 in the construction we have just outlined. But this does not work if we want to keep the rate $n^{-1}\log n$. The problem here is that when we replace P_0 by $\overline{P}$ in (4.1), we create an error of magnitude $(n^{-1}\log\log n)^{1/2}$ since $\|\overline{P} - P_0\| = O[(n^{-1}\log\log n)^{1/2}]$. This rate overwrites the desired rate $n^{-1}\log n$.

There is another difference between our code and Pilc's. The code $\overline{B}_{P_0}$ has the stated distortion on average, i.e., it is an *expected* D-semifaithful code, but the codelength is *pointwise* $R(P_0, D) + \frac{1}{2}(1+\epsilon)n^{-1}\log n$, not in expectation. On the other hand, our code $\{B_{P,D} : P \text{ any type}\}$ is *pointwise* D-semifaithful with the *expected* codelength $R(P_0, D) + (KJ + J + 4)n^{-1}\log n$. Ignoring the issue of nonuniversality, and the different constants in front of $n^{-1}\log n$, we might say that Pilc's result is dual to ours. We doubt that there exists a universal code that is *pointwise* D-semifaithful and whose log cardinality approaches the lower bound $R(P, D)$ at the rate $n^{-1}\log n$.

A technical difference of the two results concerns the mathematical tools employed. We both use the random coding argument, but the rate $n^{-1}\log n$ came out of the method of types for us, while for Pilc it came out of the large deviation results of Shannon and Gallager [25], [26]. This is not surprising, however, since large deviation results can be obtained using the method of types in the discrete memoryless source case. Due to the elegance of the method of types, our proofs are simpler and more direct than those of Pilc. Both results rely on the assumption that the source is i.i.d., although Pilc has results on noisy channels, too. Large deviation results do exist for independent not identical distributions, but we are not aware of any result as refined as that required by Pilc's bounds.

ACKNOWLEDGMENT

The authors would like to thank two referees for very helpful comments. The authors would like to thank Dr. T. Linder for pointing out that the boundedness condition on the second derivative of $R(D, P)$ in Theorem 2 is superfluous. This is because $-R'(P, D)$ is positive and bounded by $\log J/D$, as $R(P, D)$ is a convex function of D.

REFERENCES

[1] A. R. Barron and T. M. Cover, "Minimum complexity density estimation," *IEEE Trans. Inform. Theory*, vol. 37, pp. 1034–1054, July 1991.

IEEE TRANSACTIONS ON INFORMATION THEORY, VOL. 39, NO. 3, MAY 1993

[2] T. Berger, *Rate Distortion Theory: A Mathematical Basis for Data Compression.* Englewood Cliffs, NH: Prentice-Hall, 1971.
[3] T. Berger and L. D. Davisson, *Advances in Source Coding.* New York: Springer-Verlag, 1975.
[4] R. E. Blahut, *Principles and Practice of Information Theory.* Reading, MA: Addison-Wesley, 1978.
[5] B. S. Clarke and A. R. Barron, "Information theoretic asymptotics of Bayes methods," *IEEE Trans. Inform. Theory,* vol. 36, 453–471, May 1990.
[6] I. Csiszár and J. Korner, *Information Theory: Coding Theorems for Discrete Memoryless Systems.* New York: Academic, 1981.
[7] T. M. Cover and J. A. Thomas, "Elements of information theory," Lecture Notes, Stanford Univ., CA, 1990.
[8] L. D. Davisson, "Minimax noiseless universal coding for Markov sources," *IEEE Trans. Inform. Theory,* vol. IT-29, pp. 211–215, 1983.
[9] L. Gerencser and J. Rissanen, "A prediction bound for Gaussian ARMA processes," *Proc. 25th CDC,* Athens, Greece, vol. 3, 1986, pp. 1487–1490.
[10] E. J. Hannan and L. Kavalieris, "Regression, autoregressive models," *J. Time Series Anal.,* vol. 7, pp. 27–49, 1986.
[11] E. M. Hemerly and M. H. A. Davis, "Strong consistency of the predictive least squares criterion for order determination of autoregressive processes," *Ann. Statist.,* vol. 17, pp. 941–946, 1989.
[12] J. Kieffer, "A unified approach to weak universal source coding," *IEEE Trans. Inform. Theory,* vol. IT-24, pp. 674–682, 1978.
[13] K. M. Mackenthun and M. B. Pursley, "Variable-rate universal block source coding subject to a fidelity constraint," *IEEE Trans. Inform. Theory,* vol. IT-24, pp. 349–360, 1978.
[14] D. L. Neuhoff, R. M. Gray, and L. D. Davisson, "Fixed rate universal block source coding with a fidelity criterion," *IEEE Trans. Inform. Theory,* vol. IT-21, pp. 511–523, 1975.
[15] D. S. Ornstein and P. C. Shields, "Universal almost sure data compression," *Ann. Probab.,* vol. 18, pp. 441–452, 1990.
[16] R. J. Pilc, "Coding theorems for discrete source-channel pairs," Ph.D. thesis, Dept. Elect. Eng., M.I.T., Cambridge, MA, 1967.
[17] ——, "The transmission distortion of a source as a function of the encoding block length," *Bell Syst. Tech. J.,* vol. 47, pp. 827–885, 1968.
[18] D. Pollard, *Convergence of Stochastic Processes.* New York: Springer-Verlag, 1984.
[19] J. Rissanen, "Stochastic complexity and modeling," *Ann. Statist.,* vol. 14, pp. 1080–1100, 1986.
[20] ——, "Complexity of strings in the class of Markov sources," *IEEE Trans. Inform. Theory,* vol. IT-34, pp. 526–532, July 1986.
[21] J. Rissanen, T. P. Speed, and B. Yu, "Density estimation by stochastic complexity," *IEEE Trans. Inform. Theory,* vol. 38, Pt. I, pp. 315–323, Mar. 1992.
[22] P. Shields, "Universal almost sure data compression using Markov types," *Probl. Contr. Inform. Theory,* vol. 19, pp. 269–277, 1990.
[23] C. Shannon and W. Weaver, *A Mathematical Theory of Communication.* Urbana, IL: Univ. Illinois Press, 1949.
[24] C. Shannon, "Coding theorems for a discrete source with a fidelity criterion," in *Information and Decision Processes,* R. E. Machol, Ed. New York: McGraw-Hill, 1959.
[25] C. Shannon and R. G. Gallager, "Lower bounds to error probability for coding on discrete memoryless channels I," *Inform. Contr.,* vol. 10, pp. 65–103, 1969.
[26] ——, "Lower bounds to error probability for coding on discrete memoryless channels II," *Inform. Contr.,* vol. 10, pp. 523–552, 1969.
[27] M. J. Weinberger, A. Lempel, and J. Ziv, "A sequential algorithm for the universal coding of finite memory sources," *IEEE Trans. Inform Theory,* vol. 38, pp. 1002–1014, May 1992.
[28] B. Yu and T. Speed, "Data compression and histograms," *Probability Theory Related Fields 92,* pp. 195–229, 1992.

The Annals of Statistics
1997, Vol. 25, No. 1, 169–185

INFORMATION AND THE CLONE MAPPING OF CHROMOSOMES

BY BIN YU[1] AND T. P. SPEED[2]

University of California, Berkeley

A clone map of part or all of a chromosome is the result of organizing order and overlap information concerning collections of DNA fragments called clone libraries. In this paper the expected amount of information (entropy) needed to create such a map is discussed. A number of different formalizations of the notion of a clone map are considered, and exact or approximate expressions or bounds for the associated entropy are calculated for each formalization. Based on these bounds, comparisons are made for four species of the entropies associated with the mapping of their respective cosmid clone libraries. All the entropies have the same first-order term $N \log_2 N$ (when the clone library size $N \to \infty$) as that obtained by Lehrach et al.

1. Introduction. The primary goal of the Human Genome Project is to sequence the entire human genome, which consists of about 3×10^9 base pairs (bp) of DNA. Current technology only permits sequencing of fragments of the order of a few hundred to a thousand base pairs of DNA in a single reaction. Consequently, much effort is devoted to fragmenting large DNA molecules, such as chromosomes, in such a way that the sequenced fragments can be readily assembled. Clone maps, which are one form of physical mapping, play a key role in this process, as well as providing a resource permitting the detailed study of chromosomal regions of biological interest.

A clone map of part or all of a chromosome is the result of organizing order and overlap information concerning collections of DNA fragments called clone libraries. Such libraries consist of many, typically thousands or tens of thousands, of DNA fragments from a chromosome or region of interest. Each fragment exists as an insert in an autonomously replicating DNA sequence, which resides within, and replicates with its host cells. In this manner it is possible to generate many copies of the fragment of interest, and the name clone is thus used as an abbreviation for the longer and more accurate name: cloned DNA fragment.

A large clone library might consist of 5000 cloned fragments of average length 100,000 base pairs, from a chromosome of length 100,000,000 base pairs. Assuming that the cloned fragments are randomly located along the

Received February 1995; revised April 1996.
[1]Research partially supported by ARO Grant DAAH04-94-G-0232 and NSF Grant DMS-93-22817.
[2]Research partially supported by NSF Grants DMS-91-13527 and DMS-94-04267.
AMS 1991 *subject classifications.* Primary 60K99; secondary 94A99.
Key words and phrases. Configuration variable, contig, entropy, random clone library.

chromosome, this would mean that any particular spot on the chromosome should be represented on an average of five cloned fragments, giving rise to the term fivefold coverage, or a five-hit library. We note that a library of fragments of this size is still not suitable for DNA sequencing. Typically, one or two further stages of subcloning are needed prior to sequencing, and there may be additional mapping at these stages as well. In such cases both the libraries and the fragments will be smaller, but the principles of mapping remain much the same. For details on clone mapping from the perspective of applied probability, see Lander and Waterman (1988). Nelson and Speed (1994) have a more statistical perspective, and give further references to these aspects of the topic.

In their paper comparing the relative merits of fingerprinting cloned fragments of DNA by hybridization of oligonucleotide probes and by digestion into restriction fragments, Lehrach et al. (1990) raised two interesting questions concerning the creation of clone maps of a chromosome: (1) how much information is needed? and (2) how much information is gained by the hybridization and restriction digestion methods, respectively? The answer to the first question offered by these authors was $\log_2(\frac{1}{2}N!)$ for a library of N clones. This figure corresponds to the average amount of information (the entropy, see the following discussion) required to identify the true ordering of N objects labeled $1, 2, \ldots, N$ when it is not possible to distinguish between the ordering $(i_1, i_2, \ldots, i_N)$ and its reverse $(i_N, \ldots, i_2, i_1)$, but otherwise all orderings are equally likely. However, it is not entirely clear why the ordering of objects in this way corresponds to any formal notion of a physical map, and even if there is such a correspondence, why all possible configurations should be equally likely.

To illustrate these points, let us briefly consider the cases of $N = 2$ and $N = 3$ clones, regarded mathematically as having identical length L bp and being randomly located along a chromosome of length G bp [cf. Lander and Waterman (1988)]. For two such clones we have two configurations, overlap or not, with quite unequal probabilities 2β and $1 - 2\beta$, respectively, where $\beta = L/G$. For three clones there are ten distinguishable configurations: one with no overlaps, three with exactly two clones overlapping, three with two different clone pairs overlapping, but no triple overlap and three distinguishable configurations involving a triple overlap. Again these can be seen to be far from equally probable. In practice, N will be in the hundreds or thousands.

In order to answer question (1) exactly, we would need to enumerate the set $\mathscr{X}$ of distinguishable configurations, calculate their probabilities $\{p(x): x \in \mathscr{X}\}$ and then go on to calculate the entropy $H(\mathbf{X}) = -\sum_{x \in \mathscr{X}} p(x) \log_2 p(x)$ of a random configuration $\mathbf{X}$. The first part of this program has been completed [see Newberg (1993)], but to our knowledge no one has carried the calculation of the probabilities beyond $N = 3$, although this is, in principle, possible. We do not know how to obtain the entropy $H(\mathbf{X})$ exactly, but in the following discussion we will find bounds on entropies of various configuration variables which are relevant to clone mapping.

INFORMATION AND CLONE MAPPING

The reason that the entropy $H(\mathbf{X})$ is the appropriate measure of information is explained in texts on information theory [see, e.g., Craig et al. (1990) and Rényi (1984)]. We content ourselves here with a brief informal explanation, applicable when the elements of $\mathscr{X}$ *are* equally likely, each having probability $1/|\mathscr{X}|$, in which case $H(\mathscr{X})$ achieves its upper bound $\log_2|\mathscr{X}|$. The argument goes like this: to identify any particular element $x \in \mathscr{X}$, we consider successive subdivisions of $\mathscr{X}$ into halves, quarters, eighths, and so on, and if we were told at each stage which half, quarter, eighth, and so forth contained the particular element, we would gain one bit of information each time. Clearly this process cannot finish in less than k steps, where $2^k \leq |\mathscr{X}| < 2^{k+1}$, and this k is thus a lower bound to the number of such questions, equivalently bits of information, necessary to identify the particular element in question. More refined procedures can limit the amount of information necessary to $\log_2|\mathscr{X}| + \varepsilon$, where $\varepsilon > 0$ is as small as we wish [see, e.g., Rényi (1984)]. A similar but more complicated argument applies when the elements of $\mathscr{X}$ are not equiprobable [see the discussion of the noiseless coding theorem in Cover and Thomas (1991)].

In this paper we study the entropy $H(\mathbf{X})$ of a random configuration $\mathbf{X}$ most appropriate to the clone mapping problem. The study is done through seven other random structures, $\mathbf{P}$, $\mathbf{Q}$, $\mathbf{U}$, $\mathbf{V}$, $\mathbf{W}$, $\mathbf{Y}$ and Z, each of which can be regarded as embodying a greater or lesser amount of the structure implicit in $\mathbf{X}$, but whose entropies are more accessible. We derive a variety of exact and approximate expressions and lower and upper bounds for the entropies of these quantities. We compute these bounds for clone libraries of interest and the bounds are reasonable for all configuration variables considered and very tight for some. Based on these computations, comparisons are made for four "model" species in terms of information needed for the mapping of their respective cosmid clone libraries. It is somewhat surprising that all the entropies have the same first-order term $N \log_2 N$ when $N \to \infty$, as that obtained in Lehrach et al. (1990). We end the paper with some remarks concerning the more difficult question 2.

In closing this brief introduction we note that in the analysis which follows we essentially ignore the role of distances, although we do consider the placement variable $\mathbf{W}$ in units of thousands of base pairs. Many physical mapping methods produce some information concerning distances as well as clone order, and such information can be very useful in practice, even when (as is often the case) there are large error bounds attached. In particular, it would be misleading to compare the hybridization and restriction digest methods mentioned previously, solely on the basis of the information they produce concerning clone order. The restriction digest method produces fairly precise information about distances, whereas the hybridization method does not. An analysis, which incorporates distance as well as order and overlap information, is beyond us at this time.

2. What is a clone map? We now introduce several different but related abstractions of the notion of a clone map of a chromosome, this being informally an ordering of a library of cloned fragments of the chromosome in question. As noted previously, we adopt the mathematical model for a clone library used in Lander and Waterman (1988), namely, that the N cloned fragments can be identified with N randomly located subintervals of equal length L of a genome of length G. More formally, the left-hand endpoints (say) of the N intervals corresponding to the cloned fragments are independently located uniformly along $[0, G - L]$. It will be convenient at points in the argument to take an alternative, effectively equivalent view of the left-hand endpoints as being the points on $[0, G - L]$ of a homogeneous Poisson process with rate $\lambda = N/G$ per base pair.

2.1. *Fully ordered configurations.* Following the terminology of Alizadeh, Karp, Newberg and Weisser (1993), we use the term *placement* to describe a configuration of positions of the clones along the chromosome, that is, a specification $\mathbf{W} = (W_1, W_2, \dots, W_N)$, where $W_i \in [0, G - L]$ is the location of the left-hand endpoint of the ith cloned fragment, $i = 1, 2, \dots, N$. The units here are base pairs (bp) or kilobase pairs (kb); see the following discussion. Experimental procedures exist which could precisely determine these locations for a clone library, but most clone mappings have more modest aims, seeking to single out a less completely specified configuration from among a class of a priori equivalent alternatives. Before we turn to a discussion of such "coarser" configurations, we make a connection with the work of Lehrach et al. (1990), which stimulated this research. By the *linear ordering* of a clone library, we mean the sequence $\mathbf{V} = (V_1, V_2, \dots, V_N)$ of labels of the ordered left-hand endpoints of the clones; equivalently, the vector of *ranks* of $\mathbf{W} = (W_1, W_2, \dots, W_N)$ listed in reverse order. This variable seems to be the one considered in Lehrach et al. (1990).

2.2. *Island configurations.* We turn now to a second class of clone configurations, those based on the notion of an *island*, which is either a single clone, not overlapping with any other clone in the library, or a set of clones, each pair of which is connected by a chain of overlapping pairs of clones. Islands of two or more clones are usually called *contigs*, and many clone mapping projects have as their initial objective the determination of all contigs in their library and the ordering, up to inversion, of clones within contigs. This is usually the objective of *fingerprint-based* clone mapping, which attempts to infer clone order and overlap from information concerning each of the clones in the library, such as the list of fragment lengths following digestion by restriction enzymes, or the pattern of hits and misses following hybridization with a panel of probes. Fingerprint-based clone mapping projects usually turn to quite different techniques such as radiation hybrid or fluorescence in situ hybridization (FISH) mapping [see, e.g., Cox, Burmeister, Price and Myers (1990) and Trask (1991)].

The most basic island configuration variable is Z, the *number* of islands. More informative is the variable $\mathbf{U} = (U_1, U_2, \dots, U_N)$ of *island sizes*, which is a *partition of the integer* N, that is, $\Sigma_1^N U_i = Z$, $\Sigma_1^N iU_i = N$; or, equivalently,

U_i is the number of islands containing i clones. The components of $\mathbf{U}$ are the multiplicities of the *block sizes* of the *partition* $\mathbf{Q}$ *of the set* $\{1, 2, \ldots, N\}$ of clone labels into islands. Here $\mathbf{Q}$ is the unordered list of disjoint subsets of $\{1, 2, \ldots, N\}$, usually called blocks or equivalence classes, but called islands in this context, whose union is $\{1, 2, \ldots, N\}$.

More informative again than $\mathbf{Q}$ is the configuration variable we term the *distinguishable orderings* of the clones and denote by $\mathbf{Y}$, namely, the variable which refines $\mathbf{Q}$ by including information on the ordering of clones within contigs, up to inversion. Thus $\mathbf{Y}$ tells us which clones are together in a contig and, up to a flip, the order in which they appear, but it contains no information on the relative positions of distinct islands along the genome.

There is one last refinement which we mention, namely, the configuration variable discussed in Newberg (1993), which includes information on the depth of coverage within contigs. We denote this configuration variable by $\mathbf{X}$, and note that it may be regarded as refining $\mathbf{Y}$ by containing not just information on the labels of the left-hand endpoints of the clones within each contig, up to inversion, but the labels of the interleaved sequence of the left-hand and right-hand endpoints of the clones, again up to inversion. Newberg (1990) calls two configurations of clones *topologically similar* if one can be transformed into the other by permuting the islands and/or reflecting some of the islands. An adjustment of the amount by which any pair of clones overlap leaves one with a topologically similar clone ordering, if no endpoint of a clone is moved past an endpoint of another clone. With this definition, $\mathbf{X}$ is the set of equivalence classes of topologically distinct configurations, called *interleavings* in Newberg (1993) and Alizadeh, Karp, Newberg and Weisser (1993).

2.3. *Pairwise overlaps.* Many fingerprint-based clone mapping projects take as their starting points the determination of pairwise overlaps among the clones in their library [see, e.g., Branscomb et al. (1990), Craig et al. (1990) and Fu, Timberlake and Arnold (1992)]. For this reason we define the *pairwise overlap* variable $\mathbf{P} = (P_{ij}\colon 1 \le i < j \le N)$, where $P_{ij} = 1$ if clones i and j overlap, and $P_{ij} = 0$ otherwise. It is clear that $\mathbf{P}$ can be obtained from $\mathbf{X}$ but not from $\mathbf{Y}$. In seeking to estimate $H(\mathbf{P})$ we do not mean to imply that pairwise comparisons are the best, or even an effective way to ascertain pairwise overlap information. Indeed, many of the most common clone mapping methods, such as STS-content mapping [Green and Green (1991)] and restriction mapping [Olson et al. (1986)], do not attempt to determine pairwise overlaps at all. Nevertheless, it seems to us of interest to ask just how large $H(\mathbf{P})$ is in relation to the entropies of other, more refined configuration variables.

This concludes our discussion of the different abstractions of the notion of a clone map of a chromosome based on a library of cloned DNA fragments from that chromosome. As with all mathematical idealizations, our variables all fail to account for many features of real clone mapping projects. Our hope is that the features we do retain are the important ones, and that our results are at least qualitatively correct and useful. We now illustrate the different variables just introduced in a simple case.

EXAMPLE. Suppose that $G = 150$, $L = 20$ and $N = 8$. We list the set of configuration variables refining $\mathbf{W} = (120, 50, 10, 45, 105, 55, 20, 76)$. The vector of *ranks* of these values, viewed as observations on $[0, 150]$, is $(1, 5, 8, 6, 2, 4, 7, 3)$, and so $\mathbf{V} = (3, 7, 4, 2, 6, 8, 5, 1)$. Using the values in $\mathbf{W}$, it is easy to ascertain that

$$\mathbf{X} = \{(373'7')^*, (4264'2'6')^*, (88')^*, (515'1')^*\},$$

where 3 (resp. 3′) denotes the left-hand (resp. right-hand) end of clone 3 or vice versa, and * indicates the fact that the ordering is only unique up to reversal. In a similar notation we have

$$\mathbf{Y} = \{(37)^*, (426)^*, (8), (51)^*\},$$

while $\mathbf{Q} = 15|246|37|8$, $\mathbf{U} = (1^1, 2^2, 3^1)$ and $Z = 4$.

3. Results. In this section we present our approximations to the entropy of the configurations just described. All proofs are collected in the appendices.

We have sought close nonasymptotic upper and lower bounds to the entropy expressions of interest, and have been quite successful in this regard with $H(\mathbf{Q})$ and $H(\mathbf{Y})$, and somewhat less so with $H(\mathbf{X})$ and $H(\mathbf{P})$. Exact calculations of $H(\mathbf{W})$ and $H(\mathbf{V})$ are straightforward. It is also of interest to consider our results asymptotically as $N \to \infty$. In so doing, we could keep L/G fixed and let $c = NL/G$ increase, or we could keep c fixed and let L/G decrease. A value of c in the range 3–10 is typical, with $c = 5$ being quite common, although values in the range 40–50 have been used. Our figures and tables have c fixed at 5.

The easiest entropy to evaluate is $H(\mathbf{W})$ which is just $N \log_2(G - L) \approx N \log_2 G$. This last expression can be rewritten as

$$H(\mathbf{W}) = N \log_2 N + N \log_2(L/c)$$

by making the substitution $c = NL/G$. It is clear that the leading term is $N \log_2 N$, and also that the second term depends on the units in which L is measured. The most reasonable choice would seem to be kilobase pairs (kb), in which the values G = 100,000 kb, L = 40 kb (corresponding to a cosmid library) and $c = 5$ give N = 12,500 and $H(\mathbf{W}) = 2.1 \times 10^5$, compared with $N \log_2 N = 1.7 \times 10^5$.

As pointed out in Lehrach et al. (1990), we may use Stirling's formula to get

$$\begin{aligned} H(\mathbf{V}) &= \log_2(\tfrac{1}{2}N!) \\ &\approx N \log_2 N + \tfrac{1}{2}\log_2 N - (\log_2 e)N - \log_2(\sqrt{2\pi}) - 1. \end{aligned}$$

INFORMATION AND CLONE MAPPING

Now let us define

$$\underline{L}(\mathbf{U}) = \mathbb{E}\{Z\}[\log_2 N - \log_2 e] + \tfrac{1}{2}\log_2 \mathbb{E}\{Z\} + \log_2(\sqrt{2\pi}e^{1/12}),$$

$$\overline{L}(\mathbf{U}) = \mathbb{E}\{Z\}\left(\log_2 \frac{Np^2}{q(1-q^N)} + (\log_2 q)c_N\right),$$

$$\underline{M}(\mathbf{U}) = Ne^{-c}(a_N + b_N) - (\log_2 e)N,$$

$$\overline{M}(\mathbf{U}) = Ne^{-c}(a_N + b_N) - (\log_2 e)N + \left(\log_2(\sqrt{2\pi}e^{1/12})\right)\mathbb{E}\{Z\},$$

where $a_N = \mathbb{E}\{F^N \log_2 F^N\}$, $b_N = \frac{1}{2}\mathbb{E}\{\log_2 F^N\}$ and $c_N = \mathbb{E}\{F^N\}$, and F^N is a truncated geometric random variable with $p = e^{-c}$ and truncation at N. That is, for $q = 1 - p$, $P(F^N = j) = pq^{j-1}/(1 - q^N)$, $j = 1, 2, \ldots, N$. We have the following bounds on the entropies.

RESULT A (Finite-sample entropy bounds). Let us introduce the following abbreviations:

$$\underline{H}(\mathbf{Y}) = \log_2 N! - \overline{L}(\mathbf{U}) - Np(1-p),$$

$$\overline{H}(\mathbf{Y}) = \log_2 N! - \underline{L}(\mathbf{U})^{+} - Np(1-p) + \mathbb{E}\{Z\}H(F^N) + \log_2 N,$$

$$\underline{H}(\mathbf{Q}) = \log_2 N! - \overline{L}(\mathbf{U}) - \overline{M}(\mathbf{U}),$$

$$\overline{H}(\mathbf{Q}) = \log_2 N! - \underline{L}(\mathbf{U})^{+} - \underline{M}(\mathbf{U}) + \mathbb{E}\{Z\}H(F^N) + \log_2 N,$$

$$\overline{H}(\mathbf{X}) = N\log_2 N + N\log_2(4/e) - \log_2 N,$$

$$\underline{H}(\mathbf{X}) = \underline{H}(\mathbf{Y}),$$

$$\overline{H}(\mathbf{P}) = \overline{H}(\mathbf{X}),$$

$$\underline{H}(\mathbf{P}) = \underline{H}(\mathbf{Q}),$$

where

$$H(F^N) = \sum_{j=1}^{N} -P(F^N = j)\log_2 P(F^N = j).$$

Then our main bounds may be expressed as

$$\underline{H}(S) \le H(S) \le \overline{H}(S),$$

where S may be $\mathbf{X}$, $\mathbf{Y}$, $\mathbf{Q}$ or $\mathbf{P}$.

RESULT B (Asymptotic expansions for entropies). The following expressions are valid as $N \to \infty$:

(i) $$H(\mathbf{W})/N\log_2 N = 1 + o(1),$$

(ii) $$H(\mathbf{V})/N\log_2 N = 1 + o(1),$$

(iii) $$(1 - e^{-c}) + o(1) \le H(\mathbf{X})/N\log_2 N \le 1 + o(1),$$

(iv) $$H(\mathbf{Y})/N \log_2 N = (1 - e^{-c}) + o(1),$$

(v) $$H(\mathbf{Q})/N \log_2 N = (1 - e^{-c}) + o(1),$$

(vi) $$H(\mathbf{P})/N \log_2 N = (1 - e^{-c}) + o(1).$$

The finite-sample bounds in Result A are really only useful when they are not very far apart. Fortunately, they are reasonably close for all four configuration variables considered here and very close for **Y** and **Q**. Figure 1 is the log-log plot of the entropy bounds for $c = 5$ and $N = 100, \ldots, 20{,}000$, and it is clear that the bounds are very tight for $H(\mathbf{Y})$ and $H(\mathbf{Q})$, tight for $H(\mathbf{X})$, but not so close for $H(\mathbf{P})$. It is also comforting to see that **W**, **X** and **Y**, which are all reasonable definitions of a clone map, turn out to have very similar entropies. The other interesting and useful observation is that $H(\mathbf{V})$ is numerically very close to $H(\mathbf{Y})$ for the range of N that we considered and for $c = 5$. Therefore, the simple Stirling expansion for $H(\mathbf{V})$ can be used as a valid short-hand formula for $H(\mathbf{Y})$ when $c = 5$. This shows that Lehrach et al.'s intuition works well here since the coverage is high enough that most of the randomness in the configuration variable **Y** comes from the permutation which is captured in **V**.

It is perhaps remarkable that the entropies of **W**, **V**, **X**, **Y**, **Q** and **P** all turn out to have the first-order term $N \log_2 N$, asymptotically, as obtained in Lehrach et al. (1990) (cf. Result B). Moreover, the constant for the first-order

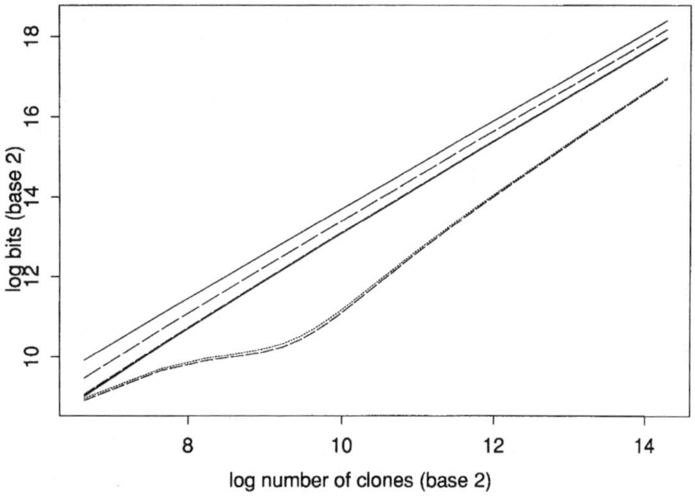

FIG. 1. $\log_2 H(\mathbf{W})$ (*top line*); $\log_2 \overline{H}(\mathbf{X})$ (*second line from top*); $\log_2 \overline{H}(\mathbf{Y})$, $\log_2 H(\mathbf{V})$ *and* $\log_2 \underline{H}(\mathbf{Y})$ *in the third line (cluster) and in that order from top*; $\log_2 \overline{H}(\mathbf{Q})$ *and* $\log_2 \underline{H}(\mathbf{Q})$ *in the bottom line (cluster) and in that order from top. Here the basic unit for* **W** *is* kb, $L = 40$ kb *and* $c = 5$. $\log_2 \underline{H}(\mathbf{Y})$ *and* $\log_2 \overline{H}(\mathbf{X})$ *serve as lower and upper bounds for* $\log_2 H(\mathbf{X})$ *and* $\log_2 \underline{H}(\mathbf{Q})$ *and* $\log_2 \overline{H}(\mathbf{X})$ *serve as lower and upper bounds for* $\log_2 H(\mathbf{P})$.

terms of $H(\mathbf{Y})$, $H(\mathbf{Q})$ and $H(\mathbf{P})$ is the same, namely, $1 - e^{-c}$. Unfortunately, this asymptotic result is not so useful for the values of N which are relevant here, because the term which makes the difference between $H(\mathbf{Y})$ and $H(\mathbf{Q})$ (cf. Figure 1) is $M(\mathbf{U})$, which is $O(N)$. The problem is that $\log_2 N$ is asymptotically larger than any constant term, but in this case it is much smaller than the corresponding constant ($\sim$ 260) in the $O(N)$ term.

An interesting fact which follows from the entropy bounds is that $H(\mathbf{P})/H(\mathbf{X}) \geq 0.20$ for $c = 5$ and $N = 100, 200, \ldots, 20{,}000$ (cf. Figure 2). (Note that the turns on the ratios for small N are probably artifacts of our bounds, not indicative of the true ratios of the entropies.) This implies that the pairwise variable **P** contains a substantial proportion of the information in the interleaving variable **X**. However, although the pairwise mapping approach is definitely a good starting point for any clone mapping effort, recovering the pairwise variable **P** efficiently may well be improved by using multiple comparisons.

Table 1 lists the entropy bounds for specific cosmid ($L = 40$ kb) clone libraries corresponding to the G for a bacterium *E. coli*, yeast *S. cerevisiae*, roundworm *C. elegans* and humans. Here we observe behavior similar to that found in the figures. Table 2 gives the bounds on $H(\mathbf{W})$, $H(\mathbf{X})$ and $H(\mathbf{Y})$ for the last three species in relation to those of the bacterium *E. coli*. The ratios are seen to be species specific rather than specific to the configuration variables. We conclude that it makes sense to say, for example, that cosmid clone mapping for the roundworm requires about 40 times as much information as that for the bacterium *E. coli*, and that such mapping for humans

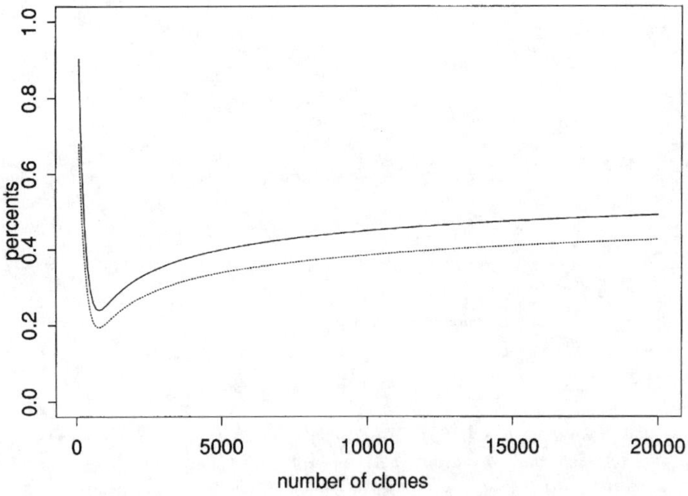

FIG. 2. *Lower bounds on* $H(\mathbf{P})/H(\mathbf{Y})$ *(upper line) and* $H(\mathbf{P})/H(\mathbf{X})$ *(lower line),* $c = 5$.

TABLE 1
Entropies and ratios for fivefold cosmid clone libraries of four species (ratios based on unrounded figures)

	Bacterium $N = 500$	**Yeast** $N = 1,875$	**Roundworm** $N = 12,500$	**Human** $N = 375,000$
$H(\mathbf{W})$	6.0×10^3	2.6×10^4	2.1×10^5	8.1×10^6
$H(\mathbf{V})$	3.8×10^3	1.8×10^4	1.5×10^5	6.4×10^6
$\underline{H}(\mathbf{X})$	3.7×10^3	1.8×10^4	1.5×10^5	6.4×10^6
$\overline{H}(\mathbf{X})$	4.8×10^3	2.1×10^4	1.8×10^5	7.2×10^6
$\underline{H}(\mathbf{X})/\overline{H}(\mathbf{X})$	0.79	0.82	0.85	0.89
$\underline{H}(\mathbf{Y})$	3.7×10^3	1.8×10^4	1.5×10^5	6.4×10^6
$\overline{H}(\mathbf{Y})$	3.8×10^3	1.8×10^4	1.5×10^5	6.4×10^6
$\underline{H}(\mathbf{Y})/\overline{H}(\mathbf{Y})$	0.99	0.99	0.99	0.99
$\underline{H}(\mathbf{Q})$	1.1×10^3	5.5×10^3	7.0×10^4	3.9×10^6
$\overline{H}(\mathbf{Q})$	1.2×10^3	5.7×10^3	7.2×10^4	4.0×10^6
$\underline{H}(\mathbf{Q})/\overline{H}(\mathbf{Q})$	0.95	0.96	0.98	0.99
$\underline{H}(\mathbf{P})$	1.1×10^3	0.6×10^4	0.7×10^5	3.9×10^6
$\overline{H}(\mathbf{P})$	4.8×10^3	2.1×10^4	1.8×10^5	7.2×10^6
$\underline{H}(\mathbf{P})/\overline{H}(\mathbf{P})$	0.23	0.26	0.40	0.55

requires about 1500 times as much information as that for the bacterium *E. coli*.

4. Final comments. We close our discussion with some brief remarks on the important question (2) raised in Section 1: how much information is gained by the hybridization and restriction digestion methods, respectively? It is not our intention to offer a thorough discussion of this topic here, as we hope to present something more complete in a future paper. Rather, our aim here is simply to point out that the situation is not quite as simple as the discussion in Lehrach et al. (1990), page 45, suggests.

Suppose that we collect data $D_1, D_2, \ldots, D_n$ on our clone library, for example, D_n could be the pattern of responses of our clones (+ or −) to the nth in a sequence of hybridization with short oligonucleotides. Each such

TABLE 2
Entropies of **W**, **X** *and* **Y** *relative to E. coli*, $c = 5$

	Yeast	**Roundworm**	**Human**
$H(\mathbf{W})$	4.3	35	1350
$\overline{H}(\mathbf{X})$	4.5	37	1505
$\underline{H}(\mathbf{X})$	4.7	40	1700
$\overline{H}(\mathbf{Y})$	4.7	40	1690
$\underline{H}(\mathbf{Y})$	4.7	40	1700

data item has an entropy $H(D_n)$, indeed the full collection has an entropy $H(D_1, D_2, \ldots, D_n)$, but if our aim is constructing a clone map using these data, the relevant entropy is $H(\mathbf{X}|D_1, D_2, \ldots, D_n)$, the conditional entropy of the library configuration $\mathbf{X}$ given the data $D_1, D_2, \ldots, D_n$. The computation of this quantity is not at all straightforward, even if the data items $D_1, D_2, \ldots, D_n$ are mutually independent and identically distributed, given $\mathbf{X}$, as might be the case with a sequence of hybridizations involving short oligonucleotides of the same length. In such a case $H(D_1, D_2, \ldots, D_n) = nH(D_1)$, but no such simplification occurs for $H(\mathbf{X}|D_1, \ldots, D_n)$, although it should be possible to determine the asymptotic behavior of this quantity as $n \to \infty$. In a future paper we hope to discuss this issue more fully.

APPENDIX A

Upper and lower bounds for H(Q) and H(Y). Let $\mathbf{u} = (1^{u_1}, 2^{u_2} \ldots)$ be a partition of the number N, and suppose that $\Sigma_1^N u_i = z$. We will use the notation $\mathbf{U}(\cdot)$ to denote the partition of N associated with the configuration in parentheses.

LEMMA A.1. *The number of configurations* $\mathbf{Q}$ *for which* $\mathbf{U}(\mathbf{Q}) = \mathbf{u}$ *is*

$$\frac{N!}{\Pi_{i=1}^N (i!)^{u_i} u_i!}. \tag{A.1}$$

PROOF. This is well known [see, e.g., Aigner (1979)].

LEMMA A.2. *The number of configurations* $\mathbf{Y}$ *for which* $\mathbf{U}(\mathbf{Y}) = \mathbf{u}$ *is*

$$\frac{N!}{\Pi_{i=1}^N u_i!} \frac{1}{2^{z-u_1}}. \tag{A.2}$$

PROOF. It is clear that the number we seek in this lemma is the number (A.1) multiplied by the number of directionless permutations of clones within islands. However, the latter is just

$$\prod_{i=2}^N \left(\tfrac{1}{2} i!\right)^{u_i}$$

and the result follows once we note that $\Sigma_{i=2}^N u_i = z - u_1$. □

LEMMA A.3. *The configurations* $\mathbf{Y}$ *with* $\mathbf{U}(\mathbf{Y}) = \mathbf{u}$ *are equally likely.*

PROOF. By symmetry.

EXAMPLE. It is easy to see that the configurations $\mathbf{y}_1 = \{(37)^*, (426)^*, (8), (51)^*\}$ and $\mathbf{y}_2 = \{(32)^*, (785)^*, (4), (16)^*\}$, for example, are equiprobable.

COROLLARY A.1.

(i) $$H(\mathbf{Q} \mid \mathbf{U}) = \log_2 N! - L(\mathbf{U}) - M(\mathbf{U}),$$

(ii) $$H(\mathbf{Y} \mid \mathbf{U}) = \log_2 N! - L(\mathbf{U}) - Np(1-p),$$

where

(A.3) $$L(\mathbf{U}) = \mathbb{E}\left\{\log_2 \prod_{i=1}^{N} U_i!\right\}$$

and

(A.4) $$M(\mathbf{U}) = \mathbb{E}\left\{\log_2 \prod_{i=1}^{N} (i!)^{U_i}\right\}.$$

PROOF. These relations are consequences of Lemmas A.1 and A.2 and the equiprobable assertion of Lemma A.3.

We turn now to obtaining upper and lower bounds $\overline{L}(\mathbf{U})$, $\overline{M}(\mathbf{U})$ and $\underline{L}(\mathbf{U})$, $\underline{M}(\mathbf{U})$ of $L(\mathbf{U})$ and $M(\mathbf{U})$. In the calculations that follow, we use upper and lower bounds for factorials easily obtained from Stirling's formula [see, e.g., Feller (1968), page 52]

(A.5) $$n^{n+1/2}e^{-n} \le n! \le n^{n+1/2}e^{-n}\sqrt{2\pi}\,e^{1/12}.$$

We also make use of the readily proved fact that the distribution of the sizes of islands is a truncated geometric with probability $p = e^{-c}$, where $c = NL/G$. More fully, the (ordered) sequence $F_1^N, F_2^N, \ldots$ of island sizes consists of identically distributed random variables with common distribution $\mathrm{pr}(F^N = i) = pq^{i-1}/(1-q^N)$, $i = 1, 2, \ldots, N$. Lander and Waterman (1988) give the proof for N large in which case F^N is approximated by a geometric. Taking the truncation into account gives more accurate results in our bounds when N is in the hundreds. It follows that $\mathbb{E}(Z - U_1) = Ne^{-c}(1 - e^{-c})$, since, for $i = 1, 2, \ldots, N$,

$$\mathbb{E}U_i = \mathbb{E}\sum_{j=1}^{Z} I_{\{F_j^N = i\}} = \mathbb{E}\{Z\}P(F^N = i)$$
$$= np^2q^{i-1}/(1-q^N).$$

[More precisely, $\mathbb{E}U_i \approx np^2q^{i-1}/(1-q^N)$, since Z is very weakly related to the sequence $\{I_{\{F_j^N = i\}}\}$ $j = 1, 2, \ldots$. Equality holds if Z is independent of this sequence.] We note that the preceding approximations are not expected to work for very small N's, but we believe they do work when N is in the hundreds, say larger than 500.

LEMMA A.4.

$$\underline{L}(\mathbf{U})^+ \le L(\mathbf{U}) \le \overline{L}(\mathbf{U}),$$

where $x^+ = \max\{x, 0\}$,

$$\overline{L}(\mathbf{U}) = \mathbb{E}\{Z\}[\log_2 N - \log_2 e] + \tfrac{1}{2}\log_2 \mathbb{E}\{Z\} + \log_2\left(\sqrt{2\pi}\,e^{1/12}\right)$$

and

$$\underline{L}(\mathbf{U}) = \mathbb{E}\{Z\}\left[\log_2 N - (2c+1)\log_2 e + (e^c - 1)\log_2(1 - e^{-c})\right].$$

PROOF. Since $\Sigma_1^N U_i = Z$, we must have

$$\begin{pmatrix} & Z & \\ U_1 & U_2 & \cdots \end{pmatrix} \geq 1,$$

in which case

$$\begin{aligned}\mathbb{E}\{\log_2 \Pi_i U_i!\} &\leq \mathbb{E}\{\log_2 Z!\}\\ &\leq \mathbb{E}\left\{\left(Z + \tfrac{1}{2}\right)\log_2 Z - (\log_2 e)Z + \log_2\left(\sqrt{2\pi}e^{1/12}\right)\right\}.\end{aligned}$$

Now $Z \leq N$, and so the right-hand side of the preceding formula is

$$\leq \mathbb{E}\{Z\}\log_2 N + \tfrac{1}{2}\mathbb{E}\{\log_2 Z\} - (\log_2 e)\mathbb{E}\{Z\} + \log_2\left(\sqrt{2\pi}e^{1/12}\right),$$

which is just the expression $\overline{L}(\mathbf{U})$.

For the lower bound $\underline{L}(\mathbf{U})$ we argue as follows:

$$\begin{aligned}L(\mathbf{U}) &= \sum_{i=1}^{N} \mathbb{E}\{\log_2 U_i!\}\\ &\geq \sum_{i=1}^{N} \mathbb{E}\{U_i \log_2 U_i - (\log_2 e)U_i\}\\ &= \sum_{i=1}^{N} \mathbb{E}\{U_i \log_2 U_i\} - (\log_2 e)\mathbb{E}\{Z\} \quad \text{since } \sum U_i = Z\\ &\geq \sum_{i=1}^{N} \mathbb{E}\{U_i\}\log_2 \mathbb{E}\{U_i\} - (\log_2 e)\mathbb{E}\{Z\} \quad \text{since } x\log_2 x \text{ is convex.}\end{aligned}$$

Now $\mathbb{E}\{U_i\} = Np^2q^{i-1}/(1-q^N)$ where $p = e^{-c}$ and $q = 1 - p$ and so, continuing the preceding sequence of inequalities,

$$\begin{aligned}L(\mathbf{U}) &\geq \sum_{i=1}^{N} Np^2q^{i-1}\left[\log_2\frac{Np^2}{1-q^N} + (i-1)\log_2 q\right] - (\log_2 e)\mathbb{E}\{Z\}\\ &= Np\log_2\frac{Np^2}{q(1-q^N)} + Np(\log_2 q)c_N - (\log_2 e)\mathbb{E}\{Z\}\\ &= \mathbb{E}\{Z\}\log_2\frac{Np^2}{q(1-q^N)} + (\log_2 q)c_N - (\log_2 e)\mathbb{E}\{Z\},\end{aligned}$$

which is seen to be $\underline{L}(\mathbf{U})$ once we recall that $\mathbb{E}\{Z\} = Ne^{-c}$ and $c_N = \mathbb{E}F^N$. This completes the proof of Lemma A.4. Note that the leading term in each case is $e^{-c}N\log_2 N$. Obviously, $U \geq 0$. Hence $L(\mathbf{U}) \geq L^+(\mathbf{U})$. □

In the following lemma a_N and b_N are moments $\mathbb{E}\{F^N \log_2 F^N\}$ and $\frac{1}{2}\mathbb{E}\{\log_2 F^N\}$, where F^N has a truncated geometric distribution with parameter $p = e^{-c}$, $c = NL/G$, and truncation at N.

LEMMA A.5.

$$\underline{M}(\mathbf{U}) \le M(\mathbf{U}) \le \overline{M}(\mathbf{U}),$$

where

$$\underline{M}(\mathbf{U}) = Ne^{-c}(a_N + b_N) - (\log_2 e)N$$

and

$$\overline{M}(\mathbf{U}) = Ne^{-c}(a_N + b_N) - (\log_2 e)N + \log_2(\sqrt{2\pi}e^{1/12}) \times \mathbb{E}\{Z\}.$$

PROOF. By definition,

$$M(\mathbf{U}) = \mathbb{E}\left\{\log_2 \prod_{i=1}^{N} (i!)^{U_i}\right\} = \mathbb{E}\left\{\sum_i U_i \log_2 i!\right\}.$$

We first use the lower bound of (A.5), obtaining

$$M(\mathbf{U}) \ge \mathbb{E}\left\{\sum_{i=1}^{N} U_i(i + \tfrac{1}{2})\log_2 i - (\log_2 e)i\right\}$$

$$= \sum_{i=1}^{N} (i \log_2 i + \tfrac{1}{2}\log_2 i)\mathbb{E}\{U_i\} - (\log_2 e)N \quad \text{since } \sum_{i=1}^{N} iU_i = N.$$

Now $\mathbb{E}(U_i) = Np^2q^{i-1}/(1 - q^N)$ as before.

To complete this, we need to recall that

$$\mathbb{E}\{F^N \log_2 F^N\} = \sum_{i=1}^{N} i(\log_2 i)\frac{pq^{i-1}}{1 - q^N}$$

and

$$\frac{1}{2}\mathbb{E}\{\log_2 F\} = \frac{1}{2}\sum_{i=1}^{N} (\log_2 i)\frac{pq^{i-1}}{1 - q^N}.$$

As mentioned in the statement of the lemma, these will be denoted by a_N and b_N, respectively, giving

$$M(\mathbf{U}) \ge Ne^{-c}(a_N + b_N) - (\log_2 e)N = \underline{M}(\mathbf{U}).$$

Turning now to the upper bound, the same reasoning leads to

$$M(\mathbf{U}) \le Ne^{-c}(a_N + b_N) - (\log_2 e)N + \big(\log_2(\sqrt{2\pi}e^{1/12})\big)\mathbb{E}\{Z\},$$

where we have used the fact that $\Sigma_i U_i = Z$. However, the right-hand side of the preceding formula is just $\overline{M}(\mathbf{U})$ and we are finished. □

LEMMA A.6.

$$0 \le H(\mathbf{U}|Z) \le \mathbb{E}\{Z\}H(F^N).$$

PROOF.

$$H(\mathbf{U}|Z) = \sum_k \mathrm{pr}(Z = k)H(\mathbf{U}|Z = k)$$

and

$$H(\mathbf{U}|Z = k) \le H(F_1^N, \dots, F_k^N) \le kH(F^N),$$

since **U** is a function of $Z = k$ identically distributed random variables with the same truncated geometric distribution as F^N and its (conditional) entropy is bounded from above by the entropy of $F_1^N, F_2^N, \dots, F_k^N$ when they are independent. The lemma now follows by substituting this second equation in the previous one. □

Corollary A.2.

$$0 \le H(\mathbf{U}) \le Ne^{-c}H(F^N) + \log_2 N.$$

Proof. The relation is an immediate consequence of the lemma, once we recall that $Z \le N$ and $\mathbb{E}Z = Ne^{-c}$. □

APPENDIX B

An upper bound for H(X). In his thesis Newberg (1993) obtained recurrence relations and asymptotic expressions for the total number $C(N)$ of interleavings involving any number of islands which can be formed from N equal-sized randomly located cloned fragments. His asymptotic expression is given in the following result.

Proposition B.1.

$$C(N) \sim \frac{e^{3/8}\sqrt{2}}{8N}\left(\frac{4N}{e}\right)^N \quad \textit{as } N \to \infty.$$

Corollary B.1.

$$\begin{aligned} H(X) &\le \log_2 C(N) \\ &= N\log_2 N + N\log_2\left(\frac{4}{e}\right) - \log_2 N \\ &\quad + \frac{3}{8}\log_2(e) - \frac{5}{2} + o(1) \quad \textit{as } N \to \infty. \end{aligned}$$

APPENDIX C

Proofs of Results A and B.

Proof of Result A. Note that, for $S = \mathbf{Y}$ or $\mathbf{Q}$,

$$H(S) = H(S \mid \mathbf{U}) + H(\mathbf{U}).$$

The bounds for $S = \mathbf{Y}$ follow from Corollaries A.1(ii) and A.2 and Lemma A.4. The bounds for $S = \mathbf{Q}$ follow from Corollaries A.1(i) and A.2 and Lemma A.5. The bounds for $S = \mathbf{X}$ follow from Corollary B.1 and the fact that

X is a function of **Y** and the lower bound on $H(\mathbf{Y})$. We dropped the constant term in the upper bound for **X** in Corollary B.1 since it makes only negligible difference. Finally, the bounds for $S = \mathbf{P}$ follow from the facts that

$$H(\mathbf{P}) \geq H(\mathbf{Q}) \geq \underline{H}(\mathbf{Q})$$

and

$$H(\mathbf{P}) \leq H(\mathbf{X}) \leq \overline{H}(\mathbf{X})$$

(because **Q** is a function of **P** and **P** is a function of **X**). □

PROOF OF RESULT B. (i) and (ii) follow directly from the finite-sample bounds on $H(\mathbf{X})$, $H(\mathbf{Y})$ and $H(\mathbf{P})$, and the exact expressions for $H(\mathbf{W})$ and $H(\mathbf{V})$, and so does

$$H(\mathbf{P}) \geq (1 - e^{-c}) N \log_2 N(1 + o(1)).$$

Because **Q** is a function of **P**,

$$H(\mathbf{P}) = H(\mathbf{Q}) + H(\mathbf{P} \mid \mathbf{Q}).$$

For any given configuration **Q**, let $\mathbf{U} = \mathbf{U}(\mathbf{Q})$. Then, for any island of i clones, **P** can only take $2^{i(i+1)/2}$ possible values. It follows that

$$\begin{aligned} H(\mathbf{P} \mid \mathbf{Q}) &\leq \mathbb{E} \log_2 \Big(\prod_i 2^{U_i \times i(i+1)/2} \Big) \\ &\leq \sum_i \mathbb{E} U_i (i^2 + i)/2 \\ &= \sum_i N p^2 q^{i-1} (i^2 + i)/2 \\ &= N e^{-c} \sum_i p q^{i-1} (i^2 + i)/2 \\ &= N e^{-c} \big(\mathbb{E}\{F_N^2\} + \mathbb{E}\{F_N\} \big)/2(1 - q^N) \\ &= O(N) \quad \text{as } N \to \infty. \end{aligned}$$

Hence

$$H(\mathbf{P}) \leq H(\mathbf{Q}) + O(N) = (1 - e^{-c}) N \log_2 N(1 + o(1)). \qquad \square$$

Acknowledgment. The authors would like to thank Dr. Phil Green for his very valuable comments and insights, which led to a greatly improved revision of their initial paper.

REFERENCES

AIGNER, M. (1979). *Combinatorial Theory*. Springer, New York.

ALIZADEH, F., KARP, R. M., NEWBERG, L. A. and WEISSER, D. C. (1993). Physical mapping of chromosomes: a combinatorial problem in molecular biology. In *Proceedings of the Fourth Annual ACM–SIAM Symposium on Discrete Algorithms*, Austin, TX. ACM, New York.

BRANSCOMB, E., SLEZAK, T., PAE, R., GALAS, D., CARRANO, A. V. and WATERMAN, M. (1990). Optimizing restriction fragment fingerprinting methods for ordering large genomic libraries. *Genomics* **8** 351–366.

INFORMATION AND CLONE MAPPING

COVER, T. and THOMAS, J. (1991). *Elements of Information Theory*. Wiley, New York.

COX, D. R., BURMEISTER, M., PRICE, E. R. and MYERS, R. M. (1990). Radiation hybrid mapping: a somatic cell genetic method for constructing high resolution maps of mammalian chromosomes. *Science* **250** 245–250.

CRAIG, A. G., NIZETIC, D., HOHEISEL, J. C., ZEHETNER, G. and LEHRACH, H. (1990). Ordering of cosmic clones covering the Herpes simplex virus type-I (HSV-I) genome—a test case for fingerprinting by hybridization. *Nucleic Acids Res.* **218** 2653–2660.

FELLER, W. (1968). *An Introduction to Probability Theory and Its Applications* **1**, 3rd ed. Wiley, New York.

FU, Y.-X., TIMBERLAKE, W. E. and ARNOLD, J. (1992). On the design of genome mapping experiments using short synthetic oligonucleotides. *Biometrics* **48** 337–359.

GREEN, E. D. and GREEN, P. (1991). Sequence-tagged sites (STS) content mapping of human chromosomes: theoretical considerations and early experiences. *PCR Methods and Applications* **1** 77–90.

LANDER, E. S. and WATERMAN, M. S. (1988). Genomic mapping by fingerprinting random clones: a mathematical analysis. *Genomics* **2** 231–239.

LEHRACH, H., DRMANAC, R., HOHEISEL, J., LARIN, Z., LENNON, G., MONACO, A. P., NIZETIC, D., ZEHETNER, G. and POUSTKA, A. (1990). Hybridization fingerprinting in genome mapping and sequencing. In *Genome Analysis. Genetic and Physical Mapping* (K. E. Davies and S. M. Tilghman, eds.) **1** 39–81. Cold Spring Harbor Laboratory Press.

NELSON, D. O. and SPEED, T. P. (1994). Statistical issues in constructing high resolution physical maps. *Statist. Sci.* **9** 334–354.

NEWBERG, L. A. (1993). Finding, evaluating and counting DNA physical maps. Ph.D. dissertation, Dept. Electrical Engineering and Computer Science, Univ. California, Berkeley.

OLSON, M. V., DUTCHIK, J. E., GRAHAM, M. Y., BROUDEUR, G. M., HELMS, C., FRANK, M., MACCOLLIN, M., SHEINMAN, R. and FRANK, T. (1986). Random-clone strategy for genomic restriction mapping in yeast. *Proc. Nat. Acad. Sci. U.S.A.* **83** 7826–7830.

RÉNYI, A. (1984). *A Diary on Information Theory*. Wiley, New York.

TRASK, B. J. (1991). Fluorescence in situ hybridization—applications in cytogenetics and gene mapping. *Trends in Genetics* **7** 149–154.

DEPARTMENT OF STATISTICS
UNIVERSITY OF CALIFORNIA
367 EVANS HALL #3860
BERKELEY, CALIFORNIA 94720-3860
E-MAIL: binyu@stat.berkeley.edu

Chapter 8
Applied Statistics and Exposition

Karl W. Broman

I particularly admire Terry for the breadth of his thinking, the depth of his understanding, the strength and openness with which he expresses his opinions, and the clarity of his writing. All of these qualities are on display in the manuscripts in this section.

But before I comment on those articles, I want to first mention that the best illustrations of Terry's admirable qualities are his regular commentaries for the IMS Bulletin, begun when he was President of the Institute of Mathematical Statistics and continued by popular demand. (To date, he has written over 70 such commentaries! They are available online at `http://bulletin.imstat.org`.) Among my favorites: In Praise of Postdocs [10], Keep Gender on the Agenda [8], A Toast to Posters! [11], Books Worth Reading [7] (his five favorite books are Feller Volume I [2], editions 1, 2 and 3, and Feller Volume II [3], editions 1 and 2), and It's Job-hunting Time! [9]. Terry has strong opinions on a wide range of topics, and we all benefit from his willingness to share them.

Terry's technical report on probabilities related to the reliability of nuclear reactors [14], commenting on probability statements in the Reactor Safety Study [4], demonstrates Terry's willingness to delve deeply and thoroughly into a problem and the clarity and insight he then gives us. (The manuscript is especially relevant today, given the problems with nuclear reactors following the tsunami in Japan.) Misuse of the addition and multiplication rules of probability is well known to statisticians. His further point, that probability statements are meaningless without an understanding of how they were derived, is obvious in retrospect yet often overlooked. It is related to a point that Terry has repeatedly emphasized: an estimate is of no value without some statement on its uncertainty (such as a standard error). I can imagine no better (or more amusing) illustration of this issue than the statement from the Chairman of the U.S. National Transportation Safety Board regarding the chance of two jets colliding on the ground, and the response to Terry's aerogramme. (The word "aero-

K.W. Broman
Department of Biostatistics & Medical Informatics, University of Wisconsin–Madison
e-mail: kbroman@biostat.wisc.edu

S. Dudoit (ed.), *Selected Works of Terry Speed*, Selected Works in Probability and Statistics, 353
DOI 10.1007/978-1-4614-1347-9_8, © Springer Science+Business Media, LLC 2012

gramme" itself makes me smile.) Also amusing is Terry's comment [14, p. 24], "The whole exercise can now be recognized as being totally without relevance to the real world." Ouch!

Terry's work on salmon populations [16] and bioassays [1] were among his last projects before he began to focus almost exclusively on genetics and genomics. The salmon work is a particularly good illustration of Terry's approach: know the science and the scientific questions and goals, and let the methods follow. I saw this in action in the bioassay project; our (or really *his*) normal-Poisson mixture model, which now seems so natural, was the last of a long sequence of preliminary models. My most vivid memory of that effort was our initial error (or really *my* initial error) in implementing an EM algorithm for the normal-Poisson mixture. The sufficient statistics include not just $\sum x_i$ but also $\sum x_i^2$, and so the E-step requires not just the expected values of the x_i, but also of the x_i^2. That one should follow the log likelihood across iterations, as a diagnostic for the correct implementation of an EM algorithm, was another important lesson.

Terry's comment [12] on Robinson's Statistical Science article on BLUP [5] is masterly. The diverse applications of BLUP that Terry provides add great emphasis to the importance of Robinson's paper. Terry's comment has been widely cited as being the first articulation of the connection between smoothing splines and BLUPs.

Finally, let me comment on the two encyclopedia entries reprinted in this section, on restricted maximum likelihood [15] and on iterative proportion fitting (IPF) [13], though I can think of little to say except that they are superb examples of the clarity of Terry's writing (and to note that Terry has an IMS Bulletin commentary [6] on IPF, too!). The breadth and depth of Terry's thinking often fills me with envy, but the effort he devotes to sharing his ideas allows me to get past the envy and simply appreciate his insights.

References

[1] K. Broman, T. Speed, and M. Tigges. Estimation of antigen-responsive T cell frequencies in PBMC from human subjects. *J. Immunol. Methods*, 198: 119–132, 1996.

[2] W. Feller. *An Introduction to Probability Theory and its Applications*, volume I. Wiley, New York, 3rd edition, 1968.

[3] W. Feller. *An Introduction to Probability Theory and its Applications*, volume II. Wiley, New York, 2nd edition, 1971.

[4] N. C. Rasmussen et al. Reactor Safety Study: An assessment of accident risk in U.S. commercial nuclear power plants. Technical report, Nuclear Regulatory Commission, Washington, D.C. (USA), October 1975. Report Number(s): WASH-1400-MR; NUREG-75/014-MR.

[5] G. K. Robinson. That BLUP is a good thing: The estimation of random effects. *Stat. Sci.*, 6:15–51, 1991.

[6] T. Speed. Terence's stuff: My favourite algorithm. *IMS Bulletin*, 37 (9):14, 2008.
[7] T. Speed. Terence's stuff: Books worth reading. *IMS Bulletin*, 34 (2):6, 2005.
[8] T. Speed. President's column: Keep gender on the agenda. *IMS Bulletin*, 33 (2):4, 2004.
[9] T. Speed. Terence's stuff: It's job-hunting time! *IMS Bulletin*, 38 (1):14, 2009.
[10] T. Speed. President's column: In praise of postdocs. *IMS Bulletin*, 32 (6):4, 2003.
[11] T. Speed. Terence's stuff: A toast to posters! *IMS Bulletin*, 34 (6):6, 2005.
[12] T. P. Speed. Comment on G. K. Robinson, "That BLUP is a good thing: The estimation of random effects". *Stat. Sci.*, 6(1):42–44, 1991.
[13] T. P. Speed. Iterative proportional fitting. In P. Armitage and T. Colton, editors, *Encyclopedia of Biostatistics*. John Wiley & Sons, New York, 1999.
[14] T. P. Speed. Negligible probabilities and nuclear reactor safety: Another misuse of probability? Technical report, Mathematics Department, University of Western Australia, 1977.
[15] T. P. Speed. Restricted maximum likelihood (REML). In S. Kotz, editor, *Encylopedia of Statistical Sciences*, volume 2, pages 472–481. John Wiley & Sons, New York, 1997.
[16] T. P. Speed. Modelling and managing a salmon population. In V. Barnett and K. T. Turkman, editors, *Statistics for the Environment: Water Related Issues*, volume 1, pages 267–292. John Wiley & Sons, 1993.

ELSEVIER

Journal of Immunological Methods 198 (1996) 119–132

JOURNAL OF IMMUNOLOGICAL METHODS

Estimation of antigen-responsive T cell frequencies in PBMC from human subjects

Karl Broman [a], Terry Speed [a,*], Michael Tigges [b]

[a] Department of Statistics, University of California, Berkeley, CA 94720, USA
[b] The Biocine Company, Chiron Corporation, Emeryville, CA 94608, USA

Received 5 April 1996; revised 10 June 1996; accepted 10 June 1996

Abstract

A new method for estimating the frequency of antigen-responsive T cells, using a cell proliferation assay, is described. In this assay, the uptake of tritiated thymidine by peripheral blood mononuclear cells which have been exposed to antigen, is measured for each well on a microtiter plate. Whereas this assay is generally used as part of a limiting dilution assay, here we estimate the frequency of responding cells using a single, carefully chosen cell density. The traditional analysis of such data uses a cut-off to separate wells which contain no responding cells and wells which contain at least one responding cell. The new method uses the scintillation count to estimate the number of responding cells for each well on the plate. We do this by fitting a two-stage model, the first stage being a Poisson model with antigen-specific frequency parameters, and the second stage a linear model with plate-specific parameters.

Keywords: Antigen-responsive T cell frequency; Limiting dilution assay; Maximum likelihood; EM algorithm; Plate-specific parameter

1. Introduction

In this paper we describe a method for estimating the frequency of antigen-responsive T cells among peripheral blood mononuclear cells (PBMC) from human subjects. It was developed in a particular context, but we believe that the approach may have wider applicability. The proliferation assay which motivates our method seeks to quantitate the response of human subjects to a herpes simplex vaccine consisting of HSV type 2 glycoproteins D and B, expressed as recombinant proteins in Chinese hamster ovary cells and administered as a vaccine combined with alum (Parr et al., 1991; Straus et al., 1993) or an oil-in-water emulsion adjuvant MF59 (Langenberg et al., 1995). A standard proliferation assay (James, 1991) utilizing the responses in triplicate wells was found to be inadequate in this context, while a full limiting dilution assay (LDA) was not feasible because of the need for more PBMC than were available from the vaccine recipients. Accordingly, our analysis was developed for estimating the number of antigen-responsive T cells based on a single carefully chosen dilution, and so we sought to make greater use of the data than is usually the case.

Abbreviations: PBMC, peripheral blood mononuclear cells; LDA, limiting dilution assay; SHPM, single-hit Poisson model; PM, Poisson model; MLE, maximum likelihood estimate; SD, standard deviation; HSV, herpes simplex virus; PHS, pooled human serum; PHA, phytohemagglutinin.

* Corresponding author. Tel.: (510) 642-2781, Fax: (510) 642-7892.

PII S0022-1759(96)00136-6

Two other factors prompted our approach, which we now describe. In the present context, a standard LDA begins with a classification of the wells in all or part of a microtiter plate into the categories positive (contains at least one responding cell) and negative (contains no responding cells). Use is then made of a statistical model, typically the single-hit Poisson model (SHPM), which relates the frequency of positive wells to the frequency of responding cells in the wells. The frequency of responding cells is then estimated from the data on wells using the method of maximum likelihood or minimum χ^2. All such analyses have to use some method of determining which wells contain responding cells, that is, of classifying wells into positives and negatives. The data with which this classification is done is generally a scintillation count, and it is usual for the assay to contain some wells which should all be negative to provide an estimate of the size of the background count. A common approach (see e.g. Langhorne and Fischer-Lindahl, 1981) is to select a threshold, often the mean plus two or three standard deviations of the background counts, and consider a well positive if its count exceeds this threshold. This approach clearly works well much of the time, as indicated by the straight lines frequently obtained when plotting the log of the frequency of negative wells against the number of cells per well. However, the background counts are usually not normally distributed about their mean, the more common situation being where there is considerable skewness, with the left-hand tail of the distribution being much shorter than its right-hand tail. Under these circumstances, reliance upon a threshold defined in terms of the mean and standard deviation of such counts can be problematic. This was the case with the data we consider in the present paper: determination of a threshold was not at all straightforward. A comparison of the counts corresponding to the wells in which no responding cells were expected with those for wells to which antigen was added revealed no clear cut-off in many cases. Efforts to develop more elaborate methods of determining the threshold for positive wells were not completely satisfactory. This led us to seek an analysis of the data which made direct use of the scintillation counts, and which did not reduce them to positive and negative well frequencies. There are good statistical reasons for avoiding the use of thresholds in situations where they are not entirely clear. In such cases, the actual threshold used can have a very great impact, indeed be the dominant contributor to the final frequency estimates, yet the very real uncertainty in the determination of the threshold is typically not reflected in the standard errors or confidence limits given for these frequency estimates. Doing so presents formidable technical statistical problems, yet ignoring this important source of uncertainty can lead to quite unrealistic impressions concerning the precision of the frequency estimates.

There was a second, independent reason why it was desirable to avoid classifying wells as positive or negative: in many cases the entire set of wells for a given antigen would be positive. This arose whenever the density of cells chosen for the assay was a poor guess, something that could not always be avoided. In such cases the standard analysis of the data, given as a frequency of 100% of wells positive, does not yield a point estimate of the frequency of responding cells, but only a lower confidence limit. This causes difficulties later, when such results are to be compared or combined with other results. Since up to 20% of our assays would yield all positive wells, however the threshold was defined, we had a strong incentive to develop a method of estimating the frequency of responding cells another way. Similar remarks apply to the less frequent cases in which all wells would have been scored negative.

As with most studies of this kind, the approach we adopt below makes use of the Poisson model (PM) for the distribution of responding cells in a well. But instead of relating frequencies of responding cells to proportions of positive wells, we relate them to averages of suitably transformed scintillation counts. Our model involves plate-specific parameters which need to be estimated, but the evidence so far suggests that this can be done well enough to permit useful estimates of the frequencies of responding T cells to be obtained from replicate pairs of microtiter plates involving cells at a single density. Details are given in Section 2.

In order to demonstrate the validity of the new method, we analyse four LDAs on cells from three subjects, each involving three antigens and a control, carried out on replicate pairs of plates at six, five and four dilutions. We then compare the results obtained from a single dilution with those obtained using the entire LDA. We also analyse a number of assays run at a single dilution. As a further demonstration of the usefulness of this approach, we reanalyse the data from a quite different type of LDA, see Langhorne and Fischer-Lindahl (1981), namely a ^{51}Cr-release assay designed to estimate the frequency of cytotoxic T lymphocyte precursor cells in mixed lymphocyte cultures. Finally, we analyse two additional proliferation assays of different designs to demonstrate the flexibility of this method: one a single density assay applied to samples pre- and post-immunization (S. Rodda, Chiron Mimotopes, Melbourne, Australia) and a set of three assays designed as limiting dilution assays (D. Koelle, University of Washington, Seattle, WA).

2. Materials and methods

2.1. Vaccine, antigens and subjects

Two subjects (#711 and #713) in a clinical trial of an HSV vaccine provided informed consent for the collection of additional blood for development of the assay. These subjects had never been infected with either HSV-1 or HSV-2 prior to vaccination with a vaccine consisting of 30 μg each of two HSV glycoproteins (D and B) expressed as a recombinant product in Chinese hamster ovary (CHO) cells. The proteins were combined with an oil-in-water emulsion adjuvant (MF59, Chiron Corporation) and administered by intramuscular injection in the deltoid muscle at 0, 1 and 6 months (Langenberg et al., 1995). A third subject (NIH 1394) who was subject to frequent recurrences of genital herpes and who is HSV-2 positive was also recruited (Kost et al., 1993). The gD and gB proteins used in the assay were from the vaccine lots or comparable lots manufactured by Chiron Biocine (Emeryville, CA). Tetanus toxoid was a gift from Wyeth Laboratories (Nutley, NJ). Phytohemagglutinin was purchased from Sigma (St. Louis, MO).

2.2. Preparation of cells

Peripheral blood mononuclear cells (PBMC) were collected from the two vaccine study subjects 60–69 days after the third immunization by three pass leukapheresis. PBMC were prepared from the third subject by one pass leukapheresis. The recovered cells were further purified by density centrifugation onto Ficoll-Hypaque (Pharmacia), washed free of separation medium and prepared for cryopreservation in RPMI 1640 medium containing 20% pooled human serum and 7.5–10% DMSO. The cells were stored in multiple aliquots in vapor-phase liquid nitrogen until assay.

2.3. Description of the assays

Six-point LDAs of PBMC from subjects #713 and #711 and a second five-point LDA of PBMC from subject #713 were set up using information obtained from a frequency analysis assay (Reece et al., 1994a; M. Tigges, unpublished observations). PBMC from #713 contain a high frequency of gD2 and gB2 specific T cells while #711 PBMC responded poorly to these two antigens. The expected frequencies were 100 and 200 responders per 10^6 PBMC to gD2 and gB2 respectively from #713 and five and two responders per 10^6 PBMC to gD2 and gB2 respectively from #711. The frequency of gB2 and gD2 responsive T cells in the NIH 1394 sample was expected to be on the order of 20–40 per 10^6 PBMC based on the results of assays with PBMC from HSV-2 seropositive subjects in other trials.

The PBMC were thawed and washed free of preservative with LGM-1 (RPMI 1640, JRH Biosciences, supplemented with 1 mM Na pyruvate, JRH Biosciences, 5 mM Hepes pH 7.2, Gibco, 2 mM glutamine, JRH Biosciences, 50 μg/ml gentamicin, Gibco, and 1% pooled human serum). The pooled human serum (PHS) was prepared from screened units of recovered plasma. The washed cells were resuspended in lymphocyte growth medium containing 10% PHS (LGM-10), the cell concentration was adjusted, and the appropriate volume of the cell suspension was placed in the culture plates. The test antigens consisted of gD2 and gB2 at a concentration of 1 μg/ml in 48 wells each distributed between two plates at each concentration. The control antigen consisted of tetanus toxoid at 2 Lf/ml in 44 wells and the control mitogen, PHA, was added to two wells on each plate. The PBMC from subject #711 were seeded into U-bottom plates and those from subjects #713 and NIH 1394 were seeded into V-bottom plates. For U-bottom plates, the test and control antigens were diluted to 10 μg/ml in LGM-10 then 20 μl were transferred to the appropriate wells. The #711 PBMC were diluted into three cell suspensions with densities of 10^6, 5×10^5 and 2×10^5 cells/ml then either 200 or 150 μl of the suspension were transferred to two plates. For PBMC from subject #713, the washed cells were adjusted to 2×10^5 cells/ml in LGM-10 then mixed with an equal volume of LGM-10 containing the test and control antigens at 2 μg/ml. The cells were then seeded into 48 replicate V-bottom wells in volumes of 150, 100, 75, 50, 35 and 25 μl. In a second assay the #713 cells were prepared similarly. The cells were resuspended at 5×10^5 cells/ml and seeded into replicate V-bottom wells in volumes of 100, 80 or 60 μl or resuspended at 2×10^5 cells/ml and seeded in volumes of 100 or 50 μl. The assay design for the NIH 1394 PBMC differed slightly in that the antigens and cells alone were seeded into 36 wells distributed over three plates (12 wells/plate); the tetanus toxoid was included in only six wells and the PHA in three wells. After the cells were diluted for dispensing into the wells, a sample was taken to determine the actual number of cells seeded per well. All of the microwell plates were cultured for four days in humidified boxes at 37°C in 7% CO^2 before being labeled for 6 h with 0.5 μCi/well of [^{3}H]thymidine. PHA was added to the appropriate wells on day 2. The plates were harvested using a Cambridge Technologies automated harvester and the filters counted in a Wallac/Pharmacia β plate scintillation counter.

The assay design for the data from S. Rodda was similar to that described above, except that 64 replicates were plated for each antigen tested and cells were plated at 200 000 cells/well. The PBMC were obtained from a single individual before and 3 weeks after immunization with tetanus toxoid (Reece et al., 1994a). The assays included preimmune cells, postimmunization cells, and a 1 : 1 mixture of the two. Test antigens included influenza ribonucleoprotein and a peptide that includes an epitope from tetanus toxoid that is recognized by $CD4^+$ T cells from many individuals (Reece et al., 1994b). The data from D. Koelle were taken from an LDA design that consisted of 10 000 irradiated autologous PBMC/well, a graded number of between 50 000 and 780 PBMC and HSV-2 viral antigen in 24 replicates. The cells alone controls were seeded in replicates of 12 or 24 on plates separate from the antigen containing wells. The cells were cultured for 5 days before being labeled with [^{3}H]thymidine and harvested on day 6.

2.4. Description of the data analysis

Our analysis of the data from a single microtiter plate makes use of all but two of the counts for the 96 wells: we do not use the counts for the two wells with PHA, which simply serve as a positive control for helping select usable data. We will describe the analysis of data from a single plate first, although our final analysis involves replicate pairs of plates.

The analysis is based upon the same PM that underlies most LDAs, but we need an extra relationship connecting the scintillation count to the number of responding cells in a well. Specifically, we suppose that there are plate-specific parameters a, b and σ, and a widely applicable power parameter p such that the pth power of the scintillation count in a well with k responding cells is approximately normally distributed with mean $a + bk$ and standard deviation σ. As with other analyses using the PM, we suppose that the number of responding cells in a well with c PBMC is Poisson distributed with mean $fc \times 10^{-6}$, where f is the frequency of responding cells per million PBMC. Since we have 24 wells with cells alone, 24 with gD2, 24 with gB2 and 22 with tetanus toxoid in any given plate, there will be four frequencies and three additional parameters for each plate. When we analyse replicate pairs of plates, there will be ten parameters: the frequencies of each of the four classes of responding cells, assumed the same in each plate, and a set of three plate-specific parameters a, b and σ for each plate.

Algebraically, our assumptions are as follows. Let y_{ij} denote the transformed scintillation count for well j of class i, and let k_{ij} denote the corresponding number of responding cells. Here $i = c, d, b, t$ corresponds to the cells alone, gD2, gB2 and tetanus toxoid classes, respectively. We assume that the (y_{ij}, k_{ij}) are mutually independent, that k_{ij} follows a Poisson distribution with mean λ_i, and that, given k_{ij}, y_{ij} follows a normal distribution with mean $a + bk_{ij}$ and standard deviation σ. The aim of our analysis is to estimate the parameters λ_i, and hence, using the estimated numbers of cells per well, the frequencies f_i of responding cells per million PBMC.

The power parameter was selected by maximum likelihood (Box and Cox, 1964) from the values 1, 1/2, 1/4 and 0 (corresponding to log). The model itself was fitted by the method of maximum likelihood, specifically, using a form of the so-called EM algorithm (Dempster et al., 1977), although we also carried out a number of confirmatory analyses using the fully calculated likelihood. Standard errors for the parameter estimates were computed using the SEM algorithm (Meng and Rubin, 1991). A more detailed description of the statistical methods can be found in Broman et al. (1996).

A by-product of the EM algorithm, which regards the 'complete' data for any well as a pair (y, k), where y is the observed scintillation count, and k the unobserved number of responding cells, is an estimate of k for each well. These can be plotted against the count, and provide an informative diagnostic for the analysis of a plate. In particular, one can usually see the effective threshold distinguishing wells with no estimated responding cells from those with one or more.

The foregoing analysis provides estimates of the frequencies λ_c, λ_d, λ_b and λ_t of responding cells *per well* for each of the four classes: cells alone, gD2, gB2 and tetanus toxoid. We next obtained MLEs of the frequencies of responding cells per well *above background*, i.e. of $\lambda_d - \lambda_c$, $\lambda_b - \lambda_c$ and $\lambda_t - \lambda_c$, and then converted the resulting figures to frequencies per 10^6 cells. This last step involved scaling by the estimated number of cells/well in each plate. We note that as long as $\hat{\lambda}_d$, $\hat{\lambda}_b$ and $\hat{\lambda}_t$ are all $\geq \hat{\lambda}_c$, the MLEs of the differences $\lambda_d - \lambda_c$, etc., are just the differences $\hat{\lambda}_d - \hat{\lambda}_c$, etc., of the MLEs. In the rare cases where one or more of $\hat{\lambda}_d$, $\hat{\lambda}_b$ and $\hat{\lambda}_t$ was $\leq \hat{\lambda}_c$, a slightly modified analysis was necessary, involving combining sets of counts. The details are straightforward, see e.g. Barlow et al. (1972), and will be omitted.

In order to analyse the ^{51}Cr-release assay, which has a more standard structure, with 24 replicate wells at each of eight concentrations, we used the same basic method, modified to correspond to only a single unknown frequency of responder cells at each concentration.

The single density assay using data from S. Rodda consisted of six plates with cells from one subject taken before and after tetanus immunization: a pair of plates with 200 000 preimmune cells per well, a pair with 200 000 post-immunization cells per well, and a pair containing a mixture of 100 000 preimmune cells per well and 100 000 post-immunization cells per well. Each plate consisted of three groups of 32 wells containing cells alone, cells treated with influenza RNP antigen, and cells treated with an epitope from tetanus toxoid, respectively. For each pair of plates, we estimate the frequency of responding cells per well for the three groups of wells (denoted λ_c, λ_t and λ_i, corresponding to cells alone, tetanus, and influenza RNP, respectively), and two sets of plate-specific parameters (a, b, σ).

The data from D. Koelle consist of three seven-point LDAs, corresponding to three different subjects. For each LDA, four plates were used. On the first plate, 24 wells were dedicated to each of four cell densities: 50 000, 25 000, 12 500 and 6250 cells per well; antigen was added to each well. On the second plate, 24 wells were dedicated to each of three cell densities: 3125, 1563 and 781 cells per

Table 1
Maximum likelihood estimates and estimated standard deviations of model parameters for the results of the first assay of subject #713 at density 11 400 cells/well

	λ_c	λ_d	λ_b	λ_t	a	b	σ
Joint							
Plate 1	0.4 (0.1)	3.5 (0.3)	3.3 (0.3)	4.7 (0.3)	16.4 (0.9)	10.3 (0.3)	3.6 (0.5)
Plate 2	0.4 (0.1)	3.5 (0.3)	3.3 (0.3)	4.7 (0.3)	14.8 (0.8)	9.4 (0.2)	2.9 (0.4)
Separate							
Plate 1	0.3 (0.1)	3.0 (0.4)	2.8 (0.4)	4.4 (0.5)	16.7 (0.9)	10.3 (0.3)	3.5 (0.4)
Plate 2	0.5 (0.1)	3.9 (0.4)	3.9 (0.4)	5.0 (0.5)	14.5 (0.7)	9.3 (0.2)	2.8 (0.3)

well; again, antigen was added to each well. The third and fourth plates in each LDA were like the first two, but with no antigen added. We analysed these data using the first and third plates together, with the parameters (a, b, σ) constrained to be equal for the two plates. The second and fourth plates were analysed similarly.

3. Results

Maximum likelihood estimates (MLEs) under the model (Finney, 1978) of the mean number of responding *cells / well* for a pair of plates from LDA #713, with 11 400 cells per well, are presented in Table 1, together with MLEs of the parameters a, b and σ and estimated standard deviations (SDs) for each estimate. We used the square root of the scintillation count, as indicated by a Box-Cox analysis. Estimates were obtained by treating each plate *separately*, and also for the *joint* analysis of the pair of plates, where the mean numbers of responding cells/well were constrained to be equal. The estimated SDs given in parentheses after each parameter estimate take into account only *within-plate* variation. No attempt has been made to include a component of variation between duplicates within duplicate pairs, and thus the SDs for the *joint* estimates are somewhat understated. The results for different members of a duplicate pair are usually quite close, and so this underestimation is unlikely to be a problem. When the two sets of results are quite different, it is usually the case that one of the pair is simply a bad plate, and the results are discarded. In any event,

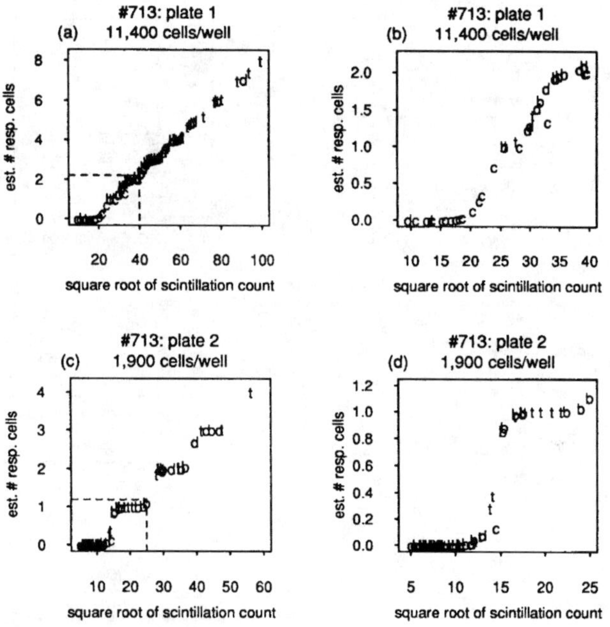

Fig. 1. Estimated number of responding cells vs. square root of scintillation count. *b* and *d* display the lower left regions (marked by dashed lines) of *a* and *c*, respectively.

K. Broman et al. / Journal of Immunological Methods 198 (1996) 119–132

the variability between duplicates within duplicate pairs is usually very much smaller than the between-assay variability, as we shall see.

After obtaining estimates of the numbers of responding *cells / well* we calculate the frequency of *responding cells / 10^6 cells* for each of the three antigens. The results for the first assay of subject #713 at density 11 400 cells/well are displayed in Table 2. In this table, these frequencies have been corrected for background using the estimated number of cells/well responding in the cells alone category. Furthermore, the estimated SDs presented incorporate variation due to cell counting and dilution errors, though they still do not reflect the between-assay variability. It can be seen that at this stage our single dilution frequency estimates have coefficients of variation around 25%.

Table 2
Maximum likelihood estimates and estimated standard deviations of frequencies of responding cells per 10^6 cells for the results of the first assay of subject #713 at density 11 400 cells/well

	f_d	f_b	f_t
Joint	271 (67)	257 (64)	378 (92)
Separate			
Plate 1	238 (60)	219 (56)	356 (87)
Plate 2	306 (75)	299 (73)	400 (97)

A by-product of the (EM) algorithm we use for maximum likelihood estimation is an estimate, for each well, of the number of responding cells in that well. These estimates are plotted against the square root of the scintillation count and exhibited in Fig. 1

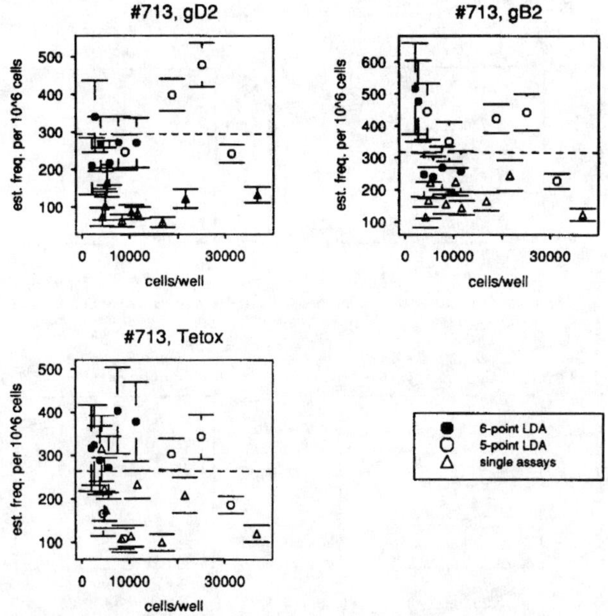

Fig. 2. Subject #713 (one six-point LDA, one five-point LDA, plus single assays). Maximum likelihood estimates of frequencies ($\times 10^6$) of responding cells using two plates at each dilution: estimates plotted against # cells/well. Error bars correspond to ± 1 SD. Dotted line corresponds to estimated frequency of responding cells ($\times 10^6$) obtained using all the data.

for two plates from LDA #713. Fig. 1b and Fig. 1d display the lower left regions (marked by dashed lines) of Fig. 1a and Fig. 1c, respectively. In Fig. 1a and Fig. 1b (#713: 11 400 cells/well, plate 1), we see that wells with counts less than 400 (20 on the square root scale) have been assigned 0 responding cells, while wells with counts greater than 625 (25 on the square root scale) have been assigned ≥ 1 responding cell. In between 400 and 625 is a grey area: there is no clear cut-off for this data. The mean and SD of the untransformed counts are 461 and 401 respectively, and so use of mean +2 SD as cut-off would lead to a figure of 1263. It is evident that at least two, perhaps three wells in the cells alone group seem to contain a responding cell. Their scintillation counts inflate the SD of the cells alone counts, which in turn can lead to an unduly large cut-off under the traditional analysis of such data. Fig. 1c and Fig. 1d (#713: 1900 cells/well, plate 2) exhibit a much sharper cut-off.

In Figs. 2–4, estimates of the frequencies of responding cells per 10^6 cells are plotted against cell density, with error bars corresponding to ± 1 SD. Here the SD incorporates both within plate variation and errors involved in counting the number of cells/well and dilution errors. The dotted line in each plot corresponds to the estimated frequency of responding cells/10^6 cells obtained by carrying out a maximum likelihood analysis using all the data from both LDAs, but not the single assays. It is

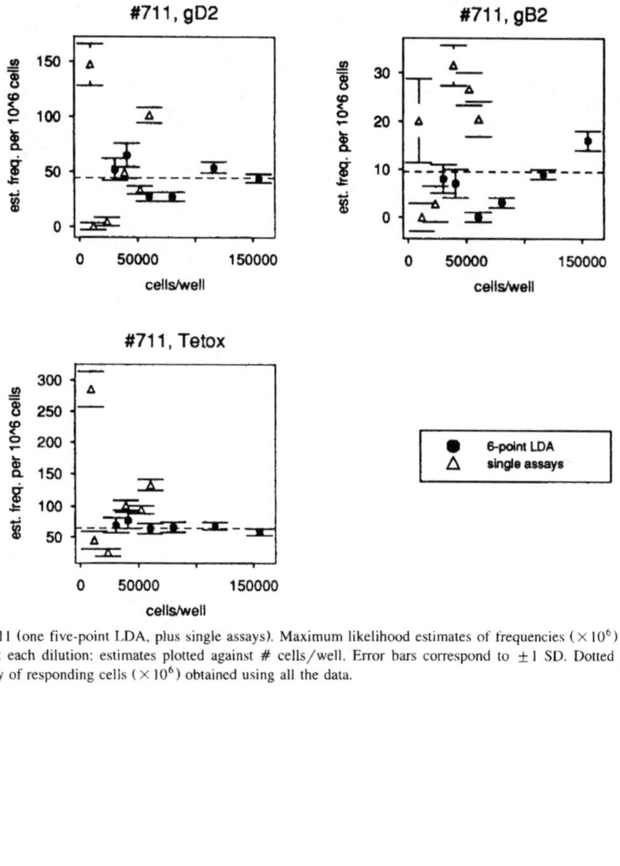

Fig. 3. Subject #711 (one five-point LDA, plus single assays). Maximum likelihood estimates of frequencies ($\times 10^6$) of responding cells using two plates at each dilution: estimates plotted against # cells/well. Error bars correspond to ± 1 SD. Dotted line corresponds to estimated frequency of responding cells ($\times 10^6$) obtained using all the data.

immediately clear from Figs. 2–4 that the error bars we have calculated understate the variability exhibited by the estimates. Roughly speaking, we would expect about 68% of these ±1 SD intervals to contain the true (but unknown) frequency, if they incorporated *all* sources of variability, and it seems evident that this is unlikely to be the case. In particular, the intervals given in the second assay for #713 at the three highest cell densities, and those for the antigens gD2 and gB2 at the four highest cell densities for #711, seem too small by perhaps a factor of 2. Despite these difficulties with the estimation of error, it is apparent that the use of these assays at a single carefully chosen density of cells will yield estimates of the frequency of responding cells with a coefficient of variation of the order of 20–25% or better. The longer error bars in the results from NIH 1394 (Fig. 4) reflect the smaller number of wells used for each antigen. The data from this assay were particularly difficult to analyse by the traditional

Table 3
Maximum likelihood estimates of CTL-precursor frequencies and plate parameters at each density for the data from Langhorne and Fischer-Lindahl (1981)

Cell density	λ	a	b	σ
0	0.0 (0.0)	5.9 (0.1)	0.0 (0.0)	0.4 (0.1)
100	0.1 (0.1)	5.7 (0.1)	1.9 (0.2)	0.4 (0.1)
300	0.5 (0.1)	6.0 (0.1)	2.2 (0.1)	0.4 (0.1)
500	1.1 (0.2)	6.3 (0.2)	2.1 (0.1)	0.5 (0.1)
750	1.7 (0.3)	6.5 (0.1)	1.8 (0.1)	0.4 (0.1)
1000	2.5 (0.3)	6.4 (0.3)	1.7 (0.1)	0.4 (0.1)
3000	3.4 (0.4)	8.2 (0.2)	1.4 (0.1)	0.3 (0.1)
10000	3.2 (0.4)	7.6 (0.5)	1.6 (0.2)	0.4 (0.1)

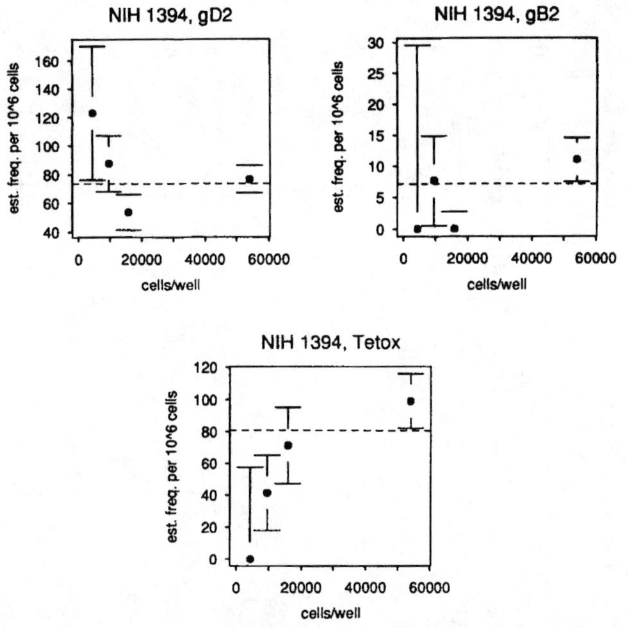

Fig. 4. Subject NIH 1394 (one four-point LDA). Maximum likelihood estimates of frequencies ($\times 10^6$) of responding cells using two plates at each dilution: estimates plotted against # cells/well. Error bars correspond to ±1 SD. Dotted line corresponds to estimated frequency of responding cells ($\times 10^6$) obtained using all the data.

K. Broman et al. / Journal of Immunological Methods 198 (1996) 119–132

Table 4
Maximum likelihood estimates of responder frequencies and plate parameters for each experimental group for the data from S. Rodda

Group	λ_c	λ_r	λ_i	a	b	σ
Post immunization	0.0 (0.0)	0.4 (0.1)	1.3 (0.2)	14.3 (0.2)	6.9 (0.2)	1.8 (0.2)
				13.9 (0.2)	4.8 (0.3)	1.5 (0.1)
1 : 1 mixture	0.0 (0.0)	0.2 (0.1)	2.9 (0.2)	16.0 (0.3)	7.1 (0.2)	2.1 (0.2)
				16.1 (0.2)	6.0 (0.2)	1.6 (0.1)
Preimmune	0.0 (0.0)	0.2 (0.1)	5.1 (0.3)	18.1 (0.4)	5.3 (0.2)	2.3 (0.3)
				19.1 (0.3)	8.2 (0.2)	2.5 (0.2)

cut-off method, as five out of 12 observed responses were all negative.

In order to re-analyse the data from Langhorne and Fischer-Lindahl (1981), it was first necessary to read the counts per minute ($\times 10^{-2}$) from their Fig. 2. In Table 3 we exhibit the MLEs of the CTL-pre-

Table 5
Maximum likelihood estimates of responder frequencies and plate parameters for the three LDAs from D. Koelle: DK2, KD and EL
(a)

# cells/well	λ_c	λ_a				
50000	1.1 (0.3)	15.1 (1.0)				
25000	0.0 (0.0)	9.3 (0.7)				
12500	0.3 (0.2)	5.4 (0.5)			DK2	
6250	0.2 (0.1)	2.4 (0.4)				
3125	1.4 (1.1)	3.5 (2.4)	Plates	a	b	σ
1563	0.0 (0.0)	1.6 (1.3)	1, 3	19.7 (0.6)	5.6 (0.2)	3.3 (0.4)
781	0.2 (0.6)	1.6 (1.2)	2, 4	20.1 (0.7)	1.7 (1.0)	3.5 (0.4)

(b)

# cells/well	λ_c	λ_a				
50000	0.2 (0.1)	16.3 (0.9)				
25000	0.1 (0.1)	11.8 (0.7)				
12500	0.0 (0.0)	8.4 (0.6)			KD	
6250	0.3 (0.1)	5.2 (0.5)				
3125	0.0 (0.0)	7.3 (1.6)	Plates	a	b	σ
1563	0.1 (0.2)	4.9 (1.1)	1, 3	17.9 (0.4)	7.7 (0.1)	2.8 (0.2)
781	0.5 (0.3)	2.5 (0.7)	2, 4	21.1 (0.7)	4.8 (0.9)	4.2 (0.4)

(c)

# cells/well	λ_c	λ_a				
50000	0.3 (0.2)	14.5 (0.9)				
25000	0.0 (0.0)	8.9 (0.7)				
12500	0.1 (0.1)	5.3 (0.5)			EL	
6250	0.0 (0.0)	2.1 (0.3)				
3125	0.4 (0.2)	2.3 (0.3)	Plates	a	b	σ
1563	0.5 (0.2)	1.7 (0.3)	1, 3	19.5 (0.5)	6.3 (0.2)	3.5 (0.3)
781	0.8 (0.2)	0.8 (0.2)	2, 4	17.4 (0.4)	4.6 (0.3)	1.8 (0.2)

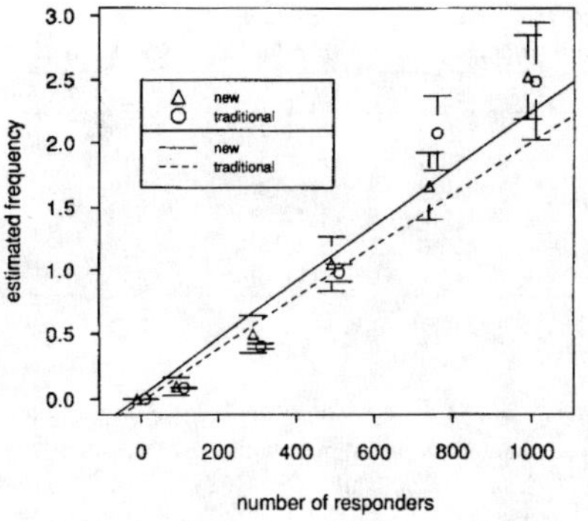

Fig. 5. Maximum likelihood estimates of frequencies of CTL-Ps plotted against number of responding cells.

cursor frequencies for each cell density, and also the associated estimates of plate-specific parameters. As before, the SDs reflect only within-plate variation. Fig. 5 shows these estimates plotted against the number of responders, where the estimates from the traditional analysis are also presented. Note that for these data, the untransformed scintillation counts were used, as indicated by a Box-Cox analysis.

The effectiveness of our analysis of the CTL-precursor assay data from Langhorne and Fischer-Lindahl (1981) is evident from Fig. 5. Not only do our estimates of the CTL-precursor frequencies give a slightly better linear relation than the traditional ones, based upon the first six frequencies, our estimates of b and σ were remarkably stable over this range, while the estimates of a increase with cell density. It is worth pointing out that we have analysed these

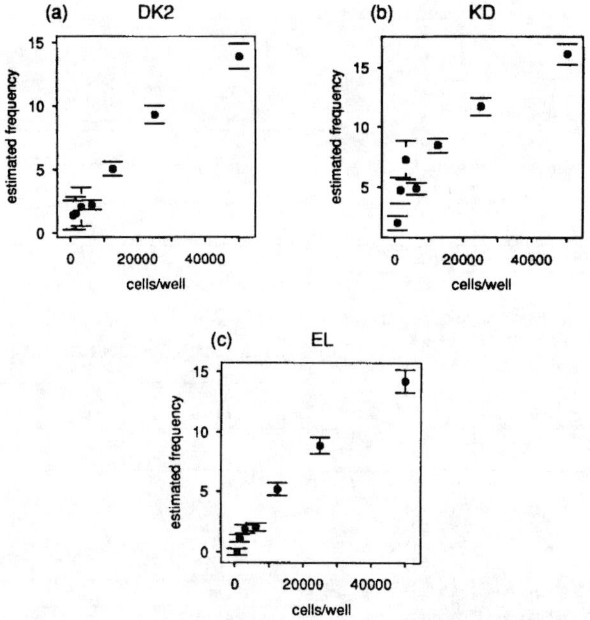

Fig. 6. Maximum likelihood estimates of frequencies of responders plotted against number of cells per well for the three LDAs from D. Koelle.

data as though the different sets of 25 counts at each precursor cell density were obtained from different microtiter plates, estimating a new a, b and σ for each set. It is not clear from the paper whether this was the case, but if not, then an analysis constraining some of the plate-specific parameters to be equal would be both more appropriate and more efficient.

Table 4 presents the results of our analysis of the data from S. Rodda. Here λ_c, λ_t and λ_i denote the estimated frequencies of responding cells per well for the cells alone, tetanus and influenza RNP groups, respectively. For these data, the square roots of the scintillation counts were used. Note that the estimated frequency for the 50:50 mixture is approximately the average of the estimated frequencies of its components, within the estimated error.

Table 5 presents the results of our analysis of the data from three LDAs from D. Koelle, while Fig. 6 gives a plot of the estimated frequencies of responding cells per well (above background) against cell density. For these data, the log scintillation counts were used. We notice that although there appears to be a reasonable linear relationship between estimated frequency of responders and number of cells per well, the estimated ratio of b to σ for two pairs of plates (DK2, #2,4; KD, #2,4) suggests difficulty fitting the model, while those for two other pairs (DK2, #1,3; EL, #1,3) are only marginally satisfactory.

4. Discussion

The main objective of our analysis was to obtain estimates of frequencies of responding cells based upon a single dilution. It is clear from the results in Figs. 2–4 that we can do this with a coefficient of variation of about 30% or lower, apart from a component of assay-to-assay variation which we discuss shortly. But before doing so, we recall the stated aim of our single dilution assay: it was intended to be a significantly more sensitive version of the standard proliferation assay, which obtains a 'stimulation index' as the ratio of the mean scintillation count from three test wells to the mean count from three wells with cells alone. Our assay was not intended to be a replacement for a standard LDA, whether analysed in the usual way by reducing the well counts to quantal responses, or under the model presented in this paper, although it does seem that our model and analysis will provide an alternative, possibly more efficient analysis of such LDAs, under certain circumstances. We have presented analyses of a variety of full LDAs simply to enable us to assess the extent to which our single dilution estimates can be relied upon.

Between assay variation is clearly an important issue, and its extent can be gauged from Figs. 2–4. Apparently it can be quite substantial, with the results for tetanus toxoid being of particular concern. It might therefore be thought essential that the SDs assigned to frequency estimates should incorporate a component of between-assay variation, so that we get a realistic impression of the true imprecision in these estimates. We feel that this topic is best studied within the context of a larger ongoing trial, in which some samples are routinely analysed in two or more different assays. For this reason we will not discuss the matter any further here, apart from noting that on the basis of the evidence presented here, it might not be unreasonable to double the estimated SDs of frequency estimates if between-assay variation is to be incorporated.

The method of analysis we have presented in this paper highlights certain issues relating to the design of assays of the kind we discuss. The most important concerns the relation of the negative control wells, our cells alone, to wells containing antigen. We have examined and analysed a number of assays in which the negative control wells and the wells containing antigen were located on different microtiter plates. Although this practice is not necessarily injudicious, there are at least three reasons why it should be avoided, particularly if our method of analysis is to be used, but even more generally. As is made explicit by our estimation of the plate-specific parameters a, b and σ, the responses of cells from the same source can differ from plate to plate, even in well-conducted assays. It is good general practice to carry out comparative analyses with the greatest possible degree of control over the conditions which could cause differences, in this case, between negative control and antigen wells, for the responses in wells with antigen will be adjusted to the extent that the negative control wells respond. If it is not possible to make such intra-plate adjustments, it becomes necessary to assume that the plate containing the negative control wells has the same parameters as that containing the wells with antigen, for otherwise the two classes of wells will not be treated similarly in their analysis.

A few comments on the interpretation of our parameters and the range of reasonable values for them are in order. The parameter a is most easily interpreted as the average or median value of the counts or transformed counts observed in the wells in the cells alone category. Of course it will also have the same relation to the counts from those wells with antigen in which there are no responding cells, but we will not usually know definitely which these are, whereas we can generally be confident that the overwhelming majority of counts from wells in the cells alone category are simply background. The parameter b is the slope of the line relating average count or transformed count to frequency of responding cells within a plate. Finally, the parameter σ corresponds to the spread of counts or transformed counts about their mean, for wells with the same number of responding cells. Under our normality assumption, about 95% of such counts would be within 2σ of the mean, which has the form $a + kb$. Thus satisfactory discrimination between adjacent values of k is only possible when σ is $b/2$ or smaller, and is really good if $\sigma \leq b/4$. Our algorithm can converge and give reasonable frequency estimates with values of σ as large as b, or even larger than b. However, we are inclined to regard such situations as failures of the model to fit the data, and include them only when there is no other way to get a frequency estimate, and a rough one is desired.

Finally, we remark that our model and analysis, originally designed for the particular assay described in Section 2, does appear to have a wider usefulness. With only relatively minor adaptations, it could be applied to the CTL-precursor assay of Langhorne and Fischer-Lindahl (1981), as well as to data from LDAs (D. Koelle), and the single density assay of S. Rodda, all of which had a structure quite different from that of the original assays. In each case analysis gave satisfactory results, and in a way which avoided the arbitrary choice of cut-offs. Assuming that the model we used is appropriate, our analysis will also be more efficient.

5. Note added in proof

A computer program that incorporates the method described here has been written for the Windows and Macintosh platforms.

Acknowledgements

Many thanks are due to David Freedman and Peter Hall for their helpful comments on the material in this paper. We also wish to thank Mario Geysen and Stuart Rodda for the initial application of the single cell density and for many helpful discussions in the early development of the frequency analysis.

References

Barlow, R.E., Bartholomew, D.J., Bremner, J.M. and Brunk, H.D. (1972) Statistical Inference under Order Restrictions; The Theory and Application of Isotonic Regression. Wiley, New York.

Box, G.E.P. and Cox D.R. (1964) The analysis of transformations. J.R. Stat. Soc. Ser. B 26, 211.

Broman, K.W., Speed, T.P. and Tigges, M. (1996) Estimation of antigen-responsive T-cell frequencies in PBMC from human subjects. Technical Report #454. Department of Statistics, University of California, Berkeley, CA.

Dempster, A., Laird, N. and Rubin, D. (1977) Maximum likelihood estimation from incomplete data via the EM algorithm. J.R. Stat. Soc. Ser. B 39, 1.

Finney, D.J. (1978) Statistical Method in Biological Assay. Academic Press, New York.

James, S.P. (1991) Measurement of basic immunologic characteristics of human mononuclear cells. In: J.E. Coligan, A.M. Kruisbeek, D.H. Margulies, E.M. Shevech and W. Stroger (Eds.), Current Protocols in Immunology. Green Publishing Company and Wiley-Interscience, New York, section 7.10.

Kost, R.G., Hill, E.L., Tigges, M. and Straus, S.E. (1993) Brief report: Recurrent acyclovir-resistant genital herpes in an immunocompetent patient. New Engl. J. Med. 329, 1777.

Langenberg, A.G.M., Burke, R.L., Adair, S.F., Sekulovich, R., Tigges, M., Dekker, C.L. and Corey, L. (1995) A recombinant glycoprotein vaccine for herpes simplex type 2: Safety and efficacy. Ann. Intern. Med. 122, 889.

Langhorne, J. and Fischer-Lindahl, K. (1981) Limiting dilution analysis of precursors of cytotoxic T-lymphocytes. In: I. Lefkovits and B. Pernis (Eds.), Immunological Methods, Vol. II, Academic Press, New York, section 12.

Meng, X.-L. and Rubin, D.B. (1991) Using EM to obtain asymptotic variance-covariance matrices: the SEM algorithm. J. Am. Stat. Assoc. 86, 899.

Parr, D., Savarese, B., Burke, R.L., Margolis, D., Meier, J., Markoff, L., Ashley, R., Corey, L., Adair, S., Dekker, C. and Straus, S.E. (1991) Ability of a recombinant herpes simplex virus type 2 glycoprotein D vaccine to induce antibody titers comparable to those following genital herpes. Clin. Res. 39, 216A.

Reece, J.C., McGregor, D.L., Geysen, H.M. and Rodda, S.J. (1994a) Scanning for T helper epitopes with human PBMC using pools of short synthetic peptides. J. Immunol. Methods 172, 241.

Reece, J.C., Geysen, H.M. and Rodda, S.J. (1994b) Mapping the major human T helper epitopes of tetanus toxoid. The emerging picture. J. Immunol. 151, 6175.

Straus, S.E., Savarese, B., Tigges, M., Freifeld, A.G., Krause, P.R., Margolis, D.M., Meier, J.L., Paar, D.P., Adair, S.F., Dina, D., Dekker, C. and Burke, R.L. (1993) Induction and enhancement of immune responses to herpes simplex virus type 2 in humans by use of a recombinant glycoprotein D vaccine. J. Infect. Dis. 167, 1045.

Comment

Terry Speed

Geoff Robinson is to be congratulated for writing this paper. It is lucidly written, it bridges a number of gulfs that have developed in our subject, and it is provocative. That he wrote it is clearly a Good Thing! I welcome the opportunity to say this and to make a few remarks that he might have made. I believe that these remarks will strengthen his already strong case for a much more explicit recognition of the role of BLUPs in our subject.

1. THE BAYESIAN DERIVATION

In Section 4.2 Robinson describes a Bayesian derivation, stating that the posterior mode is given by the BLUP estimates when β is regarded "as a parameter with a uniform, improper prior distribution and u as a parameter which has a prior distribution which has mean zero and variance $G\sigma^2$, independent of β." All this is certainly true, but it may be helpful to add that if β is given a proper prior (normal) distribution with mean zero and variance $B\sigma^2$, say, with u as before, then all of the results one could possibly want (posterior means, posterior variances, etc.) can be derived straightforwardly by the standard Bayesian formulae. Then all one has to do to derive the corresponding BLUP formulae is let $B^{-1} \to 0$. An identity which I have found useful, perhaps even indispensible, for carrying out this last step, is discussed in de Hoog, Speed and Williams (1990). Note that the approach just described is essentially that adopted in Dempster, Rubin and Tsutakawa (1981).

2. FORMULAE FOR $\hat{u}$

The only actual formulae given in the paper for $\hat{u}$ in the general case is the rather complicated one in Section 4.3. This is a pity, because there is an obvious "plug-in" expression, namely

$$\hat{u} = GZ^TV^{-1}(y - X\hat{\beta}), \tag{1}$$

where $V = ZGZ^T + R$. This may be viewed as the result of regressing u on y, with the mean $X\beta$ of y replaced by its obvious linear estimator.

A variant of (1) is

$$\hat{u} = (Z^TR^{-1}Z + G^{-1})^{-1}Z^TR^{-1}(y - X\hat{\beta}). \tag{1'}$$

Terry Speed is Professor, Department of Statistics, University of California, Berkeley, California 94720.

The simpler formulae (5.3) and (5.4) arising when there are no fixed effects also have more general analogues, namely

$$(Z^TAZ + G^{-1})\hat{u} = Z^TAy, \tag{2}$$

where $A = R^{-1}(I - S)$, $S = P^R_{\mathscr{R}(X)}$ being the projector onto $\mathscr{R}(X)$ orthogonal with respect to $\langle a, b\rangle = a^TR^{-1}b$, and for the variance-covariance matrix of $\hat{u}$:

$$\left\{G^{-1} - (Z^TAZ + G^{-1})^{-1}\right\}\sigma^2. \tag{3}$$

These expressions can be derived readily using the Bayesian approach outlined in (1) above, together with the matrix identity already referred to. I note in passing that Robinson's formulae (5.4) is in fact the variance-covariance matrix of $\hat{u} - u$, not, as stated, of $\hat{u}$.

3. SOLVING THE BLUP EQUATIONS

Perhaps in order to avoid messy algebra, Robinson has said little about the actual solution of the BLUP equations. I know that he has worked on this problem with some enormous data sets, and so I am hesitant to comment here. However, it does seem worthwhile to make one easy point, in order to connect this topic with another, closely related one. The obvious rearrangement of the first equation in (1.2),

$$X^TR^{-1}X\hat{\beta} = X^TR^{-1}(y - Z\hat{u}), \tag{4}$$

can be combined with either (1) or (1′) above, to form the basis of an iterative solution of the BLUP equations, provided, of course, that the separate problems are readily solved. Just such a strategy is recommended more generally in Green (1985) in the context of smoothing, a topic to which I shall return.

It is also worth pointing out that (1′) or (2) is to be preferred when G^{-1} has simple structure, whereas if G is simple and V is readily inverted, (1) is more useful. In many animal breeding problems it is G^{-1} which has the simpler structure, as it also does in the Kalman filter case.

4. REML AND BLUP

In Section 5.4 Robinson states that "REML is the method of estimating variance components that seems to have the best credentials from a Classical viewpoint." What he does not say, which should be of interest to readers of his paper, is that REML and BLUP are intimately connected. Indeed one view—certainly not the only one—of the REML equations for variance components is that they are simply equating *observed* with *expected* sums of squares of BLUPs. This observation goes back to the original paper by Patterson and Thompson (1971; see also Harville, 1977) and can be concisely stated within Robinson's framework as follows.

Suppose that $Z = [Z_1 \vdots \cdots \vdots Z_c]$ is blocked, corresponding to c random effects, with Z_i being $n \times q_i$, $i = 1, \ldots, c$, $u = (u_1 \vdots \cdots \vdots u_c)$ is similarly blocked into q sets of random effects, and finally $G = \text{diag}(G_1, \ldots, G_c)$ is diagonally blocked with $G_i = \gamma_i I_{q_i}$, where $\sigma_i^2 = \gamma_i \sigma^2$ is the variance of each independent component of the ith random effect u_i. It is also convenient to denote e by u_0, put $Z_0 = I_n$ and $\gamma_0 = 1$.

With this notation the REML equations take the form

$$(5) \quad y^T\left\{V^{-1}\overline{Q}\frac{\partial V}{\partial \sigma_i^2}V^{-1}\overline{Q}\right\}y = \text{tr}\left\{\frac{\partial V}{\partial \sigma_i^2}V^{-1}\overline{Q}\right\}$$

$i = 0, \ldots, c$, where $Qy = X\hat{\beta}$ and $\overline{Q} = I - Q$. (By contrast, the ML equations have no $\overline{Q}$ term in the right-hand expression.)

Turning now to BLUPs in this context, they are (in the form (1) above)

$$(6) \quad \hat{u}_i = \gamma_i Z_i^T V^{-1}\overline{Q}y$$

$i = 0, \ldots, c$, and

$$(7a) \quad \text{var}(\hat{u}_i) = (G_i - U_i)\sigma^2$$

$i = 1, \ldots, c$, where U_i is the ith diagonal block of the matrix $(Z^TAZ + G^{-1})^{-1}$. Furthermore,

$$(7b) \quad \text{var}(\hat{e}) = V^{-1}\overline{Q}\sigma^2 = (A - AZUZ^TA)\sigma^2$$

where $A = R^{-1}\overline{S}$ was defined earlier, and $U = (Z^TAZ + G^{-1})^{-1}$. If we write $p_i = \gamma_i^{-1}\text{tr}(U_i)$, $i = 1, \ldots, c$, then it follows that for $i = 1, \ldots, c$

$$(8a) \quad \mathbb{E}\,|\,\hat{u}_i\,|^2 = (q_i - p_i)\sigma_i^2$$

and

$$(8b) \quad \mathbb{E}\,|\,\hat{e}\,|^2 = \left[(n - p) - \sum_1^c (q_i - p_i)\right]\sigma^2.$$

Now the striking thing is this: the REML equations (5) can rather easily be manipulated into a form just like (8a) and (8b), with the expectation symbol $\mathbb{E}$ omitted. Although this is not necessarily the best way to solve these equations, the repeated calculation of BLUPs and then updating the variance components is one simple iterative scheme which works quite well.

5. PENALIZED LEAST SQUARES

Suppose that we regard (1.1) as an ordinary ("fixed effects") linear model, and that we wished to estimate β and u by R-weighted least squares with a "penalty" $u^TG^{-1}u$ being added to the sum of squares term being minimized. Then we would obtain just the expression given in Section 4.1, which Henderson minimized. Such penalties are added for many reasons: to smooth, to improve the condition of the matrix to be inverted, and so on, and it has long been recognized that this is a way of making one's linear model "quasi-Bayesian." More precisely, it turns the standard least squares problem into a case of BLUP. This practice has a long history, dating back at least to Whittaker (1923).

6. SMOOTHING SPLINES ARE BLUPS

Continuing with the theme of the previous remark, let us see how the smoothing splines popularized by G. Wahba (see her 1990 monograph for a comprehensive exposition) are just BLUPs. This observation corrects the terminology which has been used in the spline literature for over a decade, for the Bayesian interpretation of the smoothing spline—with a partially improper prior—is just the statement heading this section.

It is simplest to deal with cubic smoothing splines on the interval $[0, 1]$. If the observations are taken at $0 \le t_1 < \cdots < t_n \le 1$, and are

$$y_i = g(t_i) + \epsilon_i,$$

$i = 1, \ldots, n$, where g is an unknown smooth function, then the function g_λ which minimizes

$$n^{-1}\sum_1^n (y_i - g(t_i))^2 + \lambda\int_0^1 \{g''(u)\}^2\,du$$

over a suitable class $G_0 \oplus G_1$ of functions, has the values

$$\hat{y} = (g_\lambda(t_i))$$

$$(9) \quad = X(X^TV^{-1}X)^{-1}X^TV^{-1}y + Q_nV^{-1}\left(I - X(X^TV^{-1}X)^{-1}X^TV^{-1}\right)y,$$

where

$$X = (t_i^{k-1}), \qquad i = 1, \ldots, n;\ k = 1, 2;$$

$$Q_n(i, j) = Q(t_i, t_j), \qquad 1 \le i, j \le n;$$

and

$$Q(s, t) = \int_0^1 (s - w)_+(t - w)_+\,dw, 0 \le s, t \le 1.$$

It is easy to check that (9) is just the fitted value

$$\hat{y} = X\hat{\beta} + Z\hat{u},$$

where, in Robinson's notation, $\hat{\beta}$ and $\hat{u}$ are the BLUPs, X is as given above, $Z = I_n = R$, and $G = (n\lambda)^{-1}Q_n$.

Certainly there is more to smoothing splines than BLUPs; for example, estimates of the value of the function g at values of t other than those observed, but in many applications (9) and related expressions are all that is required.

By now it should come as no surprise to hear that the technique termed generalized maximum likelihood (GML) for estimating the smoothing parameter λ is no other than REML in this BLUP problem. This is readily checked by comparing formulae in this paper with ones in Wahba (1990).

With only very few changes, the identification just made to show that smoothing splines are BLUPs shows that the model robust response surface designs of Steinberg (1985) are also BLUPs. In this case the u term corresponds to sums of tensor products of orthogonal polynomials.

7. LINEAR SMOOTHERS ARE ALMOST BLUPS

There is a sense in which all linear smoothers (see Buja, Hastie and Tibshirani, 1989) are intimately related to BLUPs. A typical linear smoother S satisfies $S^n \to T$ as $n \to \infty$, where T is idempotent. This corresponds to a projector onto the subspace $\mathscr{R}(X)$ in Robinson's model (1.1), and so Ty corresponds to $X\hat{\beta}$. Thus $(S - T)y$ corresponds to $Z\hat{u}$, and in some situations it is even possible to construct a covariance matrix V such that this correspondence is precise. Furthermore, many smoothers S have form $S(\lambda)$, where λ is a parameter (bandwidth, variance ratio, smoothness penalty, etc.) that defines a family of similar smoothers. In such cases $S(\infty)$ often has the form $T + W$, where W is another projector, while $S(0) = T$. Many of the problems and the formulae in the theory of linear smoothers are analogues of ones arising in the theory of BLUPs.

8. INTERVAL ESTIMATES INVOLVING BLUPS

In Section 5.6 Robinson briefly alludes to work done on estimating the precision of BLUP estimates when uncertainty in the dispersion parameter is taken into account. This general problem, and in particular the assignment of interval estimates, has attracted a lot of attention in the literature on smoothing splines (see, e.g., Nychka, 1988, for a recent review). Much concern has been given to the question of what, if any, coverage properties can be expected of a "Bayesian" posterior interval. Making interval statements about an object which is an estimate of the sum of fixed and random effects is bound to cause problems of interpretation to many people, and I would be interested to hear Geoff Robinson's comments on this point. I know that he has studied these matters closely in the past.

9. SUMMARY

In closing these few remarks, I cannot resist paraphrasing I. J. Good's memorable aphorism: "To a Bayesian, all things are Bayesian." How does "To a non-Bayesian, all things are BLUPs" sound as a summary of this fine paper?

ACKNOWLEDGMENT

The author gratefully acknowledges partial support from NSF Grant DMS-88-02378, and the hospitality of the Department of Statistics, IAS, Australian National University whilst preparing this comment.

Iterative Proportional Fitting

Iterative proportional fitting (IPF), also known as iterative proportional scaling, is an **algorithm** for constructing tables of numbers satisfying certain constraints. In its simplest form, the algorithm enables one to construct two-way **contingency tables** with specified marginal totals and a prescribed degree of association; from a more general perspective, it may be viewed as a cyclic ascent algorithm which maximizes a specific objective function. The algorithm can also be used to construct **maximum likelihood** estimators for table entries based upon hierarchical **log-linear models** for **Poisson**, **multinomial**, or product multinomial models. We will illustrate these aspects of the algorithm and its applications by describing some simple cases.

Suppose that we are given two pairs $\mathbf{u} = (u_1, u_2)$ and $\mathbf{v} = (v_1, v_2)$ of positive numbers satisfying $u_1 + u_2 = v_1 + v_2$, and a further positive number ψ. The IPF algorithm will enable us to construct the *unique* **two-by-two table** $\mathbf{b} = (b_{ij})$ such that, for all i and j,

$$b_{i+} = u_i, \qquad b_{+j} = v_j, \qquad \frac{b_{11}b_{22}}{b_{12}b_{21}} = \psi,$$

where the subscript + denotes the result of summing over the subscript it replaces. The algorithm goes like this. Begin with the 2×2 table $\mathbf{a} = (a_{ij})$ defined by $a_{11} = \psi$, $a_{12} = a_{21} = a_{22} = 1$, noting that the cross-ratio $a_{11}a_{22}/a_{12}a_{21} = \psi$ (*see* **Odds Ratio**). Next, scale the rows of $\mathbf{a}$ to form the table $\mathbf{a}' = (a'_{ij})$:

$$a'_{ij} = a_{ij} \times \frac{u_i}{a_{i+}}, \tag{1}$$

for $i = 1, 2$ and $j = 1, 2$. It is easy to check that $\mathbf{a}'$ has the desired row sums, as well as having cross-ratio ψ. We now scale the columns of $\mathbf{a}'$ to form the table $\mathbf{a}'' = (a''_{ij})$:

$$a''_{ij} = a'_{ij} \times \frac{v_j}{a'_{+j}}. \tag{2}$$

One can check that $\mathbf{a}''$ has the desired column sums and cross-ratio, although the row sums are no longer (u_i). This completes one cycle of the IPF algorithm, beginning with the table $\mathbf{a}$.

The algorithm continues by repeatedly scaling the rows, as in (1), and then the columns, as in (2), to have the desired totals. After a number of cycles, the row totals are closer to (u_i) than they were initially, the column totals are exactly (v_j), and the cross-ratio is exactly ψ. The sequence of tables so defined converges pointwise to a 2×2 table $\mathbf{b}$ with all the desired properties; uniqueness also follows.

It is instructive to examine why these assertions are true, for in doing so we obtain further insights into the IPF algorithm. To do this, we introduce the notion of **information** (or I-) divergence between two tables $\mathbf{c} = (c_{ij})$ and $\mathbf{d} = (d_{ij})$, satisfying $c_{++} = d_{++}$, defined as follows:

$$I(\mathbf{c}|\mathbf{d}) = \sum_{ij} c_{ij} \log\left(\frac{c_{ij}}{d_{ij}}\right).$$

(A similar definition applies to singly indexed arrays.) It can be proved that $I(\mathbf{c}|\mathbf{d}) \geq 0$, and that $I(\mathbf{c}|\mathbf{d}) = 0$ if and only if $\mathbf{c} = \mathbf{d}$. Although not a symmetric function of its arguments, I behaves in many ways like a metric on tables, and it provides the basis of a proof of convergence of the IPF algorithm. We return to our construction of a table $\mathbf{b}$ having row totals $\mathbf{u}$, column totals $\mathbf{v}$, and cross-ratio ψ. First define the table $\mathbf{c} = (c_{ij})$ as follows:

$$c_{ij} = \frac{u_i v_j}{w},$$

where $w = u_+ = v_+$. The tables $\mathbf{a}, \mathbf{a}', \mathbf{a}'', \ldots$ become closer to $\mathbf{c}$ as the iterations continue, closeness here being in the sense of I-divergence. More precisely, we can check that

$$I(\mathbf{c}|\mathbf{a}) = I(\mathbf{c}|\mathbf{a}'') + I(\mathbf{v}|\mathbf{a}'_2) + I(\mathbf{u}|\mathbf{a}_1), \tag{3}$$

where $\mathbf{a}_1 = (a_{i+})$ and $\mathbf{a}'_2 = (a'_{+j})$. The convergence and uniqueness assertions above all follow from repeated use of this expansion and the stated properties of I. As long as there exists at least one table $\mathbf{c}$ with the desired marginal totals, we can begin the IPF algorithm with any table having the desired cross-ratio, and expect to converge to the stated limit. The repeated scaling gives tables closer and closer in the sense of I-divergence to the table $\mathbf{c}$, all the while retaining the original cross-ratio, and the row and column totals converge to their desired values.

All of the discussion so far applies with minimal changes to $r \times s$ tables; in the more general case, there are further cross-ratios to take into account.

Encyclopedia of Biostatistics, Online © 2005 John Wiley & Sons, Ltd.
This article is © 2005 John Wiley & Sons, Ltd.
This article was published in the *Encyclopedia of Biostatistics* in 2005 by John Wiley & Sons, Ltd.
DOI: 10.1002/0470011815.b2a10027

Iterative Proportional Fitting

Whereas in a 2×2 table there is only one cross-ratio whose value can be fixed, in an $r \times s$ table, there are $(r-1)(s-1)$ multiplicatively independent cross-ratios. A convenient set (cf. [12]) is the following:

$$\psi_{ij} = \frac{b_{ij} b_{rs}}{b_{is} b_{rj}}, \quad i = 1, \ldots, r-1; j = 1, \ldots, s-1.$$

Here we constructed our cross-ratios in relation to the index values r and s. Other choices give equivalent results; indeed there are quite different ways of defining the quantities which are preserved. This issue is addressed in the theory of **loglinear models**; see [2, 11] and [12]. Given an arbitrary set of $(r-1)(s-1)$ positive numbers (ψ_{ij}), and positive numbers $\mathbf{u} = (u_i)$ and $\mathbf{v} = (v_j)$ satisfying $u_+ = v_+$, the IPF algorithm may be initiated with the table $\mathbf{a} = (a_{ij})$ given by $a_{ij} = \psi_{ij}, i = 1, \ldots, r-1; j = 1, \ldots, s-1$, and $a_{rj} = 1 = a_{is}, i = 1, \ldots, r; j = 1, \ldots, s$. With this initial table, the steps are just as before, and the resulting sequence of tables converges to the unique table having row totals (u_i), column totals (v_j), and cross-ratios (ψ_{ij}).

We turn now to reasons for constructing such tables. One is simply to demonstrate the fact that the row totals, column totals, and cross-ratios of two-way tables may be specified independently, and to show how to obtain tables with arbitrarily specified (but consistent) values of these quantities. Historically, the algorithm was first used to adjust sample frequencies to expected marginal totals. In the examples in Deming [5], we have a table $\mathbf{n} = (n_{ij})$ based upon a **sample survey**, and marginal totals (N_{i+}) and (N_{-j}), but *not* the individual cell frequencies $\mathbf{N} = (N_{ij})$, from a census of the population. The result of applying the IPF algorithm with initial table $\mathbf{n}$, and desired marginal totals (N_{i+}) and (N_{+j}), can then be regarded as an estimate of what would have been obtained by cross-tabulating the entire population, instead of only a sample thereof. A modern treatment of these ideas can be found in [2], where the procedure is known as *raking* the table $\mathbf{n}$. The third application of the algorithm we note is to the construction of maximum likelihood estimates of table entries under loglinear models. We simply describe the results here; the reader may consult standard references such as [2, 11], or [1] for fuller details. Suppose that $\mathbf{n} = (n_{ij})$ is a two-way table of independent Poisson counts with parameters $\boldsymbol{\lambda} = (\lambda_{ij})$. Then the maximum likelihood estimate $\hat{\boldsymbol{\lambda}}$ of $\boldsymbol{\lambda}$ under the *multiplicative model* for the (λ_{ij}), has the same row and column totals as $\mathbf{n}$, and all cross-ratios equal to 1. In this case, the IPF algorithm begins with a table all of whose entries are 1, and scales the row and column totals to match those of the data $\mathbf{n}$. The algorithm converges after a single cycle to the unique maximum likelihood estimator $\hat{\boldsymbol{\lambda}}$.

Three- and Higher-Way Tables

There are a number of ways in which the IPF algorithm may be used with three-way tables. We illustrate two of these. Suppose that we have an $r \times s$ table $\mathbf{u} = (u_{ij})$ and an $s \times t$ table $\mathbf{v} = (v_{jk})$ of positive numbers satisfying $u_{+j} = v_{j+}$ for $j = 1, \ldots, s$. By analogy with our earlier construction, we might be interested in obtaining an $r \times s \times t$ table $\mathbf{b} = (b_{ijk})$ having

$$b_{ij+} = u_{ij}, \qquad b_{+jk} = v_{jk}.$$

This can be solved rather straightforwardly. For example, the table $\mathbf{c} = (c_{ijk})$ given by

$$c_{ijk} = \frac{u_{ij} v_{jk}}{w_j},$$

where $w_j = u_{+j} = v_{j+}, j = 1, \ldots, s,$ is readily checked to have ij-margin $\mathbf{u}$ and jk-margin $\mathbf{v}$.

Of course, this is not the end of the story. We may also be interested in any further structure concerning the table $\mathbf{b}$ which may be specified, in addition to these marginal totals. It turns out that we may also ask that the table has predetermined values of certain cross-ratios. In this example, and more generally, we need rules to tell us which marginal totals and which cross-ratios can be specified independently. The issue is best discussed in the language of **hierarchical** loglinear models for multiway tables, where these are commonly described in terms of the marginal subtables which constitute the **sufficient statistics** for the models (under either independent Poisson, multinomial, or independent multinomial sampling). We refer to [1, 2], and [11] for details concerning these models. In this language, the cross-ratios that we have been specifying are the antilogarithms of elements of subspaces **orthogonal** to those that define the hierarchical loglinear model corresponding to the specified marginal totals. For example, by specifying margins corresponding to the indices

Encyclopedia of Biostatistics, Online © 2005 John Wiley & Sons, Ltd.
This article is © 2005 John Wiley & Sons, Ltd.
This article was published in the *Encyclopedia of Biostatistics* in 2005 by John Wiley & Sons, Ltd.
DOI: 10.1002/0470011815.b2a10027

ij and jk, as we did in our example, we are also able to specify independently cross-ratios corresponding to the pair ik and the triple ijk – that is, all interactions other than those involved in the loglinear model defined by the prescribed marginal totals.

Now let us suppose that, in addition to **u** and **v** as above, we are given a $t \times r$ table $\mathbf{w} = (w_{ki})$ of positive numbers satisfying $w_{k+} = v_{+k}$ and $w_{+i} = u_{i+}$ for all k and i. Can we use IPF to construct a table $\mathbf{b} = (b_{ijk})$ satisfying

$$b_{ij+} = u_{ij}, \quad b_{+jk} = v_{jk}, \qquad b_{i+k} = w_{ki},$$

and having prescribed values for the ijk cross-ratios? One might think that this would be quite straightforward. Begin with a suitable initial table **a**. Then scale to achieve the ij, jk, and ki marginal totals **u**, **v**, and **w**, respectively. One cycle of the algorithm would be three such scalings, and after a few cycles, we might expect to have a table with the specified cross-ratios, and essentially the desired marginal totals.

How can this version of IPF go wrong? A clue is provided by our indication of the method used to prove that IPF converges. We made use of the existence of a table **c** satisfying the marginal constraints, and then everything followed. However, in the case of three-way tables, it is not hard to specify three consistent, positive two-way tables, for which *no* three-way table exists having positive entries, and the three specified tables as two-way marginal totals. A simple example is given by three 2×2 tables each having 1 in the diagonal cells and 2 in the off-diagonal cells. Although they are clearly consistent, it is easy to check that no $2 \times 2 \times 2$ table can exist with positive entries and these margins. Use of the IPF algorithm with an initial table whose entries are all 1, and these three marginal tables, results in a cycle through the same three tables. The tables constructed do not converge. Summarizing this discussion, we can say that only if there exists a three-way table with the given two-way tables as marginal totals is the IPF algorithm guaranteed to converge to a limiting table with the desired marginal totals and three-way cross-ratios. When it does, this table is uniquely specified by these properties.

We note that in the application of this result to maximum likelihood estimation with loglinear models, the assumption of the existence of *some* table with the given marginal totals is trivially satisfied as long as the observed table $\mathbf{n} = (n_{ijk})$ has positive entries, for in this case **n** itself suffices. If the observed table has some zero entries, but positive two-way marginal totals, the IPF algorithm still converges, but to a table with some zero entries. In a sense, this is an extended maximum likelihood estimator: one on the boundary of the natural parameter space.

The foregoing discussion applies without change to higher-way tables. For example, suppose that we have an initial four-way table $\mathbf{a} = (a_{ijkl})$, and we wish to scale it to have prescribed ij, jk, kl, and li marginal totals. What cross-ratios (equivalently, what loglinear structure) of this initial table will be preserved throughout the iterations, and could therefore be specified independently of the marginal totals? The answer is: all interactions other than those involved in the loglinear model defined by the prescribed marginal totals, that is, the $ik, jl, ijk, ijl, ikl, jkl$, and $ijkl$ interactions. Note that we still need to know that there exists a table with the specified marginal totals before the algorithm is guaranteed to converge to a limiting table with all the desired properties.

Finite Termination: Decomposable Models

Decomposable models are a class of loglinear models for complete multiway tables which possess closed-form expressions for their MLEs under the standard sampling models; see [11] and [2]. It turns out that the IPF algorithm behaves rather well for this class of models. Suppose that a set of marginal totals to be fitted via IPF defines a decomposable loglinear model. If the initial table is constant, and the margins to be fitted are taken in a suitable order, the algorithm converges after just one cycle. Furthermore, there *always* exists a table with the given set of tables as marginal subtables, when the corresponding model is decomposable. Finally, as long as the specified tables are all positive, the table whose existence has just been described has positive entries.

History

Fienberg [7] presents a discussion of the history of the IPF algorithm. Some additional references can be found in [8]. The most important early papers are [6] and [14].

This article was published in the *Encyclopedia of Biostatistics* in 2005 by John Wiley & Sons, Ltd.
DOI: 10.1002/0470011815.b2a10027

Numerical Aspects

Haberman [11] proves that tables constructed by the IPF algorithm converge to their limit at a geometric (also called first-order) rate. This means that, asymptotically, the difference between the nth iterate and the limit is bounded above by ρ^n for some ρ between zero and unity. (This compares unfavorably with the behavior of Newton or modified Newton algorithms, which typically exhibit what is known as quadratic convergence.) In many cases, ρ may be quite close to unity, and so convergence may be rather slow, giving rise to a literature concerning speeding up of the algorithm. However, at that point, the algorithm ceases to be the one we are discussing.

The great advantage of the IPF algorithm is its simplicity, stability, and economy of space. When a table is large, and the number of iterations is not a limiting factor, it is the method of choice for the problems we have discussed. For other problems, such as the calculation of MLEs under loglinear models, Newton-type methods are preferred, because of their speed of convergence and the fact that variance–**covariance matrices** are an automatic byproduct. FORTRAN IV versions of the IPF algorithm can be found in [9] and [10].

Variants and Generalizations

It is implicit in the foregoing discussion that the tables being considered are all *complete*, that is, are fully rectangular, or rectangular parallelepipeds, etc., and have no so-called **structural zeros**. This was because the algorithm is mostly used, and its properties are most easily discussed, in that context. However, variant forms of the algorithm are used successfully with tables having a variety of other structures, and preserving features corresponding to models other than hierarchical loglinear models; see [11].

For generalizations of a different kind, see [3, 4, 13]. In these papers, applications of the algorithm beyond contingency tables are given, and its connections to the information measure I and entropy are more fully explored.

References

[1] Agresti, A. (1990). *Categorical Data Analysis*. Wiley, New York.

[2] Bishop, Y.M.M., Fienberg, S.E. & Holland, P.W. (1975). *Discrete Multivariate Analysis*. MIT Press, Cambridge, Mass.

[3] Csiszar, I. (1975). I-divergence geometry of probability distributions and minimization problems, *Annals of Probability* **3**, 146–158.

[4] Darroch, J.N. & Ratcliff, D. (1972). Generalized iterative scaling for loglinear models, *Annals of Mathematical Statistics* **43**, 1470–1480.

[5] Deming, W.E. (1964). *Statistical Adjustment of Data*. Dover, New York.

[6] Deming, W.E. & Stephan, F.F. (1940). On a least squares adjustment of a sampled frequency table when the expected marginal totals are known, *Annals of Mathematical Statistics* **11**, 427–444.

[7] Fienberg, S.E. (1970). An iterative procedure for estimation in contingency tables, *Annals of Mathematical Statistics* **41**, 907–917.

[8] Fienberg, S.E. & Meyer, M.M. (1983). *Encyclopedia of Statistical Sciences*, Vol. 4, S. Kotz & N.L. Johnson, eds. Wiley, New York, p. 2275.

[9] Haberman, S.J. (1972). Loglinear fit for contingency tables, *Applied Statistics* **21**, 218–225.

[10] Haberman, S.J. (1973). Printing multidimensional tables, *Applied Statistics* **22**, 118–126.

[11] Haberman, S.J. (1974). *The Analysis of Frequency Data*. University of Chicago Press, Chicago.

[12] Plackett, R.L. (1981). *The Analysis of Categorical Data*, 2nd Ed. Griffin, London.

[13] Ruschendorf, L. (1995). Convergence of the iterative proportional fitting procedure, *Annals of Statistics* **23**, 1160–1174.

[14] Stephan, F.F. (1942). An iterative method of adjusting sample frequency tables when the expected marginal totals are known, *Annals of Mathematical Statistics* **13**, 166–178.

(*See also* **Categorical Data Analysis**)

TERRY P. SPEED

This article was published in the *Encyclopedia of Biostatistics* in 2005 by John Wiley & Sons, Ltd.
DOI: 10.1002/0470011815.b2a10027

Chapter 9
History and Teaching Statistics

Deborah Nolan

When Terry Speed arrived in Berkeley in the 1980s, I too was a new arrival. He was coming to Berkeley as a senior hire and I as a junior. It was through our connection to David Pollard that we discovered our mutual interest in teaching statistics. We first collaborated on a small project to introduce computing into the advanced undergraduate theoretical statistics course. The computing exercises we developed were aimed at students uncovering, through simulation studies, some of the rules of thumb that a practicing statistician regularly uses. This was not as successful as we had hoped because our students didn't see any reason to care about the simulation results. We had fallen into the trap Terry warns against in Speed [10]: teaching pseudo-applied statistics with context-free numbers. Subsequent attempts led us to connect the work to real applications and then to the template described in Nolan and Speed [8] and used in *Stat Labs: Mathematical Statistics through Applications* [9]. It would seem that this template should have been an immediate and obvious result of Speed [10]. It wasn't. While Terry modestly claims to be "no exception – for allowing ourselves to forget the fundamental importance of the interplay of questions, answers and statistics", I dare conjecture that one of his goals in the project was for me to gain experience through trial and error in developing an effective approach to teaching statistics.

Speed [10] successfully argues that the "whole point of statistics lies in the interplay between context and statistics." Others share this viewpoint as noted in the quotes included in the article from James, Cox, Dawid, and Tukey. However, Terry takes this assertion into the education arena and compels us to reflect this important thesis in our teaching. Following Speed [10], others have made similar arguments to change statistics education. According to Cobb and Moore [3], "The focus on variability naturally gives statistics a particular content that sets it apart from mathematics itself and from other mathematical sciences, but there is more than just content that distinguishes statistical thinking from mathematics. Statistics requires a different kind of thinking, because data are not just numbers, they are numbers

D. Nolan
Department of Statistics, University of California, Berkeley
e-mail: nolan@stat.berkeley.edu

S. Dudoit (ed.), *Selected Works of Terry Speed*, Selected Works in Probability and Statistics, 377
DOI 10.1007/978-1-4614-1347-9_9, © Springer Science+Business Media, LLC 2012

with a context." Higgins [6] and Nicholls [7] echo these statements; e.g., Higgins claims that "for the past 40 years, statistics has been doing a great job of training theoretical statisticians, but we have a more data based society and it is crucial that we identify changes to course content and delivery that need to occur." Similarly, Wild and Pfannkuch [11] note "the biggest holes in our educational fabric, limiting the ability of graduates to apply statistics, occur where methodology meets context (i.e. the real world)."

One teaching strategy offered in Speed [10] is to meet people with data by, for example, pairing the statistics teacher with a teacher in an empirical field of inquiry or pairing statistics students with students who have subject matter knowledge. Anecdotal evidence of the success of this approach appears in Field et al. [5]. There we learn of the preparation of Betty Allan, Mildred Barnard, and Helen Turner for successful careers in biometrics at CSIR in the 1930s. All three women spent significant time in Rothamsted Station where they learned statistics by designing and carrying out experiments under the guidance of Fisher, Wishart, and Yates.

Terry raised and answered in his 1986 paper two common objections to working with real problems: that these problems are too complex and the data too large to be practical in the classroom and that only the most advanced students who have a sufficiently large set of tools can successfully attack real problems. Today, we face a new version of these same concerns. Data are now free and ubiquitous. People with all sorts of backgrounds have ready access to data. This data explosion is an enormous opportunity for us to make better, more informed decisions. However, this opportunity presents challenges because people expect to be able to interact with data in new ways and the role of the statistician is changing.

As I reflected on Terry's call to change how we teach statistics, it was at first disconcerting to see that we are still asking statistics educators to consider this issue. Cobb [2] explains that "What we teach was developed a little at a time, for reasons that had a lot to do with the need to use available theory to handle problems that were essentially computational." Efron [4] describes the mathematical statistics course as "caught in a time warp" that "does not attempt to teach what we do and certainly not why we do it." Brown and Kass [1] examine statistics graduate training and warn us to break away from the view of the statistician's role as "short-term consultant" because that model "relegates the statistician to a subsidiary position, and suggests that applied statistics consists of handling well-formulated questions, so as to match an accepted method to nearly any kind of data." I have since realized that we must periodically revisit this question of how best to teach statistics and that is precisely the point. We are not aiming at a fixed target that once arrived at we will have accomplished our goal.

Efron [4] suggests starting over by imagining "a universe where computing preceded mathematics in the development of statistics" and advocates "starting more muscularly without worrying about logical order of presentation" and focusing instead on the basic kinds of reasoning and explanations that can be arrived at through randomization-based inference. Cobb [2] further develops this notion, explaining how randomization-based inference "makes a direct connection between data production and the logic of inference that deserves to be at the core of every introductory course." Cobb further posits that "Technology allows us to do more with less: more ideas, less technique. We need to recognize that the computer revolution in statistics education is far from over." Brown and Kass [1] advocate taking a "less restrictive view of what constitutes statistical training." They see a blurring of the distinction between people with data and people with statistical expertise and state "some of the most innovative and important new techniques in data analysis have come from researchers who would not identify themselves as statisticians." Brown and Kass recommend we minimize prerequisites to research, require real-world problem solving in our courses, and embrace a deeper commitment to cross-disciplinary training. Efron [4], Cobb [2], and Brown and Kass [1] advocate twenty-first century changes to statistics education that echo Terry's call to include the value of statistics in our training programs.

Terry Speed's advice from twenty-five years ago remains extremely relevant today as computational and data challenges continue to evolve and shape our field.

References

[1] E. Brown and R. Kass. What is statistics? *Am. Stat.*, 63:105–110, 2009.

[2] G. Cobb. The introductory statistics course: A Ptolemaic curriculum? *Technology Innovations in Statistics Education*, 1:1–15, 2007.

[3] G. Cobb and D. S. Moore. Mathematics, statistics, and teaching. *Am. Math. Mon.*, 104:801–823, 1997.

[4] B. Efron. Is the math stat course obsolete? In *Panel Discussion at the Joint Statistics Meeting, San Francisco, CA*, 2003. www.amstat.org/sections/educ/MathStatObsolete.pdf.

[5] J. B. F. Field, F. E. Speed, T. P. Speed, and J. M. Williams. Biometrics in the CSIR: 1930–1940. *Aust. J. Stat.*, 30B(1):54–76, 1988.

[6] J. J. Higgins. Nonmathematical statistics: A new direction for the undergraduate discipline. *Am. Stat.*, 53:1–6, 1999.

[7] D. F. Nicholls. Future directions for the teaching and training of statistics at the tertiary level. *Int. Stat. Rev.*, 69:11–15, 2001.

[8] D. Nolan and T. P. Speed. Teaching statistics: Theory through applications. *Am. Stat.*, 53(4):370–375, 1999.

[9] D. Nolan and T. P. Speed. *Stat Labs: Mathematical Statistics Through Applications*. Springer Texts in Statistics. Springer-Verlag, New York, 2000.

[10] T. P. Speed. Questions, answers and statistics. In R. Davidson and J. Swift, editors, *Proceedings: The Second International Conference on Teaching Statistics*, pages 18–28, Victoria, BC, Canada, 1986. University of Victoria.

[11] C. Wild and M. Pfannkuch. Statistical thinking in empirical enquiry. *Int. Stat. Rev.*, 67:223–248, 1999.

QUESTIONS, ANSWERS AND STATISTICS

Terry Speed
CSIRO Division of Mathematics and Statistics
Canberra, Australia

A major point, on which I cannot yet hope for universal agreement, is that our focus must be on questions, not models. . . . Models can – and will – get us in deep troubles if we expect them to tell us what the unique proper questions are.

J.W. Tukey (1977)

1. Introduction

In my view the value of statistics, by which I mean both data and the techniques we use to analyse data, stems from its use in helping us to give answers of a special type to more or less well defined questions. This is hardly a radical view, and not one with which many would disagree violently, yet I believe that much of the teaching of statistics and not a little statistical practice goes on as if something quite different was the value of statistics. Just what the other thing is I find a little hard to say, but it seems to be something like this: to summarise, display and otherwise analyse data, or to construct, fit, test and evaluate models for data, presumably in the belief that if this is done well, all (answerable) questions involving the data can then be answered. Whether this is a fair statement or not, it is certainly true that statistics and other graduates who find themselves working with statistics in government or semi-government agencies, business or industry, in areas such as health, education, welfare, economics, science and technology, are usually called upon to answer questions, not to analyse or model data, although of course the latter will in general be part of their approach to providing the answers. The interplay between questions, answers and statistics seems to me to be something which should interest teachers of statistics, for if students have a good appreciation of this interplay, they will have learned some statistical thinking, not just some statistical methods. Furthermore, I believe that a good understanding of this interplay can help resolve many of the difficulties commonly encountered in making inferences from data.

My primary aim in this paper is quite simple. I would like to encourage you to seek out or attempt to discern the main question of interest associated with any given set of data, expressing this question in the (usually non-statistical) terminology of the subject area from whence the data came, before you even think of analysing or modelling the data. Having done this, I would also like to encourage you to view analyses, models etc. simply as means towards the end of providing an answer to the question, where again the answer should be expressed in the terminology of the subject area, although there will always be the associated statement of uncertainty which characterises statistical answers. Finally, and regrettably this last point is by no means superfluous, I would then encourage you to ask yourself whether the answer you gave really did answer the question originally posed, and not some other question.

A secondary aim, which I cannot hope to achieve in the time permitted to me, would be to show you how many common difficulties experienced in attempting to draw inferences from data can be resolved by carefully framing the question of interest and the form of answer sought. A few remarks on this aspect are made in Section 6 below.

2. Why speak on this topic?

Over the years I have had many experiences which have lead me to think that the interplay between questions, answers and statistics is worthy of consideration. Let me briefly mention four, each of a different type.

The first experience is a common one for me. Someone is describing an application of statistics in some area, say biology. The speaker usually begins with an outline of the background science and goes on to give an often detailed description of the data and how they were collected. This part is new and interesting to any statisticians listening, most of whom will be unfamiliar with that particular part of biology. Sometimes the biologist who collected the data is present and contributes to the explanation, but at a certain stage the statistician starts to explain what she/he did with the data, how they were "analysed". By now the biologist is quiet, deferring to the statistician on all matters statistical, and terms like main effects, regression lines, homoscedacity, interactions, and covariates fly around the room. Sooner or later I find myself thinking "Here are the answers, but what was the question?" All too frequently in such presentations neither the statistician nor the biologist has posed the main question of biological interest in <u>non-statistical terms</u>, that is, in terms which are independent of analyses or models which may or may not be appropriate for the data, and I can certainly remember occasions when the analysis presented was seen to be inappropriate once the forgotten question was formulated. Of course many scientific questions <u>can</u> be translated into statements about parameters in a statistical model, so that I am not condemning all instances of the above practice.

A similar sort of experience is surely familiar to all who have helped people with their statistical problems. This time a scientist, say a psychologist, comes to me with a set of data and one or more questions. She/he knows some statistics, or at least some of the jargon. After being briefed on the background psychology and the mode of collection of the data 1 usually say something like "What questions do you want to answer with these data?", implicitly meaning "What <u>psychological</u> questions . . . ?" Not infrequently the answer comes back "Is the difference between such and such significant?" meaning, of course, <u>statistically</u> significant. [In my perversity I often think to myself: "Well, <u>you</u> should know; it's your data and you are the psychologist!"] Another similar query might concern interactions, or regression coefficients of covariates etc. What this has in common with the previous example is the unwillingness or inability of the psychologist to state her/his questions of interest in nonstatistical terms. We should all be familiar with the idea that scientific (e.g. psychological) significance and statistical significance are not necessarily the same thing, but how many of us keep in mind the fact that the latter involves an analysis or a statistical model, and that there may be as many answers to this question as there are analyses or models? Surely much of the blame for such thinking rests with us, the teachers of statistics, who never fail to popularize the rigid formalism of Neyman-Pearson testing theory.

My third type of experience concerns recent graduates in statistics, students I and my colleagues have taught and whom we believe should be able to operate independently as statisticians. Many of these graduates go into jobs in big public enterprises: railways, agriculture bureaux, mining companies, government departments and so on, and a few – far too many for comfort – get in touch with us when they meet a difficulty in their new job. It is not the fact that they get in touch which is discomfiting, but the questions they ask! For we then learn how little they have grasped. They have questions in abundance, often important policy questions, access to lots of data, or at least the possibility of collecting any data that they deem necessary, but they are quite unsure how to proceed, how to answer the questions. Out there in the world there are "populations" of real trains, field plots, cubic metres of ore or people, and even the simplest question relating to a mean or a proportion or a sample size can be forbidding. Perhaps they should standardize something to compare it with something else, perhaps include the variability of one factor when analysing another, or something else again, all things which we feel that a graduate of our course should be able to cope with unaided. But how well did we train them for this experience?

Finally, and briefly, let me castigate my professional colleagues – and myself, since I am no exception – for allowing ourselves to forget the fundamental importance of the interplay of questions, answers and statistics, for in so many of our professional interactions we act as if it is irrelevant. How many times have we presented new statistical techniques to one another, illustrated on sets of "real" data, drawing conclusions about those data concerning questions no one ever asked, or is ever likely to ask? And how often do we derive statistical models or demonstrate properties of models which are unrelated to any set of data collected so far, and certainly not to any questions from a substantive field of human endeavour. We are, so we tell ourselves, simply adding to the stock of statistical methods and models, for possible later use. Is it any wonder that we or our co-workers then find ourselves using these models and methods in practice, regardless of whether or not they help us to answer the main questions of interest. For a discussion of some closely related issues of great relevance to teachers of statistics, see the two excellent articles Preece (1982, 1986).

3. Why this audience?

I don't think I will be very wide off the mark if I assume that most of you – at least the active teachers of statistics amongst you – have come from a background of mathematics rather than statistics, and that few of you have actually been statisticians before you started teaching the subject. I would further guess that many of you still teach mathematics, and perhaps at the school level, statistics within a mathematics curriculum. It is on this assumption that I have chosen to focus on non-mathematical aspects of our subject, ones with which I feel you will generally be less familiar. As I said in the introduction, I hope that my talk will encourage you to give more attention to the non-mathematical aspects of statistics in your teaching, in particular to spend more time considering real questions of interest with real sets of data.

It is a curious thing that interest in the teaching of statistics in schools, colleges and universities has sprung up worldwide as an extension of mathematics teaching, because I certainly feel that the practice of statistics is no closer to mathematics than cooking is to chemistry. Both mathematics and chemistry are reasonably precise subjects in their own ways, and in general what goes on in them both is repeatable; perhaps they are true sciences. On the other hand, statistics and cooking are as much arts as they are science, although both have strong links to their corresponding science: mathematics in the case of statistics, and chemistry in the case of cooking. Who would recommend that a chemistry teacher with no cooking experience be appointed as cooking teacher as well? If I can convey to you some of the enjoyment and intellectual challenge that lies in my particular variety of cooking, and encourage you to try it yourself, I will have succeeded in my aims.

4. Two further examples

In this necessarily too brief section I offer two more concrete illustrations of interplay of the questions, answer and statistics. The first one is a very simple paraphrase of Neyman's classic illustration of hypothesis testing involving X-ray screening for tuberculosis, and I refer you to Neyman (1950, Section 5.2.1) for a fuller background and further details.

You have a single X-ray examination and, after the photograph has been read by the radiologist, you are given a clean bill of health, that is, you are told that there is no indication that you are affected by tuberculosis. With Neyman we will assume that previous experience has led to

$$\text{pr(clean bill} \mid \text{no TB)} = 0.99$$

$$\text{pr(clean bill} \mid \text{TB)} = 0.40$$

You now ask the radiologist "What are the chances that I have TB?" She says "I can't answer that question but I can say this: Of the people with TB who are examined in this way, 60% are correctly identified as having TB, and of . . . " You interrupt her. "Doctor, I know the procedure is imperfect, but you have just examined my X-ray . . . What are the chances that I have TB?"

If your radiologist is sufficiently flexible and well informed, she will answer "Well, that depends on the prevalence of TB in your population, that is, on the proportion of people affected by TB in the (a?) population from which you may be regarded as a typical individual". Indeed a simple application of Bayes' theorem yields:

$$\text{pr(TB|clean bill)} = \frac{\text{pr(clean bill|TB)pr(TB)}}{\text{pr(clean bill)}}$$

$$= \frac{0.40\text{pr(TB)}}{0.40\text{pr(TB)} + 0.99[1-\text{pr(TB)}]}$$

At last you see how to get an answer to your question. It may not be easy to obtain a value for pr(TB): your smoking habits, the location of your home, your occupation, your ancestry . . . may all play a part in defining "your population", but this is what is needed to answer the question and it is far better to recognise this than to fob you off with the answer to another question not of interest to you.

If this example smacks of Bayesian statistics it is not entirely accidental, for there are many occasions where the Bayesian view (which is certainly not necessary in this example) helps answer the question of interest, whereas classical statistics refuses, frequently answering another, unasked, question instead. For a more complex, explicitly Bayesian example, see the very fine paper Smith and West (1983) concerning the monitoring of renal transplants.

My second example concerns the determination of the age of dingos, Australia's wild native dogs. A statistician was given a large body of data relating the age of a number of dingos to a set of physical measurements including head length. The data concerned both males and females, a number of breeds and animals from a number of locations, but for this discussion we will restrict ourselves to a single combination of sex, breed and location. The question, or at least the task, to be addressed was the following: produce an age calibration curve for dingos based upon the most suitable physical measurement, that is, produce a curve so that the age of a dingo may be predicted by reading off the curve at the value of the physical measurement. This curve was for use in the field and it was taken for granted that an estimate of the precision of any age so predicted would also be obtained.

It was found that a curve of the general form $h = a + b[1 - \exp(-ct)]$, where h and t are head length and age, respectively, and a, b and c are parameters of the curve, fitted the data from each dingo extremely well over the range of ages used. This was an exercise in non-linear regression with which the statistician took great care, special concern being given to the different possible parametrizations of the curve, the convergence of the numerical algorithm used, the residuals about the fitted line and to the validity of the resulting confidence intervals for a, b and c. The parameters estimated for different dingos naturally differed, although, not surprisingly, the values of a (head length at birth) showed less variation than those of b (ultimate head length -a) and c (a growth rate parameter).

All this seems fine, and you might wonder why I am mentioning this example at all in the present context. My answer is as follows. The statistician in question knew, or knew where to find, lots of information about the fitting of individual growth curves, and so he focussed on this aspect of the problem. To answer the original question, however, his attention should have been pointed in quite a different direction, towards: the calculation of a population or group growth curve for the calibration procedure; features of the sample of dingos measured that may affect the use of their measurements as a basis for the prediction of the age of a new dingo; properties of the parameters which are relevant to this question; and, finally, towards obtaining a realistic assessment of the prediction error inherent in the use of the curve in the field.

In summary, he was willing and able to spend a lot of time on the individual animals' curves; he was less willing and less able to focus on the issues demanded by the question, those concerning population parameters, population variability, problems of selection, unrepresentativeness, and other issues including the use of normal theory, with real but not very well defined populations.

5. What is the problem?

Let me oversimplify and put my message like this. In the beginning we taught mathematics and called it statistics; much of this was probability, a quite different subject. Then, with the help of computers, we started to teach data analysis and statistical modelling; this was fine apart from one feature: it was largely context-free. The real interest (for others and many statisticians), the important difficulties and the whole point of statistics lies in the interplay between the context and the statistics, that is, in the interplay between the items of my title.

Let me offer a few similar views. A.T. James (1977, p. 157) said in the discussion of a paper on statistical inference:

> *The determination of what information in the data is relevant can only be made by a precise formulation of the question which the inference is designed to answer. . . . If one wants statistical methods to prove reliable when important practical issues are at stage, the question which the inference is to answer should be formulated in relation to these issues.*

Cox (1984, p. 309) makes the following characteristically brief contribution to our discussion:

> *It is trite that specification of the purpose of a statistical analysis is important.*

Dawid (1986) is even more to the point:

> *Fitting models is one thing; interpreting and using them is another, . . . If the model is correct and we know the parameters, how ought we to compare [schools]? . . . There is in fact no unique answer; it all depends on our purpose. . . . there remains a strong need for a careful prestatistical analysis of just what is required: following which it may well be found that it is conceptually impossible to estimate it!*

Tukey and Mosteller (1977, p. 268) offer seven purposes of regression, or, as I would paraphrase it, seven types of questions which regression analysis may help answer. Summarized, these seven purposes are:

1. to get a summary;
2. to set aside the effect of a variable;
3. as a contribution to an attempt at causal analysis;
4. to measure the size of an effect;
5. to try to discover a mathematical or empirical law;
6. for prediction;
7. to get a variable out of the way.

Similarly, Tukey (1980, pp. 10-11) gives the following six aims of time series analysis;

1. Discovery of phenomena.
2. "Modelling".
3. Preparation for further inquiry.
4. Reaching conclusions.
5. Assessment of predictability.
6. Description of variability.

Similar numbers of aims, purposes, or types of questions could be given for the analysis of variance, the analysis of contingency tables, multivariate analysis, sampling and most other major areas of statistics. Yet how often do our students meet these techniques in context with even one of these aims, much less the full range? And how else are they going to learn to cope with the special difficulties which arise when questions are asked of them in context whose answers require statistics? This is the problem.

6. Some General Comments

In this section I will mention a few difficulties which I believe can be resolved in a given case when the relation between the questions asked, the form of the answers desired and the statistical analysis to be conducted are carefully considered. A full discussion of any one of the difficulties is out of the question, and even if that had been given, there would probably remain an element of controversy, something which would be out of place in a talk like this. The section closes with some further general comments about questions.

Some elementary difficulties which I think arise include

- What is the population?
- When are population characteristics (e.g. proportions) relevant?

- What is the "correct" variance to attach to a mean or proportion?

- When should we standardize (for comparison)?

I have found that the relations between statistical models and analyses on the one hand, and populations and samples on the other, with parameters playing a role in both, are something which puzzle many students of our subject. The former play a big role in standard statistics courses whereas the latter are prominent in applications. Just how they connect is not a trivial matter.

A few somewhat more advanced difficulties include

- Which regression: y on x, x on y or some other?

- When should we use correlation and when regression analysis?

- When can/should we adjust y for x?

- Which error terms do we compare (in anova)?

- Should we regard a given effect as fixed or random?

- Which classifications (of a multiway table) correspond to factors and which to responses?

More subtle difficulties are associated with general questions such as

- Should we do a joint, marginal or conditional analysis?

I believe that in all of the above cases the difficulties arise because insufficient attention has been given to the nonstatistical context in which the discussion is taking place, and that when the question of interest is clarified and the form of answer sought understood, the difficulty either disappears completely or is readily resolved. Of course doing so takes some experience. Note that many of the difficulties listed involve, implicitly or explicitly, the notion of conditioning, or its less probabilistic forms, standardizing or adjusting. Just what we regard as being "held fixed" and what we "average over" in any given context is crucial, and here our questions and answers determine everything. The simplest form of this issue is usually: "Are we interested in just these units (the ones we have seen), or in some population of units from which these may be regarded as a (random?) sample, or both?" Models are no help here.

A simple but easy to forget aspect of the use of a statistical method is that not all questions which could be asked and answered by that method, are necessarily appropriate in a particular context. Lord's paradox, see Cox & McCullagh (1982) and references therein, provides a good example here.

7. What can/should be done?

It hardly needs saying that the best way to promote interest in the interplay between questions, answers and statistics is to put trainee statisticians into situations where they are required to provide answers to clearly stated questions on the basis of real data sets. Note that this can be a very different thing from "illustrating" a statistical technique on a set of data. In particular, much more background to the data is usually required, and this is rarely available in data sets presented in statistics texts. Indeed technical journals are now so tight with their space that it is rare to find full data sets published together with analyses and conclusions in scientific articles. This means that the best sources of suitable material of the kind being discussed, that is, of questions and data, are often one's colleagues or clients: teachers and researchers in other disciplines who make use of experimental or observational data in their work. Seeking out such material can be a way of forging links with the users of statistics and of course sandwich courses are designed with this general aim in mind.

One practice which I believe is valuable is the conduct of regular practical statistics sessions where students are asked to help answer specific questions on the basis of sets of data supplied together with background material. This is much more like the situation they will meet after their training is over. Two objections which are often expressed to me when I recommend this approach are (i) Surely it is unrealistic, except with the most advanced students, for unless they have learned a wide range of techniques, they will not be able to begin attacking "real" problems with any likelihood of success?; and (ii) Surely it is unrealistic, because real problems are so complex and real data sets so large, or even ill- defined, that nothing like what happens in practice can be presented in the classroom?

Both these objections have some validity, but let me make a few observations concerning them. Firstly, it is not necessarily a bad thing for a student (or anyone!) to attempt to answer a particular question (solve a particular problem) without knowing of the tools or techniques that may have been developed to answer just that type of question (or problem). This goes on all the time in the real world: parts of the wheel are rediscovered time and time again, and locomotion is even found to be possible without the wheel! And of course there is very seldom a single "correct" way to answer a question; an approach using less knowledge of techniques may well be better than one which uses greater knowledge. In the hands of a good teacher, such experiences can provide valuable object lessons, and, at the very least, valuable motivation for techniques not yet learned. Surely nothing could be more satisfying than hearing a student say: "What I need (to answer this question) is a way of doing such and such, under the following circumstances (e.g. errors in this variable, that factor misclassified, these observations missing or censored, that parameter chosen in a particular way, etc.)? Group discussions, where ideas are shared and knowledge pooled, are also most appropriate for this sort of work, and most enjoyable. The teacher can then play a subsidiary role, at times focussing the discussion back on the questions, perhaps at other times supplying a sought-for technique.

It would seem to me that this is just the sort of statistics which should be taught in secondary schools, not the watered-down and frequently sterile mathematical material which is often found at that level.

The second objection, that real problems are often very complex and rarely amenable to the sort of trimming that would be necessary before they could be used in a classroom, is harder to dismiss. It is certainly true that many (most?) problems are like this, but surely this highlights even more the difference between "illustrative" data sets, taken out of context, with no realistic questions or idea what would be satisfactory answers, and what we expect students to be able to cope with upon graduation. There is certainly a big gap here – between "pseudo-applied" statistics involving context-free sets of numbers, to illustrate arithmetic, and fully-fledged "warts and all" consulting problems – and I can only state that in my experience it is possible to find problem data sets which can be presented in the way I am suggesting. It certainly takes a little effort to find such material, particularly if you are not in the habit of meeting people with data and statistical problems. But as teachers of the subject, that is not such an unreasonable thing for me to expect of you is it?

A teaching strategy which could provide a means of putting these ideas into practice might be the following: pair yourself (the statistics teacher) with a teacher in an empirical field of enquiry, e.g. biology, agriculture or medicine, and also pair your statistics students with students in the corresponding class, requiring them to work together on a practical project which will enrich their understanding of both disciplines, and how statistics helps to answer questions. Many variants on this suggestion could be devised; the important thing is try something along these lines. Statistics students must meet more than mathematics and sets of numbers in their training, and it is the teachers of statistics who must arrange for this to happen.

Acknowledgement

I am very grateful for the discussions and comments on this topic offered to me by my CSIRO colleagues Peter Diggle, Geoff Eagleson and Emlyn Williams.

8. References

Cox, D.R. (1984). Present position and potential developments: Some personal views. Design of experiments and regression. J.R. Statist. Soc. Ser. A., 147, 306-315.

Cox D.R. & McCullagh P. (1982). Some aspects of the analysis of covariance. Biometrics 38, 541-553.

Dawid, A.P. (1986). Contribution to the Discussion of: "Statistical modelling issues in school effectiveness studies" by M. Aitkin and N. Longford. J. Roy. Statist. Soc. Ser. A, 149, 1-43.

James, A.T. (1977). Contribution to the Discussion of: "On resolving the controversy in statistical inference" by G.N. Wilkinson. J. Roy. Statist. Soc. Ser. B, 39, 157.

Mosteller, Frederick & Tukey, John W. (1977). Data analysis and regression. Sydney: Addison-Wesley Publishing Company.

Neyman, J. (1950). First Course in Probability and Statistics. New York: Henry Holt & Company.

Preece, D.A. (1982). t is for trouble (and textbooks): a critique of some examples of the paired samples t-test. The Statistician, 31, 169-195.

Preece, D.A. (1986). Illustrative examples: illustrative of what? The Statistician, 35, 33-44.

Smith, A.F.M. & West, M. (1983). Monitoring renal transplants: An application of the multiprocess Kalman filter. Biometrics, 39, 867-878.

Tukey, J.W. (1977). Contribution to the Discussions of "A reformulation of linear models" by J.A. Nelder. J. Roy. Statist. Soc. Ser. A, 140, 72.

Tukey, John W. (1980). Can we predict where "Time Series" should go next? Directions in Time Series Eds D.K. Brillinger & G.C. Tiao. IMS. pp.1-31.

Teacher's Corner

Teaching Statistics Theory Through Applications

D. NOLAN and T. P. SPEED

This article presents a model for developing case studies, or labs, for use in undergraduate mathematical statistics courses. The model proposed here is to design labs that are more in-depth than most examples in statistical texts by providing rich background material, investigations and analyses in the context of a scientific problem, and detailed theoretical development within the lab. An important goal of this approach is to encourage and develop statistical thinking. It is also advocated that the labs be made the centerpiece of the theoretical course. As a result, the curriculum, lectures, and assignments are significantly restructured. For example, the course work includes written assignments based on open-ended data analyses, and the lectures include group work and discussions of the case-studies.

KEY WORDS: Instruction; Quantitative literacy; Statistical thinking.

1. INTRODUCTION

Over the past 10 years we have developed a course that teaches undergraduate upper-level mathematical statistics through the use of in-depth case studies, which we call labs. It is through the labs that the theory of statistics is introduced, which leads to an integration of statistical theory and practice in a way not commonly encountered in an undergraduate course. The labs raise scientific questions that are interesting in their own right, and they contain datasets for use in addressing these questions. The context of the scientific question is the starting point for developing statistical theory.

This article presents a model that we have found to be successful for designing these labs. The model calls for the lab to be a substantial exercise with nontrivial solutions that leave room for different analyses, and for it to be a central part of the course. The lab should offer motivation and a framework for studying theoretical statistics, and it should give students experience with how statistics can be used to answer scientific questions. An important goal of this approach is to encourage and develop statistical thinking while imparting knowledge in mathematical statistics.

It often can be difficult for students to bring the mathematical statistics learned in the classroom to an independent project in school or on the job. It can also be difficult for students to make the transition from reading and understanding a critical review of a statistical analysis to successfully working on their own problem. In our attempt to address these issues, we propose here to mix theory with substantial case studies. In addition to increasing student understanding of statistical theory, through the labs students develop their quantitative reasoning and problem-solving skills in a broad, multidisciplinary setting. They become practiced in communicating their ideas orally and in writing, and they become versed in the use of statistical software.

2. BACKGROUND

The labs were designed to accompany a typical theoretical statistics course for juniors and seniors. We have used these labs for two such courses—one for mathematics and statistics majors and one for engineering and computer science majors. Both courses require students to have studied calculus for two years and probability for one semester. There is no statistics prerequisite for the course.

The material covered in the course is outlined in the following; it is divided into five broad categories. In a typical semester, roughly one week is spent on the first topic, summary statistics, and about three weeks are spent on each of the remaining four topics: sampling, estimation and testing, regression and simple linear least squares, and analysis of variance and multiple linear least squares.

1. Summarizing data
   a. *Univariate summaries*: histogram, box plots, bar charts, and quantile plots; measures of location and spread; normal approximation, skewness, kurtosis.
   b. *Bivariate summaries:* cross-tabulations, quantile-quantile plots, and scatterplots.
2. Sampling
   a. *Simple random:* sample mean, percent, and total; expectation, variance, and estimator for variance; normal approximation and confidence intervals.
   b. *Stratified:* expectation, variance; proportional and optimal allocation; bootstrap.
3. Parameter estimation, goodness-of-fit, and hypothesis testing.

D. Nolan is Professor, and T. P. Speed is Professor, Department of Statistics, University of California, 367 Evans Hall, Berkeley CA 94720-3860 (Email: nolan@stat.berkeley.edu). This work has been partially supported by NSF grant 9720490. The authors thank the reviewers for their comments which led to an improved and more informative article.

a. *Parameter estimation:* method of moments; maximum likelihood; mean square error, asymptotic bias, and variance; information and Cramer–Rao inequality.
b. *Goodness-of-fit and testing:* chi-square tests; two-sample tests; test statistic, null distribution, and Type I and II errors.

4. Regression and the simple linear model.
   a. *Regression:* correlation; least squares; residuals, confidence intervals; bivariate normal.
   b. *Simple linear least squares:* linear model, RMS error, transformations, polynomial regression, t-tests and F-tests.
5. Extensions to linear models
   a. *Analysis of variance:* one-way and two-way classification, additive model, F-test; least squares with indicators; sum of squares decomposition.
   b. *Multiple linear regression:* geometry of least squares; diagnostics, model checking.

We started to develop labs for these courses in 1989. At that time, each lab consisted of a list of exercises designed to illustrate a particular statistical concept and to be applied to the data provided. These original labs were unsatisfactory for many reasons. They lacked context; the presentation was too much like an instruction manual; and they were separate from the development of the course material. Students viewed the labs as drudge work; they did not see them as important or relevant to the course. In response to these shortcomings, we began to expand the role of the labs. We kept those labs for which we had interesting scientific problems to accompany them, and we added to them detailed descriptions of the data, background material to put the problem in context, and suggestions for how to proceed in the analysis. In a further evolution, we incorporated the statistical theory into the lab. This was an important step in the lab development, because it made the labs the centerpiece of the theoretical course, not a secondary accompaniment.

As a result of this fundamental restructure of the labs, several changes took place in the course. The format of lectures changed. More time was spent on determining how to answer general scientific questions with statistical analyses and on deriving a statistical method from its application to a specific problem. Less time was spent covering many small examples constructed to illustrate a single statistical technique. Because the application serves as the motivation for working out the theoretical material, we still cover all the material traditionally covered in the course. Also, the course work changed; now lab assignments required written reports with accompanying graphs and charts, and exams contained questions on theoretical topics related to the labs. Finally, the material distributed to the students for each lab grew from a 2-page list of instructions to a 10- to 20-page in-depth case study. More details are provided in Section 4 on how the labs are integrated into the course.

3. A LAB MODEL

Over time, we have settled on a model for the organization of a lab. Each lab is divided into five main parts: an introduction, data description, background material, investigations, and theory. Sometimes we include an extension section for a more advanced analysis and related theoretical material.

We describe the content and important features of each section in the following. The descriptions are accompanied by examples taken from a sample lab, one that introduces parameter estimation and hypothesis testing by asking students to search for unusual clusters of patterns in the DNA of a herpes virus. Despite the technical nature of the topic, this lab is one of the most popular among the students. The students derive hypothesis tests that study features of the DNA against a model for random scatter based on the homogeneous Poisson process.

3.1 Introduction

Here a clear scientific question is stated, and the motivation for answering it is given. The question is presented in the context of the scientific problem, and not as a request to perform a particular statistical method. Typically the introduction ends with the question to be addressed by the data described in the following section. We do not address questions suggested by the data, instead we consider questions raised by researchers then use statistics to seek answers.

The introduction often begins with a current newspaper article on a related subject to convey the relevance of the question at hand. The DNA lab includes an excerpt from "Feat is Milestone in Study of Evolution" by Nicholas Wade, *New York Times* (08/01/95). The article describes the sequencing of a bacterium's DNA. It includes a quote from Watson, one of the codiscoverer's of the structure of DNA, explaining the importance of the event.

> Life is a mystery, ineffable, unfathomable, the last thing on earth that might seem susceptible to exact description. Yet now, for the first time, a free-living organism has been precisely defined by the chemical identification of its complete genetic blueprint.
>
> The creature is just a humble bacterium known as *Hemophilus influenzae*, but it nonetheless possesses all the tools and tricks required for independent existence. For the first time, biologists can begin to see the entire parts list, as it were, of what a living cell needs to grow, survive and reproduce itself. . . .

Following the article, students are provided a brief introduction to the problem: how to complete the "parts list" for the human cytomegalovirus (CMV) by locating regions of DNA that may contain information on how the virus reproduces. To further motivate them, they are told that CMV is a herpes virus which is potentially life threatening for people with suppressed or deficient immune systems, and to develop strategies for combating the virus, scientists study the way in which it replicates. To study how the virus reproduces itself scientists need to find the site on the DNA that contains the instructions for starting the process of replication, called the origin of replication. DNA can be thought of as a long coded message made from the four-letter alphabet A, C, G, and T, and for other herpes viruses, the origin of replication is specially marked with a

Table 1. Locations for the First and Last 30 Palindromes in the Cytomegalovirus DNA

177	1321	1433	1477	3248	3255
3286	7263	9023	9084	9333	10884
11754	12863	14263	14719	16013	16425
16752	16812	18009	19176	19325	19415
20030	20832	22027	22739	22910	23241
...	...	204548	205503	206000	207527
207788	207898	208572	209876	210469	215802
216190	216292	216539	217076	220549	221527
221949	222159	222573	222819	223001	223544
224994	225812	226936	227238	227249	227316
228424	228953				

Source: Chee et al. (1990)

pattern of letters in their DNA code. Students are told that they are to identify any unusual patterns in the 229,354-letter-long CMV DNA. The search is restricted to one type of pattern—clusters of complementary palindromes. In the DNA alphabet, A is complementary to T, C is complementary to G, and a complementary palindrome is a sequence of letters that reads in reverse as the complement of the forward sequence—for example, GGGCATGCCC.

3.2 Data Description

Documentation for the data collected to address the question is provided. It includes a detailed description of the study protocol, as appropriate. The data for the DNA lab are quite simple to describe; the data are the locations along the CMV DNA of any palindrome longer than 10 letters. See Table 1 for a list of the first 30 locations; the entire dataset is available on the Web site www.stat.berkeley.edu/users/nolan/statlabs. Chee et. al. (1990) published the DNA sequence of CMV, and Leung, Blaisdell, Burge, and Karlin (1991) implemented a hash-coding computer program to screen the CMV DNA in search of patterns. Altogether, they found 296 palindromes that were at least 10 letters long. Palindromes shorter than 10 letters were ignored as they can occur too frequently by chance. The longest ones found were 18 letters long.

3.3 Background

Material to put the problem in context is provided in this section. Information is gathered from a variety of sources, and it is presented in nontechnical language. The idea is to present a picture of the field of interest that is accessible to a broad college audience.

For example, in the DNA lab, the material is at the level of the *Cartoon Guide to Genetics* (Gonick and Wheelis 1991). The background information provided includes: a summary of the history of the discovery of DNA; a brief introduction to the structure of DNA; the definition of a complementary palindrome in the DNA language; a description of how a virus' DNA replicates; information on the prevalence and symptoms of the herpes *Cytomegalovirus*; and mention of the burgeoning science of genomics, the study of living things through their full DNA sequences.

3.4 Investigations

Suggestions for answering the question posed in the introduction are provided next. These suggestions are written in the context of the problem, using very little statistical terminology. The ideas behind the suggestions vary in difficulty, and are grouped to enable the assignment of subsets of investigations. Also included are suggestions on how to write up the results—for example, as an article for a widely read magazine; as a memo to the head of a research group; or as a pamphlet for consumers.

For the DNA lab, the students already have had experience with histograms and quantile plots, and it is expected that they will derive diagnostic tools and statistical tests for comparing the DNA data against a benchmark homogeneous Poisson process. Following is an excerpt:

How do we find clusters of palindromes?
How do we determine whether a cluster is just a chance occurrence, or a potential replication site?
To begin, pursue the point of view that structure in the data is indicated by departures from a uniform scatter of palindromes across the DNA. Of course, a random uniform scatter does not mean that the palindromes will be equally spaced as are milestones on a freeway. There will be some gaps on the DNA where no palindromes occur, and there will be some clumping together of palindromes. To look for structure, examine the locations of the palindromes, the spacings between palindromes, and the counts of palindromes in nonoverlapping regions of the DNA.

Simulation—One starting place might be to first see how random scatter looks by using the computer to simulate it. The computer can simulate 296 palindrome sites chosen at random along a DNA sequence of 229,354 letters, using a pseudo-random number generator.

Locations and spacings—Use graphical methods to examine the spacings between palindromes that are one apart, two apart, and so on. Compare what you find for the CMV DNA to what you would expect to see in a random scatter. Consider graphical techniques for examining the locations of the palindromes.

Counts—Use graphical displays and more formal statistical tests to investigate the counts of palindromes in various regions of the DNA. Split the DNA into nonoverlapping regions of equal length to compare the number of palindromes in an interval to the number that you would expect from uniform random scatter. The counts for shorter regions will be more variable than those for longer regions. Also consider classifying the regions according to their number of counts.

The biggest cluster—Does the interval with the greatest number of palindromes indicate a potential origin of replication?

How would you advise a biologist who is about to start experimentally searching for the origin of replication? Write your recommendations in the form of a memo to the head biologist of a research team of which you are a member.

3.5 Theory

The theoretical development of the statistical concepts and methodology appear after the problem is introduced, at the end of the lab handout. The material includes information on the general topics described in Section 2 of this article, as well as topics more specific to the individual lab, such as goodness-of-fit tests for the Poisson distribution and parameter estimation for the log-normal distribution.

In the DNA lab, the homogeneous Poisson process on a line is introduced. We discuss how the Poisson process seems a reasonable model for describing the scatter of palindromes on a sequence of DNA. That is, the CMV DNA has 229,354 base pairs and 296 complementary palindromes, and given the large number of base pairs, we can think of the location of a palindrome as a point on a line segment. If the palindromes are scattered randomly across the DNA, then the number of palindromes found in any small region should be independent of the number in another nonoverlapping region, and the chance that one small region has a palindrome in it should be the same for all regions of the same length.

In addressing the main question of the lab, the analysis centers around issues of how to test the fit of the model. To help students in this task, we explore various properties of the Poisson process—for example, the uniform distribution of the locations of hits, the exponential and gamma distributions for gaps between hits, and the Poisson distribution for counts of hits in intervals. Quantile plots and simulations are used as diagnostic tools. Hypothesis testing is introduced through the chi-square goodness-of-fit test for count data. We also derive a test statistic based on the maximum number of hits (palindromes) in a set of nonoverlapping intervals (segments of DNA). To perform the tests, we need to estimate the intensity parameter, and we derive method of moments and maximum likelihood estimators for the parameter. The students also need to address practical issues of how to select interval lengths, what to do if a cluster is split across intervals, and how to reconcile various results.

4. IMPLEMENTATION

We have incorporated the labs into two theoretical statistics courses—one for statistics and mathematics majors and one for engineering students. The typical enrollment is 20 students and 40 students, respectively. The classes meet three hours a week with the instructor and one or two hours a week, respectively, with a teaching assistant. (We have also used versions of these labs with simpler investigations and theoretical material in a course for social and life science majors).

We have created more than a dozen labs for these two courses, but we use only about eight in a semester. The labs are chosen by the instructor according to the topic (theoretical or practical) and the background of the students. These eight labs are divided between those labs that are discussed primarily in lecture and those that require students to do extensive analyses outside of class and to write short papers containing their observations and solutions.

Typically we ask students to write reports for four labs, about one for each of the five main topics in the course (see Section 2). Table 2 provides two example sets of assignments. Option A could be used in a class for students in the social and life sciences, and option B could be used for engineering students. In addition to learning the core material, many labs also include special topics, which are listed in the table. In the course, we cover all of the material that we formerly covered in the theoretical course (see Section 2) plus a few of these additional topics.

4.1 Lecture

Although both types of labs feature prominently in the course, they are handled somewhat differently in class. For both types of labs we found it optimal when lectures include: discussion of the background to a particular problem, where students who have taken courses in related fields can bring their own expertise to the discussion; and motivation of the theoretical material through discussion of how to address a problem from a lab. In addition, for labs on which students are to write reports, we hold regular question-and-answer periods where students raise concerns about their work. For the other labs, we have students work in groups as described in the following.

Roughly about one class period in three is spent on these types of activities. The remainder of time is spent in a more traditional presentation of theoretical results (with reference to the current lab).

4.2 Group Work

To facilitate group work in class, we supply handouts to the students. The handouts have one of two formats. In one case, they are abbreviated lists of investigations, and students are asked to come up with a plan of attack for addressing the questions, such as a way in which to summarize and present the data. Sometimes we prepare in advance a slew of possible summaries, and we sketch results on the blackboard as students discuss their plans. In the other case, students are given a set of charts and graphs and they are asked to further summarize and interpret the output in order to answer questions from the investigation. Groups write their solutions on the blackboard, and the instructor leads the class in a discussion of the analysis. It would be ideal to lead these discussions in a multimedia classroom where output can be projected onto a screen, or where students can be seated at display stations.

An alternative format for group work that we have tried, has students sign up for a lab in groups of three to five (one of the nonreport labs). Each group meets with the instructor outside of class for assistance in their investigations. They prepare a presentation of their results for the class with their own handouts and materials, and these presentations are synchronized with the curriculum and scheduled throughout the semester.

This second type of group work has the advantage over the former in that students receive individual attention from the instructor, and they have the opportunity to work on

Table 2. Sample Sets of Lab Assignments

Lab	Option A	Option B	Topics	Additional material
Maternal smoking and infant health I		×	summary statistics	
Student use of video games	×		simple random sampling	ratio estimator
Household radon levels		×	stratified sampling	parametric bootstrap
Patterns in DNA		×	estimation and testing	Poisson process
Crab molting	×		regression	prediction
Hispanic voting behavior	×		regression	weighted regression
Snow gauge calibration		×	simple linear model	inverse regression
Maternal smoking and infant health II	×		multiple linear regression	indicator variables
Transgenic mice with down syndrome	×		analysis of variance	unbalanced design
Designing a paper helicopter		×	analysis of variance	2 × 2 factorial

NOTE: The data and investigations for these labs are available on the Web site: www.stat.berkeley.edu/users/nolan/statlabs.

the lab in greater detail than if it were only discussed in class. However, this approach exposes students to a fewer number of labs. As for the first approach, students seem to really enjoy working on problems in class where they receive immediate feedback from the instructor and see the variety of ideas their classmates have come up with.

4.3 Reports

Students work outside of class on a lab for two to three weeks. Their time is spent analyzing data and preparing reports on their findings. The datasets are complex, the analysis is open ended, and they must synthesize their findings coherently on paper. The students find this work very challenging, and we sometimes allow them to work in groups of two or three on their assignments.

It can be difficult to grade the lab reports because the investigations allow students to be creative in their solutions to the problem. We usually break down the score into four equal parts: one part for composition and presentation, including statistically sound statements; one part for a basic minimum set of analyses; one for relevant, readable, and understandable graphs and tables; and finally we make a list of several advanced analyses and look for a subset of these in the write-up. Sometimes we also request an appendix to the report for technical material.

The lab reports typically constitute 40% to 50% of the course grade, the remaining being grades for homework and exams. We also think it is important to include questions related to the labs on exams, because it maintains consistency with the approach we have taken in teaching theoretical statistics through applications.

4.4 An Alternative

M. Hahn at Tuft University teaches an undergraduate course similar to ours. The course requires three semesters of calculus and one semester of probability, and there are 15 to 20 students in the class, including statistics majors, mathematics majors, and graduate students in economics, biology, and engineering. The class meets 3 hours a week with the instructor, and there is an optional 1.5-hour problem-and-computer session led by a teaching assistant.

Last semester, she adopted some of our labs for her course, which she ran somewhat differently than discussed so far. The students worked on three labs—maternal smoking and infant health, video game usage, and patterns in DNA—plus a final project of their own choosing. Together, project and lab reports were worth about one-third of the course grade.

The course was run in a seminar style. Students were responsible for reading the background materials on their own. Class time had plenty of discussion and brainstorming on how to solve the problems in the three labs, and instruction focused on the nature of the statistical problems needed to be solved to motivate the theoretical material to be covered. Assignments were not set in advance, but made on a daily or weekly basis. This allowed the course to be somewhat flexible in meeting demands raised by the students. As a result, all the material traditionally included in the course was still covered, new mini-topics were interjected as the need arose, and students felt that class time was essential.

The inclusion of the research project has an advantage over sole reliance on the labs we have prepared in that students tend to have greater enthusiasm for projects of their own choosing, which can boost retention. Students also can present their projects in class and their projects can be the source for theoretical exam questions. Projects can be problematic, however, for it can be difficult to ensure that all students wind up with successful, challenging projects.

4.5 Computing

Bringing the computer into the theoretical course enables us to go far beyond the traditionally small, artificial examples found in textbooks. But care is taken to keep the demands made upon the student at an appropriate level. For example, the teaching assistant is mainly responsible for providing assistance in the weekly section meeting on how to use the statistical software. Often the section meets in a laboratory room, where students double up at workstations to work on the assignment and the teaching assistant provides advice as needed. Other times, the assistant distributes handouts with sample code. In a course without a teaching assistant, handouts and office hours may be able to meet the students' needs, or a mini-course on the software could be organized.

5. EVALUATION

The student feedback to our labs has been very positive. In anonymous end-of-term evaluations, students report that the labs helped them better understand statistical theory. They liked the practicality and concreteness of the course and the relevance to the real world. However, not all of the response to the laboratory exercises has been favorable. Most negative reactions center around computer anxiety, and students have requested more help with using software packages. We are considering offering a short course in the use of statistical software, for most students have no prior experience with such software.

In rating the overall effectiveness of the course, the median score was 6 with a mean of 5.6 on a scale of 1 (poor) to 7 (excellent), with 143 students from 5 classes reporting. The departmental average for courses at this level is 5.2.

We have also been pleased with feedback from students after the course. Many students report they have used their lab reports as writing samples for job applications, and that interviewers have focused on one of the lab projects in job interviews.

Hahn also met with success by including labs in her course. She found the students were motivated to learn the theoretical material necessary to understand and complete the lab assignments, and that they approached the topics with vigor since they saw them as relevant. She reports that

> The students were extremely enthusiastic about the labs. They took them as a challenge and liked the fact that they were not cookbookish. The labs generated more discussion among the students than anything I have used in the past.

The change in the course also seemed to generate an atypical interest among students in continuing their study of statistics. Students requested advanced statistics courses to be offered, they asked for advice and assistance on studying statistical applications to engineering and biology, and several decided to apply for actuarial jobs.

Although some of our evidence is anecdotal, the success of the labs is encouraging.

6. CONCLUSION

Recently Cobb and Moore (1997) called for the design of a better one-semester statistics course for mathematics majors that both strengthens their mathematical skills and integrates data analysis into the curriculum. Others have called for similar courses (Foster and Smith 1969; Hogg et. al. 1985; Kempthorne 1980; Moore and Roberts 1989; Mosteller 1988; Petruccelli, Nandram, and Chen 1995; Whitney and Urquhart 1990) and many consider training in statistical thinking important (e.g., Daisley 1979; Joiner 1989; Schuyten 1991; Riffenburgh 1995; and Nash and Quon 1996). But it is a challenge to bring statistical thinking into the advanced undergraduate mathematical statistics course. We advocate that we are better able to, as Mosteller puts it, affect broadly the minds and lives of our students if we include in the curriculum case studies that are derived from problems with important scientific questions that contain rich background information, and are integrated with development of the theoretical material.

[Received June 1998. Revised March 1999.]

REFERENCES

Chee, M. S., Bankier, A. T., Beck, S., Bohni, R., Brown, C. M., Cerny, R., Hosnell, T., Hutchinson III, C. A., Kourzarides, T., Martignetti, J. A., Preddie, E., Satchwell, S. C., Tomlinson, P., Weston, P. M., and Barell, B. G. (1990), "Analysis of the Protein Coding Content of Human Cytomegalovirus Strain ad169," *Current Topics in Microbiology and Immunology*, 154, 126–169.

Cobb, G. W., and Moore, D. S. (1997), "Mathematics, Statistics, and Teaching," *American Mathematical Monthly*, 104, 801–823.

Daisley, P. (1979), "Statistical Thinking Rather Than Statistical Methods," *The Statistician*, 28, 231–239.

Foster, F. G., and Smith, T. M. F. (1969), "The Computer as an Aid in Teaching," *Applied Statistics*, 18, 264–269.

Gonick, L., and Wheelis, M. (1991), *The Cartoon Guide to Genetics*, New York: Harper Perennial.

Hogg, R. V. (1985), "Statistical Education for Engineers: An Initial Task Force Report," *The American Statistician*, 39, 21–24.

Joiner, B. L. (1989), "Statistical Thinking: What to Teach and What Not to Teach Managers," *ASA Proceedings*, 150, Alexandria, VA: American Statistical Association, 448–461.

Kempthorne, O. (1980), "The Teaching of Statistics: Content Versus Form," *The American Statistician*, 34, 17–21.

Leung, M. Y., Blaisdell, B. E., Burge, C., and Karlin, S. (1991), "An Efficient Algorithm for Identifying Matches With Errors in Multiple Long Molecular Sequences," *Journal of Molecular Biology*, 221, 1367–1378.

Moore, T. L., and Roberts, R. A. (1989), "Statistics at Liberal Arts Colleges," *The American Statistician*, 43, 80–85.

Mosteller, F. (1988), "Broadening the Scope of Statistics and Statistical Education," *The American Statistician*, 42, 93–99.

Nash, J. C., and Quon, T. K. (1996), "Issues in Teaching Statistical Thinking With Spreadsheets," *Journal of Statistics Education* (online at http://www.amstat.org/publications/jse/), 4, 1.

Petruccelli, J. D., Nandram, B., and Chen, M.-H. (1995), "Implementation of a Modular Laboratory and Project-Based Statistics Curriculum," in *ASA Proceedings of the Section on Statistical Education*, Alexandria, VA: American Statistical Association, pp. 165–170.

Riffenburgh, R. H. (1995), Infusing Statistical Thinking into Clinical Practice," in *ASA Proceedings of the Section on Statistical Education*, Alexandria, VA: American Statistical Association, pp. 5–8.

Schuyten, G. (1991), "Statistical Thinking in Psychology and Education," in *Proceedings of the International Conference on Teaching Statistics*, pp. 486–489.

Whitney, R. E., and Urquhart, N. S. (1990), "Microcomputers in the Mathematical Sciences: Effects on Courses, Students, and Instructors," *Academic Computing*, 4, 14.

Chapter 10
Genetic Recombination

Mary Sara McPeek

Genetic recombination and genetic linkage are dual phenomena that arise in connection with observations on the joint pattern of inheritance of two or more traits or genetic markers. For example, consider two traits of the sweet pea, *Lathyrus odoratus*, an organism studied in depth by Mendel [9]: flower color, with purple (dominant) and red (recessive) phenotypes, and form of pollen, with long (dominant) and round (recessive) phenotypes. Under the Mendelian model for flower color (recast in more current terminology), each plant carries two alleles for flower color, one inherited from each parent, where each allele can be one of two types, denoted P and p. The pair of alleles carried by a plant is known as its genotype. Plants with genotype PP or Pp have purple flowers, while plants with genotype pp have red flowers. Mendel's First Law can be interpreted as specifying that a parent plant passes on a copy of one of its two alleles to each offspring, with each parental allele having an equal chance of being copied, and with this occurring independently across offspring and across parents. Similarly, each plant carries two alleles for form of pollen, where each of these can be L or l. Plants with genotype LL or Ll have long pollen, while plants with genotype ll have round pollen. Suppose one crossed a true-breeding parental line having purple flowers and long pollen (all individuals having genotype $PPLL$) with a true-breeding parental line having red flowers and round pollen (all individuals having genotype $ppll$). Then the offspring of that cross, known as the F_1 generation, would all have genotype $PpLl$, resulting in purple flowers and long pollen. Suppose a backcross were performed, in which F_1 individuals were crossed with individuals from the $ppll$ parental line. In this example, genetic **linkage** would refer to a tendency for pairs of alleles inherited from the same parent, such as the pair PL or the pair pl, to be transmitted together during meiosis, while genetic **recombination** would refer to the event that an individual transmits a pair of alleles that were inherited from different parents, such as the pair Pl or pL. If we let $0 \leq \theta \leq .5$ denote the **recombination fraction**, which is the probability of a recombination between

M.S. McPeek
Departments of Statistics and Human Genetics, University of Chicago
e-mail: mcpeek@uchicago.edu

S. Dudoit (ed.), *Selected Works of Terry Speed*, Selected Works in Probability and Statistics, 397
DOI 10.1007/978-1-4614-1347-9_10, © Springer Science+Business Media, LLC 2012

the genes for these two traits in a single meiosis, then in the backcross offspring, we expect individuals with genotypes *PpLl*, *ppll*, *Ppll* and *ppLl* to occur with relative frequencies $(1-\theta)/2$, $(1-\theta)/2$, $\theta/2$ and $\theta/2$, respectively.

A long-standing, important application of the ideas of linkage and recombination is to construction of genetic maps [15] and to subsequent localization of genes (or other genetic variants of interest) on those maps. The key observation is that the recombination fraction between a pair of genetic markers tends to increase with the chromosomal distance between them, with markers on different chromosomes having recombination fraction .5. Thus, by merely observing patterns of joint inheritance of traits, one can make inference about which trait genes lie on the same chromosome, chromosome, and make estimates of distances between them. The basic ideas of and mathematics behind linkage and recombination were developed early in the 20th century [10, 15, 5]. Notably, these problems attracted the interest of R. A. Fisher [3].

Starting in the early 1980s, there was a resurgence of interest in the problem of genetic map construction, spurred by the development of recombinant DNA technology, which resulted in the ability to collect genotype data on large numbers of neutral genetic markers throughout the human genome [1] as well as genomes of model organisms. It was not long after these technological breakthroughs occurred that Terry shifted much of his energy and interest into the field of statistical genetics, near the beginning of the explosion of new data and resulting need for new statistical models and methods. In human data, the map construction problem called for more sophisticated statistical analysis than that typically required in experimental organisms. In model organisms, experimental crosses can often be planned in such a way that it is feasible to simply observe the relative frequency of recombinants in any given interval and convert it to a distance using a "map function", an analysis method that we will call the "two-point analysis." However, in humans, crosses cannot be planned, and so any given human meiosis would typically be uninformative for some of the markers of interest. (For example, in the sweet pea example above, all meioses from an individual with genotype *Ppll* would be uninformative for recombination between these two genes, because the recombinant and non-recombinant allele pairs are indistinguishable.) When many genetic markers are considered simultaneously in each meiosis, and many meioses from different individuals (with different patterns of informativeness) are analyzed together, substantial additional information, beyond that available from a two-point analysis, can typically be obtained by a joint analysis using a suitable statistical model for joint recombination events among a collection of genetic markers.

Thus, the statistical challenges of genetic mapping in humans naturally led to consideration of probability models for the crossover process that causes the observation of recombination. In humans and other diploid eukaryotes, crossing over takes place during a phase of meiosis in which the two parental versions of a given chromosome have each been duplicated, and all four resulting strands or chromatids are lined up together, forming a tight bundle. Crossovers can be modeled as points located along this bundle, with each crossover involving exactly two of the four chromatids. It is assumed that the two chromatids involved in any particular crossover are nonsister chromatids, that is, the two chromatids cannot be the two identical copies of one of the parent's versions of the chromosome. After crossing over has occurred, the four resulting chromatids are each mosaics of the original parental chromosomes. Keeping in mind this framework, one can consider two key aspects of the model: (1) the distribution of crossover points along the bundle of four chromatids and (2) the choice of nonsister pair of chromatids involved in each crossover. Perhaps the simplest useful model is the no-interference model of Haldane [5], which models aspect (1) by assuming that the crossover points form a homogeneous Poisson process and models aspect (2) by assuming that each nonsister pair is equally likely to be chosen for each crossover, independently across crossovers. **Interference** refers to deviation from Haldane's model. Interference, in the form of local inhibition of crossover points on a resulting single chromatid, was readily apparent in early *Drosophila* data [16, 11]. It is convenient to refer to failure of assumption (1) of Haldane's model as **crossover interference** and failure of assumption (2) of Haldane's model as **chromatid interference**.

Under the assumption of no chromatid interference (NCI), Speed et al. [14] derive a set of constraints, on the multilocus recombination probabilities, that are necessary and sufficient to ensure the existence of a counting process model for the distribution of crossover points along the bundle of four chromatids. They apply these constraints to prove a consistency result for the maximum likelihood estimate of the map order of a finite number of genetic markers along a chromosome. Specifically, they show that, under the assumption of NCI, in the case of complete data, i.e. when all meioses are informative for all markers, if maximum likelihood estimation is performed assuming the Haldane model, then the MLE will converge almost surely to the true map order, even when the true crossover point process is not Poisson (it can be any counting process).

The idea that the assumption of NCI imposes constraints on multilocus recombination probabilities is developed further in Zhao et al. [18], in which the main goal is assessment of the empirical evidence for chromatid interference. This paper extends the constraints from single spore data (such as that from humans and *Drosophila*) to tetrad data (from organisms such *Neurospora crassa*, *Saccharomyces cerevisiae* and *Aspergillus nidulans*) in which data on all 4 chromatid strands are available for each meiosis, providing much more information about strand choice and, hence, allowing a more powerful test of the NCI assumption. An efficient iterative algorithm for maximum likelihood estimation under the constraints is developed, and a likelihood ratio test is proposed to assess whether there is evidence that the constraints are not satisfied by the multinomial model assumed to generate the data. An empirical bootstrap approach is used to assess significance. Some of the experiments did provide evidence for chromatid interference, but overall there was no consistent pattern. The extent and type of chromatid interference seemed to vary across organisms and across experiments. Because the loci considered in these experiments are functional genes, as opposed to neutral markers, it is possible that differential viability may play a role in the results as well. In single-spore data, in particular, the constraints imposed by NCI are rather weak, and the available data do not provide much power to contradict them. Therefore, it seemed reasonable to assume NCI and focus attention on models for the crossover process.

Because the Haldane no-interference model was so clearly contradicted by most of the available, relevant data, Terry was somewhat concerned about relying on it for map inference. If a more flexible, yet still parsimonious and tractable, model could be developed and shown to fit the data better, Terry reasoned, it could be useful for a range of applications in genetic inference. This problem is addressed by McPeek and Speed [8], in which a range of point process models, involving one or two additional parameters, are fit to *Drosophila* data by maximum likelihood. Goodness of fit of the models is assessed, and the pattern of interference generated by each model is compared to that in data. The most promising model that emerges from this study, the **gamma model**, is a stationary gamma renewal process on four strands, combined with the assumption of NCI to generate a thinned process. In addition to fitting the data better and providing a pattern of interference that mimics that in data, the gamma model is also parsimonious and, when an integer shape parameter is used, results in efficient computational methods. This promising model is further developed in Zhao et al. [19], in which the gamma model with integer shape parameter is referred to as the **chi-square model** because it results in a stationary renewal process having chi-square interarrivals (with even degrees of freedom) for the process on a single strand. The model is fit to datasets from a number of different organisms, with different datasets from the same organisms having similar estimated shape parameter. The results of the analyses suggest that it may be reasonable to use an organism-specific shape parameter to model interference.

In a closely-related line of research, Terry and colleagues sought to connect probability modeling of the crossover process with the initially mysterious-seeming map functions commonly used in two-point analysis. A **map function** is used to convert probability of recombination across an interval to genetic distance of the interval, where **genetic distance** is defined as the expected number of crossovers per strand per meiosis. A difficulty in application of map functions to multilocus analyses is that when there are more than three markers, the multilocus recombination probabilities cannot be uniquely determined from the map function [3]. Earlier work [4, 13, 7] had proposed to solve this identifiability problem by constraining the probability of an odd number of crossovers across a union of disjoint intervals to depend only on the total length of these intervals. However, this is not a biologically plausible assumption, and, as shown by Evans et al. [2], assuming NCI, the class of count-location models [6, 12] is the only class of models having map functions that satisfy this constraint. Zhao and Speed [17] remove this biologically implausible constraint, and instead solve the general problem of developing stationary renewal process models that can generate specific map functions. They show that in most cases of previously-proposed map functions, one can construct a stationary renewal process that generates the map function. Furthermore, they show that this stationary renewal process can typically be approximated quite well by the gamma or chi-square model. The useful practical consequence of this is that two-point analyses using a particular map function can easily be extended to more informative multipoint analyses, an approach that is particularly valuable in the presence of missing data.

References

[1] D. Botstein, R. L. White, M. Skolnick, and R. W. Davis. Construction of a genetic linkage map in man using restriction fragment length polymorphisms. *Am. J. Hum. Genet.*, 32:314–331, 1980.

[2] S. N. Evans, M. S. McPeek, and T. P. Speed. A characterisation of crossover models that possess map functions. *Theor. Popul. Biol.*, 43:80–90, 1993.

[3] R. A. Fisher. The theory of linkage in polysomic inheritance. *Phil. Trans. Roy. Soc. B*, 233:55–87, 1947.

[4] H. Geiringer. On genetic map functions. *Ann. Math. Statist.*, 142:1369–1377, 1944.

[5] J. B. S. Haldane. The combination of linkage values, and the calculation of distances between the loci of linked factors. *J. Genet.*, 8:299–309, 1919.

[6] S. Karlin and U. Liberman. A natural class of multilocus recombination processes and related measures of crossover interference. *Adv. Appl. Prob.*, 11: 479–501, 1979.

[7] U. Liberman and S. Karlin. Theoretical models of genetic map functions. *Theor. Popul. Biol.*, 25:331–346, 1984.

[8] M. S. McPeek and T. P. Speed. Modeling interference in genetic recombination. *Genetics*, 139:1031–1044, 1995.

[9] G. Mendel. Versuche über pflanzenhybriden. *Verh. Naturf. Ver. Brünn*, 4:3–44, 1866.

[10] T. H. Morgan. Random segregation versus coupling in Mendelian inheritance. *Science*, 34:384, 1911.

[11] H. J. Muller. The mechanism of crossing-over. *Am. Nat.*, 50:193–221, 284–305,350–366,421–434, 1916.

[12] N. Risch and K. Lange. An alternative model of recombination and interference. *Ann. Hum. Genet.*, 43:61–70, 1979.

[13] F. W. Schnell. Some general formulations of linkage effects in inbreeding. *Genetics*, 46:947–957, 1961.

[14] T. P. Speed, M. S. McPeek, and S. N. Evans. Robustness of the no-interference model for ordering genetic markers. *Proc. Natl. Acad. Sci. USA*, 89:3103–3106, 1992.

[15] A. H. Sturtevant. The linear arrangement of six sex-linked factors in *Drosophila*, as shown by their mode of association. *J. Exp. Zool.*, 14:43–59, 1913.

[16] A. H. Sturtevant. The behavior of the chromosomes as studied through linkage. *Zeit. f. ind. Abst. u. Vererb.*, 13:234–287, 1915.

[17] H. Zhao and T. P. Speed. On genetic map functions. *Genetics*, 142:1369–1377, 1996.

[18] H. Zhao, M. S. McPeek, and T. P. Speed. Statistical analysis of chromatid interference. *Genetics*, 139:1057–1065, 1995.

[19] H. Zhao, T. P. Speed, and M. S. McPeek. Statistical analysis of crossover interference using the chi-square model. *Genetics*, 139:1045–1056, 1995.

Proc. Natl. Acad. Sci. USA
Vol. 89, pp. 3103–3106, April 1992
Genetics

Robustness of the no-interference model for ordering genetic markers

(multilocus recombination/chromosome mapping/no chromatid interference)

T. P. Speed, M. S. McPeek, and S. N. Evans

Department of Statistics, University of California, Berkeley, CA 94720

Communicated by Peter J. Bickel, December 30, 1991

ABSTRACT **Under the assumption of no chromatid interference, we derive constraints on the probabilities of the different recombination patterns among $m + 1$ genetic loci. An application of these constraints is a proof that the ordering of the loci that maximizes the likelihood under the assumption of no interference is, in fact, a consistent estimator of the true order even when there is interference.**

Genetic mapping involves ordering a set of markers on a chromosome and finding distances between them. One way this may be done is through analysis of data on meiotic recombination between the markers. Meiotic recombination is believed to be the result of crossing-over between nonsister chromatids during the pachytene phase of meiosis. If a particular chromatid passed on in meiosis was involved in an odd number of crossovers between two loci, a *recombination* is said to have taken place between the two loci (see Fig. 1).

It is important to keep in mind that crossing-over takes place in the four-stranded state, when each chromosome has duplicated to form two sister chromatids. In that case, the two aspects relevant to recombination are (*i*) the distribution of crossovers along the chromosome and (*ii*) which pair of nonsister chromatids is involved in each crossover. Two simplifying assumptions are often made. The first is that the locations of different crossovers are independent and identically distributed (i.i.d.) along the chromosome. The second is that each pair of nonsister chromatids is equally likely to be involved in a crossover, independent of which were involved in other crossovers. If the occurrence of a crossover influences the probability of another's occurring nearby, in violation of the first assumption, it is termed *crossover position interference*. Following Whitehouse (1), we prefer this over the traditional term *chiasma interference*. If the second assumption is violated, it is termed *chromatid interference*. There is a considerable body of data demonstrating both kinds of interference (1), but on the whole the extent of chromatid interference seems slight. On the other hand, position interference can be substantial and can take different forms. In what follows, we will permit an arbitrary crossover location point process, but we will assume no chromatid interference.

A General Model with No Chromatid Interference

Assuming no chromatid interference, we model the occurrence of crossovers along a chromosome as a realization of a point process, with the points corresponding to the locations of the crossovers. That is, we associate the chromosome with the interval [0, 1] and require (*i*) a distribution for n = the total number of points in the interval and (*ii*) for each $n \geq 1$, the joint distribution of the positions of the points,

The publication costs of this article were defrayed in part by page charge payment. This article must therefore be hereby marked "*advertisement*" in accordance with 18 U.S.C. §1734 solely to indicate this fact.

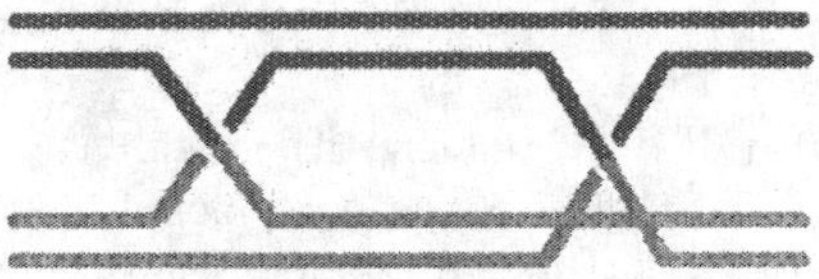

FIG. 1. Three-stand double crossover.

given that their total number is n. Furthermore, we require that the point process be *simple*: i.e., with probability 1, no two points shall occupy the same location. A simple point process on $\mathfrak{R}$ or a subset of $\mathfrak{R}$ is known as a *counting process*.

Define the *avoidance function* or *zero function* Z of the process by $Z(A) = \Pr\{\text{no points in A}\}$, for each measurable set A. It is well known that the distribution of a simple point process on a complete separable metric space is determined by the values of the avoidance function on the Borel sets (2). We shall find that the avoidance function of the crossing-over process is closely related to the recombination probabilities.

Following is a well-known derivation (see, e.g., ref. 3) of Mather's Formula (4), which expresses the chance of recombination in an interval in terms of the avoidance function, in the case of no chromatid interference. If we assume that there is no chromatid interference, then each crossover is equally likely to involve any of the four possible nonsister pairs of chromatids, independent of which pairs are involved in other crossovers. In that case, if there are $n > 0$ crossovers between loci at locations A and B, $0 \leq A < B \leq 1$, then any given chromatid has probability

$$\binom{n}{i} \times \frac{1}{2^n}$$

of being involved in exactly i of them, for $0 \leq i \leq n$. In a given meiotic product (sperm or oocyte), a recombination between A and B will have occurred if the chromatid passed on in the meiosis has been involved in an odd number of crossovers between A and B. Thus, the chance of a recombination given that $n > 0$ crossovers have occurred is

$$\frac{1}{2^n} \times \sum_{i=0}^{\left[\frac{n-1}{2}\right]} \binom{n}{2i+1} = \frac{1}{2},$$

so the chance of a recombination is $\frac{1}{2} \times \Pr\{n > 0\} = \frac{1}{2} \times (1 - Z([A, B]))$.

More generally, consider the recombination pattern among $m + 1$ ordered loci. Let A_j denote the interval between loci j and $j + 1$. Let p_x, $x = (x_1, \ldots, x_m) \in \{0, 1\}^m$, denote the probability of the event of x_j recombinations in A_j, $j = 1, \ldots, m$. Let Z_x denote $Z(\cup\{A_j : x_j = 1\})$, the probability that the point process avoids all of the intervals A_j with $x_j = 1$. In the genetics literature, p_x is known as the *crossover distribution*, and Z_x is known as the *linkage value* associated with x. The following relationship between Z_x and p_x is well known:

$$Z_x = \sum_y (-1)^{y \cdot x} p_y$$

and

$$p_x = \frac{1}{2^m} \times \sum_y (-1)^{y \cdot x} Z_y,$$

where the sum is over all $y = (y_1, \ldots, y_m) \in \{0, 1\}^m$ and $y \cdot x = \Sigma_{j=1}^m y_j x_j$. That is, (Z_x) is the $\mathbb{Z}_2^m$ Fourier transform of (p_x), and (p_x) is the inverse $\mathbb{Z}_2^m$ Fourier transform of (Z_x) (5–7).

This relationship between the distribution of recombinations among $m + 1$ loci and the avoidance function can be exploited to get necessary and sufficient conditions for the distribution of recombinations among $m + 1$ loci to be compatible with at least one underlying crossover point process. It is easy to get necessary conditions on the p_x values as follows: define q_x, $x = (x_1, \ldots, x_m) \in \{0, 1\}^m$ to be the probability of the event of no crossovers in each of the intervals A_j with $x_j = 0$ and at least one crossover in each of the intervals A_j with $x_j = 1; j = 1, \ldots, m$. Note that (q_x) and (Z_x) refer to the distribution of crossovers along the four-stranded chromosome, while (p_x) refers to the recombination distribution on a single strand. Then, clearly

$$Z_x = \sum_{y \leq x'} q_y,$$

where $y \leq x'$ denotes $y_j \leq 1 - x_j, j = 1, \ldots, m$. Plugging into the formula for p_x in terms of Z_x, we get

$$p_x = \sum_{y \geq x} \frac{1}{2^{y \cdot 1}} q_y.$$

Alternatively, the relation between the ps and the qs may be proved from *Lemma 1* of ref. 8, by induction. Inverting, we have

$$q_x = \sum_{y \geq x'} (-1)^{(y-x')\cdot 1} Z_y$$

and

$$q_x = 2^{x \cdot 1} \times \sum_{y \geq x} (-1)^{(y-x)\cdot 1} p_y.$$

Now it is clearly necessary that we have

(*i*) $\sum_x q_x = 1,$

and

(*ii*) for all x, $q_x \geq 0$.

Expressed in terms of the p_x values, these conditions are

(*i*) $\sum_x p_x = 1,$

and

(*ii*) for all x, $0 \leq \sum_{y \geq x} (-1)^{(y-x)\cdot 1} p_y.$

We further note that these conditions can be expressed in terms of the Z_x values as

(*i*) $Z(\emptyset) = 1,$

and

(*ii*) for all x, $0 \leq \sum_{y \geq x} (-1)^{(y-x)\cdot 1} Z_y.$

If we assume that any given pattern of crossovers is possible, that is, that each of the parameters q_x is nonzero, then the weak inequalities satisfied by the parameters (p_x) and (Z_x) become strict. Of course, the constraints in terms of the p_x values are of greatest interest, since the p_x values correspond to the observed data.

In fact, these conditions on the p_x values are also sufficient, under the assumption of no chromatid interference, for the existence of an underlying point process of crossovers that would be compatible with the p_x values. To show this, we need only construct, given any set of q_x values with $q_x \geq 0$ and $\Sigma_x q_x = 1$, a point process on [0, 1] that is compatible with them. This is easily done by fixing one point in each interval A_j and allowing crossovers only at those points. The pattern of crossovers among those m points is then chosen to be x with probability q_x, where for all j,

$$x_j = \begin{cases} 1 & \text{if there is a crossover in } A_j \\ 0 & \text{otherwise} \end{cases}.$$

Note that our constraints are stronger than those required by Karlin and Liberman (5) for their class of "natural" recombination distributions. They require

(*i*) $Z(\emptyset) = 1,$

and

(*ii*) for all x, $Z_x \geq 0$.

An Application of the Constraints: Robustness of the Poisson Model

PROPOSITION. *Suppose we have recombination data on* m *+ 1 loci whose true order is unknown. Assume that there is no chromatid interference, but crossover location interference of an arbitrary form may be present. That is, the positions of crossovers and the occurrence of recombinations are given by a model of the type described in the preceding section. Assume further that we are in the nondegenerate situation in which each of the parameters* q_x *is nonzero. Suppose that our data are from* n *meioses and for each meiosis we can observe whether or not a recombination occurred between each of the* ½m(m + 1) *pairs of loci. Suppose that for each possible order of the loci we fit the data by maximum likelihood under the assumption of no cross-over–location interference. Then with probability* 1 *for* n *sufficiently large, the maximized likelihood will be largest for the true order.*

If we arbitrarily choose an ordering of the loci $\mathbf{f} = (f_1, \ldots, f_{m+1})$, where $\mathbf{f}$ is a permutation of $(1, \ldots, m + 1)$ and fit the data by maximum likelihood, assuming no interference, we will be fitting m parameters $(\theta_{f_1,f_2}, \ldots, \theta_{f_m,f_{m+1}})$, where $\theta_{f_j,f_{j+1}}$ is the chance of a recombination between loci f_j and f_{j+1}. The proof (see *Appendix*) lies in showing that for any nonidentity permutation $\mathbf{f}$, with probability 1 for n sufficiently large, the collection $\{\hat{\theta}_{f_1,f_2}, \ldots, \hat{\theta}_{f_m,f_{m+1}}\}$ dominates $\{\hat{\theta}_{1,2}, \ldots, \hat{\theta}_{m,m+1}\}$. That is, the two sets can be put into a one-to-one correspondence such that each element of the first set is larger than or equal to the corresponding element of the second set, with at least one strict inequality. This is because, with probability 1 for n sufficiently large, the constraints on the p_i values imply constraints on the data. Then it follows that the likelihood will be maximized by the true order.

Discussion

As we try to increase the resolution of genetic maps, we shall find that interference plays a greater role. Although estimation of order of loci under the assumption of no crossover position interference is consistent, still the number of data points n may need to be very large in practice. It is likely that with a reasonable model for interference one could use the data to estimate order more efficiently. We note that even in the absence of a specific interference model one could compare maximized likelihoods under different orders, where the likelihood is in terms of 2^m p_x values subject to the constraints, rather than in terms of m θ values.

Another area in which the constraints may be useful is in determining when a map function has an underlying crossover point process [an extension to Liberman and Karlin (6)]. Liberman and Karlin define a map function M (which converts expected number of crossovers to chance of recombination) to be "multilocus feasible" if it satisfies

$$(i) \qquad M(0) = 0$$

and

$$(ii) \qquad \text{for all } x,\ 0 \le \sum_{y \in \{0,1\}^m} (-1)^{y \cdot x}(1 - 2M(\gamma_y)),$$

where γ_y is the map distance of $\cup_{j:y_j=1} A_j$. We could replace *ii* with the more stringent condition

$$(ii') \qquad \text{for all } x,\ 0 \le \sum_{y \ge x} (-1)^{(y-x)\cdot 1}(1 - 2M(\gamma_y)).$$

The connection between map functions and crossover point processes has been determined (unpublished work).

Appendix

Proof of Proposition: As before, let A_j denote the interval between loci j and $j + 1$ in the true order, $j = 1, \ldots, m$, and let p_x, $x = (x_1, \ldots, x_m) \in \{0, 1\}^m$ be defined as before. Assume that the constraints of the general model described above hold with *strict* inequalities.

The data consist of r_x^n, $x = (x_1, \ldots, x_m) \in \{0, 1\}^m$, where r_x^n denotes the number of meioses in which x_j recombinations occurred in A_j, $j = 1, \ldots, m$. Note that we can still calculate the set of numbers $\{r_x^n\}$ without knowing the true order. We could do the calculation assuming an arbitrary order, and the resulting set of counts $\{r_x^n\}$ would be the same but for an (unknown) permutation of indices. Assuming that the recombination patterns in different meioses are i.i.d., $(\mathbf{r}^n)$ is distributed as Multinomial(n, $\mathbf{p}$), where $\mathbf{r}^n$ is the vector of r_x^n values and $\mathbf{p}$ is the vector of p_x values.

If we arbitrarily choose an ordering of the loci $\mathbf{f} = (f_1, \ldots, f_{m+1})$, where $\mathbf{f}$ is a permutation of $(1, \ldots, m + 1)$ and fit the data by maximum likelihood, assuming no interference, we will be fitting m parameters $(\theta_{f_1,f_2}, \ldots, \theta_{f_m,f_{m+1}})$, where $\theta_{f_j,f_{j+1}}$ is the chance of a recombination between loci f_j and f_{j+1}. Suppose we maximize the likelihood under the true order and under any other order $\mathbf{f}$ and compare the maximized likelihoods. In the case in which we can observe any recombination among the loci, the maximum likelihood estimates of the θs are very simple. For the true order, we have $\hat{\theta}_j = \min(½, n^{-1} \Sigma_{x:x_j=1} r_x^n)$, where $\hat{\theta}_j$ is the maximum likelihood estimate of the chance of a recombination in the interval A_j, $j = 1, \ldots, m$. For the order $\mathbf{f}$, we have $\hat{\theta}_{f_j,f_{j+1}} = \min(½, n^{-1} \Sigma_{x \in I} r_x^n)$ where I is the set whose members x all satisfy

$$\sum_{k:A_k \text{ lies between } f_j \text{ and } f_{j+1}} x_k \quad \text{is odd.}$$

For all $x \in \{0, 1\}^m$ set

$$c_x^n = 2^{x\cdot 1} \sum_{y \ge x} (-1)^{(y-x)\cdot 1} r_y^n$$

and consider the strict constraints $c_x^n > 0$, $x \in \{0, 1\}^m$. By the law of large numbers, the probability that these constraints are satisfied is 1 for n sufficiently large because of the constraints on $\mathbf{p}$. Assume that these constraints on $\mathbf{r}^n$ hold. In terms of the c_x^n values, we have $\hat{\theta}_j = ½n^{-1} \Sigma_{x:x_j=1} c_x^n$ and $\hat{\theta}_{f_j,f_{j+1}} = ½n^{-1} \Sigma_{x \in I} c_x^n$, where $I = \{x{:}x_k = 1$ for at least one A_k lying between f_j and $f_{j+1}\}$. Note that we have $\hat{\theta}_j$ and $\hat{\theta}_{f_j,f_{j+1}}$ both $\le ½$ under the constraints. From this representation, we can see that if A_k lies between f_j and f_{j+1}, then $\hat{\theta}_k \le \hat{\theta}_{f_j,f_{j+1}}$, since $c_x^n > 0$ for all x.

Now we will match each of the m $\hat{\theta}_{f_j,f_{j+1}}$ values in one-to-one correspondence with a $\hat{\theta}_k$ that is smaller than or equal to it. This will prove the proposition, for the function $g(x) = x\log(x) + (1 - x)\log(1 - x)$ is decreasing in $0 \le x \le ½$. Thus, if $L_{\mathbf{f}}$ is the maximized likelihood under order $\mathbf{f}$, and L_{true} is the maximized likelihood under the true order, we will have

$$L_{\mathbf{f}} = n \cdot \sum_{j=1}^{m} g(\hat{\theta}_{f_j,f_{j+1}}) \le n \cdot \sum_{j=1}^{m} g(\hat{\theta}_j) = L_{\text{true}}.$$

To see that the $\{\hat{\theta}_{f_j,f_{j+1}}\}$ can be matched with the $\{\hat{\theta}_k\}$ in such a way that the $\hat{\theta}_k$ corresponding to $\theta_{f_j,f_{j+1}}$ is no larger than it, we use P. Hall's matching theorem (see p. 401 of ref. 9). We associate each interval (f_j, f_{j+1}) in the ordering $\mathbf{f}$ with the *set* of intervals A_k in the true ordering that lie between f_j and f_{j+1}. As noted above, this ensures that $\hat{\theta}_{f_j,f_{j+1}}$ is greater than or equal to $\hat{\theta}_k$ for any k of this kind. The condition of Hall's theorem that must be checked is that any set $\{(f_j, f_{j+1}): j \in J\}$ of $|J|$ distinct intervals in the order $\mathbf{f}$ must contain at least $|J|$ distinct intervals in the original order. Then there is a matching with the property stated and the proof is complete.

To show that the condition holds, argue by induction on the number of loci. The condition clearly holds for two loci. Suppose it holds for M loci, and consider the case of $M + 1$ loci. Let $\{(f_j, f_{j+1}): j \in J\}$ be any set of $|J|$ distinct intervals in some order $\mathbf{f}$. Let $i = \min\{j \in J\}$, so f_i is an endpoint of exactly one interval in the set. Now consider the M loci $\{1, 2, \ldots, M + 1\} \setminus \{f_i\}$ and the $|J| - 1$ intervals $\{(f_j, f_{j+1}): j \in J \setminus \{i\}\}$. By the induction hypothesis, these intervals cover at least $|J| - 1$ distinct intervals in the original order of $\{1, 2, \ldots, M + 1\} \setminus \{f_i\}$. Note that if the $\{(f_j, f_{j+1}): j \in J \setminus \{i\}\}$ cover exactly k distinct intervals in the original order of $\{1, 2, \ldots, M + 1\} \setminus \{f_i\}$, then they must cover either k or $k + 1$ distinct intervals in the original order of $\{1, 2, \ldots, M + 1\}$. If at least $|J|$ intervals of the original order of $\{1, \ldots, M + 1\}$ are covered by $\{(f_j, f_{j+1}): j \in J \setminus \{i\}\}$, then we are done, so assume without loss of generality that exactly $|J| - 1$ distinct intervals of the original orders of both $\{1, \ldots, M + 1\}$ and $\{1, \ldots, M + 1\} \setminus \{f_i\}$ are covered. That is, the same number of intervals are covered in the original order whether locus f_i is included or not. Thus, if we say f_i is between loci $f_i - 1$ and $f_i + 1$ in the original order, then neither $(f_i - 1, f_i)$ nor $(f_i, f_i + 1)$ could be covered. Otherwise, $(f_i - 1, f_i + 1)$ would have to be covered by an element of $\{(f_j, f_{j+1}): j \in J \setminus \{i\}\}$, but then adding in locus f_i would add in one more interval covered, which contradicts our assumption. Since the interval (f_i, f_{i+1}) must contain at least one of $(f_i - 1, f_i)$ and $(f_i, f_i + 1)$, then $\{(f_j, f_{j+1}): j \in J\}$ covers at least $|J|$ distinct intervals in the original order of the $M + 1$ loci. The argument is similar if f_i is the first or last locus in the original order.

T.P.S. was supported in part by National Science Foundation Grant DMS-880237. M.S.M. was supported in part by a National Science Foundation predoctoral research fellowship. S.N.E. was supported in part by National Science Foundation Grant DMS-9015708.

1. Whitehouse, H. L. K. (1973) *Towards an Understanding of the Mechanism of Heredity* (St. Martin's, New York), 3rd Ed., pp. 111–112.
2. Daley, D. J. & Vere-Jones, D. (1988) *An Introduction to the Theory of Point Processes* (Springer, New York), pp. 216–218.
3. Karlin, S. & Liberman, U. (1983) *Adv. Appl. Probab.* **15**, 471–487.
4. Mather, K. (1938) *Biol. Rev.* **13**, 252–292.
5. Karlin, S. & Liberman, U. (1978) *Proc. Natl. Acad. Sci. USA* **75**, 6332–6336.
6. Liberman, U. & Karlin, S. (1984) *Theor. Popul. Biol.* **25**, 331–346.
7. Risch, N. & Lange, K. (1983) *Biometrics* **39**, 949–963.
8. Lange, K. & Risch, N. (1977) *J. Math. Biol.* **5**, 55–59.
9. Aigner, M. (1979) *Combinatorial Theory* (Springer, New York), pp. 152, 401.

Modeling Interference in Genetic Recombination

Mary Sara McPeek and Terence P. Speed

Department of Statistics, University of California, Berkeley, California 94720

Manuscript received December 17, 1993
Accepted for publication October 25, 1994

ABSTRACT

In analyzing genetic linkage data it is common to assume that the locations of crossovers along a chromosome follow a Poisson process, whereas it has long been known that this assumption does not fit the data. In many organisms it appears that the presence of a crossover inhibits the formation of another nearby, a phenomenon known as "interference." We discuss several point process models for recombination that incorporate position interference but assume no chromatid interference. Using stochastic simulation, we are able to fit the models to a multilocus Drosophila dataset by the method of maximum likelihood. We find that some biologically inspired point process models incorporating one or two additional parameters provide a dramatically better fit to the data than the usual "no-interference" Poisson model.

THE phenomenon of interference in genetic recombination was noticed very early this century by *Drosophila melanogaster* geneticists in THOMAS HUNT MORGAN's lab (STURTEVANT 1915; MULLER 1916). They found that simultaneous recombination in two or three nearby chromosomal intervals occurred much less often than would be expected under independence, and that the effect appeared to decrease with distance. At present the biological nature of genetic interference is still not well understood, nor has it been adequately modeled mathematically. Virtually all multilocus linkage analyses use the assumption of no interference. Although this assumption may give consistent results for locus ordering (SPEED *et al.* 1992) and estimation of recombination fractions in some cases, it clearly does not fit the data. It is natural to ask if such an analysis could be improved by the use of a reasonable interference model. What is required is a biologically plausible point process model for crossovers along a chromosome, which should be fit to data. The Drosophila dataset of MORGAN *et al.* (1935) is ideal for this purpose. Previous attempts to fit a crossover point process to the MORGAN *et al.* (1935) Drosophila dataset (COBBS 1978; RISCH and LANGE 1983; PASCOE and MORTON 1987; GOLDGAR and FAIN 1988; FOSS *et al.* 1993) have been severely limited by the difficulty of calculating the likelihood of the data under all but the simplest models. Using a Monte Carlo method, we are able to fit a wide range of point process models to the data.

BACKGROUND

In diploid eukaryotes, crossing over takes place during the pachytene phase of meiosis, when the two homologous versions of any particular chromosome have each been duplicated and all four resulting strands or chromatids are lined up together, forming a very tight bundle. We model crossovers as being points located along this bundle, and each crossover involves exactly two of the four chromatids. We assume that the two chromatids involved in any particular crossover are nonsister chromatids, that is, the two chromatids cannot be the two copies of a single homologous chromosome. After crossing over has occurred, the four resulting chromatids are mixtures of the original parental types. In Drosophila, for each meiosis, we will have information on only one of these four resulting chromatids. If that chromatid was involved in an odd number of crossovers between two loci, a recombination is said to have taken place between the two loci.

It is important to keep in mind that crossing over takes place among four chromatids. In that case the two aspects relevant to recombination are the distribution of crossovers along the bundle of four chromatids and which pair of nonsister chromatids is involved in each crossover. The concept of interference is usually divided into two parts, corresponding to these two aspects. First, we say that there is position or chiasma interference if the number and location of crossovers in a given region are not independent of the numbers and locations of crossovers in disjoint regions. Second, we say there is chromatid interference if it is not the case that each pair of nonsister chromatids is equally likely to be involved in a crossover, independent of which were involved in other crossovers. There is little consistent evidence of chromatid interference (ZHAO *et al.* 1995b).

We assume no chromatid interference and try to find a better-fitting point process model than the Poisson for

Corresponding author: Mary Sara McPeek, Department of Mathematics, University of Southern California, Los Angeles, CA 90089-1113. E-mail: mcpeek@galton.uchicago.edu

Genetics 139: 1031–1044 (February, 1995)

M. S. McPeek and T. P. Speed

the occurrence of crossovers along the chromosome. In the case of no chromatid interference, the point process on the single chromatid can be obtained from the point process on the bundle of four by independently thinning (*i.e.*, deleting) each point with probability $^1/_2$. This is because the given chromatid has chance $^1/_2$ of being involved in a particular crossover, independent of involvement in any others. As a result of assuming no chromatid interference, the chance of recombination across an interval increases monotonically as the interval is enlarged, with an upper bound of $^1/_2$.

The traditional measure of interference is the coincidence (STURTEVANT 1915; MULLER 1916), which is expressed as a ratio. The numerator is the chance of simultaneous recombination across both of two disjoint intervals on the chromosome. The denominator is the product of the marginal probabilities of recombination across the intervals.

$$C = \frac{r_{11}}{(r_{10} + r_{11})(r_{01} + r_{11})},$$

where C is the coincidence and r_{ij} is the chance of i recombinations across the first interval and j recombinations across the second interval. If there were no position interference, the coincidence would equal one. Observed coincidences tend to be near zero for small, closely-linked intervals, increasing to one for more distant intervals.

The recombination point processes considered here are stationary in terms of some distance metric, although this distance metric is generally not equivalent to physical distance. This point will be discussed further below. For each of the point process models considered here, there is more than one such choice of a distance metric that will make the process stationary; these metrics are the same up to a multiplicative constant. The genetic distance associated with a chromosome interval is defined to be the expected number of crossovers occurring on a single chromatid within that interval. The metrics in which the models discussed here are stationary are all constant multiples of genetic distance. We shall choose one for notational convenience and call it the stationary metric. Among other things stationarity of the model means that coincidence will be a function of the distances across each of the two intervals considered and the distance between the intervals (in terms of the stationary metric) but not their actual locations. This also implies that the intensity, $\mu = \lim_{h\to 0} h^{-1}$ Pr(at least 1 point in $[t, t+h)$), where h is measured in terms of the stationary metric, is the same for all t. Furthermore, for the models considered here the conditional intensity can be defined by $\mu(z) = \lim_{\delta\to 0} \lim_{\epsilon\to 0}$ ϵ^{-1} Pr(at least 1 point in $[t+z, t+z+\epsilon)$ | at least 1 point in $[t, t+\delta)$), where ϵ is measured in terms of the stationary metric. Intuitively, this is the intensity at $t + z$ conditional on a point at t, and by stationarity this depends only on z. When the conditional intensity exists, the coincidence function can be defined as the coincidence between two intervals, in the limit when the stationary widths of the intervals are allowed to go to zero, as a function of the stationary distance between the two intervals. The coincidence function is given by the ratio of the conditional intensity to the unconditional intensity

$$C(z) = \frac{\mu(z)}{\mu}.$$

Formulae of this type appear in FOSS *et al.* (1993) and LANDE and STAHL (1993). Note that coincidence is not a complete description of interference but measures interference only between pairs of intervals.

Multilocus linkage analysis: When one has a panel of genetic markers, one may be interested in making a linkage map, that is, in ordering the markers and calculating genetic distances between them. Alternatively, one may be interested in using linkage analysis to locate a new marker or gene of interest on a previously determined marker map. In both cases the most informative kind of analysis is multilocus linkage analysis, in which one considers recombination patterns among all the marker loci simultaneously.

The Drosophila dataset of MORGAN *et al.* (1935) consists of counts of recombination events among nine marker loci on the X chromosome. The nine loci under consideration are *scute, echinus, crossveinless, cut, vermilion, sable, forked, carnation* and *bobbed,* and they correspond to observed fly phenotypes. The actual positions of the loci on the X chromosome are assumed to be fixed but unknown, and counts of recombination events are made based on the observed physical characteristics of the fly, which are associated with alleles at the nine loci. In *Drosophila melanogaster* recombination occurs only in females. Consider the X chromosome inherited by a fly from its mother. This X chromosome will be some mixture of the two maternal grandparental X chromosomes because of crossing over between them. For each of the nine loci, it can be determined whether the offspring fly inherited its maternal grandmother or maternal grandfather's allele at that locus. Thus, there are 2^9 possible observed outcomes. However, each outcome has a complementary outcome that is considered equivalent, in terms of recombination, namely the one in which all the grandmaternal and grandpaternal alleles are switched. Therefore, there are 2^8 possible recombination outcomes, and the number of times each occurs out of 16,136 meioses is recorded in the MORGAN *et al.* dataset.

If the nine loci are ordered, we let A_j denote the interval between loci j and $j + 1$. We let $x = (x_1, \ldots, x_8)$, $x_j = 0$ or 1, $j = 1, \cdots 8$, denote the event of a recombination in each interval A_j for which $x_j = 1$, and no recombination in each interval A_j for which $x_j = 0$. Then each of the 2^8 possible recombination outcomes would correspond to one of the 2^8 possible x's.

In what follows we assume the order of the loci to be known. If the order were not known, the procedures described below could be repeated for each of a small number of candidate orders, and the estimated order would be the one whose maximized likelihood was highest. We assume that each event x occurs with some fixed probability p_x. These events correspond to what is observed in the data. For each x, we wish to calculate its probability p_x under each of the point process models we consider in order to fit the models to the data by maximum likelihood.

Recall that event x occurs when a given chromatid is involved in an odd number of crossovers in each of the intervals A_j for which $x_j = 1$ and an even number of crossovers in each of the intervals A_j for which $x_j = 0$. A set of related events will be denoted $y = (y_1, \ldots, y_8)$, $y_j = 0$ or 1, $j = 1, \ldots, 8$. y is the event that, on the bundle of four chromatids, at least one crossover occurs in each of the intervals A_j for which $y_j = 1$ and no crossovers occur in each of the intervals A_j for which $y_j = 0$. Note that y is an event occurring on the bundle of four chromatids, whereas x refers to just one of the four chromatids. We let q_y denote the probability of the event y. The assumption of no chromatid interference gives a correspondence between these two sets of probabilities (WEINSTEIN 1936), namely

$$p_x = \sum_{y:1 \geq y \geq x} \frac{1}{2^{y \cdot 1}} q_y \quad \text{for all } x,$$

and inverting,

$$q_y = 2^{y \cdot 1} \times \sum_{x:1 \geq x \geq y} (-1)^{(x-y) \cdot 1} p_x \quad \text{for all } y,$$

where, for example, $y \cdot 1 = \sum_{j=1}^{8} y_j$, and $1 \geq y \geq x$ means $1 \geq y_j \geq x_j$ for all j. Thus, to get the recombination probabilities under different point process models, assuming no chromatid interference, it would suffice to calculate probabilities for the simpler events, y, corresponding to zero or nonzero crossovers in the locus intervals, on the bundle of four chromatids. Once we can calculate the q_y's in terms of the unknown parameters and from these, the p_x's, we write down the log-likelihood as

$$\text{constant} + \sum_x a_x \log(p_x),$$

where $x = (x_1, \ldots, x_8)$, $x_j = 0$ or 1, a_x is the number of times the event x occurs in the data and where the constant is irrelevant because it does not involve the unknown parameters. We would then maximize the likelihood over the unknown parameters.

MODELS

It is sometimes assumed that the problem of modeling interference is equivalent to that of formulating a map function, *i.e.*, a function from $[0, \infty)$ to $[0, 1/2]$ that maps expected number of crossovers to chance of a recombination. This is not the case. First, a map function is not a full-fledged model. By itself, a map function does not provide probabilities for multilocus recombination events when more than three loci are involved, and so it cannot be fit to data such as that described above. In an attempt to remedy this problem, LIBERMAN and KARLIN (1984) introduced multilocus map functions, which are defined as above, but with the additional property that the map function relationship between expected number of crossovers and chance of a recombination should hold on unions of disjoint intervals, as well as on intervals. Unfortunately, these multilocus map functions correspond to an extremely limited class of models under the assumption of no chromatid interference. EVANS *et al.* (1992) showed that under this assumption the only models that have multilocus map functions are the count-location models described by KARLIN and LIBERMAN (1979) and RISCH and LANGE (1979). These models have the undesirable property that the coincidence function is constant, *i.e.*, the level of interference, as measured by coincidence does not vary at all with the genetic distance between the intervals considered. In actual data coincidence seems to be close to zero for near intervals and close to one for more distant intervals (see Figure 2). For this reason the count-location model does not appear well suited to modeling interference. It is, however, easy to fit and has been previously fit to the Drosophila data of MORGAN *et al.* (RISCH and LANGE, 1983). We include it here for comparison. The model will be described in more detail below.

Poisson model: COX and ISHAM (1980) provide an introduction to point processes. The simplest crossover point process model we consider is the Poisson process. This model was proposed for recombination by HALDANE (1919) and continues to be virtually the only model used in linkage analysis. The model allows no interference at all, *i.e.*, crossing over in disjoint intervals is independent, or the presence of one crossover does not alter the chance of others occurring nearby. Thus, the coincidence function has the constant value 1. In the general formulation of the Poisson model, we let μ_t denote the intensity of the process at physical location t along the bundle of four chromatids, *i.e.*,

$$\mu_t = \lim_{h \to 0} \frac{P(\text{at least 1 point in } (t, t+h])}{h}.$$

We shall not consider cases in which this limit does not exist, nor in which μ_t is not integrable. Then the chance of no crossovers in a given interval $[a, b)$ is $\exp(-\int_a^b \mu_t dt)$, or $e^{-\mu(b-a)}$ in the homogeneous case where $\mu_t = \mu$ for all t. To get the probability of a given event y, *i.e.*, the event of at least one crossover in each interval for which $y_i = 1$ and no crossovers in each interval for which $y_i = 0$, we multiply the appropriate probabilities for each interval using independence.

If we let $l_1, \ldots, l_9$ denote the physical locations of the nine marker loci, then from recombination data the estimable quantities in the Poisson model are the genetic distances (*i.e.*, expected numbers of crossovers on a single chromatid) between the markers, namely,

$$d_1 = \frac{1}{2}\int_{l_1}^{l_2} \mu_t dt, \quad d_2 = \frac{1}{2}\int_{l_2}^{l_3} \mu_t dt, \quad \ldots,$$

$$d_8 = \frac{1}{2}\int_{l_8}^{l_9} \mu_t dt.$$

(The factor of $^1/_2$ occurs because each crossover on the bundle of four chromatids is assumed to involve a given chromatid with chance $^1/_2$.) Thus, from the data one cannot infer anything about possible inhomogeneity of the Poisson intensity, because this cannot be separated from the unknown locations of the marker loci. Similarly, one cannot estimate the actual locations of the markers nor the physical distances between them. Associating the interval $[0, 1]$ on the real line with the chromosomal segment between the outermost loci under consideration and letting μ_t and $0 = l_1 < \cdots < l_9 = 1$ be as before, let $\mu = \int_0^1 \mu_t dt$ and $M(z) = \mu^{-1} \int_0^z \mu_t dt$, for all $z \in [0, 1]$. Then M: $[0, 1] \to [0, 1]$ is monotone nondecreasing and onto (*i.e.*, for every point $w \in [0, 1]$ there is some $z \in [0, 1]$ such that $w = M(z)$), hence continuous. M can be thought of as a continuous cumulative distribution function (cdf) on $[0, 1]$. If we transform each point of the Poisson process by M, then the resulting process is homogeneous Poisson with intensity μ. The transformed homogeneous process on $[0, 1]$ with marker locations $M(l_1) < \cdots < M(l_9)$ and the original inhomogeneous process on $[0, 1]$ with marker locations $l_1 < \cdots < l_9$ would both produce recombination data with the same distribution. Thus, without loss of generality, we may consider only homogeneous Poisson processes, that is, we let $\mu_t = \mu$ for all t. The estimable parameters of the model are the genetic distances, $d_1, \ldots, d_8$, between the markers, and in terms of these the probability of the event y is

$$\prod_{i=1}^{8} e^{-2d_i y_i}(1 - e^{-2d_i})^{(1-y_i)}.$$

Note that μ can be written in terms of this parametrization: $\mu = 2\sum_{i=1}^{8} d_i$.

Gamma model: The distances between crossovers in the homogeneous Poisson model are independent exponential random variables, or equivalently, gamma with shape parameter 1. One way to generalize the Poisson model would be to consider renewal processes with general gamma interarrivals, thus adding an extra parameter to the model. On the real line the stationary renewal process with gamma interarrivals can be formulated as follows. Given a point at location w, the density of the distance to the next point to the right is

$$f_I(z) = \frac{1}{\Gamma(\gamma)} \mu^\gamma z^{\gamma-1} e^{-\mu z},$$

where I stands for interarrival, $\Gamma(\gamma) = \int_0^\infty \mu^\gamma s^{\gamma-1} e^{-\mu s} ds$, and this is independent of the occurrence of any points to the left of w. The distribution of the next point to the left of w is the same; there is no directionality. The distribution of the distance to the first point to the right (or equivalently left) of w, when it is not assumed that a point has occurred at w, has density

$$f_{1st}(z) = \frac{\mu}{\Gamma(\gamma + 1)} \int_z^\infty \mu^\gamma s^{\gamma-1} e^{-\mu s} ds.$$

The intensity of the process is then μ/γ. Letting $\gamma = 1$, we would get the Poisson model. The coincidence function for the gamma model is

$$C(z) = \frac{\gamma}{\mu} \sum_{n=1}^{\infty} f_I^n(z),$$

where $f_I^n(z) = \Gamma(\gamma n)^{-1} \mu^{\gamma n} z^{\gamma n - 1} e^{-\mu z}$.

We associate the interval $[0, 1]$ with the bundle of four chromatids, and we consider the above process restricted to $[0, 1]$. Then the chance of at least one crossover occurring on the bundle of four chromatids is $\alpha = \int_0^1 f_{1st}(z)\,dz$, and given that at least one crossover occurs, the density of the distance from the 0 end of the bundle to the first crossover is $\alpha^{-1} f_{1st}(z)$. Given a crossover at location $w \in [0, 1]$, the chance that another crossover occurs between w and 1 is $\beta(w) = \int_0^{1-w} f_I(z)\,dz$. Given a crossover at location w, and given that at least one crossover occurs between w and 1, the density of the distance from w to the next crossover between w and 1 is $\beta(w)^{-1} f_I(z)$. Note that all distances are in terms of a metric whose relationship with physical distance is unknown but which is a constant multiple of genetic distance.

We consider the gamma model on the bundle of four chromatids, and the points are then independently thinned, each with chance $^1/_2$, (*i.e.*, each point has chance $^1/_2$ to be deleted, independently of which other points are deleted) to get the crossover point process on a single chromatid. We do not restrict consideration to the particular case of gamma with integer shape parameter, but the use of the gamma renewal process with integer shape parameter to model crossover occurrence has a long history. FISHER *et al.* (1947) proposed that the crossover point process along a single chromatid be viewed as a renewal process with interarrival density $f(z) = {}^1/_2\pi \operatorname{sech}({}^1/_2\pi z) \tanh({}^1/_2\pi z)$. OWEN (1949, 1950) found that a gamma interarrival density with shape parameter 2 and scale parameter 2 was mathematically tractable and closely approximated the renewal process proposed by FISHER *et al.* CARTER and ROBERTSON (1952) used the gamma with shape parameter 2 as a four-strand model and applied various models of chromatid interference to get the chance of recombination across an interval for a single chromatid. PAYNE (1956) considered the gamma with integer shape parameter as a two-strand model. He compared coincidence curves for the gammas with shapes 2 and 3 to data. COBBS (1978) considered the gamma with integer shape parameter as a four-strand model, assuming no chromatid interference. He fit the model to Neurospora and Drosophila data by comparing the observed distribution of the number of crossovers in a segment (ignoring those that could not be observed) to the number predicted by the model. STAM (1979) considered various mathematical aspects of the same model, and FOSS *et al.* (1993) also considered this model, comparing the coincidence curves from the model to coincidence curves for Drosophila and Neurospora data. In this paper we calculate approximate probabilities of the different possible multilocus recombination events under the model, and we fit the model to the data by maximum likelihood. We are able to compare observed to expected frequencies of recombination events. This subsumes both the comparisons of distributions of number of events and of coincidence curves.

The gamma interarrival process with integer shape parameter has been used in the literature so often largely because of relative mathematical tractability. However, FOSS *et al.* (1993) propose it to explain certain empirical observations concerning recombination and gene conversion (a nonreciprocal exchange of genetic material between homologous chromosomes). First, gene conversion is associated with a high frequency of recombination of flanking markers (MORTIMER and FOGEL 1974). Second, gene conversions seem to occur independently in disjoint intervals, but gene conversions accompanied by recombination do not; rather, they appear to inhibit each other (MORTIMER and FOGEL 1974). The model proposed by FOSS *et al.* (see also STAHL 1979; MORTIMER and FOGEL 1974) is that initial precrossover events occur along the chromosome according to a Poisson process. Each such event results in a gene conversion and may or may not result in a crossover as well. Their model holds that every $(m+1)$st initial event results in a crossover. If the first event has chance $1/(m+1)$ to form a crossover, then the model is equivalent to the stationary gamma interarrival model with shape parameter $m+1$.

We note that, as in the case of the Poisson model, if we fit the stationary gamma renewal process, we are also allowing for the possibility that the true physical process may be nonstationary. The class of models covered by the analysis consists of those for which a monotone nondecreasing onto transformation M: $[0, 1] \to [0, 1]$ (*i.e.*, a continuous, cdf on $[0, 1]$) exists that maps the model to a stationary model. The models corresponding to almost-everywhere differentiable M's are those that can be formulated as follows. Given a point at w, the density of the distance to the first point to the right of w is

$$f_{\gamma}^{w}(z) = \frac{\mu_z}{\Gamma(\gamma)} \left(\int_w^z \mu_s ds\right)^{\gamma-1} e^{-\int_w^z \mu_t dt},$$

where $\mu_t = \mu \times dM(t)/dt$. The density of the distance to the first point to the right of a given location w, when it is not assumed that a point has occurred at w, is

$$f_{1st}^{w}(z) = \int_{-\infty}^{w} \frac{\mu_t}{\gamma} f_{\gamma}^{t}(z)\, dt.$$

The intensity of the process at location t is $\gamma^{-1}\mu_t$. Note that the coincidence function is identical to that for the stationary case.

The estimable parameters are the shape parameter γ and the genetic distances $d_1, \ldots, d_8$. Note that μ can be written, in terms of this parametrization, as $\mu = 2\gamma \sum_{i=1}^{8} d_i$. It has not proved possible to write down an explicit expression for the likelihood of the data under the gamma model, except in the case of integer-shape parameter (ZHAO *et al.* 1995a). Instead, the likelihood has been simulated, as described below.

Hard-core model: Another generalization of the Poisson model would be to have the points follow a Poisson process but with no two points allowed to be closer than a certain fixed distance h. This is known as a hard-core model (see *e.g.*, STOYAN *et al.* 1987). On the real line this is just a stationary renewal process with interarrivals distributed as exponential plus a constant. That is, given a point at location w, the distance to the next point to the right (equivalently, left) has density

$$f_i(z) = \begin{cases} \mu e^{-\mu(z-h)} & \text{for } z > h \\ 0 & \text{otherwise.} \end{cases}$$

The distribution of the distance to the first point to the right (equivalently, left) of w, when it is not assumed that a point has occurred at w, has density

$$f_{1st}(z) = \begin{cases} \dfrac{\mu}{1+\mu h} & \text{if } z \le h \\[2ex] \dfrac{\mu e^{-\mu(z-h)}}{1+\mu h} & \text{if } z > h. \end{cases}$$

The intensity of the process is $\mu/(1+\mu h)$. Letting h

= 0, we would get the Poisson model. The coincidence function for the hard-core model is

$$C(z) = \frac{\mu}{1 + \mu h} \sum_{n=1}^{\infty} f_I^n(z),$$

where

$$f_I^n(z) = \begin{cases} \Gamma(n)^{-1}\mu^n (z - nh)^{n-1} e^{-\mu(z-nh)} & \text{if } z \geq nh \\ 0 & \text{otherwise.} \end{cases}$$

Again, we associate the interval [0, 1] with the bundle of four chromatids. The restriction of this stationary renewal process to the interval [0, 1] works exactly as in the gamma case. As before, this hard-core model is equivalent, in terms of the data, to any model resulting from a transformation of the interval by a continuous cdf M. The models corresponding to almost-everywhere differentiable M's are those that can be formulated as follows. Given a point at w, the density of the distance to the first point to the right of w is

$$f_I^w(z) = \begin{cases} \mu_z e^{-\int_{w+h_w}^{z} \mu_s ds} & \text{if } z > w + h_w \\ 0 & \text{otherwise,} \end{cases}$$

where $\mu_t = \mu \times dM(t)/dt$, and $h_w > 0$ is any solution of $\int_w^{w+h_w} \mu_s ds = B$, where $B = h\mu$. The density of the distance to the first point to the right of a given location w, when it is not assumed that a point has occurred at w, is

$$f_{1st}^w(z) = \begin{cases} \dfrac{\mu_z}{1 + B} & \text{if } z < w + h_w \\ \mu_z e^{B - \int_0^z \mu_s ds} & \text{if } z \geq w + h_w. \end{cases}$$

The intensity of the process at location t is $\mu_t/(1 + B)$.

The estimable parameters are the hard-core genetic distance, $h_g = {}^1\!/_2\, B/(1 + B)$, and the genetic distances between the points, $d_1, \ldots, d_8$. Note that μ can be expressed in terms of this parametrization as $\mu = (1 - 2h_g)^{-1} \sum_{i=1}^{8} d_i$. As the number of loci increases, the likelihood quickly becomes very complicated, and no tractable general form has been found.

King-Mortimer model: In one version of a model suggested by KING and MORTIMER (1990), points are put down on a segment of the real line according to a Poisson process with parameter μ, just as in the original stationary Poisson model described above. Starting from time 0, each point independently waits an exponential amount of time with parameter λ before starting polymer growth, *i.e.*, the density of the time to polymer growth is $f(t) = \lambda e^{-\lambda t}$. When a polymer starts to grow from a point, it grows in both directions at a constant rate g. If a polymer from one point hits another point that has not yet started to grow, the latter point is deleted. At first glance this process appears to be nearly identical to the one-dimensional version of the Johnson-Mehl model (JOHNSON and MEHL 1939, but MEIJERING 1953 is more readable) for random tessellations. In that model points are born over time and immediately start growing polymers (called "crystals" by JOHNSON and MEHL 1939) at rate g in both directions. In the Johnson-Mehl model the rate of birth in an unpolymerized interval $[w, w + h_1)$ during an interval of time $[t, t + h_2)$ is $\rho h_1 h_2 + o(h_1 h_2)$. That is, in any infinitesimal interval of time $[t, t + h_2)$, it is a Poisson process, with parameter ρh_2, on the unpolymerized part of the line. In any small unpolymerized interval $[w, w + h_1)$ along the line, one waits an exponential amount of time, with parameter ρh_1, for a point to appear and start growing. However, in the King-Mortimer model the chance that a polymer starts to grow in a small unpolymerized interval $[w, w + h_1)$ during an interval of time $[t, t + h_2)$ can be shown to be $\mu\lambda e^{-\lambda t} h_1 h_2 + o(h_1 h_2)$, *i.e.*, it varies over time, going monotonically to zero.

In the King-Mortimer model there is an initial Poisson distribution of points, then the polymerizing process takes place over time, and then there is some resulting final distribution of points. Intuitively, it is clear that the final distribution of points would not be affected if time were slowed down or speeded up by some constant multiple c, as long as the initial Poisson distribution of points remained the same. This would be equivalent to changing λ to $c\lambda$ and changing g to cg. Therefore, it is not surprising that the final distribution of points is determined by μ and g/λ only. Thus, without loss of generality, we may take $\lambda = 1$.

KING and MORTIMER (1990) choose parameter values for which their simulations appear to resemble observed data, but they do not actually fit the model. They consider the process as defined on a line segment. In that case points are more likely to occur near the ends of the segment, because there are no points outside the segment to interfere with them. This is also true in the case of the hard-core model defined on a segment. We instead consider the process as defined on the real line and then restricted to [0, 1]. In that case the points near the ends of the segment behave as if other points lying outside the segment could interfere with them, and the model is stationary. As before, a nonstationary model resulting from a transformation of the interval [0, 1] by a continuous cdf M is indistinguishable, in terms of the data, from the stationary model. This nonstationarity is of a different type from that introduced by considering the process only on an interval, as KING and MORTIMER (1990) have done. The nonstationary models with almost-everywhere differentiable M's covered by this analysis are those in which the initial Poisson distribution has inhomogeneous parameter $\mu_t = \mu\, dM(t)/dt$, and the growth rate of a polymer that has

Modeling Interference

reached position t is $g_t = g\,dM(t)/dt$. We note that the stationary process is not a renewal process; given a point at $w \in [0, 1]$, the distribution of the distance to the next point to the right of w is not independent of the occurrence of other points to the left of w.

To specify the stationary King-Mortimer model on $[0, 1]$, it is convenient to determine the distribution of the time when a polymer growing from somewhere outside $[0, 1]$ would reach that interval. To do this, we ignore, for the moment, points inside $[0, 1]$ that could potentially polymerize and prevent any outside polymers from ever reaching $[0, 1]$. To calculate the distribution of the time when a polymer from $(1, \infty)$ reaches 1, we let X_i, for each $i > 1$, be the distance between the ith and $(i-1)$st initial points to the right of 1. If X_1 is the distance from 1 to the first initial point to the right of 1, then the X_i are independent exponential(μ). Let Z_i be the waiting time for polymer growth of the ith initial point to the right of 1. The Z_i are independent exponential(1). Let $S_i = g^{-1}\sum_{j=1}^{i} X_j$. Then the chance that no polymer arrives at 1 from the right before time t is

$$P(S_1 > t) + \sum_{i=1}^{\infty} P(S_i \leq t < S_{i+1},\ S_j + Z_j > t, \text{ for all } j \leq i).$$

The ith term of the sum is found to be

$$e^{-\mu g t}\frac{1}{i!}[\mu g(1 - e^{-t})]^i,$$

and so the required probability is

$$e^{\mu g(1-t-e^{-t})}.$$

One minus this quantity is the cdf of the time that a polymer first arrives from the right (equivalently, left) at any given point. This cdf is useful for simulating the process. Substituting $2g$ in for g gives the cdf of the time that a polymer first arrives at any given point from either the right or the left, ignoring the possibility that the point itself grows first and prevents the polymer's arrival. From this we can calculate the intensity of the point process, which is just the intensity, μ, of the initial Poisson process multiplied by the chance that a given initial point starts to grow before another polymer arrives. This is

$$\mu\int_0^{\infty} e^{-t}e^{2\mu g(1-t-e^{-t})}dt = \mu\left[\left(\frac{e}{a}\right)^a \Gamma(a, a) - \frac{1}{a}\right],$$

where $a = 2\mu g$ and $\Gamma(x, y) = \int_0^y t^{x-1}e^{-t}dt$, known as the incomplete gamma function. The estimable parameters under this version of the King-Mortimer model are the genetic distances, $d_1, \ldots, d_8$, and the expected time for a polymer to traverse the distance between two adjacent points in the initial Poisson process, as a multiple of expected waiting time for a polymer to start growing: $T = (\mu g)^{-1} = 2a^{-1}$. Note that μ can be expressed in terms of this parametrization:

$$\mu = 2\sum_{i=1}^{8} d_i \Big/ \left[\left(\frac{eT}{2}\right)^{2/T}\Gamma(2/T, 2/T) - T/2\right].$$

No closed-form expression is known for the likelihood of recombination data under this model, nor for the coincidence function. Both of these are obtained here by simulation.

K-M II model: The second King-Mortimer model (KING and MORTIMER 1990), which shall be called here K-M II, is the same as the previous model except that a parameter is added for termination of polymer growth, once it has started. That is, under the set-up of the previous model, each polymer now grows for an independent exponential$(1/\nu)$ amount of time, or until it hits another polymer, whichever happens first. The possibility of early termination of a polymer has the effect of allowing interference to operate over a smaller range than before. Interference can then be made more intense, yet more localized, than in the ordinary King-Mortimer model. For this model there is no known expression for the intensity, the likelihood of recombination data nor for the coincidence function. The parameters estimated are the genetic distances, $d_1, \ldots, d_8$, the expected time for a polymer to traverse the distance between two adjacent points in the initial Poisson configuration (given that growth does not terminate up to that time), T, and the expected time to terminate polymer growth, ν, where each of these times is expressed as a multiple of the expected waiting time for a polymer to start growing.

Count-location model: In the count-location model (KARLIN and LIBERMAN 1979, called the "generalized no interference model" by RISCH and LANGE 1979) the number of crossovers is chosen according to some count distribution c, where c is given by $c_0, c_1, \ldots,$ and $c_i = P(i \text{ crossovers})$. Given the number of crossovers, their locations are independent and identically distributed (i.i.d.) along the bundle of chromatids, according to a location distribution ν that does not vary with the count. The count-location model can be thought of as a modification of the Poisson process model, in which the number of crossovers occurring is no longer Poisson, but their locations are again i.i.d. as in the Poisson model. In the count-location model the coincidence function is constant:

$$C(z) = \sum_{i=2}^{\infty} i(i-1)c_i \Big/ \left(\sum_{j=1}^{\infty} jc_j\right)^2.$$

The version of the count-location model considered here allows no more than three crossovers to occur

M. S. McPeek and T. P. Speed

TABLE 1

Observed and expected counts under the fitted models for five loci with chi-square statistics

Recombination event	Observed data	Expected Poisson	Expected gamma	Expected hard-core	Expected King-Mortimer	Expected count-location	Expected K-M II
0000	10431	11014.1	10497.1	10837.0	10783.6	10434.0	10443.1
1000	771	596.6	738.9	682.7	686.1	777.6	767.4
0100	1579	1246.9	1537.6	1360.6	1390.0	1556.1	1560.4
0010	1221	931.3	1214.0	1058.7	1064.2	1184.7	1210.7
0001	1994	1664.1	1979.7	1685.2	1811.7	2036.4	2018.0
1100	4	67.5	1.5	13.5	25.2	16.0	7.3
1010	7	50.4	12.0	69.3	35.8	11.8	12.4
0110	4	105.4	4.0	35.1	43.1	23.4	13.2
1001	46	90.1	73.0	106.3	85.2	20.2	32.9
0101	53	188.4	68.0	208.1	138.5	42.1	48.6
0011	25	140.7	10.1	71.7	64.8	31.5	20.8
1110	1	40.3	0.1	7.9	7.8	2.3	1.3
1101	1	40.3	0.1	7.9	7.8	2.3	1.3
1011	1	40.3	0.1	7.9	7.8	2.3	1.3
0111	1	40.3	0.1	7.9	7.8	2.3	1.3
1111	1	40.3	0.1	7.9	7.8	2.3	1.3
Chi-square	—	773	51	420	267	67	17
df	—	8	7	7	7	6	6

Triple and quadruple recombination events are quite rate. These events were pooled in the analysis because accurate probabilities for the events could not be computed by simulation.

simultaneously, *i.e.*, $\Sigma_{i=1}^{3} c_i = 1$ and $c_i = 0$ for $i > 3$. The location distribution ν may be taken to be uniform without loss of generality, because the class of models covered by this analysis will include all those with location distributions having a continuous cdf. The intensity of the process is $\Sigma_{i=0}^{3} ic_i$. The model can be parametrized in terms of the genetic distances, $d_1, \ldots, d_8$, and c_0 and c_1. The likelihood can be calculated explicitly in terms of these parameters.

To fit any of these models to recombination data, it is necessary to be able to calculate q_y for any y, that is, the probability of any combination of zero and nonzero crossover counts in the eight marker intervals, as described above. This is trivial for the Poisson model and the count-location model. For the other models this involves integrating the density over all possible realizations compatible with y, assuming that the density is known or can be computed. So far, these integrals appear virtually impossible to do exactly, except in the special case of gamma with integer shape parameter. As a result, they have been calculated here by Monte Carlo methods. For the Poisson, gamma, count-location and K-M II models, the recombination probabilities were computed for events among all nine loci, whereas for the other models only a subset of five of the nine loci were used.

For a given model the probability q_y depends not only on the model parameters but also on the relative locations of the loci. For each choice of model parameters, a large number n of realizations of the point process are simulated. Using these simulations, the desired probability q_y can be estimated, for any set of locus locations, by the observed frequency of the event y. From the q's the estimates of the probabilities of the observed recombination events can be computed, and from these, the estimated likelihood of the observed data is computed. Recombination events in which three or more recombinations took place were so rare in this data that their probabilities are very difficult to estimate. These rare events were pooled in this analysis. Holding the point process parameters fixed, the likelihood can then be numerically maximized over the locus locations by the Nelder-Mead search algorithm (NELDER and MEAD 1965). This maximization is done

TABLE 2

Fitted point process parameters for five loci

Model	Parameter(s)[a]
Gamma	$\gamma = 6.41$ (0.280)
Hard-core	$h_g = 0.116$ (0.004)
King-Mortimer I	$T = 0.038$ (0.005)
Count-location	$c_0 = 0.311$ (0.005), $c_1 = 0.641$ (0.007) (redundant parameters: $c_2 =$ 0.0448 (0.005), $c_3 = 0.0032$ (0.002))
K-M II	$T = 1.01 \times 10^{-4}$ (2.44×10^{-5}), $v =$ 0.244 (0.187)

[a] Estimated standard deviations in parentheses.

Modeling Interference

TABLE 3

Fitted genetic distances between adjacent markers for five loci

Genetic distances	Poisson	Gamma	Hard-core	K-M I	Count-location	K-M II
(*sc*)-(*ec*)	0.054 (0.002)	0.051 (0.002)	0.055 (0.002)	0.052 (0.002)	0.052 (0.002)	0.050 (0.003)
(*ec*)-(*cv*)	0.114 (0.003)	0.100 (0.003)	0.100 (0.003)	0.101 (0.002)	0.104 (0.002)	0.102 (0.005)
(*cv*)-(*ct*)	0.085 (0.003)	0.077 (0.003)	0.077 (0.002)	0.076 (0.002)	0.079 (0.002)	0.079 (0.003)
(*ct*)-(*v*)	0.152 (0.004)	0.132 (0.004)	0.129 (0.003)	0.135 (0.003)	0.136 (0.003)	0.132 (0.007)

Estimated standard deviations in parentheses.

using the same set of n simulated realizations. For each choice of model parameters, the process can be repeated, generating a new set of n realizations and maximizing the estimated likelihood over the locus locations. The Nelder-Mead search algorithm can then be used to maximize over the model parameters. This is an extension of a method described in DIGGLE and GRATTON (1984). The number of realizations simulated, n, was taken to be 160,000 initially and was increased to 960,000 as a maximum was approached.

There are standard algorithms for simulating from exponential and gamma distributions (see RIPLEY 1987). In addition to these, we must be able to simulate from the first-point distributions of the gamma and the hard-core models, and also, in the King-Mortimer model, the distribution of the first time that a polymer from outside the [0, 1] interval first hits 0 from below (or equivalently, hits 1 from above).

In the gamma case if the shape parameter is an integer, say γ, then a random variable from the first-point distribution can be realized by simulating the sum of k independent exponential random variables, where k is chosen uniformly on $1, \ldots, \gamma$. For noninteger γ's one can do rejection sampling using the next highest integer (see RIPLEY 1987 for a discussion of rejection sampling). For the hard-core model with hard-core distance h and Poisson parameter μ, the first point will be uniform on $[0, h]$ with probability $h/(1/\mu + h)$ and exponentially distributed on $[h, \infty)$ with probability $(1/\mu)/(1/\mu + h)$.

To simulate from the King-Mortimer model on an interval, one must be able to simulate the distribution of the first time that a polymer from outside the interval reaches the interval from the left, or identically, from the right. The distribution, which was derived in the last section, can be simulated by rejection sampling using exponential(μ) for $\mu g \le 1$ and exponential($1/g$) for $\mu g \ge 1$. For the K-M II model even this distribution is not known but is simulated by looking at the process over a very large interval and then restricting to a small interval, where edge effects may be assumed to be negligible.

RESULTS

SPEED *et al.* (1992) found that as a result of assuming no chromatid interference, certain constraints must be satisfied by the probabilities of recombination events. Namely,

$$0 \le \sum_{y \ge x} (-1)^{(y-x)\cdot 1} p_y \quad \text{for all } x.$$

Furthermore, for any set of p's that satisfies these constraints, there is a no-chromatid-interference model that is compatible with them. It is interesting to note that for the five loci considered, the observed frequencies of the recombination events satisfy all eight nontrivial constraints of the ones required above. Only a single observation prevents this from being true for the 37 nontrivial constraints on all nine loci. Of course, this is not proof that the assumption of no chromatid interference is correct but simply that the data are not incompatible with this assumption.

TABLE 4

Point-process parameters and chi-square statistics for nine-locus fit

Model	Parameter(s)[a]	Chi-square	d.f.
Poisson	None	1672	30
Gamma	$\gamma = 4.94$ (0.124)	107	29
Count-location	$c_0 = 0.057$ (0.008), $c_1 = 0.434$ (0.017)	889	28
	redundant parameters: $c_2 = 0.434$ (0.017), $c_3 = 0.075$ (0.008)		
K-M II	$T = 1.84 \times 10^{-4}$ (7.30×10^{-6}), $v = 1.08$ (0.337)	294	28

[a] Estimated standard deviations in parentheses.

TABLE 5

Genetic distances for nine-locus fit

Genetic distances	Poisson	Gamma	Count-location	K-M II
(*sc*)-(*ec*)	0.054 (0.002)	0.053 (0.002)	0.053 (0.002)	0.046 (0.002)
(*ec*)-(*cv*)	0.114 (0.003)	0.098 (0.002)	0.107 (0.003)	0.096 (0.004)
(*cv*)-(*ct*)	0.085 (0.003)	0.075 (0.002)	0.082 (0.002)	0.078 (0.003)
(*ct*)-(*v*)	0.152 (0.004)	0.133 (0.002)	0.141 (0.003)	0.141 (0.005)
(*v*)-(*s*)	0.089 (0.003)	0.085 (0.002)	0.086 (0.002)	0.088 (0.004)
(*s*)-(*f*)	0.184 (0.004)	0.157 (0.003)	0.171 (0.004)	0.163 (0.007)
(*f*)-(*car*)	0.081 (0.002)	0.075 (0.003)	0.078 (0.002)	0.071 (0.003)
(*car*)-(*bb*)	0.046 (0.002)	0.044 (0.002)	0.044 (0.002)	0.039 (0.002)

Table 1 shows Pearson chi-square statistics for the five-locus models, and Table 4 shows them for the nine-locus models. In all cases the p values are smaller than 0.01, implying significant misfit of all models. However, with the addition of a single parameter, the chi-square statistic in the five-locus case is brought down from 773 (Poisson model) to 51 (gamma model), and in the nine-locus case from 1672 (Poisson model) to 107 (gamma model). This represents a tremendous improvement. In the five-locus case the best-fitting model is the K-M II model, which adds two parameters to the Poisson model and has a chi-square statistic of 17, compared to 773 for the Poisson model. The K-M II model is outperformed by the gamma model in the nine-locus case. One interpretation is that K-M II does a reasonably good job of modeling interference between nearby markers but that it is not flexible enough to simultaneously model both close and medium-range interference, as required in the nine-locus dataset.

Tables 2 and 3 contain fitted parameter values, with estimated standard deviations, for the five-locus models, and Tables 4 and 5 show them for the nine-locus models.

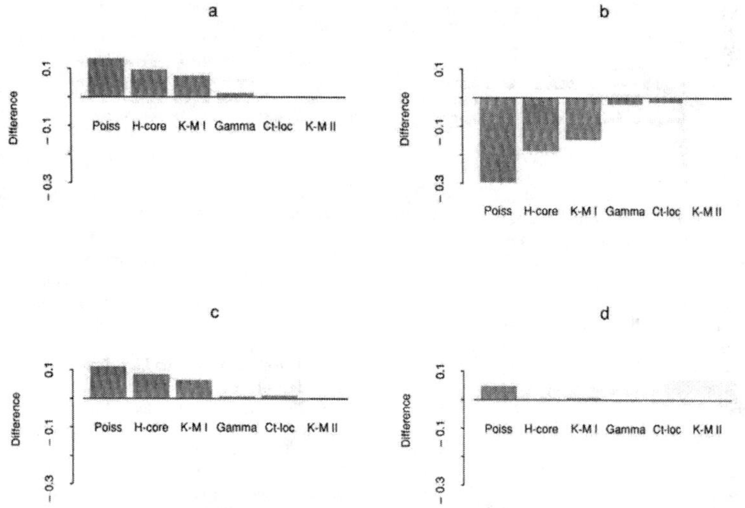

FIGURE 1.—Distribution of the number of crossovers among the first five loci of the MORGAN *et al.* (1935) dataset, for each of the models under consideration, relative to the data distribution. (a) The chance of zero crossovers for each of the models minus 0.310, the proportion of cases in which zero recombinations occurred among the first five loci in the data. (b–d) Similar bar graphs for the probabilities of 1, 2 and 3 or more crossovers, respectively. The data proportions to which these numbers are compared are 0.656, 0.0337 and 0.000496, respectively.

Modeling Interference

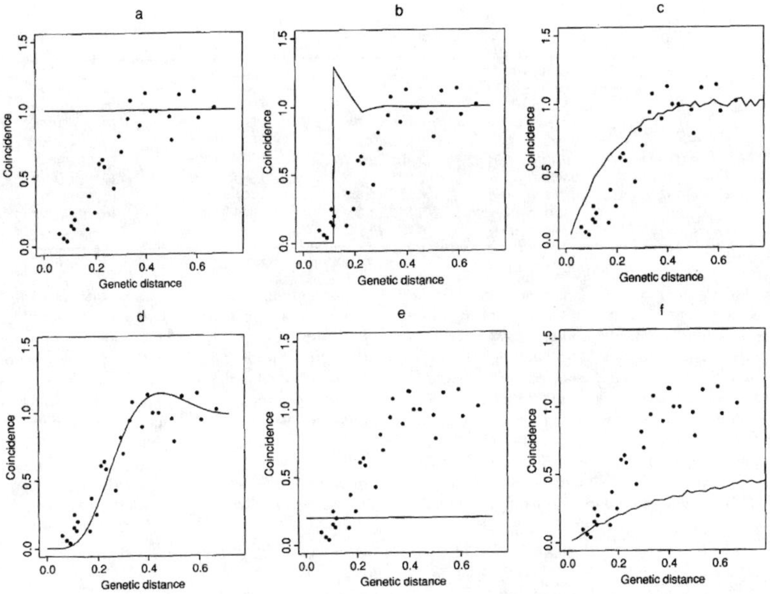

FIGURE 2.—Plots of coincidence *vs.* genetic distance for the five-locus models. The points represent coincidence calculated from the data for each pair of atomic intervals among the nine loci. The corresponding genetic distance for each pair is calculated between the midpoints of the two intervals. The curves represent five-locus model coincidences. a is the Poisson model, b is the hard-core model, c the King-Mortimer model, d the gamma model, e the count-location model and f the K-M II model.

The results shown for the Poisson are from a calculation rather than from simulation, but the simulated results are virtually identical. In the case of the gamma model, the likelihood can actually be calculated in the case of integer-shape parameter (ZHAO *et al.* 1995a). In the nine-locus case using this likelihood calculation, the best fit is obtained using shape parameter 5, which agrees with the results given here for the nine-locus case. This would correspond to four gene conversion events between each crossover in the framework described in FOSS *et al.* (1993). The count-location model has previously been fit to the MORGAN *et al.* (1935) dataset by RISCH and LANGE (1983), using a calculated, rather than simulated, likelihood. In their calculation they did not pool events in which three or more recombinations occurred. Thus, they maximized a slightly different likelihood than the one maximized here. Still, the results are very close to those obtained here. Note that the estimates of genetic distances and their standard deviations were fairly similar for all models. The Poisson model gives very similar estimates to the other models for small genetic distances (<10 cM) but overestimates larger genetic distances relative to the other models.

The SDs were estimated by simply taking numerical derivatives of the recombination probabilities (p_x's in the notation given above) with respect to the estimated parameters and using these to calculate the Fisher information matrix. A better approach, not taken here, would be to smooth the likelihood surface before calculating derivatives.

The Pearson chi-square statistic measures the overall goodness-of-fit of each model. We consider two additional criteria that do not indicate goodness-of-fit as directly as the chi-square does but that are easy to interpret and may be useful for understanding the performance of the models. First, we consider the number of crossovers predicted by the models. Figure 1 shows that the five-locus Poisson model predicts too few single crossover events and too many of all other types of events. This shows up in Table 1 as an underestimate

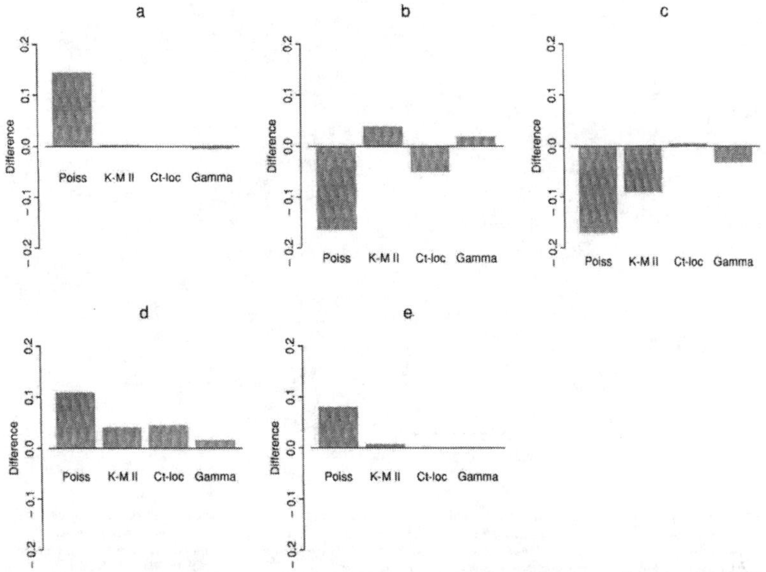

FIGURE 3.—Distribution of the number of crossovers among all nine loci of the MORGAN *et al.* (1935) dataset for each of the models under consideration relative to the data distribution. (a) The chance of zero crossovers for each of the models minus 0.056, the proportion of cases in which zero recombinations occurred in the data. (b–e) Similar bar graphs for the probabilities of 1, 2, 3 and 4 or more crossovers, respectively. The data proportions to which these numbers are compared are 0.485, 0.429, 0.030 and 0, respectively.

of the number of single recombination events among the five loci and an overestimate of all other types of recombination events. Figure 3 shows that in the nine-locus case, the Poisson model predicts too few single and double crossover events and too many of all other types. The gamma, count-location and K-M II models all do well at matching the distribution of the number of crossovers in the data, although the K-M II model does not do nearly as well with nine loci and with five.

The other aspect of the models which we shall consider is the coincidence function which is discussed above. Figure 2 shows coincidence *vs.* genetic distance for the five-locus models, and Figure 4 shows it for the nine-locus models. Recall that coincidence is a measure of the degree to which one crossover inhibits the formation of another nearby. For the Poisson model coincidence is one, reflecting no interference. The coincidence curves for the models have been calculated (or simulated, in the case of the King-Mortimer and K-M II models) using infinitesimal intervals. The coincidence curves of the gamma model and the King-Mortimer model give the best approximation to the data in the five-locus case, whereas the gamma and K-M II models do best in the nine-locus case.

The strange shape of the coincidence curve for the hard-core model can be explained as follows: in the hard-core model, no points are allowed within a fixed distance of the crossover point that is assumed to lie at 0 (for the purpose of making the coincidence curve). Immediately after that fixed distance has been surpassed, one is waiting an exponential amount of time for the next point, whereas in an interval chosen at random the intensity is less than exponential because there is a chance of another crossover already having occurred nearby. This explains the spike in the coincidence curve for the hard-core model, and the hump in the coincidence curve for the gamma model can be explained similarly. As the distance from the given crossover gets close to the mode of the gamma interarrival distribution, one is more likely to have another crossover than one would be to have a crossover in a given interval chosen at random.

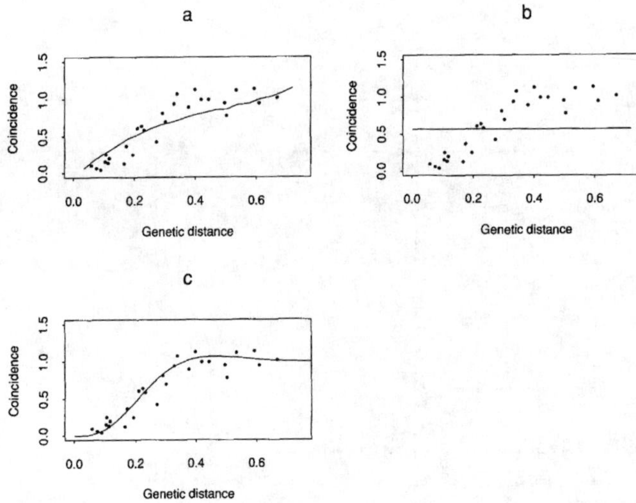

FIGURE 4.—Plots of coincidence *vs.* genetic distance for the nine-locus models. The points represent coincidence calculated from the data for each pair of atomic intervals among the nine loci. The corresponding genetic distance for each pair is calculated between the midpoints of the two intervals. The curves represent nine-locus model coincidences. a is the K-M II model, b is the count-location model and c is the gamma model.

It should be noted that the coincidence curve for the count-location model is constant, even if the distribution of points is i.i.d. but nonuniform. The reason is that the *x* axis represents genetic, not physical distance, that is, it represents the expected number of crossovers and so is automatically rescaled to be uniform. Note also the anomaly that the K-M II model fits the data better (according to chi-square) in the five-locus case than in the nine-locus case, but the coincidence curve resembles the data much more closely in the nine-locus case.

DISCUSSION

Although none of the models fit the data, several of the models were able to capture certain aspects of the data. The best overall model of the ones considered was the renewal model with gamma interarrivals. It gave reasonable estimates of the numbers of crossovers and matched the interference pattern in the data as measured by the coincidence. In terms of the chi-square statistic, the gamma model represents a huge improvement over the Poisson model. Furthermore, in the case of integer shape parameter, the gamma model may prove tractable enough to use in mapping applications.

This work represents a first attempt to systematically fit a range of point process models to recombination data. A suitable model, such as the gamma model, may be of great use in assessing the impact of interference on conventional linkage analyses and genetic mapping. It may even be possible to apply such a model in genetic mapping, thereby making more efficient use of the data.

We are grateful to B. D. RIPLEY for his advice on the Monte Carlo method and to F. W. STAHL for helpful discussions. This work was supported by National Science Foundation grants DMS-880237 and DMS-9113527.

LITERATURE CITED

CARTER, T. C., and A. ROBERTSON, 1952 A mathematical treatment of genetical recombination using a four-strand model. Proc. R. Soc. Lond. Ser. B **139:** 410–426.

COBBS, G., 1978 Renewal process approach to the theory of genetic linkage: case of no chromatid interference. Genetics **89:** 563–581.

COX, D. R., and V. ISHAM, 1980 *Point Processes.* Chapman and Hall, New York.

DIGGLE, P. J., and R. J. GRATTON, 1984 Monte Carlo methods of inference for implicit statistical models. J. Roy. Statist. Soc. B **46:** 193–227.

EVANS, S. N., M. S. MCPEEK and T. P. SPEED, 1992 A characterisation of crossover models that possess map functions. Theor. Popul. Biol. **43:** 80–90.

FISHER, R. A., M. F. LYON and A. R. G. OWEN, 1947 The sex chromosome in the house mouse. Heredity **1:** 335–365.

Foss, E., R. Lande, F. W. Stahl and C. M. Steinberg, 1993 Chiasma interference as a function of genetic distance. Genetics **133:** 681–691.

Goldgar, D. E., and P. R. Fain, 1988 Models of multilocus recombination: nonrandomness in chiasma number and crossover positions. Am. J. Hum. Genet. **43:** 38–45.

Haldane, J. B. S., 1919 The combination of linkage values, and the calculation of distances between the loci of linked factors. J. Genet. **8:** 299–309.

Johnson, W. A., and R. F. Mehl, 1939 Reaction kinetics in processes of nucleation and growth. Trans. A.I.M.M.E. **135:** 416–458.

Karlin, S., and U. Liberman, 1979 A natural class of multilocus recombination processes and related measures of crossover interference. Adv. Appl. Probab. **11:** 479–501.

Lande, R., and F. W. Stahl, 1993 Chiasma interference and the distribution of exchanges in *Drosophila melanogaster*. Cold Spring Harbor Symp. Quant. Biol. **58:** 543–552.

Liberman, U., and S. Karlin, 1984 Theoretical models of genetic map functions. Theor. Popul. Biol. **25:** 331–346.

King, J. S., and R. K. Mortimer, 1990 A polymerization model of chiasma interference and corresponding computer simulation. Genetics **126:** 1127–1138.

Meijering, J. L., 1953 Interface area, edge length, and number of vertices in crystal aggregates with random nucleation. Philips Res. Rep. **8:** 270–290.

Morgan, T. H., C. B. Bridges and J. Schultz, 1935 Constitution of the germinal material in relation to heredity. Carnegie Instit. Washington Publ. **34:** 284–291.

Mortimer, R. K., and S. Fogel, 1974 Genetical interference and gene conversion, pp. 263–275 in *Mechanisms in Recombination*, edited by R. F. Grell. Plenum, New York.

Muller, H. J., 1916 The mechanism of crossing-over. Am. Nat. **50:** 193–221, 284–305, 350–366, 421–434.

Nelder, J. A., and R. Mead, 1965 A simplex method for function minimisation. Comput. J. **7:** 303–313.

Owen, A. R. G., 1949 The theory of genetical recombination, I. Long-chromosome arms. Proc. R. Soc. Lond. Ser. B **136:** 67–94.

Owen, A. R. G., 1950 The theory of genetical recombination. Adv. Genet. **3:** 117–157.

Pascoe, L., and N. E. Morton, 1987 The use of map functions in multipoint mapping. Am. J. Hum. Genet. **40:** 174–183.

Payne, L. C., 1956 The theory of genetical recombination: a general formulation for a certain class of intercept length distributions appropriate to the discussion of multiple linkage. Proc. R. Soc. Lond. Ser. B **144:** 528–544.

Ripley, B. D. R., 1987 *Stochastic Simulation*. John Wiley and Sons, New York.

Risch, N., and K. Lange, 1979 An alternative model of recombination and interference. Ann. Hum. Genet. **43:** 61–70.

Risch, N., and K. Lange, 1983 Statistical analysis of multilocus recombination. Biometrics **39:** 949–963.

Speed, T. P., M. S. McPeek and S. N. Evans, 1992 Robustness of the no-interference model for ordering genetic markers. Proc. Natl. Acad. Sci. USA **89:** 3103–3106.

Stahl, F. W., 1979 *Genetic Recombination. Thinking About it in Phage and Fungi.* W. H. Freeman, San Francisco.

Stam, P., 1979 Interference in genetic crossing over and chromosome mapping. Genetics **92:** 573–594.

Stoyan, D., W. S. Kendall and J. Mecke, 1987 *Stochastic Geometry and its Applications*, Wiley, New York.

Sturtevant, A. H., 1915 The behavior of the chromosomes as studied through linkage. Z. Indukt. Abstammungs. Vererbungsl. **13:** 234–287.

Weinstein, A., 1936 The theory of multiple-strand crossing over. Genetics **21:** 155–199.

Zhao, H., T. P. Speed and M. S. McPeek, 1995a Statistical analysis of crossover interference using the chi-square model. Genetics **139:** 000–000.

Zhao, H., M. S. McPeek and T. P. Speed, 1995b Statistical analysis of chromatid interference. Genetics **139:** 000–000.

Communicating editor: B. S. Weir

Statistical Analysis of Crossover Interference Using the Chi-Square Model

Hongyu Zhao, Terence P. Speed and Mary Sara McPeek[1]

Department of Statistics, University of California, Berkeley, California 94720

Manuscript received December 17, 1993
Accepted for publication October 19, 1994

ABSTRACT

The chi-square model (also known as the gamma model with integer shape parameter) for the occurrence of crossovers along a chromosome was first proposed in the 1940's as a description of interference that was mathematically tractable but without biological basis. Recently, the chi-square model has been reintroduced into the literature from a biological perspective. It arises as a result of certain hypothesized constraints on the resolution of randomly distributed crossover intermediates. In this paper under the assumption of no chromatid interference, the probability for any single spore or tetrad joint recombination pattern is derived under the chi-square model. The method of maximum likelihood is then used to estimate the chi-square parameter m and genetic distances among marker loci. We discuss how to interpret the goodness-of-fit statistics appropriately when there are some recombination classes that have only a small number of observations. Finally, comparisons are made between the chi-square model and some other tractable models in the literature.

CROSSOVER interference has been observed in almost all organisms studied, although there is little consistent evidence of chromatid interference even within the same organism (ZHAO *et al.* 1995). In what follows we assume no chromatid interference (NCI).

Information on the distribution of crossovers along a chromosome generally comes from genetic experiments in which only recombinations, not crossovers, can actually be observed. In some organisms, such as Drosophila, the results of such experiments are in the form of *single spore data,* in which the products of a single meiosis are recovered separately. Other organisms, such as yeast, yield *tetrad data,* in which all four meiotic products are recovered together. It is easy to see that there are 2^n distinct recombination patterns for single spore data involving $n + 1$ markers. For tetrad data involving $n + 1$ markers, $n > 1$, there are more than 3^n distinguishable tetrad patterns, but under the assumption of NCI, there are only 3^n different probabilities among these patterns, *i.e.,* some distinct tetrad patterns have the same probability of being observed. Each different probability corresponds to one of the types $(i_1 i_2 \cdots i_n)$, where $i_j = 0, 1, 2$ corresponds to parental ditype, tetratype and nonparental ditype, respectively, between l_j and l_{j+1}.

Both single spore data and tetrad data record recombination events among a set of markers. As the underlying crossovers occurring during meiosis are not directly observable from the data, any model about interference must relate the observable recombination or tetrad patterns to the underlying unobservable crossover events. Crossing over occurs among four strands after each homologous chromosome has duplicated. A model relating crossovers to recombination should specify the distribution of crossover points along the bundle of four chromatids and the choice of nonsister chromatids to be involved in each crossover.

The chi-square model for crossovers has a long history; see BAILEY (1961). MCPEEK and SPEED (1995) briefly review the history and fit a more general class of models, renewal processes with gamma interarrivals, which includes the class of chi-square models, to Drosophila data by maximum likelihood using a Monte Carlo method. Whereas it has generally been of interest due to its mathematical tractability, the chi-square model has also been suggested as a plausible biological model by FOSS *et al.* (1993), motivated by observations from experiments on gene conversion. There the model is represented in the form $Cx(Co)^m$ as follows: assume that crossover intermediates (C events) are randomly distributed along the four-strand bundle, and every C event will either resolve in a crossover (Cx) or not (Co). When a C resolves as a Cx, the next m C's must resolve as Co events, and after m Co's the next C must resolve as a Cx, *i.e.,* the C's resolve in a sequence $\cdots Cx(Co)^m\ Cx(Co)^m \cdots$. To make the process stationary given a set of C events, the leftmost C has an equal chance to be one of $Cx(Co)^m$. In their paper FOSS *et al.* estimate the parameter m in $Cx(Co)^m$ from the observed ratio of Co to Cx. Here we perform a full maximum-likelihood estimation procedure to estimate m and genetic distances between markers from both kinds of recombination data.

Corresponding author: Hongyu Zhao, Department of Statistics, University of California, Berkeley, CA 94720.
E-mail: zhao@stat.berkeley.edu

[1] *Present address:* Department of Statistics, University of Chicago, 5734 University Ave., Chicago, IL 60637.
E-mail: mcpeek@galton.uchicago.edu

ESTIMATION UNDER THE CHI-SQUARE MODEL

Given a set of markers $\mathcal{M}_1, \ldots, \mathcal{M}_{n+1}$ along a chromosome, under the chi-square model $Cx(Co)^m$ $n+1$ parameters need to be specified, namely, m and the genetic distances between each consecutive pair of markers, $x_1, x_2, \ldots, x_n$, so that the probability of each single spore or tetrad recombination pattern can be calculated. Suppose these parameters are given, let $p = m+1$, $y_j = 2px_j$ and let $\mathbf{D_k}(y)$ be the matrix whose i, jth entry is $e^{-y}y^{pk+j-i}/(pk+j-i)!$. Then the probability of k_j crossovers between $\mathcal{M}_j$ and $\mathcal{M}_{j+1}$, $j = 1, \cdots n$, is (see APPENDIX, Lemma):

$$\frac{1}{p}\,\mathbf{1D_{k_1}}(y_1)\mathbf{D_{k_2}}(y_2)\cdots\mathbf{D_{k_n}}(y_n)\mathbf{1}', \quad \text{where}$$

$$\mathbf{1} = (1, 1, \ldots, 1).$$

Note that when $p = 1$, the above expression reduces to the Poisson case, *i.e.*, the no-interference model of HALDANE (1919). Using the above formula, we can calculate the probability of any single spore or tetrad recombination pattern $(i_1 i_2 \cdots i_n)$. We consider the two cases separately.

For single spore data, given two consecutive markers $\mathcal{M}_j$ and $\mathcal{M}_{j+1}$, we can observe a recombination or nonrecombination between them. If no crossovers occur between $\mathcal{M}_j$ and $\mathcal{M}_{j+1}$, no strand in the bundle will show any recombination between these markers. MATHER (1935) proved that under the assumption of NCI, if there are $k \geq 1$ crossovers between two markers, then the probability that these two markers recombine on any given single strand is $1/2$. Recall that for single spore data any recombination pattern can be represented as $(i_1 i_2 \cdots i_n)$, where $i_j = 0$ or 1. Define

$$\mathbf{N_j} = \mathbf{D_0}(y_j) + {}^1\!/_2 \sum_{s\geq 1} \mathbf{D_s}(y_j)$$

$$\mathbf{R_j} = {}^1\!/_2 \sum_{s\geq 1} \mathbf{D_s}(y_j).$$

Then the probability of recombination pattern $(i_1 i_2 \cdots i_n)$ is (see APPENDIX, Theorem 1)

$$P(i_1 i_2 \cdots i_n) = \frac{1}{p}\,\mathbf{1M_1M_2}\cdots\mathbf{M_n1}',$$

where $\mathbf{M_j} = \mathbf{N_j}$ when $i_j = 0$, and $\mathbf{M_j} = \mathbf{R_j}$ when $i_j = 1$.

For tetrad data recall that there are three different possible tetrad patterns between two markers. We let p_0, p_1 and p_2 denote the probabilities of parental ditype, tetratype and nonparental ditype, respectively, between a fixed pair of markers. Given $k \geq 1$ crossovers between two loci, under the assumption of NCI, the conditional probabilities $p_0^{(k)}$, $p_1^{(k)}$ and $p_2^{(k)}$ of a tetrad being of parental ditype, tetratype and nonparental ditype, respectively, are given by MATHER (1935):

$$p_0^{(k)} = {}^1\!/_3({}^1\!/_2 + (-{}^1\!/_2)^k)$$

$$p_1^{(k)} = {}^2\!/_3(1 - (-{}^1\!/_2)^k)$$

$$p_2^{(k)} = {}^1\!/_3({}^1\!/_2 + (-{}^1\!/_2)^k).$$

We can calculate the probability of any tetrad pattern $(i_1 i_2 \cdots i_n)$, where $i_j = 0$, 1 or 2. Define

$$\mathbf{P_j} = \mathbf{D_0}(y_j) + \sum_{s\geq 2} {}^1\!/_3({}^1\!/_2 + (-{}^1\!/_2)^k)\mathbf{D_s}(y_j)$$

$$\mathbf{T_j} = \mathbf{D_1}(y_j) + \sum_{s\geq 2} {}^2\!/_3(1 - (-{}^1\!/_2)^k)\mathbf{D_s}(y_j)$$

$$\mathbf{N_j} = \sum_{s\geq 2} {}^1\!/_3({}^1\!/_2 + (-{}^1\!/_2)^k)\mathbf{D_s}(y_j).$$

Then the probability of the tetrad pattern $(i_1 i_2 \cdots i_n)$ can be written as (see APPENDIX, Theorem 2)

$$P(i_1 i_2 \cdots i_n) = \frac{1}{p}\,\mathbf{1M_1M_2}\cdots\mathbf{M_n1}',$$

where $\mathbf{M_j} = \mathbf{P_j}$ if $i_j = 0$, $\mathbf{M_j} = \mathbf{T_j}$ if $i_j = 1$, and $\mathbf{M_j} = \mathbf{N_j}$ if $i_j = 2$.

Given a set of single spore or tetrad data and based upon the above formulae, the likelihood of the observations, up to a constant factor, can be calculated in terms of the parameters as $\prod P(i_1 i_2 \cdots i_n)^{x_{i_1 i_2 \cdots i_n}}$, where $x_{i_1 i_2 \cdots i_n}$ is the observed frequency of single spores or tetrads with pattern $(i_1 i_2 \cdots i_n)$. The maximum likelihood estimates of the parameters are those that maximize the likelihood among all possible parameter values. The numerical method used to find the maximum likelihood estimates used in our analysis is the downhill simplex method, see PRESS *et al.* (1988). The standard error for each estimate is approximated using the fact that as $n \to \infty$,

$$\sqrt{n}(\hat{\theta}_{jn} - \theta_j) \to N(0, [I(\theta)]^{-1}_{jj}),$$

where $I(\theta)$ is the Fisher information matrix.

APPLICATIONS TO VARIOUS ORGANISMS

In this section the $Cx(Co)^m$ model is fitted to data from various organisms via the method of maximum likelihood. Data are of tetrad form except *Drosophila melanogaster* and human recombination data that are of single spore type.

Drosophila melanogaster: Many valuable recombination datasets for this organism have appeared in the literature since it was first studied by geneticists early in this century. Among these, two large, well-known datasets, namely WEINSTEIN (1936) and MORGAN *et al.* (1935), have drawn much attention and have frequently been used as a basis upon which to compare different models.

Seven loci that cover most of the *X*-chromosome of *D. melanogaster* were used in WEINSTEIN's study. A total of 28,239 offspring genotypes were determined. Among the $Cx(Co)^m$ models $Cx(Co)^4$ fits the data best, *i.e.*, the

Chi-Square Model for Interference

TABLE 1

Observed and expected counts under the $Cx(Co)^4$ model

sc-e	*e-cv*	*cv-ct*	*ct-v*	*v-g*	*g-f*	Expected	Observed
0	0	0	0	0	0	12934	12776
1	0	0	0	0	0	1266	1407
0	1	0	0	0	0	1909	2018
0	0	1	0	0	0	1831	1976
0	0	0	1	0	0	3420	3378
0	0	0	0	1	0	2454	2356
0	0	0	0	0	1	2119	2067
1	1	0	0	0	0	5	9
1	0	1	0	0	0	34	16
1	0	0	1	0	0	205	142
1	0	0	0	1	0	240	198
1	0	0	0	0	1	226	206
0	1	1	0	0	0	8	11
0	1	0	1	0	0	146	136
0	1	0	0	1	0	280	261
0	1	0	0	0	1	327	318
0	0	1	1	0	0	33	42
0	0	1	0	1	0	150	148
0	0	1	0	0	1	258	212
0	0	0	1	1	0	65	123
0	0	0	1	0	1	252	315
0	0	0	0	1	1	30	59
1	1	0	1	0	0	0	3
1	1	0	1	0	0	0	1
1	1	0	0	0	1	1	2
1	0	1	0	1	0	3	3
1	0	1	0	0	1	5	3
1	0	0	1	1	0	3	10
1	0	0	1	0	1	14	15
1	0	0	0	1	1	3	1
0	1	1	1	0	0	0	1
0	1	0	1	1	0	2	2
0	1	0	1	0	1	9	10
0	1	0	0	1	1	3	1
0	0	1	1	0	1	2	5
0	0	1	0	1	1	1	5
0	0	0	1	1	1	0	1
1	1	1	1	0	0	0	1
1	1	1	0	0	1	0	1

$Cx(Co)^4$ gives the best fit among $Cz(Co)^m$ models. The estimated genetic distances and their standard errors between these markers are 7.13 ± 0.14, 9.55 ± 0.17, 8.28 ± 0.16, 14.75 ± 0.20, 11.45 ± 0.18 and 11.47 ± 0.19 cM. 0, no recombination; 1, recombination. The estimated P value is <0.001. Data of WEINSTEIN (1936).

largest likelihood is achieved when the $Cx(Co)^4$ model is used. The estimated genetic distances and their associated standard errors are given in Table 1. The optimal $m = 4$, estimated here by statistical analysis, is the same as that in FOSS *et al.* (1993), where they determine m from the observed proportion of gene conversions associated with crossovers.

MORGAN *et al.*'s dataset also contains markers on the *X*-chromosome of Drosophila. There are 16,136 observations on nine loci in this dataset, including six of the same loci as in WEINSTEIN (1936). The $Cx(Co)^4$ model

TABLE 2

Estimated genetic distances with standard errors under the $Cx(Co)^4$ model

Interval	Distance (cM)	SE
sc-e	5.13	0.17
e-cv	9.83	0.23
cv-ct	7.49	0.20
ct-v	13.29	0.26
v-s	8.42	0.21
s-f	15.55	0.28
f-ca	7.47	0.20
ca-b	4.42	0.16

$Cx(Co)^4$ model gives the best fit among $Cx(Xo)^m$ models. The estimated P value is <0.001. Data of MORGAN *et al.* (1935).

again gives the best fit to the data among the $Cx(Co)^m$ class. The results are given in Table 2. MCPEEK and SPEED (1995) fit a broader class of models, in which m is allowed to be noninteger, to the MORGAN *et al.* dataset and estimated $m = 3.94$, which agrees very well with the integer value $m = 4$.

Among the loci that appear in both datasets, the genetic distances estimated from the two different datasets appear rather similar, yet the differences are large compared to the standard errors. This difference probably reflects nonhomogeneity across Drosophila individuals. It is well known that recombination values are different for different individuals and can be affected by factors such as temperature (PERKINS 1962). However, in the $Cx(Co)^m$ model we assume that the crossover process follows the same distribution across the whole population. Thus, it is not surprising that we underestimate the variation in genetic distances.

Neurospora crassa: PERKINS (1962) contains data involving six markers on the right arm of linkage group I in *N. crassa*. PERKINS' data were gathered from six different experiments. This set of data was previously analyzed by COBBS (1978) and RISCH and LANGE (1983). In his paper PERKINS observes that there are significant differences between recombination values for offspring from different parents and from the same set of parents when the temperature is varied. We estimated the interference parameter m for each experiment separately and also for the data when all six experiments are combined. In all cases the best model is $Cx(Co)^2$, which has less crossover interference than $Cx(Co)^4$. This suggests crossover interference is weaker in Neurospora than in Drosophila. The results are given in Table 3.

There is another large multilocus Neurospora dataset in STRICKLAND (1961). He accumulated data from four experiments involving four markers at the end of the right arm of linkage group *V*. A total of 10,269 completely analyzable asci were recovered. We fit the

TABLE 3

Observed and expected counts under the $Cx(Co)^2$ model

cr-th	*th-ni*	*ni-au*	*au-ni*	*ni-os*	Expected	Observed
0	0	0	0	0	106	103
1	0	0	0	0	57	65
0	1	0	0	0	189	201
0	0	1	0	0	109	108
0	0	0	1	0	74	52
0	0	0	0	1	141	126
1	1	0	0	0	26	19
1	0	1	0	0	32	24
1	0	0	1	0	27	32
1	0	0	0	1	61	79
0	2	0	0	0	3	5
0	1	1	0	0	41	36
0	1	0	1	0	50	40
0	1	0	0	1	152	188
0	0	2	0	0	0	1
0	0	1	1	0	8	15
0	0	1	0	1	52	47
0	0	0	1	1	19	22
1	1	1	0	0	4	2
1	1	0	1	0	6	8
1	1	0	0	1	20	14
1	0	1	1	0	2	5
1	0	1	0	1	15	10
1	0	0	1	1	7	6
0	1	1	1	0	2	7
0	1	1	0	1	19	18
0	1	0	1	1	13	11
0	0	1	1	1	2	2
1	1	1	1	0	0	1
1	1	1	0	1	2	1
1	0	1	1	1	0	1
0	1	1	1	1	1	3

$Cx(Co)^2$ gives the best fit among $Cx(Co)^m$ models. The estimated genetic distances with standard errors between these markers are 10.66 (0.59), 22.78 (0.82), 11.81 (0.60), 8.57 (0.55) and 21.69 (0.80) cM. 0, parental ditype; 1, tetratype; 2, nonparental ditype. The estimated P value is 0.69. Data of PERKINS (1962).

$Cx(Co)^m$ model to the data from each of the four experiments separately, and the $Cx(Co)^2$ model gives the highest likelihood in all cases. The results are summarized in Table 4. As in the case of the Drosophila data, some of the estimated genetic distances are significantly different from one experiment to another.

BOLE-GOWDA *et al.* (1962) consider seven markers on linkage group I of *Neurospora crassa*, three on the left arm and four on the right arm. Altogether 2920 offspring were observed. When all markers are used in the analysis, the best model turns out to be Cx, *i.e.*, m is estimated as zero. This estimate is inconsistent with the estimates from the data of PERKINS (1962) and STRICKLAND (1961). This discrepancy might be due to the fact that no positive interference was observed across the centromere, and in that case the chi-square model may not be applicable to data which span the centromere.

The estimated $m = 2$ for Neurospora is consistent with the observation of the ratio of gene conversions to crossovers as described in FOSS *et al.* (1993). Moreover, from the fact that $Cx(Co)^2$ is the best $Cx(Co)^m$ model for data from both linkage groups I and V, we might suspect that the degree of interference is similar within the entire Neurospora genome but with no interference across the centromere.

Saccharomyces cerevisiae: There are abundant two-point cross data for *S. cerevisiae*, but, perhaps because of the high frequency of gene conversion, published multilocus tetrad data are rare. We analyze two-point cross data from a series of papers by MORTIMER and HAWTHORNE (1960, 1966, 1968, 1973). They were analyzed by SNOW (1979) using the model proposed by BARRATT *et al.* (1954). In BARRATT's model, there is a parameter k that measures the degree of crossover interference, similar to the role m plays in the $Cx(Co)^m$ model. (For a full description, see BARRATT *et al.* 1954.) $k = 1$ implies no interference, whereas $k > 1$ and $k < 1$ correspond to negative and positive interference, respectively. We say that there is positive (negative) interference if the probability of double recombinations in two intervals is less (bigger) than the product of the probabilities of recombination in each interval, *i.e.*, interference is defined through a quantity called S_3 by FOSS *et al.* (1993). SNOW fits BARRATT *et al.*'s model to tetrad data involving 34 pairs of markers on 12 chromosomes in *S. cerevisiae*. SNOW's results, along with our estimated optimal m, genetic distances and associated standard errors from the $Cx(Co)^m$ model, are given in Table 5.

Professor J. HABER kindly provided us with a multilo-

TABLE 4

Estimated genetic distances with standard errors and estimated P values

	Experiment 1	Experiment 2	Experiment 3	Experiment 4
hist1-inos	4.33 (0.33)	7.07 (0.40)	6.50 (0.35)	5.40 (0.29)
inos-bis	4.97 (0.36)	5.27 (0.35)	6.02 (0.36)	6.23 (0.32)
bis-pab2	10.1 (0.47)	9.18 (0.45)	10.7 (0.45)	10.6 (0.40)
P value	0.50	0.02	0.02	0.13

$Cx(Co)^2$ is the best model among $Cx(Co)^m$ models for all four experiments. Data of STRICKLAND (1961).

Chi-Square Model for Interference

TABLE 5

Data of *S. cerevisiae* analyzed in SNOW (1979)

Chromosome	Gene pair	m	n	x_{χ^2}	SE	p	x_s	k
2	*cyh1-gal1*	1	146	19.9	2.6	0.98	23.0	0.337
	gal1-lys2	2	383	52.7	3.5	0.71	79.5	0.488
	lys2-try1	3	335	34.8	1.9	0.77	50.0	0.245
	try1-his7	5	127	42.2	3.5	0.90	70.5	0.194
	SUP45-lys2	2	104	25.7	3.1	0.89	32.7	0.258
	SUP45-tyr1	0	105	17.4	3.4	0.64	17.4	1.817
3	*his4-mat1*	2	278	39.8	2.8	0.77	56.3	0.294
	his4-leu2	1	521	17.4	1.2	0.85	19.8	0.370
	leu2-mat1	1	481	35.8	2.2	0.73	44.6	0.429
	mat1-thr4	1	434	21.3	1.6	0.90	24.8	0.404
	thr4-MAL2	2	286	29.2	2.2	0.82	38.2	0.294
4	*SUP35-aro1*	5	101	46.1	7.5	0.99	80.9	0.204
	trp1-cdc2	1	205	18.2	2.1	0.92	20.8	0.284
5	*his1-trp2*	2	215	24.5	2.2	0.79	30.8	0.277
	ura3-hom3	5	206	34.7	2.3	0.99	53.2	0.115
6	*SUP11-his2*	1	105	22.0	3.2	0.97	25.6	0.386
7	*trp5-ade6*	0	106	82.1	4.9	0.62	83.3	1.333
	ade5-tyr3	1	166	73.3	10.	0.87	101.	0.764
	try3-lys5	0	162	8.0	1.7	0.89	0.80	2.355
	cyh2-trp5	3	160	44.8	3.9	0.88	70.5	0.265
	leu1-ade6	2	507	34.3	1.8	0.89	46.7	0.271
	MAL1-ade3	1	138	52.0	6.4	0.78	68.5	0.495
8	*pet1-CUP1*	3	232	46.7	3.4	0.80	74.7	0.271
	thr1-CUP1	3	486	24.4	1.3	0.96	31.8	0.112
	CUP1-pet1	3	240	36.0	2.6	0.98	52.4	0.210
9	*his6-lys1*	1	411	47.1	3.3	0.55	61.1	0.461
10	*SUP4-SUP7*	2	179	53.1	5.4	0.78	80.1	0.498
11	*met14-met1*	1	109	46.2	6.1	0.96	59.7	0.545
	met1-MAL4	1	133	28.0	3.4	0.70	33.5	0.633
15	*ser1-ade2*	4	210	27.4	2.0	0.98	37.5	0.098
	ade2-cyh4	1	236	33.9	3.0	0.78	41.9	0.397
	pet17-ade2	2	232	48.8	3.9	0.79	72.8	0.367
17	*met2-pha2*	2	146	36.8	3.5	0.75	50.6	0.386
	pet2-pha2	1	151	50.7	5.9	0.93	66.4	0.551

m, estimated m in the $Cx(Co)^m$ model; n, sample size; x_{χ^2}, estimated genetic distance from the $Cx(Co)^m$ model; SE, standard error for the estimated genetic distance; p, estimated P value; x_s, estimated genetic distance in SNOW (1979); k, estimated k in BARRATT *et al.*'s model in SNOW (1979).

cus *S. cerevisiae* dataset involving the five markers, *met13, cyh2, trp5, cyh3* and *leu1* on chromosome VII. The markers *met13, cyh2, trp5* and *leu1* were used in three of the 14 experiments, and *met13, trp5, cyh3* and *leu1* were used in the other 11 experiments. We grouped the data across the experiments having the same set of markers and fitted chi-square models. $Cx(Co)^6$ gives the best fit to the data from crosses involving *met13, cyh2, trp5, cyh3* and *leu1;* however, the best model for the other group is *CxCo*. The estimated P values are 0.48 and 0.89, respectively.

Genetic experiments (FOSS *et al.* 1993) have shown that in *S. cerevisiae* the ratio of gene conversions to crossovers is ~2, so we might expect the model $Cx(Co)^2$ to fit best. From Table 5 and results for HABER's data, we can see that unlike Drosophila and Neurospora, where the optimal m does not change from one experiment to another, in Saccharomyces the optimal m varies for different pairs of genes even within the same chromosome. With such a small sample size (usually ~200), it may be that there is simply not sufficient information to clearly distinguish between different $Cx(Co)^m$ models. In these Saccharomyces crosses the differences between the likelihoods under different $Cx(Co)^m$ models are usually small. For example, for the second group in HABER's data, the −log(likelihood)'s are rather close for different m's: they are 125.6, 125.3, 125.1, 125.2 and 125.3 for m is 4, 5, 6, 7 and 8, respectively. A high rate of conversions might also create a problem here. Instead of looking for the optimal m, we could take the $Cx(Co)^2$ model as our hypothesis and test if it is consistent with the data we have.

Schizosaccharomyces pombe: We analyze data from two different sources: those analyzed by SNOW (1979) that were from KOHLI *et al.* (1977) and those provided by Dr. P. MUNZ (personal communication). Two-point

TABLE 6

Data of *S. pombe* analyzed in SNOW (1979)

Chromosome	Gene pair	m	n	x_{χ^2}	SE	p	x_s	k
1	*cyh1-cdc1*	1	142	58.0	7.5	0.75	77.7	0.545
	cdc1-leu2	1	124	51.7	7.0	0.73	67.7	0.698
	his1-leu2	0	102	11.5	2.7	0.22	11.5	4.691
	sup3-aro3	0	170	75.1	10.	0.56	76.2	1.350
	ura2-ade2	0	364	35.8	3.1	0.75	35.7	0.888
	ade2-ade4	0	290	122.	20.	0.72	121.	0.869
	lys3-ura1	0	692	19.9	1.4	0.17	19.9	1.771
	ura1-lys5	0	131	38.4	5.5	0.63	32.8	0.716
	pro1-ade3	0	100	49.6	8.1	0.92	49.7	1.087
	ade3-pro2	0	589	96.9	8.5	0.94	96.7	0.985
2	*ade7-ura5*	1	556	12.3	1.0	0.94	13.4	0.229
	ade7-his3	0	392	69.5	6.1	0.72	69.8	1.117
	glu1-his3	0	212	66.8	8.0	0.81	66.5	0.911
	his3-mat1	1	728	80.1	6.0	0.49	111.	0.837
	tsl24-mat1	0	451	96.2	9.8	0.48	97.6	1.261
	leu1-his5	0	371	28.2	2.5	0.91	28.2	0.951
	his5-leu3	0	100	62.1	11.	0.94	61.8	0.921
	ade1-his4	0	498	42.5	3.0	0.48	42.7	1.249
	his4-trp1	0	128	109.	23.	0.75	111.	1.227
	ade8-arg4	0	185	31.3	4.0	0.87	31.2	0.894
3	*ade10-fur1*	0	142	17.6	3.0	0.70	17.5	0.581
	ade10-ade6	0	202	31.8	3.8	0.98	31.7	0.974
	fur1-sin2	0	206	25.4	3.1	0.90	25.3	0.892
	fur1-min5	0	199	22.4	2.9	0.85	22.4	1.177
	ade6-min5	0	337	4.5	0.8	0.05	4.4	7.693
	tsl5-arg1	5	48	5.16	7.3	0.98	42.3	0.302
	arg1-ade5	1	157	6.14	7.8	0.99	82.7	0.645
	arg1-aro4	0	67	56.3	12.	0.34	58.0	2.523
	trp3-aro4	0	126	19.7	3.4	0.90	19.6	1.151
	ade5-wee1	0	67	69.1	15.	0.85	68.5	0.850

Symbols in this table are the same as those of Table 5.

cross tetrad data on 30 pairs of markers from all three *S. pombe* chromosomes were analyzed by SNOW. Here we fit the $Cx(Co)^m$ model to all the data analyzed by SNOW. The results are given in Table 6. Overall, among the class of $Cx(Co)^m$ models, the best fitting model is the *Cx* model (where $m = 0$), which is equivalent to the no-interference model of HALDANE (1919). This suggests that there is no positive crossover interference in *S. pombe*. Recall that in BARRATT's model $k > 1$ corresponds to negative interference so the fact that the estimated k in BARRATT's model is sometimes >1 suggests that there may be negative crossover interference in *S. pombe*. Among the $Cx(Co)^m$ models slight negative interference may occur at large distances, but at near distances interference cannot be negative. However, the class of $Cx(Co)^m$ models can be extended to allow for negative interference. Instead of assuming that the distance between two crossovers is a χ^2 distribution, one may assume it is a gamma distribution of which the χ^2 is a special case. It is proved in KARLIN and LIBERMAN (1983) that when the gamma shape parameter is <1, the model thus proposed has negative interference. However, there is no explicit expression for any single spore or tetrad recombination pattern unless the shape parameter is an integer. One way to overcome this difficulty would be to use the simulation method that is described by MCPEEK and SPEED (1995).

A multilocus *S. pombe* dataset was kindly provided by Dr. P. MUNZ, who used seven markers (*ura, his, tps, h, leu, ade, lys*) in his experiment with sample size 458. As for the two-point cross data mentioned above, the *Cx* model, *i.e.*, the no-interference model, fits the data best. The estimated P value is 0.23. All the data suggest that there is no positive crossover interference in *S. pombe*, although there may be negative interference.

Aspergillus nidulans: STRICKLAND (1958) published 1231 fully classifiable asci of *A. nidulans* from three separate experiments. Crosses number 1 and 3 cover the same three intervals in the right arm of the *BI* chromosome, whereas cross 2 covers six intervals (the same three in the right arm, two in the left arm and the sixth spanning the centromere). The data from the first two experiments are fitted by the chi-square model. As in the case of *S. pombe*, the *Cx* model fits the data best. There appears to be no positive crossover interference present in this organism.

TABLE 7

Observed and expected counts under the $Cx(Co)^2$ model

Crossovers	Expected (male)	Observed (male)	Expected (female)	Observed (female)
0	206	196	123	130
1	115	131	102	93
2	11	4	20	20
3	0	1	1	3

When m is assumed to be the same for both male and female, $Cx(Co)^2$ gives the best fit. Data of McInnis *et al.* (1993).

Humans: For humans there are not yet available data of the quality and large sample size one finds for experimental organisms. We have analyzed data from McInnis *et al.* (1993) consisting of the number of crossovers inferred over a region of ~60 cM in each of 664 meioses. We estimate $m = 2$ overall, but when males and females are considered separately, we estimate $m = 1$ for females and $m = 4$ for males. The results when male and female are assumed to have the same interference parameter m are given in Table 7.

THE INTERPRETATION OF GOODNESS-OF-FIT STATISTICS

As multiple recombination tends to be rare, there are many possible multilocus recombination events that each occur only a small number of times in the datasets considered here. As a result the asymptotic χ^2 distribution of a test statistic such as Pearson's chi-square statistic or the likelihood ratio statistic may not be a good approximation to the actual distribution. There are two ways to get around with this difficulty: by using a Monte Carlo method to approximate the distribution or by grouping some classes with small expected counts together to form a bigger class. Suppose the sample size is N, the model is of the form $f(\theta_1, \ldots, \theta_k)$, and the test statistic for the sample is T. A Monte Carlo approximation can be carried out as follows: (1) Estimate parameters $\hat{\theta}_1, \ldots, \hat{\theta}_k$ in the model by the method of maximum likelihood, (2) generate N observations according to the distribution $f(\hat{\theta}_1, \ldots, \hat{\theta}_k)$ and calculate the test statistic t_i for this sample and (3) repeat step 2 M times, then the P value of T can be estimated as the proportion of times when t_i is bigger than T.

We did a simulation study to explore which test statistic should be used and whether grouping small cells gives a more powerful test when both the hypothesis and alternative are of the $Cx(Co)^m$ form. The simulation is carried out as follows. Suppose there are five equally spaced markers on the chromosome with genetic distance 10 cM between each consecutive pair of markers. The null hypothesis is that the crossover process follows the $Cx(Co)^m$ model; the alternative is that crossover process follows the $Cx(Co)^k$ model, $k \neq m$. A sample of size 500 is generated from the alternative model. The hypothesis model is used to fit the simulated data, and the P value of the test statistic is calculated by the Monte Carlo approximation as described above. Two thousand such samples are generated, and the P values are calculated. For a level-α test we reject the null hypothesis if the calculated P value is $<\alpha$. The power of the test thus can be approximated by the percentage of times that the null hypothesis is rejected. For each dataset several test statistics from the so-called *power divergence family* (Read and Cressie 1988) are used. The test statistics in this family have the form

$$\frac{2}{\lambda(\lambda+1)} \sum_{i=1}^{n} X_i\left[\left(\frac{X_i}{E_i}\right)^{\lambda} - 1\right].$$

This family includes several well-known test statistics. For example, $\lambda = 1$ gives Pearson's chi-square statistic, and $\lambda = 0$ gives the log-likelihood ratio test statistic. For each sample each test is applied on both ungrouped and grouped classes. Grouping is done in such a way that all offspring types with more than three recombinations are put together to form a larger class, whereas the other types are kept separate. Two pairs of null and alternative hypotheses are considered. In both cases the null hypothesis is set to be the model $Cx(Co)^2$. The alternative is $Cx(Co)^3$ in the first pair and $CxCo$ in the second pair. The results are summarized in Table 8 and Table 9.

TABLE 8

Power comparison between grouped and ungrouped tests

λ	$-1/2$	0	$1/2$	1	$5/2$
Ungrouped test					
$\alpha = 0.01$	0.020	0.010	0.001	0.001	0.003
$\alpha = 0.05$	0.099	0.055	0.021	0.014	0.014
$\alpha = 0.10$	0.200	0.121	0.060	0.039	0.041
Grouped test					
$\alpha = 0.01$	0.045	0.042	0.037	0.031	0.019
$\alpha = 0.05$	0.144	0.139	0.127	0.124	0.107
$\alpha = 0.10$	0.223	0.221	0.217	0.213	0.196

Simulated data are generated from the $Cx(Co)^3$ model then tested against the hypothesis that the data are from the $Cx(Co)^2$ model. Different λ's correspond to different test statistics in power divergence family. α is the significance level at which the test is carried out.

From Tables 8 and 9 we can see that on average tests based on grouped data are more powerful than those based on the original classes, and the power varies more among different test statistics when the ungrouped classes are used. For tests based on grouped classes, there are no big differences between the power of different test statistics. Actually, when the data are grouped, χ^2 is a good approximation to the real distribution of all test statistics, so they are very similar to each other. We think the reason is that large classes are more informative than small classes. Tests of goodness-of-fit should give more weight to these larger classes.

TABLE 9

Power comparison between grouped and ungrouped tests

λ	$-{}^1/_2$	0	${}^1/_2$	1	${}^5/_2$
Ungrouped test					
$\alpha = 0.01$	0.009	0.086	0.123	0.077	0.048
$\alpha = 0.05$	0.069	0.213	0.278	0.235	0.017
$\alpha = 0.10$	0.139	0.321	0.407	0.393	0.308
Grouped test					
$\alpha = 0.01$	0.099	0.111	0.115	0.115	0.125
$\alpha = 0.05$	0.250	0.254	0.251	0.256	0.271
$\alpha = 0.10$	0.359	0.363	0.366	0.372	0.379

Simulated data are generated from the *CxCo* model then tested against the hypothesis that the data are from the $Cx(Co)^2$ model.

COMPARISON WITH THE MODEL OF GOLDGAR AND FAIN

A thorough comparison of different models is made by MCPEEK and SPEED (1995). In this section we focus on the comparison of the $Cx(Co)^m$ model with a model proposed by GOLDGAR and FAIN (1988). In that model, which is similar to the count-location model (KARLIN and LIBERMAN 1979 and RISCH and LANGE 1979), GOLDGAR and FAIN (1988) assume that the number of crossovers follows a distribution that has to be estimated from the data. Their model differs from the count-location model in two respects: (1) given the number of crossovers, their locations are not independent, but they follow a specified joint distribution in which some parameters have to be estimated; (2) instead of putting the distribution on the four-strand bundle, these distributions are put on the single meiotic product, *i.e.*, it is a two-strand model. In fact, it is not possible to construct a four-strand NCI model that is consistent with their model on a single meiotic product (see D. GOLDSTEIN, H. ZHAO and T. P. SPEED unpublished results). In their paper GOLDGAR and FAIN show that their model fits data much better than the count-location model and several two-strand models based on map functions. Estimates of genetic distances among markers from WEINSTEIN's and MORGAN's data by their model (GOLDGAR and FAIN 1988), as well as the estimates based on the $Cx(Co)^m$ model, are given in Table 10 and Table 11.

Besides genetic distances the parameters used by GOLDGAR and FAIN are as follows: d_i, $i = 0, 1, 2$ and 3, the probabilities of 0, 1, 2 and 3 crossovers, respectively; k, which measures the degree of interference, and x_0, the genetic distance between the centromere and the marker closest to it. So when $n + 1$ markers are involved in the experiment, a total of $n + 5$ parameters have to be estimated. On the other hand for the $Cx(Co)^m$ model, $n + 1$ parameters are used, including n genetic distances between each pair of consecutive markers and the parameter m that measures interference. Thus, in general, four fewer parameters are needed for the $Cx(Co)^m$ model than for GOLDGAR and FAIN's model. For some organisms it is reasonable to assume there are no more than three crossovers. For example, among 28,239 offspring in MORGAN's *D. melanogaster* data, only two offspring showed recombination in four intervals at the same time, and no such individual was recorded in WEINSTEIN's data. On the other hand for those organisms that have a large number of crossovers during meiosis, *e.g.*, *S. pombe*, probabilities of 4, 5 or even more crossovers on the four-strand bundle must be estimated when GOLDGAR and FAIN's model is used. One should also specify the joint distribution of these crossover locations. In this case the model loses its simplicity and credibility when many joint distributions must be assumed based on empirical observations.

A good model should both fit the data and be biologically reasonable. Recall that crossovers occur among four-chromatid strands, so under the assumption of no chromatid interference, we can relate the probabilities of crossover patterns on the four-strand bundle and those on a single strand. Under GOLDGAR and FAIN's model when the probabilities of crossover patterns on a single strand are specified, some crossover patterns on the four-strand bundle will have negative probabilities under the assumption of NCI. Thus, the model they describe is incompatible with the assumption of NCI. We tried a variation of GOLDGAR and FAIN's model in which we put the distribution on the four-strand bundle instead of on a single strand. Under the assumption of NCI, we derived the probability of each recombination pattern for single spore data. This slightly modified version of GOLDGAR and FAIN's model fits the Drosophila data as well as the original one.

Coincidence curves: The traditional measure of interference is coincidence (STURTEVANT 1915; MULLER

TABLE 10

Comparison with GOLDGAR and FAIN's model (WEINSTEIN's data)

Interval	*sc-ec*	*ec-cv*	*cv-ct*	*ct-v*	*v-g*	*g-f*	LR
GOLDGAR and FAIN	7.2	9.9	8.6	15.0	11.4	11.4	132.4
$Cx(Co)^4$	7.1	9.6	8.3	14.8	11.5	11.5	219.1

Estimated genetic distances based on GOLDGAR and FAIN's model and the $Cx(Co)^4$ model, together with likelihood ratio statistics (LR).

Chi-Square Model for Interference

TABLE 11

Comparison with GOLDGAR and FAIN's model (MORGAN's data)

Interval	*sc-ec*	*ec-cv*	*cv-ct*	*ct-v*	*v-s*	*s-f*	*f-ca*	*ca-b*	LR
GOLDGAR and FAIN	5.2	10.1	7.8	13.5	8.3	15.7	7.3	4.5	159.5
$Cx(Co)^4$	5.1	9.8	7.5	13.3	8.4	15.6	7.5	4.4	174.8

Estimated genetic distances bas:.' on GOLDGAR and FAIN's model and the $Cx(Co)^4$ model, together with likelihood ratio statistics (LR).

1916), which is expressed as a ratio. The numerator is the chance of simultaneous recombination across both of two disjoint intervals on the chromosome. The denominator is the product of the marginal probabilities of recombination across the intervals.

$$S = \frac{r_{11}}{(r_{10} + r_{11})(r_{01} + r_{11})},$$

where S is the coincidence and r_{ij} is the chance of i recombinations across the first interval and j recombinations across the second interval. The coincidence curve for a model is a plot of the coincidence against the genetic distance between two intervals, where the widths of the two intervals are taken to be infinitesimal. (FOSS *et al.* call this quantity S_4.) FOSS *et al.* (1993) compare the coincidence curves (S) for the $Cx(Co)^m$ model with empirical coincidence curves estimated from data. They find that the theoretical curves are very close to the empirical ones. Similarly, we draw the S curves based on the modified version of GOLDGAR and FAIN's model (Figure 1). In GOLDGAR and FAIN's model S is not only a function of the genetic distance between the two regions under study but also depends on how far these regions are from the centromere, so the S curve cannot be uniquely drawn on the graph. Instead, for a given genetic distance between two regions, S can vary according to the distance to the centromere. It is clear from the graph that GOLDGAR and FAIN's model predicts that S will be >1.5 when the distance between the two regions is bigger than 60 cM, no matter where they are located on the chromosome. But this prediction is not consistent with the empirical results in which S is always smaller than 1.2.

DISCUSSION

Based on the derived probabilities for each single spore or tetrad recombination pattern, we use the method of maximum likelihood to fit the $Cx(Co)^m$ model to a variety of organisms. The estimated m's based on statistical analyses of *D. melanogaster* and *N. crassa* data agree with those given by FOSS *et al.* (1993), where m was estimated by the ratio of gene conversions

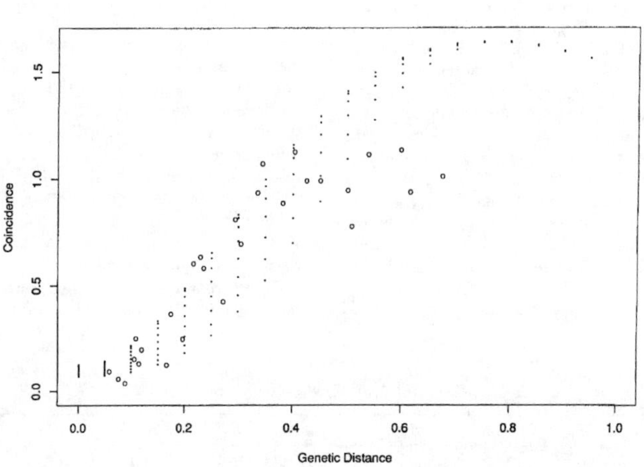

FIGURE 1.—Comparison between predicted and observed S values for GOLDGAR and FAIN's model. The dots above any given genetic distance represent the range of possible predicted S values, which vary according to location on the chromosome. O, the observed S value.

to crossovers in genetic experiments. In humans we are currently able to provide only a preliminary estimate of *m*, although we hope that more extensive human datasets will soon become available. Whereas some amount of positive interference is shown in the above organisms, it is not present in two other organisms we analyzed, *S. pombe* and *A. nidulans*.

The *m* estimated from different experiments using genes on different chromosomes within the same organism turn out to be rather similar. This implies that as a measure of interference, *m* does not change across different chromosomes, thus, the degree of interference might be determined by factors specific to each organism.

We have discussed how to interpret the goodness-of-fit test statistic appropriately after fitting a model to a multilocus dataset. In multilocus data because many single spore or tetrad recombination patterns have rather small expected numbers of observations, even in an experiment of moderate size, the χ^2 approximation to goodness-of-fit statistics often fails. Two ways of avoiding this difficulty are proposed: (1) simulating the distribution of the test statistic by a Monte Carlo method or (2) grouping small classes into a larger class. Our simulation study shows that the tests based on grouped classes usually have larger power than the tests based on ungrouped classes.

Among the four-strand models considered here, the four-strand version of the model of GOLDGAR and FAIN (1988) gave the smallest likelihood ratio statistics (see Table 10 and Table 11), but considering that this model has four more parameters than the chi-square model, the difference in likelihood ratio statistics is not impressive. The count-location model has two parameters more than the chi-square model, yet it performed worse. In comparing the $Cx(Co)^m$ model with the count-location model and GOLDGAR and FAIN's model, we consider that a good model should not only yield a small test statistic but should also be parsimonious (*i.e.*, have few parameters), biologically reasonable and generalizable to many organisms. In these respects the $Cx(Co)^m$ model seems superior. It has a biological basis and is computationally tractable. Because of its simple structure, it can be applied to a broad range of organisms.

Although the $Cx(Co)^m$ model discussed in this paper applies well to data of different organisms and gives some insight into the underlying crossover process, there is a lot of room left for improvement. First of all, the parameter *m* need not be restricted to be an integer, but when *m* is not an integer, there are no explicit expressions for the probabilities of single spore or tetrad recombination patterns. In this case M. S. MCPEEK and T. P. SPEED (unpublished results) use a simulation method to estimate the parameters. Second, we might suspect that the amount of interference varies in different regions within the same chromosome. A local *m* rather than a global *m* might be fitted in the model. Finally, for some organisms with a high proportion of conversion data observed, we need to develop a model to include both gene conversions and crossovers. A good model of this kind should help us understand more about the crossing-over process.

The no-interference model is widely used in human genome mapping. Although it has been shown by SPEED *et al.* (1992) that the no-interference model is asymptotically robust for gene ordering, we do lose some efficiency in ordering and in excluding a test locus when there is interference in the underlying crossover process. D. GOLDSTEIN, H. ZHAO and T. P. SPEED (unpublished results) study the loss in efficiency using the no-interference model when the actual crossover process follows the $Cx(Co)^m$ model. They find that the number of gametes required for these tasks is 10–50% smaller for the $Cx(Co)^m$ model than for the no interference model, depending on the degree of interference and the distances between the markers.

We thank Professor JAMES HABER and Dr. PETER MUNZ for kindly supplying the data. This work was supported by National Science Foundation grant DMS-9113527.

LITERATURE CITED

BAILEY, N. T. J., 1961 *Introduction to the Mathematical Theory of Genetic Linkage*. Oxford University Press, London.

BARRATT, R. W., D. NEWMEYER, D. D. PERKINS and L. GARNJOBST, 1954 Map construction in *Neurospora crassa*. Adv. Genet. **6:** 1–93.

BOLE-GOWDA, B. N., D. D. PERKINS and W. N. STRICKLAND, 1962 Crossing-over and interference in the centromere region of linkage group I of Neurospora. Genetics **47:** 1243–1252.

COBBS, G., 1978 Renewal process approach to the theory of genetic linkage: case of no chromatid interference. Genetics **89:** 563–581.

FOSS, E., R. LANDE, F. W. STAHL and C. M. STEINBERG, 1993 Chiasma interference as a function of genetic distance. Genetics **133:** 681–691.

GOLDGAR, D. E., and P. R. FAIN, 1988 Models of multilocus recombination: nonrandomness in chiasma number and crossover positions. Am. J. Hum. Genet. **43:** 38–45.

HALDANE, J. B. S., 1919 The combination of linkage values, and the calculation of distances between the loci of linked factors. J. Genet. **8:** 299–309.

HAWTHORNE, D. C., and R. K. MORTIMER, 1960 Chromosome mapping in Saccharomyces: centromere-linked genes. Genetics **45:** 1085–1110.

HAWTHORNE, D. C., and R. K. MORTIMER, 1968 Genetic mapping of nonsense suppressors in yeast. Genetics **60:** 735–742.

KARLIN, S., and U. LIBERMAN, 1979 A natural class of multilocus recombination processes and related measure of crossover interference. Adv. Appl. Probab. **11:** 479–501.

KARLIN, S., and U. LIBERMAN, 1983 Measuring interference in the chiasma renewal formation process. Adv. Appl. Probab. **15:** 471–487.

KOHLI, J., H. HOTTINGER, P. MUNZ, A. STRAUSS and P. THURIAUX, 1977 Genetic mapping in *Schizosaccharomyces pombe* by mitotic and meiotic analysis and induced haploidization. Genetics **87:** 471–489.

MATHER, K., 1935 Reduction and equational separation of the chromosomes in bivalents and multivalents. J. Genet. **30:** 53–78.

MCINNIS, M. G., A. CHAKRAVARTI, J. BLASCHAK, M. B. PETERSEN, V. SHARMA *et al.*, 1993 A linkage map of human chromosome 21: 43 PCR markers at average intervals of 2.5 cM. Genomics **16:** 562–571.

MCPEEK, M. S., and T. P. SPEED, 1995 Modeling interference in genetic recombination. Genetics **139:** 000–000.

MORGAN, T. H., C. B. BRIDGES and J. SCHULTZ, 1935 Report of

investigations on the constitution of the germinal material in relation to heredity. Carnegie Instit. Washington **34:** 284–291.
MORTIMER, R. K., and D. C. HAWTHORNE, 1966 Genetic mapping in Saccharomyces. Genetics **53:** 165–173.
MORTIMER, R. K., and D. C. HAWTHORNE, 1973 Genetic mapping in Saccharomyces IV. Mapping of temperature-sensitive genes and use of disomic strains in localizing genes. Genetics **74:** 33–54.
MULLER, H. J., 1916 The mechanism of crossing-over. Am. Nat. **50:** 193–434.
PERKINS, D. D., 1962 Crossing-over and interference in a multiply marked chromosome arm of Neurospora. Genetics **47:** 1253–1274.
PRESS, W. H., B. P. FLANNERY, S. A. TEUKOLSKY and W. T. VETTERLING, 1988 *Numerical Recipes in C.* Cambridge University Press, Cambridge, UK.
READ, T. R. C., and N. A. C. CRESSIE, 1988 *Goodness-of-fit Statistics for Discrete Multivariate Data.* Springer-Verlag, New York.
RISCH, N., and K. LANGE, 1979 An alternative model of recombination and interference. *Ann. Hum. Genet.* **43:** 61–70.
RISCH, N., and K. LANGE, 1983 Statistical analysis of multilocus recombination. Biometrics **39:** 949–963.
SNOW, R., 1979 Maximum likelihood estimation of linkage and interference from tetrad data. Genetics **92:** 231–245.
SPEED, T. P., M. S. MCPEEK and S. N. EVANS, 1992 Robustness of the no-interference model for ordering genetic markers. Proc. Natl. Acad. Sci. USA **89:** 3103–3106.
STRICKLAND, W. N., 1958 An analysis of interference in *Aspergillus nidulans.* Proc. R. Soc. Lond. Ser. B **149:** 82–101.
STRICKLAND, W. N., 1961 Tetrad analysis of short chromosome regions of *Neurospora crassa.* Genetics **46:** 1125–1141.
STURTEVANT, A. H., 1915 The behavior of the chromosomes as studied through linkage. *Z. Indukt. Abstammungs-Vererbungsl.* **13:** 234–287.
WEINSTEIN, A., 1936 The theory of multiple-strand crossing over. Genetics **21:** 155–199.
ZHAO, H., M. S. MCPEEK and T. P. SPEED, 1995 Statistical analysis of chromatid interference. Genetics **139:** 1057–1065.

Communicating editor: B. S. WEIR

APPENDIX

Lemma: Under the $Cx(Co)^m$ model the probability of k_j crossovers between l_j and l_{j+1}, $j = 1, \ldots, n$ is

$$\frac{1}{p}\,\mathbf{1}\mathbf{D}_{\mathbf{k}_1}(y_1)\,\mathbf{D}_{\mathbf{k}_2}(y_2)\cdots\mathbf{D}_{\mathbf{k}_n}(y_n)\,\mathbf{1}',$$

where $p = m + 1$, $y_j = 2px_j$ and $\mathbf{D}_\mathbf{k}(y)$ has i, jth entry $e^{-y}y^{pk+j-i}/(pk + j - i)!$.

Proof: We start with the simplest case when there are only two markers. Because the C events are randomly distributed on the four-strand bundle and the number of C's follows the Poisson distribution with parameter, say y, the chance of s C's is $e^{-y}y^s/s!$. $1/p$ of these C's will resolve as crossover event, and under the assumption of no chromatid interference, each strand has chance $1/2$ of being involved in each crossover. So on average each strand has $s/2p$ crossovers, given s C events. Recall that the genetic distance is defined to be the expected number of crossovers on a single strand, so the genetic distance x and the Poisson parameter y are related by $x = y/2p$, *i.e.*, $y = 2px$.

In the following discussion suppose markers l_1, l_2, $\ldots$, l_n are laid out from left to right, and the C events occur also from left to right. The $Cx(Co)^m$ model assumes that the crossover intermediate (C) events resolve in sequence like $CxCoCo\cdots CoCxCo\cdots$ and that the process is stationary, so the first C event to the right of l_1 has an equal chance of resolving as any of the $m + 1$ elements of $Cx(Co)^m$. The occurrence of k crossovers between l_1 and l_2 might be the result of p^2 possible situations, depending on the number of Co's before the first Cx to right of l_1 and the number of Co's between l_2 and the nearest Cx left to it. The number can vary from 0 to $p - 1$. Therefore, the chance of k_1 Cx's between l_1 and l_2 can be computed as

$$\frac{e^{-y_1}}{p}\sum_{i=1}^{p}\sum_{j=0}^{p-1}\frac{y_1^{pk_1-p+i+j}}{(pk_1 - p + i + j)!}.$$

The case $i = 1$ corresponds to the situation where the leftmost C between l_1 and l_2 is a Cx, and the rightmost C could be either one Cx ($pk_1 - p + 1$ C's altogether between l_1 and l_2), the first Co after a Cx ($pk_1 - p + 2$ C's) or the second Co after a Cx, *etc.* $i - 1$ corresponds to the number of Co's between l_1 and the first Cx.

We can write the sum in a matrix product form:

$$\frac{1}{p}(1 \quad 1 \quad \cdots \quad 1)\,\mathbf{D}_{\mathbf{k}_1}(y_1)\begin{pmatrix}1\\1\\\vdots\\1\end{pmatrix}.$$

Each element in the first column of the matrix corresponds to the last C event between l_1 and l_2 being a Cx; the second column corresponds to the last C being the first Co after the k_1th Cx, the jth column to the jth C after the k_1th Cx. Therefore, the sum of the jth ($j > 0$) column multiplied by $1/p$ is the probability that there are k_1 crossovers between l_1 and l_2, and the last C event is the $(j - 1)$th Co after the k_1th Cx. Therefore if we define

$$(p_{k_1}^1 p_{k_1}^2 \cdots p_{k_1}^p) = \frac{1}{p}(1 \quad 1 \quad \cdots \quad 1)\,\mathbf{D}_{\mathbf{k}_1}(y_1),$$

then $p_{k_1}^j$ is the probability that the last C between l_1 and l_2 is the $(j - 1)$th Co after the k_1th Cx with the exception that $p_{k_1}^0$ is the probability of the last C being the k_1th Cx.

Now we consider the case for three markers l_1, l_2 and l_3. Given that the first C to the right of l_2 is the lth Co after a Cx, the probability of k_2 crossovers between l_2 and l_3 is

$$e^{-y_2}\sum_{i=1}^{p}\frac{y_2^{pk_2+l-i}}{(pk_2 + l - i)!}.$$

The chance that there are l Co's between l_2 and the first Cx after l_2 is the same as the chance that the last C between l_1 and l_2 is the $(p - l - 1)$th C after a Cx

which is $p_{k_1}^{p-l-1}$. Therefore the chance of k_1 crossovers between ℓ_1, ℓ_2, and k_2 crossovers between ℓ_2 and ℓ_3 is

$$e^{-y_2} \sum_{i=1}^{p} p_{k_1}^{p+1-i} \sum_{j=0}^{p-1} \frac{y_2^{pk_2-p+i+j}}{(pk_2 - p + i + j)!}.$$

Rewriting the above relation in matrix form, we get

$$(p_{k_1}^1 \quad p_{k_1}^2 \quad \cdots \quad p_{k_1}^p)\mathbf{D_{k_2}}(y_2)\begin{pmatrix} 1 \\ 1 \\ \vdots \\ 1 \end{pmatrix}.$$

Recall that $(p_{k_1}^1 \quad p_{k_1}^2 \quad \cdots \quad p_{k_1}^p) = (1 \quad 1 \quad \cdots \quad 1)$ $\mathbf{D_{k_1}}(y_1)$, thus the probability of k_1 crossovers between ℓ_1 and ℓ_2, k_2 crossovers between ℓ_2 and ℓ_3 is

$$\mathbf{1}\mathbf{D_{k_1}}(y_1)\mathbf{D_{k_2}}(y_2)\mathbf{1}'.$$

The general result involving n intervals can be proved by the same method.

Theorem 1: Define

$$\mathbf{N_j} = \mathbf{D_0}(y_j) + {}^1/_2 \sum_{s\geq 1} \mathbf{D_s}(y_j),$$

$$\mathbf{R_j} = {}^1/_2 \sum_{s\geq 1} \mathbf{D_s}(y_j),$$

then the probability of recombination pattern $(i_1 i_2 \cdots i_n)$ is

$$P(i_1 i_2 \cdots i_n) = \frac{1}{p}\mathbf{1}\mathbf{M_1}\mathbf{M_2}\cdots\mathbf{M_n}\mathbf{1}',$$

where $\mathbf{M_j} = \mathbf{N_j}$ when $i_j = 0$, and $\mathbf{M_j} = \mathbf{R_j}$ when $i_j = 1$.

Proof: It is well known that given $k \geq 1$ crossovers between two markers, the chance of a recombination on a single strand is $^1/_2$, and there can be no recombination if no crossovers occur. We can write $p_0^{(k)}$ for the probability of no recombination and $p_1^{(k)}$ for the probability of recombination given k crossovers occurring. So $p_0^{(0)} = 1$, $p_1^{(0)} = 0$ and $p_1^{(k)} = p_0^{(k)} = {}^1/_2$ when $k \geq 1$. Write $P_{(i_1 i_2 \cdots i_n)}^{(k_1, k_2, \ldots, k_n)}$ for the probability of observing recombination pattern $(i_1 i_2 \cdots i_n)$ when there are k_j crossovers in the jth interval, $j = 1, 2, \ldots, n$. Then we have

$$\begin{aligned} P(i_1 i_2 \cdots i_n) &= \sum_{k_1, k_2, \ldots, k_n} P_{(i_1 i_2 \cdots i_n)}^{(k_1, k_2, \ldots, k_n)} \\ &= \sum_{k_1}\sum_{k_2}\cdots\sum_{k_n} p_{i_1}^{(k_1)} p_{i_2}^{(k_2)} \cdots p_{i_n}^{(k_n)} \frac{1}{p} \\ &\quad \times \mathbf{1}\mathbf{D_{k_1}}(y_1)\mathbf{D_{k_2}}(y_2)\cdots\mathbf{D_{k_n}}(y_n)\mathbf{1}' \\ &= \frac{1}{p}\mathbf{1}\Big(\sum_{k_1}\sum_{k_2}\cdots\sum_{k_n} p_{i_1}^{(k_1)} p_{i_2}^{(k_2)} \cdots p_{i_n}^{(k_n)} \\ &\quad \times \mathbf{D_{k_1}}(y_1)\mathbf{D_{k_2}}(y_2)\cdots\mathbf{D_{k_n}}(y_n)\Big)\mathbf{1}' \\ &= \frac{1}{p}\mathbf{1}\Big(\sum_{k_1} p_{i_1}^{(k_1)}\mathbf{D_{k_1}}(y_1)\sum_{k_2} p_{i_2}^{(k_2)} \\ &\quad \times \mathbf{D_{k_2}}(y_2)\cdots\sum_{k_n} p_{i_n}^{(k_n)}\mathbf{D_{k_n}}(y_n)\Big)\mathbf{1}' \\ &= \frac{1}{p}\mathbf{1}\mathbf{M_1}\mathbf{M_2}\cdots\mathbf{M_n}\mathbf{1}'. \end{aligned}$$

Theorem 2: Define

$$\mathbf{P_j} = \mathbf{D_0}(y_j) + \sum_{s\geq 2} {}^1/_3({}^1/_2 + (-{}^1/_2)^k)\mathbf{D_s}(y_j)$$

$$\mathbf{T_j} = \mathbf{D_1}(y_j) + \sum_{s\geq 2} {}^2/_3(1 - (-{}^1/_2)^k)\mathbf{D_s}(y_j)$$

$$\mathbf{N_j} = \sum_{s\geq 2} {}^1/_3({}^1/_2 + (-{}^1/_2)^k)\mathbf{D_s}(y_j).$$

Then the probability of tetrad pattern $(i_1 i_2 \cdots i_n)$ can be written as

$$P(i_1 i_2 \cdots i_n) = \frac{1}{p}\mathbf{1}\mathbf{M_1}\mathbf{M_2}\cdots\mathbf{M_n}\mathbf{1}',$$

where $\mathbf{M_j} = \mathbf{P_j}$ if $i_j = 0$, $\mathbf{M_j} = \mathbf{T_j}$ if $i_j = 1$, and $\mathbf{M_j} = \mathbf{N_j}$ if $i_j = 2$.

Proof: Notice that given $k \geq 1$ crossovers between two markers, the probabilities of parental ditype, tetratype and nonparental ditype are ${}^1/_3({}^1/_2 + (-{}^1/_2)^k)$, ${}^2/_3(1 - (-{}^1/_2)^k)$ and ${}^1/_3({}^1/_2 + (-{}^1/_2)^k)$, respectively. Using the same method as Theorem 1, the conclusion follows.

On Genetic Map Functions

Hongyu Zhao* and Terence P. Speed†

**Department of Biostatistics, University of California, Los Angeles, California 90024 and †Department of Statistics, University of California, Berkeley, California 94720*

Manuscript received June 12, 1995
Accepted for publication December 20, 1995

ABSTRACT

Various genetic map functions have been proposed to infer the unobservable genetic distance between two loci from the observable recombination fraction between them. Some map functions were found to fit data better than others. When there are more than three markers, multilocus recombination probabilities cannot be uniquely determined by the defining property of map functions, and different methods have been proposed to permit the use of map functions to analyze multilocus data. If for a given map function, there is a probability model for recombination that can give rise to it, then joint recombination probabilities can be deduced from this model. This provides another way to use map functions in multilocus analysis. In this paper we show that stationary renewal processes give rise to most of the map functions in the literature. Furthermore, we show that the interevent distributions of these renewal processes can all be approximated quite well by gamma distributions.

GENETIC maps consist of different kinds of markers positioned along a chromosome, with their relative locations measured in the map units known as centiMorgans. During meiosis, crossing over occurs after homologous chromosomes pair and duplicate, resulting in a four-strand structure. Each crossover involves two strands of different origins (nonsister pairs). The genetic map distance in Morgans between a pair of markers on the same chromosome is the average number of crossovers occurring between these markers during meiosis on one chromatid. Because crossovers are not observable, a genetic map function $r = M(d)$ is often used to infer genetic distance (d) from the observable recombination fraction (r).

In his 1919 paper, HALDANE made three contributions to the study of genetic map functions, as well as defining them. By assuming no *crossover interference* (STURTEVANT 1915; MULLER 1916), he derived the map function $r = {}^1\!/_2(1 - e^{-2d})$ with inverse $d = -{}^1\!/_2 \log(1 - 2r)$; he also proposed the empirical inverse map function $d = 0.7r + 0.3(-{}^1\!/_2 \log(1 - 2r))$ to account for crossover interference in the data then available, and he introduced a differential equation method that permitted the construction of a variety of map functions.

A commonly used measure of the degree of crossover interference is the coincidence coefficient C involving three markers. Letting $p_{i_1 i_2}$ be the probability of i_1 recombinations in the first interval and i_2 recombinations in the second interval, $i_1, i_2 = 0, 1$, C is defined by

$$C = \frac{p_{11}}{(p_{10} + p_{11})(p_{01} + p_{11})}.$$

Corresponding author: Terence P. Speed, Department of Statistics, University of California, Berkeley, CA 94720.
E-mail: terry@stat.berkeley.edu

Genetics 142: 1369–1377 (April, 1996)

The case of $C = 1$ corresponds to no crossover interference, while $C < 1$ and $C > 1$ correspond to *positive* and *negative* crossover interference, respectively. Because the $p_{i_1 i_2}$ can be directly estimated from recombination data in most experimental organisms, C is often used as an empirical index of the degree of crossover interference.

One implicit assumption underlying the use of a genetic map function for one organism is that the functional relationship between genetic distances and recombination fractions does not vary across the genome in this organism. This assumption implies that C should only depend on the map lengths of the intervals between three markers of interest, say d and h. Noting that $p_{10} + p_{11} = M(d)$, $p_{01} + p_{11} = M(h)$, and $p_{11} = \{M(d) + M(h) - M(d + h)\}/2$, we have

$$C(d,h) = \frac{M(d) + M(h) - M(d + h)}{2M(d)M(h)}.$$

Letting $h \to 0$, and assuming $\lim_{h\to 0} M(h)/h = 1$, we obtain the differential equation

$$M'(d) = 1 - 2C(d)M(d). \quad (1)$$

Different choices of $C(d)$ lead to different map functions, which will be discussed later.

One of the difficulties in using genetic map functions to construct genetic maps is that when there are more than three markers, the multilocus recombination probabilities cannot be uniquely determined from the map function, a point that was made by FISHER (1947). This identifiability problem can be solved by postulating that the probability that there is an odd number of crossovers across a set of (disjoint) intervals only depends on the total length of these intervals

(GEIRINGER 1944; SCHNELL 1961). The underlying assumption is that the distance between two disjoint intervals is irrelevant to the joint recombination probabilities in these two intervals. However experimental results suggest that the degree of interference varies with the distance between two intervals: the smaller the distance, the stronger the interference. Thus, the above approach is not consistent with experimental results. Nevertheless, this assumption can be and has been used to calculate joint recombination probabilities from map functions. LIBERMAN and KARLIN (1984) proved that a necessary condition to guarantee that the recombination probabilities obtained in this way are nonnegative is that $(-1)^k M^{(k)}(0) \leq 0$ for all k. They (inappropriately) termed map functions that always give rise to nonnegative recombination probabilities "multilocus feasible" and showed that many map functions that were found to fit data well did not satisfy these criteria.

If crossovers are viewed as a stochastic point process along the chromatid and a given map function can be realized from a crossover process, multi-locus recombination probabilities compatible with the map function can be obtained by assuming crossovers are generated from this point process. Two classes of point processes have been studied extensively in the literature in the context of modeling crossover interference: renewal processes, reviewed in BAILEY (1961) and MCPEEK and SPEED (1995), and count-location processes, studied by KARLIN and LIBERMAN (1978) and RISCH and LANGE (1979).

In this paper, we show that for most map functions in the literature there exist stationary renewal processes that give rise to them, and so these map functions are compatible with the analysis of multilocus data via this approach. Moreover, the interevent distributions of the stationary renewal processes corresponding to most map functions can be closely approximated by gamma distributions.

A special class of stationary renewal processes, called chi-square models, which have chi-square interevent distributions with even degrees of freedom, was found to give good fit to data from a variety of organisms (ZHAO *et al.* 1995b). This class of models, which evolved from an ordinary renewal process model for crossovers on a single meiotic product proposed by FISHER *et al.* (1947), was mainly studied because of its mathematical tractability (MCPEEK and SPEED 1995). It was reintroduced by FOSS *et al.* (1993) from a biological perspective, motivated by observations from experiments on gene conversion although there are now serious doubts concerning the appropriateness of this motivation (see FOSS and STAHL 1995). The chi-square model was conveniently denoted as $Cx(Co)^m$, which corresponds to a chi-square renewal density with $2(m + 1)$ degrees of freedom. Using the method of maximum likelihood, chi-square models were fitted to different organisms by ZHAO *et al.* (1995b). The estimates of m for Drosophila ($m = 4$) and Neurospora ($m = 2$) were the same as those obtained by FOSS *et al.* (1993), who estimated m from the observed ratio of *Cx* to *Co*. S. LIN and T. P. SPEED (unpublished results) fitted chi-square models to data on six loci from the CEPH consortium map of human chromosome *10*, which was analyzed by WEEKS *et al.* (1994) using other models, and estimated the parameter m to be 3. Their results suggested the presence of crossover interference during human meiosis. MCPEEK and SPEED (1995) compared the fit of the stationary renewal process model with gamma interevent distributions (the gamma model) with that of other models using one large data set of Drosophila and found that the gamma model gives a better fit than all other models. This fact, together with the observation that for most map functions the interevent distributions for the corresponding stationary renewal processes can be closely approximated by gamma distributions, suggest that to some degree the chi-square model, and more generally, the gamma model, is able to capture the important features of the unobservable crossover process.

After raising the question "What is a genetic map function?", SPEED (1995) discussed in detail many issues related to map functions. It should be pointed out that other methods have also been proposed to extend map functions to handle multilocus data (OWEN 1953; MORTON and MACLEAN 1984).

MAP FUNCTIONS AND STATIONARY RENEWAL PROCESSES

When modeling crossovers as a point process, a distinction should be made between the point process on the four strand bundle, *the chiasma process*, and the point process on a single strand, *the crossover process*. Modeling the crossover process as a renewal process has a long history. Following JENNINGS (1923) and MATHER (1936, 1937), FISHER *et al.* (1947) modeled crossing over as a renewal process; that is, crossovers along a single strand were assumed to be formed as a regular sequence starting from the centromere, with the length between two adjacent crossovers always following the same distribution. Although it was known that crossovers take place when four strands (chromatids) are present during meiosis, FISHER *et al.* (1947) only modeled the crossover process on a single strand, without relating it to the chiasma process on the four-strand bundle. On the other hand, most later work modeled the chiasma process as a renewal process and related the crossover process to the chiasma process (CARTER and ROBERTSON 1952; COBBS 1978; STAM 1979). The assumption that crossovers occur starting from the centromere and progressing toward the telomere is now known not to be true (WHITEHOUSE 1982), but it is still a convenience. To connect the chiasma process to the crossover process, assumptions have to be made concerning the way nonsister pairs are involved in crossovers. Two types of

Map Functions

interference can be distinguished: *chiasma interference,* where chiasmata on the four-strand bundle do not occur independently of each other, and *chromatid interference,* where the choices of nonsister pairs involved in different chiasmata are not independent.

The observation of crossover interference on the meiotic products (single strands) can be the result of chiasma interference alone, the result of chromatid interference alone, or the result of both types of interference. It is interesting to note that the operation of two types of interference can lead to no apparent crossover interference. Consider the case when the chiasma process follows a stationary renewal process with inter-event distribution being the gamma distribution with shape parameter $^1/_2$, and there is complete positive chromatid interference, *i.e.,* the strands involved in one chiasma are never involved in the closest chiasmata to its left and to its right. It is easy to see that the distance between two crossovers on a single meiotic product from this process follows the gamma distribution with shape parameter 1, *i.e.,* the exponential distribution. Therefore crossovers on a single strand appear to occur independently of each other. This example shows that two types of interference cannot be separated based on single-strand recombination data, where meiotic products from a single meiosis are recovered separately. Therefore tetrad data, where all meiotic products from a single meiosis are recovered together, are often used to detect chromatid interference. Since there is no strong and consistent evidence of chromatid interference (WHITEHOUSE 1982), it is generally assumed that there is no chromatid interference (NCI) in the models proposed in the literature.

The assumption of NCI imposes certain constraints on both recombination and tetrad probabilities (SPEED *et al.* 1992; ZHAO *et al.* 1995a). These constraints further impose constraints on map functions (SPEED 1995):

$$0 \le M(d) \le \tfrac{1}{2},$$

$$M'(d) \ge 0,$$

$$M''(d) \le 0.$$

It is also obvious that if the chiasma point process is simple and stationary in the map distance metric, then $M(0) = 0$ and $M'(0) = 1$ (DALEY and VERE-JONES 1988, Section 3.3). We will consider map functions that are defined on a finite interval $[0, L]$ and those defined $[0, \infty)$ separately. By imposing one more condition, we say a function M defined on $[0, \infty)$ satisfies condition (A) if

$$M(0) = 0, \qquad (A1)$$

$$M'(d) \ge 0, \quad \text{for all } d, \qquad (A2)$$

$$M'(0) = 1, \qquad (A3)$$

$$M''(d) \le 0, \quad \text{for all } d, \qquad (A4)$$

$$\lim_{d\to\infty} M(d) = {}^1/_2. \qquad (A5)$$

Apart from (*A*5), these conditions are necessary under the assumptions of NCI and that the chiasma process is a simple stationary point process. Condition (*A*5) postulates that two markers that are very far apart on the same chromosome can be considered very loosely linked, effectively behaving like markers on different chromosomes and segregating independently. Moreover, we have the following theorem to characterize this class of map functions. The proof is given in the APPENDIX.

Theorem 1: Let M be the map function for a stationary renewal chiasma process satisfying NCI on a chromosome arm of infinite length. Then M satisfies (A). Conversely, suppose that a function M from $[0, \infty)$ into $[0, {}^1/_2]$ satisfies (A). Then there is a stationary renewal chiasma process whose map function is M and whose renewal density is $-M''$.

For a map function M defined on $[0, L]$ where $L < \infty$, we say that M satisfies condition (B) if

$$M(0) = 0, \qquad (B1)$$

$$M'(d) \ge 0, \quad \text{for all } d, \qquad (B2)$$

$$M'(0) = 1, \qquad (B3)$$

$$M''(d) \le 0, \quad \text{for all } d, \qquad (B4)$$

$$M'(L) = 0, \qquad (B5)$$

$$M(L) = {}^1/_2. \qquad (B6)$$

We say that M satisfies condition (B)′ if it satisfies (B1)–(B4) and

$$M'(L) > 0, \qquad (B5)'$$

$$M(L) < {}^1/_2. \qquad (B6)'$$

For map functions defined on $[0, L]$, we have the following analogue of Theorem 1 with the proof given in the APPENDIX.

Theorem 2: Let M be the map function for a stationary renewal chiasma process satisfying NCI on a chromosome arm of finite length. Then M satisfies (B) or (B)′ for any L. Conversely, suppose that a function M from $[0, L]$ into $[0, {}^1/_2]$ satisfies (B) or (B)′. Then there is a stationary renewal chiasma process whose map function is M and whose renewal density is $-M''$ when $d \le L$.

VARIOUS MAP FUNCTIONS

In this section, we apply our two theorems to some map functions in the literature to see if there are stationary renewal processes that can give rise to them.

HALDANE (1919): $M_{H_2}^{-1}(r) = 0.7r + 0.3(-{}^1/_2 \log(1 - 2r))$. It is easy to see that $M^{-1}(0) = 0$, $\lim_{r\to 1/2} M^{-1}(r) = \infty$, $(M^{-1})' > 0$, $(M^{-1})'(0) = 1$, $\lim_{r\to 1/2} (M^{-1})' = \infty$, and $(M^{-1})'' \ge 0$. So M_{H_2} satisfies condition (A), and there is a stationary renewal process giving rise to M_{H_2}.

LUDWIG (1934): $M_L(d) = {}^1/_2 \sin(2d)$. Clearly this should only be considered a possible map function in

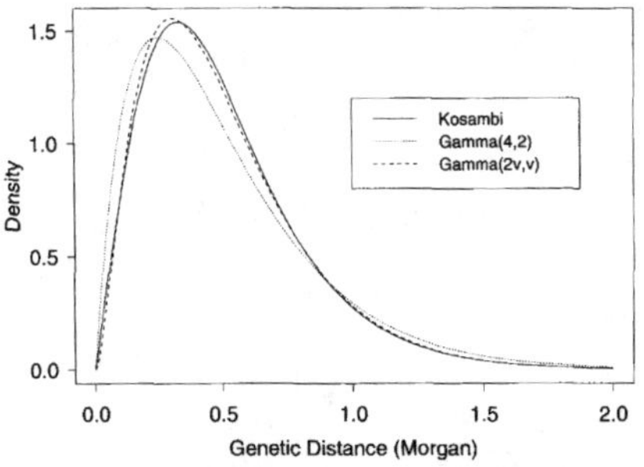

FIGURE 1.—Comparison between the interevent density ($-M_K''$) of the stationary renewal process corresponding to the Kosambi map function and two gamma densities: gamma(4,2) and gamma(2v,v), where v is $1/(2\log(2) - 1) \approx 2.6$.

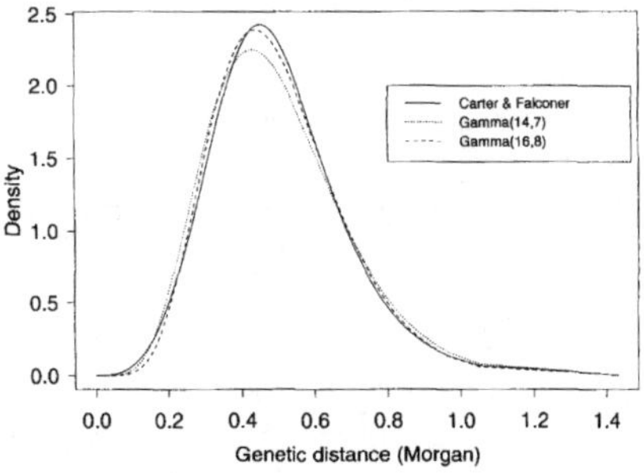

FIGURE 2.—Comparison between the interevent density ($-M_{CF}''$) of the stationary renewal process corresponding to the Carter and Falconer map function and two gamma densities: gamma(14,7) and gamma(16,8).

the interval $[0, \frac{1}{4}\pi]$, and it is easy to check condition (B) in this case, with $L = \frac{1}{4}\pi$. Thus there is a stationary renewal process having it as a map function, although the chromosome arm is rather short.

KOSAMBI (1944): $M_K(d) = \frac{1}{2}\tanh(2d)$. Since $M_K'(d) = 4(e^{2d} + e^{-2d})^{-2}$, and $M_K''(d) = -16(e^{2d} - e^{-2d})/(e^{2d} + e^{-2d})^3$, it is easy to check that (A) is satisfied. The interevent distribution for the corresponding stationary renewal process is $16(e^{2d} - e^{-2d})/(e^{2d} + e^{-2d})^3$. KOSAMBI (1944) found that this map function gave good fit to Drosophila data. FISHER *et al.* (1947) noticed that the Kosambi map function is very close to the map function from a renewal process with interevent distribution being chi-square with four degrees of freedom. Both $-M_K''$ and the density of gamma(4,2) are plotted in Figure 1. The density of a gamma(b,g) variable is $[b(bx)^{g-1}e^{-bx}]/\Gamma(g)$. The mean and variance of $-M_K''$ are $\frac{1}{2}$ and $\log(2)/2 - \frac{1}{4}$. The density of gamma(2v, v), where $v = 1/(2\log(2) - 1)$, which has the same mean and variance as $-M_K''$, is also plotted in Figure 1.

CARTER and FALCONER (1951): $M_{CF}^{-1}(r) = \frac{1}{4}(\tan^{-1}(2r) + \tanh^{-1}(2r))$. Here $M_{CF}(d)$ is the solution of the differential equation $M'(d) = 1 - 16M^4(d)$. This map function was found to fit mouse data better than other map functions (CARTER and FALCONER 1951). Since $M_{CF}''(d) = -64M_{CF}^3(d)\,(1 - 16M_{CF}^4(d))$, and $M_{CF} \leq \frac{1}{2}$, (A) is satisfied. The corresponding stationary renewal process has interevent distribution $64M_{CF}^3(d)\,(1 - 16M_{CF}^4(d))$. The Carter and Falconer map function $-M_{CF}''(d)$ together with the density of gamma(14,7) and gamma(16,8) are plotted in Figure 2. It was found that $Cx(Co)^6$ and $Cx(Co)^4$ give the best fit for two mouse data sets in BLANK *et al.* (1988) and TODD *et al.* (1991) (ZHAO 1995); the choice of the Carter and Falconer map function would be equivalent to using the $Cx(Co)^6$ model.

STURT (1976): $M_S(d) = \frac{1}{2}(1 - (1 - \frac{d}{L})e^{-d(2L-1)/L})$, $0 \leq d \leq L$. This map function arises via a count-location chiasma process that begins with an obligatory crossover event on the arm, followed by a Poisson-distributed number of crossover events having mean $2L - 1$. The total genetic length is thus L. This map function fails (B5) but satisfies (B6), and so no stationary renewal process can give rise to it.

RAO *et al.* (1977): $M_R^{-1}(r) = w_H M_H^{-1}(2r) + w_K M_K^{-1}(2r) + w_{CF} M_{CF}^{-1}(2r) + w_M M_M^{-1}(2r)$, where $w_H = p\,(1 - 2p)(1 - 4p)/3$, $w_K = -4p(1 - p)(1 - 4p)$, $w_{CF} = 32p(1 - p)(1 - 2p)/3$, $w_M = (1 - p)(1 - 2p)(1 - 4p)$, and M_H^{-1}, M_K^{-1}, M_{CF}^{-1}, and M_M^{-1} are the inverse of the Haldane, Kosambi, Carter-Falconer, and Morgan map functions. It is easy to check that for any p, w_H, w_K, w_{CF}, and w_M cannot all be positive. Indeed M_R does not satisfy our necessary conditions (WEEKS 1994). Following RAO *et al.*'s idea, we might try to define a map function from a set of $n > 2$ map functions by letting $M^{-1}(r) = \sum_{i=1}^{n} w_i(p)\, M_i^{-1}(r)$, where $w_i(p)$ is a polynomial in p of order $n - 1$, and $M(d)$ reduces to $M_i(d)$ when $p = p_i$ for given $0 < p_1 < p_2 < \cdots p_n < 1$. But it can be shown that for no p can the $w_i(p)$, $i = 1, 2, \ldots, n$, so defined all be positive. Therefore this approach to obtaining empirical map functions from existing map functions would seem questionable.

FELSENSTEIN (1979): $M_F(d) = \frac{1}{2}(1 - e^{2(K-2)d})/(1 - (K-1)e^{2(K-2)d})$. When $K > 2$, $\lim_{d\to\infty} M_F(d) = \frac{1}{2}(K - 1) \neq \frac{1}{2}$, and so no stationary renewal process exists with this mapping function. Because $M_F'(d) = (K - 2)^2 e^{2(K-2)d}/(1 - (K - 1)e^{2(K-2)d})^2$, $M_F''(d) = 2\,(K - 2)^3 e^{2(K-2)d}\,(1 + (K-1)e^{2(K-2)d})/(1 - (K-1)e^{2(K-2)d})^3$, it is easy to verify that $M_F(d)$ satisfies (A) when $0 \leq K < 2$. In Felsenstein's map functions, K is a measure of interference, with $K > 1$ corresponding to negative interference, and $K < 1$ to positive interference. The mean of $-M_F''(d)$ is $\frac{1}{2}$, and the variance is $\log(2 - K)/(2(1 - K)) - \frac{1}{4}$ ($\frac{1}{2}$ when $K = 1$). Both $-M_F''(d)$ and the gamma distribution with the same mean and variance are plotted in Figure 3 for different Ks. Note the

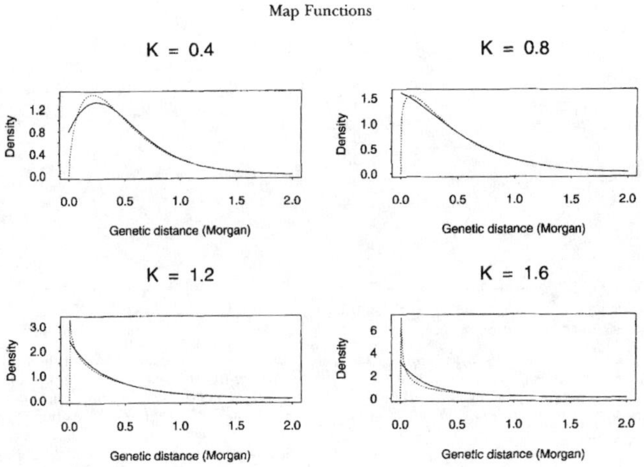

FIGURE 3.—Comparison between the interevent density ($-M_F''$, solid curve) of the stationary renewal process corresponding to the Felsenstein map function and the gamma density (dotted curve) having the same mean and variance for $K = 0.4, 0.8, 1.2$, and 1.6.

close agreement between Felsenstein's family and the gamma family.

KARLIN and LIBERMAN (1978, 1979): $M_{CL}(d) = \frac{1}{2}[1 - c(1 - d/L)]$. This class of map functions arise from the count-location process, where $c(s) = \sum_{k\geq 0} c_k s^k$ is a probability generating function of a count variable **c** with distribution (c_k), and $c'(1) = 2L$. M_{CL} is only well defined for finite L, It is easy to check that (B1)–(B4) are satisfied. Because $M_{CL}(L) = \frac{1}{2}(1 - c_0)$ and $M'_{CL}(L) = c_1/2L$, from Theorem 2, there is a corresponding stationary renewal process for M_{CL} only if (1) $c_0 = 0$ and $c_1 = 0$, or (2) $c_0 > 0$ and $c_1 > 0$.

DISCUSSION

In constructing genetic maps, map functions have been used to infer the unobservable genetic distance between two markers from the observable recombination fraction between these markers. Different genetic map functions embody different degrees of crossover interference among the crossovers. It has been observed that different organisms have different degrees of crossover interference, it so is not surprising that different map functions have been found suitable for different organisms. The major disadvantage of using map functions is that joint recombination probabilities cannot be uniquely determined in terms of them when there are more than three markers. Various approaches have been proposed to extend map functions to handle multilocus data.

One widely adopted approach, which was suggested by GEIRINGER (1944) and SCHNELL (1961) and thoroughly studied by LIBERMAN and KARLIN (1984), embodies the assumption that for a pair of noncontiguous intervals, the probabilities for joint recombination patterns across these intervals do not depend on the distance between these two intervals, something that is not consistent with observations. Those map functions that can be extended to multilocus data through this approach have been (inappropriately) called "multilocus feasible" (LIBERMAN and KARLIN 1984). This criterion excludes many functions that were found to fit well to recombination data, such as the Kosambi map function.

In this paper, another approach is proposed to extend map functions for the analysis of multilocus data. If for any given map function, we can find a point process model that gives rise to this map function, then multilocus joint recombination probabilities can be obtained in a way that is completely compatible with the map function. Stationary renewal processes can give rise to many map functions. From this perspective, most map functions that are not multilocus feasible according to KARLIN and LIBERMAN can in fact be extended to permit the analysis of multilocus data.

Another measure of interference, called S_4 by FOSS *et al.* (1993), is formally defined as

$$S_4(d) = \lim_{h\to 0}\lim_{k\to 0} \frac{p_{11}}{(p_{10} + p_{11})(p_{01} + p_{11})},$$

where the $p_{i_1 i_2}$ are as in the definition of C, with one interval having map length h, the other interval map length k, and the two intervals being separated by a map distance d. It seems that S_4 captures more important aspects of crossover interference than does C (MCPEEK and SPEED 1995). Though S_4 cannot be deter-

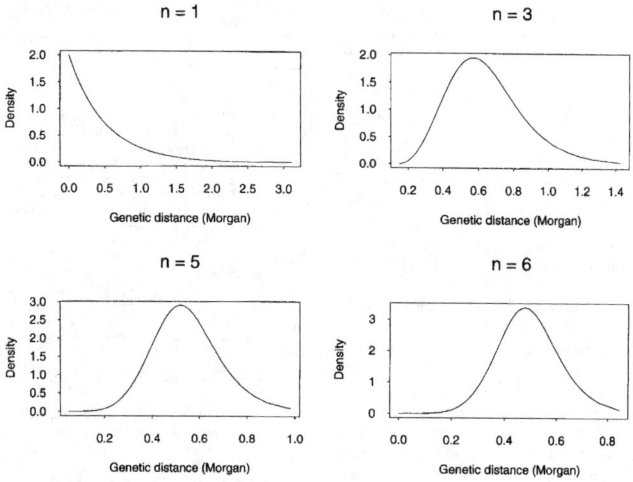

FIGURE 4.—The interevent density of the stationary renewal process corresponding to the map function M that satisfies $M' = 1 - (2M)^n$ for $n = 1, 3, 5, 6$.

mined from a map function, it can be calculated given a particular way of generalizing map functions to multilocus data and compared to empirical estimates. For the count-location processes, S_4 is constant for all d, whereas S_4 has various forms for the stationary renewal processes. The values of S_4 as a function of d were estimated from a large Drosophila data set and were very close to the S_4 values under the chi-square model (FOSS *et al.* 1993, MCPEEK and SPEED 1995).

Map functions cannot and should not be expected to reflect chiasma or crossover interference in anything but the most superficial way. Indeed, a map function could arise from both the stationary renewal process and the count-location process, although the interference differs greatly between these two classes of models. For example, the map function $M(d) = d/(1 + 2d)$ could arise from a stationary renewal process with interevent distribution $4(1 + 2d)^{-3}$, and it could also arise from a count-location model with $c_k = {}^1/_2{}^k$, where $k \geq 0$ (LIBERMAN and KARLIN 1984). Under the count-location model, the length of the chromosome defined by $\mathbf{c} = (c_k)$ is $^1/_2$, so $M(d)$ is only defined on $[0, {}^1/_2]$, while d can range from 0 to ∞ for the stationary renewal process model.

Note that for the stationary renewal processes corresponding to most map functions discussed, the interevent distributions can often be well approximated by gamma distributions. Recall that the Haldane, Kosambi, and Carter and Falconer map functions are all solutions to the differential Equation 1 with $C(d) = (2M)^{n-1}$. It can be shown that map functions M obtained via this approach always satisfy (A). Figure 4 displays plots of $-M''(d)$, the interevent distribution of the corresponding stationary renewal process for $n = 1$ (Haldane map function), $n = 3$, 5, and 6. The cases $n = 2$ and 4, which correspond to the Kosambi map function and the Carter and Falconer map function, were plotted in Figure 1 and Figure 2. When n is large, the corresponding density becomes close to a normal distribution. This suggests that if we take the approach proposed in this paper to generalize map functions to the multilocus situation, the stationary renewal processes corresponding to most genetic map functions can be well approximated by the gamma model, or the chi-square model. This provides, to some extent, an explanation of the fact that chi-square models (and hence gamma models) are flexible enough to give good fit to data from various organisms exhibiting different degrees of interference.

It has been assumed throughout this paper that there is no chromatid interference. Under this assumption, the relation between the chiasma process and the crossover process and the relation between the chiasma process and the map function are simple [see SPEED (1995) for a comprehensive review of this and related matters]. It was shown earlier that there are situations where the presence of both chiasma interference and chromatid interference could lead to no apparent crossover interference, from which we concluded that it is impossible to separate two types of interference from single-strand recombination data. Because there is no strong and consistent evidence of chromatid interference from the study of tetrad data, the assumption of no chromatid interference is generally considered valid. The use of map functions and stationary renewal processes requires that the degree of interference be the same across the chromosome, which is an obvious simplification. But a large amount of data will probably be necessary to permit the detection of nonstationarity of the underlying process.

In summary, we have shown that for most genetic map functions, there is a corresponding stationary renewal process, and that these map functions can be extended to permit the analysis of multilocus data by calculating joint recombination probabilities from their corresponding renewal processes. This provides another way of generalizing a given map function to the multilocus situation, although it seems unlikely that there are efficient methods to estimate genetic distances and other parameters from multilocus recombination data under completely general renewal processes. The calculation of multilocus recombination probabilities for stationary renewal chiasma processes is discussed in the APPENDIX. However, comparisons between the interevent distribution of general stationary renewal processes and that of chi-square distributions suggest that this class of renewal processes provides satisfactory approximations to the renewal processes corresponding to most genetic map functions in the literature. With the limited amounts of data currently available, it will probably be hard to distinguish these models, although we can look forward to many refinements in the future.

This work was supported by National Institutes of Health grant HG-01093-01. The authors thank two referees for their helpful comments.

LITERATURE CITED

BAILEY, N. T. J., 1961 *Introduction to the Mathematical Theory of Genetic Linkage*. Oxford University Press, London.

BLANK, R. D., G. R. CAMPBELL, A. CALABRO and P. D. EUSTACHIO, 1988 A linkage map of mouse chromosome 12: localization of *Igh* and effects of sex and interference on recombination. Genetics **120:** 1073–1083.

CARTER, T. C., and D. S. FALCONER, 1951 Stocks for detecting linkage in the mouse and the theory of their design. J. Genet. **50:** 307–323.

CARTER, T. C., and A. ROBERTSON, A., 1952 A mathematical treatment of genetical recombination using a four-strand model. Proc. Roy. Soc. B **139:** 410–426.

COBBS, G., 1978 Renewal process approach to the theory of genetic linkage: case of no chromatid interference. Genetics **89:** 563–581.

DALEY, D. J., and D. VERE-JONES, 1988 *An Introduction to the Theory of Point Processes*. Springer-Verlag, New York.

FELLER, W., 1971 *An Introduction to Probability Theory and Its Applications*. Vol. 2, Ed. 2. John Wiley, New York.

FELSENSTEIN, J., 1979 A mathematically tractable family of genetic mapping functions with differents amount of interference. Genetics **91:** 769–775.

FISHER, R. A., 1947 The theory of linkage in polysomic inheritance. Phil. Trans. Roy. Soc. B. **233:** 55–87.

FISHER, R. A., M. F. LYON and A. R. G. OWEN, 1947 The sex chromosome in the house mouse. Heredity **1:** 335–365.

FOSS, E., and F. W. STAHL, 1995 A test of a counting model for chiasma interference. Genetics **139:** 1201–1209.

FOSS, E., R. LANDE, F. W. STAHL and C. M. STEINBERG, 1993 Chiasma interference as a function of genetic distance. Genetics **133:** 681–691.

GEIRINGER, H., 1944 On the probability theory of linkage in Mendelian heredity. Ann. Math. Statist. **15:** 25–57.

HALDANE, J. B. S., 1919 The combination of linkage values, and the calculation of distances between the loci of linked factors. J. Genetics **8:** 299–309.

JENNINGS, H. S., 1923 The numerical relations in the crossing over of the genes with a critical examination of the theory that the genes are arranged in a linear series. Genetics **8:** 393–457.

KARLIN, S., and U. LIBERMAN, 1978 Classification and comparisons of multilocus recombination distributions. Proc. Natl. Acad. Sci. USA **75:** 6332–6336.

KARLIN, S., and U. LIBERMAN, 1979 A natural class of multilocus recombination processes and related measure of crossover interference. Adv. Appl. Prob. **11:** 479–501.

KOSAMBI, D. D., 1944 The estimation of the map distance from recombination values. Ann. Eugen. **12:** 172–175.

LIBERMAN, U., and S. KARLIN, 1984 Theoretical models of genetic map functions. Theor. Pop. Bio. **25:** 331–346.

LUDWIG, W., 1934 Uber numerische beziehungen der crossoverwerte untereinander. Z. indukt. Abatamm. Vereb. **67:** 58–95.

MATHER, K., 1935 Reductional and equational separation of the chromosomes in bivalents and multivalents. J. Genet. **30:** 53–78.

MATHER, K., 1936 The determination of position in crossing over. J. Genet. **33:** 207–235.

MATHER, K., 1937 The determination of position in crossing over. II. The chromosome length-chiasma frequency relation. Cytologia, Jub. Vol., 514–526.

MCPEEK, M. S., and T. P. SPEED, 1995 Modeling interference in genetic recombination. Genetics **139:** 1031–1044.

MORGAN, T. H., and C. B. BRIDGES, 1916 Sex-linked inheritance in Drosophila. Carniegie Institute of Washington.

MORTON, N. E., and C. J. MACLEAN, 1984 Multilocus recombination frequencies. Genet. Res. **44:** 99–108.

MULLER, H. J., 1916 The mechanism of crossing over. Am. Nat. **50:** 193–221, 284–305, 350–366, 421–434.

OWEN, A. R. G., 1953 The analysis of multiple linkage data. Heredity **7:** 247–264.

RAO, D. C., N. E. MORTON, J. LINDSTEN, M. HULTEN, and S. YEE, 1977 A mapping function for man. Hum. Hered. **27:** 99–104.

RISCH, N., and K. LANGE, 1979 An alternative model of recombination and interference. Ann. Hum. Genet. **43:** 61–70.

SCHNELL, F. W., 1961 Some general formulations of linkage effects in inbreeding. Genetics **46:** 947–957.

SPEED, T. P., 1995 What is a genetic map function? in *Genetic Mapping and DNA Sequencing*, edited by T. P. SPEED and M. S. WATERMAN. Springer-Verlag, New York.

SPEED, T. P., M. S. MCPEEK and S. N. EVANS, 1992 Robustness of the no-interference model for ordering genetic markers. Proc. Natl. Acad. Sci. USA **89:** 3103–3106.

STAM, P., 1979 Interference in genetic crossing over and chromosome mapping. Genetics **92:** 573–594.

STURT, E., 1975 A mapping function for human chromosomes. Ann. Hum. Genet. **40:** 147–163.

STURTEVANT, A. H., 1915 The behavior of chromosomes as studied through linkage. Z. Indukt. Abstammungs. Vererbugsl. **13:** 234–287.

TODD, J. A., T. J. AITMAN, R. J. CORNALL, S. GHOSH, J. R. HALL *et al.*, 1991 Genetic analysis of autoimmune type 1 diabetes mellitus in mice. Nature **351:** 542–547.

WEEKS, D. E., 1994 Invalidity of the Rao map function for three loci. Hum. Hered. **44:** 178–180.

WEEKS, D. E., G. M. LATHROP and J. OTT, 1993 Multipoint mapping under genetic interference. Hum. Hered. **43:** 86–97.

WHITEHOUSE, H. L. K., 1982 *Genetic Recombination: Understanding the Mechanisms*. John Wiley, New York.

ZHAO, H., 1995 *Statistical Analysis of Genetical Interference*. PhD thesis, University of California at Berkeley, Berkeley.

ZHAO, H., M. S., MCPEEK and T. P. SPEED, 1995a Statistical analysis of chromatid interference. Genetics **139:** 1057–1065.

ZHAO, H., T. P. SPEED and M. S. MCPEEK, 1995b Statistical analysis of crossover interference using the chi-square model. Genetics **139:** 1045–1056.

Communicating editor: B. S. WEIR

H. Zhao and T. P. Speed

APPENDIX

Proof of Theorem 1: Suppose the crossovers are from a stationary renewal chiasma process with interevent density f. Without loss of generality, we may assume the mean interevent distance is $\mu = {}^1/_2$, so the metric is simply genetic distance. For any point $\mathcal{M}_1$, say, on the chromosome, the chance that the first crossover after $\mathcal{M}_1$ occurs in the small interval $(y, y + dy)$ is

$$\int_y^\infty \frac{f(t)}{\mu}\,dtdy = 2\int_y^\infty f(t)\,dtdy.$$

The probability of no crossovers occurring before $\mathcal{M}_2$, which is map distance d from $\mathcal{M}_1$, is

$$p_0 = \int_d^\infty \left\{2\int_y^\infty f(t)\,dt\right\}dy.$$

Using Mather's formula (MATHER 1935), which asserts that the recombination fraction r between any two markers is

$$r = \frac{1}{2}(1 - p_0),$$

where p_0 is the probability of zero crossovers occurring between these markers, we have

$$M(d) = r = \frac{1}{2}\left\{1 - 2\int_d^\infty\int_y^\infty f(t)\,dtdy\right\}.$$

It is easy to verify that M so defined satisfies (A).

Conversely, if M satisfies (A), then

$$\int_0^\infty (-M''(t))\,dt = M'(0) = 1.$$

Thus $-M''$ is a probability density function on $[0, \infty)$. If the interevent distribution in a stationary renewal process is $-M''$, then their mean is

$$\int_0^\infty (-tM''(t))\,dt = \int_0^\infty\int_t^\infty (-M''(y))\,dydt$$

$$= \int_0^\infty M'(t)\,dt = {}^1/_2.$$

Thus, the map function generated from the stationary renewal chiasma process with interevent distribution $-M''$ is

$$\frac{1}{2}\left(1 - 2\int_d^\infty\int_y^\infty (-M''(t)\,dt)\,dy\right) = M(d).$$

Proof of Theorem 2: Note that

$$M(L) = \frac{1}{2}\left\{1 - 2\int_L^\infty\int_y^\infty f(t)\,dtdy\right\},$$

and

$$M'(L) = \int_L^\infty f(t)\,dt.$$

It follows that either (B5) and (B6) are true, or (B5)$'$ and (B6)$'$ are true. The first part of this theorem can then be proved as in Theorem 1.

If $M(L) < {}^1/_2$ and $M'(L) > 0$, we may define an extended map function $M_E(d)$ on $[0, \infty)$ as follows:

$$M_E(d) = \begin{cases} M(d) & \text{if } d < L \\ M(L) + \alpha(1 - e^{-\beta(d-L)}) & \text{if } d \geq L, \end{cases}$$

where $\alpha = {}^1/_2 - M(L)$ and $\alpha\beta = M'(L)$. It can be verified that $M_E(d)$ so defined satisfies (A) in Theorem 1. So there is a stationary renewal process whose corresponding map function is $M_E(d)$, which coincides with $M(d)$ on $[0, L]$. If $M(L) = {}^1/_2$ and $M'(L) = 0$, it can be easily shown that the stationary renewal process with renewal density $-M''$ gives rise to M.

Calculating multilocus recombination probabilities for stationary renewal chiasma processes: Suppose that $\mathcal{M}_0, \mathcal{M}_1, \ldots, \mathcal{M}_n$ are $n + 1$ consecutive loci along a chromosome, defining n genomic intervals $I_1 = [\mathcal{M}_0, \mathcal{M}_1), \ldots, I_n = [\mathcal{M}_{n-1}, \mathcal{M}_n)$ of map lengths $d_1, d_2, \ldots, d_n$, respectively. Extending the notation introduced earlier, we denote by $p_{i_1 \ldots i_n}$ the joint recombination probability of having $i_j = 0$ or 1 recombination across interval I_j, $j = 1, \ldots, n$. The question we address here is the calculation of all such probabilities $\mathbf{p} = (p_{i_1 \ldots i_n})$ when the underlying chiasmata form a renewal process stationary in the genetic distance metric and NCI is assumed. To do so we make use of the so-called *linkage values*, denoted by $\mathbf{z} = (z_{i_1 \ldots i_n})$, where $z_{i_1 \ldots i_n}$ is the probability of finding no chiasmata in the union $\cup\{I_j : i_j = 1\}$ of those intervals for which $i_j = 1$, see SPEED *et al.* (1992) for fuller details. We also make use of some well known facts from renewal theory and refer to FELLER (1971) for derivations. Suppose that we have a stationary renewal process with interevent density f and mean interval length μ. The distance of an arbitrary but fixed point on the chromosome to the next chiasma to its left (respectively right) is called the backward (respectively forward) recurrence length (traditionally called "time"), and if these are denoted by β and ϕ, then

$$P(\beta > u, \phi > v) = \int_{u+v}^\infty \frac{1 - F(y)}{\mu}\,dy, \qquad (2)$$

where F is the cumulative distribution function (*c.d.f.*) corresponding to f.

Now let us consider the calculation of the linkage values $\mathbf{z} = (z_{i_1 \ldots i_n})$. For $n = 2$ this is quite straightforward. Suppose that we want to calculate z_{10}, the probability of no crossovers in I_1. We regard $\mathcal{M}_1$ as the arbitrary but fixed point in the preceding discussion, and put $u = d_1$ and $v = 0$ in (2), obtaining the formula $z_{10} = 1 - F^*(d_1)$, where F^* is the *c.d.f.* corresponding to $f^* = \mu^{-1}(1 - F)$. Similarly, $z_{01} = 1 - F^*(d_2)$, $z_{11} = 1 - F^*(d_1 + d_2)$, and $z_{00} = 1$. All of these expressions are easily computed as long as F^* is tractable.

We now turn to $n = 3$ intervals and the calculation of

Map Functions

$\mathbf{z} = (z_{i_1 i_3})$. A quick run through all eight possibilities reveals that all but z_{101} can be obtained in the manner just illustrated with $n = 2$. For example, $z_{100} = 1 - F^*(d_1)$, $z_{011} = 1 - F^*(d_2 + d_3)$, etc. It turns out that the calculation of z_{101}, in general requires summing a series of multiple integrals, and that the only known cases in which these integrals simplify into something tractable are variants on thinned Poisson processes. Let us see why.

First we recall that z_{101} is the probability of no chiasmata in either I_1 or I_3; there may be zero, one or more in I_2, where the count is not constrained. Thus an initial reduction of z_{101} is as follows:

$$z_{101} = z_{111} + \sum_{k=1}^{\infty} \zeta_k,$$

where z_{111} is the probability of no chiasmata in $I_1 \cup I_2 \cup I_3$ $(= 1 - F^*(d_1 + d_2 + d_3))$, and ζ_k is the probability of k chiasmata in I_2 and none in I_1 or I_3. We now give an expression for ζ_k that, in general, does not simplify, and we remark that we know of no substantially simpler alternative expressions in the literature on renewal processes.

If there are to be $k \geq 1$ chiasmata in I_2, we may denote the forward recurrence interval length from $\mathcal{M}_1$ to the first event by y_1, and the k subsequent interevent distances by $y_2, y_3, \ldots, y_k, y_{k+1}$. Further we may denote by y_0 the backward recurrence distance to the first event to the left of $\mathcal{M}_1$. With this notation we can readily check that the probability ζ_k is the $(k + 2)$-fold integral of the joint density of $(y_0, y_1, \ldots, y_k, y_{k+1})$ over the range $\{y_0 > d_1\} \cap \{y_1 + \cdots + y_k < d_2\} \cap \{y_1 + \cdots + y_k + y_{k+1} > d_2 + d_3\}$. The joint density of $y_0, \ldots, y_{k+1}$ is the product

$$\mu^{-1} f^*(y_0 + y_1) \times \prod_{i=2}^{k+1} f(y_i),$$

and so our assertion is demonstrated: z_{101} is an infinite sum of multiple integrals and will have a tractable expression only when these sums and integrals simplify. For each $n \geq 3$ there is one or more $z_{i_1 \ldots i_n}$ requiring such expressions, and so far it is only the class of chi-square renewal processes (ZHAO *et al.* 1995b) and a slightly more general class termed Poisson-skip processes (H. ZHAO, K. LANGE and T. P. SPEED, personal communication) that are known to yield simplifications.

Chapter 11
Molecular Evolution

Steven N. Evans

Although the Department of Statistics at Berkeley decided they wanted to hire me in 1987, I didn't take up my position there until 1989. I don't have any recollection of meeting Terry when I interviewed, but, due in part to our shared Australian nationality, we became good friends shortly after I moved to Berkeley. Two years later, I jumped at the chance to move from my gloomy, north-facing office to one next to Terry's. Its corner location with a view across the San Francisco Bay through the Golden Gate was merely an added inducement.

The resulting proximity led us to discuss scientific matters to a much greater extent. I was soon meeting with Terry and his students, serving on his students' dissertation committees, and attending Terry's weekly "statistics and biology" seminar. The thing that got me irredeemably hooked on the applications of statistics and probability to biology arose out of Terry's work with his student Trang Nguyen on phylogenetics, the enterprise that seeks to reconstruct the evolutionary family tree of some collection of *taxa* (typically, species) using data such as DNA sequences. Phylogenetics was already a huge field in the early 1990s with a variety of statistical and non-statistical methods, and it has expanded greatly since then. Some idea of its scope may be gleaned from Semple and Steel [43], Felsenstein [19], Gascuel [20], and Lemey et al. [28].

Phylogenetic inference can be viewed as a standard statistical estimation problem [22]. One has a probability model for the observed DNA sequences that involves two kinds of parameters: those that define the mechanism by which DNA changes over time down a lineage and those that define the tree. The latter can be thought of as being further divided into a discrete parameter, the shape of the tree, and a set of numerical parameters, the lengths of the various branches (which represent either chronological time or evolutionary distance). In principle, the problem is therefore amenable to standard inferential techniques such as maximum likelihood or Bayesian methods. Unfortunately, likelihoods are somewhat expensive to compute for large numbers of taxa because they consist of large sums of products –

S.N. Evans
Department of Statistics, University of California, Berkeley
e-mail: evans@stat.berkeley.edu

S. Dudoit (ed.), *Selected Works of Terry Speed*, Selected Works in Probability and Statistics, 441
DOI 10.1007/978-1-4614-1347-9_11, © Springer Science+Business Media, LLC 2012

essentially, one has to sum over all the possibilities for the genetic types of the unobserved ancestors at each of the internal nodes of the tree. Even more forbidding is the fact that the number of possible trees for even a modest number taxa is enormous, so any approach that involves naively searching over tree space for the tree with maximal likelihood or summing and integrating over tree space to compute a posterior distribution will quickly become computationally infeasible, although there are widely used software packages that incorporate effective heuristics for maximizing the likelihood [21, 46, 45] and MCMC methods to computing posterior distributions [23, 24]. This computational difficulty is particularly galling because a significant proportion of the effort is expended to estimate the edge lengths of the tree and the parameters of the DNA substitution model, while the main object of interest is often just the shape of the tree.

Trang and Terry had come across an intriguing alternative approach to phylogenetic inference, the use of *phylogenetic invariants*, that had been proposed in Lake [27] and Cavender and Felsenstein [12] and developed further in Cavender [10] and Cavender [11]. The idea behind this approach is the following. Assume we have N taxa. At any site there are 4^N possibilities for the nucleotides exhibited by the taxa. Each of these possibilities has an associated probability that is a function of the parameters in our model. It is usual to assume that these probabilities are the same for each site and that different sites behave independently. Suppose that for a given tree there is a collection of polynomial functions of these probabilities such that each function has the property it has value zero regardless of the values of the numerical parameters. Such functions are called phylogenetic invariants. Moreover, suppose that the values of these polynomials are typically non-zero when they are evaluated on the corresponding probabilities associated with other trees for generic values of the numerical parameters. The hope is that one can find enough invariants to distinguish between any pair of trees, estimate their values using the observed empirical frequencies of vectors of nucleotides across many sites, and then determine which tree appears to have the estimates of the values of "its" invariants close to zero and hence is a suitable estimate of the underlying phylogenetic tree.

In order to implement this strategy, one needs ideally a procedure for finding all the invariants for a given tree. Because a linear combination of invariants is an invariant and the product of an invariant and an arbitrary polynomial is an invariant, the invariants form an *ideal* in the ring of polynomials, and so one actually wants to characterize an algebraically independent generating set. When Terry and I discussed this problem, we realized that the models of DNA substitution for which others had been successful in finding specific examples of invariants by *ad hoc* means were ones in which there is an underlying group structure. More specifically, if one identifies the nucleotides $\{A,G,C,T\}$ with the elements of the abelian group $\mathbb{Z}_2 \otimes \mathbb{Z}_2$ in an appropriate manner, then the substitution dynamics are just those of a continuous time random walk (that is, a processes with stationary independent increments) on this group. This suggested that we should attack the problem with Fourier theory for abelian groups – I should note that similar observations about the substitution models were made by others such as Székely et al. [50] around the same time. We found in our joint paper reproduced in this volume that the algebraic structure of the likelihoods looks much simpler in "Fourier coordinates" and that one can determine a generating set for the family of invariants of a given tree using essentially linear algebra. We also proposed some conjectures on the number of "independent" invariants for various models that were verified subsequently in Evans and Zhou [17] and Evans [18].

It turned out that Terry and I had been like Molière's Monsieur Jourdain in *Le Bourgeois Gentilhomme* who "for more than forty years" had been "speaking prose without knowing it." The simple structure we observed after the passage to Fourier coordinates is an instance of a *toric ideal*, and we had unwittingly reproduced some of the elementary theory related to such objects. This connection was made in Sturmfels and Sullivant [47] and it led to a large amount of work using tools from commutative algebra to construct and analyze phylogenetic invariants in a number of different settings [1, 2, 8, 15, 6, 3, 5, 4, 9, 14]. Even tools from the representation theory of non-abelian groups have turned out to be useful in this context [49, 48]. Moreover, the investigation of phylogenetic invariants led in part to an appreciation of the extent to which many statistical models could be profitably studied from the perspective of commutative algebra and algebraic geometry, and this point of view is the basis of the field of *algebraic statistics* [37, 38, 41, 16].

An extremely important observation in phylogenetics is that evolution occurs at the level of genes and that different genes can have evolutionary family trees that disagree with the associated species tree. For example, genes can be duplicated and the duplicate can mutate to take on a new function, sometimes resulting in the loss of another gene that originally performed that function. Also, the lineages of orthologous genes (that is, genes descended from a common ancestral gene) in two taxa will split some time before the corresponding split in the species tree, and if this difference is sufficiently great the shape of the tree for a given gene will differ from that of the species tree. This means that in constructing a species tree one needs to resolve the incompatibilities observed between the trees constructed for various genes. On the other hand, if one has an accepted species tree, then it is desirable to reconcile a discordant gene tree with the species tree by describing how the above phenomena might have conspired to produce the differences between the two. This general problem is discussed in Pamilo and Nei [40], Page and Charleston [39], Nichols [36], and Maddison [32].

The papers by Bourgon et al. [7] and Wilkinson et al. [51] carry out the task of clarifying the connection between a gene tree and a species tree in two important instances, the evolution of the serine repeat antigen in various *Plasmodium* species (including *P. falciparum*, the parasite responsible for the most acute form of malaria in humans) and the evolution of relaxin-like peptides across species ranging from humans to the zebra fish and the African clawed frog.

There has been considerable fascinating theoretical research on the problem of constructing species trees from gene trees, some of it showing quite paradoxical behavior; for example, the most likely gene tree can differ from the species tree and inferring a species tree by concatenating the sequences of several genes and treating the result as one gene can lead to an incorrect species tree with high probability [42, 13, 31, 25, 33, 34]. Some recent approaches to constructing well-behaved estimates of species trees using data from several genes are Liu and Pearl [30], Liu [29], Kubatko et al. [26], and Mossel and Roch [35].

The last of Terry's work on molecular evolution is his paper with Sidow and Nguyen [44] on estimating invariable codons using capture-recapture methods. Invariable codons are those that are conserved across different species because of structural or functional constraints. In essence, they are codons that are prevented from changing because any change would have fatal biochemical consequences. It is not possible to observe which codons are invariable by simply looking at sequence data because some codons might be conserved by chance across all species even though there is no biochemical reason preventing a change, and so the invariable codons form some unknown fraction of the conserved ones. This paper is another example of Terry at his best: it provides answers of genuine scientific importance using simple, sensible statistical ideas that are normally not associated with the analysis of molecular data and that he probably learned from his extensive teaching and consulting experience.

Working with Terry has been one of the high points of my career at Berkeley. He has affected deeply the areas in which I have worked and my general attitude to research. Perhaps more importantly, by my good fortune of being his neighbor for around twenty years I have had an unrivaled opportunity to witness the humanity, dedication and commitment that he always shows to his students and collaborators. I may not have always lived up to the wonderful example he continues to set, but that does not make me any the less grateful for it.

References

[1] E. S. Allman and J. A. Rhodes. Phylogenetic invariants for the general Markov model of sequence mutation. *Math. Biosci.*, 186:113–144, 2003.

[2] E. S. Allman and J. A. Rhodes. Phylogenetic invariants for stationary base composition. *J. Symbolic Comput.*, 41:138–150, 2006.

[3] E. S. Allman and J. A. Rhodes. Phylogenetic invariants. In *Reconstructing Evolution*, pages 108–146. Oxford University Press, 2007.

[4] E. S. Allman and J. A. Rhodes. Molecular phylogenetics from an algebraic viewpoint. *Stat. Sinica*, 17:1299–1316, 2007.

[5] E. S. Allman and J. A. Rhodes. Phylogenetic ideals and varieties for the general Markov model. *Adv. in Appl. Math.*, 40:127–148, 2008.

[6] C. Bocci. Topics on phylogenetic algebraic geometry. *Expo. Math.*, 25:235–259, 2007.

[7] R. Bourgon, M. Delorenzi, T. Sargeant, A. N. Hodder, B. S. Crabb, and T. P. Speed. The serine repeat antigen (SERA) gene family phylogeny in Plasmodium: The impact of gc content and reconciliation of gene and species trees. *Mol. Biol. Evol.*, 21:2161–2171, 2004.

[8] W. Buczyńska and J. A. Wiśniewski. On geometry of binary symmetric models of phylogenetic trees. *J. Eur. Math. Soc. (JEMS)*, 9:609–635, 2007.

[9] M. Casanellas and J. Fernández-Sánchez. Geometry of the Kimura 3-parameter model. *Adv. in Appl. Math.*, 41:265–292, 2008.
[10] J. Cavender. Mechanized derivation of linear invariants. *Mol. Biol. Evol.*, 6:301–316, 1989.
[11] J. Cavender. Necessary conditions for the method of inferring phylogeny by linear invariants. *Math. Biosci.*, 103:69–75, 1991.
[12] J. Cavender and J. Felsenstein. Invariants of phylogenies in a simple case with discrete states. *J. Class.*, 4:57–71, 1987.
[13] J. H. Degnan and N. A. Rosenberg. Discordance of species trees with their most likely gene trees. *PLoS Genetics*, 2, 2006.
[14] J. Draisma and J. Kuttler. On the ideals of equivariant tree models. *Math. Ann.*, 344:619–644, 2009.
[15] A. Dress and M. Steel. Phylogenetic diversity over an abelian group. *Ann. Comb.*, 11:143–160, 2007.
[16] M. Drton, B. Sturmfels, and S. Sullivant. *Lectures on Algebraic Statistics*, volume 39 of *Oberwolfach Seminars*. Birkhäuser Verlag, 2009.
[17] S. Evans and X. Zhou. Constructing and counting phylogenetic invariants. *J. Comput. Biol*, 5:713–724, 1998.
[18] S. N. Evans. Fourier analysis and phylogenetic trees. In D. Healy, Jr. and D. Rockmore, editors, *Modern Signal Processing (Lecture notes from an MSRI Summer School)*. Cambridge University Press, 2004.
[19] J. Felsenstein. *Inferring Phylogenies*. Sinauer, 2004.
[20] O. Gascuel, editor. *Mathematics of Evolution and Phylogeny*. Oxford University Press, 2007.
[21] S. Guindon and O. Gascuel. A simple, fast, and accurate algorithm to estimate large phylogenies by maximum likelihood. *Syst. Biol.*, 52:696–704, 2003.
[22] S. Holmes. Statistics for phylogenetic trees. *Theor. Popul. Biol.*, 63:17–32, 2003.
[23] J. P. Huelsenbeck and F. Ronquist. MrBayes: Bayesian inference of phylogenetic trees. *Bioinformatics*, 17:754–755, 2001.
[24] J. P. Huelsenbeck and F. Ronquist. MrBayes 3: Bayesian phylogenetic inference under mixed models. *Bioinformatics*, 19:1572–1574, 2003.
[25] L. S. Kubatko and J. H. Degnan. Inconsistency of phylogenetic estimates from concatenated data under coalescence. *Syst. Biol.*, 56:17–24, 2007.
[26] L. S. Kubatko, B. C. Carstens, and L. L. Knowles. STEM: Species tree estimation using maximum likelihood for gene trees under coalescence. *Bioinformatics*, 25:971–973, 2009.
[27] J. Lake. A rate-independent technique for analysis of nucleic acid sequences: Evolutionary parsimony. *Mol. Biol. Evol.*, 4:167–191, 1987.
[28] P. Lemey, M. Salemi, and A.-M. Vandamme, editors. *The Phylogenetic Handbook: A Practical Approach to Phylogenetic Analysis and Hypothesis Testing*. Cambridge University Press, 2nd edition, 2009.
[29] L. Liu. BEST: Bayesian estimation of species trees under the coalescent model. *Bioinformatics*, 24:2542–2543, 2008.

[30] L. Liu and D. K. Pearl. Species trees from gene trees: Reconstructing Bayesian posterior distributions of a species phylogeny using estimated gene tree distributions. *Syst. Biol.*, 56:504–514, 2007.
[31] W. Maddison and L. Knowles. Inferring phylogeny despite incomplete lineage sorting. *Syst. Biol.*, 55:21–30, 2006.
[32] W. P. Maddison. Gene trees in species trees. *Syst. Biol.*, 46:523–536, 1997.
[33] F. A. Matsen and M. Steel. Phylogenetic mixtures on a single tree can mimic a tree of another topology. *Syst. Biol.*, 56:767–775, 2007.
[34] F. A. Matsen, E. Mossel, and M. Steel. Mixed-up trees: The structure of phylogenetic mixtures. *Bull. Math. Biol.*, 70:1115–1139, 2008.
[35] E. Mossel and S. Roch. Incomplete lineage sorting: Consistent phylogeny estimation from multiple loci. *IEEE Comp. Bio. and Bioinformatics*, 7:166–171, 2010.
[36] R. Nichols. Gene trees and species trees are not the same. *Trends Ecol. Evol.*, 16:358–364, 2001.
[37] L. Pachter and B. Sturmfels, editors. *Algebraic Statistics for Computational Biology*. Cambridge University Press, 2005.
[38] L. Pachter and B. Sturmfels. The mathematics of phylogenomics. *SIAM Rev.*, 49:3–31, 2007.
[39] R. D. M. Page and M. A. Charleston. From gene to organismal phylogeny: Reconciled trees and the gene tree/species tree problem. *Mol. Phylogenet. Evol.*, 7:231–240, 1997.
[40] P. Pamilo and M. Nei. Relationships between gene trees and species trees. *Mol. Biol. Evol.*, 5:568–583, 1988.
[41] G. Pistone, E. Riccomagno, and H. P. Wynn. *Algebraic Statistics*, volume 89 of *Monographs on Statistics and Applied Probability*. Chapman & Hall/CRC, 2001.
[42] N. A. Rosenberg. The probability of topological concordance of gene trees and species trees. *Theor. Popul. Biol.*, 61:225–247, 2002.
[43] C. Semple and M. Steel. *Phylogenetics*, volume 22 of *Mathematics and its Applications*. Oxford University Press, 2003.
[44] A. Sidow, T. Nguyen, and T. P. Speed. Estimating the fraction of invariable codons with a capture-recapture method. *J. Mol. Evol.*, 35:253–260, 1992.
[45] A. Stamatakis. RAxML-VI-HPC: Maximum likelihood-based phylogenetic analyses with thousands of taxa and mixed models. *Bioinformatics*, 22:2688–2690, 2006.
[46] A. Stamatakis, T. Ludwig, and H. Meier. RAxML-III: A fast program for maximum likelihood-based inference of large phylogenetic trees. *Bioinformatics*, 21:456–463, 2005.
[47] B. Sturmfels and S. Sullivant. Toric ideals of phylogenetic invariants. *J. Comput. Biol.*, 12:204–228, 2005.
[48] J. Sumner and P. Jarvis. Markov invariants and the isotropy subgroup of a quartet tree. *J. Theoret. Biol.*, 258:302–310, 2009.
[49] J. Sumner, M. Charleston, L. Jermiin, and P. Jarvis. Markov invariants, plethysms, and phylogenetics. *J. Theoret. Biol.*, 253:601–615, 2008.

[50] L. A. Székely, M. A. Steel, and P. L. Erdős. Fourier calculus on evolutionary trees. *Adv. in Appl. Math.*, 14:200–210, 1993.

[51] T. N. Wilkinson, T. P. Speed, G. W. Tregear, and R. A. Bathgate. Evolution of the relaxin-like peptide family from neuropeptide to reproduction. *Ann. N.Y. Acad. Sci.*, 1041:530–533, 2005.

The Annals of Statistics
1993, Vol. 21, No. 1, 355–377

INVARIANTS OF SOME PROBABILITY MODELS USED IN PHYLOGENETIC INFERENCE

BY STEVEN N. EVANS[1] AND T. P. SPEED[2]

University of California, Berkeley

The so-called method of invariants is a technique in the field of molecular evolution for inferring phylogenetic relations among a number of species on the basis of nucleotide sequence data. An invariant is a polynomial function of the probability distribution defined by a stochastic model for the observed nucleotide sequence. This function has the special property that it is identically zero for one possible phylogeny and typically nonzero for another possible phylogeny. Thus it is possible to discriminate statistically between two competing phylogenies using an estimate of the invariant. The advantage of this technique is that it enables such inferences to be made without the need for estimating nuisance parameters that are related to the specific mechanisms by which the molecular evolution occurs. For a wide class of models found in the literature, we present a simple algebraic formalism for recognising whether or not a function is an invariant and for generating all possible invariants. Our work is based on recognising an underlying group structure and using discrete Fourier analysis.

1. Introduction. The problem of inferring phylogenetic relations among a group of species using nucleotide sequence data is one of continuing interest to researchers in the field of molecular evolution. There are a variety of approaches to the problem in current use, see Swofford and Olsen (1990) for a recent review, and our concern is with methods based upon simple probabilistic models for nucleotide substitution. Such models have been in use for some time now, but interest in them heightened following the revelation by Felsenstein (1978) that the popular parsimony criterion can give rise to serious biases when the rates of evolutionary change in the true phylogenetic tree differ greatly from one branch to another.

In our view, the use of statistical models fitted by maximum likelihood is currently the best method of inferring phylogenies [see, e.g., Felsenstein (1981), Tavaré (1986), Barry and Hartigan (1987), and Navidi, Churchill and von Haeseler (1992)]. However, in recent years much interest has focussed on a simpler approach using functions of the data which permit inferences

---

Received April 1991; revised December 1991.
[1]Partially supported by NSF Grant DMS-90-15708.
[2]Partially supported by NSF Grant DMS-88-02378.
AMS 1991 *subject classifications.* Primary 62H05; secondary 60K99, 62F99.
Key words and phrases. Invariant, phylogenetic inference, discrete Fourier analysis, random walk on a group.

concerning the phylogeny without requiring the estimation of other parameters describing the nucleotide substitution mechanism. This approach has been called the *method of invariants* and we may describe it informally as follows (a full description is given in Section 2).

Suppose that we have aligned DNA sequence data for a number of taxa. For a given position in the sequence (typically, the third, second or first codon position of a DNA sequence coding for a common protein such as cytochrome c) we have a stochastic model for the observed base. This model is built using two ingredients. The first ingredient is a dependence structure reflecting the putative phylogeny. The second ingredient is a collection of stochastic mechanisms describing the occurrence of base substitution events along the branches of the phylogenetic tree. An *invariant* is a polynomial function that has as its argument the probability distribution of the observed bases and that, for a particular phylogeny, is zero for all choices of the substitution mechanisms. If it is assumed that the bases at different positions are i.i.d., then it is easy to estimate such an invariant without estimating the parameters describing the base substitution mechanism; and, if the invariant is typically nonzero for another specification of the phylogeny, then it is possible to discriminate statistically between the two competing phylogenies. Moreover, one of the hopes for the method of invariants is that the assumption of identical distribution for different sites can be weakened—a generalisation that does not seem as feasible with maximum likelihood methods.

Invariants were first defined by Cavender and Felsenstein (1987) and Lake (1987) for models involving four taxa. These and subsequent attempts at finding invariants have been, to a certain extent, ad hoc. In order to fully exploit the potential of the method of invariants it is necessary to have techniques for generating all possible invariants and for recognising when a given function is an invariant. The purpose of the present paper is to describe simple algebraic procedures that achieve these ends.

The outline of the remainder of the paper is as follows. In Section 2 we formally describe the models we will be dealing with and formally define what we mean by an invariant. Having developed the relevant nomenclature, we give a comparison of our work and previous work in the area. We also make the key observation that there is a group structure inherent in the models we are considering. With this in mind, we digress in Sections 3 and 4 to give a brief overview of discrete Fourier theory and random walks on finite groups, respectively. In Section 5 we give another description of the models in group language. In Section 6 we present our procedures for constructing and recognising invariants. We discuss examples involving two, three and four taxa in Section 7. A noteworthy feature of these examples is that the number of algebraically independent invariants always coincides with the number of "degrees of freedom" obtained by an informal parameter counting argument. (Some care needs to be taken when doing this counting due to issues of over-parametrisation and parameter identifiability.) We conjecture that the equality of these two numbers is a general phenomenon, but we do not as yet have a proof.

INVARIANTS IN PHYLOGENETIC INFERENCE

2. Definitions and notation. Suppose that we have aligned DNA sequence data for m taxa. We may construct a general class of stochastic models for the bases observed at a given position in the following manner.

Consider a finite rooted tree $\mathbf{T}$ with m leaves. Let $\mathbf{V}$ denote the set of vertices of $\mathbf{T}$. Write ρ for the root of $\mathbf{T}$ and $\mathbf{L}$ for the set of leaves of $\mathbf{T}$. For each vertex $v \in \mathbf{V} \setminus \{\rho\}$, there is a unique vertex $\sigma(v)$ which is connected to v by an edge and is closer to ρ in the usual graph-theoretic distance. Write $(\sigma(v),v)$ for the unique edge which connects $\sigma(v)$ and v.

Label the taxa with the elements of $\mathbf{L}$ and think of the collection of observed bases as a realisation of a $\{A, G, C, T\}^{\mathbf{L}}$-valued random variable $(Y_l)_{l \in \mathbf{L}}$ with a distribution defined as follows. Let π be a probability distribution on $\{A, G, C, T\}$. We will refer to π as the *root distribution*. For each vertex $v \in \mathbf{V} \setminus \{\rho\}$, let $P^{(v)}$ be a stochastic matrix on $\{A, G, C, T\}$. We will refer to $P^{(v)}$ as the *substitution matrix* associated with the edge $(\sigma(v), v)$. Define a probability distribution μ on $\{A, G, C, T\}^{\mathbf{V}}$ by setting

$$\mu((i_v)_{v \in \mathbf{V}}) = \pi(i_\rho) \prod_{v \in \mathbf{V} \setminus \{\rho\}} P^{(v)}(i_{\sigma(v)}, i_v).$$

Finally, let $(Y_l)_{l \in \mathbf{L}}$ have the marginal distribution

$$\mathbb{P}\{(Y_l)_{l \in \mathbf{L}} = (i_l)_{l \in \mathbf{L}}\} = \sum_{v \in \mathbf{V} \setminus \mathbf{L}} \sum_{i_v} \mu(((i_v)_{v \in \mathbf{V} \setminus \mathbf{L}}, (i_l)_{l \in \mathbf{L}})),$$

where each of the dummy variables i_v, $v \in \mathbf{V} \setminus \mathbf{L}$, is summed over the set $\{A, G, C, T\}$.

The various elements appearing in these models have the following interpretations. The tree $\mathbf{T}$ is a candidate for the true phylogenetic tree describing the evolution of the observed present-day species corresponding to the leaves of the tree, insofar as this evolution is indicated by the evolution of the aligned sequence of nucleotides under study. The root of the tree ρ corresponds to an unobserved common ancestor of all of the observed present-day species, whilst the vertices other than the root and the leaves correspond to unobserved species intermediate in the evolutionary process, being common ancestors of pairs, triples, and so on of the observed present-day species. The root distribution π is thought of as the relative frequency of bases in the common ancestor's sequence, whilst the substitution matrices $P^{(v)}$ give a tractable and plausible probability model for the substitution process. We remark that the distribution μ on $\{A, G, C, T\}^{\mathbf{V}}$ satisfies a Markov property which may be stated as follows: for any two vertices v_1 and v_2, the base at v_1 and the base at v_2 are conditionally μ-independent given the base at any vertex v_3 on the unique path connecting v_1 and v_2.

The models of this form which appear in the literature usually take each substitution matrix to be the transition matrix at some point in time of a continuous time Markov chain on the state space $\{A, G, C, T\}$ (which particular point in time is possibly different for each edge, and these variables constitute "unknown parameters" in the model). Moreover, the Markov chain is usually taken to be from some subfamily of the possible chains on

$\{A, G, C, T\}$. The subfamilies we will be particularly interested in are described most easily in terms of the infinitesimal generator matrix of the chain. Kimura (1981) presents a model in which the infinitesimal generator matrix is of the form

$$\begin{array}{c} \\ A \\ G \\ C \\ T \end{array} \begin{array}{c} \begin{array}{cccc} A & G & C & T \end{array} \\ \begin{pmatrix} -(\alpha+\beta+\gamma) & \alpha & \beta & \gamma \\ \alpha & -(\alpha+\beta+\gamma) & \gamma & \beta \\ \beta & \gamma & -(\alpha+\beta+\gamma) & \alpha \\ \gamma & \beta & \alpha & -(\alpha+\beta+\gamma) \end{pmatrix} \end{array}$$

where $\alpha, \beta, \gamma \geq 0$. The value of the triple (α, β, γ) is possibly different for each edge, and these variables also constitute "unknown parameters" in the model. We will refer to this model as the *Kimura three-parameter model*. If we further restrict the class of allowable infinitesimal generator matrices by imposing the extra condition that $\beta = \gamma$, then we obtain the model considered by Kimura (1980). We will refer to this model as the *Kimura two-parameter model*. Finally, if we require that $\alpha = \beta = \gamma$ we obtain the model considered in Jukes and Cantor (1969) and more explicitly in Neyman (1971), which we will refer to as the *Jukes–Cantor model*.

As yet we have not said anything about the choice of the root distribution π. We will take π to be either the uniform distribution on $\{A, G, C, T\}$ or otherwise some arbitrary (and "unknown") probability distribution on $\{A, G, C, T\}$. Note that all the Markov chains described in the previous paragraph are reversible with the uniform distribution as the symmetrising stationary measure. We do not make explicit use of reversibility in this paper.

Let F be a polynomial in the dummy variables t_i, $i \in \{A, G, C, T\}^{\mathbf{L}}$. We say that F is an *invariant* for one of the models defined above if $F((\mathbb{P}\{Y = i\})_{i \in \{A, G, C, T\}^{\mathbf{L}}}) = 0$ for all choices of parameters in the model. We described the statistical uses of invariants in Section 1. The concept was introduced by Cavender and Felsenstein (1987) and Lake (1987). Cavender and Felsenstein (1987) and later Drolet and Sankoff (1990), Sankoff (1990) and Felsenstein (1991) derived invariants for Jukes–Cantor models with uniform root distribution and at most five taxa. Lake (1987) and later Cavender (1989, 1991) obtained linear invariants for a four taxa model based on a parametric family of substitution matrices that contains the Kimura two-parameter and Jukes–Cantor families. [We will show in Section 7 that, contrary to a claim made in Cavender (1991), there can be strictly fewer linear invariants for the Cavender–Lake model than there are for the Kimura two-parameter model.]

EXAMPLE. Consider the tree in Figure 1 with the leaves labelled as $1, 2, 3, 4$ and the root labelled as 0. Suppose that we have a Jukes–Cantor model with uniform root distribution constructed from this tree. In general there are only 15 distinct probabilities $\mathbb{P}\{Y = (i_1, i_2, i_3, i_4)\}$, corresponding to the possible partitions of $\{1, 2, 3, 4\}$ defined by the equalities and inequalities amongst

INVARIANTS IN PHYLOGENETIC INFERENCE

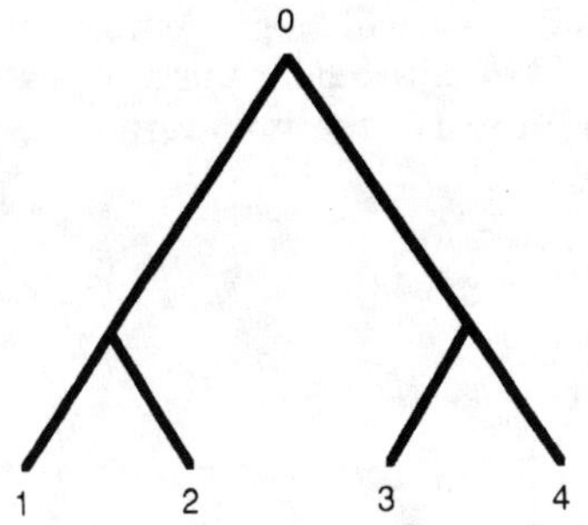

FIG. 1.

i_1, i_2, i_3, i_4. Write these 15 values as f_{1234}, $f_{1|234}$ and three similar, $f_{12|34}$ and two similar, $f_{12|3|4}$ and five similar, and $f_{1|2|3|4}$, where, for example, $f_{1234} = \mathbb{P}\{Y = (A, A, A, A)\} = \cdots = \mathbb{P}\{Y = (T, T, T, T)\}$. Lake (1987) shows that

$$f_{13|24} - f_{13|2|4} - f_{24|1|3} + f_{1|2|3|4} = 0$$

and

$$f_{14|23} - f_{14|2|3} - f_{1|4|23} + f_{1|2|3|4} = 0,$$

and these observations can be used to construct linear invariants for this model.

As we stated in Section 1, our aim in this paper is to describe a relatively simple algebraic formalism for generating/recognising all invariants for any of the models considered above when there is an arbitrary number of taxa. The key to our approach is the following observation. Suppose that we think of the bases $\{A, G, C, T\}$ as the elements of an Abelian (that is, commutative) group with the group operation defined by the following addition table:

$$\begin{array}{c} + \\ A \\ G \\ C \\ T \end{array} \begin{array}{c} \begin{array}{cccc} A & G & C & T \end{array} \\ \begin{pmatrix} A & G & C & T \\ G & A & T & C \\ C & T & A & G \\ T & C & G & A \end{pmatrix} \end{array}.$$

This group is isomorphic to the *Klein four-group* $\mathbb{Z}_2 \oplus \mathbb{Z}_2$ [i.e., the group consisting of the elements $\{(0,0),(0,1),(1,0),(1,1)\}$ with the group operation being coordinatewise addition modulo 2]. One possible isomorphism is given by $A \leftrightarrow (0,0)$, $G \leftrightarrow (0,1)$, $C \leftrightarrow (1,0)$ and $T \leftrightarrow (1,1)$. It is straightforward to check that the infinitesimal generator matrices appearing in the Kimura three-parameter model have the property that the entry corresponding to the pair of bases (i, j) is a function of the base $i - j$. The same is also true a fortiori for the Kimura two-parameter model and the Jukes–Cantor model. Thus such a matrix is nothing other than the infinitesimal generator matrix for a random walk on the group (see Section 4 for a discussion of continuous time random walks on finite Abelian groups).

3. Some discrete Fourier analysis. For the sake of reference, we review some elementary facts regarding Fourier analysis on finite Abelian groups. There seems to be no good reference to Fourier analysis and group characters which treats the Abelian case in isolation and contains all the theory we need. However, Chapter 104 of Körner (1988) is a good introduction, and all of what we have to say here may be deduced fairly easily from the more general material presented in Chapter 2 of Diaconis (1988). In the same spirit as our work, Diaconis (1990) investigates how general Fourier theory may be used to analyse the properties of patterned matrices when the pattern reflects invariance under the action of some group.

Let $\mathbb{G}$ be a finite Abelian group, with the group operation written as $+$. Let $\mathbb{T} = \{z \in \mathbb{C}: |z| = 1\}$ denote the unit circle in the complex plane, and regard $\mathbb{T}$ as an Abelian group with the group operation being ordinary complex multiplication. The *characters* of $\mathbb{G}$ are the group homomorphisms mapping $\mathbb{G}$ into $\mathbb{T}$. That is, $\chi: \mathbb{G} \to \mathbb{T}$ is a character if $\chi(g_1 + g_2) = \chi(g_1)\chi(g_2)$ for all $g_1, g_2 \in \mathbb{G}$. The characters form an Abelian group under the operation of pointwise multiplication of functions. This group is called the *dual group* of $\mathbb{G}$ and is denoted by $\hat{\mathbb{G}}$. The groups $\mathbb{G}$ and $\hat{\mathbb{G}}$ are isomorphic. Given $g \in \mathbb{G}$ and $\chi \in \hat{\mathbb{G}}$, write $\langle g, \chi \rangle$ for $\chi(g)$. The dual of the direct sum $\mathbb{G}^m = \oplus_{i=1}^m \mathbb{G}$ is isomorphic to $\hat{\mathbb{G}}^m$ under the isomorphism given by $\langle (g_1, \ldots, g_m), (\chi_1, \ldots, \chi_m) \rangle = \prod_{i=1}^m \langle g_i, \chi_i \rangle$.

EXAMPLE. Suppose that $\mathbb{G} = \mathbb{Z}_2 \oplus \mathbb{Z}_2$. Then one may write $\hat{\mathbb{G}} = \{1, \phi, \psi, \phi\psi\}$, where the following table gives the value of $\langle g, \chi \rangle$ for $g \in \mathbb{G}$ and $\chi \in \hat{\mathbb{G}}$:

	(0,0)	(0,1)	(1,0)	(1,1)
1	1	1	1	1
ϕ	1	−1	1	−1
ψ	1	1	−1	−1
$\phi\psi$	1	−1	−1	1

Given a function $f: \mathbb{G} \to \mathbb{C}$, the function $\hat{f}: \hat{\mathbb{G}} \to \mathbb{C}$ defined by

$$\hat{f}(\chi) = \sum_{g \in \mathbb{G}} \langle g, \chi \rangle f(g)$$

is called the *Fourier transform* of f. If f is a discrete probability mass function on $\mathbb{G}$ and Z is a $\mathbb{G}$-valued random variable with distribution f, then

INVARIANTS IN PHYLOGENETIC INFERENCE

$\hat{f}(\chi) = \mathbb{E}[\langle Z, \chi\rangle]$. The Fourier transform has the following properties:

$$\hat{1}(\chi) = \begin{cases} |\mathbb{G}|, & \text{if } \chi = 1, \\ 0, & \text{otherwise,} \end{cases}$$

where $|A|$ denotes the cardinality of a set A; and

$$\widehat{f_1 * f_2} = \hat{f}_1 \hat{f}_2,$$

where

$$f_1 * f_2(g) = \sum_{h \in \mathbb{G}} f_1(g-h) f_2(h), \qquad g \in \mathbb{G},$$

is the convolution of the functions f_1 and f_2. Moreover, a function may be recovered from its Fourier transform by the process of *Fourier inversion*; namely, if f has Fourier transform $\hat{f}$ then $f(g) = |\mathbb{G}|^{-1}\Sigma_{\chi \in \hat{\mathbb{G}}}\overline{\langle g, \chi\rangle}\hat{f}(\chi)$ for all $g \in \mathbb{G}$.

4. Random walks. Suppose that $\mathbb{G}$ is a finite Abelian group and $X = (X_t, \mathbb{P}^g)$ is a continuous time Markov chain on $\mathbb{G}$ (here, $\mathbb{P}^g$, $g \in \mathbb{G}$, is the probability measure on path-space corresponding to starting the chain off at the initial point g). Let P_t denote the corresponding semigroup of transition matrices (i.e., $P_t(i,j) = \mathbb{P}^i\{X_t = j\}$) and let Q denote the corresponding infinitesimal generator matrix. We say that the process X is a *random walk* if, for all $t \geq 0$ and all $i, j \in \mathbb{G}$, $P_t(i,j) = p_t(j-i)$ for some probability distribution p_t on $\mathbb{G}$ or, equivalently, that $Q(i,j) = q(i-j)$ for some function q: $\mathbb{G} \to \mathbb{R}$ such that $\Sigma_g q(g) = 0$.

We can describe such a process probabilistically as follows. Let N be a simple, homogeneous Poisson process with rate $-q(0)$ and let $\{J_n\}_{n=1}^{\infty}$ be an independent sequence of i.i.d. $\mathbb{G}$-valued random variables with common distribution given by

$$\mathbb{P}\{J_n = j\} = \begin{cases} q(j) \Big/ \left(\sum_{g \neq 0} q(g)\right), & \text{if } j \neq 0, \\ 0, & \text{if } j = 0. \end{cases}$$

Then the distribution of $\{X_t: t \geq 0\}$ under $\mathbb{P}^g$ is the same as the distribution of $\{g + \Sigma_{n=1}^{N_t} J_n: t \geq 0\}$, where we define the empty sum to be 0. More generally, the distribution of $\{X_t: t \geq 0\}$ under $\mathbb{P}^{\nu}$, where ν is some arbitrary initial distribution, is the same as the distribution of $\{J_0 + \Sigma_{n=1}^{N_t} J_n: t \geq 0\}$, where J_0 is a $\mathbb{G}$-valued random variable with distribution ν and J_0 is independent of N and $\{J_n\}_{n=1}^{\infty}$.

From this description of X it is easy to see that we have the *Lévy–Hinčin formula*

$$\hat{p}_t(\chi) = \exp\left(t \sum_{g \in \mathbb{G}\setminus 0} (\langle g, \chi\rangle - 1) q(g)\right) = \exp\left(t \sum_{g \in \mathbb{G}} \langle g, \chi\rangle q(g)\right).$$

We are particularly interested in the case when $\mathbb{G}$ is the Klein four-group $\mathbb{Z}_2 \oplus \mathbb{Z}_2$. Here the matrix Q will be of the form

$$Q = \begin{array}{c} \\ (0,0) \\ (0,1) \\ (1,0) \\ (1,1) \end{array} \begin{array}{c} \begin{array}{cccc} (0,0) & (0,1) & (1,0) & (1,1) \end{array} \\ \begin{pmatrix} -(\alpha+\beta+\gamma) & \alpha & \beta & \gamma \\ \alpha & -(\alpha+\beta+\gamma) & \gamma & \beta \\ \beta & \gamma & -(\alpha+\beta+\gamma) & \alpha \\ \gamma & \beta & \alpha & -(\alpha+\beta+\gamma) \end{pmatrix} \end{array},$$

for some parameters $\alpha, \beta, \gamma \geq 0$. If we label the characters of $\mathbb{G}$ in the same way as we did in Section 3, then we see from the Lévy–Hinčin formula that

$$\hat{p}_t(1) = 1,$$
$$\hat{p}_t(\phi) = \exp(-2t(\alpha+\gamma)),$$
$$\hat{p}_t(\psi) = \exp(-2t(\beta+\gamma))$$

and

$$\hat{p}_t(\phi\psi) = \exp(-2t(\alpha+\beta)).$$

Applying Fourier inversion, we see that

$$p_t((0,0)) = \tfrac{1}{4}[1 + \exp(-2t(\alpha+\gamma)) + \exp(-2t(\beta+\gamma)) + \exp(-2t(\alpha+\beta))],$$
$$p_t((0,1)) = \tfrac{1}{4}[1 - \exp(-2t(\alpha+\gamma)) + \exp(-2t(\beta+\gamma)) - \exp(-2t(\alpha+\beta))],$$
$$p_t((1,0)) = \tfrac{1}{4}[1 + \exp(-2t(\alpha+\gamma)) - \exp(-2t(\beta+\gamma)) - \exp(-2t(\alpha+\beta))]$$

and

$$p_t((1,1)) = \tfrac{1}{4}[1 - \exp(-2t(\alpha+\gamma)) - \exp(-2t(\beta+\gamma)) + \exp(-2t(\alpha+\beta))].$$

Define R_3 to be the set of all the probability distributions on $\mathbb{G}$ which can occur as the distributions p_t if we let α, β, γ and t range over all possible values. Define R_2 (resp., R_1) similarly, but with the restriction that $\beta = \gamma$ (resp., $\alpha = \beta = \gamma$). The following lemmas are trivial given the above calculations, but they will be crucial ingredients in our procedure for constructing all possible invariants.

LEMMA 4.1. *The set* $\{(\hat{r}(\phi), \hat{r}(\psi), \hat{r}(\phi\psi)) \colon r \in R_3\} \subset \mathbb{R}^3$ *has a nonempty interior.*

LEMMA 4.2. *The equality* $\hat{r}(\phi) = \hat{r}(\phi\psi)$ *holds for all* $r \in R_2$ *and the set* $\{(\hat{r}(\phi), \hat{r}(\psi)) \colon r \in R_2\} \subset \mathbb{R}^2$ *has a nonempty interior.*

LEMMA 4.3. *The equality* $\hat{r}(\phi) = \hat{r}(\psi) = \hat{r}(\phi\psi)$ *holds for all* $r \in R_1$ *and the set* $\{\hat{r}(\phi) \colon r \in R_1\} \subset \mathbb{R}$ *has a nonempty interior.*

INVARIANTS IN PHYLOGENETIC INFERENCE

5. Another description of the models. Identify the four bases $\{A, G, C, T\}$ with the elements of the Klein four-group $\mathbb{G} = \mathbb{Z}_2 \oplus \mathbb{Z}_2$ as we did in Section 2. Each substitution matrix appearing in the description of the Kimura three-parameter model is thus of the form $P^{(v)}(i, j) = r^{(v)}(j - i)$ for some probability distribution $r^{(v)} \in R_3$. The same is true of the Kimura two-parameter model and the Jukes–Cantor model if we replace R_3 by R_2 and R_1, respectively.

Construct independent $\mathbb{G}$-valued random variables $(Z_v)_{v \in \mathbf{V}}$ such that Z_v has the distribution $r^{(v)}$ for each $v \in \mathbf{V} \setminus \{\rho\}$, and Z_ρ has the root distribution π. For each vertex $v \in \mathbf{V}$, let $\delta(v)$ denote the sequence of states along the unique path through the tree connecting ρ and v [we include both ρ and v in $\delta(v)$]. Then it is clear that the probability distribution μ from Section 1 is the distribution of the random variables $(\Sigma_{u \in \delta(v)} Z_u)_{v \in \mathbf{V}}$, and hence the random variables $(Y_l)_{l \in \mathbf{L}}$ have the same distribution as $(\Sigma_{u \in \delta(l)} Z_u)_{l \in \mathbf{L}}$. In the future we will suppose that the random variables $(Y_l)_{l \in \mathbf{L}}$ have actually been constructed as $(\Sigma_{u \in \delta(l)} Z_u)_{l \in \mathbf{L}}$. Thus, if we set $Y = (Y_l)_{l \in \mathbf{L}}$ and $Z = (Z_v)_{v \in \mathbf{V}}$, then we have an "additive random effects model" $Y = DZ$, where D is an appropriate "design" matrix of 0's and 1's. Here, of course, we are using the usual $\mathbb{Z}$-module notation $kg = \Sigma_{i=1}^{k} g$ for $k \in \mathbb{Z}$, $k \geq 0$ and $g \in \mathbb{G}$.

EXAMPLE. Suppose that $m = 4$ and $\mathbf{T}$ is the tree in Figure 2 with the vertices labelled as shown. If we take the vertex 0 as the root then the design matrix will be

$$D = \begin{array}{c} \\ 1 \\ 2 \\ 3 \\ 4 \end{array} \begin{array}{c} \begin{array}{ccccccc} 0 & 1 & 2 & 3 & 4 & 5 & 6 \end{array} \\ \begin{pmatrix} 1 & 1 & 0 & 0 & 0 & 0 & 0 \\ 1 & 0 & 1 & 0 & 0 & 1 & 0 \\ 1 & 0 & 0 & 1 & 0 & 1 & 1 \\ 1 & 0 & 0 & 0 & 1 & 1 & 1 \end{pmatrix} \end{array}.$$

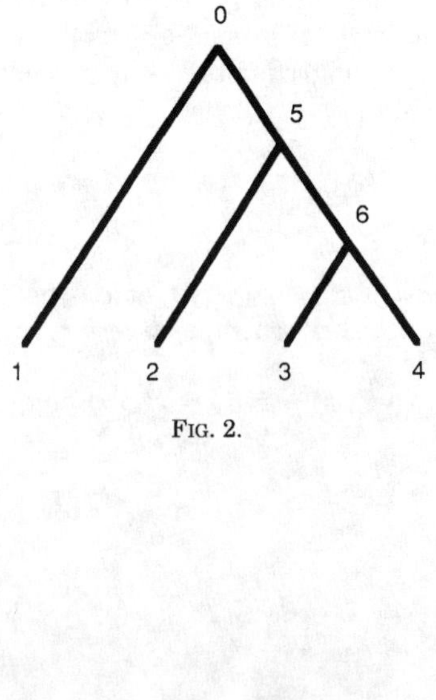

FIG. 2.

6. Constructing and classifying invariants. Let us begin by considering the Kimura three-parameter model with the uniform root distribution. Before presenting our general method for constructing invariants, we show how it works in a simple example. Suppose that $m = 3$ and $\mathbf{T}$ is the tree in Figure 3 with the vertices labelled as shown.

If we take the vertex 0 as the root then the model is

$$\begin{aligned} Y_1 &= Z_0 + Z_1, \\ Y_2 &= Z_0 + Z_2, \\ Y_3 &= Z_0 + Z_3, \end{aligned} \tag{6.1}$$

where $(Z_i)_{i=0}^3$ are independent, Z_0 has the uniform distribution on $\mathbb{G}$ and each Z_i, $i \in \{1, 2, 3\}$, has a distribution belonging to R_3.

Observe that

$$\begin{aligned} &\mathbb{E}[\langle Y_1, \phi\rangle\langle Y_2, \psi\rangle\langle Y_3, \phi\psi\rangle] \\ &\quad = \mathbb{E}[\langle Z_0 + Z_1, \phi\rangle\langle Z_0 + Z_2, \psi\rangle\langle Z_0 + Z_3, \phi\psi\rangle] \\ &\quad = \mathbb{E}[\langle Z_0, \phi\rangle\langle Z_0, \psi\rangle\langle Z_0, \phi\psi\rangle\langle Z_1, \phi\rangle\langle Z_2, \psi\rangle\langle Z_3, \phi\psi\rangle] \\ &\quad = \mathbb{E}[\langle Z_0 + Z_0, \phi\psi\rangle\langle Z_1, \phi\rangle\langle Z_2, \psi\rangle\langle Z_3, \phi\psi\rangle] \\ &\quad = \mathbb{E}[\langle Z_1, \phi\rangle]\mathbb{E}[\langle Z_2, \psi\rangle]\mathbb{E}[\langle Z_3, \phi\psi\rangle], \end{aligned}$$

where the second and third lines follow from the fact that the characters are homomorphisms and the last line follows from this fact, the fact that each element of $\mathbb{G}$ is its own inverse and the independence of $(Z_i)_{i=1}^3$. Similarly,

$$\mathbb{E}[\langle Y_1, \phi\psi\rangle\langle Y_2, \phi\rangle\langle Y_3, \psi\rangle] = \mathbb{E}[\langle Z_1, \phi\psi\rangle]\mathbb{E}[\langle Z_2, \phi\rangle]\mathbb{E}[\langle Z_3, \psi\rangle]$$

and

$$\mathbb{E}[\langle Y_1, \psi\rangle\langle Y_2, \phi\psi\rangle\langle Y_3, \phi\rangle] = \mathbb{E}[\langle Z_1, \psi\rangle]\mathbb{E}[\langle Z_2, \phi\psi\rangle]\mathbb{E}[\langle Z_3, \phi\rangle].$$

Also observe, by similar reasoning, that

$$\mathbb{E}[\langle Y_i, \theta\rangle\langle Y_j, \theta\rangle] = \mathbb{E}[\langle Z_i, \theta\rangle]\mathbb{E}[\langle Z_j, \theta\rangle]$$

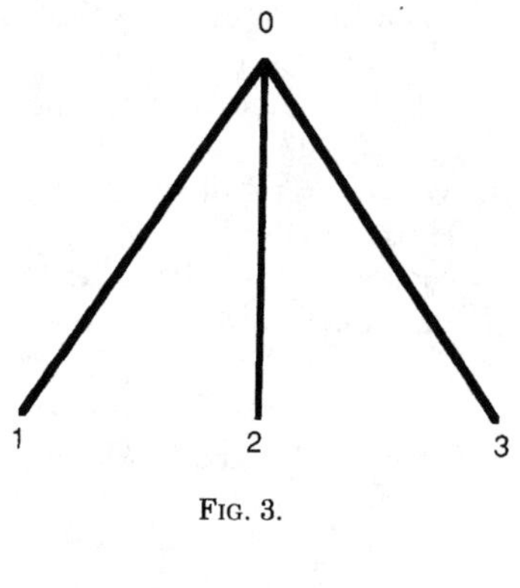

FIG. 3.

for $1 \le i < j \le 3$ and $\theta \in \{\phi, \psi, \phi\psi\}$. Therefore

$$\Big(\mathbb{E}\big[\langle Y_1, \phi\rangle\langle Y_2, \psi\rangle\langle Y_3, \phi\psi\rangle\big]\mathbb{E}\big[\langle Y_1, \phi\psi\rangle\langle Y_2, \phi\rangle\langle Y_3, \psi\rangle\big] \times \mathbb{E}\big[\langle Y_1, \psi\rangle\langle Y_2, \phi\psi\rangle\langle Y_3, \phi\rangle\big]\Big)^2 - \prod_{1 \le i < j \le 3} \prod_{\theta \in \{\phi, \psi, \phi\psi\}} \mathbb{E}\big[\langle Y_i, \theta\rangle\langle Y_j, \theta\rangle\big] = 0.$$

Thus, if we express each of the expectations appearing above in terms of the variables $\mathbb{P}\{Y = g\}$, $g \in \mathbb{G}^3$, we see that we can construct a ninth degree polynomial F in the dummy variables $(t_g)_{g \in \mathbb{G}^3}$ such that $F[(\mathbb{P}\{Y = g\})_{g \in \mathbb{G}^3}]$ is identically zero for all possible choices of parameters in the model and hence we have found an invariant.

The rationale behind what we did here is to take two expressions of the form

$$\prod_{\chi} (\mathbb{E}[\langle Y, \chi\rangle])^{k(\chi; j)},$$

for $j = 1, 2$, and show that they will be equal for all possible choices of parameters in the model by showing that they may be reexpressed as a common product of powers of the quantities $\{\mathbb{E}[\langle Z_i, \theta\rangle]\colon\ 0 \le i \le 3,\ \theta \in \{\phi, \psi, \phi\psi\}\}$. Nowhere in constructing this invariant did we make use of the fact that π, the distribution of Z_0, is uniform on $\mathbb{G}$. In fact, no term of the form $(\mathbb{E}[\langle Z_0, \theta\rangle])^l$, $l \ge 1$, appeared in the common product.

Suppose now that if we express some multinomial $\prod_{\chi}(\mathbb{E}[\langle Y, \chi\rangle])^{k(\chi)}$ as a product of powers of the quantities $\{\mathbb{E}[\langle Z_i, \theta\rangle]\colon 0 \le i \le 3, \theta \in \{\phi, \psi, \phi\psi\}\}$, then a term of the form $(\mathbb{E}[\langle Z_0, \theta\rangle])^l$, $l \ge 1$, does appear. We know from Section 3 that $\mathbb{E}[\langle Z_0, \theta\rangle] = 0$ for $\theta \in \{\phi, \psi, \phi\psi\}$. Thus $\prod_{\chi}(\mathbb{E}[\langle Y, \chi\rangle])^{k(\chi)}$ is identically zero for all choices of parameters in the model and we have an invariant. For example, we have

$$\mathbb{E}\big[\langle Y_1, \phi\rangle\langle Y_2, \phi\rangle\langle Y_3, \phi\rangle\big] = \mathbb{E}\big[\langle Z_0, \phi\rangle\big]\mathbb{E}\big[\langle Z_1, \phi\rangle\big]\mathbb{E}\big[\langle Z_2, \phi\rangle\big]\mathbb{E}\big[\langle Z_3, \phi\rangle\big] = 0.$$

We will show below in Theorem 6.1 that these two ways of constructing invariants lead to all possible invariants for a general Kimura three-parameter model with uniform root distribution. First, however, we need some notation.

Given a character $\chi \in \hat{\mathbb{G}}^{\mathbf{L}}$, we can represent the function $z \mapsto \langle Dz, \chi\rangle$, $z \in \mathbb{G}^{\mathbf{V}}$, as $z \mapsto \prod_{v \in \mathbf{V}}\langle z_v, \eta(v, \chi)\rangle$, where the characters $\eta(v, \chi) \in \hat{\mathbb{G}}$, $v \in \mathbf{V}$, are defined by $\eta(v, \chi) = \prod_{l \in \mathbb{L}} \chi_l^{D_{l,v}}$. Moreover, this is the unique such representation. Let $\mathbf{H}$ denote the collection of multinomials in the dummy variables u_{χ}, $\chi \in \hat{\mathbb{G}}^{\mathbf{L}}$. That is, $h \in \mathbf{H}$ is of the form $h((u_{\chi})) = \prod_{\chi} u_{\chi}^{k_{\chi}}$, where $k_{\chi} \in \{0, 1, 2, \ldots\}$, $\chi \in \hat{\mathbb{G}}^{\mathbf{L}}$. Given such a multinomial $h \in \mathbf{H}$, we can uniquely define another multinomial $S_3 h$ in the dummy variables $w_{v,\theta}$, $v \in \mathbf{V}$ and $\theta \in \{\phi, \psi, \phi\psi\}$, by

$$S_3 h((w_{v,\theta})) = \prod_{\chi}\Big(\prod_{v} W_3(v, \chi)\Big)^{k_{\chi}},$$

where we set

$$W_3(v,\chi) = \begin{cases} w_{v,\eta(v,\chi)}, & \text{if } \eta(v,\chi) \neq 1, \\ 1, & \text{if } \eta(v,\chi) = 1. \end{cases}$$

Observe that

$$h\big((\mathbb{E}[\langle Y,\chi\rangle])_\chi\big) = S_3 h\big((\mathbb{E}[\langle Z_v,\theta\rangle])_{v,\theta}\big).$$

We now define an equivalence relation $\sim_3$ on $\mathbf{H}$ as follows. Suppose that we have two multinomials $f, g \in \mathbf{H}$ with

$$S_3 f\big((w_{v,\theta})\big) = \prod_{v,\theta} w_{v,\theta^{a(v,\theta)}}$$

and

$$S_3 g\big((w_{v,\theta})\big) = \prod_{v,\theta} w_{v,\theta^{b(v,\theta)}}.$$

We declare that $f \sim_3 g$ if *either* $\Sigma_\theta a(\rho,\theta) \neq 0$ and $\Sigma_\theta b(\rho,\theta) \neq 0$, *or* $\Sigma_\theta a(\rho,\theta) = \Sigma_\theta b(\rho,\theta) = 0$ and $a(v,\theta) = b(v,\theta)$ for all $v \in \mathbf{V}\setminus\{\rho\}$ and all $\theta \in \{\phi,\psi,\phi\psi\}$. Write $\mathscr{H}_3$ for the family of equivalence classes in $\mathbf{H}$ under $\sim_3$, and let $\mathbf{H}_{3,\rho}$ denote the equivalence class consisting of multinomials h such that $S_3 h((w_{v,\theta}))$ is divisible by $w_{\rho,\xi}$ for some $\xi \in \{\phi,\psi,\phi\psi\}$. Observe that if $f, g \in \mathbf{H}$ with $f \sim_3 g$ then

$$f((\mathbb{E}[\langle Y,\chi\rangle])_\chi) = g((\mathbb{E}[\langle Y,\chi\rangle])_\chi),$$

for all choices of parameters in the model, with the common value being identically zero if $f, g \in \mathbf{H}_{3,\rho}$. Moreover, from Lemma 4.1 we see that the converse to this statement also holds.

Using this notation, we can describe the structure of the most general invariant as follows.

THEOREM 6.1. *Consider a Kimura three-parameter model with uniform root distribution. A polynomial F in the dummy variables $(t_g)_{g\in\mathbb{G}^{\mathbf{L}}}$ will be such that $F((\mathbb{P}\{Y=g\})_{g\in\mathbb{G}^{\mathbf{L}}}) = 0$ for all choices of parameters in the model if and only if F is of the form*

$$F\big((t_g)\big) = \sum_{h\in\mathbf{H}} c_h h\left(\left(\sum_{g\in\mathbb{G}^{\mathbf{L}}} \langle g,\chi\rangle t_g\right)_{\chi\in\hat{\mathbb{G}}^{\mathbf{L}}}\right),$$

where only finitely many of the coefficients c_h are non-zero and $\Sigma_{h\in\mathbf{K}} c_h = 0$ for all $\mathbf{K} \in \mathscr{H}_3 \setminus \{\mathbf{H}_{3,\rho}\}$.

PROOF. The sufficiency of the stated condition is already obvious from the observations we have made above.

Consider the question of necessity. Using Fourier inversion, we can express each variable t_g as a linear combination of the terms $\Sigma_{g\in\mathbb{G}^{\mathbf{L}}}\langle g,\chi\rangle t_g$, $\chi \in \hat{\mathbb{G}}^{\mathbf{L}}$, and so we can certainly write

$$F((t_g)) = \sum_{h \in \mathbf{H}} c_h h\left(\left(\sum_{g \in \mathbb{G}^{\mathbf{L}}} \langle g, \chi \rangle t_g\right)_{\chi \in \hat{\mathbb{G}}^{\mathbf{L}}}\right)$$

for some coefficients (c_h), where only finitely many of the c_h are nonzero. For each $\mathbf{K} \in \mathscr{H}_3 \setminus \{\mathbf{H}_{3,\rho}\}$ let $k_{\mathbf{K}}$ denote the common value of $S_3 h$ for $h \in \mathbf{K}$. Recall that $w_{\rho,\xi}$ does not divide $k_{\mathbf{K}}((w_{v,\theta}))$ for any $\xi \in \{\phi, \psi, \phi\psi\}$; and so, from Lemma 4.1, the collection of functions on the space of parameters given by $k_{\mathbf{K}}((\mathbb{E}[\langle Z_v, \theta \rangle]))$, $\mathbf{K} \in \mathscr{H}_3 \setminus \{\mathbf{H}_{3,\rho}\}$, is linearly independent. As

$$\begin{aligned} F((\mathbb{P}\{Y = g\})) &= \sum_{h \in \mathbf{H}} c_h h((\mathbb{E}[\langle Y, \chi \rangle])) \\ &= \sum_{\mathbf{K} \in \mathscr{H}_3 \setminus \{\mathbf{H}_{3,\rho}\}} \left(\sum_{h \in \mathbf{K}} c_h\right) k_{\mathbf{K}}((\mathbb{E}[\langle Z_v, \theta \rangle])), \end{aligned}$$

the result follows. □

Given Theorem 6.1, we see that the problem of generating all invariants for the Kimura three-parameter model with uniform root distribution reduces to the two problems of:

(i) describing all multinomials $h \in \mathbf{H}_{3,\rho}$, and
(ii) describing all pairs of multinomials h' and h'' such that $h' \notin \mathbf{H}_{3,\rho}$, $h'' \notin \mathbf{H}_{3,\rho}$ and $h' \sim_3 h''$.

Regarding problem (i), observe that h belongs to $\mathbf{H}_{3,\rho}$ if and only if $h((u_\chi))$ is divisible by some u_{χ^*}, $\chi^* \in \hat{\mathbb{G}}^{\mathbf{L}}$, such that $\eta(\rho, \chi^*) \in \{\phi, \psi, \phi\psi\}$. Thus problem (i) reduces to computing and inspecting $\eta(\rho, \chi)$ for each $\chi \in \hat{\mathbb{G}}^{\mathbf{L}}$.

Problem (ii) is a little more involved. Let $\chi^{(1)}, \ldots, \chi^{(M)}$ be a list of the characters $\chi \in \hat{\mathbb{G}}^{\mathbf{L}}$ such that $\eta(\rho, \chi) = 1$. We can write two multinomials $h' \notin \mathbf{H}_{3,\rho}$ and $h'' \notin \mathbf{H}_{3,\rho}$ as

$$h'((u_\chi)) = \prod_{i=1}^{M} u_{\chi^{(i)}}^{a(i)}$$

and

$$h''((u_\chi)) = \prod_{i=1}^{M} u_{\chi^{(i)}}^{b(i)}.$$

Associate each character $\chi^{(i)}$ with a vector $x^{(i)}$ of 0's and 1's indexed by $(\mathbf{V} \setminus \{\rho\}) \times \{\phi, \psi, \phi\psi\}$ by setting

$$x_{v,\theta}^{(i)} = \begin{cases} 1, & \text{if } \eta(v, \chi^{(i)}) = \theta, \\ 0, & \text{otherwise.} \end{cases}$$

Equivalently, $x_{v,\theta}^{(i)} = 1$ if and only if $W_3(v, \chi^{(i)}) = w_{v,\theta}$. Then $h' \sim_3 h''$ if and only if $\sum_{i=1}^{M} a(i) x^{(i)} = \sum_{i=1}^{M} b(i) x^{(i)}$. Observe that h' and h'' will be equivalent under $\sim_3$ if and only if the two multinomials obtained by removing common factors are also equivalent; so, without loss of generality, it suffices to describe all pairs of nonnegative integer M-tuples $a = (a(i))_{i=1}^M$ and $b = (b(i))_{i=1}^M$ such that $\sum_{i=1}^M a(i)x^{(i)} = \sum_{i=1}^M b(i)x^{(i)}$ and at most one of the integers $a(i)$ and $b(i)$ is nonzero for each $i \in \{1, \ldots, M\}$. This latter problem is equivalent to describing the set

$$E_3 = \left\{ e = (e(i))_{i=1}^M \in \mathbb{Z}^M : \sum_{i=1}^M e(i)x^{(i)} = 0 \right\}.$$

It is clear that E_3 is a *lattice* in the sense of Section IV.3 in Cohn (1980). Therefore, by Theorem IV.5.1 of Cohn (1980), there exists a *minimal basis* for E_3 when $E_3 \neq \{0\}$. That is, there exist vectors $e^{(1)}, \ldots, e^{(N)} \in E_3$ such that E_3 consists of all vectors of the form $e = \sum_{j=1}^N n(j)e^{(j)}$ for some $(n(j))_{j=1}^N \in \mathbb{Z}^N$; and, moreover, the coefficients appearing in this representation are unique. The number N is intrinsic to E_3, that is, it is the same for all possible minimal bases. Here N coincides with the dimension of the real vector space

$$\left\{ f = (f(i))_{i=1}^M \in \mathbb{R}^M : \sum_{i=1}^M f(i)x^{(i)} = 0 \right\}.$$

Thus, $N = M - \operatorname{rank}(C)$, where C is the matrix with ith row $x^{(i)}$. Cohn (1980) describes a procedure for constructing such a minimal basis.

The analogues of all of the above for the Kimura two-parameter model with uniform root distribution and the Jukes–Cantor model with uniform root distribution follow along similar lines, with Lemma 4.1 replaced by Lemma 4.2 and Lemma 4.3, respectively. We give a brief sketch of the main ideas and leave the details to the reader.

Given a multinomial $h \in \mathbf{H}$ of the form $h((u_\chi)) = \prod_\chi u_\chi^{k_\chi}$, where $k_\chi \in \{0, 1, 2 \ldots\}$, we can uniquely define another multinomial $S_2 h$ (resp., $S_1 h$) in the dummy variables $w_{v,\theta}$, $v \in \mathbf{V}$ and $\theta \in \{\phi, \psi\}$, (resp., in the dummy variables w_v, $v \in \mathbf{V}$) by $S_2 h((w_{v,\theta})) = \prod_\chi (\prod_v W_2(v, \chi))^{k_\chi}$ (resp., $S_1 h((w_v)) = \prod_\chi (\prod_v W_1(v, \chi))^{k_\chi}$), where we set

$$W_2(v, \chi) = \begin{cases} w_{v,\phi}, & \text{if } \eta(v, \chi) \in \{\phi, \phi\psi\}, \\ w_{v,\psi}, & \text{if } \eta(v, \chi) = \psi, \\ 1, & \text{if } \eta(v, \chi) = 1 \end{cases}$$

(resp.,

$$W_1(v, \chi) = \begin{cases} w_v, & \text{if } \eta(v, \chi) \neq 1, \\ 1, & \text{if } \eta(v, \chi) = 1). \end{cases}$$

Observe that for the Kimura two-parameter model

$$h\big((\mathbb{E}[\langle Y, \chi \rangle])_\chi\big) = S_2 h\big((\mathbb{E}[\langle Z_v, \theta \rangle])_{v,\theta}\big)$$

and that for the Jukes–Cantor model

$$h\big((\mathbb{E}[\langle Y, \chi \rangle])_\chi\big) = S_1 h\big((\mathbb{E}[\langle Z_v, \phi \rangle])_v\big).$$

INVARIANTS IN PHYLOGENETIC INFERENCE

We define equivalence relations $\sim_2$ and $\sim_1$ on $\mathbf{H}$ in the same manner as we defined $\sim_3$, but with S_3 replaced by S_2 and S_1, respectively.

Write $\mathscr{H}_2$ (resp., $\mathscr{H}_1$) for the family of equivalence classes in $\mathbf{H}$ under $\sim_2$ (resp., $\sim_1$). Let $\mathbf{H}_{2,\rho}$ denote the equivalence class in $\mathscr{H}_2$ consisting of multinomials h such that $S_2 h((w_{v,\theta}))$ is divisible by $w_{\rho,\xi}$ for some $\xi \in \{\phi, \psi\}$ and let $\mathbf{H}_{1,\rho}$ denote the equivalence class in $\mathscr{H}_1$ consisting of multinomials h such that $S_1 h((w_v))$ is divisible by w_ρ. Observe for the Kimura two-parameter model with uniform root distribution that if $f, g \in \mathbf{H}$ with $f \sim_2 g$ then

$$f((\mathbb{E}[\langle Y, \chi\rangle])_\chi) = g((\mathbb{E}[\langle Y, \chi\rangle])_\chi),$$

for all choices of parameters in the model, with the common value being identically zero if $f, g \in \mathbf{H}_{2,\rho}$. Similarly, for the Jukes–Cantor model with uniform root distribution observe that if $f, g \in \mathbf{H}$ with $f \sim_1 g$ then

$$f((\mathbb{E}[\langle Y, \chi\rangle])_\chi) = g((\mathbb{E}[\langle Y, \chi\rangle])_\chi),$$

for all choices of parameters in the model, with the common value being identically zero if $f, g \in \mathbf{H}_{1,\rho}$. Moreover, from Lemma 4.2 and Lemma 4.3 we see that the converses to the last two statements also hold.

For example, consider the Jukes–Cantor model with uniform root distribution which has the design specified by (6.1). It is easy to check that

$$\left(\mathbb{E}[\langle Y_1, \phi\rangle\langle Y_2, \psi\rangle\langle Y_3, \phi\psi\rangle]\right)^2 - \prod_{1 \le i < j \le 3} \mathbb{E}\left[\langle Y_i, \phi\rangle\langle Y_j, \phi\rangle\right] = 0,$$

for all choices of parameters in the model.

The analogues of Theorem 6.1 for the Kimura two-parameter model and the Jukes–Cantor model are combined in the following result.

THEOREM 6.2. *Consider a Kimura two-parameter model (resp., a Jukes–Cantor model) with uniform root distribution. A polynomial F in the dummy variables $(t_g)_{g \in \mathbb{G}^{\mathbf{L}}}$ will be such that $F((\mathbb{P}\{Y = g\})_{g \in \mathbb{G}^{\mathbf{L}}}) = 0$ for all choices of parameters in the model if and only if F is of the form*

$$F((t_g)) = \sum_{h \in \mathbf{H}} c_h h\left(\left(\sum_{g \in \mathbb{G}^{\mathbf{L}}} \langle g, \chi\rangle t_g\right)_{\chi \in \hat{\mathbb{G}}^{\mathbf{L}}}\right),$$

where only finitely many of the coefficients c_h are non-zero and $\sum_{h \in \mathbf{K}} c_h = 0$ for all $\mathbf{K} \in \mathscr{H}_2 \setminus \{\mathbf{H}_{2,\rho}\}$ (respectively, for all $\mathbf{K} \in \mathscr{H}_1 \setminus \{\mathbf{H}_{1,\rho}\}$).

The analogues of the discussion following Theorem 6.1 about how to generate equivalent multinomials are straightforward and are left to the reader.

Finally, we remark that the results for models which have an arbitrary root distribution are very similar to those above. The only difference is that the root is treated like a typical vertex in a Kimura three-parameter model. That is, instead of imposing the constraints $\mathbb{E}[\langle Z_\rho, \theta\rangle] = 0$, $\theta \in \{\phi, \psi, \phi\psi\}$, we use the fact (which follows a fortiori from Lemma 4.1) that the image of

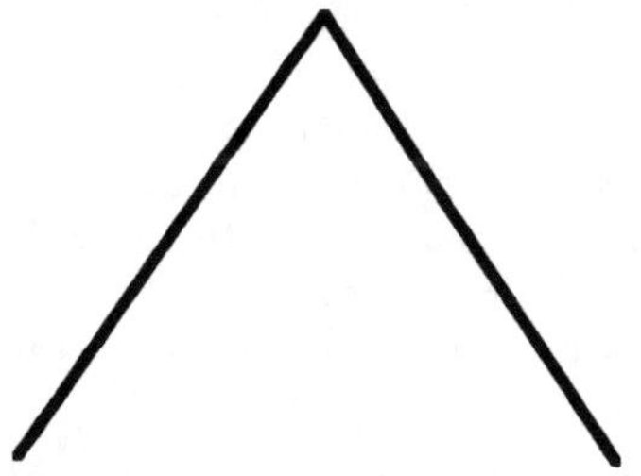

FIG. 4.

$(\mathbb{E}[\langle Z_\rho, \phi \rangle], \mathbb{E}[\langle Z_\rho, \psi \rangle], \mathbb{E}[\langle Z_\rho, \phi\psi \rangle])$ as the distribution of Z_ρ varies over all possible distributions has a nonempty interior. We once again leave the details to the reader. As an exercise, we invite the reader to show that there are no linear invariants for any Kimura three-parameter model with arbitrary root distribution.

7. Examples. In this section we present some explicit results for rooted trees with 2, 3 and 4 leaves, respectively, describing invariants under both the Kimura three-parameter and two-parameter models, with either a uniform or an arbitrary root distribution.

Two taxa. We begin with the simple two-leaf tree of Figure 4. Suppose (cf. Section 4) that Z_0, Z_1 and Z_2 are mutually independent $\mathbb{G}$-valued random variables (where, as before, $\mathbb{G} = \{A, G, C, T\}$) with distributions π, r_1 and r_2, respectively. Suppose that r_1 and r_2 are elements of R_3 with parameters $(\alpha_1, \beta_1, \gamma_1)$ and $(\alpha_2, \beta_2, \gamma_2)$ and a common value of t. Set $(Y_1, Y_2) = (Z_0 + Z_1, Z_0 + Z_2)$, so that we have a Kimura three-parameter model. The distribution $f = (f_{g_1, g_2})$ of (Y_1, Y_2) has Fourier transform

$$\hat{f}(\chi_1, \chi_2) = \hat{\pi}(\chi_1\chi_2)\hat{r}_1(\chi_1)\hat{r}_2(\chi_2),$$

where $\chi_1, \chi_2 \in \hat{\mathbb{G}}$. Expressions for $\hat{r}_1$ and $\hat{r}_2$ can be found in Section 4. When π is uniform, it follows that $\hat{f}(\chi_1, \chi_2) = 0$ unless $\chi_1 = \chi_2$, and the values of $\hat{f}(\chi, \chi)$ are just those of the Fourier transform of the convolution $r_1 * r_2$. It is easy to check that in this case only the parameters $t(\alpha_1 + \alpha_2)$, $t(\beta_1 + \beta_2)$ and $t(\gamma_1 + \gamma_2)$ are identifiable and so these, together with the requirement that the four distinct values of f_{g_1, g_2} must sum to $1/4$, account for all the degrees of freedom in this simple case. In what follows we will repeatedly use this informal counting of degrees of freedom, because in all cases we describe, the numbers match.

We turn now to the case where the initial distribution is arbitrary. The class of all such f is described by nine parameters: three for the root distribution, and three each for the two edges. Given that $\Sigma f_{g_1, g_2} = 1$, we need to find a further six constraints and in this case they are all *nonlinear*. Following the construction outlined after the proof of Theorem 6.1, we can calculate the relevant 15×9 matrix of 0's and 1's, and it is easily checked that this matrix has row rank 9, that is, there are six linearly independent relationships defining invariants. There are many ways to describe six independent invariants, although, as explained in the proof, a minimal basis of the lattice could be constructed. We content ourselves here with presenting a typical cubic invariant:

$$\mathbb{E}\langle Y_1, \phi\rangle \mathbb{E}\langle Y_2, \psi\rangle \mathbb{E}[\langle Y_1, \psi\rangle\langle Y_2, \phi\rangle] = \mathbb{E}\langle Y_1, \psi\rangle \mathbb{E}\langle Y_2, \phi\rangle \mathbb{E}[\langle Y_1, \phi\rangle\langle Y_2, \psi\rangle]$$

and there are two others like this one.

Turning to the Kimura two-parameter model for this tree (still with an arbitrary root distribution), suppose that $\beta_1 = \gamma_1$ and $\beta_2 = \gamma_2$, that is, that r_1 and r_2 are elements of R_2. Recall that $\hat{r}_i(\phi) = \hat{r}_i(\theta)$, $i = 1, 2$, where we write θ for $\phi\psi \in \hat{\mathbb{G}}$. We can then see that

$$\hat{f}(\phi, \phi) = \hat{f}(\theta, \theta)$$

and

$$\hat{f}(\phi, \theta) = \hat{f}(\theta, \phi),$$

and these identities define two *linear* invariants. A simple counting argument would suggest that there are $16 - 1 - (3 + 2 \times 2) = 8$ independent invariants in all, leaving six *nonlinear* invariants to be found, and this can be verified by following the construction given in the proof of Theorem 6.2 and computing the row rank of the appropriate 15×7 matrix of 0's and 1's. In constructing a collection of six independent *nonlinear* invariants we find that some may be taken to be the same as those described earlier, whilst others result from the fact that $\hat{r}_i(\phi) = \hat{r}_i(\theta)$, $i = 1, 2$. This latter class includes the quadratic invariant

$$\mathbb{E}\langle Y_1, \theta\rangle \mathbb{E}\langle Y_2, \phi\rangle = \mathbb{E}\langle Y_1, \phi\rangle \mathbb{E}\langle Y_2, \theta\rangle.$$

Before closing this discussion of the invariants of the Kimura two-parameter model for the two-leaf tree, it is both of independent interest and convenient for later examples to relate our notation and results to those of Cavender (1989, 1991). Following Cavender, we let A (resp. G, C, T) double as the *function* on $\mathbb{G} = \{A, G, C, T\}$ which takes the value 1 on A (resp., G, C, T) and 0 elsewhere. We then see that $A - G = (1/2)(\phi + \theta)$ and $C - T = (1/2)(\phi - \theta)$; and, letting $\otimes$ denote the tensor product of functions on $\mathbb{G}$, we see that

$$\begin{aligned}(A - G) \otimes (C - T) &= \tfrac{1}{4}(\phi + \theta) \otimes (\phi - \theta) \\ &= \tfrac{1}{4}(\phi \otimes \phi - \phi \otimes \theta + \theta \otimes \phi - \theta \otimes \theta)\end{aligned}$$

and

$$\begin{aligned}(C - T) \otimes (A - G) &= \tfrac{1}{4}(\phi - \theta) \otimes (\phi + \theta) \\ &= \tfrac{1}{4}(\phi \otimes \phi - \theta \otimes \phi + \phi \otimes \theta - \theta \otimes \theta).\end{aligned}$$

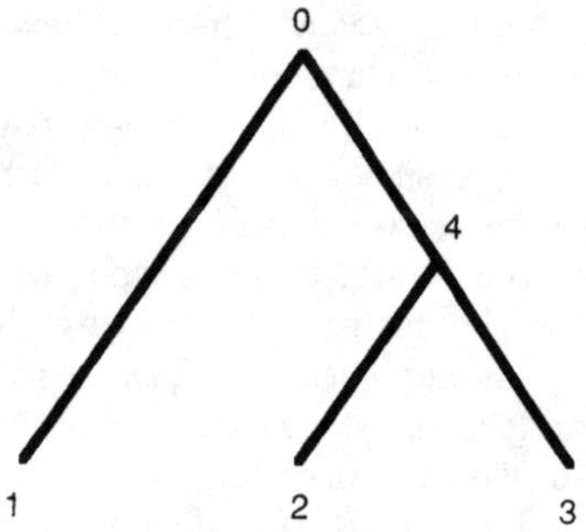

FIG. 5.

Letting f_{ij}, $i, j \in \{A, C, G, T\}$, denote the probability of observing base i in the first species and base j in the second species [cf. Cavender (1991)], we have

$$\begin{aligned} f_{AC} - f_{GC} - f_{AT} + f_{GT} &= \sum_{g_1, g_2} (A - G) \otimes (C - T)_{g_1, g_2} f_{g_1, g_2} \\ &= \sum_{g_1, g_2} \tfrac{1}{4}(\phi + \theta) \otimes (\phi - \theta)(g_1, g_2) f_{g_1, g_2} \\ &= \tfrac{1}{4}\big[\hat{f}(\phi, \phi) - \hat{f}(\phi, \theta) + \hat{f}(\theta, \phi) - \hat{f}(\theta, \theta)\big]. \end{aligned}$$

Similarly,

$$f_{CA} - f_{TA} - f_{CG} + f_{TG} = \tfrac{1}{4}\big[\hat{f}(\phi, \phi) - \hat{f}(\theta, \phi) + \hat{f}(\phi, \theta) - \hat{f}(\theta, \theta)\big].$$

Three taxa. We now discuss the rooted tree with 3 leaves, see Figure 5.

Again we will consider the Kimura three-parameter model first. Let $Z_0, \dots, Z_4$ be mutually independent $\mathbb{G}$-valued random variables with Z_0 having distribution π, and Z_i having distribution $r_i \in R_3$, $1 \le i \le 4$. If we set $Y_1 = Z_0 + Z_1$, $Y_2 = Z_0 + Z_4 + Z_2$ and $Y_3 = Z_0 + Z_4 + Z_3$, then the distribution $f = (f_{g_1, g_2, g_3})$ of (Y_1, Y_2, Y_3) has Fourier transform

$$\hat{f}(\chi_1, \chi_2, \chi_3) = \hat{\pi}(\chi_1\chi_2\chi_3)\left[\prod_{i=1}^{3} \hat{r}_i(\chi_i)\right]\hat{r}_4(\chi_2\chi_3),$$

where $\chi_1, \chi_2, \chi_3 \in \hat{\mathbb{G}}$.

Suppose that π is uniform. Then the only nonzero values of the Fourier transform occur when $\chi_1\chi_2\chi_3 = 1$, that is, when $\chi_3 = \chi_1\chi_2$. There are 16 distinct probabilities with a single sum constraint, depending upon six parameters, and so we might expect to find nine independent invariants. Calculation of the rank of the appropriate matrix reveals this to be the case. All of these invariants are *nonlinear*, a typical one being

$$\begin{aligned} &\mathbb{E}[\langle Y_1, \theta\rangle\langle Y_2, \phi\rangle\langle Y_3, \psi\rangle]\mathbb{E}\langle Y_1 + Y_2, \psi\rangle\mathbb{E}\langle Y_1 + Y_3, \phi\rangle \\ &\quad = \mathbb{E}[\langle Y_1, \theta\rangle\langle Y_2, \psi\rangle\langle Y_3, \phi\rangle]\mathbb{E}\langle Y_1 + Y_2, \phi\rangle\mathbb{E}\langle Y_1 + Y_3, \psi\rangle. \end{aligned}$$

INVARIANTS IN PHYLOGENETIC INFERENCE

Now let us suppose π to be arbitrary. Then we have 64 probabilities with a single sum constraint, given in terms of 15 parameters: three for the root distribution and three for each of the four edges of the tree. We note at this point a difference between the situation where π is uniform and π is arbitrary. In the former, the edge between the root and vertex 4 does not count in the parametrisation; the tree is effectively unrooted or, equivalently, a star phylogeny. This is because the result of convolving the uniform distribution with any distribution on $\mathbb{G}$ is again the uniform distribution, and hence the distribution r_4 gets "lost," that is, its parameters are unidentifiable. When π is general, however, we do need to count the parameters of r_4. Doing so would suggest that there are $64 - 1 - 15 = 48$ independent invariants, and again a check of the appropriate matrix rank shows this to be the case. An example of an invariant in this case is

$$\mathbb{E}[\langle Y_1, \chi\rangle\langle Y_2 + Y_3, \chi'\rangle] = \mathbb{E}[\langle Y_1, \chi\rangle]\mathbb{E}[\langle Y_2 + Y_3, \chi'\rangle]$$

for all $\chi, \chi' \in \hat{\mathbb{G}}$, expressing the obvious independence of Y_1 and $Y_2 + Y_3$.

Turning to the Kimura two-parameter for this tree (still with an arbitrary root distribution), we impose the constraints $\beta_i = \gamma_i$, $i = 1, \ldots, 4$. We now find that there are 18 *linear* invariants, arising from equalities of the form $\hat{f}(\chi_1, \chi_2, \chi_3) = \hat{f}(\chi_1', \chi_2', \chi_3')$, where for $i = 1, 2, 3$, either $\chi_i = \chi_i'$, $(\chi_i, \chi_i') = (\phi, \theta)$ or $(\chi_i, \chi_i') = (\theta, \phi)$. In essentially the same notation as Cavender (1989, 1991) these invariants may be written as $X \otimes (A - G) \otimes (C - T)$, $X \otimes (C - T) \otimes (A - G)$, where $X \in \mathbb{G}$ is arbitrary, and two similar sets of pairs, with X occupying the second and third position in the triple. This identification is easily obtained using the relations given at the end of the discussion of the two taxa case.

Let us note here a difference between our analysis and a result stated by Cavender (1991). In that paper it is asserted that the space of linear invariants of the six-parameter Cavender–Lake model coincides with that of the Kimura two-parameter model. However, we can only find 12 linearly independent linear invariants for the Cavender–Lake model, compared with the 18 obtained above for the Kimura two-parameter model. Indeed it is not hard to check that the linear function with coefficients $(\phi + \theta) \otimes (\phi - \theta) \otimes \psi$ is simultaneously an invariant for the Kimura model, and an element of the "expected spectrum" for the Cavender–Lake model (i.e., a linear combination of joint probabilities under the model).

To confirm this last assertion, set

$$\begin{aligned} D &= \tfrac{1}{4}(\phi + \theta) \otimes (\phi - \theta) \otimes \psi \\ &= (A - G) \otimes (C - T) \otimes ((A + G) - (C + T)) = E - 2F, \end{aligned}$$

where

$$E = (A - G) \otimes (C + T) \otimes ((A + G) + (C + T))$$

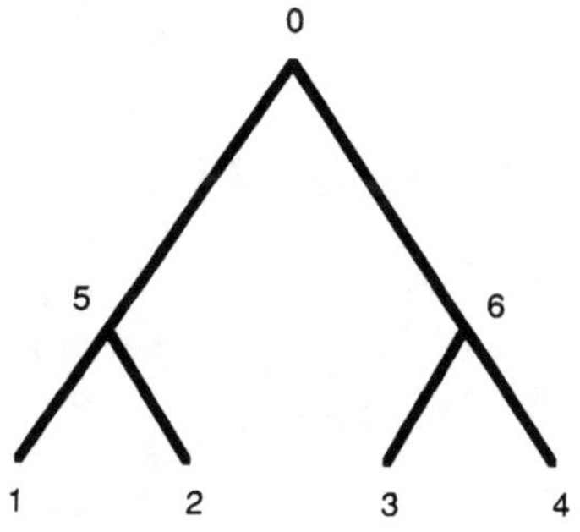

FIG. 6.

and

$$F = (A - G) \otimes C \otimes (C + T) + (A - G) \otimes T \otimes (A + C).$$

It is easy to check that D is an invariant for the two-parameter Kimura model, and we now show that E and F both belong to the expected spectrum of the Cavender–Lake model.

Firstly, consider $A \otimes (C + T) \otimes (A + G)$. Put $\pi = \delta_A$, the unit point mass at A. In the notation of Section 2, put $P^{(1)} = P^{(4)} = I$, the identity matrix. Recalling the basis $\{p_i\}$ given in Cavender (1989), put $P^{(2)} = p_3$ and $P^{(3)} = p_2$. We find that up to a scalar the result is $A \otimes (C + T) \otimes (A + C)$. Similarly, we can get $A \otimes (C + T) \otimes (C + T)$, $G \otimes (C + T) \otimes (A + G)$ and $G \otimes (C + T) \otimes (C + T)$, which implies that E is indeed in the expected spectrum of the Cavender–Lake model.

On the other hand, putting $\pi = \delta_A$, $P^{(1)} = P^{(2)} = I$, $P^{(4)} = p_3$ and $P^{(3)} = p_6$, shows that $A \otimes G \otimes (C + T) + A \otimes T \otimes (A + C)$ belongs to the expected spectrum. Similarly, replacing p_3 by p_2 shows that $G \otimes C \otimes (C + T) + G \otimes T \otimes (A + G)$ belongs to the expected spectrum. Thus F belongs to the expected spectrum and the assertion follows.

Four taxa. Our final example concerns the four-leaf tree given in Figure 6.

Yet again we consider the Kimura three-parameter model for this tree first. Let $(Z_i)_{i=0}^6$ be mutually independent $\mathbb{G}$-valued random variables, with Z_0 having distribution π and Z_i having distribution $r_i \in R_3$, $1 \le i \le 6$. Writing

$$Y_1 = Z_0 + Z_5 + Z_1,$$

$$Y_2 = Z_0 + Z_5 + Z_2,$$

$$Y_3 = Z_0 + Z_6 + Z_3,$$

$$Y_4 = Z_0 + Z_6 + Z_4,$$

we readily check that the Fourier transform of the distribution $f = (f_{g_1, g_2, g_3, g_4})$ of $(Y_i)_{i=1}^4$ factorises as follows:

$$\hat{f}(\chi_1, \chi_2, \chi_3, \chi_4) = \hat{\pi}(\chi_1\chi_2\chi_3\chi_4)\left[\prod_{i=1}^{4} \hat{r}_i(\chi_i)\right]\hat{r}_5(\chi_1\chi_2)\hat{r}_6(\chi_3\chi_4),$$

where $\chi_1, \chi_2\chi_3, \chi_4 \in \hat{\mathbb{G}}$.

When π is uniform, there are only 64 distinct probabilities in the 4^4 array f, corresponding to the 64 nonzero values of $\hat{f}$ which arise from quadruples $(\chi_1, \chi_2, \chi_3, \chi_4)$ such that $\chi_1\chi_2\chi_3\chi_4 = 1$. These quadruples are readily enumerated and it is easy to check that in this case there are 48 *nonlinear* independent invariants. We can write some of the invariants for this case in the following intuitively appealing forms:

(a) independence statements such as

$$\mathbb{E}[\langle Y_1 + Y_2, \chi\rangle\langle Y_3 + Y_4, \chi'\rangle] = \mathbb{E}[\langle Y_1 + Y_2, \chi\rangle]\mathbb{E}[\langle Y_3 + Y_4, \chi'\rangle]$$

for each of the *ordered* pair of nontrivial characters (χ, χ') (there are nine such equalities corresponding to the nine degrees of freedom in the usual chi-squared test of independence for a 4×4 contingency table);

(b) independence-like statements such as

$$\mathbb{E}\left\langle \sum_{i=1}^{4} Y_i, \chi\right\rangle\mathbb{E}\left\langle \sum_{i=1}^{4} Y_i, \chi'\right\rangle = \mathbb{E}[\langle Y_1 + Y_2, \chi\rangle\langle Y_3 + Y_4, \chi'\rangle] \times\mathbb{E}[\langle Y_1 + Y_2, \chi'\rangle\langle Y_3 + Y_4, \chi\rangle]$$

for each of the *unordered* pair of distinct nontrivial characters $\chi \neq \chi'$;

(c) equalities reminiscent of the determinantal identities on page 62 of Cavender and Felsenstein (1987) such as

$$\mathbb{E}[\langle Y_1 + Y_3, \chi\rangle\langle Y_2 + Y_4, \chi'\rangle]\mathbb{E}[\langle Y_1 + Y_3, \chi'\rangle\langle Y_2 + Y_4, \chi\rangle] = \mathbb{E}[\langle Y_1 + Y_4, \chi\rangle\langle Y_2 + Y_3, \chi'\rangle]\mathbb{E}[\langle Y_1 + Y_4, \chi'\rangle\langle Y_2 + Y_3, \chi\rangle]$$

for each of the *unordered* pairs of distinct characters $\chi \neq \chi'$ with not both χ and χ' trivial;

(d) cubic invariants obtained by considering three leaves at a time and using the invariants found in the three taxa case.

Once more the simple counting rules described earlier apply. There are $48 = 64 - 1 - 3 \times 5$ independent invariants, and we note as before that the uniform root distribution renders one of the distributions r_4 or r_5 superfluous.

Next we suppose π to be arbitrary. Then all six of the edge distributions contribute three parameters, as does the root distribution, and so there should be $256 - 1 - 3 - 6 \times 3 = 234$ independent invariants. The row rank of the appropriate 255×21 matrix is indeed 21 and so the counting rules continue to apply.

Finally, suppose that we have the Kimura two-parameter model (still with an arbitrary root distribution) obtained by fixing $\beta_i = \gamma_i$, $1 \leq i \leq 6$. There are 92 linearly independent *linear* invariants, in contrast with the 68 found by Cavender (1989) for the six-parameter Cavender–Lake model. These 92 linear invariants all arise from equalities of the form $\hat{f}(\chi_1, \ldots, \chi_4) = \hat{f}(\chi'_1, \ldots, \chi'_4)$ that occur when $(\chi'_1, \ldots, \chi'_4)$ is obtained from $(\chi_1, \ldots, \chi_4)$ by switching ϕ and θ, and such equalities are readily enumerated. Similarly, it is easy to check from the appropriate matrix that there are a total of $240 = 256 - 1 - 3 - 6 \times 2$ independent invariants, as expected.

Acknowledgments. We would like to thank Barbara Bowman, Arend Sidow and Allan Wilson for arousing our interest in this field, and for being generous with their time in explaining the area to us. A more immediate stimulus for this paper was the one-day workshop on invariants organised by Mike Waterman at the University of Southern California in July 1990. Thanks are due to him and James Lake, James Cavender and Joe Felsenstein, whose presentations and earlier work constitute the foundations on which we have built. Finally, we would like to thank Trang Nguyen for her assistance with the rank computations of Section 7.

REFERENCES

BARRY, D. and HARTIGAN, J. A. (1987). Statistical analysis of hominoid molecular evolution. *Statist. Sci.* **2** 191–210.

CAVENDER, J. A. (1989). Mechanized derivation of linear invariants. *Molecular Biology and Evolution* **6** 301–316.

CAVENDER, J. A. (1991). Necessary conditions for the method of inferring phylogeny by linear invariants. *Math. Biosci.* **103** 69–75.

CAVENDER, J. A. and FELSENSTEIN, J. (1987). Invariants of phylogenies in a simple case with discrete states. *J. Classification* **4** 57–71.

COHN, H. (1980). *Advanced Number Theory*. Dover, New York.

DIACONIS, P. (1988). *Group Representations in Probability and Statistics*. IMS, Hayward, CA.

DIACONIS, P. (1990). Patterned matrices. In *Matrix Theory and Applications—Proceedings of Symposia in Applied Mathematics* **40** (C. R. Johnson, ed.) 37–58. Amer. Math. Soc., Providence, RI.

DROLET, S. and SANKOFF, D. (1990). Quadratic tree invariants for multivalued characters. *Journal of Theoretical Biology* **144** 117–129.

FELSENSTEIN, J. (1978). Cases in which parsimony or compatibility methods will be positively misleading. *Systematic Zoology* **27** 401–410.

FELSENSTEIN, J. (1981). Evolutionary trees from DNA sequences: A maximum likelihood approach. *Journal of Molecular Evolution* **17** 368–376.

FELSENSTEIN, J. (1991). Counting phylogenetic invariants in some simple cases. *Journal of Theoretical Biology* **152** 357–376.

JUKES, T. H. and CANTOR, C. (1969). Evolution of protein molecules. In *Mammalian Protein Metabolism* (H. N. Munro, ed.) 21–132. Academic, New York.

KIMURA, M. (1980). A simple method for estimating evolutionary rates of base substitution through comparative studies of nucleotide sequences. *Journal of Molecular Evolution* **16** 111–120.

KIMURA, M. (1981). Estimation of evolutionary sequences between homologous nucleotide sequences. *Proc. Nat. Acad. Sci. USA* **78** 454–458.

KIMURA, M. (1983). *The Neutral Theory of Molecular Evolution*. Cambridge Univ. Press.

INVARIANTS IN PHYLOGENETIC INFERENCE

KÖRNER, T. W. (1988). *Fourier Analysis*. Cambridge Univ. Press.

LAKE, J. A. (1987). A rate-independent technique for analysis of nucleic acid sequences: Evolutionary parsimony. *Molecular Biology and Evolution* **4** 167–191.

NAVIDI, W., CHURCHILL, G. A. and VON HAESELER, A. (1992). Phylogenetic inference: Invariants and maximum likelihood. *Biometrics*. To appear.

NEYMAN, J. (1971). Molecular studies of evolution: A source of novel statistical problems. In *Statistical Decision Theory and Related Topics* (S. S. Gupta and J. Yackel, eds.) 1–27. Academic, New York.

SANKOFF, D. (1990). Designer invariants for large phylogenies. *Molecular Biology and Evolution* **7** 255–269.

SWOFFORD, D. L. and OLSEN, G. J. (1990). Phylogeny reconstruction. In *Molecular Systematics* (D. M. Hillis and C. Moritz, eds.) 411–501. Sinauer, Sunderland, MA.

TAVARÉ, S. (1986). Some probabilistic and statistical problems in the analysis of DNA sequences. In *Lectures on Mathematics in the Life Sciences* (R. Miura, ed.) 57–86. Amer. Math. Soc., Providence, RI.

DEPARTMENT OF STATISTICS
567 EVANS HALL
UNIVERSITY OF CALIFORNIA, BERKELEY
BERKELEY, CALIFORNIA 94720

Chapter 12
Statistical Genetics

Darlene R. Goldstein

Terry Speed has produced many interesting and important contributions to the field of statistical genetics, with work encompassing both experimental crosses and human pedigrees. He has been instrumental in uncovering and elucidating algebraic structure underlying a diverse range of statistical problems, providing new and unifying insights.

Here, I provide a brief commentary and introduction to some of the key building blocks for an understanding of the papers. Some readers may also find useful a refresher on group action (see e.g. Fraleigh [6]) and hidden Markov models [13].

Linkage mapping

Linkage analysis studies inheritance of traits in families, with the aim of determining the chromosomal location of genes influencing the trait. The analysis proceeds by tracking patterns of coinheritance of the trait of interest and other traits or genetic markers, relying on the varying degree of recombination between trait and marker loci to map the loci relative to one another.

A measure of the degree of linkage is the recombination fraction θ, the chance of recombination occurring between two loci. For unlinked genes, $\theta = 1/2$; for linked genes, $0 \leq \theta < 1/2$.

D.R. Goldstein
Institut de mathématiques d'analyse et applications, École Polytechnique Fédérale de Lausanne, and Swiss Institute of Bioinformatics, Switzerland
e-mail: darlene.goldstein@epfl.ch

S. Dudoit (ed.), *Selected Works of Terry Speed*, Selected Works in Probability and Statistics, DOI 10.1007/978-1-4614-1347-9_12, © Springer Science+Business Media, LLC 2012

Human pedigrees

S. Dudoit and T. P. Speed (1999). A score test for linkage using identity by descent data from sibships. *Annals of Statistics* 27:943–986.

This paper offers a novel and comprehensive algebraic view of sib-pair methods, fundamentally unifying a large collection of apparently ad hoc procedures and providing powerful insights into the methods.

Identical by descent allele sharing

Data for linkage analysis consist of sets of related individuals (pedigrees) and information on the genetic marker and/or trait genotypes or phenotypes. The recombination fraction is most commonly estimated by maximum likelihood for an appropriate genetic model for the coinheritance of the loci.

However, likelihood-based linkage analysis can be difficult to carry out due to the problem that the mode of inheritance may be complex and in any case is usually unknown. Nonparametric approaches are thus appealing, since they do not require a genetic inheritance model to be specified. Such methods typically focus on identical by descent (IBD) allele sharing at a locus between a pair of relatives. DNA at a locus is shared by two relatives *identical by descent* if it originated from the same ancestral chromosome. In families of individuals possessing the trait of interest, there is association between the trait and allele sharing at loci linked to trait susceptibility loci, which can be used to localize trait susceptibility genes.

Testing for linkage with IBD data has developed differently, depending on the type of trait. For qualitative traits, tests are based on *IBD sharing conditional on phenotypes*. Affected sib-pair methods are a popular choice; these are often described as nonparametric since the mode of inheritance does not need to be specified (see Hauser and Boehnke [10] for a review). On the other hand, for quantitative trait loci (QTL), linkage analysis is based on examination of *phenotypes conditional on sharing* (for example, the method of Haseman and Elston [9] or one of its many extensions).

Inheritance vector

The pattern of IBD sharing at a locus within a pedigree is summarized by an inheritance vector, which completely specifies the ancestral source of DNA. For sibships of size k, label locus (1, 2) and maternally derived alleles (3, 4). The inheritance vector at a given locus is the vector $x = (x_1, x_2, \ldots, x_{2k-1}, x_{2k})$, where for sib i, x_{2i-1} is the label of the paternally inherited allele (1 or 2) and x_{2i} is that of the maternally inherited allele (3 or 4) at the locus.

For a pair of sibs, when paternal and maternal allele sharing are not distinguished, the 16 possible inheritance vectors give rise to three IBD configurations C_j: the sibs may share 0, 1, or 2 alleles IBD at the locus (Table 12.1). The IBD configurations can be thought of as orbits of groups acting on the set of possible inheritance vectors $\mathscr{X}$ [2].

Table 12.1 Sib-pair IBD configurations

Alleles IBD		Inheritance vectors	$\mid C_j \mid$
0 IBD		(1, 3, 2, 4), (1, 4, 2, 3), (2, 3, 1, 4), (2, 4, 1, 3)	4
1 IBD	(paternal)	(1, 3, 1, 4), (1, 4, 1, 3), (2, 3, 2, 4), (2, 4, 2, 3)	8
	(maternal)	(1, 3, 2, 3), (1, 4, 2, 4), (2, 3, 1, 3), (2, 4, 1, 4)	
2 IBD		(1, 3, 1, 3), (1, 4, 1, 4), (2, 3, 2, 3), (2, 4, 2, 4)	4

Score test for linkage

The literature contains several proposed tests of the null hypothesis of no linkage ($H : \theta = 1/2$) based on score functions of IBD configurations for sibships and other pedigrees, with scores chosen to yield good power against a particular alternative. The score test of Dudoit and Speed to detect linkage represents a major breakthrough in that it creates a coherent, unified based approach to the linkage analysis of qualitative and quantitative traits using IBD data. The likelihood for the recombination fraction θ, conditional on the phenotypes of the relatives, is used to form a score test of the null hypothesis of no linkage ($\theta = 1/2$).

The probability vector of IBD configurations, conditional on pedigree phenotypes, at a *marker* locus linked to a trait susceptibility locus at recombination fraction θ can be written as $\rho(\theta,\pi)_{1\times m} = \pi_{1\times m}T(\theta)_{m\times m}$, where π represents the conditional probability vector for IBD configurations at the *trait* locus and the number of IBD configurations is m. $T(\theta)$ denotes the transition matrix between IBD configurations at loci separated by recombination fraction θ.

In general, the probability vector π depends on unknown genetic parameters. However, using their formulation of the problem, Dudoit and Speed [4] show rigorously that for affected sibships of a given size, the second-order Taylor series expansion of the log likelihood around the null of no linkage is independent of the genetic inheritance model. They thus provide a mathematically justified basis for affected sib-pair methods, which do not require an assumed mode of inheritance.

Practical advantages of the score test include: it is locally most powerful for alternatives close to the null; any genotype distribution can be used (i.e., Hardy-Weinberg equilibrium is not required); conditioning on phenotypes eliminates selection bias introduced by nonrandom ascertainment; and combining differently ascertained pairs is straightforward, providing the important benefit of allowing us to avoid discarding any data. For many realistic simulation scenarios [7, 8], the score test proves to be robust and shows large power gains over commonly used nonparametric tests.

Although the paper focuses on pairs of sibs, the same score test approach is also applicable to any set of relatives [3].

Experimental crosses

N. J. Armstrong, M. S. McPeek and T. P. Speed (2006). Incorporating interference into linkage analysis for experimental crosses. *Biostatistics* 7:374–386.

This paper improves multilocus linkage analysis of experimental crosses by incorporating a realistic model of crossover interference, and implementing it by extending the Lander-Green algorithm for genetic reconstruction. It represents the culmination of a series of studies of the modeling of crossover interference.

χ^2 *model of crossover interference*

During the (four-strand) process of crossing over in meiosis, two types of interference (nonindependence) are distinguished: chromatid interference, a situation in which the occurrence of a crossover between any pair of nonsister chromatids affects the probability of those chromatids being involved in other crossovers in the same meiosis; and crossover interference, which refers to nonrandom location of chiasmata along a chromosome.

Most genetic mapping is carried out assuming independence; that is, no chromatid interference and no crossover interference. This assumption simplifies likelihood calculations. Although there is little empirical evidence for chromatid interference, there is substantial evidence of crossover interference. Thus, more a more realistic model incorporating crossover interference should be able to provide more accurately estimated genetic maps.

The χ^2 model of crossover interference [5] provides a dramatically improved fit over a wide range of models [12, 14]. This model assumes that recombination intermediates (structures formed after initiation of recombination) are resolved in one of two ways: either with or without crossing over. Recombination initiation events are assumed to occur according to a Poisson distribution, but constraints on the resolution of intermediates creates interference. The χ^2 model assumes m unobserved intermediates between each crossover. For $m = 1$, the model reduces to the no (crossover) interference model. This model is a special case of the more general gamma model, but has the advantage of being computationally more feasible.

Genetic reconstruction and the Lander-Green algorithm

Genetic mapping in humans can be viewed as a missing data problem, since we are typically unable to observe the complete data (the number of recombinant and non-recombinant meioses for each interval). If complete data were available, maximum likelihood estimates of a set of recombination fractions θ_i, $i = 1,\ldots,T-1$, for adjacent markers $\mathscr{M}_1,\ldots,\mathscr{M}_T$ would just be the observed proportion of recombinants in an interval.

The genetic reconstruction problem is to determine the expected number of recombinations that occurred in intervals of adjacent markers, given genotypes at multiple marker loci in a pedigree and the recombination fraction for each interval. Construction is straightforward when there is complete genotype information, including the ancestral origin (paternal or maternal).

More commonly this information is not known, so a different strategy for likelihood calculation is needed to obtain recombination fraction estimates. Lander and Green [11] proposed an approach based on the use of inheritance vectors. They showed that the probability of the observed data can be calculated for any particular inheritance vector and that under no crossover interference, the inheritance vectors form a Markov chain along the chromosome. They model the pedigree and data as a hidden Markov model, where the hidden states are the (unobserved) inheritance vectors. The complexity of their algorithm for calculating likelihoods increases linearly with the number of markers but exponentially in pedigree size, making it appropriate for analysis of many markers on small to moderately sized pedigrees.

In experimental crosses, mapping is generally more straightforward since investigators can arrange crosses to produce complete data. However, the presence of unobserved recombination initiation points creates a new kind of missing data when the no interference model is not assumed. The creative insight of Armstrong et al. [1] is to model the crossover interference process as a hidden Markov model. This step works because even though crossovers resulting from initiation events do not occur independently (in the presence of crossover interference), the initiation events themselves *are* assumed to be independent. Armstrong et al. [1] are thus able to extend the Lander-Green algorithm to incorporate interference according to the χ^2 model, thereby providing more accurately estimated genetic maps.

Conclusion

Terry's work in statistical genetics has identified underlying commonalities across seemingly disparate procedures, contributing meaningful theoretical and practical improvements. An impressive aspect of these works is the fresh perspective offered by viewing the problems at a stripped-down, fundamental level. Applying an exceptional combination of extensive mathematical expertise and pragmatic sensibility, Terry provides inventive solutions and a richer structural understanding of significant questions in statistical genetics.

References

[1] N. J. Armstrong, M. S. McPeek, and T. P. Speed. Incorporating interference into linkage analysis for experimental crosses. *Biostatistics*, 7(3):374–386, 2006.
[2] K. P. Donnelly. The probability that related individuals share some section of genome identical by descent. *Theor. Popul. Biol.*, 23:34–63, 1983.
[3] S. Dudoit. *Linkage Analysis of Complex Human Traits Using Identity by Descent Data*. PhD thesis, Department of Statistics, University of California, Berkeley, 1999.
[4] S. Dudoit and T. P. Speed. A score test for linkage using identity by descent data from sibships. *Ann. Stat.*, 27(3):943–986, 1999.
[5] E. Foss, R. Lande, F. Stahl, and C. Steinberg. Chiasma interference as a function of genetic distance. *Genetics*, 133:681–691, 1993.
[6] J. B. Fraleigh. *A First Course in Abstract Algebra*. Addison-Wesley Pub. Co., Reading, Mass., 7th edition, 2002.
[7] D. R. Goldstein, S. Dudoit, and T. P. Speed. Power of a score test for quantitative trait linkage analysis of relative pairs. *Genet. Epidemiol.*, 19(Suppl. 1): S85–S91, 2000.
[8] D. R. Goldstein, S. Dudoit, and T. P. Speed. Power and robustness of a score test for linkage analysis of quantitative trait based on identity by descent data on sib pairs. *Genet. Epidemiol.*, 20(4):415–431, 2001.
[9] J. K. Haseman and R. C. Elston. The investigation of linkage between a quantitative trait and a marker locus. *Behav. Genet.*, 2:3–19, 1972.
[10] E. R. Hauser and M. Boehnke. Genetic linkage analysis of complex genetic traits by using affected sibling pairs. *Biometrics*, 54:1238–1246, 1998.
[11] E. S. Lander and P. Green. Construction of multilocus genetic maps in humans. *Proc. Natl. Acad. Sci. USA*, 84:2363–2367, 1987.
[12] M. S. McPeek and T. P. Speed. Modeling interference in genetic recombination. *Genetics*, 139:1031–1044, 1995.
[13] L. R. Rabiner. A tutorial on hidden Markov models and selected applications in speech recognition. *Proceedings of the IEEE*, 77(2):257–286, 1989.
[14] H. Zhao, T. P. Speed, and M. S. McPeek. Statistical analysis of crossover interference using the chi-square model. *Genetics*, 139:1045–1056, 1995.

The Annals of Statistics
1999, Vol. 27, No. 3, 943–986

A SCORE TEST FOR LINKAGE USING IDENTITY BY DESCENT DATA FROM SIBSHIPS

BY SANDRINE DUDOIT[1] AND TERENCE P. SPEED[2]

University of California, Berkeley

We consider score tests of the null hypothesis H_0: $\theta = \frac{1}{2}$ against the alternative hypothesis H_1: $0 \le \theta < \frac{1}{2}$, based upon counts multinomially distributed with parameters n and $\rho(\theta, \pi)_{1\times m} = \pi_{1\times m} T(\theta)_{m\times m}$, where $T(\theta)$ is a transition matrix with $T(0) = I$, the identity matrix, and $T(\frac{1}{2}) = (1, \dots, 1)^T(\alpha_1, \dots, \alpha_m)$. This type of testing problem arises in human genetics when testing the null hypothesis of no linkage between a marker and a disease susceptibility gene, using identity by descent data from families with affected members. In important cases in this genetic context, the score test is independent of the nuisance parameter π and based on a widely used test statistic in linkage analysis. The proof of this result involves embedding the states of the multinomial distribution into a continuous-time Markov chain with infinitesimal generator Q. The second largest eigenvalue of Q and its multiplicity are key in determining the form of the score statistic. We relate Q to the adjacency matrix of a quotient graph in order to derive its eigenvalues and eigenvectors.

1. Introduction. This paper concerns a class of score tests which arise naturally in human genetics. However, their essence can be described quite efficiently without any of the genetic background, and we now do so. Let $\alpha = (\alpha_1, \dots, \alpha_m)$ and $\pi = (\pi_1, \dots, \pi_m)$ be two multinomial distributions, viewed as points in a simplex, and let $\{T(\theta): 0 \le \theta \le \frac{1}{2}\}$ be a one-parameter family of transition matrices such that $T(0) = I$, the identity matrix, and $T(\frac{1}{2}) = \mathbf{1}^T\alpha$, where $\mathbf{1} = (1, \dots, 1)$. These objects allow us to define the curve $\mathscr{C}_\pi(\theta)$ of distributions $\rho(\theta, \pi) = \pi T(\theta)$, $0 \le \theta \le \frac{1}{2}$, connecting $\pi = \rho(0, \pi)$ to $\alpha = \rho(\frac{1}{2}, \pi)$. Our interest is a score test for the null hypothesis H_0: $\theta = \frac{1}{2}$ against the alternative H_1: $0 \le \theta < \frac{1}{2}$, that is, for testing H_0: $\rho = \alpha$ against alternatives along the curve $\mathscr{C}_\pi(\theta)$, based upon counts $N = (N_1, \dots, N_m)$ multinomially distributed with parameters $n = \sum_i N_i$ and $\rho(\theta, \pi)$. The associated log-likelihood is $l(\theta, \pi) = \sum_i N_i \ln(\rho_i(\theta, \pi))$, and the score test in question should be based on $l'(\frac{1}{2}, \pi) = \sum_i N_i \rho_i'(\frac{1}{2}, \pi)/\alpha_i$, where $'$ denotes differentiation in θ. It turns out in our genetic context that $l'(\frac{1}{2}, \pi) \equiv 0$, and so we consider the second derivative, obtaining $l''(\frac{1}{2}, \pi) = \sum_i N_i \rho_i''(\frac{1}{2}, \pi)/\alpha_i = \sum_i N_i(\sum_j \pi_j u_{ji})/\alpha_i$, where $U = (u_{ij}) = T''(\frac{1}{2})$. Now, we would normally need

Received August 1998; revised March 1999.

[1]Supported in part by a PMMB Burroughs-Wellcome Fellowship.

[2]Supported in part by NIH Grant R01 HG01093.

AMS 1991 *subject classifications.* Primary 62F03; secondary 92D30, 60J20, 15A18, 05C20, 05C30.

Key words and phrases. Score test, Markov chain, infinitesimal generator, quotient graph, adjacency matrix, eigenvalues, orbits, Pólya's theory, linkage analysis.

to deal with the nuisance parameter π in this score test. This study was motivated by the observation that in some important cases in linkage analysis, U has rank 1, that is, $u_{ij} = a_i b_j$, for suitable vectors (a_i) and (b_i). In such cases, $l''(\frac{1}{2}, \pi) = (\sum_j a_j \pi_j)(\sum_i b_i N_i/\alpha_i)$, and the score test is independent of the nuisance parameter π. Moreover, for our genetic problem, the score test is based on a widely used nonparametric statistic in linkage analysis, S_{pairs} [Whittemore and Halpern (1994a), Kruglyak, Daly, Reeve-Daly and Lander (1996)]. We thought it would be of interest to understand the origins of this property and to learn just how far it extended.

In Section 2, we present the genetic problem which motivated our study, the linkage analysis of disease susceptibility genes using identity by descent (IBD) data from sets of siblings (sibships). This involves describing how IBD patterns in pedigrees (i.e., collections of related individuals) may be summarized by inheritance vectors which correspond to the vertices of a hypercube. The inheritance vectors along a chromosome are embeddable in a continuous-time random walk on the vertices of the hypercube, with time parameter the genetic distance along the chromosome. For our purpose, the inheritance vectors may be partitioned into so-called IBD configurations, which are orbits of groups acting on the set of inheritance vectors. In Section 3, we derive a semigroup property for the IBD configuration transition matrix $T(\theta)$ and present a spectral decomposition of $T(\theta)$ in terms of the eigenvalues and eigenvectors of its infinitesimal generator Q. The second largest eigenvalue of Q and its multiplicity are key in determining the form of the linkage score statistic. In order to derive the eigenvalues and eigenvectors of the infinitesimal generator, we relate it to the adjacency matrix of a quotient graph. Finally, in Section 4, we derive score statistics for testing linkage in sibships and illustrate the results with sib-pairs and sib-trios in Section 5. Remarkably, in an affected-only analysis, where siblings not affected by the disease are ignored, the score test for sibships of a given size does not depend on the nuisance parameter π and is based on a well-known statistic in linkage analysis, S_{pairs} [Whittemore and Halpern (1994a), Kruglyak, Daly, Reeve-Daly and Lander (1996)].

2. Testing linkage using identity by descent data. Genetic mapping is based upon the phenomenon of *crossing-over*, which is the exchange of corresponding DNA between chromosomes from the same pair during gamete (egg/sperm) formation. The human genome is distributed along 23 pairs of chromosomes, 22 autosomal pairs and the sex chromosome pair (XX for females and XY for males). Each pair consists of a paternally inherited chromosome and a maternally inherited chromosome. As a result of crossovers, chromosomes passed from parent to offspring are combinations of the two grand-parental chromosomes (see Figures 1 and 2). In general, the DNA variants (*alleles*) passed from parent to offspring at two nearby chromosomal locations (*loci*) have the same grand-parental origin (e.g., at both loci, the maternally inherited alleles are from the maternal grandfather). This is sometimes called *cosegregation*, as *segregation* is the process leading to the choice of one of a parent's two variants (maternal or paternal) at any given locus for transmis-

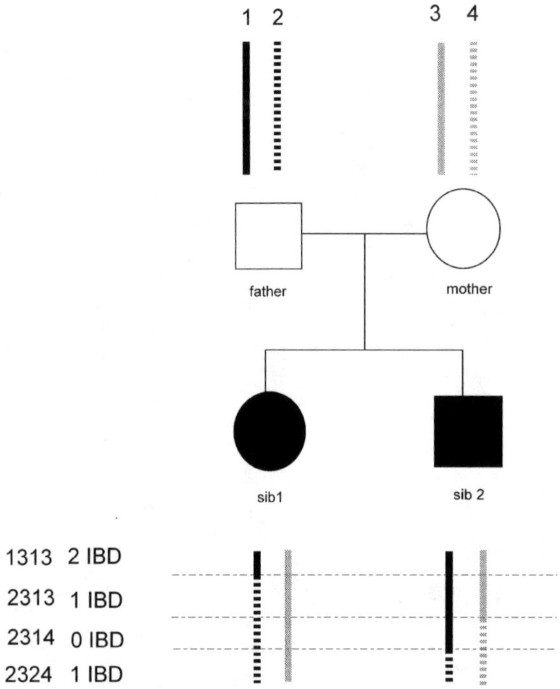

FIG. 1. *Segregation products for a sibship of size 2 and a single chromosome pair. Male and female individuals are represented by squares and circles, respectively, and colored symbols indicate affectedness by the disease under study. The paternal and maternal chromosome pairs are labeled by* $(1, 2)$ *and* $(3, 4)$*, respectively. The inheritance vectors and IBD configurations of the sib-pair are indicated on the left.*

sion to a child. Exceptions to cosegregation occur due to crossovers; then, the variants passed on to the child have different grand-parental origins at the two loci and the chromosome is said to be *recombinant* (e.g., for the maternally inherited chromosome, the variant from the maternal grandfather was inherited at one locus and that from the maternal grandmother was inherited at the other locus). The frequency with which this occurs is the *recombination fraction* between the two loci, conventionally denoted by θ. [Recombination fracctions are assumed to be constant across conditions (e.g., age and temperature) and individuals. Under general models for crossovers, the recombination fraction between two loci belongs to the interval $[0, 1/2]$.] In general, two loci are said to be *linked* if their recombination fraction is less than $\frac{1}{2}$, and *unlinked* if it is $\frac{1}{2}$. Thus, unlinked loci may be widely separated on the same chromosome, or on different chromosomes. Loci are said to be *tightly linked* if the recombination fraction θ is close to 0, for example, $\theta < 0.05$ [see Ott (1991), McPeek (1996) and Speed (1996) for an introduction to linkage analysis].

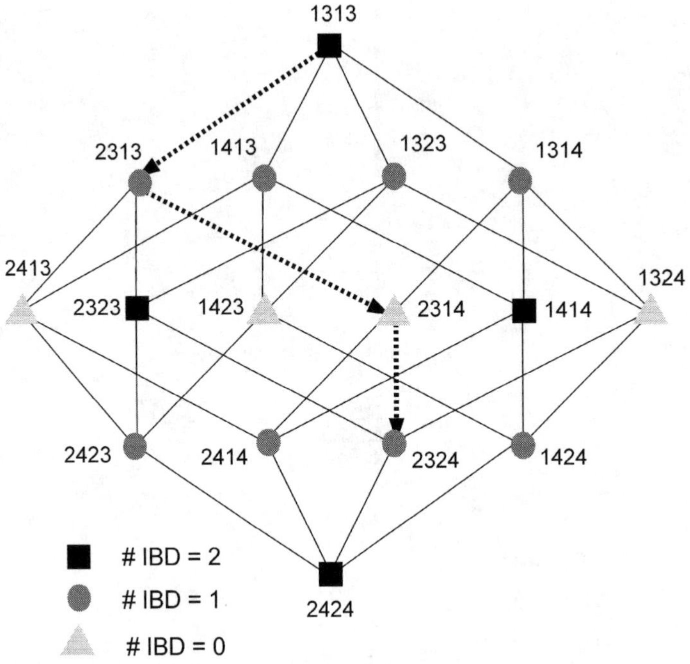

FIG. 2. *Four-dimensional hypercube whose vertices correspond to the* 16 *possible inheritance vectors for a sib-pair and whose edges represent permissible transitions. The arrows indicate the transitions for the segregation products represented in Figure* 1.

When mapping disease susceptibility (DS) genes, we are interested in testing whether genetic markers with known location are linked or not to DS genes, that is, in testing a null hypothesis of the form H_0: $\theta = \frac{1}{2}$, where θ is the recombination fraction between a genetic marker and a putative DS gene. This could be done by studying the cosegregation of variants of the DS genes with those of other mapped genes or markers. (By now, there are hundreds of well-mapped markers along each human, mouse and many other chromosomes.) Frequent cosegregation of a DS locus with a mapped marker would imply a small recombination fraction between the two loci, and hence an accurate placement of the DS locus. However, for most diseases of interest, we do not in general know, and are unable to determine, the alleles present at the DS loci prior to their being mapped. Indeed, much of the interest in mapping DS loci is to determine the variants segregating in populations. Thus, a direct approach to mapping DS loci is generally not available. Many ingenious methods have been developed by geneticists to circumvent this problem, and this paper concerns one such which studies marker identity by descent in sibships with affected members. DNA at the same locus on two chromosomes from the same pair is said to be *identical by descent* (*IBD*) if it originated from the same ancestral chromosome. This is by contrast to *identity by state* (*IBS*), where the same DNA variant in two individuals may have entered the family under study through different ancestors and hence may not be IBD. Linkage analysis methods based on IBD data seek to exploit the association between the *sharing of DNA identical by descent at loci linked to DS loci* and *disease status* (*phenotype*) in families with affected individuals. At loci unlinked to DS loci, IBD sharing is independent of phenotype. For example, for sibships, this association arises from the fact that full sibs get all their genes from the same source, their parents. Consequently, if susceptibility to the disease under study has a genetic component, the disease status (affected or not) of the sibs should be associated with their IBD status (identical or not) at the DS loci [see Tables 3 and 4 in Dudoit and Speed (1999) for a simple example of this association in sib-pairs]. This approach is effective to the extent that disease susceptibility is affected by genes rather than, say, a shared environment, or other "random" factors. Determining IBD status at or near a DS locus is usually feasible, where determining the gene variant is not, because one can readily determine IBD status at so-called marker loci, one of which may be tightly linked to the DS locus [cf. Kruglyak and Lander (1995) and Kruglyak, Daly, Reeve-Daly and Lander (1996) for a treatment of incomplete IBD data]. IBD-based methods for detecting linkage to DS loci will thus be successful if (and only if) (1) there is a noticeable association between phenotype and IBD status of relatives at the DS loci, and (2) this association is strong enough to remain detectable when IBD status at an (unknown) DS locus is replaced by observed IBD status at a marker locus. Recombination between a DS locus and a marker locus will attenuate the association between phenotype and IBD status. If we have a dense enough set of marker loci, problem (2) would appear to be solved, but in truth there is always a close connection between the magnitude of the association in (1) and the density of the marker set necessary for its detection. These issues were addressed by Thompson (1997), who refers to the two components (1) and (2) as the *specificity* of the DS loci and the *scale* of the genetic distance, respectively. We refer the reader to Whittemore and Halpern (1994a, b), Whittemore (1996), Kruglyak, Daly, Reeve-Daly and Lander (1996), Kong and Cox (1997), Feingold and Siegmund (1997), Teng and Siegmund (1997), Dudoit (1999), and Dudoit and Speed (1999), McPeek (1999) for recent discussions of linkage analysis using IBD data.

The IBD pattern within a pedigree may be summarized at any chromosomal locus by the inheritance vector. Consider a sibship of $k \geq 2$ sibs and suppose we wish to identify the parental origin of the DNA inherited by each sib at a particular autosomal locus, $\mathscr{L}$ say (for loci on sex chromosomes, males and females need to be treated differently). Arbitrarily label the paternal chromosomes containing the locus of interest by $(1, 2)$, and similarly label the maternal chromosomes by $(3, 4)$. The *inheritance vector* of the sibship at the locus $\mathscr{L}$ is the $2k$-vector $x = (x_1, x_2, \ldots, x_{2k-1}, x_{2k})$, indicating the outcome of each of the $2k$ segregations giving rise to the sibship. More precisely, for $i = 1, \ldots, k$, x_{2i-1} is the label of the paternal chromosome from which sib i inherited DNA at $\mathscr{L}$, 1 or 2, and x_{2i} is the label of the maternal chromosome from which sib i inherited DNA at $\mathscr{L}$, 3 or 4 (see Figures 1 and 2). Denote by $\mathscr{X}$ the set of all 2^{2k} inheritance vectors.

Consider now two loci, $\mathscr{L}_1$ and $\mathscr{L}_2$, separated by a recombination fraction θ, and denote the inheritance vectors at the two loci by x and y, respectively. If these two inheritance vectors differ at a particular entry, this indicates the occurrence of a recombination between $\mathscr{L}_1$ and $\mathscr{L}_2$ in the corresponding segregation. The chance of a recombination between the two loci is the recombination fraction θ, taken to be constant across conditions and individuals. the transition matrix $R(\theta)$ between inheritance vectors at loci separated by a recombination fraction θ has entries

$$r_{xy}(\theta) = \theta^{\Delta(x,y)}(1-\theta)^{2k-\Delta(x,y)}, \tag{2.1}$$

where $\Delta(x, y)$ is the number of coordinates at which the inheritance vectors x and y differ, that is, the number of recombination events between the two loci. The matrix $R(\theta)$ may be expressed as the Kronecker power of 2×2 transition matrices corresponding to transitions in each of the $2k$ coordinates,

$$R(\theta) = \begin{bmatrix} 1-\theta & \theta \\ \theta & 1-\theta \end{bmatrix}^{\otimes 2k}. \tag{2.2}$$

The notion of inheritance vector generalizes to arbitrary pedigrees, where pedigree members are separated into founders (individuals whose parents are not in the pedigree) and nonfounders (individuals whose parents are in the pedigree). In the case of sibships, the parents are founders and the sibs are nonfounders. For a pedigree with k nonfounders, the inheritance vector at a particular locus is defined to be a $2k$-vector with coordinates describing the outcome of the paternal and maternal segregations giving rise to the k nonfounders [Lander and Green (1987); Kruglyak and Lander (1995) and Kruglyak, Daly, Reeve-Daly and Lander (1996)]. The $(2i-1)$st coordinate is 0 or 1 according to whether the grand-paternal or grand-maternal DNA was transmitted in the paternal segregation giving rise to the ith nonfounder. The $(2i)$th coordinate contains the same information for the maternal segregation. For the purpose of this paper, we prefer the definition introduced earlier for sibship inheritance vectors (with labels $1, 2, 3, 4$ for parental chromosomes) to the more common definition with binary labels, since this facilitates the presentation in Section 3 of group action on inheritance vectors. It is easy to show that for general pedigrees the transition matrix $R(\theta)$ between inheritance vectors also has the form given in (2.2). Although this paper is primarily concerned with IBD data from sibships, the general setup and some of the results presented here (Propositions 1, 2 and 3) apply to arbitrary pedigree types, and we discuss generalizations where appropriate.

For the purpose of linkage analysis of disease genes, certain inheritance vectors are equivalent to each other, in that they have the same probability of arising at DS genes in pedigrees with given phenotypes. Although not needed for an understanding of this paper, a discussion of these probabilities for sibships and the genetic model under which they are calculated may be found in

Dudoit and Speed (1999) and Dudoit (1999). For an arbitrary pedigree type, we define *IBD configurations* as any partitioning of the set of inheritance vectors. For example, for k affected sibs, Ethier and Hodge (1985) partition the 2^{2k} inheritance vectors into a much smaller number of equivalence classes as follows. Two inheritance vectors belong to the same IBD configuration if one may be obtained from the other by applying any combination of the following four operations: (1) interchange the paternal labels 1 and 2, (2) interchange the maternal labels 3 and 4, (3) interchange the parental origin of the DNA by interchanging 1 and 3 and 2 and 4, and (4) permute the sibs. With this definition, the 16 possible inheritance vectors for a sib-pair are partitioned into three IBD configurations, corresponding to the number of chromosomes sharing DNA IBD at the locus of interest (cf. Table 1). In general, inheritance vectors may be partitioned in various ways for different purposes, and we address this question in greater detail for sibships in Section 3.

For a pedigree with given phenotypes, the conditional probability vector of IBD configurations at a genetic marker $\mathscr{M}$ linked to a DS locus $\mathscr{D}$ at recombination fraction θ is given by

$$\rho(\theta, \pi)_{1\times m} = \pi_{1\times m} T(\theta)_{m\times m},$$

where π is the conditional probability vector of IBD configurations at the DS locus (possibly one of several unlinked DS loci), m is the number of IBD configurations and $T(\theta)$ is the transition matrix between IBD configurations θ apart. In general, π depends on unknown and numerous genetic parameters, such as penetrances and genotype frequencies. In this paper, we consider a general genetic model with multiple genes unlinked to each other, arbitrary penetrances, and no population genetic assumptions such as random mating or Hardy–Weinberg equilibrium. Under the null hypothesis of no linkage between the marker and the DS locus, the IBD distribution at the marker is

$$\alpha = \rho\left(\frac{1}{2}, \pi\right) = \frac{1}{2^{2k}}(|\mathscr{C}_1|, \ldots, |\mathscr{C}_m|),$$

where $|\mathscr{C}_i|$ is the number of inheritance vectors in $\mathscr{C}_i$, the ith IBD configuration.

Thus, the IBD probabilities at the marker have two separate components: one component involving the recombination fraction θ between the marker and the DS locus (*scale*), the other depending on the mode of inheritance of

TABLE 1
Sib-pair IBD configurations

Orbits of $S_2 \times D_4$	**Orbits of $S_2 \times (C_2 \times C_2)$**	**Inheritance vectors**
0 IBD	0 IBD	(1, 3, 2, 4), (1, 4, 2, 3), (2, 3, 1, 4), (2, 4, 1, 3)
1 IBD	1 paternal IBD	(1, 3, 1, 4), (1, 4, 1, 3), (2, 3, 2, 4), (2, 4, 2, 3)
	1 maternal IBD	(1, 3, 2, 3), (1, 4, 2, 4), (2, 3, 1, 3), (2, 4, 1, 4)
2 IBD	2 IBD	(1, 3, 1, 3), (1, 4, 1, 4), (2, 3, 2, 3), (2, 4, 2, 4)

the disease (*specificity*). Our score test in the recombination fraction θ focuses on the scale component and seems to achieve some robustness with respect to the specificity (π). Examples of the transition matrix $T(\theta)$ are given in Section 5 for sib-pairs and sib-trios.

Suppose we collect data on n sibships of given size and phenotypes and wish to test the null hypothesis of no linkage between a genetic marker and a DS locus. Denote by N_i the number of sibships with IBD configuration i, $i = 1, \dots, m$, at the genetic marker. Under certain sampling assumptions [Dudoit (1990)], $(N_1, \dots, N_m)$ have a Multinomial$(n, \rho(\theta, \pi))$ distribution. There is no uniformly most powerful test of H_0: $\theta = \frac{1}{2}$, however, the *score test* is *locally most powerful*. Moreover, for affected-only sibships of a given size, the score statistic does not involve the nuisance parameter π and reduces to a widely used statistic in linkage analysis, S_{pairs}, which is obtained by forming all possible pairs of sibs and averaging the proportions of chromosomes on which they share DNA IBD at the marker [Whittemore and Halpern (1994a) and Kruglyak, Daly, Reeve-Daly and Lander (1996)]. This result is a corollary to Theorem 2 in Section 4.

COROLLARY 1. *For affected sib-k-tuples, using the IBD configurations of Ethier and Hodge, the score test of* H_0: $\theta = \frac{1}{2}$ *is based on* S_{pairs}, *regardless of the model for disease susceptibility, that is, regardless of* π. *For one affected sib-k-tuple,*

$$S_{\text{pairs}} = \frac{\sum_{i<j} S_{ij}}{k(k-1)},$$

where S_{ij} *is the number of chromosomes on which the ijth sib-pair shares DNA IBD. Under the null hypothesis of no linkage, the* S_{ij}*'s are pairwise independent* Binomial$(2, \frac{1}{2})$ *random variables, and thus*

$$E_0[S_{\text{pairs}}] = \frac{1}{2}, \qquad \text{Var}_0[S_{\text{pairs}}] = \frac{1}{4k(k-1)}.$$

For a collection of affected sib-k-tuples, S_{pairs} *is summed over all sibships.*

Thus, for affected-only sibships of a *given* size, S_{pairs} is locally most powerful (in θ), and may be calculated easily by considering each sib-pair one at a time and without the need for assigning sibships to IBD configurations. This finding extends the work of Knapp, Seuchter and Baur (1994a) to sibships of any size and to general genetic models with multiple unlinked DS loci and no population genetic assumptions such as random mating or Hardy–Weinberg equilibrium. Unfortunately, this simple property does not hold with all types of sibships, and we consider examples where it fails in Sections 4 and 5. Furthermore, we show that the linkage score statistic combining IBD data from sibships of *different* sizes assigns different weights to the various sibship types and these weights depend on the genetic model.

The remainder of this paper is concerned with the proof of Corollary 1 and with deriving score statistics for general sibships, with any number of affected and unaffected sibs, and distinguishing the parental origin of the DNA. In general, the form of the score statistic is based on properties of the transition matrix $T(\theta)$, which in turn are determined by the pedigree type and the choice of IBD configurations. Thus, we first describe how sibship inheritance patterns may be summarized by IBD configurations which are orbits of groups acting on the set of inheritance vectors.

3. Transition matrix for IBD configurations.

3.1. *Sibship IBD configurations.* Let $a = (1, 3)$, $b = (1, 4)$, $c = (2, 3)$ and $d = (2, 4)$ denote all four possible segregation outcomes at a particular locus for a given sib. Then we may think of the set of inheritance vectors $\mathscr{X}$ as the set of mappings $x\colon \{1, \dots, k\} \to \{a, b, c, d\}$. In this setting, the IBD configurations are *orbits* of groups acting on $\mathscr{X}$, where the groups are determined by the type of operations allowed within IBD configurations [cf. Fraleigh (1989), Section 3.2 for an introduction to group action]. Let

$$\begin{aligned} \alpha &= (ac)(bd) && \text{interchange labels 1 and 2 of paternal chromosomes,} \\ \beta &= (ab)(cd) && \text{interchange labels 3 and 4 of maternal chromosomes,} \\ \gamma &= (bc) && \text{interchange parental origin of DNA.} \end{aligned}$$

The group of permutations generated by α, β and γ is actually the *dihedral group*, D_4 (α and γ are sufficient to generate D_4), and the group generated by α and β is the *Klein four-group*, $C_2 \times C_2$. Figure 3 displays a square with vertices a, b, c, and d. Permutations α and β correspond to mirror images in the perpendicular bisectors of the sides, and permutations γ corresponds to a diagonal flip. The diherdral group D_4 is the group of symmetries of the square. The IBD configurations of Ethier and Hodge (1985) for affected-only sibships are the orbits of the direct product $S_k \times D_4$, of the *symmetric group* S_k on k letters and the dihedral group D_4, acting on $\mathscr{X}$. In some situations (e.g., parental

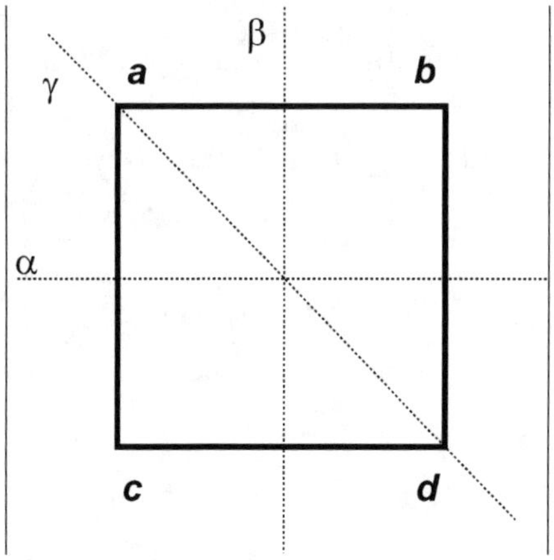

FIG. 3. *Permutations α, β and γ of the vertices of the square.*

imprinting, when disease susceptibility is different for maternally and paternally inherited disease alleles), it may be appropriate to distinguish between sharing of maternal and paternal DNA and exclude the group operation γ. For example, for a sib-pair, it may be appropriate to distinguish between the two inheritance vectors $(1, 3, 1, 4)$ and $(1, 3, 2, 3)$; for $(1, 3, 1, 4)$ the sibs share DNA IBD on the paternal chromosome, whereas for $(1, 3, 2, 3)$ the sibs share DNA IBD on the maternal chromosome (cf. Table 1). For sibships with both affected and unaffected individuals, similar configurations may be defined, but the sibs are permuted only among affecteds or unaffecteds. The different types of group action are listed in Table 2.

For a group $G \times H$ acting on the set $\mathscr{X}$, let m denote the number of orbits and $\mathscr{C}_i$ denote the ith orbit. In general, we may use the *Pólya theory of counting* [van Lint and Wilson (1992) and deBruijn (1964)] to find the number of orbits of groups acting on mappings, and hence determine the number of IBD configurations of each type (see Appendix, Section A). Ethier and Hodge derived the number of IBD configurations of affected sib-k-tuples, as well as the number of inheritance vectors in each IBD configuration, without reference to the group $S_k \times D_4$. Instead, they based their calculations on labels for the equivalence classes which are triples of integers [cf. pages 264, 265 in Ethier and Hodge (1985) and Appendix, Section A].

3.2. *Properties of transition matrix.* In this section, we derive properties of transition matrices between IBD configurations. While Theorems 1 and 2 are specific to the sibship IBD configurations defined in the previous section, Propositions 1 and 2 hold more generally for any type of pedigree, with IBD configurations defined as orbits of groups acting on the set of inheritance vectors. The transition matrix $T(\theta)$ between IBD configurations at loci separated by a recombination fraction θ is the $m \times m$ matrix with entries

$$t_{ij}(\theta) = \frac{1}{|\mathscr{C}_i|} \sum_{x \in \mathscr{C}_i} \sum_{y \in \mathscr{C}_j} r_{xy}(\theta) = \frac{1}{|\mathscr{C}_i|} \sum_{x \in \mathscr{C}_i} \sum_{y \in \mathscr{C}_j} \theta^{\Delta(x,y)}(1-\theta)^{2k-\Delta(x,y)}.$$

However, given any two inheritance vectors in $\mathscr{C}_i$, the probability of a transition to $\mathscr{C}_j$ is the same, that is,

$$\sum_{y \in \mathscr{C}_j} r_{xy}(\theta) = \sum_{y \in \mathscr{C}_j} r_{\tilde{x}y}(\theta) \quad \text{for any } x,\ \tilde{x} \in \mathscr{C}_i. \tag{3.1}$$

TABLE 2
Sibship IBD configurations

# affected, # unaffected	Distinguish maternal from paternal sharing	Group $G \times H$
$k, 0$	NO	$S_k \times D_4$
	YES	$S_k \times (C_2 \times C_2)$
$h, k-h$	NO	$(S_h \times S_{k-h}) \times D_4$
	YES	$(S_h \times S_{k-h}) \times (C_2 \times C_2)$

This result follows by observing that if $\sim$ denotes the operation applied to x to obtain $\tilde{x}$, then $\Delta(\tilde{x}, y) = \Delta(\tilde{\tilde{x}}, \tilde{y}) = \Delta(x, \tilde{y})$ and $\tilde{y} \in \mathscr{C}_j$. Consequently,

$$(3.2)\qquad \begin{aligned} t_{ij}(\theta) &= \sum_{y \in \mathscr{C}_j} \theta^{\Delta(x, y)}(1-\theta)^{2k-\Delta(x, y)} \quad \text{where } x \text{ is any } x \in \mathscr{C}_i \\ &= \frac{|\mathscr{C}_j|}{|\mathscr{C}_i|} \sum_{x \in \mathscr{C}_i} \theta^{\Delta(x, y)}(1-\theta)^{2k-\Delta(x, y)} \quad \text{where } y \text{ is any } y \in \mathscr{C}_j. \end{aligned}$$

The next two propositions relate the transition matrix $T(\theta)$ to the adjacency matrix of a quotient graph, whose eigenvalues are key in determining the form of the score statistic (see Appendix, Section B for proofs).

PROPOSITION 1. *Let $T(\theta)$ denote the transition matrix between IBD configurations which are orbits of groups acting on the set of inheritance vectors $\mathscr{X}$, and let $\theta_1 * \theta_2 = \theta_1(1-\theta_2) + \theta_2(1-\theta_1)$. Then, $T(\theta)$ satisfies the semigroup property*

$$T(\theta_1 * \theta_2) = T(\theta_1)T(\theta_2).$$

Thus, $T(\theta)$ may be written as

$$T(\theta) = e^{d(\theta)Q},$$

where $d(\theta) = -\frac{1}{2}\ln(1-2\theta)$ is the inverse of the Haldane map function and Q is the infinitesimal generator. The infinitesimal generator is given by

$$Q = B - 2kI,$$

where B is the $m \times m$ matrix with entries

$$b_{ij} = \sum_{y \in \mathscr{C}_j} I(\Delta(x, y) = 1) \quad \text{for any } x \in \mathscr{C}_i$$

$\Delta(x, y)$ is the number of coordinates at which x and y differ and $I(\ )$ denotes the indicator function. The stationary distribution of $T(\theta)$ is

$$\alpha = (\alpha_1, \ldots, \alpha_m) = \frac{1}{2^{2k}}(|\mathscr{C}_1|, \ldots, |\mathscr{C}_m|)$$

and $T(\theta)$ is reversible, that is,

$$\alpha_i t_{ij}(\theta) = \alpha_j t_{ji}(\theta).$$

Hence, for one segregation, the crossover process is embeddable in a continuous-time random walk on $\{0, 1\}$, where 0 and 1 denote, respectively, the transmission of paternal and maternal DNA to one's child, and the time parameter is $d(\theta) = -\frac{1}{2}\ln(1-2\theta)$. Jointly, the crossover processes are i.i.d. and hence embeddable in a continuous-time random walk on the vertices of the hypercube $\{0, 1\}^{2k}$ (cf. Donnelly (1983) and Figures 1 and 2). The random walk model for the crossover process is widely used and is referred to in the genetics literature as the *no interference model*. The *Haldane map function* relates the recombination fraction to the map distance under the no interference model. The *map distance* between two loci is the expected number of crossover events, that is, of changes of the grand-parental origin of the DNA, occurring on a segregation product in the chromosomal interval between the two loci. Under the no interference model, the crossover process on individual segregation products is a Poisson process with intensity 1 [cf. Ott (1991), McPeek (1996) and Speed (1996) for an introduction to map functions and a more detailed discussion of crossover processes]. Note that, if we have three ordered loci, and θ_1 and θ_2 are the recombination fractions between the first and second and second and third locus, respectively, then $\theta_1 * \theta_2$ is the recombination fraction between the first and third locus, under the assumption that recombination events in disjoint intervals are independent, that is, there is no crossover interference. Also, note that we did not need to assume no crossover interference to derive the semigroup property. If, however, we do assume no crossover interference, then the inheritance vectors along a chromosome form a continuous-time Markov chain with time parameter the genetic distance along a chromosome. From (3.1) and condition (15), page 63 in Rosenblatt (1974), it follows that the IBD configurations also form a continuous-time Markov chain.

In order to compute score statistics, we need derivatives of the transition matrix at $\theta = \frac{1}{2}$. These may be computed by differentiating (3.2); however, we gain more knowledge on the transition matrix $T(\theta)$ and on the form of the score statistic by using the following spectral decomposition of $T(\theta)$.

PROPOSITION 2. *Let $T(\theta)$ denote the transition matrix between IBD configurations which are orbits of groups acting on the set of inheritance vectors $\mathscr{X}$. Then $T(\theta)$ may be written as*

$$T(\theta) = \sum_h \exp(\lambda_h d(\theta))\, P_h = \sum_h (1-2\theta)^{-\lambda_h/2}\, P_h, \tag{3.3}$$

where λ_h are the m real eigenvalues of the infinitesimal generator Q, and P_h are projection matrices satisfying $P_h^2 = P_h = P_h^*$, $P_h P_l = 0$, $h \neq l$, and $\sum_h P_h = I$. Here P_h^* is the adjoint of P_h with respect to the inner product $\langle x, y\rangle_\alpha = \sum_i \alpha_i x_i y_i$. The ijth entry of P_h is $\alpha_j v_{ih} v_{jh}$, where v_{ih} is the ith entry of the right eigenvector $\mathbf{v}_h$ of Q corresponding to λ_h, and the eigenvectors $\mathbf{v}_h$ are orthonormal with respect to the inner product $\langle\ ,\ \rangle_\alpha$. In particular, the first two derivatives of the transition matrix with respect to θ are

$$T'(\theta) = \sum_h \lambda_h (1-2\theta)^{-(\lambda_h+2)/2} P_h \tag{3.4}$$

and

$$T''(\theta) = \sum_h \lambda_h(\lambda_h+2)(1-2\theta)^{-(\lambda_h+4)/2} P_h. \tag{3.5}$$

A SCORE TEST FOR LINKAGE

Thus, eigenvalues of Q and their multiplicity give us information regarding the derivatives of the transition matrix $T(\theta)$ and hence, the form of the score statistic. In particular, powers of θ in $T(\theta)$ are determined by the eigenvalues of Q, and the first nonzero derivative of $T(\theta)$ at $\theta = \frac{1}{2}$ and its rank are determined by the second largest eigenvalue of Q and its multiplicity. We will relate Q to the adjacency matrix of a quotient graph in order to derive its eigenvalues. Consider the graph $\mathscr{X}$ with vertex set the set of all inheritance vectors of length $2k$ and adjacency matrix $A(\mathscr{X}) = A$ with (x, y)-entry,

$$a_{xy} = \begin{cases} 1, & \text{if } \Delta(x, y) = 1, \\ 0, & \text{otherwise.} \end{cases}$$

Here $\mathscr{X}$ is the graph defined by the first associates in the Hamming scheme $H(2k, 2)$ [Chapter 30 in van Lint and Wilson (1992)]. Consider any of the four groups $G \times H$ described in Table 2. The matrix B, defined in Proposition 1, is the *adjacency matrix of the quotient graph* $\mathscr{X}/(G \times H)$, which is the multidigraph with the orbits of $G \times H$ as its vertices and with b_{ij} arcs going from $\mathscr{C}_i$ to $\mathscr{C}_j$. Recall that $Q = B - 2kI$; consequently, we may work with B to derive the eigenvalues of Q. The following theorem relies on general facts concerning eigenvectors and eigenvalues of adjacency matrices of quotient graphs, as well as specific facts regarding the behavior of eigenvectors of A on the orbits of the four groups $G \times H$ described in Section 3.1 (see Appendix, Section C for proof).

THEOREM 1 (Eigenvalues of infinitesimal generator Q for sibship IBD configurations). *The largest eigenvalue of Q is 0, with multiplicity one, and the second largest eigenvalue of Q is -4, with multiplicity depending on the group $G \times H$ defining the IBD configurations.*

(a) $S_k \times D_4$: *-4 has multiplicity one;*
(b) $S_k \times (C_2 \times C_2)$: *$-4$ has multiplicity two;*

and for $k \geq 3$

(c) $(S_h \times S_{k-h}) \times D_4$: *$-4$ has multiplicity two if $h = 1$ or $h = k - 1$, and three if $2 \leq h \leq k - 2$;*
(d) $(S_h \times S_{k-h}) \times (C_2 \times C_2)$: *$-4$ has multiplicity four if $h = 1$ or $h = k - 1$, and six if $2 \leq h \leq k - 2$.*

Furthermore, all other eigenvalues of Q belong to the set $\{-2i_{\binom{2k}{i}}: i = 3, \ldots, 2k\}$, where the subscript $\binom{2k}{i}$ is the largest possible multiplicity of the eigenvalue $-2i$. Thus, from (3.4) *and* (3.5),

$$T'\left(\tfrac{1}{2}\right) = 0 \tag{3.6}$$

and

$$U = T''\left(\tfrac{1}{2}\right) = 8P_{-4}, \tag{3.7}$$

where P_{-4} is the projection matrix for the second largest eigenvalue, -4, with rank the multiplicity of -4. In general, the ijth entry of P_{-4} is $\alpha_j \sum v_i v_j$, *where the v's are the right orthonormal (with respect to the inner product* $\langle\ ,\ \rangle_\alpha$*) eigenvectors of Q corresponding to −4, and the sum is over all such eigenvectors.*

Note that we may also show that $T'(\frac{1}{2}) = 0$ by simple algebra, but this approach does not yield any particular insight into other general properties of $T(\theta)$.

The projection matrix for the largest eigenvalue of Q, $\lambda_1 = 0$, is $T(\frac{1}{2})$, the matrix whose rows are equal to the stationary distribution α. From Proposition 2 and Theorem 1, transition matrices for sibship IBD configurations have the form

$$T(\theta) = T(\tfrac{1}{2}) + (1-2\theta)^2 P_{-4} + o((1-2\theta)^2),$$

and the rate of convergence of $T(\theta)$ to $T(\frac{1}{2})$ as $\theta \to \frac{1}{2}$ is $O((1-2\theta)^2)$. Under the no interference model, the rate of convergence in terms of the map distance $d = -\ln(1-2\theta)/2$ is $O(e^{-4d})$.

More generally, since the matrix $R(\theta)$ has the same form for any type of pedigree, Propositions 1, 2 and 3 apply to arbitrary pedigrees, as long as the IBD configurations are defined as orbits of groups. Thus, for $2k$ segregations, the transition matrix for IBD configurations has the general form

$$T(\theta) = \sum_h (1-2\theta)^{-\lambda_h/2} P_h,$$

where the eigenvalues λ_h of Q belong to the set $\{-2i_{\binom{2k}{i}}: i = 0, \ldots, 2k\}$. The ith derivative, $i = 0, \ldots, 2k$, of $T(\theta)$ is given by

$$T^{(i)}(\theta) = \sum_h \left\{ \prod_{j=0}^{i-1} (\lambda_h + 2j) \right\} (1-2\theta)^{-(\lambda_h+2i)/2} P_h.$$

The first nonzero derivative of $T(\theta)$ at $\theta = \frac{1}{2}$ and its rank are determined by the second largest eigenvalue of Q, λ_2, and its multiplicity. If $\lambda_2 = -2i$, the first nonzero derivative is the ith derivative,

$$T^{(i)}(\tfrac{1}{2}) = (-2)^i i! P_{-2i},$$

and the rank of this ith derivative is the multiplicity of $\lambda_2 = -2i$. Furthermore, the rate of convergence to $T(\frac{1}{2})$ is determined by the second largest eigenvalue,

$$T(\theta) = T(\tfrac{1}{2}) + (1-2\theta)^{-\lambda_2/2} P_{\lambda_2} + o((1-2\theta)^{-\lambda_2/2}).$$

The smaller the second largest eigenvalue, the faster the convergence to $T(\frac{1}{2})$. We proved that for sibships, the second largest eigenvalue is −4, but this is not the case for all types of relatives. In turns out that for grandparent/grandchild pairs, who can share DNA IBD on either 0 or 1 chromosome, the second largest eigenvalue is −2 [Dudoit (1999)]. It is noteworthy that for some types of relative pairs (e.g., cousin pairs), the usual IBD configurations (0 or 1 IBD) are not orbits of groups and the transition matrix for these usual configurations does not satisfy the semigroup property [Donnelly (1983) and Dudoit (1999)].

In the next section, we will explore the implications of Theorem 1 on score tests of the null hypothesis of no linkage between a marker and a gene using IBD data from sibships.

4. Linkage score test for sibships. Suppose we have data on n sibships of a given type (e.g., affected sib-k-tuples with orbits of $S_k \times D_4$), in the form of multinomial counts N_i, $i = 1, \dots, m$, for the number of sibships with IBD configuration i at a marker $\mathscr{M}$. We wish to test the null hypothesis of no linkage between the marker $\mathscr{M}$ and a DS locus $\mathscr{D}$, which could be one of several DS loci unlinked to each other; that is, we wish to test

$$\mathrm{H}_0\colon \theta = \tfrac{1}{2} \text{ (no linkage)} \quad \text{versus} \quad \mathrm{H}_1\colon 0 \le \theta < \tfrac{1}{2} \text{ (linkage)},$$

where θ denotes the recombination fraction between $\mathscr{M}$ and $\mathscr{D}$. Note that $\theta = \frac{1}{2}$ and $\pi = \alpha$ are not identifiable.

The log-likelihood of the IBD data, conditional on the phenotypes, is

$$l(\theta, \pi) = \sum_i N_i \ln(\rho_i(\theta, \pi)),$$

where

$$\rho(\theta, \pi)_{1\times m} = \pi_{1\times m} T(\theta)_{m\times m}.$$

The *score test* is based on the first nonzero derivative in the Taylor series expansion of the log-likelihood about $\theta = \frac{1}{2}$. In our problem, the first derivative vanishes, so we turn to the second derivative of the log-likelihood with respect to θ, which yields a test that maximizes the second derivative of the power function at the null. We find the score statistic for the given sibship type to be

$$S = \left.\frac{\partial^2 l(\theta, \pi)}{\partial \theta^2}\right|_{\theta=1/2} = \sum_{i=1}^{m} N_i \left.\frac{\partial^2 \rho_i(\theta, \pi)/\partial\theta^2}{\rho_i(\theta, \pi)}\right|_{\theta=1/2} = \sum_{i=1}^{m} N_i \frac{\sum_{j=1}^{m} \pi_j u_{ji}}{\alpha_i},$$

where $U = T''(\frac{1}{2}) = 8P_{-4} = (8\alpha_j \sum v_i v_j)$, the v's are right orthonormal (with respect to $\langle\ ,\ \rangle_\alpha$) eigenvectors of Q corresponding to the eigenvalue -4, and the sum in U is over all such eigenvectors. The null hypothesis of no linkage is rejected for large values of the score statistic S. We show next that for affected sib-k-tuples, with the orbits of $S_k \times D_4$, the second largest eigenvalue has multiplicity one, and as a result, the score statistic for affected-only sibships of a given size is independent of the nuisance parameter π.

4.1. *Affected sib-k-tuples, orbits of $S_k \times D_4$.* A very widely used statistic in linkage analysis is S_{pairs} [cf. Kruglyak, Daly, Reeve-Daly and Lander (1996), S_P of Whittemore and Halpern (1994a), and *PAIRS* and *WP* of Suarez and van Eerdewegh (1984)]. For a sibship of size k, S_{pairs} is obtained by forming all possible pairs of sibs and averaging the proportions of chromosomes on which they share DNA IBD at the locus of interest; that is,

$$S_{\text{pairs}} = \frac{\sum_{i<j} S_{ij}}{k(k-1)},$$

where S_{ij} is the number of chromosomes on which the ijth sib-pair shares DNA IBD. The corollary in Section 2 results from the following theorem.

THEOREM 2 (Affected sib-k-tuple score statistic, orbits of $S_k \times D_4$). *Suppose we have IBD data on n affected sib-k-tuples, with IBD configurations defined as the orbits of $S_k \times D_4$. Then the score test of the null hypothesis of no linkage between a marker and a DS locus, H_0: $\theta = \frac{1}{2}$, is based on S_{pairs}. The contribution of n affected sib-k-tuples to the overall score statistic is*

$$\begin{aligned} S &= 8\Big(\sum_{j=1}^{m} v_j \pi_j\Big)\Big(\sum_{i=1}^{m} v_i N_i\Big) \\ &= 2^{2k-2}\Big(\sum_{j=1}^{m} u_{j1}\pi_j\Big)(2S_{\text{pairs}} - n) \qquad (4.1) \\ &= 8\sqrt{k(k-1)}\Big(\sum_{j=1}^{m} v_j\pi_j\Big)(2S_{\text{pairs}} - n), \end{aligned}$$

where u_{i1} is the ith entry of the first column of $U = T''(\frac{1}{2})$, v_i is the ith entry of the right eigenvector of Q corresponding to the eigenvalue -4 and $u_{i1} = 2^{5-2k}\sqrt{k(k-1)}v_i$. For n sibships, S_{pairs} is summed over all sibships.

The proof of Theorem 2 may be found in the Appendix, Section D, and relies on Theorem 1 and the following identity. For a sibship with inheritance vector x,

$$\begin{aligned} S_{\text{pairs}} &= \frac{\sum_{i=1}^{4} a_i(x)(a_i(x)-1)}{2k(k-1)} \qquad (4.2) \\ &= \frac{a_1(x)^2 + a_2(x)^2 + a_3(x)^2 + a_4(x)^2 - 2k}{2k(k-1)}, \end{aligned}$$

where $a_i(x)$ is the number of i labels in the inheritance vector x of the sibship, $i = 1, 2, 3, 4$, and $a_1(x) + a_2(x) + a_3(x) + a_4(x) = 2k$. Without loss of generality, we let the first IBD configuration be the one for which all sibs inherited the same maternal and paternal DNA, that is, with representative inheritance vector $(1, 3, 1, 3, \ldots, 1, 3)$ and label $(0, 0, 0)$ in the notation of Ethier and Hodge (1985). The entries of the first column of U are easily computed, as seen in the proof.

Thus, for affected sib-k-tuples and without distinguishing between sharing of maternal and paternal DNA, the score test is based on S_{pairs}, regardless of the genetic model, and may be calculated easily by considering each sib-pair one at a time and without the need for assigning sibships to IBD configurations. For commonly studied genetic models, the affected sib-pair "possible triangle" constraints hold [Dudoit and Speed (1999)], and as a result, $\sum_j v_j \pi_j \geq 0$. This follows by noting that $n \sum_j v_j \pi_j$ is the expected value of $\sum_j v_j N_j = \sqrt{k(k-1)}(2S_{\text{pairs}} - n)$ when $\theta = 0$. Also, for each sibship, S_{pairs} is an average of sib-pair statistics, each with expected value $\pi_2 + \pi_1/2 \geq 1/2$, where in this expression π_i, $i = 0, 1, 2$, is the probability that an affected sib-pair shares DNA IBD on i chromosomes at the DS locus. For IBD data from sibships of a *given* size, it is thus appropriate to reject the null hypothesis of no linkage for large values of S_{pairs}. However, the score statistic for combining IBD data from sibships of *different* sizes involves weights which do depend on the genetic model ($\sum_i v_i \pi_i$).

4.2. *Affected sib-k-tuples, orbits of* $S_k \times (C_2 \times C_2)$. For affected sib-$k$-tuples and distinguishing between sharing of maternal and paternal DNA, the second largest eigenvalue of the infinitesimal generator Q, -4, has multiplicity two (Theorem 1). Hence, the second derivative of the transition matrix at $\theta = \frac{1}{2}$ has rank 2 and entries

$$u_{ij} = 8\alpha_j(v_i v_j + \tilde{v}_i \tilde{v}_j),$$

where $v = (v_1, \ldots, v_m)^T$ and $\tilde{v} = (\tilde{v}_1, \ldots, \tilde{v}_m)^T$ are the right orthonormal (with respect to the inner product $\langle\ ,\ \rangle_\alpha$) eigenvectors of Q corresponding to the second largest eigenvalue. These eigenvectors are based on V_e and V_o, respectively (see Appendix, Section C). The score statistic is given by

$$\begin{aligned} S &= \sum_{i=1}^{m} N_i \frac{\sum_{j=1}^{m} \pi_j 8\alpha_i(v_i v_j + \tilde{v}_i \tilde{v}_j)}{\alpha_i} \\ &= 8\Bigl(\sum_{j=1}^{m} v_j \pi_j\Bigr)\Bigl(\sum_{i=1}^{m} v_i N_i\Bigr) + 8\Bigl(\sum_{j=1}^{m} \tilde{v}_j \pi_j\Bigr)\Bigl(\sum_{i=1}^{m} \tilde{v}_i N_i\Bigr). \end{aligned}$$

Thus, in general, the score test depends on the parameters of the genetic model through π. However, in some situations (e.g., no imprinting), this score statistic reduces to S_{pairs}.

4.3. *Discordant sib-k-tuples.* For sibships of size at least 3, with both affected and unaffected individuals [orbits of $(S_h \times S_{k-h}) \times D_4$ or of $(S_h \times S_{k-h}) \times (C_2 \times C_2)$], the second largest eigenvalue of the infinitesimal generator has multiplicity at least two (Theorem 1). Hence, in general, the score statistic depends on the genetic model and is a sum of terms of the form $(\sum_{j=1}^{m} v_j \pi_j)(\sum_{i=1}^{m} v_i N_i)$, where the v's are the right orthonormal eigenvectors of Q corresponding to the second largest eigenvalue.

In the next section, we consider the examples of sib-pairs and sib-trios and present the transition matrix $T(\theta)$, the infinitesimal generator Q and the score statistic for these sibship types.

5. Examples.

5.1. *Sib-pairs, orbits of $S_2 \times D_4$.* For sib-pairs with either 0, 1 or 2 affected individuals and without distinguishing between sharing of maternal and paternal DNA, there are three distinct IBD configurations, labeled 0, 1, 2, according to the number of chromosomes sharing DNA IBD at the locus of interest. The transition matrix is

$$T(\theta) = \begin{bmatrix} \psi^2 & 2\psi\bar{\psi} & \bar{\psi}^2 \\ \psi\bar{\psi} & \psi^2 + \bar{\psi}^2 & \psi\bar{\psi} \\ \bar{\psi}^2 & 2\psi\bar{\psi} & \psi^2 \end{bmatrix},$$

where $\psi = \theta^2 + (1-\theta)^2$ and $\bar{\psi} = 1 - \psi$. Figure 4 is a barycentric representation of curves $\mathscr{C}_\pi(\theta) = \{\rho(\theta, \pi) = \pi T(\theta) \colon 0 \le \theta \le \frac{1}{2}\}$ for the sib-pair transition matrix $T(\theta)$ and for $\pi = (\pi_0, \pi_1, \pi_2)$ on the boundaries of the simplex. The infinitesimal generator is

$$Q = \begin{bmatrix} -4 & 4 & 0 \\ 2 & -4 & 2 \\ 0 & 4 & -4 \end{bmatrix}.$$

Q has eigenvalues $\lambda = 0, -4$ and -8. The left and right eigenvectors of Q corresponding to $\lambda_2 = -4$ are $(1/2\sqrt{2}, 0, -1/2\sqrt{2})$ and $(\sqrt{2}, 0, -\sqrt{2})$, respectively

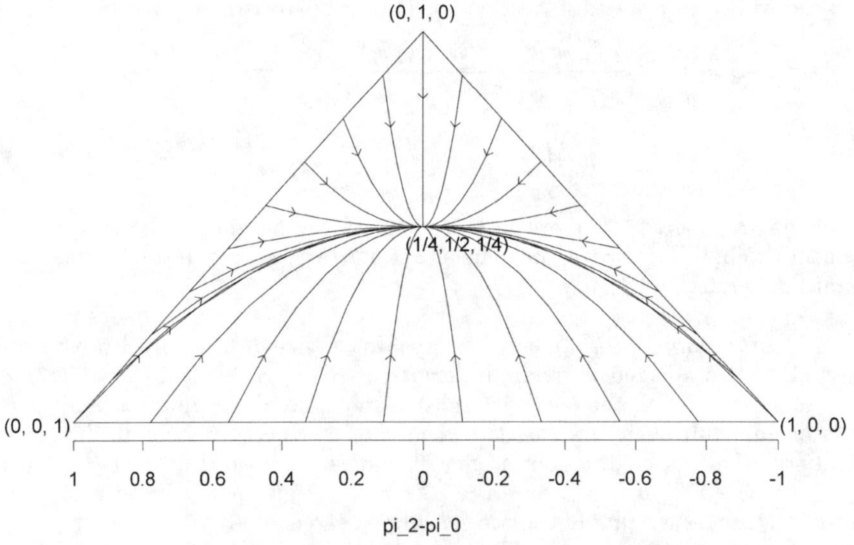

FIG. 4. *$S_2 \times D_4$-Barycentric representation of curves $\mathscr{C}_\pi(\theta)$, $0 \le \theta \le \frac{1}{2}$, for π on boundaries of simplex.*

(right eigenvector has unit norm with respect to the inner product $\langle\ ,\ \rangle_\alpha$). Hence

$$U = 8P_{-4} = 8\begin{bmatrix} \sqrt{2} \\ 0 \\ -\sqrt{2} \end{bmatrix}\begin{bmatrix} \frac{1}{2\sqrt{2}}, 0, -\frac{1}{2\sqrt{2}} \end{bmatrix} = \begin{bmatrix} 4 & 0 & -4 \\ 0 & 0 & 0 \\ -4 & 0 & 4 \end{bmatrix}.$$

If we let N_i denote the number of affected sib-pairs sharing DNA IBD on i chromosomes at the marker, $i = 0, 1, 2$, then the score statistic for affected sib-pairs is

$$16(\pi_2 - \pi_0)(N_2 - N_0),$$

similarly for discordant and unaffected sib-pairs. Note that $N_2 - N_0$ may be rewritten as $N_2 - (n - N_1 - N_2) = 2(N_2 + \frac{1}{2}N_1) - n$. Thus, the score test is based on $S_{\text{pairs}} = N_2 + \frac{1}{2}N_1$, also known as the *mean IBD statistic*. These findings extend the work of Knapp, Seuchter and Baur (1994a), who proved the local optimality of the mean IBD statistic for affected sib-pairs under a single DS locus model with random mating and Hardy–Weinberg equilibrium. The score statistic for combining IBD data from affected, discordant and unaffected sib-pairs is a linear combination of the mean IBD statistics for each type of sib-pair, with weights $\pi_2 - \pi_0$ depending on the genetic model.

Note that this setup also applies to testing linkage to quantitative trait loci. In this case, each sib-pair may have different continuous phenotypes, and the score statistic is given by

$$16\sum_i(\pi_{2i} - \pi_{0i})(N_{2i} - N_{0i}),$$

where $N_{ji} = 1$ if the ith sib-pair shares DNA IBD on $j = 0, 1, 2$ chromosomes at the marker and $N_{ji} = 0$ otherwise, and π_{ji} is the conditional probability that the ith sib-pair shares DNA IBD on j chromosomes at the trait locus given the phenotypes of the sibs [Dudoit (1999)].

5.2. *Sib-pairs, orbits of $S_2 \times (C_2 \times C_2)$.* For sib-pairs with any number of affecteds and distinguishing between sharing of maternal and paternal DNA, there are four distinct IBD configurations, conveniently labeled by the pair (i, j), $i, j = 0, 1$, where i and j denote the number of paternally and maternally inherited chromosomes sharing DNA IBD, respectively. The transition matrix is

$$T(\theta) = \begin{bmatrix} \psi^2 & \psi\bar{\psi} & \psi\bar{\psi} & \bar{\psi}^2 \\ \psi\bar{\psi} & \psi^2 & \bar{\psi}^2 & \psi\bar{\psi} \\ \psi\bar{\psi} & \bar{\psi}^2 & \psi^2 & \psi\bar{\psi} \\ \bar{\psi}^2 & \psi\bar{\psi} & \psi\bar{\psi} & \psi^2 \end{bmatrix}$$

and the infinitesimal generator is

$$Q = \begin{bmatrix} -4 & 2 & 2 & 0 \\ 2 & -4 & 0 & 2 \\ 2 & 0 & -4 & 2 \\ 0 & 2 & 2 & -4 \end{bmatrix}.$$

Q has eigenvalues $\lambda = 0, -4, -4$ and -8. The two orthonormal right eigenvectors corresponding to $\lambda_2 = -4$ are $(\sqrt{2}, 0, 0, -\sqrt{2})$ and $(0, \sqrt{2}, -\sqrt{2}, 0)$, hence

$$U = 8P_{-4} = 8\begin{bmatrix} \sqrt{2} \\ 0 \\ 0 \\ -\sqrt{2} \end{bmatrix}\left[\frac{1}{2\sqrt{2}}, 0, 0, -\frac{1}{2\sqrt{2}}\right] + 8\begin{bmatrix} 0 \\ \sqrt{2} \\ -\sqrt{2} \\ 0 \end{bmatrix}\left[0, \frac{1}{2\sqrt{2}}, -\frac{1}{2\sqrt{2}}, 0\right]$$

$$= \begin{bmatrix} 4 & 0 & 0 & -4 \\ 0 & 0 & 0 & 0 \\ 0 & 0 & 0 & 0 \\ -4 & 0 & 0 & 4 \end{bmatrix} + \begin{bmatrix} 0 & 0 & 0 & 0 \\ 0 & 4 & -4 & 0 \\ 0 & -4 & 4 & 0 \\ 0 & 0 & 0 & 0 \end{bmatrix}.$$

Let N_{ij} denote the number of affected sib-pairs sharing DNA IBD on i paternal and j maternal chromosome at the marker, $i, j = 0, 1$. The score statistic is given by

$$16(\pi_{11} - \pi_{00})(N_{11} - N_{00}) + 16(\pi_{10} - \pi_{01})(N_{10} - N_{01}),$$

and in general depends on the genetic model, similarly for discordant and unaffected sib-pairs. When $\pi_{10} = \pi_{01}$, the score test is based on $N_{11} - N_{00}$, that is, $N_2 - N_0$ in the more usual notation.

5.3. *Affected sib-trios, orbits of* $S_3 \times D_4$. For affected sib-trios (ASTs), there are four IBD configurations, with representative inheritance vectors and labels [defined as in Ethier and Hodge (1985)] listed in Table 3.

TABLE 3
IBD configurations for affected sib-trios

IBD configuration i	**Representative inheritance vector**	**Label**	$\|\mathscr{C}_i\|$
1	(1, 3, 1, 3, 1, 3)	(0, 0, 0)	4
2	(1, 3, 1, 3, 1, 4)	(0, 0, 1)	24
3	(1, 3, 1, 4, 2, 3)	(0, 1, 1)	24
4	(1, 3, 1, 3, 2, 4)	(1, 1, 1)	12

The transition matrix $T(\theta)$ is given by

$$\begin{bmatrix} (1-3\theta+3\theta^2)^2 & 6\theta\bar{\theta}(1-3\theta+3\theta^2) & 6\theta^2\bar{\theta}^2 & 3\theta^2\bar{\theta}^2 \\ \theta\bar{\theta}(1-3\theta+3\theta^2) & 1-4\theta+10\theta^2-12\theta^3+6\theta^4 & 2\theta\bar{\theta}(1-\theta+\theta^2) & \theta\bar{\theta}(1-\theta+\theta^2) \\ \theta^2\bar{\theta}^2 & 2\theta\bar{\theta}(1-\theta+\theta^2) & 1-4\theta+10\theta^2-12\theta^3+6\theta^4 & \theta\bar{\theta}(2-5\theta+5\theta^2) \\ \theta^2\bar{\theta}^2 & 2\theta\bar{\theta}(1-\theta+\theta^2) & 2\theta\bar{\theta}(2-5\theta+5\theta^2) & 1-6\theta+17\theta^2-22\theta^3+11\theta^4 \end{bmatrix}$$

and the infinitesimal generator is

$$Q = \begin{bmatrix} -6 & 6 & 0 & 0 \\ 1 & -4 & 2 & 1 \\ 0 & 2 & -4 & 2 \\ 0 & 2 & 4 & -6 \end{bmatrix}.$$

Q has eigenvalues $\lambda = 0, -4, -8, -8$, and the left and right eigenvectors corresponding to $\lambda_2 = -4$ are $\frac{1}{16}\sqrt{\frac{2}{3}}(3, 6, -6, -3)$ and $\sqrt{\frac{2}{3}}(3, 1, -1, -1)$, respectively. Hence

$$U = 8P_{-4} = 8\sqrt{\frac{2}{3}}\begin{bmatrix} 3 \\ 1 \\ -1 \\ -1 \end{bmatrix}\frac{1}{16}\sqrt{\frac{2}{3}}[3, 6, -6, -3] = \begin{bmatrix} 3 & 6 & -6 & -3 \\ 1 & 2 & -2 & -1 \\ -1 & -2 & 2 & 1 \\ -1 & -2 & 2 & 1 \end{bmatrix}.$$

Let N_i denote the number of ASTs with IBD configuration i at the marker, $i = 1, 2, 3, 4$. Then the score statistic for testing linkage is

$$S = \tfrac{16}{3}(3\pi_1 + \pi_2 - \pi_3 - \pi_4)(3N_1 + N_2 - N_3 - N_4).$$

Note that $3N_1 + N_2 - N_3 - N_4$ may be rewritten as $2(3N_1 + 2N_2 + N_3 + N_4) - 3n = 6S_{\text{pairs}} - 3n$.

5.4. *Discordant sib-trios, orbits of* $(S_1 \times S_2) \times D_4$. For discordant sib-trios (DSTs), where the first sib is the "odd" sib (i.e., the only affected sib or the only unaffected sib), there are seven IBD configurations, with representative inheritance vectors listed in Table 4.

TABLE 4
IBD configurations for discordant sib-trios

IBD configuration i	Representative inheritance vector	$\lvert\mathscr{C}_i\rvert$
1	$(1, 3, 1, 3, 1, 3)$	4
2	$(1, 3, 1, 3, 1, 4)$	16
3	$(1, 3, 1, 3, 2, 4)$	8
4	$(1, 3, 1, 4, 2, 3)$	8
5	$(1, 3, 1, 4, 2, 4)$	16
6	$(1, 3, 1, 4, 1, 4)$	8
7	$(1, 3, 2, 4, 2, 4)$	4

The infinitesimal generator is

$$Q = \begin{bmatrix} -6 & 4 & 0 & 0 & 0 & 2 & 0 \\ 1 & -5 & 1 & 1 & 1 & 1 & 0 \\ 0 & 2 & -6 & 2 & 2 & 0 & 0 \\ 0 & 2 & 2 & -6 & 2 & 0 & 0 \\ 0 & 1 & 1 & 1 & -5 & 1 & 1 \\ 1 & 2 & 0 & 0 & 2 & -6 & 1 \\ 0 & 0 & 0 & 0 & 4 & 2 & -6 \end{bmatrix}.$$

Q has eigenvalues $\lambda = 0, -4, -4, -8, -8, -8, -8$, and the two orthonormal right eigenvectors corresponding to $\lambda_2 = -4$ are $v = \sqrt{\frac{2}{3}}(-1, -1, -1, -1, 1, 1, 3)$ and $\tilde{v} = (1/\sqrt{3})(4, 1, -2, -2, -1, 2, 0)$. Hence

$$U = 8P_{-4} = \begin{bmatrix} 3 & 4 & -2 & -2 & -4 & 2 & -1 \\ 1 & 2 & 0 & 0 & -2 & 0 & -1 \\ -1 & 0 & 2 & 2 & 0 & -2 & -1 \\ -1 & 0 & 2 & 2 & 0 & -2 & -1 \\ -1 & -2 & 0 & 0 & 2 & 0 & 1 \\ 1 & 0 & -2 & -2 & 0 & 2 & 1 \\ -1 & -4 & -2 & -2 & 4 & 2 & 3 \end{bmatrix}.$$

Denote the ith column of U by u_i, then $u_1 + u_7 = -u_3 = -u_4 = u_6$ and $u_1 - u_7 = u_2 = -u_5$. Let N_i denote the number of DSTs with IBD configuration i at the marker, $i = 1, \ldots, 7$. Then, the score statistic for testing linkage is

$$\begin{aligned} S &= 8\Big(\sum_{j=1}^{7} v_j \pi_j\Big)\Big(\sum_{i=1}^{7} v_i N_i\Big) + 8\Big(\sum_{j=1}^{7} \tilde{v}_j \pi_j\Big)\Big(\sum_{i=1}^{7} \tilde{v}_i N_i\Big) \\ &= \frac{8}{3}\Big(2(-\pi_1 - \pi_2 - \pi_3 - \pi_4 + \pi_5 + \pi_6 + 3\pi_7) \\ &\qquad \times (-N_1 - N_2 - N_3 - N_4 + N_5 + N_6 + 3N_7) \\ &\qquad + (4\pi_1 + \pi_2 - 2\pi_3 - 2\pi_4 - \pi_5 + 2\pi_6) \\ &\qquad \times (4N_1 + N_2 - 2N_3 - 2N_4 - N_5 + 2N_6)\Big). \end{aligned}$$

6. Discussion. In this paper, we have derived score statistics for testing the null hypothesis of no linkage between a marker and a disease gene using identity by descent (IBD) data from sibships. We considered IBD configurations which are orbits of groups acting on the set of inheritance vectors, and proved that the transition matrix between IBD configurations satisfies a semigroup property (Proposition 1). For general pedigree types, we derived the following spectral decomposition for the IBD configuration transition matrix $T(\theta)$:

$$T(\theta) = \sum_h (1 - 2\theta)^{-\lambda_h/2} P_h,$$

where λ_h and P_h are the eigenvalues and projection matrices of the infinitesimal generator Q , respectively, and λ_h are negative even integers (Propositions 2 and 3). By relating Q to the adjacency matrix of a quotient graph, we derived properties of its eigenvalues and eigenvectors. In general, the second largest eigenvalue of Q and its multiplicity determine the form of the score statistic for a given pedigree type. If the second largest eigenvalue of Q is $-2i$, the score test is based on the ith derivative of the log-likelihood, and if it has multiplicity one, the score statistic is independent of the genetic model (i.e., of the nuisance parameter π). For affected-only sibships of a given size, the second largest eigenvalue is -4 and has multiplicity one. As a result, the score test is based on the second derivative of the log-likelihood and is independent of the genetic model. Furthermore, the score statistic for affected-only sibships of a given size reduces to a well-known nonparametric statistic in linkage analysis, S_{pairs}.

Testing the null hypothesis of no linkage using IBD data, as we do here, is an instance of a general class of testing problems in which the null hypothesis is that a Markov chain has reached its stationary distribution [Diaconis (1988)]. The second largest eigenvalue of the infinitesimal generator not only determines the rate of convergence to the stationary distribution, but also plays an important role in hypothesis testing, as illustrated by our study.

Since the mid-1970's, linkage statistics have been the subject of numerous studies in the genetics literature. Earlier research focussed on IBD data from sib-pairs, and the sib-pair case still remains popular for theoretical work on linkage analysis and in practice for genetic studies. Several test statistics have been examined to test the null hypothesis of no linkage between a marker and a DS locus, based on the numbers (N_0, N_1, N_2) of affected sib-pairs sharing DNA IBD on 0, 1 and 2 chromosomes, respectively. These include the mean IBD statistic, that is, $S_{\text{pairs}} = N_2 + \frac{1}{2}N_1$ [Blackwelder and Elston (1985), Schaid and Nick (1990), Knapp, Seuchter and Baur (1994a, b)], N_2 [Day and Simons (1976), Blackwelder and Elston (1985) and Schaid and Nick (1990)], $N_2 + \frac{1}{4}N_1$ [Feingold and Siegmund (1997)], likelihood ratio statistics and χ^2 goodness-of-fit statistics, either unrestricted or restricted to the "possible triangle" [Risch (1990a, b), Holmans (1993), Faraway (1993), Feingold, Brown and Siegmund (1993), Holmans and Clayton (1995), Knapp, Seuchter and Baur (1994c) and Kruglyak and Lander (1995)]. Knapp, Seuchter and Baur (1994a) first pointed out the relationship of S_{pairs} to likelihood-based tests, in the special case of affected sib-pairs and genetic models with a single disease locus with random mating and Hardy–Weinberg equilibrium. In the past five years, nonparametric allele-sharing statistics for small pedigrees have been actively studied. Current methods for analyzing IBD data from small pedigrees rely on "scoring functions" which are defined as functions of the IBD configurations of affected pedigree members. [Whittemore and Halpern

(1994a), Kruglyak, Daly, Reeve-Daly and Lander (1996), Kong and Cox (1997), Teng and Siegmund (1997) and McPeek (1999)]. A few scoring functions have been suggested on empirical grounds, such as S_{pairs} and S_{all} [Whittemore and Halpern (1994a)], but the optimality and relationship of these scoring functions to likelihood analysis deserve further study. In addition, while it is clear that different pedigree types differ in their linkage information, IBD data from different pedigrees are typically combined by assigning equal weights to the standardized allele-sharing statistics of the various pedigrees. Different standardization and weighing schemes have been suggested, but the issue of combining IBD data from different pedigree types has not yet been thoroughly examined in the literature and remains an open problem. Whittemore (1996) proposed a likelihood-based approach to linkage analysis using multipoint marker data from general pedigrees. In the context of a genome scan, Whittemore considered the null hypothesis of no DS gene linked to the candidate locus against the alternative hypothesis of a DS gene *at* the candidate locus ($\theta = 0$). The null and alternative hypotheses were expressed in terms of a specificity parameter β, which is a function of genetic parameters, such as penetrances and allele frequencies. Whittemore proposed testing the null hypothesis of no DS gene linked to the candidate locus using a score test in β, that is, in the *specificity* parameter. Like our score test in the recombination fraction θ, this is a likelihood-based test which simultaneously treats the IBD data from all pedigrees. However, this type of score test depends intrinsically on the parameterization of the genetic model. This is a potentially serious shortcoming for complex diseases, for which there typically is no established knowledge regarding the number of DS loci, the penetrances, or the frequency of genotypes in the study population. Similarly, McPeek (1999) derived allele-sharing statistics for different types of alternative genetic models by focusing on specificity parameters.

It is important to note that in genome scans, where marker density is high, one expects to be testing near $\theta = 0$ rather than near $\theta = \frac{1}{2}$. If one has some knowledge concerning the mode of inheritance of the disease (e.g., rare recessive disease), then making use of this information in deriving allele-sharing statistics as in McPeek (1999) is desirable. However, the main obstacle in obtaining optimal statistics for an alternative hypothesis where $\theta = 0$ is precisely their dependence on the genetic model, which is usually unknown. By contrast, our score test in the recombination fraction θ focuses on the *scale* component and seems to achieve some robustness with respect to specificity. Although our score statistic in θ is locally most powerful near $\theta = \frac{1}{2}$, S_{pairs} was also found to be optimal for some classes of genetic models when $\theta = 0$ by McPeek (1999). A recent study by Davis and Weeks (1997) on affected sib-pairs found S_{pairs} to perform well for a variety of two-locus genetic models. Our preliminary work on applying the sib-pair score test in θ to the linkage analysis of quantitative traits also demonstrates the power of this type of score test for nonlocal alternatives (Dudoit (1999)). However, when the alternative hypothesis is such that $\theta = 0$, the optimal statistic depends on the genetic model, and there are inevitably classes of models for which S_{pairs} is outperformed by other statistics. For example, Feingold and Siegmund (1997) reported that for sib-pairs N_2 performed better than S_{pairs} for rare heterogeneous traits. For larger sibships, the performance of S_{pairs} needs to be further investigated.

As mentioned earlier, an important issue is the combination of IBD data from different pedigree types. The score test in θ presented in this paper produces weights for combining IBD data from different types of sibships ($\sum_j v_j \pi_j$), but these weights are linear combinations of the IBD probabilities, π, which depend on the usually unknown genetic model. Our work on quantitative traits revealed that the sib-pair score test was robust to the choice of model for computing the weights $\pi_2 - \pi_0$. For sib-pairs sampled randomly with respect to their phenotypes, the score test was found to be similar in power to the widely used Haseman–Elston test, and for sib-pairs selected with extreme concordant or discordant phenotypes, the score test was found to be far more powerful than the Haseman–Elston test [cf. Haseman and Elston (1972), Risch and Zhang (1995, 1996), Dudoit (1999)].

We believe that score statistics in θ as described in this paper have the potential to be useful in practice for linkage analysis using IBD data from sibships and relative pairs, when the mode of inheritance is unknown. We have derived score statistics in θ for IBD data from other types of relative pairs (e.g., cousin pairs), using the IBD configurations considered by Donnelly (1983) and the general framework of Propositions 1 and 2 [Dudoit (1999)]. These score statistics may readily be extended to accommodate incomplete IBD data by replacing the IBD counts by their expected value given multipoint marker data [cf. inheritance distribution of Kruglyak and Lander (1995) and Kruglyak, Daly, Reeve-Daly and Lander (1996)]. We were very encouraged by our preliminary studies with sib-pairs and will further explore the issue of combining IBD data from different sibship types and relative pairs.

APPENDIX

A. Pólya theory of counting. Let A and B be finite sets, $|A| = n$, and let G and H be finite groups, G acting on A and H on B. By Theorem 35.3 in van Lint and Wilson (1992), the number of orbits of $G \times H$ acting on B^A, the set of mappings from A to B, is given by

$$\frac{1}{|H|} \sum_{\tau \in H} Z_G(m_1(\tau), \dots, m_n(\tau)),$$

where

$$m_i(\tau) := \sum_{j|i} j z_j(\tau), \qquad i = 1, \dots, n,$$

$$z_j(\tau) := \text{number of cycles of } \tau \text{ having length } j, \qquad j = 1, \dots, |B|.$$

For a group G acting on a set of n elements, the *cycle index* Z_G is a polynomial in n letters, $X_1, \dots, X_n$, defined by

$$Z_G(X_1, \dots, X_n) := \frac{1}{|G|} \sum_{\sigma \in G} X_1^{z_1(\sigma)} \cdots X_n^{z_n(\sigma)}.$$

The cycle index for the symmetric group on n letters is

$$Z_{S_n}(X_1, \ldots, X_n) = \sum_{(1^{k_1} \cdots n^{k_n})} \frac{1}{1^{k_1} \cdots n^{k_n} k_1! \cdots k_n!} X_1^{k_1} \cdots X_n^{k_n},$$

where $(1^{k_1} \cdots n^{k_n})$ denotes a partition of n with k_i parts of size i, $i = 1, \ldots, n$. $Z_{S_n}(X_1, \ldots, X_n)$ is also the coefficient of z^n in the expansion of $\exp(\sum_{i=1}^{\infty}(z^i/i)X_i)$ [cf. deBruijn (1964), page 147]; this is the formula which is most appropriate for our problem. In our problem, we wish to determine the number of orbits of $G \times H$ acting on the set of mappings $\{a, b, c, d\}^{\{1,\ldots,k\}}$ (i.e., the set of inheritance vectors), where $H = D_4$ or $C_2 \times C_2$ and $G = S_k$ or $S_h \times S_{k-h}$. Table 5 lists for each permutation in D_4 the number of cycles of length j, $j = 1, 2, 3, 4$. The m_i's of Table 6 are calculated using Table 5, and the fact that for $\tau \in D_4$ and $i \geq 0$,

$$\begin{aligned} m_4(\tau) &= m_{4i+4}(\tau), \\ m_1(\tau) &= m_{4i+1}(\tau) = m_{4i+3}(\tau), \\ m_2(\tau) &= m_{4i+2}(\tau). \end{aligned}$$

Table 7 lists for each permutation in D_4 the cycle index $Z_{S_k}(m_1(\tau), \ldots, m_k(\tau))$. When $H = C_2 \times C_2$, only the first and third rows of Table 7 are used. When $G = S_h \times S_{k-h}$, we note that the cycle index polynomial of the direct product of two groups is simply the product of the cycle indices of the two groups and Table 7 may be used again. The number of IBD configurations for the four groups are listed below.

$S_k \times D_4$:

$$m = \begin{cases} (k+1)(k+3)(k+5)/48, & k \text{ odd}, \\ (k+2)(k^2+7k+18)/48, & k \text{ even and } k/2 \text{ odd}, \\ (k+4)(k^2+5k+12)/48, & k \text{ even and } k/2 \text{ even}, \end{cases}$$

which agrees with equation (5) of Ethier and Hodge (1985).

TABLE 5

Cycles of D_4. $z_j(\tau)$ denotes the number of cycles of τ having length j *

Permutation τ	$z_1(\tau)$	$z_2(\tau)$	$z_3(\tau)$	$z_4(\tau)$
ι	4	0	0	0
$\rho_1 = (cadb)$	0	0	0	1
$\rho_2 = (ad)(bc)$	0	2	0	0
$\rho_3 = (bdac)$	0	0	0	1
$\mu_1 = (ab)(cd)$	0	2	0	0
$\mu_2 = (ac)(bd)$	0	2	0	0
$\delta_1 = (bc)$	2	1	0	0
$\delta_2 = (ad)$	2	1	0	0

*The elements of D_4 are listed according to the notation of Fraleigh [(1989), page 70].

A SCORE TEST FOR LINKAGE

TABLE 6
$m_i(\tau)$ *for* $\tau \in D_4$

Permutation τ	$m_{4i+1}(\tau)=m_{4i+3}(\tau)$	$m_{4i+2}(\tau)$	$m_{4i+4}(\tau)$
ι	4	4	4
ρ_1, ρ_3	0	0	4
ρ_2, μ_1, μ_2	0	4	4
δ_1, δ_2	2	4	4

$S_k \times (C_2 \times C_2)$:

$$m = \begin{cases} (k+3)(k+2)(k+1)/24, & k \text{ odd}, \\ (k+2)(k^2+4k+12)/24, & k \text{ even}. \end{cases}$$

$(S_h \times S_{k-h}) \times D_4$:

$$\begin{aligned} m = \frac{1}{8}\Bigg[&\binom{h+3}{3}\binom{k-h+3}{3} + 2I(4|h)I(4|k-h) \\ &+ \frac{3}{4}I(2|h)I(2|k-h)(h+2)(k-h+2) \\ &+ \frac{1}{8}(I(2|h)+(h+1)(h+3))(I(2|k-h)+(k-h+1)(k-h+3))\Bigg], \end{aligned}$$

where $I(j|i) = 1$ if j divides i and 0 otherwise.

$(S_h \times S_{k-h}) \times (C_2 \times C_2)$:

$$m = \frac{1}{4}\left[\binom{h+3}{3}\binom{k-h+3}{3} + \frac{3}{4}I(2|h)I(2|k-h)(h+2)(k-h+2)\right].$$

For any inheritance vector x, let pat denote the less frequent of the paternal labels 1 and 2, and similarly let mat denote the less frequent of the maternal labels 3 and 4. The number of paternal labels pat in the inheritance vector x is denoted by $|pat|$, similarly for the maternal labels. The number of sibs with a pair of labels (pat, mat) is denoted by $|(pat, mat)|$. Ethier and Hodge (1985) define the label of the particular inheritance vector x to be the triple (l_1, l_2, l_3) where

$$l_1 = |(pat, mat)|, \qquad l_2 = \min(|pat|, |mat|), \qquad l_3 = \max(|pat|, |mat|).$$

TABLE 7
$Z_{S_k}(m_1(\tau), \ldots, m_k(\tau))$ *for* $\tau \in D_4$. $I(\ )$ *is the indicator function*

Permutation τ	$\exp\left(\sum_{i=1}^{\infty} m_i(\tau)\frac{z^i}{i}\right)$	$Z_{S_k}(m_1(\tau), \ldots, m_k(\tau))$
ι	$(1-z)^{-4}$	$\binom{k+3}{3}$
ρ_1, ρ_3	$(1-z^4)^{-1}$	$I(4\|k)$
ρ_2, μ_1, μ_2	$(1-z^2)^{-2}$	$I(2\|k)(k+2)/2$
δ_1, δ_2	$(1-z)^{-2}(1-z^2)^{-1}$	$\frac{1}{4}(I(2\|k)+(k+1)(k+3))$

For example, if $x = (1, 3, 1, 4, 2, 3)$, the less frequent of the paternal labels is 2, thus $pat = 2$ and $|pat| = 1$. Similarly, $mat = 4$ and $|mat| = 1$. $(mat, pat) = (2, 4)$ and the number of sibs with pair of labels $(2, 4)$ is 0. Thus $l_1 = 0$, $l_2 = 1$ and $l_3 = 1$. In ambiguous cases such as $|1| = |2| = k/2$ or $|3| = |4| = k/2$, Ethier and Hodge suggest making a choice that results in $l_1 \geq l_2/2$. Then the triple satisfies

$$0 \leq l_1 \leq l_2 \leq l_3 \leq k/2, \quad \text{and} \quad l_1 \geq l_2/2 \text{ if } l_3 = k/2.$$

We can modify the labeling of Ethier and Hodge for the orbits of $S_k \times (C_2 \times C_2)$ and let

$$l_1 = |(pat, mat)|, \qquad l_2 = |pat|, \qquad l_3 = |mat|.$$

B. Transition matrix for IBD configurations.

B.1. *Proof of Proposition* 1. Let $\bar{\theta} = 1 - \theta$. We first prove the semigroup property for the transition matrix $R(\theta)$ of inheritance vectors. Let $\Delta = \Delta(x, y)$, then

$$\begin{aligned} r_{xy}(\theta_1 * \theta_2) &= (\theta_1 * \theta_2)^{\Delta}(1 - (\theta_1 * \theta_2))^{2k-\Delta} \\ &= (\theta_1\bar{\theta}_2 + \bar{\theta}_1\theta_2)^{\Delta}(\theta_1\theta_2 + \bar{\theta}_1\bar{\theta}_2)^{2k-\Delta} \\ &= \sum_{i=0}^{\Delta}\sum_{j=0}^{2k-\Delta}\binom{\Delta}{i}\binom{2k-\Delta}{j}\theta_1^{i+j}\bar{\theta}_1^{2k-(i+j)}\theta_2^{\Delta-i+j}\bar{\theta}_2^{2k-(\Delta-i+j)}. \end{aligned}$$

Also,

$$\sum_z r_{xz}(\theta_1)r_{zy}(\theta_2) = \sum_z \theta_1^{\Delta(x,z)}\bar{\theta}_1^{2k-\Delta(x,z)}\theta_2^{\Delta(y,z)}\bar{\theta}_2^{2k-\Delta(y,z)}.$$

Now, for $i = 0, \ldots, \Delta$, $j = 0, \ldots, 2k - \Delta$, divide the set of all 2^{2k} inheritance vectors into groups of $\binom{\Delta}{i}\binom{2k-\Delta}{j}$ inheritance vectors z, such that z differs from x at i of the Δ positions at which x and y differ and z differs from x at j of the $2k - \Delta$ positions at which x and y agree. Then, $\Delta(x, z) = i + j$, $\Delta(y, z) = (\Delta - i) + j$, and

$$\begin{aligned} &\sum_z \theta_1^{\Delta(x,z)}\bar{\theta}_1^{2k-\Delta(x,z)}\theta_2^{\Delta(y,z)}\bar{\theta}_2^{2k-\Delta(y,z)} \\ &\quad = \sum_{i=0}^{\Delta}\sum_{j=0}^{2k-\Delta}\binom{\Delta}{i}\binom{2k-\Delta}{j}\theta_1^{i+j}\bar{\theta}_1^{2k-(i+j)}\theta_2^{\Delta-i+j}\bar{\theta}_2^{2k-(\Delta-i+j)}. \end{aligned}$$

Therefore,

$$r_{xy}(\theta_1 * \theta_2) = \sum_z r_{xz}(\theta_1)r_{zy}(\theta_2).$$

A SCORE TEST FOR LINKAGE

Consider now the transition matrix for IBD configurations. From (3.2),

$$\begin{aligned} t_{ij}(\theta_1 * \theta_2) &= \sum_{y \in \mathscr{C}_j} r_{xy}(\theta_1 * \theta_2) \quad \text{(where } x \text{ is any } x \in \mathscr{C}_i) \\ &= \sum_{y \in \mathscr{C}_j} \sum_{z} r_{xz}(\theta_1) r_{zy}(\theta_2) = \sum_{y \in \mathscr{C}_j} \sum_{l} \sum_{z \in \mathscr{C}_l} r_{xz}(\theta_1) r_{zy}(\theta_2) \\ &= \sum_{l} \sum_{z \in \mathscr{C}_l} r_{xz}(\theta_1) \sum_{y \in \mathscr{C}_j} r_{zy}(\theta_2) \\ &= \sum_{l} t_{lj}(\theta_2) \sum_{z \in \mathscr{C}_l} r_{xz}(\theta_1) \\ &= \sum_{l} t_{il}(\theta_1) t_{lj}(\theta_2). \end{aligned}$$

Hence, $T(\theta)$ satisfies the semigroup property $T(\theta_1 * \theta_2) = T(\theta_1)T(\theta_2)$. Now $T(\theta)$ is differentiable and for $\theta \neq \frac{1}{2}$,

$$\begin{aligned} \frac{T(\theta + h(1-2\theta)) - T(\theta)}{h(1-2\theta)} &= \frac{T(\theta * h) - T(\theta)}{h(1-2\theta)} \\ &= \left(\frac{T(\theta)}{1-2\theta}\right)\left(\frac{T(h)-I}{h}\right) = \left(\frac{T(h)-I}{h}\right)\left(\frac{T(\theta)}{1-2\theta}\right). \end{aligned}$$

Thus $T'(\theta)$, the matrix of first derivatives of the transition probabilities, is given by

$$T'(\theta) = \lim_{h \to 0} \frac{T(\theta + h) - T(\theta)}{h} = \lim_{h \to 0} \frac{T(\theta + h(1-2\theta)) - T(\theta)}{h(1-2\theta)},$$

that is,

$$T'(\theta) = \frac{T(\theta)}{1-2\theta} T'(0) = T'(0) \frac{T(\theta)}{1-2\theta},$$

and hence

$$T(\theta) = e^{d(\theta)Q},$$

where $d(\theta) = -\frac{1}{2}\ln(1-2\theta)$ is the inverse of the Haldane map function and $Q = T'(0)$ is the infinitesimal generator. Q has entries,

$$q_{ij} = \sum_{y \in \mathscr{C}_j} (-2k\, I(\Delta(x, y) = 0) + I(\Delta(x, y) = 1)) = \sum_{y \in \mathscr{C}_j} I(\Delta(x, y) = 1) - 2k\delta_{ij},$$

where x is any inheritance vector in $\mathscr{C}_i$ and $\delta_{ij} = 1$ if $i = j$ and 0 otherwise. Then Q may be written as $Q = B - 2kI$, where B is the $m \times m$ matrix with entries

$$b_{ij} = \sum_{y \in \mathscr{C}_j} I(\Delta(x, y) = 1) \quad \text{for any } x \in \mathscr{C}_i.$$

$T(\theta)$ satisfies

$$|\mathscr{C}_i|t_{ij}(\theta) = |\mathscr{C}_j|t_{ji}(\theta);$$

hence, the stationary distribution of $T(\theta)$ is

$$\alpha = (\alpha_1, \ldots, \alpha_m) = \frac{1}{2^{2k}}(|\mathscr{C}_1|, \ldots, |\mathscr{C}_m|),$$

since

$$\sum_i \alpha_i t_{ij}(\theta) = \sum_i \alpha_j t_{ji}(\theta) = \alpha_j. \qquad \square$$

B.2. *Proof of Proposition* 2. Q satisfies the reversibility condition $\alpha_i q_{ij} = \alpha_j q_{ji}$, hence Q is self-adjoint with respect to the real inner product $\langle x, y\rangle_\alpha = \sum_i \alpha_i x_i y_i$ on $\mathbb{R}^m$. Hence, from the principal axis theorem [cf. Jacob (1990), page 288], Q has an orthonormal basis of eigenvectors with only real eigenvalues, λ_h, $h = 1, \ldots, m$ (not necessarily distinct). Denote the hth right eigenvector by $\mathbf{v}_h$ and its ith entry by v_{ih}. Then $\langle \mathbf{v}_h, \mathbf{v}_l\rangle_\alpha = \sum_i \alpha_i v_{ih} v_{il} = \delta_{hl}$. Since Q is reversible, the row vector $\mathbf{w}_h$ with ith entry $w_{hi} = \alpha_i v_{ih}$ is the left eigenvector of Q corresponding to the hth eigenvalue. Hence Q may be written as

$$Q = \sum_h \lambda_h P_h,$$

where

$$(P_h)_{ij} = v_{ih} w_{hj} = \alpha_j v_{ih} v_{jh},$$

that is,

$$P_h = \mathbf{v}_h \mathbf{w}_h.$$

The projection matrices satisfy $P_h^2 = P_h = P_h^*$, $P_h P_l = 0$, $h \neq l$ and $\sum_h P_h = I$, where P_h^* is the adjoint of P_h with respect to $\langle\ ,\ \rangle_\alpha$. It follows that

$$T(\theta) = \sum_h \exp(\lambda_h d(\theta))\, P_h = \sum_h (1-2\theta)^{-\lambda_h/2}\, P_h. \qquad \square$$

C. Adjacency matrix of quotient graph $\mathscr{X}/(G\times H)$. Consider the graph $\mathscr{X}$ with vertex set the set of all inheritance vectors of length $2k$ and adjacency matrix $A(\mathscr{X}) = A$ with (x, y)-entry,

$$a_{xy} = \begin{cases} 1, & \text{if } \Delta(x, y) = 1, \\ 0, & \text{otherwise.} \end{cases}$$

To describe the eigenvectors of A it is convenient to code the inheritance vectors $x = (x_1, x_2, \ldots, x_{2k})$ as in a 2^{2k} factorial experiment, where $x_{2i-1} = 1$ when factor $2i - 1$ is absent and 2 when it is present, and $x_{2i} = 3$ when factor $2i$ is absent and 4 when it is present. The eigenvectors of A have the following patterns.

PROPOSITION 3 (Eigenvalues and eigenvectors of adjacency matrix A). *The eigenvector corresponding to the eigenvalue* $\lambda = 2k$ *is the grand mean term* $V_0 = (1, 1, \ldots, 1)^T$. *The eigenvectors corresponding to the eigenvalue* $\lambda = 2k-2$ *are the* $2k$ *main effect terms,* $V_1, V_2, \ldots, V_{2k}$, *where*

$$V_{2i-1}(x) = I(x_{2i-1} = 2) - I(x_{2i-1} = 1),$$
$$V_{2i}(x) = I(x_{2i} = 4) - I(x_{2i} = 3).$$

The eigenvectors corresponding to the eigenvalue $\lambda = 2k-4$ *are the* $\binom{2k}{2}$ *2-factor interactions,* V_{ij}, $1 \le i < j \le 2k$, *where*

$$V_{ij}(x) = V_i(x)V_j(x).$$

In general, the eigenvectors corresponding to the eigenvalue $\lambda = 2(k-i)$, $i = 0, \ldots, 2k$, *are the* $\binom{2k}{i}$ *i-factor interactions,* $V_{j_1, j_2, \ldots, j_i}$, $1 \le j_1 < j_2 < \cdots < j_i \le 2k$, *where*

$$V_{j_1, j_2, \ldots, j_i}(x) = V_{j_1}(x)V_{j_2}(x) \cdots V_{j_i}(x).$$

Let H *denote the matrix with rows the* 2^{2k} *eigenvectors of* A *described above. Then,* H *is an Hadamard matrix; that is, its entries are* 1 *and* -1 *and* $HH^T = 2^{2k} I$.

PROOF (Partial). We need not distinguish the parental origin of the DNA, hence, for simplicity, denote 1's and 3's by 0's and 2's and 4's by 1's. Then

$$V_i(x) = I(x_i = 1) - I(x_i = 0) = 2\,I(x_i = 1) - 1.$$

$\lambda = 2k$: the rows of A sum to $2k$ hence $\lambda = 2k$ is an eigenvalue of A with eigenvector V_0.

$\lambda = 2k - 2$:

$$\begin{aligned}\sum_y a_{xy} V_i(y) &= \sum_y I(\Delta(x, y) = 1)(2\,I(y_i = 1) - 1)\\ &= 2\sum_y I(\Delta(x, y) = 1, y_i = 1) - 2k\\ &= 2(I(x_i = 1)(2k-1) + I(x_i = 0)) - 2k\\ &= 2((2k-2)\,I(x_i = 1) + 1) - 2k\\ &= (2k-2)(2\,I(x_i = 1) - 1) = (2k-2)V_i(x).\end{aligned}$$

Hence $\lambda = 2k - 2$ is an eigenvalue of A with eigenvectors V_i, $i = 1, \ldots, 2k$. It is easy to show that $\langle V_i, V_j \rangle = 2^{2k}\delta_{ij}$.

$\lambda = 2k - 4$:

$$\begin{aligned}
\sum_y a_{xy} V_i(y) V_j(y)
&= \sum_y I(\Delta(x, y) = 1)(2\,I(y_i = 1) - 1)(2\,I(y_j = 1) - 1) \\
&= \sum_y I(\Delta(x, y) = 1)(4\,I(y_i = 1,\ y_j = 1) - 2\,I(y_i = 1) - 2\,I(y_j = 1) + 1) \\
&= 4(I(x_i = 1,\ x_j = 1)(2k - 2) + I(x_i = 1,\ x_j = 0) + I(x_i = 0,\ x_j = 1)) \\
&\quad - 2(I(x_i = 1)(2k - 1) + I(x_i = 0)) \\
&\quad - 2(I(x_j = 1)(2k - 1) + I(x_j = 0)) + 2k \\
&= 4I(x_i = 1,\ x_j = 1)(2k - 2) \\
&\quad + 4(I(x_i = 1) - I(x_i = 1,\ x_j = 1)) \\
&\quad + 4(I(x_j = 1) - I(x_i = 1,\ x_j = 1)) \\
&\quad - 2((2k - 2)I(x_i = 1) + 1) - 2((2k - 2)I(x_j = 1) + 1) + 2k \\
&= (2k - 4)(4I(x_i = 1,\ x_j = 1) - 2I(x_i = 1) - 2I(x_j = 1) + 1) \\
&= (2k - 4)V_i(x)V_j(x).
\end{aligned}$$

Hence $\lambda = 2k - 4$ is an eigenvalue of A with eigenvectors V_{ij}. □

In order to prove Theorem 1, we rely on the following general facts concerning quotient graphs [Chapter 5 in Godsil (1993)]. Consider a group $G \times H$ acting on the vertices of $\mathscr{X}$, as described in Table 2. Then, by the same argument as that leading to (3.1), the orbits of $G \times H$, $\mathscr{C}_i$, $i = 1, \ldots, m$, form an equitable partition of the vertex set of $\mathscr{X}$. The matrix B defined in Proposition 1 is the *adjacency matrix of the quotient graph* $\mathscr{X}/(G \times H)$, which is the multidigraph with the orbits of $G \times H$ as its vertices and with b_{ij} arcs going from $\mathscr{C}_i$ to $\mathscr{C}_j$. Let C denote the *characteristic matrix* of the partition $(\mathscr{C}_i)$; C is a $2^{2k} \times m$ matrix, with ijth entry 1 or 0 according as the ith vertex of $\mathscr{X}$ is contained in the orbit $\mathscr{C}_j$ or not.

FACT 1 [Based on Lemma 2.2 in Godsil (1993)]. The eigenvalues of B are a subset of the eigenvalues of A.

FACT 2 [Based on Lemma 2.2 in Godsil (1993)]. If v is an eigenvector of B, then Cv is an eigenvector of A which is constant over the orbits of $G \times H$, with entry v_i on $\mathscr{C}_i$.

FACT 3. If V is an eigenvector of A which is constant over the orbits of $G \times H$, with $V(x) = v_i\ \forall\ x \in \mathscr{C}_i$, then the vector v, with ith entry v_i, is an eigenvector of B.

PROOF. For any $x \in \mathscr{C}_i$

$$\lambda v_i = \lambda V(x) = \sum_y a_{xy} V(y) = \sum_j v_j \sum_{y \in \mathscr{C}_j} a_{xy} = \sum_j v_j b_{ij}. \qquad \square$$

The proof of Theorem 1 also relies on the following specific properties of the eigenvectors of A on the orbits of $G \times H$.

C.1. *Quotient graph* $\mathscr{X}/(S_k \times D_4)$.

FACT 4. The $2k$ eigenvectors of A corresponding to the eigenvalue $2k - 2$ sum to 0 over the orbits of $S_k \times D_4$, that is, $\forall\, i = 1, \ldots, 2k$, and any orbit $\mathscr{C}$,

$$\sum_{x \in \mathscr{C}} V_i(x) = 0.$$

PROOF. Let $\iota \in S_k$ denote the identity permutation and as before, let $\alpha = (ac)(bd)$ denote the permutation of D_4 which corresponds to interchanging the paternal labels 1 and 2. Let $\tilde{x} = (\iota, \alpha)(x)$ denote the inheritance vector obtained from x by interchanging the paternal labels. Then, for $1 \leq i \leq k$,

$$\begin{aligned} V_{2i-1}(x) &= (I(x_{2i-1} = 2) - I(x_{2i-1} = 1)) \\ &= (I(\tilde{x}_{2i-1} = 1) - I(\tilde{x}_{2i-1} = 2)) = -V_{2i-1}(\tilde{x}), \end{aligned}$$

and since applying (ι, α) to the elements of $\mathscr{C}$ results in a permutation of the inheritance vectors in $\mathscr{C}$, then

$$\sum_{x \in \mathscr{C}} V_{2i-1}(x) = -\sum_{x \in \mathscr{C}} V_{2i-1}(\tilde{x}) = -\sum_{x \in \mathscr{C}} V_{2i-1}(x).$$

Consequently,

$$\sum_{x \in \mathscr{C}} V_{2i-1}(x) = 0.$$

The proof for V_{2i} is similar, but uses the permutation β instead of α. $\square$

FACT 5. The k^2 eigenvectors of A corresponding to the eigenvalue $2k - 4$ and involving "odd" and "even" factors sum to 0 over the orbits of $S_k \times D_4$, that is, $\forall\, i, j = 1, \ldots, k$, and any orbit $\mathscr{C}$,

$$\sum_{x \in \mathscr{C}} V_{2i-1}(x) V_{2j}(x) = 0.$$

PROOF. Here again, let $\tilde{x} = (\iota, \alpha)(x)$. Then

$$V_{2i-1}(x) V_{2j}(x) = (-V_{2i-1}(\tilde{x})) V_{2j}(\tilde{x})$$

and

$$\sum_{x \in \mathscr{C}} V_{2i-1}(x) V_{2j}(x) = -\sum_{x \in \mathscr{C}} V_{2i-1}(\tilde{x}) V_{2j}(\tilde{x}) = -\sum_{x \in \mathscr{C}} V_{2i-1}(x) V_{2j}(x).$$

Hence

$$\sum_{x \in \mathscr{C}} V_{2i-1}(x) V_{2j}(x) = 0. \qquad \square$$

FACT 6. Let

$$V(x) = \sum_{(i,j)} \{V_{2i-1,2j-1}(x) + V_{2i,2j}(x)\},$$

where the sum is over all $\binom{k}{2}$ unordered pairs (i, j) of distinct integers ranging from 1 to k. Then V is an eigenvector of A corresponding to the eigenvalue $2k - 4$. Furthermore, V is constant over the orbits of $S_k \times D_4$, that is, for any orbit $\mathscr{C}$,

$$V(x) = V(\tilde{x}) \quad \text{whenever } x, \tilde{x} \in \mathscr{C}.$$

PROOF. Members of the same orbit are obtained by a combination of any of the following three operations: a permutation $\sigma \in S_k$ of the sibs, and permutations α and γ of the pairs of labels of all sibs simultaneously. We consider a particular configuration $\mathscr{C}$ and the effect of each operation separately on $x \in \mathscr{C}$.

$\tilde{x} = (\iota, \alpha)(x)$, where ι is the identity in S_k and $\alpha = (ac)(bd)$: for each pair (i, j),

$$V_{2i-1}(\tilde{x})V_{2j-1}(\tilde{x}) + V_{2i}(\tilde{x})V_{2j}(\tilde{x}) = (-V_{2i-1}(x))(-V_{2j-1}(x)) + V_{2i}(x)V_{2j}(x),$$

hence $V(\tilde{x}) = V(x)$.

$\tilde{x} = (\iota, \gamma)(x)$, where ι is the identity in S_k and $\gamma = (bc)$: for each $1 \le i \le k$,

$$\begin{aligned} I(\tilde{x}_{2i-1} = 1) &= I(\tilde{x}_{2i-1} = 1,\ \tilde{x}_{2i} = 3) + I(\tilde{x}_{2i-1} = 1,\ \tilde{x}_{2i} = 4) \\ &= I(x_{2i-1} = 1,\ x_{2i} = 3) + I(x_{2i-1} = 2,\ x_{2i} = 3) \\ &= I(x_{2i} = 3), \end{aligned}$$

and similarly

$$I(\tilde{x}_{2i-1} = 2) = I(x_{2i} = 4).$$

Hence, for $1 \le i \le k$,

$$V_{2i-1}(\tilde{x}) = V_{2i}(x), \tag{C.1}$$

and consequently $V(x) = V(\tilde{x})$.

$\tilde{x} = (\sigma, \iota)(x)$, where $\sigma \in S_k$ and ι is the identity in D_4: For $1 \le i \le k$, $\tilde{x}_{2i-1} = x_{2\sigma^{-1}(i)-1}$ and $\tilde{x}_{2i} = x_{2\sigma^{-1}(i)}$, thus

$$\begin{aligned} V(\tilde{x}) &= \sum_{(i,j)} \{V_{2\sigma^{-1}(i)-1}(x)V_{2\sigma^{-1}(j)-1}(x) + V_{2\sigma^{-1}(i)}(x)V_{2\sigma^{-1}(j)}(x)\} \\ &= \sum_{(i,j)} \{V_{2i-1}(x)V_{2j-1}(x) + V_{2i}(x)V_{2j}(x)\} = V(x). \end{aligned}$$

In particular, for $k > 1$,

$$V(1,3,1,3,\ldots,1,3) = \sum_{(i,j)} (-1)(-1) + (-1)(-1) = 2\binom{k}{2} = k(k-1) \neq 0.$$

Hence, since V is a linear combination of eigenvectors of A which is nonzero, then V is an eigenvector of A corresponding to the eigenvalue $2k - 4$. Furthermore, V is constant on the orbits of $S_k \times D_4$. □

FACT 7. The $k(k-1)$ 2-factor eigenvectors $\{V_{2i-1,2j-1}, V_{2i,2j}: 1 \leq i < j \leq k\}$ have the same sums over the orbits of $S_k \times D_4$, that is, for any orbit $\mathscr{C}$ and $1 \leq i_1 < j_1 \leq k$, $1 \leq i_2 < j_2 \leq k$,

$$\sum_{x\in\mathscr{C}} V_{2i_2-1,2j_2-1}(x) = \sum_{x\in\mathscr{C}} V_{2i_1-1,2j_1-1}(x) = \sum_{x\in\mathscr{C}} V_{2i_1,2j_1}(x) = \sum_{x\in\mathscr{C}} V_{2i_2,2j_2}(x).$$

PROOF. Let $x \in \mathscr{C}$, then $\tilde{x} = (\iota, \gamma)(x) \in \mathscr{C}$, and by (C.1), for each $1 \leq i < j \leq k$,

$$\sum_{x\in\mathscr{C}} V_{2i-1}(x)V_{2j-1}(x) = \sum_{x\in\mathscr{C}} V_{2i}(\tilde{x})V_{2j}(\tilde{x}) = \sum_{x\in\mathscr{C}} V_{2i}(x)V_{2j}(x).$$

Also, consider any permutation $\sigma \in S_k$, then $\tilde{x} = (\sigma, \iota)(x) \in \mathscr{C}$ and

$$\sum_{x\in\mathscr{C}} V_{2i}(x)V_{2j}(x) = \sum_{x\in\mathscr{C}} V_{2i}(\tilde{x})V_{2j}(\tilde{x}) = \sum_{x\in\mathscr{C}} V_{2\sigma^{-1}(i)}(x)V_{2\sigma^{-1}(j)}(x).$$

Similarly for $V_{2i-1,2j-1}$.

PROPOSITON 4 [Eigenvalues of adjacency matrix B of quotient graph $\mathscr{X}/(S_k \times D_4)$]. *The two largest eigenvalues of Bare $2k$ and $2k-4$, and have multiplicity one. All other eigenvalues of B are strictly less than $2k-4$ and belong to the set $\{2(k-i)_{\binom{2k}{i}}: i = 3, \ldots, 2k\}$, where $\binom{2k}{i}$ is the largest possible multiplicity of the eigenvalue $2(k-i)$. The eigenvector v corresponding to $2k-4$ may be obtained from*

$$V(x) = \sum_{(i,j)} \{V_{2i-1,2j-1}(x) + V_{2i,2j}(x)\},$$

by letting

$$v_i = V(x) \quad \textit{where } x \textit{ is any } x \in \mathscr{C}_i.$$

PROOF. From Proposition 3 and Fact 1, the eigenvalues of B belong to the set $\{2(k-i)_{\binom{2k}{i}}: i = 0, \ldots, 2k\}$.

$\lambda = 2k$: the rows of B sum to $2k$, hence $2k$ is an eigenvalue of B with corresponding eigenvector $\mathbf{1} = (1, 1, \ldots, 1)^T$.

$\lambda = 2k - 2$: from Fact 4, eigenvectors of A corresponding to the eigenvalue $2k-2$ sum to 0 over the orbits of $S_k \times D_4$, hence no eigenvector of A can be constant and nonzero over the orbits. Hence, from Fact 2, $2k-2$ is not an eigenvalue of B.

$\lambda = 2k-4$: we have shown with Fact 6 that V is an eigenvector of A, corresponding to the eigenvalue $2k-4$, which is constant over the orbits. Hence, by Fact 3, V yields an eigenvector of B. It remains to show that B has no other eigenvector, that is, V is the only eigenvector of A which is constant over the orbits. The orthogonal complement of V in the eigenspace of A for $\lambda = 2k-4$ is spanned by the following $2k^2-k$ vectors:

$$\begin{aligned} W_{2i-1,2j} &= V_{2i-1,2j} - \frac{\langle V_{2i-1,2j}, V\rangle}{|V|^2} V \\ &= V_{2i-1,2j}, \qquad 1 \le i, j \le k, \\ W_{2i-1,2j-1} &= V_{2i-1,2j-1} - \frac{\langle V_{2i-1,2j-1}, V\rangle}{|V|^2} V \\ &= V_{2i-1,2j-1} - \frac{1}{k(k-1)} V, \qquad 1 \le i < j \le k, \\ W_{2i,2j} &= V_{2i,2j} - \frac{\langle V_{2i,2j}, V\rangle}{|V|^2} V \\ &= V_{2i,2j} - \frac{1}{k(k-1)} V, \qquad 1 \le i < j \le k. \end{aligned}$$

By Fact 5, for any orbit $\mathscr{C}$,

$$\sum_{x\in\mathscr{C}} W_{2i-1,2j}(x) = 0.$$

Also, by Fact 7,

$$\begin{aligned} &\sum_{x\in\mathscr{C}} W_{2i-1,2j-1}(x) \\ &\quad = \sum_{x\in\mathscr{C}} V_{2i-1,2j-1}(x) - \frac{1}{k(k-1)} \sum_{(i,j)} \sum_{x\in\mathscr{C}} \{V_{2i-1,2j-1}(x) + V_{2i,2j}(x)\} = 0, \end{aligned}$$

and similarly

$$\sum_{x\in\mathscr{C}} W_{2i,2j}(x) = 0.$$

Hence, no eigenvector in the orthogonal complement of V in the eigenspace of A for $\lambda = 2k-4$ is constant over the orbits of $S_k \times D_4$. Consequently, by Fact 2, $2k-4$ is an eigenvalue of B with multiplicity 1. □

C.2. *Quotient graph* $\mathscr{X}/(S_k \times (C_2 \times C_2))$. Facts 4 and 5 also apply to the orbits of $S_k \times (C_2 \times C_2)$. Facts 6 and 7 may be modified as follows.

A SCORE TEST FOR LINKAGE

FACT 8. Let

$$V_o(x) = \sum_{(i,j)} V_{2i-1,2j-1}(x)$$

and

$$V_e(x) = \sum_{(i,j)} V_{2i,2j}(x),$$

where the sums are over all $\binom{k}{2}$ unordered pairs (i, j) of distinct integers ranging from 1 to k. Then, V_e and V_o are two eigenvectors of A corresponding to the eigenvalue $2k-4$. Furthermore, V_e and V_o are constant over the orbits of $S_k \times (C_2 \times C_2)$.

FACT 9. The $k(k-1)/2$ 2-factor eigenvectors $\{V_{2i-1,2j-1}: 1 \le i < j \le k\}$ have the same sums over the orbits of $S_k \times (C_2 \times C_2)$, that is, for any orbit $\mathscr{C}$ and $1 \le i_1 < j_1 \le k$, $1 \le i_2 < j_2 \le k$,

$$\sum_{x \in \mathscr{C}} V_{2i_1-1,2j_1-1}(x) = \sum_{x \in \mathscr{C}} V_{2i_2-1,2j_2-1}(x).$$

Similarly for the $k(k-1)/2$ 2-factor eigenvectors $\{V_{2i,2j}: 1 \le i < j \le k\}$.

PROPOSITION 5 [Eigenvalues of adjacency matrix B of quotient graph $\mathscr{X}/(S_k \times (C_2 \times C_2))$]. *The largest eigenvalue of B is $2k$, with multiplicity one, and the second largest eigenvalue is $2k-4$, with multiplicity two. All other eigenvalues of B are strictly less than $2k-4$ and belong to the set $\{2(k-i)_{\binom{2k}{i}}: i = 3, \ldots, 2k\}$. The eigenvectors corresponding to $2k-4$ may be obtained from V_e and V_o.*

PROOF. From Proposition 3 and Fact 1, the eigenvalues of B belong to the set $\{2(k-i)_{\binom{2k}{i}}: i = 0, \ldots, 2k\}$.

$\lambda = 2k$: the rows of B sum to $2k$, hence $2k$ is an eigenvalue of B with corresponding eigenvector $\mathbf{1} = (1, 1, \ldots, 1)^T$.

$\lambda = 2k-2$: from Fact 4, eigenvectors of A corresponding to the eigenvalue $2k-2$ sum to 0 over the orbits of $S_k \times (C_2 \times C_2)$, hence no eigenvector of A can be constant and nonzero over the orbits. Hence, from Fact 2, $2k-2$ is not an eigenvalue of B.

$\lambda = 2k-4$: from Fact 8, V_o and V_e are eigenvectors of A, corresponding to the eigenvalue $2k-4$, which are constant over the orbits. Hence, by Fact 3, V_e and V_o yield two eigenvectors of B. It remains to show that B has only two eigenvectors, that is, V_e and V_o are the only eigenvectors of A which are constant over the orbits. The orthogonal complement of Span$\{V_o, V_e\}$ in the eigenspace of A for $\lambda = 2k-4$ is spanned by the following $2k^2-k$ vectors:

$$\begin{aligned} W_{2i-1,2j} &= V_{2i-1,2j} - \frac{\langle V_{2i-1,2j}, V_e\rangle}{|V_e|^2} V_e - \frac{\langle V_{2i-1,2j}, V_o\rangle}{|V_o|^2} V_o \\ &= V_{2i-1,2j}, \qquad 1 \le i,\ j \le k, \end{aligned}$$

$$W_{2i-1,2j-1} = V_{2i-1,2j-1} - \frac{\langle V_{2i-1,2j-1}, V_e\rangle}{|V_e|^2} V_e - \frac{\langle V_{2i-1,2j-1}, V_o\rangle}{|V_o|^2} V_o$$
$$= V_{2i-1,2j-1} - \frac{2}{k(k-1)} V_o, \qquad 1 \le i < j \le k,$$
$$W_{2i,2j} = V_{2i,2j} - \frac{\langle V_{2i,2j}, V_e\rangle}{|V_e|^2} V_e - \frac{\langle V_{2i,2j}, V_o\rangle}{|V_o|^2} V_o$$
$$= V_{2i,2j} - \frac{2}{k(k-1)} V_e, \qquad 1 \le i < j \le k.$$

By Fact 5, for any orbit $\mathscr{C}$,

$$\sum_{x\in\mathscr{C}} W_{2i-1,2j}(x) = 0.$$

Also, by Fact 9,

$$\sum_{x\in\mathscr{C}} W_{2i-1,2j-1}(x) = \sum_{x\in\mathscr{C}} V_{2i-1,2j-1}(x) - \frac{2}{k(k-1)} \sum_{(i,j)} \sum_{x\in\mathscr{C}} V_{2i-1,2j-1}(x) = 0,$$

and similarly,

$$\sum_{x\in\mathscr{C}} W_{2i,2j}(x) = 0.$$

Hence, no eigenvector in the orthogonal complement of Span$\{V_o, V_e\}$ in the eigenspace of A for $\lambda = 2k-4$ is constant over the orbits of $S_k \times (C_2 \times C_2)$. Consequently, by Fact 2, $2k-4$ is an eigenvalue of B with multiplicity 2. □

C.3. *Quotient graph* $\mathscr{X}/((S_h \times S_{k-h}) \times D_4)$. Facts 4 and 5 also apply to the orbits of $\mathscr{X}/((S_h \times S_{k-h}) \times D_4)$. The proof for sibships with both affected and unaffected sibs is similar to that for affected-only sibships, but involves new combinations of eigenvectors. Without loss of generality, order the sibs such that the first h are affected and the last $k-h$ unaffected. For $k \ge 3$, define

$$V^a(x) = \sum_{1\le i<j\le h} \{V_{2i-1,2j-1}(x) + V_{2i,2j}(x)\}, \qquad h \ge 2,$$
$$V^u(x) = \sum_{h+1\le i<j\le k} \{V_{2i-1,2j-1}(x) + V_{2i,2j}(x)\}, \qquad h \le k-2,$$
$$V^{au}(x) = \sum_{1\le i\le h,\, h+1\le j\le k} \{V_{2i-1,2j-1}(x) + V_{2i,2j}(x)\}.$$

Facts 6 and 7 are then modified as follows.

FACT 10. For $k \ge 3$, V^a ($h \ge 2$), V^u ($h \le k-2$) and V^{au} are eigenvectors of A corresponding to the eigenvalue $2k-4$. Furthermore, these are constant over the orbits of $\mathscr{X}/((S_h \times S_{k-h}) \times D_4)$.

FACT 11. For any orbit $\mathscr{C}$ of $\mathscr{X}/((S_h \times S_{k-h}) \times D_4)$ and $1 \le i_1 < j_1 \le h$, $1 \le i_2 < j_2 \le h$,

$$\sum_{x\in\mathscr{C}} V_{2i_2-1,\,2j_2-1}(x) = \sum_{x\in\mathscr{C}} V_{2i_1-1,\,2j_1-1}(x) = \sum_{x\in\mathscr{C}} V_{2i_1,\,2j_1}(x) = \sum_{x\in\mathscr{C}} V_{2i_2,\,2j_2}(x).$$

Similarly for $h+1 \le i_1 < j_1 \le k$, $h+1 \le i_2 < j_2 \le k$ and $1 \le i_1, i_2 \le h$, $h+1 \le j_1, j_2 \le k$.

PROPOSITION 6 [Eigenvalues of adjacency matrix B of quotient graph $\mathscr{X}/((S_h \times S_{k-h}) \times D_4)$]. *The largest eigenvalue of B is $2k$, with multiplicity one, and the second largest eigenvalue is $2k-4$, with multiplicity three if $2 \le h \le k-2$ and two otherwise. All other eigenvalues of B are strictly less than $2k-4$ and belong to the set $\{2(k-i)_{\binom{2k}{i}}: i = 3, \dots, 2k\}$. The eigenvectors corresponding to $2k-4$ may be obtained from V^a, V^u and V^{au}.*

PROOF. From Proposition 3 and Fact 1, the eigenvalues of B belong to the set $\{2(k-i)_{\binom{2k}{i}}: i = 0, \dots, 2k\}$. We give the proof for $\lambda = 2k-4$; for the other eigenvalues, the proof is as for Propositions 4 and 5. From Fact 10, V^a $(h \ge 2)$, V^u $(h \le k-2)$ and V^{au} are eigenvectors of A, corresponding to the eigenvalue $2k-4$, which are constant over the orbits. Hence, by Fact 3, they yield eigenvectors of B. It remains to show that these are the only eigenvectors of B, that is, V^a, V^u and V^{au} are the only eigenvectors of A which are constant over the orbits. The orthogonal complement of Span$\{V^a, V^u, V^{au}\}$ in the eigenspace of A for $\lambda = 2k-4$ is spanned by the $2k^2 - k$ vectors $W_{i,\,j}$, $1 \le i, j \le 2k$, $i \ne j$, defined as follows

$$W_{i,\,j} = V_{i,\,j} - \frac{\langle V_{i,\,j}, V^a\rangle}{|V^a|^2} V^a - \frac{\langle V_{i,\,j}, V^u\rangle}{|V^u|^2} V^u - \frac{\langle V_{i,\,j}, V^{au}\rangle}{|V^{au}|^2} V^{au}.$$

The $W_{i,\,j}$'s simplify to

$$W_{2i-1,\,2j} = V_{2i-1,\,2j}, \qquad i, j = 1, \dots, k,$$

$$\begin{aligned} W_{2i-1,\,2j-1} &= V_{2i-1,\,2j-1} - \frac{1}{h(h-1)} V^a, \qquad 1 \le i < j \le h,\\ &= V_{2i-1,\,2j-1} - \frac{1}{(k-h)(k-h-1)} V^u, \qquad h+1 \le i < j \le k,\\ &= V_{2i-1,2j-1} - \frac{1}{2h(k-h)} V^{au}, \qquad 1 \le i \le h,\ h+1 \le j \le k,\end{aligned}$$

$$\begin{aligned} W_{2i,\,2j} &= V_{2i,\,2j} - \frac{1}{h(h-1)} V^a, \qquad 1 \le i < j \le h,\\ &= V_{2i,2j} - \frac{1}{(k-h)(k-h-1)} V^u, \qquad h+1 \le i < j \le k,\\ &= V_{2i,2j} - \frac{1}{2h(k-h)} V^{au}, \qquad 1 \le i \le h,\ h+1 \le j \le k.\end{aligned}$$

By Fact 5, for any orbit $\mathscr{C}$,

$$\sum_{x\in\mathscr{C}} W_{2i-1,\,2j}(x) = 0.$$

Also, by Fact 11,

$$\sum_{x\in\mathscr{C}} W_{2i-1,\,2j-1}(x) = 0$$

and

$$\sum_{x\in\mathscr{C}} W_{2i,\,2j}(x) = 0.$$

Hence, no eigenvector in the orthogonal complement of Span$\{V^a, V^u, V^{au}\}$ in the eigenspace of A for $\lambda = 2k - 4$ is constant over the orbits of $\mathscr{X}/((S_h \times S_{k-h}) \times D_4)$. Consequently, by Fact 2, $2k - 4$ is an eigenvalue of B with multiplicity three if $2 \le h \le k - 2$ and two otherwise. □

C.4. *Quotient graph* $\mathscr{X}/((S_h \times S_{k-h}) \times (C_2 \times C_2))$. For $(S_h \times S_{k-h}) \times (C_2 \times C_2)$ we again separate "even" and "odd" eigenvectors and consider six new combinations of eigenvectors,

$$V_e^a(x) = \sum_{1\le i<j\le h} V_{2i,\,2j}(x), \qquad h \ge 2,$$

$$V_o^a(x) = \sum_{1\le i<j\le h} V_{2i-1,\,2j-1}(x), \qquad h \ge 2,$$

$$V_e^u(x) = \sum_{h+1\le i<j\le k} V_{2i,\,2j}(x), \qquad h \le k-2,$$

$$V_o^u(x) = \sum_{h+1\le i<j\le k} V_{2i-1,\,2j-1}(x), \qquad h \le k-2,$$

$$V_e^{au}(x) = \sum_{1\le i\le h,\,h+1\le j\le k} V_{2i,\,2j}(x),$$

$$V_o^{au}(x) = \sum_{1\le i\le h,\,h+1\le j\le k} V_{2i-1,\,2j-1}(x).$$

Facts 6 and 7 may then be suitably modified.

D. Score statistic: Proof of Theorem 2. From Theorem 1, -4 is an eigenvalue of the infinitesimal generator Q with multiplicity 1. Hence, the second derivative of the transition matrix at $\theta = \frac{1}{2}$ has rank 1 and entries

$$u_{ij} = 8\alpha_j v_i v_j,$$

where $v = (v_1, \ldots, v_m)^T$ is the right eigenvector of Q with unit norm with respect to the inner product $\langle \cdot, \cdot \rangle_\alpha$. The score statistic is given by

$$S = \sum_{i=1}^{m} N_i \frac{\sum_{j=1}^{m} \pi_j 8\alpha_i v_i v_j}{\alpha_i}$$
$$= 8\Big(\sum_{j=1}^{m} v_j \pi_j \Big)\Big(\sum_{i=1}^{m} v_i N_i \Big).$$

It is convenient to express the score statistic in terms of the first column of U, $8\alpha_1 v_1 v$. Without loss of generality, we let the first IBD configuration be the one for which all sibs inherited the same maternal and paternal DNA, that is, with representative inheritance vector $(1, 3, 1, 3, \ldots, 1, 3)$ and label $(0, 0, 0)$ in the notation of Ethier and Hodge (1985):

$$S = \frac{8}{8(8\alpha_1 v_1^2)\alpha_1} \Big(\sum_{j=1}^{m} u_{j1}\pi_j \Big)\Big(\sum_{i=1}^{m} u_{i1} N_i \Big)$$
$$= \frac{2^{2k}}{u_{11}|\mathscr{C}_1|} \Big(\sum_{j=1}^{m} u_{j1}\pi_j \Big)\Big(\sum_{i=1}^{m} u_{i1} N_i \Big).$$

By differentiating (3.2) we find that

$$u_{ij} = 2^{4-2k} \sum_{y \in \mathscr{C}_j} ((\Delta(x, y) - k)^2 - k/2), \text{ where } x \text{ is any inheritance vector in } \mathscr{C}_i$$

and in particular, $u_{11} = 2^{5-2k} k(k-1)$. The contribution of an affected sib-k-tuple with inheritance vector $x \in \mathscr{C}_i$ to the score statistic is based on

$$u_{i1} = 2^{4-2k} \sum_{y \in \mathscr{C}_1} ((\Delta(x, y) - k)^2 - k/2)$$
$$= 2^{4-2k}((a_2(x) + a_4(x) - k)^2 + (a_2(x) + a_3(x) - k)^2$$
$$+ (a_1(x) + a_4(x) - k)^2 + (a_1(x) + a_3(x) - k)^2 - 2k)$$
$$= 2^{4-2k}(2(a_1(x)^2 + a_2(x)^2 + a_3(x)^2 + a_4(x)^2)$$
$$+ 2(a_2(x)a_4(x) + a_2(x)a_3(x) + a_1(x)a_4(x) + a_1(x)a_3(x))$$
$$- 2k(2a_1(x) + 2a_2(x) + 2a_3(x) + 2a_4(x)) + 4k^2 - 2k)$$
$$= 2^{5-2k}(a_1(x)^2 + a_2(x)^2 + a_3(x)^2 + a_4(x)^2$$
$$+ (a_1(x) + a_2(x))(a_3(x) + a_4(x)) - 4k^2 + 2k^2 - k)$$
$$= 2^{5-2k}(a_1(x)^2 + a_2(x)^2 + a_3(x)^2 + a_4(x)^2 + k^2 - 2k^2 - k)$$
$$= 2^{5-2k}(a_1(x)^2 + a_2(x)^2 + a_3(x)^2 + a_4(x)^2 - k(k+1)).$$

Hence, from (4.2),

$$u_{i1} = 2^{5-2k}k(k-1)(2S_{\text{pairs}} - 1).$$

Now, $u_{i1} = 8\alpha_1 v_1 v_i$, $u_{11} = 2^{5-2k}k(k-1)$ and $\alpha_1 = 2^{2-2k}$. Thus

$$v_1 = \sqrt{k(k-1)} \quad \text{and} \quad v_i = \sqrt{k(k-1)}(2S_{\text{pairs}} - 1).$$

Hence

$$S = 2^{2k-2}\Big(\sum_j u_{j1}\pi_j\Big)(2S_{\text{pairs}} - n) = 8\sqrt{k(k-1)}\Big(\sum_j v_j\pi_j\Big)(2S_{\text{pairs}} - n),$$

where S_{pairs} is summed over all sibships with k affected sibs. □

Acknowledgments. We are grateful to Steve Evans for many helpful discussions on Markov chains and to Cheryl Praeger, Alice Niemeyer and Nick Wormald for graph-related ideas. We also thank the referees for their valuable comments and suggestions on a previous version of this paper.

REFERENCES

BLACKWELDER, W. C. and ELSTON, R. C. (1985). A comparison of sib-pair linkage tests for disease susceptibility loci. *Genet. Epidemiol.* **2** 85–97.

DAVIS, S. and WEEKS, D. E. (1997). Comparison of nonparametric statistics for detection of linkage in nuclear families: single-marker evaluation. *Amer. J. Hum. Genet.* **61** 1431–1444.

DAY, N. E. and SIMONS, M. J. (1976). Disease-susceptibility genes—their identification by multiple case family studies. *Tissue Antigens* **8** 109–119.

DEBRUIJN, N. G. (1964). Pólya's theory of counting. In *Applied Combinatorial Mathematics* (E. F. Beckenbach, ed.) 144–184. Wiley, New York.

DIACONIS, P. (1988). *Group Representations in Probability and Statistics.* IMS, Hayward, CA.

DONNELLY, K. P. (1983). The probability that related individuals share some section of genome identical by descent. *Theoret. Population Biol.* **23** 34–63.

DUDOIT, S. (1999). Linkage analysis of complex human traits using identity by descent data. Ph.D. dissertation, Univ. California, Berkeley.

DUDOIT, S. and SPEED, T. P. (1999). Triangle constraints for sib-pair identity by descent probabilities under a general multilocus model for disease susceptibility. In *Statistics in Genetics* (M. E. Halloran and S. Geisser, eds.). Springer, New York.

ETHIER, S. N. and HODGE, S. E. (1985). Identity-by-descent analysis of sibship configurations. *Amer. J. Med. Genet.* **22** 263–272.

FARAWAY, J. J. (1993). Improved sib-pair linkage test for disease susceptibility loci. *Genet. Epidemiol.* **10** 225–233.

FEINGOLD, E., BROWN, P. O. and SIEGMUND, D. (1993). Gaussian models for genetic linkage analysis using complete high-resolution maps of identity by descent. *Amer. J. Hum. Genet.* **53** 234–251.

FEINGOLD, E. and SIEGMUND, D. (1997). Strategies for mapping heterogeneous recessive traits by allele sharing methods. *Amer. J. Hum. Genet.* **60** 965–978.

FRALEIGH, J. B. (1989). *A First Course in Abstract Algebra*, 4th ed. Addison-Wesley, Reading, MA.

GODSIL, C. D. (1993). *Algebraic Combinatorics.* Chapman and Hall, New York.

A SCORE TEST FOR LINKAGE

HASEMAN, J. K. and ELSTON, R. C. (1972). The investigation of linkage between a quantitative trait and a marker locus. *Behavior Genetics* **2** 3–19.

HOLMANS, P. (1993). Asymptotic properties of affected sib-pair linkage analysis. *Amer. J. Hum. Genet.* **52** 362–374.

HOLMANS, P. and CLAYTON, D. (1995). Efficiency of typing unaffected relatives in an affected-sib-pair linkage study with single-locus and multiple tightly linked markers. *Amer. J. Hum. Genet.* **57** 1221–1232.

JACOB, B. (1990). *Linear Algebra*. Freeman, New York.

KNAPP, M., SEUCHTER, S. A. and BAUR, M. P. (1994a). Linkage analysis in nuclear families I: optimality criteria for affected sib-pair tests. *Hum. Hered.* **44** 37–43.

KNAPP, M., SEUCHTER, S. A. and BAUR, M. P. (1994b). Linkage analysis in nuclear families II: relationship between affected sib-pair tests and lod score analysis. *Hum. Hered.* **44** 44–51.

KNAPP, M., SEUCHTER, S. A. and BAUR, M. P. (1994c). Two-locus disease models with two marker loci: the power of affected-sib-pair tests. *Amer. J. Hum. Genet.* **55** 1030–1041.

KONG, A. and COX, N. J. (1997). Allele-sharing models: lod scores and accurate linkage tests. *Amer. J. Hum. Genet.* **61** 1179–1188.

KRUGLYAK, L., DALY, M. J., REEVE-DALY, M. P. and LANDER, E. S. (1996). Parametric and non-parametric linkage analysis: a unified multipoint approach. *Amer. J. Hum. Genet.* **58** 1347–1363.

KRUGLYAK, L. and LANDER, E. S. (1995). Complete multipoint sib-pair analysis of qualitative and quantitative traits. *Amer. J. Hum. Genet.* **57** 439–454.

LANDER, E. S. and GREEN, P. (1987). Construction of multilocus genetic maps in humans. *Proc. Nat. Acad. Sci. U.S.A.* **84** 2363–2367.

MCPEEK, M. S. (1996). An introduction to recombination and linkage analysis. In *Genetic Mapping and DNA Sequencing* (T. P. Speed and M. S. Waterman, eds.) 1–14. Springer, New York.

MCPEEK, M. S. (1999). Optimal allele-sharing statistics for genetic mapping using affected relatives. *Genet. Epidemiol.* **16** 225–249.

OTT, J. (1991). *Analysis of Human Genetic Linkage*, rev. eds. Johns Hopkins Univ. Press, Baltimore.

RISCH, N. (1990a). Linkage strategies for genetically complex traits II. The power of affected relative pairs. *Amer. J. Hum. Genet.* **46** 229–241.

RISCH, N. (1990b). Linkage strategies for genetically complex traits III. The effect of marker polymorphism on analysis of affected relative pairs. *Amer. J. Hum. Genet.* **46** 242–253.

RISCH, N. and ZHANG, H. (1995). Extreme discordant sib pairs for mapping quantitative trait loci in humans. *Science* **268** 1584–1589.

RISCH, N. and ZHANG, H. (1996). Mapping quantitative trait loci with extreme discordant sib pairs: sampling considerations. *Amer. J. Hum. Genet.* **58** 836–843.

ROSENBLATT, M. (1974). *Random Processes. Graduate Texts in Mathematics* **17**, 2nd ed. Springer, New York.

SCHAID, D. J. and NICK, T. G. (1990). Sib-pair linkage tests for disease suscpetibility loci: common tests vs. the asymptotically most powerful test. *Genet. Epidemiol.* **7** 359–370.

SPEED, T. P. (1996). What is a genetic map function? In *Genetic Mapping and DNA Sequencing* (T. P. Speed and M. S. Waterman, eds.) 65–88. Springer, New York.

SUAREZ, B. K. and VAN EERDEWEGH, P. (1984). A comparison of three affected-sib-pair scoring methods to detect HLA-linked disease susceptibility genes. *Amer. J. Med. Genet.* **18** 135–146.

TENG, J. and SIEGMUND, D. O. (1997). Combining information within and between pedigrees for mapping complex traits. *Amer. J. Hum. Genet.* **60** 979–992.

THOMPSON, E. A. (1997). Conditional gene identity in affected individuals. In *Genetic Mapping of Disease Genes* (I.-H. Pawlowitzki, J. H. Edwards and E. A. Thompson, eds.) 137–146. Academic Press, San Diego.

VAN LINT, J. H. and WILSON, R. M. (1992). *A Course in Combinatorics*. Cambridge Univ. Press.

WHITTEMORE, A. S. (1996). Genome scanning for linkage: an overview. *Amer. J. Hum. Genet.* **59** 704–716.

WHITTEMORE, A. S. and HALPERN, J. (1994a). A class of tests for linkage using affected pedigree members. *Biometrics* **50** 118–127.

WHITTEMORE, A. S. and HALPERN, J. (1994b). Probability of gene identity by descent: computation and applications. *Biometrics* **50** 109–117.

DEPARTMENT OF STATISTICS
UNIVERSITY OF CALIFORNIA, BERKELEY
BERKELEY, CALIFORNIA 94720-3860
E-MAIL: sandrine@stat.berkeley.edu
terry@stat.berkeley.edu

Biostatistics (2006), **7**, 3, *pp.* 374–386
doi:10.1093/biostatistics/kxj012
Advance Access publication on December 14, 2005

Incorporating interference into linkage analysis for experimental crosses

NICOLA J. ARMSTRONG*
Division of Radiotherapy, Netherlands Cancer Institute, Plesmanlaan 121, 1066 CX Amsterdam, The Netherlands
n.armstrong@nki.nl

MARY SARA MCPEEK
Department of Statistics, University of Chicago, Chicago, IL 60637, USA

TERENCE P. SPEED
Department of Statistics, University of California at Berkeley, Berkeley, CA 94720, USA and Division of Genetics and Bioinformatics, Walter and Eliza Hall Institute of Medical Research, Melbourne, Australia

SUMMARY

The phenomenon of interference in genetic recombination is well-known and studied in a wide variety of organisms. Multilocus linkage analysis, which makes use of recombination patterns among all genetic markers simultaneously, is routinely used with data on humans and experimental organisms to build genetic maps. It is also used to try to determine the genes involved in traits of interest, such as common diseases. Most linkage analyses performed today ignore the occurrence of genetical interference. We present an extension to the Lander–Green algorithm for experimental crosses (backcross and intercross) to incorporate crossover interference according to the χ^2 model. Simulation results show the impact of using this model on the accuracy of estimated genetic maps.

Keywords: Crossover interference; Hidden Markov model; Linkage analysis.

1. INTRODUCTION

Genetical interference was first observed in Thomas Hunt Morgan's *Drosophila* laboratory early last century (Muller, 1916). During meiosis, chromosomes replicate, producing sister chromatids. The homologous chromosomes then pair and synapse, forming a four-strand bundle. Once pairing is complete, crossing over, the reciprocal exchange of chromosomal segments among nonsister chromatids, begins. As the chromosomes separate, the crossover positions become visible as chiasmata. Two types of interference are distinguished in this crossover process: chromatid interference, the nonrandom choice of chromatids involved in adjacent chiasma, and crossover interference, the nonrandom placement of chiasma along a chromosome. Experimental evidence in various organisms has shown that crossovers do not occur at

*To whom correspondence should be addressed.

Incorporating interference into linkage analysis

random, rather they are more evenly spaced along a chromosome (e.g. Hultén, 1974; Blank *et al.*, 1988). Chromatid interference can only be detected if all four products of meiosis can be recovered. However, to date there has been little evidence of this type of interference in experimental organisms.

For experimental crosses, the most common method of linkage analysis in use is the Lander–Green (LG) algorithm (Lander and Green, 1987). This algorithm makes use of a hidden Markov model (HMM) to construct a genetic map using multilocus linkage analysis. The LG approach is based on the assumption that there is no interference, either crossover or chromatid (NI model). The crossover positions are assumed to follow a Poisson distribution and the nonsister chromatids are chosen at random for each crossover.

Various models for the crossover process have been introduced over the years (e.g. Karlin and Liberman, 1979; Risch and Lange, 1979; King and Mortimer, 1990; McPeek and Speed, 1995). The χ^2 distribution was first used in a model for crossing over in Fisher *et al.* (1947). We consider here a four-strand version of this model. Foss *et al.* (1993) gave a simple biological motivation and interpretation for the model based on gene conversions. They proposed that gene conversions are intermediate events that can resolve in reciprocal crossing over but do not need to do so. The χ^2 model assumes that m noncrossover resolutions of gene conversion occur between each crossover. The gene conversion events follow a Poisson distribution, and the distance between actual crossovers follows a χ^2 distribution with $2(m+1)$ degrees of freedom. This model has been found to fit observed data from a variety of organisms as well (Foss *et al.*, 1993; Zhao *et al.*, 1995; Broman *et al.*, 2002), indeed better than other models (McPeek and Speed, 1995; Broman and Weber, 2000). Yet, despite the literature on alternative crossover models, in practice the phenomenon of interference is routinely ignored in genetic-mapping exercises. At most, the Kosambi map function is used after recombination fractions have been estimated under the NI model.

Geneticists have also known for many years that the rate of recombination differs, often substantially, between males and females for organisms such as humans and the mouse. In other species, such as the Lepidoptera *Bombyx mori* and *Drosophila melanogaster*, recombination does not occur at all in one of the sexes. This is equivalent to the genetic distances between markers being zero since no recombination is observed between them. Unfortunately, due to lack of parameter identifiability it is not possible to estimate sex-specific distances from an F_2 cross. The two heterozygous genotypes are indistinguishable, so that if both recombination fractions are greater than zero, the chromosome on which the crossover occurred cannot be determined. In the case of no recombination in one sex, genetic distances can be estimated from a group of F_2 progeny. However, if a model of equal recombination between the sexes is used, the genetic distance estimates will be too small as the model produces, in effect, sex-averaged genetic distance estimates. When recombination is absent in one sex, these distances will be much smaller than the true distances in the recombinant sex.

In this paper, we present an extension of the LG algorithm to incorporate crossover interference according to the χ^2 model. The following section presents the algorithm for both the backcross (BC) and F_2 cases. The special case of no recombination in one sex is also considered for the F_2 progeny. In the results section, the accuracy of this extended algorithm in estimating genetic maps is evaluated through simulation studies and compared to the original LG algorithm and the Kosambi transformation. The extra time required to obtain genetic maps when incorporating crossover interference is also considered.

2. Model

Let $\mathcal{M}_1, \ldots, \mathcal{M}_T$ denote T genetic loci listed in order along a chromosome. The observed genotypic information can be denoted $O = (O_1, \ldots, O_T)$, where O_t contains the genotypes of all individuals at $\mathcal{M}_t$. Define $\lambda_t = 2(m+1)d_t$, where d_t is the genetic distance, in Morgans, between $\mathcal{M}_t$ and $\mathcal{M}_{t+1}$ and m is a defined constant. An important difference between the LG and χ^2 algorithms is that LG estimates the recombination fraction while the χ^2 algorithm estimates λ_t. If $m = 0$, this gives a direct estimate of the genetic distance and no transformation is necessary. For $m > 0$, it estimates a fixed multiple of the genetic distance.

Under the χ^2 model, there are assumed to be m unobserved crossover intermediates between each crossover. The hidden chain y_t for the χ^2 model keeps track of both the number of crossover intermediates that have occurred since the last crossover and the origin of the DNA, grandmaternal (gm) or grandpaternal (gp). Specifically, the states $i = 0, \ldots, m$ represent the event of i crossover intermediates and gm origin of the DNA, while states $i = m+1, \ldots, 2m+1$ represent the event of $i-(m+1)$ crossover intermediates and gp origin of the DNA. The process along a chromosome can be described by the embedded chain in Figure 1. Every $(m+1)$st event results in a crossover that involves the strand of interest with probability $\frac{1}{2}$. For example, starting in state 0 we must progress through m steps to state m, the next intermediate which will result in a crossover. When in state m, there is either a crossover and switch to gp DNA (state $m+1$) or there is no crossover and we stay on gm DNA, returning to state 0. Note that if $m = 0$, the chain is that of the NI model with the hidden states simply representing regions of gm and gp DNA on the chromosome.

Conditional on parental genotypes, the genotypes of each offspring are independent. Hence, the likelihood may be calculated for each individual separately and then combined. In the discussion that follows, the HMM is defined for one individual. The superscript k is used when necessary to denote the kth offspring and omitted otherwise. It is also assumed, without loss of generality, that the F_1 parent is crossed with individuals from the grandmaternal strain to produce BC progeny.

2.1 *Backcross*

Given the inbred strains AA and aa, at any locus the BC offspring will have genotype Aa or the genotype of the homozygous parent (either aa or AA depending on which line the homozygous parent came from). We define the observed genotypes at a given locus of a BC animal to be A if homozygous and H if

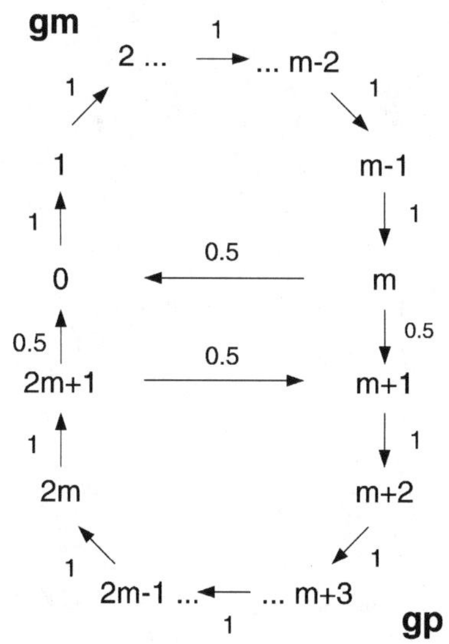

Fig. 1. Embedded chain along a chromosome for the χ^2 model with parameter m.

heterozygous. Hence, for a BC experiment, the elements of O are A, H and $-$, the three possible observed genotypes, where '$-$' denotes missing. In this case, we need consider only the chromosome inherited from the F_1 parent.

The initial-state distribution is uniform over all states. The transition probability matrices, $A(\lambda_t)$, at each marker locus have $2(m+1) \times 2(m+1)$ entries of which only $3m+2$ are distinct. The entries $a_{ij}(t)$ can be related to the distinct elements $a_l(t)$ by grouping the pairs (i, j) into sets I_l:

1. If $0 \leqslant i, j \leqslant m$ or $i, j > m$, $(i, j) \in I_l$, where $l = j - i$.
2. If $0 \leqslant i \leqslant m$, $j > m$, then for $j > i + m$, $(i, j) \in I_l$, where $l = j - i$, else $(i, j) \in I_l$, where $l = j - i - (m+1)$.
3. If $i > m$, $0 \leqslant j \leqslant m$, then for $j + (m+1) > i - 1$, $(i, j) \in I_l$, where $l = j - i + 2m + 2$, else $(i, j) \in I_l$, where $l = j - i + (m+1)$.

For example, consider $a_{12}(t)$ when $m = 2$. The chain progresses from state 1 to state 2 if there has been a single intermediate event, if there have been four intermediate events but no crossover has occurred ($1 \to 2 \to 0 \to 1 \to 2$), or if any multiple of four intermediate events and an even number of crossovers have occurred. This is also the same set of circumstances under which the chain could move from state 4 to state 5. In this case, the transition matrix is given by:

$$A(\lambda_t) = \begin{pmatrix} a_0 & a_1 & a_2 & a_3 & a_4 & a_5 \\ a_{-1} & a_0 & a_1 & a_{-1} & a_3 & a_4 \\ a_{-2} & a_{-1} & a_0 & a_{-2} & a_{-1} & a_3 \\ a_3 & a_4 & a_5 & a_0 & a_1 & a_2 \\ a_{-1} & a_3 & a_4 & a_{-1} & a_0 & a_1 \\ a_{-2} & a_{-1} & a_3 & a_{-2} & a_{-1} & a_0 \end{pmatrix}.$$

Defining

$$f_{m,s} = \frac{1}{2} e^{-\lambda_t} \sum_{k=1}^{\infty} \frac{\lambda_t^{k(m+1)+s}}{(k(m+1)+s)!},$$

the distinct $a_l(t)$ are

$$a_l(t) = \begin{cases} f_{m,l}(\lambda_t), & -m \leqslant l \leqslant -1, \\ \frac{e^{-\lambda_t}\lambda_t^l}{l!} + f_{m,l}(\lambda_t), & 0 \leqslant l \leqslant m, \\ f_{m,l-(m+1)}(\lambda_t), & m+1 \leqslant l \leqslant 2m+1. \end{cases}$$

The probability that a particular genotype is observed at a given marker is conditional on the hidden state at that marker, that is $b_i(O_t) = P(O_t | y_t = i)$. Random errors in the genotyping process of rate ϵ are allowed. If there is no genotyping error, $0 \leqslant y_t \leqslant m$ implies O_t = A and $m + 1 \leqslant y_t \leqslant 2m + 1$ implies O_t = H.

Likelihood calculations are done by way of the 'forward variables':

$$\alpha_t(i) = P(O_1, \ldots, O_t, y_t = i | \lambda).$$

For n offspring, the likelihood is given by

$$P(O|\lambda) = \prod_{k=1}^{n} \sum_{i=0}^{2m+1} \alpha_T^k(i), \tag{2.1}$$

where $\alpha_T^k(i)$ denotes the value of $\alpha_T(i)$ for the kth offspring. The forward variables give the probability of observing the partial genotype sequence up to $\mathcal{M}_t$ and being in hidden state i at $\mathcal{M}_t$. Likewise, there are corresponding 'backward variables', $\beta_t(i)$, giving the probability of observing the partial genotype sequence from $\mathcal{M}_{t+1}$ to $\mathcal{M}_T$ given the hidden state i at $\mathcal{M}_t$. Both the forward and backward variables are computed recursively.

The expectation maximization (EM) algorithm is employed to determine the maximum likelihood estimates (MLEs) of the intermarker genetic distances, d_t, $t = 1, \ldots, T-1$. In particular, Baum's lemma is used, so that $Q(\lambda, \lambda')$ is explicitly maximized, where

$$Q(\lambda, \lambda') = \sum_y P(y, O|\lambda) \log P(y, O|\lambda')$$

and $y = (y_1, \ldots, y_T)$. In terms of λ_t' this is equivalent to maximizing, for n BC offspring,

$$\sum_{k=1}^{n} \sum_{y_t^k, y_{t+1}^k} P(y_t^k, y_{t+1}^k | O^k, \lambda) \log P(y_{t+1}^k | y_t^k, \lambda_t') = \sum_{k=1}^{n} \sum_{l=-m}^{2m+1} E(n_l^k(t)|O, \lambda) \log a_l(\lambda_t'),$$

where $n_l^k(t) \in \{0, 1\}$ is defined to be the number of transitions from state i to state j for a pair $(i, j) \in I_l$ for the kth offspring. The E step then involves calculating

$$E(n_l^k(t)|O, \lambda) = \sum_{i,j \in I_l} \xi_t^k(i, j),$$

where $\xi_t(i, j) = \frac{\alpha_t(i) a_{ij}(t) b_j(O_{t+1}) \beta_{t+1}(j)}{\sum_{i,j} \alpha_t(i) a_{ij}(t) b_j(O_{t+1}) \beta_{t+1}(j)}$.

For general m there is no closed-form solution to the maximization step. Instead, we must search for the maximum,

$$\underset{\lambda_t'}{\operatorname{argmax}} \sum_{k=1}^{n} \sum_{l=-m}^{2m+1} \log a_l(\lambda_t') E(n_l^k(t)|O, \lambda).$$

Brent's (1973) method is used to find a solution numerically, iterating the E and M steps until the scaled genetic distance estimates, λ_t, change by less than a prespecified tolerance level.

2.2 F_2 *intercross*

Given the two inbred strains AA and aa, at a given locus each F_2 offspring will be homozygous AA, homozygous aa, or heterozygous Aa. The observed genotypes are defined as A or B (homozygous AA or aa) and H if heterozygous. We can further define two incomplete genotypes C and D to be not B (i.e. AA or Aa) and not A (i.e. aa or Aa), respectively. For an F_2 intercross, both chromosomes must be considered because it is impossible to determine, at each heterozygous locus, which allele came from which parent. There are $4(m+1)^2$ hidden states, since on each chromosome the chain could be in any one of the states $0, \ldots, 2m+1$. For example, if $m = 1$, there are 16 hidden states, shown here with their corresponding

Incorporating interference into linkage analysis

states on each individual chromosome:

$0 \leftrightarrow 0, 0$	$4 \leftrightarrow 1, 0$	$8 \leftrightarrow 2, 0$	$12 \leftrightarrow 3, 0$
$1 \leftrightarrow 0, 1$	$5 \leftrightarrow 1, 1$	$9 \leftrightarrow 2, 1$	$13 \leftrightarrow 3, 1$
$2 \leftrightarrow 0, 2$	$6 \leftrightarrow 1, 2$	$10 \leftrightarrow 2, 2$	$14 \leftrightarrow 3, 2$
$3 \leftrightarrow 0, 3$	$7 \leftrightarrow 1, 3$	$11 \leftrightarrow 2, 3$	$15 \leftrightarrow 3, 3.$

Again, the convention that states $0, \ldots, m$ represent gm DNA and $m+1, \ldots, 2m+1$ gp DNA on a single chromosome is used. The initial-state distribution is assumed uniform. The conditional distribution of the observed genotypes given the hidden state i at $\mathcal{M}_t$ allows for the incomplete marker genotypes C and D, and incorporates a random error rate of ϵ.

The transitions on the maternal and paternal chromosomes are independent; hence, the joint probability is simply the product of the marginal probabilities of the transitions on each chromosome. Formally, the transition probability matrix for the states is given by

$$C(\lambda_t^f, \lambda_t^p) = A(\lambda_t^f) \otimes A(\lambda_t^p),$$

where $A(\lambda_t^f)$ and $A(\lambda_t^p)$ are the transition matrices on the maternal and paternal chromosomes, respectively, and $\otimes$ the Kronecker product. It is generally assumed that the recombination rates for the two sexes are equal, i.e. $\lambda_t^f = \lambda_t^p = \lambda_t$ and $C(\lambda_t) = A(\lambda_t) \otimes A(\lambda_t)$. In general, the (nondistinct) $c_{ij}(t)$ are obtained by translating the indices i and j into the corresponding indices for the transition matrix $A(\lambda_t)$ for each individual chromosome. The algorithm detailed in the previous section is used to determine the distinct quantities in both cases ($a_{s_1}(t)$ and $a_{s_2}(t)$), giving

$$c_{ij}(t) = a_{s_1}(t) a_{s_2}(t), \qquad -m \leqslant s_1 \leqslant s_2 \leqslant 2m+1.$$

For example, when $m = 1$ the states are given above and the probability of moving from state 1 to state 4 involves a transition from states 0 to 1 on the first chromosome and 1 to 0 on the second chromosome. Using the notation from Section 2.1, the marginal probabilities for these transitions are $a_1(t)$ and $a_{-1}(t)$, respectively. Therefore, in this case, $c_{14}(t) = a_1(t) a_{-1}(t)$.

The special case of no recombination in one sex only affects the transition matrix in the model. Without loss of generality, if $\lambda_t^f = 0$ for $t = 1, \ldots, T-1$ (i.e. no female recombination), then:

$$A(\lambda_t^f) = \mathcal{I}_{2m+1},$$

where $\mathcal{I}_{2m+1}$ is the $(2m+1) \times (2m+1)$ identity matrix. The joint transition matrix then simplifies to

$$C(\lambda_t^f, \lambda_t^p) = \begin{pmatrix} A(\lambda_t^p) & 0 & \cdots & 0 \\ 0 & A(\lambda_t^p) & \cdots & 0 \\ \vdots & & \ddots & \vdots \\ 0 & \cdots & 0 & A(\lambda_t^p) \end{pmatrix}.$$

In other words, only the block diagonal elements of $C(\lambda_t^f, \lambda_t^p)$ are nonzero.

Having defined the elements of the HMM, calculation of the forward, backward, and ξ variables can then be carried out as in Section 2.1, substituting $c_{ij}(t)$ in place of $a_{ij}(t)$. In order to find the MLEs of genetic distance, we invoke the EM algorithm once more. For n F_2 offspring and equal recombination rates for the sexes, i.e. $\lambda^f = \lambda^p$, $Q(\lambda, \lambda')$ in terms of λ'_t becomes, in the general case,

$$\sum_{k=1}^{n} \sum_{y_t^k, y_{t+1}^k} P(y_t^k, y_{t+1}^k | O^k, \lambda) \log P(y_{t+1}^k | y_t^k, \lambda'_t)$$

$$= \sum_{k=1}^{n} \sum_{l_1, l_2=-m}^{2m+1} \log[a_{l_1}(\lambda'_t) a_{l_2}(\lambda'_t)] E(n_{l_1,l_2}^k(t) | O, \lambda).$$

$n_{l_1,l_2}^k(t)$ is the number of transitions from i to j for a pair (i, j) related to I_{l_1} and I_{l_2} on the individual chromosomes. The E step consists then of calculating $E(n_{l_1,l_2}^k(t)|O, \lambda)$ and the M step is carried out numerically using Brent's (1973) method. To avoid underflow problems, the forward and backward variables are scaled by way of logarithms for all values of m and for all progeny numbers, when there are more than 100 markers in the data set for the BC HMM and, similarly, when there are more than 70 markers in the F_2 case.

3. RESULTS

To test the performance of the χ^2 HMM, genotype data were simulated for both F_2 and BC breeding schemes. The data were simulated under the assumption of crossover interference. To mimic biological reality, the χ^2 model with $m = 6$ was used. This model was previously estimated to provide the best fit for mouse data (Lin and Speed, 1996).

The programs were first verified by comparing all results (estimated genetic distances and likelihood values) obtained when $m = 0$ against both MapMaker and R/qtl (Broman *et al.*, 2003) which assume no crossover interference. Furthermore, for the BC case, the theoretical and empirical map functions were compared for a variety of m values. For fixed m, 1000 BC mice were simulated and genotyped every centimorgan along a single chromosome. Using this genotype data, the empirical recombination fractions could then be computed. Plotting the empirical recombination fractions against the true genetic distance and the theoretical map function, given by

$$\theta = \frac{1}{2}\left[1 - e^{-2d} \sum_{i=0}^{m} \frac{(2d)^i}{i!}\left(1 - \frac{i}{m+1}\right)\right],$$

on the same set of axes found close agreement. Figure 2 depicts the results for $m = 0, 1, 2,$ and 3.

3.1 *Bombyx mori*

A set of F_2 genotype data for *B. mori*, the domesticated silkworm, was obtained from D. G. Heckel (Shi *et al.*, 1995). In Lepidoptera, females are the heterogametic sex and have achiasmatic meiosis (no crossing over). The purpose of the original study was to create a genetic linkage map for *B. mori*. The data consist of the genotypes of 52 F_2 progeny at a total of 58 markers. The markers are both dominant and codominant. Fifty of the autosomal loci were grouped into a total of 15 primary linkage groups (PLGs), ranging from 2 to 8 loci each, by Shi *et al.* (1995). The PLGs were taken as originally assigned and the genetic maps were re-estimated using the HMM with $m = 0$ in order to verify the published map. In addition to estimating the intermarker genetic distances, the log-likelihood values for all possible map orders were calculated conditional on their PLG assignment to verify the published orders for all PLGs. The results, both estimated intermarker genetic distances and map orders, matched the published maps for all PLGs except PLG 9 and 15 (Figure 3).

Incorporating interference into linkage analysis

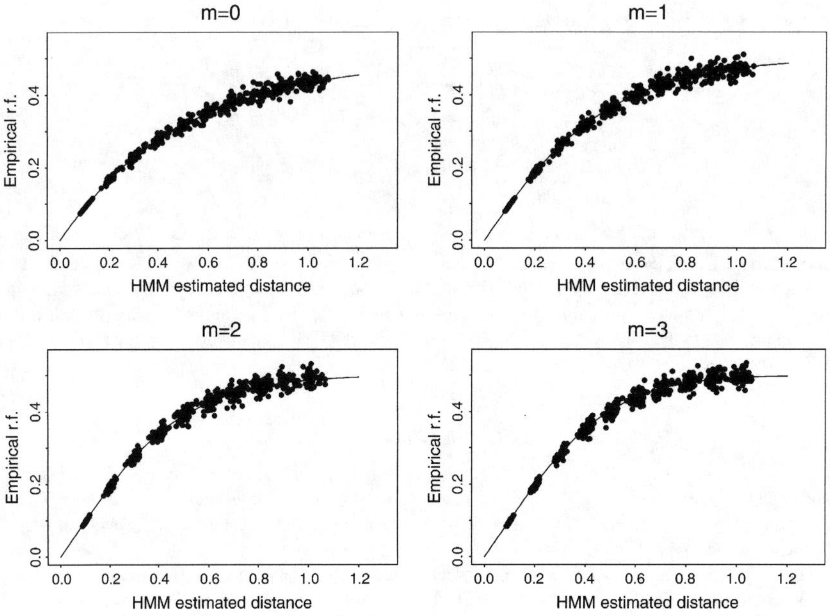

Fig. 2. Empirical and theoretical map functions for various values of m.

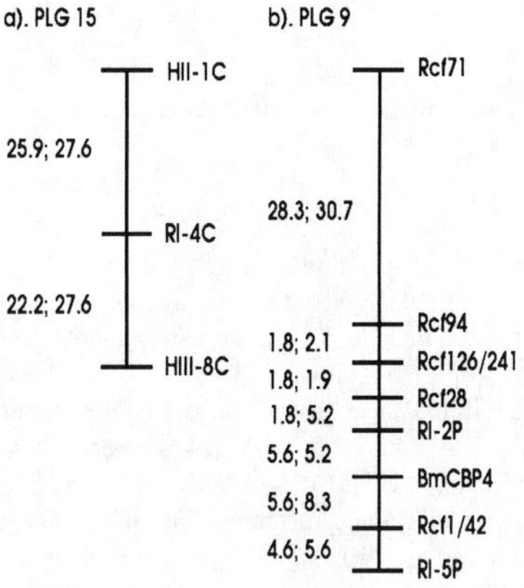

Fig. 3. Genetic maps for PLG 9 and 15, as estimated by Shi *et al.* (1995) (listed first) and by using the HMM (second).

PLG 15 consists of three dominant markers. The slight difference in estimated distances can be explained by the different estimation methods used. Shi *et al.* (1995) analyzed the linkage data under the assumptions of no crossing over in the female and no crossover interference, using maximum likelihood estimation and the EM algorithm. For some triple genotype combinations of dominant markers the Markov property of the observed genotype information O is lost, e.g. $P(O_3 = C|O_1 = C, O_2 = C) \neq P(O_3 = C|O_1 = A, O_2 = C)$, making it incorrect to calculate the likelihood using a Markov chain, as was done by Shi *et al.* Hence, multilocus probabilities must be calculated some other way, such as by using an HMM. The inclusion of the hidden chain allows for the correct calculation of multilocus probabilities when there is dominance or missing data at some loci. Our method therefore makes more efficient use of the data for this PLG.

The largest PLG was PLG 9, consisting of eight loci. For this PLG, Shi *et al.* (1995) used MapMaker to estimate an approximation to the maximum likelihood map, as their program was too slow to calculate the maps for all possible orders in a reasonable time. Furthermore, the recombination fractions between loci were estimated from the MapMaker sex-averaged estimates using a linear regression (for details see Shi *et al.*, 1995). In contrast, using the HMM approach, the likelihood values for all possible orders were calculated and the genetic distances estimated under the assumption of no crossing over in the female. We concluded that the order of loci was the same as the published order but that the genetic distance estimates were different. Given their ad hoc method of estimating the distances for this PLG, it is not surprising that the HMM distance estimates differ from their published ones (Figure 3).

3.2 *Timing*

As m increases, the number of states in the HMM also increases. Recall there are $2(m + 1)$ states for BC progeny and $4(m + 1)^2$ states for F_2 progeny. The increase in the number of states leads, necessarily, to an increase in the time taken to converge to the maximum likelihood estimates due to more computations being required to calculate the forward, backward, and ξ variables. The time required to perform the E step also increases with m. However, finding the values of λ_t which maximize $Q(\lambda, \lambda')$ at each iteration (the M step) requires only a slight increase in time. An important issue with regard to the time taken to find the estimated genetic distances is that λ_t, not d_t, is estimated. The shape of $Q(\lambda_t, \lambda'_t)$ rapidly becomes very flat as the value of λ_t approaches the maximum likelihood estimate, so that in order to get an accurate estimate of d_t many iterations are often required. It is the increased number of iterations of the EM algorithm coupled with the increase in E step time which gives rise to the large overall increase in time.

The time taken to find the maximum likelihood estimates for varying m values was investigated for both BC and F_2 cases. The impact of increasing the number of mice in the data set and, separately, the impact of increasing the number of genetic markers were investigated. The value of m used for simulating the genotype data was equal to the value of m used in the HMM. Initial values for the distances were taken to be the NI ($m = 0$) estimates.

All programs were run on 400-MHz UltraSparc machines with 256 MB RAM. The HMM does not require huge amounts of memory, but the time taken depends quite heavily on the speed of the machine. Under each set of conditions investigated, the differences in the time taken for each data set were directly attributable to the number of iterations of the EM algorithm required before the convergence tolerance was satisfied. The time taken rises linearly as both the number of markers and the number of mice rise, with times also increasing with m. Kinks in Figure 4 can be attributed to the fact that the machines used were not available solely for our purposes. However, the graphs give a rough indication of the increase in time required.

The relationship between number of mice, time taken, and m is very similar for BC (Figure 4) and F_2 mice (not shown), except that the times involved are much larger in the F_2 case. For 1600 mice, it took

Incorporating interference into linkage analysis

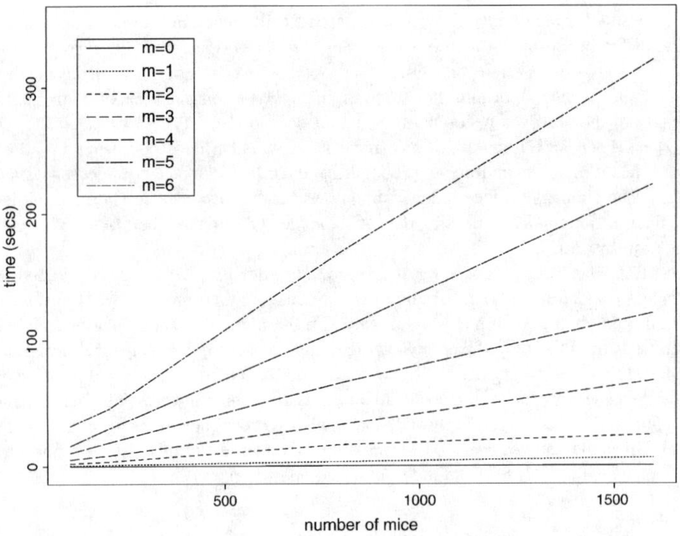

Fig. 4. Impact of increasing the number of mice on the time taken to find the distance MLEs for 10 intervals in the BC case.

the BC HMM 0.4 s, on average, to compute distance estimates for 10 intervals when $m = 0$, whereas when $m = 6$ the time taken rose to 321 s (5.5 min), an 800-fold increase. In the F_2 case, it took 12.2 s for $m = 0$, and 235 880 s (65.5 h) for $m = 6$, an almost 20 000-fold increase. When investigating the impact of the number of markers in a data set on time taken to find the distance MLEs, the number of mice was fixed at 400 in the BC case but reduced to 100 for F_2 data. Again, the relationship between the number of markers, m, and the amount of time taken to calculate the MLEs was very similar for F_2 and BC mice (data not shown). The time taken to calculate distance estimates rises linearly with an exponential increase in markers, similar to that shown in Figure 4 for the number of mice. For example, for 32 markers, the time taken to produce distance estimates when $m = 0$ was, on average, 4.4 s and increased to approximately 13 h if $m = 6$, for F_2 data, a 10 000-fold increase. In the BC case, under the no-interference model, estimates were obtained in 0.18 s compared to 163 s (2.7 min) if $m = 6$.

3.3 *Accuracy*

The most important gain that we hope to make by extending the LG algorithm to incorporate crossover interference is in the accuracy of the estimates. In order to investigate this, both BC and F_2 mice were simulated under the χ^2 model with $m = 6$. In each instance, 1000 sets of 300 BC mice and 100 sets of 300 F_2 mice were simulated and the distance estimates obtained using both the NI model and the χ^2 model recorded. The accuracies of distance estimates obtained using the Kosambi map function to transform the recombination fractions obtained under the NI model were also compared.

The entries in Table 1 demonstrate that as the distance between markers increases, the gains in precision from using the χ^2 model also increase when the data are fully informative, even if the wrong level

Table 1. *Root mean squared error in centimorgans of the distance estimates obtained under various models for fully informative BC and F_2 data simulated under the χ^2 model with $m = 6$*

Model used	True distance (cM)					
	BC data			F_2 data		
	5	10	20	5	10	20
NI	1.38	2.28	3.90	0.87	1.58	3.04
$\chi^2_{10}(m = 4)$	1.24	1.82	2.46	0.78	1.27	1.93
$\chi^2_{14}(m = 6)$	1.23	1.81	2.38	0.78	1.26	1.87
$\chi^2_{18}(m = 8)$	1.23	1.81	2.34			
NI + Kosambi	1.25	1.89	2.78	0.79	1.32	2.17

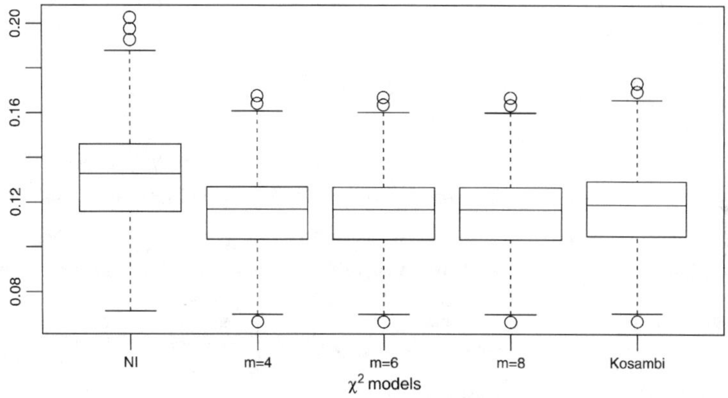

Fig. 5. Box plot of distance estimates obtained when true distance is 10 cM. In this case 300 BC meioses were simulated with 1% random genotyping error.

of interference is used in the HMM. Introducing a random genotyping error into the genotype data when either no error or the wrong amount of error is allowed for in the HMM produces similar results. For all distances the χ^2 model is more accurate, but the gains in accuracy are sometimes slight, especially if the genetic distances involved are small. Figure 5 shows the results found under one scenario, the results being typical of those found under other scenarios also. Simulations with 5% and 10% missing genotype data, and both 2% and 4% incomplete genotypes for the F_2 case were also performed. We found that this conclusion holds true in those cases also (data not shown). The results presented in Figure 6 are representative of those found under the different conditions investigated, for both F_2 and BC data.

4. CONCLUSIONS

We have developed and tested an HMM for estimating genetic distances that incorporates crossover interference, specifically under the χ^2 model. The HMM for F_2 data was extended to incorporate the situation where there is no crossing over in one of the sexes. The χ^2 HMM produces more accurate estimates of

Incorporating interference into linkage analysis

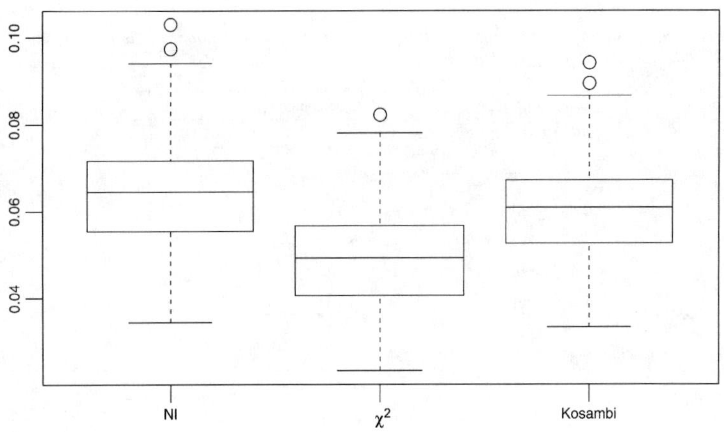

Fig. 6. Box plot of distance estimates obtained when true distance is 5 cM. In this case, 300 F_2 meioses were simulated with 5% missing and 4% incomplete genotyping data.

genetic distance, in the presence of crossover interference, than the NI model which is commonly implemented.

The gains in accuracy achieved by using the χ^2 model, with $m = 6$, instead of the Kosambi map function, are small for distances under 10 cM. With the ever-increasing number of genetic markers available and the falling costs of genotyping, in future we can expect many markers to be typed for a particular experimental cross, leading to a reduction in the typical intermarker genetic distance. The use of single-nucleotide polymorphisms and microarray technology for genotyping purposes will also mean smaller distances between markers. Using a lower level of interference than true ($m = 4$ instead of $m = 6$) resulted in very little difference in the level of accuracy obtained, likewise if a slightly higher m is used in the HMM. The problem of estimation of m has been treated in detail by Zhao *et al.* (1995) and Goldstein *et al.* (1997). The method they use is to compute the maximum likelihood for different values of m and compare them. The HMM presented in this paper may also be used to estimate the level of interference using the same approach of comparing likelihood values for different m values (NJ Armstrong and TP Speed, in preparation). The computational cost is equal to the sum of the computational costs of maximizing the likelihood for the individual values of m. The accuracy of the estimate obtained is in large part dependent on the number of individuals in the cohort and the number of markers genotyped. A more realistic model for the distribution of crossovers along a chromosome is a mixture of the χ^2 and NI models (Copenhaver *et al.*, 2002). For small-mixture probabilities, the results on accuracy of distance estimates are likely to be similar to those presented here.

The time taken to find estimates of genetic distance for F_2 data with either a large number of markers or progeny, for large levels of interference ($m = 6$), is at least 1000 times more than if no interference is assumed, and can take as much as 20 000 times longer. The increase in time required in conjunction with the small increases in accuracy suggests that this model will not become widely implemented in practice, even with the rapid increase in computing power. Development of the HMM allowing for crossing over in only one sex, for both the χ^2 model and the NI model, will hopefully be of use for researchers involved with organisms such as those in the order Lepidoptera.

The software used in this paper is available on request from the first author.

ACKNOWLEDGMENTS

We thank David Heckel for kindly allowing us access to the *Bombyx mori* data. This work was supported by National Institutes of Health grant GM59506-01 (to Speed).

REFERENCES

BLANK, R., CAMPBELL, G., CALABRO, A. AND D'EUSTACHIO, P. (1988). A linkage map of mouse chromosome 12: localization of Igh and effects of sex and interference on recombination. *Genetics* **120**, 1073–1083.

BRENT, R. (1973). *Algorithms for Minimization Without Derivatives*. Englewood Cliffs, NJ: Prentice-Hall.

BROMAN, K., ROWE, L., CHURCHILL, G. AND PAIGEN, K. (2002). Crossover interference in the mouse. *Genetics* **160**, 1123–1131.

BROMAN, K. AND WEBER, J. (2000). Characterization of human crossover interference. *American Journal of Human Genetics* **66**, 1911–1926.

BROMAN, K., WU, H., SEN, S. AND CHURCHILL, G. (2003). R/qtl: Qtl mapping in experimental crosses. *Bioinformatics* **19**, 889–890.

COPENHAVER, G., HOUSWORTH, E. AND STAHL, F. (2002). Crossover interference in Arabidopsis. *Genetics* **160**, 1631–1639.

FISHER, R., LYON, M. AND OWEN, A. (1947). The sex chromosome in the house mouse. *Heredity* **1**, 335–365.

FOSS, E., LANDE, R., STAHL, F. AND STEINBERG, C. (1993). Chiasma interference as a function of genetic distance. *Genetics* **133**, 681–691.

GOLDSTEIN, D., ZHAO, H. AND SPEED, T. (1997). The effects of genotyping errors and interference on estimation of genetic distance. *Human Heredity* **47**, 86–100.

HULTÉN, M. (1974). Chiasma distribution at diakinesis in the normal human male. *Hereditas* **76**, 55–78.

KARLIN, S. AND LIBERMAN, U. (1979). A natural class of multilocus recombination processes and related measures of crossover interference. *Advances in Applied Probability* **11**, 479–501.

KING, J. AND MORTIMER, R. (1990). A polymerization model of chiasma interference and corresponding computer simulation. *Genetics* **126**, 1127–1138.

LANDER, E. AND GREEN, P. (1987). Construction of multilocus genetic linkage maps in humans. *Proceedings of the National Academy of Science of the United States of America* **84**, 2363–2367.

LIN, S. AND SPEED, T. (1996). Incorporating crossover interference into pedigree analysis using the χ^2 model. *Human Heredity* **46**, 315–322.

MCPEEK, M. AND SPEED, T. (1995). Modeling interference in genetic recombination. *Genetics* **139**, 1031–1044.

MULLER, H. (1916). The mechanism of crossing-over. *The American Naturalist* **50**, 193–221, 284–305, 350–366, 421–434.

RISCH, N. AND LANGE, K. (1979). An alternative model of recombination and interference. *Annals of Human Genetics* **43**, 61–70.

SHI, J., HECKEL, D. AND GOLDSMITH, M. (1995). A genetic linkage map for the domesticated silkworm, Bombyx mori, based on restriction fragment length polymorphisms. *Genetic Research (Cambridge)* **66**, 109–126.

ZHAO, H., SPEED, T. AND MCPEEK, M. (1995). Statistical analysis of crossover interference using the chi-square model. *Genetics* **139**, 1045–1056.

[*Received April 19, 2005; first revision August 12, 2005; second revision October 7, 2005; third revision November 21, 2005; accepted for publication December 7, 2005*]

Chapter 13
DNA Sequencing

Lei M. Li

DNA sequencing is one great leap in the advance of life sciences. The research in these two sequencing articles came respectively from two chapters of my PhD thesis at Berkeley, and it is my luck to connect my life with DNA sequencing through my wonderful thesis advisor, Professor Terry Speed.

Several factors motivated the selection of DNA sequencing as my thesis topic. First, I had an ambition to be an applied mathematician when I was young. In my last year of college, however, I was tortured by fatigue and infection. My mood was very low then, and I had no appetite for more mathematics at all. With the help of my family, I gradually recovered with a therapy of Chinese herbs, and I continued to study hard mathematics. From then on, I had a vague yet deep thought in my mind that someday I should apply my mathematical knowledge to the understanding of life and medicine. This was one reason why in graduate school I looked for some applied topic related to life sciences. Second, Terry had been working on statistical genetics and was very enthusiastic about new statistical problems in genomics. He gave me a physical mapping problem as a start-up project. I quickly made some progress that helped me pass the oral exam. Third, in 1995 the Human Genome Project accelerated and researchers from many disciplines such as chemistry, engineering, computer science, mathematics, statistics jumped into the field. And Terry brought me into the adventure with a good will.

Among the interesting mathematical problems associated with genomics, I picked DNA sequencing, or more exactly, DNA base-calling as my thesis topic, as Terry suggested. In 1994-95, DNA sequencing was based on Sanger's dideoxy DNA amplification, fluorescence dye technique and electrophoresis. In the beginning, I knew nothing about molecular biology, and Terry helped me understand the basic ideas with a great patience. His former student, David Nelson, participated in DNA sequencing research at that time too, and provided us with fairly complete background on electrophoresis [8, 9, 10]. Another source of collaboration came from Professor

L.M. Li
Academy of Mathematics and Systems Science, Chinese Academy of Sciences,
and Computational Biology and Bioinformatics, University of Southern California
e-mail: lilei@amss.ac.cn

S. Dudoit (ed.), *Selected Works of Terry Speed*, Selected Works in Probability and Statistics, 535
DOI 10.1007/978-1-4614-1347-9_13, © Springer Science+Business Media, LLC 2012

Richard Mathies' group in the chemistry department at Berkeley, who were conducting research on capillary DNA sequencing. In a statistical consulting service, of which Terry was in charge for the Statistics Department during one semester in 1995, Dr. Indu Kheterpal, who was a PhD graduate student in Professor Mathies' group, brought in an interesting estimation problem in fluorescence dye technique. Terry set up a good collaboration with them and I learned a lot of chemistry related to DNA sequencing through the interaction.

Sanger DNA sequencing generates a signal trace from each template DNA, and base-calling is the data analysis part of DNA sequencing, aimed at reconstructing the nucleotide sequence with a fair fidelity. We decomposed the problem into three parts: color correction, deconvolution, and base-calling. Then we tried to work out solutions to each of them. In my opinion, the work on color correction and deconvolution is mathematically and statistically more elegant and original, and we put a lot of effort into publishing it. In comparison, the solution to the last step of base-calling is more engineering-like in flavor. Terry introduced me to the technique of the hidden Markov model (HMM), which was not so widely known then as it is now. I was intrigued by the idea and we designed an HMM for base-calling. In genome research, a good idea alone is not sufficient to have an impact, and a good implementation is equally important, if not more so. The implementation of the HMM base-calling requires model-training and a lot of serious software programming. Due to graduation and my limited programming strength, I only tested the idea and did not develop a real software solution. A little later, Dr. Green and his team published their famous work on base-calling. In the meantime, microarray technology gradually caught people's attention. And our HMM base-calling idea was not pursued further [3].

By now our most influential contribution to DNA sequencing is color correction [4]. A few years ago, Terry told me that Solexa, now owned by Illumina, one major next generation sequencing platform, adopted our scheme. This is encouraging and yet not surprising because we have shown, at least in one important perspective, that the color correction scheme we proposed is optimal. In capillary Sanger sequencing, four dyes, which emit different colors as excited by laser, are used to distinguish four kinds of nucleotides. The purpose of color correction is to remove the cross-talk phenomenon of the four dyes' emission spectra. One key idea of our work is that we need to estimate the cross-talk phenomenon adaptively from each experiment. Another key idea of our estimation is that we make use of the "canonical" distribution of data without any cross-talk. As a PhD student, I was enthusiastic about the solution when it was first discovered. In a late afternoon, we walked home down Hearst Avenue, and Terry asked me a serious question, "how do we know our solution is right?" I gave him an answer, "If we estimate the cross-talk matrix properly, the distribution of the corrected data should match the nominal one." Terry agreed. After I graduated, I went through several interesting problems in engineering and science, and realized that they share a common nature with the color correction problem. I wrote an article about this class of blind inversion problems in the festschrift for Professor

Terry Speed's 60th birthday [6], because Terry's question partially inspired the formulation of this notion.

In usual DNA sequencing light intensities at four wavebands are measured, since four dyes are used. Interestingly, Dr. Kheterpal and Professor Mathies asked us if we could instead use only three light intensities for base-calling. After some struggles, we designed a procedure consisting of a series of nonnegative least squares and a model selection scheme [1]. Professor Mathies was very pleased with the result.

The work on deconvolution is also motivated by Sanger sequencing and is more technical than color correction. Each base in a Sanger sequencing trace can roughly be represented by a Gaussian-shaped peak on a continuous scale, and the four kinds of nucleotides, namely A, G, C and T, are represented by four different colors respectively. The motion of DNA molecules in capillary is usually explained by the reptation theory. The aggregation of the molecules of the same size can approximately be described by a Brownian motion. That explains why each peak looks like a normal distribution. In Sanger sequencing, most base-calling errors come from the regions with runs of the same kind of nucleotides, and lead to insertions and deletions, or simply indels. Once an error of this type occurs in base-calling, it often causes more trouble than a substitution error does in an alignment. How to separate these peaks, or in other words, how to count the bases in a run correctly, is a problem that we solved with the deconvolution technique.

The parametric deconvolution [5] was something we worked out without much prior literature knowledge on the topic. Terry suggested that we do a literature survey. In 1995, I searched the key word, deconvolution, on Yahoo (I am sure it was not Google then), and got over one thousand hits, and the early work went back to the nineteenth century. Obviously deconvolution is a common problem in many areas. I read almost all relevant papers I could find, and discussed them with Terry over a long period of time till 2000 in Melbourne. One issue that puzzled us was whether deconvolution is an ill-posed problem — a notion postulated by Hadamard in 1902. Without any constraint on the solution space, deconvolution is an ill-posed problem, and had been classified so in applied mathematics. Nevertheless, in many cases, the signals to be reconstructed are positive and "sparse". In parametric deconvolution, we formulate the unknowns by a mixture of finite Dirac spikes, and we can estimate them well in a regular sense, see Theorem 4.1 and 4.2 in Li and Speed [5], although the dimension of the solution space needs to be estimated too by model selection, see Algorithm 5.2 in Li and Speed [5] and Proposition 3.3 in Li [2]. Thus Terry and I came to the conclusion: if the signal to be reconstructed is positive and sparse, then deconvolution is well-posed.

The well-posedness explains why historically some nonparametric deconvolvers such as the Jansson's method and the folk iteration (5.2) in Li and Speed [7], obtained in different scenarios by either EM algorithms or Bayesian methods in the literature, work quite well in their respective applications. Furthermore, Terry and I did an investigation on the general linear inverse problem with positive constraints (LININPOS) that underlies the folk iteration. We discovered that the iteration in fact minimizes the Kullback-Leibler divergence between the target and the fit, and this result clarifies the core structure of the LININPOS solution.

The work described here has been a source of both enlightenment and enjoyment to me. When I was writing down these words, those scenes when Terry and I walked down Hearst Avenue and chatted on various issues came upon my mind like yesterday. I am sure that Terry's other students and colleagues had their own pleasant study and work experiences with him as well. His spirit is no doubt the source of many good things. In addition to his passion for science and mathematics, his respect of the interests and talents of each student, each collaborator and his own may partially explain his wide research spectrum.

References

[1] I. Kheterpal, L. Li, T. P. Speed, and R. A. Mathies. A three-color labeling approach for DNA sequencing using energy transfer primers and capillary electrophoresis. *Electrophoresis*, 19:1403–1414, 1999.

[2] L. Li. DNA sequencing and parametric deconvolution. *Stat. Sinica*, 12(1): 179–202, 2002.

[3] L. Li. *Statistical Models of DNA Base-calling*. PhD thesis, University of California, Berkeley, 1998.

[4] L. Li and T. P. Speed. An estimate of the crosstalk matrix in four-dye fluorescence-based DNA sequencing. *Electrophoresis*, 20(7):1433–1442, 1999.

[5] L. Li and T. P. Speed. Parametric deconvolution of positive spike trains. *Ann. Stat.*, 28(5):1279–1301, 2000.

[6] L. M. Li. Blind inversion needs distribution (BIND): General notation and case studies. In D. Goldstein, editor, *Science and Statistics: A Festschrift for Terry Speed*, volume 40 of *Lecture Notes – Monograph Series*, pages 275–293. Institute of Mathematical Statistics, Beachwood, OH, 2003.

[7] L. M. Li and T. P. Speed. Deconvolution of sparse positive spikes. *J. Comput. Graph. Stat.*, 13(4):853–870, 2004.

[8] D. O. Nelson. Improving DNA sequence accuracy and throughput. In T. P. Speed and M. S. Waterman, editors, *Genetic Mapping and DNA Sequencing*, volume 81 of *The IMA Volumes in Mathematics and its Applications*, pages 183–206. Springer, 1996.

[9] D. O. Nelson. Introduction of reptation. Technical report, Lawrence Livermore National Laboratory, 1995.

[10] D. O. Nelson and T. P. Speed. Recovering DNA sequences from electrophoresis data. In S. E. Levinson and L. Shepp, editors, *Image Models (and their Speech Model Cousins)*, pages 141–152. Springer-Verlag, 1996.

The Annals of Statistics
2000, Vol. 28, No. 5, 1279–1301

PARAMETRIC DECONVOLUTION OF POSITIVE SPIKE TRAINS

BY LEI LI[1] AND TERENCE P. SPEED[2]

Florida State University and University of California, Berkeley

This paper describes a parametric deconvolution method (PDPS) appropriate for a particular class of signals which we call spike-convolution models. These models arise when a sparse spike train—Dirac deltas according to our mathematical treatment—is convolved with a fixed point-spread function, and additive noise or measurement error is superimposed. We view deconvolution as an estimation problem, regarding the locations and heights of the underlying spikes, as well as the baseline and the measurement error variance as unknown parameters. Our estimation scheme consists of two parts: model fitting and model selection. To fit a spike-convolution model of a specific order, we estimate peak locations by trigonometric moments, and heights and the baseline by least squares. The model selection procedure has two stages. Its first stage is so designed that we expect a model of a somewhat larger order than the truth to be selected. In the second stage, the final model is obtained using backwards deletion. This results in not only an estimate of the model order, but also an estimate of peak locations and heights with much smaller bias and variation than that found in a direct trigonometric moment estimate. A more efficient maximum likelihood estimate can be calculated from these estimates using a Gauss–Newton algorithm. We also present some relevant results concerning the spectral structure of Toeplitz matrices which play a key role in the estimation. Finally, we illustrate the behavior of these estimates using simulated and real DNA sequencing data.

1. Introduction and background. Deconvolution is used in many scientific disciplines, including geophysics, spectroscopy, chromatography and pharmacokinetics. In abstract form, an unobserved signal $x(t)$ is blurred by a known point spread function w, resulting in an observed signal $y(t)$. Mathematically, this can be represented in terms of the convolution operation $*$: $y = w * x$. The task of deconvolution is to reconstruct the unobserved signal x from the observed signal y. The point spread function is assumed to be known throughout this paper. An additive measurement error is also assumed in y in all discrete situations discussed.

Our motivation is the problem of base-calling in DNA sequencing. The Sanger sequencing technique is a combination of enzymatic reactions, electrophoresis and fluorescence-based detection; see [1]. This procedure produces a four-component vector time series. Base-calling is the analysis part of DNA

Received June 1998; revised May 2000.
[1]Supported by NSF Grant DMS-99-71698.
[2]Supported by DOE Grant DE-FG03-97-ER62387.
AMS 1991 *subject classifications*. Primary 62F10; secondary 62F12, 86A22.
Key words and phrases. Deconvolution, spike train, model selection, DNA sequencing, Toeplitz matrix.

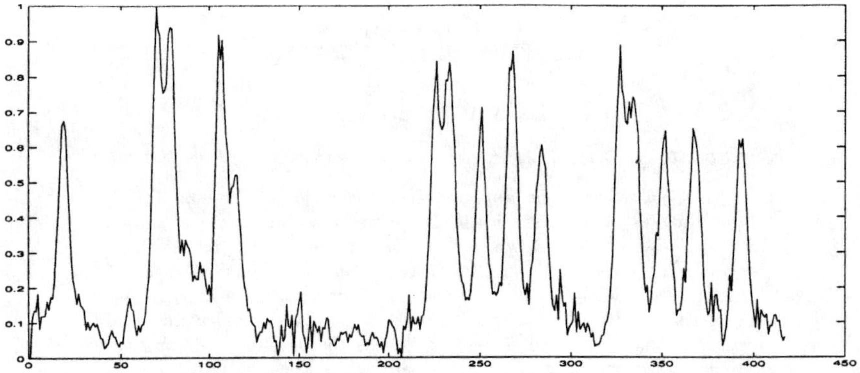

FIG. 1. *A segment slab gel electrophoresis sequencing data (provided by the engineering group at Lawrence Berkeley National Laboratory).*

sequencing, which attempts to recover the underlying DNA sequence from the vector time series. Figure 1 shows a segment of one channel of such a series. (This series is different from the original sequencing data because of the "cross talk" phenomenon. We here pass over this issue, and refer to [21] for details.) Typically, each major peak in the series corresponds to one base. As sequencing progresses, electrophoretic diffusion spreads peaks more and more. In regions where there are multiple occurrences of the same base, several successive peaks may merge into one large block. In this situation, base-calling is far more difficult. A number of studies exist of the errors made by one widely used base-calling system, see [3, 17, 18]. These reports show that errors associated with runs of the same base constitute more than half of the total errors. Furthermore, this kind of error causes more serious difficulties in later analysis than other kinds. For this reason, it seems important to try and do better in resolving peaks, and this is the motivation of the research we report here.

The literature on deconvolution is rich and scattered across a wide variety of fields. Frequently, deconvolution is an ill-posed inverse problem; see [13]. Two techniques, regularization and exploiting bound or nonnegativity constraints on the unknown functions, have been proved to be useful in dealing with ill-posedness. Regularization was introduced by Tikhonov [30], and has been well studied since then. For example, the long-standing iterative deconvolution method of Van Cittert (see [13]) can be viewed as a regularization method. Jansson [13] adjusted this regularization method, added nonnegativity constraints to the unknown function and applied it successfully to problems in spectroscopy.

Maximum entropy deconvolution can also be regarded as a regularization procedure, one which only applies to nonnegative signals; see [10, 11]. Donoho, Johnstone, Hoch and Stern [5] showed this procedure has certain advantages such as signal-to-noise enhancement and superresolution when the signal is nearly black. Stark and Parker [29] proposed some new algorithms to solve this type of constrained minimization problems.

The deblurring method introduced by Shepp and Vardi [27], and Vardi and Lee [32] uses maximum likelihood estimation under a Poisson or a multinomial model. Again it is assumed that the unknown function is nonnegative. Snyder, Schulz and O'Sullivan [28] obtained a similar algorithm as a solution to a general Fredholm integral equation of the first kind. They derive the formula by minimizing Csiszár's I-divergence, which is closely related to the concept of likelihood. Richardson [26], Kennett, Prestwich and Robertson [14, 15, 16] and Di Gesù and Maccarone [4] obtained the same result from a more intuitive Bayesian point of view and term it as "Bayesian deconvolution." All of these methods could be used in the base-calling problem, and their behavior on DNA sequencing data can be found in [20].

Poskitt, Dogancay and Chung [25] described a double-blind deconvolution method to analyze postsynaptic currents in nerve cells. This analysis is based on an elegant statistical model, and the estimator is derived by minimizing the quasi-profile-likelihood. Though the data from postsynaptic currents in nerve cells are different from DNA seqnencing data, we find that the likelihood structures of the two models proposed for each problem have some similarities. Therefore, we can borrow some ideas from [25] to study the deconvolution problem in DNA sequencing.

Motivated by the DNA sequencing data, we define what we call the spike-convolution model, in which the unknown function is represented as a linear combination of a finite number of positive spikes (Dirac functions) together with a constant baseline. In this model, deconvolution is nothing but a standard parameter estimation problem, where the parameters include the number, locations and heights of the underlying spikes, the baseline and the measurement error variance. Our estimation procedure uses the spectral properties of Toeplitz matrices, least squares and statistical model selection techniques, and we call it parametric deconvolution of positive spikes (PDPS).

The paper is arranged as follows. In Section 2 we introduce the spike-convolution model and some notation, discuss several aspects such as identifiability, and outline the estimation procedure. In Section 3 we study spectral structures of Toeplitz matrices constructed from trigonometric moments and present algorithms and asymptotics of the trigonometric moments estimates. In Section 4 we describe the algorithms and asymptotics of the maximum likelihood estimates. In Section 5 we propose our ultimate method, which is a package including a model fitting procedure and a two-stage model selection procedure. Section 6 contains a simulation study of the proposed methodologies. Finally, we apply PDPS to real DNA sequencing data, and compare the result with that using another nonparametric deconvolution method. Several other practical issues of PDPS are also discussed. The Appendix contains the proofs and the relevant mathematical details.

2. The spike-convolution model. Throughout the paper, the point spread function $w(\cdot)$ is assumed to be known, and to satisfy the following conditions, where $v_k = (1/2\pi)\int_{-\pi}^{\pi} w(t)e^{ikt}\,dt$ are w's Fourier coefficients:

1. It has finite support $(-\kappa_1, \kappa_2)$, where $0 < \kappa_1, \kappa_2 < \pi$;
2. $w(\cdot) \in C^2[-\pi, \pi]$;
3. $v_k \neq 0$, for $k = 0, \pm 1, \ldots, \pm K_0$.

The last assumption is described as requiring that there be no hole in the Fourier transform of the point spread function. It worth mentioning that $w(\cdot)$ is not necessarily nonnegative or causal. There exist studies of the shape of the point spread function in DNA sequencing (see [23] for references), and the determination of the width of the point spread function is another matter worthy of attention. However, these issues are not critical to the present work, and so we pass over them here. The signal to be estimated is assumed to have the following form:

$$x(t) = A_0 + \sum_{j=1}^{p} A_j \delta(t - \tau_j), \tag{1}$$

where $\delta(\cdot)$ is the Dirac delta function. We assume the coefficients A_j of the Dirac functions are positive and refer to them as "heights" of the spikes. Thus the underlying signal $x(t)$ is a linear combination of a finite number of spikes with positive heights, together with a constant baseline. We also assume $-\pi + \kappa_1 < \tau_1 < \cdots < \tau_p < \pi - \kappa_2$. Hence the support of the convolution y of w and x will stay in the range $[-\pi, \pi]$. Explicitly, we have

$$y(t) = A_0 + \sum_{j=1}^{p} A_j w(t - \tau_j) = (x * w)(t), \qquad t \in [-\pi, \pi], \tag{2}$$

where the time range has been scaled to $[-\pi, \pi]$ for convenience. We have assumed there are no peaks near the two ends. In real DNA sequencing, we can always cut the raw data into pieces at valley points and apply the deconvolution to each piece separately. The observations $\{z(t_l)\}$ are a sample of the above model, corrupted by measurement errors which are assumed to be additive:

$$z(t_l) = y(t_l) + \varepsilon(t_l) = A_0 + \sum_{j=1}^{p} A_j w(t_l - \tau_j) + \varepsilon(t_l), \tag{3}$$

where $t_l = 2\pi l/n$, $l = -[n/2], \ldots, 0, \ldots, [n/2] - 1$ if n is even, or $l = -[n/2], \ldots, 0, \ldots, [n/2]$ if n is odd. The $\{\varepsilon(t_l)\}$ are supposed to be i.i.d. with $E(\varepsilon(t_l)) = 0$, $\mathrm{Var}(\varepsilon(t_l)) = \sigma^2$ and a finite third moment.

Before we proceed, we introduce some notation. We denote the signal in the spike-convolution model (2) by $SC(w; p; \mathbf{A}; \tau)$ where $\mathbf{A}$ and τ, respectively, represent $\{A_j\}$ and $\{\tau_j\}$, and formally denote the signal x in (1) by $SC(\delta; p; \mathbf{A}; \tau)$. We define the inner product of two functions $y_1(t)$ and $y_2(t)$, belonging to $L^2[-\pi, \pi]$, by $\langle y_1, y_2 \rangle = \frac{1}{2\pi}\int_{-\pi}^{\pi} y_1(t) y_2(t)\,dt$. For functions $z_1(t)$ and $z_2(t)$

well defined at the lattice points $t_l = 2\pi l/n$, we also define the following inner product:

$$\langle z_1, z_2\rangle_n = \begin{cases} \dfrac{1}{n} \sum_{l=-[n/2]}^{[n/2]} z_1(t_l)z_2(t_l), & \text{if } n \text{ is odd,} \\ \dfrac{1}{n} \sum_{l=-[n/2]}^{[n/2]-1} z_1(t_l)z_2(t_l), & \text{if } n \text{ is even.} \end{cases}$$

The norms induced by $\langle\cdot,\cdot\rangle$ and $\langle\cdot,\cdot\rangle_n$ are denoted by $\|\cdot\|$ and $\|\cdot\|_n$, respectively.

The Hilbert–Schmidt theory for Fredholm integral equations of the first kind assumes that the signal $x(t)$ is in $L^2[\pi, \pi]$, and thus excludes the Dirac functions. Consequently, the signal reconstructed by methods within that framework will not contain any Dirac functions. This means that only incomplete deconvolution would be achieved if (2) were the truth. We prefer to regard deconvolution as a problem of parameter estimation, and this can result in a complete deconvolution.

The parameters in a $SC(w; p; \mathbf{A}; \tau)$ model include the number, locations and heights of the peaks, and the baseline. Although these parameters are closely related, they play quite different roles from the perspective of statistical estimation, and it seems very difficult to estimate them all, simultaneously and efficiently, in one step. The difficulty lies in the irregular structure of the the parameter space. For example, suppose we have a spike-convolution model with three positive peaks. If we let the height of one peak tend to zero, then the limiting model can only be regarded as a spike-convolution model with two peaks, for we require peak heights to be positive. Following this limiting process, the dimension of the parameter space changes. However, we have the following result saying that a spike-convolution model of order p cannot arbitrarily well be approximated by models of smaller orders.

THEOREM 2.1. *Let* $y(t)$ *be a* $SC(w; p; \mathbf{A}; \tau)$ *model. Then we have:* $\inf_{\bar{y}} \|y - \bar{y}\| = d > 0$, *where the infimum is taken over all* $\bar{y} \in SC(w; l; \bar{\mathbf{A}}; \bar{\tau})$, $l < p$.

A relating problem in signal processing is to estimate the so-called hidden frequencies ω_j in the following model:

$$s_t = \sum_{j=1}^{p} A_j \cos(t\omega_j + \theta_j) + \varepsilon_t, \tag{4}$$

where the A_j are positive and ε_t is white noise. In fact, the spectral density function of the above process, in the generalized sense, is a constant plus jumps occurring at the ω_j, with corresponding heights A_j. This is exactly the signal to be reconstructed in Model (2). Apart from the point spread function, the two problems are Fourier duals of each other. In order to estimate the hidden frequencies, Pisarenko [24] suggested a method using the Toeplitz matrices constructed from autocovariance functions. We exploit a similar idea as part of our estimation procedure.

Our estimation procedure consists of several steps. In Algorithm 3.1, we estimate the peak locations by connecting deconvolution with the spectral structure of Toeplitz matrices constructed from the Fourier coefficients of the observations. The peak heights of the estimated locations can be estimated either by a trigonometric moment method, Algorithm 3.2, or by least squares, where the latter exploits the connection between spike-convolution models and hypothetical linear regressions; see Algorithm 5.1. This leaves us with the task of estimating the number of peaks, usually regarded as a model selection problem. In our method, models of each candidate order have to be fitted before model selection takes place. The model selection strategy described in Algorithm 5.2 consists of two stages. First we choose a model which should come close to including the true model as a submodel, as overfitting is not completely suppressed at this step. We then use a modified GIC criterion together with a backward deletion procedure to obtain our final model. The resulting estimate can further be tuned by an optional step: maximizing likelihood if the distribution of the measurement errors is assumed known; see Section 4. Under the assumption of normal errors, we calculate the Fisher information matrix of the spike-convolution model, whose inverse gives the nominal standard errors of estimates. Note that these standard errors will not have taken into account the model selection process. The computation of the maximum likelihood estimate or one-step estimate can be carried out by Gauss–Newton algorithm. Please keep in mind that Sections 3 and 4 are about inference given the model order m, the number of spikes included in the spike-convolution model. This is a brief description of PDPS, which will be explained more fully in the rest of the paper.

3. The trigonometric moment estimates. Throughout this section, we assume the model order m is given. The connection between the trigonometric moments and the parameters in a spike-convolution model can be given using the spectral structure of Toeplitz matrices as follows.

THEOREM 3.1. *Let the Fourier coefficients of $y(t)$, which follows $SC(w; m; \mathbf{A}; \tau)$ with $m \leq K_0$, be $f_k = \langle y(t), e^{ikt}\rangle$, for $k = 0, \pm 1, \ldots$. First, write $g_0 = f_0$ and $g_k = f_k v_0 / v_k$, $g_{-k} = \bar{g}_k$, for $0 < k \leq K_0$. Next, form the Toeplitz matrices $G_m = (g_{j-1})_{i,\, j=0,\ldots,m}$. Finally, write $U^{(m)}(z) = \prod_{j=1}^{m}(z - \exp(i\tau_j)) = \sum_{j=0}^{m} \alpha_j z^j$. Then we have*

(i)

$$\begin{cases} f_0 = A_0 + \left(\sum\limits_{j=1}^{m} A_j\right) v_0, \\ f_k = \left(\sum\limits_{j=1}^{m} A_j e^{ik\tau_j}\right) v_k, \qquad k \neq 0. \end{cases} \tag{5}$$

(ii) *A_0 is the smallest eigenvalue of G_m with multiplicity one and eigenvector $(\alpha_0, \ldots, \alpha_m)^T$.*

(iii) *The* $\{A_j\}$ *satisfy the following linear system:*

$$v_0 \begin{pmatrix} 1 & 1 & \cdots & 1 \\ \exp(i\tau_1) & \exp(i\tau_2) & \cdots & \exp(i\tau_m) \\ \vdots & \vdots & \ddots & \vdots \\ \exp(i(m-1)\tau_1) & \exp(i(m-1)\tau_2) & \cdots & \exp(i(m-1)\tau_m) \end{pmatrix} \begin{pmatrix} A_1 \\ A_2 \\ \vdots \\ A_m \end{pmatrix} = \begin{pmatrix} g_0 - A_0 \\ g_1 \\ \vdots \\ g_{m-1} \end{pmatrix} \tag{6}$$

Note. When a Toeplitz matrix has distinct eigenvalues, the relation between eigenvalues and eigenvectors is one-to-one. For simplicity, we will use expressions such as "smallest eigenvector" to refer to the eigenvector corresponding to the smallest eigenvalue. The converse of the above theorem is also true in the following sense.

THEOREM 3.2. *Suppose we are given* $2m+1$ *complex numbers* $\{f_j, -m \le j \le m\}$, *where* $m \le K_0$, $\bar{f}_j = f_{-j}$. *Let* $g_0 = f_0$, *and* $g_k = f_k v_0/v_k$, $g_{-k} = \bar{g}_k$, *for* $0 < k \le K_0$. *Assume that the smallest eigenvalue* A_0 *of the Toeplitz matrix* $G_m = (g_{j-i})_{i,\, j=0,\ldots,m}$ *has multiplicity* 1. *Let the smallest eigenvector be* $\alpha = (\alpha_0, \ldots, \alpha_m)^T$, *and* $U^{(m)}(z) = \sum_{j=0}^{m} \alpha_j z^j$. *Then*:

(i) $U^{(m)}(z)$ *has m distinct roots exactly on the unit circle, which are denoted by* $\{\exp(i\tau_j)\}$.

(ii) *Furthermore, if* $-\pi + \kappa_1 < \tau_1 < \cdots < \tau_m < \pi - \kappa_2$, *then there exists a* $SC(w; m; \mathbf{A}; \tau)$ *whose first* $m+1$ *Fourier coefficients are* $\{f_j, 0 \le j \le m\}$. *Its baseline and heights* $\{A_j\}$ *are determined by the linear system (6), and the resulting heights are positive.*

This result is of great significance for practical model fitting from the computational point of view, since the peak locations could be found by restricting the search of roots of $U^{(m)}(z)$ to the unit circle. Starting from data $z(t_l)$, we estimate the trigonometric moments by $\hat{f}_k = \langle z, e^{ikt} \rangle_n$. Based on Theorem 3.2, for any given nonnegative integer $m \le K_0$, we can input these empirical trigonometric moments into the following two algorithms.

ALGORITHM 3.1 (Trigonometric moment estimates of peak locations).

(i) *Deconvolution: let* $\hat{g}_0 = \hat{f}_0$, $\hat{g}_k = \hat{f}_k v_0/v_k$, *for* $k = \pm 1, \ldots, \pm m$.

(ii) *Computing the smallest eigenvalue-vector of the Toeplitz matrix: construct the Toeplitz matrix* $\widehat{G}_m = (\hat{g}_{j-i})$ *and compute its smallest eigenvalue* $\widehat{A}_0^{(m)}$ *(assuming its multiplicity is one) and corresponding eigenvector* $\hat{\alpha}^{(m)} = (\hat{\alpha}_0^{(m)}, \ldots, \hat{\alpha}_m^{(m)})$.

(iii) *Solving a polynomial: on the unit circle, find the* m *distinct roots of* $\widehat{U}^{(m)}(z) = \sum_{j=0}^{m} \hat{\alpha}_j^{(m)} z^j$, *which we denote by* $\{\exp(i\hat{\tau}^{(m)})\}$, $j = 1, \ldots, m$.

ALGORITHM 3.2 (Trigonometric moment estimates of heights). *Solve the following Vandermonde linear system:*

$$v_0 \begin{pmatrix} 1 & 1 & \cdots & 1 \\ \exp(i\hat{\tau}_1^{(m)}) & \exp(i\hat{\tau}_2^{(m)}) & \cdots & \exp(i\hat{\tau}_m^{(m)}) \\ \vdots & \vdots & \ddots & \vdots \\ \exp(i(m-1)\hat{\tau}_1^{(m)}) & \exp(i(m-1)\hat{\tau}_2^{(m)}) & \cdots & \exp(i(m-1)\hat{\tau}_m^{(m)}) \end{pmatrix} \begin{pmatrix} \widehat{A}_1^{(m)} \\ \widehat{A}_2^{(m)} \\ \vdots \\ \widehat{A}_m^{(m)} \end{pmatrix}$$
$$= \begin{pmatrix} \hat{g}_0 - \widehat{A}_0^{(m)} \\ \hat{g}_1 \\ \vdots \\ \hat{g}_{m-1} \end{pmatrix}.$$

The output of these two algorithms is a $SC(w; m; \widehat{\mathbf{A}}^{(m)}; \hat{\tau}^{(m)})$ whose first $m+1$ Fourier coefficients are $\hat{f}_k$, $k = 0, \ldots, m$. We make some remarks here. First, we have ignored the case when the multiplicity of A_0 is greater than 1, since the Lebesgue measure of this singular case is zero. Second, strictly speaking, the fitted model makes sense only when $-\pi + \kappa_1 < \hat{\tau}_1^{(m)} < \cdots < \hat{\tau}_m^{(m)} < \pi - \kappa_2$. We thus delete those peaks outside the legitimate range in the regression stage discussed later, and then estimate the heights of the remaining peaks by least squares; see Algorithm 5.1 for more details. Third, our numerical experiments carried out in MATLAB show these two algorithms are robust to round off and noise in the data. For example, the roots of $\widehat{U}^{(m)}(z)$ do indeed lie on the unit circle to the necessary accuracy. Finally, in the case that the observations are generated from a $SC(w; p; \mathbf{A}; \tau)$, if we take $m = p$, then $\hat{\tau}_j$ and $\widehat{A}_j$ are the trigonometric moment estimates, which are consistent. Indeed, we have the following central limit theorem.

THEOREM 3.3.

$$\begin{aligned} &\sqrt{n}[(\widehat{A}_0, \widehat{A}_1, \ldots, \widehat{A}_p, \hat{\tau}_1, \ldots, \hat{\tau}_p)^T - (A_0, A_1, \ldots, A_p, \tau_1, \ldots, \tau_p)^T] \\ &\quad \overset{d}{\rightarrow} N(0, V), \end{aligned} \tag{7}$$

where $V = 4\pi^2 Q^{-1} P Q^{-T}$,

$$P = \frac{\sigma^2}{2} \mathrm{diag}\left(2, \left|\frac{v_0}{v_1}\right|^2, \ldots, \left|\frac{v_0}{v_p}\right|^2, \left|\frac{v_0}{v_1}\right|^2, \ldots, \left|\frac{v_0}{v_p}\right|^2\right) \tag{8}$$

and

$$
(9)\qquad Q = \left(\begin{array}{c|cccc|cccc}
1 & 1 & 1 & \cdots & 1 & 0 & 0 & \cdots & 0 \\ \hline
0 & \cos\tau_1 & \cos\tau_2 & \cdots & \cos\tau_p & -A_1\sin\tau_1 & -A_2\sin\tau_2 & \cdots & -A_p\sin\tau_p \\
\vdots & \vdots & \vdots & \ddots & \vdots & \vdots & \vdots & \ddots & \vdots \\
0 & \cos p\tau_1 & \cos p\tau_2 & \cdots & \cos p\tau_p & -pA_1\sin p\tau_1 & -pA_2\sin p\tau_2 & \cdots & -pA_p\sin p\tau_p \\ \hline
0 & \sin\tau_1 & \sin\tau_2 & \cdots & \sin\tau_p & A_1\cos\tau_1 & A_2\cos\tau_2 & \cdots & A_p\cos\tau_p \\
\vdots & \vdots & \vdots & \ddots & \vdots & \vdots & \vdots & \ddots & \vdots \\
0 & \sin p\tau_1 & \sin p\tau_2 & \cdots & \sin p\tau_p & pA_1\cos p\tau_1 & pA_2\cos p\tau_2 & \cdots & pA_p\cos p\tau_p
\end{array}\right)
$$

More algebra shows that the asymptotic variances of the $\{A_j\}$ depend only on the $\{\tau_j\}$, while the asymptotic variances of the $\{\tau_j\}$ depend not only on the configuration of the $\{\tau_j\}$, but also on the heights $\{A_j\}$. In fact, if we define A/σ to be the local signal-to-noise ratio, then the asymptotic standard deviation of $\hat{\tau}_j$ is proportional to the reciprocal of its local signal-to-noise ratio.

4. The maximum likelihood estimates. Throughout this section, we assume the model order m is given. In general, trigonometric moment estimates are not as efficient as maximum likelihood estimates. However, starting from trigonometric moment estimates, which are $\sqrt{n}$-consistent, we can construct one-step estimates or find maximum likelihood estimates using Fisher scoring. In either case, we need to specify the error distribution to calculate the Fisher information matrix. Under the assumption of normal errors, the -2 *loglikelihood* of the observations generated from Model (3) is

$$
(10)\qquad n\log(2\pi\sigma^2) + \frac{1}{\sigma^2}\sum_l \left\{ z(t_l) - A_0 - \sum_{j=1}^{p} A_j w(t_l - \tau_j) \right\}^2 .
$$

More notation is needed in this section. We write $\theta = (A_0, A_1, \ldots, A_p, \tau_1, \ldots, \tau_p)^T$, and sometimes we use $y_\theta(t)$ to denote $SC(w; p; \mathbf{A}; \tau)$. Denote the gradient vector by $\nabla_\theta = (\partial \log L/\partial\theta)^T$, and $\nabla_{(\theta,\sigma^2)} = (\nabla_\theta^T, \partial\log L/\partial\sigma^2)^T$. As usual, the Fisher information matrix is defined by $I_{(\theta,\sigma^2)} = \frac{1}{n}E[\nabla_{(\theta,\sigma^2)}\nabla^T_{(\theta,\sigma^2)}]$ and $I_\theta = \frac{1}{n}E[\nabla_\theta\nabla_\theta^T]$.

PROPOSITION 4.1. *Let* $\Psi_\theta = (\psi_{A_0}, \psi_{A_1}, \ldots, \psi_{A_p}, \psi_{\tau_1}, \ldots, \psi_{\tau_p})^T$, *where* $\psi_{A_0} = 1$, $\psi_{A_j} = w(t-\tau_j)$, $\psi_{\tau_j} = -A_j w'(t - t_j)$; $j = 1, \ldots, p$. *Then*

$$
(11)\qquad I_{(\theta,\sigma^2)} = \begin{pmatrix} I_\theta & 0_p^T \\ 0_p & \frac{1}{2\sigma^4} \end{pmatrix},
$$

where 0_p *is a vector with* p *zeros, and*

$$
(12)\qquad I_\theta = \frac{1}{\sigma^2}\langle \Psi_\theta, \Psi_\theta^T \rangle_n \to \frac{1}{\sigma^2}\langle \Psi_\theta, \Psi_\theta^T\rangle.
$$

Here $\langle \Psi_\theta, \Psi_\theta^T \rangle$ *is defined by*

$$\left(\begin{array}{c|ccc|ccc}
\langle \psi_{A_0}, \psi_{A_0}\rangle & \langle \psi_{A_0}, \psi_{A_1}\rangle & \cdots & \langle \psi_{A_0}, \psi_{A_p}\rangle & \langle \psi_{A_0}, \psi_{\tau_1}\rangle & \cdots & \langle \psi_{A_0}, \psi_{\tau_p}\rangle \\ \hline
\langle \psi_{A_1}, \psi_{A_0}\rangle & \langle \psi_{A_1}, \psi_{A_1}\rangle & \cdots & \langle \psi_{A_1}, \psi_{A_p}\rangle & \langle \psi_{A_1}, \psi_{\tau_1}\rangle & \cdots & \langle \psi_{A_1}, \psi_{\tau_p}\rangle \\
\vdots & \vdots & \ddots & \vdots & \vdots & \ddots & \vdots \\
\langle \psi_{A_p}, \psi_{A_0}\rangle & \langle \psi_{A_p}, \psi_{A_1}\rangle & \cdots & \langle \psi_{A_p}, \psi_{A_p}\rangle & \langle \psi_{A_p}, \psi_{\tau_1}\rangle & \cdots & \langle \psi_{\tau_1}, \psi_{\tau_p}\rangle \\ \hline
\langle \psi_{\tau_1}, \psi_{A_0}\rangle & \langle \psi_{\tau_1}, \psi_{A_1}\rangle & \cdots & \langle \psi_{\tau_1}, \psi_{A_p}\rangle & \langle \psi_{\tau_1}, \psi_{\tau_1}\rangle & \cdots & \langle \psi_{\tau_1}, \psi_{\tau_p}\rangle \\
\vdots & \vdots & \ddots & \vdots & \vdots & \ddots & \vdots \\
\langle \psi_{\tau_p}, \psi_{A_0}\rangle & \langle \psi_{\tau_p}, \psi_{A_1}\rangle & \cdots & \langle \psi_{\tau_p}, \psi_{A_p}\rangle & \langle \psi_{\tau_p}, \psi_{\tau_1}\rangle & \cdots & \langle \psi_{\tau_p}, \psi_{\tau_p}\rangle
\end{array}\right),$$

and similarly for $\langle \Psi_\theta, \Psi_\theta^T \rangle_n$. *Using a similar notation, we compute the gradient vector as follows:*

$$\begin{aligned}
\frac{1}{n}\nabla_\theta &= \frac{1}{\sigma^2}\langle \Psi_\theta, \varepsilon\rangle_n \\
&= \frac{1}{\sigma^2}(\langle \psi_{A_0}, \varepsilon_\theta\rangle_n, \langle \psi_{A_1}, \varepsilon_\theta\rangle_n, \ldots, \langle \psi_{A_p}, \varepsilon_\theta\rangle_n, \langle \psi_{\tau_1}, \varepsilon_\theta\rangle_n, \ldots, \langle \psi_{\tau_p}, \varepsilon_\theta\rangle_n)^T,
\end{aligned}$$

where $\varepsilon_\theta(t) = z(t) - y_\theta(t)$.

Just as with i.i.d. observations, the MLEs are both consistent and asymptotically efficient under the assumption of normal errors.

THEOREM 4.1. *The maximum likelihood estimates are consistent; indeed, as* $n \to \infty$, *we have*

$$\sqrt{n}[(\tilde{\theta}, \tilde{\sigma}^2)^T - (\theta, \sigma^2)^T] \xrightarrow{d} N(0, I^{-1}_{(\theta,\sigma^2)}).$$

In order to compute the maximum likelihood estimate, we can use Fisher scoring as follows, taking the trigonometric moment estimates as the starting value.

$$\begin{pmatrix} \theta_{\text{new}} \\ \sigma^2_{\text{new}} \end{pmatrix} = \begin{pmatrix} \theta_{\text{old}} \\ \sigma^2_{\text{old}} \end{pmatrix} + \frac{1}{n} I^{-1}_{(\theta,\sigma^2)} \nabla_{(\theta,\sigma^2)}\big|_{(\theta_{\text{old}},\sigma^2_{\text{old}})}.$$

Because of the orthogonality of θ and σ^2, we can first use Fisher scoring method to improve the estimate of θ. This leads to the well-known Gauss–Newton algorithm.

ALGORITHM 4.1 (Gauss–Newton).

(i) *Let* θ_{old} *be a* $\sqrt{n}$*-consistent estimate of* θ.
(ii) *Calculate* θ_{new} *by the following:*

$$\begin{aligned}
\theta_{\text{new}} - \theta_{\text{old}} &= \frac{1}{n} I_\theta^{-1} \nabla_\theta\big|_{\theta_{\text{old}}} = [\langle \Psi_{\theta_{\text{old}}}, \Psi^T_{\theta_{\text{old}}}\rangle_n]^{-1} \langle \Psi_{\theta_{\text{old}}}, z - y_{\theta_{\text{old}}}\rangle_n \\
&= \left[\sum_l \Psi_{\theta_{\text{old}}}(t_l)\Psi_{\theta_{\text{old}}}(t_l)^T\right]^{-1} \left[\sum_l \Psi_{\theta_{\text{old}}}(t_l)(z(t_l) - y_{\theta_{\text{old}}}(t_l))\right]. \quad (13)
\end{aligned}$$

Although we can iterate the above procedure, the following result shows one step is enough for the consideration of efficiency.

THEOREM 4.2.

$$\sqrt{n}(\theta_{\text{new}} - \theta)^T \xrightarrow{d} N(0, I_\theta^{-1}).$$

Finally, σ^2 can be estimated as

$$\sigma_{\text{new}}^2 = \left\| z(t) - A_{0,\,\text{new}} - \sum_{j=1}^{p} A_{j,\,\text{new}} w(t - \tau_{j,\,\text{new}}) \right\|_n^2 .$$

5. Hypothetical regressions and model selection. In this section, we deal with the case of m unknown.

5.1. *The least squares estimates and model fitting.* Because of the linear dependence of the signal on the baseline and heights in a $SC(w; p; \mathbf{A}; \tau)$, once the number and locations of the peaks are obtained, we can use least squares to estimate the baseline and peak heights. Combining this idea with Algorithm 3.1, we are led to the following model fitting algorithm.

ALGORITHM 5.1 (Model-fitting). *Starting with the empirical trigonometric moments $\hat{f}_k$ for any given nonnegative integer $m \le K_0$:*

(i) *Use the method of trigonometric moments to estimate the peak locations $\{\bar{\tau}_j\}$ using Algorithm* 3.1.

(ii) *Eliminate those peaks falling outside $[-\pi + \kappa_1, \pi - \kappa_2]$, and denote the locations of the remaining peaks by $\{\bar{\tau}_j, j = 1, \ldots, \bar{m}\}$, where $\bar{m} \le m$.*

(iii) *Estimate the heights $\overline{A}_j$ corresponding to these peaks by minimizing the following:*

$$\left\| z(t) - \overline{A}_0 - \sum_{j=1}^{\bar{m}} \overline{A}_j w(t - \bar{\tau}_j) \right\|_n^2 . \tag{14}$$

This results in the least squares estimates of the baseline and heights.

The output of this algorithm is a $SC(w; \bar{m}; \overline{\mathbf{A}}^{(\bar{m})}; \bar{\tau}^{(\bar{m})})$. Next the problem of model selection arises because we can fit a spike-convolution model for each nonnegative integer in a given range.

5.2. *Model selection.* Suppose that the data is generated from a $SC(w; p; \mathbf{A}; \tau)$, and noise is added. In light of Theorem 2.1, models $SC(w; \bar{m}; \overline{\mathbf{A}}^{(\bar{m})}; \bar{\tau}^{(\bar{m})})$ with $\bar{m} < p$ are not good. When $\bar{m} \ge p$, we might expect a subset of $\{\bar{\tau}_j\}$ will be close to the real peak locations $\{\tau_j\}$. This is the basis of our model selection procedure, whose motivation will be clearer following the next result.

PROPOSITION 5.1. *Suppose that a Toeplitz matrix G_m has smallest eigenvalue Λ_0 with multiplicity $r > 1$.*

(i) *Let $m-r+1 = p$. Then the Toeplitz submatrix G_p has the same smallest eigenvalue Λ_0 with multiplicity 1.*

(ii) *Let $\{e^{i\tau_j}, j = 1, \dots, p\}$ be the roots corresponding to the eigenvector of G_p associated with the eigenvector Λ_0. Then any polynomial whose coefficients form an eigenvector in the invariant space of G_m corresponding to Λ_0 has $\{e^{i\tau_j}, j = 1, \dots, p\}$ as a subset of its roots.*

Let us consider Algorithm 3.1 in light of this result. Supposing that $m \geq p$, we explain the situation from two perspectives. From the computational perspective, in the absence of errors, any polynomial (not unique) whose coefficients form an eigenvector in the invariant space of G_m corresponding to the smallest eigenvalue always has $\{e^{i\tau_j}, j = 1, \dots, p\}$ as a subset of its roots, where τ_j are the real peak locations. But in general it is not true that all its zeros are on the unit circle. In the presence of errors, Theorem 3.2 implies that the multiplicity of the smallest eigenvalue is 1 (apart from a set of Lebesgue measure zero), and the resulting (unique) polynomial has all its roots on the unit circle. This attractive feature, in the computational sense, is the "positive" aspect of noise. From the perspective of spectral structures, the situation here has a close connection to the perturbation theory of matrices, see [8]. In the absence of errors, the distance between the invariant space $\mathscr{S}$ of G_m corresponding to the smallest eigenvalue and other invariant spaces is positive. In other words, it is well separated from other invariant spaces. In the presence of errors and when the sample size is large enough, on the one hand, the last $m - p + 1$ eigenvalues could be close to each other, and their eigenvectors could be "wobbly" (see [8]) under small perturbation of the noise. On the other hand the invariant space $\widehat{\mathscr{S}}$ defined by these possibly "wobbly" eigenvectors is stable. This is true because when the sample size is large enough, $\widehat{\mathscr{S}}$ is close to $\mathscr{S}$, which is well separated from other invariant spaces. Therefore we expect a subset of the roots obtained from Algorithm 3.1 to be close to $\{e^{i\tau_j}, j = 1, \dots, p\}$ when the sample size is large enough. It is so because the eigenvector from which these roots are obtained belongs to $\widehat{\mathscr{S}}$, and is close to one eigenvector belonging to $\mathscr{S}$ and having the property as shown in Proposition 5.1. Now let us return to model fitting. The above argument means the regressors $1, w(t-\bar{\tau}_1), \dots, w(t-\bar{\tau}_{\bar{m}})$ will include a subset of "explanatory variables" close to the true regressors $1, w(t-\tau_1), \dots, w(t-\tau_p)$. For large enough n we can therefore expect model selection criteria to behave as they do in the context of variable selection in regression. Our model selection procedure has two stages. We assume the model order has an upper bound $M(\leq K_0)$.

ALGORITHM 5.2 (Two-stage model selection). (i) *First stage. Among all the $SC(w; \bar{m}; \bar{\mathbf{A}}^{(\bar{m})}; \bar{\tau}^{(\bar{m})})$ models fitted by Algorithm 5.1, choose the one that* *minimizes the following*:

$$MGIC_1(r) = \bar{\sigma}(r)^2 + \frac{c_1(n)\log n}{n} r, \tag{15}$$

where $\bar{\sigma}(r)^2$ *is the quantity in* (14), *and* $c_1(n) \geq 0$ *is a penalty coefficient. Denote this model by* $SC(w; \bar{m}_0; \overline{\mathbf{A}}^{(\bar{m}_0)}; \bar{\tau}^{(\bar{m}_0)})$.

(ii) *Second stage. We regard the model selected in the first stage as a hypothetical regression model, and use a backward deletion procedure to select the final model. That is, starting from* $SC(w; \bar{m}_0; \overline{\mathbf{A}}^{(\bar{m}_0)}; \bar{\tau}^{(\bar{m}_0)})$, *we delete the peak that is least significant in terms of sum of squares. Compare the two models according to the following statistic*:

$$MGIC_2(r) = \check{\sigma}(r)^2 + \frac{c_2(n)\log n}{n} r, \tag{16}$$

where $\check{\sigma}^{(r)2}$ *is the sum of squares fitted by a model with r peaks, and* $c_2(n) > 0$ *is another penalty coefficient possibly depending on n. Choose the one that minimizes* $MGIC_2$. *If one peak can be deleted according to this criterion, then we iterate this procedure until we cannot delete any more peaks.*

We make some remarks about this procedure. First, existing model selection procedures such as AIC and BIC cannot be applied here, for the parameter estimates are not maximum likelihood ones. Second, the penalty term $c_1(n)$ is used in the first stage to compare all models obtained from Algorithm 5.1, and overfitting is not suppressed but encouraged to some extent. In fact, we would like to find the "best overfitting" model in this stage. The penalty term $c_2(n)$ is used in the second stage to eliminate those false peaks in the model obtained in the first stage. This suggests that we impose another restriction $c_1(n) < c_2(n)$. Third, the purpose of this two-stage model selection procedure is not only estimating the model order, but also producing a parameter estimate with much smaller bias and variance than that of the trigonometric moment estimate if the order could be assumed to be known; see the numerical example in Section 6 for details. Fourth, the determination of the two penalty terms needs more investigation in both theory and implementation, though some experience has been gained for the dataset we have been working on. Use of the bootstrap or cross validation is possible. For example, in the analogous problem of estimating the number of hidden frequencies in Model (4), Ulrych and Sacchi [31] chose the number using Kullback divergence as the risk, and a bootstrap method to estimate the risk of each model. Fifth, when applying this methodology to DNA sequencing data, we can set a lower bound as well as an upper bound on the model order, since the numbers of the four kinds of DNA bases in a given range can be estimated. Our experience shows moderate overfitting is not an issue for DNA traces. Finally, in the sequel, we refer to the procedures in Algorithm 3.1, 5.1, 5.2 as PDPS (parametric deconvolution of positive spikes). If the error distribution is known to be normal, Algorithm 4.1 (Gauss–Newton) could be included in PDPS to improve the accuracy of the estimate.

6. Examples and discussion.

6.1. *A simulated example.* Our simulation study is based the following model.

EXAMPLE 6.1.

$$z(t_l) = 0.5 + w(t_l + 1.9) + 1.25w(t_l + 1.6) + 1.25w(t_l + 1.3) + w(t_l) \\ + 1.25w(t_l - 0.5) + 1.1w(t_l - 1.0) + 1.25w(t_l - 2.5) + \varepsilon(t_l),$$

where the sample size $n = 1024, w(t)$ is a Gaussian function $b/2\pi \times \exp\{-b^2t^2/2\}$ with the scale parameter $b = 8$ being truncated at ± 4 SD. Errors are normally distributed with mean 0 and standard deviation 0.3. Figure 2 shows a simulated sample from this model. The signal contains seven peaks, and the three on the left are quite close to one another. The peak heights are generally similar, which is typical for sequencing data. Simulations in this section are carried out in MATLAB, and are repeated 1000 times for each method we have studied. We apply three estimation procedures to this example. First, we assume the model order is known, and use the method of trigonometric moments. Second, we use PDPS to estimate the parameter. The penalty coefficients $c_1(n)$ and $c_2(n)$ are taken to be 2 and 10, respectively. The upper bound of the model order is taken to be 20. Out of the 1000 replications, all but one of the final-fitted models had order 7, which is the truth. Finally, this result was further tuned by the Gauss–Newton algorithm. Two iterations were used. The statistics of estimates of the peak locations and heights are summarized in Table 1. The estimate tuned by the Gauss–Newton algorithm is almost unbiased, and its standard errors are close to the nominal ones. It

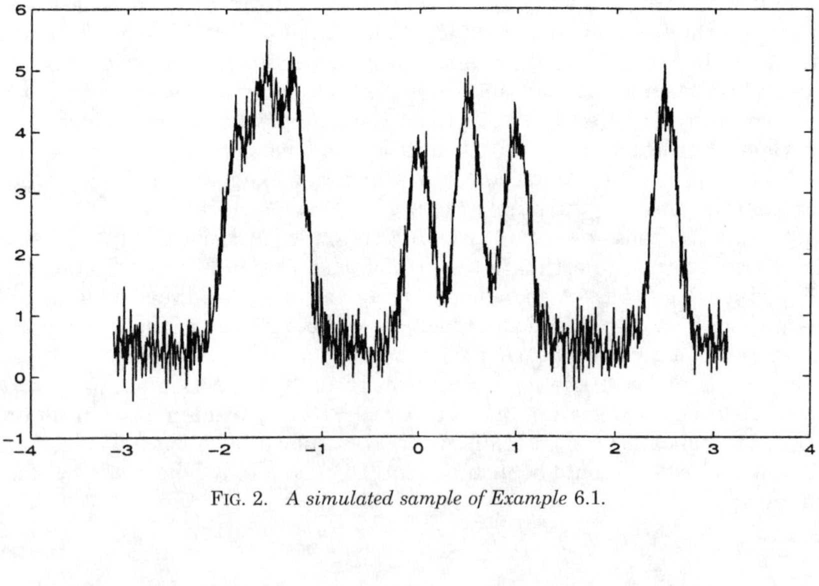

FIG. 2. *A simulated sample of Example* 6.1.

PARAMETRIC DECONVOLUTION

TABLE 1
Statistics of the three estimates of the parameters in Example 6.1 (1000 *replications*)

Parameter truth		Method of trigonometric moments			PDPS without MLE tuning			Gauss–Newton algorithm			
		Bias ($\times 10^3$)	**SD** ($\times 10^2$)	**CV** (%)	**Bias** ($\times 10^3$)	**SD** ($\times 10^2$)	**CV** (%)	**Bias** ($\times 10^3$)	**SD** ($\times 10^2$)	**CV** (%)	**Nominal SE** ($\times 10^2$)
τ_1	−1.9	−211	32		−8	2		0	0.4		0.4
τ_2	−1.6	0	19		3	4		0	0.4		0.4
τ_3	−1.3	315	48		11	2		0	0.3		0.3
τ_4	0.0	64	14		−2	1		0	0.3		0.3
τ_5	0.5	78	16		−2	1		0	0.2		0.2
τ_6	1.0	97	26		−1	1		0	0.3		0.3
τ_7	2.5	2	4		−6	1		0	0.2		0.2
A_1	1.0	−216	67	85	−8	11	11	0	2.1	2	2.1
A_2	1.25	506	20	11	52	6	5	0	2.1	2	2.1
A_3	1.25	−158	68	62	−30	11	9	−1	2.1	2	2.1
A_4	1.0	62	18	17	−1	2	2	0	1.7	2	1.7
A_5	1.25	21	8	6	−1	2	2	−1	1.7	1	1.7
A_6	1.1	−170	36	38	0	2	2	0	1.7	2	1.7
A_7	1.25	−9	13	10	−1	2	1	−1	1.7	1	1.7
A_0	0.5	−6	2	4	−1	2	3	1	1.3	3	1.2

is quite surprising that the accuracy of the trigonometric moment estimate is so poor, even though the order is assumed to be known. In comparison, PDPS has achieved an accuracy much closer to that of the maximum likelihood estimate. This means that modest overfitting in the first stage can greatly control the bias and variance of the estimate. In other words, even when the number of peaks is known, PDPS is better than the direct trigonometric moment estimate in terms of bias and variance. In this case, we set a lower bound by the known order at the first stage, and stop the backward deletion when the number of left peaks equals the known order at the second stage. The frequency of the model orders selected at stage one is shown in Table 2. We see that most models selected at stage one have orders ranging from 9 to 12.

6.2. *Real trace data.* Next we apply PDPS to the sequencing trace shown in Figure 1. The point spread function is taken to be a truncated Gaussian function. The result is shown in Figure 3. (Maximum likelihood estimation was not used.) For a comparison, a nonnegative least squares deconvolution was carried out; see [19]. The results are displayed in Figure 4. The two methods

TABLE 2
Frequency of model orders selected at the first stage

Model orders	8	9	10	11	12	13	14
Frequency	48	248	212	202	244	45	1

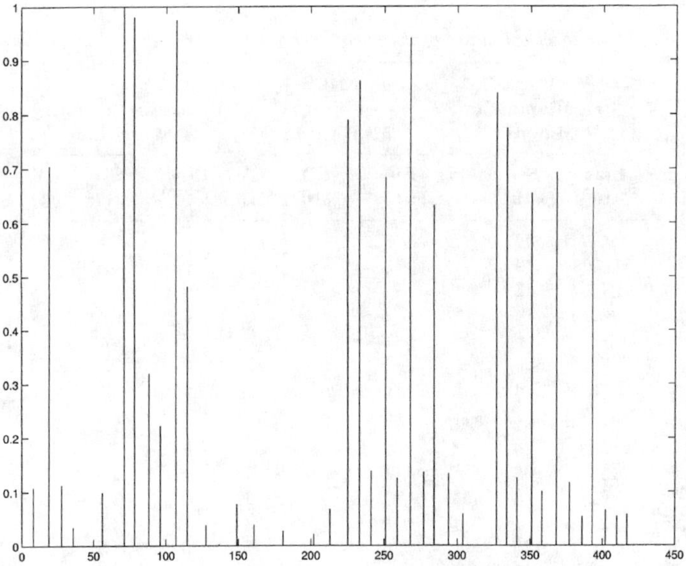

FIG. 3. *Parametric deconvolution.*

yield quite different results. First, that obtained from the parametric deconvolution is cleaner. Second, the relative heights of the major spikes following the parametric deconvolution are more similar to those in the original data. Third, parametric deconvolution is more efficient from the computational point of view. On a Sun Ultra-2 workstation, the Lawson and Hanson algorithm took more than one hour while the parametric method took only two minutes. A more systematic comparison of this parametric deconvolution method with others can be found in [20].

6.3. *Colored noise and reblurring.* The result in this paper can be generalized to situations when the errors are serially correlated. We might approximate such errors by an autoregressive process. That is, we could assume that the errors ε in (2) can be modeled by

$$\varepsilon_t + \sum_{k=1}^{p} \phi_k \varepsilon_{t-k} = \xi_t, \tag{17}$$

where the ξ_t is i.i.d. $N(0, \sigma^2)$. Then we could prewhiten the signal as follows:

$$(\phi * z)(t) = (\phi * w * x)(t) + \xi(t). \tag{18}$$

By replacing $z(t)$ by $(\phi * z)(t)$, we could use our original scheme PDPS to do the deconvolution. This reblurring idea has been used in other similar situations; see [7,13].

PARAMETRIC DECONVOLUTION

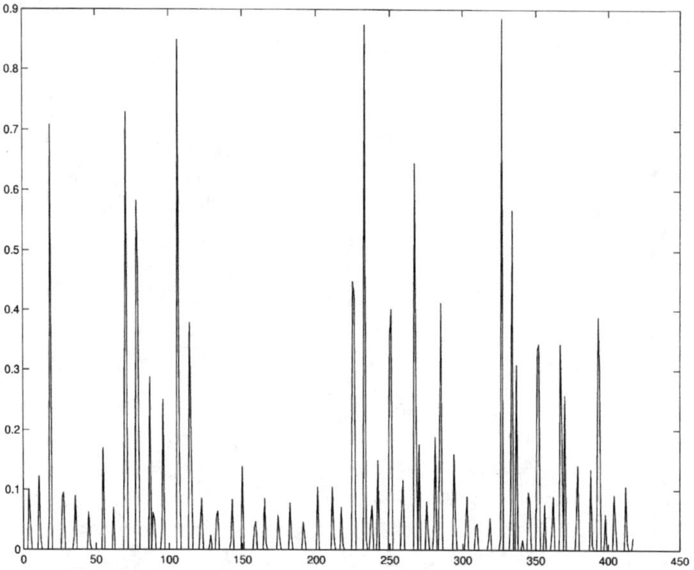

FIG. 4. *NNLS deconvolution (carried out by the Lawson and Hanson algorithm).*

6.4. *Implementations.* The numerical implementation of PDPS hinges on linear regression and computation of smallest eigenvalues and their eigenvectors of Toeplitz matrices. Because of the special structure of Toeplitz matrices, efficient algorithms do exist, and we refer to [12, 22]. Though we have discussed maximum likelihood estimates within the assumption of Gaussian errors in this paper, we may skip the MLE tuning in applications, since either the assumption of normal errors may not be appropriate, or highly accurate estimates of the peak positions and heights may not be necessary.

APPENDIX

This section contains the proofs and relevant mathematical facts. The following theorem is in essence the so-called trigonometric moment problem solved by Carathéodory; see [9]. We present a version under the framework of the spike-convolution model.

THEOREM A.1.

(i) *Given a* $SC(\delta; m; \mathbf{A}; \tau)$, *let* G_m *be the Toeplitz matrix constructed from its Fourier coefficients* $\{g_k\}$. *Then* A_0 *is the smallest eigenvalue of* G_m *with multiplicity one and eigenvector* $(\alpha_0, \ldots, \alpha_m)^T$. *The* $\{A_j\}$ *satisfy the following linear system:*

$$
(19) \qquad \frac{1}{2\pi}\begin{pmatrix} 1 & 1 & \cdots & 1 \\ \exp(i\tau_1) & \exp(i\tau_2) & \cdots & \exp(i\tau_m) \\ \vdots & \vdots & \ddots & \vdots \\ \exp(i(m-1)\tau_1) & \exp(i(m-1)\tau_2) & \cdots & \exp(i(m-1)\tau_m) \end{pmatrix}\begin{pmatrix} A_1 \\ A_2 \\ \vdots \\ A_m \end{pmatrix} = \begin{pmatrix} g_0 - A_0 \\ g_1 \\ \vdots \\ g_{m-1} \end{pmatrix}.
$$

(ii) *Conversely, suppose we are given* $2m+1$ *complex numbers* $\{g_j, -m \le j \le m\}$, *where* $\bar{g}_j = g_{-j}$. *Assume that the smallest eigenvalue* A_0 *of the Toeplitz matrix* $G_m = (g_{j-i})_{i,\, j=0,\dots,m}$ *has multiplicity* 1. *Let the smallest eigenvector be* $\alpha = (\alpha_0, \dots, \alpha_m)^T$, *and* $U^{(m)}(z) = \sum_{j=0}^{m} \alpha_j z^j$. *Then there exists a unique* $SC(\delta; m; \mathbf{A}; \tau)$ *whose first* $m+1$ *Fourier coefficients are* $\{g_j, 0 \le j \le m\}$. *The* $\{\tau_j\}$ *are determined from the m distinct roots* $\{e^{i\tau_j}\}$ *of* $U^{(m)}(z)$ *lying exactly on the unit circle. The* $\{A_j\}$ *are determined by the linear system* (19), *and the resulting heights are positive.*

For the proof, the first part is easy to check. As for the second part, an algebraic proof can be found in [9]. Li [20] gave a measure-theoretic proof.

PROOF OF THEOREM 2.1. Let the Fourier coefficients of the corresponding $x(t)$ and $\bar{x}(t)$ be $\{g_k\}, \{\bar{g}_k\}$, respectively. According to the Parseval identity, $\|y - \bar{y}\|^2 = \sum_{k=-\infty}^{\infty} |v_k(g_k - \bar{g}_k)|^2 \ge (\min_{0 \le k \le p} |v_k|^2) \sum_{k=0}^{p} |g_k - \bar{g}_k|^2$. Thus if there were a sequence $SC(w; l; \overline{\mathbf{A}}^{(l)}; \tau^{(l)})$, $l < p$ which could approach $SC(w; p; \mathbf{A}; \tau)$, then their Fourier coefficients $\bar{g}_k^{(l)} \to g_k$, for $0 \le k \le p$. Therefore their Toeplitz matrices $\overline{T}_p^{(l)}$ will converge to T_p as do their characteristic polynomials. But this is impossible because T_p's smallest eigenvalue has multiplicity 1 according to Theorem 7.1, while $\overline{T}_p^{(l)}$'s smallest eigenvalue has multiplicity larger than 1, as can be easily checked.

PROOF OF THEOREM 3.1. This can be checked by direct calculation.

PROOF OF THEOREM 3.2. By Theorem 7.1, we can find a function x of the form (1) with Fourier coefficients $\{g_k\}$. Next the proof is completed by convolving x with w and using the convolution theorem.

PROOF OF THEOREM 3.3. Without loss of generality, assume n is even. A trigonometric moment $\hat{f}_k$ can be decomposed into three parts,

$$
\begin{aligned}
\hat{f}_k &= f_k + \sum_{l=-[n/2]}^{[n/2]-1} \left[\frac{2\pi}{n} y(t_l)\exp(ikt_l) - \int_{t_l}^{t_{l+1}} y(t)\exp(ikt)dt \right] \\
&\quad + \frac{1}{n} \sum_{l=-[n/2]}^{[n/2]-1} \varepsilon(t_l)\exp(ikt_l),
\end{aligned} \tag{20}
$$

for $k = 0, \pm 1, \ldots$. Denote the second and third terms by $\check{f}_k$ and $\tilde{f}_k$, respectively. Suppose initially the order p of the model is known in advance, and let $\hat{g}_0 = \hat{f}_0$, $\hat{g}_k = \hat{f}_k v_0/v_k$, for $0 < k \le K_0$. Similarly, decompose $\hat{g}_k$ into three parts, g_k, $\check{g}_k$ and $\tilde{g}_k$ corresponding to f_k, $\check{f}_k$ and $\tilde{f}_k$, respectively. Under the smoothness condition on $w(t)$, the effect of $\check{f}_k$ is $O(1/n)$ using Taylor expansion and the bounded property of $w'(\cdot)$ and is thus negligible compared with that of $\tilde{f}_k$. Next, notice that $\tilde{k} = \langle \varepsilon(t), e^{ikt} \rangle_n$. According to the Lyapunov central limit theorem, $\tilde{f}_0, \tilde{f}_1, \ldots, \tilde{f}_p$ are asymptotically normally distributed,

$$\sqrt{n}(\tilde{f}_0, \tilde{f}_1, \ldots, \tilde{f}_p)^T \xrightarrow{d} N(0, \sigma^2 I_{p+1}).$$

Consequently, we have a CLT for the $\hat{f}_k$,

$$\sqrt{n}[(\hat{f}_0, \hat{f}_1, \ldots, \hat{f}_p)^T - (f_0, f_1, \ldots, f_p)^T] \xrightarrow{d} N(0, \sigma^2 I_{p+1}).$$

Thus

$$\sqrt{n}[(\hat{g}_0, \hat{g}_1, \ldots, \hat{g}_p)^T - (g_0, g_1, \ldots, g_p)^T] \xrightarrow{d} N\left(0, \sigma^2 \mathrm{diag}\left(\left|\frac{v_0}{v_k}\right|^2\right)\right).$$

For $k \neq 0$, we can split each term g_k into its real part $g_{k,r}$ and its imaginary part $g_{k,i}$. Then another form of the above CLT is given by

$$
\begin{aligned}
&\sqrt{n}[(\hat{g}_0, \hat{g}_{1,r}, \ldots, \hat{g}_{p,r}, \hat{g}_{1,i}, \ldots, \hat{g}_{p,i})^T - (g_0, g_1, r, \ldots, g_{p,r}, g_{1,i}, \ldots, g_{p,i})^T] \\
&\xrightarrow{d} N(0, P),
\end{aligned}
$$

where P is given by (8). The mapping between the trigonometric moments and the parameters is continuous. Thus we can find the CLT for the trigonometric moment estimates using the delta method. Now we calculate the Jacobian matrix. From

$$
\begin{pmatrix} g_0 \\ g_1 \\ \vdots \\ g_p \end{pmatrix}
= \frac{1}{2\pi}
\begin{pmatrix}
1 & 1 & 1 & \cdots & 1 \\
0 & \exp(i\tau_1) & \exp(i\tau_2) & \cdots & \exp(i\tau_p) \\
\vdots & \vdots & \vdots & \ddots & \vdots \\
0 & \exp(ip\tau_1) & \exp(ip\tau_2) & \cdots & \exp(ip\tau_p)
\end{pmatrix}
\begin{pmatrix} A_0 \\ A_1 \\ A_2 \\ \vdots \\ A_p \end{pmatrix},
$$

we have

$$\frac{\partial(g_0, g_1, \dots, g_p)}{\partial(A_0, A_1, \dots, A_p, \tau_1, \dots, \tau_p)} = \frac{1}{2\pi}$$

$$\times \left(\begin{array}{c|cccc|cccc} 1 & 1 & 1 & \cdots & 1 & 0 & 0 & \cdots & 0 \\ \hline 0 & \exp(i\tau_1) & \exp(i\tau_2) & \cdots & \exp(i\tau_p) & iA_1\exp(i\tau_1) & iA_2\exp(i\tau_2) & \cdots & iA_p\exp(i\tau_p) \\ \vdots & \vdots & \vdots & \ddots & \vdots & \vdots & \vdots & \ddots & \vdots \\ 0 & \exp(ip\tau_1) & \exp(ip\tau_2) & \cdots & \exp(ip\tau_p) & ipA_1\exp(ip\tau_1) & ipA_2\exp(ip\tau_2) & \cdots & ipA_p\exp(ip\tau_p) \end{array} \right).$$

Rewriting this as a square matrix by splitting the derivative into real and imaginary parts, we get

$$\frac{\partial(g_0, g_{1,r}, \dots, g_{p,r}, g_{1,i}, \dots, g_{p,i})}{\partial(A_0, A_1, \dots, A_p, \tau_1, \dots, \tau_p)} = \frac{1}{2\pi} Q.$$

Therefore $V = 4\pi^2 Q^{-1} P (Q^{-1})^T$. We refer to [24] for the invertibility of Q.

PROOF OF PROPOSITION 4.1. This can be checked by direct calculation.

PROOF OF THEOREM 4.1. (i) Let $l(\theta, \sigma^2)$ be the log likelihood evaluated at (θ, σ^2). To prove the consistency, we need to show that for any fixed $(\theta_0, \sigma_0^2) \neq (\theta, \sigma^2)$, where (θ, σ^2) is the truth,

$$P(l(\theta_0, \sigma_0^2) < l(\theta, \sigma^2)) \to 1 \quad \text{as } n \to \infty. \tag{21}$$

Notice that

$$l(\theta, \sigma^2) - l(\theta_0, \sigma_0^2) = [l(\theta, \sigma^2) - l(\theta, \sigma_0^2)] + [l(\theta, \sigma_0^2) - l(\theta_0, \sigma_0^2)]. \tag{22}$$

Denote $\theta_0 = (A_{0,0}, A_{1,0}, \dots, A_{p,0}, \tau_{1,0}, \dots, \tau_{p,0})^T$, and let $\eta(t) = y(t) - [A_{0,0} + \sum_{k=1}^{p} A_{k,0} w(t - \tau_{k,0})]$. Then $\int_{-\pi}^{\pi} \eta(t)^2 dt > 0$ for $\theta_0 \neq \theta$ because of the identifiability of the parameterization. By the law of large numbers, (cf. [2, 6]), the second term in (22) is

$$\begin{aligned} \frac{1}{n}[l(\theta, \sigma_0^2) - l(\theta_0, \sigma_0^2)] &= \frac{1}{2n\sigma_0^2}\left[\sum_l (\varepsilon(t_l) - \eta(t_l))^2 - \sum_l \varepsilon(t_l)^2\right] \\ &= \frac{1}{2n\sigma_0^2}\left[\sum_l (\eta(t_l))^2 - 2\sum_l \eta(t_l)\varepsilon(t_l)\right] \\ &\overrightarrow{p} \; \frac{1}{2\sigma_0^2}\int_{-\pi}^{\pi} \eta(t)^2\, dt \geq 0. \end{aligned}$$

The first term is

$$\begin{aligned} \frac{1}{n}[l(\theta, \sigma^2) - l(\theta, \sigma_0^2)] &= \left[-\frac{1}{2}\log\sigma^2 - \frac{1}{2n\sigma^2}\sum_l \varepsilon(t_l)^2\right] \\ &\quad - \left[-\frac{1}{2}\log\sigma_0^2 - \frac{1}{2n\sigma_0^2}\sum_l \varepsilon(t_l)^2\right] \overrightarrow{p} \left[-\frac{1}{2}\log\sigma^2 - \frac{\sigma^2}{2\sigma^2}\right] \end{aligned}$$

PARAMETRIC DECONVOLUTION

$$-\left[-\frac{1}{2}\log\sigma_0^2-\frac{\sigma^2}{2\sigma_0^2}\right]\geq 0,$$

since the function $x(s) = -\log s - \sigma^2/s$ has its unique maximum at σ^2. Hence (21) is true.

(ii) Let D be the matrix with entries of second derivatives of the log-likelihood. Since the MLE$(\tilde{\theta}, \tilde{\sigma}^2)$ is consistent, the following Taylor expansion can be checked straightforwardly:

$$\frac{1}{\sqrt{n}}\left[\nabla(\tilde{\theta},\tilde{\sigma}^2)-\nabla(\theta,\sigma^2)\right]=\frac{1}{\sqrt{n}}D(\theta,\sigma^2)(\tilde{\theta}-\theta,\tilde{\sigma}^2-\sigma^2)^T + O_p\|(\tilde{\theta}-\theta,\tilde{\sigma}^2-\sigma^2)\|^2. \tag{23}$$

Thus

$$\sqrt{n}(\tilde{\theta}-\theta,\tilde{\sigma}^2-\sigma^2)^T\stackrel{d}{\to}-\left[\frac{D(\theta,\sigma^2)}{n}\right]^{-1}\frac{\nabla(\theta,\sigma^2)}{\sqrt{n}},$$

where the first term converges to $I_{(\theta,\sigma^2)}^{-1}$ in probability and the second term converges to $N(0, I_{(\theta,\sigma^2)})$ in distribution. The remainder is an application of the Slutsky theorem.

Proof of Theorem 4.2. Notice that by the Taylor expansion,

$$y_{\theta\text{old}}-y=\Psi_\theta^T(\theta_{\text{old}}-\theta)+o_p(\|\theta_{\text{old}}-\theta\|),$$

where the second term is uniform with respect to the variable t, and so

$$z-y_{\theta\text{old}}=\varepsilon-(y_{\theta\text{old}}-y)=\varepsilon-\Psi_\theta^T(\theta_{\text{old}}-\theta)+o_p(\|\theta_{\text{old}}-\theta\|).$$

Inserting this into line 2 of the following, we obtain

$$\begin{aligned}
\sqrt{n}(\theta_{\text{new}}-\theta) &= \sqrt{n}(\theta_{\text{new}}-\theta_{\text{old}})+\sqrt{n}(\theta_{\text{old}}-\theta)\\
&= \sqrt{n}[\langle\Psi_{\theta_{\text{old}}},\Psi_{\theta_{\text{old}}}^T\rangle_n]^{-1}\langle\Psi_{\theta_{\text{old}}},z-y_{\theta_{\text{old}}}\rangle_n+\sqrt{n}(\theta_{\text{old}}-\theta)\\
&= \sqrt{n}[\langle\Psi_{\theta_{\text{old}}},\Psi_{\theta_{\text{old}}}^T\rangle_n]^{-1}\langle\Psi_{\theta_{\text{old}}},\varepsilon\rangle_n\\
&\quad -[\langle\Psi_{\theta_{\text{old}}},\Psi_{\theta_{\text{old}}}^T\rangle_n]^{-1}\langle\Psi_{\theta_{\text{old}}},\Psi_\theta\rangle_n\sqrt{n}(\theta_{\text{old}}-\theta)\\
&\quad +\sqrt{n}(\theta_{\text{old}}-\theta)+o_p(\|\sqrt{n}(\theta_{\text{old}}-\theta\|)\\
&\overrightarrow{p}\ \sqrt{n}[\langle\Psi_\theta,\Psi_\theta^T\rangle_n]^{-1}\langle\Psi_{\theta,}\varepsilon\rangle_n+\sqrt{n}[\langle\Psi_\theta,\Psi_\theta^T\rangle_n]^{-1}\\
&\quad\times\langle\Psi_{\theta_{\text{old}}}-\Psi_\theta,\varepsilon\rangle_n+o_p(1).
\end{aligned}$$

Here we need the $\sqrt{n}$-consistency of θ_{old}. The second term in the last line is $o_p(1)$, since we can apply a Taylor expansion to $\Psi_{\theta_{\text{old}}}-\Psi_\theta$ at θ. Then we complete the proof by applying the central limit theorem to the first term and using the Slutsky theorem.

PROOF OF PROPOSITION 5.1. The dimension of the invariant space corresponding to the smallest eigenvalue A_0 is r. By Gaussian elimination, we can always find a vector in this space with the form $\alpha = (\alpha_0, \dots, \alpha_p, 0, \dots, 0)^T$. Of course, $(\alpha_0, \dots, \alpha_p)^T$ is a eigenvector of G_p with eigenvalue A_0. A_0 is the smallest eigenvalue of G_p because of the monotone property of eigenvalues of nested matrices. Its multiplicity is 1. Otherwise, by repeating the above reasoning, we can find an eigenvector of G_{m-r} with the form of $(\beta_0, \dots, \beta_{m-r})^T$. Then by extending this vector into $m+1$ Euclidean space by adding zeros in the beginning and the end, we can construct $r+1$ linear independent eigenvectors of G_m corresponding to A_0, which would imply that the multiplicity of A_0 is greater than r, a contradiction. In order to prove the second part, notice that $(\alpha_0, \dots, \alpha_p, 0, \dots, 0)^T, (0, \alpha_0, \dots, \alpha_p, 0, \dots, 0)^T, \dots, (0, \dots, 0, \alpha_0, \dots, \alpha_p)^T$ is a basis of the invariant space corresponding to A_0. Consequently, any polynomial whose coefficients are an eigenvector corresponding to A_0 is a linear combination of polynomials with common roots $\{e^{i\tau_j}, j = 1, \dots, p\}$. The rest is obvious.

Acknowledgments. We give our grateful thanks to Prof. J. Rice for his very helpful comments. We owe many debts to the two referees. Their constructive suggestions have greatly enhanced this research and improved the organization of the material.

REFERENCES

[1] ADAMS, M. D., FIELDS, C. and VENTOR, J. C. (eds.). (1994). *Automated DNA Sequencing and Analysis*. Academic Press, London.

[2] BILLINGSLEY. P. (1986). *Probability and Measure*. Wiley, New York.

[3] CHEN, W.-Q. and HUNKAPILLER, T. (1992). Sequence accuracy of larger DNA sequencing projects. *J. DNA Sequencing and Mapping* **2** 335–342.

[4] DI GES, V. and MACCARONE, M. C. (1984). The Bayesian direct deconvolution method: properties and and applications. *Signal Processing* **6** 201–211.

[5] DONOHO, D. L., JOHNSTONE, I. M., HOCH, J. C. and STERN, A. S. (1992). Maximum entropy and the nearly black object. *J. Roy. Statist. Soc. Ser. B* **54** 41–81.

[6] DURRETT, R. (1991). *Probability: Theory and Examples*. Wadsworth and Brooks Cole, Belmont, CA.

[7] FREDKIN, D. R. and RICE, J. A. (1997). Fast evaluation of the likelihood of an HMM: ion channel currents with filetering and colored noise. Dept. Statistics, Univ. California, Berkeley.

[8] GOLUB, G. H. and VAN LOAN, C. F. (1996). *Matrix Computations*, 3rd ed. John Hopkins Univ. Press.

[9] GRENANDER, U. and SZEGÖ, G. (1958). *Toeplitz Forms and Their Applications*. Univ. California Press, Berkeley.

[10] GULL, S. F. (1989). Developments in maximum entropy data analysis. In *Maximum Entropy and Bayesian Methods* (J. Skilling, ed.) Kluwer, Boston.

[11] GULL, S. F. and DANIELL, G. J. (1978). Image reconstruction from incomplete and noisy data. *Nature* **272** 686–690.

[12] HUANG, D. (1992). Symmetric solutions and eigenvalue problems of Toeplitz systems. *IEEE Trans. Acoust. Speech Signal Processing* **40** 3069–3074.

[13] JANSSON, P. A. (ed.). (1997). *Deconvolution of Images and Spectra*. Academic Press, New York.

[14] KENNETT, T. J., PRESTWICH, W. V. and ROBERTSON, A. (1978). Bayesian deconvolution 1. convergence properties. *Nuclear Instrument and Methods* **151** 285–292.

[15] KENNETT, T. J., PRESTWICH, W. V. and ROBERTSON, A. (1978). Bayesian deconvolution 2. noise properties. *Nuclear Instrument and Methods* **151** 293–301.

[16] KENNETT, T. J., PRESTWICH, W. V. and ROBERTSON, A. (1978). Bayesian deconvolution 3. application and algorithm implementation. *Nuclear Instrument and Methods* **153** 125–135.

[17] KOOP, B. F., ROWEN, L., CHEN, W.-Q., DESHPANDE, P., LEE, H. and HOOD, L. (1993). Sequence length and error analysis of sequence and automated *taq* cycle seqeuncing methods. *BioTechniques* **14** 442–447.

[18] LAWRENCE, C. B. and SOLOVYEV, V. V. (1994). Assignment of position-specific error probability to primary DNA sequence data. *Nucleic Acid Research* **22** 1272–1280.

[19] LAWSON, C. L. and HANSON, R. J. (1974). *Solving Least Squares Problems*. Prentice Hall, Englewood Cliff, NJ.

[20] LI, L. (1998). Statistical models of DNA base-calling. Ph.D. dissertation, Univ. California, Berkeley.

[21] LI, L. and SPEED, T. P. (1999). An estimate of the color separation matrix in four-dye fluorescence-based DNA sequencing. *Electrophoresis* **20** 1433–1442.

[22] MAKHOUL, J. (1981). On the eigenvectors of symmetric Toeplitz matrices. *IEEE Trans. Acoust. Speech. Signal Processing* **29** 868–872.

[23] NELSON, D. O. (1995). Introduction of reptation. Technical report, Lawrence Livermore National Lab.

[24] PISARENKO, V. F. (1973). The retrieval of harmonics from a convariance function. *Geophys. J. Roy. Astrophys. Soc.* **33** 347–366.

[25] POSKITT, D. S., DOGANCAY, K. and CHUNG, S.-H. (1999). Double-blind deconvolution: the analysis of post-synaptic currents in nerve cells. *J. Roy. Statist. Soc. Ser. B* **61** 191–212.

[26] RICHARDSON, W. H. (1972). Bayesian-based iterative method of image restoration. *J. Opt. Soc. Amer. A* **62** 55–59.

[27] SHEPP, L. A. and VARDI, Y. (1982). Maximum-likelihood reconstruction for emission tomography. *IEEE Trans. Medical Imaging* **MI-1** 113–121.

[28] SNYDER, D. L., SCHULZ, T. J. and O'SULLIVAN, J. A. (1992). Deblurring subject to nonnegativity constraints. *IEEE Trans. Signal Processing* **40** 1143–1150.

[29] STARK, P. B. and PARKER, R. L. (1995). Bounded-variable least-squares: an algorithm and applications. *Comput. Statist.* **10** 129–141.

[30] TIKHONOV, A. (1963). Solution of incorrectly formulated problems and the regularization method. *Soviet Math. Dokl.* **5** 1035–1038.

[31] ULRYCH, T. J. and SACCHI, M. D. (1995). Sompi, Pisarenko and the extended information criterion. *Geophysical J.* **122** 719–724.

[32] VARDI, Y. and LEE, D. (1993). From image deblurring to optimal investment: maximum likelihood solutions for positive linear inverse problems. *J. Roy. Statist. Soc. Ser. B* **55** 569–612.

DEPARTMENT OF STATISTICS
FLORIDA STATE UNIVERSITY
TALLAHASSEE, FLORIDA 32306-4330
E-MAIL: lilei@stat.fsu.edu

DEPARTMENT OF STATISTICS
UNIVERSITY OF CALIFORNIA
BERKELEY, CALIFORNIA

Chapter 14
Biological Sequence Analysis

Simon E. Cawley

Shortly after the start of my graduate studies at the U.C. Berkeley Statistics department in 1995, I had the good fortune to meet Terry and learn about some of his work in the area of the application of statistics to genetics and molecular biology. Not having thought about biology since high school, I was very impressed by the large impact statistical approaches were making in a field I had naively considered as one that had little to do with quantitative analysis. I eagerly dove in to a collaboration that Terry had put in place with the Human and Drosophila Genome Projects at Lawrence Berkeley National Laboratories and spent the next few years having a great time working on interesting and practical statistical problems that arose in the context of the ongoing genome sequencing efforts.

In this section we present some of Terry's contributions in the area of Sequence Analysis – generally speaking, the area of analysis of biological sequences such as DNA or protein sequences. The papers presented here relate to the interpretation of DNA sequences.

DNA sequence analysis has been an area of growing importance since DNA sequencing techniques started to emerge in the early 1970s. The chain-terminator method developed by Frederick Sanger at the University of Cambridge [7] was a pivotal moment, enabling the first rapid scaling up of DNA sequencing capabilities. The rate of sequencing was further accelerated through the 1980s and 1990s as ever-greater levels of automation were brought to bear on Sangers original concept.

As the level of automation increased, it became possible to sequence entire genomes of successively more complex organisms with larger genomes, ranging from bacteriophage phiX174 in the late 1970s, various microbial genomes in the early 1990s through to the draft of the human genome sequence published in 2001. The Sanger method showed remarkable longevity and was at the core of the vast majority of sequencing efforts through to the early 2000s.

The dominance of Sanger sequencing finally ended in the early 2000s with the advent of a renaissance of sorts as multiple new massively parallel technologies such

S.E. Cawley
Ion Torrent
e-mail: simon.cawley@lifetech.com

S. Dudoit (ed.), *Selected Works of Terry Speed*, Selected Works in Probability and Statistics, 563
DOI 10.1007/978-1-4614-1347-9_14,

as 454 pyrosequencing, followed soon after by Solexa (Illumina), SOLiD, polony, DNA nanoball and Ion Torrent sequencing.

As DNA sequencing technologies scaled up, huge opportunities arose along the way for the application of statistics, both in the area of analysis of the signals generated from each of the various instruments and technologies to improve DNA sequencing accuracy (the subject of Chapter 13), and in the downstream analysis of the DNA sequence collected. In particular, as the volumes of sequence generated started to exceed what an expert molecular biologist could manually browse and interpret, it became crucial to develop statistical models for assembling and interpreting the sequences.

The papers presented in this chapter cover two important areas in the interpretation of DNA sequences. The first, Cawley et al. [3], addresses the problem of analyzing stretches of DNA to search for the collections of sub-sequences that correspond to gene transcripts. The model presented was not the first of its kind; similar Hidden Markov Models (HMMs) had been published before [2, 4, 5]. Its novel contributions were various observations about computational shortcuts that can be made, at no cost to accuracy, taking advantage of some of the structure of the problem of applying HMMs to gene finding. This paper was also the first instance where the probabilistic formulation of the HMM gene finder was used to derive posterior probabilities of bases being part of the gene; previous attempts focused exclusively on the use of the Viterbi algorithm to predict gene structures. The software implementing the gene finder was also the first HMM gene finder made available as open-source software, something of value given the rate at which new organisms were then being sequenced.

As an interesting side note, while doing some of the work that was described in the publication, I had a near-death experience with the very Malaria parasite that was the subject of the work. A pure coincidence – the work had involved nothing more than electronic interaction with the parasite!

The second paper, Zhao et al. [8], introduced the novel concept of a Permuted Variable Length Markov Model (PVLMM), a generalization of the VLMM [1, 6]. VLMMs themselves are a generalization of Markov models. When applied to sequence analysis, they have the advantage of allowing for modeling of long context dependencies without necessarily coming at the cost of an exponential increase in the number of parameters to estimate. However, the dependencies that VLMMs best model are still relatively local dependencies and they are ill-suited to describe long-range dependencies between particular positions in a sequence as sometimes occurs. PVLMMs offer a way around that limitation by providing a framework in which the modeled sequence can be permuted to bring dependent positions together, turning long-range dependencies into local ones.

The paper provides some impressive work, putting the new theory into practice in two substantial applications: modeling of splice sites, a sub-component of gene sequences; and modeling of Transcription Factor Binding Sites (TFBS), important regions of DNA to which regulatory molecules known as transcription factors bind as part of the regulation mechanism for gene expression. By showing effective performance in two different sequence analysis problems, a strong case is made for the PVLMM as a general tool that will be well suited to a broad range of applications.

These papers, along with the diverse range of publications reviewed in the other chapters, provide a sense of the amazing breadth of Terry's work. I am a direct beneficiary of his diverse interests – when he introduced me to the field of statistics applied to molecular biology, I enjoyed it so much that it ended up being the basis of my career to-date. I will always be grateful to him for how selflessly he shared his time and insights, and for the patient guidance he provided during my graduate years and beyond.

References

[1] P. Bühlmann and A. J. Wyner. Model selection for variable length Markov chains and tuning the context algorithm. *Ann. Inst. Stat. Math.*, 52(2):287–315, 2000.

[2] C. Burge and S. Karlin. Prediction of complete gene structures in human genomic DNA. *J. Mol. Biol.*, 268:78–94, 1997.

[3] S. Cawley, A. Wirth, and T. P. Speed. Phat—a gene finding program for *Plasmodium falciparum*. *Mol. Biochem. Parasit.*, 118:167–174, 2001.

[4] A. Krogh. Two methods for improving performance of an HMM and their application for gene finding. In *Proc. Int. Conf. Intell. Syst. Mol. Biol.*, volume 5, pages 179–186. ISMB, 1997.

[5] D. Kulp, D. Haussler, M. G. Reese, and F. H. Eeckman. A generalized hidden Markov model for the recognition of human genes in DNA. In *Proc. Int. Conf. Intell. Syst. Mol. Biol.*, volume 4, pages 134–142. ISMB, 1996.

[6] J. Rissanen. Complexity of strings in the class of Markov sources. *IEEE Trans. Inform. Theory*, 32:526–532, 1986.

[7] F. Sanger and A. R. Coulson. A rapid method for determining sequences in DNA by primed synthesis with DNA polymerase. *J. Mol. Biol.*, 94(3):441–448, 1975.

[8] X. Zhao, H. Huang, and T. P. Speed. Finding short DNA motifs using permuted Markov models. *J. Comput. Biol.*, 12(6):894–906, 2005.

ELSEVIER

Molecular & Biochemical Parasitology 118 (2001) 167–174

MOLECULAR
& BIOCHEMICAL
PARASITOLOGY

www.parasitology-online.com.

Phat—a gene finding program for *Plasmodium falciparum*

Simon E. Cawley [a,b,*], Anthony I. Wirth [c], Terence P. Speed [a]

[a] *Department of Statistics, U.C. Berkeley, Berkeley, CA 94720, USA*
[b] *Affymetrix, Emeryville, CA 94608, USA*
[c] *Department of Computer Science, Princeton University, Princeton, NJ 08544, USA*

Abstract

We describe and assess the performance of the gene finding program pretty handy annotation tool (Phat) on sequence from the malaria parasite *Plasmodium falciparum*. Phat is based on a generalized hidden Markov model (GHMM) similar to the models used in GENSCAN, Genie and HMMgene. In a test set of 44 confirmed gene structures Phat achieves nucleotide-level sensitivity and specificity of greater than 95%, performing as well as the other *P. falciparum* gene finding programs Hexamer and GlimmerM. Phat is particularly useful for *P. falciparum* and other eukaryotes for which there are few gene finding programs available as it is distributed with code for retraining it on new organisms. Moreover, the full source code is freely available under the GNU General Public License, allowing for users to further develop and customize it.

Keywords: Plasmodium falciparum; Gene-finding; Generalized hidden Markov model; Viterbi algorithm

1. Introduction

Sequencing of the *Plasmodium falciparum* genome is proceeding apace. Two completely sequenced chromosomes have been published [1,2] as well as the mitochondrion, and substantial amounts of the sequence of other chromosomes are already available [3–6]. The two published chromosomes have been annotated extensively, in each case making use of a gene-finding program. GlimmerM [7,8], a eukaryotic gene-finding program based on Glimmer [9], was used in the analysis of chromosome 2, while chromosome 3 was annotated with the help of Hexamer [10] and Genefinder [11]. Furthermore, chromosome 3 was revisited later with GlimmerM [12].

Before either of these chromosome sequences was published, there was no publicly available gene-finding program trained on *P. falciparum* sequence, which is known to have a base composition different enough from other organisms to preclude simply using an existing program. Since some of our colleagues had a desire to analyze the sequence then available for genes, one of us wrote a gene-finding program [13]. This paper is about a descendent of that original program which we call pretty handy annotation tool (Phat).

Broadly speaking, there are now four publicly available *Plasmodium* gene-finding programs: Genefinder, GlimmerM, Hexamer and Phat. They each differ somewhat in the way in which they seek to exploit sequence features to find genes, in their availability, and in the extent to which they can be re-trained on new data and used by people other than their authors. As well as introducing Phat, we compare and contrast it with the other programs.

2. Methods

2.1. The model

Phat models genomic DNA with a generalized hidden Markov model (GHMM), similar to existing GHMM gene models such as GENSCAN [14] Genie [15,16] and HMMgene [17]. There is an underlying state space consisting of three main types of states: exons, introns and intergenic regions (Fig. 1). Introns are classified as phase 0, 1 or 2 according to the number of bases of the final codon generated in the previous exon (where previous means the last exon in the 5′ direction, on the coding strand). Exons are classified into four

* Corresponding author. Tel.: +1-510-428-8534; fax: +1-510-428-8585.
E-mail address: simon_cawley@affymetrix.com (S.E. Cawley).

0166-6851/01/$ - see front matter
PII: S0166-6851(01)00363-2

S.E. Cawley et al. / Molecular & Biochemical Parasitology 118 (2001) 167–174

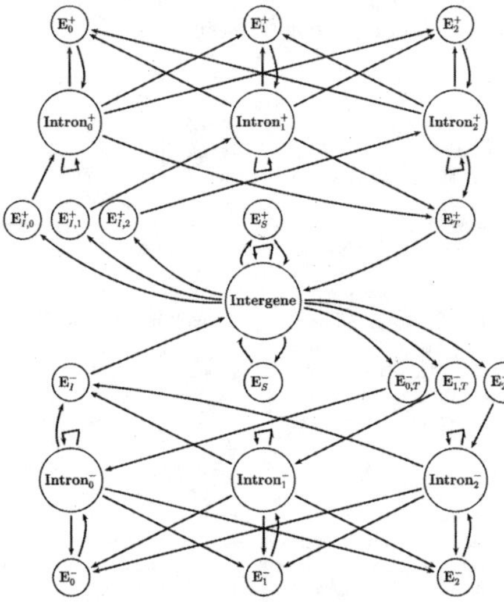

Fig. 1. Markovian state space of the Phat GHMM. States labeled '+' model genes on the forward strand, those labeled '−' model genes on the reverse. There are three intron states on each strand, one for each possible intron phase. E_S denotes a single exon gene and exons labeled E_I are internal exons located just upstream of a phase i intron. $E^+_{I,i}$ denotes an initial exon on the forward strand located just upstream of a phase i intron, and $E^-_{T,i}$ denotes a terminal exon on the reverse strand located just downstream of a phase i intron (note that since the DNA sequence is modeled left to right the reverse-strand genes are modeled 3′–5′).

types, i.e. single, initial, internal and terminal, and each exon state is composed of three parts—a pair of exon boundary state sites flanking a coding region. The possible exon boundary states are translation start, donor, acceptor and translation stop.

The most direct way to understand the model is to consider how it generates data (though in practice it is not used for data generation). A start state is chosen from some initial probability distribution. Say we start off in the intergene state. A single nucleotide is generated from an intergenic output distribution and the next state is selected. The Markov property specifies that the next state chosen depends only on the current state. Since intergenic regions tend to be reasonably long, the most likely choice for the next state will again be the intergenic state, but with some positive probability it could be an initial exon on the forward strand, a terminal exon on the reverse strand or a single exon on either strand (as indicated by the arrows leading from the intergenic state in Fig. 2).

The procedure is slightly different in exon states. First, the length of the exon is generated from an exon length distribution. This distribution is specific to both the type of the exon (single/initial/internal/terminal) and to the previous state. The corresponding number of nucleotides for the two exon boundaries and for the internal coding region are then generated and the next state is chosen.

Note that introns, internal exons, initial exons (on the forward strand) and terminal exons (on the reverse strand) are each represented by three states. The tripli-

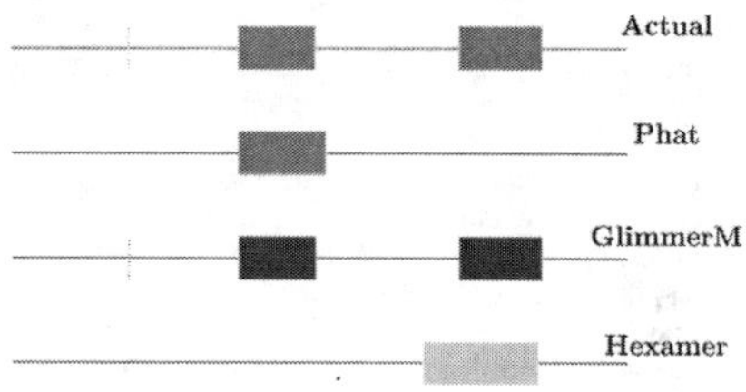

Fig. 2. Gene finding performance on a 15-kDa vesicular-like antigen gene (GenBank accession M94732). The solid blocks represent coding exons (untranslated regions are not presented since none of the gene prediction methods tries to identify non-coding exons). The tiers represent the actual structure (green), Phat (red), GlimmerM (blue) and Hexamer (yellow). The coding part of the first exon is so small (3 bp) that it does not show up in the plot. Most of the exons containing the translation start codon is untranslated and its last three bases form the start codon.

cate representation is to keep track of the frame. When arriving in a given exon state from a particular intron state, the next state is fully determined and the length of the exon must have the appropriate remainder after division by three.

The possibility of the length of exon output sequences following an arbitrary distribution makes the model slightly more flexible than a regular HMM (in which the output will always have length 1 for each state in the hidden state space). Hence the name 'generalized' HMM. For intron and intergene sequences we only allow the generation of one base of output at a time (so the output length is always fixed at 1) but unlike the exon states we allow self-transitions. Accordingly, intron and intergene lengths will follow a geometric distribution. This restriction of the model allows for large decreases in both running time and memory requirements, and turns out to be a reasonable approximation for many organisms.

2.2. Gene predictions

In practice the aim is to use the model to predict the location of genes in sequence data, which involves the estimation of the hidden states and their duration given an observed genomic sequence. A reasonable approach is to determine the sequence of states and duration that maximizes the joint probability of the hidden and observed data. This approach is the one usually adopted in HMM gene finders [14–16] (though see [17] for a nice alternative) and is the one we use in Phat. The approach is popular not only because it is effective, but also because it can be implemented in an efficient manner using the *Viterbi* algorithm [18].

The key idea in the Viterbi algorithm is to record, for each hidden state and sequence position, the maximum joint probability of hidden and observed data up to that position. The actual algorithm is a dynamic programming procedure that computes recursively the single most likely sequence.

The standard Viterbi algorithm applies to any GHMM, but there are certain features of the state space of the Phat model that can be exploited to yield savings in time and memory. Firstly, the state space can be decomposed into the set of exon states and the set of intron and intergene states. This decomposition has the special property that no exon state can jump directly to another exon state, which implies that an intron or intergene state must be preceded by another state of the same kind either one or two states back.

By modifying the Viterbi algorithm to maximize over the previous two states rather than just the previous one we can achieve a 70% reduction in the storage space required. Other features that help in reducing computation include the fact that exon states have only one possibility for the next state, that intron or intergenic states always have a duration of 1, and that the only way for a state of the latter kind to be followed directly by another such state is via a self-transition.

There is a trick that can be used to further reduce the number of computations by a significant factor, at the expense of a modest extra storage requirement. The distributions we use for exons all consist of three parts, i.e. a pair of exon boundary distributions and a distribution for the coding portion. For the last we use a three-periodic Markov model of order five. There are thus three sets of probabilities used for coding sequence, one corresponding to each codon position. Certain quantities related to these probabilities can be computed in advance and can be stored in a look-up table, after which exon probabilities can be computed when needed with only a single division. Such a look-up table approach reduces the runtime of the program by orders of magnitude.

One of the attractive features of using a GHMM for gene prediction is that it provides a natural way of computing the probability of a predicted exon, given the observed data. In addition we are often interested in the probability of a particular base or region being part of an exon. The probability that a particular sequence position is non-coding can be calculated, and subtracting it from 1, we get the probability that the position is part of some coding state. While useful in predicting potential exons that the Viterbi reconstruction may have missed, this probability says nothing about the possible strand of the coding region.

2.3. Training the model

There is a number of standard techniques for training hidden Markov models. Perhaps the best-known is the Baum–Welch method (also known as the expectation maximization, or EM algorithm) presented by Rabiner [18]. Given a collection of training sequences and initial values for the model parameters a single iteration of the Baum–Welch method provides new model parameters under which the training sequences have greater or equal likelihood. Repeated iterations yield maximum likelihood parameter estimates.

Though reasonably straightforward to write down, actual implementation of a maximum likelihood training method is a tricky and time-consuming task. The common approach, and the one we adopt, is to obtain parameter estimates independently from categorized training data. See [17] for an alternative approach, where parameters are estimated by a conditional maximum likelihood approach. We now describe our training in a little more detail.

First consider the transition probabilities between the underlying states. Any pair of states not connected in Fig. 1 has a transition probability of zero. As we have constructed the state space so that every exon state is followed by a unique state, we can now restrict our

S.E. Cawley et al. / Molecular & Biochemical Parasitology 118 (2001) 167–174

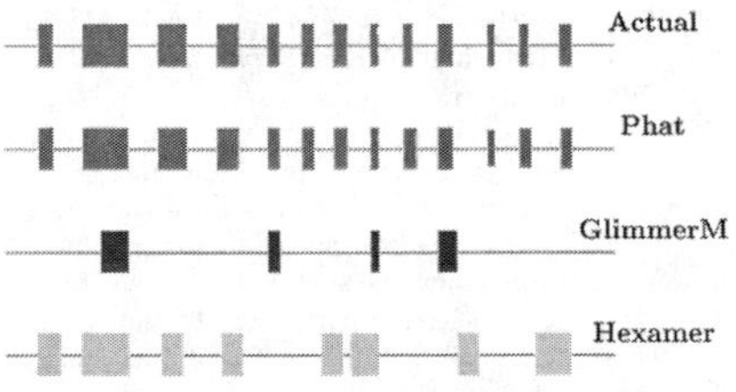

Fig. 3. Gene finding performance on a chloroquine resistance transporter (GenBank accession AF030694). The scheme is as in Fig. 2. The actual structure (green), Phat (red), GlimmerM (blue) and Hexamer (yellow). The complete coding region is shown 5′–3′.

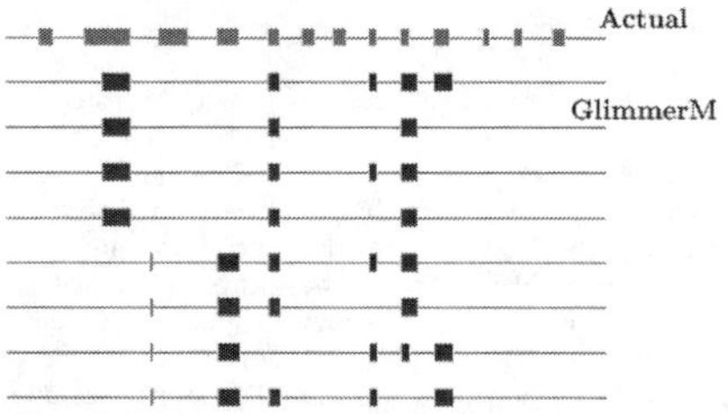

Fig. 5. GlimmerM's alternative predictions for the chloroquine resistance transporter from Fig. 3. Note that some of the predictions include the fourth and ninth exons, which are among the exons missing from the original prediction.

attention to transitions from non-coding states. In fact, the previous discussion implies that we can consider effectively such transitions as being from non-coding to non-coding states. Since we assume that all non-coding states emit one nucleotide at a time, the self-transition probabilities parameterize the length of the non-coding sections. Using our training data, we can obtain frequency counts for each of these transitions, which can be used to compute maximum likelihood estimates of the transition probabilities. One slight complication is that transitions from the Intergene state could go to either the forward or reverse strand states. For convenience, we assume that the next gene is equally likely to be on either strand.

We assume that the initial state is not an exon state and set the initial probability of the intergene state to be the fraction of all non-coding sequence that is intergenic in the training set. The initial probability of each type of intron is set to its relative frequency and we assume that each intron type is equally likely on both strands. We use different initial probabilities for the three intron phases to allow for the observed fact that phase 2 introns in *P. falciparum* are relatively rare.

Non-coding states have geometric length distributions, a result of the model's restriction that their state durations are always 1. The mean lengths are fully determined by the transition probabilities, which are estimated from a training data set of intron/intergene lengths. For coding state lengths we use a shifted γ-distribution, whose three parameters allow for a reasonable fit to observed data. Given the characteristically different types of distributions for single, initial, internal and terminal exons in *P. falciparum*, a different distribution is estimated for each. The reliance on both the previous and current states in the term length distribution is a feature of the frame constraints. A single exon must have a length that is a multiple of

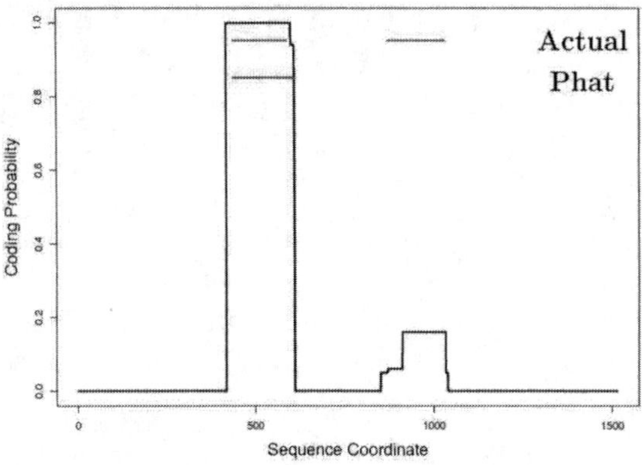

Fig. 4. Phat's coding probability plot for the vesicular-like antigen of Fig. 2. The green bars near the top represent the actual structure (the first 3 bp coding exon is invisible) and the red bar represents the single exon predicted by Phat. Note that even though the terminal exon was not predicted, its presence is suggested by the coding probability plot.

three, while there are similar constraints on the lengths of other exons.

As stated above, we model each exon by three sequential components: an exon boundary, followed by an internal coding model, terminated by another exon boundary. The internal coding model is a three-periodic Markov model. Introns and intergenic regions are modeled by a regular Markov model. In the case of *P. falciparum* we have enough previously annotated data to use a fifth order model for coding regions and a second order model for introns and intergenic regions. Maximum likelihood estimates for the Markovian probabilities can be obtained from coding Hexamer frequencies and trimer frequencies for introns and intergenic regions. One slight problem with frequency-based estimation is that some observed frequencies may be zero, which we get around by adding a prior frequency count of one to all values. The probabilities for the reverse strand are also calculated from the observed frequency counts, with a few modifications. Appropriate adjustment also has to be made for the codon phase. If we define the first nucleotide of a codon to be in codon phase 0 and the last to be in codon phase 2, then for a fifth order model phase 0 forwards is equivalent to phase 1 backwards and vice versa, while codon phase 2 forwards is equivalent to phase 2 backwards.

We use very simple models for translation start and translation stop sites. On the forward strand the translation start site produces ATG with probability one and the translation stop site produces one of the three stop codons TAA, TAG or TGA according to probabilities estimated from stop codon frequencies in the training set. The reverse strand uses the same probabilities for the reverse complements.

Splice sites are modeled with variable length Markov chains (VLMCs) [19], a generalization of Markov chains. For donor sites we use three bases upstream and ten bases downstream of the actual site, for acceptor sites we use 20 bases upstream and three bases downstream. The model is that the base at each position of the site follows a distribution that is conditional on some of the previous bases. In a Markov model the number of previous bases upon which the next is dependent is a fixed value, but for a VLMC the number of previous bases influencing the next depends on the sequence context. The advantage is the ability to model longer-range interactions without having to deal with an exponential increase in the number of parameters to estimate.

2.4. Measures of prediction accuracy

We compare the accuracy of predictions at two levels, i.e. nucleotide and exon. At the nucleotide level, we measure the accuracy of a prediction by comparing the predicted coding value (coding or non-coding) with the true coding value along the test sequence. This is the approach adopted by most of the authors (see [20] for a comprehensive discussion of the issues involved and references to earlier research). Sensitivity (Sn) and specificity (Sp) are widely used measures of prediction quality, each being defined in terms of the quantities TP, TN, FP and FN. Here TP denotes the number of coding nucleotides that are predicted to be coding, called true positives, while TN are the non-coding nucleotides predicted to be non-coding, called true negatives and similarly for false positives and false negatives. We write $Sn = TP/(TP + FN)$ for the proportion

Table 1
A gene finding comparison between Phat and GlimmerM on the 25-gene test set

Test set	Nucleotide-level			Exon-level			
Program	Sn	Sp1	Sp2	Correct	Partial	Wrong	Missing
GlimmerM	89.3	93.1	97.6	57.8	42.2	0.0	22.6
Phat	99.0	98.9	99.6	77.2	22.8	0.0	6.0

The set contains 84 exons and three of the genes are single-exon genes. For the exon-level results, each predicted exon is classified as 'correct' if both boundaries are precisely correct, as 'wrong' if the prediction has no overlap with a true exon, and as 'partial' otherwise. The column labeled 'missing' shows the percentage of true exons for which there is no overlapping prediction. All reported values are percentages.

Table 2
A gene finding comparison between Phat and GlimmerM on the 19-gene training set

Train Set	Nucleotide-level			Exon-level			
Program	Sn	Sp1	Sp2	Correct	Partial	Wrong	Missing
GlimmerM	90.2	96.2	97.1	79.7	16.9	3.4	30.6
Phat	95.8	95.1	96.5	80.3	19.7	0.0	16.5

The set contains 85 exons and all but one of the genes are multi-exon. Notation is the same as in Table 1.

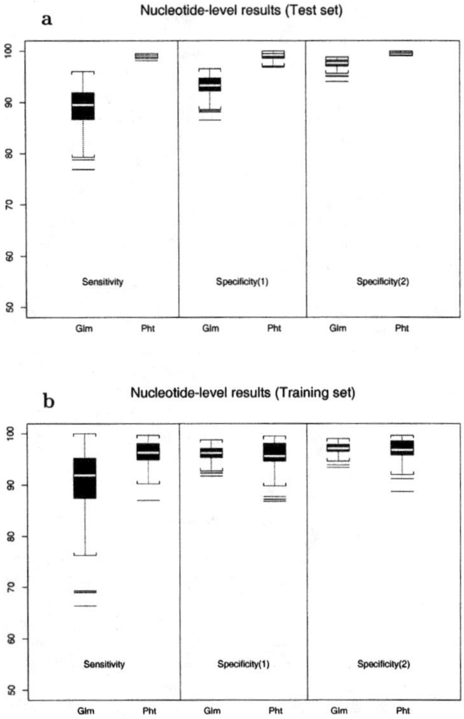

Fig. 6. Box plots of bootstrapped sensitivities and specificities (as defined in Table 1) for GlimmerM (Glm) and Phat (Pht) on the test set (a) and the training set (b). For each box plot, the solid box covers the inter-quartile range of the data, with the line within the box representing the median. The 'whiskers' extend to the nearest values not beyond 1.5 times the inter-quartile range from the quartiles. Remaining points are represented by isolated dashes.

of positives that are correctly annotated, Sp1 = TN/(TN + FP) for the proportion of negatives correctly annotated, and Sp2 = TP/(TP + FP) for the proportion of coding predictions which are correct (also called the positive predictive value).

In many contexts Sp1 is the more natural measure of specificity, but in discussion of the accuracy of genefinders, it has generally been replaced by Sp2. This is because the typically high proportion of non-coding sequence predicted readily as such can dominate Sp1, and thus make the measure less sensitive. We present both quantities below.

Exon level results are also reported. Predicted exons are classified as either correct (both boundaries correct), partial (overlapping a true exon), or wrong. A missing exon is one which the genefinder did not detect at all.

3. Results

We have conducted a study to compare the performance of Phat [21,13] with other gene finding programs on *P. falciparum* sequence. Currently the other main programs are Hexamer [10], Genefinder [11] and GlimmerM [7,8]. Hexamer operates quite differently to the others using only Hexamer frequencies to predict individual coding regions. It does not attempt to detect exon boundaries, nor does it assemble its predicted coding regions together into whole genes.

The remaining three programs all attempt to predict whole gene structures where possible, and can be applied to large sequences containing multiple genes. GlimmerM has models for coding regions and splice sites. Genefinder models coding regions, splice sites, introns, intergenic regions and has a model for the transcription start site. Phat models all the aforementioned features, save for the transcription start site, also using explicit state length distributions to model feature lengths.

For the purpose of comparing gene finding programs it is important to train the programs on a common data set. Phat and Hexamer are distributed along with code for retraining on new data sets. The GlimmerM version used here (obtained from the authors in August 2000) comes pre-trained on a set of around 300 genes and there are no means available to re-train GlimmerM on new data sets. Genefinder also comes with code for retraining on new data, however, we experienced technical difficulties getting the retraining code to work. The end result is that the only way to do a fair comparison is to drop Genefinder out of the analysis and train the others on GlimmerM's training set.

P. falciparum researchers from the Sanger Centre and from the Walter and Eliza Hall Institute (WEHI) were asked to provide a list of genes, which have been confirmed biologically by reverse transcription-polymerase chain reaction (RT-PCR) experiments, leading to an evaluation data set of 44 genes. Of these 44 genes it turned out that 19 were already in the GlimmerM training set. In what follows we refer to these 19 genes as the training set, and to the remaining 25 in the evaluation set as the test set.

Comparing the gene finders on the evaluation set, it is clear that Phat and GlimmerM often provide accurate predictions. Hexamer is a very simple model and while it does a reasonable job of generally indicating regions of coding potential it has no model for splice sites nor for how to join the regions together as genes, so performs much worse than Phat and GlimmerM.

Looking at Figs. 2 and 3 there are cases when Phat outperforms GlimmerM, and vice versa. Each program also provides some useful features for detecting possibly missed exons. Fig. 4 is a plot of the coding probability computed by Phat (as earlier) for the gene in 2. Phat missed two exons in the optimal prediction, but the coding probability plot is suggestive of the larger of the two exons missed. For each gene predicted by GlimmerM, a list of alternative gene predictions is also provided—these are genes achieving high scores in GlimmerM's model. Fig. 5 shows GlimmerM's alternative predictions for the same gene in Fig. 3, the alternative predictions suggest an extra two exons missing from the original prediction.

Tables 1 and 2 present the nucleotide-level and exon-level results for the test and training sets. An understanding of the variability of these estimates can be helpful, and we address this using a bootstrap study. For one bootstrap iteration, we draw with replacement a new test data set from the original, then evaluate the Genefinder on this bootstrapped data set and compute the performance measures. This is repeated many times and the results are collected. Fig. 6 presents results for the test and training data sets. Across both sets GlimmerM's performance is clearly more variable. It is important to note that the extent to which the results on this evaluation set can be extrapolated to the set of all *P. falciparum* genes will depend on the extent to which the evaluation set is a representative sample.

The time and memory requirements of the programs are important, particularly if they are to be used in a high throughput environment. We compared the time and memory requirements of the two programs on a 700 MHz Pentium III processor running under Linux. For a sequence of 100 kbp Phat requires 39 Mb of memory and takes 20 s of CPU time. For the same sequence GlimmerM requires 12 Mb of memory and takes 34 s. Phat runs faster than GlimmerM, its chief gain probably coming from the use of look-up tables for fast computation of exon probabilities, but at the expense of increased memory requirements. Both programs scale roughly linearly in the length of the sequence being analyzed and have been used to analyze sequences of up to 1 Mbp.

4. Discussion

Both genefinders displayed relatively high sensitivity and specificity on both the training and test sets of genes. It is a little surprising that both gene finders performed better on the examples on which they had not been trained, perhaps the genes in the training set are in some sense more difficult to predict accurately. A reviewer with extensive experience in the field has found that GlimmerM tends consistently to under-annotate while Phat tends consistently to over-annotate. He also found Phat sometimes returned abnormally short introns and abnormally long exons. These aspects of Phat are perhaps general features of the GHMM approach, and all we can say is that they have their advantages as well, the example of Fig. 3 being a clear example.

As mentioned earlier, the use of our specificity measure Sp1 has all but ceased in the assessment of genefinding algorithms, due to its being dominated by large values of TN. Nonetheless, we have included it along with Sp2, because in our case the two measures are comparable, indeed Sp2 is slightly larger than Sp1. The reason for this is that *P. falciparum* has a relatively high coding content, (chromosomes 2 and 3 are about 50% coding), so the values of TP and TN are much more similar than in other organisms (e.g. human, Drosophila) for which these measures have been calculated previously.

In conclusion, we have demonstrated that Phat performs well upon *P. falciparum* sequence, and compares favorably with GlimmerM. There is thus a good case for making use of both GlimmerM and Phat for new *P.*

falciparum data as there are examples where one predicts correctly what the other misses. Each program also has useful functionality to try to detect exons that may have been missed in the single best prediction (Figs. 4 and 5). These gene finders should prove useful as the *P. falciparum* genome approaches completion.

Acknowledgements

This work was made possible with advice and data from many others. The authors would like to thank Mauro De Lorenzi, Alan Cowman and Tony Triglia at WEHI, Mihaela Pertea and Steven Salzberg at TIGR, Allan Saul and Robert Heustis at QIMR, Sharen Bowman and Neil Hall at the Sanger Centre, Winston Hide and Ralhston Muller at SANBI and Jane Carlton at NCBI. S.C. was supported by DOE grant DE-FGO3-97ER62387.

References

[1] Gardner MJ, Tettelin H, Carucci DJ, Cummings LM, Aravind L, Koonin EV, Shallom S, et al. Chromosome 2 sequence of the human malaria parasite *Plasmodium falciparum*. Science 1998;282:1126–32.

[2] Bowman S, Lawson D, Basham D, Brown D, Chillingworth T, Churcher CM, Craig A, et al. The complete nucleotide sequence of chromosome 3 of *plasmodium falciparum*. Nature 1999;400:532–8.

[3] The Malaria genome project at Stanford University, URL: http://sequence-www.stanford.edu/group/malaria/.

[4] The *Plasmodium falciparum* Genome Database at the University of Pennsylvania, URL: http://PlasmodiumDB.cis.upenn.edu/.

[5] The *Plasmodium falciparum* genome database at The Institute for Genomic Research, URL: http://www.tigr.org/tdb/edb2/pfa1/htmls/.

[6] The *Plasmodium falciparum* genome project at the Sanger Centre, URL: http://www.sanger.ac.uk/Projects/P_falciparum/.

[7] GlimmerM, http://www.tigr.org/softlab/glimmerm.

[8] Salzberg S, Pertea M, Deicher A, Gardner M, Tettelin H. Interpolated Markov models for eukaryotic gene finding. Genomics 1999;59:24–31.

[9] Salzberg S. Decision trees and Markov chains for gene finding. In: Salzberg S, Searls D, Kasif S, editors. Computational methods in molecular biology. Amsterdam: Elsevier, 1998:187–203.

[10] Durbin R. Hexamer. 1995. Source code available at ftp://ftp.sanger.ac.uk/pub/pathogens/software/hexamer.

[11] Green P. Genefinder. 1994. Contact phg@u.washington.edu for details.

[12] Pertea M, Salzberg S, Gardner M. Finding genes in *Plasmodium falciparum*. Nature 2000;404:34–5.

[13] Wirth A. A *Plasmodium falciparum* genefinder. Honours thesis, University of Melbourne; 1998.

[14] Burge C, Karlin S. Prediction of complete gene structures in human genomic DNA. J Mol Biol 1997;268:78–94.

[15] Kulp D, Haussler D, Reese MG, Eeckman FH. A generalized hidden Markov model for the recognition of human genes in DNA. In: States D, Agarwal P, Gaasterland T, Hunter L, Smith RF, editors. ISMB-96: proceedings of the fourth international conference on intelligent systems for molecular biology. AAI Press, 1996:134–41.

[16] Reese MG, Kulp D, Tammana H, Haussler D. Genie—gene finding in *Drosophila melanogaster*. Genome Res 2000;10:529–38.

[17] Krogh A. Using database matches with HMMGene for automated gene detection in *Drosophila*. Genome Res 2000;10:523–8.

[18] Rabiner LR. A tutorial on hidden Markov models and selected applications in speech recognition. Proc IEEE 1989;77:257–86.

[19] Buhlmann P, Wyner AJ. Variable length Markov models. Ann Statistics 1999;27:480–513.

[20] Burset M, Guigo R. Evaluation of gene structure prediction programs. Genomics 1996;34:353–67.

[21] Wirth A, Cawley S, Speed T. Phat. 1998. Source code available at http://www.stat.berkeley.edu/users/scawley/Phat.

ICM 2002 · Vol. III · 1–3

Biological Sequence Analysis

T. P. Speed*

Abstract

This talk will review a little over a decade's research on applying certain stochastic models to biological sequence analysis. The models themselves have a longer history, going back over 30 years, although many novel variants have arisen since that time. The function of the models in biological sequence analysis is to summarize the information concerning what is known as a motif or a domain in bioinformatics, and to provide a tool for discovering instances of that motif or domain in a separate sequence segment. We will introduce the motif models in stages, beginning from very simple, non-stochastic versions, progressively becoming more complex, until we reach modern profile HMMs for motifs. A second example will come from gene finding using sequence data from one or two species, where generalized HMMs or generalized pair HMMs have proved to be very effective.

2000 Mathematics Subject Classification: 60J20, 92C40.
Keywords and Phrases: Motif, Regular expression, Profile, Hidden Markov model.

1. Introduction

DNA (deoxyribonucleic acid), RNA (ribonucleic acid), and proteins are macromolecules which are unbranched polymers built up from smaller units. In the case of DNA these units are the 4 nucleotide residues A (adenine), C (cytosine), G (guanine) and T (thymine) while for RNA the units are the 4 nucleotide residues A, C, G and U (uracil). For proteins the units are the 20 amino acid residues A (alanine), C (cysteine) D (aspartic acid), E (glutamic acid), F (phenylalanine), G (glycine), H (histidine), I (isoleucine), K (lysine), L (leucine), M (methionine), N (asparagine), P (proline), Q (glutamine), R (arginine), S (serine), T (threonine), V (valine), W (tryptophan) and Y (tyrosine). To a considerable extent, the chemical properties of DNA, RNA and protein molecules are encoded in the linear sequence of these basic units: their primary structure.

*Department of Statistics, University of California, Berkeley, CA 94720, USA; Division of Genetics and Bioinformatics, Walter and Eliza Hall Institute of Medical Research, VIC 3050, Australia. E-mail: terry@stat.berkeley.edu

The use of statistics to study linear sequences of biomolecular units can be descriptive or it can be predictive. A very wide range of statistical techniques has been used in this context, and while statistical models can be extremely useful, the underlying stochastic mechanisms should never be taken literally. A model or method can break down at any time without notice. Further, biological confirmation of predictions is almost always necessary.

The statistics of biological sequences can be global or it can be local. For example, we might consider the global base composition of genomes: *E. coli* has 25% A, 25% C, 25% G, 25% T, while *P. falciparum* has 82%A+T. At the very local, the triple ATG is the near universal motif indicating the start of translation in DNA coding sequence. A major role of statistics in this context is to characterize individual sequences or classes of biological sequences using probability models, and to make use of these models to identify them against a background of other sequences. Needless to say, the models and the tools vary greatly in complexity.

Extensive use is made in biological sequence analysis of the notions of motif or domain in proteins, and site in DNA. We shall use these terms interchangeably to describe the recurring elements of interest to us. It is important to note that while we focus on the sequence characteristics of motifs, domains or sites, in practice they also embody (biochemical) structural significance.

2. Deterministic models

The C2H2 (cysteine-cysteine histidine-histidine) zinc-finger DNA binding domain is composed of 25-30 amino acid residues including two conserved cysteines and two conserved histidines spaced in a particular way, with some restrictions on the residues in between and nearby. Of course the arrangement reflects the three-dimensional molecular structure into which the amino-acid sequence folds, for it is the structure which has the real biochemical significance, see Figure 1, which was obtained from `http://www.rcsb.org/pdb/`. An example of this motif is the 27-

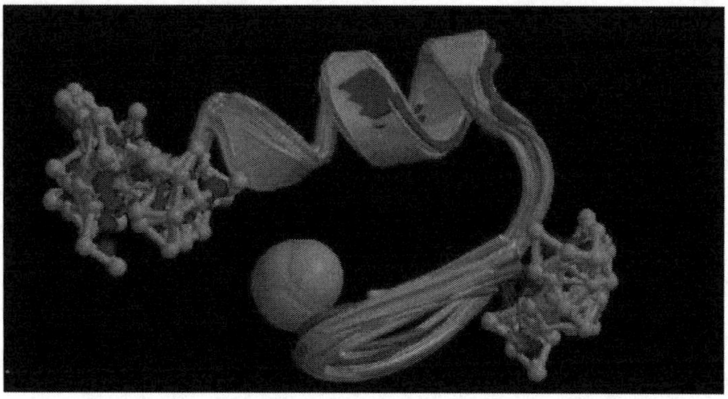

Figure 1: A C2H2 zinc finger DNA binding domain

letter sequence known as 1ZNF, this being a Protein Data Bank identifier for the structure XFIN-31 of *X. laevis.* Its amino acid sequence is

```
1ZNF:   XYKCGLCERSFVEKSALSRHQRVHKNX
```

Note the presence of the two Cs separated by 2 other residues, and the two Hs separated by 3 other residues. Here and elsewhere, X denotes an arbitrary amino acid residue. A popular and useful summary description of C2H2 zinc fingers which clearly includes our example, is the regular expression

$$C - X(2,4) - C - X(3) - [LIVMFYWC] - X(8) - H - X(3,5) - H$$

where $X(m)$ denotes a sequence of n unspecified amino acids, while $X(m,n)$ denotes from m to n such, and the brackets enclose mutually exclusive alternatives. There is a richer set of notation for *regular expressions* of this kind, but for our purposes it is enough to note that this representation is essentially deterministic, with uncertainty included only through mutually exclusive possibilities (e.g. length or residue) which are not otherwise distinguished.

Simple and efficient algorithms exist for searching query sequences of residues to find every instance of the regular expression above. In so doing with sequence in which all instances of the motif are known, we may identify some sub-sequences of the query sequence which are not C2H2 zinc finger DNA binding domains, i.e. which are false positives, and we may miss some sub-sequences which are C2H2 zinc fingers, i.e. which are false negatives. Thus we have essentially deterministic descriptions and search algorithms for the C2H2 motifs using regular expressions. Their performance can be described by the frequency of false positives and false negatives, equivalently, their complements, the specificity and sensitivity of the regular expression. We do not have space for an extensive bibliography, so for more on regular expressions and on most of the other concepts we introduce below, see [2].

3. Regular expressions can be limiting

Most protein binding sites are characterized by some degree of sequence specificity, but seeking a consensus DNA sequence is often an inadequate way to recognize their motifs. Simply listing the alternatives seen at a position may not be very informative, but keeping track of the frequencies with which the different alternatives appear can be very valuable. Thus position-specific nucleotide or amino acid distributions came to represent the variability in DNA or protein motif composition. This is just the set of marginal distribution of letters at each position. Rather than present an extensive tabulation of frequencies for our C2H2 zinc finger example, we present a pictorial representation: a sequence logo coming from `http://blocks.fhcrc.org`.

Sequence logos are scaled representation of position-specific nucleotide or amino acid distributions. The overall height at a given position is proportional to information content, which is a constant minus the entropy of the distribution at that

T. P. Speed

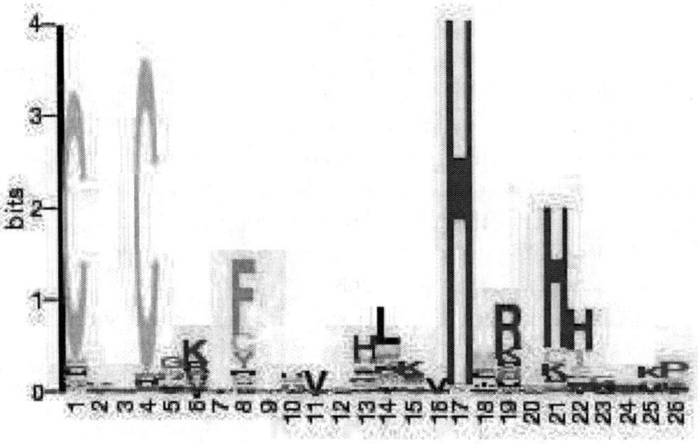

Figure 2: Sequence logo for C2H2 zinc finger

position. The proportions of each nucleotide or amino acid at a position are in relation to their observed frequency at that position, with the most frequent on top, the next most frequent below, etc.

4. Profiles

It is convenient for our present purposes to define a profile as a set of position-specific distributions describing a motif. (Traditionally the term has been used for the derived scores.) How would we use a set of such distributions to search a query sequence for instances of the motif? The answer from bioinformatics is that we *score* the query sequence, and for suitably large scores, declare that a candidate subsequence is an instance of our motif.

There are a number of approaches for deriving profile scores, but the easiest to explain here is this: scores are *log-likelihood ratio test statistics*, for discriminating between a probability model M for the motif and a model B for the background. The model M will be the direct product of the position-specific distributions, (i.e. the independent but not identical distribution model), while the background model B will be the direct product of a set of relevant background frequencies (i.e. the independent and identical distribution model). Thus, if f_{al} is the frequency of residue a at position l of the motif, and f_a background frequency of the same residue, then the profile score assigned to residue a at position l in a possible instance of the motif will be $s_{al} = \log f_{al}/f_a$. These scores are then summed across the positions in the motif, and compared to a suitably defined threshold. Note that proper setting of the threshold requires a set of data in which all instances of the motif are known. The false positive and false negative rate could then be determined for various thresholds, and a suitable choice made.

We briefly discuss variants of the log-likelihood ratio scores. In many contexts, it will matter little whether a position is occupied by a leucine (L) rather than an isoleucine (I), as each can evolve in time to or from the other rather more readily than from other residues. Thus it might make sense to modify the scores to take this and similar evolutionary patterns into account. Indeed the first use of profiles involved scores of this kind, using the position specific amino acid distribution of an alignment of instances of the motif and entries from what are known as PAM matrices, which embody patterns of molecular evolution. In addition, the background distribution of residues may be modelled more detailed manner, e.g. using the so-called Dirichlet mixture models.

It is also possible to include position-specific scores for insertion and deletion of residues, relative to a consensus pattern. When these are used, the scoring becomes a little more subtle, as the problem is then quite analogous to pairwise sequence alignment, but with position dependent scoring parameters for matches, mismatches, insertions and deletions.

We summarise this section by noting that probability has entered into our description through the use of frequencies, and scores based on them, but so far we do not have global statistical models, at least not ones embodying insertions and deletions, on which we base our estimation and testing. These are all part of the use of profile HMMs, but first we introduce HMMs.

5. Hidden Markov models

Hidden Markov models (HMMs) are processes $(S_t, O_t), t = 1, \ldots, T$, where S_t is the hidden state and O_t the observation at time t. Their probabilistic evolution is constrained by the equations

$$\begin{aligned} pr(S_t|S_{t-1}, O_{t-1}, S_{t-2}, O_{t-2}, \ldots) &= pr(S_t|S_{t-1}), \\ pr(O_t|S_{t-1}, O_{t-1}, S_{t-2}, O_{t-2}, \ldots) &= pr(O_t|S_t, S_{t-1}). \end{aligned}$$

The definitions and basic facts concerning HMMs were laid out in a series of beautiful papers by L. E. Baum and colleagues around 1970, see [2] for references. Much of their formulation has been used almost unchanged to this day. Many variants are now used. For example, the distribution of O may not depend on previous S, or it may also depend on previous O values,

$$\begin{aligned} pr(O_t|S_t, S_{t-1}, O_{t-1}, \ldots) &= pr(O_t|S_t), \quad \text{or} \\ pr(O_t|S_t, S_{t-1}, O_{t-1}, \ldots) &= pr(O_t|S_t, S_{t-1}, O_{t-1}). \end{aligned}$$

Most importantly for us below, the times of S and O may be decoupled, permitting the observation corresponding to state time t to be a string whose length and composition depends on S_t (and possibly S_{t-1} and part or all of the previous observations). This is called a hidden semi-Markov or generalized hidden Markov model.

T. P. Speed

Early applications of HMMs were to finance, but these were never published, to speech recognition, and to modelling ion channels. In the mid-late 1980s HMMs entered genetics and molecular biology, where they are now firmly entrenched. One of the major reasons for the success of HMMs as stochastic models is the fact that although they are substantial generalizations of Markov chains, there are elegant dynamic programming algorithms which permit full likelihood calculations in many cases of interest. Specifically, there are algorithms which permit the efficient calculation of a) $pr(sequence|M)$, where *sequence* is a sequence of observations and M is an HMM; b) the maximum over *states* of $pr(states|sequence, M)$, where *states* is the unobserved state sequence underlying the observation *sequence*; and c) the maximum likelihood estimates of parameters in M based on the observation *sequence*. Step c) is carried out by an iterative procedure which in the case of independent states was later termed the EM algorithm.

6. Profile HMMs

In a landmark paper A. Krogh, D. Haussler and co-workers introduced profile HMMs into bioinformatics. An illustrative form of their profile HMM architecture is given in Figure 3. There we depict the underlying state space of the hidden

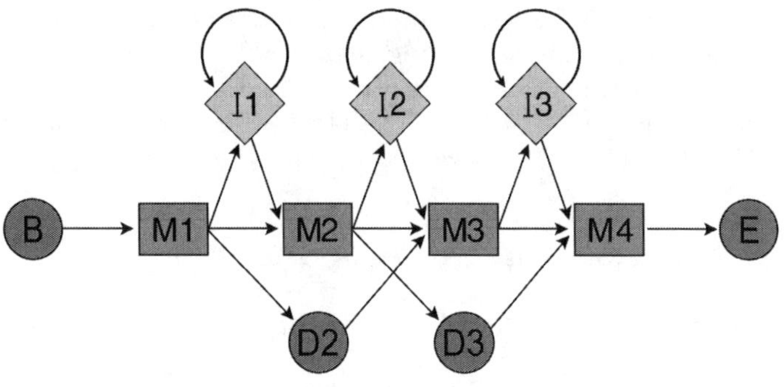

Figure 3: State space of a simple profile HMM

Markov chain of a profile HMM of length 4, with M denoting *match* states, I *insert* states and D *delete* states, while B and E are *begin* and *end* states, respectively. Encircled states (D, B and E) do not emit observations, while each of the match and insert states will have position-specific observation or emission distributions. Finally, each arrow will have associated transition probabilities, with the expectation being that the horizontal transition probabilities are typically near unity. This the chain proceeds from left to right, and if it remains within match states, its output will be an amino acid sequence of length 4. Deviation to the insert or delete states will modify the output accordingly. The similarity with a direct product of a sequence of position-specific distributions should be unmistakeable. The profile HMMs in use now have considerably more features, while sharing the basic M, I and D architecture.

Why was the introduction of the HMM formalism such an advance? The answer is simple: it permitted the construction and application of profiles to be conducted entirely within a formal statistical framework, and that really helped. Instances of the motif embodied in an HMM could be identified by calculating $pr(sequence|M)/pr(sequence|B)$ as was done with profiles, using the algorithm for problem a) in X above. Instances of the motif could be aligned to the HMM by calculating the most probable state sequence giving rise to the motif sequence, in essence finding the most probable sequence of matches, insertions, deletions which align the given sequence to the others which gave rise to the HMM, cf. problem b) above. And finally, the parameters in the HMMs could be estimated from data comprising known instances of the motif by using maximum likelihood, an important step for many reasons, one being that it put insert and delete scores on precisely the same footing as match and mismatch scores. Although the estimation of HMM parameters is easiest if the example sequences are properly aligned, the EM algorithm (problem c) above) does not require aligned sequences.

In the years since the introduction of profile HMMs, they have been become the standard approach to representing motifs and protein domains. The database Pfam (`http://pfam.wustl.edu`) now has 3,849 hidden Markov models (May 2002) representing recognized protein or DNA domains or motifs. Profile HMMs have essentially replaced the use of regular expressions and the original profiles for searching other databases to find novel instances of a motif, for finding a motif or domain match to an input sequence, and for aligning a motif or domain to a an existing family. There is considerable evidence that the HMM-based searches are more powerful than the older profile based ones, though they are slower computationally, and at times that is an important consideration.

7. Finding genes in DNA sequence

Identifying genes in DNA sequence is one of the most challenging, interesting and important problems in bioinformatics today. With so many genomes being sequenced so rapidly, and the experimental verification of genes lagging far behind, it is necessary to rely on computationally derived genes in order to make immediate use of the sequence.

What is a gene? Most readers will have heard of the famous *central dogma* of molecular biology, in which the hereditary material of an organism resides in its genome, usually DNA, and where genes are expressed in a two-stage process: first DNA is *transcribed* into a messenger RNA (mRNA) sequence, and later a processed form of this sequence is *translated* into an amino acid sequence, i.e. a protein. In general the transcribed sequence is longer than the translated portion: parts called introns (intervening sequence) are removed, leaving exons (expressed sequence), of which only some are expressed, while the rest remain untranslated. The translated sequence comes in triples called codons, beginning and ending with a unique start (ATG) and one of three stop (TAA, TAG, TGA) codons. There are also characteristic intron-exon boundaries called splice donor and acceptor sites, and a variety of other motifs: promoters, transcription start sites, polyA sites, branching sites, and so on.

All of the foregoing have statistical characterizations, and in principle they can all help identify genes in long otherwise unannotated DNA sequence segments. To get an idea of the magnitude of the task with the human genome, consider the following facts about human gene sequences [5]: the coding regions comprise about 1.5% of the entire genome; the average gene length is about 27,000 bp (base pair); the average total coding region is 1,340 bp; and the average intron length is about 3,300 bp. Further, only about 8% of genes have a single exon. We see that the information in human genes is very dispersed along the genome, and that in general the parts of primary interest, the coding exons, are a relatively small fraction of the gene, on average about $\frac{1}{20}$.

8. Generalized HMMs for finding genes

The HMMs which are effective in finding genes are the generalized HMMs (GHMMs) described in section 5. above. Space does not permit our giving an adequate description here, so we simply outline the architecture of Genscan [1] one of the most widely used human genefinders. States represent the gene features we mentioned above: exon, intron, and of course intergenic region, and a variety of other features (promotor, untranslated region, polyA site, and so on. Output observations embody state-dependent nucleotide composition, dependence, and specific signal features (such as stop codons). In a GHMM the state *duration* needs to be modelled, as well as two other important features of genes in DNA: the *reading frame*, which corresponds to the triples along the mRNA sequence which are sequentially translated, and the *strand*, as DNA is double stranded, and genes can be on either strand, i.e. they can point in either direction. These features can be seen in Figure 4, which was kindly supplied by Lior Pachter.

The output of a GHMM genefinder after processing a genomic segment is broadly similar to that from a profile HMM after processing an amino acid sequence: the most probable state sequence given an observation sequence is a best gene annotation of that sequence, and a variety of probabilities can be calculated to indicate the support in the observation sequence for various specific gene features.

9. Comparative sequence analysis using HMMs

The large number of sequenced genomes now available, and the observation that functionally important regions are evolutionarily conserved, has led to efforts to incorporate conservation into the models and methods of biological sequence analysis. Pair HMMs were introduced in [2] as a way of including alignment problems under the HMM framework, and recently [4] they were combined with GHMMs (forming GPHMMs) to carry out alignment and genefinding with homologous segments of the mouse and human genomes. Use of the program SLAM on the whole

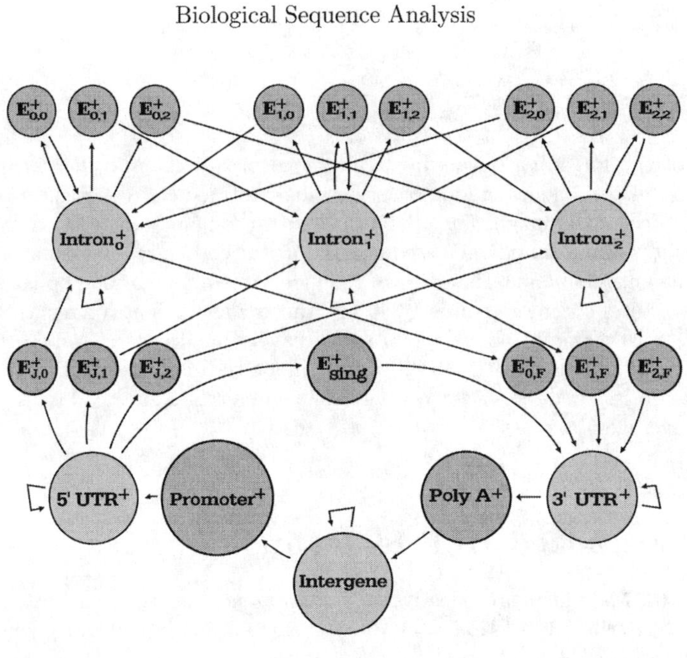

Figure 4: Forward half of the Genscan GHMM state space

mouse genome (`http://bio.math.berkeley.edu/slam/mouse/`) demonstrated the value of GPHMMs in this context.

10. Challenges in biological sequence analysis

The first challenge is to understand the biology well enough to begin biological sequence analysis. This part will frequently involve collaborations with biologists. With HMMs, GHMMs and GPHMMs, designing the underlying architecture, and carrying out the modelling for the components parts, e.g. for splice sites in genefinding GHMM is perhaps the next major challenge. Undoubtedly the hardest and most important task of all is the implementation: coding up the algorithms and making it all work with error-prone and incomplete sequence data. Finally, it is usually a real challenge to find good data sets for calibrating and evaluating the algorithms, and for carrying out studies of competing algorithms.

For a recent example of this process, which is a model of its kind, see [3]. There an HMM is presented for the so-called σ^A recognition sites, which involve two DNA motifs separated by a variable number of base pairs. In addition to the examples mentioned so far, there are many more HMMs in the bioinformatics literature, see p. 79 of [2] for ones published before 1998.

11. Closing remarks

T. P. Speed

In this short survey of biological sequence analysis, I have simply touched on some of the major ideas. A much more comprehensive treatment of material covered here can be found in the book [2], whose title not coincidentally is the same as that of this paper. Many important ideas from biological sequence analysis have not been mentioned here, including molecular evolution and phylogenetic inference, and the use of stochastic context-free grammars, a form of generalization of HMMs suited to the analysis of RNA sequence data.

At this Congress I have talked (and am now writing) on the research of others, in an area in which my own contributions have been negligible. I chose to do so upon being honoured by the invitation to speak at this Congress because I believe this topic – HMMs – to be one of the great success stories of applying mathematics to bioinformatics. In my view it is the one most worthy of a wider mathematical audience. I hope that the fact that there are many others better suited than me to speak on this topic will not prevent readers from appreciating it and following it up through the bibliography.

I owe what understanding I have of this field to collaborations and discussions with a number of people, and I would like to acknowledge them here. Firstly, Tony Wirth, Simon Cawley and Mauro Delorenzi, with whom I have worked on HHMMs. Next, it has been an honour and pleasure to observe from close by the development of SLAM, by Simon Cawley, Lior Pachter and Marina Alexandersson. Finally I'd like to thank Xiaoyue Zhou and Ken Simpson for their kind help to me when I was preparing my talk and this paper.

References

[1] C. Burge & S. Karlin, Prediction of complete gene structures in human genomic DNA, *J. Mol. Biol.* 268 (1997) 78–94.

[2] R. Durbin, S. Eddy, A. Krogh & G. Mitchison, *Biological Sequence Analysis. Probabilistic models of proteins and nucleic acids*, Cambridge University Press, 1998.

[3] H. Jarmer, T. S. Larsen, A. Krogh, H. H. Saxild, S. Brunak & S. Knudsen, Sigma A recognition sites in the *Bacillus subtilis* genome, *Microbiology* 147 (2001), 2417–2424.

[4] L. Pachter, M. Alexandersson & S. Cawley, Applications of generalized pair hidden Markov models to alignment and gene finding problems, *J. Comp. Biol.* 9 (2002), 389–399.

[5] The Genome Sequencing Consortium, Initial sequencing and analysis of the human genome, *Nature* 409 (2001), 860–921.

Chapter 15
Microarray Data Analysis

Jane Fridlyand

I met Terry when I was a beginning graduate student in the Department of Statistics at UC Berkeley. The first year is the time when bright-eyed and idealistic graduate students start thinking about what they want to do for the next 30 years of their lives, or at least until they are handed their PhD diploma and a job offer from Wall Street. I was in awe of Terry, but gathered my courage to approach him with that crucial question: "Would you work with me?" Now that I write this, I find that it sounds rather like a marriage proposal. And indeed, it becomes one: a covenant of commitment between a student and the mentor, with all the ups and downs, for better or for worse, lasting a lifetime. I wanted to work with Terry because he inspired me, as a scientist and as a person, and his interests in biological and medical applications were close to my heart. I also had reason to hope for a positive response – I was told in confidence by several people that in his 20 years of working with students Terry had never turned anyone down. So, here I was asking "Would you work with me?". His reply was immediate and crushing: "Why?". I did not know what to say – with all the mental rehearsals I had done, I was not prepared for this comeback. I must have blushed, mumbled something and run away. I guess there is always a first one to be turned down, unfortunately it just happened to be me!

My despair did not last long. The next day I found a thick stack of papers on statistical genetics and schizophrenia research in my mailbox with a note asking me to read them and meet Terry the next day at a specific time to discuss. And this is how our work together began. Although I transitioned from schizophrenia research, our working relationship had been established, and Terry has remained a very important part of my life since (15 years and counting!).

J. Fridlyand
Department of Biostatistics, Genentech
e-mail: fridlyand.jane@gene.com

S. Dudoit (ed.), *Selected Works of Terry Speed*, Selected Works in Probability and Statistics, DOI 10.1007/978-1-4614-1347-9_15,

Microarrays and high-dimensional data

Starting in the late nineties, the field of applied statistics in biomedical research has transformed from the traditional paradigm of many samples and few variables to a situation that had not been greatly considered before by statisticians outside of the machine learning community – one of few samples and an enormous number of variables, also known as the "small *n*, large *p*" problem. Unlike the past when existing theory and methods foreshadowed (or even dictated) data types that would occur in practice, this time the technology came first along with excitement and great promise. cDNA microarrays and high-density oligonucleotide chips allowed measurement of many thousands, and eventually, millions, of gene products simultaneously. High-density SNP (Single Nucleotide Polymorphism) arrays enabled high-throughput genetic profiling of living organisms. Taken together and occurring in parallel with the ongoing human and other genome projects, these breakthroughs in technology generated exciting possibilities in biomedical research: human disease prognosis and classification, new drug targets, mammalian models, basic research, and, finally, the ability to conduct discovery experiments on a scale previously unimaginable. As new technologies were quickly adopted by researchers and clinicians, questions encompassing a broad spectrum of statistical issues arose, including:

- *"How reliable and reproducible are the measurements?"* (quality assessment and control)
- *"Can I really find a needle in a haystack?"* (experimental design, estimation, testing)
- *"What can one do with so many variables at a time?"* (modeling, prediction techniques)
- *"How can I minimize the false leads?"* (testing)

Re-formulating and addressing such questions falls in the purview of statisticians, who are able to draw on their knowledge of experimental design, prediction techniques, modeling, estimation and testing, and adapt and expand existing concepts to work with these new and unprecedentedly large datasets.

When I think of Terry's approach to statistics and mentorship, a few quotes from Albert Einstein come to mind – *In theory, practice and theory are the same. In reality, they are not* and *Everything should be made as simple as possible, but not simpler.* These points could not have been more appropriate or timely than when excited statisticians, physicists, and computer scientists started working on high-dimensional biomedical problems.

It is difficult to overemphasize Terry's contributions to the field of high-dimensional data analysis in biomedical research. He stepped in at the very start and, with vigor, rigor and great enthusiasm, began to transform the analytical methods used in the field. Generically, we can consider two levels of microarray data analysis: low-level analysis, which is concerned with preprocessing the raw data into meaningful and analyzable measures; and high-level analysis, which is the statistical analysis of the resulting data matrix. Most methodological researchers tend to specialize in one or the other of these. Terry has made major, fundamental, and very widely used contributions across the analysis spectrum.

The May 8th 2011 PubMed search for "TP Speed" reveals that Terry has co-authored in excess of 150 peer-reviewed publications, a large number of which focus on the analysis of high-dimensional biological data. Here, I provide a historical commentary to only a few of the most ground breaking of those.

Your results will only be as good as the information you put in (more commonly known as "garbage in, garbage out")

Perhaps the most widely cited microarray contributions from Terry have been from his work on low-level preprocessing of the measurements. Early on, it was recognized that there are many sources of systematic variation in both cDNA and oligonucleotide microarrays. Although understanding the underlying physical reasons for the observed variation is useful, it is not always feasible. Terry recognized that simple empirical normalization approaches may be competitive with more complex biophysical models. Terry also proposed a number of what are now among the most commonly used quality control visualization tools assuring that appropriate preprocessing has been done (e.g. MA-plots, chip pseudo-image plots). Finally, for a formal evaluation of preprocessing methods, relevant biological calibration experiments had to be designed and conducted.

Yang et al. [10] and Irizarry et al. [4] represent some of the papers describing revolutionary microarray normalization (preprocessing) techniques for cDNA arrays (lowess) and short oligonucleotide chips (RMA), respectively. RMA, or some subsequent variant of it, is the most frequently used and cited preprocessing technique for short oligonucleotide chips. Rabbee and Speed [7] describe a multi-chip, multi-SNP approach to genotype calling for Affymetrix SNP chips, providing the first such alternative to the standard (at the time) genotype calling procedures, which processed all the features associated with one chip and one SNP at a time.

Microarray data analysis $\neq$ clustering

In the very early days of microarray data analysis, probably due to the high dimensionality of the data, virtually all analyses included a cluster analysis – regardless of the scientific question under study (for which clustering may or may not be appropriate).

For the problem of identifying genes that are differentially expressed in two conditions, a more natural, statistically based approach would be to use the mean difference or standardized mean difference of the expression levels, separately for each gene. However, these statistics are problematic. A large mean difference may be due to large variability or an outlier. But taking account of the variability by using the standardized difference is also problematic because when the number of replicates is small the estimate of variance is less reliable, and in particular may be artificially small. In this case, a small average difference can be highly statistically significant, yet biologically meaningless.

Lönnstedt and Speed [6] address this issue using an empirical Bayes approach that avoids these problems. They use a Bayes log posterior odds for differential versus equal expression to select differentially expressed genes. Tai and Speed [9] extend the model to allow for analysis of time-course microarray data.

Do complex datasets require complex methods?

A new laboratory technology without an established methodology for analysis of the resulting data may be attractive to aspiring quantitative analysts eager to apply new sophisticated analytical methods "brewing" in their labs yet lacking an exciting application. This situation violates a firm rule that Terry had for his students: it is the real life problems that motivate methodological research, not the reverse. Thus, when human cancer microarray datasets were first publicly released for re-analysis by other groups, Terry questioned whether the complex, state-of-the-art prediction methods that were being published with the aim of addressing biomedical research problems (e.g. prediction of a patient's tumor subtype or treatment outcome) do indeed outperform more simplistic approaches that place tight restrictions on the parameter space, such as a linear discriminant analysis with diagonal covariance matrix. Another question that came up was how to measure the relative performance of multiple candidate predictors in the absence of true, independent datasets, particularly when many parameters are estimated.

These two issues are discussed in-depth in Dudoit et al. [2]. Somewhat surprisingly, the main conclusion of this work (later supported theoretically by Levina and Bickel [5]) was that for small n, large p problems the simplest methods, with the most restrictive assumptions, perform as well as or better than the latest machine learning approaches. Moreover, unbiased assessment of performance can be challenging and must be done through a careful and valid cross-validation – an important caveat ignored by several groups in early publications. In view of these results, it is not surprising that in high-dimensional genomics, the rate of independently validated predictions remains low. Nevertheless, much of the progress that has been made is due to Terry's work on formulating and disseminating the appropriate message.

If you torture data enough, it will confess

Testing many thousands of genes for association with the phenotype of interest invariably presents an issue. From the statistical point of view, testing must be performed at an exceedingly stringent alpha level to control the overall number of false positive findings. At the time of this writing, this idea seems obvious; however, even 5 years ago, it was not – a change of mindset was required as a great majority of papers reported the significance of individual tests without regard to the number of the comparisons performed. The initial discussion of permutation-based adjusted (rather than nominal) p-values took place in Callow et al. [1]; an extensive review of approaches to multiple testing was presented in Ge et al. [3].

And finally...

On a very personal note, I would like to conclude with the story of a paper that Terry and I have never written but work on which manifests in my mind many of the wonderful qualities that Terry possesses: inspiration, mentorship and willingness to always give one more chance. In 1998, I hit a creativity wall, a not uncommon occurrence in the life of a PhD student. Terry had many PhD students, but each of us was important to him as an individual. Terry brought me to Australia with him and there I was able to stumble on a topic that excited and reinvigorated me – the search for interactions in high-dimensional SNP studies. There we were able to utilize binary tree partitioning techniques to discover epistatic genes without prominent independent (marginal) effect, while at the same time illuminating an underlying interaction structure. The results summarizing the application of our approach are described in Symons et al. [8]; however, the methodological paper was never written. Nevertheless, this is our joint work for which I am most grateful to Terry, and that ultimately motivated me to start and finish my PhD dissertation. A lesson in this to all the mentors out there: do not give up on your students, and eventually you will be thanked in print!

References

[1] M. J. Callow, S. Dudoit, E. J. Gong, T. P. Speed, and E. M. Rubin. Microarray expression profiling identifies genes with altered expression in HDL-deficient mice. *Genome Res.*, 10(12):2022–2029, 2000.

[2] S. Dudoit, J. Fridlyand, and T. P. Speed. Comparison of discrimination methods for the classification of tumors using gene expression data. *J. Am. Stat. Assoc.*, 97(457):77–87, 2002.

[3] Y. Ge, S. Dudoit, and T. P. Speed. Resampling-based multiple testing for microarray data analysis. *TEST*, 12(1):1–44, 2003.

[4] R. A. Irizarry, B. Hobbs, F. Collin, Y. D. Beazer-Barclay, K. J. Antonellis, U. Scherf, and T. P. Speed. Exploration, normalization, and summaries of high density oligonucleotide array probe level data. *Biostatistics*, 4(2):249–264, 2003.

[5] E. Levina and P. J. Bickel. Some theory for Fisher's linear discriminant function, "naive Bayes", and some alternatives when there are many more variables than observations. *Bernoulli*, 10(6):989–1010, 2004.

[6] I. Lönnstedt and T. P. Speed. Replicated microarray data. *Stat. Sinica*, 12: 31–46, 2001.

[7] N. Rabbee and T. P. Speed. A genotype calling algorithm for Affymetrix SNP arrays. *Bioinformatics*, 22(1):7–12, 2006.

[8] R. C. A. Symons, M. J. Daly, J. Fridlyand, T. P. Speed, W. D. Cook, S. Gerondakis, A. W. Harris, and S. J. Foote. Multiple genetic loci modify susceptibility to plasmacytoma-related morbidity in Eμ–v–*abl* transgenic mice. *Proc. Natl. Acad. Sci. USA*, 99(17):11299–11304, 2002.

[9] Y. C. Tai and T. P. Speed. A multivariate empirical Bayes statistic for replicated microarray time course data. *Ann. Stat.*, 34(5):2387–2412, 2006.

[10] Y. H. Yang, S. Dudoit, P. Luu, D. M. Lin, V. Peng, J. Ngai, and T. P. Speed. Normalization for cDNA microarray data: A robust composite method addressing single and multiple slide systematic variation. *Nucleic Acids Res.*, 30(4):e15, 2002.

 Nucleic Acids Research, 2002, Vol. 30, No. 4 e15

Normalization for cDNA microarray data: a robust composite method addressing single and multiple slide systematic variation

Yee Hwa Yang[1], Sandrine Dudoit[2], Percy Luu[3], David M. Lin[3], Vivian Peng[3,4], John Ngai[3,4,5] and Terence P. Speed[1,6,*]

[1]Department of Statistics, [2]Division of Biostatistics, [3]Department of Molecular and Cell Biology, [4]Functional Genomics Laboratory and [5]Neurogenomics Center, Helen Wills Neuroscience Institute, University of California, Berkeley, CA 94720-3860, USA and [6]Division of Genetics and Bioinformatics, The Walter and Eliza Hall Institute, Melbourne, Australia

Received October 3, 2001; Revised November 20, 2001; Accepted December 1, 2001

ABSTRACT

There are many sources of systematic variation in cDNA microarray experiments which affect the measured gene expression levels (e.g. differences in labeling efficiency between the two fluorescent dyes). The term normalization refers to the process of removing such variation. A constant adjustment is often used to force the distribution of the intensity log ratios to have a median of zero for each slide. However, such global normalization approaches are not adequate in situations where dye biases can depend on spot overall intensity and/or spatial location within the array. This article proposes normalization methods that are based on robust local regression and account for intensity and spatial dependence in dye biases for different types of cDNA microarray experiments. The selection of appropriate controls for normalization is discussed and a novel set of controls (microarray sample pool, MSP) is introduced to aid in intensity-dependent normalization. Lastly, to allow for comparisons of expression levels across slides, a robust method based on maximum likelihood estimation is proposed to adjust for scale differences among slides.

INTRODUCTION

DNA microarrays are part of a new class of biotechnologies that allow the monitoring of expression levels in cells for thousands of genes simultaneously. In a typical microarray experiment utilizing 'spotted arrays', the two mRNA samples to be compared are reverse transcribed into cDNA, labeled using two different fluorophores (usually a red fluorescent dye, Cy5, and a green fluorescent dye, Cy3) and then hybridized simultaneously to the glass slide. Intensity values generated from hybridization to individual DNA spots are indicative of gene expression levels, and comparisons in gene expression levels between the two samples are derived from the resulting intensity ratios (1). Applications of microarrays range from the study of gene expression in yeast under different environmental stress conditions (2,3) to the comparison of gene expression profiles for tumors from cancer patients (4–9).

In order to accurately and precisely measure gene expression changes, it is important to take into account the random (experimental) and systematic variations that occur in every microarray experiment. For example, a well-known source of systematic variation arises from biases associated with the different fluorescent dyes. This can most easily be seen in an experiment where two identical mRNA samples are labeled with different dyes and subsequently hybridized to the same slide (10). In this instance, it is rare to have the dye intensities equal across all spots between the two samples. Even though such systematic biases may be comparatively small, they may be confounding when searching for subtle biological differences. Dye biases can stem from a variety of factors, including physical properties of the dyes (heat and light sensitivity, relative half-life), efficiency of dye incorporation, experimental variability in hybridization and processing procedures, or scanner settings at the data collection step. Furthermore, the relative gene expression levels from replicate experiments may have different sample variances due to differences in experimental conditions. Many of these factors, whether internal or external to the target samples, make distinctions between differentially and constantly expressed genes difficult [in this article we adopt the definitions of 'probe' and 'target' from the January 1999 supplement to *Nature Genetics* (11), whereby the term target refers to the samples hybridized to the array and the term probe refers to the DNA sequences spotted on the array].

The purpose of normalization is to minimize systematic variations in the measured gene expression levels of two co-hybridized mRNA samples, so that biological differences can be more easily distinguished, as well as to allow the comparison of expression levels across slides. Current methods of normalization

*To whom correspondence should be addressed at: Department of Statistics, University of California, 367 Evans Hall, #3860 Berkeley, CA 94720-3860, USA. Tel: +1 510 642 2781; Fax: +1 510 642 7892; Email: terry@stat.berkeley.edu

The authors wish it to be known that, in their opinion, the first three authors should be regarded as joint First Authors

e15 Nucleic Acids Research, 2002, Vol. 30, No. 4

fail to account for important sources of systematic variation (e.g. intensity- or spatially-dependent dye biases). In this article we propose a composite normalization procedure, based on robust local regression, to accommodate different types of dye biases and the use of control sequences spotted on the array. The selection of a suitable set of control spots for use in the normalization procedure is critical for proper normalization. To this end, we introduce a novel control sample (microarray sample pool, MSP), with minimal sample-specific bias over a large intensity range, and show that it is effective in many types of microarray experiments.

MATERIALS AND METHODS

Biological samples

Preparation of RNA samples and microarray analysis. Tissues were dissected, solubilized in Trizol (Gibco BRL) and total RNA was prepared according to the manufacturer's suggested protocol. Prior to reverse transcription and labeling, total RNA samples were treated with DNase using RQ RNase-free DNase (Promega) for 20 min at 37°C. RNA samples were reverse transcribed and labeled for microarray analysis using standard techniques (6,12). Briefly, RNA samples were reverse transcribed with Superscript II reverse transcriptase in the presence of 2-aminoallyl-dUTP. Samples were purified and coupled to Cy3 or Cy5 as described (6,12,13). Labeled targets were resuspended in hybridization buffer and applied to glass microarrays. Hybridizations were performed overnight at 50–55°C. Washed and dried slides were imaged in an Axon GenePix 4000A scanner.

Experiment A: apolipoprotein AI (apo AI) experiment. The treatment group consisted of eight mice with the apo AI gene knocked out and the control group consisted of eight control C57Bl/6 mice. For each of these 16 mice, target cDNA was obtained from mRNA by reverse transcription and labeled using a red fluorescent dye, Cy5. The reference sample used in all hybridizations was prepared by pooling cDNA from the eight control mice and was labeled with a green fluorescent dye, Cy3. Target cDNA was hybridized to microarrays containing 6384 cDNA probes, which included 257 genes thought to be related to lipid metabolism. Probes were spotted onto the glass slides using a 4×4 print head and each of the corresponding 16 print tip groups was laid out in a 19×21 array or sub-grid. For further details the reader is referred to Callow *et al.* (14).

Experiment B: olfactory bulb experiment. In this experiment, comparisons were made between different spatial regions of the mouse olfactory bulb to screen for possible region-specific differences in gene expression (D.M.Lin, Y.H.Yang, J.Scolnick, L.Brunet, V.Peng, T.Speed and J.Ngai, submitted for publication). The target cDNA was hybridized to glass microarrays containing ~18 000 isolated expressed sequence tags (ESTs) from the RIKEN Release 1 mouse cDNA library (15). The olfactory bulb is an ellipsoidal structure, so in order to make a 3-dimensional representation using binary comparisons, bulbs were separately sub-dissected into three sections along each of the three orthogonal axes. RNA was collected from a number of different mice and samples from the same anatomical domains were harvested and pooled (D.M.Lin, Y.H.Yang, J.Scolnick, L.Brunet, V.Peng, T.Speed and J.Ngai, submitted for publication). Comparisons were made between maximally separated regions: anterior versus posterior, medial versus lateral and dorsal versus ventral regions.

MSP titration series. Total EST collections were generated from amplification of PCR products for microarray fabrication. Samples corresponding to all 18 816 ESTs from the RIKEN Release 1 cDNA library were pooled and precipitated. An MSP was also made from a randomly picked non-normalized plasmid library generated from mouse cerebellum (A.Finn and T.Serafini, unpublished results). Precipitated samples were resuspended and serially diluted in preparation for printing. Six steps were used in the dilution series and the samples were then spotted in the middle of the first or last row of each of the print tip groups. Microarrays were prepared as discussed previously (6).

Image processing

Each hybridization produced a pair of 16-bit images, which were processed using the software package Spot (16). The main quantities of interest produced by the image analysis methods (segmentation and background correction) are the (R,G) fluorescence intensity pairs for each gene on each array (where R = red for Cy5 and G = green for Cy3). Note that we call the spotted DNA sequences 'genes', whether they correspond to actual genes, ESTs or DNA sequences from other sources.

Statistical methods

An 'MA-plot', as described in Dudoit *et al.* (10), is used to represent the (R,G) data, where $M = \log_2 R/G$ and $A = \log_2\sqrt{(R \times G)}$. We have found MA-plots to be helpful in terms of identifying spot artifacts and detecting intensity-dependent patterns in the log ratios M. They are also very useful for the purpose of normalization, as illustrated next with several location normalization procedures. Within-slide normalization for location consists of subtracting a function $c(.)$ from individual intensity log ratios, where the function $c(.)$ is computed separately for each slide, using only data from that hybridization.

Global normalization. Global methods assume that the red and green intensities are related by a constant factor, i.e. $R = kG$, and the center of the distribution of log ratios is shifted to zero

$$\log_2 R/G \rightarrow \log_2 R/G - c = \log_2 R/(kG)$$

A common choice for the location parameter $c = \log_2 k$ is the median or mean of the intensity log ratios M for a particular gene set.

Intensity-dependent normalization. We use the robust scatter plot smoother 'lowess', implemented in the statistical software package R (17), to perform a local A-dependent normalization

$$\log_2 R/G \rightarrow \log_2 R/G - c(A) = \log_2 R/[k(A)G]$$

where $c(A)$ is the lowess fit to the MA-plot. The lowess scatter plot smoother performs robust locally linear fits. In particular, it will not be affected by a small percentage of differentially expressed genes, which will appear as outliers in the MA-plot. The user-defined parameter f is the fraction of the data used for smoothing at each point; the larger the f value, the smoother the fit. We typically use $f = 40\%$.

Within-print tip group normalization. Within-print tip group normalization is simply a (print tip + A)-dependent normalization, i.e.

$$\log_2 R/G \rightarrow \log_2 R/G - c_i(A) = \log_2 R/[k_i(A)G]$$

where $c_i(A)$ is the lowess fit to the MA-plot for the ith grid only (i.e. for the ith print tip group), $i = 1, \ldots, I$, and I denotes the number of print tips.

Scale normalization. Starting from data which have been location normalized as just described, we suppose that the log ratios from the ith print tip group follow a normal distribution with mean zero and variance $a_i^2 \sigma^2$, where σ^2 is the variance of the true log ratios and a_i^2 is the scale factor for the ith print tip group. In order to perform scale normalization, the scale factors a_i for the different print tip groups are estimated and then eliminated. Enforcing the natural constraint $\sum_{i=1}^{I} \log a_i^2 = 0$, with I denoting the total number of print tip groups on the array (or the number of slides, for multiple slide normalization discussed below), the maximum likelihood estimate for a_i is

$$\hat{a}_i^2 = (\sum_{j=1}^{n_i} M_{ij}^2)/[\sqrt[I]{(\prod_{k=1}^{I} \sum_{j=1}^{n_k} M_{kj}^2)}]$$

where M_{ij} denotes the jth log ratio in the ith print tip group, $j = 1, \ldots, n_i$. A robust alternative to this estimate, which we find preferable, is

$$\hat{a}_i = (MAD_i)/[\sqrt[I]{(\prod_{i=1}^{I} MAD_i)}$$

where the median absolute deviation MAD is defined by

$$MAD_i = \text{median}_j\{\,|M_{ij} - \text{median}_j(M_{ij})|\,\}$$

Composite normalization. For a given print tip group the composite normalization curve is a weighted average of the MSP lowess curve and the lowess curve based on all genes in the print tip group. The weights are dependent on the cumulative number of genes at different intensity levels A. An outline of this procedure for a spot in the ith print tip group is as follows. (i) Estimate $\hat{f}_i(A)$, the lowess fit to the MA-plot for the ith print tip group. (ii) Estimate $\hat{g}(A)$, the lowess fit to the MA-plot using only spots from the MSP titration series. (iii) Calculate the weighted average, $c_i(A) = \alpha_A \hat{g}(A) + (1 - \alpha_A)\hat{f}_i(A)$, where α_A is defined as the proportion of genes less than a given intensity A.

Comparison between different normalization methods. After image processing and normalization, the gene expression data can be summarized by a matrix X of intensity log ratios $M = \log_2 R/G$, with p rows corresponding to the genes being studied and n columns corresponding to the different hybridizations. In the apo AI experiment $p = 6384$ and there were $n_1 = 8$ control (C57Bl/6 mice) and $n_2 = 8$ treatment (apo AI knockout mice) hybridizations. Differentially expressed genes were identified by computing two-sample Welch t-statistics. For gene j the t-statistic comparing gene expression in the control and treatment groups is

$$t_j = (\bar{x}_{2j} - \bar{x}_{1j})/\sqrt{[(s_{1j}^2/n_1) + (s_{2j}^2/n_2)]}$$

where $\bar{x}_{1j}$ and $\bar{x}_{2j}$ denote the average background-corrected and normalized expression level of gene j in the eight control and eight treatment hybridizations, respectively. Similarly, s_{1j}^2 and s_{2j}^2 denote the variances of gene j expression level in the control and treatment hybridizations, respectively. Large absolute t-statistics suggest that the corresponding genes have different expression levels in the control and treatment groups. The statistical significance of the results was assessed based on P-values adjusted for multiple comparisons. These adjusted P-values were estimated by permutation, using Westfall and Young's step-down adjusted P-value procedure in algorithm 4.1 (18). The analysis of the apo AI experiment is described in detail in Dudoit *et al.* (10).

In order to compare the different within-slide normalization procedures, we considered their effect on the location and scale of the log ratios M using box plots. A Gaussian kernel density estimator (the 'density' function from the statistical software package R, bandwidth size 0.17) is also used to produce density plots of the log ratios for each of the normalization methods. For experiment A we considered the effect of the normalization procedures on the t-statistics for the knockout gene.

RESULTS

Within-slide normalization: intensity- and spatially-dependent systematic error

We first address within-slide normalization, i.e. normalization issues associated with data obtained from a single slide. A well-known source of error can be attributed to biases linked to the different dyes used at the labeling step. Current methods of global normalization assume a uniform grading of systematic error across all variables in an experiment. Two major assumptions are usually made: (i) all cDNA species within a sample will incorporate an equivalent amount of dye per mole cDNA; (ii) there are no other variables (e.g. spatial location, overall intensity, plate) that contribute to dye biases across the slide. These assumptions are too simplistic to account for the multiple sources of systematic error typically encountered in microarray experiments. The problem is best illustrated in an experiment where identical mRNA samples are labeled with Cy3 and Cy5 and subsequently hybridized to the same slide [self–self comparison; described in Dudoit *et al.* (10)]. In a 'perfect' self–self hybridization the intensity log ratios M in an MA-plot should be evenly distributed around zero across all intensity values A. However, this is rarely the case, and systematic error often manifests itself in terms of non-zero log ratios M. Furthermore, the imbalance in the red and green intensities is usually not constant across the spots and can vary according to overall spot intensity A (indicated by a curvature in the MA-plot), location on the array, plate origin and possibly other variables.

Intensity-dependent dye bias can be seen in the apo AI experiment (14). Apo AI is a gene known to play a pivotal role in high-density lipoprotein metabolism. The goal of the experiment was to identify genes with altered expression in the livers of mice with the apo AI gene knocked out compared with inbred C57Bl/6 control mice. In this instance, it was found that the vast majority of genes examined on the microarray showed no difference in expression level. The clear curvature in the MA-plot in Figure 1A strongly suggests the existence of an intensity-dependent dye bias.

Some systematic differences may exist between the print tips, such as slight differences in the length or in the opening of the tips, and deformation after many hours of printing. We therefore also performed individual lowess fits within each print tip group. The arrays in the apo AI experiment were printed with a 4 × 4 print head, so each lowess fit in Figure 1 corresponds to spots printed with a single print tip. Four

e15 Nucleic Acids Research, 2002, Vol. 30, No. 4

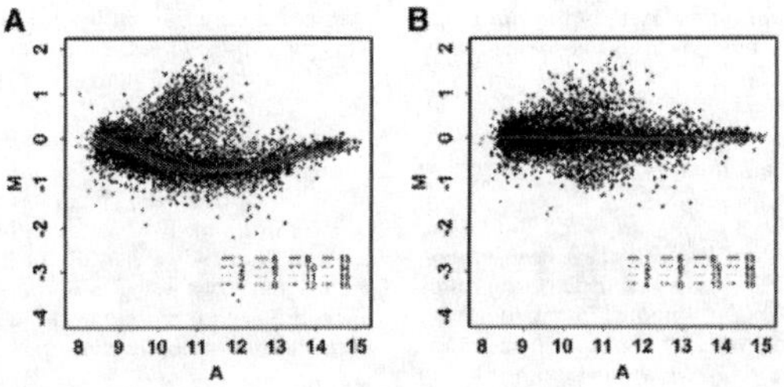

Figure 1. Within-slide normalization. (**A**) MA-plot demonstrating the need for within-print tip group location normalization. (**B**) MA-plot after within-print tip group location normalization. Both panels display the lowess fits ($f = 40\%$) for each of the 16 print tip groups (data from apo AI knockout mouse number 8 in experiment A).

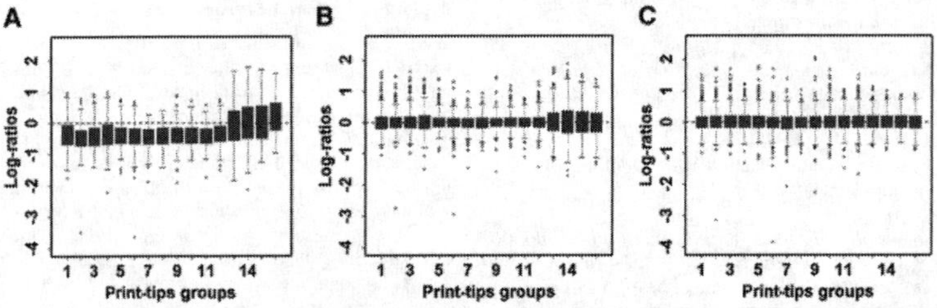

Figure 2. Within-slide normalization: box plots displaying the intensity log ratio distribution, for each of the 16 print tip groups before and after different normalization procedures. The array was printed using a 4 × 4 print head and the print tip groups are numbered first from left to right, then from top to bottom, starting from the top left corner (data from apo AI knockout mouse number 8 in experiment A). (**A**) Before normalization. (**B**) After within-print tip group location normalization, but before scale adjustment. (**C**) After within-print tip group location and scale normalization.

within-print tip group lowess curves stand out from the remaining 12 curves, indicating strong print tip or spatial effects. These four curves correspond to the last row of print tips in the 4 × 4 print head (print tips 13–16). This pattern was visible in the raw images, where the bottom four grids tended to have a higher red signal. We further examined the spatial effects by considering box plots of the log ratios M for each print tip group. Figure 2 shows that print tip groups 13–16 have a larger spread in their log ratios than any of the other 12 print tip groups. Such a difference in spread may result in misidentification of genes that are differentially expressed in the knockout mice compared to the control mice. Thus, normalization for scale across print tip groups seems desirable here.

Within-slide normalization using the majority of genes on the microarray

For the apo AI experiment considered in Figures 1 and 2, global normalization, in which a constant adjustment is used to force the distribution of the log ratios to have a median zero within each slide, would result in a vertical translation of the MA-plot. It would not correct for intensity- or spatially-dependent effects, including local differences in the spread of the log ratios M. As a first pass towards eliminating intensity and spatial biases, we considered a normalization procedure in which the majority of genes on the array are used for normalization. This is a reasonable assumption when there are good reasons to expect that (i) only a relatively small proportion of the genes will vary significantly in expression between the two co-hybridized mRNA samples or (ii) there is symmetry in the expression levels of the up/down-regulated genes. The data shown in Figures 1 and 2 are good examples of this situation.

To address both intensity and spatial normalization issues, we first incorporated an intensity modifier into our normalization procedures. We used the scatter plot smoother lowess to produce robust location estimates of the intensity log ratios M for various intensity levels A and to adjust each gene with a different normalization value depending on its overall intensity. Other variables that may contribute to systematic bias include differences in print tips and spatial location. Because every grid in an array is printed using the same print tip, print tip groups can also be used as proxies for spatial effects on the slide. Thus, we also incorporated a print tip modifier into the intensity-dependent normalization. It might be thought that the layout of genes on the slide could lead to one or more print tip groups being enriched for differentially expressed genes and, hence, invalidate the assumption underlying print tip group normalization. While we cannot rule out chance imbalances in the spatial distribution of differentially expressed genes, the mechanics of spotting cDNA onto the slide makes a large effect of this kind unlikely. Even if one had a collection of

Nucleic Acids Research, 2002, Vol. 30, No. 4 **e15**

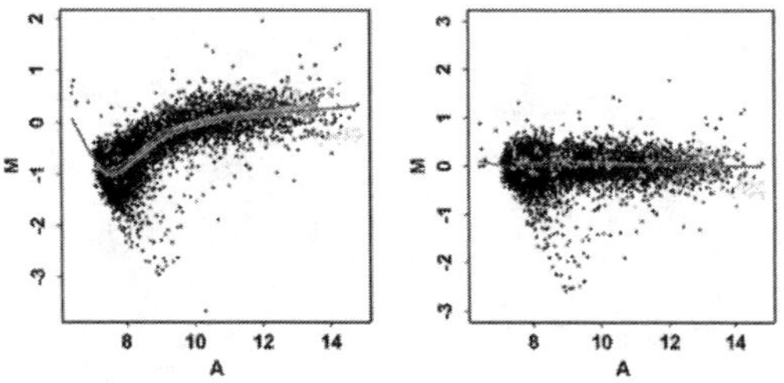

Figure 3. Within-slide normalization: MA-plot for comparison of the anterior versus posterior portion of the olfactory bulb. These samples are very similar and we do not expect many genes to change. The cyan dots represent the MSP titration series and the cyan curve represents the corresponding lowess fit. The red curve corresponds to the lowess fit for the entire dataset. Control genes are highlighted in yellow (tubulin and GAPDH), green (mouse genomic DNA) and orange (an approximate rank-invariant set of genes with $P = 0.01$ and $l = 25$). (Left) MA-plot before normalization. (Right) MA-plot after within-print tip group location normalization.

genes known or expected to be differentially expressed in one or more microtiter plates, they would be spotted evenly across the slide by the printer.

In principle, after within-print tip group location normalization, the log ratios from the different print tip groups should be centered around zero (Fig. 2B). However, it is possible that the log ratios from the various print tip groups have different spreads; if this is the case, a scale adjustment may be required. Figure 2 displays box plots of the intensity log ratios M for a slide in experiment A, before normalization (Fig. 2A), after within-print tip group location normalization (Fig. 2B) and after within-print tip group location and scale normalization (Fig. 2C). In Figure 2B there is a disproportionately large number of extreme log ratios in the lower four grids. After scale normalization, the extreme log ratios are evenly distributed on the array (Fig. 2C). Again, this procedure assumes that a relatively small proportion of the genes vary significantly in expression between the two co-hybridized mRNA samples, as would be expected when comparing samples from wild-type mice versus mice harboring a mutation in a single gene. In addition, it is assumed that the spread of the distribution of the log ratios should be roughly the same for all print tip groups. The robust statistic *MAD*, like the robust lowess smoother, will not be affected by a small percentage of differentially expressed genes, which will appear as outliers in the MA-plots.

In another example of within-slide location normalization, Figure 3 shows an MA-plot from experiment B, for a comparison of mRNA levels in the anterior and posterior portions of the mouse olfactory bulb. These mRNA samples are biologically very similar and very few genes are expected to be differentially expressed. Indeed, the MA-plot shows very little divergence from the lowess fit to all genes and the scatter plot is roughly symmetrical about the lowess curve. We thus performed a print tip group normalization using all genes. Figure 3 displays the intensity data before (Fig. 3, left) and after normalization (Fig. 3, right). The normalization procedure resulted in a scatter plot centered around an M value of zero across the A intensity range, thus indicating that the types of systematic errors we have identified have been minimized.

Within-slide normalization using MSP

Frequently, the expression profiles in biological samples are more divergent in nature than in the examples investigated above. Thus, normalization based upon all genes may be inaccurate. A control sample that spans the intensity range and exhibits a relatively constant expression level across biological samples is desirable. Yeast genomic DNA has been used for normalization in that system. Since all species within an mRNA sample can hybridize to this control, sample-specific bias is reduced. The genomic DNA approach does not, however, directly extend to more complex metazoan systems, where the high ratio of non-coding to coding DNA effectively reduces the signal from such a control below the detection threshold in a microarray experiment.

We therefore constructed a novel control sample ensemble, MSP, inclusive of all genes present on the microarray. This sample should be analogous to genomic DNA without the intervening sequences and, thus, provides a potential probe for every species within a labeled cDNA target. We titrated this sample over the intensity range of a typical microarray experiment in order to account for all levels of intensity-dependent bias. The utility of this control is demonstrated in Figure 3, which highlights the MSP titration series (cyan dots) and the corresponding lowess fit to the MSP spots (cyan curve). Notice that the MSP curve is the same as the lowess fit to the MA-plot based on all genes (red curve). An intensity-dependent normalization using the MSP control as a reference would thus be similar, in this case, to that using all genes.

In experiment B we made a more divergent comparison between mRNA samples from the medial and lateral portions of the olfactory bulb. Due to the presence of vascular tissue near the medial bulb, medial samples have a higher representation of blood tissue. Figure 4A displays the MA-plot for the medial versus lateral comparison. The genetic divergence between the samples is evident in the increased spread of the log ratios, particularly in the high intensity range. The lowess curve based on all genes (red) and the lowess curve based on the MSP titration series (cyan) are different at high intensity values. In such a case, where samples are widely divergent at high intensities, normalization based on the MSP titration series appears to be more accurate. However, whereas the MSP

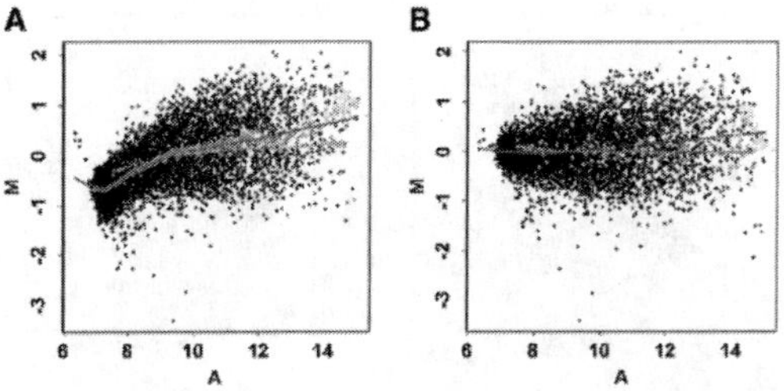

Figure 4. Within-slide normalization: MA-plot for comparison of the medial versus lateral portion of the olfactory bulb. The cyan dots represent the MSP titration series and the cyan curve represents the corresponding lowess fit. The red curve corresponds to the lowess fit for the entire dataset. The green curve represents the composite normalization curve. Control genes are highlighted in yellow (tubulin and GAPDH), green (mouse genomic DNA) and orange (an approximate rank-invariant set of genes). (**A**) MA-plot before normalization. (**B**) MA-plot after composite normalization.

spots produce more accurate estimates of expression levels, these estimates may be less stable in the context of spatial normalization, due to the small number of MSP spots per print tip group.

Comparisons of MSP to other control samples

In some instances a small number of known genes, for example housekeeping genes whose expression is expected to be constant across samples, are utilized for microarray normalization. Such genes are often highly expressed, as illustrated in Figure 3 for tubulin and glyceraldehyde-3-phosphate dehydrogenase (GAPDH) (yellow). Typically, housekeeping genes are not representative of all intensity values *A* and are therefore limited in their utility for intensity-dependent normalization. In addition, there is a sample-specific bias for many genes which may not be predictable; this is again non-ideal for use as a control.

Another approach is to select a rank-invariant set of genes. A set of genes is said to be rank-invariant if their ranks are the same for the red and green intensities. In practice, a maximal invariant set tends to be too small and an iterative procedure for finding an approximately invariant set of genes has been proposed (19,20). These genes are highlighted in orange in Figure 3 and were obtained using the method described in Tseng *et al.* (20) with $P = 0.01$ and $l = 25$. The value P is chosen such that a conserved set of genes is selected. Notice that this set of spots overlaps the lowess fit to the MA-plot based on all genes.

Composite within-slide normalization

We propose a composite normalization method to address the limitations of using all genes or only the MSP titration series for normalization. The composite normalization curve is a weighted average of the MSP curve and the within-print tip group lowess curve based on all genes. The weights are dependent on the cumulative number of genes at different intensity levels. Figure 4B shows the MA-plot after within-print tip group normalization. In this figure the green composite normalization curve is a weighted average of the red and cyan colored curves. Note that the divergence of the red from the green curve at high intensity values still persists after normalization. In practice, we find composite normalization necessary in the case of divergent samples. Biologically significant outliers in experiment B were more consistently identified when composite normalization was incorporated in the analysis (data not shown).

Comparison between different normalization methods

In order to compare the different within-slide normalization methods, we considered their effect on the location and scale of the log ratios *M*. Figure 5A shows density plots of the log ratios for different normalization methods. Without normalization (black curve) the log ratios are centered around –1, indicating a bias towards the green (Cy3) dye. A global median normalization (red curve) shifts the center of the log ratio distribution to zero, but does not affect the spread. The dependence of the log ratio *M* on the overall intensity *A* is also still present (see Fig. 1). Both the intensity-dependent (green curve) and within-print tip group (blue curve) location normalization methods reduce the spread of the log ratios compared to a global normalization. A within-print tip group scale normalization (cyan curve) further reduces the spread slightly.

The different methods were also evaluated based on their ability to identify genes which are known to be differentially expressed. For experiment A the apo AI gene is knocked out in the eight treatment mice, so one expects the *t*-statistics to take on very large negative values for this gene. Figure 5B shows a truncated plot of the extreme *t*-statistics for each of the methods. The global median, intensity-dependent and within-print tip group location normalization methods seem to perform best in terms of their ability to detect the three copies of the knocked out apo AI gene. A good method should enable a clear distinction between differentially and constantly expressed genes as reflected by the *t*-statistic, i.e. one expects a large jump in the *t*-statistic between the least extreme of the differentially expressed genes and the most extreme of the remaining genes. The largest jump in *P*-values is observed for within-print tip group location normalization. Thus, in the situation presented by experiment A, where log ratios from the different arrays have fairly similar spreads (see Fig. 2), within-print tip group location normalization enables the best separation between differentially expressed genes and noise.

Multiple slide normalization

Having addressed location and scale normalization issues within a slide, all normalized log ratios should be centered

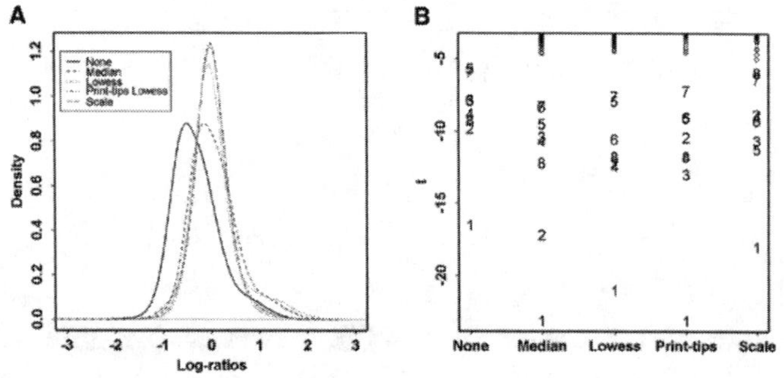

Figure 5. Within-slide normalization. (**A**) Density plots of the log ratios *M* before and after different normalization procedures. The solid black curve represents the density of the log ratios before normalization. The red, green, blue and cyan curves represent the densities after global median normalization, intensity-dependent location normalization, within-print tip group location normalization and within-print tip group scale normalization, respectively (data from apo AI knockout mouse number 8 in experiment A). (**B**) Plot of *t*-statistics for different normalization methods. The numbers 1–8 represent the differentially expressed genes identified in Dudoit *et al.* (10) and confirmed using RT–PCR: indices 1–3 represent the three apo AI genes spotted on the array. Empty circles represent the remaining 6376 genes where no effect is expected. Only *t* values less than –4 are shown.

around zero. However, in many experiments expression levels must be compared across different slides. It is important to note that individual slides in a multiple slide comparison may need to be adjusted for scale when the different slides have substantially different spreads in their intensity log ratios. Failing to perform a scale normalization could lead to one or more slides having undue weight when averaging log ratios across slides. We can apply the principles used for within-slide print tip group scale normalization to multiple slide scale adjustment.

In practice, the need for scale normalization between slides will be determined empirically. Figure 6 displays box plots of the log ratios for each of the 16 slides in experiment A, after within-print tip group location and scale normalization. The box plots are centered at zero and have fairly similar spreads. In this instance we chose not to adjust for scale, as the noise introduced by a scale normalization of the different slides may be more detrimental than a small difference in scale.

DISCUSSION

Intensity data from microarray experiments are subject to a variety of random and systematic errors. This paper has introduced location and scale normalization methods for different types of cDNA microarray experiments and discussed different sets of control spots utilized in normalization. The location normalization procedure is based on robust local regression of the intensity log ratios on overall spot intensity and accounts for intensity and spatial dependence in the dye biases. A MSP titration series was constructed and used as a set of controls for normalization. The advantages of the MSP are the minimal sample-specific bias and the coverage of a wide intensity range. In addition, we have proposed a composite normalization procedure, whereby the utility of different sets of control spots and normalization methods are combined. The different normalization methods were compared using gene expression data from two experiments: the apo AI experiment (experiment A), with replicated treatment and control slides, and the mouse olfactory bulb experiment (experiment B). Normalization can be performed at three different levels: (i) within a single slide; (ii) between a pair of slides for dye-swap experiments (21); and (iii) among multiple slides.

Within-slide normalization methods

For within-slide normalization, global methods have been used as pre-processing steps in a number of papers on the identification of differentially expressed genes in single slide cDNA microarray experiments (22,23). Such procedures assume that the red and green intensities can be related by a multiplicative constant. In one of the first proposed normalization methods, Chen *et al.* (22) derived an iterative procedure for estimating normalization constants. Similar approaches have been implemented in widely used microarray software packages [e.g. GenePix (24)]. Kerr *et al.* (25) and R.D.Wolfinger, G.Gibson, E.D.Wolfinger, L.Bennett, H.Hamadeh, P.Bushel, C.Afshari and R.S.Paules (SAS Institute, unpublished data) proposed the use of ANOVA models for normalization purposes. Their methods essentially perform only a global normalization and do not correct for intensity or scale differences. We have found that the standard global median normalization can often be inadequate due to spatially- and intensity-dependent dye biases. We propose instead a within-print tip group location normalization method which is based on robust local regression of the log ratios *M* on overall spot intensity *A* (the lowess smoother for MA-plots). Compared with other normalization procedures, this approach provided a clearer distinction between the differentially and constantly expressed genes in experiment A.

Other intensity-dependent normalization methods have been proposed in recent articles. Finkelstein *et al.* (26) recommended an iterative linear regression procedure, which essentially amounts to robust linear regression. Sapir and Churchill (27) suggested using the orthogonal residuals from the robust regression of log*R* versus log*G* as the normalized log ratios.

e15 Nucleic Acids Research, 2002, Vol. 30, No. 4

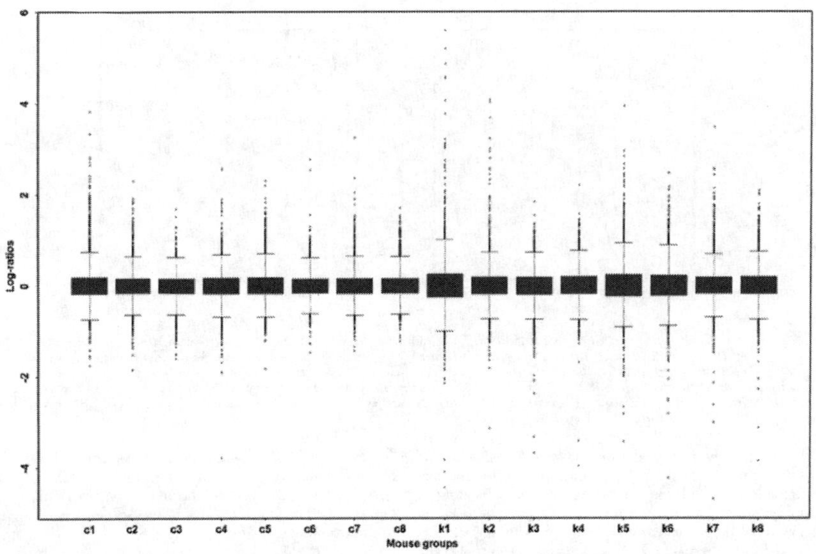

Figure 6. Multiple slide normalization: box plots displaying the intensity log ratio distribution for different slides/mice for experiment A, after within-print tip group location and scale normalization. The first eight box plots represent the data for the eight control mice and the last eight represent the data for the eight apo AI knockout mice.

Since an MA-plot amounts to a 45° counterclockwise rotation of the (log*G*,log*R*) coordinate system (up to multiplicative constants), their method is similar to fitting a robust regression line through the MA-plot, instead of a lowess curve. One can view these two linear normalizations as a more constrained version of our intensity-dependent normalization. Kepler *et al.* (28) proposed a more general intensity-dependent normalization approach, which uses a different local regression method instead of the lowess smoother. Most methods suggested thus far do not correct for spatial biases in the log ratios. As we have shown, spatial bias is also a significant source of systematic error, due to hybridization artifacts or print tip effects during printing of the microarray. Our proposed normalization procedures correct for these artifacts.

Within-slide location normalization methods adjust the intensity log ratios *M* such that they are approximately zero for genes that are constantly expressed in the two co-hybridized samples. The box plots of the location normalized log ratios in each print tip group in Figure 2 suggest that some scale adjustment may also be required within slide. However, within-print tip group scale normalization seems to have decreased our ability to identify the differentially expressed genes in experiment A. We believe that this is due to an increase in the variability (the denominator of the *t*-statistic) of the log ratios for the eight differentially expressed genes compared to the rest of the genes.

Multiple slide normalization methods

A similar approach to that described for within-slide scale normalization may also be extended to perform scale normalization across slides. In practice, multiple slide normalization aims to adjust for different sample variances in log ratios across slides. Such adjustments are required so that the relative expression levels from one particular slide do not dominate the average relative expression levels across replicate slides. In general there is a trade-off between the gains achieved by scale normalization and the possible increase in variability introduced by this additional step. In cases where the scale differences are fairly small it may thus be preferable to perform only a location normalization. Further investigations are underway to develop an improved procedure for scale adjustment and to identify better comparison criteria to assess the effectiveness of various normalization procedures.

Comparisons of MSP to commonly used control samples

In general, the set of control spots most appropriate for normalization depends on the nature of the experiment. Traditional methods based upon intensity values of housekeeping genes often show sample-specific bias and do not address the issue of intensity-dependent dye biases. Other drawbacks include the possibility that housekeeping genes may actually be regulated within an experimental sample. Housekeeping genes also tend to be highly expressed and, hence, may not be

Table 1. The various normalization methods considered in this article

		Within-slide				Multiple slide
		Global, location	Intensity-dependent, location	Print tip-dependent, location	Print tip, location and scale	Scale
		$c(.)$ constant, $a(.) = 1$	$c(.) = c(A)$, $a(.) = 1$	$c(.) = c(A$, print tip), $a(.) = 1$	$c(.) = c(A$, print tip), $a(.) = a$(print tip)	
All genes	Assumes the majority of genes in the two mRNA samples have similar overall expression levels	Yes	Yes	Yes	Yes	Yes
Housekeeping genes	Usually highly expressed and do not capture intensity-dependent structure	Yes	No	No	No	No
MSP titration series	Doesn't require any prior biological assumption, however, estimating $c(A$, print tip) based on a small number of spots may not be very stable	Yes	Yes	No	No	No
Rank-invariant set	May not span the whole intensity range	Yes	Yes	No	No	No

For within-slide normalization, the log ratios are normalized by $\log_2 R/G \rightarrow [\log_2 R/G - c(.)]/a(.)$, where $c(.)$ and $a(.)$ correspond to location and scale adjustment, respectively. The columns refer to different normalization methods and the rows correspond to different sets of control spots. The Yes or No in each cell refers to the feasibility of performing the normalization in practice. For example, it is possible in practice to perform global normalization based only on housekeeping genes, but it is not advisable to perform intensity-dependent normalization on housekeeping genes only.

representative of other genes of interest. It is clear that a less localized type of control is required to obtain accurate normalization. The other three types of control examined in this article were chosen for their representation of as many genes and intensity values as possible so as to minimize sample bias.

The MSP titration series was constructed with this specific aim in mind. In the yeast system, normalization is typically performed using yeast genomic DNA, which contains proportionately small amounts of non-coding DNA. In contrast, the genomes of higher organisms such as mice contain a much higher representation of non-coding DNA. The MSP is analogous to genomic DNA as a control, with the exception that non-coding regions are removed. Typically, a concentration titration is done to span as wide an intensity range as possible. However, due to limitations in the construction of the MSP, very high expression values cannot be represented. In practice, one could construct an MSP of lower complexity with a larger representation of highly expressed genes. Since most rare and low expression genes do not contribute significantly to an MSP signal, removing this population is analogous to further removal of non-coding DNA. Theoretically, all labeled cDNA sequences could hybridize to this mixed probe sample, so it is therefore minimally subject to any sample-specific bias.

The use of all genes for normalization offers the most stability in terms of estimating spatially- and intensity-dependent trends in the log ratios. However, in biological samples which show significant divergence, a lowess fit to the MA-plot based on all genes may not produce accurate normalized log ratios. In such instances, it would be more appropriate to normalize using the MSP spots alone. While the MSP and rank-invariant controls are effective for intensity-dependent normalization, we have found that normalization based on all genes is more reliable for spatial normalization . This is due in part to the low representation of MSP and rank-invariant spots per print tip group (6–12 spots per 400 spots) and is an example of bias variance trade-off.

Composite normalization and the MSP titration series

This article has proposed a composite normalization procedure which combines the utility of normalization methods based on all genes and those based on only the MSP titration spots. For low A intensity values, normalization is based on all genes in the corresponding intensity range. For higher A values, particularly in more divergent biological samples, normalization is based primarily on the MSP titration series. In other circumstances as they warrant, other normalization methods may be incorporated into the composite technique. For example, in cases where microarrays are printed without MSP titration spots, very high intensities may be normalized using housekeeping genes and median to low intensities may be normalized using all genes in the corresponding range.

The MSP spots were essential to validate the assumptions behind our various normalization procedures and are necessary for normalizing biologically divergent samples. The construction of the MSP titration series is important, as we observed intensity-dependent dye biases in many experiments. Efforts are still in progress to devise variants of this control set for scale normalization procedures. It is evident that no single control sample or normalization procedure is accurate or adequate for all types of microarray comparisons. However, it is becoming increasingly common for investigators to print microarrays with a large complement of control spots. This flexibility expands the opportunity to customize normalization procedures, depending on the experimental conditions.

The strengths and weaknesses of the normalization techniques and control samples discussed in this paper are summarized in Table 1. Finally, the methods described in the article are implemented in the package R (17), SMA (Statistics for Microarray Analysis), which may be downloaded from http://www.R-project.org. Supplementary analyses, figures and datasets are available at http://www.stat.berkeley.edu/users/terry/zarray/Html/index.html.

ACKNOWLEDGEMENTS

We would like to acknowledge Matthew J. Callow from the Lawrence Berkeley National Laboratory for providing the data we used to develop the various normalization approaches and Yasushi Okazaki and Yoshihide Hayashizaki of the RIKEN Genomics Sciences Center for graciously providing their normalized mouse cDNA clone set for our use. We would like to thank Elva Diaz, Andrew Finn, Jonathan Scolnick and Tito Serafini for discussions and assistance over the course of this project. We are also grateful to Eric Schadt and Wing Hung Wong for discussions as well as for providing the code for their rank-invariant normalization methods. This work was supported in part by the NIH through grants 5R01MH61665-02 (J.N. and T.P.S.) and 8RO1GM59506A (T.P.S.), by funds from the Department of Molecular and Cell Biology and Helen Wills Neuroscience Institute (University of California at Berkeley) and by a PMMB Burroughs-Wellcome postdoctoral fellowship (S.D.).

REFERENCES

1. Taniguchi,M., Miura,K., Iwao,H. and Yamanaka,S. (2001) Quantitative assessment of DNA microarrays—comparison with northern blot analyses. *Genomics*, **71**, 34–39.
2. Hughes,T.R., Marton,M.J., Jones,A.R., Roberts,C.J., Stoughton,R., Armour,C.D., Bennett,H.A., Coffey,E., Dai,H., He,Y.D., Kidd,M.J., King,A.M., Meyer,M.R., Slade,D., Lum,P.Y., Stepaniants,S.B., Shoemaker,D.D., Gachotte,D., Chakraburtty,K., Simon,J., Bard,M. and Friend,S.H. (2000) Functional discovery via a compendium of expression profiles. *Cell*, **102**, 109–126.
3. Spellman,P.T., Sherlock,G., Zhang,M.Q., Iyer,V.R., Anders,K., Eisen,M.B., Brown,P.O., Botstein,D. and Futcher,B. (1998) Comprehensive identification of cell cycle-regulated genes of the yeast *Saccharomyces cerevisiae* by microarray hybridization. *Mol. Biol. Cell*, **9**, 3273–3297.
4. Alizadeh,A.A., Eisen,M.B., Davis,R.E., Ma,C., Lossos,I.S., Rosenwald,A., Boldrick,J.C., Sabet,H., Tran,T., Yu,X., Powell,J.I., Yang,L., Marti,G.E., Moore,T., Hudson,J.,Jr, Lu,L., Lewis,D.B., Tibshirani,R., Sherlock,G., Chan,W.C., Greiner,T.C., Weisenburger,D.D., Armitage,J.O., Warnke,R., Staudt,L.M. *et al.* (2000) Distinct types of diffuse large B-cell lymphoma identified by gene expression profiling. *Nature*, **403**, 503–511.
5. Alon,U., Barkai,N., Notterman,D.A., Gish,K., Ybarra,S., Mack,D. and Levine,A.J. (1999) Broad patterns of gene expression revealed by clustering analysis of tumor and normal colon tissues probed by oligonucleotide arrays. *Proc. Natl Acad. Sci. USA*, **96**, 6745–6750.
6. DeRisi,J., Penland,L., Brown,P.O., Bittner,M.L., Meltzer,P.S., Ray,M., Chen,Y., Su,Y.A. and Trent,J.M. (1996) Use of a cDNA microarray to analyse gene expression patterns in human cancer. *Nature Genet.*, **14**, 457–460.
7. Golub,T.R., Slonim,D.K., Tamayo,P., Huard,C., Gaasenbeek,M., Mesirov,J.P., Coller,H., Loh,M.L., Downing,J.R., Caligiuri,M.A., Bloomfield,C.D. and Lander,E.S. (1999) Molecular classification of cancer: class discovery and class prediction by gene expression monitoring. *Science*, **286**, 531–537.
8. Perou,C.M., Jeffrey,S.S., van de Rijn,M., Rees,C.A., Eisen,M.B., Ross,D.T., Pergamenschikov,A., Williams,C.F., Zhu,S.X., Lee,J.C., Lashkari,D., Shalon,D., Brown,P.O. and Botstein,D. (1999) Distinctive gene expression patterns in human mammary epithelial cells and breast cancers. *Proc. Natl Acad. Sci. USA*, **96**, 9212–9217.
9. Ross,D.T., Scherf,U., Eisen,M.B., Perou,C.M., Rees,C., Spellman,P., Iyer,V., Jeffrey,S.S., Van de Rijn,M., Waltham,M., Pergamenschikov,A., Lee,J.C., Lashkari,D., Shalon,D., Myers,T.G., Weinstein,J.N., Botstein,D. and Brown,P.O. (2000) Systematic variation in gene expression patterns in human cancer cell lines. *Nature Genet.*, **24**, 227–235.
10. Dudoit,S., Yang,Y.H., Callow,M.J. and Speed,T.P. (2002) Statistical methods for identifying genes with differential expression in replicated cDNA microarray experiments. *Stat. Sin.*, in press.
11. The Chipping forecast. (1999) *Nature Genet.*, **21** (suppl.).
12. Marton,M.J., DeRisi,J.L., Bennett,H.A., Iyer,V.R., Meyer,M.R., Roberts,C.J., Stoughton,R., Burchard,J., Slade,D., Dai,H., Bassett,D.E.,Jr, Hartwell,L.H., Brown,P.O. and Friend,S.H. (1998) Drug target validation and identification of secondary drug target effects using DNA microarrays. *Nature Med.*, **4**, 1293–1301.
13. Shoemaker,D.D., Schadt,E.E., Armour,C.D., He,Y.D., Garrett-Engele,P., McDonagh,P.D., Loerch,P.M., Leonardson,A., Lum,P.Y., Cavet,G., Wu,L.F., Altschuler,S.J., Edwards,S., King,J., Tsang,J.S., Schimmack,G., Schelter,J.M., Koch,J., Ziman,M., Marton,M.J., Li,B., Cundiff,P., Ward,T., Castle,J., Krolewski,M., Meyer,M.R., Mao,M., Burchard,J., Kidd,M.J., Dai,H., Phillips,J.W., Linsley,P.S., Stoughton,R., Scherer,S. and Boguski,M.S. (2001) Experimental annotation of the human genome using microarray technology. *Nature*, **409**, 922–927.
14. Callow,M.J., Dudoit,S., Gong,E.L., Speed,T.P. and Rubin,E.M. (2000) Microarray expression profiling identifies genes with altered expression in HDL-deficient mice. *Genome Res.*, **10**, 2022–2029.
15. Miki,R., Kadota,K., Bono,H., Mizuno,Y., Tomaru,Y., Carninci,P., Itoh,M., Shibata,K., Kawai,J., Konno,H., Watanabe,S., Sato,K., Tokusumi,Y., Kikuchi,N., Ishii,Y., Hamaguchi,Y., Nishizuka,I., Goto,H., Nitanda,H., Satomi,S., Yoshiki,A., Kusakabe,M., DeRisi,J.L., Eisen,M.B., Iyer,V.R., Brown,P.O., Muramatsu,M., Shimada,H., Okazaki,Y. and Hayashizaki,Y. (2001) Delineating developmental and metabolic pathways *in vivo* by expression profiling using the RIKEN set of 18,816 full-length enriched mouse cDNA arrays. *Proc. Natl Acad. Sci. USA*, **98**, 2199–2204.
16. Buckley,M.J. (2000) *The Spot User's Guide*. CSIRO Mathematical and Information Sciences, North Ryde, NSW 1670, Australia. http://www.cmis.csiro.au/iap/spot.htm
17. Ihaka,R. and Gentleman,R. (1996) R: a language for data analysis and graphics. *J. Comput. Graph. Statist.*, **5**, 299–314.
18. Westfall,P.H. and Young,S.S. (1993) *Resampling-based Multiple Testing: Examples and Methods for P-value Adjustment*. Wiley, New York.
19. Schadt,E., Li,C., Ellis,B. and Wo,W.H. (1999) Feature extraction and normalization algorithms for high-density oligonucleotide gene expression array data. Department of Statistics, UCLA. Preprint 303, www.stat.ucla.edu
20. Tseng,G.C., Oh,M.K., Rohlin,L., Liao,J. and Wong,W.H. (2001) Issues in cDNA microarray analysis: quality filtering, channel normalization, models of variations and assessment of gene effects. *Nucleic Acids Res.*, **29**, 2549–2557.
21. Yang,Y.H., Dudoit,S., Luu,P. and Speed,T.P. (2001) Normalization for cDNA microarray. In Bittner,M.L., Chen,Y., Dorsel,A.N. and Dougherty,E.R. (eds), *Microarrays: Optical Technologies and Informatics*. SPIE, Society for Optical Engineering, San Jose, CA.
22. Chen,Y., Dougherty,E.R. and Bittner,M.L. (1997) Ratio-based decisions and the quantitative analysis of cDNA microarray images. *J. Biomed. Optics*, **2**, 364–374.
23. Newton,M.A., Kendziorski,C.M., Richmond,C.S., Blattner,F.R. and Tsui,K.W. (2001) On differential variability of expression ratios: improving statistical inference about gene expression changes from microarray data. *J. Comput. Biol.*, **8**, 37–52.
24. Axon Instruments Inc. (1999) *GenePix 4000A User's Guide*. Axon Instruments, Union City, CA.
25. Kerr,M.K., Martin,M. and Churchill,G.A. (2000) Analysis of variance for gene expression microarray data. *J. Comput. Biol.*, **7**, 819–837.
26. Finkelstein,D.B., Gollub,J., Ewing,R., Sterky,F., Somerville,S. and Cherry,J.M. (2000) Iterative linear regression by sector: renormalization of cDNA microarray data and cluster analysis weighted by cross homology. In *CAMDA*. http://afgc.stanford.edu/afgc_html/site2Stat.htm
27. Sapir,M. and Churchill,G.A. (2000) Estimating the posterior probability of differential gene expression from microarray data, Poster. The Jackson Laboratory. www.jax.org/research/churchill/pubs
28. Kepler,T.B., Crosby,L. and Morgan,K.T. (2000) Normalization and analysis of DNA microarray data by self-consistency and local regression. Santa Fe Institute. 00-99-055, www.santafe.edu/sfi/publications/00wplist.html

Biostatistics (2003), **4**, 2, *pp*. 249–264
Printed in Great Britain

Exploration, normalization, and summaries of high density oligonucleotide array probe level data

RAFAEL A. IRIZARRY
Department of Biostatistics, Johns Hopkins University, Baltimore MD 21205, USA
rafa@jhu.edu

BRIDGET HOBBS
Division of Genetics and Bioinformatics, WEHI, Melbourne, Australia

FRANCOIS COLLIN
Gene Logic Inc., Berkeley, CA, USA

YASMIN D. BEAZER-BARCLAY, KRISTEN J. ANTONELLIS, UWE SCHERF
Gene Logic Inc., Gaithersburg, MD, USA

TERENCE P. SPEED
Division of Genetics and Bioinformatics, WEHI, Melbourne, Australia. Department of Statistics, University of California at Berkeley

SUMMARY

In this paper we report exploratory analyses of high-density oligonucleotide array data from the Affymetrix GeneChip R system with the objective of improving upon currently used measures of gene expression. Our analyses make use of three data sets: a small experimental study consisting of five MGU74A mouse GeneChip R arrays, part of the data from an extensive spike-in study conducted by Gene Logic and Wyeth's Genetics Institute involving 95 HG-U95A human GeneChip R arrays; and part of a dilution study conducted by Gene Logic involving 75 HG-U95A GeneChip R arrays. We display some familiar features of the perfect match and mismatch probe (PM and MM) values of these data, and examine the variance–mean relationship with probe-level data from probes believed to be defective, and so delivering noise only. We explain why we need to normalize the arrays to one another using probe level intensities. We then examine the behavior of the PM and MM using spike-in data and assess three commonly used summary measures: Affymetrix's (i) average difference (AvDiff) and (ii) MAS 5.0 signal, and (iii) the Li and Wong multiplicative model-based expression index (MBEI). The exploratory data analyses of the probe level data motivate a new summary measure that is a robust multi-array average (RMA) of background-adjusted, normalized, and log-transformed PM values. We evaluate the four expression summary measures using the dilution study data, assessing their behavior in terms of bias, variance and (for MBEI and RMA) model fit. Finally, we evaluate the algorithms in terms of their ability to detect known levels of differential expression using the spike-in data. We conclude that there is no obvious downside to using RMA and attaching a standard error (SE) to this quantity using a linear model which removes probe-specific affinities.

To whom correspondence should be addressed

An R package with the functions used for the analyses in this paper is part of the Bioconductor project and can be downloaded (http://www.bioconductor.org). Supplemental material, such as color versions of the figures, is available on the web (http://www.biostat.jhsph.edu/ ririzarr/affy).

1. INTRODUCTION

High-density oligonucleotide expression array technology is now widely used in many areas of biomedical research. The system (Lockhart *et al.*, 1996) uses oligonucleotides with length of 25 base pairs that are used to probe genes. Typically, each gene will be represented by 16–20 pairs of oligonucleotides referred to as *probe sets*. The first component of these pairs is referred to as a perfect match (PM) probe. Each PM probe is paired with a mismatch (MM) probe that is created by changing the middle (13th) base with the intention of measuring non-specific binding. The PM and MM are referred to as a *probe pair*. See the Affymetrix Microarray Suite User Guide (1999) for details. RNA samples are prepared, labeled and hybridized with arrays. Arrays are scanned and images are produced and analysed to obtain an intensity value for each probe. These intensities represent how much hybridization occurred for each oligonucleotide probe. Of interest is finding a way to combine the 16–20 probe pair intensities for a given gene to define a measure of expression that represents the amount of the corresponding mRNA species.

We denote the intensities obtained for each probe as

$$PM_{ijn} \text{ and } MM_{ijn}, i = 1, \ldots, I, j = 1, \ldots, J_n, \text{ and } n = 1, \ldots, N$$

with n representing the different genes, i representing different RNA samples, and j representing the probe pair number (this number is related to the physical position of the oligonucleotide in the gene). The number of genes N usually ranges from 8 000 to 20 000, the number of arrays I ranges from one to hundreds, and the number of probe pairs within each gene J_n usually ranges from 16 to 20. Throughout the text indices are suppressed when there is no ambiguity.

Section 2 describes the three data sets used in this paper. Section 3 explores various interesting features of the data with the objective of defining an effective measure of gene expression using the probe level data. Section 4 describes normalization. Some expression measures, for example AvDiff and Li and Wong's MBEI, are based on $PM - MM$. Other measures, for example Affymetrix's Average Log Ratio, are based on $\log(PM/MM)$. In Sections 3 and 4 we also explore the behavior of these quantities. Section 5 describes four measures of expression. Section 6 assesses the four expression measures in terms of bias, variance, and model fit. Section 7 examines the ability of the four methods at detecting differentially expressed probe sets. Section 8 presents our conclusions.

2. DESCRIPTION OF DATA

To properly compare summary measures of expression in terms of bias, variance, sensitivity, and specificity, data for which we know the 'truth' is required. In this paper we examine three data sets for which assessments can be performed where specific results are expected. Data set A provides probes for which we can assume the measurements are entirely due to non-specific binding. This permits us to study the variance–mean relationship for intensity measures. Data set B provides the results of a spike-in experiment where gene fragments have been added at known concentrations. These data can be used to assess bias, sensitivity and specificity. Data set C provides the results from a study in which samples were hybridized at different dilutions. This permits us to assess bias and variance in a more 'realistic' scenario than with data set B.

Exploration, normalization, and summaries of high density oligonucleotide array probe level data

Data sets B and C are available from the web at http://qolotus02.genelogic.com/datasets.nsf/. In this section we describe them in detail for readers interested in using them. We also explain which specific subsets of the data were used for the analyses presented in this paper.

2.1 *Mouse data set A*

Data set A comes from an experiment where five MG-U74A mouse GeneChip ^{R} arrays were used. These were hybridized with samples of lung tissue mRNA obtained from five mice exposed to different experimental conditions. About 1/5 of the probe pairs in the MG-U74A array were incorrectly sequenced. We therefore assume that the measurements read for most of these probes are entirely due to non-specific binding.

2.2 *Spike-in data sets B*

Data set B consists of experiments where 11 different cRNA fragments were added to the hybridization mixture of the GeneChip ^{R} arrays at different picomolar (pM) concentrations. The 11 control cRNAs were BioB-5, BioB-M, BioB-3, BioC-5, BioC-3, BioDn-5 (all *E. coli*), CreX-5, CreX-3 (phage P1), and DapX-5, DapX-M, DapX-3 (*B. subtilis*) (Hill *et al.*, 2000, 2001; Baugh *et al.*, 2001). The cRNA were chosen to match the target sequence for each of the Affymetrix control probe sets. For example, for DapX (a *B. subtilis* gene), the 5 , middle and 3 target sequences (identified by DapX-5, DapX-M, DapX-3) were each synthesized separately and spiked-in at a specific concentration. Thus, for example, on one of the arrays DapX-3 target sequence was added to the total hybridization solution of 200 μl to give a final concentration of 0.5 pM.

There are two series of spike-in experiments. The experiments were originally carried out for the development of normalization procedures (Hill *et al.*, 2001). In this paper we use the data in a different way, mainly for the comparison of expression measures.

2.2.1 *The varying concentration series data set, B1.* For an individual array, all of the 11 control cRNAs were spiked-in at the same concentration and this concentration was varied across arrays, taking the values 0.0, 0.5, 0.75, 1, 1.5, 2, 3, 5, 12.5, 25, 50, and 150 pM. For example, array 1 had all control cRNAs spiked with 0.0 pM and array 2 had all control cRNAs spiked with 0.5 pM, etc. Of these 12 concentrations, 0, 0.5, 0.75, 1, 1.5, 2, 3 were represented on just one array, 5 and 100 on two arrays, and the rest were in triplicate, i.e. on three arrays for a total of 27 arrays. All arrays have a common background cRNA from an acute myeloid leukemia (AML) tumor cell line. In this paper we use only 12 arrays, one replicate for each of the 12 concentrations. One of the probe set spike-in combinations (CreX-3) failed to respond adequately, and data from that probe set is entirely omitted from the analysis. Thus we analyse data from 10 spiked-in probe-sets.

2.2.2 *Latin square series data set, B2.* In this series each of the 11 control cRNAs were spiked-in at a different concentration on each array (apart from replicates). The 12 concentrations used were 0.5, 1, 1.5, 2, 3, 5, 12.5, 25, 37.5, 50, 75, and 100 pM, and these were arranged in a 12 × 12 cyclic Latin square, with each concentration appearing once in each row and column. The 12 combinations of concentrations used on the arrays were taken from the first 11 entries of the 12 rows of this Latin square. Of the 12 combinations used, 11 were done on three arrays and one on just one array. All of these arrays had the same AML background as in data set B1.

The analysis in this paper makes use of data from six arrays that are a pair of triplicates. The spike-in concentrations for each of the 11 control RNAs on the two sets of triplicates is shown in Table 1.

Table 1. *Concentrations and observed ranks of each spiked-in gene in a comparison of two sets of triplicates from the Latin square series spike-in data set*

	Concentration		Expected Rank	Observed Rank			
Probe set	Set of triplicates 1	set of triplicates 2		AvDiff	MAS 5.0	Li & Wong	RMA
BioB-5	100.0	0.5	1	6	2	1	1
BioB-3	0.5	25.0	2	16	1	3	2
BioC-5	2.0	75.0	4	74	6	2	3
BioB-M	1.0	37.5	4	30	3	7	5
BioDn-3	1.5	50.0	5	44	5	6	4
DapX-3	35.7	3.0	6	239	24	24	7
CreX-3	50.0	5.0	7	333	73	36	9
CreX-5	12.5	2.0	8	3276	33	3128	8
BioC-3	25.0	100.0	9	2709	8579	681	6431
DapX-5	5.0	1.5	10	4598	102	12203	10
DapX-M	3.0	1.0	11	165	19	13	6

Notice that relative concentrations of the spike-ins are three fold or more, which permits us to check the sensitivity of expression indices.

2.3 *Dilution data set C*

Two sources of cRNA, A (human liver tissue) and B (central nervous system cell line), were hybridized to human array (HG-U95A) in a range of proportions and dilutions. In this publication, we study data from arrays hybridized to source A starting with 1.25 μg cRNA, and rising through 2.5, 5.0, 7.5, 10.0 to 20.0 μg. There were five replicate arrays for each tissue: that is, each generated cRNA was hybridized on five HG-U95 GeneChip^{R} arrays. Five scanners were used in this study. Each array replicate was processed in a different scanner.

3. FEATURES OF PROBE LEVEL DATA

Figure 1(a) shows histograms of log ratio, $\log_2(PM/MM)$, stratified by quantiles of abundance, $\log_2 \sqrt{PM \times MM}$, with gray scale representing height of histogram (light is high and dark is low) for one array from data set A. The histograms have been scaled so that the mode of each histogram is represented with the same gray scale. This figure shows that, in general, MM grows with PM. Furthermore, for larger values of abundance the differences have a bimodal distribution with the second mode occurring for negative differences. The same bimodal effect is seen when we stratify by $\log_2(PM)$, thus it is not an artifact of conditioning on sums. In Figures 1(b)–1(e), four histograms with a broader stratification clearly show this effect. The figure also displays (in darker grays) the histograms of the defective probes where the bimodal distribution is also seen. Notice, there are many probe pairs with $MM >> PM$. Finally, notice that for about 1/3 of the probes $MM > PM$. The number of probe pairs within probe sets for which $MM > PM$ varies from 0 to 14. The distribution across probe sets is the following:

# of times $MM > PM$	0	1	2	3	4	5	6	7	8	9	10	11	12	13	14
# of probe sets	7401	481	628	819	1123	1461	1759	1906	1555	1200	760	345	152	50	14

All these effects have been seen in many arrays.

The defective probes are used to assess the variance–mean relationship. Intensities obtained from probe j in arrays $i = 1, \ldots, I$, PM_{ijn}, are expected to have the same mean and variance. If standard deviations (SDs) $\sqrt{\{(I-1)^{-1} \sum_i (PM_{ijn} - \bar{PM}_{\cdot jn})^2\}}$ and averages $\bar{PM}_{\cdot jn} = I^{-1} \sum_i PM_{ijn}$ are computed

Exploration, normalization, and summaries of high density oligonucleotide array probe level data

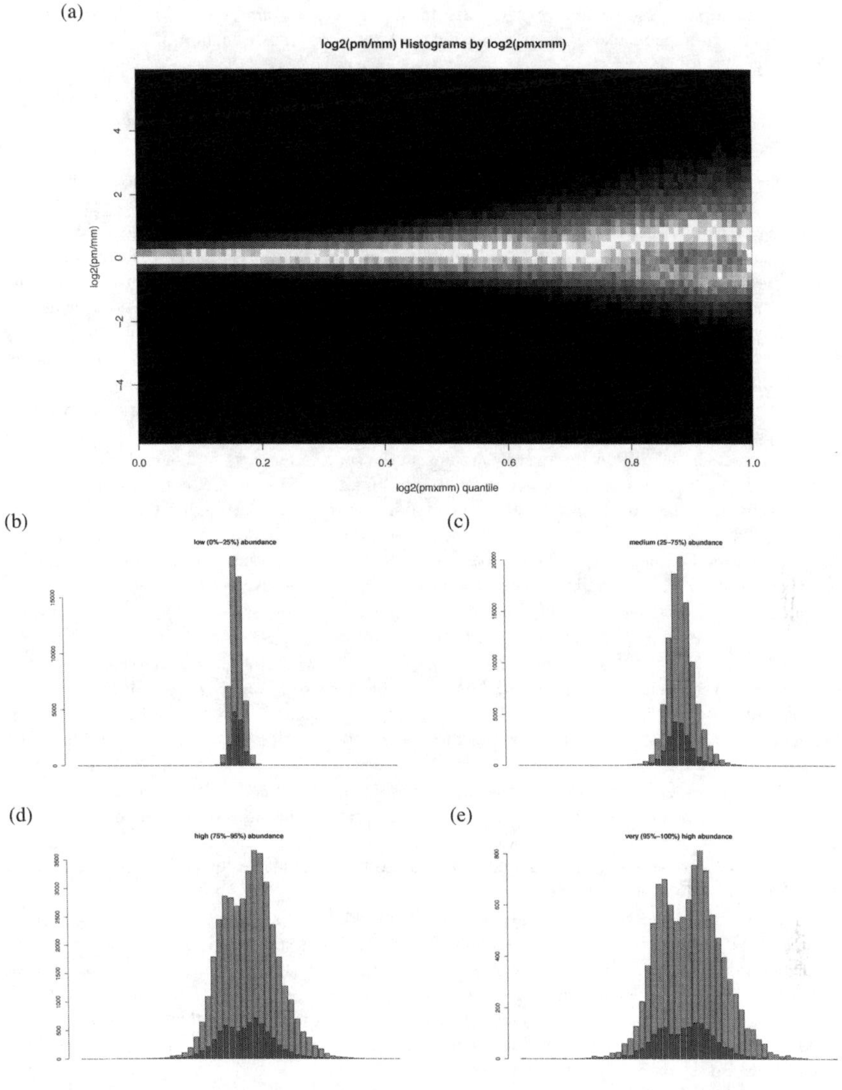

Fig. 1. (a) Histograms of log ratio $\log_2(PM/MM)$, stratified by quantiles of abundance, $\log_2 \sqrt{PM \times MM}$, with gray scale representing height of histogram (light grays are high and dark grays are low) for one array from the mouse data set. The histograms have been scaled so that the mode of each histogram is represented with the same gray scale. (b) Histogram of log ratios for first quartile of abundance with the histogram for the defective probes represented by a darker gray. (c) Like (b) for abundance values between first and third quartile. (d) Like (b) for abundance values in the last quartile excluding the highest 5 percent. (e) Like (b) for the highest 5% of abundance.

for a random sample of 2000 defective probe sets, the SD increases from roughly 50 to 5000, a factor of 100-fold, as the average increases on its entire range. After a log transformation of the PM intensities there is only a 1.5-fold increase.

4. NORMALIZATION

In many of the applications of high-density oligonucleotide arrays, the goal is to learn how RNA populations differ in expression in response to genetic and environmental differences. For example, large expression of a particular gene or genes may cause an illness resulting in variation between diseased and normal tissue. These sources of variation are referred to as *interesting variation*. Observed expression levels also include variation introduced during the sample preparation, manufacture of the arrays, and the processing of the arrays (labeling, hybridization, and scanning). These are referred to as sources of 'obscuring variation'. See (Hartemink *et al.*, 2001) for a more detailed discussion. The obscuring sources of variation can have many different effects on data. Unless arrays are appropriatly *normalized*, comparing data from different arrays can lead to misleading results.

Dudoit *et al.* (2001) describe the need for normalization procedures for cDNA microarray data. Similar issues are present with GeneChip ^{R} arrays. Figures 2(a) and 2(b) show box plots of $\log_2(PM)$ and $PM - MM$ for data set C. The different gray scales represent the six different sets of five replicates processed on scanners 1 to 5. The scanner effect is clearly seen in Figure 2. For example, note that the $\log_2(PM)$ boxplot intensities obtained using scanner/fluidic station 1 were in general higher than those obtained from scanner/fluidic station 5. For the replicate arrays we expect no genes to be differentially expressed. This figure shows direct array to array comparison of PM values warrants normalization. Figure 2(b) boxplot shows that further normalization is needed for the $PM - MM$ as well.

Figures 3(a) and 3(b) show log ratios, $M = \log_2(y/x)$ versus abundance $A = \log_2 \sqrt{x \times y}$, (MVA) plots for $x = PM_1, y = PM_2$ and $x = PM_1 - MM_1, y = PM_2 - MM_2$ for two arrays (denoted with 1 and 2) in which the BioDn-3 gene has been spiked at 5 pM and 2 pM respectively. These plots have been used by, for example, Dudoit *et al.* (2002) to explore intensity related biases. Because the same RNA background was hybridized to arrays 1 and 2, we do not expect any of the non-spiked-in genes to be differentially expressed and therefore these plots to scatter around 0. It is clear from Figure 3 that these data need normalization.

For cDNA arrays the normalization procedure presented in Dudoit *et al.* (2002) has worked well in practice. For each array, a loess curve is fitted to the MVA plot of intensities of the red and green labels and the residuals are considered the normalized log ratios. However, this approach is not appropriate for GeneChip ^{R} arrays because only one sample is hybridized to each array instead of two (red and green). A procedure that normalizes each array against all others is needed.

Various methods have been proposed for normalizing GeneChip ^{R} arrays. Bolstad *et al.* (2002) present a review of these methods and find *quantile normalization* to perform best. The goal of quantile normalization is to make the distribution of probe intensities the same for arrays $i = 1, \ldots, I$. The normalization maps probe level data from all arrays, $i = 1, \ldots, I$, so that an I-dimensional quantile–quantile plot follows the I-dimensional identity line. A possible problem with this approach is that we risk removing some of the signal in the tails. However, empirical evidence suggest this is not a problem in practice: see Bolstad *et al.* (2002) for details.

In Figures 3(c) and 3(d) the MVA plots of the normalized arrays are shown. Notice how the normalization has removed the bias seen in Figures 3(a) and 3(b). The large points represent the 20 spiked-in probes and the small black dots represent a random sample of non-spiked-in probes. Notice that in all plots, normalization helps identify the spiked-in probes as differentially expressed. The benefits of this normalization at the probe level are also seen in Figures 2(c) and 2(d).

Exploration, normalization, and summaries of high density oligonucleotide array probe level data

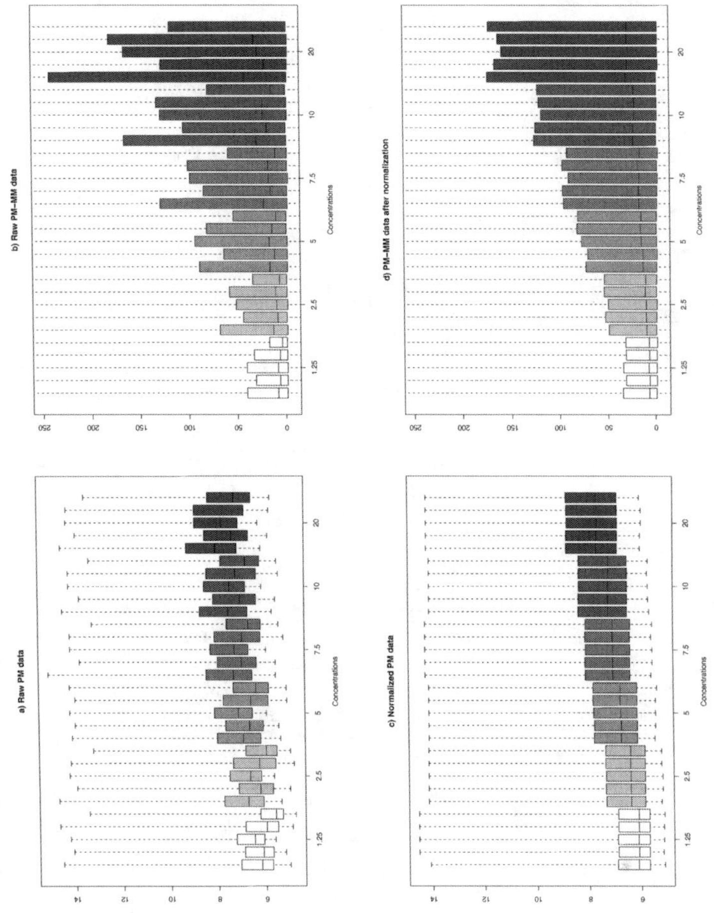

Fig. 2. Boxplots of $\log_2(PM)$ and $PM - MM$ for the 30 arrays from data set C. Because $PM - MM$ values are usually between −2000 and 10 000, a reduced range is used to get a better view of the interquartile range. The bottom row are the after quantile normalization boxplots. The y-axis scale can be deduced from the plot titles.

5. Measures of expression

Various measures of expression have been proposed: for example see Li and Wong (2001), Naef *et al.* (2001), and Holder *et al.* (2001). The most commonly used (at the time this paper was written) is AvDiff,

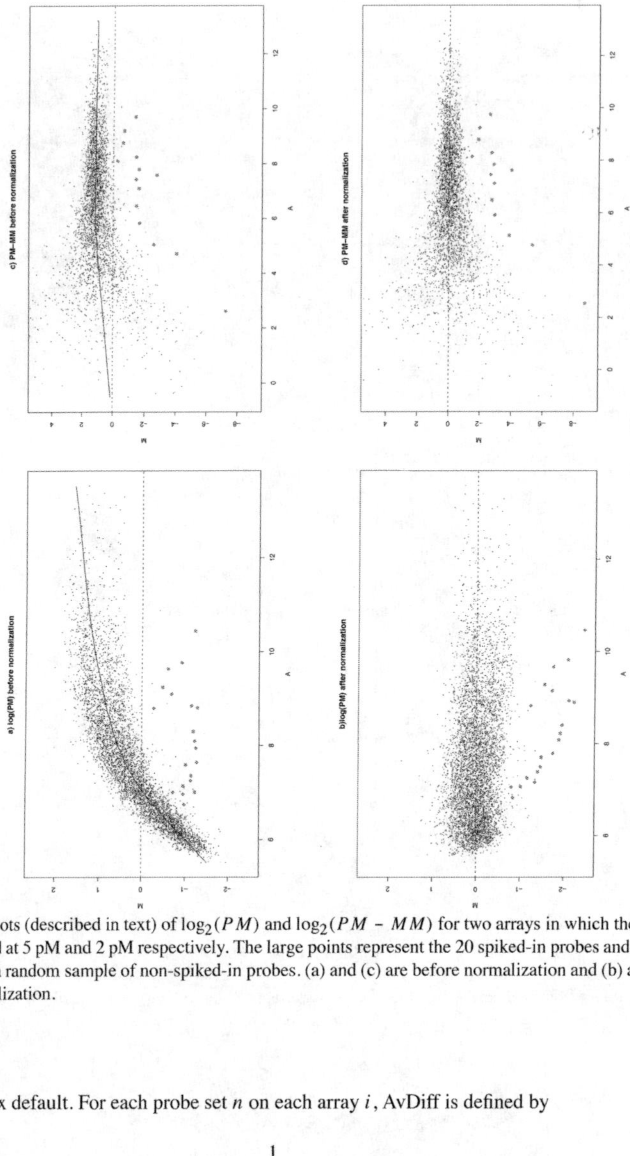

Fig. 3. MVA plots (described in text) of $\log_2(PM)$ and $\log_2(PM - MM)$ for two arrays in which the BioDn-3 gene has been spiked at 5 pM and 2 pM respectively. The large points represent the 20 spiked-in probes and the small black dots represent a random sample of non-spiked-in probes. (a) and (c) are before normalization and (b) and (d) are after quantile normalization.

the Affymetrix default. For each probe set n on each array i, AvDiff is defined by

$$\text{AvDiff} = \frac{1}{\#A} \sum_{j \in A} (PM_j - MM_j)$$

with A the subset of probes for which $d_j = PM_j - MM_j$ are within 3 SDs away from the average of $d_{(2)}, .., d_{(J-1)}$ with $d_{(j)}$ the jth smallest difference. $\#A$ represents the cardinality of A. Many of the other expression measures are versions of AvDiff with different ways of removing outliers and different ways of dealing with small values.

We have observed that linear scale measures, such as AvDiff, are not optimal. Li and Wong (2001) observed this and proposed an alternative model based expression index. For each probe set n, Li and Wong's measure is defined as the maximum likelihood estimates of the $\theta_i, i = 1, \ldots, I$ obtained from fitting

$$PM_{ij} - MM_{ij} = \theta_i \phi_j + \epsilon_{ij} \tag{1}$$

with ϕ_j representing probe-specific affinities and the ϵ_{ijn} are assumed to be independent normally distributed errors. The estimation procedure includes rules for outlier removal.

Affymetrix also appears to have noticed that the linear scale is not appropriate and, in the new version of their analysis algorithm MAS 5.0, are now using a log scale measure. Specifically the MAS 5.0 signal (measure) is defined as

$$\text{signal} = \text{Tukey Biweight}\{\log(PM_j - CT_j)\}$$

with CT_j a quantity derived from the MMs that is never bigger than its PM pair. See Hubbell (2001) for more details.

Each of these measures rely upon the difference $PM - MM$ with the intention of correcting for non-specific binding. However, the exploratory analysis presented in Section 3 suggests that the MM may be detecting signal as well as non-specific binding. Some researchers (Naef *et al.*, 2001) propose expression measures based only on the PM. In Figure 4 we show the PM, MM, PM/MM and $PM - MM$ values for each of the 20 probes representing BioB-5 in the 12 spiked-in arrays, from data set B1, plotted against spike-in concentration. The 20 different probe pairs are represented with different symbols and line types. As expected, the PM values are growing in proportion to the concentration. Notice also that the lines representing the 20 probes are close to being parallel, showing there is a strong additive (in the log scale) probe-specific effect. As evident in Figure 4(c), the additive probe-specific effect is also detected by the MM motivating their subtraction from the PM. However, in Figure 4(d) the parallel lines are still seen in $PM - MM$, demonstrating that subtracting is not enough to remove the probe effect. The fact that parallel lines are not as obvious in Figure 4(c) shows that dividing by MM removes, to some degree, the probe effect. However, the MM also grow with concentrations, because they detect signal as well as non-specific binding, hence the signal in PM/MM is attenuated. Notice, in particular, that PM/MM is unable to distinguish between concentrations of 25 and 150. Since subtracting probe-specific MM adds noise with no obvious gain in bias and because PM/MM results in a biased signal, in this paper we propose an alternative measure to those based on $PM - MM$ or PM/MM.

Figure 4(a) shows that on a log scale (i) the PMs grow roughly linearly with respect to concentrations, (ii) the variances are roughly constant and (iii) the probe-specific affinity is approximately additive. This suggests an additive linear model for the probe set data and the average $J^{-1} \sum_{j=1}^{J} \log(PM_{ij})$ as a log scale measure of expression. However, this measure does not account for non-specific binding. Because, in Figure 4, the log-scale slope of the PM is less than 1, particularly for small concentrations, the PM values should be adjusted to account for non-specific binding. To see this consider a hypothetical case with two arrays where the signal of a probe set is twice as big in one of the arrays, but an additive signal of 100 units occurs due to non-specific binding and/or background noise in both arrays. In this case the observed difference in the signals would be about $\log_2(100 + 2s) - \log_2(100 + s)$ instead of $\log_2(2s) - \log_2(s)$. For small values of s the incorrect difference would be close to 0 instead of 1.

Figure 5 shows histograms of $\log_2(MM)$ for an array in which no probe-set was spiked along with the three arrays in which BioB-5 was spiked-in at concentrations of 0.5, 0.75, and 1 pM (from data set

R. A. IRIZARRY ET AL.

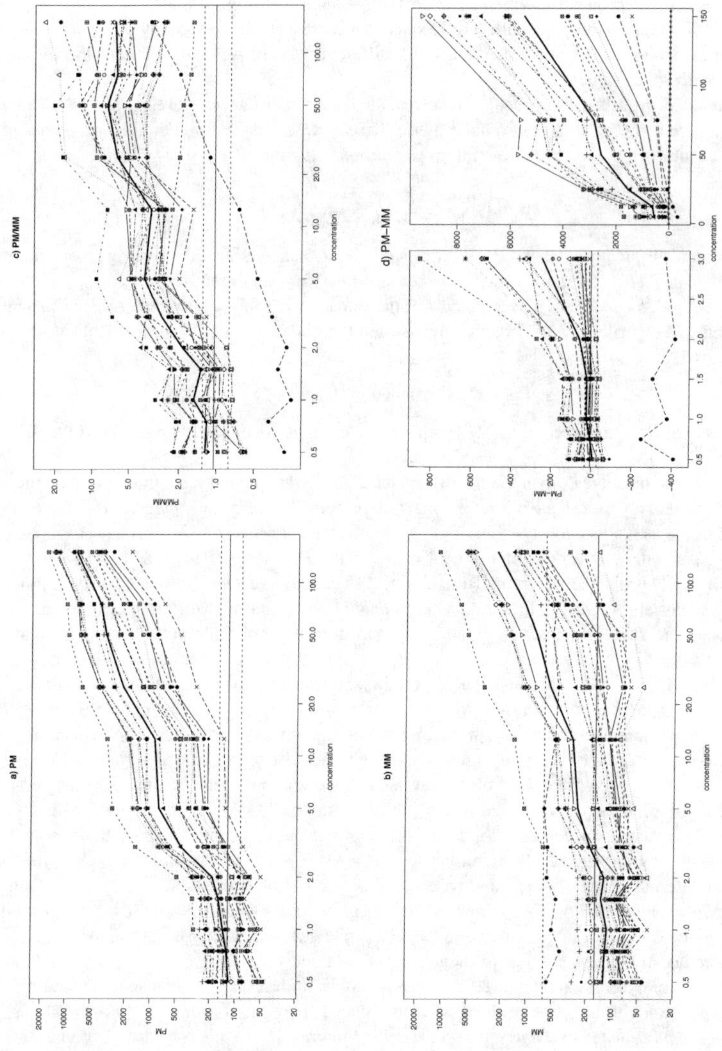

Fig. 4. PM, MM, PM/MM, and $PM - MM$ values for each of the 20 probes representing BioB-5 (with the exception of CreX-3, all other spike-in genes behaved similarly to BioB-5) in the 12 spiked-in arrays from the varying concentration experiment plotted against concentration. The different probes are represented by the different line types and symbols. The horizontal line represents the median of the 20 BioB-5 probes for the non-spiked-in array. The dashed lines are the 25th and 75th quantiles.

Exploration, normalization, and summaries of high density oligonucleotide array probe level data

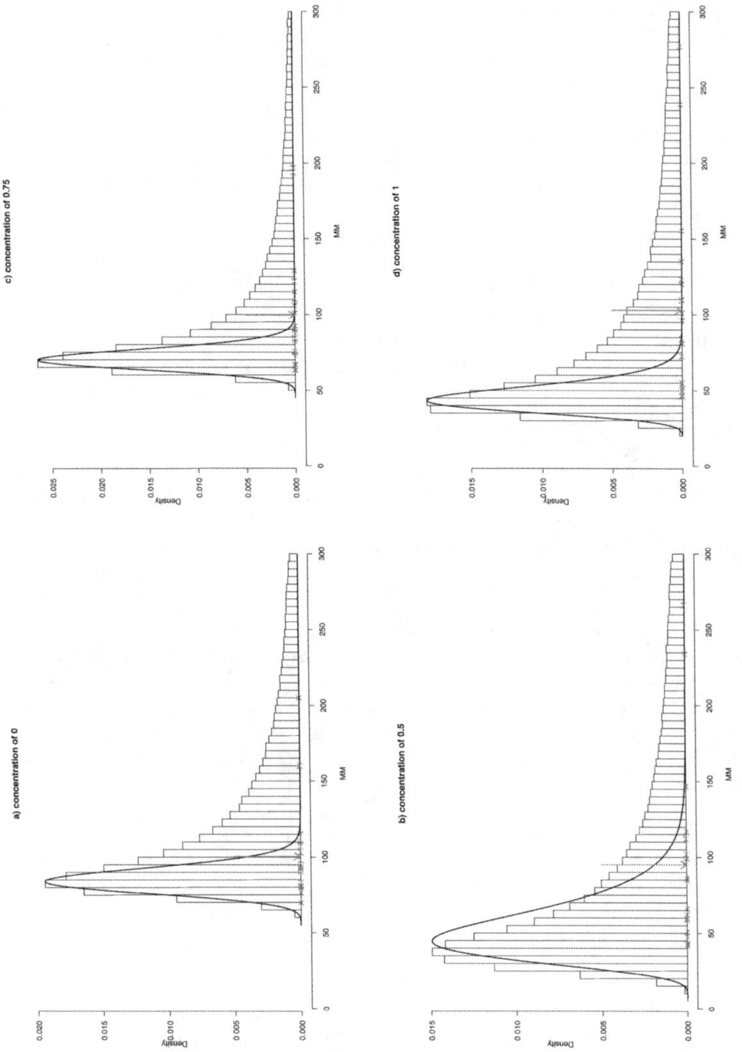

Fig. 5. Histograms of $\log_2(MM)$ for a array in which no probe-set was spiked along with the three arrays in which BioB-5 was spiked-in at concentrations of 0.5, 0.75, and 1 pM. The observed PM values for the 20 probes associated with BioB-5 are marked with crosses and the average with an arrow. The black curve represents the log normal distribution obtained from left-of-the-mode data.

B1). All arrays in all data sets had similar shaped $\log_2(MM)$ histograms. Furthermore, the $\log_2(MM)$ histograms for the spiked-in probe set had similar histograms as well. The MMs to the left of the mode of the histogram can be approximated with the left-hand tail of a log-normal distribution. This suggests that the MMs are a mixture of probes for which (i) the intensities are largely due to non-specific binding and background noise and (ii) the intensities include transcript signal just like the PMs. The mode of the histogram is a natural estimate of the mean background level. The observed PM values for the 20 probes associated with BioB-5 are marked with crosses and the average with an arrow. All the average PM values are close to 100. Thus, judging solely on the average, a difference would be hard to detect. However, distance of the average PM from the average background noise does in fact increase with concentration.

Figure 5 motivates a background plus signal model of the form $PM_{ijn} = bg_{ijn} + s_{ijn}$. Here bg_{ijn} represents background signal in array i caused by optical noise and non-specific binding. We assume each array has a common mean background level, $\mathrm{E}(bg_{ijn}) = \beta_i$. We want to adjust the PM intensities to remove the background effect. A naive approach is to consider $PM_{ijn} - \hat{\beta}_i$, with $\log_2(\hat{\beta}_i)$ the mode of the $\log_2(MM)$ distribution. An estimate of this distribution can be obtained using a density kernel estimate. In practice, a problem with this measure is that for a small percentage of probes $PM_{ijn} \leqslant \hat{\beta}_i$ and log transforming $PM_{ijn} - \hat{\beta}_i$ becomes a problem. An alternative background correction is to consider $B(PM_{ijn}) \quad \mathrm{E}(s_{ijn}|PM_{ijn})$. If we impose a strictly positive distribution on s_{ijn}, then $B(PM_{ijn}) > 0$. To obtain a computationally feasible $B(\cdot)$ we consider the closed-form transformation obtained when assuming s_{ijn} is exponential and bg_{ijn} is normal. Although the data suggest that this model can be improved, the results obtained using $B(\cdot)$ work well in practice, as is demonstrated in the next section.

To obtain an expression measure we assume that for each probe set n, the background-adjusted, normalized, and log-transformed PM intensities, denoted with Y, follow a linear additive model

$$Y_{ijn} = \mu_{in} + \alpha_{jn} + \epsilon_{ijn}, i = 1, \ldots, I, j = 1, \ldots, J, n = 1, \ldots, n \tag{2}$$

with α_j a probe affinity effect, μ_i representing the log scale expression level for array i, and ϵ_{ij} representing an independent identically distributed error term with mean 0. For identifiability of the parameters we assume that $\quad_j \alpha_j = 0$ for all probe sets. This assumption is saying that Affymetrix technology has chosen probes with intensities that on average are representative of the associated genes expression. The estimate of μ_i gives the expression measures for probe set n on array i.

To summarize, in this paper we consider a new expression measure that (i) background-corrects the arrays using the transformation $B(\cdot)$, (ii) normalizes the arrays using quantile normalization, and (iii) for each probe set n, fits a linear model (2) to the background-corrected, normalized and log (base 2) transformed probe intensities denoted here with $Y_{ij}, i = 1, \ldots, I, j = 1, \ldots, J$. To protect against outlier probes we use a robust procedure, such as median polish (Holder *et al.*, 2001), to estimate model parameters. We use the estimate of μ_i as the log scale measure of expression which we refer to as robust multi-array average (RMA).

6. BIAS, VARIANCE, AND GOODNESS OF FIT COMPARISONS

Plots of log observed expression versus known concentration (not shown) demonstrate that the expression measures perform similarly in detecting the spiked-in probe sets. However, for the highest concentration, AvDiff and MBEI sometimes underestimate the predicted value from the known concentrations. This results from the attenuation caused by subtracting MM. We also notice that RMA is less noisy than all other measures at lower concentrations.

It is possible that the control genes used in data set B1 provide a stronger than usual signal. Therefore, a comparison based on all probe sets of the HG-U95A arrays is conducted using data set C. For these data

Exploration, normalization, and summaries of high density oligonucleotide array probe level data

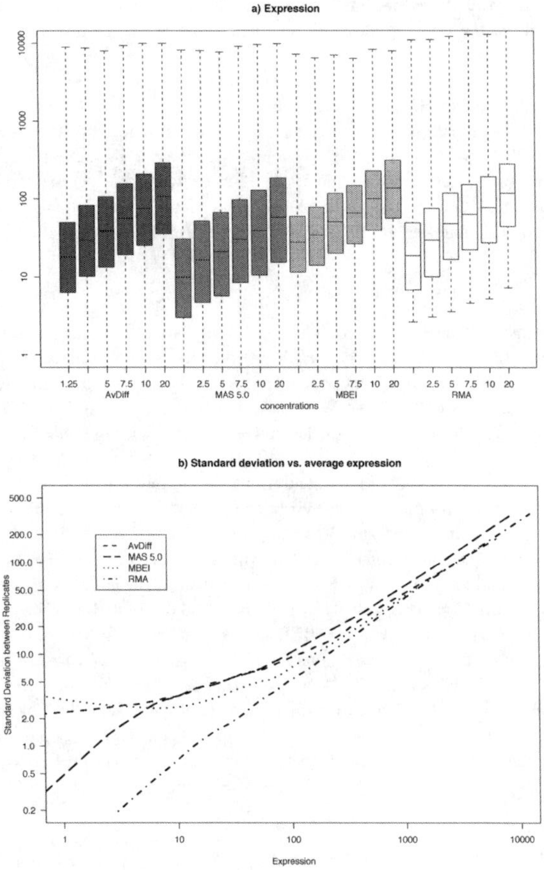

Fig. 6. Data set C boxplots. (a) Averages over replicates for each gene in (b). (b) Loess curves fitted to standard deviation versus average expression scatter-plots.

the amount of hybridization of probe sets representing expressed genes is expected to double when the amount of RNA hybridized to the array is double. Furthermore, the difference in gene expression across replicate arrays should be small.

For each of the four measures, we denote the expression values with $E_{ik}, i = 1, \ldots, 6, k = 1, \ldots, 5$ with i representing the dilution concentration level and k the replicate (which also identifies scanner). The averages are denoted with $E_{i\cdot} = (1/5) \sum_{k=1}^{5} E_{ik}$ and the SDs with $SD_i = \sqrt{(1/4) \sum_{k=1}^{5}(E_{ik} - E_{i\cdot})^2}$. Figure 6(a) shows boxplots of the $E_{i\cdot}$ for each dilution concentration i. Notice that all measures have roughly the same ability to detect signal. Figure 6(b) shows loess curves fitted to the scatter plot (on the log scale) of SD_i vs $E_{i\cdot}$. Clearly, RMA has the smallest SD across replicates. The advantage of RMA is especially noticeable in the low expression values where the SD is 10 times smaller than the other measures.

Li and Wong's method provides not only an estimate of θ_i but a nominal SE for this estimate, denoted here with $\hat{\sigma}_i$. Under (2) one can obtain a naive nominal estimate for the SE of $\hat{\mu}$ using an analysis of variance approach. Because there are five replicates one can also obtain an observed SE of any estimate by simply considering the SD_i defined above. If the model is close to the actual mechanism giving rise to the data, the nominal and observed SE should agree. Plots of nominal to observed SE log ratios versus expression (not shown) show that in general, RMA is closer to 0 than Li and Wong's MBEI showing that the observed and nominal standard error methods are, in general, closer when using (2) instead of (1).

7. Detection of differential expression

Data set B2 was used to assess how well the different expression measures perform at detecting differentially expressed probe sets. For each of the six arrays studied expression measures $E_{11n}, E_{12n}, E_{13n}, E_{21n}, E_{22n}, E_{23n}$ were obtained in their respective scale (log for MAS 5.0 and RMA) for each probe set $n = 1, \ldots, N$. We then computed the averages over triplicates $E_{i\cdot n} = (1/3) \sum_{k=1}^{3} E_{ikn}, i = 1, 2, n = 1, \ldots, N$. For the probe sets representing spike-in RNAs the observed ratios or 'fold changes' ($E_{2\cdot n}/E_{1\cdot n}$ for AvDiff and MBEI or $2^{E_{1\cdot n} - E_{2\cdot n}}$ for MAS 5.0 and RMA) should coincide with the true ratio of the spike-in concentrations shown in Table 1. Recall that apart from the spiked-in probe sets, the background samples hybridized to the six arrays are the same. We therefore expect only the 11 probe sets shown in Table 1 to be differentially expressed. In the left side of Figure 7 MVA plots of the average expressions obtained are shown. Probe sets with negative expression measures were left out for AvDiff and Li and Wong's MBEI. Notice that all measures separate 10 out of the 11 spiked-in probe sets from the cloud of points. However, the cloud of points for probe sets with small total intensity has a much larger spread for AvDiff, MBEI, and MAS 5.0 than for RMA. For this reason, many of the probe sets with high differential expressions for AvDiff, MBEI, and MAS 5.0 are not actually the spiked-in probe-sets. The smaller spread of RMA results in better detection of differentially expressed probe-sets. In the right side of Figure 7, quantile–quantile plots of the observed ratios are shown. RMA is the only measure to perfectly differentiate the spiked-in probe sets (with the exception BioC-3, which no measure was able to detect) from the rest. Table 1 shows the observed rank of the spiked-in probe sets.

8. Conclusion

In this paper we have developed a novel measure of gene expression and compared it to other standard measures. Through the analyses of three data sets, we have shown that expression is better measured using log-transformed PM values, after carrying out a global background adjustment and across-array normalization. We studied the performance of a version of the Affymetrix summary measures AvDiff and MAS 5.0, the Li and Wong model-based expression index, and the new measure RMA. We evaluated the four expression summary measures using spike-in and dilution study data, assessing their behavior in terms of bias, variance, the ability to detect known differential expression levels, and (for MBEI and RMA) model fit. We conclude that there is no obvious downside to summarizing the expression level of a probe set with RMA, and attaching an SE to this quantity using a linear model that removes probe-specific affinities. The greater sensitivity and specificity of RMA in detection of differential expression provides a useful improvement for researchers using the GeneChip R technology. We expect marginal though worthwhile gains to be achievable by using a more carefully designed and tested background correction procedure.

Exploration, normalization, and summaries of high density oligonucleotide array probe level data

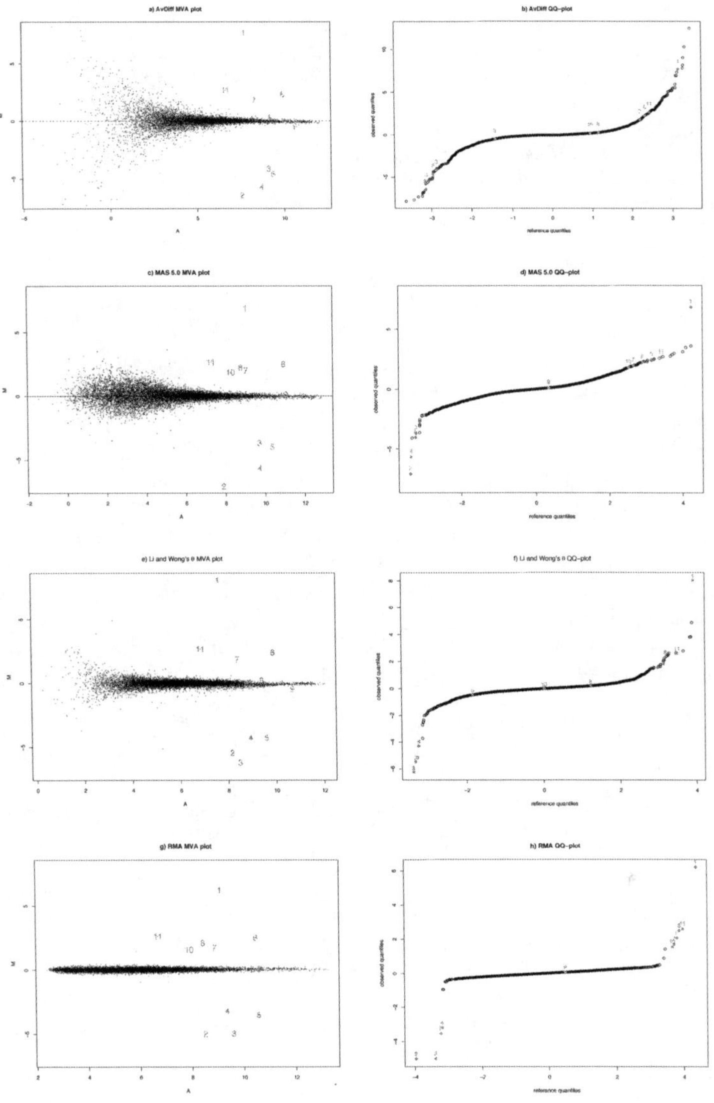

Fig. 7. MVA (described in text) and qq-plots indicating the positions of differentially expressed genes ranked by their absolute log relative expression values.

ACKNOWLEDGEMENTS

The work of Rafael A. Irizarry was supported by the PGA U01 HL66583. We would like to thank Skip Garcia, Tom Cappola and Joshua M. Hare from Johns Hopkins University for the mouse data and Gene Brown's group at Wyeth/Genetics Institute for helpful suggestions in the design of the spike-in experiment. We would like to thank Rehannah Borup and Eric Hoffman from the Children's National Medical Center Microarray Center for help obtaining the raw mouse data. We would like to thank Laurent Gautier from the Technical University of Denmark, Ben Bolstad from UC Berkeley and Magnus strand from Astra Zeneca Mölndal for developing and coding up the normalization routines. Finally, we thank Earl Hubbell (Affymetrix), Cheng Li (Harvard), the Associate Editor, and the referee for suggestions that have improved this paper.

REFERENCES

AFFYMETRIX (1999). *Affymetrix Microarray Suite User Guide, version 4 edition*. Santa Clara, CA: Affymetrix.

BAUGH, L., HILL, A., BROWN, E. AND HUNTER, C. P. (2001). Quantitative analysis of mRNA amplification by *in vitro* transcription. *Nucleic Acids Research* **29**, 1–9.

BOLSTAD, B., IRIZARRY, R., STRAND, M. AND SPEED, T. (2002). A comparison of normalization methods for high density oligonucleotide array data based on variance and bias. *Bioinformatics,* to appear.

DUDOIT, S., YANG, Y. H., CALLOW, M. J. AND SPEED, T. P. (2001). Statistical methods for identifying genes with differentialexpression in replicated cDNA microarray experiments. *Statistica Sinica,* **12**, 111–139.

HARTEMINK, A. J., GIFFORD, D. K., JAAKOLA, T. S. AND YOUNG, R. A. (2001). Maximum likelihood estimation of optimal scaling factors for expression array normalization. *SPIE BiOS*.

HILL, A., HUNTER, C., TSUNG, B., TUCKER-KELLOGG, G. AND BROWN, E. (2000). Genomic analysis of gene expression in c. elegans. *Science* **290**, 809–812.

HILL, A. A., BROWN, E. L., WHITLEY, M. Z., TUCKER-KELLOGG, G., HUNTER, C. P. AND SLONIM, D. K. (2001). Evaluation of normalization procedures for oligonucleotide array data based on spiked cRNA controls. *Genomebiology* **2**, 1–13.

HOLDER, D., RAUBERTAS, R. F., PIKOUNIS, V. B., SVETNIK, V. AND SOPER, K. (2001). Statistical analysis of high density oligonucleotide arrays: a SAFER approach. *Proceedings of the ASA Annual Meeting 2001*. Atlanta, GA.

HUBBELL, E. (2001). Estimating signal with next generation Affymetrix software. *Gene Logic Workshop on Low Level Analysis of AffymetrixGeneChip* ^{R} *data*.
http://www.stat.berkeley.edu/users/terry/zarray/Affy/GL_Workshop/genelogic2001.html.

LI, C. AND WONG, W. (2001). Model-based analysis of oligonucleotide arrays: Expression index computation and outlier detection. *Proceedings of the National Academy of Science U S A* **98**, 31–36.

LOCKHART, D. J., DONG, H., BYRNE, M. C., FOLLETTIE, M. T., GALLO, M. V., CHEE, M. S., MITTMANN, M., WANG, C., KOBAYASHI, M., HORTON, H. AND BROWN, E. L. (1996). Expression monitoring by hybridization to high-density oligonucleotide arrays. *Nature Biotechnology* **14**, 1675–1680.

NAEF, F., LIM, D. A., PATIL, N. AND MAGNASCO, M. O. (2001). From features to expression: High density oligonucleotide array analysis revisited. *Tech Report* **1**, 1–9.

[*Received June 3, 2002; revised July 8, 2002; accepted for publication July 22, 2002*]

The Annals of Statistics
2006, Vol. 34, No. 5, 2387–2412
DOI: 10.1214/009053606000000759

A MULTIVARIATE EMPIRICAL BAYES STATISTIC FOR REPLICATED MICROARRAY TIME COURSE DATA

BY YU CHUAN TAI[1] AND TERENCE P. SPEED[2]

University of California, Berkeley and Walter and Eliza Hall Institute of Medical Research, Australia

In this paper we derive one- and two-sample multivariate empirical Bayes statistics (the *MB*-statistics) to rank genes in order of interest from longitudinal replicated developmental microarray time course experiments. We first use conjugate priors to develop our one-sample multivariate empirical Bayes framework for the null hypothesis that the expected temporal profile stays at 0. This leads to our one-sample *MB*-statistic and a one-sample $\widetilde{T}^2$-statistic, a variant of the one-sample Hotelling T^2-statistic. Both the *MB*-statistic and $\widetilde{T}^2$-statistic can be used to rank genes in the order of evidence of nonzero mean, incorporating the correlation structure across time points, moderation and replication. We also derive the corresponding *MB*-statistics and $\widetilde{T}^2$-statistics for the one-sample problem where the null hypothesis states that the expected temporal profile is constant, and for the two-sample problem where the null hypothesis is that two expected temporal profiles are the same.

1. Introduction. Microarray time course experiments differ from other microarray experiments in that gene expression values at different time points can be correlated. This may happen when the design is longitudinal, that is, where the mRNA samples at successive time points are taken from the same *units*. Such longitudinal experiments make it possible to monitor and study the temporal *changes* within units of biological processes of interest for thousands of genes simultaneously. Two major categories of time course experiments are those involving periodic and developmental phenomena. Periodic time courses typically concern natural biological processes such as cell cycles or circadian rhythms, where the temporal profiles follow regular patterns [7, 8, 25, 26]. On the other hand, developmental time course experiments measure gene expression levels at a series of times in a developmental process, or after applying a treatment such as a drug to the organism, tissue or cells [9, 32, 34]. In this case, we typically have few prior expectations concerning the temporal patterns of gene expression. The gene

Received August 2004; revised August 2005.
[1]Supported in part by NIH Grant HG00047.
[2]Supported in part by NIH Grant R01 Lmo7609-01.
AMS 2000 subject classifications. Primary 62M10; secondary 62C12, 92D10.
Key words and phrases. Microarray time course, longitudinal, multivariate empirical Bayes, moderation, gene ranking, replication.

ranking methods we develop in this paper are mainly for longitudinal replicated developmental time course data.

A typical microarray time course dataset consists of expression measurements of G genes across k time points, under one or more biological conditions (e.g., wildtype versus mutant). The number of genes G (1,000–40,000) is very much larger than the number of time points k, which can be 5–10 for shorter and 11–20 for longer time courses. Many such experiments are unreplicated due to cost or other limitations, and when replicates are done, the number n is typically quite small, say, 2–5. We refer the reader to [29] for a fuller review of microarray time course experiments.

One of the statistical challenges here is to identify genes of interest. In what we call the one-sample problem, these are genes whose patterns of expression change over time, perhaps in some specific way. In the two-sample problem we seek genes whose temporal patterns differ across two biological conditions. Such genes are of interest to biologists because they are often involved in the biological processes motivating the experiment. The challenge arises from the fact that there are very few time points, and very few replicates per gene. The series are usually so short that we cannot consider using standard time series methods as described in [10], such as Fourier analysis, ARMA models or wavelets. The methods proposed in this paper are for the one- and two-sample problems with longitudinal replicated microarray time course experiments of the developmental kind.

The gene ranking problem for such microarray experiments is relatively new. Few methods have been proposed to deal specifically with these problems. The most widely used method for identifying temporally changing genes in replicated microarray experiments is to carry out multiple pairwise comparisons across times, using statistics developed for comparing two independent samples, [2, 6, 13, 19, 20, 23, 24, 33]. These methods are not entirely appropriate as they do not incorporate the fact that longitudinal microarray time course samples are correlated. A simple and intuitive approach to our problem is to use classical or mixed ANOVA models; see Chapter 6 of [11] for a discussion of the latter for analyzing longitudinal data and [22] for a modified approach based on the former for use in the microarray context. However, a number of questions are not adequately addressed by the classical ANOVA methods, or the variants of [22]. As with the pairwise comparisons, the F-statistic assumes that gene expression measurements at different times are independent. The classical ANOVA models also assume normality of the gene expression measurements, which may not be a great concern when these are on the log scale. More importantly, standard F-statistics in this context will generally lead to more false positives and false negatives than is desirable, due to poorly estimated error variances in the denominator. This issue can be dealt with using the idea of moderation; see, for example, [2, 14, 20, 24, 33]. Moderation in our longitudinal context means the smoothing of gene-specific sample variance–covariance matrices toward a common matrix. When we do this, fewer genes which are not differentially expressed over time but have very small replicate variances are falsely identified as being differentially expressed, and fewer genes which are differentially expressed over time but have large replicate variances (e.g., due to outliers) are missed by the F-statistic.

A moderated gene-specific score based on the Wald statistic for the longitudinal one-sample problem was proposed in [15]. However, their method is only applicable to the situation when the number of replicates is greater than the number of time points. In [3] the expression profiles for each gene and each of two biological conditions were represented by continuous curves fitted using B-splines. A global difference between the two continuous curves and an ad hoc likelihood-based p-value was calculated for each gene. B-splines were also used in [17] to identify genes with different temporal profiles in the two-sample case. Recently, B-splines were again adopted by [27] to model the population mean, constructing the F-statistics for both longitudinal and cross-sectional data with one or more biological conditions. A major feature of this paper was a careful treatment of the multiple testing issue in this context. A novel HMM approach which incorporates the dependency in gene expression measurements across times was proposed in [36] for data with two or more biological conditions. This is one example of using HMM to identify differentially expressed genes across at least two biological conditions in this context.

The multivariate empirical Bayes model proposed in this paper was motivated by the analogous univariate model proposed in [20] for identifying differentially expressed genes in two-color comparative microarray experiments, and the more recent extensions by Smyth [24]. The B-statistic in [20] and [24], and the univariate moderates t-statistic $\tilde{t}_g$ in [24], consider just one parameter or contrast at a time in the null hypotheses. They are not for null hypotheses with two or more parameters or contrasts of interest simultaneously. However, a partly-moderated F-statistic was introduced in [24], which moderates the error variance in the denominator of the ordinary F. This partly-moderated F-statistic is useful for the simultaneous comparison of multiple uncorrelated contrasts, but as mentioned above, it is not appropriate for longitudinal experiments. Both the MB-statistics and the $\widetilde{T}^2$-statistics derived in this paper provide a degree of moderation, while retaining the temporal correlation structure. They can be used with both single- and two-channel microarray experiments.

This paper is organized as follows. After briefly introducing our notation, we formally state the null and alternative hypotheses for the gene ranking problem in Section 2. In Section 3 we present moderated versions of the standard likelihood-ratio and Hotelling T^2-statistics. We formally build up our multivariate empirical Bayes model and derive the MB- and $\widetilde{T}^2$-statistics in Section 4. A brief description of a case study is presented in Section 5. Section 6 reports results from a simulation study in which we compare the one-sample MB-statistic (the $\widetilde{T}^2$-statistic) of Section 4.3 with other statistics. We discuss our models and give directions for future work in Section 7.

Y. C. TAI AND T. P. SPEED

Now we introduce some notation for one-sample problems. For two-sample problems, the notation is similar and easily conceived. For each gene g, $g = 1, \dots, G$, we have n_g independent time series, and we model these as i.i.d. $k \times 1$ random vectors $\mathbf{X}_{g1}, \dots, \mathbf{X}_{gn_g}$ with gene-specific means $\boldsymbol{\mu}_g$ and covariance matrices $\boldsymbol{\Sigma}_g$. Since relative or absolute gene expression measurements are approximately normal on the log scale, we make the multivariate normality assumption on data $\mathbf{X}_{g1}, \dots, \mathbf{X}_{gn_g}$. Our results are to be judged on their practical usefulness, not on the precise fit of our data to a multivariate normal distribution. As will be seen shortly, our final formulae involve the multivariate t distribution. Thus, a measure of robustness is built in, and our approach will probably be about as effective for elliptically distributed random vectors. We use the natural conjugate priors for $\boldsymbol{\mu}_g$ and $\boldsymbol{\Sigma}_g$, that is, an inverse Wishart prior for $\boldsymbol{\Sigma}_g$ and a dependent multivariate normal prior for $\boldsymbol{\mu}_g$. To simplify the notation, the subscript g will be dropped for the rest of this paper. The statistical models presented in the rest of this paper are for an arbitrary single gene g.

The details in this paper differ in two ways from the standard conjugate priors. First, we also have an indicator I such that $I = 1$ when the alternative K is true and $I = 0$ when the null H is true, with the priors for $\boldsymbol{\mu}$ differing in these two cases. Second, when the null hypothesis states that a gene's mean expression level is constant, in order to get a simple closed form expression for the posterior odds, we assume that the gene-specific covariance matrix $\boldsymbol{\Sigma}$ commutes with the $k \times k$ projection matrix $\mathbf{P} = k^{-1}\mathbf{1}_k\mathbf{1}_k'$, that is, $\mathbf{P}\boldsymbol{\Sigma} = \boldsymbol{\Sigma}\mathbf{P}$. In this case the $k \times k$ inverse Wishart prior for $\boldsymbol{\Sigma}$ is replaced by a $(k-1) \times (k-1)$ inverse Wishart prior for a part of $\boldsymbol{\Sigma}$ and an inverse gamma prior for the remainder. These two-part priors are independent; see Section 4.3 for details.

2. Hypothesis testing. Our gene ranking problem will be formally stated as a hypothesis testing problem. In this paper we only seek a statistic for ranking genes in the order of evidence against the null hypothesis; we do not hope to obtain raw or adjusted p-values as in [27].

Following the notation in [4], the null hypothesis is denoted by H, while the alternative hypothesis is denoted by K. The null hypothesis corresponding to a gene's mean expression level being 0 is $H : \boldsymbol{\mu} = \mathbf{0}$, $\boldsymbol{\Sigma} > 0$; the alternative is $K : \boldsymbol{\mu} \neq \mathbf{0}$, $\boldsymbol{\Sigma} > 0$. An easy extension is $H : \boldsymbol{\mu} = \boldsymbol{\mu}_0$, $\boldsymbol{\Sigma} > 0$ versus $K : \boldsymbol{\mu} \neq \boldsymbol{\mu}_0$, $\boldsymbol{\Sigma} > 0$, where $\boldsymbol{\mu}_0$ is known. Later we will consider the null hypothesis that a gene's expression is constant, against the alternative that it is not: $H : \boldsymbol{\mu} = \mu_0\mathbf{1}$, $\boldsymbol{\Sigma} > 0$, where μ_0 is a scalar representing the expected value of the gene's expression level at any time point under H, and $\mathbf{1}$ is the $k \times 1$ constant vector of $1s$; $K : \boldsymbol{\mu} \neq \mu_0\mathbf{1}$, $\boldsymbol{\Sigma} > 0$. Finally, we will consider the null hypothesis that a gene's mean expression levels are the same under two different biological conditions, versus the alternative that they are not: $H : \boldsymbol{\mu}_Z = \boldsymbol{\mu}_Y$, $\boldsymbol{\Sigma}_Z = \boldsymbol{\Sigma}_Y = \boldsymbol{\Sigma} > 0$; $K : \boldsymbol{\mu}_Z \neq \boldsymbol{\mu}_Y$, $\boldsymbol{\Sigma}_Z = \boldsymbol{\Sigma}_Y = \boldsymbol{\Sigma} > 0$.

REPLICATED MICROARRAY TIME COURSE DATA

3. The moderated *LR*-statistic.

3.1. *One-sample or paired two-sample problem.* A likelihood-ratio statistic can be used directly to test the null hypothesis H against the alternative hypothesis K when $n > k$. According to standard multivariate results (e.g., [21]), under the alternative hypothesis that there are no constraints on $\boldsymbol{\mu}$ and $\boldsymbol{\Sigma}$, their maximum likelihood estimates are

$$\hat{\boldsymbol{\mu}}_K = \overline{\mathbf{X}},$$
$$\widehat{\boldsymbol{\Sigma}}_K = \frac{n-1}{n}\mathbf{S},$$

where $\mathbf{S} = (n-1)^{-1}\sum_{i=1}^{n}(\mathbf{X}_i - \overline{\mathbf{X}})(\mathbf{X}_i - \overline{\mathbf{X}})'$ is the sample variance–covariance matrix. Also as in [21], the maximum likelihood estimate for the unconstrained $\boldsymbol{\Sigma}$ under the null that $\boldsymbol{\mu} = \boldsymbol{\mu}_0$ is

$$\widehat{\boldsymbol{\Sigma}}_H = \frac{n-1}{n}\mathbf{S} + \mathbf{d}\mathbf{d}',$$

where $\mathbf{d} = \hat{\boldsymbol{\mu}}_K - \hat{\boldsymbol{\mu}}_H$, the difference between the maximum likelihood estimates for $\boldsymbol{\mu}_0$ (if unknown) under H and K. If $\boldsymbol{\mu}_0$ is known, then $\mathbf{d} = \hat{\boldsymbol{\mu}}_K - \boldsymbol{\mu}_0$. The likelihood ratio statistic for testing any such H against the above K is

$$(3.1)\qquad \begin{aligned} LR &= 2(l_K^{\max} - l_H^{\max}) \\ &= n\log\left(1 + \frac{n}{n-1}\mathbf{d}'\mathbf{S}^{-1}\mathbf{d}\right). \end{aligned}$$

If the null hypothesis states that $\boldsymbol{\mu}_0 = \mathbf{0}$, then $\mathbf{d}$ reduces to $\overline{\mathbf{X}}$, hence, the likelihood-ratio statistic is equation (3.1) with $\mathbf{d}$ replaced by $\overline{\mathbf{X}}$. Similarly, if $\boldsymbol{\mu}_0$ is known, then $\mathbf{d} = \overline{\mathbf{X}} - \boldsymbol{\mu}_0$. On the other hand, if the null hypothesis states that $\boldsymbol{\mu}_0 = \mu_0\mathbf{1}$, then the maximum likelihood estimate for $\boldsymbol{\mu}$ is

$$\hat{\boldsymbol{\mu}}_H = \left(\frac{\mathbf{1}'\mathbf{S}^{-1}\overline{\mathbf{X}}}{\mathbf{1}'\mathbf{S}^{-1}\mathbf{1}}\right)\mathbf{1}.$$

The statistic $n\mathbf{d}'\mathbf{S}^{-1}\mathbf{d}$ is the one-sample Hotelling T^2-statistic, and by Section 5.3.1b in [21], it is distributed as a Hotelling $T^2(k-1, n-1)$ under H, that is, $((n-k+1)n\mathbf{d}'\mathbf{S}^{-1}\mathbf{d})/((n-1)(k-1))$ has an F-distribution with degrees of freedom $(k-1, n-k+1)$.

In the microarray time course context, the number of replicates n is typically smaller than the number of time points k, and so $\mathbf{S}$ has less than full rank. Furthermore, as discussed in [29], we wish to moderate the sample variance–covariance matrix. Our moderated $\mathbf{S}$ will take the form

$$\widetilde{\mathbf{S}} = \frac{\nu\mathbf{\Lambda} + (n-1)\mathbf{S}}{\nu + n - 1},$$

where $\nu > 0$ controls the degree of moderation, and $\mathbf{\Lambda} > 0$ is the common $k \times k$ matrix toward which $\mathbf{S}$ is smoothed. In Section 4.1 we give the theoretical reason for choosing this moderated variance–covariance matrix $\widetilde{\mathbf{S}}$ and explain how we estimate ν and $\mathbf{\Lambda}$. Replacing $\mathbf{S}$ with $\widetilde{\mathbf{S}}$ in the *LR*-statistic, our moderated *LR*-statistic is

$$\widetilde{LR} = 2(\boldsymbol{l}_K^{\max} - \boldsymbol{l}_H^{\max}) = n \log\left(1 + \frac{n}{n-1}\tilde{\mathbf{d}}'\widetilde{\mathbf{S}}^{-1}\tilde{\mathbf{d}}\right). \tag{3.2}$$

When all the genes have an equal number of replicates n, equation (3.2) is a monotonic increasing function of $n\tilde{\mathbf{d}}'\widetilde{\mathbf{S}}^{-1}\tilde{\mathbf{d}}$. We define the quadratic form $n\tilde{\mathbf{d}}'\widetilde{\mathbf{S}}^{-1}\tilde{\mathbf{d}} = \|n^{1/2}\widetilde{\mathbf{S}}^{-1/2}\tilde{\mathbf{d}}\|^2$ to be the moderated one-sample Hotelling T^2-statistic. In the case of the null $H: \boldsymbol{\mu} = \mathbf{0}, \mathbf{\Sigma} > 0$, this is identical to the $\widetilde{T}^2$-statistic we derive in Section 4.1. The one-sample moderated *LR*-statistic and the moderated Hotelling T^2-statistic are hybrids of likelihood and Bayesian statistics, since $\widetilde{\mathbf{S}}$ is estimated using the multivariate empirical Bayes procedure we describe below.

3.2. *Unpaired two-sample problem.* Similarly, in the unpaired two-sample case, the moderated *LR*-statistic can be written as a function of the moderated two-sample Hotelling T^2-statistic

$$\widetilde{LR} = (m+n)\log\left(1 + \frac{1}{m+n-2}\frac{mn}{m+n}\mathbf{d}'\widetilde{\mathbf{S}}^{-1}\mathbf{d}\right),$$

where $\mathbf{d} = \overline{\mathbf{Z}} - \overline{\mathbf{Y}}$ is the difference between sample averages and $\mathbf{S} = (m+n-2)^{-1}((m-1)\mathbf{S}_Z + (n-1)\mathbf{S}_Y)$ is the pooled sample variance–covariance matrix, and

$$\widetilde{\mathbf{S}} = \frac{(m+n-2)\mathbf{S} + \nu\mathbf{\Lambda}}{m+n-2+\nu}.$$

The term $(m+n)^{-1}mn\mathbf{d}'\widetilde{\mathbf{S}}^{-1}\mathbf{d}$ is our moderated two-sample Hotelling T^2-statistic. We use the same approach to estimate $\widetilde{\mathbf{S}}$ here as that for our two-sample multivariate empirical Bayes model described in Section 4.2.

4. The multivariate empirical Bayes model.

4.1. *One-sample or paired two-sample problem.*

4.1.1. *Models and priors.* The data $\mathbf{X}_1, \ldots, \mathbf{X}_n$ are multivariate normal with mean $\boldsymbol{\mu}$ and covariance matrix $\mathbf{\Sigma}$, denoted by $N_k(\boldsymbol{\mu}, \mathbf{\Sigma})$. We define an indicator random variable I to reflect the status of the gene,

$$I = \begin{cases} 1, & \text{if } K \text{ is true,} \\ 0, & \text{if } H \text{ is true.} \end{cases}$$

We suppose that I has a Bernoulli distribution with success probability p, $0 < p < 1$. Now we build up our multivariate hierarchical Bayesian model by first assigning independent and identical inverse Wishart priors to the gene-specific covariance matrices $\mathbf{\Sigma}$:

$$\mathbf{\Sigma} \sim \text{Inv-Wishart}_\nu((\nu\mathbf{\Lambda})^{-1}), \tag{4.1}$$

where $\nu > 0$ and $\nu\mathbf{\Lambda} > 0$ are the degrees of freedom and scale matrix, respectively. Given $\mathbf{\Sigma}$, we assign multivariate normal priors for the gene-specific mean $\boldsymbol{\mu}$ for the two cases $(I = 1)$ and $(I = 0)$:

$$\boldsymbol{\mu}|\mathbf{\Sigma}, I = 1 \sim N_k(\mathbf{0}, \eta^{-1}\mathbf{\Sigma}),$$
$$\boldsymbol{\mu}|\mathbf{\Sigma}, I = 0 \equiv 0,$$

where $\eta > 0$ is a scale parameter.

The posterior odds are the probability that the expected time course $\boldsymbol{\mu}$ does not stay at 0 (i.e., $I = 1$) over the probability that $\boldsymbol{\mu}$ stays at 0 (i.e., $I = 0$), given the data $\mathbf{X}_1, \ldots, \mathbf{X}_n$. Following [24]'s notation, we write

$$\begin{aligned} \mathbf{O} &= \frac{P(I = 1|\mathbf{X}_1, \ldots, \mathbf{X}_n)}{P(I = 0|\mathbf{X}_1, \ldots, \mathbf{X}_n)} \\ &= \frac{p}{1-p} \frac{P(\mathbf{X}_1, \ldots, \mathbf{X}_n|I = 1)}{P(\mathbf{X}_1, \ldots, \mathbf{X}_n|I = 0)}. \end{aligned} \tag{4.2}$$

The distribution of the data given I can be written as

$$P(\mathbf{X}_1, \ldots, \mathbf{X}_n|I) = \int P(\overline{\mathbf{X}}|\mathbf{\Sigma}, I) P(\mathbf{S}|\mathbf{\Sigma}, I) P(\mathbf{\Sigma}|I)\, d\mathbf{\Sigma}. \tag{4.3}$$

4.1.2. *Multivariate joint distributions.* Once the priors and the models are set, the joint distributions of the data can be determined given I. We omit the standard calculations leading to

$$\begin{aligned} &P(\mathbf{X}_1, \ldots, \mathbf{X}_n|I = 1) \\ &\quad = \frac{\mathbf{\Gamma}_k((n+\nu)/2)}{\mathbf{\Gamma}_k((n-1)/2)\mathbf{\Gamma}_k(\nu/2)} \\ &\qquad \times (n-1)^{(1/2)k(n-1)} \nu^{-(1/2)kn} \pi^{-(1/2)k} (n^{-1} + \eta^{-1})^{-(1/2)k} \\ &\qquad \times \frac{|\mathbf{\Lambda}|^{-(1/2)n} |\mathbf{S}|^{(1/2)(n-k-2)}}{|\mathbf{I}_k + ((n^{-1} + \eta^{-1})\nu\mathbf{\Lambda})^{-1}\overline{\mathbf{X}}\overline{\mathbf{X}}' + (\nu\mathbf{\Lambda}/(n-1))^{-1}\mathbf{S}|^{(1/2)(n+\nu)}}. \end{aligned} \tag{4.4}$$

Thus, given $I = 1$, the probability density function of the data is a function of $\overline{\mathbf{X}}$ and $\mathbf{S}$ only, which follows a Student–Siegel distribution [1]. Following [1]'s notation, this distribution is denoted by $StSi_k(\nu, \mathbf{0}, (n^{-1} + \eta^{-1})\mathbf{\Lambda}, n-1, (n-1)^{-1}\nu\mathbf{\Lambda})$. For the case of $I = 0$, we get the same distribution with different parameters, namely, $StSi_k(\nu, \mathbf{0}, n^{-1}\mathbf{\Lambda}, n-1, (n-1)^{-1}\nu\mathbf{\Lambda})$.

4.1.3. *MB-statistic and $\widetilde{T}^2$-statistic.* Define our moderated gene-specific sample variance–covariance matrix $\widetilde{\mathbf{S}}$ to be the inverse of the posterior mean of $\mathbf{\Sigma}^{-1}$ given $\mathbf{S}$,

$$\widetilde{\mathbf{S}} = [E(\mathbf{\Sigma}^{-1}|\mathbf{S})]^{-1} = \frac{(n-1)\mathbf{S} + \nu\mathbf{\Lambda}}{n-1+\nu}. \tag{4.5}$$

The posterior odds $\mathbf{O}$ we defined earlier can be derived using the distributions of the data given I and is

$$\begin{aligned}\mathbf{O} &= \frac{p}{1-p}\left(\frac{\eta}{n+\eta}\right)^{(1/2)k} \\ &\quad\times \left(\frac{n-1+\nu+\widetilde{T}^2}{n-1+\nu+(\eta/(n+\eta))\widetilde{T}^2}\right)^{(1/2)(n+\nu)},\end{aligned} \tag{4.6}$$

where $\widetilde{T}^2 = \tilde{\mathbf{t}}'\tilde{\mathbf{t}}$ and $\tilde{\mathbf{t}}$ is the moderated multivariate t-statistic defined by

$$\tilde{\mathbf{t}} = n^{1/2}\widetilde{\mathbf{S}}^{-1/2}\overline{\mathbf{X}}. \tag{4.7}$$

Following the tradition in genetics, the log base 10 of $\mathbf{O}$ is called the LOD score. To distinguish it from the LOD score (also called the B-statistic) in the univariate model of [20] and [24], the multivariate LOD score in this paper is called the MB-statistic,

$$MB = \log_{10}\mathbf{O}. \tag{4.8}$$

When all genes have the same number of replicates n, equation (4.8) is a monotonic increasing function of $\widetilde{T}^2$. This shows that the MB-statistic is equivalent to the $\widetilde{T}^2$-statistic when n is the same across genes, and therefore, one is encouraged to use the $\widetilde{T}^2$-statistic in this case since it does not require the estimation of η and leads to the same rankings as equation (4.8). We now derive the distribution for $\widetilde{T}^2$.

By Gupta and Nagar [16], the Jacobian transformation from $\overline{\mathbf{X}}$ to $\tilde{\mathbf{t}}$ is $J(\overline{\mathbf{X}} \to \tilde{\mathbf{t}}) = |n^{-1/2}\widetilde{\mathbf{S}}^{1/2}|$. Since equation (4.4) is a function of $\overline{\mathbf{X}}$ and $\mathbf{S}$ only, it is the joint probability density function for these two random variables. Substituting for $\overline{\mathbf{X}}$ in terms of $\tilde{\mathbf{t}}$ in equation (4.4), and multiplying the resulting expression by $J(\overline{\mathbf{X}} \to \tilde{\mathbf{t}})$, the joint probability density function for $\tilde{\mathbf{t}}$ and $\mathbf{S}$ given $I = 1$ is

$$\begin{aligned} & P(\tilde{\mathbf{t}}, \mathbf{S}|I = 1) \\ & = \pi^{-(1/2)k} \frac{\Gamma((n+\nu)/2)}{\Gamma((n+\nu-k)/2)} \left(\frac{n+\eta}{\eta}\right)^{-(1/2)k} (n-1+\nu)^{-(1/2)k} \\ & \times \left(1 + \frac{1}{n-1+\nu}\left(\frac{\eta}{n+\eta}\right)\tilde{\mathbf{t}}'\tilde{\mathbf{t}}\right)^{-(1/2)(n+\nu)} \\ & \times \frac{1}{\boldsymbol{\beta}_k((n-1)/2, \nu/2)} \\ & \times \frac{|\mathbf{S}|^{(1/2)(n-k-2)}}{|\nu\boldsymbol{\Lambda}/(n-1)|^{(1/2)(n-1)}|\mathbf{I}_k + (\nu\boldsymbol{\Lambda}/(n-1))^{-1}\mathbf{S}|^{(1/2)(n+\nu-1)}}. \end{aligned} \tag{4.9}$$

The above expression is factorized into parts involving $\mathbf{S}$ only and $\tilde{\mathbf{t}}$ only, proving that $\tilde{\mathbf{t}}$ and $\mathbf{S}$ are independent. It is apparent that $\tilde{\mathbf{t}}$ has a multivariate t distribution with degrees of freedom $n + \nu - k$, scale parameter $n + \nu - 1$, mean vector $\mathbf{0}$ and covariance matrix $\eta^{-1}(n + \eta)\mathbf{I}_k$. This distribution is denoted by $\tilde{\mathbf{t}}|I = 1 \sim \mathbf{t}_k(n + \nu - k, n + \nu - 1, \mathbf{0}, \eta^{-1}(n + \eta)\mathbf{I}_k)$ [16]. It is straightforward to see that $\tilde{\mathbf{t}}|I = 0 \sim \mathbf{t}_k(n + \nu - k, n + \nu - 1, \mathbf{0}, \mathbf{I}_k)$. Given $I = 1$, $\mathbf{S}$ is distributed as a generalized type-II beta distribution with parameters $(n - 1)/2$, $\nu/2$, scale matrix $\nu\boldsymbol{\Lambda}/(n - 1)$ and location matrix $\mathbf{0}$. The distribution is denoted by $GB_k^{II}((n - 1)/2, \nu/2, \nu\boldsymbol{\Lambda}/(n - 1), \mathbf{0})$ [16]. The marginal distribution of $\mathbf{S}$ does not depend on I so that $P(\mathbf{S}|I = 0) = P(\mathbf{S}|I = 1)$. This distributional result is used to estimate the hyperparameter $\boldsymbol{\Lambda}$. The distribution for $\widetilde{T}^2$ under the null follows immediately. Under H, $k^{-1}\widetilde{T}^2$ has an F distribution with degrees of freedom $(k, n + \nu - k)$; equivalently, $(n + \nu - k)^{-1}(n + \nu - 1)\widetilde{T}^2$ has the Hotelling T^2-distribution $T^2(k, n + \nu - 1)$.

The $\widetilde{T}^2$-statistic is identical to the one-sample moderated Hotelling T^2-statistic in Section 3.1 with the same null hypothesis.

For the easy extension to the above model, $H : \boldsymbol{\mu} = \boldsymbol{\mu}_0, \boldsymbol{\Sigma} > 0$ and $K : \boldsymbol{\mu} \neq \boldsymbol{\mu}_0, \boldsymbol{\Sigma} > 0$, where $\boldsymbol{\mu}_0$ is known, all the results above hold with $\overline{\mathbf{X}}$ replaced by $\overline{\mathbf{X}} - \boldsymbol{\mu}_0$.

4.1.4. *Special cases.*

1. $\boldsymbol{\Sigma} = \sigma^2\mathbf{I}_k$. By constraining $\boldsymbol{\Sigma} = \sigma^2\mathbf{I}_k$, we ignore the correlations among gene expression values at different times, and assume the variances at different times are equal. Suppose that the prior for σ^2 is

$$\sigma^2 \sim \text{inv-gamma}(\tfrac{1}{2}\nu, \tfrac{1}{2}\nu\lambda^2).$$

Define

$$s_j^2 = (n-1)^{-1}\sum_{i=1}^{n}(\mathbf{X}_{ij} - \overline{\mathbf{X}}_j)^2,$$

Y. C. TAI AND T. P. SPEED

$$\tilde{s}_j^2 = (n-1+\nu)^{-1}((n-1)s_j^2 + \nu\lambda^2)$$

and

$$\tilde{t}_j = n^{1/2}\overline{\mathbf{X}}_j \tilde{s}_j^{-1}, \qquad j = 1, \ldots, k.$$

In this case, the posterior odds are equivalent to a product of k independent univariate odds,

$$\text{(4.10)} \quad \mathbf{O} = \frac{p}{1-p}\left(\frac{\eta}{n+\eta}\right)^{(1/2)k} \prod_{j=1}^{k}\left(\frac{n-1+\nu+\tilde{t}_j^{\,2}}{n-1+\nu+(\eta/(n+\eta))\tilde{t}_j^{\,2}}\right)^{(1/2)(n+\nu)},$$

and the MB-statistic is equivalent to the sum of k univariate B-statistics.

2. $n = 1$. When $n = 1$, that is, when there is no replication at all, each gene has its own unknown variability. The moderated multivariate t-statistic becomes $\tilde{\mathbf{t}} = \mathbf{\Lambda}^{-1/2}\mathbf{X}$. The posterior odds are obtained by plugging in $n = 1$ in equation (4.6), and are found to be a function of $\mathbf{X}$ only. Since there is no replication, our hyperparameters must be assigned values, for example, from previous experiments.

3. $k = 1$. When $k = 1$, that is, when there is only one time point, the alternative hypothesis states that there is differential expression at this single time point. Our multivariate model should and does reduce to the univariate model in [20] and [24].

4.1.5. *Limiting cases.*

1. $\nu \to \infty$. In this case, the gene-specific variance–covariance matrices are totally ignored. The moderated multivariate t-statistic above becomes $\tilde{\mathbf{t}}_\infty = n^{1/2}\mathbf{\Lambda}^{-1/2}\overline{\mathbf{X}}$, and $\tilde{T}_\infty^2 = \tilde{\mathbf{t}}'_\infty \tilde{\mathbf{t}}_\infty$. The posterior odds become

$$\mathbf{O} = \frac{p}{1-p}\left(\frac{\eta}{n+\eta}\right)^{(1/2)k} \exp\left(\frac{1}{2}\left(\frac{n}{n+\eta}\right)\tilde{T}_\infty^2\right).$$

2. $\nu \to 0$. In this case, there is no moderation at all. The posterior odds are just equation (4.6) with ν replaced by 0. If $n < k$, then $\mathbf{S}^{-1/2}$ should be calculated by a g-inverse.

3. $\nu \to \infty$ *and* $\mathbf{\Sigma} = \sigma^2\mathbf{I}_k$. Define $\tilde{t}_{\infty j} = n^{1/2}\lambda^{-1}\overline{\mathbf{X}}_j$, $j = 1, \ldots, k$. The posterior odds become

$$\mathbf{O} = \frac{p}{1-p}\left(\frac{\eta}{n+\eta}\right)^{(1/2)k} \exp\left(\frac{1}{2}\left(\frac{n}{n+\eta}\right)\sum_{j=1}^{k}\tilde{t}_{\infty j}^2\right).$$

4. $\nu \to 0$ *and* $\mathbf{\Sigma} = \sigma^2\mathbf{I}_k$. In this case, the posterior odds are just equation (4.10) with ν replaced with 0.

4.1.6. *Hyperparameter estimation.* We have shown that the *MB*-statistic for assessing whether or not a time course has mean $\mathbf{0}$ depends on $(k^2 + k + 6)/2$ hyperparameters: ν, $\mathbf{\Lambda}$, η and p. In practice, we need to estimate these hyperparameters, and plug in our estimates into the formulae for $\widetilde{\mathbf{S}}$, $\tilde{\mathbf{t}}$, $\mathbf{O}, \ldots$ and so on. Slightly abusing our notation, we will use the same symbols for these estimates, relying on context to make it clear whether we are assuming the hyperparameters to be known or not. In our multivariate model, many more hyperparameters need to be estimated, compared to the univariate models in [20] and [24], both of which have four hyperparameters. Closed form estimators for the hyperparameters in the univariate linear model setting are derived in [24], using the marginal sampling distributions of the statistic $\tilde{t}$ and the sample variance s^2, and are shown to be better than the simple estimators in [20]. Following [24], the aim of this section is to derive estimators for the hyperparameters in our multivariate model. In general, the hyperparameter η associated with the case $I = 1$ is estimated based on only a small subset of genes, while ν and $\mathbf{\Lambda}$ are estimated using the whole gene set. Instead of estimating the proportion of differentially expressed genes p, we plug in a user-defined value, since the choice of p does not affect the rankings of genes based on the *MB*-statistic.

EB estimation of ν *and* $\mathbf{\Lambda}$. The hyperparameter ν determines the degree of smoothing between $\mathbf{S}$ and $\mathbf{\Lambda}$. The method we use to estimate ν builds on that used to estimate d_0 in Section 6.2 in [24]. However, unlike d_0 in [24], ν is associated with the $k \times k$ matrix $\mathbf{\Sigma}$. Therefore, a method appropriate to this multivariate framework is needed. Let $\hat{\nu}_j$ be the estimated prior degrees of freedom based on the jth diagonal elements of the gene-specific sample variance–covariance matrices (i.e., the replicate variances for the jth time point from the whole gene set) using the method proposed in Section 6.2 in [24]. Our estimation of ν is based on the following two-step strategy. As the first step, set ν as $\hat{\nu} = \max(\text{mean}(\hat{\nu}_j), k + 6)$, $j = 1, \ldots, k$. This estimated $\hat{\nu}$ is used to estimate $\mathbf{\Lambda}$. Once $\mathbf{\Lambda}$ is estimated, $\hat{\nu}$ is reset to be $\hat{\nu} = \text{mean}(\hat{\nu}_j)$. In practice, one can even just plug in a user-defined value ν_0 which gives the desired amount of smoothing. In such a case, the first step sets $\hat{\nu} = \max(\nu_0, k + 6)$. This $\hat{\nu}$ is used to estimate $\mathbf{\Lambda}$. After $\mathbf{\Lambda}$ is estimated, $\hat{\nu}$ can be reset to the user-defined value ν_0.

Our estimate of $\mathbf{\Lambda}$ comes after the first step in the estimation of ν. We showed that, under our model, $\mathbf{S}$ follows the generalized type-II beta distribution with expectation $(\nu - k - 1)^{-1}\nu\mathbf{\Lambda}$. By the weak law of large numbers, $\overline{\mathbf{S}}$ converges in probability to $(\nu - k - 1)^{-1}\nu\mathbf{\Lambda}$. We can thus estimate $\mathbf{\Lambda}$ by $\hat{\nu}^{-1}(\hat{\nu} - k - 1)\overline{\mathbf{S}}$. If $\hat{\nu} \to \infty$, then $\mathbf{\Lambda}$ is estimated by $\overline{\mathbf{S}}$. The above estimates give quite satisfactory results on real data. A theoretical analysis of the estimation of our hyperparameters will be given later. For the moment, we content ourselves with obtaining reasonable estimates.

Y. C. TAI AND T. P. SPEED

EB estimation of η. The hyperparameter η is related to the moderated multivariate t-statistic $\tilde{\mathbf{t}}$ of nonzero genes. The method we use to estimate η builds on that of estimating v_0 in [24], except that we now need to deal with the multivariate case. Let $\tilde{t}_j$ be the jth element of the moderated multivariate t-statistic $\tilde{\mathbf{t}}$, $j = 1, \dots, k$. As in Section 6.3 in [24], each $\tilde{t}_j$ gives an estimate of η, call it $\hat{\eta}_j$, based on the top $p/2$ portion of genes with the largest $|\tilde{t}_j|$. We set $\hat{\eta}$ to be the mean of the $\hat{\eta}_j$.

4.2. *Unpaired two-sample problem.* Suppose there are two independent biological conditions Z and Y with sample sizes m and n, respectively. We can also derive the *MB*-statistic for testing the null $H : \boldsymbol{\mu}_Z = \boldsymbol{\mu}_Y, \boldsymbol{\Sigma}_Z = \boldsymbol{\Sigma}_Y > 0$. The null hypothesis turns out to be the same as that in Section 4.1: $H : \boldsymbol{\mu} = \mathbf{0}, \boldsymbol{\Sigma} > 0$, if we write $\boldsymbol{\mu} = \boldsymbol{\mu}_Z - \boldsymbol{\mu}_Y$ and $\boldsymbol{\Sigma} = \boldsymbol{\Sigma}_Z = \boldsymbol{\Sigma}_Y$. That is, we solve this two-sample problem using the one-sample approach in Section 4.1. We denote the m i.i.d. time course vectors for biological condition Z by $\mathbf{Z}_1, \dots, \mathbf{Z}_m$, each from a multivariate normal distribution with mean $\boldsymbol{\mu}_Z$ and variance–covariance matrix $\boldsymbol{\Sigma}$. Similarly, those for biological condition Y are denoted by $\mathbf{Y}_1, \dots, \mathbf{Y}_n$, each with mean $\boldsymbol{\mu}_Y$ and variance–covariance matrix $\boldsymbol{\Sigma}$. Since the null hypothesis here is identical to that in Section 4.1, the priors for $\boldsymbol{\mu}$ and $\boldsymbol{\Sigma}$ are exactly the same as those in Section 4.1, and we omit the details here. In a later paper we will attack this problem by assigning independent priors for $\boldsymbol{\mu}_Y$ and $\boldsymbol{\mu}_Z$ separately.

All the results follow immediately. The moderated multivariate t-statistic $\tilde{\mathbf{t}}$ here is defined as equation (4.7) with n replaced by $(m^{-1} + n^{-1})^{-1}$ and $\overline{\mathbf{X}}$ replaced by $\overline{\mathbf{Z}} - \overline{\mathbf{Y}}$. $\tilde{\mathbf{S}}$ here is the same as that defined in Section 3.2. The posterior odds $\mathbf{O}$ against the null hypothesis that the expected time courses are the same are

$$\mathbf{O} = \frac{p}{1-p}\left(\frac{m^{-1}+n^{-1}}{m^{-1}+n^{-1}+\eta^{-1}}\right)^{(1/2)k} \times \left(\frac{m+n-2+\nu+\widetilde{T}^2}{m+n-2+\nu+((m^{-1}+n^{-1})/(m^{-1}+n^{-1}+\eta^{-1}))\widetilde{T}^2}\right)^{(1/2)(m+n+\nu-1)}.$$

The log base 10 of $\mathbf{O}$ is our two-sample *MB*-statistic. Again, when all genes have the same sample sizes m and n, the two-sample *MB*-statistic is equivalent to the $\widetilde{T}^2 = \tilde{\mathbf{t}}'\tilde{\mathbf{t}}$. Under H, $k^{-1}\widetilde{T}^2$ has an F distribution with degrees of freedom $(k, m + n + \nu - k - 1)$; equivalently, $(m + n + \nu - k - 1)^{-1}(m + n + \nu - 2)\widetilde{T}^2$ has the Hotelling T^2 distribution $T^2(k, m + n + \nu - 2)$.

The *MB*-statistic described in this section involves hyperparameters ν, $\boldsymbol{\Lambda}$, η and p. The estimation procedures for these hyperparameters are very similar to those in Section 4.1, except that we have to use the gene-specific pooled sample variance–covariance matrices when estimating ν and $\boldsymbol{\Lambda}$, and use the $k \times 1$ moderated multivariate t-statistic $\tilde{\mathbf{t}}$ defined here to estimate η. We omit the details here.

It should be noted that the *MB*-statistic derived in this section has a slightly different definition: instead of using all the data we observe, we only use the difference in sample averages and the pooled sample variance–covariance matrix. The

$\widetilde{T}^2$-statistic here is identical to the moderated two-sample Hotelling T^2-statistic in Section 3.2.

4.3. *One-sample problem of constancy.* In this section we derive the posterior odds against the null that a gene's mean expression level stays constant over time. We obtain a closed form solution similar to that in the preceding sections, but only under a constraint on the variance–covariance matrix $\boldsymbol{\Sigma}$.

4.3.1. *Transformation.* For each gene, let I be the indicator variable defined in Section 4.1. Let $\mathbf{P} = k^{-1}\mathbf{1}_k\mathbf{1}_k'$ be the $k \times k$ projection matrix onto the rank 1 space of constant vectors, where $\mathbf{1}_k' = (1, \ldots, 1)$ is a $k \times 1$ vector of 1s. Let $\mathbf{P}^c = \mathbf{I}_k - \mathbf{P}$ be the projection onto the orthogonal complement of $R(\mathbf{P})$. We can write any vector $\boldsymbol{\mu} \in R^k$ as $\boldsymbol{\mu} = \mathbf{P}\boldsymbol{\mu} + \mathbf{P}^c\boldsymbol{\mu}$, and in the case $I = 0$, the second term $\mathbf{P}^c\boldsymbol{\mu}$ vanishes. As in Section 4.1, we build up our multivariate model by first assigning independent inverse Wishart priors to the gene-specific covariance matrices $\boldsymbol{\Sigma}$; see equation (4.1). Given $\boldsymbol{\Sigma}$, we next assign multivariate normal priors to the gene-specific mean parameters $\boldsymbol{\mu}$ for the case of nonconstant ($I = 1$) and constant genes ($I = 0$), respectively:

$$(4.11)\qquad \begin{cases} \boldsymbol{\mu}|\boldsymbol{\Sigma}, I = 1 \sim N(0, \tau^{-1}\mathbf{P}\boldsymbol{\Sigma}\mathbf{P} + \kappa^{-1}\mathbf{P}^c\boldsymbol{\Sigma}\mathbf{P}^c), \\ \boldsymbol{\mu}|\boldsymbol{\Sigma}, I = 0 \sim N(0, \tau^{-1}\mathbf{P}\boldsymbol{\Sigma}\mathbf{P}). \end{cases}$$

Given $\boldsymbol{\Sigma}$ and $I = 0$, the covariance matrix $\mathbf{P}\boldsymbol{\Sigma}\mathbf{P}$ guarantees that $\boldsymbol{\mu}$ is a constant vector, while when $I = 1$, the extra component $\mathbf{P}^c\boldsymbol{\Sigma}\mathbf{P}^c$ adds further variance to $\boldsymbol{\mu}$ so that it becomes a nonconstant vector. Again, in order to obtain the full expression for the posterior odds $\mathbf{O}$, we need to derive $P(\mathbf{X}_1, \ldots, \mathbf{X}_n|I)$ using equation (4.3). To get a closed-form expression for the posterior odds, we find it necessary to make an additional assumption, namely, that $\mathbf{P}\boldsymbol{\Sigma} = \boldsymbol{\Sigma}\mathbf{P}$. With this assumption, given $\boldsymbol{\Sigma}$ and $I = 0$, $\overline{\mathbf{X}}$ is a multivariate normal distribution with mean $\mathbf{0}$ and covariance matrix $(n^{-1}\boldsymbol{\Sigma} + \tau^{-1}\boldsymbol{\Sigma}\mathbf{P})$. Similarly, given $\boldsymbol{\Sigma}$ and $I = 1$, $\overline{\mathbf{X}}$ is a multivariate normal distribution with mean $\mathbf{0}$ and the covariance matrix $(n^{-1}\boldsymbol{\Sigma} + \tau^{-1}\boldsymbol{\Sigma}\mathbf{P} + \kappa^{-1}\boldsymbol{\Sigma}\mathbf{P}^c)$.

For the rest of this section, unless stated otherwise, we assume $\mathbf{P}\boldsymbol{\Sigma} = \boldsymbol{\Sigma}\mathbf{P}$, and we make use of the following lemma, whose proof is omitted.

LEMMA 4.1. *Suppose $\mathbf{T}$ is any $k \times k$ nonsingular matrix whose first row is constant c and the remaining rows have row sums equal to 0. Write $\mathbf{T} = (\mathbf{T}_0', \mathbf{T}_1')'$, where $\mathbf{T}_0$ is the first row of $\mathbf{T}$, and $\mathbf{T}_1$ is the remainder. Then, for any $\boldsymbol{\Sigma} > 0$ satisfying $\mathbf{P}\boldsymbol{\Sigma} = \boldsymbol{\Sigma}\mathbf{P}$, $\mathbf{T}\boldsymbol{\Sigma}\mathbf{T}' = \widetilde{\boldsymbol{\Sigma}}$ is a $k \times k$ block diagonal matrix with the scalar $\tilde{\sigma}^2 > 0$ as the first block and $(k-1) \times (k-1)$ matrix $\widetilde{\boldsymbol{\Sigma}}_1 > 0$ as the second block: that is,*

$$\mathbf{T}\boldsymbol{\Sigma}\mathbf{T}' = \widetilde{\boldsymbol{\Sigma}} = \begin{pmatrix} \tilde{\sigma}^2 & \mathbf{0} \\ \mathbf{0} & \widetilde{\boldsymbol{\Sigma}}_1 \end{pmatrix}.$$

As the first example, let $\mathbf{T}$ be the Helmert matrix, where the jith element of $\mathbf{T}$ is defined as

$$\begin{cases} t_{ji} = 1/\sqrt{k}, & \text{for } j = 1, i = 1, \ldots, k, \\ t_{ji} = 1/\sqrt{j(j-1)}, & \text{for } 2 \le j \le k, 1 \le i \le j-1, \\ t_{ji} = -(j-1)/\sqrt{j(j-1)}, & \text{for } 2 \le j \le k, i = j, \\ t_{ji} = 0, & \text{for } 2 \le j \le k-1, j+1 \le i \le k. \end{cases}$$

$\mathbf{T}$ can also be the following matrix, where the jith element of $\mathbf{T}$ is defined as

$$\begin{cases} t_{ji} = 1, & \text{for } j = 1, i = 1, \ldots, k, \\ t_{ji} = 1, & \text{for } 2 \le j \le k, i = j-1, \\ t_{ji} = -1, & \text{for } 2 \le j \le k, i = j, \\ t_{ji} = 0, & \text{otherwise.} \end{cases}$$

For our multivariate empirical Bayes model in this section, we use the Helmert matrix $\mathbf{T}$ to proceed with our calculations. The results can be applied to other $\mathbf{T}$ immediately.

4.3.2. *Models and priors.* Here $\mathbf{T}$ is partitioned into its first row $\mathbf{T}_0$ $(1 \times k)$ and its last $k-1$ rows $\mathbf{T}_1$ $((k-1) \times k)$. Since $\mathbf{X}_1, \ldots, \mathbf{X}_n$ are i.i.d. $N_k(\boldsymbol{\mu}, \boldsymbol{\Sigma})$, the transformed random vectors $\mathbf{T}\mathbf{X}_i$ are also multivariate normally distributed with mean $\mathbf{T}\boldsymbol{\mu}$ and covariance matrix $\widetilde{\boldsymbol{\Sigma}}$. By Lemma 4.1, the matrix $\widetilde{\boldsymbol{\Sigma}}$ is a block diagonal matrix with $\tilde{\sigma}^2$ as the first block, and $\widetilde{\boldsymbol{\Sigma}}_1$ as the second block. Defining $\bar{x}_i = k^{-1} \sum_{j=1}^{k} X_{ij}$, then $\sqrt{k}\bar{x}_i$ and the random vectors $\mathbf{T}_1\mathbf{X}_i$ are independent and normally distributed, with distributions

$$\sqrt{k}\bar{x}_i | \mathbf{T}_0\mu, \tilde{\sigma}^2 \sim N(\mathbf{T}_0\boldsymbol{\mu}, \tilde{\sigma}^2),$$
$$\mathbf{T}_1\mathbf{X}_i | \mathbf{T}_1\boldsymbol{\mu}, \widetilde{\boldsymbol{\Sigma}}_1 \sim N(\mathbf{T}_1\boldsymbol{\mu}, \widetilde{\boldsymbol{\Sigma}}_1).$$

This transformation allows us to separate the gene expression changes into constant and nonconstant changes.

As we have seen in Section 4.1, the joint distributions of data given I can be fully described using the sufficient statistics $\bar{\bar{x}}$, $\mathbf{T}_1\overline{\mathbf{X}}$, s^2 and $\mathbf{S}_1$, where $\bar{\bar{x}} = n^{-1} \sum_{i=1}^{n} \bar{x}_i$, $\mathbf{T}_1\overline{\mathbf{X}} = n^{-1} \sum_{i=1}^{n} \mathbf{T}_1\mathbf{X}_i$, $s^2 = (n-1)^{-1} \sum_{i=1}^{n} (\bar{x}_i - \bar{\bar{x}})^2$ and $\mathbf{S}_1 = (n-1)^{-1} \sum_{i=1}^{n} (\mathbf{T}_1\mathbf{X}_i - \mathbf{T}_1\overline{\mathbf{X}})(\mathbf{T}_1\mathbf{X}_i - \mathbf{T}_1\overline{\mathbf{X}})'$. The prior for $\widetilde{\boldsymbol{\Sigma}}$ is first set through the independent priors for $\tilde{\sigma}^2$ and $\widetilde{\boldsymbol{\Sigma}}_1$. We suppose that $\tilde{\sigma}^2$ and $\widetilde{\boldsymbol{\Sigma}}_1$ are independently distributed, with an inverse gamma distribution with shape parameter $\xi/2$ and scale parameter $\xi\lambda^2/2$, and an inverse Wishart distribution with degrees of freedom ν and scale matrix $\nu\boldsymbol{\Lambda}_1$, respectively, that is,

$$\begin{cases} \tilde{\sigma}^2 \sim \text{inv-gamma}(\frac{1}{2}\xi, \frac{1}{2}\xi\lambda^2), \\ \widetilde{\boldsymbol{\Sigma}}_1 \sim \text{inv-Wishart}_\nu((\nu\boldsymbol{\Lambda}_1)^{-1}). \end{cases} \tag{4.12}$$

The prior for $\mathbf{T}\boldsymbol{\mu}$ has four parts. We assign independent priors to $\mathbf{T}_0\boldsymbol{\mu}$ and $\mathbf{T}_1\boldsymbol{\mu}$ separately for the cases $I = 1$ and $I = 0$. For the case $I = 1$, priors are

$$\begin{cases} \mathbf{T}_0\boldsymbol{\mu}|\tilde{\sigma}^2, I = 1 \sim N(\theta, \kappa^{-1}\tilde{\sigma}^2), \\ \mathbf{T}_1\boldsymbol{\mu}|\widetilde{\boldsymbol{\Sigma}}_1, I = 1 \sim N(\mathbf{0}, \eta^{-1}\widetilde{\boldsymbol{\Sigma}}_1), \end{cases} \tag{4.13}$$

where $\theta \geq 0$ is the mean, and $\kappa > 0$ and $\eta > 0$ are scale parameters. When $I = 0$, $\mathbf{T}_1\boldsymbol{\mu} = \mathbf{0}$ with probability 1. Thus, the priors in this case are

$$\begin{cases} \mathbf{T}_0\boldsymbol{\mu}|\tilde{\sigma}^2, I = 0 \sim N(\theta, \kappa^{-1}\tilde{\sigma}^2), \\ \mathbf{T}_1\boldsymbol{\mu}|\widetilde{\boldsymbol{\Sigma}}_1, I = 0 \equiv \mathbf{0}. \end{cases} \tag{4.14}$$

It is reasonable to assume $P(\mathbf{T}_0\boldsymbol{\mu}|\tilde{\sigma}^2, I = 0) = P(\mathbf{T}_0\boldsymbol{\mu}|\tilde{\sigma}^2, I = 1)$ for large genome-wide arrays since there is no obvious reason why the expected grand mean of the expression levels for nonconstant genes should differ from that of constant genes. For two-color comparative microarray experiments, it is also reasonable to assume $\theta = 0$.

4.3.3. *Multivariate joint distributions.* The joint distributions can be derived quite readily using a previously established formula, and so we omit the details. Given $I = 1$, $\mathbf{T}_1\overline{\mathbf{X}}$ and $\mathbf{S}_1$ follow the Student–Siegel distribution $StSi_{k-1}(\nu, \mathbf{0}, (n^{-1} + \eta^{-1})\boldsymbol{\Lambda}_1, n - 1, (n - 1)^{-1}\nu\boldsymbol{\Lambda}_1)$. Similarly, the joint distribution of $\mathbf{T}_1\overline{\mathbf{X}}$ and $\mathbf{S}_1$ given $I = 0$ is $StSi_{k-1}(\nu, \mathbf{0}, n^{-1}\boldsymbol{\Lambda}_1, n - 1, (n - 1)^{-1}\nu\boldsymbol{\Lambda}_1)$.

4.3.4. *MB-statistic and $\widetilde{T}^2$-statistic.* The posterior odds against the null that a gene's mean expression level stays constant over time are equation (4.6) in Section 4.1 with k replaced by $k - 1$, $\tilde{\mathbf{t}}$ expressed by equation (4.7) with $\widetilde{\mathbf{S}}$ replaced by $\widetilde{\mathbf{S}}_1$ and $\overline{\mathbf{X}}$ replaced by $\mathbf{T}_1\overline{\mathbf{X}}$. $\widetilde{\mathbf{S}}_1$ is just equation (4.5) with $\mathbf{S}$ replaced by $\mathbf{S}_1$ and $\boldsymbol{\Lambda}$ replaced by $\boldsymbol{\Lambda}_1$. As in Section 4.1, the *MB*-statistic is a monotonic increasing function of $\widetilde{T}^2 = \tilde{\mathbf{t}}'\tilde{\mathbf{t}}$ when all genes have the same sample size n.

Under H, $(k - 1)^{-1}\widetilde{T}^2$ has an F distribution with degrees of freedom $(k - 1, n + \nu - k + 1)$, or, equivalently, $(n + \nu - k + 1)^{-1}(n + \nu - 1)\widetilde{T}^2$ has a Hotelling T^2-distribution $T^2(k - 1, n + \nu - 1)$. The hyperparameter estimation procedures here are similar to those described in Section 4.1, except that all the estimations are performed based on transformed data.

5. Case study. In this section we illustrate our results with a paired two-sample problem, using the *Arabidopsis thaliana* dataset in [35]. Here we only give a very brief description of the data and the results. We refer the reader to [35] and to Chapter 5 of [28] for more thorough discussions.

A. thaliana wildtype (Columbia) and ics1-2 null mutant plants were evenly positioned, intermixed and grown in growth chambers under controlled conditions. When the plants were four weeks old, they were infected with a moderately heavy innoculum of the powdery mildew *G. orontii*. Each pair of mRNA samples from wildtype and mutant plants was harvested and collected at six time points post-infection. Plants could not be resampled, so mRNA samples at one time point were from different plants than those of any other time point. We report here on the analysis of three replicate experiments under similar environmental conditions which contribute four biological replicates: one from the first and third experiments (1–3, 3–1) and two from the second experiment (2–1, 2–2). These mRNA samples were hybridized onto Affymetrix *Arabidopsis* ATH1 GeneChips, yielding 22,810 probesets and 48 arrays in our analysis. The array preprocessing were done using the Robust Multi-array Analysis (RMA) algorithm described in [18, 5] which is implemented in the Bioconductor package `affy`.

This study is longitudinal if we treat *experiments* as units, while it is cross-sectional if we treat *plants* as units. We believe it is worthwhile to treat it as a paired longitudinal study, since samples within the same experiment are more similar than those from different experiments. We thus have a paired two-sample problem, with the genes of interest being those whose wildtype and mutant temporal profiles are different. We subtracted the $\log_2$ intensities of the wildtype from those of the paired mutant at each time within each replicate, yielding the $\log_2$ ratios of mutant relative to wildtype. Since the number of time course replicates is the same ($n = 4$) across genes for this dataset, we used the $\widetilde{T}^2$-statistic instead of the MB-statistic to rank genes, so that we did not have to estimate the hyperparameter η.

For comparison purposes, we fitted a linear model to the log ratios for each gene, with time and replicate effects, and calculated the F-statistic for the time effect.

5.1. *Results.* The extent of moderation from $\hat{\nu} = 5$ was 63%. The left panel of Figure 1 presents three genes falling into different ranges of ranks (rank = 1, 175, 859) by $\widetilde{T}^2$, while the ones on the right panel have the same ranks by F. The gene ranked most highly by $\widetilde{T}^2$ exhibits much greater differences between the wildtype and mutant temporal profiles than the one ranked most highly by F. The magnitude of the difference, as measured by $\widetilde{T}^2$, decreases as the rank goes down. The gene ranked 1 by $\widetilde{T}^2$ is well known: pathogenesis-related protein 1 (PR1). Other known pathogenesis-related genes also ranked highly by $\widetilde{T}^2$ and were less highly ranked by the F-statistic, as detailed in [35].

To investigate the effect of the amount of moderation on gene ranking, we kept $\mathbf{\Lambda}$ fixed, and re-calculated the $\widetilde{T}^2$-statistic with several different ν values. We then computed the Spearman rank correlation between the different sets of $\widetilde{T}^2$-statistics for all genes and for the top 859 genes only. Table 1 gives the results. The correlations are lower in the two extremes. All of these sets have the same number one gene. This comparison shows that the gene ranks are reasonably stable when the extent of moderation varies within a certain window, and that moderation seems to have more effect on the top genes relative to the whole gene set.

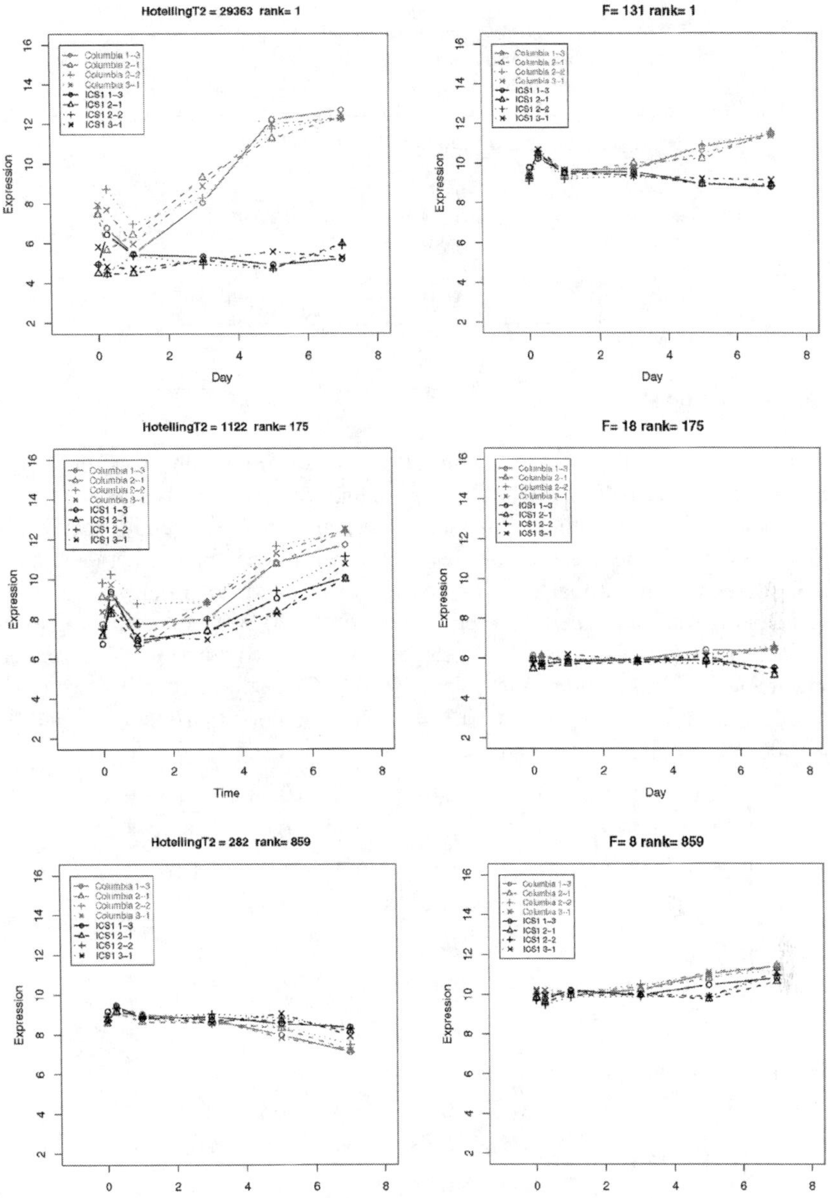

FIG. 1. *Genes of ranks* 1, 175, 859 *by the paired two-sample* $\widetilde{T}^2$*-statistic in Section* 4.1 (*left panel*) *and the F-statistic* (*right panel*). *The temporal difference between wildtype and mutant decreases as the rank by* $\widetilde{T}^2$ *goes down. The genes on the left panel all show larger differences between the wildtype and mutant than the corresponding ones on the right panel.*

TABLE 1
Spearman rank correlations between $\widetilde{T}^2$ with different ν and the estimated ν for all and the top 859 genes. The percent moderation is defined by $(\nu + n - 1)^{-1}\nu \times 100$

% moderation	Correlation (all)	Correlation (top 859)
97 ($\nu = 100$)	0.97	0.90
80 ($\nu = 12$)	0.99	0.98
40 ($\nu = 2$)	0.99	0.98
25 ($\nu = 1$)	0.98	0.96
0 ($\nu = 0.01$)	0.93	0.90

6. Simulation study.

6.1. *Method.* In this section we report on a small simulation study for the null hypothesis $H : \boldsymbol{\mu} = \mu_0\mathbf{1}, \boldsymbol{\Sigma} > 0$ based on an actual example we have met. We simulate 100 data sets, each with 20,000 genes. The genes are simulated independently, which we regard as an assumption that makes sense to compare methods, but it should be kept in mind that gene expression measures in real data can be quite dependent. In each simulated data set, 400 out of the 20,000 genes are assigned to be nonconstant. That is, $p = 0.02$. Each gene is simulated with three independent replicates ($n = 3$) and eight time points ($k = 8$). The other hyperparameters are the following: $\nu = 13$, $\xi = 3$, $\lambda^2 = 0.3$, $\theta = 0$ (two-color experiments), $\kappa = 0.02$, $\eta = 0.08$, and

$$\boldsymbol{\Lambda} = \begin{pmatrix} 14.69 & 0.57 & 0.99 & 0.40 & 0.55 & 0.51 & -0.23 \\ 0.57 & 15.36 & 1.22 & 0.84 & 1.19 & 0.91 & 0.86 \\ 0.99 & 1.22 & 14.41 & 2.47 & 1.81 & 1.51 & 1.07 \\ 0.40 & 0.84 & 2.47 & 17.05 & 2.40 & 2.32 & 1.33 \\ 0.55 & 1.19 & 1.81 & 2.40 & 15.63 & 3.31 & 2.75 \\ 0.51 & 0.91 & 1.51 & 2.32 & 3.31 & 13.38 & 3.15 \\ -0.23 & 0.86 & 1.07 & 1.33 & 2.75 & 3.15 & 12.90 \end{pmatrix} \times 10^{-3}.$$

The correlation matrix of $\boldsymbol{\Lambda}$ is

$$\begin{pmatrix} 1 & 0.04 & 0.07 & 0.03 & 0.04 & 0.04 & -0.02 \\ 0.04 & 1 & 0.08 & 0.05 & 0.08 & 0.06 & 0.06 \\ 0.07 & 0.08 & 1 & 0.16 & 0.12 & 0.11 & 0.08 \\ 0.03 & 0.05 & 0.16 & 1 & 0.15 & 0.15 & 0.09 \\ 0.04 & 0.08 & 0.12 & 0.15 & 1 & 0.23 & 0.20 \\ 0.04 & 0.06 & 0.11 & 0.15 & 0.23 & 1 & 0.24 \\ -0.02 & 0.06 & 0.08 & 0.09 & 0.19 & 0.24 & 1 \end{pmatrix},$$

and we see clear evidence of serial correlation. Note that, in the real world, although often the case, the correlation does not always decrease with time lag. The statistics compared in our study are the following:

(1) *MB*-statistic, or equivalently, the $\widetilde{T}^2$-statistic;

(2) *MB*-statistic using first differences: take the differences in gene expression values at consecutive time points within replicates, and use them to test the null hypothesis $H : \boldsymbol{\mu} = \mathbf{0}$, $\boldsymbol{\Sigma} > 0$ (Section 4.1), where $\boldsymbol{\mu}$ is the mean of the differences;

(3) *MB*-statistic in the special case $\boldsymbol{\Sigma} = \sigma^2 \mathbf{I}_k$;

(4) *MB*-statistic in the limiting case $\nu \to \infty$;

(5) *MB*-statistic in the limiting case $\nu \to 0$;

(6) ordinary F-statistic from an ANOVA model with time and replicate effects;

(7) partly-moderated F-statistic proposed in [24] from an ANOVA model with time and replicate effects;

(8) one-sample moderated Hotelling T^2-statistic $\|n^{1/2}\tilde{\mathbf{S}}^{-1/2}\tilde{\mathbf{d}}\|^2$ derived in Section 3, equivalently, the moderated *LR*-statistic, where the degree of moderation and the common matrix toward which each sample covariance matrix moves is estimated by the method given in Section 4.1;

(9) the variance across time course replicates $(nk-1)^{-1}\sum_{i=1}^{n}\sum_{j=1}^{k}(X_{ij}-\bar{\bar{x}})^2$.

Here each of the nine statistics incorporates either none (e.g., variance) or one (ordinary F-statistic) or more of the following: moderation, correlation structure and replicate variances, and thus can be used to show the importance of the above properties. It is not appropriate to set the prior degrees of freedom ν to be a very small number, since we have the constraint that $\nu \geq k - 1$. We choose ν to be $k + 5 = 13$ because it simulates more stable $\boldsymbol{\Sigma}$'s across genes.

6.2. *Results.* Table 2 compares the means and standard deviations of the hyperparameter estimates of the diagonal elements of $\boldsymbol{\Lambda}_1$ (λ_j^2), $j = 1, \ldots, k-1$, with their true values. The mean estimate of $\boldsymbol{\Lambda}_1$ is very close to the true $\boldsymbol{\Lambda}_1$, and the standard deviations are very small. The hyperparameter η is always underestimated (mean $= 0.026$, SD $= 0.002$), which agrees with Section 8 in [24], where v_0 was usually over-estimated. The hyperparameter ν is also always underestimated (mean $= 7.0$, SD $= 0.2$). In Section 5 we observed that the amount of moderation ν does not greatly affect gene ranking except at the two extremes. One can even choose a user-defined ν which gives reasonable results. Although not well estimated, η only affects the rankings when the number of replicates is different across genes. However, this does not happen often in the real world. Even when that happens, the effect is very small. To investigate the effects of η on gene rankings, we tried a couple of η values from different ranges, while keeping the remaining hyperparameters fixed, and calculated the *MB*-statistics. The rank correlations between rankings of the *MB*-statistics with the user-defined η's and the estimated η for one set of simulated data are the following: 0.91, 0.94, 0.99, 0.99, 0.99 for η =2, 1, 0.08 (true value), 0.05, 0.001, respectively.

TABLE 2
The means and standard deviations (SD) of the diagonal elements of the estimated $\mathbf{\Lambda}_1$

Hyperparameters	True value $\times 10^3$	Mean $\times 10^3$	SD $\times 10^3$
λ_1^2	14.69	14.71	0.16
λ_2^2	15.36	15.37	0.17
λ_3^2	14.41	14.43	0.15
λ_4^2	17.05	17.04	0.19
λ_5^2	15.63	15.63	0.15
λ_6^2	13.38	13.40	0.15
λ_7^2	12.90	12.92	0.17

To examine the relationship between the $\widetilde{T}^2$-statistic and the true deviation from constancy, the $\log_{10}$ transformed $\widetilde{T}^2$-statistic from one simulated dataset is plotted against the Mahalanobis distance between the expected time course vector $\boldsymbol{\mu}$ and its projection onto the rank 1 constant space $\bar{\boldsymbol{\mu}} = \mathbf{P}\boldsymbol{\mu}$ (Figure 2). The squared Mahalanobis distance is defined by $d(\boldsymbol{\mu}, \bar{\boldsymbol{\mu}})^2 = (\boldsymbol{\mu} - \bar{\boldsymbol{\mu}})'\boldsymbol{\Sigma}^{-1}(\boldsymbol{\mu} - \bar{\boldsymbol{\mu}})$. Figure 2 clearly

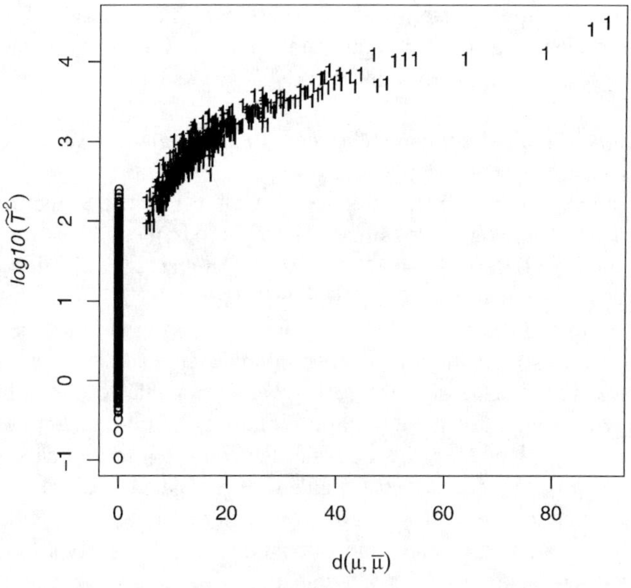

FIG. 2. *The* $\log_{10}\widetilde{T}^2$ *statistic versus the true deviation from constancy* $d(\boldsymbol{\mu}, \bar{\boldsymbol{\mu}})$ *for one simulated dataset. Here* 1 *denotes nonconstant, and* o *constant genes.*

REPLICATED MICROARRAY TIME COURSE DATA

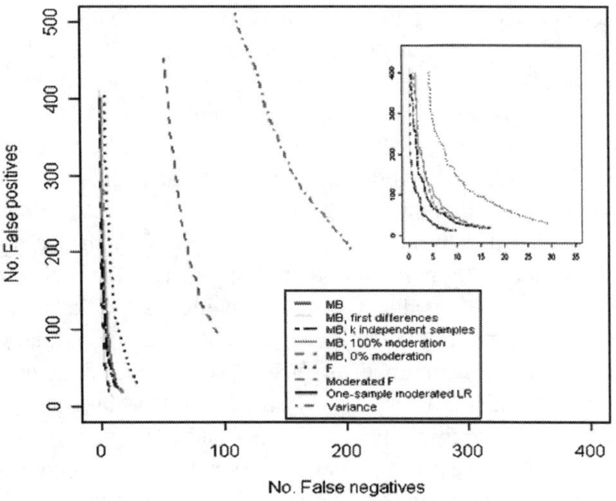

FIG. 3. *Average number of false positives versus number of false negatives of all the nine statistics. The subplot presents the curves for the best seven statistics.*

shows that $\log_{10} \widetilde{T}^2$ is positively correlated with $d(\boldsymbol{\mu}, \bar{\boldsymbol{\mu}})$, and most of the 400 true nonconstant genes achieve higher $\widetilde{T}^2$-statistics than the constant genes.

Figure 3 plots the average number of false positives against average number of false negatives at different cutoffs. By different cutoffs, we mean choosing the top x genes and calculating the numbers of false positives and false negatives, where x varies across the integers from 400 to 800. The lines in Figure 3 from left to right represent the following: *MB*-statistic ($\widetilde{T}^2$), *MB*-statistic with first differences (indistinguishable from the *MB*-statistic), one-sample moderated Hotelling T^2-statistic (indistinguishable from the *MB*-statistic), *MB*-statistic with $\boldsymbol{\Sigma} = \sigma^2 \mathbf{I}_k$, *MB*-statistic with $\nu \to \infty$, partly-moderated F-statistic [24], ordinary F-statistic, *MB*-statistic with $\nu \to 0$ and variance. The *MB*-statistic ($\widetilde{T}^2$) attains almost the same numbers of false positives and false negatives as *MB* with first differences and the one-sample moderated Hotelling T^2-statistic. The effectiveness of moderation is demonstrated by comparing the lines for the *MB*-statistic, the *MB*-statistic in the limiting case $\nu \to \infty$ and the *MB*-statistic in the limiting case that $\nu \to 0$. Both of these limiting cases lead to higher aggregate false positives and false negatives. This result supports the view stated in [29] that moderation is useful. In particular, the case $\nu \to 0$ (no moderation at all) produces much higher numbers of false positives and false negatives. This is likely due to the poor estimation of sample variance–covariance matrices with a small number of replicates. Indeed, the ordinary unmoderated F-statistic which ignores the correlation structure achieves smaller numbers of false positives and false negatives than the unmoderated *MB*-statistic. A similar situation also arises in the microarray discrimination context; see Section 7 of [12]. The partly-moderated F-statistic [24] which ignores the dependence among times behaves like the *MB*-statistic in the special case $\mathbf{\Sigma} = \sigma^2\mathbf{I}_k$. Moreover, it achieves fewer false positives and false negatives than the ordinary F-statistic. Figure 3 also demonstrates the importance of incorporating the correlation structure among time points. The *MB*-statistic ($\widetilde{T}^2$) and the one-sample moderated Hotelling T^2-statistic perform better than the partly-moderated F-statistic in [24] and the ordinary F-statistic; the former incorporates the correlation structure among time points, whereas the latter does not. However, we observe that the amount of moderation given by the partly-moderated F-statistic in [24] is usually much less than that given by the *MB*-statistic. When there are a large number of residual degrees of freedom from the linear model, the partly-moderated F-statistic [24] behaves very much like the ordinary F-statistic. This suggests that the lower number of false positives and number of false negatives from the *MB*-statistic than the partly-moderated F-statistic in [24] involve both the incorporation of correlation structures and the amounts of moderation. As expected, the simple variance statistic across replicates, which totally ignores the replicate variances, performs the worst. This demonstrates the importance of incorporating the replicate variances into any statistic.

7. Discussion. In this paper we introduced the *MB*- and $\widetilde{T}^2$-statistics for one- and two-sample longitudinal replicated developmental microarray time course experiments. Our main focus was the one-sample or paired two-sample problem with the null hypothesis $H: \boldsymbol{\mu} = \mathbf{0}, \mathbf{\Sigma} > \mathbf{0}$. This *MB*-statistic can be used when there are two biological conditions and the samples are paired across conditions, and it is shown to perform better than the classical F-statistic on a problem briefly described in Section 5. In addition, we also derive the *MB*-statistics and $\widetilde{T}^2$-statistics for the two-sample problem with the null $H: \boldsymbol{\mu}_Z = \boldsymbol{\mu}_Y, \mathbf{\Sigma}_Z = \mathbf{\Sigma}_Y = \mathbf{\Sigma} > 0$, and the one-sample problem with the null $H: \boldsymbol{\mu} = \mu_0\mathbf{1}, \ \mathbf{\Sigma} > \mathbf{0}$ using similar approaches. The latter situation requires a slight assumption on $\mathbf{\Sigma}$ in order to get a simple closed-form solution for the posterior odds against the null. All the *MB*-statistics and $\widetilde{T}^2$-statistics incorporate the correlation structure, replication and moderation. The moderated versions of some standard likelihood-ratio test statistics are also described. When all genes have the same sample size(s), our $\widetilde{T}^2$-statistics are not only equivalent to the *MB*-statistics, but also are identical to their corresponding moderated Hotelling T^2-statistics, apart from the one-sample problem with the null $H: \boldsymbol{\mu} = \mu_0\mathbf{1}, \ \mathbf{\Sigma} > \mathbf{0}$, where there is an additional constraint on $\mathbf{\Sigma}$. In this case the $\widetilde{T}^2$-statistic performed as well as the moderated Hotelling T^2-statistic in our simulation study, and also on several real datasets we have encountered. We have shown in the simulation study that, with this null, the *MB*-statistic ($\widetilde{T}^2$), the *MB*-statistic using first differences and the one-sample moderated Hotelling T^2-statistic perform best among all the nine statistics compared. This is not entirely surprising given that we simulated data under our model, but the comparisons are still informative. In practice, we regard the *MB*-statistic, the *MB*-statistic with first differences and the moderated *LR*-statistic as performing equally well, giving very similar (or identical) results. However, to use the *LR*-statistic (the moderated Hotelling T^2-statistic), we still need to insert moderated sample variance–covariance matrices, and these come from our multivariate empirical Bayes framework. In other words, our models provide a natural way to estimate the gene-specific moderated variance–covariance matrices (Sections 4.1–4.3), while the likelihood-based approach alone does not.

The assumption of $\mathbf{P\Sigma} = \mathbf{\Sigma P}$ with the null $H : \boldsymbol{\mu} = \mu_0 \mathbf{1}, \mathbf{\Sigma} > 0$ allows the mathematical calculations in Section 4.3, and leads to a closed-form formula for the *MB*-statistic. One question which naturally arises to be the impact of this constraint on the rankings of genes. From the practical point of view, the impact of this constraint on gene rankings is very *slight*. The rank correlations between the one-sample *MB*-statistic *with* the commuting assumption and the moderated *LR*-statistic from likelihood-based approach *without* the constraint, from the actual examples we have met, are typically very high (over 0.99). The rank correlations from our simulated data are also over 0.99. On the other hand, using the *MB*-statistic with first differences produces very similar or even identical results to the *MB*-statistic in Section 4.3. Indeed, instead of using the Helmert matrix, if we choose $\mathbf{T}$ to be the second transformation matrix of Section 4.3, we get identical results. Even so, we do not consider the first differences approach to be the solution to this commuting constraint, since the inference drawn is based on reduced, not the original data. In other words, the null hypotheses are not equivalent. The likelihood-based approach with moderation and the first differences approach support the fact that this constraint does not have much effect on the results. In addition, the former is a good way to avoid this assumption, since it performs as well as our *MB*-statistic in Section 4.3.

The statistics proposed in this paper are for one- and two-sample longitudinal data. We should be aware that many experiments in the real world exhibit some features from both longitudinal and cross-sectional experiments (e.g., Section 5).

One thing we plan to investigate in the future is the effect of assuming the same variance–covariance matrix $\mathbf{\Sigma}$ for both $I = 1$ and $I = 0$. Another issue which interests us is the effect of assuming the same $\mathbf{\Sigma}$ across biological conditions in the unpaired two-sample model in Section 4.2. The proposed methods may be extended in several ways, for example, identifying genes of some *specific* pattern, rather than *any* pattern. The statistics for a longitudinal multi-sample problem when there are at least three biological conditions and genes of interest are those with different temporal profiles across conditions derived in [30]. The corresponding statistics for cross-sectional data are also presented in [31].

The proposed methods focus on gene ranking, but not assessing the *significance* using p-values. However, if desired, we believe that generating p-values from a bootstrap analysis should be successful in this context.

We constructed our models using conjugate priors for the multivariate normal likelihoods, so that we got simple closed-form solutions for the posteriors odds.

Finding a closed-form statistic when the priors on $\boldsymbol{\mu}$ and $\boldsymbol{\Sigma}$ are independent seems to be an open and probably hard problem; that problem probably needs to be dealt with using MCMC.

Acknowledgments. We thank the Editor Jianqing Fan, an Associate Editor and the two referees for their valuable comments on this paper. We are grateful to Mary Wildermuth and her colleagues for sharing their *A. thaliana* dataset prior to publication of Wildermuth et al. [35]. Thanks are also due to Gordon Smyth, Ingrid Lönnstedt, Darlene Goldstein, Greg Hather, Avner Bar-Hen, Alfred Hero, Cavan Reilly and Christina Kendziorski for their valuable comments on an earlier draft, and to Ben Bolstad for his assistance with our simulations. Finally, we would like to acknowledge Mary Wildermuth, John Ngai and their lab members for their helpful discussions on the biological background of time course experiments and access to their data to test the methods developed in this paper.

REFERENCES

[1] AITCHISON, J. and DUNSMORE, I. R. (1975). *Statistical Prediction Analysis*. Cambridge Univ. Press. MR0408097

[2] BALDI, P. and LONG, A. D. (2001). A Bayesian framework for the analysis of microarray expression data: Regularized t-test and statistical inferences of gene changes. *Bioinformatics* **17** 509–519.

[3] BAR-JOSEPH, Z., GERBER, G., SIMON, I., GIFFORD, D. K. and JAAKKOLA, T. S. (2003). Comparing the continuous representation of time-series expression profiles to identify differentially expressed genes. *Proc. Natl. Acad. Sci. USA* **100** 10,146–10,151. MR1998142

[4] BICKEL, P. J. and DOKSUM, K. A. (2001). *Mathematical Statistics: Basic Ideas and Selected Topics*, 2nd ed. **1**. Prentice Hall, Upper Saddle River, NJ.

[5] BOLSTAD, B., IRIZARRY, R., ÂSTRAND, M. and SPEED, T. (2003). A comparison of normalization methods for high density oligonucleotide array data based on bias and variance. *Bioinformatics* **19** 185–193.

[6] BROBERG, P. (2003). Statistical methods for ranking differentially expressed genes. *Genome Biology* **4** R41.

[7] CHO, R., CAMPBELL, M., WINZELER, E., STEINMETZ, L., CONWAY, A., WODICKA, L., WOLFSBERG, T., GABRIELIAN, A., LANDSMAN, D., LOCKHART, D. and DAVIS, R. (1998). A genome-wide transcriptional analysis of the mitotic cell cycle. *Molecular Cell* **2** 65–73.

[8] CHO, R., HUANG, M., CAMPBELL, M., DONG, H., STEINMETZ, L., SAPINOSO, L., HAMPTON, G., ELLEDGE, S., DAVIS, R. and LOCKHART, D. (2001). Transcriptional regulation and function during the human cell cycle. *Nature Genetics* **27** 48–54.

[9] CHU, S., DERISI, J., EISEN, M., MULHOLLAND, J., BOTSTEIN, D., BROWN, P. O. and HERSKOWITZ, I. (1998). The transcriptional program of sporulation in budding yeast. *Science* **282** 699–705.

[10] DIGGLE, P. J. (1990). *Time Series: A Biostatistical Introduction*. Oxford Univ. Press, New York. MR1055357

[11] DIGGLE, P. J., HEAGERTY, P., LIANG, K.-Y. and ZEGER, S. L. (2002). *Analysis of Longitudinal Data*, 2nd ed. Oxford Univ. Press, New York. MR2049007

[12] DUDOIT, S., FRIDLYAND, J. and SPEED, T. (2002). Comparison of discrimination methods for the classification of tumors using gene expression data. *J. Amer. Statist. Assoc.* **97** 77–87. MR1963389

[13] DUDOIT, S., YANG, Y. H., CALLOW, M. and SPEED, T. (2002). Statistical methods for identifying differentially expressed genes in replicated cDNA microarray experiments. *Statist. Sinica* **12** 111–139. MR1894191

[14] EFRON, B., TIBSHIRANI, R., STOREY, J. D. and TUSHER, V. (2001). Empirical Bayes analysis of a microarray experiment. *J. Amer. Statist. Assoc.* **96** 1151–1160. MR1946571

[15] GUO, X., QI, H., VERFAILLIE, C. M. and PAN, W. (2003). Statistical significance analysis of longitudinal gene expression data. *Bioinformatics* **19** 1628–1635.

[16] GUPTA, A. and NAGAR, D. (2000). *Matrix Variate Distributions*. Chapman and Hall/CRC, Boca Raton, FL. MR1738933

[17] HONG, F. and LI, H. (2006). Functional hierarchical models for identifying genes with different time-course expression profiles. *Biometrics* **62** 534–544.

[18] IRIZARRY, R. A., BOLSTAD, B. M., COLLIN, F., COPE, L. M., HOBBS, B. and SPEED, T. P. (2003). Summaries of Affymetrix GeneChip probe level data. *Nucleic Acids Res.* **31** e15.

[19] KENDZIORSKI, C., NEWTON, M., LAN, H. and GOULD, M. (2003). On parametric empirical Bayes methods for comparing multiple groups using replicated gene expression profiles. *Statistics in Medicine* **22** 3899–3914.

[20] LÖNNSTEDT, I. and SPEED, T. P. (2002). Replicated microarray data. *Statist. Sinica* **12** 31–46. MR1894187

[21] MARDIA, K., KENT, J. and BIBBY, J. (1979). *Multivariate Analysis*. Academic Press, New York. MR0560319

[22] PARK, T., YI, S.-G., LEE, S., LEE, S. Y., YOO, D.-H., AHN, J.-I. and LEE, Y.-S. (2003). Statistical tests for identifying differentially expressed genes in time-course microarray experiments. *Bioinformatics* **19** 694–703.

[23] REINER, A., YEKUTIELI, D. and BENJAMINI, Y. (2003). Identifying differentially expressed genes using false discovery rate controlling procedures. *Bioinformatics* **19** 368–375.

[24] SMYTH, G. K. (2004). Linear models and empirical Bayes methods for assessing differential expression in microarray experiments. *Stat. Appl. Genet. Mol. Biol.* **3** article 3. MR2101454

[25] SPELLMAN, P. T., SHERLOCK, G., ZHANG, M. Q., IYER, V. R., ANDERS, K., EISEN, M. B., BROWN, P. O., BOTSTEIN, D. and FUTCHER, B. (1998). Comprehensive identification of cell cycle-regulated genes of the yeast Saccharomyces cerevisiae by microarray hybridization. *Molecular Biology of the Cell* **9** 3273–3297.

[26] STORCH, K.-F., LIPAN, O., LEYKIN, I., VISWANATHAN, N., DAVIS, F. C., WONG, W. H. and WEITZ, C. J. (2002). Extensive and divergent circadian gene expression in liver and heart. *Nature* **417** 78–83.

[27] STOREY, J., XIAO, W., LEEK, J. T., TOMPKINS, R. G. and DAVIS, R. W. (2005). Significance analysis of time course microarray experiments. *Proc. Natl. Acad. Sci. USA* **102** 12,837–12,842.

[28] TAI, Y. C. (2005). Multivariate empirical Bayes models for replicated microarray time course data. Ph.D. dissertation, Div. Biostatistics, Univ. California, Berkeley.

[29] TAI, Y. C. and SPEED, T. P. (2005). Statistical analysis of microarray time course data. In *DNA Microarrays* (U. Nuber, ed.) Chapter 20. Chapman and Hall/CRC, New York.

[30] TAI, Y. C. and SPEED, T. P. (2005). Longitudinal microarray time course *MB*-statistic for multiple biological conditions. Dept. Statistics, Univ. California, Berkeley. In preparation.

[31] TAI, Y. C. and SPEED, T. P. (2005). Cross-sectional microarray time course *MB*-statistic. Dept. Statistics, Univ. California, Berkeley. In preparation.

[32] TAMAYO, P., SLONIM, D., MESIROV, J., ZHU, Q., KITAREEWAN, S., DMITROVSKY, E., LANDER, E. S. and GOLUB, T. R. (1999). Interpreting patterns of gene expression with self-organizing maps: Methods and application to hematopoietic differentiation. *Proc. Natl. Acad. Sci. USA* **96** 2907–2912.

[33] TUSHER, V. G., TIBSHIRANI, R. and CHU, G. (2001). Significance analysis of microarrays applied to the ionizing radiation response. *Proc. Natl. Acad. Sci. USA* **98** 5116–5121.

[34] WEN, X., FUHRMAN, S., MICHAELS, G. S., CARR, D. B., SMITH, S., BARKER, J. L. and SOMOGYI, R. (1998). Large-scale temporal gene expression mapping of central nervous system development. *Proc. Natl. Acad. Sci. USA* **95** 334–339.

[35] WILDERMUTH, M. C., TAI, Y. C., DEWDNEY, J., DENOUX, C., HATHER, G., SPEED, T. P. and AUSUBEL, F. M. (2006). Application of $\widetilde{T}^2$ statistic to temporal global Arabidopsis expression data reveals known and novel salicylate-impacted processes. To appear.

[36] YUAN, M. and KENDZIORSKI, C. (2006). Hidden Markov models for microarray time course data in multiple biological conditions (with discussion). *J. Amer. Statist. Assoc.* **101** 1323–1340.

DIVISION OF BIOSTATISTICS
367 EVANS HALL 3860
UNIVERSITY OF CALIFORNIA, BERKELEY
BERKELEY, CALIFORNIA 94720-3860
USA
E-MAIL: yuchuan@stat.berkeley.edu
terry@stat.berkeley.edu

Bibliography of Terry Speed

[1] T. P. Speed. A note on commutative semigroups. *J. Aust. Math. Soc.*, 8:731–736, 1968.

[2] T. P. Speed. On rings of sets. *J. Aust. Math. Soc.*, 8:723–730, 1968.

[3] T. P. Speed. On Stone lattices. *J. Aust. Math. Soc.*, 9:297–307, 1969.

[4] T. P. Speed. Some remarks on a class of distributive lattices. *J. Aust. Math. Soc.*, 9:289–296, 1969.

[5] T. P. Speed. Spaces of ideals of distributive lattices I. Prime ideals. *Bull. Soc. Roy. Sci. Liege*, 38(11–12):610–628, 1969.

[6] T. P. Speed. Two congruences on distributive lattices. *Bull. Soc. Roy. Sci. Liege*, 38(3–4):86–95, 1969.

[7] T. P. Speed. A note on commutative semigroups II. *J. Lond. Math. Soc. (2)*, 2(1):80–82, 1970.

[8] R. M. Phatarfod, T. P. Speed, and A. W. Walker. A note on random walks. *J. Appl. Probab.*, 8(1):198–201, 1971.

[9] T. P. Speed. A note on Post algebras. *Colloq. Math.*, 14:37–44, 1971.

[10] T. P. Speed. A note on Stone lattices. *Can. Math. Bull.*, 14(1):81–86, 1971.

[11] T. P Speed and M. W. Evans. A note on commutative Baer rings. *J. Aust. Math. Soc.*, 13(1):1–6, 1971.

[12] T. P. Speed and E. Strzelecki. A note on commutative l-groups. *J. Aust. Math. Soc.*, 12(1):69–74, 1971.

[13] T. P. Speed. A note on commutative Baer rings. *J. Aust. Math. Soc.*, 14:257–263, 1972.

[14] T. P. Speed. On the order of prime ideals. *Algebra Univers.*, 2:85–87, 1972.

[15] T. P. Speed. Profinite posets. *Bull. Aust. Math. Soc.*, 6:177–183, 1972.

[16] E. Arjas and T. P. Speed. An extension of Cramér's estimate for the absorption probability of a random walk. *Math. Proc. Camb. Phil. Soc.*, 73(2):355–359, 1973.

[17] E. Arjas and T. P. Speed. A stopping problem in Markov additive processes. *Adv. Appl. Prob.*, 5:2–3, 1973.

S. Dudoit (ed.), *Selected Works of Terry Speed*, Selected Works in Probability and Statistics, DOI 10.1007/978-1-4614-1347-9,

[18] E. Arjas and T. P. Speed. Symmetric Wiener–Hopf factorisations in Markov additive processes. *Z. Wahrscheinlichkeitstheorie und verw. Geb.*, 26(2):105–118, 1973.

[19] E. Arjas and T. P. Speed. Topics in Markov additive processes. *Math. Scand.*, 33:171–192, 1973.

[20] J. W. Pitman and T. P. Speed. A note on random times. *Stoch. Proc. Appl.*, 1(4):369–374, 1973.

[21] T. P. Speed. A note on commutative Baer rings III. *J. Aust. Math. Soc.*, 15: 15–21, 1973.

[22] T. P. Speed. A note on random walks, II. *J. Appl. Probab.*, 10(1):218–222, 1973.

[23] T. P. Speed. On rings of sets. II: Zero-sets. *J. Aust. Math. Soc.*, 16:185–199, 1973.

[24] T. P. Speed. Some remarks on a result of Blomqvist. *J. Appl. Probab.*, 10(1):229–232, 1973.

[25] T. P. Speed. Statistics in school and society. *Math. Spectrum*, 6(1):7–11, 1973.

[26] E. Arjas and T. P. Speed. A note on the second factorization identity of A. A. Borovkov. *Theor. Probab. Appl.*, 18(3):576–578, 1974.

[27] J. D. Brodrick, I. M. Strachan, and T. P. Speed. Cytological changes in the conjunctiva in the megaloblastic anemias. *Invest. Ophth. Vis. Sci.*, 13(11):870–872, 1974.

[28] T. P. Speed. Spaces of ideals of distributive lattices, II. Minimal prime ideals. *J. Aust. Math. Soc.*, 18:54–72, 1974.

[29] E. Arjas and T. P. Speed. Markov chains with replacement. *Stoch. Proc. Appl.*, 3(2):175–184, 1975.

[30] J. Neveu. *Discrete-Parameter Martingales*, Volume 10 of *North-Holland Mathematical Library*. North-Holland, Amsterdam, 1975. Translated from French by T. P. Speed.

[31] T. P. Speed. Geometric and probabilistic aspects of some combinatorial identities. *J. Aust. Math. Soc. A*, 22:462–468, 1976.

[32] T. P. Speed. A note on pairwise sufficiency and completions. *Sankhyā Ser. A*, 38(2):194–196, 1976.

[33] W. W. Barker, J. Graham, T. C. Parks, and T. P. Speed. Electrostatic energy of disordered distributions of vacancies or altervalent ions. *J. Solid State Chem.*, 22(3):321–329, 1977.

[34] A. G. Pakes and T. P. Speed. Lagrange distributions and their limit theorems. *SIAM J. Appl. Math.*, 32(4):745–754, 1977.

[35] T. P. Speed. Negligible probabilities and nuclear reactor safety: Another misuse of probability? Technical report, Mathematics Department, University of Western Australia, 1977. Published in Danish in [40].

[36] V. Burke, A. Malajczuk, M. Gracey, T. P. Speed, and M. L. Thornett. Intestinal transport of monosaccharide after biliary diversion in the rat. *Aust. J. Exp. Biol. Med.*, 56(3):253–263, 1978.

[37] T. P. Speed. Decompositions of graphs and hypergraphs. In D. A. Holton and J. Seberry, editors, *Combinatorial Mathematics: Proceedings of the International Conference on Combinational Theory*, Volume 686 of *Lecture Notes in Mathematics*, pages 300–307. Springer, 1978.

[38] T. P. Speed. A factorisation theorem for adequate statistics. *Aust. J. Stat.*, 20:240–249, 1978.

[39] T. P. Speed. Relations between models for spatial data, contingency tables and Markov fields on graphs. *Adv. Appl. Prob.*, 10 (Supplement: Proceedings of the Conference on Spatial Patterns and Processes, March 1978):111–122, 1978.

[40] T. P. Speed. Forsvindende sandsynligheder og atomkraftsikkerhed: Et nyt misbrug af sandsynlighedsregningen? *RAMA*, 1:1–33, 1979. Danish version of [35].

[41] T. P. Speed. A note on nearest-neighbour Gibbs and Markov probabilities. *Sankhyā Ser. A*, 41(3–4):184–197, 1979.

[42] J. N. Darroch, S. L. Lauritzen, and T. P. Speed. Markov fields and log-linear interaction models for contingency tables. *Ann. Stat.*, 8(3):522–539, 1980.

[43] T. Hida. *Brownian Motion*, Volume 11 of *Applications of Mathematics*. Springer-Verlag, New York, 1980. Translation from Japanese by the author and T. P. Speed.

[44] D. Ratcliff, E. R. Williams, and T. P. Speed. Estimating missing values in multi-stratum experiments. *J. Roy. Stat. Soc. C*, 30(1):71–72, 1981.

[45] R. A. Bailey and T. P. Speed. On a class of association schemes derived from lattices of equivalence relations. In P. Schultz, C. E. Praeger, and R. P. Sullivan, editors, *Algebraic Structures and their Applications*, Volume 74 of *Lecture Notes in Pure and Applied Mathematics*, pages 55–74. Marcel Dekker, New York, 1982.

[46] R. P. Hart, J. B. Iveson, S. D. Bradshaw, and T. P. Speed. A study of isolation procedures for multiple infections of *Salmonella* and *Arizona* in a wild marsupial, the quokka (*Setonix brachyurus*). *J. Appl. Bacteriol.*, 53(3):395–406, 1982.

[47] H. Kiiveri and T. P. Speed. The structural analysis of multivariate data: A review. In S. Leinhardt, editor, *Sociological Methodology*, Volume 32, Chapter 6, pages 209–290. Jossey-Bass, San Francisco, CA, 1982.

[48] A. A. Landauer, S. E. McDonald, N. Papandreou, S. Skates, and T. P. Speed. Factors influencing the pre-sentence report: An analysis of hypothetical responses. *Aust. N. Z. J. Criminol.*, 15:207–218, 1982.

[49] R. A. Bailey, C. E. Praeger, C. A. Rowley, and T. P. Speed. Generalized wreath products of permutation groups. *Proc. Lond. Math. Soc. (3)*, 47: 69–82, 1983.

[50] J. N. Darroch and T. P. Speed. Additive and multiplicative models and interactions. *Ann. Stat.*, 11(3):724–738, 1983.

[51] A. M. Houtman and T. P. Speed. Balance in designed experiments with orthogonal block structure. *Ann. Stat.*, 11(4):1069–1085, 1983.

[52] T. P. Speed. Cumulants and partition lattices. *Aust. J. Stat.*, 25(2):378–388, 1983.

[53] T. P. Speed. General balance. In S. Kotz and N. L. Johnson, editors, *Encyclopedia of Statistical Sciences*, Volume 3. John Wiley & Sons, 1983.

[54] D. J. Best, M. Cameron, G. Eagleson, D. E. Shaw, and T. P. Speed. Downwind and long-term effects of cloud seeding in southeastern Australia. *Search*, 15:154–157, 1984.

[55] A. Houtman and T. P. Speed. The analysis of multistrata designed experiments with incomplete data. *Aust. J. Stat.*, 26(3):227–246, 1984.

[56] H. Kiiveri, T. P. Speed, and J. B. Carlin. Recursive causal models. *J. Aust. Math. Soc. A*, 36:30–52, 1984.

[57] S. L. Lauritzen, T. P. Speed, and K. Vijayan. Decomposable graphs and hypergraphs. *J. Aust. Math. Soc. A*, 36:12–29, 1984.

[58] D. Radcliff, E. R. Williams, and T. P. Speed. A note on the analysis of covariance in balanced incomplete block designs. *Aust. J. Stat.*, 26(3):337–341, 1984.

[59] T. P. Speed. On the Möbius function of $Hom(P,Q)$. *Bull. Aust. Math. Soc.*, 29(1):39–46, 1984.

[60] T. Tjur, R. A. Bailey, T. P. Speed, and H. P. Wynn. Analysis of variance models in orthogonal designs. *Int. Stat. Rev.*, 52(1):33–81, 1984.

[61] L. Råde and T. P. Speed, editors. *Teaching Statistics in the Computer Age*, Proceedings of the Round Table Conference on the Impact of Calculators and Computers on Teaching Statistics. Studentlitteratur, Lund, Sweden, 1985.

[62] H. D. Patterson, E. R. Williams, and T. P. Speed. A note on the analysis of resolvable block designs. *J. Roy. Stat. Soc. B*, 47(2):357–361, 1985.

[63] C. E. Praeger, C. A. Rowley, and T. P. Speed. A note on generalized wreath product groups. *J. Aust. Math. Soc. A.*, 39:415–420, 1985.

[64] T. P. Speed. Dispersion models for factorial experiments. *Bull. Int. Stat. Inst.*, 51(4):24.1, 16, 1985.

[65] T. P. Speed. Probabilistic risk assessment in the nuclear industry: WASH-1400 and beyond. In L. M. Le Cam and R. A. Olshen, editors, *Proceedings of the Berkeley Conference in Honor of Jerzy Neyman and Jack Kiefer*, Volume 2 of *Statistics/Probability Series*, pages 173–200. Wadsworth, Pacific Grove, CA, 1985.

[66] T. P. Speed. Some practical and statistical aspects of filtering and spectrum estimation. In J. F. Price, editor, *Fourier Techniques and Applications*, pages 101–120. Plenum Press, New York and London, 1985.

[67] T. P. Speed. Teaching of statistics at University level: How computers can help us find realistic models for real data and reasonably assess their reliability. In L. Råde and T. P. Speed, editors, *Teaching Statistics in the Computer Age*, Proceedings of the Round Table Conference on the Impact of Calculators and Computers on Teaching Statistics, pages 184–195. Studentlitteratur, Lund, Sweden, 1985.

[68] R. A. Bailey and T. P. Speed. Rectangular lattice designs: Efficiency factors and analysis. *Ann. Stat.*, 14(3):874–895, 1986.

[69] D. Culpin and T. P. Speed. The role of statistics in nuclear materials accounting: Issues and problems. *J. Roy. Stat. Soc. A*, 149(4):281–313, 1986.

[70] T. P. Speed and H. T. Kiiveri. Gaussian Markov distributions over finite graphs. *Ann. Stat.*, 14(1):138–150, 1986.

[71] T. P. Speed. ANOVA models with random effects: An approach via symmetry. In J. Gani and M. B. Priestley, editors, *Essays in Time Series and Allied Processes: Papers in Honour of E. J. Hannan*, pages 355–368. Applied Probability Trust, Sheffield, UK, 1986.

[72] T. P. Speed. Applications of cumulants and their generalisations. In I. S. Francis, B. F. J. Manly, and F. C. Lam, editors, *Proceedings of the Pacific Statistical Congress – 1985*, pages 12–20. North-Holland, Amsterdam, 1986.

[73] T. P. Speed. Cumulants and partition lattices II: Generalized k-statistics. *J. Aust. Math. Soc. A*, 40:34–53, 1986.

[74] T. P. Speed. Cumulants and partition lattices III: Multiply-indexed arrays. *J. Aust. Math. Soc. A*, 40:161–182, 1986.

[75] T. P. Speed. Cumulants and partition lattices IV: A.S. convergence of generalized k-statistics. *J. Aust. Math. Soc. A*, 41:79–94, 1986.

[76] T. P. Speed. Questions, answers and statistics. In R. Davidson and J. Swift, editors, *Proceedings: The Second International Conference on Teaching Statistics*, pages 18–28. University of Victoria, Victoria, BC, Canada, 1986.

[77] T. P. Speed and R. A. Bailey. Factorial dispersion models. *Int. Stat. Rev.*, 55(3):261–277, 1987.

[78] M. W. Knuiman and T. P. Speed. Incorporating previous results into the analysis of general balanced designed experiments. *Aust. J. Stat.*, 29(3):317–333, 1987.

[79] T. P. Speed. Generalized variance component models. In T. Pukkila and S. Puntanen, editors, *Proceedings of the Second Tampere International Conference on Statistics*. Dept. of Mathematical Sciences/Statistics, University of Tampere, Tampere, Finland, 1987.

[80] T. P. Speed. What is an analysis of variance? *Ann. Stat.*, 15:885–910, 1987.

[81] J. N. Darroch, G. F. V. Glonek, and T. P. Speed. On the existence of maximum likelihood estimators for hierarchical loglinear models. *Scand. J. Statist.*, 15(3):187–193, 1988.

[82] J. N. Darroch, M. Jirina, and T. P. Speed. Sampling without replacement: Approximation to the probability distribution. *J. Aust. Math. Soc. A*, 44:197–213, 1988.

[83] A. W. Davis and T. P. Speed. An Edgeworth expansion for the distribution of the F-ratio under a randomization model for the randomized block design. In S. S. Gupta and J. O. Berger, editors, *Statistical Decision Theory and Related Topics IV*. Springer-Verlag, New York, 1988.

[84] J. B. F. Field, F. E. Speed, T. P. Speed, and J. M. Williams. Biometrics in the CSIR: 1930–1940. *Aust. J. Stat.*, 30B(1):54–76, 1988.

[85] M. W. Knuiman and T. P. Speed. Incorporating prior information into the analysis of contingency tables. *Biometrics*, 44(4):1061–1071, 1988.

[86] T. P. Speed. The role of statisticians in CSIRO: Past, present and future. *Aust. J. Stat.*, 30(1):15–34, 1988.

[87] T. P. Speed and H. L. Silcock. Cumulants and partition lattices V. Calculating generalized k-statistics. *J. Aust. Math. Soc. A*, 44(2):171–196, 1988.

[88] T. P. Speed and H. L. Silcock. Cumulants and partition lattices VI. Variances and covariances of mean squares. *J. Aust. Math. Soc. A*, 44(3):362–388, 1988.

[89] E. R. Williams, J. T. Wood, and T. P. Speed. Non-orthogonal block structure in two-phase designs. *Aust. J. Stat.*, 30A(1):225–237, 1988.

[90] D. M. Dabrowska and T. P. Speed. Translation of "On the application of probability theory to agricultural experiments. Essay on principles. Section 9." by J. Splawa-Neyman. *Stat. Sci.*, 5(4):465–472, 1990.

[91] F. R. de Hoog, T. P. Speed, and E. R. Williams. On a matrix identity associated with generalized least squares. *Linear Algebra Appl.*, 127:449–456, 1990.

[92] T. P. Speed. Complexity, calibration and causality in influence diagrams. In R. M. Oliver and J. Q. Smith, editors, *Influence Diagrams, Belief Nets, and Decision Analysis*, Wiley Series in Probability and Mathematical Statistics, pages 49–63. John Wiley & Sons, Hoboken, NJ, 1990.

[93] T. P. Speed. Invariant moments and cumulants. In D. R. Chaudhuri, editor, *Coding Theory and Design Theory. Part II. Design Theory*, Volume 20 of *The IMA Volumes in Mathematics and its Applications*, pages 319–335. Springer-Verlag, New York, 1990.

[94] A. E. Sutherland, T. P. Speed, and P. G. Calarco. Inner cell allocation in the mouse morula: The role of oriented division during fourth cleavage. *Dev. Biol.*, 137(1):13–25, 1990.

[95] T. P. Speed. Comment on G. K. Robinson, "That BLUP is a good thing: The estimation of random effects". *Stat. Sci.*, 6(1):42–44, 1991.

[96] T. P. Speed. Review of "Stochastic complexity in statistical inquiry" (J. Rissanen, 1989). *IEEE Trans. Inform. Theory*, 37(6):1739–1740, 1991.

[97] R. Guerra, M. S. McPeek, T. P. Speed, and P. M. Stewart. A Bayesian analysis for mapping from radiation hybrid data. *Cytogenet. Cell Genet.*, 59(2–3):104–106, 1992.

[98] T. Nguyen and T. P. Speed. A derivation of all linear invariants for a nonbalanced transversion model. *J. Mol. Evol.*, 35:60–76, 1992.

[99] J. Rissanen, T. P. Speed, and B. Yu. Density estimation by stochastic complexity. *IEEE Trans. Inform. Theory*, 38(2):315–323, 1992.

[100] A. Sidow, T. Nguyen, and T. P. Speed. Estimating the fraction of invariable codons with a capture-recapture method. *J. Mol. Evol.*, 35:253–260, 1992.

[101] T. P. Speed. Introduction to Fisher (1926) The arrangement of field experiments. In S. Kotz and N. L. Johnson, editors, *Breakthroughs in Statistics*, Volume 2 of *Springer Series in Statistics: Perspectives in Statistics*, pages 71–81. Springer-Verlag, New York, 1992.

[102] T. P. Speed, M. S. McPeek, and S. N. Evans. Robustness of the no-interference model for ordering genetic markers. *Proc. Natl. Acad. Sci. USA*, 89(7):3103–3106, 1992.

[103] B. Yu and T. P. Speed. Data compression and histograms. *Probab. Theory Relat. Fields*, 92(2):195–229, 1992.

[104] S. N. Evans, M. S. McPeek, and T. P. Speed. A characterization of crossover models that possess map functions. *Theor. Popul. Biol.*, 43:80–90, 1993.

[105] S. N. Evans and T. P. Speed. Invariants of some probability models used in phylogenetic inference. *Ann. Stat.*, 21(1):355–377, 1993.

[106] A. Gelman and T. P. Speed. Characterizing a joint probability density by conditionals. *J. Roy. Stat. Soc. B*, 55(1):185–188, 1993.

[107] N. Hengartner and T. P. Speed. Assessing between-block heterogeneity within the post-strata of the 1990 Post-Enumeration Survey. *J. Am. Stat. Assoc.*, 88(423):119–129, 1993.

[108] A. Karu, A. McClatchie, A. Perman, S. J. Richman, and T. P. Speed. AutoElisa: A data management system for regulatory and diagnostic immunoassays, using parallel fitting for data evaluation. In D. Kurtz, editor, *Proceedings of the Symposium "Immunoassay: An emerging analytical chemistry technology"*. Association of Official Analytical Chemists, 1993.

[109] M. C. Samuel, N. Hessol, S. Shiboski, R. R. Engel, T. P. Speed, and W. Winkelstein. Factors associated with human immunodeficiency virus seroconversion in homosexual men in three San Francisco cohort studies, 1984–1989. *J. Acq. Immun. Def. Synd.*, 6(3):303–312, 1993.

[110] M. C. Samuel, M. S. Mohr, T. P. Speed, and W. Winkelstein. Infectivity of human immunodeficiency virus by anal and oral intercourse among male homosexuals: Estimates from a prospective study in San Francisco. In E. Kaplan and P. Brandau, editors, *Modelling the AIDS Epidemic: Planning, Policy, and Prediction*, pages 19–24. Raven Press, New York, 1993.

[111] T. P. Speed. Modelling and managing a salmon population. In V. Barnett and K. T. Turkman, editors, *Statistics for the Environment: Water Related Issues*, Volume 1, pages 267–292. John Wiley & Sons, 1993.

[112] T. P. Speed and B. Yu. Model selection and prediction: Normal regression. *Ann. Inst. Stat. Math.*, 45(1):35–54, 1993.

[113] B. Yu and T. P. Speed. A rate of convergence result for a universal D-semifaithful code. *IEEE Trans. Inform. Theory*, 39(3):813–820, 1993.

[114] K. Jin, T. P. Speed, W. Klitz, and G. Thomson. Testing for segregation distortion in the HLA complex. *Biometrics*, 50(4):1189–1198, 1994.

[115] D. O. Nelson and T. P. Speed. Predicting progress in directed mapping projects. *Genomics*, 24(1):41–52, 1994.

[116] D. O. Nelson and T. P. Speed. Statistical issues in constructing high resolution physical maps. *Stat. Sci.*, 9(3):334–354, 1994.

[117] S. Scherer, M. S. McPeek, and T. P. Speed. Atypical regions in large genomic DNA sequences. *Proc. Natl. Acad. Sci. USA*, 91(15):7134–7138, 1994.

[118] P. F. Baker, T. P. Speed, and F. K. Ligon. Estimating the influence of temperature on the survival of chinook salmon smolts (*Oncorhynchus tshawytscha*) migrating through the Sacramento–San Joaquin River Delta of California. *Can. J. Fish. Aquat. Sci.*, 52(4):855–863, 1995.

[119] J. Bastacky, C. Y. Lee, J. Goerke, H. Koushafar, D. Yager, L. Kenaga, T. P. Speed, Y. Chen, and J. A. Clements. Alveolar lining layer is thin and continuous: Low-temperature scanning electron microscopy of rat lung. *J. Appl. Physiol.*, 79(5):1615–1620, 1995.

[120] D. R. Goldstein, H. Zhao, and T. P. Speed. Relative efficiencies of χ^2 models of recombination for exclusion mapping and gene ordering. *Genomics*, 27(2):265–273, 1995.

[121] K. Jin, H.-N. Ho, T. P. Speed, and T. J. Gill, 3rd. Reproductive failure and the major histocompatibility complex. *Am. J. Hum. Genet.*, 56(6):1456–1467, 1995.

[122] K. Jin, T. P. Speed, and G. Thomson. Tests of random mating for a highly polymorphic locus: Application to HLA data. *Biometrics*, 51:1064–1076, 1995.

[123] M. S. McPeek and T. P. Speed. Modeling interference in genetic recombination. *Genetics*, 139:1031–1044, 1995.

[124] H. Zhao, M. S. McPeek, and T. P. Speed. Statistical analysis of chromatid interference. *Genetics*, 139:1057–1065, 1995.

[125] H. Zhao, T. P. Speed, and M. S. McPeek. Statistical analysis of crossover interference using the chi-square model. *Genetics*, 139:1045–1056, 1995.

[126] S. Zhou, R. A. Maller, and T. P. Speed. On a shared allele test of random mating. *Aust. J. Stat.*, 37(1):61–72, 1995.

[127] T. P. Speed and M. S. Waterman, editors. *Genetic Mapping and DNA Sequencing*, Volume 81 of *The IMA Volumes in Mathematics and its Applications*. Springer-Verlag, New York, 1996.

[128] K. Broman, T. P. Speed, and M. Tigges. Estimating antigen-responsive T cell frequencies in PBMC from human subjects. *J. Immunol. Methods*, 198:119–132, 1996.

[129] R. Guerra and T. P. Speed. Statistical issues arising in the analysis of DNA-DNA hybridization data. *Syst. Biol.*, 45(4):586–595, 1996.

[130] M.-Y. Leung, G. M. Marsh, and T. P. Speed. Over- and underrepresentation of short DNA words in herpesvirus genomes. *J. Comput. Biol.*, 3(3):345–360, 1996.

[131] S. Lin and T. P. Speed. Incorporating crossover interference into pedigree analysis using the χ^2 model. *Hum. Hered.*, 46:315–322, 1996.

[132] S. Lin and T. P. Speed. A note on the combination of estimates of a recombination fraction. *Ann. Hum. Genet.*, 60(3):251–257, 1996.

[133] D. O. Nelson and T. P. Speed. Recovering DNA sequences from electrophoresis data. In S. E. Levinson and L. Shepp, editors, *Image Models (and their Speech Model Cousins)*. Springer-Verlag, New York, 1996.

[134] T. P. Speed. What is a genetic map function? In T. P. Speed and M. S. Waterman, editors, *Genetic Mapping and DNA Sequencing*, Volume 81 of *The IMA Volumes in Mathematics and its Applications*, pages 65–88. Springer-Verlag, New York, 1996.

[135] H. Zhao and T. P. Speed. On genetic map functions. *Genetics*, 142(4):1369–1377, 1996.

[136] D. R. Goldstein, H. Zhao, and T. P. Speed. The effects of genotyping errors and interference on estimation of genetic distance. *Hum. Hered.*, 47(2): 86–100, 1997.

[137] K. Lange, T. P. Speed, and H. Zhao. The Poisson-skip model of crossing-over. *Ann. Appl. Probab.*, 7(2):299–313, 1997.

[138] S. Lin and T. P. Speed. An algorithm for haplotype analysis. *J. Comput. Biol.*, 4(4):535–546, 1997.

[139] D. O. Nelson, T. P. Speed, and B. Yu. The limits of random fingerprinting. *Genomics*, 40(1):1–12, 1997.

[140] T. P. Speed. Restricted maximum likelihood (REML). In S. Kotz, editor, *Encylopedia of Statistical Sciences*, Volume 2, pages 472–481. John Wiley & Sons, New York, 1997.

[141] B. Yu and T. P. Speed. Information and the clone mapping of chromosomes. *Ann. Stat.*, 25(1):169–185, 1997.

[142] K. Broman, T. P. Speed, and M. Tigges. Estimation of antigen-responsive T cell frequencies in PBMC from human subjects. *Stat. Sci.*, 13(1):4–8, 1998.

[143] S. Foote, T. P. Speed, and E. Handman. What can bioinformatics do for parasitology research? *Parasitol. Today*, 14(9):346–347, 1998.

[144] I. Kheterpal, L. Li, T. P. Speed, and R. A. Mathies. A three-wavelength labeling approach for DNA sequencing using energy transfer primers and capillary electrophoresis. *Electrophoresis*, 19(8–9):1403–1414, 1998.

[145] T. P. Speed. Genetic map functions. In P. Armitage and T. Colton, editors, *Encyclopedia of Biostatistics*. John Wiley & Sons, 1998.

[146] H. Zhao and T. P. Speed. Statistical analysis of half-tetrads. *Genetics*, 150(1):473–485, 1998.

[147] H. Zhao and T. P. Speed. Statistical analysis of ordered tetrads. *Genetics*, 150(1):459–472, 1998.

[148] H. Zhao and T. P. Speed. Stochastic modeling of the crossover process during meiosis. *Comm. Stat. A - Theor.*, 27(6):1557–1580, 1998.

[149] K. Broman and T. P. Speed. A review of methods for identifying QTLs in experimental crosses. In F. Seillier-Moisewitsch, editor, *Statistics in Molecular Biology and Genetics*, Volume 33 of *Lecture Notes–Monograph Series*, pages 114–142. Institute of Mathematical Statistics, Hayward, CA, 1999.

[150] S. Dudoit and T. P. Speed. A score test for linkage using identity by descent data from sibships. *Ann. Stat.*, 27(3):943–986, 1999.

[151] S. Dudoit and T. P. Speed. Triangle constraints for sib-pair identity by descent probabilities under a general multilocus model for disease susceptibility. In M. E. Halloran and S. Geisser, editors, *Statistics in Genetics*, Volume 112 of *The IMA Volumes in Mathematics and its Applications*, pages 181–221. Springer-Verlag, New York, 1999.

[152] A. Gelman and T. P. Speed. Corrigendum: Characterizing a joint probability distribution by conditionals. *J. Roy. Stat. Soc. B*, 61(2):483, 1999.

[153] L. Li and T. P. Speed. An estimate of the crosstalk matrix in four-dye fluorescence-based DNA sequencing. *Electrophoresis*, 20(7):1433–1442, 1999.

[154] S. Lin and T. P. Speed. Relative efficiencies of the Chi-square recombination models for gene mapping with human pedigree data. *Ann. Hum. Genet.*, 63:81–95, 1999.

[155] D. Nolan and T. P. Speed. Teaching statistics: Theory through applications. *Am. Stat.*, 53(4):370–375, 1999.

[156] L. J. Roberts, T. M. Baldwin, T. P. Speed, E. Handman, and S. J. Foote. Chromosomes X, 9, and the H2 locus interact epistatically to control *Leishmania major* infection. *Eur. J. Immunol.*, 29(9):3047–3050, 1999.

[157] T. P. Speed. Iterative proportional fitting. In P. Armitage and T. Colton, editors, *Encyclopedia of Biostatistics*. John Wiley & Sons, New York, 1999.

[158] T. P. Speed. Linkage analysis, Multipoint. In P. Armitage and T. Colton, editors, *Encyclopedia of Biostatistics*. John Wiley & Sons, New York, 1999.

[159] H. Zhao and T. P. Speed. On a Markov model for chromatid interference. In F. Seillier-Moiseiwitsch, editor, *Statistics in Molecular Biology and Genetics*, Volume 33 of *Lecture Notes–Monograph Series*, pages 1–20. Institute of Mathematical Statistics, Hayward, CA, 1999.

[160] M. J. Callow, S. Dudoit, E. L. Gong, T. P. Speed, and E. M. Rubin. Microarray expression profiling identifies genes with altered expression in HDL-deficient mice. *Genome Res.*, 10(12):2022–2029, 2000.

[161] S. Cawley and T. P. Speed. DNA sequencing with transposons. *J. Comput. Biol.*, 7(5):717–729, 2000.

[162] S. Dudoit and T. P. Speed. A score test for the linkage analysis of qualitative and quantitative traits based on identity by descent data from sib-pairs. *Biostatistics*, 1:1–26, 2000.

[163] D. R. Goldstein, S. Dudoit, and T. P. Speed. Power of a score test for quantitative trait linkage analysis of relative pairs. *Genet. Epidemiol.*, 19(Suppl. 1):S85–S91, 2000.

[164] A. Kavner, T. P. Speed, and R. Jeanloz. Statistical analysis of phase-boundary observations. In H. Aoki, Y. Syono, and R. J. Hemley, editors, *Mineralogy Meets Physics: Condensed-Matter Physics in Geosciences*, pages 71–79. Cambridge University Press, Cambridge, UK, 2000.

[165] L. Li and T. P. Speed. Parametric deconvolution of positive spike trains. *Ann. Stat.*, 28(5):1279–1301, 2000.

[166] D. Nolan and T. P. Speed. *Stat Labs: Mathematical Statistics Through Applications*. Springer Texts in Statistics. Springer-Verlag, New York, 2000.

[167] J. P. Rubio, T. P. Speed, M. Bahlo, T. J. Kilpatrick, and S. J. Foote. The current state of multiple sclerosis genetic research. *Ann. Acad. Med. Singapore*, 29(3):322–330, 2000.

[168] H. Bengtsson, B. Calder, I. S. Mian, M. Callow, E. Rubin, and T. P. Speed. Identifying differentially expressed genes in cDNA microarray experiments authors. *Sci. Aging Knowl. Environ.*, 19(12):vp8, 2001.

[169] S. Cawley, A. Wirth, and T. P. Speed. Phat—a gene finding program for *Plasmodium falciparum*. *Mol. Biochem. Parasit.*, 118:167–174, 2001.

[170] D. R. Goldstein, S. Dudoit, and T. P. Speed. Power and robustness of a score test for linkage analysis of quantitative trait based on identity by descent data on sib pairs. *Genet. Epidemiol.*, 20(4):415–431, 2001.

[171] T. Triglia, J. Thompson, S. R. Carvana, M. Delorenzi, T. P. Speed, and A. F. Cowman. Identification of proteins from *Plasmodium falciparum* that are homologous to the reticulocyte binding proteins in *Plasmodium vivax*. *Infect. Immun.*, 69(2):1084–1092, 2001.

[172] Y. H. Yang, M. J. Buckley, and T. P. Speed. Analysis of cDNA microarray images. *Brief. Bioinform.*, 2(4):341–349, 2001.

[173] Y. H. Yang, S. Dudoit, P. Luu, and T. P. Speed. Normalization of cDNA microarray data. In M. L. Bittner, Y. Chen, A. N. Dorsel, and E. R. Dougherty, editors, *Microarrays: Optical Technologies and Informatics*, Volume 4266 of *Proceedings of SPIE*, pages 141–152. Society of Photo-optical Instrumentation Engineers, 2001.

[174] P. Banerjee, M. Bahlo, J. R. Schwartz, G. G. Loots, K. A. Houston, I. Dubchak, T. P. Speed, and E. M. Rubin. SNPs in putative regulatory regions identified by human mouse comparative sequencing and transcription factor binding site data. *Mamm. Genome*, 13(10):554–557, 2002.

[175] K. W. Broman and T. P. Speed. A model selection approach for the identification of quantitative trait loci in experimental crosses. *J. Roy. Stat. Soc. B*, 64:641–656, 2002.

[176] P. A. Butcher, E. R. Williams, D. Whitaker, S. Ling, T. P. Speed, and G. F. Moran. Improving linkage analysis in outcrossed forest trees – an example from *Acacia mangium*. *Theor. Appl. Genet.*, 104(6–7):1185–1191, 2002.

[177] M. Delorenzi, A. Sexton, H. Shams-Eldin, R. T. Schwartz, T. P. Speed, and L. Schofield. Genes for glycosylphosphatidylinositol toxin biosynthesis in *Plasmodium falciparum*. *Infect. Immun.*, 70(8):4510–4522, 2002.

[178] M. Delorenzi and T. P. Speed. An HMM model for coiled-coil domains and a comparison with PSSM-based predictions. *Bioinformatics*, 18(4):617–625, 2002.

[179] E. Díaz, Y. Ge, Y. H. Yang, K. C. Loh, T. A. Serafini, Y. Okazaki, Y. Hayashizaki, T. P. Speed, J. Ngai, and P. Scheiffele. Molecular analysis of gene expression in the developing pontocerebellar projection system. *Neuron*, 36(3):417–434, 2002.

[180] S. Dudoit, J. Fridlyand, and T. P. Speed. Comparison of discrimination methods for the classification of tumors using gene expression data. *J. Am. Stat. Assoc.*, 97(457):77–87, 2002.

[181] S. Dudoit, Y. H. Yang, M. J. Callow, and T. P. Speed. Statistical methods for identifying differentially expressed genes in replicated cDNA microarray experiments. *Stat. Sinica*, 12(1):111–139, 2002.

[182] M. Grote and T. P. Speed. Approximate Ewens formulae for symmetric overdominance selection. *Ann. Appl. Probab.*, 12(2):637–663, 2002.

[183] I. Lönnstedt and T. P. Speed. Replicated microarray data. *Stat. Sinica*, 12:31–46, 2002.

[184] J. P. Rubio, M. Bahlo, H. Butzkueven, I. A. F. van Der Mei, M. M. Sale, J. L. Dickinson, P. Groom, L. J. Johnson, R. D. Simmons, B. Tait, M. Varney, B. Taylor *et al.* Genetic dissection of the human leukocyte antigen region by use of haplotypes of Tasmanians with multiple sclerosis. *Am. J. Hum. Genet.*, 70(5):1125–1137, 2002.

[185] T. P. Speed. Biological sequence analysis. In D. Li, editor, *Proceedings of the International Congress of Mathematicians*, Volume III, pages 97–106. Higher Education Press, 2002.

[186] T. P. Speed. Gene expression analysis. In R. C. Elston, J. M. Olson, and L. Palmer, editors, *Biostatistical Genetics and Genetic Epidemiology*, Wiley Reference Series in Biostatistics, pages 286–293. Wiley, London, 2002.

[187] T. P. Speed. John Tukey's contributions to analysis of variance. *Ann. Stat.*, 30(6):1649–1665, 2002.

[188] R. C. A. Symons, M. J. Daly, J. Fridlyand, T. P. Speed, W. D. Cook, S. Gerondakis, A. W. Harris, and S. J. Foote. Multiple genetic loci modify susceptibility to plasmacytoma-related morbidity in Eμ–v–*abl* transgenic mice. *Proc. Natl. Acad. Sci. USA*, 99(17):11299–11304, 2002.

[189] A. S. Wilson, B. G. Hobbs, T. P. Speed, and P. E. Rakoczy. The microarray: Potential applications for ophthalmic research. *Mol. Vis.*, 8:259–270, 2002.

[190] Y. H. Yang, M. J. Buckley, S. Dudoit, and T. P. Speed. Comparison of methods for image analysis on cDNA microarray data. *J. Comput. Graph. Stat.*, 11(1):108–136, 2002.

[191] Y. H. Yang, S. Dudoit, P. Luu, D. M. Lin, V. Peng, J. Ngai, and T. P. Speed. Normalization for cDNA microarray data: A robust composite method addressing single and multiple slide systematic variation. *Nucleic Acids Res.*, 30(4):e15, 2002.

[192] Y. H. Yang and T. P. Speed. Design issues for cDNA microarray experiments. *Nat. Rev. Genet.*, 3(8):579–588, 2002.

[193] Y. H. Yang and T. P. Speed. Direct and indirect hybridizations for cDNA microarray experiments. *Sankhyā Ser. A*, 3:706–720, 2002.

[194] Y. H. Yang and T. P. Speed. An introduction to microarray bioinformatics (parts 1–4). In D. Bowtell and J. Sambrook, editors, *DNA Microarrays: A Molecular Cloning Manual*, pages 509–551. Cold Spring Harbor Laboratory Press, Cold Spring Harbor, NY, 1st edition, 2002.

[195] V. B. Yap and T. P. Speed. Modeling genomic DNA base substitution. In *JSM 2002: Statistics in an Era of Technological Change*, pages 3855–3864. Proceedings of Joint Statistical Meetings, American Statistical Association, Alexandria, VA, 2002.

[196] R.-F. Yeh, T. P. Speed, M. S. Waterman, and X. Li. Predicting progress in shotgun sequencing with paired ends. Unpublished, 2002.

[197] T. P. Speed, editor. *Statistical Analysis of Gene Expression Microarray Data*. Interdisciplinary Statistics. CRC Press, Boca Raton, FL, 2003.

[198] A. Barczak, M. V. Rodriguez, K. Hanspers, L. Koth, Y. C. Tai, B. M. Bolstad, T. P. Speed, and D. J. Erle. Spotted long oligonucleotide arrays for human gene expression analysis. *Genome Res.*, 13(7):1775–1785, 2003.

[199] B. M. Bolstad, R. A. Irizarry, M. Åstrand, and T. P. Speed. A comparison of normalization methods for high density oligonucleotide array data based on variance and bias. *Bioinformatics*, 19(2):185–193, 2003.

[200] T. F. de Koning-Ward, R. A. O'Donnell, D. R. Drew, R. Thomson, T. P. Speed, and B. S. Crabb. A new rodent model to assess blood stage immunity to the *Plasmodium falciparum* antigen merozoite surface protein 1_19 reveals a protective role for invasion inhibitory antibodies. *J. Exp. Med.*, 198(6):869–875, 2003.

[201] E. Díaz, Y. H. Yang, T. Ferreira, K. C. Loh, Y. Okazaki, Y. Hayashizaki, M. Tessier-Lavigne, T. P. Speed, and J. Ngai. Analysis of gene expression in the developing mouse retina. *Proc. Natl. Acad. Sci. USA*, 100(9):5491–5496, 2003.

[202] Y. Ge, S. Dudoit, and T. P. Speed. Resampling-based multiple testing for microarray data analysis. *TEST*, 12(1):1–44, 2003.

[203] A. N. Hodder, D. R. Drew, V. C. Epa, M. Delorenzi, R. Bourgon, S. K. Miller, R. L. Moritz, D. F. Frecklington, R. J. Simpson, T. P. Speed, R. N. Pike, and B. S. Crabb. Enzymic, phylogenetic and structural characterisation of the unusual papain-like protease domain of *Plasmodium falciparum* SERA5. *J. Biol. Chem.*, 278(48):48169–48177, 2003.

[204] R. A. Irizarry, B. M. Bolstad, F. Collin, L. M. Cope, B. Hobbs, and T. P. Speed. Summaries of Affymetrix GeneChip probe level data. *Nucleic Acids Res.*, 31(4):e15, 2003.

[205] R. A. Irizarry, B. Hobbs, F. Collin, Y. D. Beazer-Barclay, K. J. Antonellis, U. Scherf, and T. P. Speed. Exploration, normalization and summaries of high density oligonucleotide array probe level data. *Biostatistics*, 4(2):249–264, 2003.

[206] E. A. Kapp, F. Schütz, G. E. Reid, J. S. Eddes, R. L. Moritz, R. A. J. O'Hair, T. P. Speed, and R. J. Simpson. Mining a tandem mass spectrometry database to determine the trends and global factors influencing peptide fragmentation. *Anal. Chem.*, 75(22):6251–6264, 2003.

[207] B. I. P. Rubinstein, M. McAuliffe, S. Cawley, M. Palaniswami, K. Ramamohanarao, and T. P. Speed. Machine learning in low-level microarray analysis. *ACM SIGKDD Expl.*, 5(2):130–139, 2003.

[208] F. Schütz, E. A. Kapp, R. J. Simpson, and T. P. Speed. Deriving statistical models for predicting peptide tandem MS product ion intensities. *Biochem. Soc. Trans.*, 31(6):1479–1483, 2003.

[209] G. K. Smyth and T. P. Speed. Normalization of cDNA microarray data. *Methods*, 31(4):265–273, 2003.

[210] G. K. Smyth, Y. H. Yang, and T. P. Speed. Statistical issues in cDNA microarray data analysis. In M. J. Brownstein and A. Khodursky, editors, *Functional Genomics: Methods and Protocols*, Volume 224 of *Methods in Molecular Biology*, pages 111–136. Humana Press, Totowa, NJ, 2003.

[211] A. S. Wilson, B. G. Hobbs, W. Y. Shen, T. P. Speed, A. S. Schmidt, C. G. Begley, and P. E. Rakoczy. Argon laser photocoagulation–induced modification of gene expression in the retina. *Invest. Ophth. Vis. Sci.*, 44(4):1426–1434, 2003.

[212] Y. H. Yang and T. P. Speed. Design and analysis of comparative microarray experiments. In T. P. Speed, editor, *Statistical Analysis of Gene Expression Microarray Data*, Interdisciplinary Statistics, pages 35–92. CRC Press, Boca Raton, FL, 2003.

[213] T. Beißbarth, L. Hyde, G. K. Smyth, C. Job, W.-M. Boon, S.-S. Tan, H. S. Scott, and T. P. Speed. Statistical modeling of sequencing errors in SAGE libraries. *Bioinformatics*, 20(Suppl. 1):i31–i39, 2004.

[214] T. Beißbarth and T. P. Speed. GOstat: Find statistically overrepresented Gene Ontologies within a group of genes. *Bioinformatics*, 20(9):1464–1465, 2004.

[215] B. M. Bolstad, F. Collin, K. M. Simpson, R. A. Irizarry, and T. P. Speed. Experimental design and low-level analysis of microarray data. *Int. Rev. Neurobiol.*, 60:25–58, 2004.

[216] R. Bourgon, M. Delorenzi, T. Sargeant, A. N. Hodder, B. S. Crabb, and T. P. Speed. The serine repeat antigen (SERA) gene family phylogeny in *Plasmodium*: The impact of GC content and reconciliation of gene and species trees. *Mol. Biol. Evol.*, 21(11):2161–2171, 2004.

[217] L. M. Cope, R. A. Irizarry, H. A. Jaffee, Z. Wu, and T. P. Speed. A benchmark for Affymetrix GeneChip expression measures. *Bioinformatics*, 20(3):323–331, 2004.

[218] L. Li and T. P. Speed. Deconvolution of sparse positive spikes. *J. Comput. Graph. Stat.*, 13(4):853–870, 2004.

[219] D. M. Lin, Y. H. Yang, J. A. Scolnick, L. J. Brunet, H. Marsh, V. Peng, Y. Okazaki, Y. Hayashizaki, T. P. Speed, and J. Ngai. Spatial patterns of gene expression in the olfactory bulb. *Proc. Natl. Acad. Sci. USA*, 101(34):hbox12718–12723, 2004.

[220] P. G. Marciano, J. Brettschneider, E. Manduchi, J. E. Davis, S. Eastman, R. Raghupathi, K. E. Saatman, T. P. Speed, C. J. Stoeckert, Jr., J. H. Eberwine, and T. K. McIntosh. Neuron-specific mRNA complexity responses during hippocampal apoptosis after traumatic brain injury. *J. Neurosci.*, 24(12):2866–2876, 2004.

[221] R. L. Moritz, H. Ji, F. Schütz, L. M. Connolly, E. A. Kapp, T. P. Speed, and R. J. Simpson. A proteome strategy for fractionating proteins and peptides using continuous free-flow electrophoresis coupled off-line to reversed-phase high-performance liquid chromatography. *Anal. Chem.*, 76(16):4811–4824, 2004.

[222] N. Rabbee, D. Speca, N. J. Armstrong, and T. P. Speed. Power calculations for selective genotyping in QTL mapping in backcross mice. *Genet. Res.*, 84(2):103–108, 2004.

[223] J. P. Rubio, M. Bahlo, N. Tubridy, J. Stankovich, R. Burfoot, H. Butzkueven, C. Chapman, L. Johnson, M. Marriott, G. Mraz, B. Tait, C. Wilkinson *et al.* Extended haplotype analysis in the HLA complex reveals an increased frequency of the HFE-C282Y mutation in individuals with multiple sclerosis. *Hum. Genet.*, 114(6):573–580, 2004.

[224] V. B. Yap and T. P. Speed. Modeling DNA base substitution in large genomic regions from two organisms. *J. Mol. Evol.*, 58(1):12–18, 2004.

[225] X. Zhao, H. Huang, and T. P. Speed. Finding short DNA motifs using permuted Markov models. In D. Gusfield, P. Bourne, S. Istrail, P. Pevzner, and M. Waterman, editors, *RECOMB 2004: Proceedings of the Eighth Annual Conference on Research in Computational Molecular Biology*, pages 68–75. Association of Computing Machinery, New York, 2004.

[226] M. Bahlo and T. P. Speed. How many genes? Mapping mouse traits. In R. Peck, G. Casella, G. Cobb, R. Hoerl, and D. Nolan, editors, *Statistics: A Guide to the Unknown*, Chapter 17. Duxbury Press, Pacific Grove, CA, 4th edition, 2005.

[227] P. N. Baird, S. J. Foote, D. A. Mackey, J. Craig, T. P. Speed, and A. Bureau. Evidence for a novel glaucoma locus at chromosome 3p21-22. *Hum. Genet.*, 117(2–3):249–257, 2005.

[228] A. Barrier, A. Lemoine, P.-Y. Boelle, C. Tse, D. Brault, F. Chiappini, J. Brettschneider, F. Lacaine, S. Houry, M. Huguier, M. J. van der Laan, T. P. Speed *et al.* Colon cancer prognosis prediction by gene expression profiling. *Oncogene*, 24(40):6155–6164, 2005.

[229] T. Beißbarth, J. A. Tye-Din, G. K. Smyth, T. P. Speed, and R. P. Anderson. A systematic approach for comprehensive T-cell epitope discovery using peptide libraries. *Bioinformatics*, 21(Suppl. 1):i29–i37, 2005.

[230] B. M. Bolstad, F. Collin, J. Brettschneider, K. Simpson, L. Cope, R. A. Irizarry, and T. P. Speed. Quality control of Affymetrix GeneChip data. In R. Gentleman, V. J. Carey, W. Huber, R. A. Irizarry, and S. Dudoit, editors, *Bioinformatics and Computational Biology Solutions using R and Bioconductor*, Statistics for Biology and Health, Chapter 3, pages 33–48. Springer, New York, 2005.

[231] M. Dai, P. Wang, A. D. Boyd, G. Kostov, B. Athey, E. G. Jones, W. E. Bunney, R. M. Myers, T. P. Speed, H. Akil, S. J. Watson, and F. Meng. Evolving gene/transcript definitions significantly alter the interpretation of GeneChip data. *Nucleic Acids Res.*, 33(20):e175, 2005.

[232] S. J. Foote, J. P. Rubio, M. Bahlo, T. J. Kilpatrick, T. P. Speed, J. Stankovich, R. Burfoot, H. Butzkueven, L. Johnson, C. Wilkinson, B. Taylor, M. Sale *et al.* Multiple sclerosis: A haplotype association study. In G. Bock and J. Goode, editors, *Genetics of Autoimmunity*, Volume 267 of *Novartis Foundation Symposia*, pages 31–39. John Wiley & Sons, Chichester, UK, 2005.

[233] V. A. Likic, V. A. Gooley, T. P. Speed, and E. E. Strehler. A statistical approach to the interpretation of molecular dynamics simulations of calmodulin equilibrium dynamics. *Protein Sci.*, 14(12):2955–2963, 2005.

[234] A. Sakthianandeswaren, C. M. Elso, K. Simpson, J. M. Curtis, B. Kumar, T. P. Speed, E. Handman, and S. J. Foote. The wound repair response controls outcome to cutaneous leishmaniasis. *Proc. Natl. Acad. Sci. USA*, 102(43):15551–15556, 2005.

[235] K. M. Simpson, J. Baum, R. T. Good, E. A. Winzeler, A. F. Cowman, and T. P. Speed. A comparison of match-only algorithms for the analysis of *Plasmodium falciparum* oligonucleotide arrays. *Int. J. Parasitol.*, 35(5):523–531, 2005.

[236] J. Stankovich, M. Bahlo, J. P. Rubio, C. R. Wilkinson, R. Thomson, A. Banks, M. Ring, S. J. Foote, and T. P. Speed. Identifying nineteenth century genealogical links from genotypes. *Hum. Genet.*, 117(2–3):188–199, 2005.

[237] T. N. Wilkinson, T. P. Speed, G. W. Tregear, and R. A. Bathgate. Coevolution of the relaxin-like peptides and their receptors. *Ann. N. Y. Acad. Sci.*, 1041:534–539, 2005.

[238] T. N. Wilkinson, T. P. Speed, G. W. Tregear, and R. A. Bathgate. Evolution of the relaxin-like peptide family: From neuropeptide to reproduction. *Ann. N. Y. Acad. Sci.*, 1041:530–533, 2005.

[239] T. N. Wilkinson, T. P. Speed, G. W. Tregear, and R. A. D. Bathgate. Evolution of the relaxin-like peptide family. *BMC Evol. Biol.*, 5(1):14, 2005.

[240] V. B. Yap and T. P. Speed. Rooting a phylogenetic tree with nonreversible substitution models. *BMC Evol. Biol.*, 5(1):2, 2005.

[241] X. Zhao, H. Huang, and T. P. Speed. Finding short DNA motifs using permuted Markov models. *J. Comput. Biol.*, 12(6):894–906, 2005.

[242] N. J. Armstrong, T. C. Brodnicki, and T. P. Speed. Mind the gap: Analysis of marker-assisted breeding strategies for inbred mouse strains. *Mamm. Genome*, 17(4):273–287, 2006.

[243] N. J. Armstrong, M. S. McPeek, and T. P. Speed. Incorporating interference into linkage analysis for experimental crosses. *Biostatistics*, 7(3):374–386, 2006.

[244] M. Bahlo, J. Stankovich, T. P. Speed, J. P. Rubio, R. K. Burfoot, and S. J. Foote. Detecting genome wide haplotype sharing using SNP or microsatellite haplotype data. *Hum. Genet.*, 119(1–2):38–50, 2006.

[245] J. Baum, A. T. Papenfuss, B. Baum, T. P. Speed, and A. F. Cowman. Regulation of apicomplexan actin-based motility. *Nat. Rev. Microbiol.*, 4(8): 621–628, 2006.

[246] K. Belov, J. E. Deakin, A. T. Papenfuss, M. L. Baker, S. D. Melman, H. V. Siddle, N. Gouin, D. L. Goode, T. J. Sargeant, M. D. Robinson, M. J. Wakefield, S. Mahony *et al.* Reconstructing an ancestral mammalian immune supercomplex from a marsupial major histocompatibility complex. *PLoS Biology*, 4(3):e46, 2006.

[247] N. Binz, C. E. Graham, K. Simpson, Y. K. Y. Lai, W.-Y. Shen, C.-M. Lai, T. P. Speed, and P. E. Rakoczy. Long-term effect of therapeutic laser photocoagulation on gene expression in the eye. *FASEB J.*, 20(2):383–385, 2006.

[248] R. Bourgon and T. P. Speed. A model for chromatin immunoprecipitation/high density tiling array experiments: Implications for data analysis. In M. Takahashi and N. Winegarden, editors, *Profiling Transcriptional Activity with Promoter and CpG Microarrays*. DNA Press, Eagleville, PA, 2006.

[249] J. E. Deakin, A. T. Papenfuss, K. Belov, J. G. R. Cross, P. Coggill, S. Palmer, S. Sims, T. P. Speed, S. Beck, and J. A. Marshall Graves. Evolution and comparative analysis of the MHC Class III inflammatory region. *BMC Genomics*, 7:281, 2006.

[250] J. C. Dugas, Y. C. Tai, T. P. Speed, J. Ngai, and B. A. Barres. Functional genomic analysis of oligodendrocyte differentiation. *J. Neurosci.*, 26(43):10967–10983, 2006.

[251] Y. Gilad, A. Oshlack, G. K. Smyth, T. P. Speed, and K. P. White. Expression profiling in primates reveals a rapid evolution of human transcription factors. *Nature*, 440(7081):242–245, 2006.

[252] P. R. Gilson, T. Nebl, D. Vukcevic, R. L. Moritz, T. Sargeant, T. P. Speed, L. Schofield, and B. S. Crabb. Identification and stoichiometry of glycosylphosphatidylinositol-anchored membrane proteins of the human malaria parasite *Plasmodium falciparum*. *Mol. Cell. Proteomics*, 5(7):1286–1299, 2006.

[253] E. Lin, T. Pappenfuss, R. B. Tan, D. Senyschyn, M. Bahlo, T. P. Speed, and S. J. Foote. Mapping of the *Plasmodium chabaudi* resistance locus *char2*. *Infect. Immun.*, 74(10):5814–5819, 2006.

[254] R. M. Neve, K. Chin, J. Fridlyand, J. Yeh, F. L. Baehner, T. Fevr, L. Clark, N. Bayani, J. P. Coppe, F. Tong, T. P. Speed, P. T. Spellman *et al.* A collection of breast cancer cell lines for the study of functionally distinct cancer subtypes. *Cancer Cell*, 10(6):515–527, 2006.

[255] N. Rabbee and T. P. Speed. A genotype calling algorithm for affymetrix SNP arrays. *Bioinformatics*, 22(1):7–12, 2006.

[256] T. J. Sargeant, M. Marti, E. Caler, J. M. Carlton, T. P. Speed, and A. F. Cowman. Lineage-specific expansion of proteins exported to erythrocytes in malaria parasites. *Genome Biol.*, 7(2):R12, 2006.

[257] D. J. Speca, N. Rabbee, D. Chihara, T. P. Speed, and A. S. Peterson. A genetic screen for behavioral mutations that perturb dopaminergic homeostasis in mice. *Genes Brain Behav.*, 5(1):19–28, 2006.

[258] J. Stankovich, C. J. Cox, R. B. Tan, D. S. Montgomery, S. J. Huxtable, J. P. Rubio, M. G. Ehm, L. Johnson, H. Butzkueven, T. J. Kilpatrick, T. P. Speed, A. D. Roses *et al.* On the utility of data from the International HapMap project for Australian association studies. *Hum. Genet.*, 119(1–2):220–222, 2006.

[259] Y. C. Tai and T. P. Speed. A multivariate empirical Bayes statistic for replicated microarray time course data. *Ann. Stat.*, 34(5):2387–2412, 2006.

[260] M. P. Vawter, H. Tomita, F. Meng, B. Bolstad, J. Li, S. Evans, P. Choudary, M. Atz, L. Shao, C. Neal, D. M. Walsh, M. Burmeister *et al.* Mitochondrial-related gene expression changes are sensitive to agonal-pH state: Implications for brain disorders. *Mol. Psychiatr.*, 11(7):663–679, 2006.

[261] S. Wormald, D. J. Hilton, G. K. Smyth, and T. P. Speed. Proximal genomic localization of STAT1 binding and regulated transcriptional activity. *BMC Genomics*, 7:254, 2006.

[262] S. Wormald, J.-G. Zhang, D. L. Krebs, L. A. Mielke, J. Silver, W. S. Alexander, T. P. Speed, N. A. Nicola, and D. J. Hilton. The comparative roles of suppressor of cytokine signaling-1 and -3 in the inhibition and desensitization of cytokine signaling. *J. Biol. Chem.*, 281(16):11135–11143, 2006.

[263] K. Belov, C. E. Sanderson, J. E. Deakin, E. S. W. Wong, D. Assange, K. A. McColl, A. Gout, B. de Bono, A. D. Barrow, T. P. Speed, J. Trowsdale, and A. T. Papenfuss. Characterization of the opossum immune genome provides insights into the evolution of the mammalian immune system. *Genome Res.*, 17(7):982–991, 2007.

[264] J. Brettschneider, F. Collin, B. Bolstad, and T. P. Speed. Quality assessment for short oligonucleotide microarray data. *Technometrics*, 50(3):241–264, 2007.

[265] B. Carvalho, H. Bengtsson, T. P. Speed, and R. A. Irizarry. Exploration, normalization, and genotype calls of high density oligonucleotide SNP array data. *Biostatistics*, 8(2):485–499, 2007.

[266] Y. Ge, S. C. Sealfon, and T. P. Speed. Some step-down procedures controlling the false discovery rate under dependence. *Stat. Sinica*, 18(3):881–904, 2007.

[267] Y. Ge, S. C. Sealfon, C.-H. Tseng, and Speed T. P. A Holm-type procedure controlling the false discovery rate. *Stat. Prob. Lett.*, 77(18):1756–1762, 2007.

[268] D. Georlette, S. Ahn, D. M. MacAlpine, Cheung E., P. W. Lewis, E. L. Beall, S. P. Bell, T. P. Speed, J. R. Manak, and M. R. Botchan. Genomic profiling and expression studies reveal both positive and negative activities for the *Drosophila* Myb–MuvB/dREAM complex in proliferating cells. *Genes Dev.*, 21(22):2880–2896, 2007.

[269] N. Levy, X. Zhao, H. Tang, R. B. Jaffe, T. P. Speed, and D. C. Leitman. Multiple transcription factor elements collaborate with estrogen receptor α to activate an inducible estrogen response element in the *NKG2E* gene. *Endocrinology*, 148(7):3449–3458, 2007.

[270] J. E. McCoubrie, S. K. Miller, T. Sargeant, R. T. Good, A. N. Hodder, T. P. Speed, T. F. de Koning-Ward, and B. S. Crabb. Evidence for a common role for the serine-type *Plasmodium falciparum* serine repeat antigen proteases: Implications for vaccine and drug design. *Infect. Immun.*, 75(12):5565–5574, 2007.

[271] T. S. Mikkelsen, M. J. Wakefield, B. Aken, C. T. Amemiya, J. L. Chang, S. Duke, M. Garber, A. J. Gentles, L. Goodstadt, A. Heger, J. Jurka, M. Kamal *et al.* Genome of the marsupial *Monodelphis domestica* reveals innovation in non-coding sequences. *Nature*, 447(7141):167–177, 2007.

[272] X. Qin, S. Ahn, T. P. Speed, and G. M. Rubin. Global analyses of mRNA translational control during early *Drosophila* embryogenesis. *Genome Biol.*, 8(4):R63–R63, 2007.

[273] M. D. Robinson, D. P. De Souza, W. Wai Keen, E. C. Saunders, M. J. McConville, T. P. Speed, and V. A. Likic. A dynamic programming approach for the alignment of signal peaks in multiple gas chromatography-mass spectrometry experiments. *BMC Bioinformatics*, 8(1):419, 2007.

[274] M. D. Robinson and T. P. Speed. A comparison of Affymetrix gene expression arrays. *BMC Bioinformatics*, 8(1):449, 2007.

[275] J. P. Rubio, M. Bahlo, J. Stankovich, R. K. Burfoot, L. J. Johnson, S. Huxtable, H. Butzkueven, L. Lin, B. V. Taylor, T. P. Speed, T. J. Kilpatrick,

E. Mignot *et al.* Analysis of extended HLA haplotypes in multiple sclerosis and narcolepsy families confirms a predisposing effect for the class I region in Tasmanian MS patients. *Immunogenetics*, 59(3):177–186, 2007.

[276] E. Sliwerska, F. Meng, T. P. Speed, E. G. Jones, W. E. Bunney, H. Akil, S. J. Watson, and M. Burmeister. SNPs on chips: The hidden genetic code in expression arrays. *Biol. Psychiat.*, 61(1):13–16, 2007.

[277] M. L. Szpara, K. Vranizan, Y. C. Tai, C. S. Goodman, T. P. Speed, and J. Ngai. Analysis of gene expression during neurite outgrowth and regeneration. *BMC Neurosci.*, 8(1):100, 2007.

[278] N. R. Zhang, M. C. Wildermuth, and T. P. Speed. Transcription factor binding site prediction with multivariate gene expression data. *Ann. Appl. Stat.*, 2(1):332–365, 2007.

[279] H. Bengtsson, R. Irizarry, B. Carvalho, and T. P. Speed. Estimation and assessment of raw copy numbers at the single locus level. *Bioinformatics*, 24:759–767, 2008.

[280] R. K. Burfoot, C. J. Jensen, J. Field, J. Stankovich, M. D. Varney, L. J. Johnson, H. Butzkueven, D. Booth, M. Bahlo, B. D. Tait, B. V. Taylor, T. P. Speed *et al.* SNP mapping and candidate gene sequencing in the class I region of the HLA complex: Searching for multiple sclerosis susceptibility genes in Tasmanians. *Tissue Antigens*, 71(1):42–50, 2008.

[281] M. Guipponi, M. Y. Toh, J. Tan, D. Park, K. Hanson, E. Ballana, D. Kwong, P. Z. Cannon, Q. Wu, A. Gout, M. Delorenzi, T. P. Speed *et al.* An integrated genetic and functional analysis of the role of type II transmembrane serine proteases (TMPRSSs) in hearing loss. *Hum. Mutat.*, 29(1):130–141, 2008.

[282] I. Hallgrímsdóttir and T. P. Speed. The power of two-locus affected sib-pair linkage analysis to detect interacting disease loci. *Genet. Epidemiol.*, 32(1):84–88, 2008.

[283] S. Husain, C. Yildirim-Toruner, J. P. Rubio, J. Field, The Southern MS Genetics Consortium, M. Schwalb, S. Cook, M. Devoto, and E. Vitale. Variants of *ST8SIA1* are associated with risk of developing multiple sclerosis. *PLoS One*, 3(7):e2653, 2008.

[284] N. Levy, D. Tatomer, C. B. Herber, X. Zhao, H. Tang, T. Sargeant, L. J. Ball, J. Summers, T. P. Speed, and D. C. Leitman. Differential regulation of native estrogen receptor-regulatory elements by estradiol, tamoxifen, and raloxifene. *Mol. Endocrinol.*, 22(2):287–303, 2008.

[285] X.-Y. Li, S. MacArthur, R. Bourgon, D. Nix, D. A. Pollard, V. N. Iyer, A. Hechmer, L. Simirenko, M. Stapleton, C. L. Luengo Hendriks, H. C. Chu, N. Ogawa *et al.* Transcription factors bind thousands of active and inactive regions in the *Drosophila* blastoderm. *PLoS Biology*, 6(2):e27, 2008.

[286] J. Michaud, K. M. Simpson, R. Escher, K. Buchet-Poyau, T. Beißbarth, C. Carmichael, M. E. Ritchie, F. Schütz, P. Cannon, M. Liu, X. Shen, Y. Ito *et al.* Integrative analysis of RUNX1 downstream pathways and target genes. *BMC Genomics*, 9(1):363, 2008.

[287] S. Mukherjee and T. P. Speed. Network inference using informative priors. *Proc. Natl. Acad. Sci. USA*, 105(38):14313–14318, 2008.

[288] The Cancer Genome Atlas Research Network. Comprehensive genomic characterization defines human glioblastoma genes and core pathways. *Nature*, 455(7216):1061–1068, 2008.

[289] E. Purdom, K. M. Simpson, M. D. Robinson, J. G. Conboy, A. V. Lapuk, and T. P. Speed. FIRMA: A method for detection of alternative splicing from exon array data. *Bioinformatics*, 24(15):1707–1714, 2008.

[290] J. P. Rubio, J. Stankovich, J. Field, N. Tubridy, M. Marriott, C. Chapman, M. Bahlo, D. Perera, L. J. Johnson, B. D. Tait, M. D. Varney, T. P. Speed *et al.* Replication of *KIAA0350*, *IL2RA*, *RPL5* and *CD58* as multiple sclerosis susceptibility genes in Australians. *Genes Immun.*, 9:624–630, 2008.

[291] L. J. Ball, N. Levy, X. Zhao, C. Griffin, M. Tagliaferri, I. Cohen, W. A. Ricke, T. P. Speed, G. L. Firestone, and D. C. Leitman. Cell type- and estrogen receptor-subtype specific regulation of selective estrogen receptor modulator regulatory elements. *Mol. Cell. Endocrinol.*, 299(2):204–211, 2009.

[292] H. Bengtsson, A. Ray, P. Spellman, and T. P. Speed. A single-sample method for normalizing and combining full-resolution copy numbers from multiple platforms, labs and analysis methods. *Bioinformatics*, 25(7):861–867, 2009.

[293] H. Bengtsson, P. Wirapati, and T. P. Speed. A single-array preprocessing method for estimating full-resolution raw copy numbers from all Affymetrix genotyping arrays including GenomeWideSNP 5 & 6. *Bioinformatics*, 25(17):2149–2156, 2009.

[294] D. Chandran, Y. C. Tai, G. Hather, J. Dewdney, C. Denoux, D. G. Burgess, F. M. Ausubel, T. P. Speed, and M. C. Wildermuth. Temporal global expression data reveal known and novel salicylate-impacted processes and regulators mediating powdery mildew growth and reproduction on Arabidopsis. *Plant Physiol.*, 149(3):1435–1451, 2009.

[295] Y. Ge, S. C. Sealfon, and T. P. Speed. Multiple testing and its applications to microarrays. *Stat. Methods Med. Res.*, 18:543–63, 2009.

[296] R. A. Irizarry, C. Wang, Y. Zhou, and T. P. Speed. Gene set enrichment analysis made simple. *Stat. Methods Med. Res.*, 18:565–75, 2009.

[297] N. Levy, S. Paruthiyil, X. Zhao, O. I. Vivar, E. F. Saunier, C. Griffin, M. Tagliaferri, I. Cohen, T. P. Speed, and D. C. Leitman. Unliganded estrogen receptor-beta regulation of genes is inhibited by tamoxifen. *Mol. Cell. Endocrinol.*, 315:201–7, 2009.

[298] S. Loi, C. Sotiriou, B. Haibe-Kains, F. Lallemand, N. M. Conus, M. J. Piccart, T. P. Speed, and G. A. McArthur. Gene expression profiling identifies activated growth factor signaling in poor prognosis (Luminal-B) estrogen receptor positive breast cancer. *BMC Med. Genomics*, 2:37, 2009.

[299] S. Mukherjee, S. Pelech, R. M. Neve, W. L. Kuo, S. Ziyad, P. T. Spellman, J. W. Gray, and T. P. Speed. Sparse combinatorial inference with an application in cancer biology. *Bioinformatics*, 25(2):265–271, 2009.

[300] S. Paruthiyil, A. Cvoro, X. Zhao, Z. Wu, Y. Sui, R. E. Staub, S. Baggett, C. B. Herber, C. Griffin, M. Tagliaferri, H. A. Harris, I. Cohen *et al.* Drug and cell type-specific regulation of genes with different classes of estrogen receptor β-selective agonists. *PLoS One*, 4(7):e6271, 2009.

[301] A. J. Pask, A. T. Papenfuss, E. I. Ager, K. A. McColl, T. P. Speed, and M. B. Renfree. Analysis of the platypus genome suggests a transposon origin for mammalian imprinting. *Genome Biol.*, 10(1):R1, 2009.

[302] J. A. Powell, D. Thomas, E. F. Barry, C. H. Kok, B. J. McClure, A. Tsykin, L. B. To, A. Brown, I. D. Lewis, K. Herbert, G. J. Goodall, T. P. Speed *et al.* Expression profiling of a hemopoietic cell survival transcriptome implicates osteopontin as a functional prognostic factor in AML. *Blood*, 114:4859–70, 2009.

[303] M. D. Robinson and T. P. Speed. Differential splicing using whole-transcript microarrays. *BMC Bioinformatics*, 10(1):156, 2009.

[304] Y. Sui, X. Zhao, T. P. Speed, and Z. Wu. Background adjustment for DNA microarrays using a database of microarray experiments. *J. Comput. Biol.*, 16:1501–15, 2009.

[305] Y. C. Tai and T. P. Speed. On gene ranking using replicated microarray time course data. *Biometrics*, 65(1):40–51, 2009.

[306] C. J. Tonkin, C. K. Carret, M. T. Duraisingh, T. S. Voss, S. A. Ralph, M. Hommel, M. F. Duffy, L. M. Silva, A. Scherf, A. Ivens, T. P. Speed, J. G. Beeson *et al.* Sir2 paralogues cooperate to regulate virulence genes and antigenic variation in *Plasmodium falciparum*. *PLoS Biol.*, 7(4):e1000084., 2009.

[307] E. J. Atkinson, S. K. McDonnell, J. S. Witte, D. C. Crawford, Y. Fan, B. Fridley, D. Li, L. Li, A. Rodin, W. Sadee, T. Speed, S. T. Weiss *et al.* Conference scene: Lessons learned from the 5th Statistical Analysis Workshop of the Pharmacogenetics Research Network. *Pharmacogenomics*, 11(3):297–303, 2010.

[308] H. Bengtsson, P. Neuvial, and T. P. Speed. TumorBoost: Normalization of allele-specific tumor copy numbers from a single pair of tumor-normal genotyping microarrays. *BMC Bioinformatics*, 11:245, 2010.

[309] C. L. Carmichael, E. J. Wilkins, H. Bengtsson, M. S. Horwitz, T. P. Speed, P. C. Vincent, G. Young, C. N. Hahn, R. Escher, and H. S. Scott. Poor prognosis in familial acute myeloid leukaemia with combined biallelic CEBPA mutations and downstream events affecting the ATM, FLT3 and CDX2 genes. *Br. J. Haematol.*, 150:382–5, 2010.

[310] International Cancer Genome Consortium. International network of cancer genome projects. *Nature*, 464(7291):993–8, 2010.

[311] M. W. Coolen, C. Stirzaker, J. Z. Song, A. L. Statham, Z. Kassir, C. S. Moreno, A. N. Young, V. Varma, T. P. Speed, M. Cowley, P. Lacaze, W. Kaplan *et al.* Consolidation of the cancer genome into domains of repressive chromatin by long-range epigenetic silencing (LRES) reduces transcriptional plasticity. *Nat. Cell. Biol.*, 12:235–46, 2010.

[312] A. Lapuk, H. Marr, L. Jakkula, H. Pedro, S. Bhattacharya, E. Purdom, Z. Hu, K. Simpson, L. Pachter, S. Durinck, N. Wang, B. Parvin *et al.* Exon-level microarray analyses identify alternative splicing programs in breast cancer. *Mol. Cancer Res.*, 8(7):961–74, 2010.

[313] D. C. Leitman, S. Paruthiyil, O. I. Vivar, E. F. Saunier, C. B. Herber, I. Cohen, M. Tagliaferri, and T. P. Speed. Regulation of specific target genes and biological responses by estrogen receptor subtype agonists. *Curr. Opin. Pharmacol.*, 10:629–36, 2010.

[314] S. Loi, B. Haibe-Kains, S. Majjaj, F. Lallemand, V. Durbecq, D. Larsimont, A. M. Gonzalez-Angulo, L. Pusztai, W. F. Symmans, A. Bardelli, P. Ellis, A. N. Tutt *et al.* PIK3CA mutations associated with gene signature of low mTORC1 signaling and better outcomes in estrogen receptor-positive breast cancer. *Proc. Natl. Acad. Sci. USA*, 107:10208–13, 2010.

[315] M. Ramakrishna, L. H. Williams, S. E. Boyle, J. L. Bearfoot, A. Sridhar, T. P. Speed, K. L. Gorringe, and I. G. Campbell. Identification of candidate growth promoting genes in ovarian cancer through integrated copy number and expression analysis. *PLoS One*, 5(4):e9983, 2010.

[316] M. D. Robinson, A. L. Statham, T. P. Speed, and S. J. Clark. Protocol matters: Which methylome are you actually studying? *Epigenomics*, 2(4):587–598, 2010.

[317] M. D. Robinson, C. Stirzaker, A. L. Statham, M. W. Coolen, J. Z. Song, S. S. Nair, D. Strbenac, T. P. Speed, and S. J. Clark. Evaluation of affinity-based genome-wide DNA methylation data: Effects of CpG density, amplification bias, and copy number variation. *Genome Res.*, 20:1719–1729, 2010.

[318] D. J. Speca, D. Chihara, A. M. Ashique, M. S. Bowers, J. T. Pierce-Shimomura, J. Lee, N. Rabbee, T. P. Speed, R. J. Gularte, J. Chitwood, J. F. Medrano, M. Liao *et al.* Conserved role of unc-79 in ethanol responses in lightweight mutant mice. *PLoS Genet.*, 6(8):e1001057, 2010.

[319] I. K. Tan, L. Mackin, N. Wang, A. T. Papenfuss, C. M. Elso, M. P. Ashton, F. Quirk, B. Phipson, M. Bahlo, T. P. Speed, G. K. Smyth, G. Morahan *et al.* A recombination hotspot leads to sequence variability within a novel gene (AK005651) and contributes to type 1 diabetes susceptibility. *Genome Res.*, 20:1629–38, 2010.

[320] R. G. Verhaak, K. A. Hoadley, E. Purdom, V. Wang, Y. Qi, M. D. Wilkerson, C. R. Miller, L. Ding, T. Golub, J. P. Mesirov, G. Alexe, M. Lawrence *et al.* Integrated genomic analysis identifies clinically relevant subtypes of glioblastoma characterized by abnormalities in PDGFRA, IDH1, EGFR, and NF1. *Cancer Cell*, 17:98–110, 2010.

[321] O. I. Vivar, X. Zhao, E. F. Saunier, C. Griffin, O. S. Mayba, M. Tagliaferri, I. Cohen, T. P. Speed, and D. C. Leitman. Estrogen receptor beta binds to and regulates three distinct classes of target genes. *J. Biol. Chem.*, 285:22059–66, 2010.

[322] C. Y. Yu, O. Mayba, J. V. Lee, J. Tran, C. Harris, T. P. Speed, and J. C. Wang. Genome-wide analysis of glucocorticoid receptor binding regions in adipocytes reveal gene network involved in triglyceride homeostasis. *PLoS One*, 5(12):e15188, 2010.

[323] E. F. Lee, O. B. Clarke, M. Evangelista, Z. Feng, T. P. Speed, E. B. Tchoubrieva, A. Strasser, B. H. Kalinna, P. M. Colman, and W. D. Fairlie.

Discovery and molecular characterization of a Bcl-2-regulated cell death pathway in schistosomes. *Proc. Natl. Acad. Sci. USA*, 108(17):6999–7003, 2011.

[324] S. Lopaticki, A. G. Maier, J. Thompson, D. W. Wilson, W. H. Tham, T. Triglia, A. Gout, T. P. Speed, J. G. Beeson, J. Healer, and A. F. Cowman. Reticulocyte and erythrocyte binding-like proteins function cooperatively in invasion of human erythrocytes by malaria parasites. *Infect. Immun.*, 79(3):1107–17, 2011.

[325] The Cancer Genome Atlas Research Network. Integrated genomic analyses of ovarian carcinoma. *Nature*, 474(7353):609–615, 2011.

[326] P. Neuvial, H. Bengtsson, and T. P. Speed. Statistical analysis of single nucleotide polymorphism microarray in cancer studies. In H. H. S. Lu, B. Schölkopf, and H. Zhao, editors, *Handbook of Statistical Bioinformatics*. Springer, New York, 2011.

[327] P. Shen, W. Wang, S. Krishnakumar, C. Palm, A. K. Chi, G. M. Enns, R. W. Davis, T. P. Speed, N. N. Mindrinos, and C. Scharfe. High-quality DNA sequence capture of 524 disease candidate genes. *Proc. Natl. Acad. Sci. USA*, 108(16):6549–54, 2011.

[328] T. P. Speed. Commentary on D. Basu's papers on sufficiency and related topics. In A. DasGupta, editor, *Selected Works of Debrabata Basu*, pages 35–40. Springer, New York, 2011.

[329] W. Wang, P. Shen, S. Thyagarajan, S. Lin, C. Palm, R. Horvath, T. Klopstock, D. Cutler, L. Pique, I. Schrijver, R. W. Davis, M. Mindrinos *et al.* Identification of rare DNA variants in mitochondrial disorders with improved array-based sequencing. *Nucleic Acids Res.*, 39(1):44–58, 2011.

[330] X. V. Wang, R. G. Verhaak, E. Purdom, P. T. Spellman, and T. P. Speed. Unifying gene expression measures from multiple platforms using factor analysis. *PLoS One*, 6(3):e17691, 2011.

[331] S. White, T. Ohnesorg, A. Notini, K. Roeszler, J. Hewitt, H. Daggag, C. Smith, E. Turbitt, S. Gustin, J. van den Bergen, D. Miles, P. Western *et al.* Copy number variation in patients with disorders of sex development due to 46, XY gonadal dysgenesis. *PLoS One*, 6(3):e17793, 2011.